D0220615

PRECALCULUS

FUNCTIONS and GRAPHS

FRANKLIN DEMANA • BERT K. WAITS

with the assistance of
ALAN OSBORNE • GREGORY D. FOLEY
The Ohio State University

ADDISON-WESLEY PUBLISHING COMPANY

Reading, Massachusetts • Menlo Park, California • New York
Don Mills, Ontario • Wokingham, England • Amsterdam • Bonn
Sydney • Singapore • Tokyo • Madrid • San Juan

The authors dedicate this book to their wives, Christine Demana and Barbara Waits, without whose patience, love, and understanding this book would not have been possible.

Executive Editor	David F. Pallai
Associate Editor	Stephanie Botvin
Production Administrator	Catherine Felgar
Senior Electronic Production Supervisor	Mona Zeftel
Electronic Production Consultant	Laurie Petrycki
Electronic Production Assistant	Sally Simpson
TEX Consultant	Frederick H. Bartlett of The Bartlett Press, Inc.
Text Designer	Geri Davis of Quadrata, Inc.
Copyeditor	Lorraine Ferrier
Manufacturing Supervisor	Roy Logan
Cover Designer	Marshall Henrichs

Figures 9.5.7, 9.5.8, 9.5.9, 9.5.10, 9.5.11, and 9.5.12 are from *Calculus, One and Several Variables, 4th edition*, by S. L. Salas and Einar Hille. Copyright © 1982 by John Wiley and Sons, Inc. Reprinted by permission of John Wiley and Sons, Inc.

This book was produced with TEXtures.

Library of Congress Cataloging-in-Publication Data

Demana, Franklin D., 1938–
Precalculus Functions / by Franklin Demana and Bert K. Waits; with the assistance of Alan Osborne and Gregory D. Foley.

p. cm.
ISBN 0-201-52781-2
1. Algebra. 2. Trigonometry I. Waits, Bert K. II. Title.
QA154.2.D43 1990
512'.13–dc20 89-37247

ABCDEFGHIJ-MU-943210

Preface

This text is designed for a one-semester college precalculus course. It was written because the authors believe that conventional texts that do not incorporate technology do not prepare students well for further study in mathematics and science in the 1990s. For example, only about 17% of the entering freshmen at The Ohio State University with four or more years of high school college-preparatory mathematics, including a precalculus course, are ready to begin their collegiate study of mathematics with calculus. This is consistent with national evidence that performance in college calculus courses is dismal at best.

Calculator- and computer-based graphing technology is incorporated in this text to enhance the teaching and learning of precalculus mathematics. Students are expected to have *regular and frequent* access to graphing calculators or computers with appropriate software for homework outside of class and perhaps for occasional classroom laboratory activities as well. Use of computer graphing or graphing calculator technology in this text is *not* optional.

Modern technology has evolved to the stage that it should be routinely used by mathematics students at all levels. Computer- or calculator-based graphing removes the need for contrived problems and opens the door for students to explore and solve realistic and interesting applications. The teaching and learning of traditional topics can be improved with the full use of these new tools. Computer- and calculator-based technology can turn the mathematics classroom into a mathematics laboratory. Technology gives rise to interactive instructional approaches that permit a focus on problem solving and encourage generalizations based on strong geometric evidence. The new instructional approaches that are possible with today's technology make you and your students active partners in an exciting, rewarding, enjoyable, and intensive educational experience. It is in this spirit of exploration and experimentation that this text is written.

Content Features

- **Applications** Problems are used to introduce and develop much of the mathematics in this text. Students using this approach become flexible problem solvers. Using problems as a basis for discussion makes the mathematics understandable to students, and students come to value mathematics because they appreciate its power. More realistic applications are possible because of the speed and power of technology (see Example 4 of Section 2.6, Example 5 of Section 7.6, and Example 6 of Section 8.2). You can select exercises with a business andeconomics theme, a science theme, or other themes.

- **Graphs** In the past careful, time consuming numerical and analytical techniques were used to produce accurate graphs that were rarely used. Now accurate graphs are obtained quickly and used to study the numerical and analytical properties of functions (see Example 5 of Section 4.2). Important questions are often generated by students viewing graphs. Sometimes these questions are answered using algebraic manipulation, which is then often well received by students. Other times they are answered by student exploration, taking advantage of the speed and power of graphing technology (see Example 1 of Section 3.7).

- **Scale** Scale is a crucial issue. The shape of a graph depends on the viewing window or rectangle in which the graph is viewed (see Example 6 of Section 1.2). Care in selecting viewing rectangles must be exercised because familiar graphs may not be recognized when they are distorted.

- **Multiple Representations** Algebraic representations (equations, inequalities, etc.) and geometric representations (graphs of algebraic representations) are established for a given problem situation (see Figure 1.1.6 of Section 1.1). Then connections among algebraic representations, geometric representations, and the problem situation are exploited to provide understanding about mathematical concepts and to give a geometric base for the algebraic ideas. Modeling receives special attention in this text (see Example 1 of Section 2.5).

- **Foreshadowing Calculus** Maxima and minima of functions are found in this text by using graphs (see Example 5 of Section 2.7). Intervals where functions are increasing or decreasing (see Example 5 of Section 1.7) and limiting behavior of functions are determined graphically. We do *not* borrow the techniques of calculus—rather we lay the foundation for the later study of calculus by providing students with rich intuitions and understandings about functions and graphs (see Esample 6 of Section 3.7).

- **Answers** Emphasis on exact answers is reduced, and approximate answers are emphasized. Technology provides a proper balance between exact answers that are rarely needed in the real world and approximate answers. What is usually needed is an answer with prescribed accuracy (see Example 7 of Section 2.1). Graphing techniques, such as zoom-in, provide an excellent geometric vehicle for discussing error in answers. Students can read answers from graphs with accuracy up to the limits of machine precision.

- **Algebraic Manipulation** You will find a good deal of ordinary algebraic manipulation in this text. However, the algebraic techniques often arise from problem situations or are used to answer questions generated by graphs (see Example 2 of Section 6.6). Students are more willing to perform algebraic manipulations when they are developed in context but are *not* the focus of a lesson.

- **Geometric Transformations** The exploratory nature of graphing technology helps students learn how a given graph may be obtained from a basic graph using the following geometric transformations: horizontal or vertical shifting, horizontal stretching or shrinking, vertical stretching or shrinking, reflection with respect to a coordinate axis. This technology-enhanced approach develops students' abilities so that they can sketch graphs of complicated functions (see Example 1 of Section 1.7, Example 4 of Section 3.5, Example 5 of Section 3.8, Example 6 of Section 4.5, and Example 4 of Section 6.6).
- **Trigonometric Identities** We deliberately spend less time in this textbook verifying identities algebraically and believe this to be appropriate. We do use graphs to investigate the plausibility of equations that are possible identities.

Pedagogical Features

Each chapter begins with an introductory overview of the material to be covered. The first paragraph of a section gives an overview of the section. The text requires considerably more reading than conventional texts. Students are expected to read the text and work through the examples.

- **Exercises** Directions for some exercises state *not* to use graphing technology. Whenever possible, it is good practice to have students verify responses to these exercises with graphing technology. Many exercises will require the use of technology. We have *not* written an example for each type of exercise given. Some exercises that are more difficult or extend the ideas of the section are marked with an asterisk. There are more exercises than can be reasonably assigned to any one student. The exercises were designed to provide you with flexibility to accommodate students with different backgrounds and interests.
- **Intensity** The extensive examples found within a section, coupled with the interrelated explanations, provide for an intensity of topic coverage. Thus, it is usually not possible to cover a section in one class meeting. Some examples should be left as reading exercises for students. You may find it helpful sometimes to work one problem through completely even if it takes the entire class, especially early in the course. However, you may find it necessary to omit some topics.
- **The Role of the Instructor Using Technology** The role of an instructor changes from a lecturer to a facilitator of learning. Students become active partners in the learning process as a consequence of technology. They learn to explore and experiment with mathematical concepts because of the speed and power of technology. You can use a single computer or an overhead graphing calculator in the classroom to provide an interactive lecture-demonstration, or use a guided-discovery approach in a computer lab or in a classroom where each student has a graphing calculator. You can use realistic problems to motivate and teach mathematical concepts because technology makes such problems accessible. Technology provides a much richer classroom environment

that fosters student involvement in the educational experience. You should encourage students to explore and experiment with the technology.

- **Visualization** Graphing helps students learn how to see and describe graphs and their characteristics. You use the power of visualization by carefully selecting a sequence of visual experiences to help students understand or discover mathematical concepts. Graphing technology makes the addition of geometric representations to the usual numeric and algebraic representations very natural. Exploring the connections between representations and problems deepens student understanding about mathematics and helps students value mathematics.

- **Review** Each chapter ends with a list of key terms introduced in that chapter and an extensive set of exercises. These exercises can be used to prepare students for exams or used for quizzes.

Supplements for the Instructor

- **Instructor's Manual** The Instructor's Manual contains an introductory chapter that gives an extensive overview and advice about using a technology-enhanced approach. There is a chapter-by-chapter commentary including remarks about selected exercises. Two versions are given for each chapter test. The Instructor's Manual contains answers to all exercises. The text contains the answers to approximately half of the exercises—usually the odd ones.

- **Tests** The Instructor's Manual that accompanies this text contains two versions of each chapter test.

- **Software** Instructors of class-size adoptions of this text receive *Master Grapher*, a powerful, interactive computer software package designed by the authors that graphs functions, conic equations, parametric equations, polar equations, and functions of two variables. *Master Grapher* comes in versions that work on Apple II, Macintosh, IBM, and most IBM-compatible computers.

- **Graphing Calculator and Computer Graphing Laboratory Manual** This laboratory manual gives details about the use of graphing calculators and computers to support the text. Students also receive a copy of the manual. It contains a user's guide to the *Master Grapher* graphing software as well as guides to the use of a Casio or Sharp graphing calculator.

A Request

For many of you using this text, incorporating technology into the teaching of mathematics is a new venture. You are a pathfinder in the true sense of the word and, consequently, will enjoy some highs when instruction goes well but some lows when your experiences do not match your expectations.

We would like to hear from you concerning your successes and frustrations. This is partly to guide the design of a revision of this text and partly to gain understanding of what it means for instructors and institutions to shift to using technology for mathematics teaching.

Acknowledgements

The authors are indebted to a great many people who participated in the development of this text. We very much appreciate the important and constructive suggestions of our colleagues Alan Osborne and Greg Foley. Special thanks are due to Janice McDonald and Barbara Waldron, our skillful TEX typists, and to Amy Edwards, who prepared the artwork for the manuscript and worked tirelessly to help the authors meet important deadlines. We also appreciate the efforts of David New, our resident TEXpert, who facilitated production of the book; and Jill Baumer, Cindy Bernlohr, Christine Browning, Sandy Davey, Tony DeGennaro, Penny Dunham, Ann Farrell, Bishnu Naraine, Jeri Nichols, Anne Sadeghipour, Linda Taylor, and Laurie Wern, who typed drafts, prepared answers, proofread, and made numerous helpful suggestions. The staff at Addison-Wesley, especially Stephanie Botvin and David Pallai, have provided enthusiastic support and expert guidance throughout the production process, making us proud to be part of the Addison-Wesley family.

We sincerely thank our colleagues who participated in the 1988–89 field test of a preliminary version of the textbook and who made valuable contributions that helped shape this book.

1988–89 College Field Test Instructors

Chris Allgyer, *Mountain Empire Community College*
George R. Barnes, *University of Louisville*
Daniel Buchanan, *Henry Ford Community College*
James G. Carr, *Iona College*
Gloria Child, *Rollins College*
Carolyn Crandell, *Muskingum College*
Professor Philip A. DeMarois, *National College of Education*
Ann Dinkheller, *Xavier University*
Gloria Dion, *Pennsylvania State University-Ogontz*
Eunice Everett, *Seminole Community College*
Max O. Gerling, *Eastern Illinois University*
Margaret J. Greene, *Florida Community College at Jacksonville*
Thomas Gregory, *The Ohio State University–Mansfield*
William L. Grimes, *Central Missouri State University*
John G. Harvey, *University of Wisconsin-Madison*
Robert Hathway, *Illinois State University*
Ingrid Holzner, *University of Wisconsin-Milwaukee*
Spencer P. Hurd, *The Citadel*
Lynne K. Ipiña, *University of Wyoming*
Bill Jordan, *Seminole Community College*
Rose Kaplan, *The Ohio State University-Newark*
Thomas J. Kearns, *Northern Kentucky University*
Stephen C. King, *The University of South Carolina at Aiken*
David E. Kullman, *Miami University*
Larry Lance, *Columbus State Community College*
Edward Laughbaum, *Columbus State Community College*

Robert Lavelle, *Iona College*
Tommy Leavelle, *John Brown University*
Millianne Lehmann, *University of San Francisco*
Gerald Leibowitz, *The University of Connecticut*
John Long, *University of Rhode Island*
Mary E. Maxwell, *The University of Akron*
Roger B. Nelsen, *Lewis and Clark College*
Henry Nixt, *Shawnee State University*
Susan D. Penko, *Baldwin Wallace College*
Anthony Perrescini, *University of Illinois–Urbana-Champaign*
Ruth A. Pruitt, *Fort Hays State University*
William Rettig, *Indiana University of Pennsylvania*
John Savige, *St. Petersburg Junior College*
Lawrence Sher, *Manhattan Community College (CUNY)*
Donald Shriner, *Frostburg State College*
Al Stickney, *Wittenberg University*
Karen Sutherland, *The College of St. Catharine*
Frederic Tufte, *University of Wisconsin-Platteville*
Kathy Underdown, *Rollins College*
Marjie Vittum-Jones, *South Seattle Community College*
Chuck Vonder Embse, *Central Michigan University*
Ron Waite, *Blue Mountain Community College*
Suzanne Welsch, *Sierra Nevada College*
Howard L. Wilson, *Oregon State University*

F.D.
B.K.W.

Columbus, Ohio

Contents

1

Relations, Functions, and Graphs

• Introduction

In this chapter we introduce models, algebraic representations, geometric representations, and complete graphs—important concepts that will be used throughout the textbook. The definitions of these terms will not be precise, but their meanings will become clear as examples are given throughout the textbook. Roughly speaking, algebraic representations are equations, inequalities, or systems of equations or inequalities, and geometric representations are graphs. We use algebraic representations and geometric representations as models of problem situations. We introduce functions and relations and draw graphs in the Cartesian coordinate system in this chapter. We will explain what is meant by the domain, range, and graph of a relation, function, equation, algebraic representation, and problem situation. Functions, relations, equations, inequalities, systems of equations and inequalities, and graphs are examples of models. Models can be very complex, and generally there is no single, correct model for a given problem situation. The person analyzing the problem situation usually chooses a model to represent the problem situation. Ideally a model should contain all the pertinent information about the problem situation and not be unduly complicated.

In this textbook we pay a good bit of attention to the way the following topics are interrelated: problem situations, algebraic representations of problem situations, and their graphs. Textbooks often focus most of their attention on working with various algebraic representations that are common in mathematics and give too little attention to geometric representations and their connections with algebraic representations and problem situations. Applications and graphical problem solving are important themes in this textbook.

Functions are important models in mathematics. We use graphing calculators or computer graphing software to obtain graphs of functions and certain relations. Knowledge of the symmetry properties is used to help determine their complete graphs. The vertical line test is used to decide if the relation determined by a graph is a function. We will see that graphing is an extremely important tool in mathematics. The absolute value function, the greatest integer function, and piecewise-defined functions will be introduced in this chapter. We write equations for lines and determine the graphs of linear functions and linear inequalities. We will see how to obtain the graph of any function of the form $y = ax^2 + bx + c$ from the graph of $y = x^2$ using geometric transformations.

In this chapter we introduce operations on functions and investigate composition of functions. Composition of functions is a way of putting two or more functions together to create a new function. Complicated functions can be expressed as composition of simpler functions. We also study the applications involving composition of functions.

1.1 Cartesian Coordinate System and Complete Graphs

In this section we introduce the Cartesian coordinate system and explain what is meant by domain, range, and the graph of a relation, equation, and problem situation. The meaning of *complete graph* will be introduced. We also explain what is meant by an algebraic representation of a problem situation and a geometric representation of a problem situation. A geometric representation will be given for a problem situation and compared with the graph of an algebraic representation of the problem situation.

Data that arise in problem situations can be graphed. A branch of mathematics called *data analysis* is devoted to analyzing data and determining appropriate algebraic representations of data. In this textbook, data will generally be obtained from well defined algebraic relationships. In the first example we use a familiar problem situation to give an example of an algebraic representation and a geometric representation of a problem situation. In this case the algebraic representation will be an equation and the geometric representation a portion of a straight line.

- EXAMPLE 1: Quality Rent-a-Car charges \$15 plus \$0.20 a mile to rent a car. Give an algebraic representation and a geometric representation that shows the relationship between the number of miles driven and the charges.

SOLUTION: If 50 miles are driven, the charges are $0.20(50) + 15$ or \$25. If x represents the number of miles driven and y the corresponding charges in dollars, then the equation $y = 0.2x + 15$ can be used to compute additional possibilities.

x (miles driven)	50	75	100	200
y (rental charges)	25	30	35	55

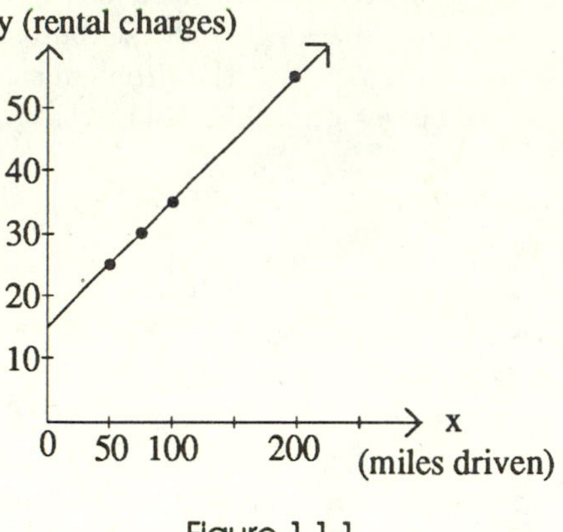

Figure 1.1.1

This equation is an algebraic representation of the problem situation. We can represent each of the possible miles driven and corresponding charges as a point on a graph. For example, the point representing 50 miles driven ($x = 50$) and charges of \$25 ($y = 25$) is located above 50 on the horizontal axis and across from 25 on the vertical axis (Figure 1.1.1). Can any real number greater than or equal to zero be the number of miles driven? Many would say that the answer is no. They would claim that miles driven should be given as positive integers, or maybe as positive numbers in tenths. Regardless of your point of view, we cannot graph all the possible points because there are infinitely many. However, if you continue to plot additional points, you will be convinced that the graph of the relationship between the number of miles driven and the charges appears to be the ray indicated in Figure 1.1.1. The ray is a geometric representation of the problem situation. The arrow on the graph indicates that the graph continues in the direction suggested by the arrow, even though we can only draw a part of it on the page. •

The collection of points on the ray indicated in Figure 1.1.1 that corresponds only to possible miles driven is a *graph of the Quality Rent-a-Car problem situation* because the graph geometrically describes the complete relationship between the number of miles driven and the charges. Because of the scale in Figure 1.1.1 the graph will appear to be the same whether you use all positive real numbers, only positive numbers in tenths, or only positive integers as possible number of miles driven.

Scale

In Figure 1.1.1 each scale mark on the horizontal axis represents 50 (miles) and each scale mark on the vertical axis represents 10 (dollars). If we had chosen the scale marks on the y-axis and x-axis to represent one dollar and one mile, respectively, then we would need a very large piece of paper to display the same information that is shown in Figure 1.1.1. The scale will often have to be different on the two axes, and should be based on the size of the numbers involved in the problem situation.

We will see that we do *not* always need a graph of a problem situation to answer questions about the problem. Often a graph that contains the graph of the problem situation will be used to answer questions about the problem situation. For example, we can use the graph in Figure 1.1.1 to answer questions about renting cars from Quality Rent-a-Car.

• EXAMPLE 2: Sarah is charged $50 for renting a car at Quality Rent-a-Car. How many miles did she drive the car?

SOLUTION: To solve this problem graphically, we draw a horizontal line from the point marked 50 ($50) on the vertical axis (rental charges) to the graph in Figure 1.1.1. Then, we draw a vertical line down to the horizontal axis (miles driven) to read the answer to this question from the graph. The result is shown in Figure 1.1.2. We estimate from the graph that Sarah drove 175 miles. In this case, 175 miles is the exact solution. This can be confirmed by direct substitution, that is, $0.2(175) + 15 = 50$, or algebraically by solving the equation $0.2x + 15 = 50$ for x. •

Next we give some of the standard terminology used in graphing.

The Real Number Line

We call $1, 2, 3, \ldots$ the **positive integers**, and $-1, -2, -3, \ldots$ the **negative integers**. The positive and negative integers together with zero form the **integers**. The positive integers together with 0 are called the **whole numbers.** A **rational number** is any number of the form $\frac{a}{b}$, where a and b are integers with $b \neq 0$. The decimal form of a rational number either terminates or is nonterminating *and* repeating. For example, $\frac{3}{4} = 0.75$ terminates and $\frac{13}{22} = 0.5909090\ldots$ is nonterminating (the digits 9 and 0 repeat forever as suggested by the pattern). A real number that is not rational is called an **irrational number**. Some examples of irrational numbers are $\sqrt{2}$, $-\sqrt{3}$, π, and $\frac{1}{\pi+1}$. Irrational numbers have infinite decimal representations that do *not* repeat; for example, $0.4404004000400004\ldots$. Notice

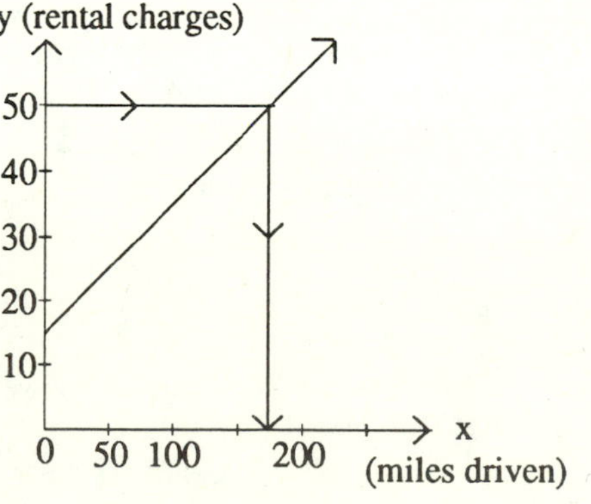

Figure 1.1.2

that the number of 0's between a pair of consecutive 4's increases by one as we move from left to right through the decimal representation.

The **real number line** provides a geometric representation of the real numbers. Each real number corresponds to one and only one point on the number line, and each point on the number line corresponds to one and only one real number. We first choose a point on the number line to be labeled zero (0). Then we mark equally spaced points on each side of 0. The points to the right of 0 are labeled $1, 2, 3, \ldots$, and the points to the left of 0 are labeled $-1, -2, -3, \ldots$ (Figure 1.1.3). Every real number, including irrational numbers, corresponds to one and only one point on the number line. The small vertical line segments used to mark the points are called **scale marks**.

The points corresponding to rational numbers are obtained by subdividing the unit line segment. For example, if the interval between 1 and 2 is divided into three equal segments, the right-hand end point of the second segment is labeled $\frac{5}{3}$ (Figure 1.1.3).

We use the notation a to represent both the real number a and the name of the point on the number line corresponding to a. We call a the **coordinate** of the point on the real number line corresponding to the real number a.

The Cartesian Coordinate System

A **Cartesian coordinate system** can be introduced in a plane by placing two real number lines (coordinate lines) in the plane so that they intersect at the zero on each line and are perpendicular. Usually one of the lines is horizontal with positive direction to the right, and the other line is vertical with positive direction upward as illustrated in Figure 1.1.4. The point O of intersection of the two lines is called the **origin**. The two lines are called **coordinate axes**, the horizontal line the ***x*-axis**, the vertical line the ***y*-axis**, and the plane a **coordinate plane**. The coordinate axes divide the plane into four parts called the **first, second, third,** and **fourth quadrants** as shown in Figure 1.1.4. The points on the coordinate axes are *not* in any quadrant.

Each point P in the coordinate plane determines a unique ordered pair of real numbers. If the vertical line through P (Figure 1.1.4) intersects the x-axis at a and the horizontal line through P intersects the y-axis at b, then P determines the **ordered pair** (a, b). We call these pairs *ordered pairs* because in general $(a, b) \neq (b, a)$; in fact $(a, b) = (b, a)$ if and

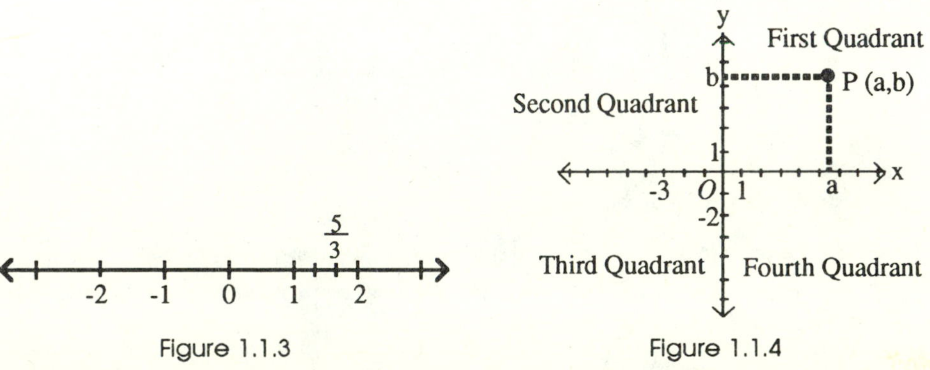

Figure 1.1.3 Figure 1.1.4

only if $a = b$. In general, $(a, b) = (c, d)$ if and only if $a = c$ and $b = d$. We call a the **x-coordinate** of P and b the **y-coordinate** of P. In addition, we say P has **coordinates (a, b)** and write $P(a, b)$. The origin O has coordinates $(0, 0)$.

Moreover, each ordered pair of real numbers (a, b) determines a unique point P in the coordinate plane. P is the point of intersection of the line perpendicular to the x-axis through a and the line perpendicular to the y-axis through b. In this way we have established a one-to-one correspondence between the set of all points in the coordinate plane and the set of all ordered pairs of real numbers. For brevity, we will refer to the point with coordinates (a, b) as the point (a, b) or simply (a, b).

Relations and Graphs

• DEFINITION Any set of ordered pairs of real numbers is called a **relation**, and the set of points in the coordinate plane corresponding to the ordered pairs is called the **graph of the relation**. The set of all first entries a of the ordered pairs (a, b) is called the **domain of the relation**, and the set of all second entries b is called the **range of the relation**. •

Example 1 determines a relation with miles driven as the first entry of each ordered pair and the rental charges as the second entry. The portion of the graph in Figure 1.1.1 that corresponds to the number of miles driven and the charges is a graph of this relation. The domain of this relation depends on your point of view. The range consists of all real numbers determined by $0.2x + 15$ as x takes all values in the domain. Now $0.2x + 15$ takes on all real numbers greater than or equal to 15 as x takes on all real numbers greater than or equal to zero. Thus, if we assume the domain of this relation (miles driven) is the set of all real numbers greater than or equal to zero then the range (rental charges) is the set of all real numbers greater than or equal to 15. If we assume that the domain is the set of positive integers, then the range is $0.2(1) + 15 = 15.2$, $0.2(2) + 15 = 15.4$, and so forth.

The ray in Figure 1.1.1 contains a graph of the Quality Rent-a-Car problem regardless of your point of view. We make the following definition.

• DEFINITION Any graph that represents *all* and *only* the data from a given problem situation is called a **graph of the problem situation**. •

• DEFINITION Any graph that contains a graph of a problem situation is called a **geometric representation** of the problem situation. •

Thus, the graph in Figure 1.1.1 is a *geometric representation* of the Quality Rent-a-Car problem. The collection of points on this graph that represent miles driven and rental charges is a *graph of the problem situation*.

Next we define what we mean by a *solution* to an equation in two variables.

• DEFINITION The pair of real numbers $x = a$ and $y = b$ is called a **solution** to an equation in the two variables x and y if we get a true statement when x and y are replaced by a and b, respectively, everywhere in the equation. •

The pair of numbers $x = 2$ and $y = 9$ is a solution to the equation $y = 2x^2 + 1$ because $9 = 2(2)^2 + 1$. The pair $x = -1$ and $y = 3$ is also a solution because $3 = 2(-1)^2 + 1$. The equation $y = 2x^2 + 1$ has *infinitely* many solutions. We focus only on real number solutions at this time. Complex number solutions will be discussed in Chapter 3.

Each pair $x = a$ and $y = b$ of real numbers that is a solution to $y = 2x^2 + 1$ determines a point (a, b) in the plane. The set of all such ordered pairs of real numbers is a *relation*, and the corresponding set of points in the plane is the *graph of this relation*.

• DEFINITION The set of all points in the coordinate plane determined by the solutions to an equation in two variables is called the **graph of the equation**. The set of all first coordinates of the points on the graph of an equation is called the **domain of the equation**, and the set of all second coordinates of the points on the graph is called the **range of the equation**. The domain variable is called the **independent variable**, and the range variable the **dependent variable**. •

The domain and range of an equation are, respectively, the domain and range of the relation determined by the set of all solutions to the equation. In Example 1 we used the algebraic equation $y = 0.2x+15$ to compute charges in the Quality Rent-a-Car problem. This equation is an example of an algebraic representation of the Quality Rent-a-Car problem. Solutions to this equation with x greater than zero represent the Rent-a-Car problem, but solutions with x less than or equal to zero do not. (Why?) We use the following *informal* definition.

• DEFINITION An **algebraic representation** of a problem situation is an equation, inequality, system of equations or inequalities, or other symbolic expression that represents the problem situation. The domain of the algebraic representation must contain the domain of the problem situation. If a is in the domain of the problem situation, then the algebraic representation must give a good approximation to the element that corresponds to a in the problem situation. •

The process of **modeling** is important in this textbook. The process involves creating an appropriate algebraic or geometric representation of a problem situation and then determining the portion of the domain of the representation that is the domain of the problem situation.

In the next example we see that a graph of an algebraic representation contains the graph of the problem situation. Thus, a graph of an algebraic representation of a problem situation *is* a geometric representation of the problem situation. However, a graph of the algebraic representation may also contain information that is not meaningful in the problem situation. Next to the actual graph of a problem situation, the graph of an algebraic repre-

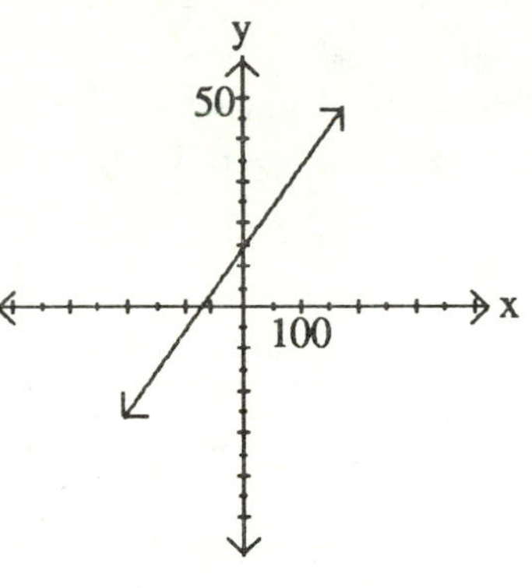

Figure 1.1.5

sentation of a problem situation is the most important and most frequently used geometric representation of a problem situation.

• EXAMPLE 3: Draw a graph of the equation $y = 0.2x + 15$ and specify the domain and range of the equation. What part of the graph represents the Quality Rent-a-Car problem?

SOLUTION: Every point on the graph in Figure 1.1.1 is a point on the graph of the equation $y = 0.2x + 15$. However, there are points on the graph of this equation with negative x-coordinates. For example, if $x = -50$, then $y = 5$. The graph of this equation consists of all points on the straight line indicated in Figure 1.1.5. Both the domain and range of the equation are the set of all real numbers.

The portion of the graph in Figure 1.1.5 to the right of the y-axis with x-coordinates that are possible miles driven is the part of the graph that represents the Quality Rent-a-Car problem. Regardless of the agreement about what numbers represent miles driven, the graph of the problem situation will appear to be all of the graph in Figure 1.1.5 to the right of the y-axis. •

We say that the graph drawn in Figure 1.1.5 is *complete* because it suggests all of the behavior of the relation determined by the equation $y = 0.2x + 15$. Loosely speaking, we say that a graph of a relation is *complete* if it suggests all of the points of the graph of the relation and all of the important features of the graph. You will need to do many examples before this idea becomes clear to you. However, determining complete graphs is an activity of most sections of this textbook. In this textbook, we use the following *informal* definition.

• DEFINITION A graph of a relation or equation is said to be **complete** if it suggests all points of the graph and all of the important features of the graph of the relation or equation. •

We say that the graph in Figure 1.1.5 is *complete* because it conveys the fact that the graph of the equation $y = 0.2x + 15$ is a straight line.

We have found an algebraic representation and two geometric representations of the Quality Rent-a-Car problem. The function defined by the equation $y = 0.2x + 15$ is the algebraic representation, and the graph of the problem situation in Figure 1.1.1 is one of the geometric representations. The complete graph of the algebraic representation $y = 0.2x+15$ given in Figure 1.1.5 is a second geometric representation. The graph of the problem situation is part of the complete graph of the algebraic representation $y = 0.2x + 15$. In general, if we have an algebraic representation of a problem situation, then a portion (maybe all) of a complete graph of the algebraic representation will give a graph of the problem situation. The diagram in Figure 1.1.6 summarizes the above important discussion about representations of problem situations.

Inverse variations. Inverse variations, or reciprocal relationships, are frequently encountered in real world situations. For example, the total time required for a trip is inversely related to the speed of travel of the trip. Pressure and volume of a gas are inversely related according to Boyle's law. Intensity of light and the square of the distance from the source of the light are inversely related. In general, y is inversely related to x if $y = \frac{K}{x}$, where K is a real number. In the next example, we investigate the most basic of all inverse relationships.

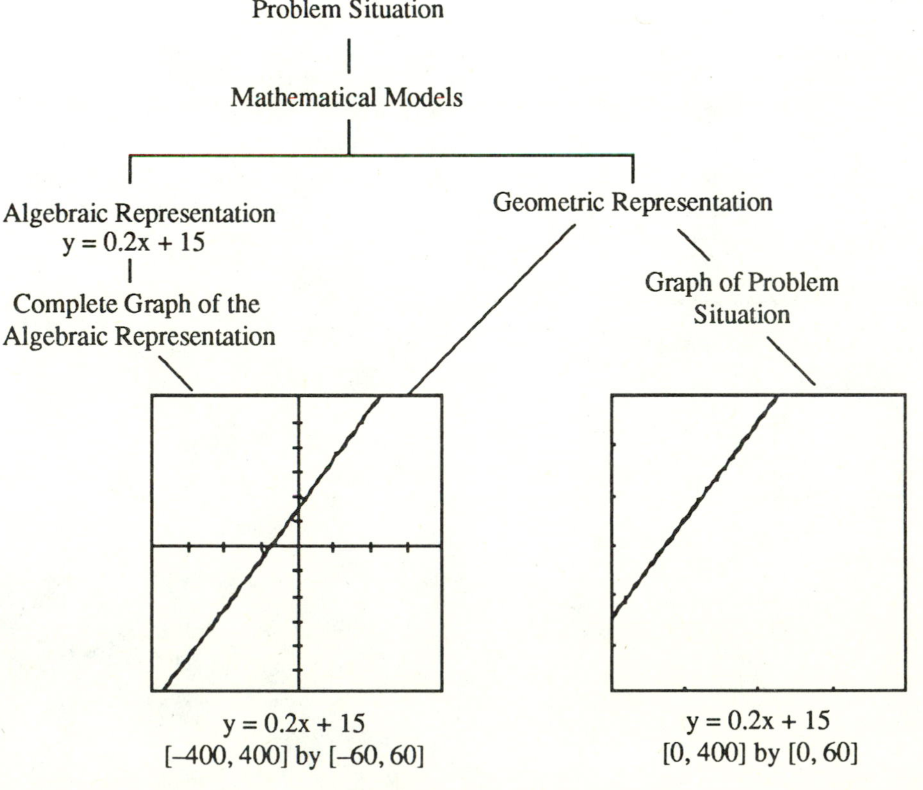

Figure 1.1.6

• EXAMPLE 4: The second of two numbers is the reciprocal of the first. Give a geometric and an algebraic representation of this problem situation.

SOLUTION: Let x represent the first number and y the second number. Then $y = \frac{1}{x}$. This equation is an *algebraic representation* of the problem situation because every solution to the problem is also a solution to the equation. A graph of the algebraic representation $y = \frac{1}{x}$ is a *geometric representation* of the problem situation. In this case, every solution to the equation is also a solution to the problem situation. Thus, a *complete graph* of the algebraic representation $y = \frac{1}{x}$ is also a *complete graph* of the problem situation.

Any real number different from zero has a reciprocal. Zero does *not* have a reciprocal. Thus, there is no second number with first number zero. Table 1 gives values of the second number y for some values of the first number x. Each value of y is rounded to hundredths. The graph of the 12 points determined by Table 1 is given in Figure 1.1.7.

You might be tempted to connect consecutive points from left to right. This gives an *incorrect* graph. We need more information about this relation in order to complete its graph. If x is 0.1, 0.01, or 0.001, then y is 10, 100, 1000, respectively. (Why?) If x is -0.1, -0.01, or -0.001, then y is -10, -100, -1000, respectively. We can see that if x is very close to zero, then y is very far from zero. Similar computations show that if x is very far from zero, then y is very close to zero.

There is a point on the graph with first coordinate a for every real number a different from zero. If we were able to plot all these points we would get the two solid curves indicated in Figure 1.1.8. Once again we use arrows on the graph to indicate that the graph continues as suggested by the arrows. The graph in Figure 1.1.8 is a *complete graph* of the equation $y = \frac{1}{x}$ because this graph suggests the complete behavior of the relation determined by the equation. It is also a geometric representation of the problem situation. Notice that the points on the graph with x-coordinates very far from zero are very close to the horizontal

x	y
-5	-0.2
-4	-0.25
-3	-0.33
-2	-0.5
-1	-1
-0.5	-2
0.5	2
1	1
2	0.5
3	0.33
4	0.25
5	0.2

Table 1

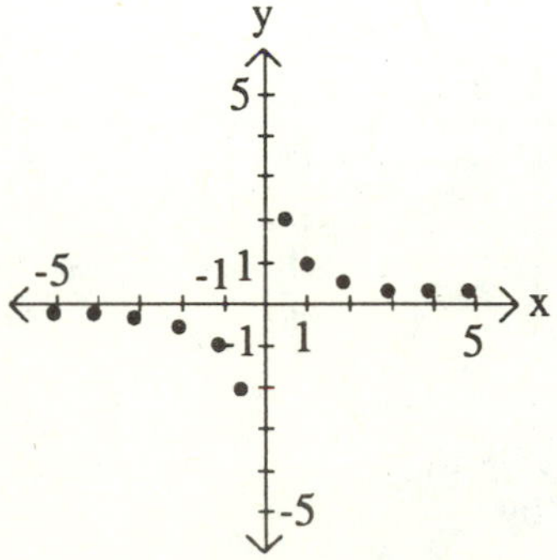

Figure 1.1.7

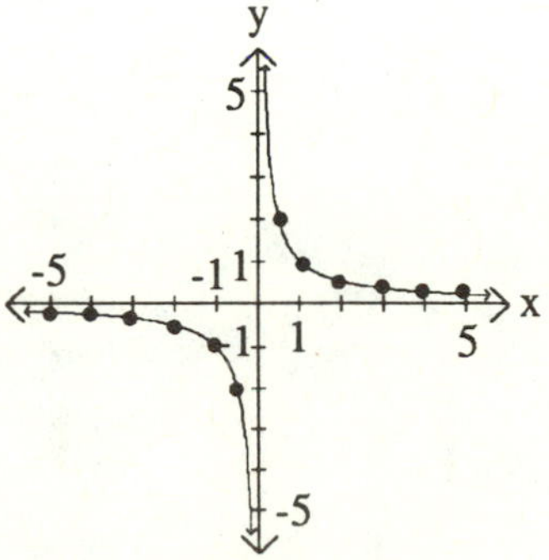

Figure 1.1.8

axis, and the points with x-coordinates very close to zero are very close to the vertical axis. •

Calculators and computers can be used to define relations. The first entry of an ordered pair is the real number that we enter into a calculator or computer, and the second entry is the corresponding real number the calculator or computer gives as an output. For example, the square key $\boxed{x^2}$ on a calculator defines a specific relation. We use the symbol $\square$ to represent a key on a calculator. If we input the number 3, then the calculator gives 9 as the output.

We agree that, when not explicitly stated, the domain of a relation defined by a calculator key is the set of all real numbers that would produce real numbers as output using the relation, and the range is the corresponding set of all output numbers obtained using the relation. For the calculator key $\boxed{\sqrt{\ }}$, the domain and range are the set of all **nonnegative numbers** (all real numbers greater than or equal to zero). For the calculator key $\boxed{x^2}$, the domain is the set of all real numbers and the range is the set of all nonnegative real numbers.

In the next two examples we explore the meaning of *complete graph*. Specifically, we will see that if a few points of a graph are known, there are infinitely many possibilities for the *complete graph* unless we have an explicit rule for the points of the graph.

• EXAMPLE 5: Table 2 gives some values of a certain relation. Graph the points determined by the table. Then give three possibilities for a complete graph of the relation, and specify the domain and range of each.

SOLUTION: The graph of the points determined by Table 2 is given in Figure 1.1.9. Here the first entries (x values) are called inputs, and the second entries (y values) corresponding to x are called outputs. There are many possibilities for a complete graph of this relation. One possibility is that these are all of the values of the relation so that the graph in Figure

Input	Output
−3	−15
−2.5	−5.625
−2	0
−1.5	2.625
−1	3
−0.5	1.875
0	0
0.5	−1.875
1	−3
1.5	−2.625
2	0
2.5	5.625
3	15

Table 2

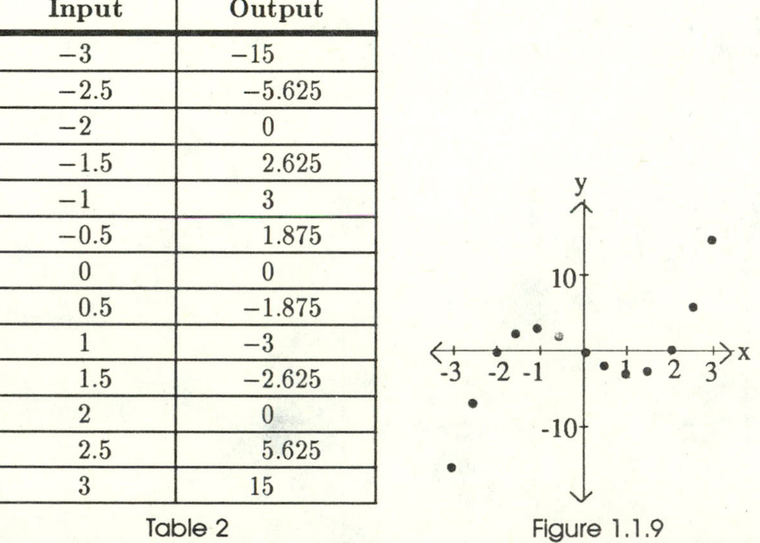

Figure 1.1.9

1.1.9 is one possible *complete* graph. In this case the domain is the set of all numbers in the input column of the table, and the range is the set of all numbers in the output column. Two other possible complete graphs of the relation are given in Figures 1.1.10 and 1.1.11.

The solid dots in Figure 1.1.10 are used to indicate that the graph starts at the point $(-3, -15)$ and stops at the point $(3, 15)$. The domain of this relation is the set of all real numbers between -3 and 3, including -3 and 3, which can be denoted by $-3 \le$ Input ≤ 3; the range is the set of real numbers between -15 and 15, including -15 and 15, or $-15 \le$ Output ≤ 15.

Figure 1.1.11 indicates that the graph starts at the point $(-3, -15)$ and continues in the direction suggested by the *arrow* above $(3, 15)$. The domain of this relation is the set of all real numbers greater than or equal to -3, or Input ≥ -3; the range is the set of all real numbers greater than or equal to -15, or Output ≥ -15. •

Generally it is difficult to graph accurately points such as the ones determined by Table 2. For example, to display carefully both 1.875 and 15 would require a very large piece of graph paper. The reason we used these numbers will become clear in the next example.

• EXAMPLE 6: Complete the graph of the relation started in Figure 1.1.9 assuming that the output y is obtained from the input x using the equation $y = x^3 - 4x$.

SOLUTION: You can verify that the entries in Table 2 can be obtained using the equation $y = x^3 - 4x$ with x as input and y as output. For example, $(2.5)^3 - 4(2.5) = 5.625$. Now there is only *one* possible *complete graph*, and it is indicated in Figure 1.1.12. Arrows are drawn on the graph to indicate that it continues in both directions. •

If we are given a complete graph of a relation, then we can use that graph to estimate the domain and range of the corresponding relation.

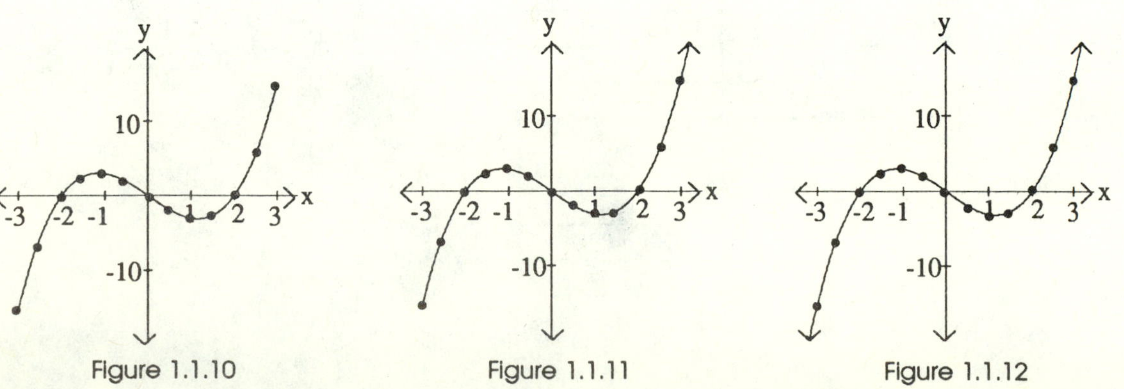

Figure 1.1.10 Figure 1.1.11 Figure 1.1.12

• EXAMPLE 7: Assume the graph given in Figure 1.1.13 is complete.

(a) Complete the table for the relation determined by this graph.

x	y
0	
1	
	5

(b) What are the domain and range of this relation?

SOLUTION:

(a) We can read from the graph in Figure 1.1.13 that y is 1 when x is 0, and y is 2 when x is 1. If we draw a horizontal line segment from 5 on the y-axis to the graph and then a vertical line segment down to the x-axis, then we can read that x is 2 when y is 5. The completed table is as follows.

x	y
0	1
1	2
2	5

(b) The domain of this relation is the set of all real numbers x greater than or equal to 0 ($x \geq 0$), and the range is the set of all real numbers y greater than or equal to 1 ($y \geq 1$). •

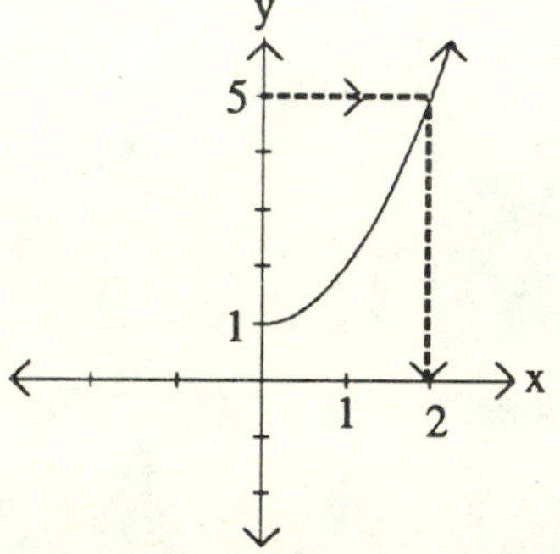

Figure 1.1.13

• EXERCISES 1-1

Exercises in this textbook that are marked with an asterisk (*) either are unusually challenging or have not received attention in the narrative.

1. Estimate the coordinates of the seven points in Figure 1.1.14.
2. Graph the points in (a) – (h) on the same coordinate plane. Identify the quadrant containing each point.

   (a) $(2,4)$ (b) $(0,3)$ (c) $(3,0)$ (d) $(-1,2)$
   (e) $(-2,3)$ (f) $(2,-\frac{5}{2})$ (g) $(3,-2)$ (h) $(-\frac{1}{2},-5)$

Graph the three points on the same coordinate system. Choose an *appropriate scale.*

3. $(5,200)$, $(3,100)$, $(7,160)$
4. $(2.01,5)$, $(2.03,8)$, $(1.98,-4)$
5. Draw and identify the polygons determined by each set of vertices. How many different polygons can be determined in each case?

   (a) $(-1,2)$, $(3,5)$, $(2,-4)$
   (b) $(0,0)$, $(2,3)$, $(-1,3)$, $(-1,0)$
   (c) $(-1,-2)$, $(6,-4)$, $(10,1)$, $(3,3)$
   (d) $(-3,0)$, $(-2,3)$, $(0,4)$, $(3,0)$, $(2,-3)$
6. The point $(a,6)$ is on the graph of $y = 2x$. What is a?
7. The point $(-2,T)$ is on the graph of $y = 5x - 3$. What is T?
8. Draw complete graphs of three different relations that include the data given in the table.

First number	Second number
−4	−2
−1	0
1	3
4	6
9	11

9. Draw complete graphs of three different relations that include the data given in the table.

x	y
1	1
2	4
3	9
4	16
5	25

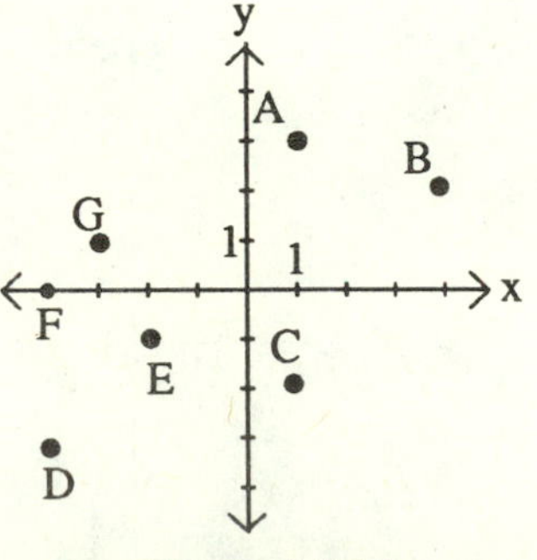

Figure 1.1.14

10. (a) Complete the table for the relation given by the complete graph in Figure 1.1.15.
    (b) What are the domain and range of this relation?

First number	Second number
0	
	0
2	
	−2

11. (a) Complete the table for the relation given by the complete graph in Figure 1.1.16.

Height (ft)	Weight (lbs)
2	
	100
4	
	200

    (b) What are the domain and range of this relation?

12. (a) Complete the table for the relation given by the complete graph in Figure 1.1.17.

Input	Output
0	
	2
1	
	0
−3	

    (b) What are the domain and range of this relation?

Second Number
1
First Number
1

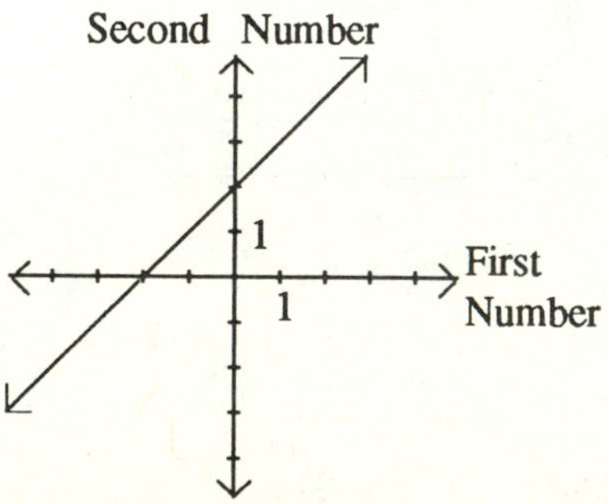

Figure 1.1.15

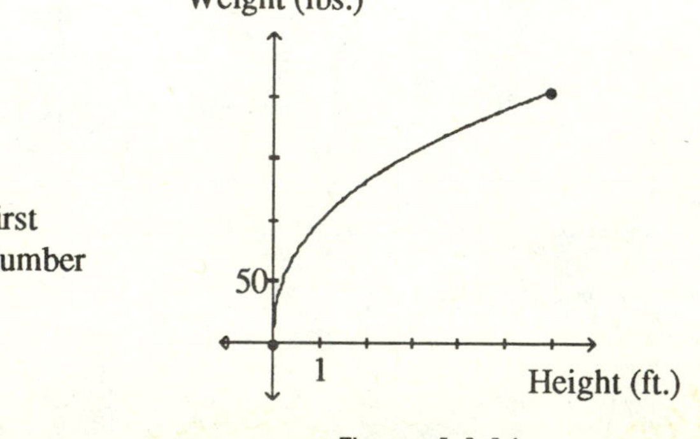

Figure 1.1.16

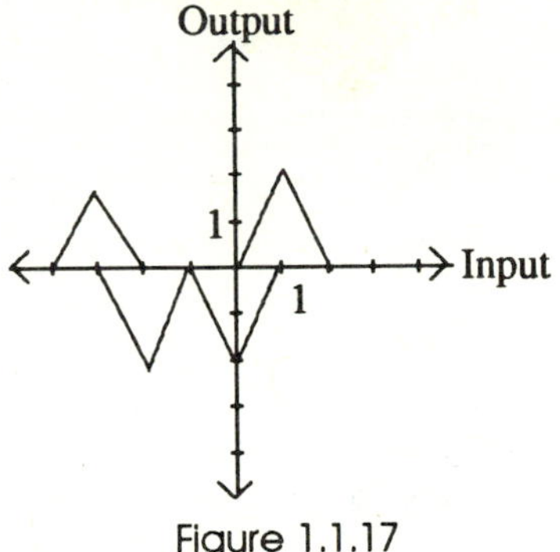

Figure 1.1.17

Draw a complete graph of each equation. Choose an appropriate scale for each axis.

13. $x + y = 6$
14. $y = 200x - 10$
15. $y = 2x^2 + 1$
16. $y = 10x^2 + 500$
17. $y = -\frac{2}{x}$
18. $y = x^3 + x$

19. (a) Use the square root key $\boxed{\sqrt{\ }}$ on a calculator to complete the table. Give your answers accurate to hundredths.

Input	Output (Square Root)
−3	
−2	
0	
1	
2	
3	
4	
5	
6	

(b) Graph the points determined by the table in (a).

(c) Draw a complete graph of the relation determined by the calculator square root key $\boxed{\sqrt{\ }}$. (The output is the square root of the input.)

(d) What are the domain and range of the square root relation?

20. (a) Use the natural logarithm key [ln] on a calculator to complete the table. Give your answers accurate to hundredths.

Input	Output (ln)
0.25	
0.5	
1	
1.5	
2	
3	
4	
5	

(b) Graph the points determined by the table in (a).

(c) Draw a complete graph of the relation determined by the calculator key [ln]. (The output is the natural logarithm of the input.)

(d) What are the domain and range of the natural logarithm relation?

21. (a) Use the sine key [sin] on a calculator to complete the tables. Use radian mode. Give your answers accurate to hundredths.

Input	Output (sin)
0	
0.52	
1.05	
1.57	
2.09	
2.62	
3.14	
3.67	
4.19	
4.71	

Input	Output (sin)
5.24	
5.76	
6.28	
6.81	
7.33	
7.85	
8.38	
8.90	
9.42	
9.95	

(b) Graph the points determined by the table in (a).

(c) Assuming the pattern in the table continues (repeats itself), draw a complete graph of the relation determined by the [sin] key.

(d) What are the domain and range of the sine relation?

22. A family takes a 100-mile trip. Let R be their average speed in miles per hour and T be the total time of the trip in hours.
    (a) Write an equation (algebraic representation) that shows how T (output) depends on R (input).
    (b) Draw a complete graph of the algebraic representation (equation) in (a).
    (c) Which part of the graph in (b) represents the problem situation, that is, the family trip?
    (d) Use the graph in (b) to determine the total time T if the average speed is 40 mph.
23. If the point (a, b) is on the graph of $T = \frac{100}{R}$, then find ab.
24. A school club pays $35.75 to rent a video cassette player to show a movie for a fund raising project. The club charges $0.25 per ticket for the movie.
    (a) Write an equation (algebraic representation) that shows how the club's profit (or loss) P (output) depends on the number n (input) of tickets sold.
    (b) Draw a complete graph of the equation in (a).
    (c) What are the domain and range of the algebraic representation (equation)?
    (d) Which part of the graph in (b) represents the problem situation? *Hint:* You cannot sell half a ticket!
    (e) Use the graph in (b) to determine the minimum number of tickets that must be sold for the club to realize a profit.
25. Let the point (z, w) be on the graph of $P = 0.25n - 35.75$. Determine a formula for w in terms of z.
26. Does the graph of the relation given in Figure 1.1.8 ever touch the horizontal axis or vertical axis? (Why?)

*27. Consider the relation *the second number is less than or equal to the first number.*
    (a) Give an algebraic representation of this relation.
    (b) Draw a complete graph of this relation.
    (c) What are the domain and range of this relation?

1.2 Functions and Graphing Utilities

In this section we begin the study of functions and use computers or graphing calculators to obtain their graphs. A function is a special type of relation. Recall that a relation is any set of ordered pairs of real numbers. A vertical line test is introduced and used to identify graphically relations that are functions. We investigate an application that has a function for an algebraic representation.

Functions are encountered routinely in our daily lives in tabular form: height–weight charts, time–temperature charts, price–sales-tax charts, income tax tables, and so forth. Functions play a very crucial role in the study of mathematics and science. They provide important models for problem situations. For example, functions can be used to model the growth of a population, the growth of money in a savings account, the force of attraction between objects, the illumination provided by a source of light, the cost of producing goods, and so forth.

Calculators can be used to give examples of functions. In fact, calculators are sometimes referred to as function machines. If a calculator computes unique output for given input, then the calculator is producing values of a function.

• DEFINITION A **function** is a relation with the property: If (a,b) and (a,c) belong to the relation, then $b = c$. The set of all first entries of the ordered pairs is called the **domain of the function**, and the set of all second entries is called the **range of the function**. •

Using an input–output analogy, a relation is a function if each input is paired with no more than one output. Suppose that $(2,b)$ and $(2,c)$ are two ordered pairs of a given relation. If the relation is a function, then b must be equal to c. That is, *only one* output is possible for the input 2. This means each second entry of an ordered pair of a function is *uniquely determined* by the corresponding first entry.

• EXAMPLE 1: Which of the following relations are functions?

(a) $R = \{(0,0),(-1,1),(1,1),(2,4)\}$ (b) $T = \{(0,0),(1,-1),(1,1),(4,2)\}$

SOLUTION: In order for a relation to be a function, there cannot be *two* different ordered pairs with the same first entry. The relation R is a function because there is one and only one pair with a given first entry: 0, -1, 1, or 2. The relation T is *not* a function because there are two ordered pairs with first entry 1, namely $(1,-1)$ and $(1,1)$. Alternatively, if we think of the first entry as input and the second entry as output, then the input 1 produces two distinct outputs: -1 and 1. Thus, the input–output relation T is not a function. •

In the next example we consider relations defined by calculator keys.

• EXAMPLE 2: Show that the relations determined by the following calculator keys are functions: $\boxed{\sqrt{\ }}$, $\boxed{x^2}$, $\boxed{\sin}$, $\boxed{\log}$.

SOLUTION: The relation determined by the calculator key $\boxed{\sqrt{\ }}$ consists of ordered pairs $(x,\sqrt{x})$ where $\sqrt{x}$ stands for the nonnegative square root of x. This relation is a function because $\sqrt{x}$ is uniquely determined by x, that is, if $a = b$ then $\sqrt{a} = \sqrt{b}$. Similarly, the key $\boxed{x^2}$ determines a function. Later in this textbook we study the functions $\sin x$ and $\log x$ in greater detail. For now, the fact that the calculator keys $\boxed{\sin}$ and $\boxed{\log}$ produce a unique output for a given input means that the relation determined by each of these keys is a function. •

• EXAMPLE 3: Show that the relation $R = \{(x,y) \mid y^2 = x\}$ determined by the equation $y^2 = x$ is not a function.

SOLUTION: The relation determined by $y^2 = x$ is the set of all solutions to $y^2 = x$ considered as ordered pairs of real numbers. The ordered pair $(1,1)$ belongs to the relation R because

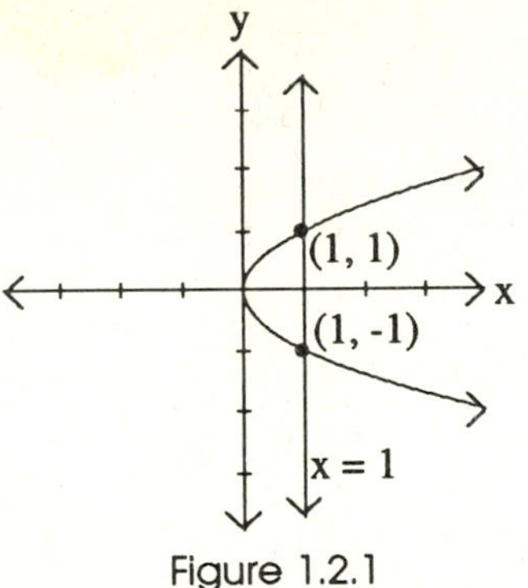

Figure 1.2.1

the pair $x = 1$ and $y = 1$ is a solution to the equation $y^2 = x$. Similarly, the ordered pair $(1, -1)$ belongs to R because the pair $x = 1$ and $y = -1$ is also a solution to the equation $y^2 = x$. R is *not* a function because the two different ordered pairs $(1, 1)$ and $(1, -1)$ have the same first entry. In other words, the first entry 1 does not uniquely determine the second entry. •

In Section 1.3 we show that the graph in Figure 1.2.1 is a complete graph of the equation $y^2 = x$. Notice that the two points $(1, 1)$ and $(1, -1)$ lie on the vertical line $x = 1$.

If a relation is *not* a function, then the relation must contain at least two ordered pairs (a, b) and (a, c) with $b \neq c$. Because both of these points lie on the vertical line $x = a$, the vertical line $x = a$ must intersect the graph of the relation in at least two points. Thus, a complete graph of a relation can be used to decide if the relation is a function.

- **Vertical Line Test for Functions.** If every vertical line intersects the graph of a relation in at most one point, then the relation is a function.

- **EXAMPLE 4:** Which of the relations represented by the graphs in Figure 1.2.2 are functions?

SOLUTION:

(a) The vertical lines $x = a$ with a greater than 2 or less than -2 do not intersect this graph. The vertical lines $x = a$ with a between -2 and 2, including 2 and -2 intersect the graph in exactly one point. Thus, the relation represented by this graph is a function.

(b) The vertical lines $x = a$ with a negative do not intersect this graph. The line $x = 0$ (y-axis) intersects the graph exactly once, but each line $x = a$ with a positive intersects the graph in two points (Figure 1.2.3). Thus, the relation represented by this graph is not a function.

(c) Every vertical line intersects this graph exactly once, so this graph is the graph of a function.

(d) Every vertical line intersects this horizontal line in exactly one point, so this graph is a graph of a function. •

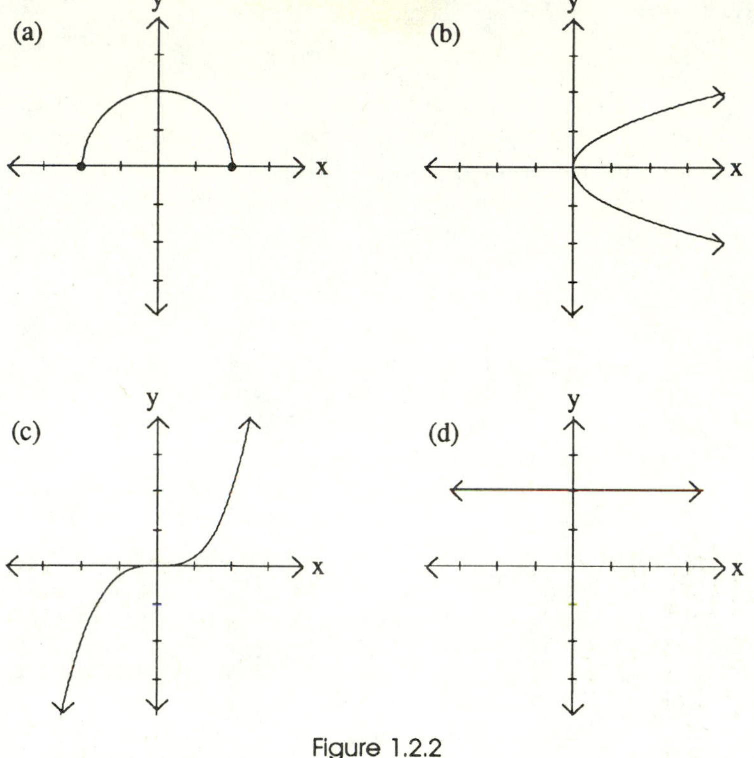

Figure 1.2.2

If f is a function and x a first entry of an ordered pair in f, we denote the second entry by $f(x)$. The notation $f(x)$ for the number paired with x by the function f is read **"f of x"** or **"f at x."** Often we call $f(x)$ the **value** of f at x or the **output** associated with **input** x. If (x, y) is an ordered pair in f, then $y = f(x)$.

• EXAMPLE 5: Let f be the relation defined by the equation $y = x^2 + 1$. Show that f is a function, and find $f(-1)$, $f(3)$, $f(a)$, $f\left(\frac{1}{a}\right)$, $f(a + h)$, and $f(x)$.

SOLUTION: First, we must show that f is a function. If f contains two ordered pairs with the same first entry, we must show that the second entries are also the same; in other words, the two ordered pairs are really the same ordered pair. The number paired with 2 by f is $2^2 + 1$ or 5. Notice that 5 is the only possible number that can be paired with 2 by f. More precisely, if (a, b) and (a, c) are ordered pairs in f, we must show that $b = c$. If (a, b) and (a, c) are ordered pairs in f, then, by the definition of f, $b = a^2 + 1$ and $c = a^2 + 1$. Thus, $b = c$, and f is a function by definition. Usually we will not take the time to verify algebraically that a given relation is a function.

The notation $f(-1)$ stands for the number paired with -1 by f, or $f(-1)$ is the output produced by f with input -1. Thus, $f(-1) = (-1)^2 + 1 = 2$. Similarly, $f(3) = 3^2 + 1 = 10$,

$f(a) = a^2+1$, and $f\left(\frac{1}{a}\right) = \left(\frac{1}{a}\right)^2+1 = \frac{1}{a^2}+1$. Next, $f(a+h) = (a+h)^2+1 = a^2+2ah+h^2+1$. Finally, $f(x) = x^2 + 1$. •

Now that we have had some practice with drawing graphs by computing values and plotting points, we will use graphing utilities to draw graphs of functions. Because a function is a relation, a **graph of a function** is the graph of the function considered as a relation. We use the term **function graphing utility** to refer to any device that draws a graph of a function. A computer with appropriate software and a graphing calculator are graphing utilities. We take advantage of the speed and power of these utilities to draw graphs of many more functions than would be reasonably possible using only pencil and paper.

We think of the display screen of a graphing utility as representing a portion of the coordinate plane. In fact, if the four corners of the screen have coordinates as illustrated in Figure 1.2.4, we are looking at a graph in the **viewing rectangle** $[L, R]$ **by** $[B, T]$ where $L \leq x \leq R$ and $B \leq y \leq T$. We use the notations $L \leq x \leq R$ to represent all the real numbers between L and R, including L and R. Similarly, the notation $B \leq y \leq T$ represents all the real numbers between B and T. It is easy to remember this choice of notation if we think of L, R, B, and T as meaning *left*, *right*, *bottom*, and *top*, respectively. Sometimes L, R, B, and T are referred to as "x min," "x max," "y min," and "y max," respectively. It is important to understand that the viewing rectangle frames a portion, usually not all, of a graph.

You are expected to have access to a graphing calculator or computer on a regular basis for homework. This utility should allow you to

- enter and obtain the graph of any function;
- enter any number between 10^{-6} and 10^6 for the value of L, R, B, or T;
- overlay the graph of a second function on the same screen with the graph of the first function.

Consult the accompanying *Graphing Calculator and Computer Graphing Laboratory Manual* to see how to use a graphing utility. All graphs in this textbook have been plotted using the Apple Macintosh™ version of a graphing utility called *Master Grapher.*

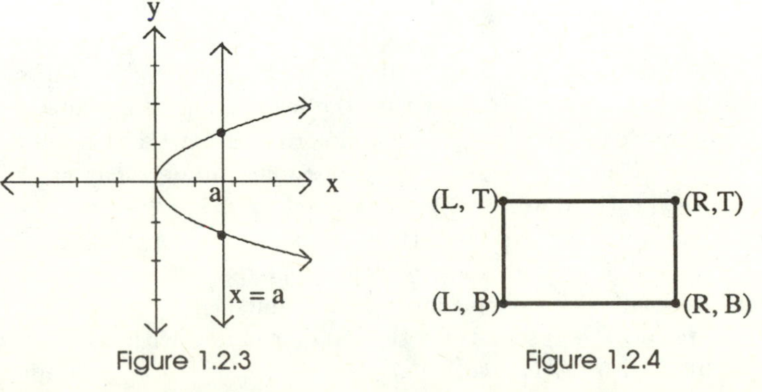

Figure 1.2.3

Figure 1.2.4

Most figures that appear in this text specify an equation for the function or relation whose graph appears in the figure together with the viewing rectangle used. In addition, the graph will be framed by a border to suggest what actually appears on the display screen of a graphing utility in the specified viewing rectangle. Scale marks that appear on the coordinate axes are equally spaced. The distance represented by the interval between consecutive scale marks on the x-axis often is different from the distance represented by the interval between consecutive scale marks on the y-axis. The shape of a given graph depends on the viewing rectangle in which it is viewed.

• DEFINITION A graph of a function is said to be **complete** if it suggests all of the points of the graph and all of the important features of the graph. •

We need to choose viewing rectangles carefully in order to investigate critical portions of a graph and to determine a complete graph of a function. Sometimes we need to display a graph in several different viewing rectangles to describe a complete graph of a given function.

• EXAMPLE 6: Draw a complete graph of the function $f(x) = x^2 + 1$ and determine its domain and range.

SOLUTION: First, we choose a viewing rectangle to frame a portion of the graph of f. Suppose we choose the viewing rectangle $[-1, 1]$ by $[-1, 1]$, that is, $-1 \leq x \leq 1$ and $-1 \leq y \leq 1$. Because $(0, 1)$ is the only point of the graph of f in this viewing rectangle, the display screen will appear to be blank except for the single point and the coordinate axes. In fact, the point will not be visible because it is covered up by the y-axis. We have positioned our viewing rectangle (framed our picture) so that only one point of the graph appears on the display screen.

If we now choose $[-1, 1]$ by $[-2, 2]$ for the viewing rectangle we get the graph in Figure 1.2.5. This time more than one point on the graph of f appears in the viewing rectangle.

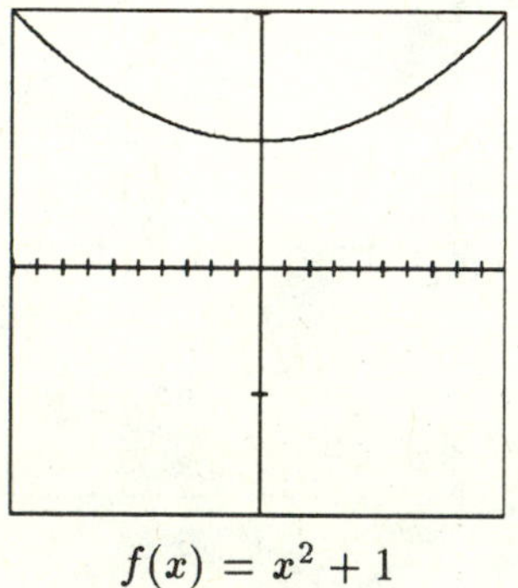

$f(x) = x^2 + 1$
$[-1, 1]$ by $[-2, 2]$

Figure 1.2.5

Notice that the equation $f(x) = x^2 + 1$ and the viewing rectangle $[-1, 1]$ by $[-2, 2]$ are displayed with the graph in Figure 1.2.5.

The graph in Figure 1.2.5 gives a clue to the complete behavior of f. However, we must be careful not to jump to conclusions too quickly. The graph of f in the **standard viewing rectangle** $[-10, 10]$ by $[-10, 10]$ is given in Figure 1.2.6. We can convince ourselves that the graph in Figure 1.2.6 is *complete* by viewing the graph in a large viewing rectangle (Figure 1.2.7). Thus, as we move from left to right through x the graph of f steadily increases. The lowest point of the graph of f is $(0, 1)$. Notice the shape of the graph of f is different in these three figures.

In this case, the graph of f in the standard viewing rectangle is a *complete graph*. Example 7 will show that this does not always happen, especially when drawing the graph of a problem situation.

Notice that the scale marks on the coordinate axes in Figure 1.2.6 are one unit apart. In Figure 1.2.7 the scale marks on the x-axis are 10 units apart and the scale marks on the y-axis are 100 units apart. These scale marks can be used to estimate coordinates of points on the graph. We can see from Figure 1.2.6 that the domain of f is the set of all real numbers, and the range of f the set of all real numbers greater than or equal to 1, or $y \geq 1$. •

It is important to understand how graphing utilities draw graphs of functions in viewing rectangles $[L, R]$ by $[B, T]$. Generally, the values of the functions are computed for a specified number of equally spaced values of x between L and R, and then the corresponding points are graphed if they fall within the rectangle $[L, R]$ by $[B, T]$. Usually, these points are joined with straight line segments to produce the graph.

In Sections 1.6 and 1.7 we study in detail the graphs of functions of the form $f(x) = ax^2 + bx + c$, where a, b, and c are real numbers with $a \neq 0$. The graph of such a function is called a **parabola**. For now you may assume that a complete graph of $f(x) = ax^2 + bx + c$ with $a \neq 0$ looks like one of the two parabolas in Figure 1.2.8. The domain of any such function is the set of all real numbers. If a is positive, the graph has a lowest point and

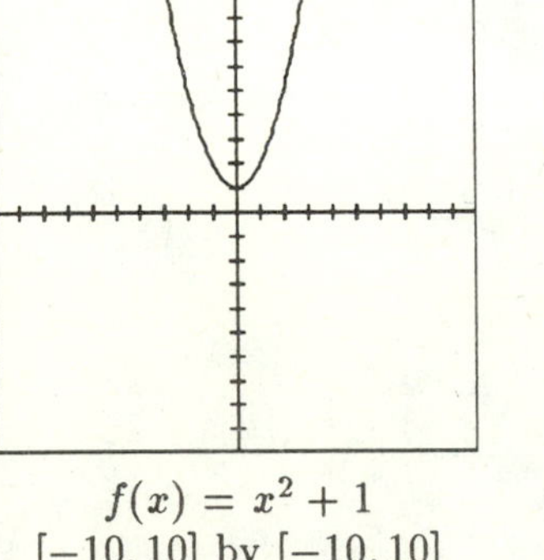

$f(x) = x^2 + 1$
$[-10, 10]$ by $[-10, 10]$

Figure 1.2.6

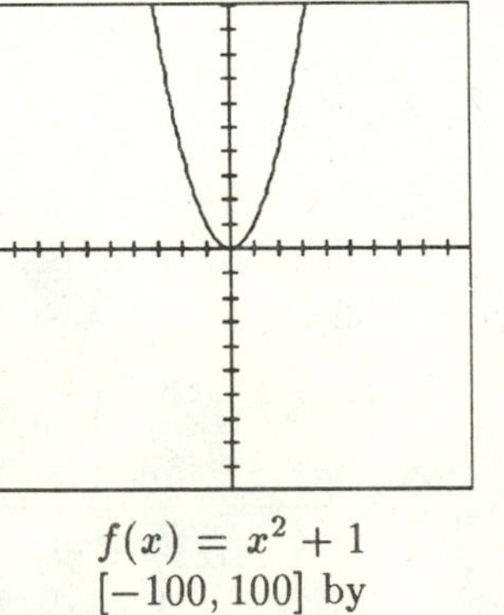

$f(x) = x^2 + 1$
$[-100, 100]$ by
$[-1000, 1000]$

Figure 1.2.7

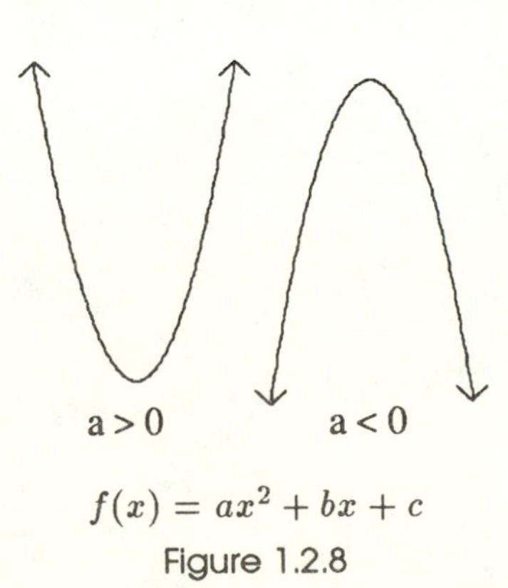

$f(x) = ax^2 + bx + c$

Figure 1.2.8

the range is the set of all real numbers greater than or equal to the smallest value of the function. If a is negative, the graph has a highest point and the range is the set of all real numbers less than or equal to the largest value of the function. The highest or lowest point is called the vertex of the parabola. In Section 1.7, we will confirm that the coordinates of the vertex are $\left(-\frac{b}{2a}, f\left(-\frac{b}{2a}\right)\right)$.

Next, the area of a sidewalk that surrounds a swimming pool will be found. A careful choice of viewing rectangles is required to obtain a complete graph of an algebraic representation of this area.

A rectangular swimming pool with dimensions 30 feet by 50 feet is surrounded by a walk of uniform width x as illustrated in Figure 1.2.9. The area of the sidewalk is the area of the outer rectangle minus the area of the inner rectangle. If A stands for the area of the sidewalk, then the following gives A as a function of x.

$$\begin{aligned} A(x) &= (30 + 2x)(50 + 2x) - (30)(50) \\ &= 4x^2 + 160x \end{aligned}$$

The function A is an algebraic representation of the sidewalk problem. In the next example, we draw a *complete* graph of this model.

• EXAMPLE 7: Let $A(x) = 4x^2 + 160x$.

(a) Draw a complete graph of the function A.
(b) What portion of the graph in (a) represents the problem situation?
(c) What are the domain and range of A?

SOLUTION:

(a) A complete graph of A must look like one of the two graphs in Figure 1.2.8. We know that $(0, 0)$ is one point of the graph of A since $A(0) = 0$. The graph of A in the standard viewing rectangle $[-10, 10]$ by $[-10, 10]$ gives very little information about A because most graphing utilities plot very few, if any, points of the graph of A in this viewing rectangle. If we look at some values of A for x near zero (Table 1), we see why this is so. Notice that the values of $A(x)$ in Table 1 are greater than 10 for $x \geq 0.5$, and the values of $A(x)$ are less than -10 for $x \leq -0.5$. Thus, most of the points of the graph of A that a graphing utility would try to produce lie outside the standard viewing rectangle $[-10, 10]$ by $[-10, 10]$. To see the graph

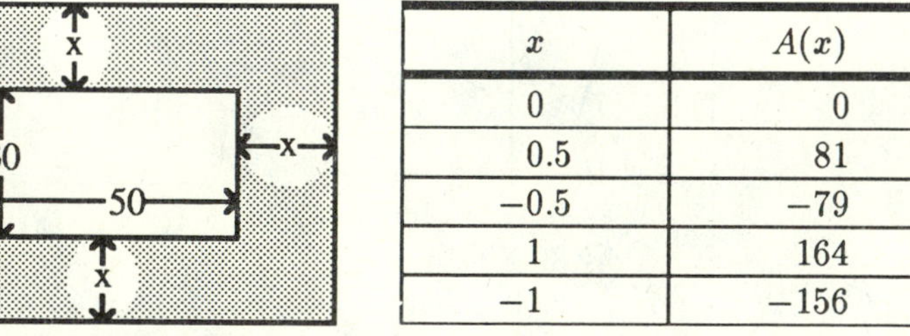

Figure 1.2.9

x	$A(x)$
0	0
0.5	81
−0.5	−79
1	164
−1	−156

Table 1

of A near $(0,0)$ we either have to choose L and R very close together or make B and T further apart. For example, the graph in $[-1,1]$ by $[-10,10]$ or in $[-10,10]$ by $[-100,100]$ gives more information about the behavior of A for x near 0 than does the standard viewing rectangle.

We obtain a complete graph of A by viewing the graph of A in the $[-80,80]$ by $[-2000,2000]$ viewing rectangle (Figure 1.2.10). We know this is a complete graph because it looks like the graph on the left in Figure 1.2.8. Notice that this graph has the same general shape as the graph in Example 6.

(b) If we are interested only in the values of x that could be possible widths of the sidewalk, then x must be positive. Thus, only the portion of the graph of A in Figure 1.2.10 to the right of the y-axis represents the sidewalk problem.

(c) The domain of A is the set of all real numbers. To find the range of A we need to determine the coordinates of the *vertex* (lowest point) of the graph of A. We can use the graph in Figure 1.2.10 to estimate that the coordinates of the vertex are $(-20,-1600)$. This estimate can be confirmed using the formula for the vertex given earlier in this section. The x-coordinate of the vertex is $-\frac{b}{2a}$, where a is the coefficient of x^2 and b is the coefficient of x. Thus, $-\frac{160}{8} = -20$ is the x-coordinate of the vertex. The y-coordinate of the vertex is $A(-20) = -1600$. Thus, the range is the set of all real numbers greater than or equal to -1600, or $y \geq -1600$. •

• EXAMPLE 8: Use a graph to approximate the width of the sidewalk in Figure 1.2.9 if its area is 600 square feet.

SOLUTION: If x is positive and $(x, A(x))$ a point on the graph in Figure 1.2.10, then x is a possible width of the sidewalk and $A(x)$ the corresponding area of the walk. Thus, Figure 1.2.11 is a graph of the problem situation. Because we want the area of the walk to be 600, we need to find a point on the graph of A with second coordinate 600. The intersection of the horizontal line $y = 600$ with the graph of A in Figure 1.2.11 gives a point with second coordinate 600. Let the first coordinate of this point be a. We can read from the graph in

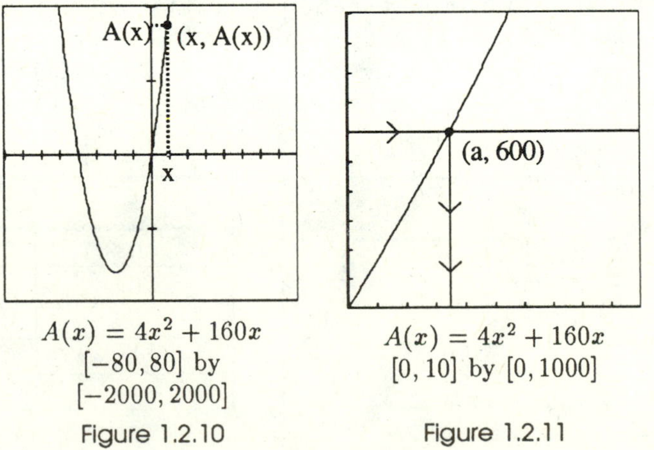

$A(x) = 4x^2 + 160x$
$[-80,80]$ by
$[-2000,2000]$
Figure 1.2.10

$A(x) = 4x^2 + 160x$
$[0,10]$ by $[0,1000]$
Figure 1.2.11

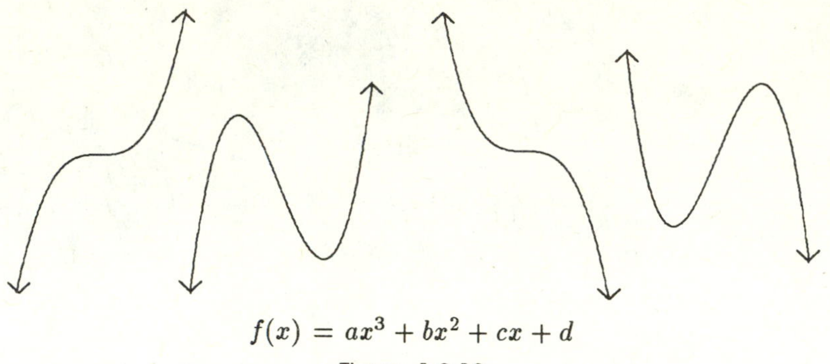

$f(x) = ax^3 + bx^2 + cx + d$

Figure 1.2.12

Figure 1.2.11 that the width of the sidewalk (the value of a) is approximately 3.5 feet. You can check that $A(3.5) = 609$. In Chapter 2 we review how to solve the quadratic equation $A(x) = 600$ algebraically and determine the exact solution. •

In the exercises you will be asked to find complete graphs of functions of the form $f(x) = ax^3 + bx^2 + cx + d$ $(a \neq 0)$. You may assume that a complete graph of such a function looks like one of the graphs in Figure 1.2.12. Each such function has both its domain and its range equal to the set of all real numbers.

• EXERCISES 1-2

1. Which of the graphs in Figure 1.2.13 are graphs of functions?

Use a graph to determine whether the relation is a function.

2. $y = x + 1$
3. $y = 2x - 4$
4. $y = \frac{2}{x}$
5. $y = x^2 - 2$
6. $y^2 = x$
7. $x = 3$
8. $y = -2$
9. $x + y = 0$
10. Let $f(x) = x^2 - 1$. Find $f(0)$, $f(1)$, $f(3)$, $f(-5)$, $f(t)$, $f(-t)$, $f\left(\frac{2}{t}\right)$, $f(a+h)$, and $\frac{f(a+h)-f(a)}{h}$.
11. Let $g(x) = \frac{1}{x+1}$. Find $g(0)$, $g(1)$, $g(3)$, $g(-5)$, $g(t)$, $g(-t)$, $g\left(\frac{2}{t}\right)$, $g(a+h)$, and $\frac{g(a+h)-g(a)}{h}$.
12. Consider the graph of the function $y = f(x)$ given in Figure 1.2.14. Estimate the indicated values.
    (a) $f(0)$, $f(-1)$, $f(-2)$, $f(4)$ (b) x if $f(x) = 0$ (c) x if $f(x) = 1$
    (d) x if $f(x) = -8$ (e) x if $f(x) = 8$
13. Consider the graph of the function $y = V(x)$ given in Figure 1.2.15. Estimate the indicated values.
    (a) $V(0)$, $V(-5)$, $V(4)$, $V(11)$ (b) x if $V(x) = 0$ (c) x if $V(x) = 100$
    (d) x if $V(x) = 1000$ (e) x if $V(x) = -100$

Draw the graph of each pair of functions in the same $[-10, 10]$ by $[-10, 10]$ viewing rectangle. Compare the graphs and explain why they are different. The caret symbol, ∧ , stands for exponenti-

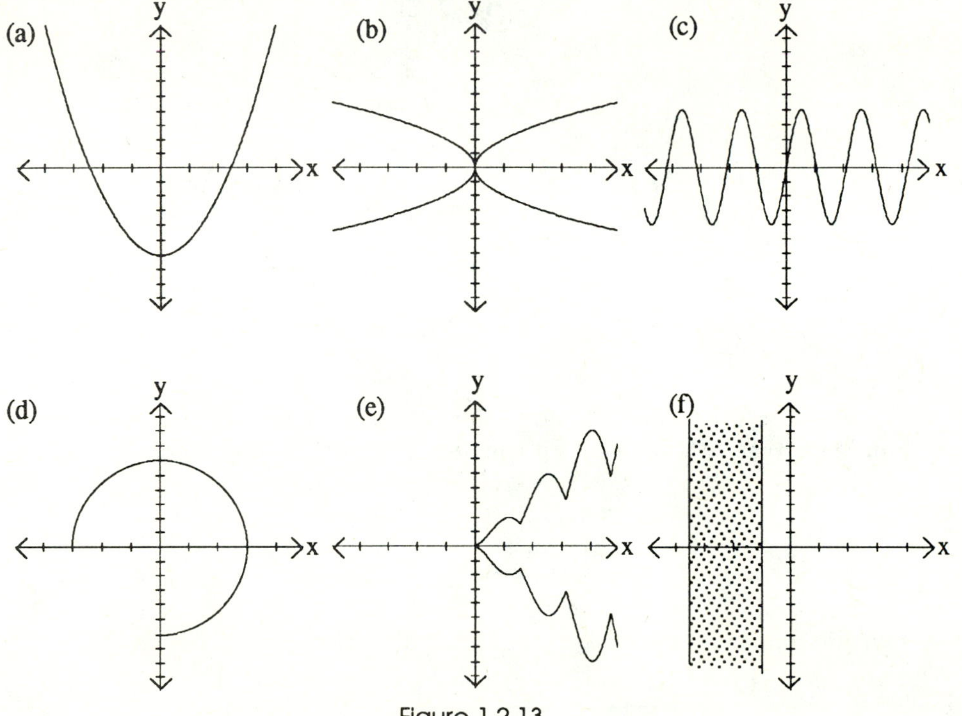

Figure 1.2.13

ation and the asterisk, $*$, for multiplication. If necessary, replace these symbols with your graphing utility's symbol for these operations.

14. (a) $y = x \wedge (1/2)$ (b) $y = x \wedge 1/2$
15. (a) $y = 1/(x-2)$ (b) $y = 1/x - 2$
16. (a) $y = 3 * x \wedge 2$ (b) $y = (3 * x) \wedge 2$

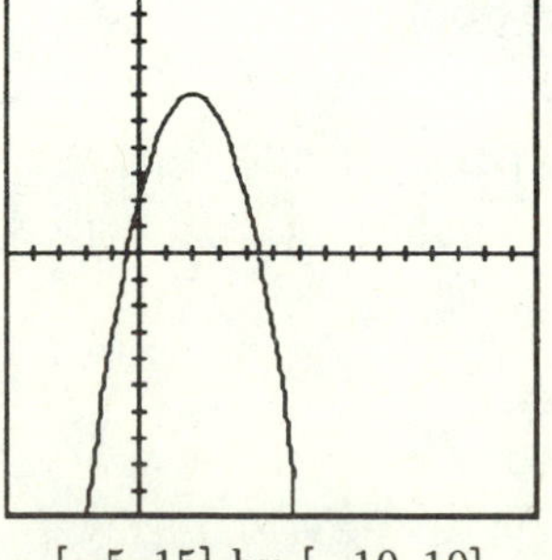

$[-5, 15]$ by $[-10, 10]$
Figure 1.2.14

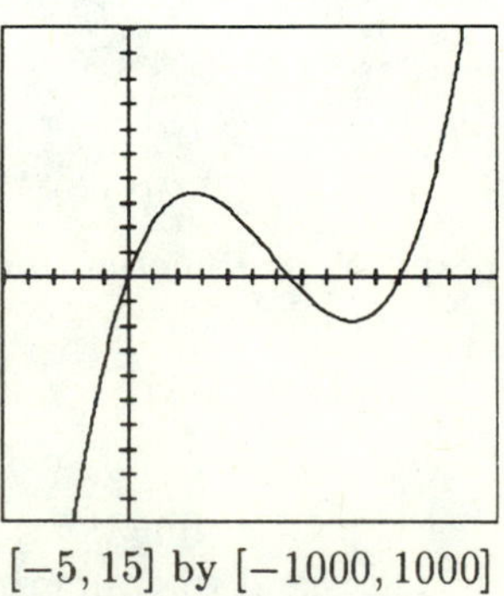

$[-5, 15]$ by $[-1000, 1000]$
Figure 1.2.15

17. Which of the following points lie in the viewing rectangle $[-2,3]$ by $[1,10]$?
    (a) $(0,0)$ (b) $(0,5)$ (c) $(5,0)$
    (d) $(3,2)$ (e) $(1,6)$ (f) $(1,-1)$
18. Choose a viewing rectangle $[L,R]$ by $[B,T]$ that includes all of the indicated points.
    (a) $(-9,10)$, $(2,8)$ and $(3,12)$
    (b) $(20,-3)$, $(13,56)$ and $(-11,2)$
19. Which of the following viewing rectangles gives the best complete graph of $f(x) = x^2 - 30x + 225$?
    (a) $[-5,5]$ by $[-5,5]$ (b) $[-10,10]$ by $[-10,10]$
    (c) $[15,30]$ by $[-50,50]$ (d) $[5,25]$ by $[-100,100]$
    (e) $[10,20]$ by $[-1000,1000]$
20. Which of the viewing rectangles gives the best complete graph of $f(x) = 20 - 30x + x^3$?
    (a) $[-10,10]$ by $[-10,10]$ (b) $[-50,50]$ by $[-10,10]$
    (c) $[-10,10]$ by $[-50,50]$ (d) $[0,10]$ by $[-100,100]$
    (e) $[-10,10]$ by $[-50,100]$

Draw a graph of the relation in several different viewing rectangles. Use this information to determine a complete graph. Determine whether the relation is a function.

21. $y = 1 + \log x$
22. $y = -\cos x$
23. $y = 3\sin x$
24. $y = 1 - \log x$

Draw a complete graph of each function. Explain why the graph is a complete graph.

25. $g(x) = \dfrac{x-4}{5}$
26. $f(x) = x^2 - 4$
27. $t(x) = 2x^2 - 3x + 5$
28. $f(x) = \left(\dfrac{x+3}{5}\right)^2 + 2x + 4$
29. $f(x) = x^3 - x + 1$
30. $h(x) = 4 - x - x^3$
31. Determine the domain and range of the function in Exercise 25.
32. Determine the domain and range of the function in Exercise 27. Hint: You need to find the coordinates of the vertex of the graph of t in order to determine the range of t.
33. Determine the domain and range of the function in Exercise 28.
34. Jerry runs at a constant speed of 4.25 mph.
    (a) Express the distance d that Jerry runs as a function of time t.
    (b) Draw a complete graph of the function $y = d(t)$.
    (c) Use the graph in (b) to estimate the time that it takes Jerry to run a 26-mile marathon.
35. A 4-inch-by-6.5-inch picture pasted on a cardboard sheet is surrounded by a region (border) of uniform width x.
    (a) Determine the area A of the border region as a function of x.
    (b) Draw a complete graph of the function $y = A(x)$ found in (a).
    (c) What are the domain and range of the function A?

(d) What values in the domain are possible widths?

(e) Use a graph to determine the width of the border if its area is 20 square inches.

36. Eighty feet of fencing is cut into two unequal lengths, and each piece is used to make a square enclosure. Let x be the perimeter of the smaller enclosure.

(a) Express the total area A of both enclosures as a function of x.

(b) Draw a complete graph of the function in (a).

(c) What are the domain and range of the function A? What numbers in the domain represent possible lengths of the smaller piece of fence?

(d) If $A = 300$ square feet use the graph in (b) to approximate the smaller length of fencing.

Draw a complete graph of each function. Record each series of graphs on the *same* coordinate system.

37. $y = x^2$, $y = (x-1)^2$, $y = (x-2)^2$, $y = (x+3)^2$.

38. $y = -1 + x^2$, $y = 2 + x^2$, $y = 4 - x^2$, $y = x^2 - 4$.

39. Consider the two relations in Example 1. Determine an equation for each with domain that contains the domain of the given relation.

40. Consider the relation $R = \{(-2,-1), (-1,-4), (0,-5), (1,-4)\}$.

(a) Graph this relation.

(b) Determine an equation with domain that contains the domain of this relation.

41. Use the method in Example 5 to show that the relation $y = 3x - 1$ is a function.

42. Use the method in Example 5 to show that the relation $y = 2x^2 - 1$ is a function.

1.3 Graphs and Symmetry

In this section we describe what it means for a graph to be symmetric with respect to the x-axis, the y-axis, or the origin. Both algebraic and geometric tests are discussed. We use symmetry properties to obtain graphs without a graphing utility and then use a graphing utility to verify our computations.

Some graphs have symmetry properties that can help us better understand the relation determined by the graph. Additionally, certain relations have symmetry properties that can be used to speed up the drawing of their graphs. Points P and Q are said to be **symmetric with respect to the line L** if L is the perpendicular bisector of the line segment PQ (Figure 1.3.1). For example, the points $(-3, y)$ and $(3, y)$ are symmetric with respect to the y-axis for all values of y; the points $(x, 2)$ and $(x, -2)$ are symmetric with respect to the x-axis for all values of x (Figure 1.3.2).

Points P and Q are said to be **symmetric with respect to the point M** if M is the midpoint of the line segment PQ (Figure 1.3.3). The points $(3, 4)$ and $(-3, -4)$ are symmetric with respect to the origin (Figure 1.3.4).

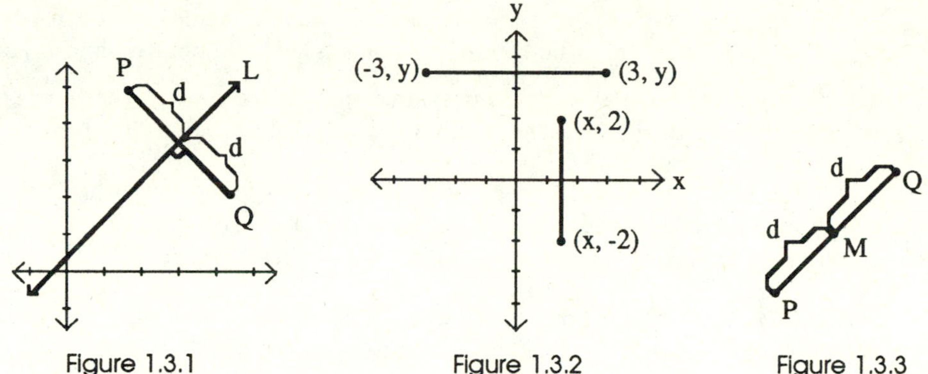

Figure 1.3.1 Figure 1.3.2 Figure 1.3.3

Let (a,b) be a point in the coordinate plane. The point $(-a,b)$ is symmetric to (a,b) with respect to the y-axis (Figure 1.3.5). The point $(a,-b)$ is symmetric to (a,b) with respect to the x-axis. The point $(-a,-b)$ is symmetric to (a,b) with respect to the origin. We use this information to describe what it means for a graph to be symmetric with respect to the y-axis, x-axis, and origin.

Suppose we fold the coordinate plane along the y-axis. If the part of the graph that lies on the left of the y-axis coincides with the part that lies on the right, the graph is *symmetric with respect to the y-axis*. More precisely, we have the following algebraic definition.

• DEFINITION A graph is said to be **symmetric with respect to the y-axis** if the point $(-a,b)$ is on the graph whenever the point (a,b) is on the graph. •

In general, we can use the following geometric test to determine symmetry of a graph with respect to a line.

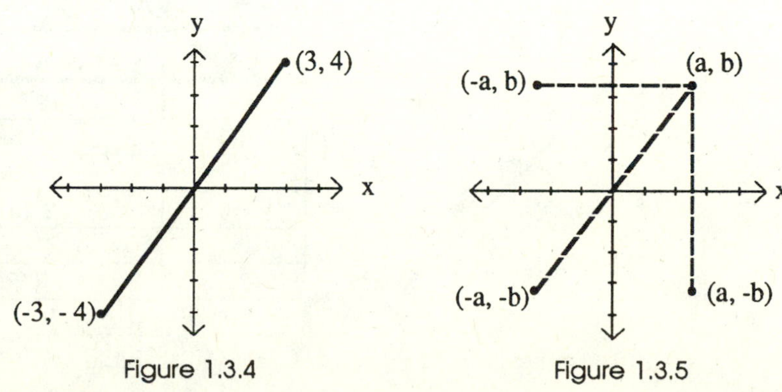

Figure 1.3.4 Figure 1.3.5

- **Geometric Test for Line Symmetry.** A graph is symmetric with respect to the line L if the part of the graph on one side of L coincides with the part on the other side of L when the coordinate plane is folded along the line L (Figure 1.3.6).

If we determine that a graph is symmetric with respect to the y-axis, we reduce the number of computations needed to obtain a complete graph.

- EXAMPLE 1: Show that the graph of $f(x) = x^2$ is symmetric with respect to the y-axis and draw a complete graph of the function without using a graphing utility. Then, use the graph to determine the domain and range of f. Check the graph with a graphing utility.

SOLUTION: Notice that both $(2,4)$ and $(-2,4)$ are on the graph of $f(x) = x^2$ because $f(2) = 4$ and $f(-2) = 4$. Suppose that (a,b) is on the graph of $f(x) = x^2$. Then, $f(a) = b$ or $a^2 = b$. Now, observe that $(-a,b)$ is also on the graph of $f(x) = x^2$ because $f(-a) = b$. This follows because $f(-a) = (-a)^2 = a^2 = b = f(a)$. Thus, the point $(-a,b)$ is on the graph of $f(x) = x^2$ whenever the point (a,b) is on this graph. The graph is symmetric with respect to the y-axis. Therefore, we need only compute the values of $f(x)$ corresponding to nonnegative values of x. Table 1 gives some of these values, and Table 2 gives the corresponding ones obtained by using the symmetry of the graph.

These two tables indicate that the values of $f(x)$ steadily increase if x is positive and increasing, or if x is negative and decreasing. The graph of the points determined by Tables 1 and 2 and a complete graph of f are given in Figure 1.3.7.

We can infer from this graph that the domain of f (values of x) is the set of all real numbers, and the range (values of y) is the set of all nonnegative real numbers. (Why?) The computer generated graph in Figure 1.3.8 confirms our computations. •

Notice if we fold the coordinate plane along the y-axis in Figure 1.3.7 or 1.3.8, then the part of the graph that lies to the left of the y-axis coincides with the part that lies to the right of the y-axis. We can use the symmetry with respect to the y-axis to help draw the graph. We first draw the portion of the graph on one side of the y-axis and then use the folding symmetry with respect to the y-axis to draw the portion on the other side of the

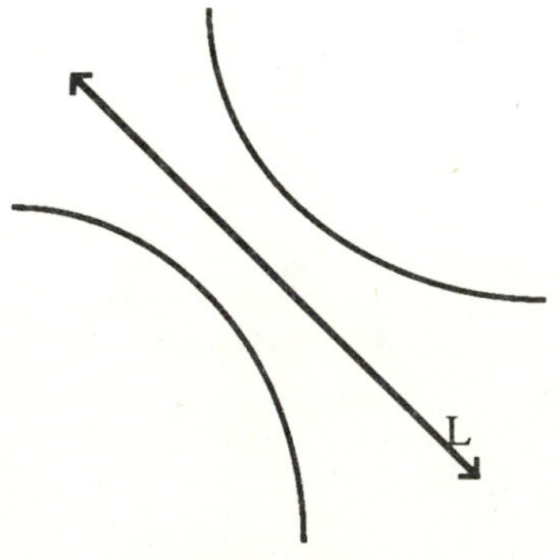

Figure 1.3.6

Computed Values	
x	$f(x)$
0	0
0.5	0.25
1	1
1.5	2.25
2	4
2.5	6.25
3	9

Table 1

y-axis. This geometric folding principle gives a complete graph with half the calculations. Symmetry helps both in computing values and in drawing the graph.

The definition of symmetry with respect to the x-axis is similar to the definition of symmetry with respect to the y-axis. Suppose we fold the coordinate plane along the x-axis. If the parts of the graph on opposite sides of the x-axis coincide, the graph is *symmetric with respect to the x-axis.* More precisely, we have the following algebraic definition.

• DEFINITION A graph is said to be **symmetric with respect to the x-axis** if the point $(a, -b)$ is on the graph whenever the point (a, b) is on the graph. •

• EXAMPLE 2: Show that the graph of $y^2 = x$ is symmetric with respect to the x-axis and draw a complete graph of $R = \{(x, y)|y^2 = x\}$ determined by the equation R. Then, use the graph to determine the domain and range of R. Check the graph with a graphing utility.

SOLUTION: Notice that both $(4, 2)$ and $(4, -2)$ are on the graph of R because $2^2 = 4$ and $(-2)^2 = 4$. Suppose that (a, b) is on the graph of R. Then, $b^2 = a$. Now, observe that $(a, -b)$ is also on the graph of the equation R because $(-b)^2 = b^2 = a$. Thus, the point $(a, -b)$ is on the graph of R whenever the point (a, b) is on this graph. The graph is symmetric with respect to the x-axis.

Each positive value of x is paired with two values of y; one is positive, and the other is negative. Therefore, two points on the graph are determined. One point is above the x-axis and the other, symmetric point is below the x-axis. Table 3 gives some values of x and the corresponding values of y rounded to tenths. This table indicates that for x positive and increasing the positive values of y steadily increase while the negative values of y steadily decrease. The graph of the points determined by Table 3 and a complete graph of R are given in Figure 1.3.9. We infer from this graph that the domain of R is the set of all nonnegative real numbers, and the range is the set of all real numbers.

Values from Symmetry	
x	$f(x)$
0	0
-0.5	0.25
-1	1
-1.5	2.25
-2	4
-2.5	6.25
-3	9

Table 2

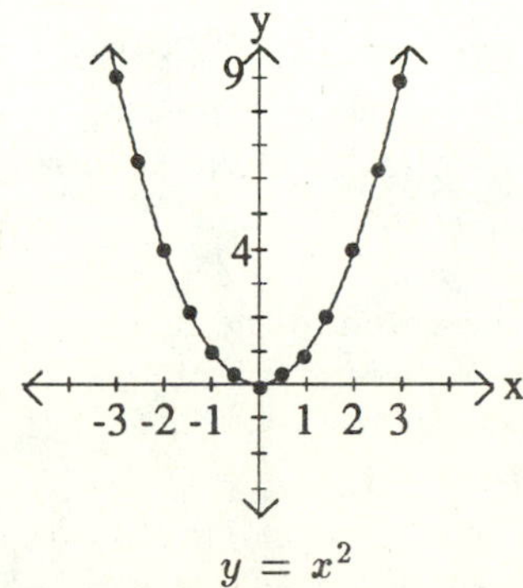

$y = x^2$

Figure 1.3.7

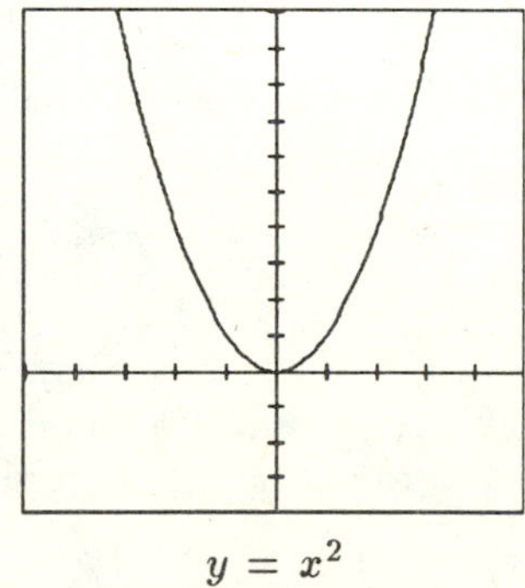
$y = x^2$

$[-5, 5]$ by $[-4, 10]$

Figure 1.3.8

x	y
0.5	± 0.7
1	± 1
1.5	± 1.2
2	± 1.4
2.5	± 1.6
3	± 1.7
4	± 2
5	± 2.2

Table 3

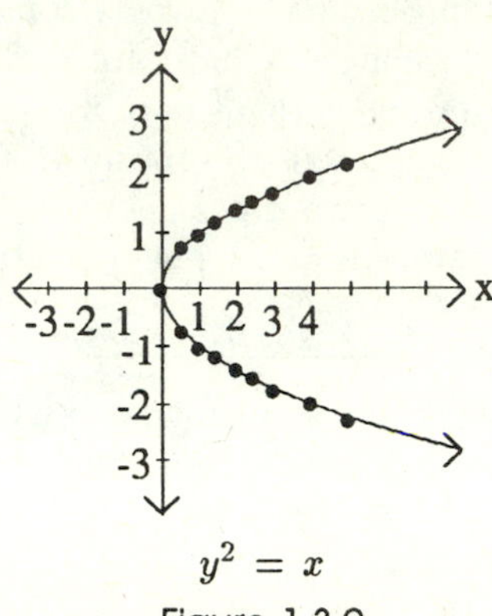

$y^2 = x$

Figure 1.3.9

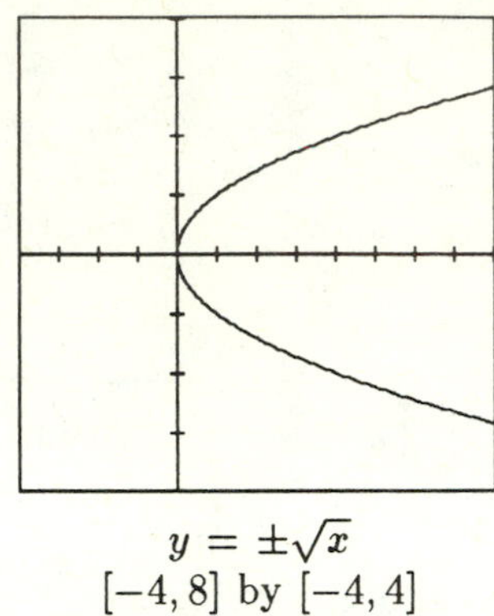

$y = \pm\sqrt{x}$

$[-4, 8]$ by $[-4, 4]$

Figure 1.3.10

We *cannot* verify directly with a function graphing utility that the graph in Figure 1.3.9 is correct because it is not the graph of a function. (Why?) However, if we first solve for y in terms of x to obtain $y = \pm\sqrt{x}$, then we can use a function graphing utility. By drawing complete graphs of $y = \sqrt{x}$ *and* $y = -\sqrt{x}$ in the same viewing rectangle, we obtain a complete graph of $y^2 = x$ (Figure 1.3.10). •

If we fold the coordinate plane along the x-axis in Figure 1.3.9 or 1.3.10, then the part of the graph above the x-axis would coincide with the part that lies below the x-axis. This is another way to describe symmetry with respect to the x-axis.

Suppose that a graph is symmetric with respect to a line L. We can draw the portion of the graph on one side of L, and then use folding symmetry with respect to the line L to draw the portion of the graph on the other side of L. The graphs of the equations in Examples 1 and 2 were symmetric with respect to a line. In the next example we draw a complete graph of an equation that is symmetric with respect to the origin.

• DEFINITION A graph is said to be **symmetric with respect to the origin** if the point $(-a, -b)$ is on the graph whenever the point (a, b) is on the graph. •

• **Geometric Test for Symmetry with Respect to the Origin.** A graph is symmetric with respect to the origin if the graph coincides with itself when the coordinate plane is folded along the x-axis followed by folding along the y-axis. The order of folding may be interchanged.

• EXAMPLE 3: Show that the graph of the function $f(x) = x^3 + x$ is symmetric with respect to the origin and draw a complete graph of f. Then use the graph to determine the domain and the range of the function.

SOLUTION: If (a, b) is on the graph of $f(x) = x^3 + x$, then $b = f(a) = a^3 + a$. Notice that $f(-a) = (-a)^3 + (-a) = -a^3 - a = -(a^3 + a) = -b$. This means that the point $(-a, -b)$ is also on the graph of $f(x) = x^3 + x$. Thus, the graph is symmetric with respect to the origin,

Computed Values	
x	$f(x)$
0	0
0.5	0.6
1	2
1.5	4.9
2	10
2.5	18.1
3	30

Table 4

Values from Symmetry	
x	$f(x)$
0	0
−0.5	−0.6
−1	−2
−1.5	−4.9
−2	−10
−2.5	−18.1
−3	−30

Table 5

and we need only compute values of f corresponding to nonnegative values of x. Tables 4 and 5 give some values of x and the corresponding values of f rounded to tenths.

These two tables indicate that the values of f steadily increase for x positive and increasing, and the values of f steadily decrease for x negative and decreasing. The graph of the points determined by Tables 4 and 5 and a complete graph of f are given in Figure 1.3.11. We can see from this graph that both the domain and the range of f are the set of all real numbers. The computer drawn graph of f in Figure 1.3.12 confirms our computations. •

The graph in Figure 1.3.11 or 1.3.12 can be made to coincide with itself using geometric folding along the axes. This can be accomplished in two distinct ways. Fold the coordinate plane along the x-axis and then along the y-axis or vice versa (Figure 1.3.13). This two-step geometric folding process illustrates the geometric test for symmetry with respect to the origin.

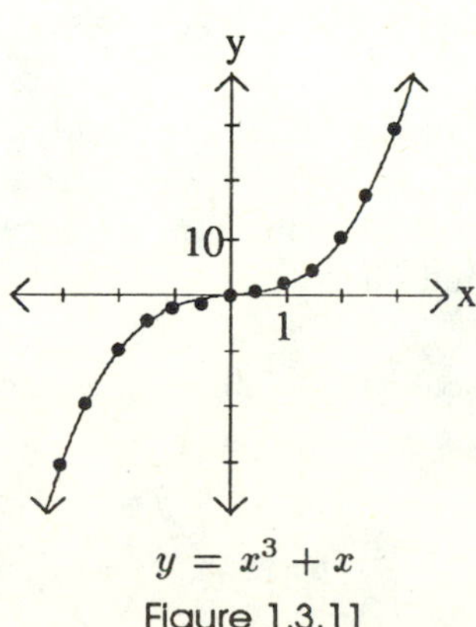

$y = x^3 + x$

Figure 1.3.11

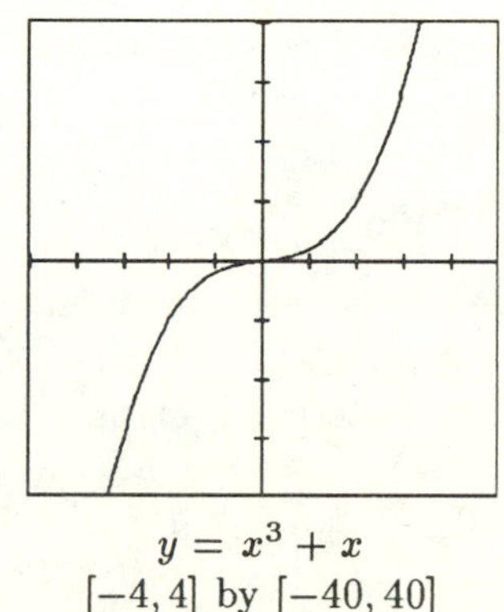

$y = x^3 + x$

$[-4, 4]$ by $[-40, 40]$

Figure 1.3.12

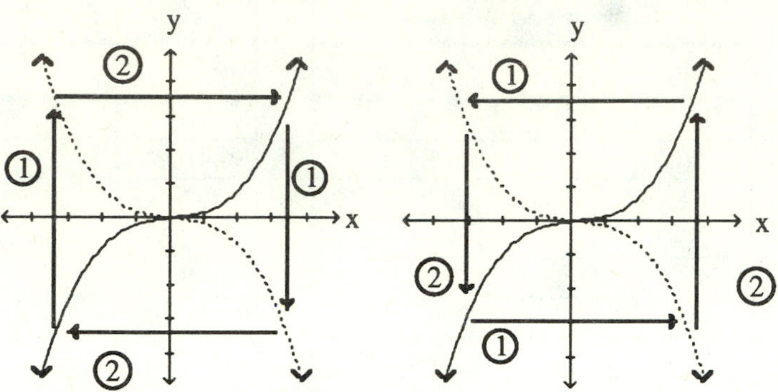

Figure 1.3.13

Folding the coordinate plane first along the x-axis and then along the y-axis has the same effect on the coordinate plane as first folding along the y-axis and then along the x-axis. (Try it!) These two folding processes satisfy the commutative property; the order in which these two operations are performed does not matter.

Next, we want to determine an algebraic test for symmetry. In Example 1 we showed that $y = x^2$ was symmetric with respect to the y-axis. Notice if x is replaced by $-x$ in the equation $y = x^2$ the resulting equation is the same as the original equation.

$$y = (-x)^2$$
$$y = x^2$$

This is the algebraic test for symmetry with respect to the y-axis. Similar observations about symmetry with respect to the x-axis and the origin lead to the following algebraic test for symmetry.

- **Algebraic Test for Symmetry.** The graph of an equation is symmetric with respect to the

  (a) y-axis, if and only if replacement of x by $-x$ everywhere in the equation leads to the same equation.

  (b) x-axis, if and only if replacement of y by $-y$ everywhere in the equation leads to the same equation.

  (c) origin, if and only if replacement of x by $-x$ and y by $-y$ everywhere in the equation leads to the same equation.

- EXAMPLE 4: Determine whether the graphs of the following equations are symmetric with respect to the x-axis, the y-axis, or the origin.

  (a) $x^4 - y^2 = 9$ (b) $y^2 + y = x$

  (c) $x^2 + y^3 = 1$ (d) $y = x^3$

SOLUTION:

(a) Replace y by $-y$ in the original equation $x^4 - y^2 = 9$.

$$x^4 - y^2 = 9$$
$$x^4 - (-y)^2 = 9$$
$$x^4 - y^2 = 9$$

The equation is unchanged. Thus, the graph of $x^4 - y^2 = 9$ is symmetric with respect to the x-axis. Next, replace x by $-x$ in the original equation $x^4 - y^2 = 9$.

$$x^4 - y^2 = 9$$
$$(-x)^4 - y^2 = 9$$
$$x^4 - y^2 = 9$$

Again, the equation is unchanged. Thus, this graph is also symmetric with respect to the y-axis. Finally, replace x by $-x$ and y by $-y$ in the original equation $x^4 - y^2 = 9$.

$$x^4 - y^2 = 9$$
$$(-x)^4 - (-y)^2 = 9$$
$$x^4 - y^2 = 9$$

The equation is unchanged. This graph is also symmetric with respect to the origin.

In the exercises you will be asked to show that any graph symmetric with respect to both the x-axis and the y-axis is also symmetric with respect to the origin.

(b) Replace y by $-y$ in the equation $y^2 + y = x$.

$$y^2 + y = x$$
$$(-y)^2 + (-y) = x$$
$$y^2 - y = x$$

The new equation is different from the original equation. The equation also changes if we replace x by $-x$ or x and y by $-x$ and $-y$, respectively. Thus the graph of this equation is not symmetric with respect to the x-axis, the y-axis, or the origin.

(c) The equation $x^2 + y^3 = 1$ does not change if x is replaced by $-x$. Thus, the graph is symmetric with respect to the y-axis. However, the equation does change if y is replaced by $-y$, or if x and y are replaced by $-x$ and $-y$, respectively. Thus, the graph of $x^2 + y^3 = 1$ is not symmetric with respect to the x-axis or the origin.

(d) Replacing x by $-x$ or y by $-y$ in this equation shows that this graph is not symmetric with respect to the x-axis or y-axis. Suppose we replace x and y by $-x$ and $-y$, respectively.

$$\begin{aligned} y &= x^3 \\ -y &= (-x)^3 \\ -y &= -x^3 \\ y &= x^3 \end{aligned}$$

The equation is unchanged. Thus, the graph of the function $y = x^3$ is symmetric with respect to the origin. A computer drawn graph of $y = x^3$ could be used to confirm this result. •

In the Exercises we will ask you to use a function graphing utility to confirm the graphs of the relations of Example 4(a,b).

• EXAMPLE 5: Show that the graph of $y = |x|$ is symmetric with respect to the y-axis.

SOLUTION: This equation is not changed if x is replaced by $-x$ because $|-x| = |x|$. Thus, the graph of $y = |x|$ is symmetric with respect to the y-axis. •

• EXAMPLE 6: Draw a complete graph of $f(x) = |x|$.

SOLUTION: Let $y = f(x)$. In Example 5 we showed that this graph is symmetric with respect to the y-axis. Thus, we need only to compute values of y for nonnegative values of x. Table 6 gives the values of y corresponding to some nonnegative values of x, and Table 7 gives the values obtained using symmetry.

These tables indicate that the values of y steadily increase if x is positive and increasing or if x is negative and decreasing. A complete graph is given in Figure 1.3.14. If we fold the coordinate plane in Figure 1.3.14 along the y-axis, the parts of the graph on either side of the y-axis coincide. We can first draw the part of the graph to the right of the y-axis

Computed Values	
x	y
0	0
5	5
10	10
15	15

Table 6

Values from Symmetry	
x	y
0	0
-5	5
-10	10
-15	15

Table 7

and then obtain the rest using folding symmetry. A computer drawn graph of $f(x) = |x|$ in Figure 1.3.15 confirms our computation. Most computers require that $f(x) = |x|$ be entered as $f(x) = \text{ABS}(x)$. •

• EXAMPLE 7: Draw a complete graph of $x = |y|$ and show that it is symmetric with respect to the x-axis. Compare the domain and range of $y = |x|$ and $x = |y|$.

SOLUTION: The equation $x = |y|$ is unchanged if y is replaced by $-y$. Thus, the graph of $x = |y|$ is symmetric with respect to the x-axis. There are no points on the graph of $x = |y|$ with x negative. (Why?) Each positive value of x is paired with two values of y that produce a pair of points symmetric with respect to the x-axis. For example, if $x = 3$, then $y = 3$ or $y = -3$. A complete graph of $x = |y|$ is shown in Figure 1.3.16. From Figure 1.3.14, we can see that the domain of $y = |x|$ is the set of all real numbers, and its range is the set of all real numbers greater than or equal to zero. By contrast, Figure 1.3.16 shows that the domain of $x = |y|$ is the set of all real numbers greater than or equal to zero, and its range is the set of all real numbers. The domain and range of $x = |y|$ are obtained by interchanging the domain and range of $y = |x|$. Notice that if we interchange x and y in the equation $y = |x|$, we obtain the equation $x = |y|$. •

Graphs that involve absolute value can have a line of symmetry that is parallel to the x-axis or the y-axis. The graph in the next example is symmetric with respect to the vertical line $x = 3$.

• EXAMPLE 8: Draw a complete graph of $f(x) = |x - 3|$. Give an equation for the line of symmetry.

SOLUTION: First, $f(3) = |3 - 3| = |0| = 0$ and the corresponding point $(3, 0)$ is the lowest point, or vertex, of the V-shaped absolute value graph (Figure 1.3.17). (Why?) Notice that $f(2) = |2 - 3| = |-1| = 1$ and $f(4) = |4 - 3| = |1| = 1$. The corresponding points on the graph, $(2, 1)$ and $(4, 1)$, are symmetric with respect to the vertical line $x = 3$.

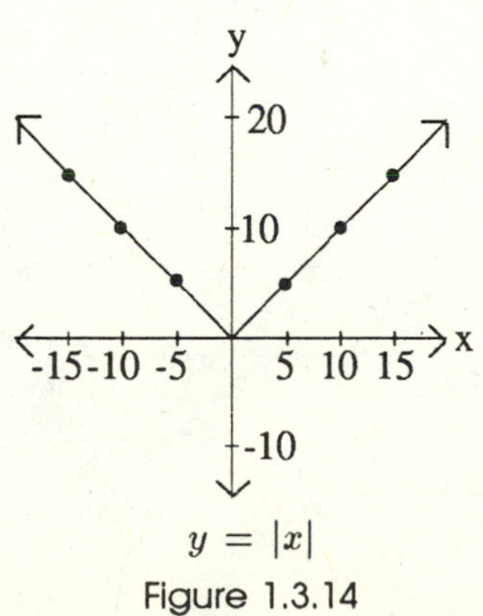

$y = |x|$

Figure 1.3.14

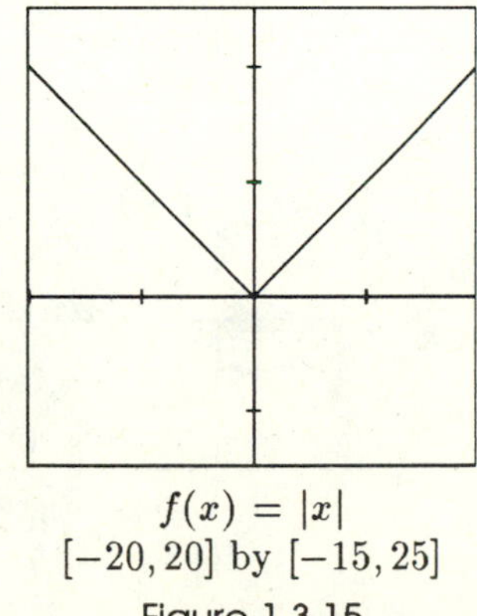
$f(x) = |x|$

$[-20, 20]$ by $[-15, 25]$

Figure 1.3.15

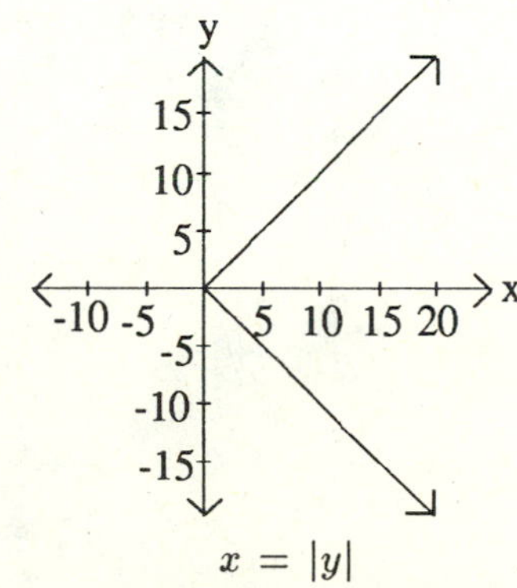

$x = |y|$

Figure 1.3.16

Similarly, $f(1) = 2$ and $f(5) = 2$. Again, the corresponding points on the graph, $(1,2)$ and $(5,2)$, are symmetric with respect to the vertical line $x = 3$. The value of f is the same for any two values of x equally distant from 3. These two values of x produce a pair of points on the graph of $f(x) = |x - 3|$ that are symmetric with respect to the line $x = 3$. Thus, the vertical line $x = 3$ is a line of symmetry for the graph of $f(x) = |x - 3|$.

The computer drawn graph in Figure 1.3.18 confirms our analysis and is a complete graph. Notice if we fold the coordinate plane along the line $x = 3$, then the two parts of the graph on opposite sides of this line coincide. •

We will see that the graphs of $y = f(x)$ and $y = f(x - a)$ are always closely related. In the next example, the line of symmetry is a horizontal line.

• EXAMPLE 9: Draw a complete graph of $x = |y+3|$. Give an equation for the line of symmetry.

SOLUTION: This time it is easier to assign values to y and compute the corresponding values of x. If $y = -3$, then $x = 0$. The corresponding point $(0, -3)$ is the vertex of the V-shaped graph in Figure 1.3.19. (Why?) Notice that $x = 1$ for both $y = -2$ and $y = -4$. The corresponding points, $(1, -2)$ and $(1, -4)$, are symmetric with respect to the horizontal line $y = -3$. Similarly, $x = 2$ for both $y = -1$ and $y = -5$. Again the corresponding points, $(2, -1)$ and $(2, -5)$, are symmetric with respect to the line $y = -3$. The graph in Figure 1.3.19 is complete and is symmetric with respect to the horizontal line $y = -3$. •

In Chapter 5, we will see that $f(x) = \sin x$ and $g(x) = \cos x$ have the same domain and range. The domain of each function is the set of all real numbers, and the range of each function is $-1 \leq y \leq 1$. The graphs of f and g in Figures 1.3.20 and 1.3.21 will be shown to be complete graphs.

• EXAMPLE 10: Determine whether the graphs of $f(x) = \sin x$ and $g(x) = \cos x$ are symmetric with respect to the x-axis, the y-axis, or the origin.

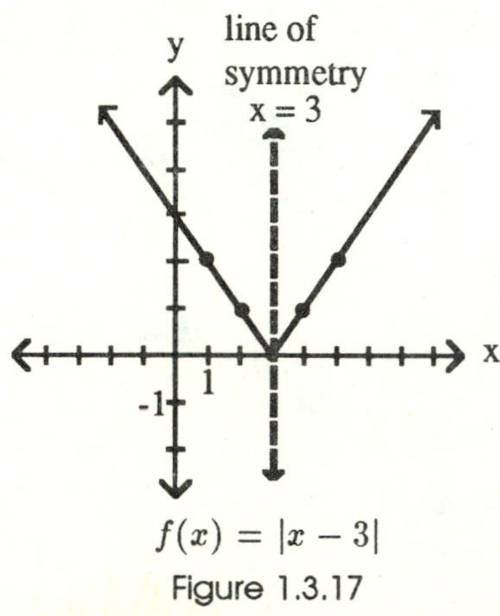

$f(x) = |x - 3|$

Figure 1.3.17

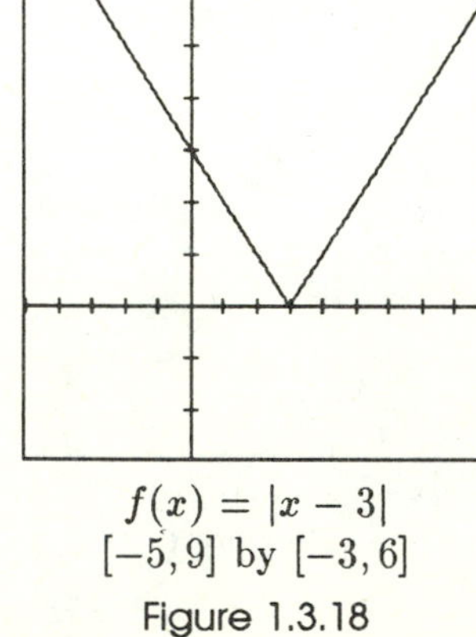

$f(x) = |x - 3|$

$[-5, 9]$ by $[-3, 6]$

Figure 1.3.18

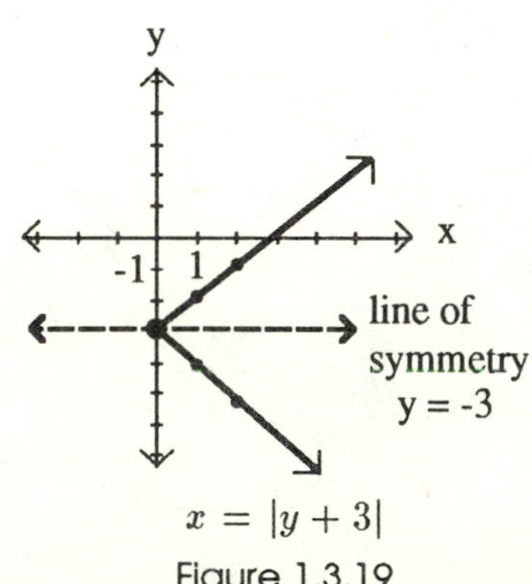

$x = |y + 3|$

Figure 1.3.19

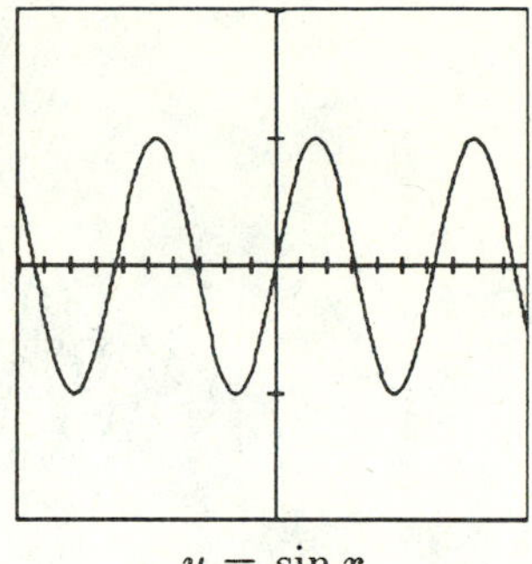

$y = \sin x$
$[-10, 10]$ by $[-2, 2]$
Figure 1.3.20

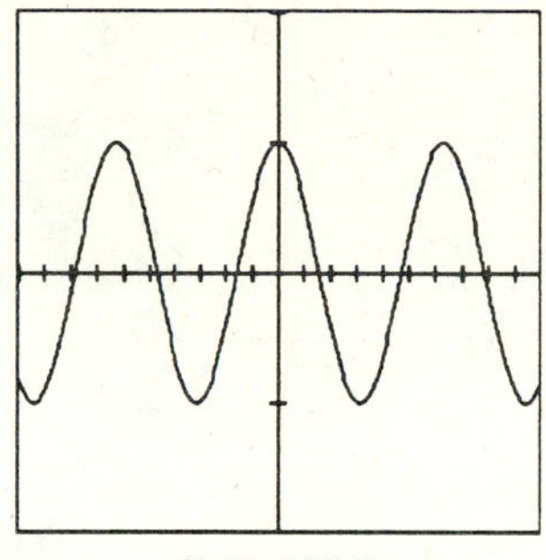

$y = \cos x$
$[-10, 10]$ by $[-2, 2]$
Figure 1.3.21

SOLUTION: The graph of $f(x) = \sin x$ given in Figure 1.3.20 is *not* symmetric with respect to the x-axis or y-axis. However, we can use folding symmetry to see that the graph of $f(x) = \sin x$ is symmetric with respect to the origin. The graph of $g(x) = \cos x$ given in Figure 1.3.21 is symmetric with respect to the y-axis. It is *not* symmetric with respect to the x-axis or the origin. •

EXERCISES 1-3

1. Let $(2, 3)$ be on a graph. Determine a second point on the graph if we know that the graph is symmetric with respect to

   (a) the x-axis. (b) the y-axis. (c) the origin.

2. Let $(-1, 4)$ be on a graph. Determine a second point on the graph if we know that the graph is symmetric with respect to

   (a) the x-axis. (b) the y-axis. (c) the origin.

3. Let $(a, -b)$ be on a graph. Determine a second point on the graph if we know that the graph is symmetric with respect to

   (a) the x-axis. (b) the y-axis. (c) the origin.

4. Determine whether the graphs in Figure 1.3.22 are symmetric with respect to the x-axis, the y-axis, or the origin. Justify your answers.

5. Determine whether the graphs in Figure 1.3.23 are symmetric with respect to the x-axis, y-axis, or the origin. Justify your answers.

Draw a complete graph of each equation. If possible, use symmetry. Then, use the graph to determine the domain and range of the equation.

6. $y = 4 - x^2$
7. $x^2 - y + 6 = 0$
8. $y = (x + 3)^2$
9. $y = x^3 - x$
10. $y = 3 \sin x$
11. $y = -2 \cos x$

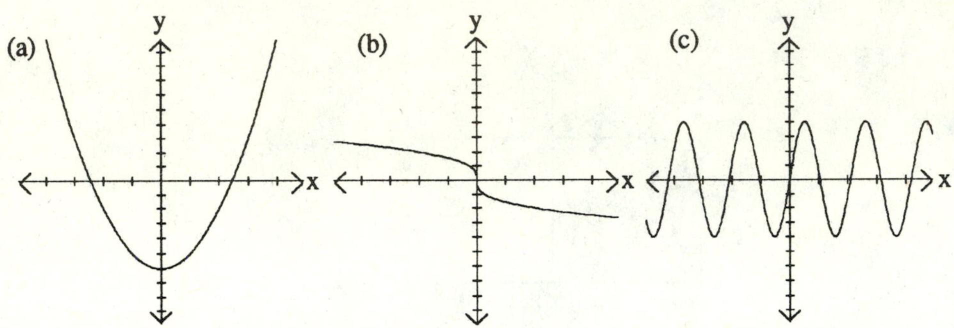

Figure 1.3.22

12. Complete the graph shown in Figure 1.3.24, if a complete graph is symmetric with respect to
    (a) the x-axis only. (b) the y-axis only.
    (c) the origin only. (d) the x-axis and the origin.
13. Complete the graph shown in Figure 1.3.25 if a complete graph is symmetric with respect to
    (a) the x-axis only. (b) the y-axis only.
    (c) the origin only. (d) the y-axis and the origin.

Determine whether the graph of the equation is symmetric with respect to the x-axis, y-axis or the origin.

14. $y = x^2 + 1$
15. $y^2 = x + 1$
16. $y^3 = x$
17. $x^2 y = 1$
18. $x^2 - y^4 = 8$
19. $y = (x - 2)^2$

Draw a complete graph of each equation and specify its domain and range. Give an equation for any line of symmetry and indicate which equations define y as a function of x.

20. $y = |x + 2|$
21. $y = |x - 5|$
22. $|y + 1| = x$
23. $|y - 3| = x$
24. $y = |3 - x|$
25. $y = |x| + 2$

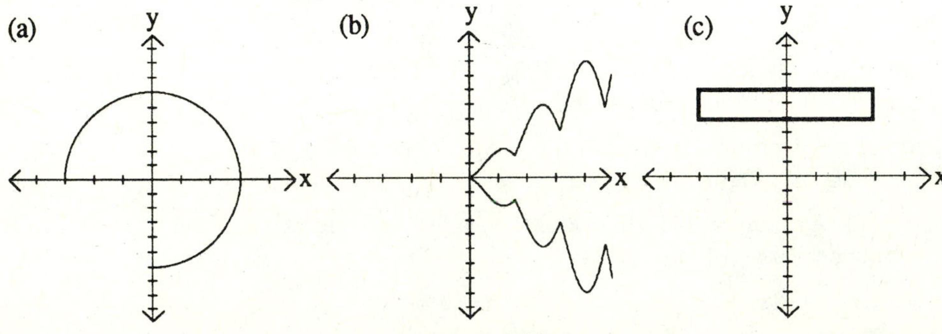

Figure 1.3.23

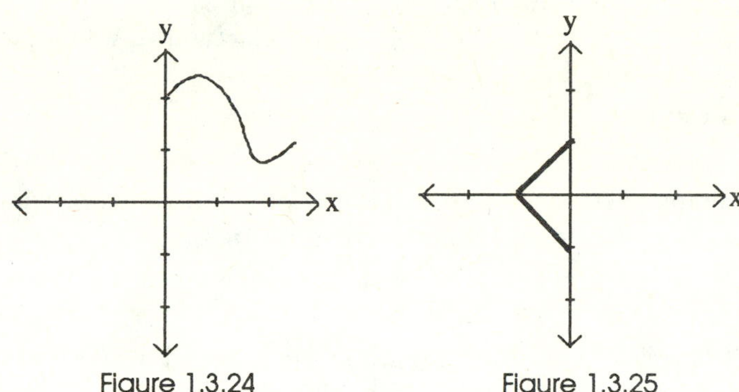

Figure 1.3.24 Figure 1.3.25

26. $y = |x| - 3$ 27. $|y| + 1 = x$ 28. $|y| - 3 = x$

Use computer drawn graphs to determine whether the graphs are symmetric with respect to the x-axis, the y-axis or the origin.

29. $y = -\sin x$ 30. $y = 2\cos x$

31. Explain how you would use a function graphing utility to draw a complete graph of $y^2 = 2x^2+1$.

32. A hat manufacturer determines the annual cost of making x hats to be \$7 per hat plus \$83,000 in fixed overhead costs.
    (a) Write the total cost C as a function of the number of hats made.
    (b) Draw a complete graph of the function in (a).
    (c) What values of x make sense in this problem situation?
    (d) Use the graph in (b) to determine the total cost of manufacturing 150 hats.

33. A three-sided rectangular fence is constructed against a barn. Let x be the length of the fence parallel to the barn using 128 feet of fencing material.
    (a) Write the area enclosed by the fence as a function of x.
    (b) What are the domain and range of the algebraic representation in (a)?
    (c) Draw a complete graph of the algebraic representation in (a).
    (d) What values of x make sense in this problem situation? Draw a graph of the problem situation.
    (e) What is the largest area that can be enclosed by the fence, and what are the dimensions of this maximum enclosure?

34. The hat manufacturer in Exercise 32 receives \$12.25 for each hat sold. Assume x hats are sold in one year.
    (a) Write the the total revenue R as a function of x.
    (b) What are the domain and range of the function in (a)?
    (c) Draw a complete graph of the function in (a).
    (d) Use the graph in (c) to determine the number of hats that are sold if the total revenue is \$97,853.

35. Consider the hat manufacturer in Exercises 32 and 34.
   (a) Write the total annual profit P of the manufacturer as a function of the number of hats made and sold.
   (b) Draw a complete graph of the function in (a).
   (c) Use the graph in (b) to determine the annual profit if 23,500 hats are made and sold.
36. Show that a graph possessing any two of the following three symmetries must possess the third: symmetry with respect to the x-axis, y-axis, or origin.
37. Draw a complete graph of $y = (x - 2)^2$. Is the graph symmetric about some line? If so, what is an equation of the line of symmetry?
38. A function f is called **even** if its graph is symmetric with respect to the y-axis. A function is called **odd** if its graph is symmetric with respect to the origin. Classify the function as even, odd, or neither.
   (a) $y = x^2 + 1$ (b) $f(x) = x^3$ (c) $y = \sin x$
   (d) $g(x) = x^3 - 4$ (e) $f(x) = (x - 2)^2$ (f) $y = \cos x$
*39. Draw a complete graph of $y = x^3 + 2$. Is the graph symmetric about some point? If so, what is the point of symmetry?
40. Confirm that the graphs of the equations of Example 4(a,b) are correct by drawing in the same viewing rectangle the graphs of the two functions obtained by solving for y in terms of x.

1.4 More on Functions

Another way to define a function is to say that it is a *domain* together with a *rule* that assigns an element of the *range* to each element of the domain. Sometimes a separate rule is used for various portions of the domain. Such functions are said to be **piecewise defined** and are introduced in this section. We find the domain of a function by analyzing the rule that defines the function. The graphs of some of the functions that are built-in on most computers and graphing calculators are studied in this section.

Finding the domain and range of a relation is often a difficult task. We have found the domain and range of a relation given by an equation by reading this information from a complete graph of the relation. To obtain this complete graph, we first found all the real solutions to the equation. Now we want to turn our attention to a method of finding the domain of a function by analyzing the equation or rule that defines the function.

Domain of a Function

- Understood Domain of a Function. Let $y = f(x)$ be a function. The **understood domain** of f is the largest subset of the real numbers for which the equation $y = f(x)$ makes sense. That is, the real number a is in the *understood domain* of f if and only if $f(a)$ is a real number.

Unless otherwise specified, the *domain* of a function is its *understood domain*.

• EXAMPLE 1: Find the domain of the function $f(x) = x^2$.

SOLUTION: For every real number a, $f(a) = a^2$ is also a real number. Thus, the domain of f is the set of all real numbers. •

• EXAMPLE 2: Find the domain and range of the function $f(x) = \sqrt{x-3}$.

SOLUTION: Notice that $f(2) = \sqrt{2-3} = \sqrt{-1}$ is not a real number. Thus, 2 is not in the domain of f. If a is any real number less than 3, then $f(a)$ is *not* a real number. If a is greater than or equal to 3, then $f(a)$ is a real number. Therefore, the domain of f is the set of all real numbers greater than or equal to 3, or $x \geq 3$. The computer drawn graph of f in Figure 1.4.1 confirms the domain determined algebraically. We can read from this graph that the range is $y \geq 0$. •

We can algebraically find the domain of a function by analyzing its rule. Finding the range of the equation by analyzing the rule algebraically will often be difficult. Most of the time we will still read the range from a complete graph.

Built-In Functions

Computers have special built-in functions that can be accessed by typing a special code. For example, typing SQR(x) (or SQRT(x) on some computers) will usually produce the **square root function** $f(x) = \sqrt{x}$. This function can also be accessed by entering it in exponential form: $x \wedge (1/2)$ or $x \wedge 0.5$. The parentheses around the 1/2 are needed because computers follow the conventional rules for order of operations. Entering $x \wedge 1/2$ produces the function $\frac{x^1}{2}$, that is, $\frac{x}{2}$.

Inequalities and interval notation will be studied in detail in Chapter 2. For the time being we use the following notation.

$x \geq a$ The set of all real numbers greater than or equal to a.

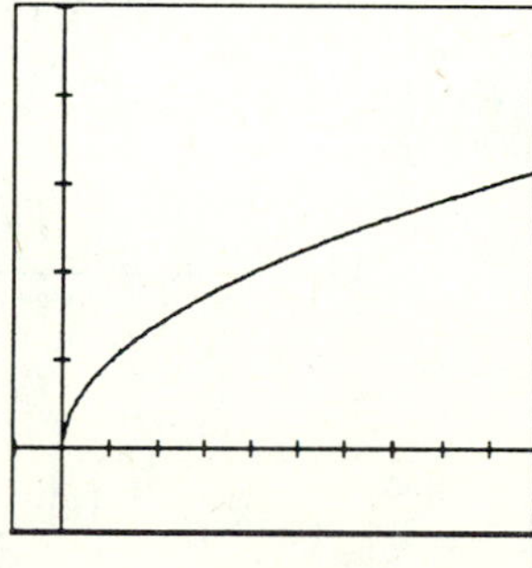

$y = \sqrt{x-3}$

$[-100, 1000]$ by $[-10, 50]$

Figure 1.4.1

$x > a$ The set of all real numbers greater than a.
$x \leq a$ The set of all real numbers less than or equal to a.
$x < a$ The set of all real numbers less than a.
$a \leq x \leq b$ The set of all real numbers between a and b, including a and b.
$a \leq x < b$ The set of all real numbers between a and b, including a but *not* b.
$a < x \leq b$ The set of all real numbers between a and b, including b but *not* a.
$a < x < b$ The set of all real numbers between a and b, excluding both a and b.

• EXAMPLE 3: Find the domain and range of the function $f(x) = \frac{\sqrt{x}}{2x-4}$.

SOLUTION: The denominator $2x - 4$ of the rule for f is zero if x is 2. Division by zero does *not* produce a real number. Thus, 2 is *not* in the domain of f. The numerator $\sqrt{x}$ of the rule for f is real if and only if x is nonnegative. Thus, the domain of f consists of all nonnegative real numbers different from 2, or $0 \leq x < 2$ together with $x > 2$. The computer drawn graph of f in Figure 1.4.2 confirms the domain found algebraically. This figure suggests that the range of f is the set of all real numbers. Notice that $f(0) = 0$ so that 0 is in the range of f. •

Another function that is built-in on most computers is the greatest integer function.

• DEFINITION The **greatest integer function** pairs a number with the largest integer that is less than or equal to the number. This function is denoted by INT(x). Another notation often used for this function is $[\![x]\!]$. •

• EXAMPLE 4: Compute:

(a) INT(2), INT(2.01), and INT(2.999).
(b) INT(−2.01), INT(−2.999), and INT(−3).

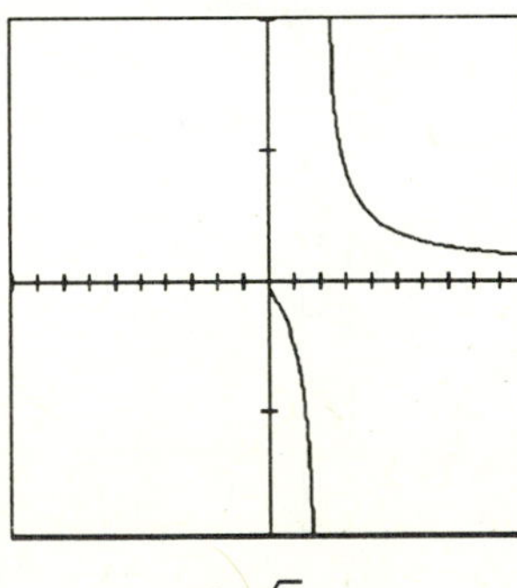

$y = \frac{\sqrt{x}}{sx-4}$
$[-10, 10]$ by $[-2, 2]$
Figure 1.4.2

SOLUTION:

(a) Notice that 2 is the largest integer that is less than or equal to 2, 2.01, and 2.999. Thus,

$$\text{INT}(2) = \text{INT}(2.01) = \text{INT}(2.999) = 2.$$

(b) The largest integer that is less than or equal to -2.01, -2.999, and -3 is -3. (Why?) Thus,

$$\text{INT}(-2.01) = \text{INT}(-2.999) = \text{INT}(-3) = -3.$$

•

• EXAMPLE 5: Draw a complete graph of $f(x) = \text{INT}(x)$ and determine its domain and range.

SOLUTION: There is a unique largest integer less than or equal to any number x. Therefore, the domain of this function is the set of all real numbers. The graph of this function in the $[-10, 10]$ by $[-10, 10]$ viewing rectangle is shown in Figure 1.4.3. The scale marks on the coordinate axes are one unit apart.

If x is between 0 and 1, then $\text{INT}(x) = 0$. Therefore, the graph of $f(x) = \text{INT}(x)$ between $x = 0$ and $x = 1$ coincides with the x-axis. This is not apparent from the computer drawn graph in Figure 1.4.3.

If x is an integer, then $\text{INT}(x) = x$. This is also not clear from the computer drawn graph in Figure 1.4.3. For example, from the graph in Figure 1.4.3 it appears that INT(3) is either 2 or 3. To clarify that INT(3) is 3 and not 2, we put a solid circle at $(3, 3)$ and an open circle at $(3, 2)$ as illustrated in Figure 1.4.4. We can, of course, do this at each integer value of x. We sometimes call this function a **step function** because its completed graph looks like a set of stair steps.

Finally, we can see that the range of $f(x)$ is the set of integers. By definition, no real number that is not an integer can be a value of this function. •

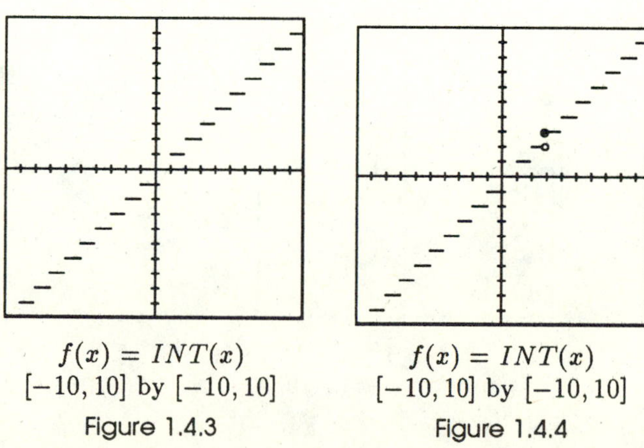

$f(x) = INT(x)$
$[-10, 10]$ by $[-10, 10]$
Figure 1.4.3

$f(x) = INT(x)$
$[-10, 10]$ by $[-10, 10]$
Figure 1.4.4

- **Warning.** There are some graphing calculators that use INT(x) for a function different from the greatest integer function. Their graphs agree for $x \geq 0$ but disagree for $x < 0$.

Another function that is built-in on a computer is the **absolute value function.**

- EXAMPLE 6: Compare the graphs of $f(x) = x^2 - 4$ and $g(x) = |x^2 - 4|$. Determine the domain and range of each function.

SOLUTION: A complete graph of $f(x) = x^2 - 4$ is given in Figure 1.4.5. You may need to rewrite g in the form $g(x) = \text{ABS}(x^2 - 4)$ in order to enter it on your graphing utility. Figure 1.4.6 gives the graph of g in the $[-10, 10]$ by $[-10, 10]$ viewing rectangle. The graphs of f and g coincide for $x \geq 2$ and for $x \leq -2$. If the coordinate plane is folded along the x-axis, then the portion of the graph of f below the x-axis coincides with the graph of g for x between -2 and 2. The graph of g can never be below the x-axis. (Why?) The effect of the absolute value in g is to bring the part of the graph of f that is below the x-axis above the x-axis to produce the graph of g. This happens because the values of f are negative if the graph of f lies below the x-axis, and the absolute value is always positive.

The domain of f is the set of all real numbers and the range of f is $y \geq -4$. The domain of g is also the set of all real numbers, but its range is $y \geq 0$. •

Example 6 illustrates how to find a complete graph of $y = |f(x)|$ whenever we have a complete graph of $y = f(x)$. The portions of the graph of $y = f(x)$ that are below the x-axis must be reflected through the x-axis using folding symmetry with respect to the x-axis. The resulting graph is a complete graph of $y = |f(x)|$. The same general strategy can be applied to determine complete graphs of $y = \sqrt{f(x)}$ and $y = \sqrt{|f(x)|}$ from a complete graph of $y = f(x)$. We will see other applications of reflections in subsequent chapters.

- EXAMPLE 7: Draw a complete graph of the function $f(x) = 16x - x^3$. Find the domain and range of f.

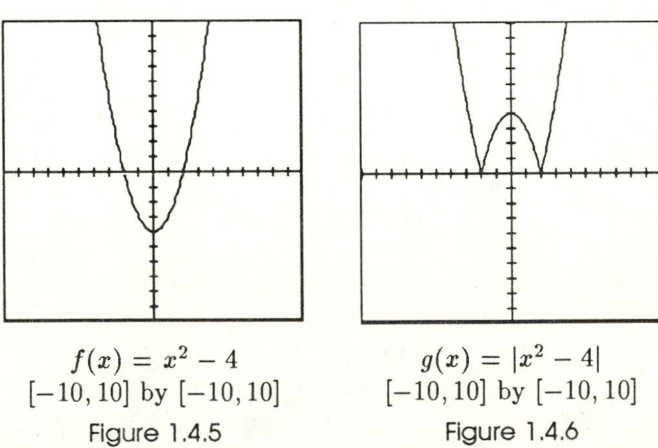

$f(x) = x^2 - 4$
$[-10, 10]$ by $[-10, 10]$
Figure 1.4.5

$g(x) = |x^2 - 4|$
$[-10, 10]$ by $[-10, 10]$
Figure 1.4.6

SOLUTION: The graph of f in the default viewing rectangle $[-10, 10]$ by $[-10, 10]$ is given in Figure 1.4.7. Careful observation while a graphing utility draws the graph in Figure 1.4.7 indicates that the graph goes off the screen and then comes back on the screen twice during the plotting process. This means that we need to see more of this graph in the vertical direction. Figure 1.4.8 shows the graph of f in the $[-10, 10]$ by $[-30, 30]$ viewing rectangle. In Section 1.2 we indicated that there are four possibilities for the complete graph of a function of the form $f(x) = ax^3 + bx^2 + cx + d$ with $a \neq 0$. We know the graph in Figure 1.4.8 is a complete graph because it looks like the second of the four graphs in Figure 1.2.12.

We can use the rule of f or the graph in Figure 1.4.8 to see that the domain of f is the set of all real numbers. It is clear from the graph that the range of f is also the set of all real numbers. •

Functions can be used to model many problem situations. Computer generated graphing can help us better understand the problem situation. One classic problem concerns the behavior of an object thrown straight up into the air with some initial velocity of v_0 feet per second. The height s of the object above ground at any time t is given by $s = -16t^2 + v_0 t$. Notice that the initial velocity is the coefficient of t. If the object is thrown from a point s_0 feet above ground with the same initial velocity, then the height s is given by $s = -16t^2 + v_0 t + s_0$ (Figure 1.4.9). The number 16 is half the acceleration due to gravity.

• EXAMPLE 8: An object is shot straight up from level ground with initial velocity of 64 feet per second.

(a) Write the height s of the object above ground as a function of time t.
(b) Draw a complete graph of the function in (a) indicating what portion of the graph represents the problem situation.
(c) Find the maximum height reached by the object and the time required to reach that height.

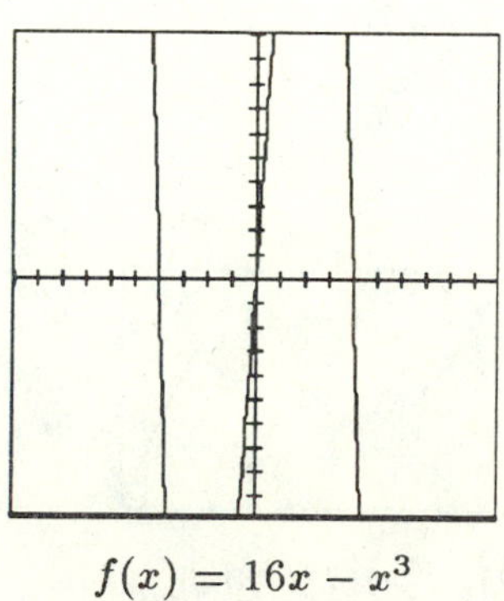

$f(x) = 16x - x^3$
$[-10, 10]$ by $[-10, 10]$
Figure 1.4.7

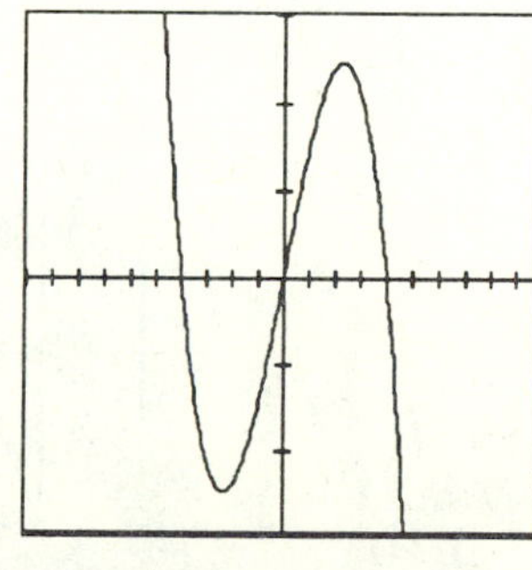

$f(x) = 16x - x^3$
$[-10, 10]$ by $[-30, 30]$
Figure 1.4.8

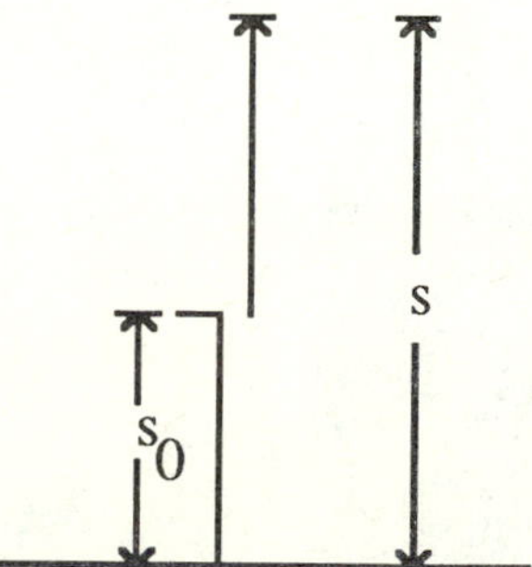

Figure 1.4.9

SOLUTION:

(a) The height above ground t seconds after launch is given by $s = -16t^2 + 64t$.

(b) You may need to view the graph of s in several viewing rectangles to find the complete graph given in Figure 1.4.10. Because distance above ground must be nonnegative, only the *first* quadrant portion of the graph in Figure 1.4.10 represents the problem situation.

(c) We can estimate from the graph in Figure 1.4.10 that the coordinates of the highest point are about $(2, 65)$. A formula for the exact coordinates of the highest point was given in Section 1.2. We know from Section 1.2 that the t coordinate of the highest point in Figure 1.4.10 is $-\frac{64}{2(-16)} = 2$. The corresponding value of s is 64. Thus, the object reaches a maximum height of 64 feet exactly 2 seconds after launch. •

• EXAMPLE 9: At what time will the object in Example 8 be 50 feet above ground?

SOLUTION: We know that only the first quadrant portion of the graph in Figure 1.4.10 represents the problem situation. Notice that s is 0 when t is 4. Figure 1.4.11 gives a graph of the problem situation and the horizontal line $s = 50$. The *two* points of intersection in Figure 1.4.11 have second coordinate 50. The first coordinates give values of t for which $s(t) = -16t^2 + 64t = 50$. These two values of t give the times when the object is 50 feet above ground; once on the way up and once on the way down. We can estimate from Figure 1.4.11 that the two times are 1.1 and 2.9 seconds after launch. •

The graph in Figure 1.4.11 represents the height above ground of the object at any time t. It does *not* represent the path of the object. The object goes straight up and comes straight down. The actual path of the object lies on a vertical line. In the next chapter, we review algebraic methods for solving equations and introduce a powerful geometric method for solving equations.

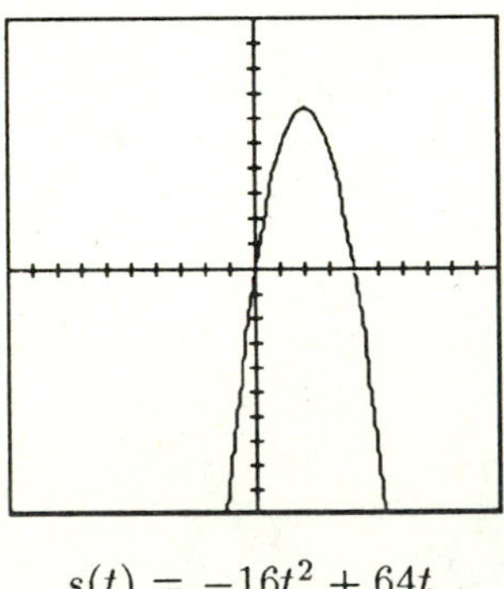

$s(t) = -16t^2 + 64t$
$[-10, 10]$ by $[-100, 100]$
Figure 1.4.10

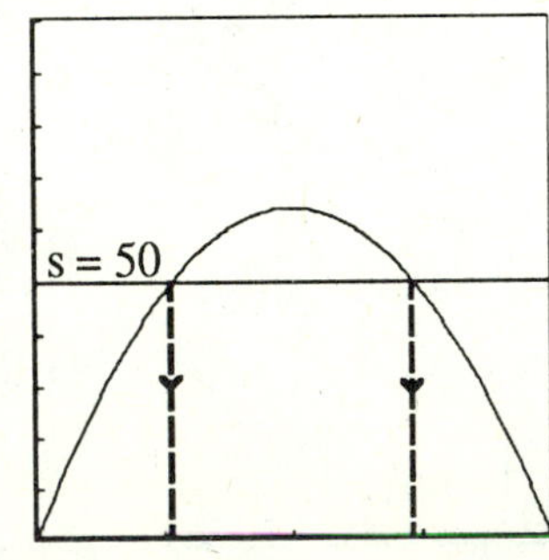

$s(t) = -16t^2 + 64t$
$[0, 4]$ by $[0, 100]$
Figure 1.4.11

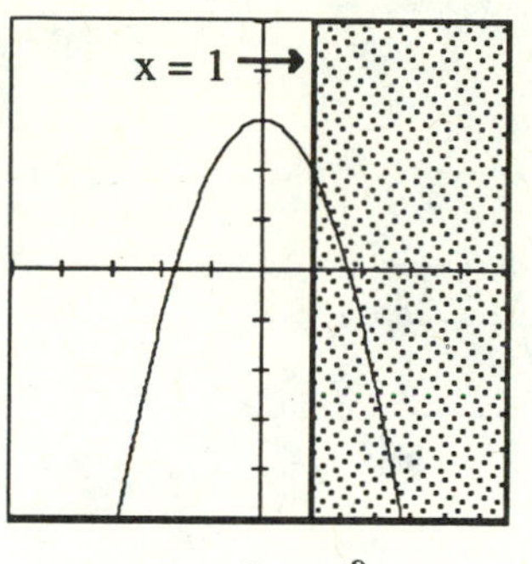

$y = 3 - x^2$
$[-5, 5]$ by $[-5, 5]$
Figure 1.4.12

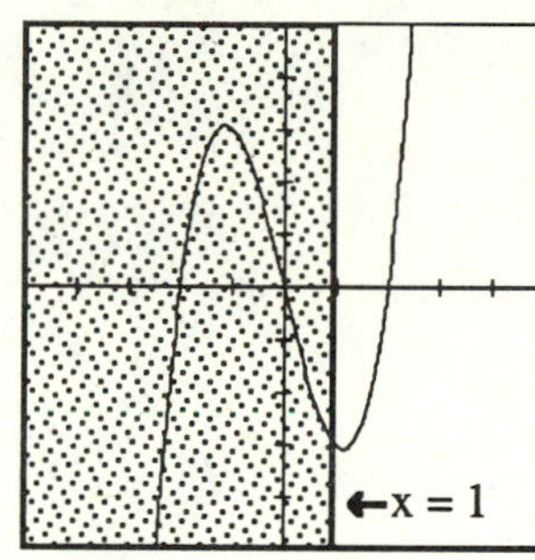

$y = x^3 - 4x$
$[-5, 5]$ by $[-5, 5]$
Figure 1.4.13

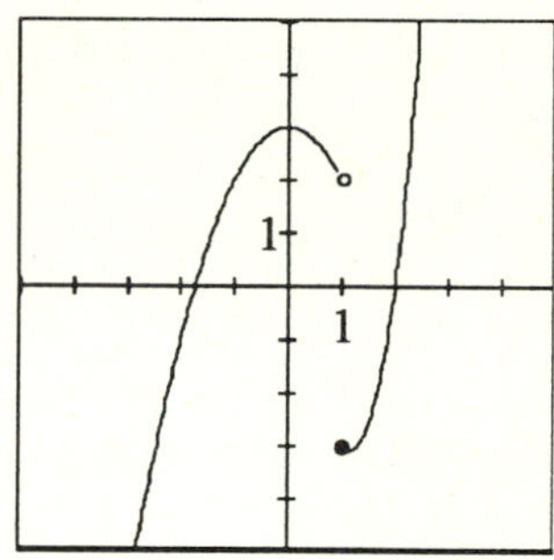

$$y = \begin{cases} 3 - x^2, & \text{for } x < 1 \\ x^3 - 4x, & \text{for} x \geq 1 \end{cases}$$
Figure 1.4.14

Piecewise-Defined Functions

Piecewise-defined functions are often used to provide examples of functions with specified properties or to give counterexamples to certain conjectures. We can use a graphing utility to help us draw a complete graph of a piecewise-defined function.

- EXAMPLE 10: Draw a complete graph of the function

$$f(x) = \begin{cases} 3 - x^2, & \text{for } x < 1 \\ x^3 - 4x, & \text{for } x \geq 1 \end{cases}.$$

Determine the domain and range of f.

SOLUTION: The computer drawn graph of $y = 3 - x^2$ gives the portion of the graph of f to the left of the line $x = 1$ (Figure 1.4.12). The computer drawn graph of $y = x^3 - 4x$ gives the portion of the graph of f to the right of the line $x = 1$ (Figure 1.4.13). Notice that $f(1) = -3$. Combining the unshaded portions of the graphs in Figures 1.4.12 and 1.4.13 and putting an open circle at $(1, 2)$ and a closed circle at $(1, -3)$ gives the complete graph of f shown in Figure 1.4.14. The domain and range of f are the set of all real numbers. •

• EXERCISES 1-4

Draw a complete graph of each function.

1. $y = \sqrt{x + 1}$
2. $y = -1 - \sqrt{x - 1}$
3. $y = 1 + |x - 2|$
4. $y = -|2x - 3|$
5. $y = \text{INT}(x + 2)$
6. $y = x^2 - 5x - 10$
7. $y = (x - 30)(x + 20)$
8. $y = \left(\frac{x - 20}{2}\right)^2 - 50x$
9. $y = |x^2 - 6x - 12|$
10. $y = 2x^3 - x + 3$
11. $y = x^3 - 8x$
12. $y = 10x^3 - 20x^2 + 5x - 30$
13. What are the domain and range of the function in Exercise 1?

14. What are the domain and range of the function in Exercise 6?

Algebraically determine the domain of the function. Check with a graphing utility. Resolve conflicting results.

15. $f(x) = \dfrac{x^2 - 5}{3}$

16. $g(x) = \sqrt{8 - x}$

17. $f(x) = \dfrac{x - 2}{x^2 + 5}$

18. $f(x) = \sqrt{\dfrac{1}{x+1}}$

19. $g(x) = \dfrac{\sqrt{x - 5}}{x^2 - 5}$

20. $T(x) = \sqrt{x^3 - 8x}$

21. Compare complete graphs of $f(x) = \frac{|x|}{x}$ and $g(x) = \frac{x}{|x|}$.

22. Which of the computer expressions are equal to $3x^{1/2}$?

(a) $3 * x \wedge (1/2)$ (b) $3 * (x \wedge (1/2))$

(c) $(3 * x) \wedge (1/2)$ (d) $3 * x \wedge 1/2$

23. Which of the viewing rectangles gives the best complete graph of $f(x) = x^3 + 10x + 100$?

(a) $[-10, 10]$ by $[-10, 10]$ (b) $[-10, 10]$ by $[-100, 100]$

(c) $[-10, 10]$ by $[0, 100]$ (d) $[-10, 10]$ by $[-100, 0]$

(e) $[-10, 10]$ by $[-1000, 1000]$

24. Which of the viewing rectangles gives the best complete graph of $y = 12x^2 - 5x + 19$?

(a) $[-10, 10]$ by $[-10, 10]$ (b) $[-10, 10]$ by $[-100, 100]$

(c) $[-50, 50]$ by $[-5, 5]$ (d) $[-5, 5]$ by $[-50, 50]$

(e) $[-10, 0]$ by $[0, 50]$

25. Which of the viewing rectangles gives the best complete graph of $f(x) = 0.005x^3 - 5x^2 - 60$?

(a) $[-10, 10]$ by $[-100, 100]$ (b) $[-100, 100]$ by $[-100{,}000, 100{,}000]$

(c) $[-1000, 2000]$ by $[-1{,}000{,}000, 1{,}000{,}000]$ (d) $[0, 2000]$ by $[0, 1{,}000{,}000]$

(e) $[-1000, 1000]$ by $[-1000, 1000]$

26. Determine a viewing rectangle that gives a complete graph of $f(x) = 50 - 25x + 100x^2 - x^3$.

Sketch a complete graph of the function. Determine the domain and range.

27. $f(x) = \begin{cases} x^2, & x < 3 \\ x - 2, & x \geq 3 \end{cases}$

28. $g(x) = \begin{cases} \dfrac{1}{x}, & x < -2 \\ 2x, & x \geq -2 \end{cases}$

29. $f(x) = \begin{cases} 4 - x^2, & |x| \leq 5 \\ 6, & |x| > 5 \end{cases}$

30. $g(x) = \begin{cases} -x, & 0 \leq x < 4 \\ \sqrt{x - 3}, & x \geq 4 \end{cases}$

31. $f(x) = \begin{cases} 4 - x + x^2, & x < 2 \\ 1 - 2x + 3x^2, & x \geq 2 \end{cases}$

32. $g(x) = \begin{cases} 2x^3 - 4x^2 + x - 6, & x < 0 \\ \sqrt{2x + 3}, & x \geq 0 \end{cases}$

33. Sue charges \$3 per hour for babysitting plus \$5 to cover her transportation costs.

(a) Express Sue's babysitting fee F as a function of time t.

(b) Draw a complete graph of the function determined in (a).

(c) What are the domain and range of the function in (a)?

(d) What part of the domain of the function represents the problem situation?

(e) Use the graph in (b) to estimate the time Sue spent babysitting if her charges amounted to \$14.50.

34. A trucker averages 42 mph on a cross country trip from New York City to San Francisco, California. Let t be the time elapsed since the trucker left New York.

(a) Express the total distance the trucker travels as a function of time t.

(b) Draw a complete graph of the function determined in (a).

(c) What are the domain and range of the function in (a)?

(d) What part of the domain of the function represents the problem situation?

(e) Use the graph in (b) to determine how many hours have elapsed since the trucker left New York if the distance traveled is 850 miles.

35. The annual profit P of a candy manufacturer is determined by the formula $P = R - C$, where R is the total revenue generated from selling x pounds of candy and C is the total cost of making and selling x pounds of candy. Each pound of candy sells for \$1.50 and costs \$1.25 to make. The fixed costs of making and selling the candy are \$20,000 annually.

(a) Express the company's annual profit as a function of x.

(b) Draw a complete graph of the function determined in (a).

(c) How many pounds of candy must be sold for the company to "break even?"

(d) How many pounds of candy must be sold for the company to make a profit of \$10,000?

(e) Suppose the company gives some buyers a reduced price of \$1.40 a pound when large quantities of candy are purchased. If exactly 300 pounds of the candy are sold at the reduced price, what is the answer to (c)?

36. An object 10 feet above level ground is shot straight up with an initial velocity of 150 feet per second.

(a) Determine the height s of the object above the ground as a function of time t.

(b) Draw a complete graph of the function in (a) indicating what portion of the graph represents the problem situation.

(c) At what time will the object reach its maximum height? What is the maximum height?

(d) At what time t will the object be 300 feet above the ground?

*37. The regular long distance telephone charges C from Columbus, Ohio, to Lancaster, Pennsylvania, are \$0.48 for the first minute or fraction of a minute, and \$0.28 for each additional minute or fraction of a minute.

(a) Write an equation relating the cost of calling Lancaster from Columbus in terms of time t.

(b) Draw a complete graph of the function determined in (a).

(c) What are the domain and range of the function in (a)?

(d) What part of the domain of the function represents the problem situation?

(e) Bill can spend \$2.00 for a call from Columbus to his girlfriend in Lancaster. Use the graph in (b) to determine how long Bill can talk to his girlfriend.

*38. A parking garage in downtown Houston, Texas, charges \$2 for the first hour and \$1 for each additional hour or fraction of an hour, up to a maximum of \$12 per 24-hour day. Let t represent the number of hours parked at the garage.

(a) Express the total possible parking charges for one day as a function of time t. *Hint:* You will have to consider two cases: One case for maximum charges of \$12 and the other for charges less than or equal to the maximum charge of \$12.

(b) Draw a complete graph of the function determined in (a).

(c) Use the graph in (b) to determine the time parked if the parking charges are \$8.

39. Let the point P be on the line $x + y = 1$. Let the point Q have coordinates $(3, 5)$.

(a) Write the distance D between P and Q as a function of x, where x is the x-coordinate of the point P.

(b) What is the domain of the function D?

(c) Draw a complete graph of the function $y = D(x)$.

(d) Use your graph in (c) to determine x if the distance is 10. What are the coordinates of the associated point P?

40. A function f is **even** if $f(-x) = f(x)$ for every x in the domain of f. Show that the graph of an even function is symmetric with respect to the y-axis.

41. Which of the functions are even? Use a graphical argument.

(a) $f(x) = x^2 + 1$ (b) $g(x) = 1 - x^3$

(c) $f(x) = |x|$ (d) $f(x) = (x - 1)^2$

(e) $f(x) = \sin x$ (f) $g(x) = \cos x$

42. A function f is **odd** if $f(-x) = -f(x)$ for every x in the domain of f. Show that the graph of an odd function is symmetric with respect to the origin.

43. Which of the functions are odd? Use a graphical argument.

(a) $f(x) = x^3$ (b) $f(x) = (x - 1)^3$

(c) $g(x) = \sqrt{x}$ (d) $g(x) = 1/x$

(e) $f(x) = 2\sin x$ (f) $g(x) = -3\cos x$

Draw a complete graph of the function. Determine the domain, range, and equation for any lines of symmetry. Which graphs are symmetric with respect to the origin?

44. $f(x) = |x^2 - 3|$ 45. $g(x) = |x^3 - 1|$ 46. $h(x) = \sqrt{|x + 5|}$

47. $f(x) = |x^3 - 8x|$ 48. $g(x) = |x^2 - 5x + 8|$ 49. $h(x) = \left|\dfrac{2}{x - 3}\right|$

1.5 Linear Functions and Linear Inequalities

In this section we draw graphs of linear functions and graphs of inequalities that are linear in two variables. The graph of a linear function is a straight line. The graph of a linear inequality in two variables consists of a portion of the coordinate plane that lies on one side or the other of a straight line, and may or may not include the line itself.

• DEFINITION A function f is called a **linear function** if it can be written in the form $f(x) = mx + b$, where m and b are real numbers and $m \neq 0$. •

The first example gives information about the coefficient m of x in a linear function.

• EXAMPLE 1: Draw complete graphs of the functions

$$y = \frac{1}{2}x\ , \quad y = 3x\ , \quad y = -2x\ , \quad \text{and} \quad y = -3x$$

on the same coordinate system.

SOLUTION: The four graphs are shown in Figure 1.5.1 and are complete. •

The graphs of $y = \frac{1}{2}x$ and $y = 3x$ each rise as we move from left to right across the viewing rectangle of Figure 1.5.1. The graph of $y = 3x$ rises more rapidly than the graph of $y = \frac{1}{2}x$. The graph of $y = 3x$ is steeper because the coefficient 3 is greater than the coefficient $\frac{1}{2}$.

The graphs of $y = -2x$ and $y = -3x$ fall as we move from left to right across the viewing rectangle. This happens when the coefficient of x in a linear function is negative. If $m > 0$, the graph of $y = mx$ rises as we move from left to right, that is, as x increases so does y. If $m < 0$, the graph of $y = mx$ falls as we move from left to right; that is, as x increases, y decreases. Further, if m_1 and m_2 are positive with $m_1 > m_2$, then the graph of $y = m_1x$ is steeper than the graph of $y = m_2x$. Finally, if m_1 and m_2 are negative with $m_1 < m_2$, then both graphs fall, but the graph of $y = m_1x$ falls more rapidly than the graph of $y = m_2x$. Use the graphs of $y = -2x$ and $y = -3x$ in Figure 1.5.1 to check the last statement.

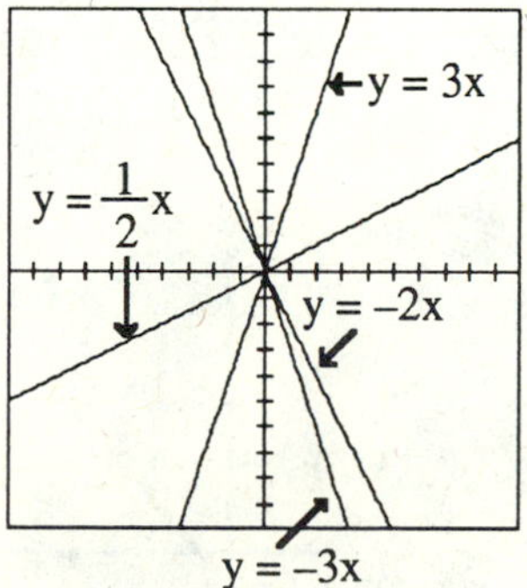

[−10, 10] by [−10, 10]

Figure 1.5.1

The next example gives more information about the coefficient m in the linear function $y = mx + b$.

• EXAMPLE 2: Let (x_1, y_1) and (x_2, y_2) be any two points on the graph of $y = mx + b$ with $x_1 \neq x_2$. Show that $m = \frac{y_2 - y_1}{x_2 - x_1}$ (Figure 1.5.2).

SOLUTION: If (x_1, y_1) and (x_2, y_2) lie on the graph of $y = mx + b$, then $y_1 = mx_1 + b$ and $y_2 = mx_2 + b$. (Why?) Subtracting y_1 from y_2 allows us to solve for m, the coefficient of x in $y = mx + b$.

$$y_2 - y_1 = (mx_2 + b) - (mx_1 + b)$$
$$y_2 - y_1 = m(x_2 - x_1)$$
$$m = \frac{y_2 - y_1}{x_2 - x_1}$$ •

• DEFINITION If (x_1, y_1) and (x_2, y_2) are any two points on the graph of $y = mx + b$ with $x_1 \neq x_2$, then $m = \frac{y_2 - y_1}{x_2 - x_1}$ is called the **slope** of the graph of $y = mx + b$. •

The sign of m (positive or negative) determines whether the graph rises or falls as we move from left to right, and the absolute value of m determines the steepness of the rise or fall. Further, we can view m as the "change in y" divided by the "change in x" as we move from the point (x_1, y_1) to the point (x_2, y_2).

• EXAMPLE 3: Determine the slope of the line passing through the points $(-1, 2)$ and $(4, -2)$.

SOLUTION: Let $(x_1, y_1) = (-1, 2)$ and $(x_2, y_2) = (4, -2)$; then

$$m = \frac{y_2 - y_1}{x_2 - x_1} = \frac{-2 - 2}{4 - (-1)} = -\frac{4}{5}.$$ •

$$m = \frac{y_2 - y_1}{x_2 - x_1} = \frac{\text{change in y}}{\text{change in x}}$$

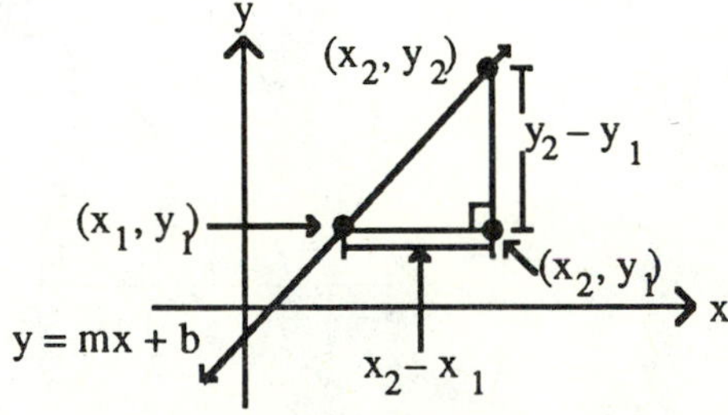

Figure 1.5.2

If (x_1, y_1) and (x_2, y_2) lie on a horizontal line, then the slope of the line is zero because $y_1 = y_2$. If (x_1, y_1) and (x_2, y_2) lie on a vertical line, then the line has no slope because $x_1 = x_2$ and division by zero is not permitted.

Next, we establish a geometric interpretation of the coefficient b in $f(x) = mx + b$.

• EXAMPLE 4: Draw and compare the graphs of the functions $y = x$, $y = x + 3$, and $y = x - 2$ on the same coordinate system.

SOLUTION: The three graphs are parallel straight lines as shown in Figure 1.5.3. The graph of $y = x$ goes through the origin $(0, 0)$. The graph of $y = x + 3$ lies above the graph of $y = x$ and crosses the y-axis at $(0, 3)$. The graph of $y = x - 2$ lies below the graph of $y = x$ and crosses the y-axis at $(0, -2)$. •

• DEFINITION If a graph crosses the y-axis at $(0, b)$, then b is called a **y-intercept** of the graph. •

The graph of $y = x + 3$ has y-intercept 3 and the graph of $y = x - 2$ has y-intercept -2. In general, the graph of $y = mx + b$ has y-intercept b. Every linear function has exactly one y-intercept. The graphs of $y = mx + b$ for fixed m and different b's are *parallel.* (Why?)

• DEFINITION If a graph crosses the x-axis at $(a, 0)$, then a is called an **x-intercept** of the graph. •

Graphs of Linear Functions

The graph of $y = mx + b$ is a straight line with slope m and y-intercept b. This information can be used to draw a quick sketch of the graph of a linear function.

• EXAMPLE 5: Determine the slope, the y-intercept, the x-intercept, and draw a complete graph of $2x - 3y = 9$.

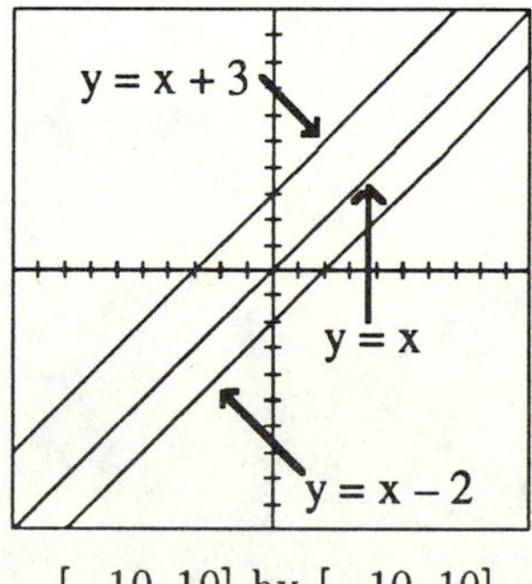

$[-10, 10]$ by $[-10, 10]$

Figure 1.5.3

SOLUTION: We rewrite the equation in the form $y = mx + b$ to determine the slope and y-intercept.

$$\begin{aligned} 2x - 3y &= 9 \\ -3y &= -2x + 9 \\ y &= \frac{2}{3}x - 3 \end{aligned}$$

The slope is $\frac{2}{3}$ and the y-intercept is -3. A second point on the graph of y can be found by starting with the y-intercept and moving 2 units up and 3 units to the right. (Why?) This second point has coordinates $(3, -1)$. A complete graph of $2x - 3y = 9$ is sketched in Figure 1.5.4. •

If we set $y = 0$ in $2x - 3y = 9$ we find $x = 4.5$. Thus, 4.5 is the x-intercept. We can also determine a complete graph by drawing the straight line through the points $(0, -3)$ and $(4.5, 0)$ determined by the y- and x-intercepts, respectively. (Why?)

Graphs of Linear Inequalities in Two Variables

• DEFINITION An inequality in two variables is called a **linear inequality** in x and y if it can be written in one of the following forms: $y < mx + b$, $y \leq mx + b$, $y > mx + b$, or $y \geq mx + b$. •

• EXAMPLE 6: Draw a complete graph of each inequality.

(a) $y < 2x + 3$ (b) $y \geq 2x + 3$

SOLUTION: The graph of the straight line $y = 2x + 3$ is given in Figure 1.5.5.

(a) The point $(1, 5)$ is on the graph of $y = 2x + 3$ because $(1, 5)$ is a solution to the equation $y = 2x + 3$. If $(1, b)$ is on the graph of $y < 2x + 3$, then $b < 2(1) + 3$ or

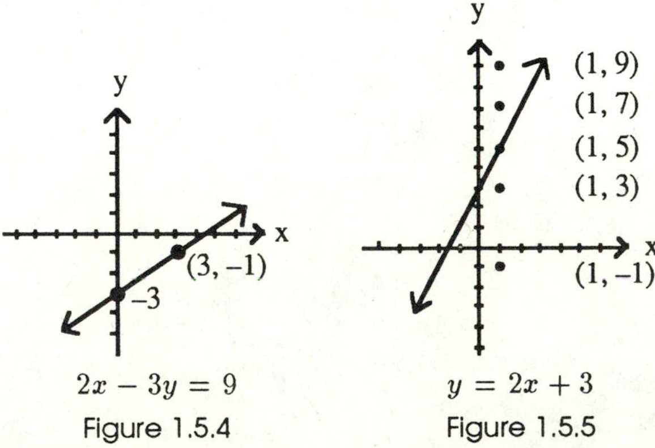

$2x - 3y = 9$

Figure 1.5.4

$y = 2x + 3$

Figure 1.5.5

$b < 5$. For example, $(1,3)$ and $(1,-1)$ are on the graph of $y < 2x + 3$, but $(1,7)$ and $(1,9)$ are not. (Why?) Thus, among the points on the vertical line $x = 1$, the points *below* the graph of $y = 2x + 3$ are *solutions* to $y < 2x + 3$ and the points *above* the graph of $y = 2x + 3$ are *not solutions* to $y < 2x + 3$. The same is true for every vertical line $x = c$. Therefore, a complete graph of $y < 2x + 3$ consists of all the points *below* the graph of $y = 2x + 3$ (Figure 1.5.6). The graph of the line $y = 2x + 3$ in Figure 1.5.6 is dashed to indicate it is not part of the solution to $y < 2x + 3$. The region below the graph of $y = 2x + 3$ is shaded to indicate it is part of the solution to $y < 2x + 3$.

(b) The solution to $y \geq 2x+3$ consists of the solution to $y > 2x+3$ together with the solution to $y = 2x + 3$. Thus, the graph of the line $y \geq 2x + 3$ consists of those points in the coordinate plane that lie *above* or *on* the line $y = 2x + 3$ (Figure 1.5.7). The graph of the line $y = 2x + 3$ in Figure 1.5.7 is solid to indicate it *is* part of the solution to $y \geq 2x + 3$. The region above the line $y = 2x + 3$ is shaded to indicate it is part of the solution to $y \geq 2x + 3$. •

The graph of a linear inequality in two variables is a region of the plane. The *boundary* of the region is a straight line that may or may not be part of the solution to the inequality.

The graph of $y < mx + b$ consists of the half plane below the line $y = mx + b$, and the graph of $y > mx + b$ consists of the half plane above the line $y = mx + b$. In each case, the graph of $y = mx + b$ is drawn dashed to indicate it is not part of the graph of the inequality.

The graph of $y \leq mx + b$ consists of the half plane below the line $y = mx + b$ together with the graph of the line $y = mx + b$. The graph of $y \geq mx + b$ consists of the half plane above the line $y = mx + b$ together with the graph of the line $y = mx + b$. In each case, the graph of $y = mx + b$ is drawn solid to indicate it is part of the graph of the inequality.

• EXAMPLE 7: Draw a complete graph of $2x + 3y < 4$.

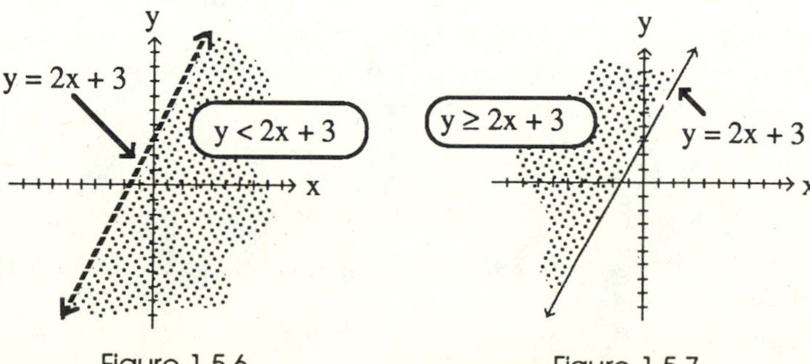

Figure 1.5.6   Figure 1.5.7

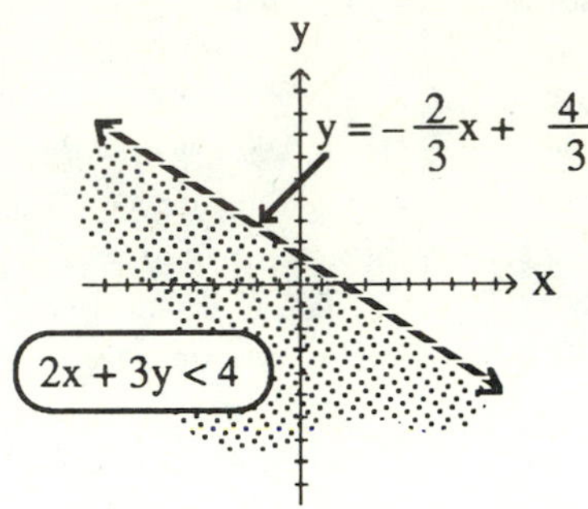

Figure 1.5.8

SOLUTION: First, we rewrite $2x+3y<4$ as an equivalent inequality in the form $y<mx+b$.

$$2x+3y<4$$
$$3y<-2x+4$$
$$y<-\frac{2}{3}x+\frac{4}{3}$$

Next, we draw a complete graph of $y=-\frac{2}{3}x+\frac{4}{3}$. A complete graph of $2x+3y<4$ consists of all points below the line $y=-\frac{2}{3}x+\frac{4}{3}$ (Figure 1.5.8).

Another way to obtain the graph of $2x+3y<4$ is based on the fact that we know the graph of $2x+3y<4$ consists of either the points above the line $2x+3y=4$ or the points below the line $2x+3y=4$. Instead of solving the inequality $2x+3y<4$ for y, we select one point not on the line and check if it is a solution to $2x+3y<4$. If the selected point is a solution to the inequality, then the solution to the inequality consists of all points on the same side of the line as the selected point. If the selected point is not a solution to the inequality, then the solution to the inequality consists of all points on the side of the line opposite to the side containing the selected point. For example, the point $(0,0)$ is below the line $2x+3y=4$ and is a solution to $2x+3y<4$. (Why?) Therefore, the solution to $2x+3y<4$ consists of all the points below the line $2x+3y=4$. •

Problem situations often have linear equations or inequalities for an algebraic representation. The next example involves a problem situation with a linear equation for a model. Answers to questions about the problem situation will be found algebraically. In the exercises, you will give a geometric representation for the problem situation and find the answers to the same questions geometrically.

• EXAMPLE 8: A house was purchased ten years ago for \$60,000. This year it was appraised at \$85,000. Assume the value of the house is a linear function of time.

(a) Write an algebraic representation that gives the value of the house at any time after the purchase.
(b) When was the house worth \$71,250?
(c) When was the house worth at least \$80,000?

SOLUTION:

(a) Let V stand for the value of the house and t for the time in years since its purchase. It is reasonable to let time t be 0 when the house was purchased. Thus, $V = 60{,}000$ when $t = 0$ and $V = 85{,}000$ when $t = 10$. Because V is a linear function of t we can write $V(t) = mt + b$ for some real numbers m and b. Now, substitute $t = 0$ and $V = 60{,}000$ in the equation.

$$\begin{aligned} V(t) &= mt + b \\ 60{,}000 &= m(0) + b \\ 60{,}000 &= b \end{aligned}$$

Thus, $V(t) = mt + 60{,}000$. Next, substitute $V = 85{,}000$ and $t = 10$ in the equation.

$$\begin{aligned} V(t) &= m(t) + 60{,}000 \\ 85{,}000 &= m(10) + 60{,}000 \\ 25{,}000 &= 10m \\ 2500 &= m \end{aligned}$$

Therefore, $V(t) = 2500t + 60{,}000$.

(b) To determine when the house was worth \$71,250, we set $V = 71{,}250$ and solve for t.

$$\begin{aligned} V(t) &= 2500t + 60{,}000 \\ 71{,}250 &= 2500t + 60{,}000 \\ 11{,}250 &= 2500t \\ 4.5 &= t \end{aligned}$$

Thus, the house was worth \$71,250 four and one-half years after its purchase.

(c) To find when the house was worth at least \$80,000 we need to solve the inequality $2500t + 60{,}000 \geq 80{,}000$.

$$\begin{aligned} 2500t + 60{,}000 &\geq 80{,}000 \\ 2500t &\geq 20{,}000 \\ t &\geq 8 \end{aligned}$$

Thus, the house was worth at least \$80,000 eight or more years after its purchase. •

• EXERCISES 1-5

1. Draw the line through the point $(1, 2)$ with the given slope.

   (a) 2 (b) $\frac{1}{3}$ (c) 0 (d) no slope

2. Draw the line through the point $(-3, 4)$ with the given slope.

   (a) -1 (b) -3 (c) $-\frac{1}{4}$ (d) 0

Draw complete graphs of the three functions in the same coordinate system without using a graphing utility. Check your answers with a graphing utility.

3. $y = 3x$, $y = \frac{1}{3}x$, $y = -3x$
4. $y = -3x$, $y = -\frac{1}{3}x$, $y = -2x$
5. $y = x - 1$, $y = x + 3$, $y = x + 5$
6. $y = 2x + 3$, $y = 2x - 1$, $y = 2x - 3$

Determine the slope of the line through each pair of points.

7. $(-1, 2)$, $(2, -6)$
8. $(0, -6)$, $(-4, 7)$

Determine the y-intercept of the graph of each function. Draw a complete graph of each function.

9. $y = 3 - 2x$
10. $2x - 5y + 6 = 0$

Determine the x-intercept of the graph of each function. Draw a complete graph of each function.

11. $x + 2y = 3$
12. $-x - 3y = 8$

Assume f is a linear function. Determine a formula for f that satisfies the given conditions.

13. $f(0) = 3$ and $f(2) = 6$
14. $f(1) = 1$ and $f(0) = 2$
15. $f(0) = 300$ and $f(10) = 365$
16. $f(5) = 200$ and $f(35) = 1050$

Draw a complete graph of each equation or inequality.

17. $y = 3x - 6$
18. $y < x - 2$
19. $x - 2y \geq 6$
20. $2x + y = 8$
21. $2x - 6y = 8$
22. $y \geq x$
23. $5x - 2y < 8$
24. $2x + 3y - 5 \geq 0$
25. (a) Draw a complete graph of the algebraic representation of the value of the house in Example 8. What portion of the graph is the graph of the problem situation?
    (b) Use a graphing method to solve (b) and (c) of Example 8.
26. Explain why either point can be labeled (x_1, y_1) or (x_2, y_2) when applying the slope formula in Example 3.
27. What do a and b represent in $x/a + y/b = 1$? *Hint:* Draw a complete graph of the equation for a few specific values of a and b.
28. Use the concept of slope to determine if the three points $(-1, 2)$, $(2, 4)$, and $(6, 9)$ are collinear, that is, all lie on the same line.
29. A house was purchased eight years ago for \$42,000. This year it was appraised at \$67,500. Assume the value of the house is a linear function of time.
    (a) Determine an algebraic representation of the value of the house at any time t after the purchase.
    (b) Draw a complete graph of the algebraic representation in (a). What values of t make sense in the problem situation?
    (c) Use a graph to estimate when the house will be worth \$90,000. Confirm your estimate algebraically.
30. A hardware store is marking all items down 15% for its autumn sale.
    (a) Draw a complete graph of the algebraic representation of the relationship between the original price and the sale price.
    (b) What portion of the graph in (a) is the graph of the problem situation?
    (c) If the sale price of an item is \$76.50, use a graph to determine its original price.

31. Hank intends to invest \$18,000, putting part of the money in one savings account that pays 5% annually and the rest in another account that pays 8% annually.

    (a) Determine an algebraic representation of the total interest received at the end of one year as a function of the amount invested at 8%.

    (b) Draw a complete graph of the algebraic representation in (a).

    (c) Use a graph to determine how the total interest received at the end of one year depends on the amount invested at 8%.

    (d) If Hank's annual interest is \$1020, how much of his \$18,000 did he invest at 8%?

32. Sue invests \$15,000 in two accounts, one part at 5.5% annual interest and the remainder at 7.5% annual interest.

    (a) Determine an algebraic representation of total interest received at the end of one year as a function of the amount invested at 5.5%.

    (b) Draw a complete graph of the algebraic representation in (a).

    (c) What portion of the graph in (b) is the graph of the problem situation?

    (d) Use a graph to find the amount invested at each rate if the annual interest is \$1000.

33. Tickets to a concert cost \$2.50 for students and \$3.75 for adults. Determine an algebraic representation expressing the requirement that the total receipts should exceed \$1500, and draw its graph.

34. Candy worth \$0.95 per pound is mixed with candy worth \$1.35 per pound. Determine an algebraic representation expressing the requirement that the value of the new mixture should not exceed \$1.15 per pound, and draw its graph.

35. Jim has some money invested in stocks that yield 10.5% annually and the rest in an account that pays 16.7% annually. Determine an algebraic representation expressing the requirement that his investments should yield at least 13% annually, and draw its graph.

36. An alcohol solution is formed by mixing a 34%-alcohol solution with a 62%-alcohol solution. Determine an algebraic representation expressing the requirement that the percent of alcohol in the new mixture should not exceed 44%, and draw its graph.

37. A fast-food restaurant makes \$0.25 profit on each hamburger, \$0.15 profit on each order of French fries, and breaks even on the other items. If the weekly overhead is \$600, determine

an algebraic representation expressing the requirement that the weekly profits should exceed \$400, and draw its graph.

Consider the problem situation in Example 8.

38. Complete the following table.

t	$V(t)$	**Annual Percentage Increase in Value of Home**
0	60,000	
1	62,500	$\dfrac{V(1)-V(0)}{V(0)} = \dfrac{2500}{60{,}000} = 4.17\%$
2	65,000	$\dfrac{V(2)-V(1)}{V(1)} = \dfrac{2500}{62{,}500} = 4\%$
3	67,500	$\dfrac{V(3)-V(2)}{V(2)} = \dfrac{2500}{65{,}000} = 3.85\%$
4	70,000	
5	72,500	
10	85,000	
30	135,000	

39. What is the annual percentage increase in the value of the house from year 49 to year 50? Write an algebraic representation of the annual percentage increase in the value of the house from year $t-1$ to year t.

40. Draw a complete graph of the algebraic representation in Exercise 39. What portion of the graph represents the problem situation?

41. Is the model of home appreciation given in Example 8 a realistic model? Why?

Long Distance Charges

A certain long distance phone company charges the following amounts for calls after 6 P.M. and before 8 A.M. from Columbus, Ohio to San Francisco, California. For calls up to 5 minutes in length the charge is \$0.72 per minute or fraction of a minute. For calls greater than 5 and up to 15 minutes in length the charge is \$0.63 per minute or fraction of a minute. For calls greater than 15 minutes in length, the charge is \$0.51 per minute or fraction of a minute.

42. Determine a piecewise-defined function that is an algebraic representation of the problem situation.

43. Determine the domain and range of this function.
44. Sketch a graph of this problem situation. Do not use a graphing utility.

Depreciation and Appreciation

The following notation and definitions will be used in connection with depreciation and appreciation of an asset.

C = original value (cost) of the asset

n = length of term (useful life under consideration)

S = value of the asset at end of term (in case of depreciation, S is called the *salvage value*)

$D = C - S$ is the *total depreciation* if $S < C$ and is the *total appreciation* if $S > C$

B = *book value*, the value for accounting or tax purposes at any time t

$\dfrac{C-S}{n}$ = the annual depreciation or appreciation amount

A continuous model for *book value* at any time t using the *straight line* method is

$$B = C - t\left(\frac{C-S}{n}\right) = C - \frac{Dt}{n}$$

where $0 \le t \le n$.

45. A machine costs \$17,000, has a useful life of 6 years, and has salvage value of \$1200. Draw the graph of the "book value" model $y = B(t)$. Why is the model called straight line depreciation?
46. What is the book value of the machine in 4 years, 3 months?
47. Assume a home valued today at \$78,000 has a "useful life" of 25 years and *appreciates* at \$5000 per year using the straight line method. Draw a complete graph of this problem situation.
48. What is the value of the home in Exercise 52 after 10 years? 15 years? 25 years?

A precision computer-assisted design (CAD) machine purchased from Japan costs \$22,600. It appreciates in value for 3 years to \$26,400 (due to the fall in the value of the dollar) and then depreciates in value for 4 years (when the age of the machine overshadows the fall in the value of the dollar) and has a salvage value of \$4700. Assume straight line appreciation and depreciation for computing book value.

49. Write a piecewise-defined function of time that is a model of the book value for this problem situation.
50. Draw a complete graph of the model of the book value of the machine.
51. What is the book value of the machine after 5 years, 6 months?

1.6 Quadratic Functions and Geometric Transformations

In Sections 1.6 and 1.7 we show how the graph of any function of the form $f(x) = ax^2+bx+c$ can be obtained from the graph of $y = x^2$ by applying a sequence of geometric transformations. Three of the transformations are rigid-motion transformations that produce **congruent graphs**, that is, graphs that coincide when superimposed. The other transformation stretches or shrinks the graph producing a new graph *not* congruent to the original graph. These geometric transformations are also applied to other functions in these sections.

• DEFINITION A function f is called a **quadratic function** if it can be written in the form $f(x) = ax^2 + bx + c$, where a, b, and c are real numbers and $a \neq 0$. •

• EXAMPLE 1: Draw complete graphs of the functions $y = x^2$, $y = 3x^2$, and $y = \frac{1}{2}x^2$ on the same coordinate system.

SOLUTION: The three graphs are shown in Figure 1.6.1 and are complete. •

Each graph in Figure 1.6.1 passes through the point $(0, 0)$. Excluding the origin, the graph of $y = 3x^2$ lies above the graph of $y = x^2$, and the graph of $y = \frac{1}{2}x^2$ lies below the graph of $y = x^2$. This happens because, for each nonzero real number x, the value of $3x^2$ is greater than the value of x^2, and the value of $\frac{1}{2}x^2$ is less than the value of x^2. In fact, for $x \neq 0$, the graph of $y = ax^2$ will always lie above the graph of $y = x^2$ if $a > 1$, or below the graph of $y = x^2$ if $0 < a < 1$.

• DEFINITION If $a > 1$, the graph of $y = ax^2$ is said to be obtained from the graph of $y = x^2$ by **vertically stretching** the graph of $y = x^2$ by the factor a. If $0 < a < 1$, the

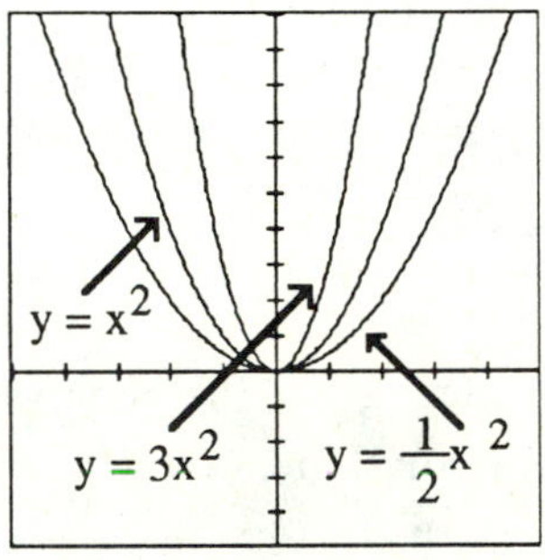

[−5, 5] by [−5, 10]

Figure 1.6.1

graph of $y = ax^2$ is said to be obtained from the graph of $y = x^2$ by **vertically shrinking** the graph of $y = x^2$ by the factor a. Vertically stretching or shrinking a given graph by the positive factor a, with $a \neq 1$, does not produce a congruent graph. •

• EXAMPLE 2: Draw complete graphs of the functions $y = -x^2$, $y = -3x^2$, and $y = -\frac{1}{2}x^2$ on the same coordinate system.

SOLUTION: The three graphs are shown in Figure 1.6.2 and are complete. •

If we fold the coordinate plane in Figure 1.6.1 along the x-axis, the resulting graphs would coincide with the graphs in Figure 1.6.2. For $x \neq 0$, the graph of $y = ax^2$ lies above the x-axis if $a > 0$, or below the x-axis if $a < 0$.

• DEFINITION The graph of $y = -ax^2$ is said to be obtained from the graph of $y = ax^2$ by **reflecting** the graph of $y = ax^2$ through the x-axis. A reflection produces a congruent graph. •

If $a < -1$, the graph of $y = ax^2$ is obtained by vertically stretching the graph of $y = x^2$ by the factor $|a|$ and then reflecting the resulting graph $y = |a|x^2$ through the x-axis; or by reflecting the graph of $y = x^2$ through the x-axis and then vertically stretching the resulting graph $y = -x^2$ by the factor $|a|$. If $-1 < a < 0$, the graph of $y = ax^2$ is obtained by vertically shrinking the graph of $y = x^2$ by the factor $|a|$ and then reflecting the resulting graph through the x-axis or vice versa. Alternatively, the graph of $y = ax^2$ can be obtained from the graph of $y = x^2$ by multiplying each y-coordinate of the graph of $y = x^2$ by a.

Notice that points on the x-axis are not moved under vertical stretching, vertical shrinking, or reflection through the x-axis. As we graph other classes of functions throughout this textbook you will be convinced that for any function f, the graph of $y = af(x)$ can be

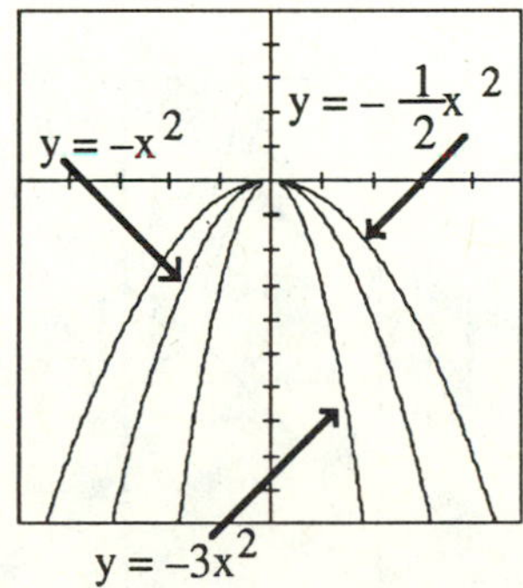

$[-5, 5]$ by $[-10, 5]$

Figure 1.6.2

obtained from the graph of $y = f(x)$ in the same way that the graph of $y = ax^2$ is obtained from the graph of $y = x$. The next example illustrates this.

• EXAMPLE 3: Draw complete graphs of $g(x) = -2(2x^2 - 3x - 2)$ and $f(x) = 2x^2 - 3x - 2$, and explain geometrically how a complete graph of g can be obtained from a complete graph of f.

SOLUTION: Figure 1.6.3 gives complete graphs of f and g. We can confirm by direct substitution that -1.5 and 2 are x-intercepts of both f and g; that is $f(-1.5) = f(2) = g(-1.5) = g(2) = 0$. Notice that these points are not moved under the transformation that multiplies each y-coordinate of the graph of f by -2. Figure 1.6.3 suggests that the graph of g can be obtained from the graph of f by applying a vertical stretch by a factor of 2 followed by a reflection through the x-axis or vice versa. Notice that the point $(1, -3)$ of f corresponds to the point $(1, 6)$ of g, and the point $(-2, 4)$ of f corresponds to the point $(-2, -8)$ of f under a vertical stretch by a factor of 2 followed by a reflection through the x-axis. Additional points can be traced to convince yourself that the graph of g is obtained from the graph of f as stated above. •

The x-intercepts of the graph of f in Example 3 are sometimes called "pivot points" because they are not moved by the geometric transformations that send the graph of f to the graph of g.

• EXAMPLE 4: Explain how complete graphs of the functions $y = x^2 + 2$ and $y = x^2 - 3$ are obtained from a complete graph of $y = x^2$.

SOLUTION: Complete graphs of $y = x^2 + 2$, $y = x^2 - 3$, and $y = x^2$ are given in Figure 1.6.4. The graph of $y = x^2 + 2$ is obtained from the graph of $y = x^2$ by moving each point vertically upward 2 units. The graph of $y = x^2 - 3$ is obtained from the graph of $y = x^2$ by moving each point vertically downward 3 units. •

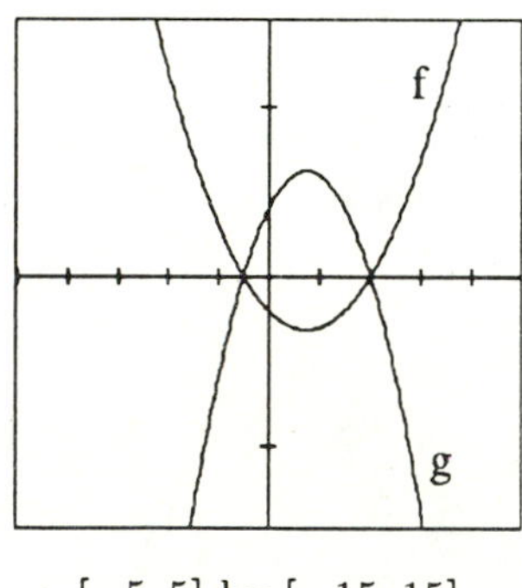

[−5, 5] by [−15, 15]

Figure 1.6.3

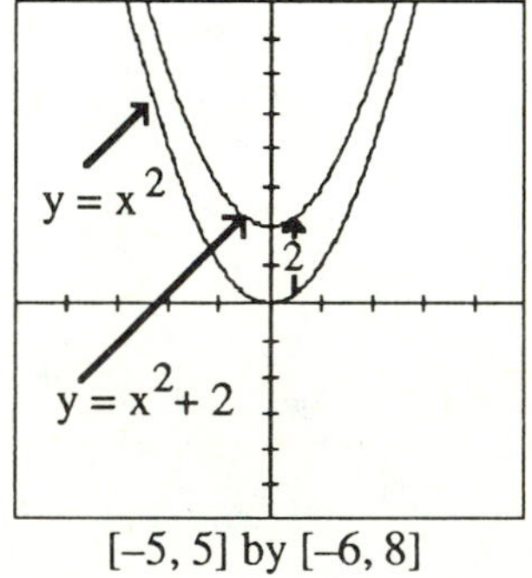

[−5, 5] by [−6, 8]

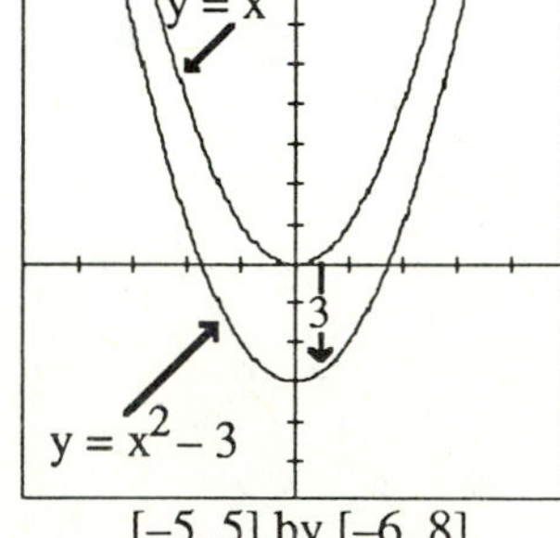

[−5, 5] by [−6, 8]

Figure 1.6.4

• DEFINITION The graph of $y = x^2 + k$ is said to be obtained from the graph of $y = x^2$ by a **vertical shift** or **slide**. We shift the graph of $y = x^2$ up k units if $k > 0$, or we shift the graph of $y = x^2$ down $|k|$ units if $k < 0$. A vertical shift of a given graph produces a congruent graph. •

In Example 4, the graph of $y = x^2 + 2$ can be obtained by vertically shifting the graph of $y = x^2$ up 2 units, and the graph of $y = x^2 - 3$ can be obtained by vertically shifting the graph of $y = x^2$ down 3 units. You will also discover that the graph of $y = f(x) + k$ can be obtained from the graph of $y = f(x)$ by a vertical shift. If $k \neq 0$, then every point in the coordinate plane is moved under the vertical shift determined by k.

• EXAMPLE 5: Explain how complete graphs of the functions $y = (x - 3)^2$ and $y = (x + 2)^2$ are obtained from a complete graph of $y = x^2$.-

SOLUTION: Complete graphs of $y = (x - 3)^2$, $y = (x + 2)^2$, and $y = x^2$ are given in Figure 1.6.5. This figure shows that the graph of $y = (x - 3)^2$ is obtained from the graph of $y = x^2$ by moving each point horizontally to the right 3 units. The graph of $y = (x+2)^2$ is obtained from the graph of $y = x^2$ by moving each point horizontally to the left 2 units. Be careful because the correct shift may be the opposite of what you expect. •

• DEFINITION The graph of $y = (x - h)^2$ is said to be obtained from the graph of $y = x^2$ by a **horizontal shift** or **slide**. We shift the graph of $y = x^2$ to the right h units if $h > 0$, and we shift the graph of $y = x^2$ to the left $|h|$ units if $h < 0$. A horizontal shift of a given graph produces a congruent graph. •

In Example 5, the graph of $y = (x - 3)^2$ can be obtained by horizontally shifting the graph of $y = x^2$ right 3 units and the graph of $y = (x + 2)^2$ can be obtained by horizontally shifting the graph of $y = x^2$ left 2 units. We will see that the graph of $y = f(x + h)$ can

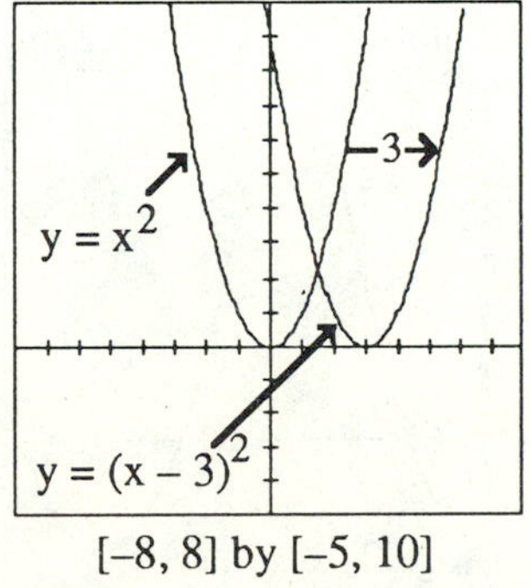

[–8, 8] by [–5, 10]

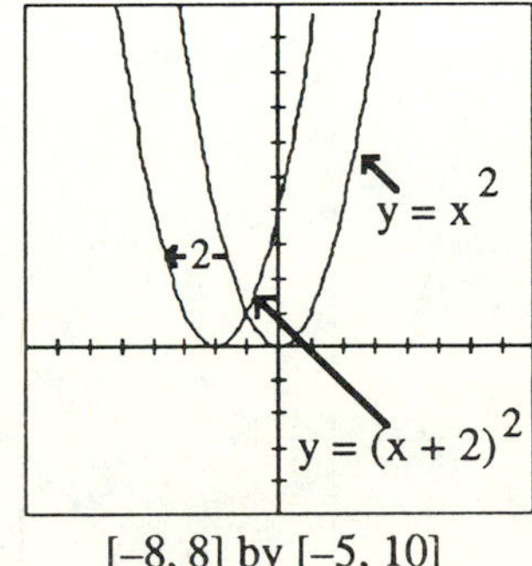

[–8, 8] by [–5, 10]

Figure 1.6.5

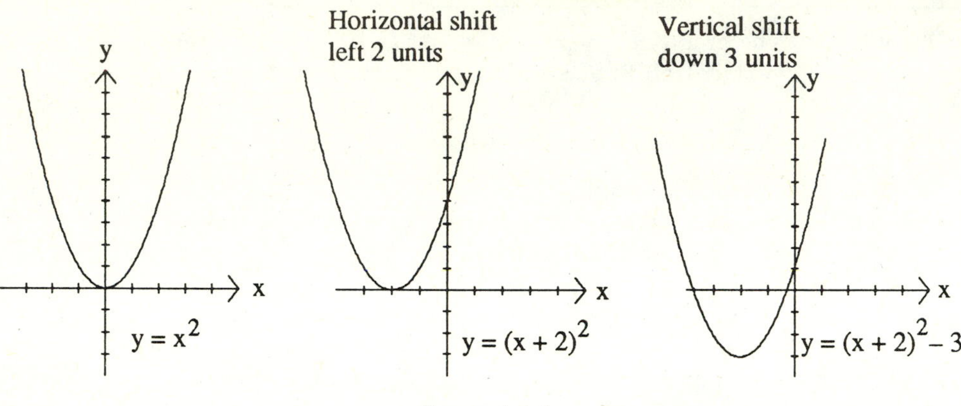

Figure 1.6.6

be obtained from the graph of f by a horizontal shift. If $h \neq 0$, then every point in the coordinate plane is moved by the horizontal shift determined by h.

• EXAMPLE 6: Describe how the graph of $y = (x + 2)^2 - 3$ can be obtained from the graph of $y = x^2$.

SOLUTION: We can rewrite the expression for y as $y = (x - (-2))^2 - 3$. The following two transformations performed in the given order produce a graph of $y = (x + 2)^2 - 3$.

1. Horizontal shift left 2 units.
2. Vertical shift down 3 units.

The effect of the two transformations is illustrated in Figure 1.6.6.
The same graph results if we apply the steps in reverse order (Figure 1.6.7). •

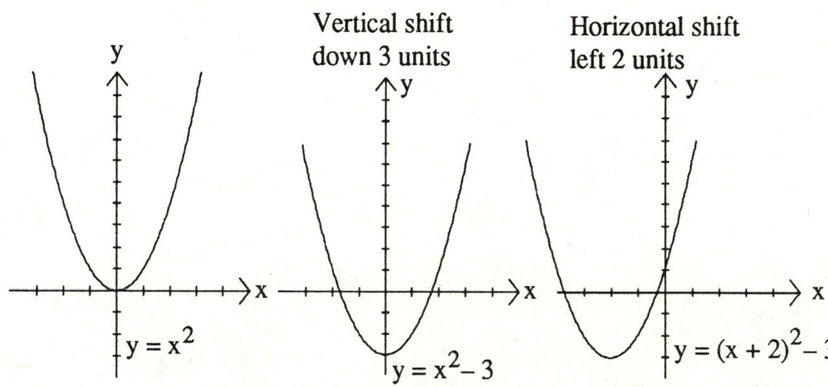

Figure 1.6.7

In Example 6, the order in which the transformations were applied made no difference. However, the next example shows that the order in which geometric transformations are applied does make a difference.

• EXAMPLE 7: Describe how the graph of $y = 5x^2 + 4$ can be obtained from the graph of $y = x^2$. Then show that the order cannot be reversed.

SOLUTION: The following two transformations performed in the given order produce the graph of $y = 5x^2 + 4$.

1. Vertical stretch by a factor of 5.
2. Vertical shift up 4 units.

The effect of the two transformations is illustrated in Figure 1.6.8. However, if we apply the transformations in reverse order we get the graph of $y = 5(x^2 + 4)$ (Figure 1.6.9). The last function in Figure 1.6.9 results because a vertical stretch by a factor of 5 is the same as multiplying the function values by 5. Thus, a vertical stretch of $y = x^2 + 4$ by a factor of 5 gives

$$y = 5(x^2 + 4) = 5x^2 + 20\,. \quad \bullet$$

The order in which the transformations described in this section are applied is important. Reversing the order may produce a different result. Example 7 shows that these operations are *not* necessarily commutative.

• EXAMPLE 8: Describe how the graph of $y = -3(x - 1)^2 + 4$ can be obtained from the graph of $y = x^2$.

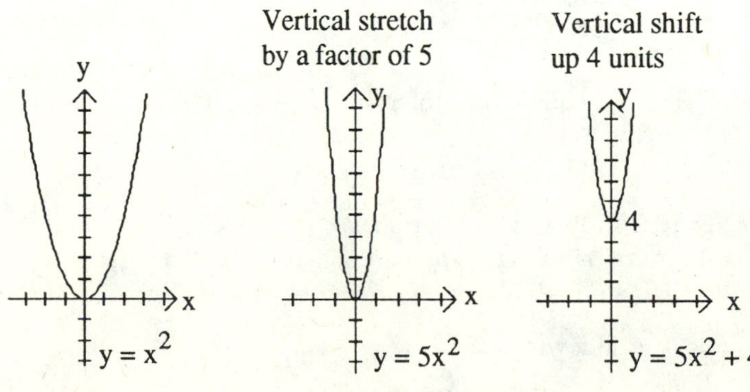

Figure 1.6.8

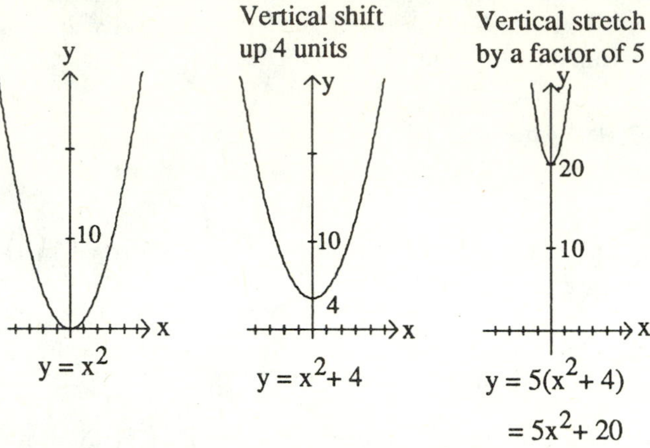

Figure 1.6.9

SOLUTION: The following four transformations performed in the given order produce a graph of $y = -3(x-1)^2 + 4$.

1. Vertical stretch by a factor of 3 to get $y = 3x^2$.
2. Reflection through the x-axis to get $y = -3x^2$.
3. Horizontal shift right 1 unit to get $y = -3(x-1)^2$.
4. Vertical shift up 4 units to get $y = -3(x-1)^2 + 4$.

The effect of the four transformations is illustrated in Figure 1.6.10. The first three transformations can be rearranged in any order and still produce the same graph. Moreover, the horizontal shift can be performed at any time. However, the vertical shift must follow the vertical stretch and the reflection. (Why?) •

In general, the graph of $y = a(x-h)^2+k$ can be obtained from the graph of $y = x^2$ using the geometric transformations described in this section. The following summary explains how to obtain the graph.

SUMMARY The graph of $y = a(x-h)^2 + k$ can be obtained from the graph of $y = x^2$ as follows.
If $a > 0$, the following steps are needed.

1. Vertical stretch ($a > 1$) or shrink ($0 < a < 1$) by a factor of a to get $y = ax^2$.
2. Horizontal shift right ($h > 0$) or left ($h < 0$) to get $y = a(x-h)^2$.
3. Vertical shift up ($k > 0$) or down ($k < 0$) to get $y = a(x-h)^2 + k$.

If $a < 0$, the following steps are needed.

1. Vertical stretch ($|a| > 1$) or shrink ($0 < |a| < 1$) by a factor of $|a|$ to get $y = |a|x^2$.
2. Reflection through the x-axis to get $y = ax^2$ (because $-|a| = a$).

3. Horizontal shift right ($h > 0$) or left ($h < 0$) to get $y = a(x - h)^2$.
4. Vertical shift up ($k > 0$) or down ($k < 0$) to get $y = a(x - h)^2 + k$.

The techniques of this section can be applied to other functions.

• EXAMPLE 9: Explain how the graph of the second function can be obtained from the graph of the first function.

(a) $f(x) = \sin x$; $g(x) = \sin(x - 2)$ (b) $f(x) = \cos x$; $g(x) = 3\cos x$

(c) $f(x) = \sin x$; $g(x) = 2 + \sin x$

SOLUTION:

(a) Figures 1.6.11 and 1.6.12 give the graphs of $f(x) = \sin x$ and $g(x) = \sin(x - 2)$ respectively, in the $[-10, 10]$ by $[-2, 2]$ viewing rectangle. Notice that the graph of g is the graph of f shifted 2 units to the right.

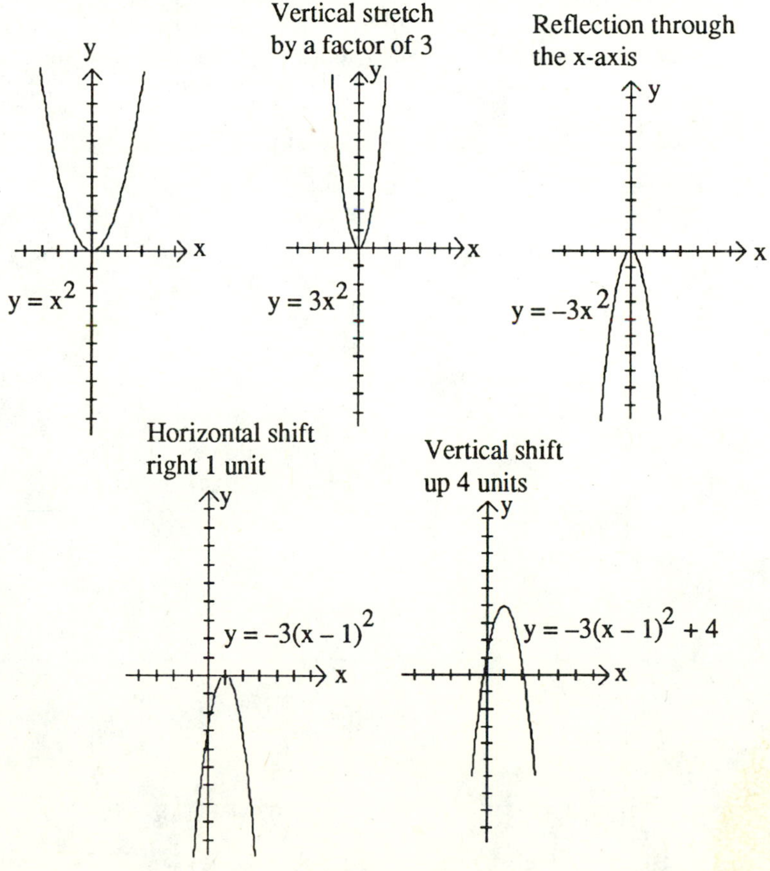

Figure 1.6.10

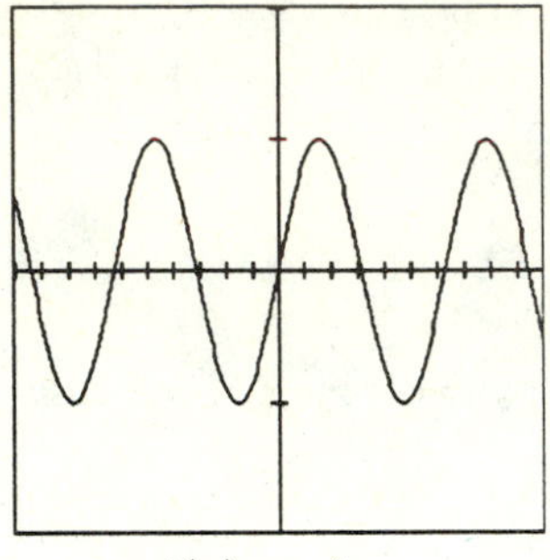

$f(x) = \sin x$
$[-10, 10]$ by $[-2, 2]$
Figure 1.6.11

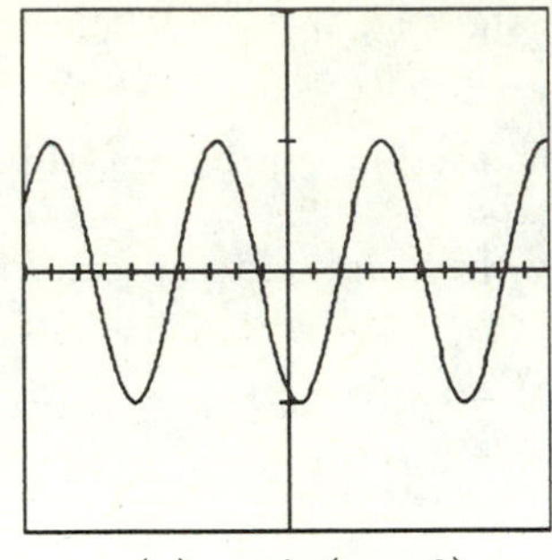

$g(x) = \sin(x-2)$
$[-10, 10]$ by $[-2, 2]$
Figure 1.6.12

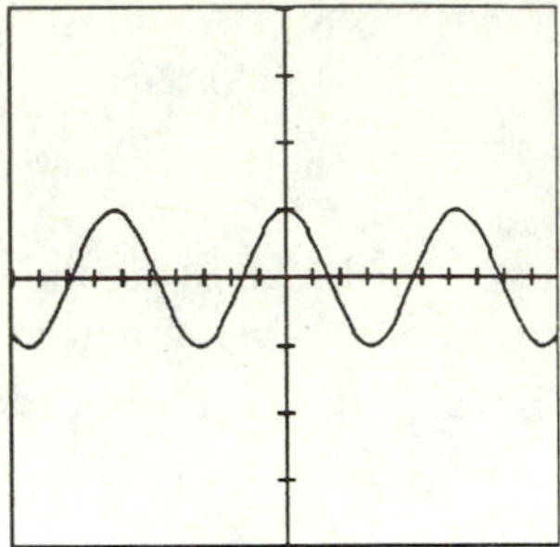

$f(x) = \cos x$
$[-10, 10]$ by $[-4, 4]$
Figure 1.6.13

(b) Figures 1.6.13 and 1.6.14 give the graphs of $f(x) = \cos x$ and $g(x) = 3\cos x$ respectively, in the $[-10, 10]$ by $[-4, 4]$ viewing rectangle. Notice that the graph of g can be obtained from the graph of f by applying a vertical stretch by a factor of 3.

(c) Figures 1.6.15 and 1.6.16 give the graphs of $f(x) = \sin x$ and $g(x) = 2 + \sin x$, respectively, in the $[-10, 10]$ by $[-4, 4]$ viewing rectangle. Notice that the graph of g is the graph of f shifted 2 units up. •

• EXAMPLE 10: A graph of a function f is given in Figure 1.6.17. Sketch the graph of $y = 2f(x+1) - 3$.

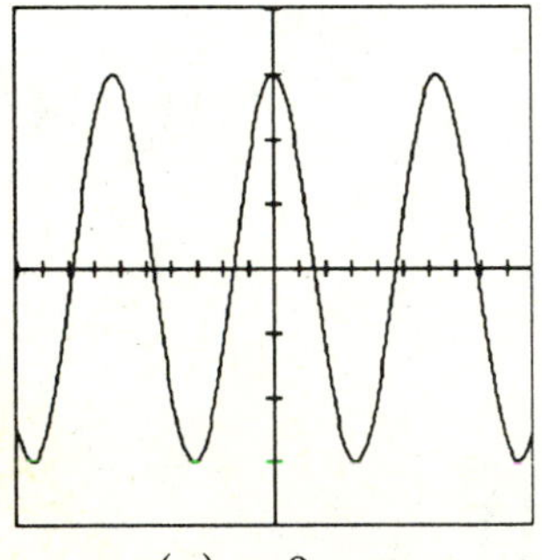

$g(x) = 3\cos x$
$[-10, 10]$ by $[-4, 4]$
Figure 1.6.14

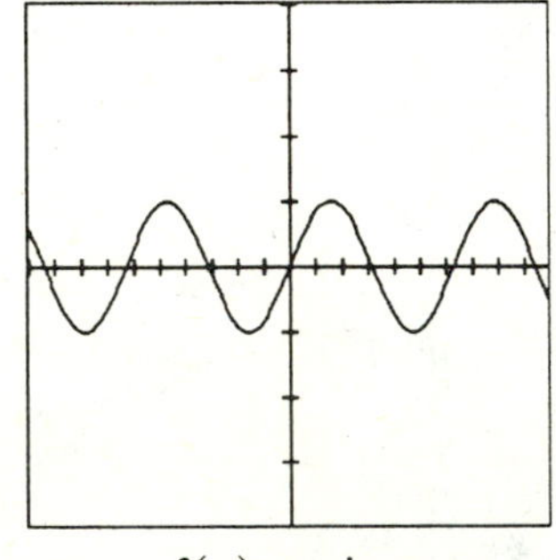

$f(x) = \sin x$
$[-10, 10]$ by $[-4, 4]$
Figure 1.6.15

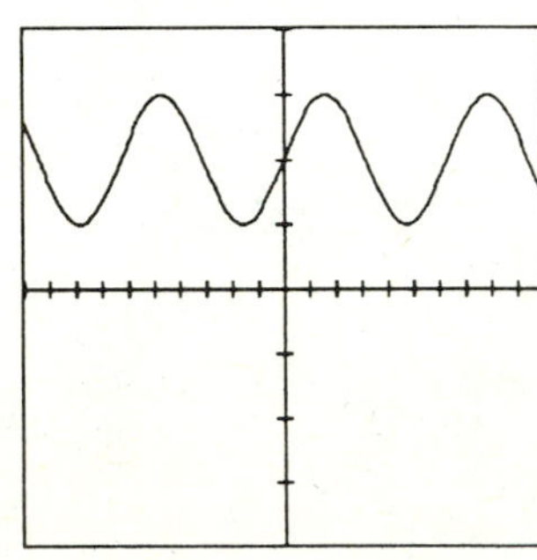

$g(x) = 2 + \sin x$
$[-10, 10]$ by $[-4, 4]$
Figure 1.6.16

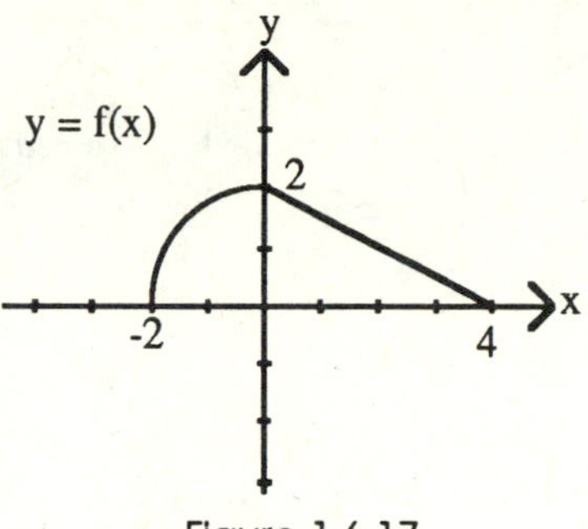

Figure 1.6.17

SOLUTION: The following three transformations performed in the given order produce the graph of $y = 2f(x+1) - 3$.

1. Vertical stretch by a factor of 2 to get $y = 2f(x)$.
2. Horizontal shift left 1 unit to get $y = 2f(x+1)$.
3. Vertical shift down 3 units to get $y = 2f(x+1) - 3$. •

The effect of the three transformations described above is illustrated in Figure 1.6.18.

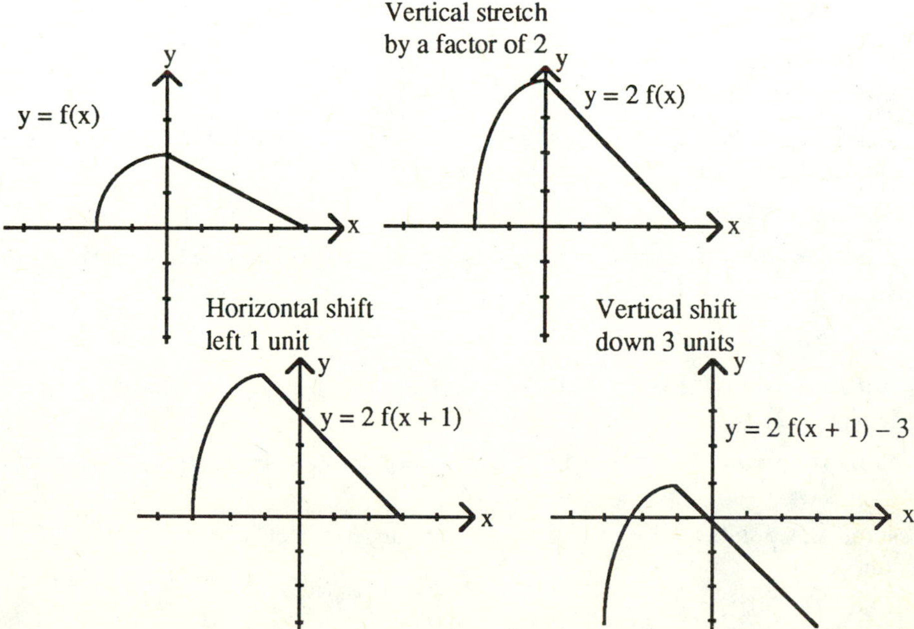

Figure 1.6.18

• EXERCISES 1-6

Use a graphing utility to draw complete graphs of the three equations on the same coordinate system.

1. $y = 3x^2$, $y = 3(x-2)^2$, $y = 3(x+2)^2$
2. $y = -x^2$, $y = -(x+4)^2$, $y = -(x-4)^2$
3. $y = 2x^2$, $y = 2(x-1)^2$, $y = 2(x+1)^2$
4. $y = -x^2$, $y = -x^2 - 4$, $y = -x^2 + 4$
5. $y = x^2 + 3$, $y = (x+2)^2 + 3$, $y = (x-3)^2 + 3$
6. $y = x^2 - 2$, $y = 3x^2 - 2$, $y = -2x^2 - 2$

Without using a graphing utility, draw a complete graph of each function. Describe how each graph can be obtained from the graph of $y = x^2$.

7. $y = 4x^2$
8. $y = -3x^2$
9. $y = (x-5)^2$
10. $y = (x+1)^2$
11. $y = 2x^2 - 3$
12. $y = -3x^2 + 2$
13. $y = (x-4)^2 + 3$
14. $y = -(x+3)^2 - 2$
15. $y = 2(x-1)^2 + 3$
16. $y = -3(x+4)^2 - 5$

Draw complete graphs of the three equations on the same coordinate system.

17. $y = \sin x$, $y = \sin(x-3)$, $y = \sin(x+2)$
18. $y = \sin x$, $y = 2\sin x$, $y = 4\sin x$
19. $y = \cos x$, $y = -2\cos x$, $y = -3\cos x$
20. $y = \cos x$, $y = 2\cos(x-2)$, $y = -\cos(x+3)$

Without using a graphing utility, draw a complete graph of each function. Describe how each graph can be obtained from the graph of $y = \sin x$.

21. $y = 2\sin(x-3)$
22. $y = 1 - \sin(x+2)$
23. Assume the point $(3, 4)$ is on the graph of $y = f(x)$. What point is on the graph of
    (a) $y = 2f(x)$?
    (b) $y = 4f(x)$?
    (c) $y = 2 + f(x)$?
    (d) $y = -2 - 3f(x)$?
24. Assume the point $(4, 3)$ is on the graph of $y = f(x)$. Find b so that $(2, b)$ is on the graph of $y = f(x+2)$.
25. Assume the point $(-1, 5)$ is on the graph of $y = f(x)$. Find b so that $(2, b)$ is on the graph of $y = f(x-3)$.
26. Assume the points $(2, 5)$ and $(0, 3)$ are on the graph of $y = f(x)$. What points are on the graph of
    (a) $y = f(x-2)$?
    (b) $y = f(x-2)$?
    (c) $y = -3f(x-2)$?
    (d) $y = 3 + 2f(x-2)$?

27. (a) What is an equation of the graph that is obtained from the graph of $y = x^2$ after applying a vertical stretch by a factor of 3 followed by a horizontal shift right 4 units? Sketch a complete graph of the equation.
    (b) What is an equation of the graph if the order of the two transformations in (a) is interchanged? Sketch a complete graph of the equation.
    (c) Are the graphs in (a) and (b) the same? Explain any differences.
28. (a) What is an equation of the graph that is obtained from the graph of $y = x^2$ after applying a vertical stretch by a factor of 3 followed by a vertical shift up 4 units? Sketch a complete graph of the equation.
    (b) What is an equation if the order of the two transformations in (a) is interchanged? Sketch a complete graph of the equation.
    (c) Are the graphs in (a) and (b) the same? Explain any differences.
29. What is an equation of the graph that is obtained from the graph of $y = x^2$ by a horizontal shift left 2 units, followed by a vertical stretch by a factor of 3, followed by a vertical shift down 4 units?
30. What is an equation of the graph that is obtained from the graph of $y = x^2$ by a horizontal shift right 4 units, followed by a vertical stretch by a factor of 2, followed by a reflection through the x-axis, followed by a vertical shift up 3 units?
31. Let $f(x) = 2x^2 - 3x + 5$. Apply a vertical stretch by a factor of 2 followed by a horizontal shift right 3 units to the graph of f. Show what happens to the points $(-1, f(-1)), (0, f(0))$, and $(2, f(2))$.
32. Let f be given by the graph in Figure 1.6.19. Determine the points on the graph of $y = 2 + 3f(x + 1)$ corresponding to the points $(-2, f(-2))$, $(0, f(0))$ and $(4, f(4))$.
33. Consider the graph in Figure 1.6.20.
    (a) Sketch a complete graph of $y = -1 + f(x)$.
    (b) Sketch a complete graph of $y = f(x + 1)$.
    (c) Sketch a complete graph of $y = -2f(x)$.
    (d) Sketch a complete graph of $y = 2f(x - 1)$.

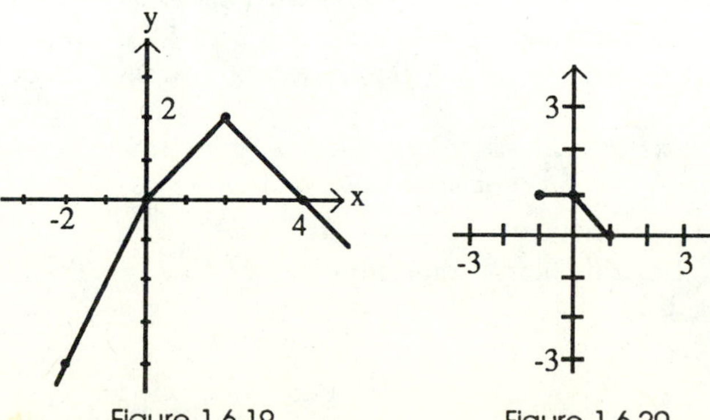

Figure 1.6.19 Figure 1.6.20

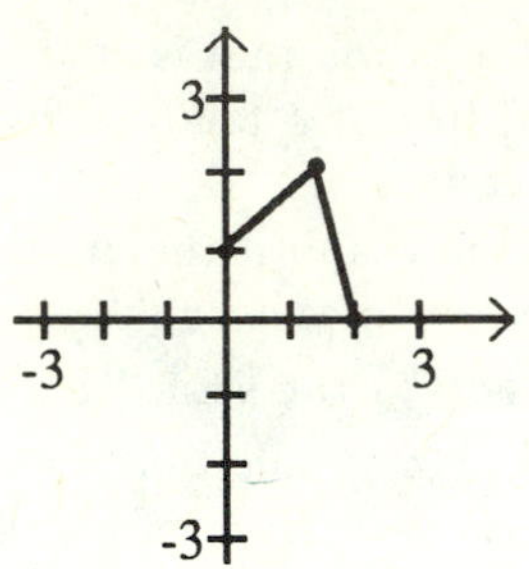

Figure 1.6.21

34. Consider the graph in Figure 1.6.21.
   (a) Sketch a complete graph of $y = 1 + f(x)$.
   (b) Sketch a complete graph of $y = -2 + f(x)$.
   (c) Sketch a complete graph of $y = 2f(x)$.
   (d) Sketch a complete graph of $y = -2f(x)$.
   (e) Sketch a complete graph of $y = f(x + 2)$.
   (f) Sketch a complete graph of $y = -1 + 2f(x - 1)$.

1.7 More on Quadratic Functions and Geometric Transformations

In this section we show that the graph of any quadratic function can be obtained from the graph of $y = x^2$ using the geometric transformations described in the previous section. These graphs have either a highest point or a lowest point. We also investigate applications with solutions that involve identifying the highest or lowest point of the graph of a quadratic function.

The first example illustrates how to determine transformations needed to obtain the graph of a quadratic function from the graph of $y = x^2$.

• EXAMPLE 1: Describe how a complete graph of $f(x) = -2x^2 - 12x - 13$ can be obtained from the graph of $y = x^2$.

SOLUTION: We rewrite $f(x) = -2x^2 - 12x - 13$ in the form $y = a(x - h)^2 + k$ in order to identify the needed transformations. This can be accomplished by completing the square.

The values of a, h, and k give us the transformations we need.

$$f(x) = -2x^2 - 12x - 13$$
$$f(x) = -2(x^2 + 6x) - 13$$
$$f(x) = -2(x^2 + 6x + 9) - 13 + 18$$
$$f(x) = -2(x + 3)^2 + 5$$

The graph of $f(x) = -2x^2 - 12x - 13 = -2(x+3)^2 + 5$ can be obtained from the graph of $y = x^2$ by applying, in order, the following transformations.

1. Vertical stretch by a factor of 2 to get $y = 2x^2$.
2. Reflection through the x-axis to get $y = -2x^2$.
3. Horizontal shift left 3 units to get $y = -2(x+3)^2$.
4. Vertical shift up 5 units to get $f(x) = -2(x+3)^2 + 5$.

The effect of the four transformations described above is illustrated in Figure 1.7.1. •

The algebraic computations in Example 1 can be checked using a graphing utility. We can see from Figure 1.7.1 that the standard viewing rectangle $[-10, 10]$ by $[-10, 10]$ should display a complete graph of $f(x) = -2x^2 - 12x - 13$. If we draw the graphs of $f(x) = -2x^2 - 12x - 13$ and $g(x) = -2(x+3)^2 + 5$ in the viewing rectangle $[-10, 10]$ by $[-10, 10]$, then the two graphs will coincide if no algebraic mistake was made.

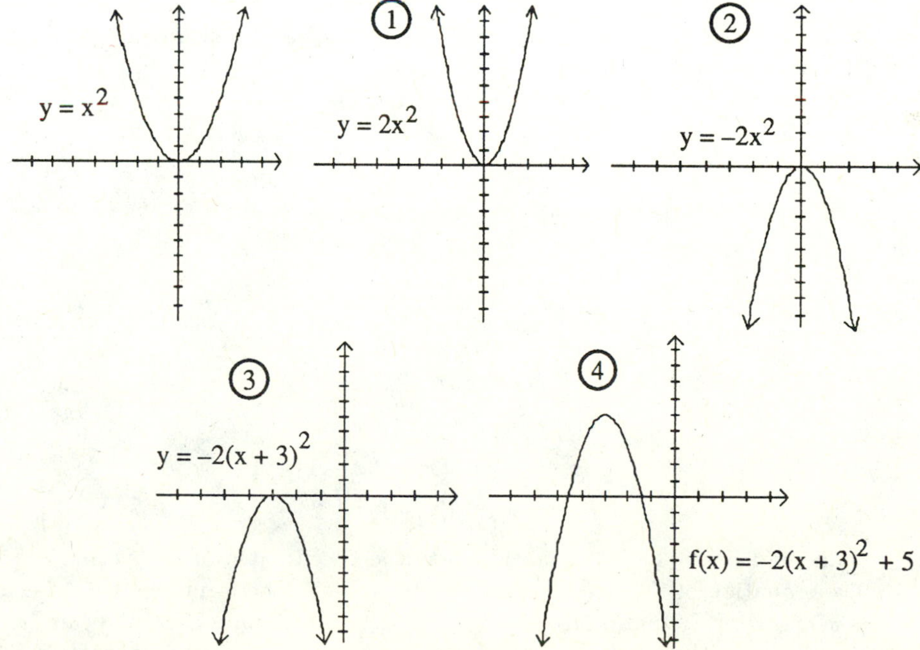

Figure 1.7.1

Recall that the graph of a quadratic function is called a *parabola*. The graph of $y = x^2$ has a lowest point at $(0,0)$ and is symmetric with respect to the line $x = 0$ (y-axis). From Figure 1.7.1, we can see that the highest point on the graph of $f(x) = -2(x+3)^2 + 5$ has coordinates $(-3,5)$. If the coordinate plane is folded along the line $x = -3$, then the portion of the graph of $f(x) = -2(x+3)^2 + 5$ to the left of the line $x = -3$ will coincide with the portion of the graph to the right of the line $x = -3$. Thus, the graph of $f(x) = -2(x+3)^2+5$ is symmetric with respect to the vertical line $x = -3$. If we keep track of how the parabola is obtained from $y = x^2$, the equation of the line of symmetry and the coordinates of the highest or lowest point can be determined. In Example 1, the coordinates of the highest point are $(-3,5)$ and the line of symmetry is $x = -3$ because we have shifted the graph of $y = -2x^2$ left 3 units and up 5 units.

We can also find the coordinates of the vertex and the line of symmetry for the graph of $f(x) = -2(x+3)^2 + 5$ by reasoning numerically. The largest value f can have is 5 because $-2(x+3)^2$ is *not* positive for any value of x. The largest value occurs when $-2(x+3)^2 = 0$. Thus, $x = -3$ produces the largest value of f and $(-3,5)$ is the vertex of the parabola. The value of $-2(x+3)^2$ is the same for any two values of x the same distance from $x = -3$. Therefore, the value of f is the same for these two values of x, and the line of symmetry is $x = -3$.

• **Quadratic Function** Every quadratic function f can be written in the form $f(x) = a(x-h)^2 + k$. The point (h,k) is called the **vertex** of this parabola. The graph of $f(x) = a(x-h)^2 + k$ is symmetric with respect to the vertical line $x = h$. This line is called the **line of symmetry** of the parabola.

The vertex (h,k) of a parabola will be the highest point on the graph if $a < 0$, and it will be the lowest point on the graph if $a > 0$.

The vertex of $y = x^2$ is $(0,0)$ and the line of symmetry is $x = 0$. The vertex of $f(x) = -2(x+3)^2 + 5$ is $(-3,5)$ and the line of symmetry is $x = -3$.

• EXAMPLE 2: Find the coordinates of the vertex, an equation of the line of symmetry, the y-intercept, and sketch a complete graph of the parabola $f(x) = 3x^2 + 6x + 1$.

SOLUTION: First we complete the square to rewrite f as follows.

$$f(x) = 3x^2 + 6x + 1$$
$$f(x) = 3(x^2 + 2x) + 1$$
$$f(x) = 3(x^2 + 2x + 1) + 1 - 3$$
$$f(x) = 3(x+1)^2 - 2$$

We can read from the last equation that the coordinates of the vertex are $(-1,-2)$ and that an equation of the line of symmetry is $x = -1$. The y-intercept is 1 because $f(0) = 1$. This is enough information to produce the sketch in Figure 1.7.2. We can draw the portion of the graph to the right of $x = -1$ and then use symmetry to complete the graph. The computer graph given in Figure 1.7.3 confirms our sketch. •

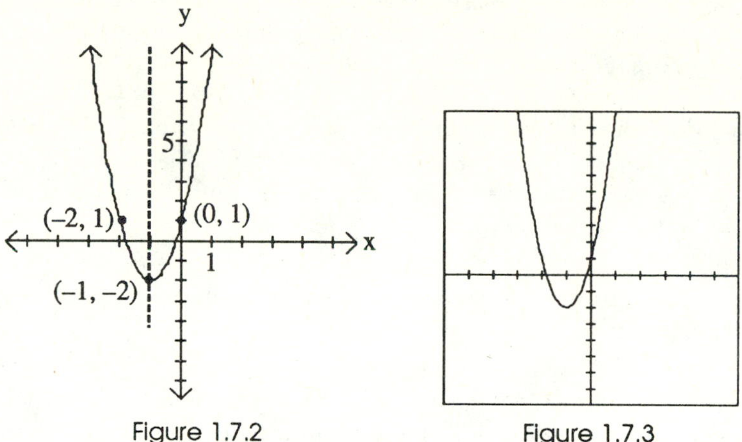

Figure 1.7.2 Figure 1.7.3

We are now in a position to prove a claim we made in Section 1.2 about the coordinates of the vertex of a parabola.

• EXAMPLE 3: Show that the line of symmetry of $f(x) = ax^2 + bx + c$ is $x = -\frac{b}{2a}$ and the vertex is $\left(-\frac{b}{2a}, c - \frac{b^2}{4a}\right)$ where $f\left(-\frac{b}{2a}\right) = c - \frac{b^2}{4a}$.

SOLUTION: We complete the square to rewrite f as follows.

$$f(x) = ax^2 + bx + c$$

$$f(x) = a(x^2 + \frac{b}{a}x) + c$$

$$f(x) = a(x^2 + \frac{b}{a}x + \frac{b^2}{4a^2}) + c - \frac{b^2}{4a}$$

$$f(x) = a(x + \frac{b}{2a})^2 + c - \frac{b^2}{4a}$$

Now we can see that the line of symmetry is $x = -\frac{b}{2a}$. Thus, the x-coordinate of the vertex is $-\frac{b}{2a}$, the y-coordinate of the vertex is $f\left(-\frac{b}{2a}\right) = c - \frac{b^2}{4a}$, and the vertex is $\left(-\frac{b}{2a}, c - \frac{b^2}{4a}\right)$. •

Using Example 3, we find that the coordinates of the vertex of the parabola $f(x) = 3x^2 + 6x + 1$ of Example 2 are (h, k), where $h = -\frac{b}{2a} = -\frac{6}{2(3)} = -1$ and $k = c - \frac{b^2}{4a} = 1 - \frac{6^2}{4(3)} = -2$. This is consistent with the coordinates of the vertex found in Example 2 by completing the square.

• EXAMPLE 4: Determine the possible number of real zeros that a quadratic function can have.

SOLUTION: Let $f(x) = ax^2 + bx + c$ be a quadratic function. A zero of f is a value of x for which $f(x) = 0$. The real zeros of a function are the same as the x-intercepts of its graph.

There are three possibilities for the graph of f: the graph may *not* cross the x-axis; the graph may be tangent to the x-axis; or the graph may cross the x-axis twice (Figure 1.7.4). Thus, the number of real zeros can be 0, 1, or 2. •

Using the formulas of Example 3 is much easier than using completing the square to find the vertex and line of symmetry of the graph of a quadratic function. If a quadratic function is given in factored form, then we can find the equation of the line of symmetry and the vertex without expanding the product.

• EXAMPLE 5: Find the coordinates of the vertex, an equation of the line of symmetry, the x- and y-intercepts, and sketch a complete graph of $f(x) = (x-1)(x+5)$.

SOLUTION: The line of symmetry of the parabola $f(x) = (x-1)(x+5)$ is a vertical line. The two real zeros -5 and 1 are the x-intercepts, must lie on opposite sides of the line of symmetry, and must be equally distant from the line of symmetry. Thus, the line of symmetry must be halfway between $x = -5$ and $x = 1$. Therefore, an equation for the line of symmetry is

$$x = \frac{-5+1}{2} = -2\,.$$

Notice that -2 is also the midpoint of the line segment on the x-axis from $x = -5$ to $x = 1$.

Now $f(-2)$ is the y-coordinate of the vertex. The vertex is $(-2, -9)$ because $f(-2) = (-2-1)(-2+5) = -9$. The y-intercept is -5 because $f(0) = -5$ when $x = 0$. This is enough information to draw the sketch in Figure 1.7.5. The computer-drawn graph of f in Figure 1.7.6 confirms our sketch. •

• EXAMPLE 6: A graph of a function f is given in Figure 1.7.7. Sketch a graph of

(a) $y = 3f(x)$. (b) $y = -2f(x)$.

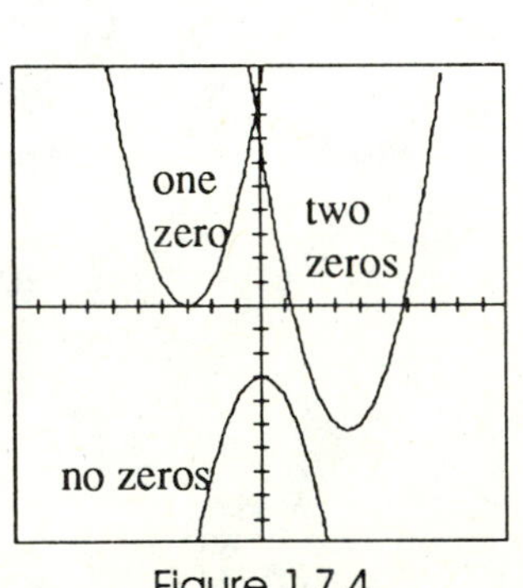

Figure 1.7.4

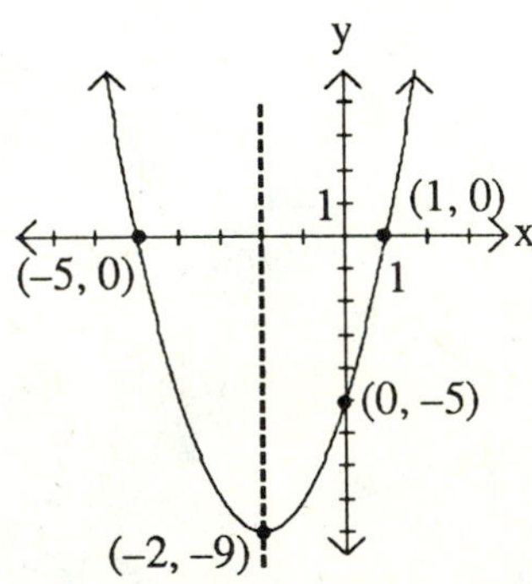

Figure 1.7.5

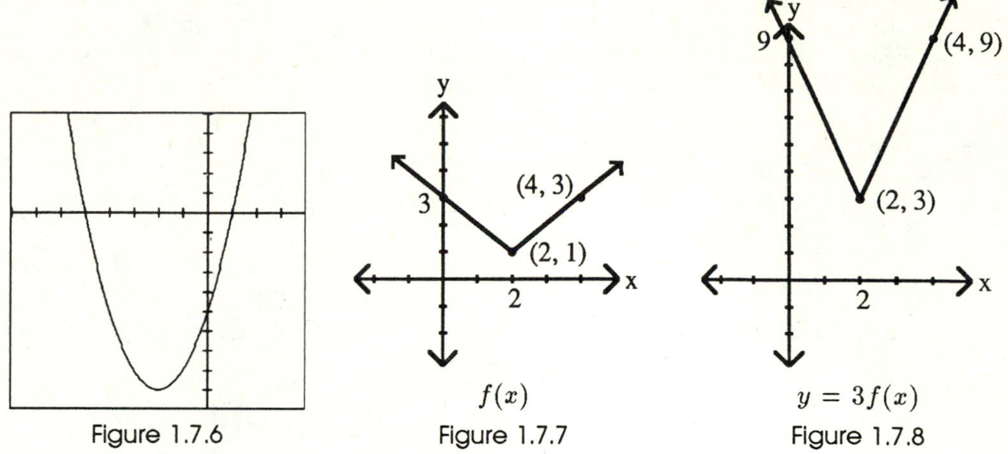

Figure 1.7.6 Figure 1.7.7 Figure 1.7.8

SOLUTION:

(a) Under a vertical stretch by a factor of 3 the three points $(0,3)$, $(2,1)$, and $(4,3)$ of f are moved to $(0,9)$, $(2,3)$, and $(4,9)$. (Why?) A complete graph of $y = 3f(x)$ is given in Figure 1.7.8.

(b) Under a vertical stretch by a factor of 2 followed by a reflection through the x-axis, the three points $(0,3)$, $(2,1)$ and $(4,3)$ of f are moved to $(0,-6)$, $(2,-2)$, and $(4,-6)$. (Why?) A complete graph of $y = -2f(x)$ is given in Figure 1.7.9. •

Discriminant

In Section 2.1 we will show that the solutions (zeros) of $ax^2 + bx + c = 0$ are given by the quadratic formula

$$x = \frac{-b \pm \sqrt{b^2 - 4ac}}{2a}.$$

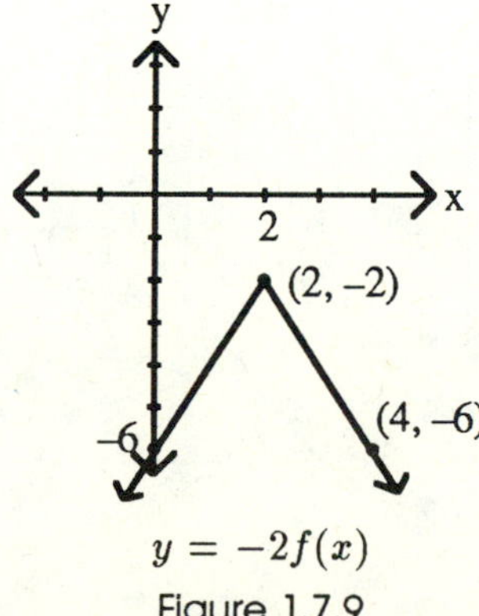

Figure 1.7.9

The expression $b^2 - 4ac$ under the radical sign is called the **discriminant** of the quadratic equation $ax^2 + bx + c = 0$. The discriminant determines the number of real zeros of the quadratic function $f(x) = ax^2 + bx + c$. The discriminant together with Example 4 above suggests that the following theorem is true.

• THEOREM 1 Let $f(x) = ax^2 + bx + c$.

(i) If $b^2 - 4ac < 0$, then $f(x) = ax^2 + bx + c$ has no real zeros and the graph of f lies entirely above or entirely below the x-axis.

(ii) If $b^2 - 4ac = 0$, then $f(x) = ax^2 + bx + c$ has exactly one real zero and the graph of f is tangent to the x-axis.

(iii) If $b^2 - 4ac > 0$, then $f(x) = ax^2 + bx + c$ has two real zeros and the graph of f crosses the x-axis twice.

• EXAMPLE 7: Determine the number of real zeros of $f(x) = 4x^2 - 4x + 1$.

SOLUTION: The number of real zeros of f can be determined by the discriminant,

$$b^2 - 4ac = (-4)^2 - 4(4)(1) = 0.$$

Because the discriminant is 0, $f = 4x^2 - 4x + 1$ has exactly one real zero. The graph in Figure 1.7.10 confirms this information. •

• EXAMPLE 8: A long, rectangular sheet of metal 10 inches wide is to be made into a gutter by turning up sides of equal length perpendicular to the sheet (Figure 1.7.11). Find the length that must be turned up to give a gutter with maximum cross-sectional area.

SOLUTION: If x is the length of a side turned up, then the bottom of the gutter is $10 - 2x$ inches wide. The cross-sectional area is $A(x) = x(10 - 2x)$. The graph of A is a parabola that opens downward (Figure 1.7.12). Only the portion of the graph in the first quadrant is a graph of the problem situation. (Why?)

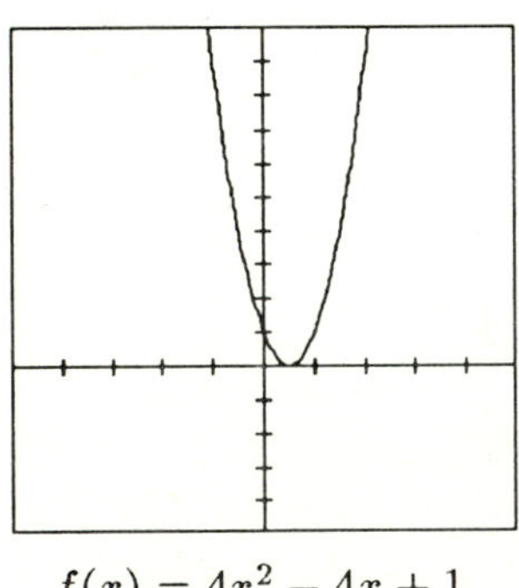

$f(x) = 4x^2 - 4x + 1$
$[-5, 5]$ by $[-5, 10]$
Figure 1.7.10

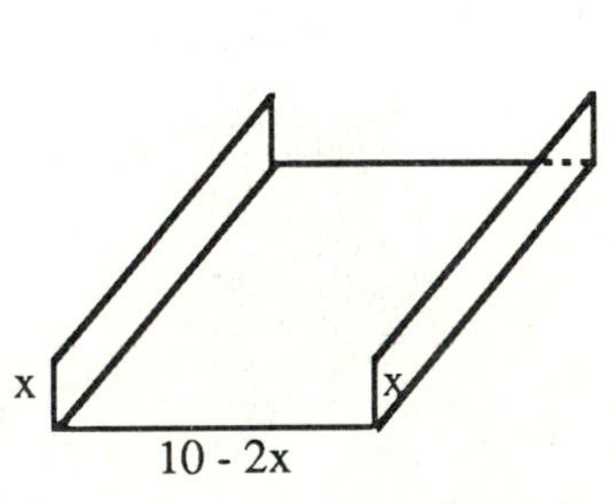

Figure 1.7.11

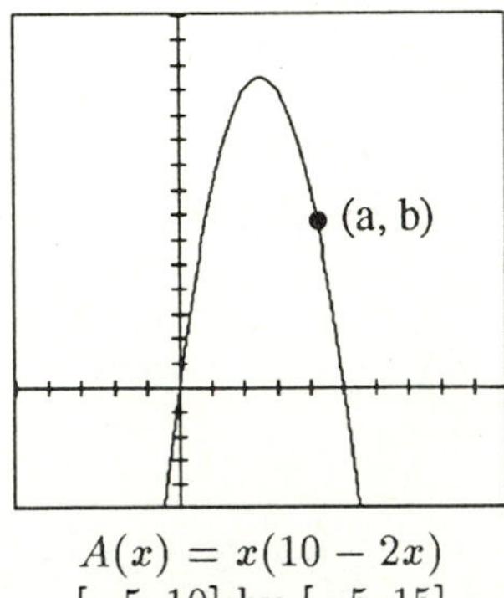

$A(x) = x(10 - 2x)$
$[-5, 10]$ by $[-5, 15]$
Figure 1.7.12

If (a, b) is a point in the first quadrant on the graph of A, then a is a possible length of a side turned up and b the corresponding cross-sectional area of the resulting gutter. (Why?) The maximum value of the area function A occurs at the vertex of the parabola in Figure 1.7.12. The vertex appears to have coordinates $(2.5, 12.5)$ so that x is 2.5 inches and the corresponding cross-sectional area is 12.5 square inches.

The exact solution can be determined algebraically. The x-intercepts (zeros) of $A(x) = x(10 - 2x)$ are 0 and 5. The x-coordinate of the vertex is $\frac{0+5}{2} = 2.5$. The y-coordinate of the vertex is 12.5 because $A(2.5) = 12.5$. Because the vertex is the highest point of $A(x) = x(10 - 2x)$, turning up sides of length 2.5 inches produces a gutter of maximum possible cross-sectional area, namely 12.5 square inches. •

Suppose we have 200 feet of fence and want to enclose a rectangular plot of land using an existing wall as one side of the rectangle (Figure 1.7.13). If the length of a side of the rectangle perpendicular to the wall is x, then the length of the side parallel to the wall is $200 - 2x$. The area of the rectangular plot enclosed by the fence is $x(200 - 2x)$.

• EXAMPLE 9: If 200 feet of fence are used to enclose a rectangular plot of land using an existing wall, find the dimensions of the rectangle with maximum enclosed area.

SOLUTION: The function $A(x) = x(200 - 2x)$ is an algebraic representation for the area of the rectangular plot. The graph of this function in the $[-10, 110]$ by $[-1000, 6000]$ viewing rectangle is given in Figure 1.7.14 and suggests that the maximum enclosed area is about 5000 when x is about 50. Several applications of zoom-in lead to the graph in Figure 1.7.15. This figure permits us to read that the maximum area is 5000 and occurs for $x = 50$ with error at most 0.01.

The zeros of $A(x) = x(200 - 2x)$ are 0 and 100. Thus, the x-coordinate of the vertex is 50 and the y-coordinate of the vertex is $A(50) = 5000$. Therefore, the exact coordinates of the vertex are $(50, 5000)$. This means the maximum enclosed area is 5000 square feet, and

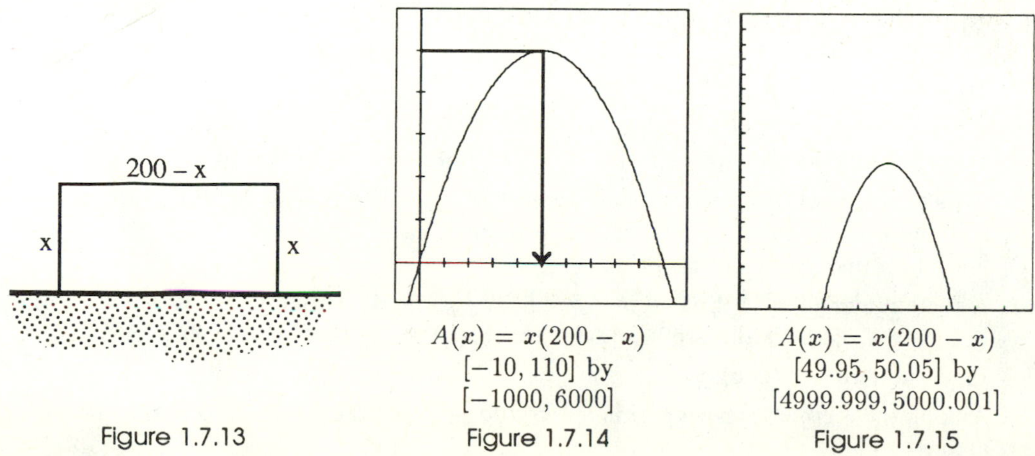

Figure 1.7.13

$A(x) = x(200 - x)$
$[-10, 110]$ by
$[-1000, 6000]$

Figure 1.7.14

$A(x) = x(200 - x)$
$[49.95, 50.05]$ by
$[4999.999, 5000.001]$

Figure 1.7.15

it occurs when the length of a side of the rectangle perpendicular to the wall is 50 feet. The dimensions of the maxium enclosed area are 50 feet by 100 feet. •

• EXERCISES 1-7

Write a sequence of transformations that will produce the graph of each function from the graph of $y = x^2$. Specify an order in which the transformations could be applied.

1. $f(x) = 3(x-2)^2$
2. $f(x) = -2(x+1)^2 + 3$
3. $f(x) = x^2 + 3$
4. $f(x) = x^2 - 4x + 6$
5. $f(x) = x^2 - 6x + 12$
6. $f(x) = 2x^2 - 8x + 20$
7. $f(x) = 10 - 16x - x^2$
8. $f(x) = 6 - 10x + 5x^2$

9–16. Use the information from Exercises 1–8 to graph each function *without* the aid of a graphing utility. Indicate the vertex, line of symmetry, y-intercept, and algebraically determine the real zeros (if any). Check your answers using a graphing utility.

Without using a graphing utility, draw a complete graph of $y = f(x)$. List transformations that can be applied, in order, to obtain the graph from the graph of $y = |x|$

17. $f(x) = |1x - 3|$
18. $f(x) = -2|x+3|$
19. $f(x) = 2 + 3|x-4|$
20. $f(x) = 3 - |x-2|$
21. Suppose that a parabola has line of symmetry $x = 2$ and contains the points $(1, -1)$, $(4, 8)$, and $(-1, 23)$.
    (a) Find three additional points on the parabola using symmetry.
    (b) Find an equation of the parabola and draw its graph. *Hint:* An equation is of the form $y = a(x-2)^2 + k$. Use two points to determine a and k by solving two equations simultaneously.
22. Let T_1 be a vertical shift of 3 units upward and T_2 be a vertical stretch by a factor of 2. Show that applying T_1 to $y = x^2$ followed by applying T_2 to the resulting graph is *not* the same as applying T_2 to $y = x^2$ followed by applying T_1 to the resulting graph.
23. Determine the midpoint M of the real zeros of each quadratic function. Find the equation of the line of symmetry, the coordinates of the vertex, x-intercepts, y-intercept, and sketch a complete graph.
    (a) $f(x) = (x-8)(x+2)$ (b) $f(x) = (x+8)(2x-6)$
24. A rectangle is 3 feet longer than it is wide. If each side is increased by 1 foot, the area of the new rectangle is 208 square feet. Find the dimensions of the original rectangle.
25. A page of a book is 4 inches longer than it is wide. The margin at top and bottom is 2 inches, and the margin on each side is 1 inch. If the area of the printed matter is 35 square inches, find the dimensions of the page.
26. A rectangular pool of dimensions 25 feet by 40 feet is surrounded by a walk of uniform width. If the area of the walk is 504 square feet, find the width of the walk.
27. Find the dimensions of a rectangle with a perimeter of 37 feet and an area of 85 square feet.
28. Among the rectangles with perimeters of 100 feet, find the dimensions of the one with maximum area.

29. A piece of wire 20 feet long is cut into two pieces so that the sum of the squares of the length of each piece is 202 square feet. Find the length of each piece.

30. A rectangular fence is to be constructed so that one side is against the wall of a large building. Determine the maximum area that can be enclosed if the total length of fencing to be used is 500 feet.

In physics, it can be shown that the trajectory (that is, the path) of an object projected with constant acceleration at a given angle with the horizontal is given by a quadratic function $y = ax^2 + bx + c$. For each of the given values of a, b, and c, determine the greatest height H attained by the object and the horizontal range R of the object (Figure 1.7.16).

31. $a = -2$, $b = 48$, $c = 22$

32. $a = -4$, $b = 79$, $c = 38$

33. A large apartment rental company has 1600 units available, and 800 are currently rented at \$300 per month. A market survey indicates that each \$5 decrease in monthly rent will result in 20 new tenants.
    (a) Determine a function R that represents the total rental income realized by the company where x is the number of \$5 decreases in monthly rent (that is, if $x = 3$, then the monthly rent is \$285). Assume that all rents are adjusted to the new figure.
    (b) Draw a complete graph of the function $y = R(x)$.
    (c) What is the domain of the function R? What values of x make sense in the problem situation?
    (d) Determine the rent that will yield the rental company the greatest monthly income.

34. Determine a, b, and c so the graph of $y = ax^2 + bx + c$ contains the points $(1, 8)$, $(-1, 2)$, and $(3, 6)$. *Hint:* Write and solve three equations simultaneously.

35. If one solution of $2x^2 + 11x + k = 0$ is $x = -5$, find the other solution.

36. The sum of the solutions of a quadratic equation is -2, and their product is -15. Find the solutions and write the quadratic equation.

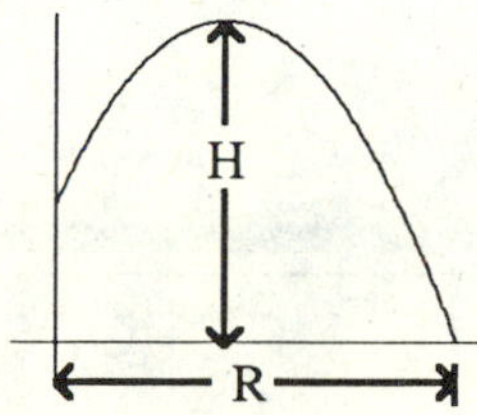

Figure 1.7.16

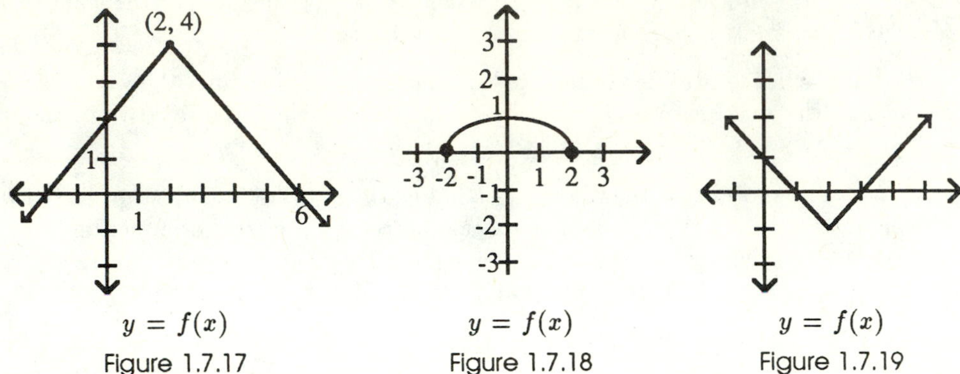

$y = f(x)$
Figure 1.7.17

$y = f(x)$
Figure 1.7.18

$y = f(x)$
Figure 1.7.19

37. Show that if x_1 and x_2 are the real zeros of a quadratic function $f(x) = ax^2 + bx + c$, then
    (a) the x-coordinate of the vertex of the graph is the *midpoint* of the two points labeled x_1 and x_2 on the x-axis.
    (b) the vertical line
$$x = \frac{x_1 + x_2}{2}$$
    is the line of symmetry of the graph of $y = f(x)$.
38. A graph of a function f is given by Figure 1.7.17. Sketch the graph of
    (a) $y = 4f(x)$ (b) $y = f(x-2)$
    (c) $y = -f(x)$ (d) $y = -2 + 3f(x)$
39. Consider the graph of a function f in Figure 1.7.18. Sketch the graph of
    (a) $g(x) = 3 + f(x)$. (b) $g(x) = 3 \cdot f(x)$.
    (c) $g(x) = f(x-3)$. (d) $g(x) = f(x+3)$.
40. Consider the graph of a function f in Figure 1.7.19. Sketch the graph of
    (a) $g(x) = -3 \cdot f(x)$. (b) $g(x) = -1 + 2 \cdot f(x)$.
    (c) $g(x) = 1 - 2 \cdot f(x-1)$. (d) $g(x) = 1 + 2 \cdot f(-x)$.
41. What are the x-intercepts of $f(x) = a(x-h)^2 + k$ in terms of a, h, and k? What is the discriminant in terms of a, h, and k?

1.8 Operations on Functions

The sum, difference, product, and quotient of two functions are defined in this section. We have been implicitly using these definitions in this textbook.

• DEFINITION Let f and g be two functions. We define the **sum** $f+g$, the **difference** $f-g$, and the **product** fg to be the functions with domains the set of all numbers common to the domains of f and g and with rules given by the following formulas:

$$(f+g)(x) = f(x) + g(x),$$
$$(f-g)(x) = f(x) - g(x),$$
$$(fg)(x) = f(x)g(x).$$ •

• EXAMPLE 1: Let $f(x) = \sqrt{2-x}$ and $g(x) = \frac{1}{x}$. Find the domains and rules of $f+g$, $f-g$, and fg.

SOLUTION: The domain of f is the set of all real numbers for which $2-x \geq 0$. (Why?) Thus, the domain of f is $(-\infty, 2]$. The domain of g is the set of all nonzero real numbers. The set of elements common to the domains of f and g is $(-\infty, 0) \cup (0, 2]$. Therefore, the domains of $f+g$, $f-g$, and fg are $(-\infty, 0) \cup (0, 2]$. The rules for these functions are given by the following formulas:

$$(f+g)(x) = f(x) + g(x) = \sqrt{2-x} + \frac{1}{x},$$
$$(f-g)(x) = f(x) - g(x) = \sqrt{2-x} - \frac{1}{x},$$
$$(fg)(x) = f(x)g(x) = \frac{1}{x}\sqrt{2-x}.$$ •

• EXAMPLE 2: Use complete graphs of $f(x) = x$ and $g(x) = \sin x$ to obtain a complete graph of $h(x) = x + \sin x$.

SOLUTION: Complete graphs of f and g are given in Figure 1.8.1. For each value of x we add the corresponding values of $f(x)$ and $g(x)$ to obtain the value of $h(x)$. Thus, if (a, b) and (a, c) are points on the graphs of f and g, respectively, then $(a, b+c)$ is a point on the graph of h (Figure 1.8.2). Each point on the graph of h can be obtained this way.

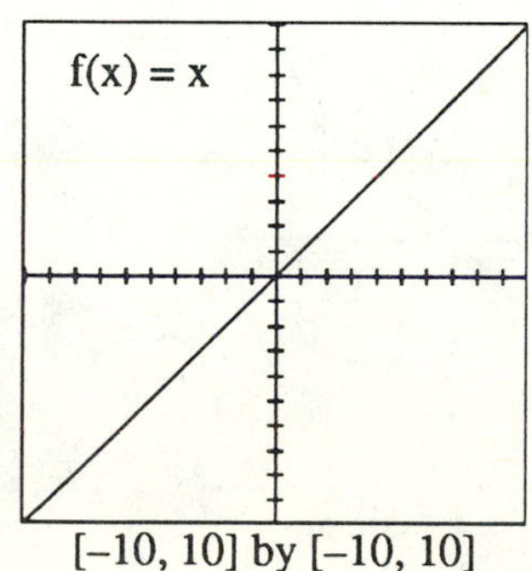

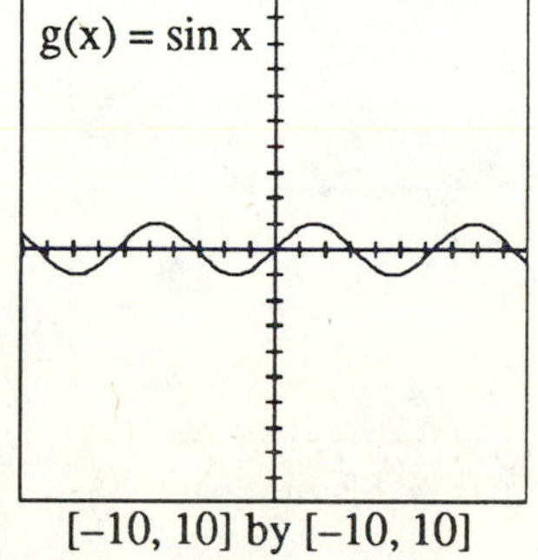

Figure 1.8.1

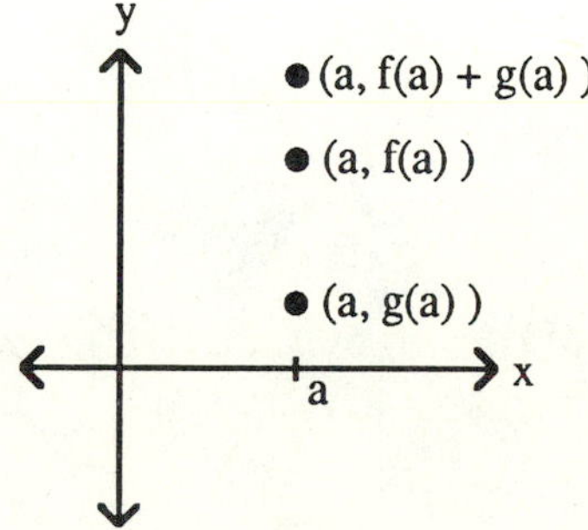

Figure 1.8.2

Notice that the values of $g(x)$ vary between -1 and 1 for all x, but that $f(x) \to \infty$ as $x \to \infty$ and $f(x) \to -\infty$ as $x \to -\infty$. For values of x far away from 0, the values of $h(x) = x + \sin x$ are dominated by the values of $f(x) = x$. In fact, for $|x|$ large, $f(x) = x$ is a very good approximation to $h(x) = x + \sin x$ because the values of $g(x) = \sin x$ are relatively small. For each value of x, the value of $h(x)$ is within one unit of the value of $f(x)$. Figure 1.8.3 gives a complete graph of h. Compare the graphs of $h(x) = x + \sin x$ and $f(x) = x$ that are shown together in Figure 1.8.4. •

An additional restriction is needed to define the quotient of two functions.

• DEFINITION Let f and g be two functions. We define the **quotient** $\frac{f}{g}$ to be the function with domain the set of all numbers x common to the domains of f and g for which $g(x) \neq 0$, and with rule given by the following formula:

$$\frac{f}{g}(x) = \frac{f(x)}{g(x)}, \quad \text{provided} \quad g(x) \neq 0.$$ •

• EXAMPLE 3: Let $f(x) = x^2 - 1$ and $g(x) = x^2 + 1$. Find the domain and rule of $\frac{f}{g}$ and $\frac{g}{f}$.

SOLUTION: The domains of f and g are the set of all real numbers. Thus, the domain of $\frac{f}{g}$ is the set of all real numbers x for which $g(x) \neq 0$. Because g has no real zeros, the domain of $\frac{f}{g}$ is the set of all real numbers, and its rule is given by

$$\frac{f}{g}(x) = \frac{f(x)}{g(x)} = \frac{x^2 - 1}{x^2 + 1}.$$

The domain of $\frac{g}{f}$ is the set of all real numbers x for which $f(x) \neq 0$. The real zeros of f are -1 and 1. Therefore, the domain of $\frac{g}{f}$ is $(-\infty, -1) \cup (-1, 1) \cup (1, \infty)$ and its rule is given by

$$\frac{g}{f}(x) = \frac{g(x)}{f(x)} = \frac{x^2 + 1}{x^2 - 1}.$$ •

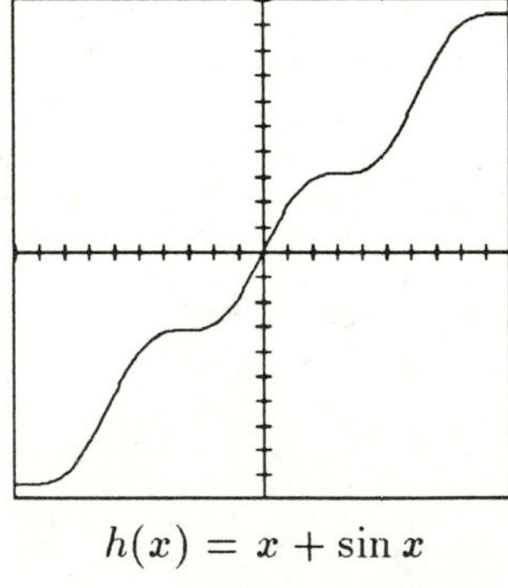

$h(x) = x + \sin x$

$[-10, 10]$ by $[-10, 10]$

Figure 1.8.3

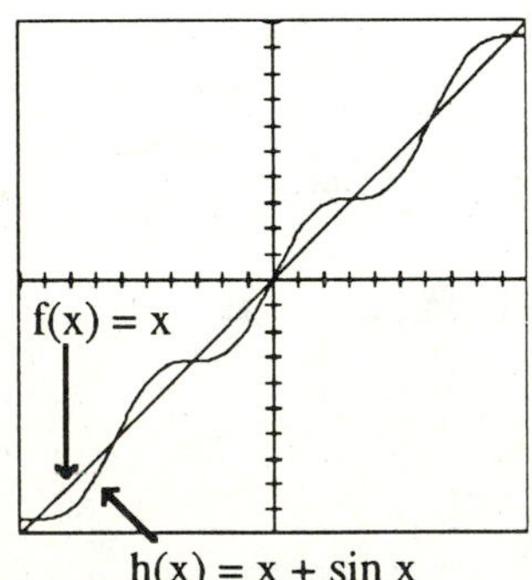

$[-10, 10]$ by $[-10, 10]$

Figure 1.8.4

Example 3 suggests that every rational function is a quotient of functions. Recall that f is a rational function if it can be expressed in the form $\frac{p(x)}{h(x)}$ where p and h are polynomial functions. Thus, f is the quotient of the polynomial functions p and h, that is, $f = \frac{p}{h}$.

Composition of Functions

Let f and g be functions and a an element in the domain of f. If $f(a)$ is in the domain of g, then we can determine $g(f(a))$.

- EXAMPLE 4: Let $f(x) = x + 1$ and $g(x) = \sqrt{x}$. If possible, determine $g(f(1))$ and $g(f(-2))$.

SOLUTION: We can determine $g(f(1))$ because $f(1) = 2$ and 2 is in the domain of g:

$$g(f(1)) = g(2) = \sqrt{2}.$$

Because $f(-2) = -1$ and -1 is not in the domain of g, $g(f(-2))$ is not defined. •

• DEFINITION Let f and g be functions. The **composition of f by g** is the function defined by $g \circ f(a) = g(f(a))$ for all values a in the domain of f for which $f(a)$ is in the domain of g. •

Notice that the domain of $g \circ f$ consists of all real numbers x for which $f(x)$ lies in the domain of g. Figure 1.8.5 gives a geometric illustration of $g \circ f$ in the case that the range of f is a subset of the domain of g. Let D be the domain of f, E the range of f, S the domain of g, and T the range of g. The domain of $g \circ f$ is D and the range of $g \circ f$ is a subset B of T. Loosely speaking, the function $g \circ f$ is the function obtained by applying f first and then applying g to the result. We also say that f *maps* D to E and g *maps* S to T. The composite function $g \circ f$ *maps* D to B.

- EXAMPLE 5: Let $f(x) = x^2+1$ and $g(x) = \sqrt{x}$. Determine $g \circ f(2)$, $f \circ g(9)$, and a rule for $f \circ g$.

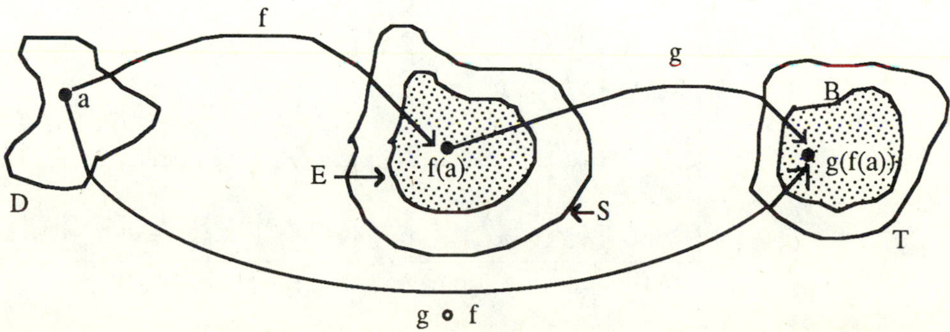

Figure 1.8.5

SOLUTION: The graphs of f and g are given in Figure 1.8.6. The range of f is $[1,\infty)$, and the domain of g is $[0,\infty)$. Thus, the range of f is a subset of the domain of g, and $g \circ f$ is defined for every a in the domain of f.

Because $g \circ f(2) = g(f(2))$ and $f(2) = 2^2 + 1 = 5$, we obtain:

$$\begin{aligned} g \circ f(2) &= g(f(2)) \\ &= g(5) \\ &= \sqrt{5}. \end{aligned}$$

Notice $f \circ g$ means to first apply g and then apply f to the result. The range of g is $[0,\infty)$ and the domain of f is $(-\infty,\infty)$. Thus, the range of g is a subset of the domain of f, and $f \circ g$ is defined for every a in the domain of g. A computation similar to the one above gives the following result:

$$\begin{aligned} f \circ g(9) &= f(g(9)) \\ &= f(\sqrt{9}) \\ &= f(3) \\ &= 10. \end{aligned}$$

Let a be any element in the domain of g. Then $a \geq 0$ and $g(a) = \sqrt{a}$ is in the domain of f. Thus, the domain of $f \circ g$ is $[0,\infty)$, the entire domain of g. We obtain the following rule for $f \circ g$:

$$\begin{aligned} (f \circ g)(x) &= f(g(x)) \\ &= f(\sqrt{x}) \\ &= (\sqrt{x})^2 + 1 \\ &= x + 1. \end{aligned}$$

Thus, $f \circ g(x) = x + 1$ with $x \geq 0$ because the domain of $f \circ g$ is $[0,\infty)$. •

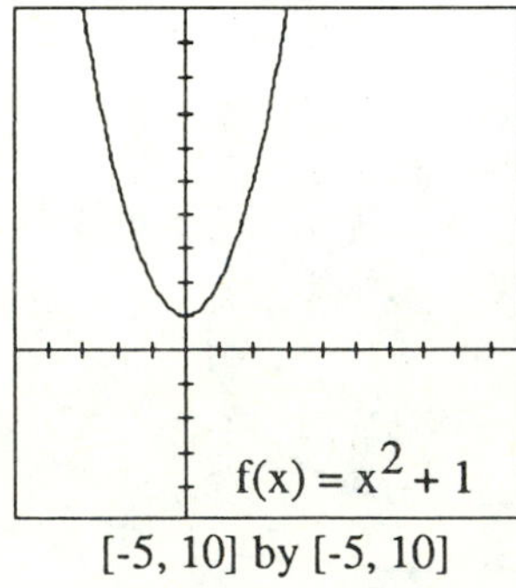

[-5, 10] by [-5, 10]

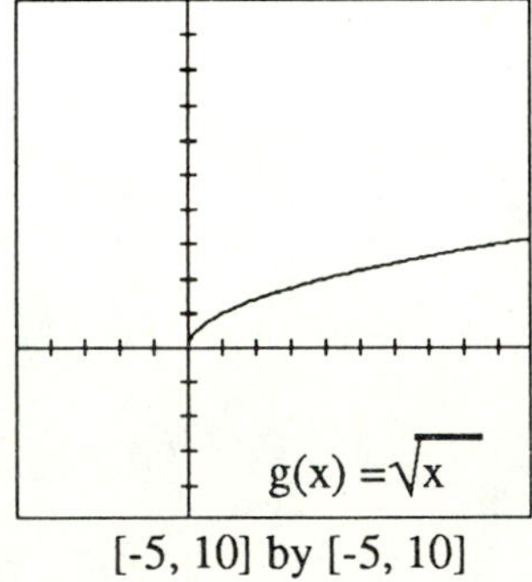

[-5, 10] by [-5, 10]

Figure 1.8.6

Notice in the above example we cannot simply write $f \circ g(x) = x + 1$ as a rule for $f \circ g$. The restriction $x \geq 0$ must be stated because the expression $x + 1$ is valid for x negative, but the domain of $f \circ g$ is $[0, \infty)$.

• EXAMPLE 6: Let $f(x) = \sqrt{x}$ and $g(x) = x - 3$. Compute $f \circ g(5)$ and $g \circ f(5)$. Determine the domain and range of f, g, $f \circ g$ and $g \circ f$, and then give a rule for $f \circ g$ and $g \circ f$.

SOLUTION: We use the definitions of $f \circ g$ and $g \circ f$ to complete the following. Notice that $f \circ g(5) \neq g \circ f(5)$.

$$\begin{aligned} g \circ f(5) &= g(f(5)) \\ &= g(\sqrt{5}) \\ &= \sqrt{5} - 3 \end{aligned} \qquad \begin{aligned} f \circ g(5) &= f(g(5)) \\ &= f(5-3) \\ &= f(2) \\ &= \sqrt{2} \end{aligned}$$

The rules or graphs of f and g can be used to establish the following:

$$\text{domain of } f = [0, \infty) \qquad \text{domain of } g = (-\infty, \infty)$$

$$\text{range of } f = [0, \infty) \qquad \text{range of } g = (-\infty, \infty)$$

Thus, the range of f is a subset of the domain of g so that the domain of $g \circ f$ is the entire domain of f, namely $[0, \infty)$. We determine a rule for $g \circ f$ in the following way. Let x be any element in the domain of $g \circ f$. Then $x \geq 0$ and

$$\begin{aligned} g \circ f(x) &= g(f(x)) \\ &= g(\sqrt{x}) \\ &= \sqrt{x} - 3. \end{aligned}$$

Notice that no additional restriction on x in $g \circ f(x) = \sqrt{x} - 3$ is needed because the expression $\sqrt{x} - 3$ is real if and only if x is in the domain of $g \circ f$. We can use the rule for $g \circ f$ or the graph of $g \circ f$ in Figure 1.8.7 to see that the range of $g \circ f$ is $[-3, \infty)$.

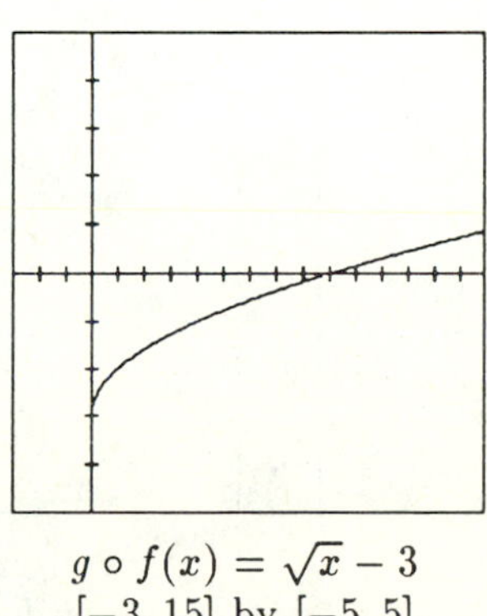

$g \circ f(x) = \sqrt{x} - 3$
$[-3, 15]$ by $[-5, 5]$
Figure 1.8.7

The range of g is not a subset of the domain of f. Thus, the domain of $f \circ g$ consists of those values of x for which $g(x)$ lies in the domain of f, namely, those values of x for which $x - 3 \geq 0$. Therefore, the domain of $f \circ g$ is $[3, \infty)$. We determine a rule for $f \circ g$ in the following way. Let x be any element in the domain of $f \circ g$. Then $x \geq 3$ and

$$\begin{aligned} f \circ g(x) &= f(g(x)) \\ &= f(x-3) \\ &= \sqrt{x-3}. \end{aligned}$$

Again, the expression $\sqrt{x-3}$ is real if and only if x is in the domain of $f \circ g$ (Figure 1.8.8). Thus, $f \circ g(x) = \sqrt{x-3}$ with no additional restriction on x gives a rule for $f \circ g$.

Summarizing, we have established the following information:

$g \circ f(x) = \sqrt{x} - 3$	$f \circ g(x) = \sqrt{x-3}$
domain of $g \circ f = [0, \infty)$	domain of $f \circ g = [3, \infty)$
range of $g \circ f = [-3, \infty)$	range of $f \circ g = [0, \infty)$

•

It follows from Example 3 that $f \circ g \neq g \circ f$. This shows that the operation of composition of functions is *not commutative.*

In the next two examples, we illustrate how a complicated function can be decomposed into simpler functions using composition of functions.

• EXAMPLE 7: Let $f(x) = (x-3)^{1/3}$. Find functions g and h so that $f = h \circ g$.

SOLUTION: You should not be surprised to see that the *cube root* function $y = x^{1/3}$ and the *linear* function $y = x - 3$ are involved. Let $g(x) = x - 3$, $h(x) = x^{1/3}$, and determine $h \circ g$:

$$\begin{aligned} h \circ g(x) &= h(g(x)) \\ &= h(x-3) \\ &= (x-3)^{1/3}. \end{aligned}$$

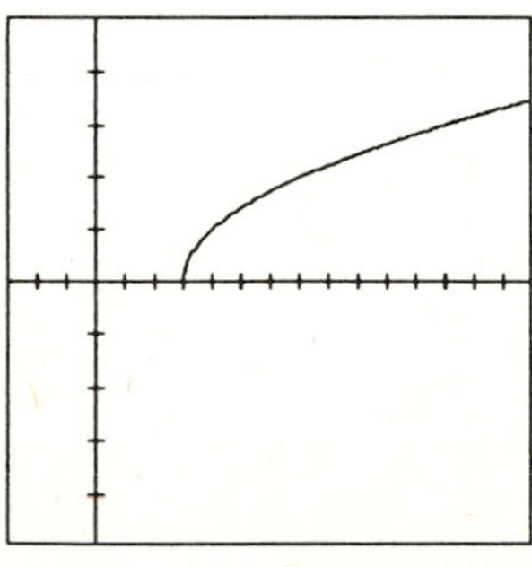

$f \circ g(x) = \sqrt{x-3}$
$[-3, 15]$ by $[-5, 5]$
Figure 1.8.8

Thus, $f = h \circ g$. •

Example 8 illustrates that a given function can be expressed as a composition of functions in more than one way.

• EXAMPLE 8: Express $f(x) = \frac{1}{x^2-4x+7}$ as a composition of simpler functions.

SOLUTION: There are numerous possibilities. We give two of them. Let $g(x) = \frac{1}{x}$, $h(x) = x^2 - 4x + 7$, and determine $g \circ h$:

$$\begin{aligned} g \circ h(x) &= g(h(x)) \\ &= g(x^2 - 4x + 7) \\ &= \frac{1}{x^2 - 4x + 7}. \end{aligned}$$

Thus, $f = g \circ h$.

We can obtain another decomposition of f by writing h as a composition of functions. First we complete the square to rewrite h as $h(x) = (x-2)^2 + 3$. Now, let $h_1(x) = x - 2$, $h_2(x) = x^2$ and $h_3(x) = x + 3$. Then, $h = h_3 \circ h_2 \circ h_1$ so that $f = g \circ h_3 \circ h_2 \circ h_1$. •

Composition of functions is important in applications. Suppose that a balloon in the shape of a sphere is being inflated with a gas (Figure 1.8.9). This means that the *radius* of the sphere is increasing as a function of time, say $r = f(t)$. Then, we can determine an algebraic representation for the *volume* of the sphere as a function of time.

The volume of a sphere is a function of its radius r given by $V = g(r) = \frac{4}{3}\pi r^3$. Substituting $r = f(t)$, we obtain the following form for V:

$$V = g(f(t)) = \frac{4}{3}\pi(f(t))^3,$$

$$V = g \circ f(t) = \frac{4}{3}\pi(f(t))^3.$$

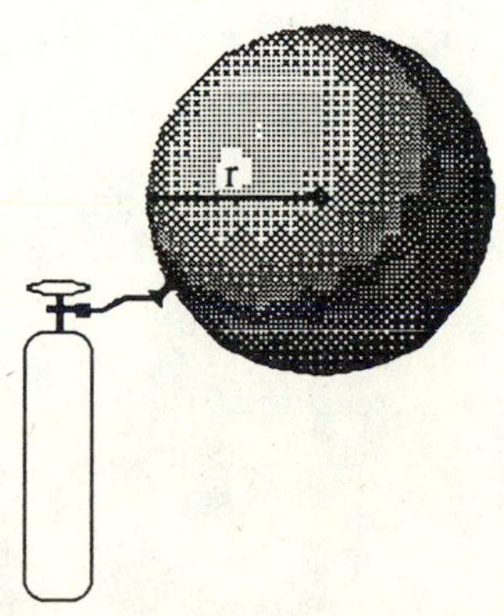

Figure 1.8.9

Thus, the volume of the sphere is also a function of time given by the algebraic model $V = g \circ f(t)$.

• EXAMPLE 9: A balloon in the shape of a sphere is being inflated with a gas (Figure 1.8.9). Assume that the radius r of the balloon is increasing at the constant rate of 2 inches per second and is zero when $t = 0$.

(a) Find an algebraic representation for the volume V of the balloon as a function of time t.
(b) Determine the volume of the balloon at $t = 5$ seconds.
(c) Suppose that the balloon will burst when its volume is 10,000 cubic inches. When will the balloon burst?

SOLUTION:

(a) The radius r of the balloon depends on time t. Thus, we can write $r = f(t)$ for some function f to be determined. Because the radius of the balloon is increasing at the constant rate of 2 inches per second, and $r = 0$ when $t = 0$, we have $r = 2t$, where t is the number of seconds. Therefore, $f(t) = 2t$. Notice, for example, that at $t = 3.5$ seconds, the radius is 7 inches. The volume of the sphere as a function of the radius is given by $V = g(r) = \frac{4}{3}\pi r^3$. We use composition of functions to write the volume of the sphere as a function of time:

$$V = g(r) = g(f(t)) = g \circ f(t) = \frac{4}{3}\pi(f(t))^3.$$

Using $f(t) = 2t$ we can express the volume V in the following form:

$$\begin{aligned} V &= \frac{4}{3}\pi(2t)^3 \\ &= \frac{32}{3}\pi t^3. \end{aligned}$$

This equation gives an algebraic representation for the value V as a function of time t.

(b) Let $t = 5$ in the equation $V = \frac{32}{3}\pi t^3$ of (a). Then $V = \frac{32}{3}\pi(5^3)$. Thus, the volume V is approximately 4188.79 cubic inches after five seconds of inflation.

(c) The balloon will burst when $V = 10{,}000$. We must find the value of t for which $10{,}000 = \frac{32}{3}\pi t^3$. Solving for t^3, we find $t^3 = \frac{30{,}000}{32\pi}$. Thus $t = \left(\frac{30{,}000}{32\pi}\right)^{1/3}$. Using a calculator we can determine that $t = 6.68$ seconds. •

• EXAMPLE 10: The initial dimensions of a rectangle are 3 cm by 4 cm and the length and width of the rectangle are increasing at the rate of 1 cm per second. How long will it take for the area to be at least ten times its initial size?

SOLUTION: We can write the length and width as functions of time t: $\ell = 3 + t$, $w = 4 + t$. Thus, the area $A = \ell w$ of the rectangle as a function of t is $A(t) = (3 + t)(4 + t)$. To determine when the area of the rectangle will be at least ten times its initial size, we need

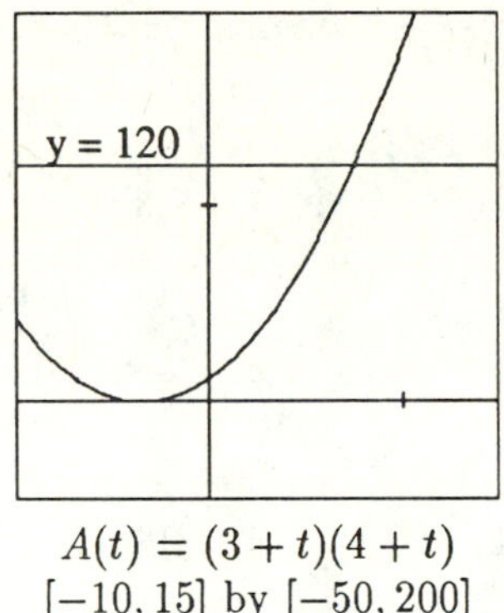

$A(t) = (3+t)(4+t)$
$[-10, 15]$ by $[-50, 200]$
Figure 1.8.10

to solve the inequality $A(t) \geq 120$. The graph of $y = A(t)$ and $y = 120$ is given in Figure 1.8.10. We need to determine the positive values of t for which the graph of $y = A(t)$ lies above or coincides with the graph of $y = 120$. The solution to the problem consists of those values of t greater than or equal to the value of t at the point of intersection of $y = A(t)$ and $y = 120$ in the first quadrant.

We can use zoom-in to determine that the value of t at this point of intersection is $t = 7.66$ with error at most 0.01. The solution to the problem is $[7.66, \infty)$. Alternatively, we could solve the inequality $(3+t)(4+t) \geq 120$ algebraically for t. Thus, after approximately 7.66 seconds the rectangle will have area ten times the original area of 12cm^2. •

We will return to the study of composition of functions when we study inverses of functions. Composition of functions plays an important role in the study of calculus, advanced mathematics, and science.

• EXERCISES 1-8

Determine the domain and a rule for each of $f+g$, $f-g$, fg, and f/g.

1. $f(x) = 2x - 1$; $g(x) = x^2$
2. $f(x) = (x-1)^2$; $g(x) = 3 - x$
3. $f(x) = x^2$; $g(x) = 2x$
4. $f(x) = \sqrt{x}$; $g(x) = x - 2$
5. $f(x) = x + 3$; $g(x) = \dfrac{2x-1}{3}$
6. $f(x) = \dfrac{1}{x-1}$; $g(x) = (x+2)^2$

7–8. Find the range of the functions $f+g$, $f-g$, fg, and f/g with f and g defined as in Exercises 5 and 6. You may need to use a graphing utility.

9. Explain why fg and $f \circ g$ are different in general. Give an example.

10. Determine if *addition* of polynomial functions is commutative, is associative, has an identity, or has an inverse.

11. Determine if *multiplication* of polynomial functions is commutative, is associative, has an identity, or has an inverse.

Draw complete graphs of f, g, and $f+g$ in the same viewing rectangle.

12. $f(x) = -x;\ g(x) = \cos x$
13. $f(x) = \sin x;\ g(x) = \cos x$
14. $f(x) = \dfrac{1}{x};\ g(x) = \sin x$
15. $f(x) = -x;\ g(x) = \dfrac{1}{x}.$
16. Which function in Exercise 12 will most closely approximate the sum function $f+g$ for x large in absolute value? Why?

Determine $f \circ g(3)$ and $g \circ f(-2)$.

17. $f(x) = 2x - 3;\ g(x) = x + 1$
18. $f(x) = x^2;\ g(x) = \sqrt{x-1}$
19. $f(x) = x^2 - 1;\ g(x) = 2x - 3$
20. $f(x) = 2x - 3;\ g(x) = x^2 - 2x + 3$

Determine a rule for $f \circ g$ and $g \circ f$. Find the domain and range of f, g, $f \circ g$, and $g \circ f$.

21. $f(x) = 3x + 2;\ g(x) = x - 1$
22. $f(x) = 2x - 5;\ g(x) = \dfrac{x+3}{2}$
23. $f(x) = x^2 - 1;\ g(x) = \dfrac{1}{x-1}$
24. $f(x) = x^2 - 2;\ g(x) = \sqrt{x+1}$
25. $f(x) = \dfrac{1}{x-1};\ g(x) = (x+1)^2$
26. $f(x) = x^2 - 3;\ g(x) = \sqrt{x+2}$

Draw a complete graph of f, g, $f \circ g$, and $g \circ f$ in the same viewing rectangle.

27. $f(x) = \dfrac{2x-3}{4};\ g(x) = \dfrac{4x+3}{2}$
28. $f(x) = \sqrt{x+3};\ g(x) = x^2 - 3$

Determine functions g and h so that $f(x) = h \circ g(x)$.

29. $f(x) = x^2 + 3$
30. $f(x) = (x+1)^{1/4}$
31. $f(x) = (x+3)^2$
32. $f(x) = \left(\dfrac{1}{x+1}\right)^3$
33. $f(x) = \sqrt[3]{x-2}$
34. $f(x) = \dfrac{2}{(x-3)^2}$

Express f as a composition of simpler functions in two different ways.

35. $f(x) = \dfrac{1}{(x-1)^2}$
36. $f(x) = \dfrac{1}{\sqrt{x+1}}$
37. $f(x) = \dfrac{1}{x^2 + 2x + 1}$
38. $f(x) = (x^2 - 1)^3$
39. The initial dimensions of a rectangle are 5 cm by 7 cm, and the length and width of the rectangle are increasing at the rate of 2 cm per second. How long will it take for the area to be at least five times its initial size?
40. Paul is 6 feet 8 inches tall and walks at the rate of 5 feet per second away from a street light with a lamp 15 feet above level ground.
    (a) Determine an algebraic representation for the length of Paul's shadow.
    (b) At what rate is the length of Paul's shadow increasing?
    (c) Express the distance D between the lamp and the tip of Paul's shadow as a function of time t.
    (d) When will the distance D be 150 feet?

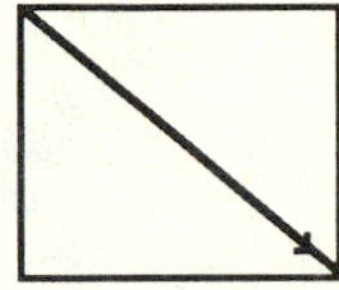

Figure 1.8.11

41. A balloon in the shape of a sphere is being inflated with a gas. Assume that the radius r of the balloon is increasing at the rate of 3 inches per second and is zero when $t = 0$.
    (a) Determine an algebraic representation for the volume of the balloon.
    (b) Express the volume V of the balloon as a function of time t.
    (c) Determine the volume of the balloon at $t = 3$ seconds.
    (d) Suppose that the balloon will burst when its volume is 12,000 cubic inches. When will the balloon burst?
42. A rock is tossed into a pond. The radius of the first circular ripple (wave) increases at the rate of 2.3 feet/second.
    (a) Express the area A enclosed by the circular ripple as a composite function $A = f \circ g$.
    (b) What is the area enclosed by the ripple after 6 seconds have elapsed?
43. The surface area of a sphere of radius r is $S = 4\pi r^2$.
    (a) When a hard candy ball of radius 1.6 cm is dropped into a glass of water, the radius of the ball decreases at the rate of 0.0027 cm/sec. Express the surface area of the candy ball as a function of t by writing S as a composite function $S = f \circ g$.
    (b) When will the candy be completely dissolved?
44. A rectangle is formed on a computer screen by dragging a mouse pointer from one vertex to the opposite vertex along a diagonal. (See Figure 1.8.11.)
    (a) Assume the rectangle's diagonal increases in length at the rate of 1.3 cm/sec. Further assume the length of the rectangle is always twice the width. Express the area A of the rectangle as a function of time t by writing A as a composite function $A = f \circ g$.
    (b) At what time t will the area be 18 centimeters?
45. The initial dimensions of a box are 5 feet by 7 feet by 3 feet. If each of the three side lengths are increasing at a rate of 2 feet per second, how long will it take for the volume of the box to be at least 5 times its initial size?

• KEY TERMS

Listed below are the key terms, vocabulary, and concepts in this chapter. You should be familiar with their definitions and meanings as well as be able to illustrate each of them.

Absolute value
Algebraic representation
Algebraic test for symmetry
Axes
Cartesian coordinate system
Complete graph

Composition of functions
Congruent graph
Discriminant
Domain
Folding symmetry
Function
Geometric representation
Geometric test for symmetry
Graph of an equation
Graph of a problem situation
Graph of a function
Graph of a relation
Graphing utility
Greatest integer function
Horizontal shift
Integer
Irrational number
Line of symmetry
Model
Piecewise-defined function
Point-slope form
Quadrant
Range
Rational number
Real number
Real number line
Reflection through the x-axis
Relation
Slope
Slope-intercept form
Solution to an equation
Square root function
Standard viewing rectangle
Sum, difference, product, and quotient of functions
Symmetry with respect to a line (axis)
Symmetry with respect to a point (origin)
Understood domain
x-intercept
Vertex of a parabola
Vertical line test
Vertical shift
Vertical stretch
Vertical shrink
Viewing rectangle

• REVIEW EXERCISES

This section contains representative problems from each of the previous sections. You should use these problems to test your understanding of the material covered in this chapter.

1. Graph the points on the same coordinate plane. Identify the quadrant containing each point.
   (a) $(1, 4)$ (b) $(0, -4)$ (c) $(-\frac{1}{3}, -2)$ (d) $(-3, \frac{1}{2})$
2. Graph the three points on the *same* coordinate system. Choose an appropriate scale.
   (a) $(5.1, 2000)$ (b) $(5.25, 3500)$ (c) $(5.45, 6000)$
3. The point $(a, 50)$ is on the graph of $y = 2x^2$. What is a?
4. The point $(-3, T)$ is on the graph of $y = 5x^2 - 8$. What is T?
5. Draw complete graphs of two different relations that include the data given in Table 1.
6. (a) Complete Table 2 for the graph of the relation in Figure R.1.
   (b) What are the domain and range of this relation?

Input	Output
−2	3
−1	0
1	0
2	3
3	8

Table 1

t	D
0	
1	
2	
3	
	2
	0

Table 2

Draw a complete graph of each equation. Do not use a graphing utility. Choose an appropriate scale for each axis.

7. $y - 5 = \frac{1}{2}x$

8. $x - y = 6$

9. $y = 200x - 10$

10. $y = 2x^2 + 1000$

Draw a complete graph of each equation.

11. $D = 3N^2 - 30$

12. $V = L^2(L - 20)$

13. What are the domain and range of the equation given in Exercise 12?

14. Let $(\frac{1}{2}, -3)$ be on a graph. Determine a second point on the graph if we know that the graph is symmetric with respect to

   (a) the x-axis. (b) the y-axis. (c) the origin.

15. Determine whether the graphs in Figure R.2 are symmetric with respect to the x-axis, the y-axis, or the origin. Justify your answers.

Draw a complete graph of each equation. (If possible, use symmetry.) Determine the domain and range of the equation.

16. $y = 3 + x^2$

17. $y^2 - 3 = x$

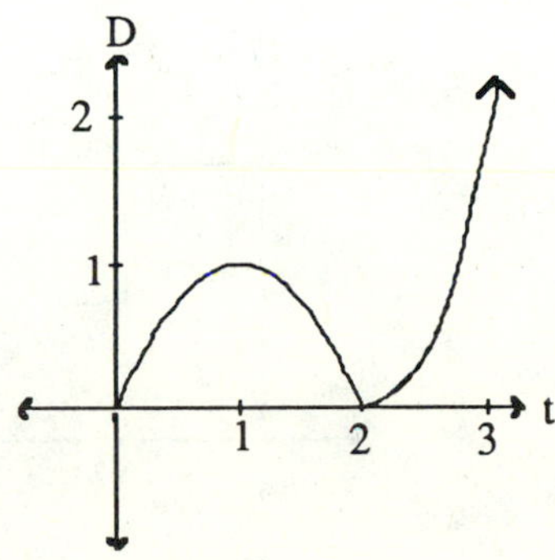

Figure R.1

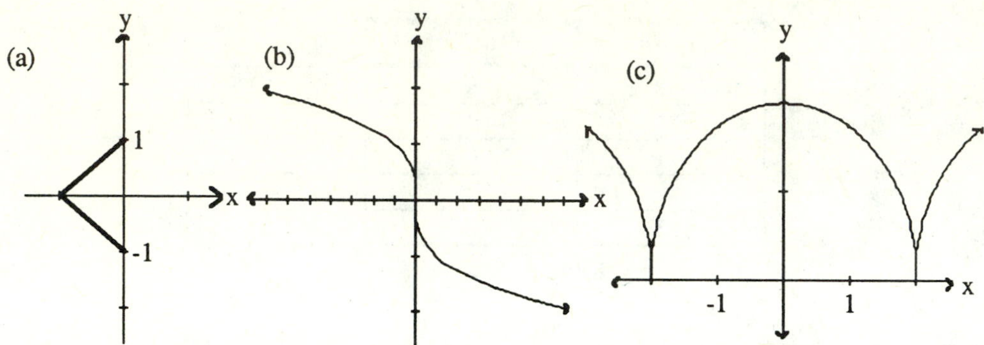

Figure R.2

18. Complete the graph in Figure R.3 if a complete graph is symmetric with respect to
    (a) the x-axis. (b) the y-axis. (c) the origin.

Determine whether the graph of the equation is symmetric with respect to the x-axis, the y-axis, or the origin.

19. $x^2y = 1$
20. $x^2 - xy^4 = 2$
21. Draw a complete graph of $y = (x + 5)^2$. Is the graph symmetric about some line? If so, what is an equation of the line of symmetry?
22. Compute each value.
    (a) $|-2.5|$ (b) $|2 - 6|$ (c) $|-(5 - 2)|$
23. Write the exact value of the indicated expression without using absolute value notation.
    (a) $|\sqrt{7} - 2.6|$ (b) $|2.6 - \sqrt{7}|$
    (c) $|\pi - 3|$ (d) $|x - 5|$ where $x > 5$

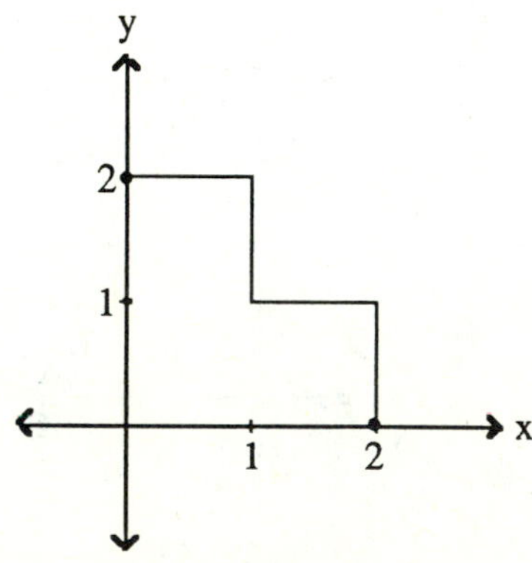

Figure R.3

24. Find the distance between the two points on the number line with the given coordinates.

(a) $-2, 4$ (b) $5, -2$ (c) $-\pi, -1$

Find the distance between the points in the plane with the given coordinates.

25. $(1,1)$, $(3,4)$

26. $(-3,-8)$, $(2,3)$

27. Write an expression that gives the distance in the plane between each pair of points.

(a) $(0,y)$, $(-2,3)$ (b) $(x,0)$, $(0,-4)$

28. Let $f(x) = x^2 + 5$ and $g(x) = \frac{1}{x-3}$. Find the distance between the points.

(a) $(2, f(2))$ and $(6, g(6))$ (b) $(-1, f(-1))$ and $(-5, f(-5))$

29. Consider the graph in Figure R.4 of an equation in two variables x and y.

(a) Use the graph to complete Table 3.

(b) Draw the line of symmetry of the graph. Give an equation for the line of symmetry.

(c) What are the domain and range of the equation represented by this graph?

Draw a complete graph of each equation and specify its domain and range. Give an equation of the line of symmetry and indicate which equations define y as a function of x.

30. $y = |x - 3|$

31. $y = 5 - |x|$

32. $|y - 4| = x$

33. $y = |x^3 - 8x|$

34. $x|y| = 3$

35. $y = \left|\frac{3}{x-2}\right|$

36. Determine the equation of the circle with given center and radius.

(a) $(-1,3)$, 6 (b) $(2,-5)$, 4

37. Which of the graphs in Figure R.5 are graphs of functions?

Use a graph to determine whether each relation is a function.

38. $y = |x - 1|$

39. $|y + 2| = x$

40. $y = 2\cos x$

41. $y = \sin x + 3$

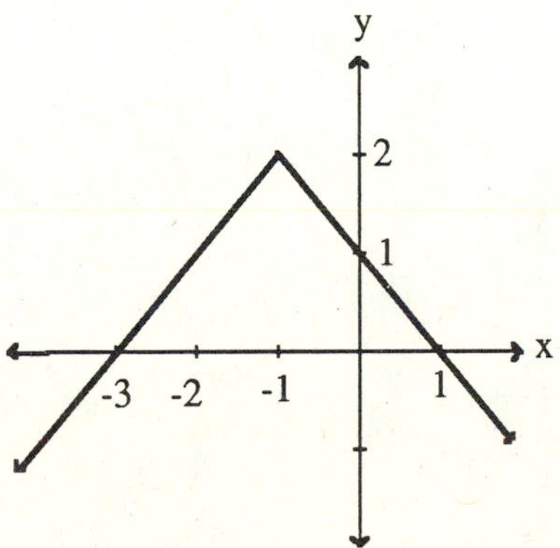

Figure R.4

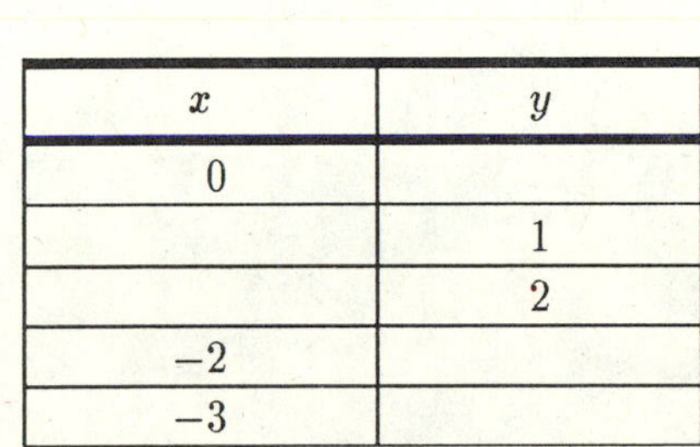

x	y
0	
	1
	2
-2	
-3	

Table 3

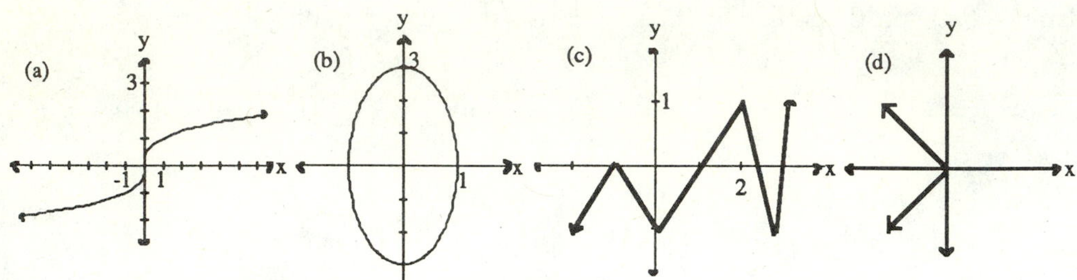

Figure R.5

42. (a) Let $f(x) = x^2 + 2$. Find $f(0)$, $f(1)$, $f(3)$, $f(-5)$, $f(t)$, $f(-t)$, $f\left(-\frac{1}{t}\right)$, $f(a+h)$, and $\frac{f(a+h)-f(a)}{h}$.

    (b) Let $g(x) = \frac{1}{2-x}$. Find $g(2)$, $g(0)$, $g(-2)$, $g(a)$, $g\left(\frac{1}{a}\right)$ and $g(a+h)$.

43. Consider the graph of the function $y = f(x)$ given in Figure R.6. Estimate the indicated values.

    (a) $f(0)$, $f(-1)$, $f(2)$

    (b) x if $f(x) = 0$

    (c) x if $f(x) = 2$

44. Choose a viewing rectangle [L, R] by [B, T] that will include all of the indicated points.

    (a) $(-9, 12)$, $(2, -8)$ and $(3, 20)$

    (b) $(20, -11)$, $(18, 156)$ and $(-11, 2)$

45. Which of the following viewing rectangles gives the best complete graph of $y = x^2 - 30x - 100$?

    (a) $[-10, 5]$ by $[-10, 5]$

    (b) $[-10, 10]$ by $[-10, 10]$

    (c) $[25, 50]$ by $[-500, 1000]$

    (d) $[-10, 50]$ by $[-400, 800]$

    (e) $[-100, 100]$ by $[-10, 10]$

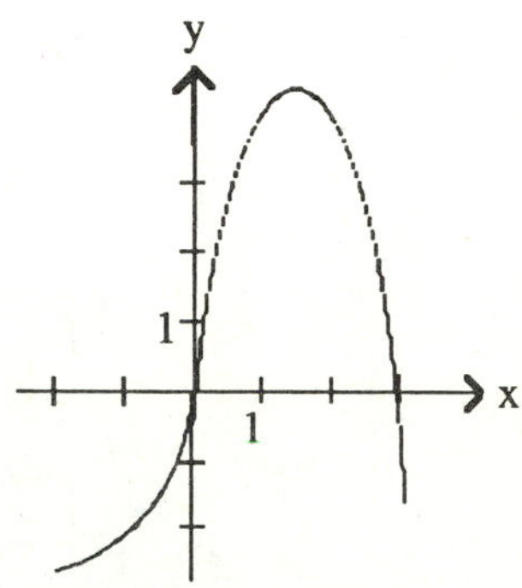

Figure R.6

46. Which of the following viewing rectangles gives the best complete graph of $y = 200 - 10x - x^3$?
(a) $[-10, 10]$ by $[-10, 10]$ (b) $[-50, 50]$ by $[-10, 10]$
(c) $[-10, 10]$ by $[-50, 50]$ (d) $[-10, 10]$ by $[-1000, 1000]$
(e) $[-100, 100]$ by $[-10, 10]$

Draw a complete graph of each function. Explain why the graph is a complete graph.

47. $g(x) = x^2 - 40$ 48. $f(x) = 10x^3 - 400x$ 49. $t(x) = \left(\frac{x}{4}\right)^2 + \left(\frac{200 - x}{4}\right)^2$

50. $y = \sqrt{x - 3}$ 51. $y = -2 - \sqrt{x + 3}$ 52. $y = \frac{1}{2 - x}$

53. Determine the domain and range of the function in Exercise 47.

54. Determine the domain and range of the function in Exercise 48.

55. Determine the domain and range of the function in Exercise 50.

56. Determine a viewing rectangle which gives a complete graph of $f(x) = 50 - 25x + 100x^2 - x^3$. Sketch a complete graph of the function. Do not use a graphing utility. Determine the domain and range.

57. $f(x) = \begin{cases} x + 7, & x \leq -3 \\ x^2 + 1, & x > -3 \end{cases}$ 58. $f(x) = \begin{cases} 4 - x^2, & x < 1 \\ \sqrt{x + 2}, & x \geq 1 \end{cases}$

59. Draw a line through the point $(3, 4)$ with slope $-\frac{1}{2}$.

60. Draw a line through the point $(-1, 3)$ with slope 2.

61. Draw the graphs of the functions $y = 2x^2 + 3$, $y = 2x^2 - 1$, and $y = 2x^2 - 3$ on the same coordinate system. *Do not* use a graphing utility.

62. Determine the slope of the line through the points $(0, 3)$ and $(-2, 6)$.

63. Determine the y-intercept of the graph of the function $3x - 2y + 8 = 0$. Graph the function.

Assume f is a linear function. Determine a formula for f that satisfies the given conditions.

64. $f(2) = 0$ and $f(0) = 4$ 65. $f(0) = 250$ and $f(-10) = 200$

Draw a complete graph of each equation or inequality. Do not use a graphing utility.

66. $2x + y \leq 3$ 67. $3x - 4y = 12$

68. $x + 3y = 9$ 69. $y \leq x$

70. Write an equation for the line with slope $-\frac{2}{3}$ and y-intercept 4.

71. Write an equation for the line with slope $\frac{3}{4}$ that contains the point $(1, 2)$.

72. Write an equation for the line determined by the points $(-3, 4)$ and $(2, 5)$.

73. Given the point $A = (5, 7)$, write an equation for the
(a) vertical line through A. (b) horizontal line through A.

Draw complete graphs of the three equations in the same viewing rectangle.

74. $y = 4x^2$, $y = 4(x - 2)^2$, $y = 4(x + 4)^2$ 75. $y = -2x^2$, $y = -2(x + 3)^2$, $y = 2(x - 3)^2$

Without using a graphing utility, draw a complete graph of $y = f(x)$. Describe how each graph is obtained from the graph of $y = x^2$.

76. $y = -x^2$ 77. $y = (x - 3)^2$

78. $y = -2(x+3)^2 + 4$

79. $y = (x+1)^2 - 4$

Using a graphing utility, draw a complete graph of $y = f(x)$. Describe how each graph is obtained from the graph of $y = |x|$.

80. $y = 2|x-3|$

81. $y = -3 - |x+4|$

82. Assume the point $(4, 3)$ is on the graph of $y = f(x)$. Find b so that $(1, b)$ is on the graph of $y = f(x+3)$.

83. Assume the points $(-3, 2)$ and $(1, 0)$ are on the graph of $y = f(x)$. Find b so that:
    (a) $(-1, b)$ is on the graph of $y = f(x-2)$.
    (b) $(3, b)$ is on the graph of $y = f(x-2)$.
    (c) $(-1, b)$ is on the graph of $y = -3f(x-2)$.
    (d) $(3, b)$ is on the graph of $y = 2 + 2f(x-2)$.

84. (a) What is an equation of the graph that is obtained by vertically stretching the graph of $y = x^2$ by a factor of 2 followed by a vertical shift up 1 unit? Sketch a complete graph.
    (b) What is an equation if the order of the two transformations in (a) is interchanged?
    (c) Are the two graphs the same? Explain the effect of reversing the order of the transformations.

Write a sequence of transformations that will produce a complete graph of each function from the graph of $y = x^2$. Specify an order in which the transformations should be applied.

85. $f(x) = 2x^2 - 12x + 4$

86. $f(x) = 14 - 6x - x^2$

87–88. Use the information from Exercises 85 and 86 to graph each function without the aid of a graphing utility. Indicate the vertex, line of symmetry, x-intercepts, and y-intercept. Check your answers with a graphing utility.

89–90. Algebraically determine the real zeros (if any) of each of the functions in Exercises 85 and 86. Check your answers with a graphing utility.

91. Determine the number of real solutions of $3x^2 - 2x + 1 = 0$ by computing the discriminant. Check your answers with a graphing utility.

92. Determine the midpoint M of the real zeros of each of the indicated quadratic functions. Find an equation of the line of symmetry, the coordinates of the vertex, the x-intercepts, the y-intercept, and sketch a complete graph.
    (a) $f(x) = (x-2)(x+4)$
    (b) $f(x) = (2x+2)(x-3)$

List the transformations that will produce a complete graph of each function from the graph of $y = x^3$. Specify the order in which the transformations should be applied and then sketch a complete graph of the function without using a graphing utility. Use a graphing utility to check your answer.

93. $y = (x+2)^3$

94. $y = -2(x-3)^3 - 5$

Determine $f \circ g(-3)$ and $g \circ f(2)$.

95. $f(x) = 5x + 7$; $g(x) = x - 4$

96. $f(x) = x^2 + 4$; $g(x) = \sqrt{1-x}$

97. Determine $f \circ g$ and $g \circ f$ if $f(x) = x^2 + 2$ and $g(x) = \frac{x+3}{2}$.

98. Find the domain and range of f, g, $f \circ g$, and $g \circ f$ if $f(x) = \frac{2x-3}{4}$ and $g(x) = x^2 - 2$.

99. Determine the domain and a rule for each of $f+g$, $f-g$, fg, and f/g where $f(x) = 1/(x-1)$ and $g(x) = (x+2)^2$.

Find functions g and h so that $f(x) = h \circ g(x)$.

100. $f(x) = (x+5)^2$

101. $f(x) = \dfrac{3}{(x-2)^2}$

Sketch a complete graph of $y = f(x)$ without using a graphing utility. Check your answer with a graphing utility.

102. $y = 2(x-2)^2 + 3$

103. $y = 4 - 3\sqrt{x+1}$

104. If the point (a, b) is on the graph of $t = \frac{450}{R}$, then determine ab.

105. A school club buys a scientific calculator for \$18.25 to use as a raffle prize for a fund raising project. The club charges \$0.50 per raffle ticket.
   (a) Determine an algebraic representation that gives the club's profit (or loss) P as a function of the number n of tickets sold.
   (b) Draw a complete graph of the algebraic representation.
   (c) What are the domain and range of the algebraic representation?
   (d) What part of the graph in (b) represents the problem situation? *Hint:* You cannot sell half a ticket!
   (e) Use the graph in (b) to determine the minimum number of tickets that must be sold for the club to realize a profit.

106. The sum of the length and the width of some rectangles is 150 inches. Let x be the length of such a rectangle.
   (a) If the length is 25 inches, then what is the area of the rectangle?
   (b) Show that $A = x(150 - x)$ is an algebraic representation that gives the area of such rectangles in terms of their length.
   (c) Draw a complete graph of the algebraic representation and specify its domain.
   (d) Verify that $(10, 1400)$ is a point on the graph in (c). What meaning do these coordinates have in this problem situation?
   (e) Verify that $(-10, -1600)$ is a point on the graph in (c). What meaning do these coordinates have in this problem situation?
   (f) What values of x make sense in this problem situation? Draw a graph of the problem situation.

107. The perimeter of some rectangles is 500 inches. Let x be the length of such a rectangle.
   (a) If the length is 50 inches, then what is the area of such a rectangle?
   (b) Show that $A = x(250 - x)$ is an algebraic representation that gives the area of such rectangles in terms of their length.
   (c) Draw a complete graph of the algebraic representation.
   (d) Verify that (100, 15,000) is a point on the graph in (c). What meaning do these coordinates have in this problem situation?

(e) Verify that $(-50, -15{,}000)$ is a point on the graph in (c). What meaning do these coordinates have in this problem situation?

(f) Draw a graph of the problem situation.

108. $A = \pi r^2$ is an equation that gives the area A of a circle with radius r units.

(a) Draw a complete graph of the algebraic representation.

(b) What are the domain and range of the algebraic representation?

(c) Draw a graph of the problem situation. How is its domain different from the domain of the graph of the algebraic representation?

(d) Use the graph in (a) to determine the radius of a circle with area 150 square units.

109. The volume V of a box is given by $V = LWH$ where L is the length, W is the width, and H is the height of the box.

(a) Write the volume as a function of the height if the base of the box has area 40 square units.

(b) Draw a complete graph of the algebraic representation.

(c) What is the domain of the algebraic representation?

(d) What is the range of the algebraic representation?

(e) Use the graph in (b) to determine the height of the box if the volume is 300 square units.

110. Sherri invests \$20,000, part at 6% simple interest and the remainder at 7.5% simple interest.

(a) Write an equation that shows how the total interest Sherri receives in one year depends on the amount she invests at 6%.

(b) Draw a complete graph of the algebraic representation in (a).

(c) What portion of the graph in (b) represents the problem situation?

(d) Use the graph in (b) to estimate the amount invested at each rate if Sherri receives \$1300 interest in one year.

(e) Is there a maximum amount of interest Sherri can receive in one year?

111. A bike manufacturer determines the annual cost C of making x bikes to be \$85 per bike plus \$75,000 in fixed overhead costs.

(a) Write the total annual cost as a function of the number of bikes made.

(b) Draw a complete graph of the algebraic representation in (a).

(c) What are the domain and range of the function?

(d) What portion of the graph in (b) represents possible total annual costs of producing the bikes?

(e) Use the graph in (b) to determine the number of bikes made if the total cost is \$143,000.

(f) If (A, B) is a point on the graph representing the problem situation, then what does A represent in the problem situation?

112. The bike manufacturer in Exercise 111 determines that he can sell each bike for \$98.00.

(a) Write an equation that gives the total annual revenue R as a function of the number of bikes produced and sold.

(b) Draw a complete graph of the function in (a).

(c) What are the domain and range of the function?

(d) What portion of the graph in (b) represents possible total annual revenue from selling the bikes?

(e) Use the graph in (b) to determine the number of bikes made if the total revenue is \$107,800.

(f) If (A, B) is a point on the graph representing the problem situation, then what does B represent in the problem situation?

113. A three-sided rectangular fence with a perimeter of 72 feet is constructed against a barn. Let x be the length of the fence parallel to the barn.

(a) Write an equation for the area enclosed by the fence as a function of x.

(b) What are the domain and range of the function in (a)?

(c) Draw a complete graph of the function in (a).

(d) What values of x make sense in this problem situation? Draw a graph of the problem situation.

(e) Use the graph in (c) to determine the length of the fence parallel to the barn if the area enclosed is 200 square feet.

(f) Is there a largest area that can be enclosed by the fence?

114. One hundred twenty feet of fencing is cut into two unequal lengths, and each piece is used to make a square enclosure. Let x be the perimeter of the smaller enclosure.

(a) Express the total area A of both enclosures as a function of x.

(b) Draw a complete graph of the function in (a).

(c) What is the domain of the function A? What numbers in the domain represent possible lengths of the smaller piece of fence?

(d) If $A = 6500$ square feet, use the graph in (b) to approximate the smaller length of fencing.

115. A trucker averages 48 mph on a cross country trip from Boston to Seattle. Let t be the time since the trucker left Boston.

(a) Express the total distance the trucker travels as a function of time.

(b) Draw a complete graph of the function determined in (a).

(c) What are the domain and range of the function in (a)?

(d) What part of the domain of the function represents the problem situation?

(e) Use the graph in (b) to determine how many hours have elapsed since the trucker left Boston if the distance traveled is 1200 miles.

116. The annual profit P of a baby food manufacturer is determined by the formula $P = R - C$, where R is the total revenue generated from selling x jars of baby food and C is the total cost of making and selling x jars of baby food. Each jar sells for \$0.60 and costs \$0.45 to make. The fixed costs of making and selling the baby food are \$83,000 annually.

(a) Express the company's annual profit as a function of x.

(b) Draw a complete graph of the function determined in (a).

(c) How many jars of baby food must be sold for the company to "break even?"

(d) How many jars of baby food must be sold for the company to make a profit of \$10,000?

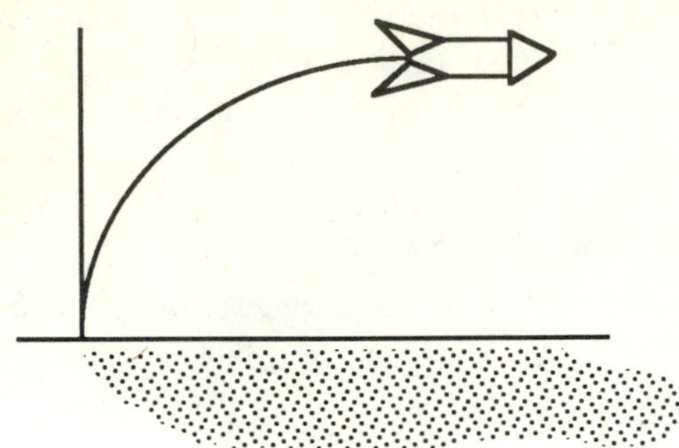

Figure R.7

117. An object 30 feet above level ground is shot straight up with an initial velocity of 250 feet per second.
    (a) Determine the height s of the object above the ground as a function of time t.
    (b) Draw a complete graph of the function in (a) indicating what portion of the graph represents the problem situation.
    (c) At what time t will the object be 550 feet above the ground?
118. Judy is 5 feet 6 inches tall and walks at the rate of 4 feet per second away from a street light with a lamp 14.5 feet above level ground.
    (a) Determine an algebraic representation of the length of Judy's shadow.
    (b) At what rate is the length of Judy's shadow increasing?
    (c) Express the distance D between the lamp and the tip of Judy's shadow as a function of time t.
    (d) When will the distance D be 100 feet?
119. A balloon in the shape of a sphere is being inflated. Assume the radius r of the balloon is increasing at the rate of 3 inches per second and is zero when $t = 0$.
    (a) Express the volume V of the balloon as a function of time t.
    (b) Determine the volume of the balloon at $t = 5$ seconds.
    (c) Suppose the balloon will burst when its volume is 15,000 cubic inches. When will the balloon burst?

A rocket launched from ground level is d miles above the earth t minutes after launch (Figure R.7). While the rocket engine is on, d is given by

$$d = 6t^2 + 185t\,.$$

When the engines are off, d is given by

$$d = 1728 + 25t - t^2\,.$$

120. Assume the rocket engine is on for only the first 8 minutes. Determine a piecewise-defined function that is a model for this problem situation.

2 Polynomials and Solving Equations and Inequalities

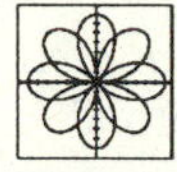

• Introduction

In Chapter 1 we introduced both algebraic and geometric representations of problem situations. An equation, inequality, or system of equations or inequalities is often an algebraic representation of a problem situation. In this chapter we find real number solutions to equations and inequalities using both algebraic and graphical techniques. We introduce the important graphical problem-solving technique called *zoom-in*. Computer based graphing techniques are more general and more powerful than algebraic techniques. In this chapter we concentrate on equations and inequalities that involve *polynomials*. Absolute value is introduced, and equations and inequalities involving absolute value are solved both algebraically and graphically.

In the following definition we use the symbol a_i, where i is a positive integer. The i in a_i is called a **subscript**, and a_i is read "a sub i." There are 26 letters in the alphabet. Subscripting letters of the alphabet is a common way to expand the list of letters or variables for use in mathematics.

• DEFINITION Any expression of the form

$$a_n x^n + a_{n-1} x^{n-1} + \cdots + a_1 x + a_0,$$

where n is a nonnegative integer and $a_n, a_{n-1}, \ldots, a_0$ are real numbers is called a **polynomial in x**. If $a_n \neq 0$, then n is called the **degree** of the polynomial. The a_i's are called **coefficients**, a_n is often called the **leading coefficient** of the polynomial and a_0 the **constant term**. If a function f is a polynomial in x, then f is called a **polynomial function**. •

A polynomial of degree 1 is called a **linear polynomial**, and a polynomial of degree 2 is called a **quadratic polynomial**. For example, $\frac{1}{2}x-3$ is a linear polynomial and $2x^2-x+1$ is a quadratic polynomial. The expression $3x^4-2x^2-x+15$ is a polynomial of degree 4. Be sure you understand why $\frac{2x-1}{x^2-1}$ and $3\sqrt{x}+5$ are *not* polynomials.

• DEFINITION Any equation of the form

$$a_nx^n+a_{n-1}x^{n-1}+\cdots+a_1x+a_0=0,$$

is called a **polynomial equation**. If $a_n\neq 0$, the equation is called a **polynomial equation of degree n**. •

The equation $\frac{1}{2}x-3=0$ is a polynomial equation of degree 1, and the equation $2x^2-x+1=0$ is a polynomial equation of degree 2. The equations $\frac{2x-1}{x^2-1}=0$ and $3\sqrt{x}+5=0$ are *not* polynomial equations because $\frac{2x-1}{x^2-1}$ and $3\sqrt{x}+5$ are *not* polynomials.

You have used algebraic methods and formulas in previous mathematics courses to solve polynomial equations of degrees 1 and 2. There are very complicated algebraic formulas that give the exact solutions to all polynomial equations of degrees 3 and 4. There is *no* formula that gives the exact solutions to all polynomial equations of degree 5 or higher. Solving polynomial equations graphically is a general solution method that does not depend on the degree. A graphical solution is quick and accurate and leads to important geometric intuition about the meaning of solutions to an equation, inequality, or system of equations.

We establish the connection between finding solutions algebraically and geometrically. Solving equations graphically is equivalent to finding the *x-intercepts* of a related graph or points *where two graphs intersect.* Solving inequalities graphically involves determining when a graph is above or below the x-axis or when one graph is above or below another. When solving problems graphically, we must concern ourselves with the *accuracy* of the solutions found. A discussion about error is included in this chapter. As in Chapter 1 we must be careful to determine which solutions to an algebraic or geometric representation are also solutions to the problem situation.

We find the intervals on which a function is increasing or decreasing, and we determine the coordinates of local maximum and local minimum values of functions. Graphs of inequalities of the form $y>f(x)$, $y\geq f(x)$, $y<f(x)$, and $y\leq f(x)$, where f is a polynomial, are drawn.

2.1 Solving Equations

The main purpose of this section is to illustrate how to use a graphing utility to solve equations. This method produces accurate approximate solutions rather than exact solutions. The formulas that give exact solutions to polynomial equations of degrees 2, 3, or 4 involve radicals. However, there are no formulas involving radicals that give the exact solutions to polynomial equations of degree 5 or higher. Niels Henrik Abel (1802–1829), who died when he was only 26 years old, proved that it is not possible to give a formula involving only radicals for the exact solution to a polynomial equation of degree 5 or higher. In fact, there is no known method for finding exact solutions to polynomial equations of degree 5.

The formulas for the exact solutions of polynomial equations of degree 3 or 4 are not very useful unless we approximate the radicals involved. Generally, the best we can do when solving a given equation is to find approximations to its solutions using numerical techniques. M.J. Maron puts it this way in his textbook on numerical analysis:
"Fortunately, one rarely needs *exact* numerical answers. Indeed, in the 'real world' the problems themselves are generally inexact because they are posed in terms of parameters that are only approximate because they are *measured*. What one is likely to require in a realistic situation is not an exact answer, but rather one having a prescribed accuracy."

• DEFINITION The number a is a **solution to an equation** in one variable if we get a true statement when the variable is replaced by a everywhere in the equation. •

For example, -2 is a solution to the equation $x^3 - x + 6 = 0$ because we get the following true statement when x is replaced by -2 in this equation: $(-2)^3 - (-2) + 6 = 0$. Similarly, $\sqrt{3}$ is a solution to $y^2 - 3 = 0$ because $(\sqrt{3})^2 - 3 = 0$. When we have found *all* solutions to an equation we say that we have **solved the equation**. The study of complex numbers will be delayed until Chapter 3. Thus, in this chapter we restrict our attention to real number solutions to equations and inequalities. Often we refer to real number solutions more simply as **real solutions**.

One important algebraic procedure to solve an equation is to transform it into an equivalent equation with solutions that are obvious.

• DEFINITION Two equations are said to be **equivalent** if they have exactly the same set of solutions. •

For example, the equations $x - 2 = 0$ and $2x - 4 = 0$ are equivalent. Similarly, $x = 3$ and $x + 2 = 5$ are equivalent.

Equivalent Equations

An equivalent equation is obtained if any one of the following operations is performed.

(1) Add the same real number to, or subtract it from, each side of a given equation.
(2) Add the same polynomial to, or subtract it from, each side of a given equation.
(3) Multiply each side of an equation, or divide each side of an equation, by the same nonzero real number.

Solutions and x-intercepts

Every equation can be rewritten in an equivalent form with one side equal to zero. For example, an equivalent form of $x^3 + 2x = 1$ is $x^3 + 2x - 1 = 0$. Any equation in one variable

x can be rewritten in the form $f(x) = 0$. For the equation $x^3 + 2x = 1$, $f(x)$ is $x^3 + 2x - 1$. Notice that f is a *polynomial function* because $x^3 + 2x - 1$ is a polynomial.

The *graph* of the function $y = f(x)$ can be used to solve the equation $f(x) = 0$. If $x = a$ is a real solution to $f(x) = 0$, then $(a, 0)$ is a point on the graph of $y = f(x)$. Thus, the solution a to $f(x) = 0$ corresponds to the point $(a, 0)$ where the graph of $y = f(x)$ crosses the x-axis (Figure 2.1.1). Such points $(a, 0)$ are called the **x-intercepts** of $y = f(x)$. We also call a a **zero**, or a **root**, of $f(x)$. The value a is called a zero because it is an x-value that makes $f(x)$ equal to zero.

Conversely, if $(a, 0)$ is an x-intercept of $y = f(x)$, then $x = a$ is a real solution of $f(x) = 0$. Therefore, the set of all real solutions of $f(x) = 0$ is the same as the set of all first coordinates of the x-intercepts of the graph of $y = f(x)$.

Notation for x-intercept. If $(a, 0)$ is an x-intercept of $y = f(x)$, we often call a an x-intercept. With this agreement, the set of all real solutions of $f(x) = 0$ is the same as the set of all x-intercepts of the graph of $y = f(x)$.

Linear Equations

• DEFINITION An equation is **linear** in x if it can be written in the form $ax + b = 0$, where a and b are real numbers and $a \neq 0$. •

Once we transform a linear equation into the form $ax + b = 0$ it is quite easy to solve the equation. We subtract b from both sides of $ax + b = 0$. Then we divide both sides of the resulting equation by a. We can divide both sides by a because, by the definition of linear, $a \neq 0$. The linear equation $ax + b = 0$ has exactly one solution, namely, $-\frac{b}{a}$. Notice that the graph of the equation $y = ax + b$ is a *straight line* that crosses the x-axis at $(-\frac{b}{a}, 0)$. (Figure 2.1.2.)

• EXAMPLE 1: Solve $2x + 3 = 0$.

SOLUTION: Algebraically, we know that $-\frac{3}{2}$ is the solution. Indeed, the graph of $y = 2x + 3$ shown in Figure 2.1.3 appears to have x-intercept $-\frac{3}{2}$. Later in this section we will see

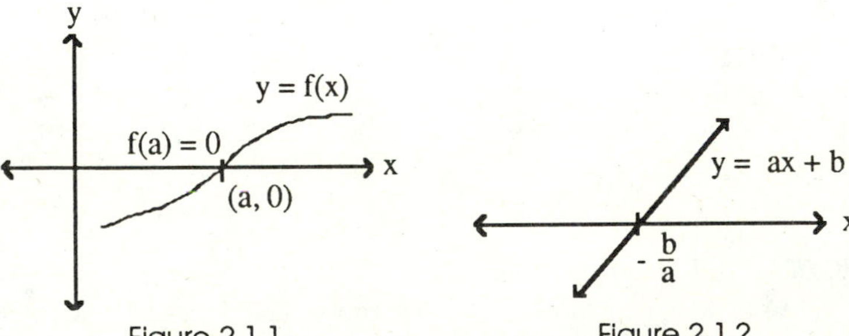

Figure 2.1.1 Figure 2.1.2

how to find an x-intercept and, therefore, a solution to an equation with a high degree of accuracy. •

The diagram in Figure 2.1.4 demonstrates the correspondence between solutions of equations and x-intercepts of graphs.

Quadratic Equations

• DEFINITION An equation is **quadratic** in x if it can be written in the form $ax^2 + bx + c = 0$, where a, b, and c are real numbers and $a \neq 0$. •

Transforming a quadratic equation into an equivalent form with solutions that are obvious may require more work than transforming linear equations. We use a technique called **completing the square** to derive the **quadratic formula** which gives the exact solutions to $ax^2 + bx + c = 0$. The following equations are equivalent.

$$ax^2 + bx + c = 0$$

$$ax^2 + bx = -c \quad \text{(Subtract } c \text{ from each side.)}$$

$$x^2 + \frac{b}{a}x = \frac{-c}{a} \quad \text{(Divide each side by } a.\text{)}$$

We can divide each side by a because, by definition of quadratic, $a \neq 0$. Next we want to add an expression to each side that makes the left side of $x^2 + \frac{b}{a}x = -\frac{c}{a}$ the square

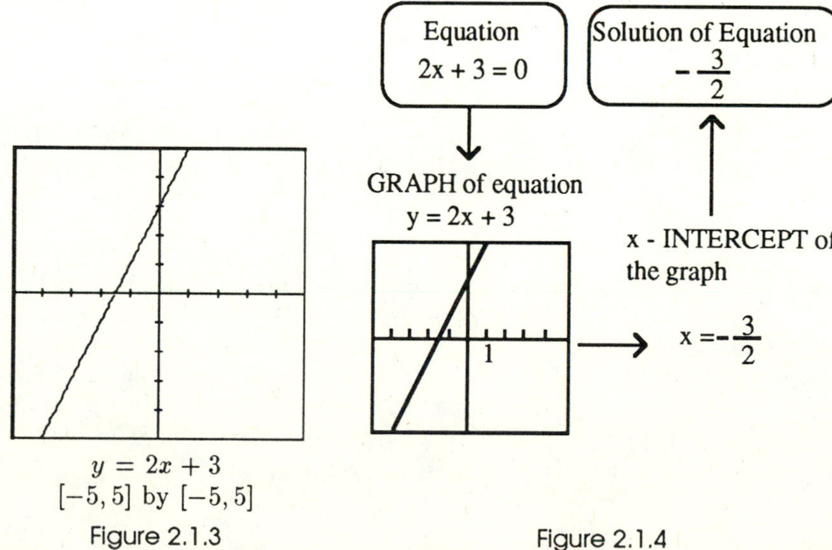

$y = 2x + 3$
$[-5, 5]$ by $[-5, 5]$
Figure 2.1.3

Figure 2.1.4

of a binomial. Because $(x+t)^2 = x^2 + 2tx + t^2$, we must add the square of half of the coefficient of x.

$$x^2 + \frac{b}{a}x = \frac{-c}{a}$$

$$x^2 + \frac{b}{a}x + \left(\frac{b}{2a}\right)^2 = \frac{-c}{a} + \left(\frac{b}{2a}\right)^2 \quad \text{(Add } \left(\frac{b}{2a}\right)^2 \text{ to each side.)}$$

$$\left(x + \frac{b}{2a}\right)^2 = \frac{-c}{a} + \left(\frac{b}{2a}\right)^2 \quad \text{(Factor left side.)}$$

$$\left(x + \frac{b}{2a}\right)^2 = \frac{b^2 - 4ac}{4a^2} \quad \text{(Combine fractions on right side.)}$$

Because the square of $x + \frac{b}{2a}$ equals $\frac{b^2-4ac}{4a^2}$, we must have

$$x + \frac{b}{2a} = \frac{\sqrt{b^2 - 4ac}}{2a} \quad \text{or} \quad x + \frac{b}{2a} = -\frac{\sqrt{b^2 - 4ac}}{2a}.$$

Solving for x we have,

$$x = -\frac{b}{2a} + \frac{\sqrt{b^2 - 4ac}}{2a} \quad \text{or} \quad x = -\frac{b}{2a} - \frac{\sqrt{b^2 - 4ac}}{2a}.$$

Generally, the two solutions above are combined into the single expression

$$x = \frac{-b \pm \sqrt{b^2 - 4ac}}{2a}.$$

This last expression is called the **quadratic formula** and gives the solutions to the quadratic equation $ax^2 + bx + c = 0$ in terms of the coefficients a, b, and c.

• DEFINITION The expression $b^2 - 4ac$ in the quadratic formula is called the **discriminant** of the quadratic equation $ax^2 + bx + c = 0$. •

Number of Real Solutions. If $b^2 - 4ac$ is positive, the quadratic equation has exactly two real solutions. If $b^2 - 4ac$ is zero, the quadratic equation has exactly one real solution. If $b^2 - 4ac$ is negative, the quadratic equation has no real solutions.

Figure 1.2.8 from Section 1.2 gives the possible graphs of $y = ax^2 + bx + c$. The x-intercepts of $y = ax^2+bx+c$ give the real solutions to the quadratic equation $ax^2+bx+c = 0$. Figure 2.1.5 illustrates that the number of x-two. We get the same possibilities if the graph of $y = ax^2 + bx + c$ opens downward ($a < 0$). The discriminant $b^2 - 4ac$ of the quadratic equation will be negative, zero, or positive, respectively. (Why?)

• EXAMPLE 2: Solve $2x^2 = 1 - 2x$.

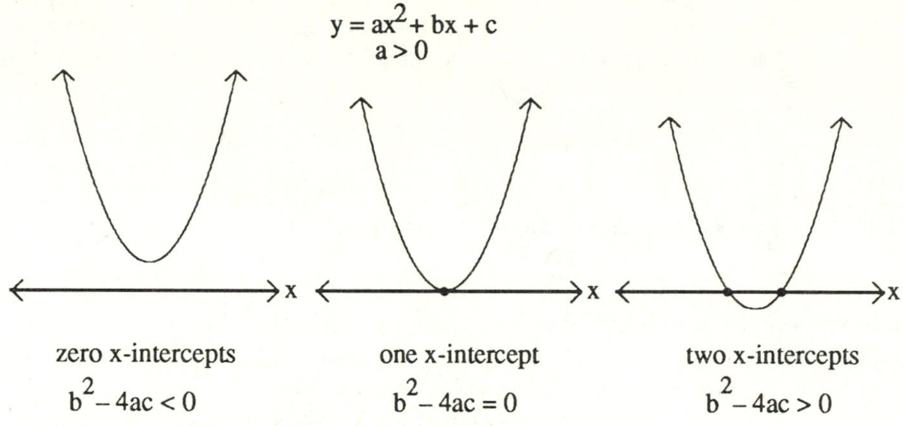

Figure 2.1.5

SOLUTION: First we rewrite the equation in the form $f(x) = 0$. The desired equivalent equation is $2x^2 + 2x - 1 = 0$, where $f(x) = 2x^2 + 2x - 1$. The quadratic formula gives

$$x = \frac{-2 \pm \sqrt{2^2 - 4(2)(-1)}}{2(2)} = \frac{-2 \pm \sqrt{12}}{4} = \frac{-1 \pm \sqrt{3}}{2}.$$

These exact solutions are generally not as useful as decimal approximations. We can use a calculator to show that, accurate to seven decimal places, $\frac{-1+\sqrt{3}}{2}$ is 0.3660254 and $\frac{-1-\sqrt{3}}{2}$ is -1.3660254. The graph of $f(x) = 2x^2 + 2x - 1$ in the $[-5, 5]$ by $[-5, 10]$ viewing rectangle in Figure 2.1.6 suggests that these two values of x are reasonable approximations to the solutions because they are good approximations to the x-intercepts. •

• EXAMPLE 3: Solve $x^2 + 2x - 15 = 0$.

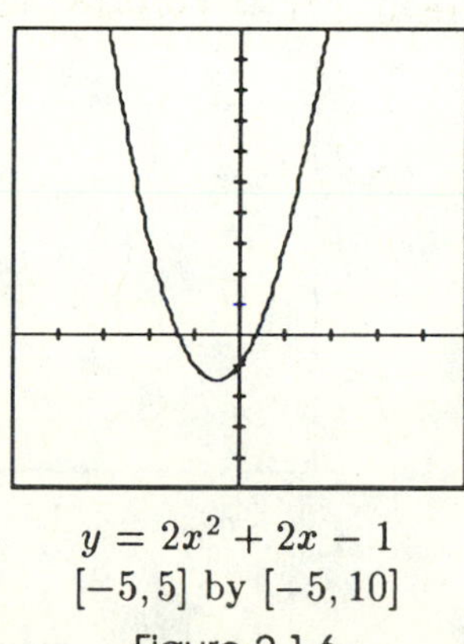

$y = 2x^2 + 2x - 1$
$[-5, 5]$ by $[-5, 10]$
Figure 2.1.6

SOLUTION: The graph of $y = x^2 + 2x - 15$ in Figure 2.1.7 strongly suggests that the x-intercepts are -5 and 3. This means that -5 and 3 are likely the solutions to the equation. We can confirm this conjecture made graphically either by direct substitution or by factoring the equation.

$$x^2 + 2x - 15 = 0$$
$$(x+5)(x-3) = 0$$

Thus, -5 and 3 are exact solutions to this equation. •

The techniques used to solve quadratic equations can occasionally be applied to polynomial equations of higher degree.

• EXAMPLE 4: Solve $x^4 - 5x^2 - 36 = 0$.

SOLUTION: The easiest way to solve this equation is to notice that it can be factored; it is quadratic in x^2.

$$x^4 - 5x^2 - 36 = 0$$
$$(x^2+4)(x^2-9) = 0$$

Thus, either $x^2 + 4 = 0$ or $x^2 - 9 = 0$. The equation $x^2 + 4 = 0$ has no real solutions, and the solutions to $x^2 - 9 = 0$ are -3 and 3. Therefore, the original equation has two real solutions: -3 and 3.

Alternatively, the equation $x^4 - 5x^2 - 36 = 0$ can be viewed as a quadratic equation in the variable x^2 by rewriting it in the form

$$(x^2)^2 - 5(x^2) - 36 = 0\,.$$

Then, the quadratic formula can be applied with $a = 1$, $b = -5$, and $c = -36$, to obtain

$$x^2 = \frac{5 \pm \sqrt{25+144}}{2} = \frac{5 \pm 13}{2}\,.$$

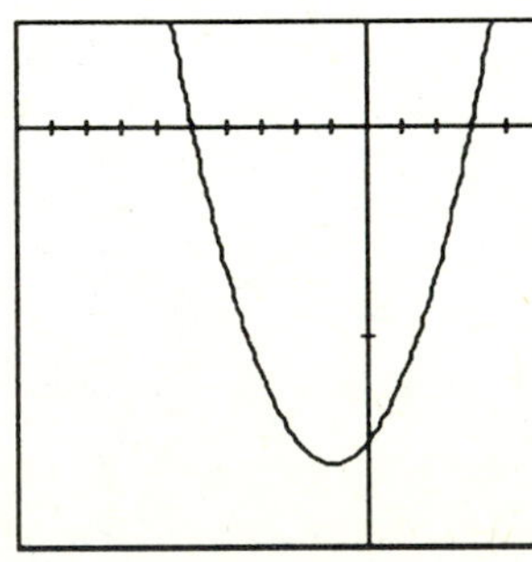

$y = x^2 + 2x - 15$
$[-10, 5]$ by $[-20, 5]$
Figure 2.1.7

Thus, $x^2 = 9$ or $x^2 = -4$. The equation $x^2 = -4$ has no real solutions, and -3 and 3 are the solutions to $x^2 = 9$. The graph of $f(x) = x^4 - 5x^2 - 36$ in Figure 2.1.8 confirms the real solutions. •

Cubic Equations

• DEFINITION A polynomial equation of degree 3 is called a **cubic equation**. •

The quadratic formula always gives exact solutions, which can then be approximated if desired or needed. There are formulas for the exact solutions to polynomial equations of degrees 3 and 4. However, these formulas are quite complicated, and solving equations graphically is so easy that we will *not* use these formulas in this course. To illustrate their complexity we give the solutions to

$$x^3 + a_1x^2 + a_2x + a_3 = 0\,. \tag{1}$$

First, we set

$$p = a_2 - \frac{{a_1}^2}{3}, \quad q = \frac{2{a_1}^3}{27} - \frac{a_1a_2}{3} + a_3\,,$$

$$P = \sqrt[3]{-\frac{q}{2} + \sqrt{\frac{p^3}{27} + \frac{q^2}{4}}}\,, \quad \text{and} \quad Q = \sqrt[3]{-\frac{q}{2} - \sqrt{\frac{p^3}{27} + \frac{q^2}{4}}}\,.$$

Then, the three solutions of equation (1) are

$$P + Q - \frac{a_1}{3}, \quad WP + W^2Q - \frac{a_1}{3}, \quad \text{and} \quad W^2P + WQ - \frac{a_1}{3}\,,$$

where W is a complex number with third power equal to 1. These formulas are called the **Cardan formulas**. You will see that a computer-based graphing approach makes the use of these complicated formulas unnecessary.

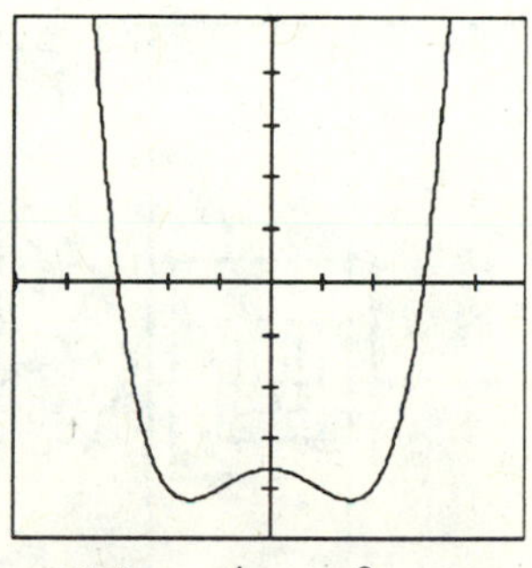

$f(x) = x^4 - 5x^2 - 36$
$[-5, 5]$ by $[-50, 50]$
Figure 2.1.8

The Cardan solutions appeared in *Ars Magna* in the year 1545. The graphical techniques in this text represent the first major practical advance in equation-solving techniques in the intervening 400 plus years that are easily available to precalculus students.

Finding Solutions Graphically Using Zoom-In

Next, we turn our attention to the use of a graphing utility to find solutions to an equation. One advantage of a graphical approach is that we are able to solve very complicated equations. We can solve any equation $f(x) = 0$ graphically provided we can enter $y = f(x)$ into our graphing utility. In particular, polynomial equations of *any* degree are no more difficult to solve than quadratic equations.

To find a solution of $f(x) = 0$ graphically we *trap the corresponding x-intercept* in a viewing rectangle that is small enough to read the value of x as accurately as desired. In fact, we actually create a nested sequence of viewing rectangles, each containing the next, that trap the x-intercept (Figure 2.1.9). However, solutions can be read only within the limits of machine precision, that is, only as accurately as the graphing utility permits. Most computers and graphing calculators allow answers to be read accurately to at least 7 or 8 significant figures. This procedure for finding the solutions to $f(x) = 0$ graphically is called **zoom-in**.

• EXAMPLE 5: Solve the equation $x^3 + 2x = 1$.

SOLUTION: First we write the equation in the form $x^3+2x-1 = 0$ and set $f(x) = x^3+2x-1$. The graph of f in the $[-10, 10]$ by $[-10, 10]$ viewing rectangle is given in Figure 2.1.10. You can check that the graph in Figure 2.1.10 is a complete graph. Thus, there is exactly one real solution to $x^3 + 2x - 1 = 0$. We can see from this figure that the graph of f crosses the x-axis between $x = 0$ and $x = 1$ because the scale marks on the x-axis are one unit apart. Remember that scale marks that appear in computer graphs produced by our graphing utility are equally spaced and depend on the viewing rectangle parameters L, R, B, and T.

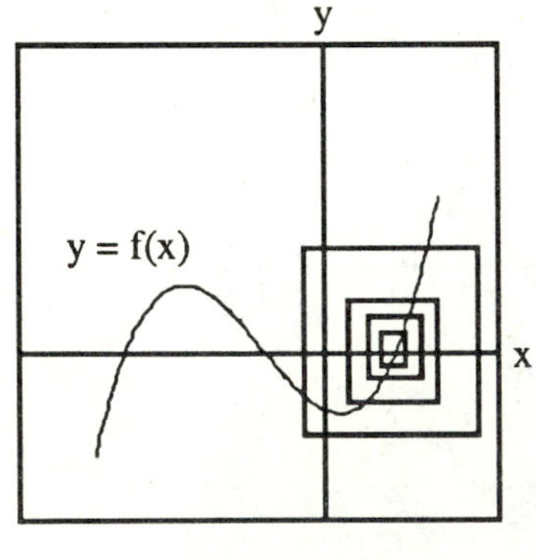

Figure 2.1.9

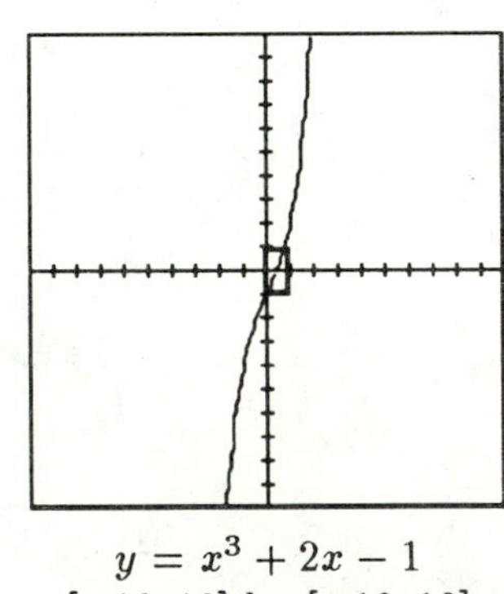

$y = x^3 + 2x - 1$
$[-10, 10]$ by $[-10, 10]$

Figure 2.1.10

Figure 2.1.10 suggests that we can choose the next viewing rectangle to be $[0, 1]$ by $[-1, 1]$. The graph of f in this viewing rectangle is shown in Figure 2.1.11.

The scale marks on the x-axis in Figure 2.1.11 are 0.1 units apart. (Why?) We can see that the graph of f crosses the x-axis between $x = 0.4$ and $x = 0.5$. Thus, $[0.4, 0.5]$ by $[-0.1, 0.1]$ is a reasonable viewing rectangle to choose, and the graph of f in this viewing rectangle is given in Figure 2.1.12.

Figure 2.1.12 indicates that the graph of f crosses the x-axis between $x = 0.45$ and $x = 0.46$. Any number between 0.45 and 0.46 that is used to approximate the actual solution is in error by at most 0.01, the distance between consecutive scale marks on the x-axis in Figure 2.1.12. For example, we could say that the solution is 0.453 with error of at most 0.01 because the graph of f in Figure 2.1.12 appears to cross the x-axis three-tenths of the way from the scale mark labeled 0.45 to the scale mark labeled 0.46. •

Error

In Example 5 we found that 0.453 was a solution to $x^3 + 2x - 1 = 0$ with error of at most 0.01. The reason the error is at most 0.01 is that Figure 2.1.12 shows that the *exact solution* lies between consecutive scale marks on the x-axis that are 0.01 units apart. In fact, *any number* in the interval $[0.45, 0.46]$, determined by these two scale marks, is a *solution to* $x^3 + 2x - 1 = 0$ *with error of at most* 0.01. For example, the left endpoint 0.45 and the right endpoint 0.46 are also solutions to $x^3 + 2x - 1 = 0$ with error of at most 0.01. Notice that we are *not* saying that the corresponding values of $f(x) = x^3 + 2x - 1$ differ from 0 by at most 0.01. In fact, Figure 2.1.12 shows that for some values of x in $[0.45, 0.46]$ the corresponding values of f differ from 0 by more than 0.01. Notice that the distance between consecutive vertical scale marks is also 0.01.

In a sufficiently small viewing rectangle a portion of the graph of a polynomial function will always appear to be a straight line. You can convince yourself that this statement is true by graphing several polynomial functions in small viewing rectangles that contain a portion of the graph. From Figure 2.1.12, we can see that the graph of f rises as we move from left to right in $0.45 \leq x \leq 0.46$. This means $f(x)$ will differ from zero by at most the larger of the two numbers $|f(0.45)|$ and $|f(0.46)|$ for x in $0.45 \leq x \leq 0.46$. Both $|f(0.45)|$

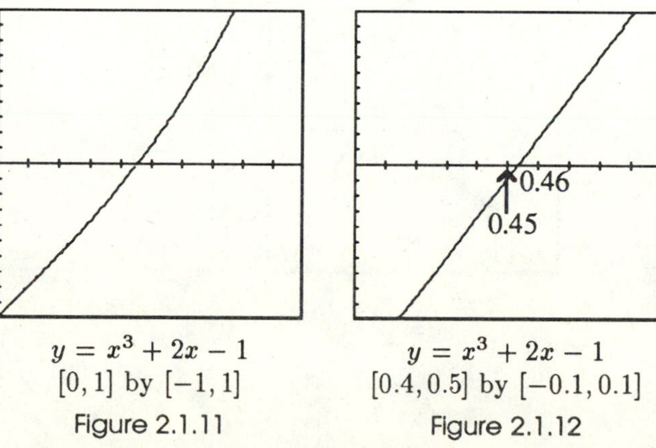

$y = x^3 + 2x - 1$
$[0, 1]$ by $[-1, 1]$
Figure 2.1.11

$y = x^3 + 2x - 1$
$[0.4, 0.5]$ by $[-0.1, 0.1]$
Figure 2.1.12

and $|f(0.46)|$ are less than 0.02 so that 0.02 is a good *estimate* for the error in the value of f when we use a value of x in $0.45 \leq x \leq 0.46$ to approximate the zero of f.

• DEFINITION Suppose that the graph of $y = f(x)$ in some viewing rectangle crosses the x-axis between two consecutive scale marks a and b that are r units apart. If c is any number in the interval determined by the scale marks, then c is said to be a **solution with error of at most r** to the equation $f(x) = 0$ (Figure 2.1.13). If the graph of f is always rising or always falling in the viewing rectangle, then the larger of $|f(a)|$ and $|f(b)|$ is a good estimate for the corresponding error bound on the function values $f(x)$. •

Suppose that the graph of $y = f(x)$ in the viewing rectangle $[L, R]$ by $[B, T]$ crosses the x-axis but displays no scale marks. Then, any number between L and R would be a solution with error of at most $r = R - L$.

• EXAMPLE 6: Solve the equation $x^3 + 2x = 1$ with error of at most 0.0000001. Estimate the error in $f(x) = x^3 + 2x - 1$.

SOLUTION: If we continue the process started in Example 5 we can look at the graph of f in smaller and smaller viewing rectangles. Figure 2.1.14 gives the graph of f in the viewing rectangle $[0.453397, 0.453398]$ by $[-0.000001, 0.000001]$. The actual solution lies between the scale marks labeled 0.4533976 and 0.4533977. Any number between 0.4533976 and 0.4533977 that is used to approximate the actual solution is in error by at most 0.0000001, the distance between horizontal scale marks in Figure 2.1.14. Figure 2.1.14 suggests we use 0.45339765 because the graph of $f(x) = x^3 + 2x - 1$ appears to cross the x-axis half-way between the scale marks labeled 0.4533976 and 0.4533977. Notice that the graph of f appears to be a straight line in this viewing rectangle. Both $|f(0.4533976)|$ and $|f(0.4533977)|$ are less than 0.0000002. Thus 0.0000002 is a good approximation to the error on $f(x)$. •

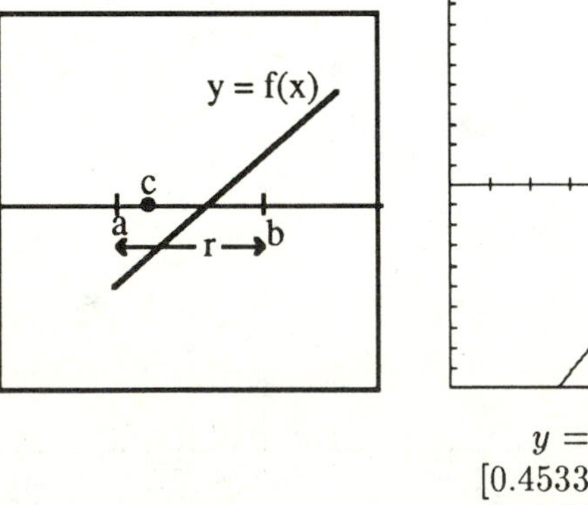

Figure 2.1.13

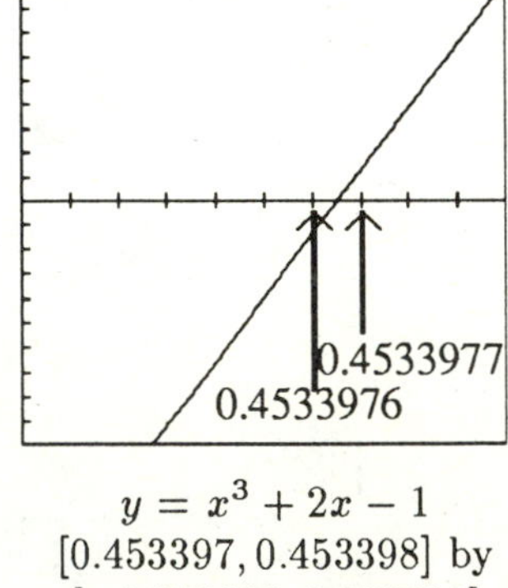

$y = x^3 + 2x - 1$
$[0.453397, 0.453398]$ by
$[-0.000001, 0.000001]$

Figure 2.1.14

It is a good idea for you to think about trying to find the real solution to the equation in Example 5 using the Cardan formulas. This should convince you that finding solutions graphically is both very quick and very accurate.

Generally, we need not approximate solutions as accurately as we did in Example 6. However, this example illustrates that a computer based graphing approach can do more than simply produce rough approximations to solutions; accuracy is limited only by machine precision.

To find all the real solutions to an equation $f(x) = 0$ graphically, we first find a complete graph of $y = f(x)$. Once we know the complete behavior of f, we can determine the number and approximate locations of the real solutions to the equation $f(x) = 0$. Then, using the technique illustrated in Examples 5 and 6, we can find each of the real solutions with a high degree of accuracy.

The procedure used in Examples 5 and 6 to trap the solutions is called *zoom-in* because this procedure is quite like using the focus adjustment on a very powerful microscope. Until recently, very specialized numerical techniques were needed to find solutions with this degree of accuracy. Using graphing utilities eliminates, or at least reduces, the need for sophisticated and complicated numerical techniques.

• EXAMPLE 7: Find a solution to $x^3 - 5x^2 + 6x - 1 = 0$ between 1 and 2 with error of at most 0.01.

SOLUTION: The graph of $f(x) = x^3 - 5x^2 + 6x - 1$ in the standard viewing rectangle (Figure 2.1.15) is complete (why?) and shows that this equation has three real solutions. One of the solutions is between 1 and 2. In order to find a solution with error of at most 0.01 we need a viewing rectangle with distance between scale marks equal to 0.01 or less. Figure 2.1.16 gives the graph of f in the $[1.5, 1.6]$ by $[-0.1, 0.1]$ viewing rectangle. Notice that the distance between scale marks in Figure 2.1.16 is 0.01. Thus, 1.555 is a solution to $x^3 - 5x^2 + 6x - 1 = 0$ that is between 1 and 2 and has error of at most 0.01. Can you give other approximations to the solution that are also in error by at most 0.01? •

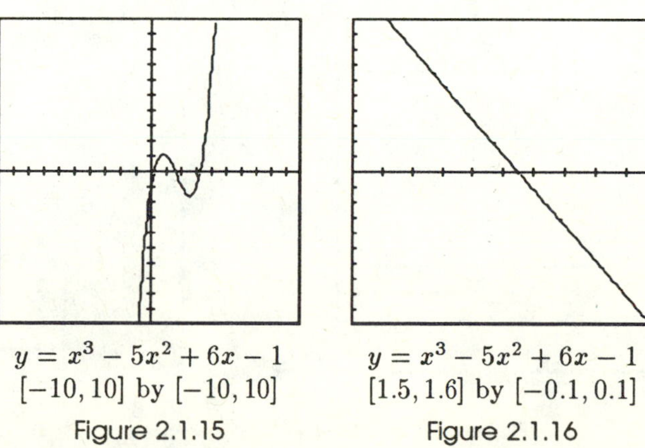

$y = x^3 - 5x^2 + 6x - 1$
$[-10, 10]$ by $[-10, 10]$
Figure 2.1.15

$y = x^3 - 5x^2 + 6x - 1$
$[1.5, 1.6]$ by $[-0.1, 0.1]$
Figure 2.1.16

Accuracy Agreement. Unless otherwise stated, from now on when we ask you to *solve an equation*, we mean for you to *approximate the real solutions with error of at most* 0.01. Furthermore, we assume that the numbers provided in Examples and Exercises are exact unless otherwise specified. Some textbooks assume that numbers like 8.7 and 7.9 are accurate only to tenths, and numbers like 6 and -3 are accurate only to units.

Application. Next, we look at an application about making a cardboard box from a rectangular sheet of cardboard. Squares of side length 5 inches are cut from each corner of a rectangular piece of cardboard with width W and length L (Figure 2.1.17). Then, we fold along the dashed lines in Figure 2.1.17 to form a box with no top (Figure 2.1.18). The height of the box is 5, the width is $W - 10$, and the length is $L - 10$.

- EXAMPLE 8: Suppose that the length of the rectangular piece of cardboard in Figure 2.1.17 is twice its width. Determine the dimensions of the cardboard if the volume of the resulting box is 2040 cubic inches.

SOLUTION: From Figure 2.1.18, the volume of the box is $5(W - 10)(L - 10)$. Because $L = 2W$ and the volume of the box is 2040 we have $5(W - 10)(2W - 10) = 2040$. The following equations are equivalent.

$$5(W - 10)(2W - 10) = 2040$$
$$5(W - 10)(2)(W - 5) = 2040$$
$$(W - 10)(W - 5) = 204$$
$$W^2 - 15W + 50 = 204$$
$$W^2 - 15W - 154 = 0$$

We can use the quadratic formula or notice that the above quadratic equation factors as

$$(W - 22)(W + 7) = 0\,.$$

Both -7 and 22 are solutions to this equation. Because width must be positive only $W = 22$ is a solution to the problem. The dimensions of the original piece of cardboard must be 22 inches by 44 inches. Verify that the resulting box has volume 2040 cubic inches. •

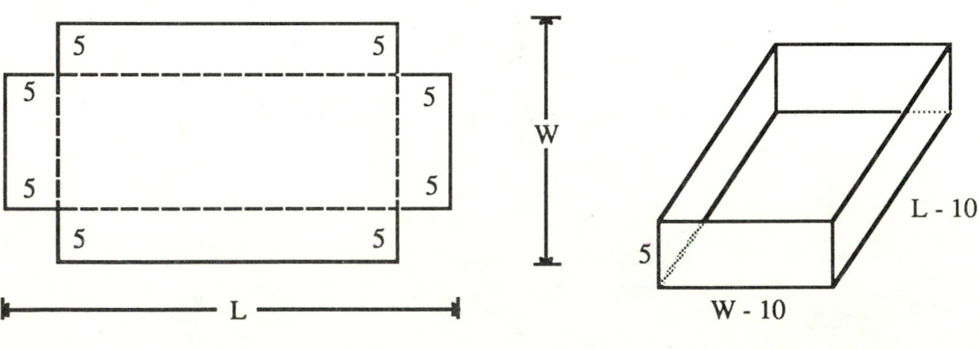

Figure 2.1.17

Figure 2.1.18

• EXERCISES 2-1

1. Solve the equation $x^2 - 3x - 10 = 0$ algebraically. Draw a complete graph of $y = x^2 - 3x - 10$ and compare the x-intercepts of the graph with the solutions to the equation.
2. Solve the equation $x^3 - 25x = 0$ algebraically. Draw a complete graph of $y = x^3 - 25x$ and compare the x-intercepts of the graph with the solutions to the equation.
3. Solve the equation $|x + 4| = 8$ algebraically. Draw a complete graph of $y = |x + 4| - 8$ and compare the x-intercepts of the graph with the solutions to the equation.

Use an *algebraic* method to solve each equation. Check your answer with a graphing utility.

4. $(x - 3)(x + 2) = 0$
5. $x^2 = 14$
6. $x^2 - 3x + 2 = 0$
7. $x^2 - 2x + 3 = 0$
8. $x^3 - x = 0$
9. $|x - 2| = 6$

Use an algebraic method to solve each equation. Check by substitution into the original equation.

10. $y^2 = 9$
11. $x^2 - 5x + 6 = 30$
12. $3y^3 - 2y^2 + y = 0$
13. Bill thinks -1, 2, and 3 are solutions to the equation $x^3 + 2x^2 - 5x - 6 = 0$. Is he right? If not, which of these numbers are solutions? Does this equation have any other real solutions? How can graphing help determine all real solutions?

Use a graphing utility to determine the number of real solutions to each equation.

14. $x^3 - 2x^2 + 3x - 5 = 0$
15. $x^3 - 65x + 10 = 0$

Use factoring to find all real solutions to the equation.

16. $x^3 - 2x^2 - 2x + 4 = 0$
17. $10 - 15x + 6x^2 - 9x^3 = 0$
18. Consider the equation $x^3 - x^2 + x - 3 = 0$. Find a sequence of four viewing rectangles containing each solution. The first viewing rectangle should permit the solution to be read with error of at most 0.1; the second 0.01; the third 0.001; and the last 0.0001.
19. Consider the equation $x^3 - 2x + 3 = 0$. It has one real solution. Find a sequence of four viewing rectangles containing the solution. The first viewing rectangle should permit the solution to be read with error of at most 0.1; the second 0.01; the third 0.001; and the last 0.0001.
20. Find the other two solutions to the equation in Example 7 with error of at most 0.01.

Determine *one positive solution* to each equation $f(x) = 0$ with error of at most 0.01. Then estimate the maximum error in the corresponding $f(x)$ value.

21. $f(x) = x^3 - x - 2$
22. $f(x) = \frac{1}{100}x^3 - x - 2$
23. $f(x) = \frac{1}{10}x^3 - x^2 - 2$
24. $f(x) = x^3 - \frac{1}{10}x$

Solve each equation.

25. $3x^2 - 15x + 8 = 0$
26. $x^3 - 2x^2 + 3x - 1 = 0$
27. $x(x - 25)(x - 35) = 3000$
28. Find three distinct approximations to the one real solution to $x^3 - 10 = 0$ with error of at most 0.01. What is the exact solution?

Solve each equation for x in the given interval with error of at most 0.01.

29. $x^4 - 3x^3 - 6x + 5 = 0$ where $0 \leq x \leq 10$
30. $\dfrac{x^3 - 10x^2 + x + 50}{x - 2} = 0$ where $-10 \leq x \leq 10$

31. $3\sin(x-5)=0$ where $0 \le x \le 10$

32. $\sqrt[3]{x^2-2x+3}=0$ where $-10 \le x \le 10$

33. Assume it is possible to find the exact solutions to $x^3+a_1x^2+a_2x+a_3=0$. Explain why it is then possible to find the exact solutions to $b_1x^3+b_2x^2+b_3x+b_4=0$, where $b_1 \neq 0$.

34. The perimeter of some rectangles is 320 inches. Let x be the length of such a rectangle.
    (a) If the length is 20 inches, then what is the area of the rectangle?
    (b) Show that $A=x(160-x)$ is an algebraic representation of the area of such rectangles.
    (c) Draw a complete graph of the algebraic representation in (b).
    (d) Verify that $(80, 6400)$ is a point on the graph in (c). What meaning do these coordinates have in this problem situation?
    (e) Verify that $(-40, -8000)$ is a point on the graph in (c). What meaning do these coordinates have in this problem situation?
    (f) What values of x make sense in this problem situation? Why? Draw a graph of the problem situation.
    (g) Use zoom-in to determine the area of a rectangle if its length is 83 inches.

35. A storage box of height 25 cm has a volume of 5400 cm^3. Let x be the length of the box in centimeters and y the width in centimeters.
    (a) Write y as a function of x.
    (b) Draw a complete graph of this equation.
    (c) What part of the graph in (b) represents the problem situation?
    (d) Use the graph in (b) to determine the width of the box if the length is 24 cm.

36. A laboratory keeps two acid solutions on hand. One is 20% acid and the other is 35% acid. An order is received for 25 liters of a 26% acid solution. How much 20% acid solution and how much 35% acid solution should be used to fill this order?

37. The formula $F=\frac{9}{5}C+32$ gives the Fahrenheit (F) temperature as a function of the Celsius (C) temperature. Solve this equation for C.

38. The formula for the area of a trapezoid (Figure 2.1.19) is $A=\frac{1}{2}h(b_1+b_2)$. Solve this equation for b_1.

39. $S=P(1+rn)$ is an algebraic representation of the value S after n years of an initial investment of P dollars that earns simple interest at rate r per year. *Note:* For a simple interest rate of 10%, r is 0.10.
    (a) Solve this equation for n.
    (b) How many years are required for an investment earning 8% simple interest to *triple* in value?

40. A box with no top is made from a rectangular sheet of cardboard that is three times as long as it is wide by removing 6-inch squares from each corner and folding up the sides as suggested by Figures 2.1.17 and 2.1.18. Determine the dimensions of the cardboard sheet if the volume of the resulting box is 3000 cubic inches.

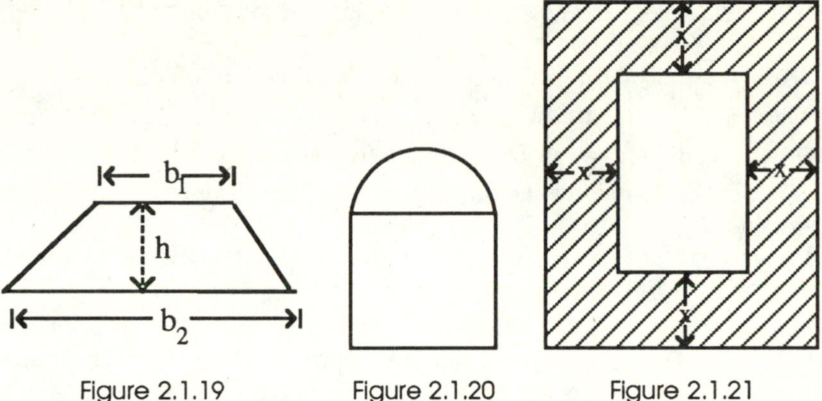

Figure 2.1.19 Figure 2.1.20 Figure 2.1.21

41. A semicircle is placed on one side of a square so that its diameter coincides with a side of the square (Figure 2.1.20). Find the side length of the square if the total area of the square plus the semicircle is 200 square units.

42. The owner of the Olde Time Ice Cream Shop pays \$1000 per month for fixed expenses such as rent, lights, and wages. Ice cream cones are sold at \$0.75 each, of which \$0.40 goes for ice cream, cone, and napkin. How many cones must be sold to break even?

43. For a certain make of car it is determined that when the car is traveling at a speed of r mph the stopping distance D in feet is given by $D = r + \frac{r^2}{19.85}$.
    (a) Draw a complete graph of $y = D(r)$.
    (b) What are the domain and range of the algebraic representation? What values of r make sense in this problem situation?
    (c) Use the graph in (a) to estimate the speed of the car if the stopping distance is 300 feet.

44. An 8 1/2 inch by 11 inch piece of paper contains a picture with uniform border. The distance from the edge of the paper to the picture is x inches on all sides (Figure 2.1.21).
    (a) Express the area A of the picture as a function of x.
    (b) Draw a complete graph of the function $y = A(x)$.
    (c) What are the domain and range of the function in (a)? What values of x make sense in this problem situation?
    (d) Use the graph in (b) to estimate the uniform border if the area of the picture is 50 square inches.

45. The distribution of velocities v (in cm/sec) in the laminar flow of arterial blood is given by the equation $v = 1.19 - (1.85 \times 10^4)r^2$ where r is the distance (in cm) of the blood layer from the center of the artery.
    (a) Draw a complete graph of $y = v(r)$.
    (b) What are the domain and range of the algebraic representation? What values of r make sense in this problem situation?
    (c) If the blood layer velocity is 0.975 cm/sec, use the graph in (a) to estimate the distance of the layer from the center of the artery.

46. A toy company makes x toys each day at a total cost of $\frac{1}{4}(x-20)^2+10$ dollars. The company receives \$0.67 for each toy sold.
    (a) Write the daily profit of the company as a function of x.
    (b) What are the fixed costs?
    (c) Draw a complete graph of the profit function in (a).
    (d) What is the domain of the function in (a)? What values of x make sense in this problem situation?
    (e) Use the graph in (c) to determine what level of daily production the company must maintain to break even.

A single-commodity open market is driven by the supply and demand principle. Economists have determined that supply curves are usually increasing (as the price increases, the sellers increase production) and demand curves are usually decreasing (as the price increases, the consumer buys less). For each demand equation, find the price p if the production level is 120 units, and find the production level if the price is \$2.30. Draw a complete graph of each equation and check your answers with a graphing utility.

47. $p = 15 - 0.023x$

48. $p = 100 - 0.0015x^2$

49. Draw a complete graph of the function $f(x) = x^3 - 2x + 3$.
    (a) How many real solutions are there to the equation $f(x) = 0$?
    (b) How many real solutions are there to the equation $f(x) = 3$?
    (c) Estimate the endpoints of the interval of x values where the values of f are decreasing as x is increasing.

50. Solve the equation: $|x| + |x-3| = 6$.

51. Solve the equation $|x| - |x-6| = 0$ using the distance interpretation of absolute value. Check with a graphing utility.

2.2 Solving Systems of Equations

Sometimes it is convenient, or even necessary, to give an algebraic representation of a problem situation that consists of a system of functions or a system of equations. In this section we focus on solving these systems graphically.

We solve systems of equations graphically by using zoom-in to trap each point of intersection in a viewing rectangle small enough to read the coordinates of the point as accurately as desired. In Section 2.1, we found solutions to an equation in one variable with error less than any prescribed number. This was accomplished by trapping the x-intercept between consecutive scale marks on the x-axis that were at a distance less than or equal to the prescribed error. If a given graphing utility did not draw scale marks, then to satisfy the error condition we had to choose the viewing rectangle $[L, R]$ by $[B, T]$ so that the difference $R - L$ was less than or equal to the prescribed error.

In an analogous way, we can prescribe a maximum error for solutions to systems of equations in two variables. Solutions of systems of equations in two variables are pairs of numbers, one for each of the two variables. Thus, it seems technically possible to prescribe the error in each variable. However, we have to be careful because these errors are generally *not independent.* The first example illustrates this point. When we ask you to give a simultaneous solution to a system of equations with error at most r, we mean that *each* coordinate should be estimated with error less than r.

The graphing utility we use to draw the graphs in this text overlays a lattice using the horizontal and vertical scale marks. For example, Figure 2.2.1 shows the lattice in the standard viewing rectangle with the distance between scale marks equal to one unit in both the horizontal and vertical direction.

The lattice in Figure 2.2.1 subdivides the viewing rectangle $[-10, 10]$ by $[-10, 10]$ into 400 unit squares. If we can determine that a specific point lies in one of these unit squares, then this lattice permits us to read the x- and y-coordinates of the point with error at most one. If scale marks but *no* lattice appear in a viewing rectangle, then it is usually possible to use the scale marks to estimate a smaller rectangle that contains a specific point. This smaller rectangle can be used to determine the error in a given approximation. If neither a lattice nor scale marks appear in a given viewing rectangle $[L, R]$ by $[B, T]$, then the x-coordinate of any point in the viewing rectangle can be read with error at most $R - L$, and the y-coordinate with error at most $T - B$. Most graphing calculators do *not* have the grid feature. You need to use the scale marks to read answers with a prescribed error on a graphing calculator. For this reason, we rarely use the grid feature of our graphing utility.

Suppose we want to find the length and width of a rectangle with perimeter 100 and area 300 (Figure 2.2.2). Let L and W stand for the length and width of such a rectangle, respectively. Then, the following system of equations is an algebraic representation of the problem.

$$\begin{aligned} 2L + 2W &= 100 \quad \text{(perimeter)} \\ LW &= 300 \quad \text{(area)} \end{aligned}$$

A solution to this system of equations is a pair of values, one for L and one for W, that is a solution to each of the equations.

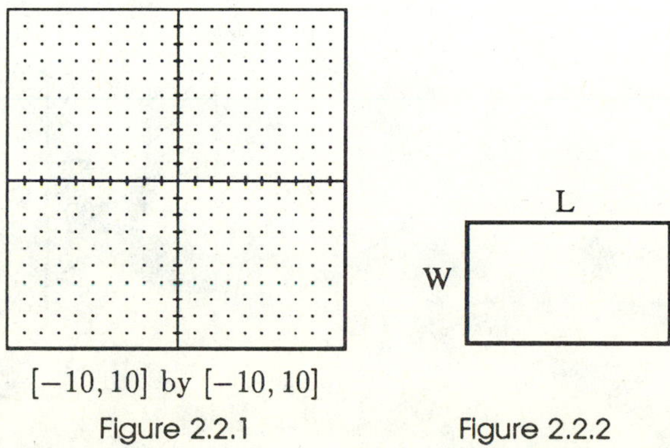

$[-10, 10]$ by $[-10, 10]$

Figure 2.2.1

Figure 2.2.2

• DEFINITION A pair of real numbers is a **solution to a system of equations in two variables** if and only if the pair of numbers is a solution to each of the equations. These solutions are called **simultaneous solutions** to the system of equations. When we have found all solutions to the system of equations we say that we have **solved the system of equations.** •

The graph of an equation in two variables is the set of solutions to the equation considered as points in the plane. A simultaneous solution to a system of equations is a common point, or point of intersection, of the graphs of the equations in the system.

We solve systems of equations graphically by finding the points of intersection of the graphs of the equations in the system. Algebraic techniques can be used to solve systems of equations.

• EXAMPLE 1: Find the simultaneous solutions to $\begin{cases} 2y + 2x = 100, \\ xy = 300. \end{cases}$

SOLUTION: To graph these equations with a graphing utility, you will need to solve each equation for y and rewrite the system as follows.

$$y = 50 - x$$
$$y = \frac{300}{x}$$

Figure 2.2.3 gives complete graphs of the two equations in the viewing rectangle $[-60, 60]$ by $[-60, 60]$. We can see from Figure 2.2.3 that the graphs have two points in common. Thus, we should expect to find two simultaneous solutions. In Example 3 we will use a graphing zoom-in procedure to find the coordinates of the two points of intersection. In this example we illustrate an algebraic procedure called the substitution method.

In the substitution method we solve one equation for a variable and then substitute the expression for that variable into the other equation. Solving the first equation for y gives

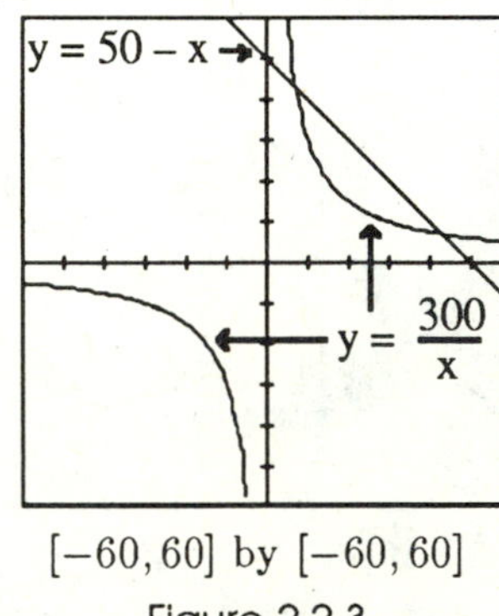

$[-60, 60]$ by $[-60, 60]$

Figure 2.2.3

$y = 50 - x$. Substituting for y in the second equation gives the following.

$$\begin{aligned} xy &= 300 \\ x(50 - x) &= 300 \\ x^2 - 50x + 300 &= 0 \end{aligned}$$

Using the quadratic formula we find the following values for x.

$$\begin{aligned} x &= \frac{50 \pm \sqrt{(-50)^2 - 4(300)}}{2} \\ &= \frac{50 \pm 10\sqrt{13}}{2} \\ &= 25 \pm 5\sqrt{13} \end{aligned}$$

Accurate to hundredths, $25 + 5\sqrt{13}$ is 43.03 and $25 - 5\sqrt{13}$ is 6.97.

The corresponding values of y can be found from the equation $y = 50 - x$. If $x = 25 + 5\sqrt{13}$, then $y = 50-(25+5\sqrt{13}) = 25-5\sqrt{13}$. If $x = 25-5\sqrt{13}$, then $y = 50-(25-5\sqrt{13}) = 25+5\sqrt{13}$. Thus, the two solutions are $(25+5\sqrt{13}, 25-5\sqrt{13})$ and $(25-5\sqrt{13}, 25+5\sqrt{13})$. In decimal form accurate to hundredths, the two solutions are $(43.03, 6.97)$ and $(6.97, 43.03)$. Each coordinate of the solution in decimal form is in error of at most 0.01. Remember, unless otherwise specified, we find solutions with error of at most 0.01. •

• EXAMPLE 2: Find the exact dimensions of a rectangle with perimeter 100 and area 300.

SOLUTION: From Example 1 we know that $L = 25 + 5\sqrt{13}$ and $W = 25 - 5\sqrt{13}$, or $L = 25 - 5\sqrt{13}$ and $W = 25 + 5\sqrt{13}$. These two solutions produce two rectangles (Figure 2.2.4).

Actually the two rectangles are congruent (really the same rectangle). Thus there is a unique (one and only one) solution and the dimensions are $25 - 5\sqrt{13}$ by $25 + 5\sqrt{13}$. In decimal form, the dimensions are 6.97 by 43.03 with error of at most 0.01. •

• EXAMPLE 3: Find the simultaneous solutions to $\begin{cases} y = -x^3 + 3x^2 + x - 3, \\ y = -2x^2 + 5. \end{cases}$

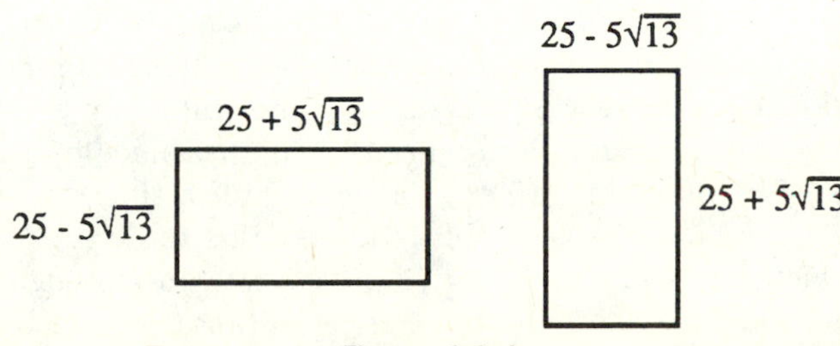

Figure 2.2.4

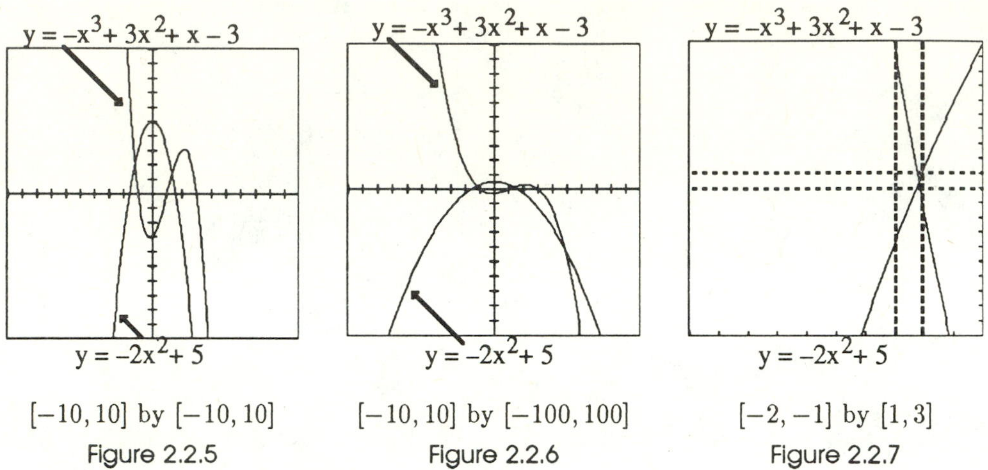

[−10, 10] by [−10, 10]
Figure 2.2.5

[−10, 10] by [−100, 100]
Figure 2.2.6

[−2, −1] by [1, 3]
Figure 2.2.7

SOLUTION: According to our agreement about error, we must determine the coordinates of each solution with error at most 0.01. Figure 2.2.5 gives the graph of these two equations in the standard viewing rectangle. This figure suggests there are two points of intersection. However, there is a third point of intersection in the fourth quadrant that is outside of this viewing rectangle. The third point of intersection shows up in the viewing rectangle $[-10, 10]$ by $[-100, 100]$ (Figure 2.2.6). Figure 2.2.6 gives a complete graph of each equation. (Why?) Thus, there are *three* solutions to this system of equations.

We can estimate from Figure 2.2.5 that the point of intersection with smallest x-coordinate lies in the $[-2, -1]$ by $[1, 3]$ viewing rectangle. The graphs of the two equations in this viewing rectangle are given in Figure 2.2.7. Figure 2.2.7 suggests we draw the graphs of the two equations in the $[-1.3, -1.2]$ by $[2.0, 2.1]$ viewing rectangle (Figure 2.2.8).

The scale marks in Figure 2.2.8 are 0.01 units apart in both the horizontal and vertical directions. Thus, any point in the $[-1.22, -1.21]$ by $[2.03, 2.04]$ rectangle gives the coordinates of this point of intersection with error at most 0.01. We read the point of intersection to be $(-1.218, 2.035)$. Similarly, we can find that the other solution in Figure 2.2.5 is $(1.350, 1.356)$ with error at most 0.01.

We were able to keep the horizontal and vertical distance between scale marks the same for the first two points of intersection. This will *not* be possible for the remaining point of intersection. Figure 2.2.6 suggests that we look at the graphs in the $[4, 5]$ by $[-50, -40]$ viewing rectangle. Notice that the two graphs drop very sharply near this third point of intersection, because small changes in x produce very large changes in y near this point. Continuing as before, the next two viewing rectangles found will likely be $[4.8, 4.9]$ by $[-43, -42]$ and $[4.86, 4.87]$ by $[-42.4, -42.3]$. You will likely find it difficult to determine the coordinates of this point of intersection with error at most 0.01 in the latter viewing rectangle. We can estimate the coordinates of the point of intersection to be $(4.868, -42.39)$ from the graphs in $[4.864, 4.869]$ by $[-42.42, -42.36]$ (Figure 2.2.9). In this figure the horizontal distance between scale marks is 0.001 and the vertical distance is 0.01. *We need error at most* 0.001 *in the x-coordinate in order to obtain error at most* 0.01 *in the y-coordinate.* If you try to

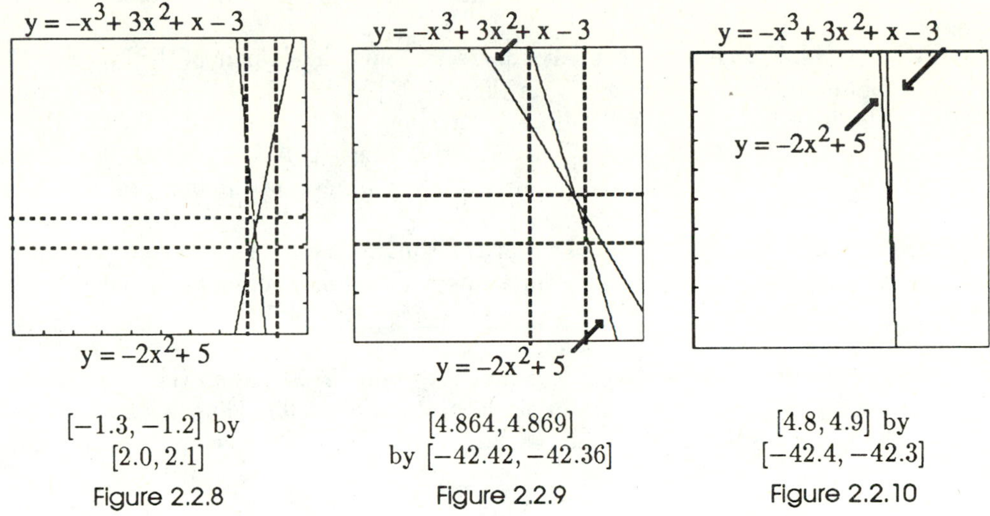

$[-1.3, -1.2]$ by $[2.0, 2.1]$

Figure 2.2.8

$[4.864, 4.869]$ by $[-42.42, -42.36]$

Figure 2.2.9

$[4.8, 4.9]$ by $[-42.4, -42.3]$

Figure 2.2.10

keep both horizontal and vertical distances between scale marks the same, you will end up with graphs like the one in Figure 2.2.10. Notice it is very hard to read the coordinates of the point of intersection from Figure 2.2.10. •

The Box Problem

Squares of side length x are cut from each corner of a rectangular piece of cardboard that is 20 inches wide by 25 inches long (Figure 2.2.11). Then the cardboard is folded along the dashed line segments in Figure 2.2.11 to form a box with no top. The height of this box is x. The base of the box has dimensions $20 - 2x$ by $25 - 2x$. Finally, the volume of the box is $V(x) = x(20 - 2x)(25 - 2x)$. V is an algebraic representation of this problem situation.

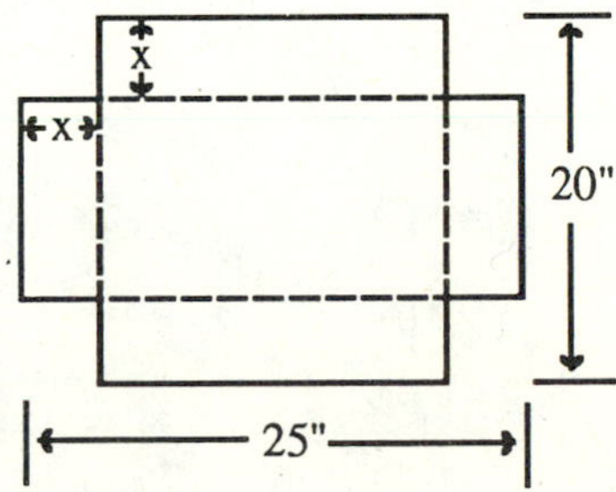

Figure 2.2.11

• EXAMPLE 4: Draw a complete graph of the function $V(x) = x(20-2x)(25-2x)$ and indicate what part of the graph represents the box problem. Determine the domain and range of V.

SOLUTION: $V(x)$ is a polynomial of degree 3. Thus, the domain of the function V is the set of all real numbers. The graph of V in the $[-5, 15]$ by $[-1000, 1000]$ viewing rectangle (Figure 2.2.12) is a complete graph. (Why?) The range of V is also the set of all real numbers. However, V represents the volume of the box constructed above only for x between 0 and 10. (Why?) The graph of V in Figure 2.2.13 is a graph of the box problem. •

• EXAMPLE 5: Find the side length of the square that must be cut out from a 20-inch by 25-inch piece of cardboard to form a box with volume 500 cubic inches.

SOLUTION: We need to find a value of x for which the volume is 500, that is, $V(x) = 500$. Remember that $V(x) = 500$ means that the point $(x, 500)$ is on the graph of V. Notice also that the point $(x, 500)$ lies on the horizontal line $y = 500$. Thus, a solution to this problem is the first coordinate of a simultaneous solution to the following system of equations.

$$y = V(x) = x(20-2x)(25-2x)$$
$$y = 500$$

The graphs of these two equations in the $[-5, 15]$ by $[-1000, 1000]$ viewing rectangle are given in Figure 2.2.14. The horizontal line $y = 500$ intersects the graph of V three times, but only two are possible values for the side length of the square. (Why?) Figure 2.2.15 is a geometric representation for only those values of x that satisfy the problem situation. Notice that these two graphs intersect twice.

We can use zoom-in to determine the two values of x. Figure 2.2.16 gives the graphs in the $[1.2, 1.3]$ by $[490, 510]$ viewing rectangle. The distance between horizontal scale marks is 0.01. The graph of $y = V(x)$ crosses the horizontal line $y = 500$ at some value of x in $1.27 \le x \le 1.28$. The graph of V appears to be a straight line in $1.27 \le x \le 1.28$. The graph in Figure 2.2.16 suggests that $V(1.27)$ is about 498 and $V(1.28)$ is about 501.

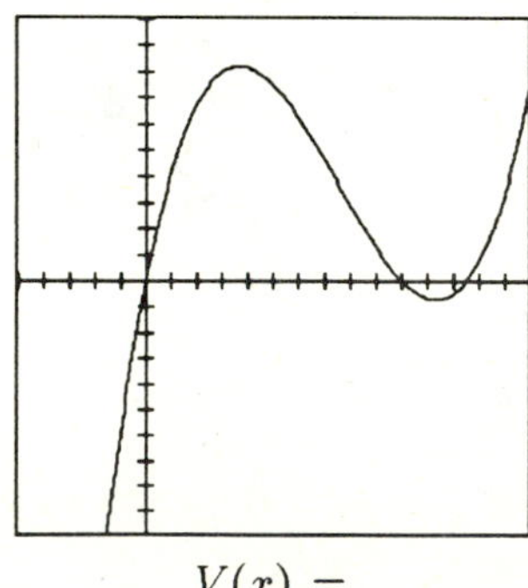

$V(x) =$
$x(20-2x)(25-2x)$
$[-5, 15]$ by $[-1000, 1000]$
Figure 2.2.12

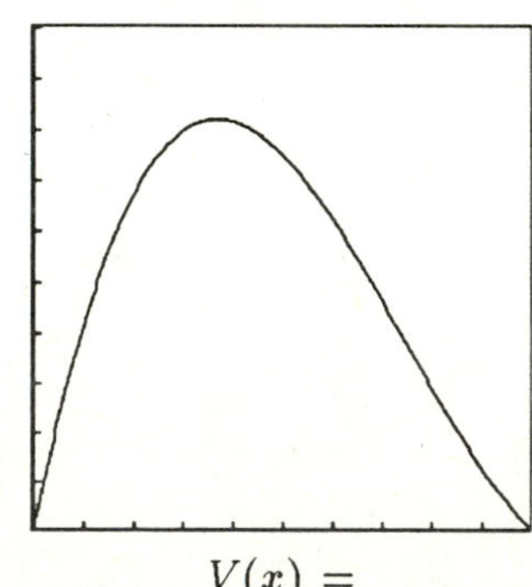

$V(x) =$
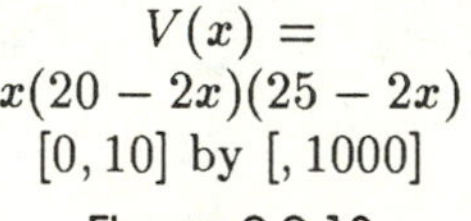
$x(20-2x)(25-2x)$
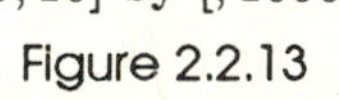
$[0, 10]$ by $[, 1000]$
Figure 2.2.13

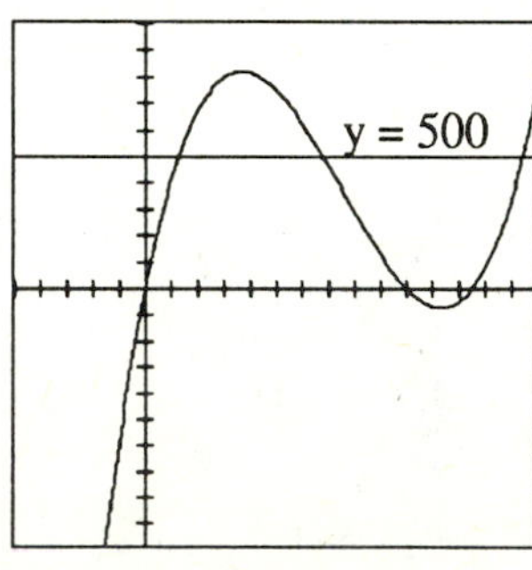

$y = x$
$(20-2x)(25-2x)$
$[-5, 15]$ by $[-1000, 1000]$
Figure 2.2.14

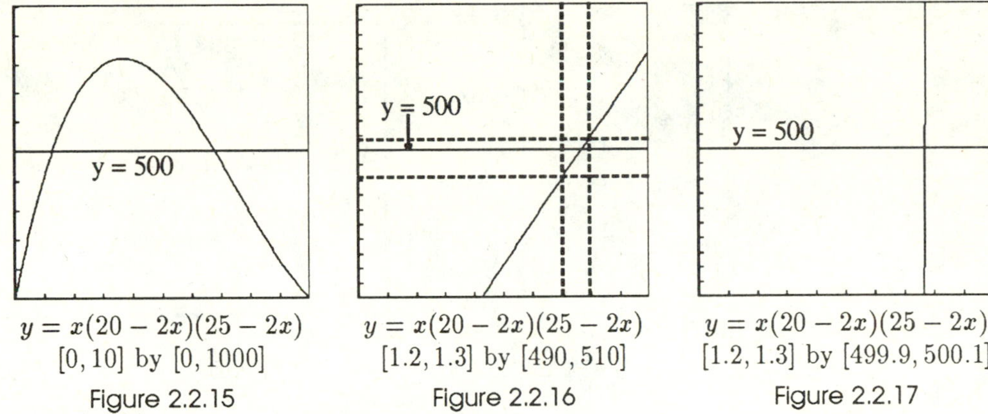

$y = x(20-2x)(25-2x)$
$[0,10]$ by $[0,1000]$

Figure 2.2.15

$y = x(20-2x)(25-2x)$
$[1.2,1.3]$ by $[490,510]$

Figure 2.2.16

$y = x(20-2x)(25-2x)$
$[1.2,1.3]$ by $[499.9,500.1]$

Figure 2.2.17

We can compute to see that $V(1.27)$ is actually a little larger than 498 and $V(1.28)$ is a little less than 501. Thus, for x between 1.27 and 1.28 the corresponding values of V range between 498 and 501. Therefore, Figure 2.2.16 permits us to read a solution to this system of equations with error at most 0.01 in the first coordinate and with error at most 2 in the second coordinate. It follows that any value of x between 1.27 and 1.28 used for the side length of the square cut out gives a solution to the problem with error of at most 0.01. Figure 2.2.16 suggests we use 1.277 as the value of x.

Suppose we also want the value of V to be in error by at most 0.01. Then, we must zoom-in more because the error in V from Figure 2.2.16 can be as much as 2 cubic inches. It doesn't help to change B and T and fix L and R in the viewing rectangle of Figure 2.2.16. For example, the graph of the system of equations in the $[1.2,1.3]$ by $[499.9,500.1]$ viewing rectangle (Figure 2.2.17) does not permit us to read the error in the second coordinate for x between 1.27 and 1.28 because the graph of V goes outside the viewing rectangle for x between 1.27 and 1.28.

If we want to increase the accuracy of V we must also increase the accuracy of x. The graph of the system of equations in the $[1.276,1.277]$ by $[499.9,500.1]$ viewing rectangle (Figure 2.2.18) permits us to read a solution with error in the second coordinate at most 0.03 because, for x between 1.2767 and 1.2768, the value of V differs from 500 by at most 0.03. Again, this can be checked by direct computation using the fact that the graph of V appears to be a straight line in $1.2767 \leq x \leq 1.2768$, or estimated from Figure 2.2.18. Thus, $x = 1.27678$ gives a solution to the problem with error in the value of V at most 0.03 and error in the value of x at most 0.0001, the distance between horizontal scale marks in Figure 2.2.18. •

Solving the equation $V(x) = 500$ in Example 5 graphically is very easy. Most of the discussion in Example 5 was concerned with the accuracy of the solution. To appreciate the power of a graphical approach, you should try to solve $V(x) = x(20-2x)(25-2x) = 500$ algebraically. You will find that the Cardan cubic equation formulas will be needed!

The next example illustrates that exact solutions can *sometimes* be found when we solve graphically.

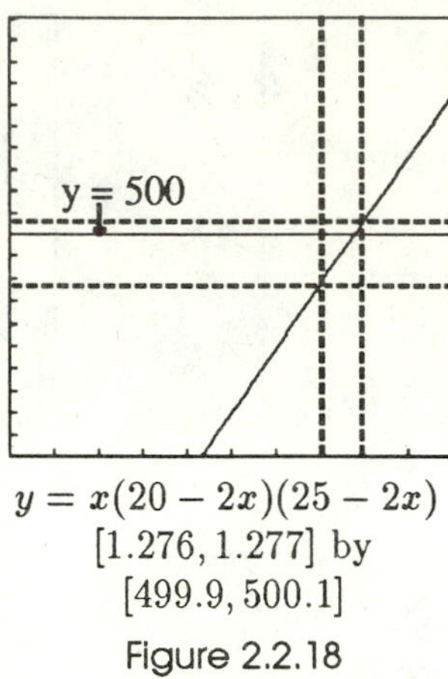

$y = x(20 - 2x)(25 - 2x)$
$[1.276, 1.277]$ by
$[499.9, 500.1]$

Figure 2.2.18

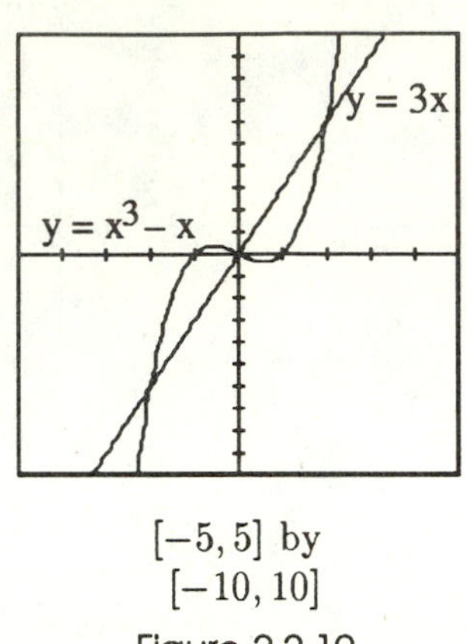

$[-5, 5]$ by
$[-10, 10]$

Figure 2.2.19

• EXAMPLE 6: Find the simultaneous solutions to $\begin{cases} y = x^3 - x \\ y = 3x \end{cases}$

SOLUTION: The graphs of these two equations in the $[-5, 5]$ by $[-10, 10]$ viewing rectangle shown in Figure 2.2.19 are complete. Figure 2.2.19 strongly suggests that the three points of intersection are: $(-2, -6)$, $(0, 0)$, and $(2, 6)$. Exactness of an answer *cannot* be established completely graphically. You can verify directly by substitution that these three pairs are solutions to the system of equations. Thus, we have found all solutions to this system of equations.

We can also find these solutions by solving the system algebraically using the substitution method. Substituting the value of y from the first equation into the second equation we obtain $x^3 - x = 3x$. Now we have one equation in one variable that we solve as follows.

$$x^3 - x = 3x$$
$$x^3 - 4x = 0$$
$$x(x^2 - 4) = 0$$
$$x(x - 2)(x + 2) = 0$$

We see that x is -2, 0, or 2. The corresponding values of y are -6, 0, and 6, respectively. Thus, the three solutions are $(-2, -6)$, $(0, 0)$, and $(2, 6)$. •

Actually, the equations in Example 6 were selected so that we could use an algebraic method of solution. Most of the time we will *not* be able to solve such systems algebraically. Frequently you will want to use the strategy of approximating a solution graphically and then substitute to verify your solution. However, a graphical approach will always work provided the equations can be graphed with a graphing utility.

• EXAMPLE 7: Determine the simultaneous solution to $\begin{cases} y = \cos x \\ y = x^2 \end{cases}$ with x between 0 and 1.

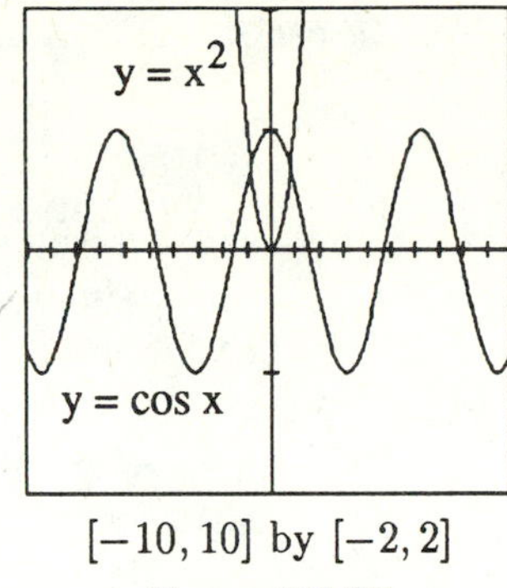

[−10, 10] by [−2, 2]

Figure 2.2.20

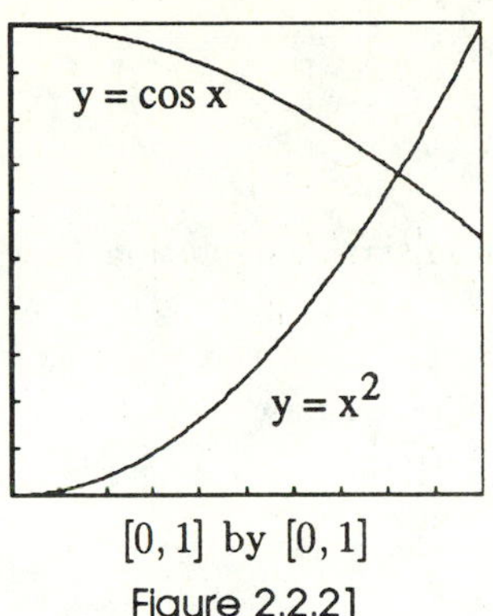

[0, 1] by [0, 1]

Figure 2.2.21

SOLUTION: The graphs of the equations in Figure 2.2.20 indicate that there are two points of intersection. When trigonometry is studied later in the book, we will find that Figure 2.2.20 gives a complete graph of each function. (If you are using a graphing calculator, be sure it is in radian mode.) Figure 2.2.21 gives a closer look at the desired point of intersection. We can use zoom-in to determine that the coordinates of the point of intersection are $(0.82, 0.68)$. Thus, the desired solution for $0 < x < 1$ is the pair of numbers $x = 0.82$ and $y = 0.68$. •

• EXERCISES 2-2

Use the substitution method to find the simultaneous solutions to each system of equations.

1. $2x - 3y = -23$

   $x + y - 6 = 0$

2. $3x + y = 20$

   $x - 2y = 10$

Use the elimination method to solve each system of equations.

3. $x - y = 10$

   $x + y = 6$

4. $2x + y = 10$

   $x - 2y = -5$

5. Graph the equations E_1, E_2, and E_3 in the standard viewing rectangle, $[-10, 10]$ by $[-10, 10]$.

$$E_1: \quad x - 2y = 4$$
$$E_2: \quad 4y - 2x = 12$$
$$E_3: \quad 4y - 2x = -8$$

   (a) Describe geometrically the simultaneous solution to the system of equations E_1 and E_2.

   (b) Describe geometrically the simultaneous solution to the system of equations E_1 and E_3.

Use an algebraic method to determine the simultaneous solution to each system. Use a graphing utility and zoom-in to check your answer.

6. $x - y = 4$

   $2x + y = 14$

7. $2x - 3y = 33$

   $5x + 2y = 35$

8. $2x + 2y = 40$

9. $y = x^3 - x$

$xy = 100$ $\qquad$ $y = x$

10. Consider the system of equations $\begin{cases} y = x^3 - 2x^2 + 3x - 5, \\ y = 5x. \end{cases}$

    Find each simultaneous solution (x, y) to the system and determine a maximum error in x to ensure the maximum error in y is at most 0.01. For each simultaneous solution, specify a viewing rectangle (give the parameters L, R, B, and T) for which the solution can be read with an error of at most 0.01.

Solve each system. (Remember *both* x and y must have an error of at most 0.01.)

11. $2x - 5y = 6$
    $y = 6 - x^2$
12. $y = x^2 - 4$
    $y = 6 - x^2$
13. $y = x^3$
    $y = 4 - x^2$
14. $y = 2x^2 - 3x - 10$
    $y = \dfrac{1}{x}$
15. $y = 16x - x^3$
    $y = 10 - x^2$
16. $y = 2x + 10 - x^2$
    $y = 7x^3 + 13x^2 - 9x + 1$
17. $y = \sin x$
    $y = 2 - x^2$
18. $y = 2 - \cos x$
    $y = x^3$
19. $y = \sin x$
    $y = x^3 - x$
20. $y = \dfrac{1}{x}$
    $y = 1 + \cos x$
21. A candy recipe calls for 2.25 times as much brown sugar as white sugar. If 23 ounces of sugar are required, how much of each type is needed?
22. Find the dimensions of a rectangular cornfield with perimeter 220 yards and area 3000 square yards.
23. Hank can row a boat one mile upstream (against the current) in 24 minutes. He can row the same distance downstream in 13 minutes. If both the rowing speed and current speed are constant, find Hank's rowing speed and the speed of the current.
24. Four hundred fifty-two tickets were sold for a high school basketball game. There were two prices of tickets: student at \$0.75 and nonstudent at \$2.00. How many tickets of each type were sold if the total proceeds from the sale of the tickets were \$429?
25. Let $H(x) = ax + b$ be a polynomial function of degree one. If $H(2) = -3$ and $H(-4) = -5$, determine a and b.
26. A function T describing the behavior of a certain physical phenomenon is known to be a polynomial function of degree two; that is, $T(x) = ax^2 + bx + c$. It is also known that $T(0) = 5$, $T(1) = -2$, and $T(3) = 6$. Determine the function T.
27. A scholarship fund earns income from three investments at simple interest rates of 5%, 6%, and 10%. The annual income available for scholarships from the three investments is \$1000. If the amount invested at 5% is five times the amount invested at 6%, and the total amount invested at 5% and 6% is equal to the amount invested at 10%, determine how much is invested at each rate.

28. The total value of 17 coins consisting of nickels, dimes, and quarters is \$2.95. There are twice as many quarters as nickels. Determine the number of nickels, dimes, and quarters.

29. The sum of the digits of a certain two-digit number is 16. If the digits are reversed, the original number is increased by 18. What is the original number?

30. A cyclist peddling against a headwind travels 22 miles down a straight road in 2 hours. The return trip with the wind takes only 1.6 hours. If the speed of the cyclist and the speed of the wind are both constant, find both the speed of the cyclist and the speed of the wind.

31. Let x be the side length of the square as illustrated in Figure 2.2.11 that must be cut out from each corner of a 30-inch-by-40-inch piece of cardboard to form a box with no top.
    (a) Express the volume of the box as a function of x.
    (b) Write two equations that can be solved simultaneously to determine the size of the square that must be cut out to produce a box with volume 1200 cubic inches.
    (c) Draw complete graphs of each equation in the same viewing rectangle.
    (d) How many simultaneous solutions are there to your system in (b)? Which of these solutions are also solutions to the problem situation? Explain.
    (e) Use zoom-in to determine all values of x that produce a box of volume 1200 cubic inches. State the maximum error in x in order to have the maximum error in the volume of at most 0.01 cubic inches.

Do the following for each pair of indicated supply and demand equations.

(i) Graph both the supply and demand equations on the same coordinate system in the *first* quadrant; p is the price (vertical axis), and x represents the number of units of the commodity produced.

(ii) Determine the equilibrium price, that is, the price where supply is equal to demand.

32. Supply: $10 + 0.1x^2$.
    Demand: $\dfrac{50}{1 + 0.1x}$.

33. Supply: $5 + 0.014x^2$.
    Demand: $\dfrac{133}{1 + 0.025x}$.

34. Sam makes two initial investments of \$1000 each. The first earns 10% simple interest and the second earns interest using 5% *simple discount*. After n years, the value of the first investment is $S = 1000(1 + 0.10n)$ and the value of the second investment is $S = 1000/(1 - 0.05n)$.
    (a) Draw the graph of each equation in the viewing rectangle $[0, 30]$ by $[0, 6000]$.
    (b) Is there a time when the future values of both investments are equal?
    (c) What happens to the simple discount investment value at the end of 19 years? 19.5 years? 19.9 years? 19.99 years? exactly 20 years? Could your results explain why the simple discount model is used for only short periods of time in actual practice? *Note:* Normally, interest is paid at the *end* of a term. In the case of simple discount, interest is paid *in advance*, at the beginning of the term.

35. Assume that the graph of $x^2 + y^2 = r^2$ is a circle of radius r centered at the origin.
    (a) Sketch the graphs of $x^2 + y^2 = 16$ and $x + y = 2$ on the same coordinate system.
    (b) Determine the number of simultaneous solutions to the system $\begin{cases} x^2 + y^2 = 16, \\ x + y = 2. \end{cases}$

(c) Determine the solutions with error at most 0.01 using zoom-in with a graphing utility. *Hint:* Write $x^2 + y^2 = 16$ as $y = \sqrt{16 - x^2}$ or $y = -\sqrt{16 - x^2}$.

(d) Determine the simultaneous solutions algebraically.

(e) Identify the possibilities for the number of simultaneous solutions to the system

$$x^2 + y^2 = r^2$$
$$ax + by + c = 0.$$

Sketch graphs illustrating each possibility.

36. Assume that the graph of $x^2 + y^2 = r^2$ is a circle of radius r centered at the origin.

(a) Sketch the graphs of $x^2 + y^2 = 25$ and $y = x^2 - 16$ on the same coordinate system.

(b) Determine the number of simultaneous solutions to the system $\begin{cases} x^2 + y^2 = 25, \\ y = x^2 - 16. \end{cases}$

(c) Determine the solutions with error at most 0.01 using zoom-in with a graphing utility.

(d) Determine the simultaneous solutions algebraically.

(e) Identify the possibilities for the number of simultaneous solutions to the system

$$x^2 + y^2 = r^2,$$
$$y = x^2 + k.$$

Sketch graphs illustrating each possibility.

*37. Determine the y-intercepts of all lines of the form $y = x + b$ that are *tangent* to the circle $x^2 + y^2 = 16$.

*38. Matrices can be used to represent and solve systems of linear equations. Compare the result of each step in the matrix solution with the corresponding algebraic result.

ALGEBRAIC	MATRIX REPRESENTATION
$3x + y = 14$ $2x - 3y = 2$	$\begin{pmatrix} 3 & 1 & 14 \\ 2 & -3 & 2 \end{pmatrix}$

Multiply Row 1 by $\frac{1}{3}$.

$x + \frac{1}{3}y = \frac{14}{3}$ $2x - 3y = 2$	$\begin{pmatrix} 1 & \frac{1}{3} & \frac{14}{3} \\ 2 & -3 & 2 \end{pmatrix}$

Add -2 times Row 1 to Row 2.

$x + \frac{1}{3}y = \frac{14}{3}$ $-\frac{11}{3}y = -\frac{22}{3}$	$\begin{pmatrix} 1 & \frac{1}{3} & \frac{14}{3} \\ 0 & -\frac{11}{3} & -\frac{22}{3} \end{pmatrix}$

Multiply Row 2 by $-\frac{3}{11}$.

$$x + \frac{1}{3}y = \frac{14}{3} \qquad \begin{pmatrix} 1 & \frac{1}{3} & \frac{14}{3} \\ 0 & 1 & 2 \end{pmatrix}$$

$$y = 2$$

Add $-\frac{1}{3}$ times Row 2 to Row 1.

$$x = 4 \qquad \begin{pmatrix} 1 & 0 & 4 \\ 0 & 1 & 2 \end{pmatrix}$$

$$y = 2$$

Use the illustrated matrix method to find the simultaneous solutions to each system.

(a) $x - 3y = 1$
$x + y = 5$

(b) $2x - 5y = 13$
$3x + 2y = -12$

(c) $2x - 6y = 8$
$5x + 3y = 15$
$3x + y - z = 10$

(d) $x - y + 2z = 3$
$2y - z = -3$
$x + 2y = 0$

2.3 Solving Inequalities

In this section we formally introduce inequalities. Basic properties of inequalities are established. Inequalities are solved both algebraically and graphically. The solutions to an inequality will often be represented by a graph on a number line. We must be careful not to confuse these number-line graphs with solving inequalities graphically, which occurs in the Cartesian plane.

Problem situations can be represented algebraically by inequalities as well as equations. Information about a problem may indicate that two expressions are equal, resulting in an equation, or that one expression is greater than another expression, resulting in an inequality. For example, in the Quality Rent-a-Car problem in Section 1.1 we could ask how many miles can be driven without exceeding \$100 in charges. The charges can be represented by the expression $0.2x + 15$ where x stands for the number of miles driven. Recall that the charges were \$15 plus \$0.20 a mile driven. To determine the number of miles that can be driven without exceeding \$100 in charges requires that the value of the expression $0.2x + 15$ be less than or equal to 100.

• DEFINITION Let a and b be real numbers. If $a - b$ is positive, we say that a **is greater than** b and write $a > b$. If $a - b$ is negative, we say that a **is less than** b and write $a < b$. The symbols $>$ and $<$ are called **inequality signs**. Each of the expressions $a > b$ and $a < b$ is called an **inequality**. •

On the number line, $a > b$ means that the point with coordinate a is to the *right* of the point with coordinate b (Figure 2.3.1). Because the inequalities $a > b$ and $b < a$ are equivalent, $b < a$ means that the point with coordinate b appears to the *left* of the point with coordinate a.

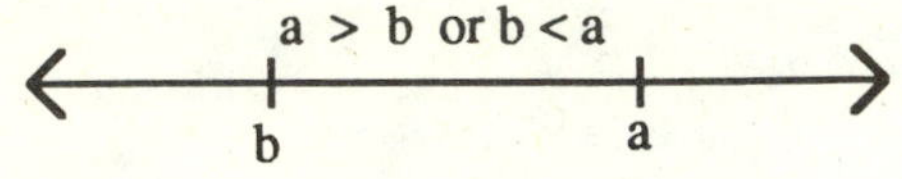

Figure 2.3.1

The inequality $a \geq b$ means that either $a > b$ or $a = b$. Thus, $a - b$ is nonnegative, that is, positive or zero. Similarly, the inequality $a \leq b$ means that either $a < b$ or $a = b$. Therefore, $a - b$ is nonpositive, that is, $a - b$ is negative or zero.

• DEFINITION Let a and b be real numbers. If $a - b$ is nonnegative, we say that a **is greater than or equal to** b and write $a \geq b$. If $a - b$ is nonpositive, we say that a **is less than or equal to** b and write $a \leq b$. •

Be sure that you understand these special cases of the above definitions.

$a > 0$ means a is positive

$a \geq 0$ means a is nonnegative

$a < 0$ means a is negative

$a \leq 0$ means a is nonpositive

• EXAMPLE 1: Use the definitions of inequalities to show that each inequality is true.

(a) $-7 < -2$ (b) $-0.5 \geq -1$

SOLUTION:

(a) Because the difference $-7 - (-2) = -5$ is negative, $-7 < -2$ is true by definition.
(b) Because the difference $-0.5 - (-1) = 0.5$ is nonnegative, $-0.5 \geq -1$ is true by definition. •

We can write the information that the charges are not to exceed \$100 in the Quality Rent-a-Car problem as an inequality, $0.2x + 15 \leq 100$.

• DEFINITION The number a is a **solution to an inequality** in one variable if we get a true statement when the variable is replaced by a everywhere in the inequality. When we have found all solutions to an inequality we say that we have **solved the inequality**. •

For example, $x = 120$ is a solution of the inequality $0.2x + 15 \leq 100$ because $0.2(120) + 15 = 39 \leq 100$. Solving an inequality algebraically is similar to solving an equation algebraically. The goal is to transform a given inequality into an equivalent one with solutions that are obvious.

• DEFINITION Two inequalities are said to be **equivalent** if they have exactly the same set of solutions. •

Equivalent Inequalities. An equivalent inequality is obtained if any one of the following operations is performed.

(1) Add the same real number to, or subtract it from, each side of the inequality.
(2) Add the same polynomial to, or subtract it from, each side of the inequality.
(3) Multiply each side of the inequality, or divide each side of the inequality, by the same positive number.
(4) Multiply each side of the inequality, or divide each side of the inequality, by the same negative number *and* reverse the inequality sign.

Notice that both $5 > 2$ and $5(3) > 2(3)$ are true statements. However, we must reverse the sign of the inequality if we multiply each side by the same negative number. Check that both $5 > 2$ and $5(-3) < 2(-3)$ are true.

We need the following properties of inequalities involving real numbers.

• THEOREM 1 Let a, b, and c be real numbers.

(i) If $a > b$, then $a + c > b + c$. (ii) If $a > b$ and $c > 0$, then $ac > bc$.

(iii) If $a > b$ and $c < 0$, then $ac < bc$. (iv) If $a > b$ and $b > c$, then $a > c$.

PROOF

(i) We need to show that $(a+c)-(b+c)$ is positive. Now, $(a+c)-(b+c) = a-b$ and $a - b$ is positive because $a > b$. Thus, by definition, $a + c > b + c$.

(ii) Because $a > b$ we know that $a - b$ is positive. Then, $(a - b)c$ is positive because both $a - b$ and c are positive. Therefore, $(a - b)c = ac - bc$ is positive and, by definition, $ac > bc$.

(iii) Because $c < 0$, $-c > 0$. Then $(a-b)(-c)$ is positive because both $a-b$ and $-c$ are positive. Thus, $(a - b)(-c) = -ac + bc$ is positive so that $bc > ac$, or $ac < bc$.

(iv) We know that $a - b$ and $b - c$ are positive. Thus, $(a - b) + (b - c) = a - c$ is positive. By definition, $a > c$.

Parts (ii) and (iii) of Theorem 1 tell us what to do when we divide an inequality by the real number d, because dividing by d is the same as multiplying by $\frac{1}{d}$. Part (iv) is called the **transitive property** of inequalities.

Properties similar to those established in Theorem 1 are true if $>$ is replaced by $\geq$, $<$, or $\leq$. The next theorem restates Theorem 1 with $>$ replaced by $<$. The proof of this theorem will be left as an exercise.

• THEOREM 2 Let a, b, and c be real numbers.

(i) If $a < b$, then $a + c < b + c$. (ii) If $a < b$ and $c > 0$, then $ac < bc$.

(iii) If $a < b$ and $c < 0$, then $ac > bc$. (iv) If $a < b$ and $b < c$, then $a < c$.

We use both algebraic and geometric techniques to solve inequalities. We begin our study with linear inequalities.

Linear Inequalities

• DEFINITION An inequality is said to be **linear in x**, or simply **linear**, if it can be written in the form $ax + b > 0$, $ax + b \geq 0$, $ax + b < 0$, or $ax + b \leq 0$, where a and b are real numbers and $a \neq 0$. •

Solving a linear inequality algebraically will seem very much like solving a linear equation algebraically.

• EXAMPLE 2: Solve $4x - 1 < 2$ algebraically.

SOLUTION: Our goal is to find an equivalent form of the inequality with the variable isolated on one side of the inequality. The following inequalities are equivalent.

$$4x - 1 < 2$$

$$4x < 3 \quad \text{(Add 1 to each side.)}$$

$$x < \frac{3}{4} \quad \text{(Divide each side by 4.)}$$

The solution consists of the set of all real numbers less than $\frac{3}{4}$. •

There are two methods we can use to solve an inequality graphically. We can graph each side of the inequality and then determine where the graph of the left side is above ($>$), above or equal to ($\geq$), below ($<$), or below or equal to ($\leq$) the graph of the right side. A second, usually easier, approach is to rewrite the inequality in an equivalent form with the left side of the form $f(x)$ and the right side 0. Then, we determine where the graph of f is above ($>$) or below ($<$) the x-axis. If $\geq$ or $\leq$ are involved, then we must also determine the x-intercepts of f, that is, the zeros of f.

Next we solve Example 2 graphically.

• EXAMPLE 3: Solve $4x - 1 < 2$ graphically.

SOLUTION: First we rewrite the inequality in the form $f(x) < 0$. The desired equivalent inequality is $4x - 3 < 0$, where $f(x) = 4x - 3$. The inequality can be solved graphically by finding the x-coordinates of those points for which the graph of $f(x) = 4x - 3$ lies *below* the x-axis (Figure 2.3.2). The graph of f appears to cross the x-axis at $x = \frac{3}{4}$. We can confirm this by computing $f\left(\frac{3}{4}\right)$. Thus, the graph of f lies below the x-axis for $x < \frac{3}{4}$.

Alternatively, we can solve this inequality graphically using the graphs of the left hand side, $g(x) = 4x - 1$, and the right hand side, $h(x) = 2$. Figure 2.3.3 shows the graphs of $g(x) = 4x - 1$ and $h(x) = 2$ in the $[-5, 5]$ by $[-5, 5]$ viewing rectangle. The solution to the original inequality $4x - 1 < 2$ consists of the x-coordinates of those points for which the graph of $g(x) = 4x - 1$ lies *below* the graph of $h(x) = 2$. The two graphs appear to intersect

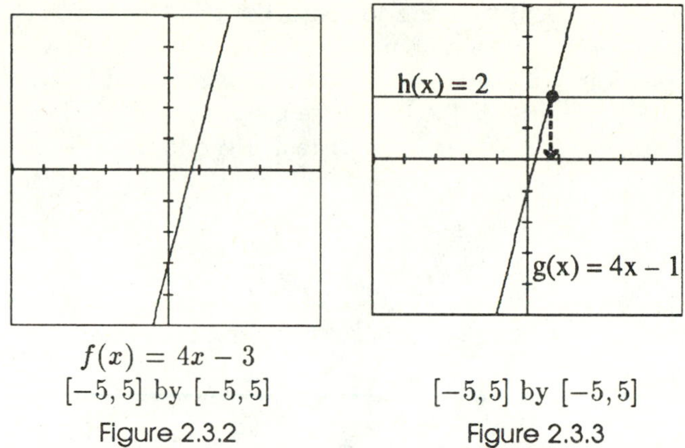

$f(x) = 4x - 3$
$[-5, 5]$ by $[-5, 5]$
Figure 2.3.2

$[-5, 5]$ by $[-5, 5]$
Figure 2.3.3

at $\left(\frac{3}{4}, 2\right)$. We can confirm the point of intersection by showing that $g\left(\frac{3}{4}\right) = 2$. For $x < \frac{3}{4}$ the graph of $g(x) = 4x - 1$ lies *below* the graph of $h(x) = 2$. •

The pair of graphs in Figure 2.3.3 or the single graph in Figure 2.3.2 is considered a geometric representation of the inequality $4x - 1 < 2$. We must be careful if we rewrite an inequality so that we do get a true geometric representation.

It is usually easier to solve a linear inequality algebraically. We give both algebraic and graphical solutions to inequalities in this section to prepare for the more difficult inequalities that occur in the next two sections. You will find a graphical solution easier when the inequalities are more complicated. In fact, a graphical solution is often the only reasonable method to use.

The algebraic steps in Example 2 and the graphs in Example 3 show that the solution to the inequality $4x - 1 \leq 2$ is $x \leq \frac{3}{4}$, the solution to $4x - 1 > 2$ is $x > \frac{3}{4}$, and the solution to $4x - 1 \geq 2$ is $x \geq \frac{3}{4}$.

Another way to describe the set of solutions to the inequality $4x - 1 < 2$ in Example 2 is to draw the graph of the solutions on a *number line* as illustrated in Figure 2.3.4.

The portion of the number line to the left of $\frac{3}{4}$ is darkened to indicate that the corresponding values of x are solutions to the inequality. We use a round parenthesis,), at $\frac{3}{4}$ to indicate that $\frac{3}{4}$ is *not* a solution to the inequality. The solution to the inequality $4x - 1 \leq 2$ *does* include $\frac{3}{4}$, and is illustrated in the number-line graph of Figure 2.3.5. We use a square bracket,], at $\frac{3}{4}$ to indicate that $\frac{3}{4}$ is a solution to the inequality $4x - 1 \leq 2$.

Solutions to inequalities often involve intervals on a number line. We use $(2, 5)$ to stand for all the numbers between 2 and 5. Neither 2 nor 5 is included in $(2, 5)$. You may think

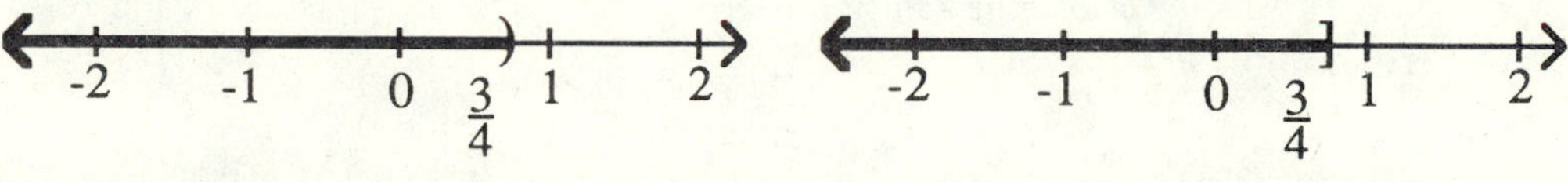

Figure 2.3.4

Figure 2.3.5

it confusing to use $(2,5)$ to stand for both the point in the plane with coordinates $x = 2$ and $y = 5$, and for the set of all numbers between 2 and 5. However, the context will clarify the meaning of the notation. The notation $2 < x < 5$ also stands for the set of numbers between 2 and 5. To include 2 or 5 we use a square bracket,] or [, next to the number. For example, the interval $2 \le x < 5$ is written in the form $[2,5)$.

With the above conventions we have the following descriptions of intervals on a number line.

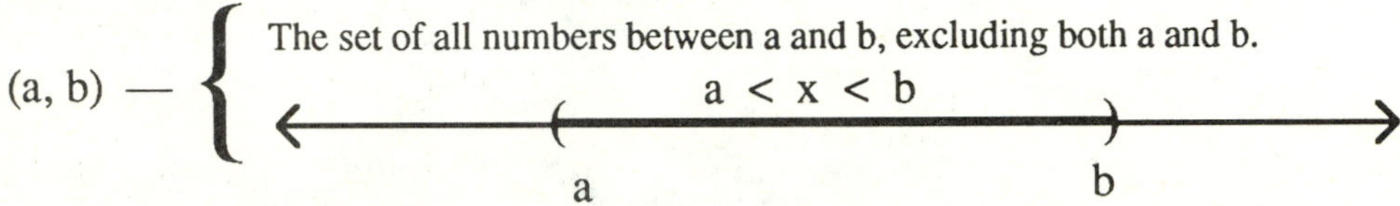
(a, b) —
The set of all numbers between a and b, excluding both a and b.
a < x < b
a
b

[a, b) — The set of all numbers between a and b, excluding b and including a.

a ≤ x < b

a b

(a, b] — The set of all numbers between a and b, excluding a and including b.

a < x ≤ b

a b

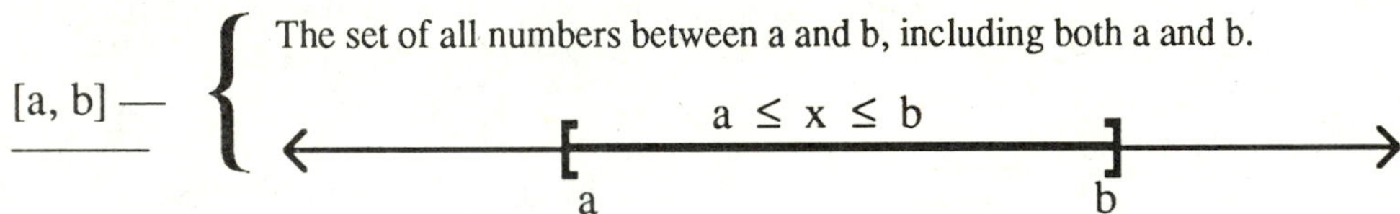
[a, b] —
The set of all numbers between a and b, including both a and b.
a ≤ x ≤ b
a
b

We also use the symbols $-\infty$ and ∞ as entries in the above ordered pairs. For example, $(-\infty, 3)$ stands for all numbers less than 3. Thus, the solution to Example 2 can be written in interval form $(-\infty, \frac{3}{4})$. We always use a round parenthesis next to the symbols $-\infty$ and ∞. With this notation, the solution to $4x - 1 \le 2$ is $(-\infty, \frac{3}{4}]$, the solution to $4x - 1 > 2$ is $(\frac{3}{4}, \infty)$, and the solution to $4x - 1 \ge 2$ is $[\frac{3}{4}, \infty)$.

• EXAMPLE 4: Draw a number-line graph of the solutions to each inequality.

(a) $x \ge 2$ (b) $x < 5$ (c) $-1 \le x < 3$

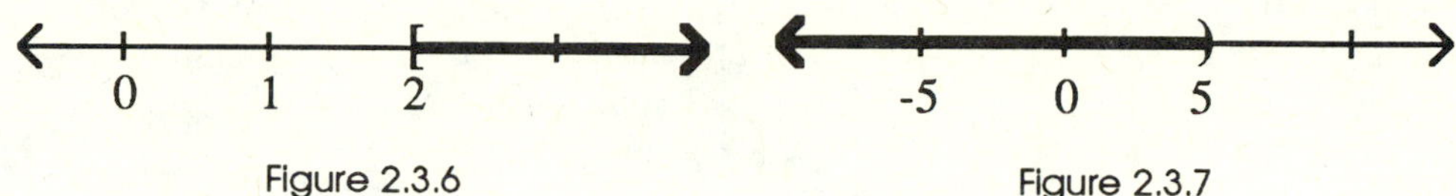

Figure 2.3.6 Figure 2.3.7

SOLUTION:

(a) $x \geq 2$ means that x is greater than or equal to 2. Because 2 is a solution we must use a square bracket at $x = 2$ (Figure 2.3.6).

(b) $x < 5$ means that x is less than 5 but not equal to 5. Because 5 is not a solution we must use a round parenthesis at $x = 5$ (Figure 2.3.7).

(c) $-1 \leq x < 3$ means that x is between -1 and 3, including -1 but excluding 3 (Figure 2.3.8). •

• EXAMPLE 5: Solve the inequality $3(x-1)+2 \leq 5x+6$ algebraically. Draw a number-line graph of the solution.

SOLUTION: The following inequalities are equivalent.

$$
\begin{aligned}
3(x-1)+2 &\leq 5x+6 \\
3x-3+2 &\leq 5x+6 \quad \text{(Distributive Property.)} \\
3x-1 &\leq 5x+6 \\
-2x &\leq 7 \quad \text{(Subtract } 5x \text{ from both sides.)} \\
x &\geq -3.5 \quad \text{(Divide both sides by } -2.\text{)}
\end{aligned}
$$

In the fourth step, we divided both sides of the inequality by -2. This is the reason the inequality sign was reversed. The solution to the inequality $3(x-1)+2 \leq 5x+6$ is $[-3.5, \infty)$, and the number-line graph of the solution is shown in Figure 2.3.9. •

The algebraic and geometric techniques used to solve linear inequalities in Examples 2 through 4 can be extended to more complicated inequalities.

• EXAMPLE 6: Solve $-3 < \frac{2x+5}{3} \leq 5$ both algebraically and graphically.

SOLUTION: The solutions to this inequality consist of those values of x for which $\frac{2x+5}{3}$ is between -3 and 5, including 5 but not including -3. Thus, we can rewrite the inequality $-3 < \frac{2x+5}{3} \leq 5$ as a pair of linear inequalities.

$$-3 < \frac{2x+5}{3} \quad \text{and} \quad \frac{2x+5}{3} \leq 5$$

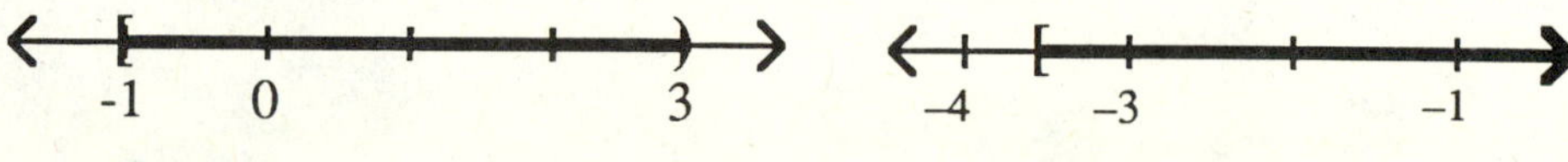

Figure 2.3.8 Figure 2.3.9

The solution to $-3 < \frac{2x+5}{3} \le 5$ consists of the common solutions to the pair of inequalities $-3 < \frac{2x+5}{3}$ and $\frac{2x+5}{3} \le 5$. Next, we solve each of these linear inequalities algebraically.

$$\begin{aligned} -3 &< \frac{2x+5}{3} \\ -9 &< 2x+5 \\ -14 &< 2x \\ -7 &< x \end{aligned} \qquad\qquad \begin{aligned} \frac{2x+5}{3} &\le 5 \\ 2x+5 &\le 15 \\ 2x &\le 10 \\ x &\le 5 \end{aligned}$$

We used exactly the same operations to solve each of these inequalities. The common solution to these two inequalities is $-7 < x \le 5$, that is, the numbers greater than -7 that are also less than or equal to 5. In interval notation, the solution to $-3 < \frac{2x+5}{3} \le 5$ is $(-7, 5]$.

It is not necessary to write the separate pair of inequalities to solve the original inequality. We can obtain the solution in a more concise manner as follows.

$$\begin{aligned} -3 < &\frac{2x+5}{3} \le 5 \\ -9 < &\; 2x+5 \le 15 \\ -14 < &\; 2x \le 10 \\ -7 < &\; x \le 5 \end{aligned}$$

Figure 2.3.10 gives graphs of $y = -3$, $y = 5$, and $y = \frac{2x+5}{3}$ and is a geometric representation of the inequality of this example. The solution to the inequality $-3 < \frac{2x+5}{3} \le 5$ consists of the x-coordinates of the points on the graph of $y = \frac{2x+5}{3}$ that lie *between* the graphs of $y = -3$ and $y = 5$, or the point of intersection of the graphs of $y = 5$ and $y = \frac{2x+5}{3}$. This graph confirms the solution, $(-7, 5]$, found algebraically. •

• DEFINITION Let $P(x, y)$ be any point in the coordinate plane. The **projection of P onto the x-axis** is x and the **projection of P onto the y-axis** is y. •

The projection of a point P onto the x-axis is determined by dropping a perpendicular from P to the x-axis. Similarly, the projection of a point P onto the y-axis is determined by dropping a perpendicular from P to the y-axis. In Example 6, x is a solution to $-3 < \frac{2x+5}{3} \le 5$ if and only if x is the projection onto the x-axis of a point on the portion of the graph of $y = \frac{2x+5}{3}$ that lies between the graph of $y = -3$ and $y = 5$ or the point of intersection of $y = 5$ and $y = \frac{2x+5}{3}$.

In the next example we divide by a negative number. Remember to reverse the inequality sign when dividing by a negative number.

• EXAMPLE 7: Solve $-1 \le \frac{3-5x}{4} \le 3$.

SOLUTION: The following inequalities are equivalent.

$$-1 \leq \frac{3-5x}{4} \leq 3$$
$$-4 \leq 3-5x \leq 12$$
$$-7 \leq \quad -5x \quad \leq 9$$
$$\frac{7}{5} \geq \quad x \quad \geq -\frac{9}{5}$$

We had to reverse the inequality signs above because we divided by a negative number. Therefore, the solution to $-1 \leq \frac{3-5x}{4} \leq 3$ is $\left[-\frac{9}{5}, \frac{7}{5}\right]$. Check the solution with your graphing utility. •

• EXAMPLE 8: Some rectangles have length one inch more than twice the width. Find the possible widths of the rectangle if its perimeter is less than 100 inches (Figure 2.3.11).

SOLUTION: If w is the width of the rectangle, then the length is $2w+1$ and the perimeter is $2w+2(2w+1)$. Because it is given that the perimeter is less than 100 inches, we have $2w+2(2w+1) < 100$. This inequality is an algebraic representation of the problem situation. The following inequalities are equivalent.

$$2w + 2(2w+1) < 100$$
$$2w + 4w + 2 < 100$$
$$6w + 2 < 100$$
$$6w < 98$$
$$w < \frac{49}{3}$$

Thus, the width can be any positive number less than $\frac{49}{3}$, or $(0, \frac{49}{3})$.

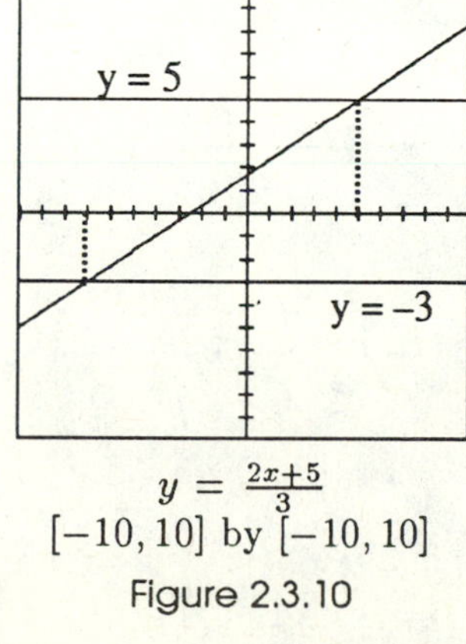

$y = \frac{2x+5}{3}$
$[-10, 10]$ by $[-10, 10]$
Figure 2.3.10

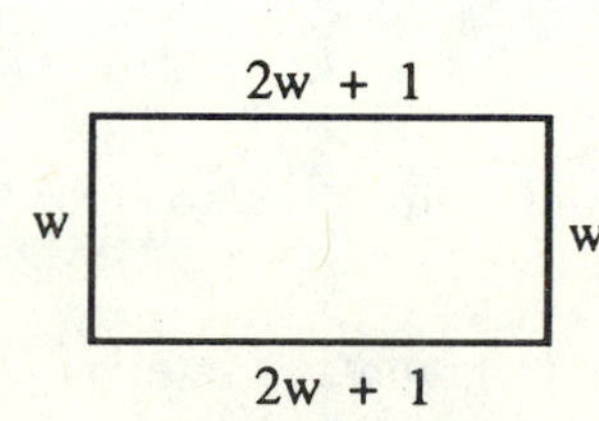

Figure 2.3.11

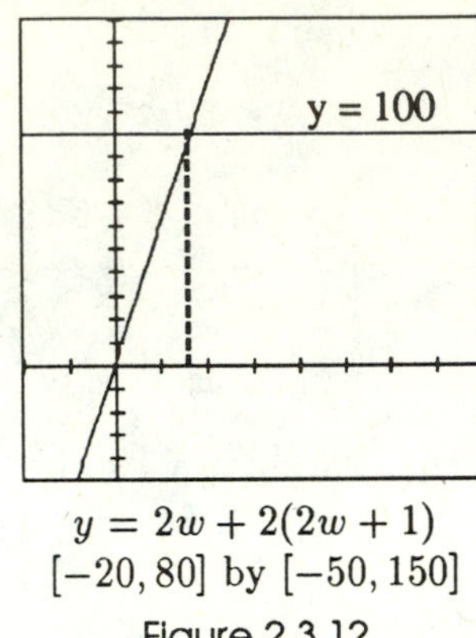

$y = 2w + 2(2w + 1)$
$[-20, 80]$ by $[-50, 150]$
Figure 2.3.12

The graphs of $y = 2w+2(2w+1)$ and $y = 100$ shown in the $[-20, 80]$ by $[-50, 150]$ viewing rectangle (Figure 2.3.12) give a geometric representation of the algebraic representation, and geometrically confirm the solution found algebraically. •

• EXERCISES 2-3

Use interval notation to name the intervals depicted graphically.

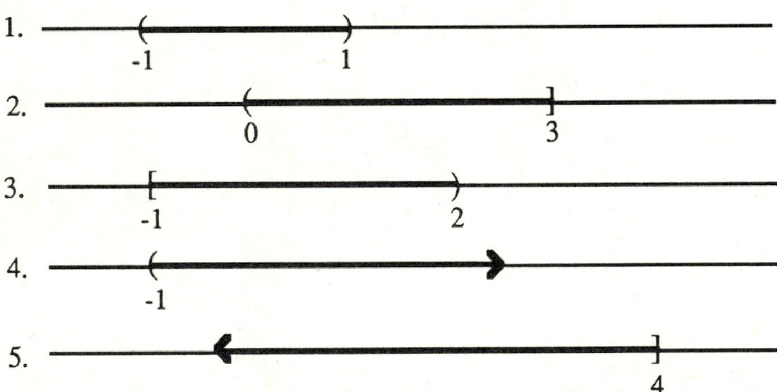

Use the definitions of inequalities to show that each inequality is true.

6. $5 > -1$

7. $-11 < -7$

Use a number line to graph the intervals.

8. $[-1, 2)$

9. $[3, \infty)$

10. $x < 8$

11. $-3 \leq x < 5$

Complete the table.

	Interval Notation	**Inequality Notation**	**English Phrase**
Example:	$(2, 5]$	$2 < x \leq 5$	The set of all real numbers less than or equal to 5 and greater than 2
12.		$7 > x > 3$	
13.	$(8, \infty)$		
14.			The set of all real numbers less than or equal to negative 2

Find the intersection of the intervals. If possible, write your answer in interval notation.

15. $(-2, \infty)$, $(-\infty, 5)$

16. $(-5, \infty)$, $(-\infty, -5)$

17. $(-\infty, 5]$, $(-2, \infty)$

Consider the graph of $y = f(x)$ in Figure 2.3.13.

18. (a) Compare $f(3)$ and 4. That is, determine if $f(3) = 4, f(3) < 4$, or $f(3) > 4$.
    (b) Is $x = 3$ a solution to $f(x) > 4$?
    (c) Use the graph to solve $f(x) > 4$.

Consider the graphs of $y = f(x)$ and $y = g(x)$ in Figure 2.3.14.

19. Use an inequality to compare
    (a) $f(-1)$ and 2.
    (b) $g(2)$ and 5.

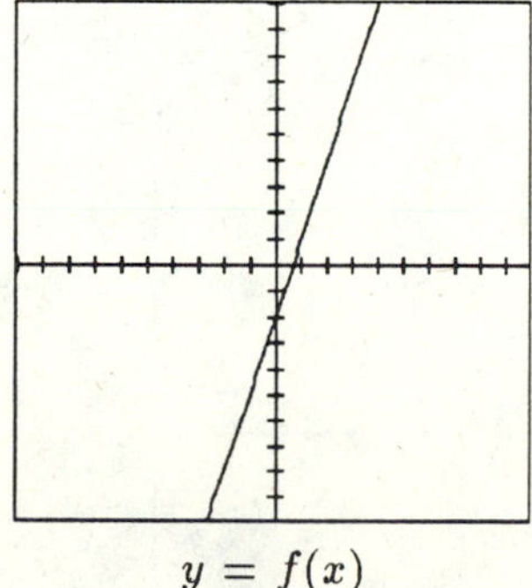

$y = f(x)$
$[-10, 10]$ by $[-10, 10]$
Figure 2.3.13

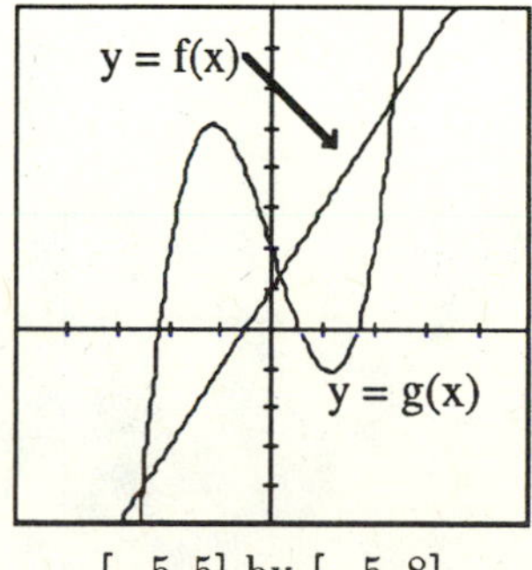

$[-5, 5]$ by $[-5, 8]$
Figure 2.3.14

20. Compare $f(a)$ and $g(a)$ if
    (a) $a = 2$. (b) $a = -1$.
21. Is $x = 1$ a solution to
    (a) $f < g$? (b) $f = g$? (c) $f > g$?
22. Is $x = -1$ a solution to
    (a) $f < g$? (b) $f = g$? (c) $f > g$?

Consider the graph of $y = f(x)$ in Figure 2.3.15.

23. Use the graph to solve $f(x) < -3$.
24. Use the graph to solve $f(x) \geq 0$.
25. Solve $4+3x < 19$ algebraically. Write the solution as an interval. Graph $y = 4+3x-19 = 3x-15$. Compare the values of x for which the graph of $y = 3x - 15$ is *below* the x-axis with the algebraic solution to $4 + 3x < 19$.
26. Solve $2x - 1 > 6$ algebraically. Write the solution as an interval. Graph $y = 2x - 1$ and $y = 6$ in the same viewing rectangle. Compare the values of x for which the graph of $y = 2x - 1$ is *above* the graph of $y = 6$ with the algebraic solution to $2x - 1 > 6$.
27. Use the graph of $y = 3(x - 1) + 2 - (5x + 6)$ to give a geometric solution to the inequality of Example 5.
28. Explain how the inequality in Exercise 26 can be solved geometrically by determining the values of x for which a graph is above or below the *x-axis*.
29. Explain how the inequality in Exercise 27 can be solved geometrically by determining the values of x for which a graph is above or below the *x-axis*.

Use an algebraic method to solve the inequalities. Write your answers in interval notation. Check your solutions with a graphing utility.

30. $2x - 1 > 4x + 3$
31. $\frac{1}{2}(x - 4) - 2x \leq 5(3 - x)$
32. $\frac{3x - 2}{5} > -1$
33. $\frac{1}{2}(x + 3) + 2(x - 4) < \frac{1}{3}(x - 3)$
34. $\frac{3 - x}{2} + \frac{5x - 2}{3} < -1$
35. $2 \leq x + 6 < 9$

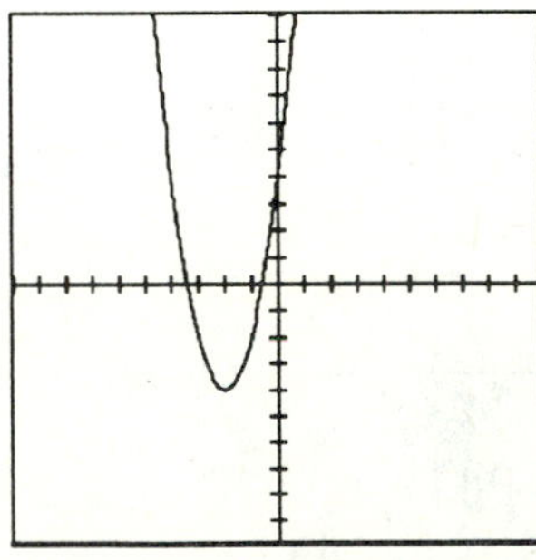

Figure 2.3.15

36. $-1 < 3x - 2 < 7$

37. $4 \geq \dfrac{2x-5}{3} \geq -2$

38. $\dfrac{1}{2} < \dfrac{5x-2}{6} \leq \dfrac{8}{3}$

39. $\dfrac{3}{x-2} > 0$

40. $-\dfrac{3}{4} < \dfrac{3-x}{2} < 8$

41. $-2(4 - \dfrac{x}{3}) < 3 + 5x$

42. $\dfrac{x+5}{3} < 2$

43. $0 < \dfrac{2}{x+5} < 6$

44. Barb wants to drive to a city 105 miles from her home in no more than 2 hours. What average speed must she drive?

45. An electrician charges \$18 per hour plus \$25 per service call for home repair work. How long did she work if her charges were less than \$100? Assume she rounds her time to the nearest quarter of an hour.

46. Sarah has \$45 to spend and wishes to take as many friends as possible to a concert. Parking is \$5.75 and concert tickets are \$7.50 each. Let x represent the number of friends Sarah takes to the concert.

    (a) Write an inequality that is an algebraic representation for this problem situation.

    (b) Solve the inequality in (a).

    (c) How many friends can Sarah take to the concert?

47. Some rectangles have length two inches less than twice the width. Find the possible widths of the rectangle if its perimeter is less than 200 inches. Solve this problem algebraically and with a graphing utility.

48. A candy company finds the cost of making a certain candy bar is \$0.23 per candy bar plus fixed costs of \$2000 per week. If the candy bar sells for \$0.25, find the minimum number of candy bars that must be made and sold in order for the company to make a profit.

49. If $a < b$ and $c < d$, does it always follow that $ac < bd$? If not, find specific numbers to give a counterexample.

50. *Boyle's law* for a certain gas states that $PV = 400$, where P is pressure and V is volume. If $20 \leq V \leq 40$, what is the corresponding range for P?

*51. A company has current assets (cash, property, inventory, and accounts receivable) of \$200,000 and current liabilities (taxes, loans, accounts payable) of \$50,000. How much can they borrow if they want their ratio of assets to liabilities to be less than 2? Assume the amount borrowed is added to both current assets and current liabilities.

2.4 Inequalities Involving Absolute Value

In this section we investigate and solve inequalities that involve absolute value both algebraically and graphically.

Every real number corresponds to a unique point on the number line. We define the **absolute value** of a real number c, denoted by $|c|$, as the distance from zero (0) of the point on the number line that corresponds to c. Because distance is nonnegative, the absolute

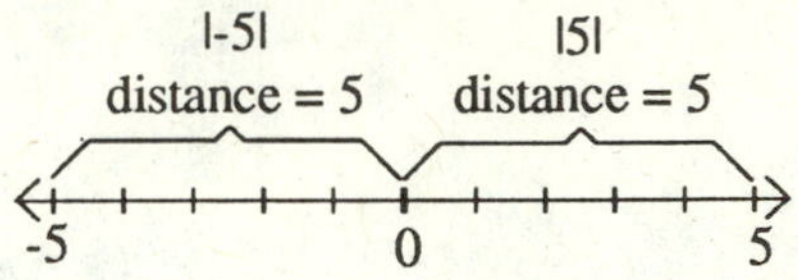

Figure 2.4.1

value of any real number is nonnegative. For example, because the distance from zero of the points that correspond to 5 and -5 is 5, we have $|5| = 5$ and $|-5| = 5$ (Figure 2.4.1).

Notice that we use the symbol x to represent both the real number x and the point on the number line with coordinate x. Thus, a real number x is a solution to the inequality $|x| < 2$ if and only if the distance from the origin of the point with coordinate x is less than 2. These are precisely the real numbers between -2 and 2, not including either 2 or -2. Therefore, the solution to the inequality $|x| < 2$ is the set of all real numbers x satisfying $-2 < x < 2$.

The inequality $|x| > 4$ has for its solution the real numbers with distance from the origin greater than 4. Thus, the solution to $|x| > 4$ consists of all real numbers x satisfying $x < -4$ or $x > 4$.

Inequalities that involve absolute value have the equivalent representations illustrated in Figure 2.4.2 ($a > 0$).

In interval notation, $|x| < a$ is $(-a, a)$ and $|x| \leq a$ is $[-a, a]$. It is not possible to write $|x| > a$, or $|x| \geq a$ as a single interval. $|x| > a$ consists of the real numbers in either the interval $(-\infty, -a)$, or the interval (a, ∞). $|x| \geq a$ consists of the real numbers in either the interval $(-\infty, -a]$ or the interval $[a, \infty)$.

• DEFINITION We use the notation $(a, b) \cup (c, d)$ to indicate the collection of all real numbers that belong to (a, b) or (c, d) or both. The symbol $\cup$ is read **union**. •

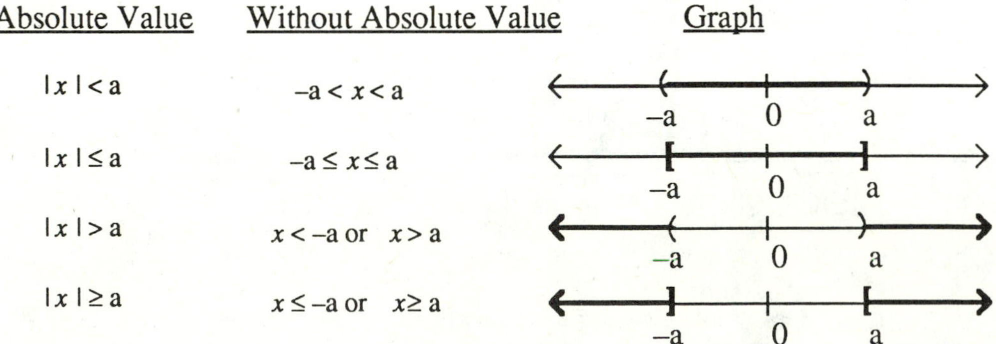

Absolute Value
Without Absolute Value
Graph
|x|<a
−a < x < a
−a 0 a
|x|≤a
−a ≤ x ≤ a
−a 0 a
|x|>a
x < −a or x > a
−a 0 a
|x|≥a
x ≤ −a or x ≥ a
−a 0 a

Figure 2.4.2

Thus, we can include interval notation among the equivalent representations of $|x|$ where $a > 0$ given in Figure 2.4.2.

Absolute Value	**Interval Notation**
$\|x\| < a$	$(-a, a)$
$\|x\| \leq a$	$[-a, a]$
$\|x\| > a$	$(-\infty, -a) \cup (a, \infty)$
$\|x\| \geq a$	$(-\infty, -a] \cup [a, \infty)$

• EXAMPLE 1: Write an inequality using absolute value for each of the following statements.

(a) x is within 4 units of the origin 0 on the number line.
(b) x is less than 3 units from the point 2 on the number line.
(c) x is at least 5 units from the point -3 on the number line.

SOLUTION:

(a) Here we are thinking of x as a point on the number line. The distance of x from 0 is less than 4. Thus, as a real number, x is a solution to the inequality $|x| < 4$.
(b) Notice that $|x-2|$ gives the distance between the points on the number line with coordinates x and 2. (Why?) As a real number, x is a solution to the inequality $|x-2| < 3$.
(c) Again we are thinking of x as a point on the number line. The distance from x to -3 is $|x-(-3)| = |x+3|$. As a real number, x is a solution to the inequality $|x+3| \geq 5$. •

• EXAMPLE 2: Solve $|x-2| < 3$ both algebraically and graphically. Give a number-line graph of the solution.

SOLUTION: The inequality $|x-2| < 3$ is equivalent to $|x-2| - 3 < 0$. We use the graph of $f(x) = |x-2| - 3$ to solve the inequality graphically. Figure 2.4.3 gives a complete graph of $f(x) = |x-2| - 3$. The graph appears to cross the x-axis at $x = -1$ and $x = 5$. This can be confirmed by direct substitution. The graph of f is *below* the x-axis for all x between -1 and 5. Thus, the solution is $-1 < x < 5$, that is, the interval $(-1, 5)$.

Next we give an algebraic solution. Remember that $|a| < 3$ is equivalent to $-3 < a < 3$. Replace a by $x - 2$ and we obtain the following equivalent inequalities.

$$|x-2| < 3$$
$$-3 < x - 2 < 3$$
$$-1 < \quad x \quad < 5$$

A number-line graph of the solution is given in Figure 2.4.4. •

• EXAMPLE 3: Solve $|3x - 2| > 1$ both algebraically and graphically.

SOLUTION: First we write the inequality in the equivalent form $|3x - 2| - 1 > 0$. Figure 2.4.5 gives a complete graph of $f(x) = |3x - 2| - 1$. You can check that the graph crosses the x-axis at $x = \frac{1}{3}$ and $x = 1$. We need to determine the values of x for which the graph of $f(x) = |3x - 2| - 1$ lies above the x-axis. The solution is $x < \frac{1}{3}$ or $x > 1$. We can also write this solution in interval form, $(-\infty, \frac{1}{3}) \cup (1, \infty)$.

To determine the solutions algebraically, remember that $|a| > 1$ is equivalent to $a < -1$ or $a > 1$. Replacing a by $3x - 2$, we obtain the following equivalent inequalities.

$$\begin{aligned} |3x - 2| &> 1 \\ 3x - 2 < -1 \quad &\text{or} \quad 3x - 2 > 1 \\ 3x < 1 \quad &\text{or} \quad 3x > 3 \\ x < \frac{1}{3} \quad &\text{or} \quad x > 1 \end{aligned}$$

The solution to $|3x - 2| > 1$ is $(-\infty, \frac{1}{3}) \cup (1, \infty)$. •

• EXAMPLE 4: Solve $|1 - 2x| \leq 4$ and draw a number-line graph of the solution.

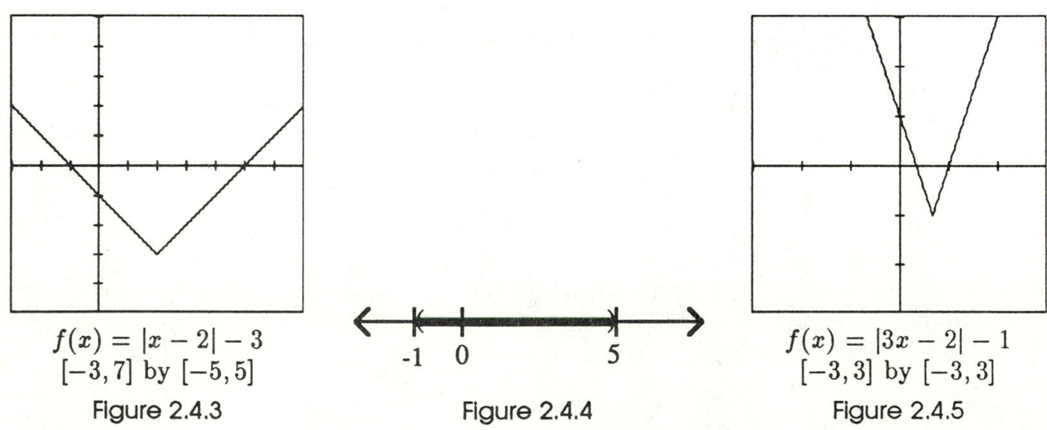

$f(x) = |x - 2| - 3$
$[-3, 7]$ by $[-5, 5]$
Figure 2.4.3

Figure 2.4.4

$f(x) = |3x - 2| - 1$
$[-3, 3]$ by $[-3, 3]$
Figure 2.4.5

SOLUTION: The previous examples should have convinced you that it is easier to solve such inequalities algebraically.

$$|1-2x| \leq 4$$
$$-4 \leq 1-2x \leq 4$$
$$-5 \leq \ -2x \ \leq 3$$
$$\frac{5}{2} \geq \quad x \quad \geq -\frac{3}{2}$$

Thus, the solution is $\left[-\frac{3}{2}, \frac{5}{2}\right]$ and a number-line graph is given in Figure 2.4.6. Use your graphing utility to check the solution. •

• EXAMPLE 5: Solve $\frac{x+3}{|x-2|} > 0$.

SOLUTION: This inequality may appear complicated at first glance because of the absolute value symbol. We can assume that $x \neq 2$ because the fraction $\frac{x+3}{|x-2|}$ is not defined for $x = 2$. Thus, to solve this inequality we need to find the values of x for which this fraction is positive. A fraction is positive if the numerator and denominator are both positive or both negative. Because this denominator is always positive, we need to find the values of x for which the numerator $x+3$ is positive; that is, the values of x for which $x+3 > 0$. Therefore, the solution consists of all real numbers greater than -3, excluding 2. The solution can be written in the form $(-3, 2) \cup (2, \infty)$. •

In Example 5 we really showed that the inequality $\frac{x+3}{|x-2|} > 0$ was equivalent to the pair of statements $x+3 > 0$ and $x \neq 2$. We must be careful when we solve this inequality graphically. The graph of $y = \frac{x+3}{|x-2|}$ in the $[-10, 10]$ by $[-1, 10]$ viewing rectangle is shown in Figure 2.4.7 and is a complete graph. To solve this inequality graphically we need to

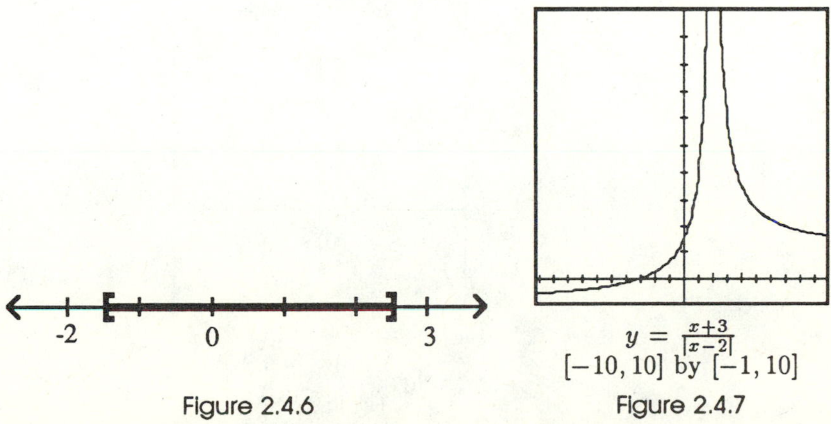

Figure 2.4.6

$y = \frac{x+3}{|x-2|}$
$[-10, 10]$ by $[-1, 10]$

Figure 2.4.7

determine the values of x for which the graph of $y = \frac{x+3}{|x-2|}$ is *above* the x-axis (the line $y = 0$).

We can see from this graph that $\frac{x+3}{|x-2|} > 0$ for $x > -3$. However, it is *not* possible to determine graphically that $x = -3$ is not a solution. You will need to check this by substituting $x = -3$ into the inequality. Finally, this graph should alert you to check that x cannot be equal to 2 because the graph suggests that the function is not defined at $x = 2$. (Why?)

• EXAMPLE 6: Solve $(x + 3)|x + 2| \geq 0$ graphically.

SOLUTION: The graph of $y = (x + 3)|x + 2|$ in the $[-5, 5]$ by $[-5, 10]$ viewing rectangle is a complete graph (Figure 2.4.8). The graph of $y = (x + 3)|x + 2|$ is above the x-axis for $-3 < x < -2$ and for $x > -2$. You can verify by substitution that $x = -3$ and $x = -2$ are also solutions to this inequality. Therefore, the solution to the inequality is $[-3, \infty)$.

To solve this inequality algebraically, first observe that $|x + 2| \geq 0$ for all x. Then, $(x + 3)|x + 2| \geq 0$ if and only if $x + 3 \geq 0$. Again we find the solution to be $[-3, \infty)$, confirming algebraically the solution we found graphically. •

Figure 2.4.8 together with the discussion in Example 6 shows that the solution to $(x + 3)|x + 2| > 0$ is $(-3, -2) \cup (-2, \infty)$. (Why?)

An algebraic approach to the next example is very tedious. However, if we use a distance interpretation of the absolute value, finding the solution is considerably easier.

• EXAMPLE 7: Solve $|x| > |x - 4|$.

SOLUTION: $|x|$ gives the distance from x to the origin, and $|x - 4|$ gives the distance from x to 4. The solution to the inequality consists of those values of x for which the distance from x to the origin is greater than the distance from x to 4. Now, 2 is half-way between

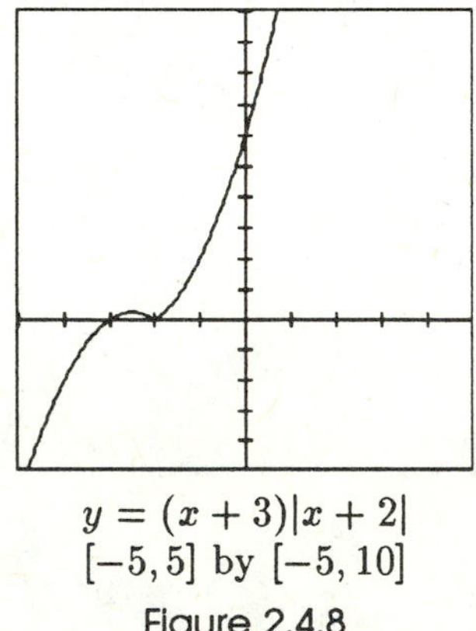

$y = (x + 3)|x + 2|$
$[-5, 5]$ by $[-5, 10]$
Figure 2.4.8

0 and 4. Points to the right of 2 have distance from the origin greater than their distance from 4. Therefore, the solution is $(2, \infty)$. You should confirm this solution using a graphing utility. •

In this section we have seen that solving some inequalities algebraically may be easier than solving them graphically. Even in cases where the algebraic approach is easier, the graphical approach can provide additional insight. In the next section we deal with problems where solving graphically is the only practical method.

• EXERCISES 2-4

Use interval notation to describe the intervals depicted graphically.

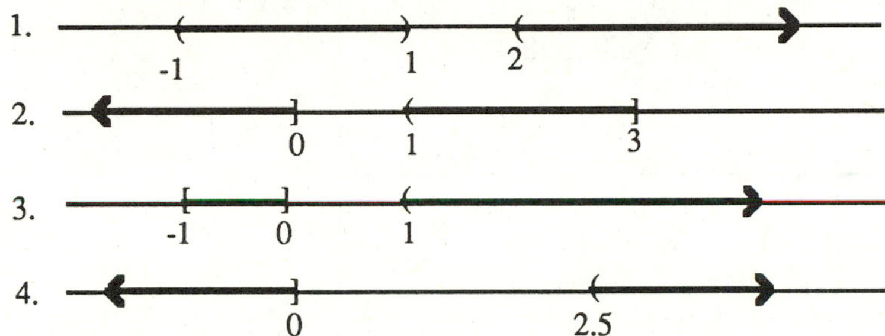

Use a number line to graph the intervals.

5. $[-4, 2)$, $[3, 7)$
6. $x \leq 2$ or $x > 5$
7. $(-\infty, 2)$, $[5, \infty)$
8. $x > -3$ or $x < -6$

Complete the table.

	Interval Notation	Inequality Notation	English Phrase
Example:	$(-\infty, 3) \cup (5, 7)$	$x < 3$ or $5 < x < 7$	The set of all real numbers less than 3 or between 5 and 7
9.		$x > 5$ or $x < -3$	
10.	$[-2, 3) \cup (6, \infty)$		
11.			The set of all real numbers between -3 and 2 or greater than 6

Complete the table.

	Interval Notation	Absolute Value Inequality Notation	English Phrase
Example:	$(-3, 3)$	$\|x\| < 3$	The set of all real numbers less than 3 units from the origin
12.		$\|x\| \geq 5$	
13.	$(-1, 7)$		
14.			The set of all real numbers greater than 2 units from -5
15.	$(-\infty, 1] \cup [6, \infty)$		

16. Solve $|x-2| < 4$ algebraically. Write the solution as an interval. Graph $y = |x-2|-4$. Compare the values of x for which the graph of $y = |x-2|-4$ is *below* the x-axis with the algebraic solution to $|x-2| < 4$.

17. Solve $|2x-6| > 8$ algebraically. Write the solution as an interval. Graph $y = |2x-6|-8$. Compare the values of x for which the graph of $y = |2x-6| = 8$ is *above* the x-axis with the algebraic solution to $|2x-6| > 8$.

Solve the inequalities algebraically. Write your answers in interval notation. Check your answers with a graphing utility.

18. $2x - 3 \leq 4$ or $x + 6 > -4$

19. $\dfrac{2x-3}{2} < 8$ or $\dfrac{2x+4}{3} \geq 2$

20. $|x-3| < 2$

21. $|x+3| \leq 5$

22. $|x-5| > 3$

23. $|x-3| \geq 5$

24. $|3-x| < 8$

25. $|2x-8| < 20$

26. $\dfrac{3x-8}{2} > 6$

27. $3|x| - 4 > 0$

28. $x|x-2| > 0$

29. $\dfrac{x-3}{|x+2|} < 0$

30. $|\dfrac{1}{x}| < 3$

Use absolute value notation to describe the intervals depicted graphically. *Hint:* Find numbers a and b so that the solution to $|x-a| < (\leq, >, \text{ or } \geq)\ b$ is the indicated interval or intervals.

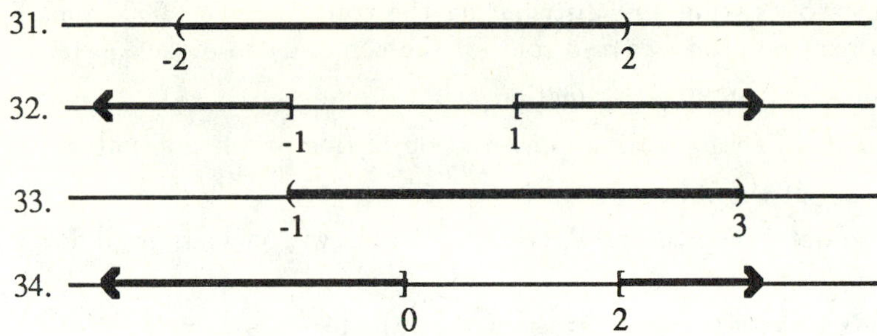

Consider the graph of $y = f(x)$ in Figure 2.4.9.

35. (a) Compare $f(2)$ and -2.
    (b) Is $x = 2$ a solution to $f(x) < -2$?
    (c) Use the graph to solve $f(x) < -2$.
36. (a) Compare $f(-3)$ and 0.
    (b) Is $x = -3$ a solution to $f(x) > 0$?
    (c) Use the graph to solve $f(x) > 0$.
37. Consider the Celsius-to-Fahrenheit temperature conversion formula $F = \frac{9}{5}C + 32$. Water boils when its temperature is greater than or equal to 212° Fahrenheit. At what temperature Celsius will water boil?
    (a) Write an inequality that is an algebraic representation of this problem situation.
    (b) Graph each side of the algebraic representation in (a) in the same viewing rectangle.
    (c) Use a graphing utility to solve this problem.
38. The annual profit P of a candy manufacturer is determined by the formula $P = R - C$, where R is the total revenue generated from selling x pounds of candy and C is the total cost of making and distributing x pounds of candy. Each pound of candy sells for \$1.80 and costs

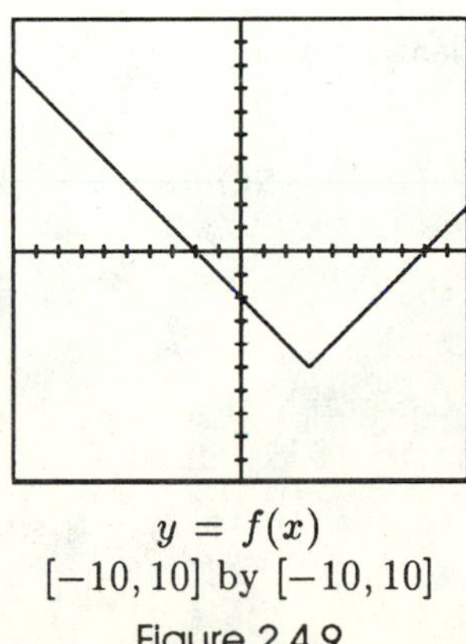

$y = f(x)$
$[-10, 10]$ by $[-10, 10]$
Figure 2.4.9

\$1.38 to make. The fixed costs of making and distributing the candy are \$20,000 annually. How many pounds of candy must be produced and sold for the company to make a profit?

(a) Express the company's annual profit as a function of x.

(b) Write an inequality that gives an algebraic representation of this problem situation.

(c) Draw a graph of each side of the inequality in (b).

(d) Use a graphing utility to determine the production level that will yield a profit for the year.

39. The annual profit P of a candy manufacturer is determined by the formula $P = R - C$, where R is the total revenue generated from selling x pounds of candy and C is the total cost of making and distributing x pounds of candy. Each pound of candy sells for \$1.80 and costs \$1.38 to make. The fixed costs of making and distributing the candy are \$20,000 annually. How many pounds of candy must be produced and sold for the company to make a profit of at least \$25,000 for the year?

(a) Express the company's annual profit as a function of x.

(b) Write an inequality that gives an algebraic representation of this problem situation.

(c) Draw a graph of each side of the inequality in (b).

(d) Use a graphing utility to solve this problem.

Solve the inequality using the distance representation of absolute value. Check your answers with a graphing utility.

40. $|x| - |x - 5| > 0$

41. $|x| - |8 - x| > 0$

42. $|x + 3| < |x|$

43. $|x + 5| > |x|$

*44. Assume $y = f(x)$ is the graph of a line and $x = 4$ is a line of symmetry of the graph of $y = |f(x)|$. Further assume that $f(6) = 8$.

(a) Compute $|f(2)|$.

(b) Solve $|f(x)| \leq 8$.

45. Show that if $|x - 2| < d$ for $d > 0$, then x is in the interval $(2 - d, 2 + d)$ which has length $2d$ and the number 2 as its midpoint, or center.

*46. Let $f(x) = 3x - 5$.

(a) Write intervals equivalent to each inequality.

(i) $|x - 2| < 0.1$

(ii) $|f(x) - 1| < 0.3$

(b) Draw a number-line graph of each interval in (a). Draw (i) on the x-axis and (ii) on the y-axis.

(c) Identify the points of the graph of $y = f(x)$ with projection onto the x-axis satisfying $|x - 2| < 0.1$.

(d) Identify the points of the graph of $y = f(x)$ with projection onto the y-axis satisfying $|f(x) - 1| < 0.3$. Compare with your answer to (c).

*47. Let $f(x) = 3x - 5$. Find all $d > 0$ so that if $|x - 2| < d$ then $|f(x) - 1| < E$ for each value of E.

(a) $E = 0.1$

(b) $E = 0.01$

(c) $E = 0.001$

(d) Find d in terms of E.

*48. Let $f(x) = mx + b$. Given any $E > 0$, find a $d > 0$ so that if $|x - a| < d$, then $|f(x) - f(a)| < E$. *Remark:* This result is a proof of the result that $f(x)$ approaches $f(a)$ as x approaches a, using the formal definition of a limit.

2.5 Solving Higher Order Inequalities Algebraically and Graphically

In this section we introduce an algebraic technique that is used to solve certain higher order inequalities. This technique is called the sign chart method and is particularly useful with inequalities involving polynomials that are easily factored. However, solving inequalities with a graphing utility is almost always the most reasonable and preferred method to use with higher order inequalities. In general, algebraic techniques, like the sign chart method, are very limited because the expressions involved in the inequality must be relatively simple.

We begin with an application studied in Section 1.4. The height s at any time t of an object shot straight up from a point s_0 feet above level ground with initial velocity v_0 is given by the model $s = -16t^2 + v_0 t + s_0$. Notice that the coefficient of t is v_0 and the constant term is s_0. The coefficient of t^2 is one-half of the directed acceleration due to gravity.

• EXAMPLE 1: Suppose that a baseball is thrown straight up from level ground with initial velocity 80 feet per second (Figure 2.5.1).

(a) Determine an algebraic representation of the problem situation.
(b) Draw a complete graph of the algebraic representation in (a). What are the domain and range of the algebraic representation?
(c) What part of the graph in (b) represents the problem situation? Draw a geometric representation of the problem situation.
(d) When will the baseball be above ground?
(e) When will the baseball be at least 64 feet above ground?

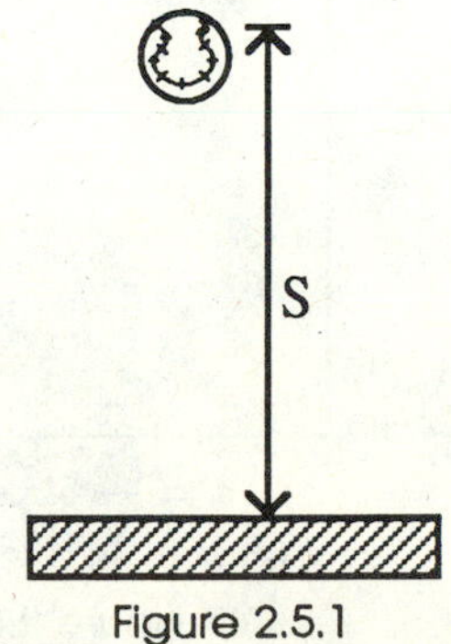

Figure 2.5.1

SOLUTION:

(a) The height above ground of the baseball at any time t is given by $s = -16t^2 + 80t$. This function is an algebraic representation of the problem situation.

(b) A complete graph of the function s is given in Figure 2.5.2. The domain of s is $(-\infty, \infty)$. To find the range of s we can use zoom-in to find the y-coordinate of the highest point of the graph of s. We must be careful when we zoom in because it is possible to produce viewing rectangles where the graph of s will appear so flat that we are unable to distinguish the high point from the rest of the points of the graph in the viewing rectangle. We need to choose viewing rectangles that are wider than they are tall. The graph of s in Figure 2.5.3 clearly highlights the high point of s. We can read from this graph that the coordinates of the high point are approximately $(2.5, 100)$. Using the formula for the coordinates of the vertex given in Section 1.2, you will find that the coordinates of the highest point of the graph of s are exactly $(2.5, 100)$. Thus, the range of s is $(-\infty, 100]$.

(c) Notice that $s = 0$ for $t = 0$ and $t = 5$. The distance s is positive for $0 < t < 5$. Thus, the graph in Figure 2.5.2 represents the problem situation only for $0 \le t \le 5$. Figure 2.5.4 gives a geometric representation that is also a graph of the problem situation.

(d) The baseball is above ground when $s > 0$. It follows from (c) that the baseball will be above ground for any time between 0 and 5 seconds.

(e) The baseball will be at least 64 feet above ground provided that $s \ge 64$. To determine the values of t for which $s \ge 64$ we overlay the graph of $y = 64$ in Figure 2.5.4. The result is given in Figure 2.5.5. The solution to the inequality $s \ge 64$ consists of those values of t for which the graph of s lies above or coincides with the graph of $y = 64$. It appears from Figure 2.5.5 that $s \ge 64$ for $1 \le t \le 4$. You should verify directly that s is 64 when t is 1 or 4. Thus, the baseball will be at least 64 feet above ground for any time from 1 to 4 seconds after it is thrown. •

• EXAMPLE 2: Solve (d) and (e) of Example 1 algebraically.

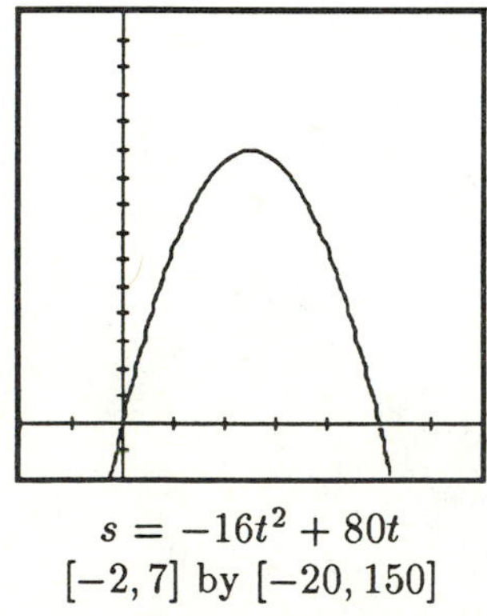

$s = -16t^2 + 80t$
$[-2, 7]$ by $[-20, 150]$
Figure 2.5.2

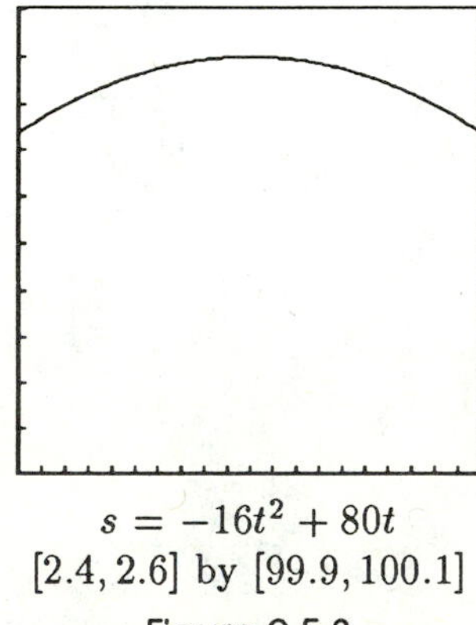

$s = -16t^2 + 80t$
$[2.4, 2.6]$ by $[99.9, 100.1]$
Figure 2.5.3

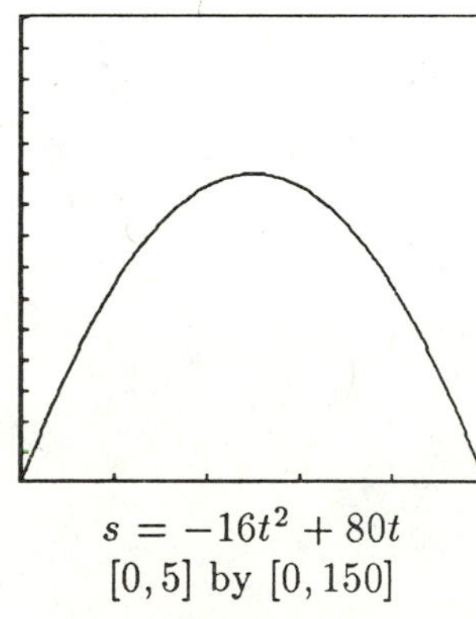

$s = -16t^2 + 80t$
$[0, 5]$ by $[0, 150]$
Figure 2.5.4

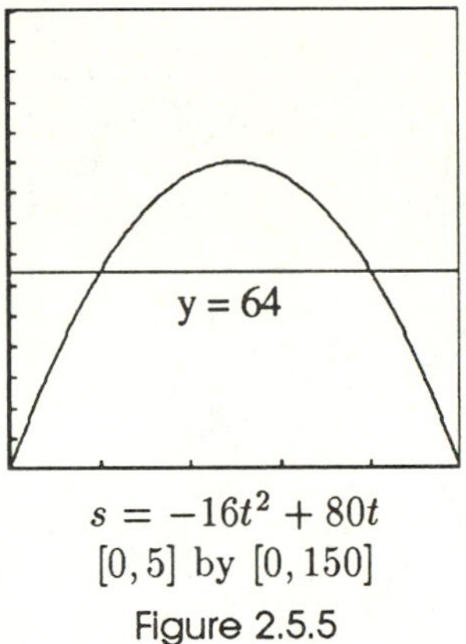

$s = -16t^2 + 80t$
$[0, 5]$ by $[0, 150]$

Figure 2.5.5

SOLUTION:

(d) To solve the inequality $-16t^2 + 80t > 0$ algebraically, first factor $-16t^2 + 80t$ to get the equivalent inequality $-16t(t-5) > 0$. The conditions of the problem situation require that $t \geq 0$. Because we are looking for values of t for which the baseball is above the ground, $-16t(t-5)$ cannot be zero, and the factors $-16t$ and $t - 5$ must be both positive or both negative. However, $-16t$ is negative because $t \geq 0$. Thus, $t - 5$ must also be negative and it follows that $0 < t < 5$.

(e) The inequality $-16t^2 + 80t \geq 64$ is equivalent to the following inequalities.

$$-16t^2 + 80t - 64 \geq 0$$
$$t^2 - 5t + 4 \leq 0$$
$$(t-1)(t-4) \leq 0$$

The product $(t-1)(t-4)$ is positive for $t > 4$ because each factor is positive for $t > 4$. Similarly, the product $(t-1)(t-4)$ is positive for $t < 1$ because each factor is negative for $t < 1$. If t is between 1 and 4, then $t - 1$ is positive and $t - 4$ is negative. Thus, the product $(t-1)(t-4)$ is negative for $1 < t < 4$. Finally, the product is zero for $t = 1$ or $t = 4$. Thus, the solution to the inequality is $1 \leq t \leq 4$, or $[1, 4]$. •

Sign Chart

We will organize the argument used to solve the inequality $(t-1)(t-4) \leq 0$ in Example 2 in the form of a **sign chart**. The linear expression $x - 2$ is zero for $x = 2$, positive for $x > 2$ and negative for $x < 2$. Every linear polynomial $ax + b$, where $a \neq 0$, is zero for exactly one value of x and *opposite* in sign to the right and left of that value. This is the fact we use to fill in the sign chart.

The solution to $(t-1)(t-4) \leq 0$ consists of those values of t for which $(t-1)(t-4) < 0$ together with the values of t for which $(t-1)(t-4) = 0$. The values of t that make each factor equal to zero determine the range for t in the following sign chart.

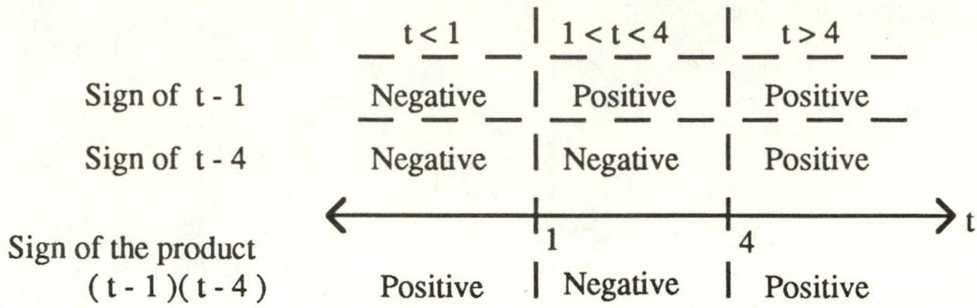

Notice that we use $t = 1$ and $t = 4$, the two values of t for which $(t-1)(t-4) = 0$, to divide the number line into the three intervals $t < 1$, $1 < t < 4$, and $t > 4$. Each row of the sign chart above the number line gives the sign of one of the two factors in each of the intervals $t < 1$, $1 < t < 4$, and $t > 4$. The first line of the sign chart tells us that $t - 1$ is negative for $t < 1$ and positive for $1 < t < 4$ and $t > 4$.

The last line of the sign chart gives the sign of the product of the two factors in these three intervals. The last line of the sign chart tells us that the product $(t-1)(t-4)$ is negative for $1 < t < 4$ and positive for $t < 1$ or $t > 4$. In addition, $(t-1)(t-4)$ is zero for $t = 1$ and $t = 4$. Thus, the solution to $(t-1)(t-4) \leq 0$ is $1 \leq t \leq 4$.

The sign of $t - 1$ is fixed in each of the intervals $t < 1$, $1 < t < 4$, and $t > 4$. Thus, the sign of $t-1$ for t in these intervals can be determined by evaluating the expression $t-1$ for exactly one value of t in each of these intervals. Now, 5 is in the interval $t > 4$ and $t - 1$ is positive if $t = 5$. Thus, $t - 1$ is positive in $t > 4$. Next, 2 is in the interval $1 < t < 4$ and $t - 1$ is positive if $t = 2$. Thus, $t - 1$ is positive in $1 < t < 4$. Finally, 0 is in the interval $t < 1$ and $t - 1$ is negative if $t = 0$. Thus, $t - 1$ is negative in $t < 1$.

We can simplify the above argument even further. To solve the inequality $f(x) \geq 0$, $f(x) > 0$, $f(x) \leq 0$, or $f(x) < 0$, where f is a polynomial, we first solve the corresponding equation $f(x) = 0$. Then the solutions are used to partition the number line into one more interval than the number of solutions. The inequality is either always true or always false in each interval. Therefore, testing the inequality with just one number in the interval identifies whether or not the inequality is true for that interval. We can summarize the above argument for $f(t) = (t-1)(t-4) \leq 0$ in another chart that is also called a **sign chart**.

Interval	$t < 1$	$1 < t < 4$	$t > 4$
Test-value of t	0	2	5
Value of $f(t) = (t-1)(t-4)$	4	-2	4
Sign of $f(t) = (t-1)(t-4)$	+	−	+

The last line of the chart tells us that $f(t) < 0$ for $1 < t < 4$ and $f(t) > 0$ for $t < 1$ or $t > 4$. Thus, the solution to $(t-1)(t-4) \leq 0$ is $1 \leq t \leq 4$.

The sign chart is useful whenever we are able to *factor* the polynomial f completely, or determine all of the zeros by other means. The *Intermediate Value Theorem* introduced in Chapter 3 is the basis for a sign chart method.

• EXAMPLE 3: Use a sign chart to solve $2x^2 + 9x - 5 > 0$, and draw a number-line graph of the solution. Check your answer with a graphing utility.

SOLUTION: First we factor $2x^2 + 9x - 5 = (2x - 1)(x + 5)$, and then we make a sign chart. Notice that $2x^2 + 9x - 5 = 0$ for $x = \frac{1}{2}$ and $x = -5$. These numbers are used to determine the intervals on the number line.

Interval	$x < -5$	$-5 < x < \frac{1}{2}$	$x > \frac{1}{2}$
Test value of x	-6	0	1
Value of $2x^2 + 9x - 5$	13	-5	6
Sign of $2x^2 + 9x - 5$	+	−	+

The values of x that make $(2x-1)(x+5)$ equal to 0 are not solutions to $(2x-1)(x+5) > 0$. The last line of the sign chart shows that the solution to the inequality $(2x - 1)(x + 5) > 0$ is $x < -5$ or $x > \frac{1}{2}$. In interval notation, the solution to the original inequality is $(-\infty, -5) \cup (\frac{1}{2}, \infty)$. A number-line graph of the solution is given in Figure 2.5.6.

A complete graph of $f(x) = 2x^2 + 9x - 5$ is given in Figure 2.5.7. It appears that the graph of f crosses the x-axis at -5 and $\frac{1}{2}$. This can be confirmed by direct substitution. Thus, the graph of $f(x) = 2x^2 + 9x - 5$ is above the x-axis for $x < -5$ or $x > \frac{1}{2}$. This geometrically confirms the solution $(-\infty, -5) \cup (\frac{1}{2}, \infty)$ found algebraically. •

• EXAMPLE 4: Solve $x(x + 3)(x - 1) \geq 0$ both algebraically and graphically.

SOLUTION: The expression $x(x + 3)(x - 1)$ is equal to 0 for $x = 0, -3,$ or 1. Be sure you take the factor x into account when making the sign chart.

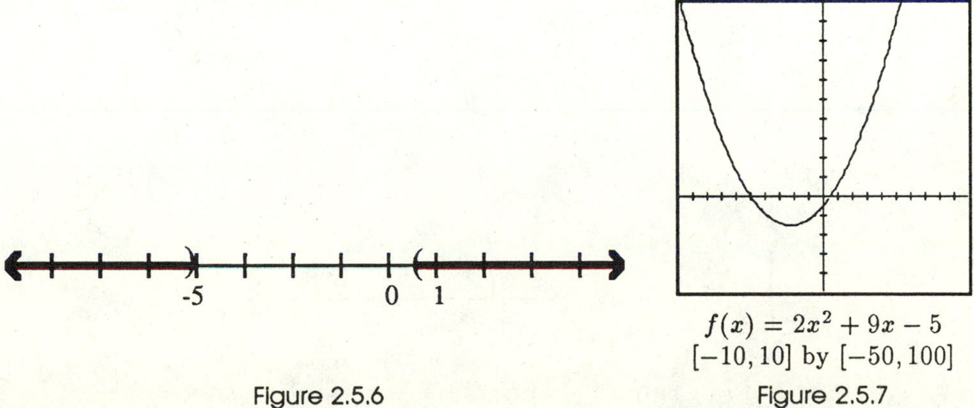

$f(x) = 2x^2 + 9x - 5$
$[-10, 10]$ by $[-50, 100]$

Figure 2.5.6

Figure 2.5.7

Interval	$x < -3$	$-3 < x < 0$	$0 < x < 1$	$x > 1$
Test value of x	-4	-1	0.5	2
Value of $x(x+3)(x-1)$	-20	4	-0.875	10
Sign of $x(x+3)(x-1)$	$-$	$+$	$-$	$+$

The solution to $x(x+3)(x-1) > 0$ is $-3 < x < 0$ or $x > 1$. The product $x(x+3)(x-1)$ is 0 for $x = -3$, 0, or 1. Therefore, the solution to $x(x+3)(x-1) \geq 0$ in interval notation is $[-3, 0] \cup [1, \infty)$.

The graph of $f(x) = x(x+3)(x-1)$ in Figure 2.5.8 is a complete graph. This graph is above or on the x-axis for $-3 \leq x \leq 0$ or $x \geq 1$, geometrically confirming the solution found algebraically. •

The sign chart method can also be used with inequalities of the form $f(x) \geq 0$, $f(x) > 0$, $f(x) \leq 0$, or $f(x) < 0$ with f the quotient of two polynomials provided that we can determine all the zeros of the two polynomials. Such functions f are called **rational functions** and will be studied in detail in Chapter 3. In this situation, the zeros of the two polynomials of f are used to partition the number line, and then we proceed as we did with polynomials. Notice that the zeros of the polynomial in the denominator of f are *not* in the domain of f. The next example illustrates the procedure for rational functions.

• EXAMPLE 5: Use a sign chart to solve

$$\frac{x-4}{2x+5} \leq 0 \; .$$

SOLUTION: The zeros of the polynomials in the numerator and denominator of the rational function $f(x) = \frac{x-4}{2x+5}$ are 4 and -2.5. These numbers are used to determine the intervals on the number line in the following sign chart.

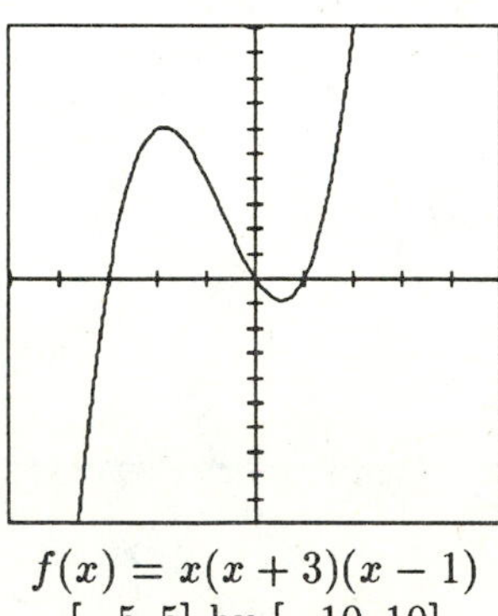

$f(x) = x(x+3)(x-1)$
$[-5, 5]$ by $[-10, 10]$
Figure 2.5.8

Interval	$x < -2.5$	$-2.5 < x < 4$	$x > 4$
Test value of x	-3	0	5
Value of $\frac{x-4}{2x+5}$	7	$-\frac{4}{5}$	$\frac{1}{15}$
Sign of $\frac{x-4}{2x+5}$	$+$	$-$	$+$

The last line of the sign chart tells us that $\frac{x-4}{2x+5} < 0$ for $-2.5 < x < 4$. Notice that $\frac{x-4}{2x+5}$ is zero for $x = 4$ and undefined for -2.5. Thus, the solution to $\frac{x-4}{2x+5} \leq 0$ is $-2.5 < x \leq 4$, or $(-2.5, 4]$. •

In Example 4, we solved an inequality that was given in factored form. The next example involves a function that cannot be factored. This example illustrates both the power of a graphical approach and the limitation of algebraic procedures. *It is extremely rare to find a real-world inequality in which the expression can be factored algebraically to apply a sign chart method.* We use zoom-in to determine the endpoints of the intervals involved in the solution. Unless we specify otherwise, these endpoints should be determined with error of at most 0.01.

• EXAMPLE 6: Solve the inequality $x^3 < 4x - 1$.

SOLUTION: This inequality can be rewritten in the form $x^3 - 4x + 1 < 0$. Let $f(x) = x^3 - 4x + 1$. The solution to the inequality consists of those values of x for which the graph of f lies below the x-axis (Figure 2.5.9). This graph is a complete graph of f. We can use zoom-in to determine that the zeros of f are -2.11, 0.25, and 1.86 with error less than 0.01. Thus, the solution to this inequality is $(-\infty, -2.11) \cup (0.25, 1.86)$. •

• **Agreement** We used round parentheses at the endpoints of the intervals in Example 6 because the inequality $<$ means strictly less than. We would use square brackets if the inequality symbol was $\leq$. We use this convention even though the endpoints

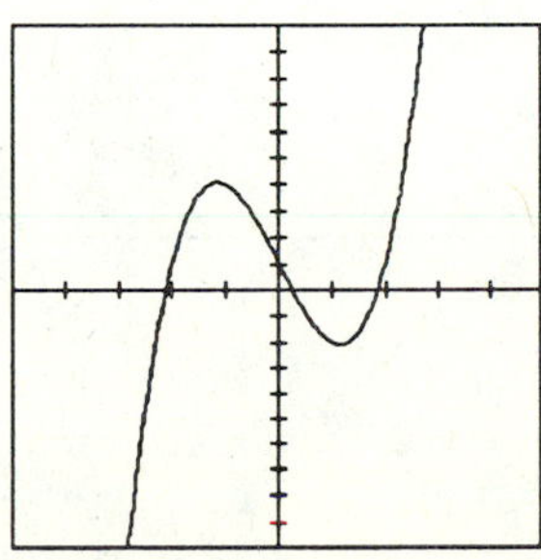

$f(x) = x^3 - 4x + 1$
$[-5, 5]$ by $[-10, 10]$
Figure 2.5.9

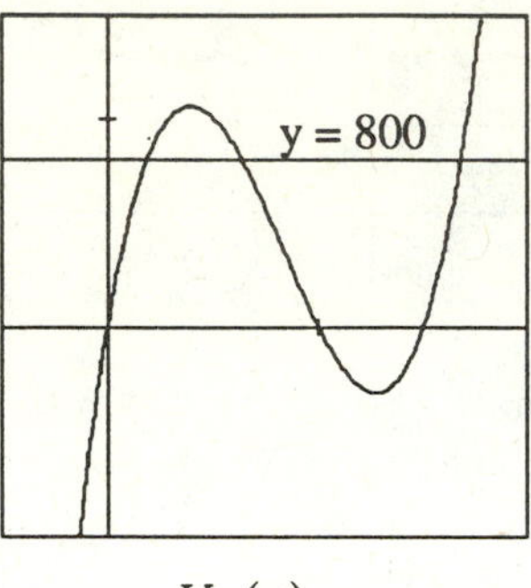

$Vx(x) =$
$x(20-2x)(30-2x)$
$[-5, 20]$ by $[-1000, 1500]$
Figure 2.5.10

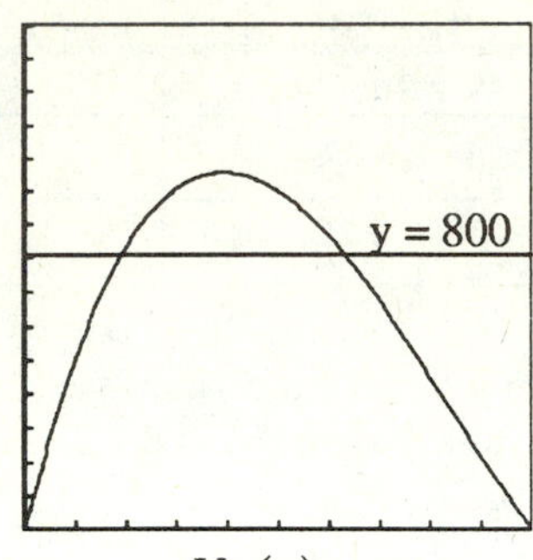

$Vx(x) =$
$x(20-2x)(30-2x)$
$[0, 10]$ by $[0, 1500]$
Figure 2.5.11

we report in solutions are approximations. That is, we act as if the endpoints are exact when deciding whether to use round parentheses or square brackets.

• EXAMPLE 7: A box is formed by removing squares of side length x from each corner of a rectangular piece of cardboard 20 inches wide by 30 inches long. Determine x so that the volume of the resulting box is at least 800 cubic inches.

SOLUTION: You may want to refer to Example 4 of Section 2.2 to see how to form the box. The volume of the box is $V(x) = x(20-2x)(30-2x)$. Figure 2.5.10 gives a complete graph of V and the horizontal line $y = 800$. The graph of V represents the problem situation only for values of x between 0 and 10. (Why?) Thus, the graph of V in Figure 2.5.11 is a geometric representation that is also a graph of the problem situation. The solution to the problem consists of those values of x for which the graph of V is on or above the graph of $y = 800$. We can use zoom-in to determine that the x-coordinates of the two points of intersection of V and $y = 800$ between $x = 0$ and $x = 10$ are 1.88 and 6.36 with error at most 0.01. The solution to the problem is the interval $[1.88, 6.36]$, that is, the volume of the resulting box is at least 800 cubic inches, provided that the side length x of the square cut out satisfies $1.88 \leq x \leq 6.36$. •

• EXERCISES 2-5

In Exercises 1 through 15 use a sign chart to solve the inequalities. Write the solution in interval notation. Use a graphing utility to check your answers.

1. $(x-1)(x+2) < 0$
2. $x(x-3) \leq 0$
3. $x^2 - 5x + 6 \geq 0$
4. $x^2 - 4x - 21 < 0$
5. $2x^2 + 5x \geq 3$
6. $x^2 < x$
7. $x^4 + 3x^2 < 4$
8. $\dfrac{x+3}{x-1} < 0$
9. $\dfrac{x+3}{x-1} \geq 0$
10. $x(x-4)(x+2) \geq 0$
11. $(x+1)(x-2)(x+4) < 0$
12. $(x-2)^3 > 0$

13. $(x-2)^2(x-1) \leq 0$

14. $\dfrac{x+9}{x+4} \leq 0$

15. $\dfrac{3x-10}{4-x} > 0$

16. Solve the inequality in Example 6 by determining the values of x for which the graph of $f(x) = x^3$ lies below the graph of $g(x) = 4x - 1$. Compare this method with that given in the example.

Solve each inequality.

17. $x^2 - 5x + 3 > 0$

18. $2 - 8x + 3x^2 \leq 0$

19. $3x^3 - 8x^2 - 5x + 6 \leq 0$

20. $x^3 - 9x^2 + 6x + 55 < 0$

21. $|x^2 - 1|(x-2) \geq 3$

22. $2x + 1 > x^3 + 2x^2 - 3x + 5$

23. $3x^2 - x + 6 \leq 2x^3 - x^2 + 3x - 4$

24. $5 \sin x < 2$ for $0 \leq x \leq 10$

25. An object is propelled upward from a tower 200 feet in height with initial velocity 100 feet per second.
    (a) Determine an algebraic representation of the problem situation.
    (b) Draw a complete graph of the algebraic representation.
    (c) What are the domain and range of the algebraic representation?
    (d) What part of the graph in (b) represents the problem situation?
    (e) When will the object be above ground?
    (f) When will the object hit the ground?
    (g) What is the maximum height attained by the object?

26. An object is propelled upward from a tower 155 feet in height with initial velocity 275 feet per second.
    (a) Determine an algebraic representation of the problem situation.
    (b) Draw a complete graph of the algebraic representation.
    (c) What are the domain and range of the algebraic representation?
    (d) What part of the graph in (b) represents the problem situation?
    (e) When will the object be above ground?
    (f) When will the object hit the ground?
    (g) What is the maximum height attained by the object?

A single-commodity open market is driven by the supply and demand principle. Economists have determined that supply curves are usually increasing (as the price increases, the sellers increase production) and the demand curves are decreasing (as the price increases, the consumers buy less). For each pair of indicated supply and demand equations do the following.

(i) Draw complete graphs of both the supply and demand equations on the same coordinate system. p is the price (vertical axis) and x represents the number of units of the commodity produced.

(ii) What portion of the graphs in (i) are complete graphs of the problem situation?

(iii) Determine the *equilibrium price*, that is, the price where supply is equal to demand.

(iv) How many items must be produced for the supply to be greater than the demand?

27. Supply: $p = 5 + \frac{2}{13}x$
Demand: $p = 10 - \frac{1}{20}x^2$

28. Supply: $p = \frac{1}{20}x + 1$
Demand: $p = \frac{10}{x}$

29. A swimming pool with dimensions 20 feet by 30 feet is surrounded by a sidewalk of uniform width x. Find the possible widths of the sidewalk if the total area of the sidewalk is to be greater than 200 square feet but less than 360 square feet.
   (a) Write an equation that gives the area of the sidewalk as a function of x.
   (b) Write an inequality that is an algebraic representation of this problem situation.
   (c) Solve the inequality in (b) algebraically, and then determine the solution to the problem situation.
   (d) Use a graphing utility to check your answer to (c).

30. Consider the simple interest formula $S = P(1 + rn)$ (Exercise 39, Section 2.1). Find the rate of simple interest necessary for an investor to accumulate at least \$30,000 in 18 years by investing \$5000 today.
   (a) Write an inequality that is an algebraic representation of this problem situation.
   (b) Draw a complete graph of each side of the inequality in (a) in the same viewing rectangle.
   (c) Determine the solution to this problem situation both algebraically and geometrically.

31. Equal squares are removed from the four corners of a 22-inch-by-29-inch rectangular sheet of cardboard. The sides are turned up to make a box with no lid. Find the possible lengths of the sides of the removed squares if the volume of the box is to be less than 2000 cubic inches.
   (a) Write an inequality that is an algebraic representation of this problem situation.
   (b) Draw a complete graph of each side of the inequality in (a) in the same viewing rectangle. Draw a complete graph of the problem situation.
   (c) Determine the solution to this problem situation geometrically.

32. Sam makes two initial investments of \$1000 each. The first earns 10% simple interest, and the second earns interest using 5% *simple discount*. After n years, the value of the first investment is $S = 1000/(1 + 0.10n)$ and the value of the second investment is $S = 1000/(1 - 0.05n)$.
   (a) Draw a complete graph of each algebraic representation in the viewing rectangle $[0, 30]$ by $[0, 6000]$.
   (b) Is there a time when the future values of both investments are equal?
   (c) Is there a period of time when the simple interest investment is less than the value of the simple discount investment?

*33. Which of the following are true for all real numbers a and b? For each part that is not true, give a counterexample.
   (a) $|ab| = |a||b|$
   (b) $|a + b| = |a| + |b|$
   (c) $|a + b| \leq |a| + |b|$
   (d) $|a + b| \geq |a| + |b|$

34. A rectangular area, with one side against an existing wall, is to be enclosed by three sides of fencing totaling 335 feet in length. Let x be side length of the fence perpendicular to the existing wall.
   (a) Determine an algebraic representation that gives the area enclosed as a function of x.

(b) Draw a complete graph of the algebraic representation.

(c) What portion of the graph of the algebraic representation represents the problem situation?

(d) Determine x so that the area is less than or equal to 11,750 square feet.

35. A 300-inch piece of wire is cut into two pieces. Each piece of wire is used to make a square. Let x be the length of one piece of the wire.

(a) Determine an algebraic representation that gives the total area of the two squares as a function of x.

(b) Draw a complete graph of the algebraic representation.

(c) Draw a complete graph of the problem situation.

(d) Determine x so that the total area of the two rectangles is less than 4000 square inches.

36. Squares of side length x are removed from a 15 inch by 60 inch piece of cardboard, and a box with no top is formed by folding (see Figure 2.2.11). Determine x so that the volume of the resulting box is at most 450 cubic inches.

37. Squares of side length x are removed from a 10 cm by 25 cm piece of cardboard, and a box with no top is formed by folding (see Figure 2.2.11). Determine x so that the volume of the resulting box is at least 175 cm^3.

38. An object is shot straight up from the top of a building 186 feet tall with initial velocity 80 feet per second. Let t be the number of seconds elapsed since the object was launched.

(a) Determine an algebraic representation that gives the height of the object above ground level as a function of t.

(b) Draw a complete graph of the algebraic representation.

(c) What portion of the graph of the model represents the problem situation?

(d) Determine when the object is at least 300 feet above ground.

2.6 Maximum and Minimum Values

Many problem situations require that we find the largest or smallest value of a given function over some portion of its domain. We have encountered such situations in some of the real-world applications in previous sections. In this section we use zoom-in to find the coordinates of points where maximum and minimum values occur. Applications whose solutions involve finding a largest or smallest value of a model are also investigated.

• EXAMPLE 1: Draw a complete graph of $f(x) = x^3 - 4x$.

SOLUTION: The graph of $f(x) = x^3 - 4x$ in the viewing rectangle $[-6, 6]$ by $[-10, 10]$ is shown in Figure 2.6.1 and is a complete graph. •

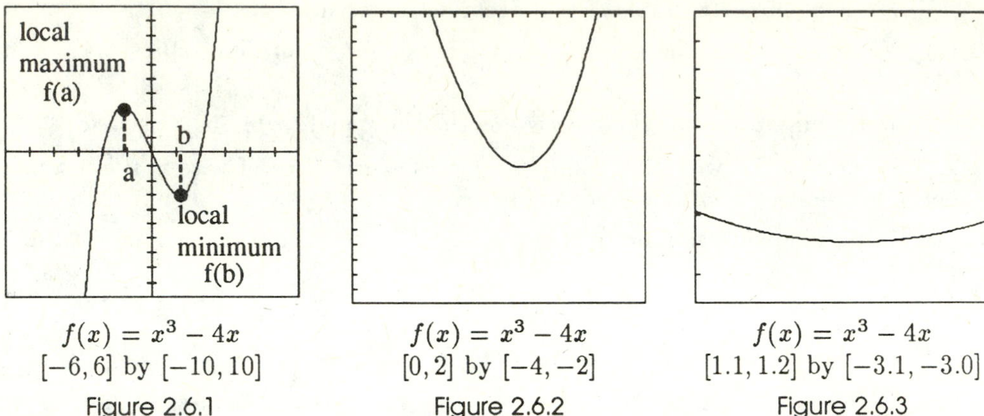

$f(x) = x^3 - 4x$
$[-6, 6]$ by $[-10, 10]$
Figure 2.6.1

$f(x) = x^3 - 4x$
$[0, 2]$ by $[-4, -2]$
Figure 2.6.2

$f(x) = x^3 - 4x$
$[1.1, 1.2]$ by $[-3.1, -3.0]$
Figure 2.6.3

The graph of $f(x) = x^3 - 4x$ has a high point (peak) between $x = -2$ and $x = 0$. The value of the function at this point is called a local maximum value of f. Similarly, the graph of $f(x) = x^3 - 4x$ has a low point (valley) between $x = 0$ and $x = 2$. The value of f at this point is called a local minimum value of f. More precisely we have the following definitions.

• DEFINITION The value of f at $x = a$ is called a **local maximum value** of f if there is an interval (c, d) with $c < a < d$ so that $f(x) \leq f(a)$ for all x in (c, d). The value of f at $x = b$ is called a **local minimum value** of f if there is an interval (c, d) with $c < b < d$ so that $f(x) \geq f(b)$ for all x in (c, d). •

When we use the phrase "for all x in (c, d)" in the above definition, it is implicitly understood that we mean those x's in (c, d) that are also in the domain of f. For example, using the above definition with this understanding about domain, the function $f(x) = \sqrt{x}$ has a local minimum value at $x = 0$.

• EXAMPLE 2: Determine the local maximum and local minimum values of the function $f(x) = x^3 - 4x$.

SOLUTION: Refer to Figure 2.6.1. We use zoom-in to trap the low point between $x = 0$ and $x = 2$ in a very small viewing rectangle. The graph in Figure 2.6.1 suggests that we next look at the graph of f in the viewing rectangle $[0, 2]$ by $[-4, -2]$ (Figure 2.6.2). Because the distance between scale marks in Figure 2.6.2 is 0.1, we can estimate that the low point has x-coordinate between 1.1 and 1.2 and y-coordinate between -3.1 and -3.0. The graph of f in the viewing rectangle $[1.1, 1.2]$ by $[-3.1, -3.0]$ is given in Figure 2.6.3. Notice that this viewing rectangle is actually a *square*.

The graph of f in Figure 2.6.3 appears relatively flat. In calculus you will see that this is exactly what should be expected when you magnify around such points. If we are not careful when we zoom in, we may produce viewing rectangles where the graph of f is so flat that we are unable to distinguish the low point (or high point) from the rest of the points in the viewing rectangle.

What we need to do is choose viewing rectangles that are much wider than they are tall. (Why?) The graph of f in the viewing rectangle $[1.1, 1.2]$ by $[-3.08, -3.07]$ (Figure 2.6.4) more clearly highlights the low point of f.

Notice that L and R are the same in Figures 2.6.3 and 2.6.4. The height of the viewing rectangle in Figure 2.6.3 is 0.1, and the height of the viewing rectangle in Figure 2.6.4 is 0.01. In fact, the width of the viewing rectangle in Figure 2.6.4 is ten times its height.

From Figure 2.6.4 we can estimate that the coordinates of the low point are $(1.155, -3.0792)$. The error in the first coordinate is at most 0.01 and the error in the second coordinate is at most 0.001. It was necessary to increase the precision on the second coordinate in order to achieve error at most 0.01 in the first coordinate. Thus, f achieves a local minimum value of -3.0792 when x is 1.155.

We can use the graph in Figure 2.6.1 or verify directly that $f(-x) = -f(x)$ to show that the graph of $f(x) = x^3 - 4x$ is symmetric with respect to the origin. (Why?) This means that the coordinates of the point where the local maximum value occurs, with error at most 0.01, are $(-1.155, 3.0792)$. (Why?) •

Polynomials of Degree 3

The graph of the function $g = -f$ where $f(x) = x^3 - 4x$ is the reflection of the graph of f through the x-axis (Figure 2.6.5). We can see from this graph that g has a local minimum value at $x = a$ and a local maximum value at $x = b$. It can be shown that every polynomial function of degree 3 has behavior like one of the following functions: $y = x^3$, $y = -x^3$, $y = x^3 - 4x$, or $y = -x^3 + 4x$. Thus, it follows that every polynomial function of degree 3 has either one local maximum value and one local minimum value like the function in Example 1 or no local maximum or minimum value like $y = x^3$.

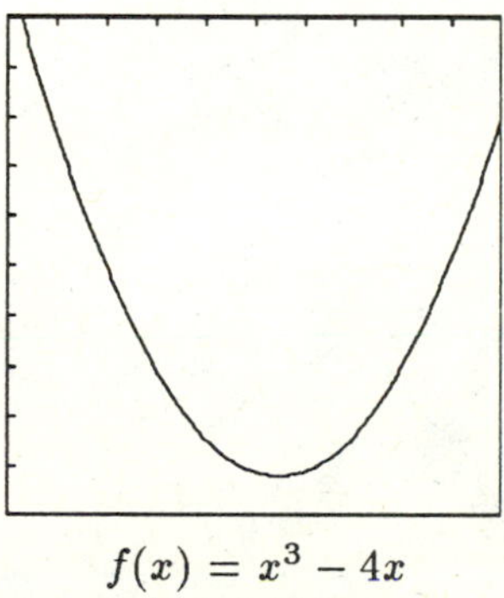

$f(x) = x^3 - 4x$
$[1.1, 1.2]$ by
$[-3.08, -3.07]$

Figure 2.6.4

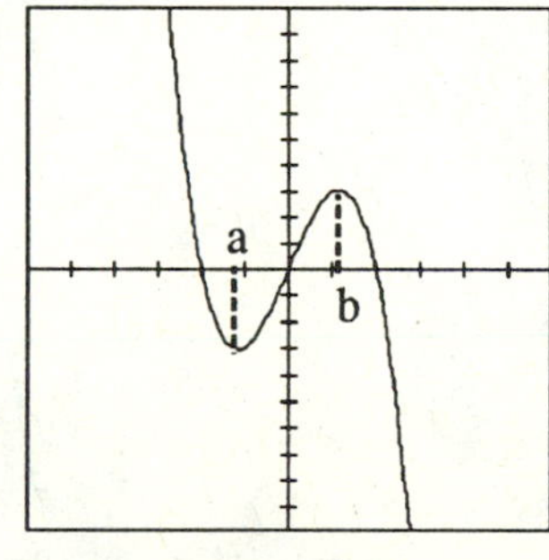

$g(x) = -x^3 + 4x$
$[-6, 6]$ by
$[-10, 10]$

Figure 2.6.5

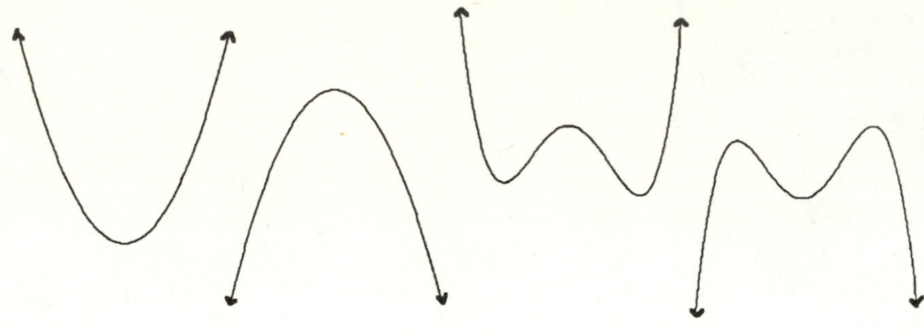

Figure 2.6.6

Polynomials of Degree 4 or 5

A polynomial of degree 4 has local maximum and minimum behavior like one of the four graphs in Figure 2.6.6. In Figure 2.6.6, the lowest point of the third graph or the highest point of the fourth graph need not be in the order suggested. A value of a function that is either a local maximum value or a local minimum value of the function is called a **local extreme value** of the function. As suggested by Figure 2.6.6, a polynomial of degree 4 has either *one* or *three* local extreme values.

A polynomial of degree 5 has local maximum and minimum behavior like one of the six graphs in Figure 2.6.7. Again, the relative position of the high or low points in the last four graphs need not be in the order suggested. As suggested by Figure 2.6.7, a polynomial of degree 5 has zero, two, or four local extreme values. You will be expected to use the information in Figures 2.6.6 and 2.6.7 in the Exercises.

- EXAMPLE 3: An object is shot straight up (launched) from the top of a building 200 feet tall with initial velocity 64 feet per second.

  (a) When will the object hit the ground?

  (b) When does the object reach its maximum height above ground, and what is this maximum height?

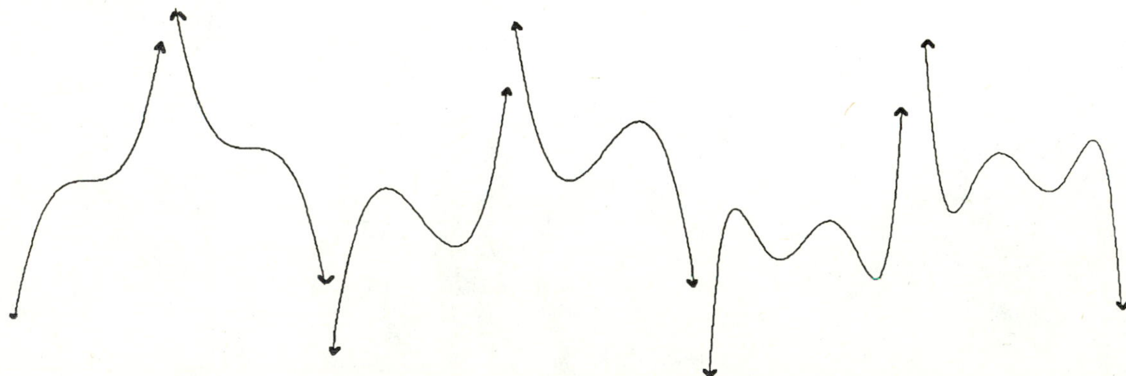

Figure 2.6.7

SOLUTION: The height of the object above ground t seconds after launch is $s = -16t^2 + 64t + 200$. The graph of s in the $[-10, 10]$ by $[-400, 400]$ viewing rectangle is shown in Figure 2.6.8. Notice that this is *not* a graph of the trajectory (flight path) of the object. (Why?)

(a) The object will hit the ground when $s = 0$. From Figure 2.6.8, it appears that the object will hit the ground when t is approximately 6. We can zoom in and find that this value of t is 6.06 with error at most 0.01. Thus, the object will hit the ground about 6.06 seconds after launch.

(b) From Figure 2.6.8 we can see that the object will reach its maximum height when t is approximately 2. We can zoom in and find that the coordinates of the highest point are $(2, 264)$ with error at most 0.01. We can complete the square to write $y = -16(t-2)^2 + 264$ and see that the coordinates of the vertex are exactly $(2, 264)$. Thus, the object reaches its maximum height of 264 feet above ground 2 seconds after launch. •

• EXAMPLE 4: Squares of side length x are removed from a 20-inch by 25-inch piece of cardboard (Figure 2.6.9). A box with no top is formed by folding along the dashed line segments in Figure 2.6.9. Determine x so that the box has maximum possible volume, and find this maximum volume.

SOLUTION: The height of the box is x, and the base of the box has dimensions $20 - 2x$ by $25 - 2x$. Thus, the volume of the box is $V(x) = x(20-2x)(25-2x)$. The graph of the function $V(x) = x(20-2x)(25-2x)$ in the viewing rectangle $[-5, 20]$ by $[-1000, 1000]$ is shown in Figure 2.6.10. The side length x must be a number between 0 and 10 because x, $20 - 2x$, and $25 - 2x$ all have to be positive. (Why?)

We can see from the graph in Figure 2.6.10 that V has a local maximum value between $x = 0$ and $x = 10$. The x-coordinate of this high point gives the value of x that will produce a box with the maximum possible volume. The y-coordinate of this high point gives the value of the maximum possible volume. We use zoom-in to find the coordinates of the high point between $x = 0$ and $x = 10$.

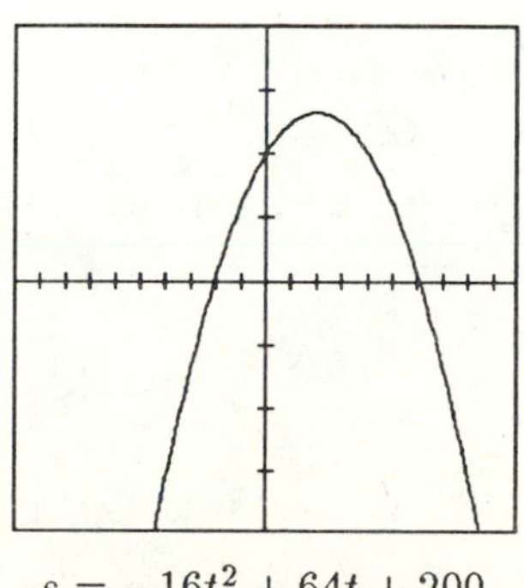

$s = -16t^2 + 64t + 200$
$[-10, 10]$ by $[-400, 400]$

Figure 2.6.8

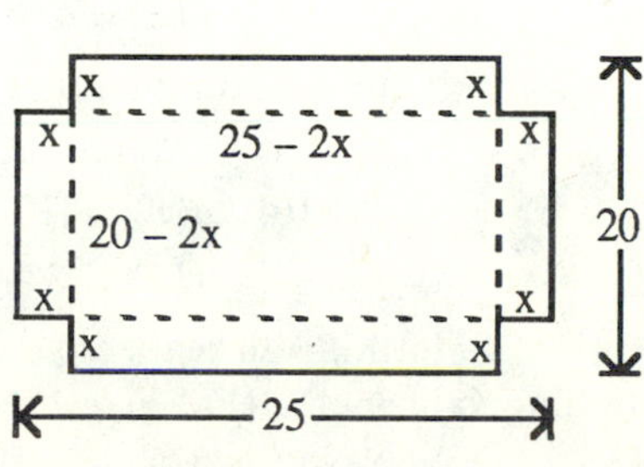

Figure 2.6.9

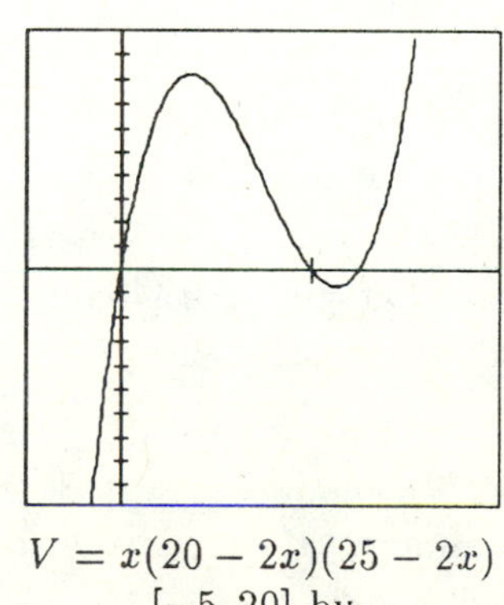

$V = x(20-2x)(25-2x)$
$[-5, 20]$ by
$[-1000, 1000]$

Figure 2.6.10

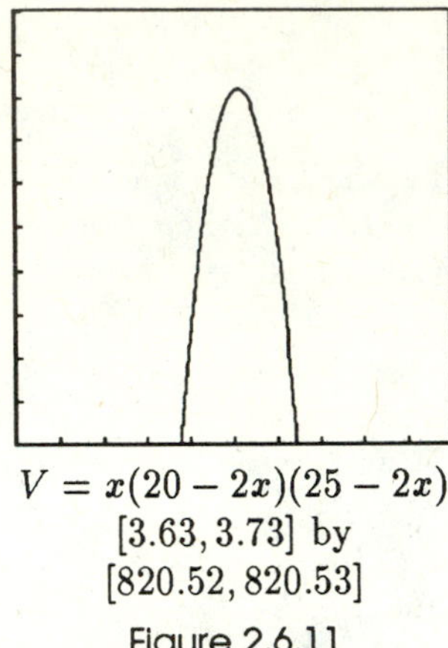

$V = x(20 - 2x)(25 - 2x)$
$[3.63, 3.73]$ by
$[820.52, 820.53]$

Figure 2.6.11

The graph of V in the $[3.63, 3.73]$ by $[820.52, 820.53]$ viewing rectangle (Figure 2.6.11) permits us to read the coordinates of the high point as $(3.68, 820.5282)$ with error at most 0.01 in either coordinate. Actually, it was necessary to find the second coordinate with error at most 0.001 in order to determine the first coordinate with error at most 0.01.

Thus, if the side length of the removed squares is 3.68 inches, we produce a box with maximum volume. The corresponding maximum value of the volume is 820.5282 cubic inches. •

• EXERCISES 2-6

Draw a complete graph of each polynomial function. Specify a viewing rectangle that displays a *complete* graph of the function.

1. $f(x) = x^2 - x + 3$
2. $y = 12x^2 - 20x + 60$
3. $A = \left(\frac{x}{4}\right)^2 + \left(\frac{500 - x}{4}\right)^2$
4. $A = x(8 - x^2)$
5. $V = x(30 - 2x)(40 - 2x)^2$
6. $g(x) = 2x^3 - 6x^2 + 3x - 5$
7. $V(x) = 200 - 20x^2 + 0.01x^3$
8. $f(x) = 2x^5 - 10x^4 + 3x^3 + 2x^2 - 10x + 5$

Determine the *exact* values of all local maximum and minimum values and the corresponding values of x that give these values.

9. $y = 30 - 3x + 7x^2$
10. $g(x) = |10 - 7x + x^2|$
11. $y = \sqrt{x - 1}$
12. $f(x) = 3 - 2|4x + 7|$

Draw a complete graph of the function. Determine all real zeros. Determine all local maximum and minimum values and the corresponding values of x that give these values.

13. $y = 10 - x^2$
14. $f(x) = 2x^2 - 6x + 10$
15. $g(x) = x^3 - 10x^2$
16. $y = 2x^3 - x^2 + x - 4$
17. $f(x) = x^4 - 4x^2 - 3x + 12$
18. $y = |2x^3 - 4x^2 + x - 3|$

19. $f(x) = x^3 - 2x^2 + x - 30$

20. $V(x) = x(34 - 2x)(53 - 2x)$

21. $T(x) = |20x^3 + 2x^2 - 10x + 5|$

22. $V(x) = x(22 - 2x)(8 - 2x)$

23. $f(x) = 12 - x + 3x^2 - 2x^3$

24. $y = x^5 - 3x^2 + 3x - 6$

25. An object is shot straight up from the top of a building 300 feet tall with initial velocity 40 feet per second.
    (a) Determine an algebraic representation of the height above ground of the object as a function of time.
    (b) When will the object hit the ground? Assume it does not land on the building.
    (c) When does the object reach its maximum height above ground, and what is this maximum height?

26. A large apartment rental company has 2500 units available, and 1900 are currently rented at \$450 per month. A market survey indicates that each \$15 decrease in monthly rent will result in 20 new tenants.
    (a) Determine a function R which represents the total rental income realized by the company where x is the number of \$15 decreases in monthly rent (that is, if $x = 3$, then the monthly rent is \$405). Assume that all rents are adjusted to the new figure.
    (b) Draw a complete graph of the function $y = R(x)$.
    (c) Determine the rent that will yield the rental company the maximum monthly income.

27. A rectangular area is to be fenced against an existing wall. The three sides of the fence must total 1050 feet in length. Find the dimensions of the maximum area that can be enclosed. What is the maximum area?

28. A 300 inch piece of wire is cut into two pieces. Each piece of wire is used to make a square. Let x be the length of one piece of the wire.
    (a) Determine an algebraic representation that gives the total area of the two squares as a function of x.
    (b) Draw a complete graph of the algebraic representation.
    (c) What portion of the graph of the algebraic representation represents the problem situation?
    (d) Find the lengths of the two pieces of wire that produce two squares of maximum total area.
    (e) Find the lengths of the two pieces of wire that produce two squares of minimum total area.

29. The total daily revenue of a lemonade stand at a state fair is given by the equation $R = x \cdot p$ where x is the number of glasses of lemonade sold daily and p is the price of one glass of lemonade. Assume the price of the lemonade is given by the "supply" equation $p = 2 + 0.002x - 0.0001x^2$.
    (a) Determine an algebraic representation that gives the total daily revenue of the lemonade stand as a function of x.
    (b) Draw a complete graph of the algebraic representation.
    (c) Draw the graph of the problem situation.
    (d) Determine the number of glasses of lemonade to be sold to produce maximum daily revenue. What is the maximum daily revenue?

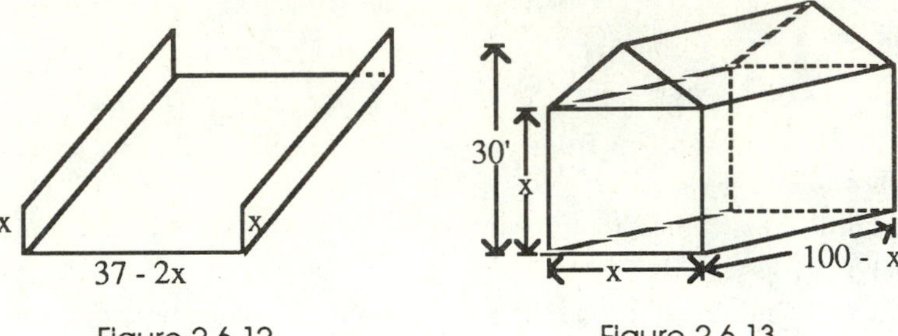

Figure 2.6.12

Figure 2.6.13

30. A rectangular sheet of metal 37 inches wide is to be made into a trough to feed hogs by turning up sides of equal length perpendicular to the sheet (Figure 2.6.12). Find the length of the turned-up sides that give the trough the maximum cross-sectional area. What is the maximum cross-sectional area?

31. A cardboard box with no lid is to be constructed from a 55 inch by 86 inch piece of flat cardboard by removing equal squares from each corner and folding up the sides. Find the dimensions of the box with maximum volume. What is the maximum volume?

32. An apartment building is built in the shape of a box with square cross section and with a triangular prism forming the roof as shown in Figure 2.6.13. The total height of the building is 30 feet and the sum of the length and width of the building is 100 feet.
    (a) Determine an algebraic representation of the volume V of the building as a function of x.
    (b) Draw a complete graph of the algebraic representation.
    (c) What values of x represent the problem situation? Draw a complete graph of the problem situation.
    (d) What are the dimensions of the building with maximum volume?
    (e) What is the maximum volume?

2.7 Increasing and Decreasing Functions

In Chapter 9, we see how polynomial functions of higher degrees are used to approximate values of complicated functions. Calculators and computers often use such approximations to calculate the values of their built-in functions. In this section we begin the formal study of polynomial functions of degree 3 or higher. We investigate the polynomial functions of degree 3 obtained by applying the geometric transformations of Sections 1.6 and 1.7 to the graph of $y = x^3$. We determine where functions are increasing and decreasing.

Polynomial functions of degree 3 or higher are more complicated than those of degree 1 or 2. For example, in Section 1.7 we found that every polynomial of degree 2 could be obtained from x^2 by appropriate use of reflection, vertical stretching or shrinking, horizontal shifting, and vertical shifting. It is *not* true that all polynomials of degree 3 can be obtained from x^3 using such techniques. A complete graph of $f(x) = x^3$ is given in Figure 2.7.1. This graph shows that the function $f(x) = x^3$ is an increasing function; that is, the function

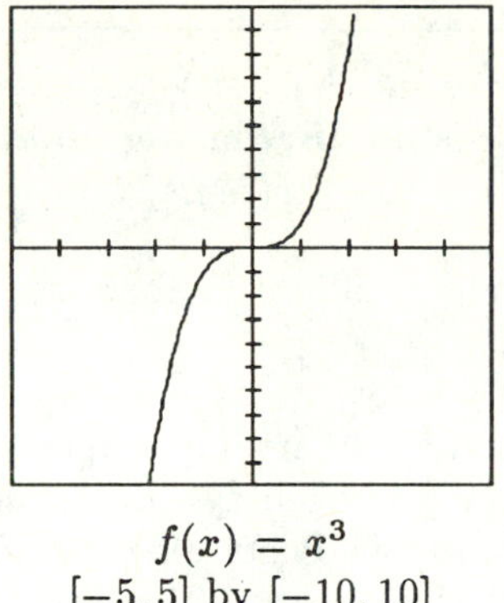

$f(x) = x^3$
$[-5, 5]$ by $[-10, 10]$
Figure 2.7.1

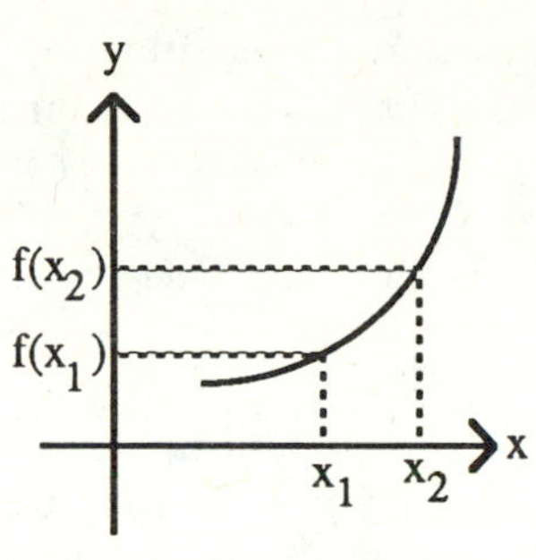

Figure 2.7.2

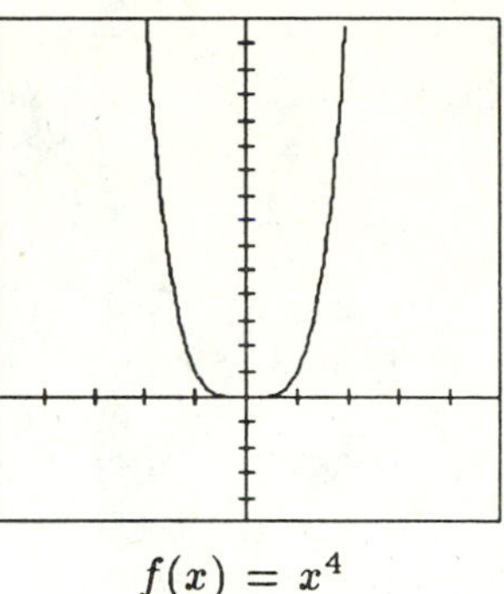

$f(x) = x^4$
$[-5, 5]$ by $[-5, 15]$
Figure 2.7.3

values $f(x)$ increase as the values of x increase. This means as we walk along the graph in the direction of increasing x, that is, from left to right, we climb. More precisely, we have the following definition.

• DEFINITION A function f is **increasing on an interval J** if for all x_1, x_2 in J with $x_2 > x_1$ we have $f(x_2) > f(x_1)$ (Figure 2.7.2). •

The function $f(x) = x^3$ is increasing on the interval $(-\infty, \infty)$. In fact, f is increasing on any interval.

A complete graph of $f(x) = x^4$ is given in Figure 2.7.3. This function is decreasing on the interval $(-\infty, 0]$ and increasing on the interval $[0, \infty)$.

• DEFINITION A function f is said to be **decreasing on an interval J** if for all x_1, x_2 in J with $x_2 > x_1$ we have $f(x_2) < f(x_1)$ (Figure 2.7.4). •

If the inequality signs between $f(x_1)$ and $f(x_2)$ in the previous two definitions are not strict inequalities, then we have the following definitions.

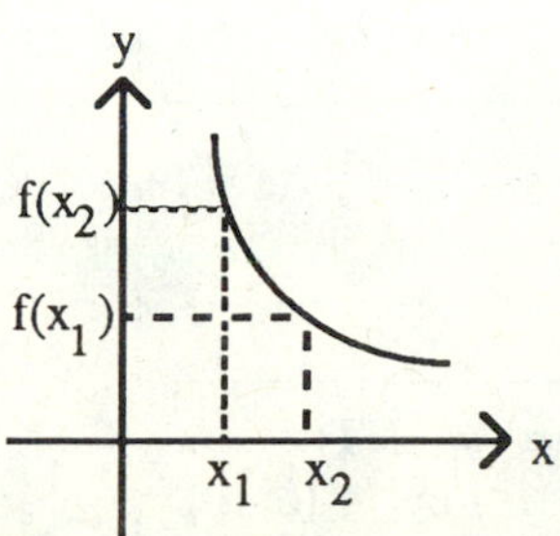

Figure 2.7.4

• DEFINITION A function f is **nondecreasing on an interval J** if for all x_1, x_2 in J with $x_2 > x_1$ we have $f(x_2) \geq f(x_1)$. A function f is **nonincreasing on an interval J** if for all x_1, x_2 in J with $x_2 > x_1$ we have $f(x_2) \leq f(x_1)$. •

• EXAMPLE 1: Let $f(x) = x^4$, and let J be any interval. What can be said about the increasing or decreasing behavior of f on J?

SOLUTION: Let a and b be the left and right endpoints of the interval J, respectively. Refer to Figure 2.7.3. If $b \leq 0$, then f is decreasing on J. If $a \geq 0$, then f is increasing on J. If $a < 0$ and $b > 0$, then f is neither increasing nor decreasing on J. For example, f is neither increasing nor decreasing on $[-1, 1]$. •

• EXAMPLE 2: Let $g(x) = \text{INT}(x)$ and let J be any interval. What can be said about the increasing or decreasing behavior of g on J?

SOLUTION: The graph of the step function $g(x) = \text{INT}(x)$ is given in Figure 2.7.5. We see from this figure that g is *nondecreasing* on every interval J. Furthermore, g is *not* an increasing function on any interval, nor is g a decreasing function on any interval. g is nonincreasing on every interval of the form $[n, n + 1)$, n an integer. •

In the previous section we found the coordinates of the points at which $f(x) = x^3 - 4x$ had local maximum and minimum values. This information can be used to determine the intervals on which f is increasing and the intervals on which f is decreasing.

• EXAMPLE 3: Determine the intervals on which $f(x) = x^3 - 4x$ is increasing and the intervals on which f is decreasing.

SOLUTION: In the previous section we showed that the graph of f in Figure 2.7.6 was complete. We also showed that the coordinates of the points at which f had local maximum and minimum values were $(-1.155, 3.0792)$ and $(1.155, -3.0792)$, respectively, with error of

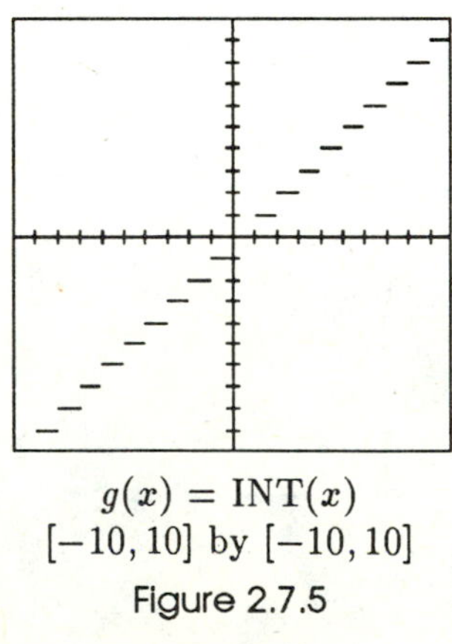

$g(x) = \text{INT}(x)$
$[-10, 10]$ by $[-10, 10]$
Figure 2.7.5

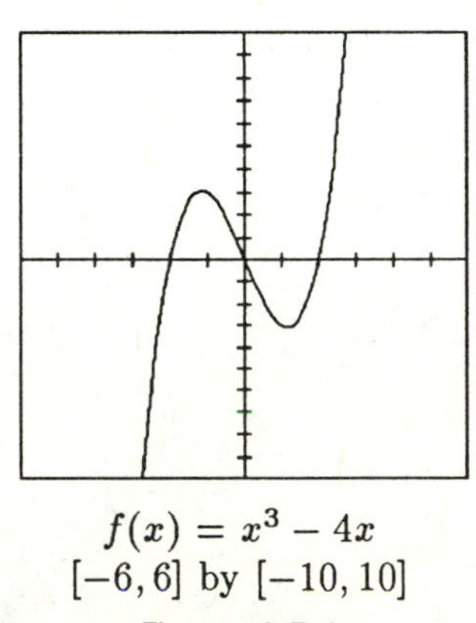

$f(x) = x^3 - 4x$
$[-6, 6]$ by $[-10, 10]$
Figure 2.7.6

at most 0.01. Therefore, f is increasing on the intervals $(-\infty, -1.155]$ and $[1.155, \infty)$ and is decreasing on $[-1.155, 1.155]$. •

• EXAMPLE 4: Draw a complete graph of each inequality.

(a) $y < x^3$ (b) $y \geq x^4$

SOLUTION: The graph of $y < x^3$ consists of all points below but not on the graph of $y = x^3$ (Figure 2.7.7). The graph of $y \geq x^4$ consists of all points above or on the graph of $y = x^4$ (Figure 2.7.8). •

• **Agreement** Unless specified otherwise, when we ask you to determine the intervals on which a function is increasing or decreasing we want the intervals to be as inclusive as possible.

The graph of a function can be used to find the intervals on which the function is increasing or decreasing. Often the endpoints of these intervals occur at points where the function has a local maximum or minimum value. In this section, we will be content to estimate the endpoints of these intervals. In the previous section we determined the endpoints with greater accuracy using zoom-in. In calculus, you will be able to learn more about the increasing and decreasing behavior of functions and how to determine the endpoints of intervals over which a function is increasing or decreasing.

• EXAMPLE 5: Estimate the x-coordinates of the points at which $f(x) = x^3 - 4x^2 - 4x + 16$ has local extremum values. Then determine the intervals on which f is increasing and on which f is decreasing.

SOLUTION: The graph of f in the $[-5, 5]$ by $[-10, 20]$ viewing rectangle is shown in Figure 2.7.9. We can use this graph to estimate that the x-coordinate of the point at which the graph of f has a local maximum value is -0.5. Similarly, the x-coordinate of the point at

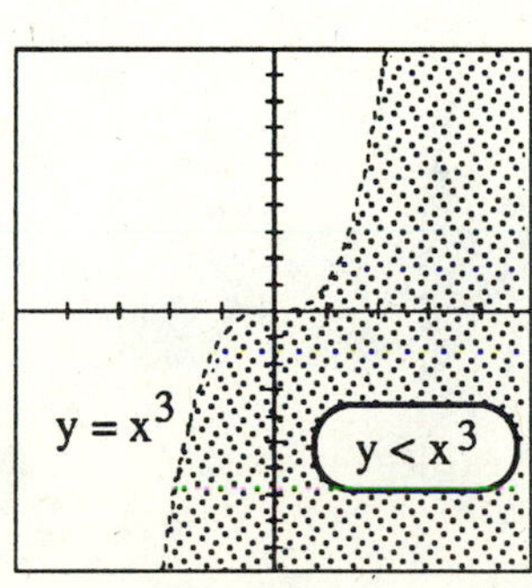

$[-5, 5]$ by $[-10, 10]$

Figure 2.7.7

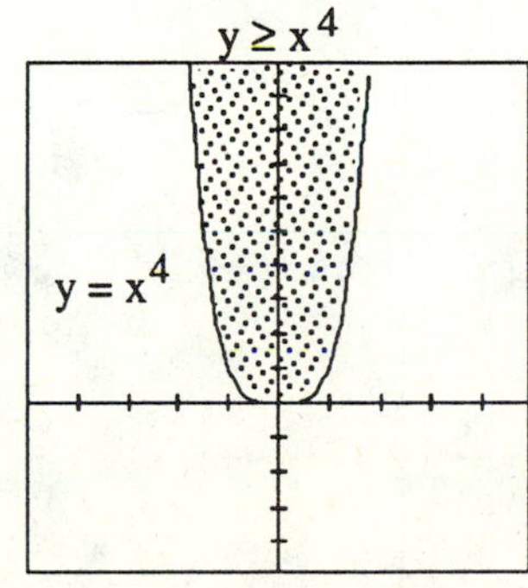

$[-5, 5]$ by $[-5, 10]$

Figure 2.7.8

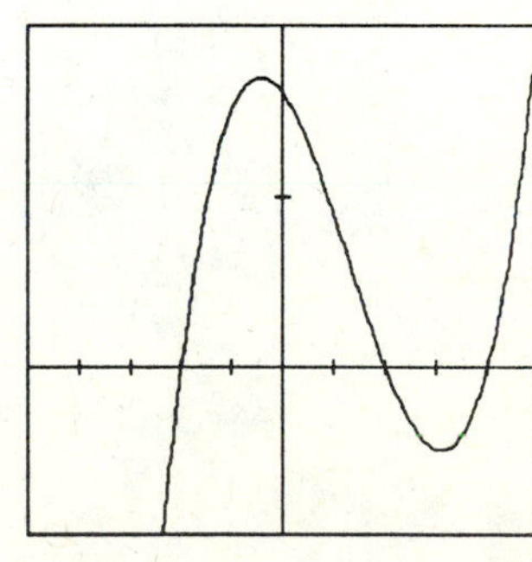

$f(x) = x^3 - 4x^2 - 4x + 16$

$[-5, 5]$ by $[-5, 10]$

Figure 2.7.9

which the graph of f has a local minimum value is 3. Then, f is increasing on $(-\infty, -0.5]$ and $[3, \infty)$ and decreasing on $[-0.5, 3]$. •

• EXAMPLE 6: Draw a complete graph of $y = -2(x+3)^3 + 1$ and determine the intervals on which the function is increasing or decreasing.

SOLUTION: The graph of this function can be obtained from the graph of $y = x^3$ by first vertically stretching by a factor of 2 to get $y = 2x^3$, then reflecting through the x-axis to get $y = -2x^3$, next shifting horizontally left 3 units to get $y = -2(x+3)^3$, and finally shifting vertically upward one unit to get $y = -2(x+3)^3 + 1$ (Figure 2.7.10). This can be confirmed by using a graphing utility to draw the graph of $y = -2(x+3)^3 + 1$ (Figure 2.7.11).

We can see from Figure 2.7.11 that $y = -2(x+3)^3 + 1$ is a decreasing function on $(-\infty, \infty)$. Alternatively, we can analyze the steps used in Figure 2.7.10 to see that $y = -2(x+3)^3 + 1$ is decreasing on $(-\infty, \infty)$. From Figure 2.7.10 we can see that both $y = x^3$ and $y = 2x^3$ are increasing on $(-\infty, \infty)$. However, the function $y = -2x^3$ that results after reflecting $y = 2x^3$ through the x-axis is decreasing on $(-\infty, \infty)$. Applying a horizontal shift 3 units left to $y = -2x^3$ followed by a vertical shift one unit up gives a congruent function with the same increasing and decreasing behavior as $y = -2x^3$. Thus, $y = -2(x+3)^3 + 1$ is also decreasing on $(-\infty, \infty)$. •

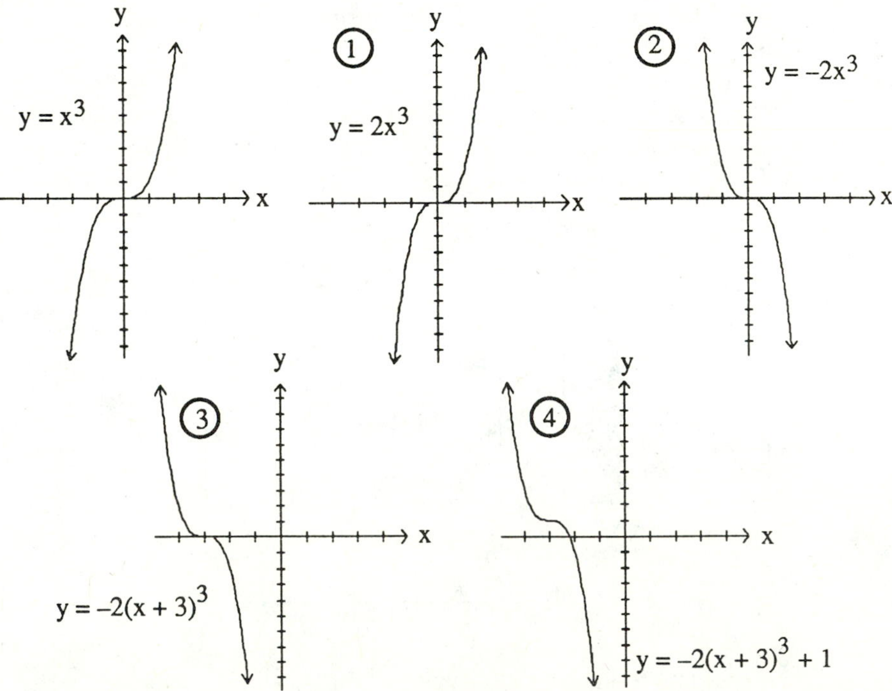

Figure 2.7.10

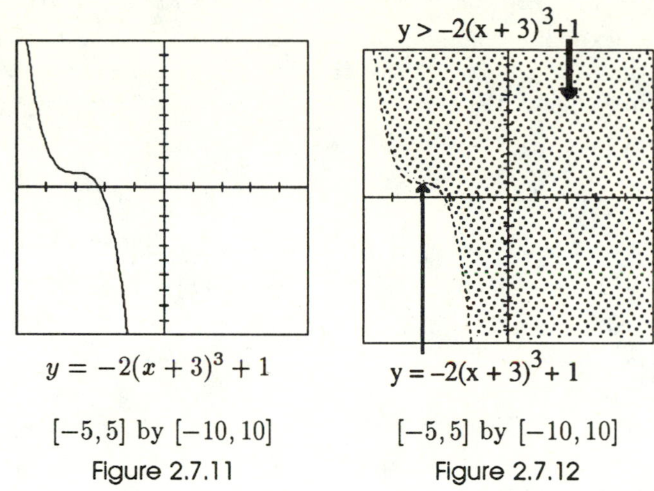

$y = -2(x+3)^3 + 1$

$[-5, 5]$ by $[-10, 10]$
Figure 2.7.11

$[-5, 5]$ by $[-10, 10]$
Figure 2.7.12

In Example 6 we used the geometric transformations needed to obtain the graph of $y = -2(x+3)^3+1$ from $y = x^3$ to reason that $y = -2(x+3)^3+1$ is decreasing on $(-\infty, \infty)$. This reasoning can be extended to obtain the following theorem.

- **THEOREM 1** The function $y = a(x-h)^3 + k$ is
  (a) increasing on $(-\infty, \infty)$ if $a > 0$. (b) decreasing on $(-\infty, \infty)$ if $a < 0$.

In Example 5 we showed that the polynomial function $f(x) = x^3 - 4x^2 - 4x + 16$ is neither increasing nor decreasing on $(-\infty, \infty)$. This fact together with Theorem 1 shows that $f(x) = x^3 - 4x^2 - 4x + 16$ *cannot* be obtained from $y = x^3$ using vertical stretching or shrinking, reflecting, horizontal shifting, or vertical shifting. Contrast this with the fact that every quadratic function can be obtained from $y = x^2$ by using these geometric transformations.

We can use the graph of $y = -2(x+3)^3 + 1$ to determine the graph that results if $=$ is replaced by $>$.

- **EXAMPLE 7:** Draw a complete graph of $y > -2(x+3)^3 + 1$.

SOLUTION: The graph of $y > -2(x+3)^3 + 1$ consists of all points above but not on the graph of $y = -2(x+3)^3 + 1$ (Figure 2.7.12). •

• EXERCISES 2-7

List the transformations that will produce the graph of each function from the graph of $y = x^3$. Specify an order in which the transformations could be applied and then sketch the graph of the function without using a graphing utility. Use a graphing utility to check your answer.

1. $y = x^3 - 4$
2. $y = (x-4)^3$
3. $y = 3(x+4)^3 - 5$
4. $y = -2(x-5)^3 + 3$

5. $y = -(x-2)^3 - 2$

6. $y = 2 + 3(x+4)^3$

Determine all local maximum and minimum values. Determine the intervals on which each function is *increasing* and the intervals on which each function is *decreasing*. *Warning:* There is *hidden* behavior in problems 11 and 12.

7. $g(x) = x(x-3)$

8. $P(x) = \dfrac{1}{2-x}$

9. $t(x) = x(x-2)(x+3)$

10. $g(x) = x^3 - 4x^2 + 4x$

11. $f(x) = x^3 - 2x^2 + x - 30$

12. $f(x) = 20x^3 + 2x^2 - 10$

13. $g(x) = \text{INT}(x-2)$

14. $T(x) = |x-5|$

Draw a complete graph of each inequality.

15. $y \le x^3 - 4x$

16. $y > \dfrac{1}{x-2}$

17. $y > |2x-1|$

18. $y \ge 2x^3 - 3x^2 + 2x - 3$

19. Draw the graph of $f(x) = 33x^3 - 100x^2 + 101x + 5$ in each viewing rectangle.
    (a) $[-10, 10]$ by $[-1000, 1000]$ (b) $[-1, 1]$ by $[-100, 100]$ (c) $[0.9, 1.1]$ by $[20, 50]$

20. The function f in Exercise 19 appears to be increasing on any interval. *It is not.* There is *hidden behavior* that can be determined by zooming in near the point $(1, f(1))$. Find a viewing rectangle that exhibits the *hidden behavior.*

21. It can be shown that $f(x) = x^3 + x + 1$ is an increasing function on any interval. Show that it is impossible to find real numbers a, b, and c so that $x^3 + x + 1 = a(x-b)^3 + c$.

Draw a complete graph. Determine all local maximum and minimum values. Determine all real zeros. Determine the intervals on which the function is increasing and the intervals on which the function is decreasing.

22. $f(x) = 2 - 3x + x^2 - x^3$

23. $f(x) = (x-1)^2 x^3$

24. $g(x) = 3x^4 - 5x^3 + 2x^2 - 3x + 6$

25. $T(x) = 2x^5 - 3x^4 + 2x^3 - 3x^2 + 7x - 4$

26. A rectangular area, with one side against an existing wall, is to be enclosed by three sides of fencing totaling 335 feet in length. Let x be the side length of the fence perpendicular to the existing wall.
    (a) Determine an algebraic representation of the enclosed area as a function of x.
    (b) Draw a complete graph of the algebraic representation.
    (c) What portion of the graph of the algebraic representation represents the problem situation?
    (d) Determine the intervals on which the algebraic representation is increasing or decreasing.

27. A 300-inch piece of wire is cut into two pieces. Each piece of wire is used to make a square. Let x be the length of one piece of the wire.
    (a) Determine an algebraic representation of the total area of the two squares as a function of x.
    (b) Draw a complete graph of the algebraic representation.
    (c) What portion of the graph of the algebraic representation represents the problem situation?
    (d) Determine the intervals on which the algebraic representation is increasing or decreasing.

28. An object is shot straight up from the top of a building 186 feet tall with initial velocity 80 feet per second. Let t be the number of seconds elapsed since the object was launched.
    (a) Determine an algebraic representation of the height of the object above ground level as a function of t.
    (b) Draw a complete graph of the algebraic representation.
    (c) What portion of the graph of the algebraic representation represents the problem situation?
    (d) Determine the intervals on which the algebraic representation is increasing or decreasing.
29. A box with no lid is made from a rectangular sheet of cardboard 50 inches by 85 inches by removing squares of equal side length from each corner and turning up the sides. Let x be the side length of the removed squares.
    (a) Determine an algebraic representation of the volume of the box as a function of x.
    (b) Draw a complete graph of the algebraic representation.
    (c) What portion of the graph of the algebraic representation represents the problem situation?
    (d) Determine the intervals on which the algebraic representation is increasing or decreasing.
30. A rectangular sheet of metal 20 inches long is to be made into a gutter by turning up sides of equal length perpendicular to the sheet. Let x be the height of the turned-up sides.
    (a) Determine an algebraic representation of the cross-sectional area of the gutter as a function of x.
    (b) Draw a complete graph of the algebraic representation.
    (c) What portion of the graph of the algebraic representation represents the problem situation?
    (d) Determine the intervals on which the algebraic representation is increasing or decreasing.
31. The total daily revenue of a lemonade stand at a state fair is given by the equation $R = x \cdot p$ where x is the number of glasses of lemonade sold daily and p is the price of one glass of lemonade. Assume the price of the lemonade is given by the supply equation $p = 1.2x - 0.01x^2$.
    (a) Determine an algebraic representation of the total daily revenue of the lemonade stand as a function of x.
    (b) Draw a complete graph of the algebraic representation.
    (c) Draw a complete graph of the problem situation.
    (d) Determine the intervals on which the algebraic representation is increasing or decreasing.

• KEY TERMS

Listed below are the key terms, vocabulary, and concepts in this chapter. You should be familiar with their definitions and meanings as well as be able to illustrate each of them.

Absolute value
Accuracy agreement
Coefficient
Completing the square
Constant term
Cubic equation

Decreasing function
Degree of a polynomial
Discriminant
Equivalent equations
Equivalent inequalities
Error
Greater than
Greater than or equal to
Increasing function
Inequality
Interval notation
Leading coefficient
Less than
Less than or equal to
Linear
Linear equation
Linear function
Linear inequality
Linear polynomial
Local maximum value
Local minimum value
Nondecreasing function
Nonincreasing function
Number of real solutions
Parabola
Polynomial
Polynomial equation
Polynomial function
Quadratic equation
Quadratic formula
Quadratic function
Quadratic polynomial
Sign chart
Simultaneous solutions
Solution to an equation
Solution to an inequality
Solution to a system of equations
Substitution method
x-intercept
Zoom-in

The Review Exercises contain representative problems from each of the sections in Chapter 2. You should use these problems to test your understanding of the material covered in this chapter.

• REVIEW EXERCISES

Use an *algebraic* method to solve each equation. Check your answers by substitution into the original equation.

1. $\frac{1}{3}x + 2 = 3x + \frac{1}{5}$
2. $3(2 - x) - 2(3x + 7) = x + 2$
3. $x^2 - 4x - 21 = 0$
4. $|2 - 3x| = 7$
5. $2x^4 - 4x^2 - 6 = 0$

Use an *algebraic* method to solve each equation. Check your answer with a graphing utility.

6. $x^2 = 6$
7. $x^2 - 4x + 7 = 0$

Solve each equation.

8. $3x^2 - 4x + 2 = 0$
9. $2x^3 - 4x + 7 = 0$
10. $4x^3 + 60x^2 - 103x - 65 = 0$

Without solving *or* graphing, find the number of real solutions to each equation.

11. $4x^2 - 6x + 3 = 0$
12. $x^2 - 4x + 4 = 0$

Use a graphing utility to determine the number of real solutions to each equation.

13. $2x^3 - 4x^2 + 3x - 9 = 0$
14. $x^3 - 31x + 2 = 0$

15. $2\cos x = 3x$

16. $x^3 - 5x = \dfrac{1}{x}$

Solve each system.

17. $y = |2x+5|$
$y = 3x^2 - 2x + 1$

18. $3x + 2y = 6$
$y = x^2 - 2x + 5$

19. $y = 16 - x^2$
$y = 9x - x^3$

20. $x^2 - 2y - 4x + 10 = 0$
$5x - 4y + 24 = 0$

21. $y = 115x - 3x^3$
$y = 50\cos x$

22. $y = 6 - x^2$
$y = 3\sin(x-4)$

Use a number line to graph the intervals.

23. $(-2, 4]$

24. $1 \le x < 7$

Complete the table.

	Interval Notation	**Inequality Notation**	**English Phrase**
25.	$[-4, 2)$		
26.		$x < 4$ or $15 \le x < 20$	
27.			The set of all real numbers between -4 and 1 or greater than 7

Consider the graphs of $y = f(x)$ and $y = g(x)$ in Figure R.1

28. Compare
(a) $f(1)$ and -3.
(b) $g(4)$ and 1.

29. Compare $f(a)$ and $g(a)$ if
(a) $a = -2$.
(b) $a = 2$.

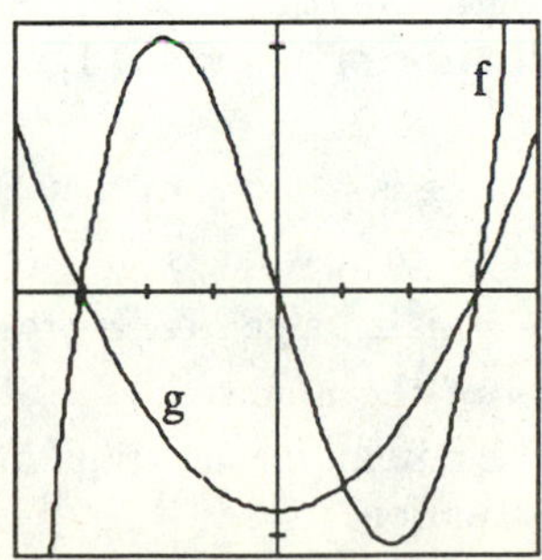

Figure R.1

30. Is $x = -1$ a solution to

(a) $f < g$? (b) $f = g$? (c) $f > g$?

(d) Use the graph to solve $f(x) > g(x)$.

Use an *algebraic* method to solve each inequality. Write your answers in interval notation. Check your answers with a graphing utility.

31. $3x + 5 \leq x - 4$

32. $\dfrac{x-1}{2} - \dfrac{2x+1}{5} > 1$

33. $\dfrac{1}{4} \leq \dfrac{4x-1}{12} < \dfrac{11}{6}$

34. $|3x - 2| \geq 7$

35. $5|x| + 3 < 2$

36. $\dfrac{3x+6}{|x+1|} \geq 0$

Use a sign chart to solve each inequality. Write the solution in interval notation. Check your answers with a graphing utility.

37. $(x-4)(x+1) \geq 0$

38. $x^2 - 5x + 6 \geq 0$

39. $x^2 + 20 < x$

40. $10 + x - 2x^2 < 0$

41. $|x-1|(2x-4) \leq 0$

42. $x(x+2)(2x-3) > 0$

43. $\dfrac{4x-8}{x+5} < 0$

44. $\dfrac{5-3x}{x-1} \geq 0$

Solve each inequality.

45. $3x^2 - 5x + 1 \geq 0$

46. $x^3 - 4x^2 - 2x + 3 < 0$

47. $2x^2 + x - 6 \leq x^3 - x^2 + 4x$

48. $3\sin x - 1 > 0$ for $0 \leq x \leq 10$

49. A storage box of height 25 cm has a volume of 10,500 cm^3. Let x be the length of the box in centimeters and y the width in centimeters.

(a) Write y as a function of x.

(b) Determine the length if the width is 12 cm.

50. The formula for the volume of a cone is $V = \frac{1}{3}\pi r^2 h$. Solve this equation for r.

51. A single-commodity open market is driven by the supply and demand principle. Economists have determined that supply curves are usually increasing (as the price increases, the sellers increase production). For the supply equation $p = 0.003x + 2$, find the price p if the production level is 2000 units.

52. Consider the equation $3x^3 + 8x^2 + 24x - 9 = 0$. It has one real solution. Find a sequence of four viewing rectangles containing the solution. The first viewing rectangle should permit the solution to be read with error of at most 0.1, the second 0.01, the third 0.001, and the last 0.0001.

53. The perimeter of some rectangles is 530 inches. Let x be the length of such a rectangle.

(a) If the length is 170 inches, then what is the area of the rectangle?

(b) Show that $A = x(265 - x)$ is an algebraic representation of the area of such rectangles.

(c) Draw a complete graph of the model.

(d) Verify that (100, 16500) is a point on the graph in (c). What meaning do these coordinates have in this problem situation?

(e) Verify that (−100, −36500) is a point on the graph in (c). What meaning do these coordinates have in this problem situation?

54. The owner of Christine's Crewel Craft Center pays \$1500 per month for fixed expenses such as rent, lights, and wages. Craft kits are sold at \$22 each and \$15 is required for material, floss, needle, and instructions for each kit. How many kits must be sold to break even?

55. A 16-inch by 20-inch frame is used to frame a picture. The picture will be surrounded by a mat with a uniform border. The distance from the edge of the frame to the picture is x inches on all sides.

 (a) Express the area A of the picture as a function of x.
 (b) Draw a complete graph of the function $y = A(x)$.
 (c) What are the domain and range of the function in (b)? What values of x make sense in this problem situation?
 (d) Use the graph in (b) to estimate the uniform border if the area of the picture is 250 square inches.

56. A publishing company prints x books each week at a total cost of $\frac{1}{2}(x - 100)^2 + 400$ dollars. The company receives \$15.25 for each book sold. Assume all books printed are sold.

 (a) Write an equation expressing the weekly profit of the company as a function of x.
 (b) What are the fixed costs?
 (c) Draw a complete graph of the profit function in (a).
 (d) What is the domain of the function in (a)? What portion of the graph in (c) is the graph of the problem situation?
 (e) Use the graph in (c) to determine what level of weekly production the company must maintain to break even.

57. Three hundred twenty-five tickets were sold for a movie. There were two ticket prices: adults at \$5.50, and children at \$3.00. How many tickets of each type were sold if the total proceeds from the sale of the tickets was \$1500?

58. Sandy can swim one mile upstream (against the current) in 20 minutes. She can swim the same distance downstream in 9 minutes. If both the swimming speed and current speed are constant, find Sandy's swimming speed and the speed of the current.

59. The total value of 23 coins consisting of pennies, nickels, and dimes is \$1.51. There are twice as many dimes as pennies. Determine the number of pennies, nickels, and dimes.

60. Chris has \$36 to spend on a pizza party for herself and her friends. Each pizza costs \$8.50 and serves two. Let x represent the number of friends Chris can invite.

 (a) Write an inequality that gives an algebraic representation for this problem situation.
 (b) What is the solution to the inequality?
 (c) How many friends can Chris invite?

61. The length of a certain rectangle is five inches greater than its width. Find the possible widths of the rectangle if its perimeter is less than 300 inches.

62. The annual profit P of a candy manufacturer is determined by the formula $P = R - C$, where R is the total revenue generated from selling x pounds of candy and C is the total cost of making and selling x pounds of candy. Each pound of candy sells for \$5.15 and costs \$1.26 to make. The

fixed costs of making and selling the candy are \$34,000 annually. How many pounds of candy must be produced and sold for the company to make a profit of at least \$42,000 for the year?

(a) Express the company's annual profit as a function of x.

(b) Write an inequality that gives an algebraic representation of this problem situation.

(c) Draw a complete graph of each side of the inequality in (b). Draw a graph of the problem situation.

(d) Solve this problem situation both algebraically and geometrically.

63. An object is propelled upward from a tower 200 feet above ground with initial velocity 80 feet per second.

(a) Determine an algebraic representation of the problem situation.

(b) Draw a complete graph of the algebraic representation.

(c) What are the domain and range of the algebraic representation?

(d) What part of the graph in (b) represents the problem situation?

(e) When will the object be above ground?

(f) When will the object hit the ground?

(g) What height above ground will the object reach?

(h) At what time will the object reach maximum height above ground?

64. A park with dimensions 125 feet by 230 feet is surrounded by a sidewalk of uniform width x. Find the possible widths of the sidewalk if the total area of the sidewalk is to be greater than 2900 square feet but less than 3900 square feet.

(a) Write an equation that gives the area of the sidewalk as a function of x.

(b) Write an inequality that is an algebraic representation of this problem situation.

(c) Solve the inequality in (b) algebraically, and then determine the solution to the problem situation.

(d) Use a graphing utility to check your answer to (c).

Determine the intervals on which each function is *increasing* and the intervals on which each function is *decreasing*. Determine all real zeros.

65. $g(x) = (x-2)(x-5)$

66. $f(x) = |x^2 - 5x + 3|$

67. $f(x) = x^3 - x^2 + 2x - 5$

68. $f(x) = 3 - 2x^3 + 2x^4$

69. List the transformations that will produce a complete graph of the function $y = (x+3)^4 + 2$ from the graph of $y = x^4$. Specify the order in which the transformations should be applied and then sketch a complete graph of the function without using a graphing utility. Use a graphing utility to check your answer.

Draw a complete graph of each polynomial function. Specify the viewing rectangle. Determine all real zeros.

70. $y = 5x^3 - 2x - 17$

71. $T(x) = 5 + 6x + 2x^3 - 2x^4$

Determine the intervals on which each function is *increasing* and the intervals on which each function is *decreasing*.

72. $y = 10 - 3x - x^2$

73. $f(x) = x^3 - 5x^2 + 2x - 4$

Determine all local maximum and minimum values, the corresponding values of x that give these values, and draw a complete graph.

74. $g(x) = |x^2 - 6x + 7|$

75. $y = 2x^3 + 4x^2 - 16$

76. $V(x) = x(22 - 2x)(30 - 2x)$

77. $f(x) = 2x^4 - 8x^2 + x - 3$

78. A dress store is marking all items down 20% for its "spring fling" sale.
   (a) Draw a complete graph that shows the relationship between the original price and the sale price.
   (b) If the sale price of an item is \$74.00, use a graph to determine its original price.

79. Ethanol worth \$0.75 per gallon is mixed with gasoline worth \$1.10 per gallon. Write an inequality expressing the requirement that the value of the new mixture cannot exceed \$0.999 (at a gas station this would be written as $0.99\frac{9}{10}$) per gallon. Draw its complete graph.

80. A fast-food restaurant makes \$0.40 profit on a medium soft drink, \$0.15 profit on a taco, and breaks even on the other items. If the weekly overhead is \$700, write an inequality that expresses the requirement that the weekly profits must exceed \$500, and draw its complete graph.

81. A rectangle is 2 feet longer than it is wide. If each side is increased by 2 feet, the area of the new rectangle is 288 square feet. Find the dimensions of the original rectangle.

82. A piece of copper tubing 8 feet long is cut into two pieces so that the sum of the squares of the length of each piece is 44.5 square feet. Find the length of each piece.

83. An object is shot straight up (launched) from the top of a building 200 feet tall with initial velocity 70 feet per second. Let t be the number of seconds since the object was launched.
   (a) Determine an algebraic representation that gives the height of the object above ground level as a function of t.
   (b) Draw a complete graph of the algebraic representation.
   (c) What portion of the graph of the algebraic representation represents the problem situation?
   (d) Determine the time when the object is 225 feet above the ground.
   (e) Determine the time when the object is more than 225 feet above the ground.

84. A cardboard box with no lid is constructed from a 40 inch by 55 inch piece of cardboard by removing equal squares of side length x from each corner and folding up the sides. Determine x so that the volume of the resulting box is at least 6750 cubic inches.

85. The total daily revenue of a cotton candy stand at a state fair is given by the equation $R = x \cdot p$ where x is the number of bags of cotton candy sold daily and p is the price of one bag of cotton candy. Assume the price of the cotton candy is given by the "supply" equation $p = 1.5x - 0.03x^2$.
   (a) Determine an algebraic representation that gives the total daily revenue of the cotton candy stand as a function of x.
   (b) Draw a complete graph of the algebraic representation.
   (c) Draw a graph of the problem situation.
   (d) Determine the number of bags of cotton candy to be sold to produce maximum daily revenue. What is the maximum daily revenue?

86. A large apartment rental company has 3000 units available, and 2643 are currently rented at \$500 per month. A market survey indicates that each \$20 decrease in monthly rent will result in ten new tenants.
   (a) Determine a function R which represents the total rental income realized by the company where x is the number of \$20 decreases in monthly rent (i.e., if $x = 3$, then the monthly rent is \$440). Assume that all rents are adjusted to the new figure.
   (b) Draw a complete graph of the function $y = R(x)$.
   (c) Determine the rent that will yield the rental company the maximum monthly income.

3 

Polynomials, Rational Functions, and Radical Functions

• Introduction

Several important concepts about functions are introduced in this chapter. We use the continuity property of polynomial functions to help study their graphs. The end behavior, or behavior of a polynomial function for x large in absolute value, is determined and used to help establish its complete graph. The concepts introduced in this chapter provide important numerical and geometric intuition that foreshadows the study of calculus, advanced mathematics, and science.

Continuous functions on closed intervals $[a, b]$ possess the important Intermediate Value Property. We made extensive, intuitive use of this property in previous chapters. This property is central to graphical problem solving. In fact, graphing utilities use this important property to draw graphs.

We begin the study of complex numbers in this chapter. Both real and complex zeros of polynomial functions are found. Extending the kinds of numbers to include complex numbers completes our study of zeros of polynomial functions; no other types of numbers are needed. We restrict our attention to real zeros and to polynomials with real number coefficients through Section 3.3. However, some of the theorems and processes introduced in these sections are valid for complex number zeros and for polynomials with complex number coefficients.

The Fundamental Theorem of Algebra is introduced and used to determine the number of zeros of a polynomial. Two important classes of functions called rational functions and radical functions are studied in this chapter. We determine the horizontal asymptotes, vertical asymptotes, and end behavior of rational functions. Rough sketches of radical functions are obtained by applying geometric transformations to the basic radical function $y = \sqrt[n]{x}$.

Equations and inequalities involving radical functions are solved both graphically and algebraically. Finally, applications involving rational functions and radical functions are studied.

3.1 Continuity and End Behavior

In this section we introduce the concepts of continuity and end behavior of functions. An important consequence of continuity called the Intermediate Value Property is studied. We have made intuitive use of the Intermediate Value Property in earlier sections to find real zeros as well as other values of functions.

It is not an accident that graphs of polynomial functions obtained so far have no breaks in them. Some functions have graphs with breaks in them. For example, the graph of $f(x) = \frac{1}{x}$ obtained in Chapter 1 has a break at $x = 0$ (Figure 3.1.1). In calculus, you will be given a precise pointwise definition of continuity. In this textbook we will be content with an intuitive idea of the meaning of continuity. We begin by giving a rough idea of what we mean by continuity on an interval.

- Understanding A function is said to be **continuous on the interval J** if, for all a and b in J, it is possible to trace the graph of the function between $x = a$ and $x = b$ with a pencil without lifting the pencil from the paper.

For example, $f(x) = \frac{1}{x}$ is continuous on the intervals $(0, \infty)$ and $(-\infty, 0)$, but it is *not* continuous on $(-\infty, \infty)$. In fact, f is not continuous on any interval containing zero. However, it is common to say that $f(x) = \frac{1}{x}$ is continuous on its domain $D = (-\infty, 0) \cup (0, \infty)$. Notice that the above understanding about continuity does not hold for $f(x) = \frac{1}{x}$ on D, but does on each interval that makes up D. In calculus, you may find that there are conventions about continuity that seem to violate our above understanding.

We give the following theorem without proof.

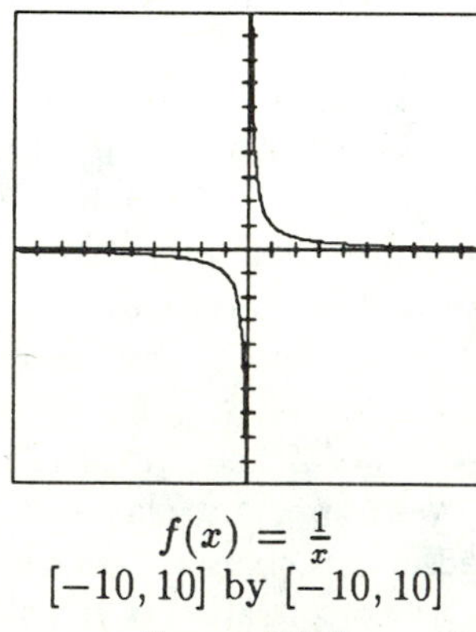

$f(x) = \frac{1}{x}$
$[-10, 10]$ by $[-10, 10]$
Figure 3.1.1

• THEOREM 1 Every polynomial function is continuous on $(-\infty, \infty)$.

In light of the above theorem we make the following definition.

• DEFINITION A function is said to be continuous if and only if it is continuous on the interval $(-\infty, \infty)$. •

Thus, we say that polynomial functions are continuous functions. In the next example we consider a function with breaks that are different than the break in $f(x) = \frac{1}{x}$ at $x = 0$.

• EXAMPLE 1: Determine the largest intervals on which the function $f(x) = \text{INT}(x)$ is continuous.

SOLUTION: Figure 3.1.2 shows the graph of $f(x) = \text{INT}(x)$ obtained in Section 1.4. Examine the graph on the interval $[3, 4)$. The open circle at the point $(3, 2)$ indicates that $f(3) \neq 2$, and the closed circle at the point $(3, 3)$ indicates that $f(3) = 3$. For all values of x in the interval $[3, 4)$, $f(x) = 3$. Similarly, for all values of x in the interval $[n, n + 1)$ where n is an integer, $f(x) = n$. We could illustrate that $f(n) = n$ and $f(n) \neq n - 1$ by using a closed circle at the point (n, n) and an open circle at the point $(n, n - 1)$. Several are illustrated in Figure 3.1.2.

The graph has a break at each integer value of x. The function $f(x) = \text{INT}(x)$ is continuous on each interval $[n, n+1)$ where n is an integer. On any interval that contains an integer as an *interior* point the function is not continuous. For example, f is not continuous on the interval $[-2.1, -1.5]$ or on the interval $[0, 2.5)$. Thus, the largest intervals on which f is continuous are of the form $[n, n + 1)$, where n is an integer. •

Notice that the union of the intervals on which $f(x) = \text{INT}(x)$ is continuous is $(-\infty, \infty)$, the domain of f, but even in calculus we do *not* consider this function continuous on $(-\infty, \infty)$.

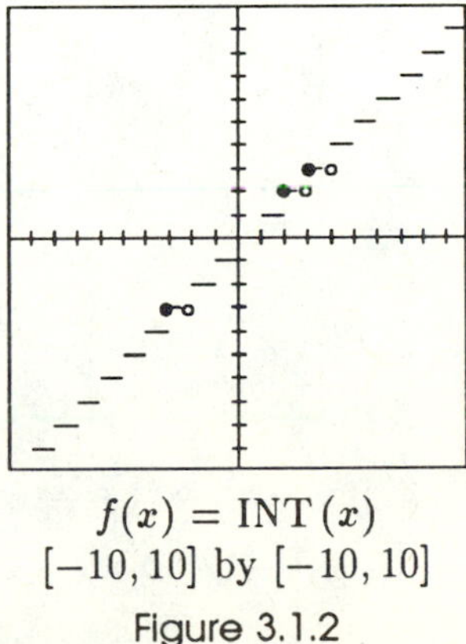

$f(x) = \text{INT}(x)$
$[-10, 10]$ by $[-10, 10]$
Figure 3.1.2

The function $y = \frac{1}{x}$ is not defined at $x = 0$, and its graph has a break in it at $x = 0$. We say that $y = \frac{1}{x}$ has a *point of discontinuity* at $x = 0$. This discontinuity is sometimes called an "infinite" discontinuity because of the behavior of the function near $x = 0$.

The graph of $y = \text{INT}(x)$ has a break at every integer value of x. These points are also called points of discontinuity. This type of discontinuity is sometimes called a "jump" discontinuity because of the type of break at integer values of x.

There is another type of point of discontinuity that will be studied in more detail in Section 3.6. This type of discontinuity is illustrated by the following piecewise defined function whose graph appears in Figure 3.1.3:

$$f(x) = \begin{cases} \frac{x^2-1}{x-1}, & \text{for } x \neq 1, \\ 3, & \text{for } x = 1. \end{cases}$$

The graph of f appears to have a hole punched out at $(1, 2)$ and the point moved up to $(1, 3)$. This is also a point of discontinuity and is sometimes called a "removable" discontinuity because the break in the graph can be repaired by replacing the missing point.

We have the following understanding about breaks in graphs of functions in this textbook.

- Understanding The function f has a **point of discontinuity** or is **discontinuous** at $x = a$ if one of the following three conditions holds:

  1. The function is not defined at $x = a$. For example, $f(x) = \frac{1}{x}$ at $x = 0$ (Figure 3.1.1).
  2. The graph of the function has a break at $x = a$. For example, $f(x) = \text{INT}(x)$ at any integer value of x (Figure 3.1.2).
  3. The graph of the function has a hole in it at $x = a$. For example,

  $$f(x) = \begin{cases} \frac{x^2-1}{x-1}, & \text{for } x \neq 1, \\ 3, & \text{for } x = 1, \end{cases}$$

  at $x = 1$ (Figure 3.1.3).

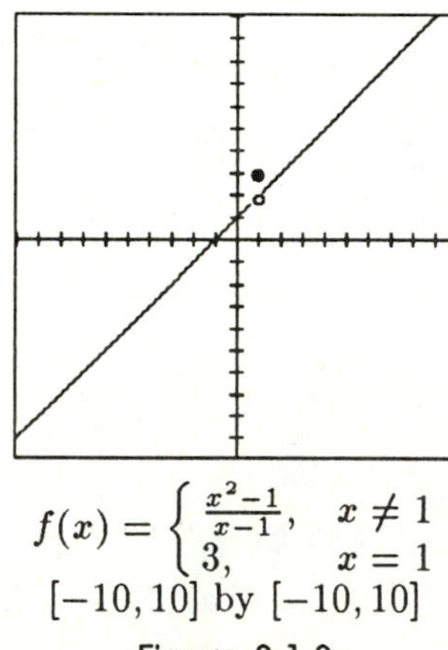

$$f(x) = \begin{cases} \frac{x^2-1}{x-1}, & x \neq 1 \\ 3, & x = 1 \end{cases}$$

$[-10, 10]$ by $[-10, 10]$

Figure 3.1.3

With the above understanding, $y = \frac{1}{x}$ has exactly one point of discontinuity, and it is at $x = 0$. Similarly, the function whose graph appears in Figure 3.1.3 has only one point of discontinuity, and it is at $x = 2$. However, $y = \text{INT}(x)$ has a point of discontinuity at and only at each integer value of x.

Piecewise-defined functions can have breaks similar to the greatest integer function.

• EXAMPLE 2: Determine the domain, the range, the points of discontinuity, and the largest intervals on which each function is continuous.

(a) $f(x) = \begin{cases} 2x+4 & \text{for } x \leq -1 \\ x^2 & \text{for } x > -1 \end{cases}$

(b) $g(x) = \sqrt{x+1}$

SOLUTION:

(a) Figure 3.1.4 gives a complete graph of f. The domain and the range are the set of all real numbers. The graph has a break, or jump, at $x = -1$. Notice f is continuous on the intervals $(-\infty, -1]$ and $(-1, \infty)$, and f is *discontinuous* at $x = -1$.

(b) Figure 3.1.5 gives a complete graph of g. The domain of g is $[-1, \infty)$, and the range of g is $[0, \infty)$. Notice that g is continuous on the interval $[-1, \infty)$, and g is discontinuous at each value of x less than -1, because g is not defined at those values. We leave the issue of continuity or discontinuity at endpoints of intervals like $x = -1$ to calculus. In this textbook, we will *not* say that g has a point of discontinuity at $x = -1$. •

By Theorem 1, polynomial functions are continuous on the set of all real numbers, that is, on $(-\infty, \infty)$. Thus, by the definition following Theorem 1, polynomial functions are called continuous functions. For example, linear functions $f(x) = ax + b$ and quadratic functions $g(x) = ax^2 + bx + c$ are continuous functions. One important consequence of continuity is the Intermediate Value Property.

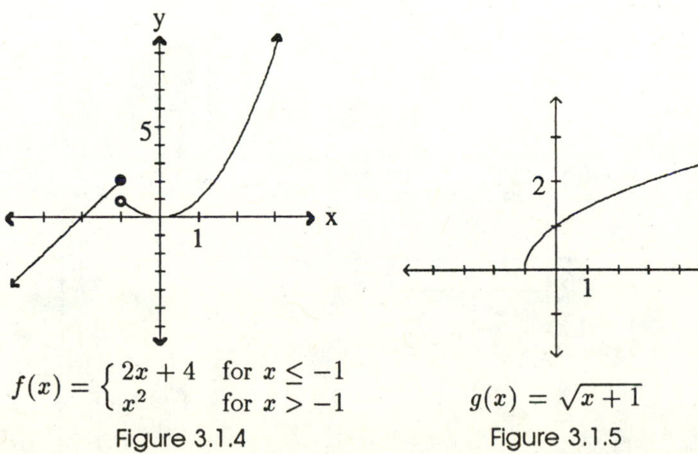

$f(x) = \begin{cases} 2x+4 & \text{for } x \leq -1 \\ x^2 & \text{for } x > -1 \end{cases}$

Figure 3.1.4

$g(x) = \sqrt{x+1}$

Figure 3.1.5

- **Intermediate Value Property** If a function f is continuous on $[a, b]$, then f assumes every value between $f(a)$ and $f(b)$. In other words, if L is any number between $f(a)$ and $f(b)$, then there is a value c between a and b such that $f(c) = L$ (Figure 3.1.6).

- EXAMPLE 3: Let $f(x) = 3x - 4$. Determine c on $[-10, 10]$ so that $f(c) = 15$.

SOLUTION: Notice that $f(-10) = -34$, $f(10) = 26$, and f is continuous on $[-10, 10]$. Furthermore, $-34 < 15 < 26$. Thus the Intermediate Value Property assures us that there must be a c satisfying $-10 \leq c \leq 10$ with $f(c) = 15$. We find c by solving the following equation:

$$\begin{aligned} f(c) &= 15 \\ 3c - 4 &= 15 \\ 3c &= 19 \\ c &= \frac{19}{3}. \end{aligned}$$

Thus $\frac{19}{3}$ is a number between -10 and 10 with $f(\frac{19}{3}) = 15$ (Figure 3.1.7). •

- EXAMPLE 4: Let $f(x) = x^2 + 3x - 4$. Determine c on $[-2, 3]$ so that $f(c) = -3$.

SOLUTION: Notice that $f(-2) = -6$, $f(3) = 14$, and f is continuous on $[-2, 3]$. Furthermore, $-6 < -3 < 14$. Thus, the Intermediate Value Property assures us that there is a c between -2 and 3 for which $f(c) = -3$. To determine c we solve the following quadratic equation:

$$\begin{aligned} f(c) &= -3 \\ c^2 + 3c - 4 &= -3 \\ c^2 + 3c - 1 &= 0. \end{aligned}$$

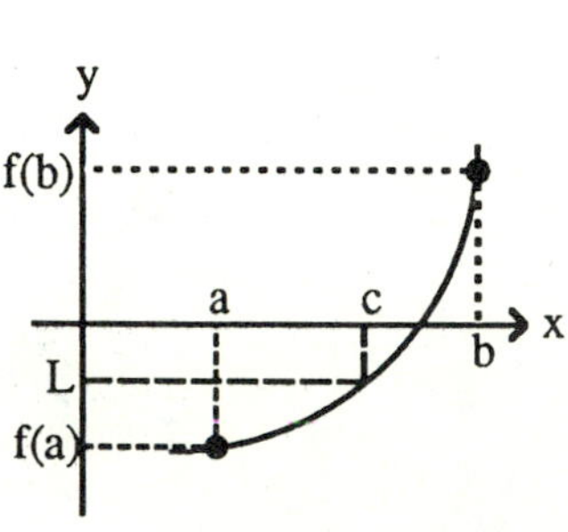

Figure 3.1.6

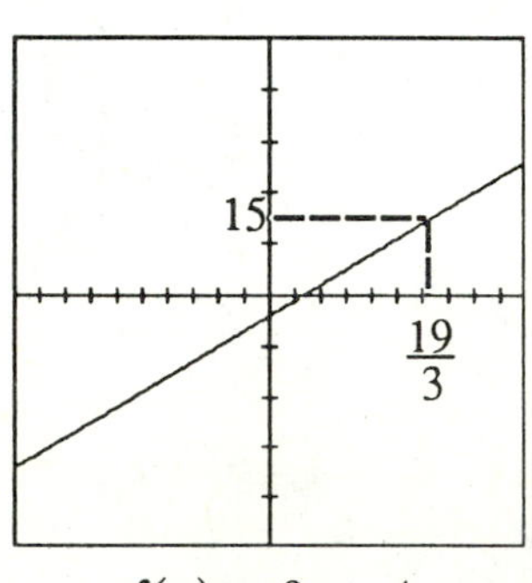

$f(x) = 3x - 4$
$[-10, 10]$ by $[-50, 50]$

Figure 3.1.7

The quadratic formula gives $c = \frac{-3\pm\sqrt{13}}{2}$. Accurate to tenths, the two possible values for c are -3.3 and 0.3. Only 0.3 is between -2 and 3. Thus, $\frac{-3+\sqrt{13}}{2}$ is a number between -2 and 3 with $f\left(\frac{-3+\sqrt{13}}{2}\right) = -3$ (Figure 3.1.8). •

The Intermediate Value Property is often the basis for approximating zeros of continuous functions. For example, consider the function $f(x) = x^5 + 2x - 1$. Because f is continuous on $(-\infty, \infty)$, it is continuous on every interval. In particular, f is continuous on $[0, 1]$, and you can check that $f(0) = -1$ and $f(1) = 2$. Applying the Intermediate Value Property to f on $[0, 1]$ we can conclude that $f(c) = 0$ for some c between 0 and 1, because 0 is between -1 and 2 (Figure 3.1.9). Next, we compute the value of f at the midpoint 0.5 of $[0, 1]$. Because $f(0.5)$ is positive, the Intermediate Value Property applied to the interval $[0, 0.5]$ shows that f has a zero between 0 and 0.5 (Figure 3.1.10). If we compute the value of f at the midpoint 0.25 of $[0, 0.5]$, we will find that this zero of f lies in $[0.25, 0.5]$. By repeated application of the Intermediate Value Property, we can find this zero as accurately as we like. This process, called the **bisection method**, is a numerical zoom-in procedure.

Computer and graphing utility techniques to find zeros and local maximum and minimum values of functions are based on the Intermediate Value Property. We have seen how to find zeros and local maximum and minimum values of functions using zoom-in. Before the easy availability of technology it was necessary to do much of this work by brute force. The power of modern technology allows today's precalculus students to investigate functions that were not accessible to precalculus students in the past.

Graphs of $y = x^n$

• EXAMPLE 5: Draw and compare complete graphs of $f(x) = x^2$ and $g(x) = x^4$ on the same coordinate system. Find the coordinates of all points of intersection of these two graphs.

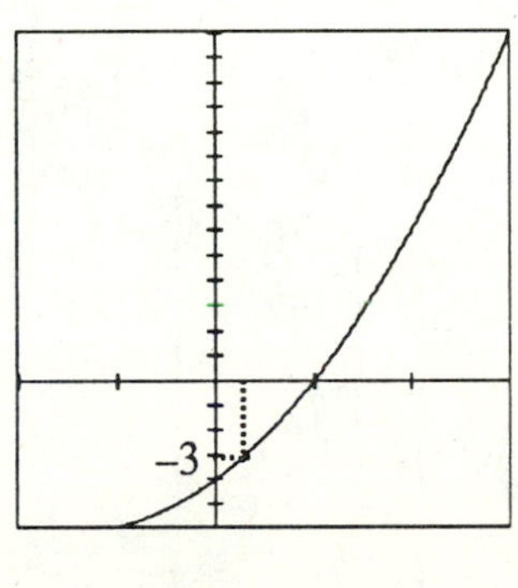

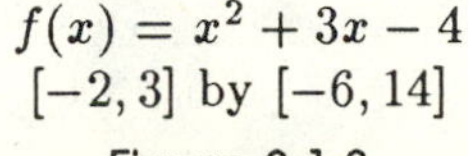

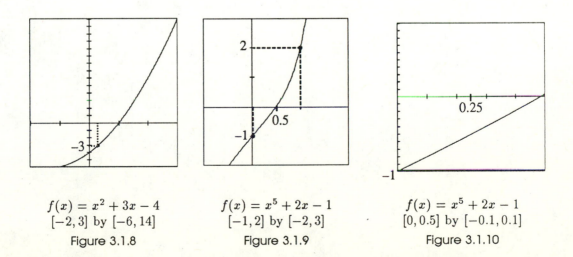

$f(x) = x^2 + 3x - 4$
$[-2, 3]$ by $[-6, 14]$

Figure 3.1.8

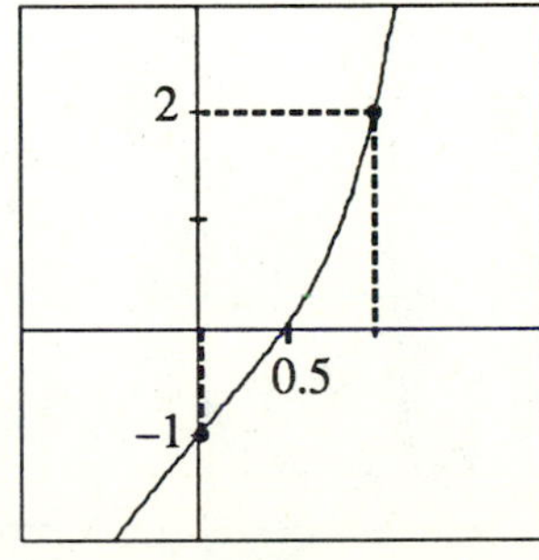

$f(x) = x^5 + 2x - 1$
$[-1, 2]$ by $[-2, 3]$

Figure 3.1.9

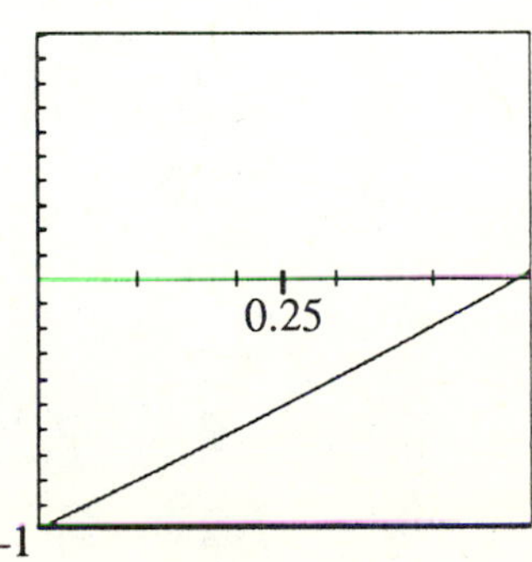

$f(x) = x^5 + 2x - 1$
$[0, 0.5]$ by $[-0.1, 0.1]$

Figure 3.1.10

SOLUTION: The graphs of f and g in the viewing rectangle $[-10,10]$ by $[-50,100]$ are given in Figure 3.1.11. This figure suggests that the graph of $g(x) = x^4$ is *above* the graph of $f(x) = x^2$ in most of $-10 \le x \le 10$. However, the behavior near $x = 0$ is not clear from this graph. Figure 3.1.12 gives the graph of these two functions in the $[-2,2]$ by $[-1,2]$ viewing rectangle.

Now we can see that the graphs of f and g intersect at the points $(-1,1)$, $(0,0)$, and $(1,1)$. This can be confirmed by substitution into the two equations, or by finding the simultaneous solutions to the pair of equations $y = x^2$ and $y = x^4$ using substitution.

$$x^4 = x^2$$
$$x^4 - x^2 = 0$$
$$x^2(x^2 - 1) = 0$$
$$x^2(x-1)(x+1) = 0$$

Thus, $x = -1$, 0, or 1, and the graphs intersect at the points $(-1,1)$, $(0,0)$, and $(1,1)$.

The graph of $f(x) = x^2$ lies *above* the graph of $g(x) = x^4$ in the intervals $(-1,0)$ and $(0,1)$, and the graph of $f(x) = x^2$ lies *below* the graph of $g(x) = x^4$ in the intervals $(-\infty,-1)$ and $(1,\infty)$. •

The graphs of $y = x^n$ for even integers n look alike and are related as suggested by the above comparison between $y = x^2$ and $y = x^4$. The graph of $y = x^n$ for n even is decreasing on $(-\infty,0]$ and increasing on $[0,\infty)$. (Why?)

Next, we investigate the graphs of $y = x^n$ for n odd.

• EXAMPLE 6: Draw and compare complete graphs of $f(x) = x^3$ and $g(x) = x^5$ on the same coordinate system. Find the coordinates of all points of intersection of the two graphs.

SOLUTION: The graphs of the two functions f and g in the viewing rectangles $[-10,10]$ by $[-1000,1000]$ and $[-1.5,1.5]$ by $[-1.5,1.5]$ given in Figures 3.1.13 and 3.1.14, respectively, are complete. The graphs of f and g intersect at the points $(-1,-1)$, $(0,0)$, and $(1,1)$.

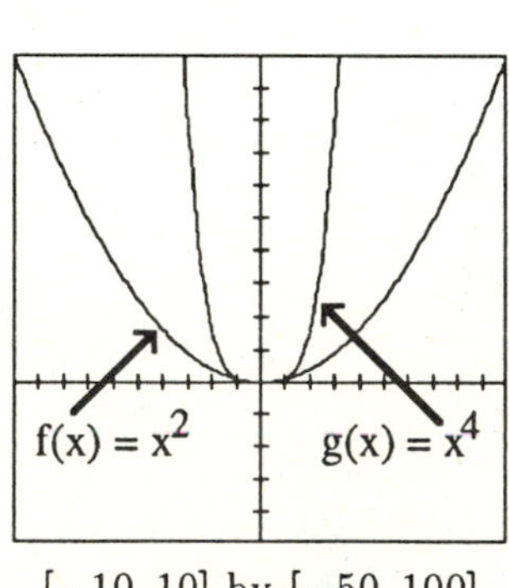

$[-10,10]$ by $[-50,100]$

Figure 3.1.11

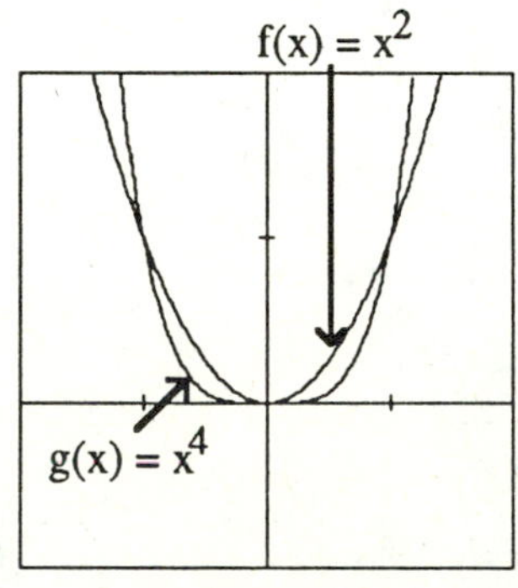

$[-2,2]$ by $[-1,2]$

Figure 3.1.12

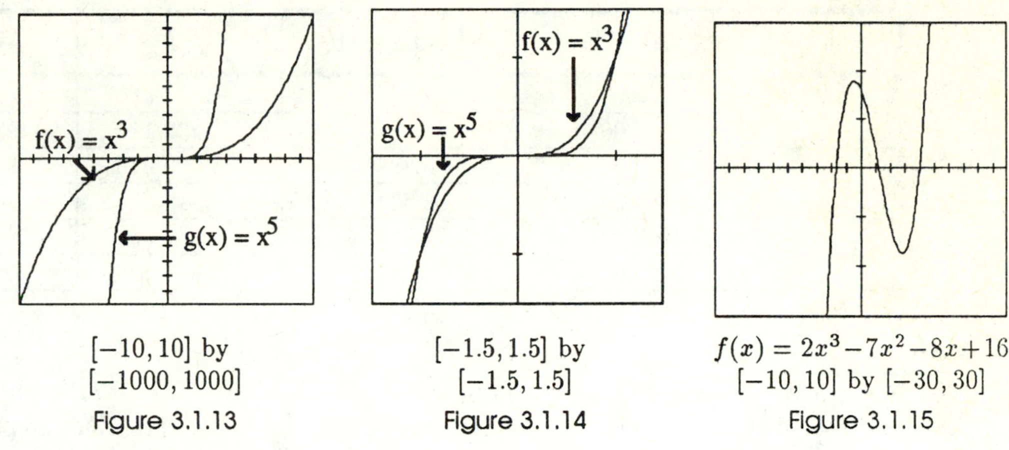

[−10, 10] by [−1000, 1000]
Figure 3.1.13

[−1.5, 1.5] by [−1.5, 1.5]
Figure 3.1.14

$f(x) = 2x^3 - 7x^2 - 8x + 16$
[−10, 10] by [−30, 30]
Figure 3.1.15

Furthermore, the graph of $f(x) = x^3$ is *above* the graph of $g(x) = x^5$ in the intervals $(-\infty, -1)$ and $(0, 1)$, and the graph of $f(x) = x^3$ is *below* the graph of $g(x) = x^5$ in the intervals $(-1, 0)$ and $(1, \infty)$. •

The graphs of $y = x^n$ for odd integers $n \geq 3$ look alike and are related as suggested by the above comparison between $y = x^3$ and $y = x^5$. The graph of $y = x^n$ for n odd is increasing on $(-\infty, \infty)$. (Why?)

End Behavior Using Zoom-Out

• DEFINITION The behavior of a function for values of x large in absolute value is called the **end behavior** of the function. •

For purposes of drawing graphs and understanding the general behavior of functions, it is often enough to know that the end behavior of a given function is very much like the end behavior of some simpler function.

• EXAMPLE 7: Show that the end behavior of $f(x) = 2x^3 - 7x^2 - 8x + 16$ is the same as the end behavior of $g(x) = 2x^3$.

SOLUTION: The graph of f in the viewing rectangle $[-10, 10]$ by $[-30, 30]$ is a complete graph and suggests the behavior of f for values of x large in absolute value (Figure 3.1.15). It appears the values of f are positive and increasing if the values of x are greater than or equal to 4 and increasing. We say $f(x)$ approaches infinity as x approaches infinity, and we write $f(x) \to \infty$ as $x \to \infty$. Figure 3.1.15 also suggests the values of f are negative and decreasing if the values of x are less than or equal to -2 and decreasing. In this case we say $f(x)$ approaches negative infinity as x approaches negative infinity, and we write $f(x) \to -\infty$ as $x \to -\infty$.

x	$2x^3$	$-7x^2 - 8x + 16$	$f(x)$	$\frac{f(x)}{2x^3}$
20	16,000	−2,944	13,056	0.816
50	250,000	−17,884	232,116	0.928
100	2,000,000	−70,884	1,929,216	0.965
−20	−16,000	−2,624	−18,624	1.164
−50	−250,000	−17,084	−267,084	1.068
−100	−2,000,000	−69,184	−2,069,184	1.035

Table 1

Table 1 demonstrates that the end behavior of $f(x) = 2x^3 - 7x^2 - 8x + 16$ is similar to the end behavior of $y = 2x^3$. Compare the values of $2x^3$ and $f(x)$ for x equal to 20, 50, 100, −20, −50, and −100.

Notice as the absolute value of x increases, the values in the third column of Table 1 become negligible in comparison to the values in the second column. This means we can approximate the value of f by the value of $2x^3$ when x is far away from zero. Another way to arrive at the same conclusion is to divide $f(x)$ by $2x^3$.

$$\frac{f(x)}{2x^3} = 1 - \frac{7}{2x} - \frac{4}{x^2} + \frac{8}{x^3}$$

Now, we can see that $\frac{f(x)}{2x^3}$ gets closer and closer to 1 as the absolute value of x becomes larger and larger, because each term of the right-hand side of the above equation after the first term approaches zero as the absolute value of x approaches infinity. We can also say $\frac{f(x)}{2x^3} \to 1$ as $|x| \to \infty$. The last column of Table 1 gives the value of $\frac{f(x)}{2x^3}$ for some values of x. Thus, $f(x)$ is approximately equal to $2x^3$ for values of x large in absolute value. •

We describe the *end behavior* of the function $f(x) = 2x^3 - 7x^2 - 8x + 16$ of Example 5 as follows: $f(x) \to \infty$ as $x \to \infty$ and $f(x) \to -\infty$ as $x \to -\infty$. We call $2x^3$ an *end behavior model* of f. It can be shown that $2x^2 + 3$ is another end behavior model of f because $\frac{f(x)}{2x^2+3} \to 1$ as $|x| \to \infty$. Thus, end behavior model of a function is not unique. We try to choose an end behavior model that is as simple as possible.

• DEFINITION The function g is an **end behavior model** for the function f if and only if

$$\frac{f(x)}{g(x)} \to 1 \quad \text{as} \quad |x| \to \infty.$$ •

Information about the end behavior of a function can be obtained by drawing its graph in very large viewing rectangles. For example, the graph of $f(x) = 2x^3 - 7x^2 - 8x + 16$ in the viewing rectangle $[-30, 30]$ by $[-3000, 3000]$ is shown in Figure 3.1.16. This graph appears to be much like the graph of $y = 2x^3$ in the same viewing rectangle (Figure 3.1.17).

The graphs in Figures 3.1.16 and 3.1.17 strongly suggest that $y = 2x^3$ is an end behavior model of $f(x) = 2x^3 - 7x^2 - 8x + 16$. This confirms the numerical evidence in Table 1 showing that the value of $\frac{f(x)}{2x^3}$ is approximately 1 for x large in absolute value.

The notation introduced in Example 7 is helpful when describing the end behavior of certain functions.

• DEFINITION Let f be a function.

1. If the values of f are positive and increase without bound for positive values of x that increase without bound, we say that **$f(x)$ approaches infinity as x approaches infinity**, and we write $f(x) \to \infty$ as $x \to \infty$.
2. If the values of f are negative and the values of $|f|$ increase without bound for positive values of x that increase without bound, we say that **$f(x)$ approaches negative infinity as x approaches infinity**, and we write $f(x) \to -\infty$ as $x \to \infty$.
3. If the values of f are positive and increase without bound for negative values of x for which $|x|$ increases without bound, we say that **$f(x)$ approaches infinity as x approaches negative infinity**, and we write $f(x) \to \infty$ as $x \to -\infty$.
4. If the values of f are negative and the values of $|f|$ increase without bound for negative values of x for which $|x|$ increases without bound, we say that **$f(x)$ approaches negative infinity as x approaches negative infinity**, and we write $f(x) \to -\infty$ as $x \to -\infty$. •

• EXAMPLE 8: Describe the end behavior of $f(x) = x^4$ and $g(x) = -x^4$.

SOLUTION: The graph of $f(x) = x^4$ in Figure 3.1.18 indicates that $f(x) \to \infty$ as $x \to \infty$ and that $f(x) \to \infty$ as $x \to -\infty$. The graph of $g(x) = -x^4$ in Figure 3.1.19 indicates that $g(x) \to -\infty$ as $x \to \infty$ and $g(x) \to -\infty$ as $x \to -\infty$. •

Notice that the end behavior of $y = x^2$ and $y = x^4$ is the same. However, $y = x^2$ is not an end behavior model for $y = x^4$ because $\frac{x^4}{x^2}$ does *not* approach 1 as $|x|$ approaches infinity.

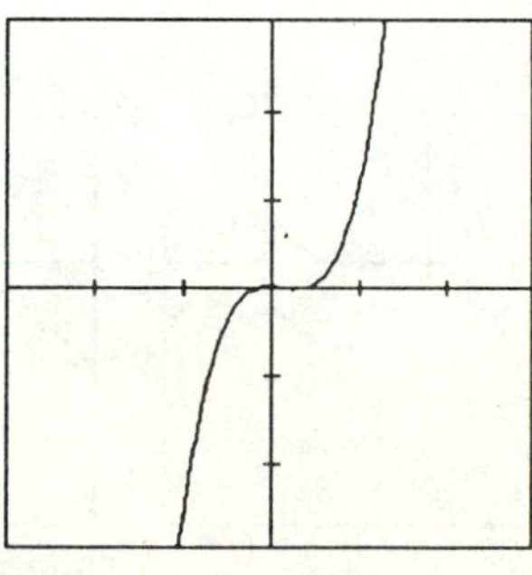

$f(x) = 2x^3 - 7x^2 - 8x + 16$
$[-30, 30]$ by $[-3000, 3000]$
Figure 3.1.16

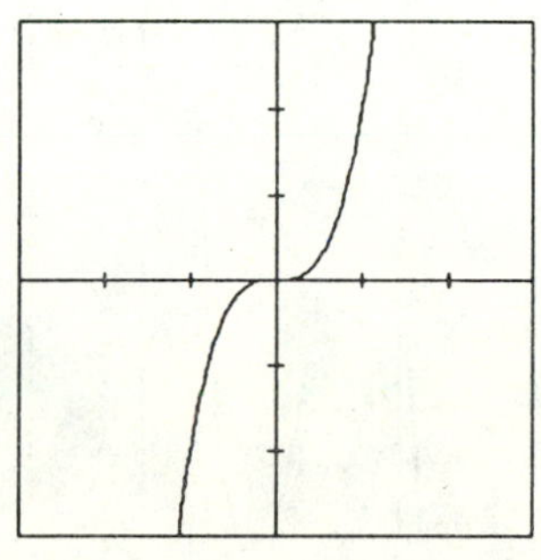

$y = 2x^3$
$[-30, 30]$ by $[-3000, 3000]$
Figure 3.1.17

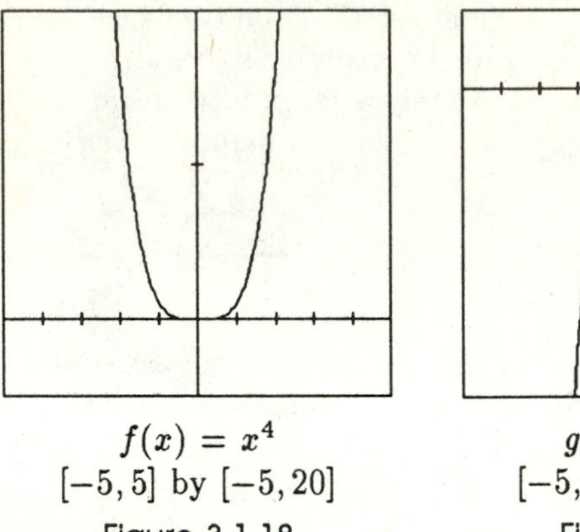

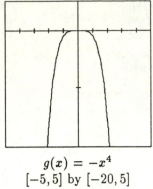

$f(x) = x^4$
$[-5, 5]$ by $[-5, 20]$
Figure 3.1.18

$g(x) = -x^4$
$[-5, 5]$ by $[-20, 5]$
Figure 3.1.19

The process of determining end behavior of a function graphically requires viewing the graph of a function in a nested sequence of viewing rectangles each contained in the next. Figure 3.1.20 shows three possible viewing rectangles beyond the standard viewing rectangle for a certain function. Notice that the end behavior appears to be like $y = x^3$. The process of viewing a graph in increasingly larger viewing rectangles is called **zoom-out**.

The viewing rectangle of Figure 3.1.16 can be viewed as the last step in a *zoom-out* process applied to $f(x) = 2x^3 - 7x^2 - 8x + 16$. This graph gives strong evidence that $y = 2x^3$ is an end behavior model for f. Generally, we only report the last viewing rectangle used when we zoom out.

• EXAMPLE 9: Graphically show that $g(x) = -0.5x^4$ is an end behavior model of $f(x) = -0.5x^4 + 5x^3 - 10x + 1$. Determine the end behavior of f.

SOLUTION: The graphs of f and g in $[-200, 200]$ by $[-100{,}000, 100{,}000]$ (Figure 3.1.21) give strong geometric evidence that these functions have the same end behavior. It can be shown that $\frac{f(x)}{g(x)} \to 1$ as $|x| \to \infty$. Notice that the end behavior of f is $f(x) \to -\infty$ as $|x| \to \infty$. •

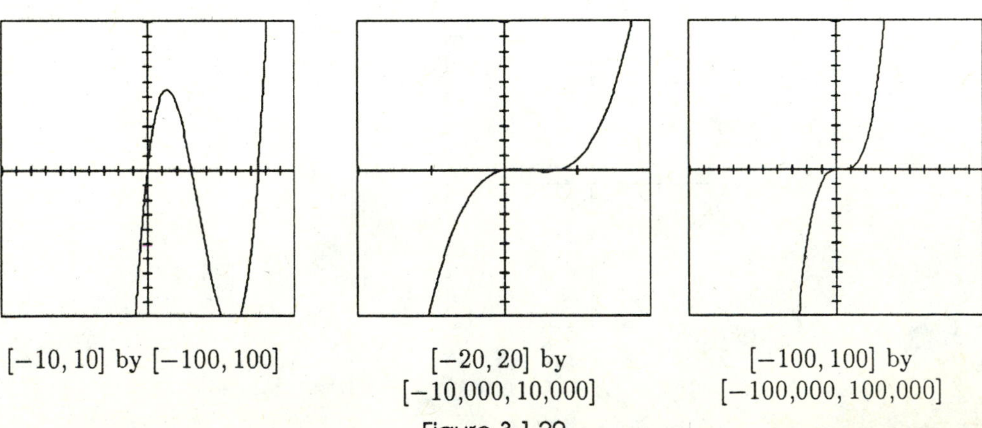

$[-10, 10]$ by $[-100, 100]$

$[-20, 20]$ by $[-10{,}000, 10{,}000]$

$[-100, 100]$ by $[-100{,}000, 100{,}000]$

Figure 3.1.20

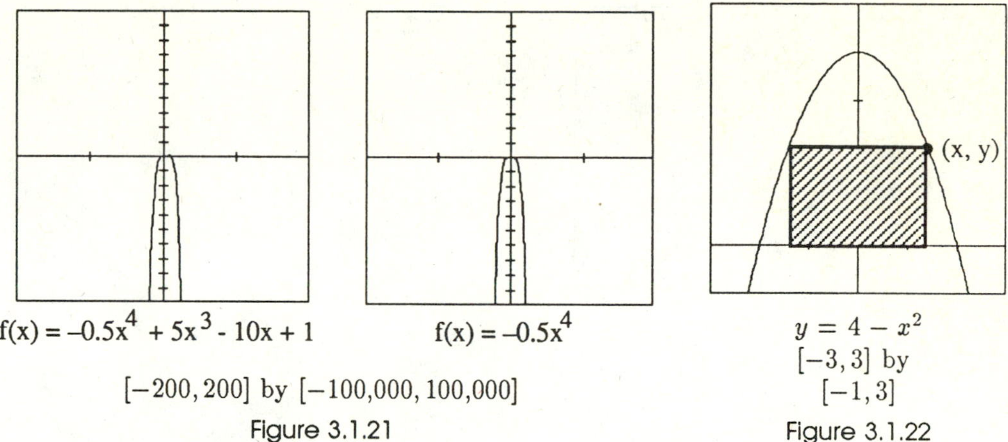

$f(x) = -0.5x^4 + 5x^3 - 10x + 1$ $\quad$ $f(x) = -0.5x^4$

$[-200, 200]$ by $[-100{,}000, 100{,}000]$

Figure 3.1.21

$y = 4 - x^2$
$[-3, 3]$ by
$[-1, 3]$

Figure 3.1.22

You might be tempted to say that $y = -x^4$ or $y = -2x^4$ appear to be end behavior models for the function $f(x) = -0.5x^4+5x^3-10x+1$ of Example 9 because their graphs seem to overlap in some viewing rectangles and they have the same end behavior. However, in the viewing rectangles $[-40, 40]$ by $[-1{,}000{,}000, 100{,}000{,}000]$ the graphs of f, $y = -x^4$, and $y = -2x^4$ are clearly separated. Notice also that $\frac{f(x)}{-x^4} \to 0.5$ and $\frac{f(x)}{-2x^4} \to 0.25$ as $|x| \to \infty$. Thus, both numerically and graphically, it is possible to determine that $y = -0.5x^4$ is an end behavior model for $f(x) = -0.5x^4 + 5x^3 - 10x + 1$ and that $y = -x^4$ or $y = -2x^4$ are *not* end behavior models for f.

It is also true that any polynomial function of degree 4 with the coefficient of x^4 negative will have the same end behavior as the function f of Example 9. However, only those polynomial functions of degree 4 with leading coefficient -0.5 are *end behavior models* for f. (Why?) The term with largest exponent is the simplest end behavior model of a polynomial function.

SUMMARY $\quad$ The polynomial function $f(x) = a_nx^n + a_{n-1}x^{n-1} + \cdots + a_0$ of degree $n \geq 1$ has end behavior model $y = a_nx^n$. This means the values of a_nx^n give good approximations to the values of f for values of x large in absolute value. Furthermore, f has the following end behavior:

(a) If $a_n > 0$ and n is even, then $f(x) \to \infty$ as $|x| \to \infty$.
(b) If $a_n > 0$ and n is odd, then $f(x) \to \infty$ as $x \to \infty$ and $f(x) \to -\infty$ as $x \to -\infty$.
(c) If $a_n < 0$ and n is even, then $f(x) \to -\infty$ as $|x| \to \infty$.
(d) If $a_n < 0$ and n is odd, then $f(x) \to -\infty$ as $x \to \infty$ and $f(x) \to \infty$ as $x \to -\infty$.

• EXAMPLE 10: Find the maximum possible area of a rectangle that has its base on the x-axis and its upper two vertices on the graph of the equation $y = 4 - x^2$ (Figure 3.1.22).

SOLUTION: Let the coordinates of the upper right-hand corner of the rectangle be (x, y). Then, the width of the rectangle is $2x$, and the height of the rectangle is y. Because the point (x, y) lies on the graph of $y = 4 - x^2$, the height y of the rectangle can also be

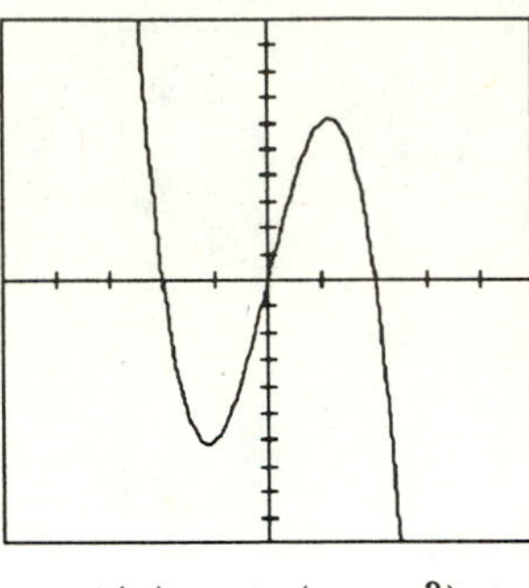

$A(x) = 2x(4 - x^2)$
$[-5, 5]$ by $[-10, 10]$
Figure 3.1.23

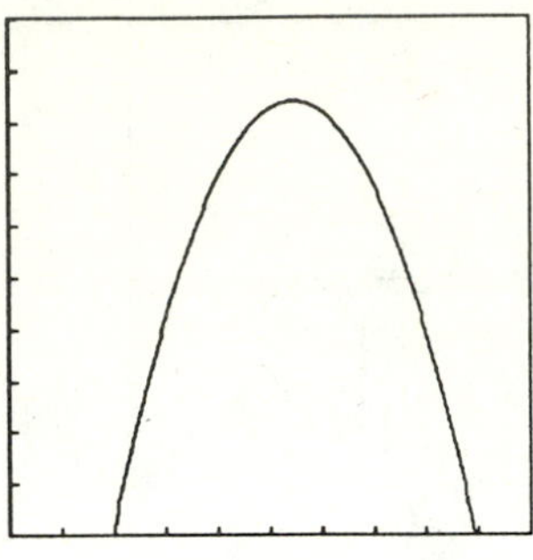

$A(x) = 2x(4 - x^2)$
$[1.1, 1.2]$ by $[6.15, 6.16]$
Figure 3.1.24

expressed as $4 - x^2$. Thus, the area of the rectangle can be written as a function of x : $A(x) = 2xy = 2x(4 - x^2)$.

From Figure 3.1.22 we can see that the only values of x that make sense in the problem situation are $0 < x < 2$. (Why?) A complete graph of A is given in Figure 3.1.23. We can see from this graph that A has a local maximum value between $x = 0$ and $x = 2$. Because the algebraic representation of A represents the problem situation only for values of x between 0 and 2, this local maximum value of A must be the solution to the problem. Using zoom-in we can determine that the value of x, with error at most 0.01, producing the largest value for the area of the rectangle is 1.155 (Figure 3.1.24). The area of the corresponding rectangle with maximum area is 6.1585. •

EXERCISES 3-1

In Exercises 1–8, determine the domain, range, and the values of x for which the function given by the graph is discontinuous. Assume each graph is complete.

1.

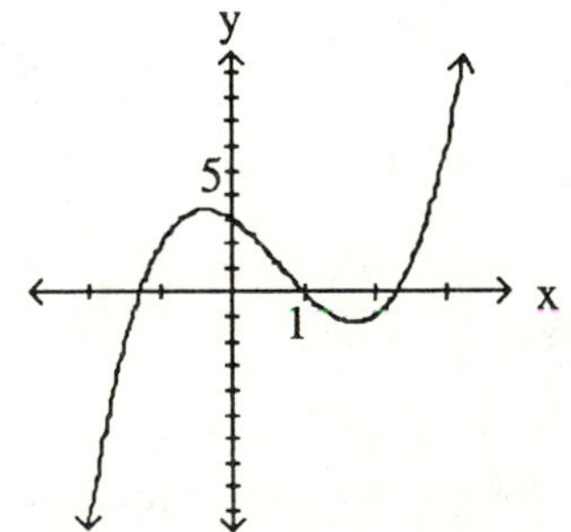

2.

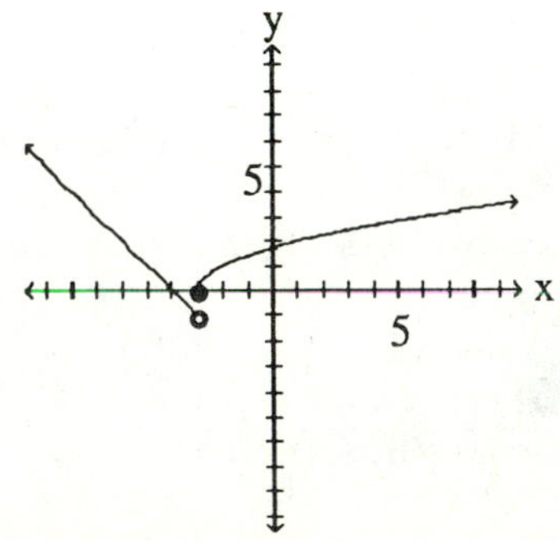

3.

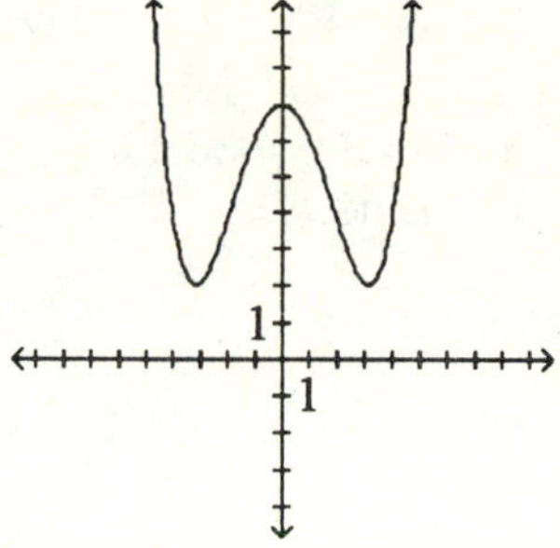

4.

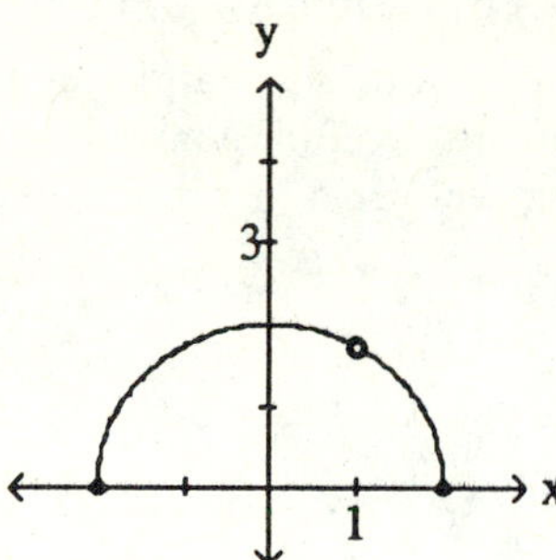

5.

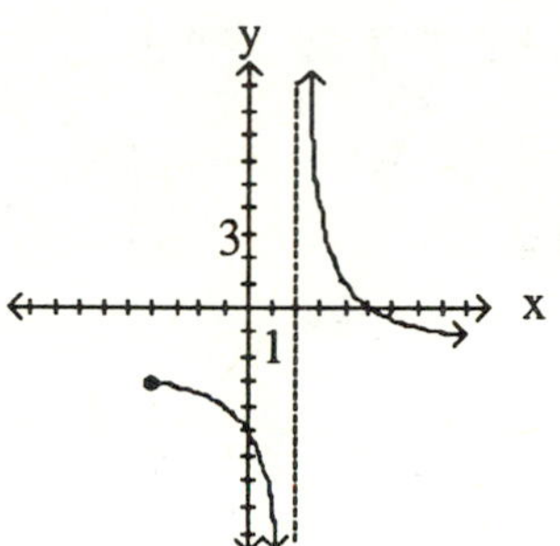

6.

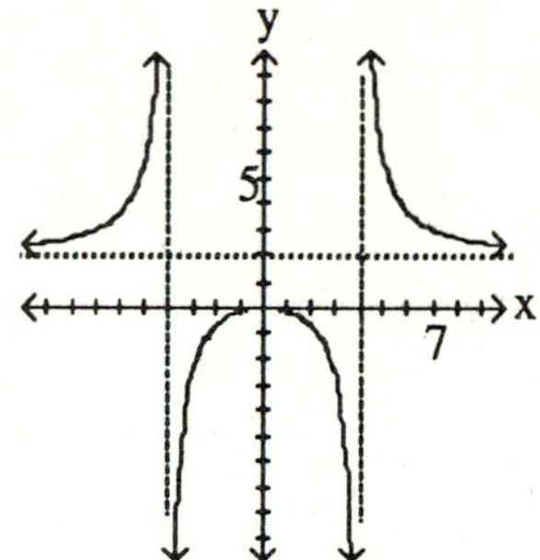

7.

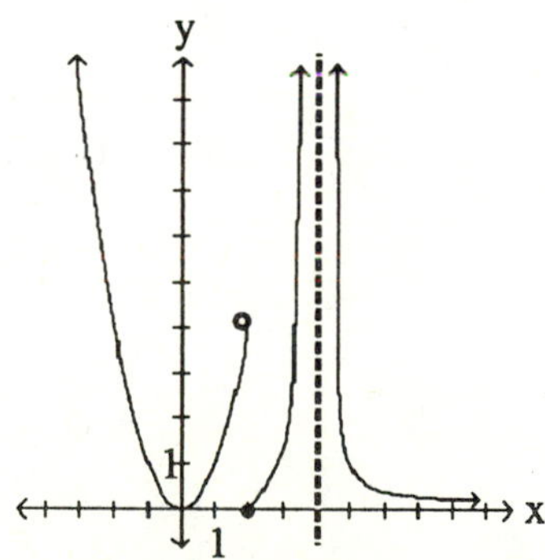

8.

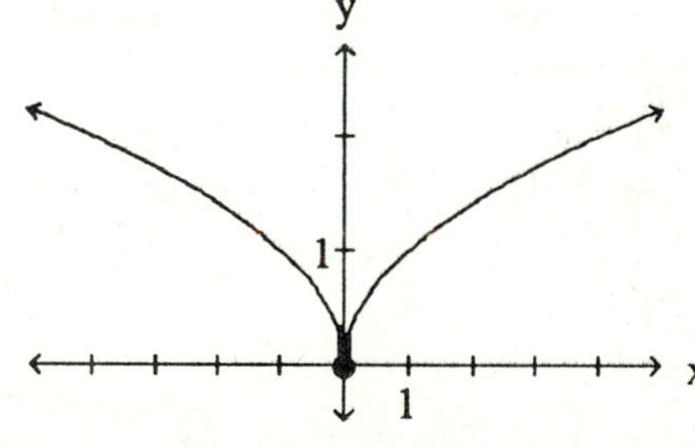

Determine the domain, range, and the points of discontinuity of each function.

9. $f(x) = \text{INT}\ (x+2)$

10. $g(x) = 2x^2 - 3x + 6$

11. $V(x) = 4x^3 - 200x^2 + 30x$

12. $f(x) = \begin{cases} |x-2| & \text{for } x < -1 \\ 2x & \text{for } x \geq -1 \end{cases}$

13. $y = \begin{cases} -\dfrac{1}{x} & \text{for } x < 0 \\ x^2 + 1 & \text{for } x \geq 0 \end{cases}$

14. $y = \dfrac{2}{x-3}$

15. Overlay the graphs of $f(x) = -5x^3 + 10x^2 - 3x + 15$ and $g(x) = -5x^3$ in each of the viewing rectangles. Which viewing rectangle *best* suggests that g is an end behavior model for f?

(a) $[-10, 10]$ by $[-100, 100]$

(b) $[-10, 10]$ by $[-1000, 1000]$

(c) $[-10, 10]$ by $[-3000, 3000]$

(d) $[-10, 10]$ by $[-5000, 5000]$

(e) $[-100, 100]$ by $[-100{,}000, 100{,}000]$ (f) $[-100, 100]$ by $[-100, 100]$

Graph the polynomial functions f and g in a viewing rectangle that both graphically illustrates that the end behavior of the two functions is the same, and that g is an end behavior model for f. That is, choose a viewing rectangle in which the graphs appear to be almost identical.

16. $f(x) = 3x^3 - 2x^2 + x - 20$; $g(x) = 3x^3$
17. $f(x) = -2x^3 - x^2 + 3x + 5$; $g(x) = -2x^3$
18. $f(x) = 3x^4 - 2x^3 + x^2 - x + 5$; $g(x) = 3x^4$
19. $f(x) = -5x^4 + 2x^3 - x + 3$; $g(x) = -5x^4$
20. For the function f in Exercise 18 write $\frac{f(x)}{3x^4}$ in the form $\frac{f(x)}{3x^4} = 1 + r_1(x) + r_2(x) + r_3(x) + r_4(x)$.
    (a) Graph $y = r_1(x)$, $y = r_2(x)$, $y = r_3(x)$, and $y = r_4(x)$ in the same viewing rectangle. Zoom out and investigate what happens to $r_i(x)$ as $x \to \infty$ or as $x \to -\infty$.
    (b) Explain how the result of (a) shows that $y = 3x^4$ is an end behavior model for f.
21. For the function f in Exercise 19 write $\frac{f(x)}{-5x^4}$ in the form $\frac{f(x)}{-5x^4} = 1 + r_1(x) + r_2(x) + r_3(x)$.
    (a) Graph $y = r_1(x)$, $y = r_2(x)$, and $y = r_3(x)$ in the same viewing rectangle. Zoom out and investigate what happens to $r_i(x)$ as $x \to \infty$ or as $x \to -\infty$.
    (b) Explain how the result of (a) shows that $y = -5x^4$ is an end behavior model for f.
22. Consider the linear function $f(x) = 3x - 12$. Verify that 5 is a number between $f(-10)$ and $f(10)$. Determine a value of c so that $f(c) = 5$ and $-10 \le c \le 10$.
23. Consider the linear function $f(x) = 4x - 7$. Verify that 0 is a number between $f(1)$ and $f(2)$. Determine a value of c so that $f(c) = 0$ and $1 \le c \le 2$.
24. Consider the quadratic function $f(x) = 2x^2 + 4x - 10$. Verify that 50 is a number between $f(0)$ and $f(10)$. Determine a value of c so that $f(c) = 50$ and $0 \le c \le 10$.
25. Consider the linear function $f(x) = ax + b$. Let L be a number between $f(-10)$ and $f(10)$. Determine a value of c so that $f(c) = L$ and $-10 \le c \le 10$.

Assume the graph of $y = f(x)$ given in Figure 3.1.25 is complete. Complete each statement.

26. As $x \to \infty$, $f(x) \to$?

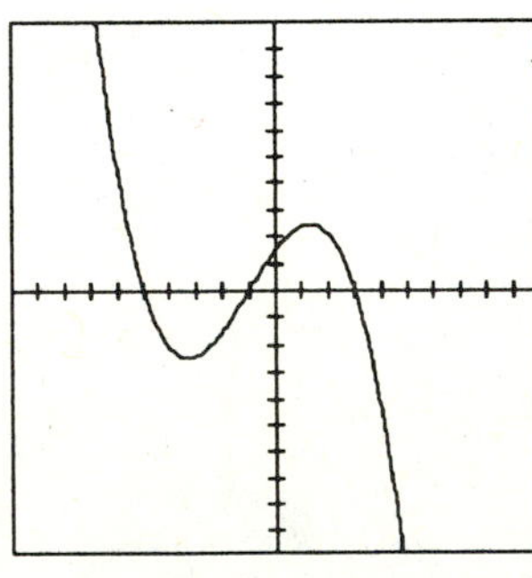

$[-10, 10]$ by $[-100, 100]$

Figure 3.1.25

27. As $x \to -\infty$, $f(x) \to$?

Assume the graph of $y = f(x)$ given in Figure 3.1.26 is complete. Complete each statement.

28. As $x \to \infty$, $f(x) \to$?

29. As $x \to -\infty$, $f(x) \to$?

*30. As $x \to 3$ through values of x satisfying $x > 3$, $f(x) \to$?

*31. As $x \to 3$ through values of x satisfying $x < 3$, $f(x) \to$?

32. A rectangle has its base on the x-axis and its upper two vertices on the graph of the equation $y = 12 - x^2$. Find the maximum possible area of such a rectangle and the dimensions of the rectangle with maximum area.

33. The total daily revenue of Chris's Cookie Shop is given by the equation $R = x \cdot p$, where x is the number of pounds of cookies sold and p is the price of one pound of cookies. Assume the price per pound of the cookies is given by the supply equation $p = 0.2 + 0.01x - 0.00001x^2$.
    (a) Determine for what positive values of x the revenue function is increasing.
    (b) Draw a complete graph of the revenue function R.
    (c) What values of x make sense in the problem situation?
    (d) Determine the maximum possible daily revenue of Chris's Cookie Shop and the number of pounds of cookies she needs to sell to achieve maximum revenue.

34. A box is made by cutting equal squares from the four corners of a 16 inch by 28 inch piece of material. Determine the size of the square that must be cut out to produce a box with maximum volume. What is the corresponding maximum volume?

*35. It is shown in numerical analysis that there is a unique polynomial of degree less than or equal to n that interpolates (the graph goes through) $n+1$ given points. These polynomials are called *Lagrange* polynomials. Such interpolating polynomials are important tools in understanding data generated from many kinds of scientific experiments and statistical analyses. Let $L(x) = 2x^3 - 11x^2 + 12x + 5$ be a Lagrange interpolating polynomial for four data points.
    (a) Draw a complete graph of $y = L(x)$.

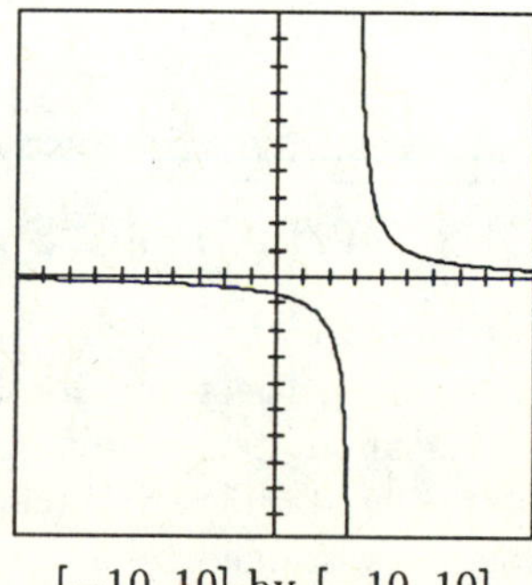

$[-10, 10]$ by $[-10, 10]$

Figure 3.1.26

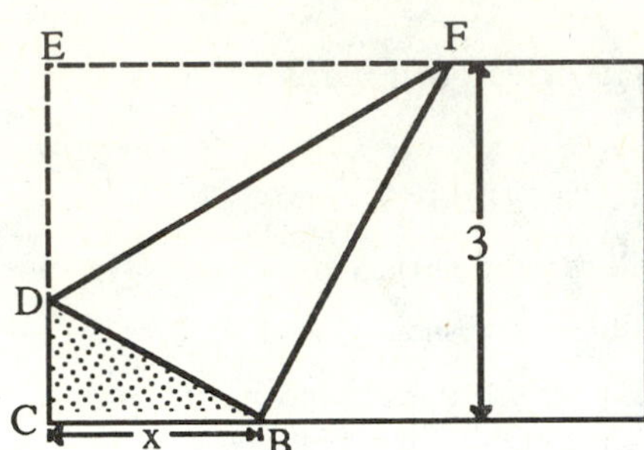

Figure 3.1.27

(b) Suppose an experiment generates the four data points $(-1,-20)$, $(0,5)$, $(1,8)$, and $(2,1)$. Show that L interpolates these four points.

(c) Suppose it is later determined that a fifth data point $(1.5,6)$ is generated by the experiment. What would have been the predicted outcome if $x = 1.5$ was used in the Lagrange interpolating polynomial $L(x)$?

36. Suppose one corner E of a long, flat, rectangular piece of material 3 feet wide is folded over along a line DF until the corner E touches the opposite edge at B (Figure 3.1.27). Let A be the area of the shaded triangle BCD. Let x be the distance between the vertex C and point B. It can be shown that the area of triangle BCD is $A = \frac{3}{4}x - \frac{1}{12}x^3$. What is the largest possible area A?

37. In Exercise 36, prove that $A = \frac{9x-x^3}{12}$. *Hint*: Observe that triangles DEF and BDF are congruent. Then, if the distance from D to C is y, the Pythagorean theorem can be used to determine y as a function of x. Finally, substitution for y in the area formula $A = \frac{1}{2}xy$ will give the desired result.

*38. Consider the function $f(x) = 2x^3 - 7x^2 - 8x + 16$ used in Table 1. How large should the absolute value of x be so that

(a) $\left|\frac{f(x)}{2x^3} - 1\right| < 0.1$? (b) $\left|\frac{f(x)}{2x^3} - 1\right| < 0.01$? (c) $\left|\frac{f(x)}{2x^3} - 1\right| < 0.001$?

3.2 Real Zeros of Polynomials

We use the division algorithm to establish the Remainder Theorem and the Factor Theorem for polynomials. Then, these theorems are used to help show the equivalence of the following important concepts about polynomials: solution of a polynomial equation, zero of a polynomial, factor of a polynomial, x-intercept of the graph of a polynomial, and the remainder when a polynomial is divided by $x - c$. These connections will be used to help find all the zeros of a polynomial function. We focus on real zeros in Sections 3.2 and 3.3 and on complex zeros in Section 3.4.

Zeros and Solutions

The words "zero" and "root" are used interchangeably with functions. The zeros, or roots, of a function f are the solutions to the equation $f(x) = 0$. To find the zeros of a linear or quadratic polynomial we must solve the corresponding linear or quadratic equation. For example, the zeros of $f(x) = 2x^2 - 3x + 1$ are found by solving the quadratic equation $2x^2 - 3x + 1 = 0$. In Section 2.1 we obtained formulas for the exact solutions of linear and quadratic equations. We also gave the Cardan formulas for the exact solutions to a polynomial equation of degree 3 in Section 2.1. However, the cubic formulas were not used because of their complexity. We have seen that finding exact solutions to equations algebraically can be difficult and often impossible. Approximations to the real solutions to any desired degree of accuracy within the limits of machine precision can be found graphically.

Zeros and x-Intercepts

Finding the zeros of a polynomial of degree 3 or higher requires finding the solutions to polynomial equations of degree 3 or higher. Because this is such a formidable algebraic task, we almost always find the solutions graphically. However, in Sections 3.2 and 3.3 we will focus on special techniques to find zeros algebraically.

In Section 2.1 we observed that finding the real zeros of a function f is equivalent to finding the x-intercepts of the graph of f. The real zeros were found graphically by using zoom-in to find the x-intercepts. We will see that a graph can sometimes help us locate exact real zeros.

Zeros and Factors

Finding zeros of polynomials and factoring polynomials are closely related. For example, if 2 is a zero of f, then $x - 2$ is a factor of f. Long division of polynomials can then be used to find the factor $g(x)$ such that $f(x) = (x - 2)g(x)$.

Long division of polynomials is a procedure for dividing one polynomial by another that is similar to long division for integers. We illustrate this procedure by dividing $2x^4 - x^3 - 2$ by $2x^2 + x + 1$. The polynomial to be divided is called the **dividend**. The polynomial that we divide by is called the **divisor**. Both polynomials must be arranged with the powers of x in descending order.

$$\begin{array}{r|l} & x^2 \\ \hline 2x^2 + x + 1 & 2x^4 - x^3 - 2 \\ & 2x^4 + x^3 + x^2 \\ \hline & -2x^3 - x^2 - 2 \end{array}$$

The partial quotient x^2 is obtained by dividing the first term $2x^4$ of the dividend $2x^4 - x^3 - 2$, by the first term $2x^2$ of the divisor $2x^2 + x + 1$. Then, the divisor is multiplied by this partial quotient and the result subtracted from the dividend.

The next term of the quotient is obtained by dividing the first term $-2x^3$ of the polynomial $-2x^3 - x^2 - 2$ obtained by subtraction by the first term $2x^2$ of the divisor. The divisor is then multiplied by the new term of the quotient and the result subtracted from the last line obtained above. The process continues until the last line is either the zero polynomial (the number 0) or a polynomial of degree less than the degree of the divisor. The completed process is given below.

$$
\begin{array}{rll}
 & x^2 - x & \longleftarrow \text{quotient} \\
\text{divisor} \longrightarrow 2x^2 + x + 1 & \overline{)\,2x^4 - x^3 - 2} & \longleftarrow \text{dividend} \\
 & 2x^4 + x^3 + x^2 & \\
 & \overline{\quad -2x^3 - x^2 - 2} & \\
 & \quad -2x^3 - x^2 - x & \\
 & \overline{\qquad\qquad x - 2} & \longleftarrow \text{remainder}
\end{array}
$$

The polynomial $x^2 - x$ in the top line is called the **quotient**. The polynomial $x - 2$ in the last line is called the **remainder**. The degree of the remainder $x - 2$ is less than the degree of the divisor $2x^2 + x + 1$. Notice we can write the dividend as the product of the divisor and the quotient plus the remainder:

$$
\begin{array}{ccccccc}
2x^4 - x^3 - 2 & = & (2x^2 + x + 1) & & (x^2 - x) & + & x - 2. \\
\downarrow & & \downarrow & & \downarrow & & \downarrow \\
\text{dividend} & = & \text{divisor} & \times & \text{quotient} & + & \text{remainder}
\end{array}
$$

For those values of x for which $2x^2 + x + 1 \neq 0$, we can write this equation in the following form:

$$
\begin{array}{ccccc}
\dfrac{2x^4 - x^3 - 2}{2x^2 + x + 1} & = & x^2 - x & + & \dfrac{x - 2}{2x^2 + x + 1} \quad (\text{if } 2x^2 + x + 1 \neq 0). \\
\downarrow & & \downarrow & & \downarrow \\
\dfrac{\text{dividend}}{\text{divisor}} & = & \text{quotient} & + & \dfrac{\text{remainder}}{\text{divisor}} \quad (\text{if divisor} \neq 0)
\end{array}
$$

The above process can be summarized as follows.

- **Division Algorithm for Polynomials** If $f(x)$ and $h(x)$ are polynomials, then there are polynomials $q(x)$ and $r(x)$ so that $f(x) = h(x)q(x) + r(x)$. Either $r(x) = 0$ or $r(x)$ has degree less than the degree of $h(x)$.

- EXAMPLE 1: Find the quotient and the remainder when $x^3 - 2x^2 - 5x + 11$ is divided by $x - 3$.

SOLUTION: The degree of the divisor $x - 3$ is 1. The long division process continues until either we obtain zero as a remainder or the degree of the remainder is 0, that is, a constant.

Applying the long division process gives the following:

$$\begin{array}{r|l} & x^2 + x - 2 \\ \hline x-3 & x^3 - 2x^2 - 5x + 11 \\ & x^3 - 3x^2 \\ \hline & \quad x^2 - 5x + 11 \\ & \quad x^2 - 3x \\ \hline & \quad\quad -2x + 11 \\ & \quad\quad -2x + 6 \\ \hline & \quad\quad\quad 5 \end{array}$$

Thus, the quotient is $x^2 + x - 2$ and the remainder is 5. •

The information in Example 1 can be summarized in the following way:

$$x^3 - 2x^2 - 5x + 11 = (x-3)(x^2 + x - 2) + 5.$$

For $x \neq 3$, we can rewrite this equation in the following form:

$$\frac{x^3 - 2x^2 - 5x + 11}{x-3} = x^2 + x - 2 + \frac{5}{x-3} \quad (x \neq 3).$$

The Division Algorithm assures us that whenever we divide a given polynomial by a linear polynomial the remainder must be either zero or some other constant. In either case, the remainder must be a number. Thus, if we divide $f(x)$ by $x - c$ we can write

$$f(x) = (x-c)q(x) + r\,, \text{ where } r \text{ is a number}\,.$$

Because this equation must be true for all real numbers x, it must be true when x is replaced by the real number c:

$$f(c) = (c-c)q(c) + r$$
$$f(c) = r.$$

Because r is a constant it follows that $r = f(c)$. This important result is called the Remainder Theorem.

• **REMAINDER THEOREM** If a polynomial $f(x)$ is divided by $x - c$, then the remainder is $f(c)$. We can write $f(x) = (x-c)q(x) + f(c)$ where $q(x)$ is the quotient.

• **EXAMPLE 2:** Find the remainder when the polynomial $f(x) = x^3 - 2x^2 + x - 5$ is divided by $x - 3$, and when it is divided by $x + 1$.

SOLUTION: Long division could be used to find the remainder when $f(x)$ is divided by $x - 3$ or $x + 1$. However, the Remainder Theorem states that $f(3)$ is the remainder when $f(x)$ is divided by $x - 3$:

$$f(3) = 3^3 - 2(3^2) + 3 - 5 = 7.$$

Thus, the remainder when $f(x)$ is divided by $x - 3$ is 7. Similarly, the remainder when $f(x)$ is divided by $x + 1$ is -9 because $x + 1 = x - (-1)$ and

$$f(-1) = (-1)^3 - 2(-1)^2 + (-1) - 5 = -9.$$

Check these results using long division. Notice that *only* the remainder (and *not* the quotient) is determined when the Remainder Theorem is used in this way. •

The Factor Theorem establishes the connection between the zeros and the factors of a polynomial.

• FACTOR THEOREM Let $f(x)$ be a polynomial. Then, $x - c$ is a factor of $f(x)$ if and only if c is a zero of $f(x)$.

PROOF If $x - c$ is a factor of $f(x)$, then $f(x) = (x - c)q(x)$. Thus, $f(c) = (c - c)q(c) = 0$, that is, c is a zero of $f(x)$. Now suppose c is a zero of $f(x)$, that is, $f(c) = 0$. Then, the Remainder Theorem gives $f(x) = (x - c)q(x) + f(c) = (x - c)q(x)$. Thus, $x - c$ is a factor of $f(x)$.

SUMMARY The following statements are equivalent for a polynomial $f(x)$ and a real number c:

1. c is a solution to the equation $f(x) = 0$.
2. c is a zero of $f(x)$.
3. c is a root of $f(x)$.
4. $x - c$ is a factor of $f(x)$.
5. The remainder when $f(x)$ is divided by $x - c$ is 0.
6. c is an x-intercept of the graph of $y = f(x)$.

• EXAMPLE 3: Find a polynomial with real coefficients of degree 3 that has the following zeros:

(a) -2, 0, and 3 (b) -3 and 1

SOLUTION:

(a) If -2 is to be a zero, then, by the Factor Theorem, $x+2$ must be a factor. Similarly, to have 0 and 3 as zeros requires that x and $x - 3$ be factors. Thus, the product $x(x + 2)(x - 3)$ must be a factor of the desired polynomial. Because the product $x(x + 2)(x - 3)$ has degree 3, the product itself is a polynomial of degree 3 with zeros -2, 0, and 3. Every polynomial of degree 3 with zeros -2, 0, and 3 must be of the form $ax(x + 2)(x - 3)$ where a is a nonzero real number.

(b) For -3 and 1 to be zeros, we must have $x+3$ and $x-1$ as factors. Thus, the product $(x - 1)(x + 3)$ must be a factor of the desired polynomial. To obtain a polynomial of degree 3 with zeros -3 and 1, we can multiply $(x - 1)(x + 3)$ by any polynomial of degree 1. For example, $(x - 1)^2(x + 3)$, $(x - 1)(x + 3)^2$, or $(x - 1)(x + 3)(x - 5)$ are each polynomials of degree 3 with zeros -3 and 1. In fact, every polynomial of degree 3 with zeros -3 and 1 must be of the form $(ax + b)(x + 3)(x - 1)$ where a and b are real numbers with $a \neq 0$. •

The next example illustrates how factoring can be used to find zeros of polynomials algebraically.

• EXAMPLE 4: Use factoring to find the real zeros of

(a) $f(x) = x^3 - 4x$. (b) $g(x) = x^3 + 2x^2 - 4x - 8$.

SOLUTION:

(a) Once we remove the common factor x the factorization can be completed:

$$x^3 - 4x = x(x^2 - 4) = x(x-2)(x+2).$$

The factor x corresponds to the root 0. The factors $x-2$ and $x+2$ correspond to the roots 2 and -2, respectively. Thus, 0, 2, and -2 are the real zeros of $f(x) = x^3 - 4x$.

(b) We can factor g by grouping the first two and the last two terms:

$$\begin{aligned} g(x) &= x^3 + 2x^2 - 4x - 8 \\ &= x^2(x+2) - 4(x+2) \\ &= (x+2)(x^2-4) \\ &= (x+2)(x-2)(x+2) \\ &= (x+2)^2(x-2). \end{aligned}$$

The real zeros of g are 2 and -2. •

Finding the zeros of a polynomial by factoring the polynomial can be very difficult. Sometimes we can discover linear factors corresponding to real zeros by looking at a graph. The graph of $f(x) = x^3 - 4x$ (Figure 3.2.1) appears to cross the x-axis at $x = 0$, 2, and -2. Thus, x, $x-2$, and $x+2$ may be good guesses of the factors of $f(x)$. The Factor Theorem can be used to confirm that these are factors by showing that $f(0) = 0$, $f(2) = 0$, and $f(-2) = 0$.

• EXAMPLE 5: Use the graph of $f(x) = 2x^3 - 4x^2 + x - 2$ to help find the real zeros of $f(x)$.

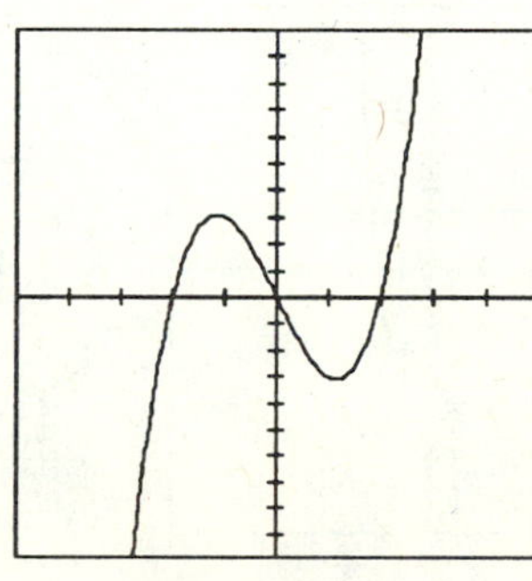

$f(x) = x^3 - 4x$
$[-5, 5]$ by $[-10, 10]$
Figure 3.2.1

SOLUTION: The graph of f in the viewing rectangle $[-10, 10]$ by $[-10, 10]$ is shown in Figure 3.2.2. It appears f has a zero at $x = 2$. This means $x - 2$ might be a factor of $f(x)$. Because $f(2) = 2(2^3) - 4(2^2) + 2 - 2 = 0$, it follows from the Factor Theorem that $x - 2$ is indeed a factor. However, if we want the other factor we need to use long division:

$$\begin{array}{r|l} & 2x^2 + \quad 1 \\ \hline x - 2 & 2x^3 - 4x^2 + x - 2 \\ & 2x^3 - 4x^2 \\ \hline & \qquad\qquad x - 2 \\ & \qquad\qquad x - 2 \\ \hline & \qquad\qquad\quad 0 \end{array}$$

Thus, $f(x) = (x - 2)(2x^2 + 1)$. Now, we can see that 2 is the only real zero because the factor $2x^2 + 1$ has no real zeros. (Why?) This is consistent with the graph in Figure 3.2.2. •

A graph may suggest that a certain real number is a zero of a polynomial. However, arithmetic or algebraic techniques must be used to confirm the graphical evidence. We will see in Section 3.4 that graphs of the polynomial do *not* help us find their complex zeros. We close this section with an application that requires finding the zeros of a polynomial of degree 4.

Ladders of 20 feet and 30 feet are placed between two buildings as shown in Figure 3.2.3. We want to find how far apart the buildings are if the ladders cross at a height of 8 feet above ground. Let D be the distance between the buildings and x the vertical distance from the top of the 20 foot ladder to the ground. In Exercise 40 you will be asked to show that

$$x^4 - 16x^3 + 500x^2 - 8000x + 32{,}000 = 0\,.$$

Once x is known, we can find the value of D because $20^2 = D^2 + x^2$.

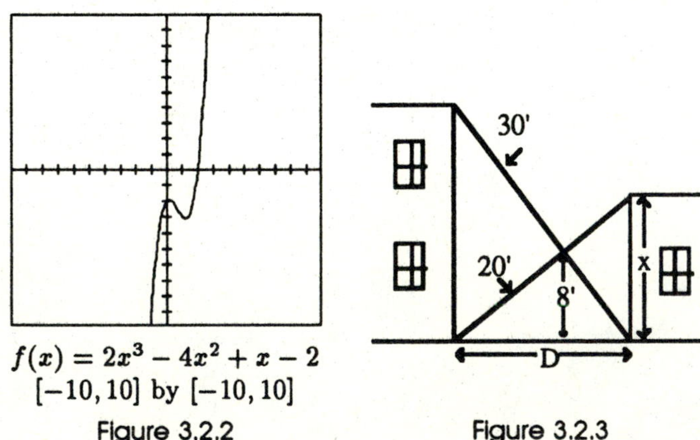

$f(x) = 2x^3 - 4x^2 + x - 2$
$[-10, 10]$ by $[-10, 10]$
Figure 3.2.2

Figure 3.2.3

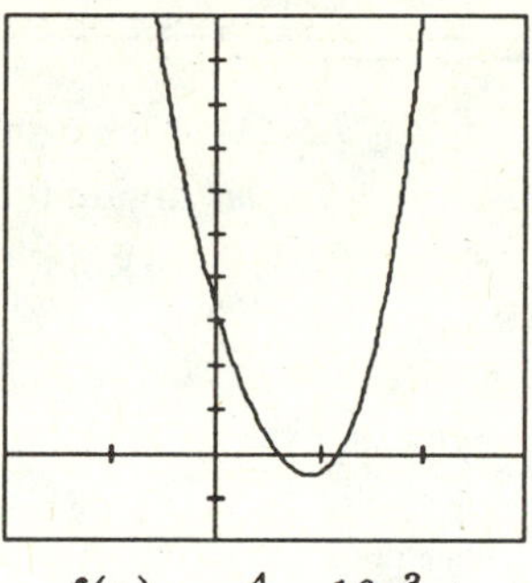

$f(x) = x^4 - 16x^3 + 500x^2 - 8000x + 32{,}000$
$[-20, 30]$ by $[-20{,}000, 100{,}000]$
Figure 3.2.4

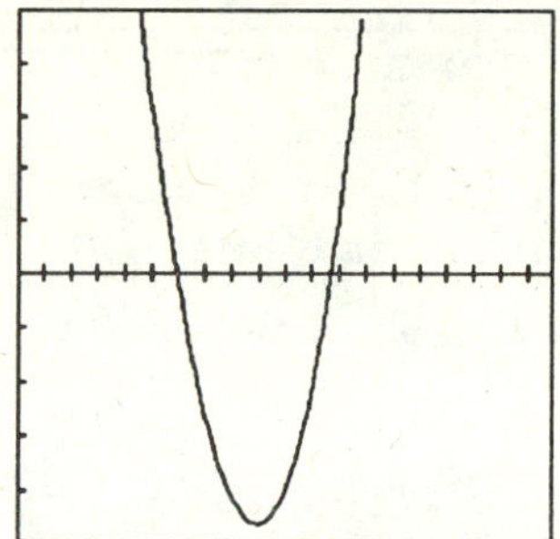

$f(x) = x^4 - 16x^3 + 500x^2 - 8000x + 32{,}000$
$[0, 20]$ by $[-5000, 5000]$
Figure 3.2.5

• EXAMPLE 6: Find all real zeros of

$$f(x) = x^4 - 16x^3 + 500x^2 - 8000x + 32{,}000\,.$$

SOLUTION: The end behavior of f is the same as that of $y = x^4$. You may need to experiment with several viewing rectangles to determine a complete graph of f. The fact that the y-intercept of this graph is 32,000 should alert you to the need for a large viewing rectangle to determine a complete graph. Figure 3.2.4 gives the graph of f in the viewing rectangle $[-20, 30]$ by $[-20{,}000, 100{,}000]$ and is a complete graph. We can see from this figure that f has two real zeros. The graph of f in the viewing rectangle $[0, 20]$ by $[-5000, 5000]$ is shown in Figure 3.2.5. We can see that there is a zero near 6 and a zero close to 12 because the horizontal scale marks in this figure are one unit apart.

Using zoom-in we can determine that the zeros are 5.944 and 11.712 with error at most 0.01. •

• EXAMPLE 7: Ladders of 20 feet and 30 feet are placed between two buildings as shown in Figure 3.2.3. Determine the distance D between the two buildings if the ladders cross at a height of 8 feet above the ground.

SOLUTION: We can see from Figure 3.2.3 that x must be greater than 8 and less than 20. Thus, only the larger of the two zeros in Example 6 actually produces a solution to the ladder problem. Substituting $x = 11.712$ into $20^2 = D^2 + x^2$ gives $D = 16.212$. Thus, the buildings are 16.212 feet apart. •

• EXERCISES 3-2

Find the quotient $q(x)$ and remainder $r(x)$ when $f(x)$ is divided by $h(x)$. Then, compute $q(x)h(x) + r(x)$ and compare with $f(x)$.

1. $f(x) = x^2 - 2x + 3$; $h(x) = x - 1$
2. $f(x) = x^3 - 1$; $h(x) = x + 1$
3. $f(x) = 4x^3 - 8x^2 + 2x - 1$; $h(x) = 2x + 1$
4. $f(x) = x^4 - 2x^3 + 3x^2 - 4x + 6$; $h(x) = x^2 + 2x - 1$

Use the Remainder Theorem to determine the remainder when $f(x)$ is divided by $x - c$. Check by using long division.

5. $f(x) = 2x^2 - 3x + 1$; $c = 2$
6. $f(x) = x^4 - 5$; $c = 1$
7. $f(x) = 2x^3 - 3x^2 + 4x - 7$; $c = 2$
8. $f(x) = x^5 - 2x^4 + 3x^2 - 20x + 3$; $c = -1$

Use the Factor Theorem to determine whether the first polynomial is a factor of the second polynomial.

9. $x + 2$; $x^2 - 4$
10. $x - 1$; $x^3 - x^2 + x - 1$
11. $x - 3$; $x^3 - x^2 - x - 15$
12. $x + 1$; $2x^{10} - x^9 + x^8 + x^7 + 2x^6 - 3$

Find all real zeros of each polynomial by factoring. Use a graphing utility to check your answer.

13. $f(x) = x^2 - 5x + 6$
14. $f(x) = 6x^2 + 8x - 8$
15. $f(x) = x^3 - 9x$
16. $g(x) = x^3 - 1$
17. $T(x) = 2x^4 + x^2 - 15$
18. $g(x) = x^3 - x^2 - 2x + 2$

Draw a complete graph of each function. Use the graph as an aid to factor the polynomial, and then determine all real zeros.

19. $f(x) = 5x^2 - 2x - 51$
20. $f(x) = x^3 - 2x^2 - 3x + 6$
21. $f(x) = x^3 - 11x^2 + x - 11$
22. $g(x) = x^4 - 16$

Determine all real zeros.

23. $f(x) = x^3 + 3x^2 - 10x - 1$
24. $g(x) = 2 - 15x + 11x^2 - 3x^3$
25. $f(x) = 100x^3 - 403x^2 + 406x + 1$
26. $f(x) = 100x^3 - 403x^2 + 406x - 1$
27. A toy rocket is shot straight up into the air with an initial velocity of 48 feet per second. Its height $s(t)$ above the ground after t seconds is given by $s(t) = -16t^2 + 48t$. What is the maximum height attained by the rocket?
28. Draw the graph of $f(x) = 100x^3 - 203x^2 + 103x - 1$ in the $[-5, 5]$ by $[-100, 100]$ viewing rectangle.
    (a) How many real zeros are evident from this graph?
    (b) How many actual real zeros exist in this case?
    (c) Find all real zeros.

29. An object is shot straight up from the top of a 260-foot-tall tower with initial velocity 35 feet per second.

   (a) When will the object hit the ground?

   (b) When does the object reach its maximum height above the ground, and what is this maximum height?

Find a polynomial with real coefficients satisfying the given conditions.

30. Degree 2, with 3 and -4 as zeros

31. Degree 2, with 2 as the only real zero

32. Degree 3, with -2, 1, and 4 as zeros

33. Degree 3, with -1 as the only real zero

NICECALC Company manufactures the only calculator with the ZAP feature. For the monthly supply function $P = S(x)$ and monthly demand function $P = D(x)$, find the equilibrium price (break even price) and the associated production level to achieve equilibrium.

34. $S(x) = 6 + 0.001x^3$; x is in thousands
    $D(x) = 80 - 0.02x^2$; x is in thousands

35. $S(x) = 20 - 0.1x + 0.00007x^4$; x is in thousands
    $D(x) = 150 - 0.004x^3$; x is in thousands

36. If $f(x) = 2x^3 - 3kx^2 + kx - 1$, find a number k so that the graph of f contains $(1, 9)$.

37. Determine a linear factor of $f(x) = (x-1)^6 - 64$.

38. (a) Determine the remainder when $x^{40} - 3$ is divided by $x + 1$

    (b) Determine the remainder when $x^{63} - 17$ is divided by $x - 1$

39. The hypotenuse of a right triangle is 2 inches longer than one of its legs and has area 50 square inches.

   (a) Show that if x denotes the length of this leg, then $10{,}000 - 4x^3 - 4x^2 = 0$.

   (b) Determine the solutions to the equation in (a).

40. Derive the fourth degree polynomial that gives the solution to the ladder problem of this section. *Hint:* Consider Figure 3.2.6. Notice that $x^2 + z^2 = 20^2$ and $y^2 + z^2 = 30^2$. Write one equation that involves only x and y. Now show that the length of $AF = 8z/x$, and the length of $BF = 8z/y$ (supply the reasons). Next, use the fact that the length of AF plus the length of BF is equal to z to determine a second equation in terms of x and y only. Solve the second equation for y and substitute in the first equation.

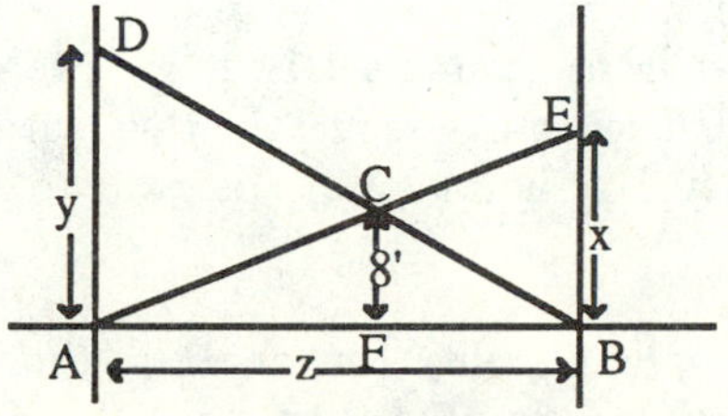

Figure 3.2.6

3.3 More on Real Zeros

Real zeros can be either rational numbers or irrational numbers. In this section we find the exact value of the rational number zeros of a polynomial. Usually the best we can do with irrational number zeros is to approximate them.

We state the following theorem about rational number zeros of a polynomial without proof.

• RATIONAL ZEROS THEOREM Suppose that all the coefficients of the polynomial $f(x) = a_n x^n + \cdots + a_1 x + a_0$ are integers with $a_n \neq 0$ and $a_0 \neq 0$. Let $\frac{u}{v}$ be a nonzero rational number in lowest terms. If $\frac{u}{v}$ is a zero of $f(x)$, then u is a factor of a_0 and v is a factor of a_n.

• EXAMPLE 1: Make a complete list of possible rational number zeros of $f(x) = 10x^5 - 3x^2 + x - 6$.

SOLUTION: We form rational numbers using the factors of the constant term -6 as numerators and the factors of the coefficient 10 of the highest power of x as denominators. The factors of -6 are ± 1, ± 2, ± 3, and ± 6, and the factors of 10 are ± 1, ± 2, ± 5, and ± 10. Thus, if we divide each and every factor of -6 by each and every factor of 10, we get the complete list of possible rational number zeros of f. Dividing all the factors of -6 by ± 1 gives ± 1, ± 2, ± 3, and ± 6 as possible rational zeros. Dividing all the factors of -6 by ± 2 gives $\pm\frac{1}{2}$, $\pm\frac{2}{2}$, $\pm\frac{3}{2}$, and $\pm\frac{6}{2}$. Notice that the possibilities ± 1 and ± 3 are repeated. Continuing, the complete list of 24 candidates is: ± 1, ± 2, ± 3, ± 6, $\pm\frac{1}{2}$, $\pm\frac{3}{2}$, $\pm\frac{1}{5}$, $\pm\frac{2}{5}$, $\pm\frac{3}{5}$, $\pm\frac{6}{5}$, $\pm\frac{1}{10}$, $\pm\frac{3}{10}$. •

It turns out that *none* of the 24 rational numbers listed in Example 1 is a zero! Checking this fact is very time consuming and tedious, even if a calculator is used. However, we can use a graph to reduce the computational task, as illustrated in the next example.

• EXAMPLE 2: Show that $f(x) = 10x^5 - 3x^2 + x - 6$ has exactly one real root and that it is an irrational number.

SOLUTION: The graph of f in the $[-5, 5]$ by $[-20, 20]$ viewing rectangle is given in Figure 3.3.1 and is a complete graph. This graph shows that f has exactly one real zero and it appears to be equal to 1. However, $f(1) \neq 0$ so 1 is not a zero. A magnified view of the graph of f (Figure 3.3.2) shows that the zero is less than 1. The largest rational number less than 1 in the list from Example 1 is $\frac{3}{5}$, and we can check that $\frac{3}{5}$ is not a zero of f. Thus, f has no rational zeros. Consequently, this real zero must be irrational. The graph of f in $[0.9, 1]$ by $[-0.1, 0.1]$ (Figure 3.3.2) allows us to read this irrational zero as 0.95 with error at most 0.01. •

Notice that a graphing utility can *never* distinguish between rational and irrational zeros. An algebraic analysis is needed to make this distinction.

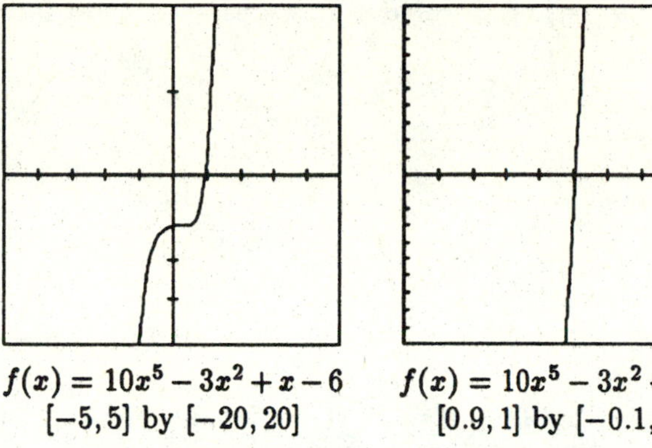

$f(x) = 10x^5 - 3x^2 + x - 6$
$[-5, 5]$ by $[-20, 20]$
Figure 3.3.1

$f(x) = 10x^5 - 3x^2 + x - 6$
$[0.9, 1]$ by $[-0.1, 0.1]$
Figure 3.3.2

Factoring Polynomials

If c is a zero of a polynomial f, then, by the Factor Theorem, $x - c$ is a factor of f. Unless we can find the exact value of c we cannot actually produce the factors of f. For example, try to factor the polynomial in Example 2!

We can find the exact value of all rational number zeros of a polynomial with integer coefficients using the Rational Zeros Theorem. Thus, rational number zeros can be used to factor polynomials.

- EXAMPLE 3: Show that $\frac{3}{2}$ is zero of $f(x) = 2x^3 - 5x^2 + x + 3$ and then factor f.

SOLUTION: $f(\frac{3}{2}) = 2(\frac{27}{8}) - 5(\frac{9}{4}) + \frac{3}{2} + 3 = 0$. Thus, $\frac{3}{2}$ is a zero and $x - \frac{3}{2}$ a factor. We can use long division to find the other factor of f:

$$
\begin{array}{r|l}
 & 2x^2 - 2x - 2 \\
\hline
x - \frac{3}{2} & 2x^3 - 5x^2 + x + 3 \\
 & 2x^3 - 3x^2 \\
\hline
 & -2x^2 + x + 3 \\
 & -2x^2 + 3x \\
\hline
 & -2x + 3 \\
 & -2x + 3 \\
\hline
 & 0
\end{array}
$$

Now, we can factor f:

$$f(x) = (x - \tfrac{3}{2})(2x^2 - 2x - 2).$$

Notice that the second factor contains 2 as a common factor so that we can rewrite f as follows:

$$\begin{aligned} f(x) &= (x - \tfrac{3}{2})(2)(x^2 - x - 1) \\ &= (2x - 3)(x^2 - x - 1). \end{aligned}$$

•

In Example 3 we found that $\frac{3}{2}$ is a zero of $f(x) = 2x^3 - 5x^2 + x + 3$ and $2x - 3$ a factor. This example is a special case of the following theorem.

• THEOREM 1 Let $\frac{u}{v}$ be a rational number in lowest terms and let f be a polynomial with integer coefficients. If $\frac{u}{v}$ is a zero of f, then $f(x) = (vx - u)q(x)$ and $q(x)$ also has integer coefficients.

If we start with a polynomial with integer coefficients, then its rational number zeros can be used to factor the polynomial using factors with integer coefficients. In previous courses, factoring probably meant to find factors with integer coefficients. We usually require you only to find factors with integer coefficients. An exception to this is when we ask you to use zeros to factor polynomials. For example, $x^2 - 3$ cannot be expressed as a product of factors with integer coefficients. However, it seems reasonable to write $x^2 - 3 = (x - \sqrt{3})(x + \sqrt{3})$ so that its zeros $\sqrt{3}$ and $-\sqrt{3}$ are displayed. Notice that $\sqrt{3}$ and $-\sqrt{3}$ are irrational numbers. Similarly, in Section 3.4 we will sometimes use the complex zeros to factor a polynomial. This can result in polynomials with complex number coefficients. Unless we are interested in displaying zeros we will not ask you to produce factors with irrational or complex numbers. Instead, we say that it does not have factors with integer coefficients, or, more simply, it does not factor.

In Example 3 we used long division to factor a polynomial. Long division of polynomials is an important process, but it can be very tedious. We introduce a simplified version called synthetic division that is valid whenever the divisor has degree 1.

Synthetic Division

This simplified long division procedure will work only for divisors that are linear. We arrange both divisor and dividend with powers of x in descending order leaving a space for any missing power of x. Suppose we divide $2x^4 - 3x^2 + x - 1$ by $x - 2$:

$$
\begin{array}{r|rrrrr}
 & 2x^3 & +4x^2 & +5x & +11 & \\
\hline
x-2 & 2x^4 & - & 3x^2 + & x - & 1 \\
 & 2x^4 & -4x^3 & & & \\
\hline
 & & 4x^3 & -3x^2 + & x - & 1 \\
 & & 4x^3 & -8x^2 & & \\
\hline
 & & & 5x^2 + & x - & 1 \\
 & & & 5x^2 & -10x & \\
\hline
 & & & & 11x - & 1 \\
 & & & & 11x & -22 \\
\hline
 & & & & & 21
\end{array}
$$

Synthetic division streamlines the above procedure. We want to record on paper as little as possible to represent the long division procedure. If terms involving a given power of x are kept in the same vertical column, we can record the coefficient and ignore recording the power of x. Further, many of the terms in the dividend are needlessly repeated several times throughout the long division procedure. In addition, we need not record the first term to be subtracted in each step. Eliminating these duplications, we obtain the following simplification:

$$
\begin{array}{r|rrrrr}
 & 2 & 4 & 5 & 11 & \\
\hline
1-2 & 2 & 0 & -3 & 1 & -1 \\
 & & -4 & & & \\
\hline
 & & 4 & & & \\
 & & & -8 & & \\
\hline
 & & & 5 & & \\
 & & & & -10 & \\
\hline
 & & & & 11 & \\
 & & & & & -22 \\
\hline
 & & & & & 21
\end{array}
$$

Next, we collapse vertically:

$$
\begin{array}{r|rrrrr}
 & 2 & 4 & 5 & 11 & \\
\hline
1-2 & 2 & 0 & -3 & 1 & -1 \\
 & & -4 & -8 & -10 & -22 \\
\hline
 & & 4 & 5 & 11 & 21
\end{array}
$$

Because the divisor will always be of the form $x - c$, we can eliminate recording the coefficient 1 of x in $x - 2$. All coefficients of the quotient, except for the first term 2, appear in the last row. If we put 2 in the last row, then we can record the same information as follows:

$$\begin{array}{l|rrrrrl} \text{(Divisor)}\quad -2 & 2 & 0 & -3 & 1 & -1 & \text{(Dividend)} \\ & & -4 & -8 & -10 & -22 & \\ \hline & 2 & 4 & 5 & 11 & 21 & \end{array}$$

$$\underbrace{2\quad 4\quad 5\quad 11}_{\text{Quotient}}\quad \underbrace{21}_{\text{Remainder}}$$

The numbers in the second row are obtained by multiplying the element in the previous column in the third row by the number -2 (from the divisor). The numbers in the third row are obtained by subtracting the corresponding number in the second row from the number in the first row.

Next, we change -2 to 2 in the divisor, and then we obtain the elements in the second row as before. Finally, we add corresponding entries in the first and second rows to get the last row. This produces the same last row as in the previous step. This is the form we use for synthetic division:

$$\begin{array}{l|rrrrrl} \text{(The } c \text{ in } x-c\text{)}\quad 2 & 2 & 0 & -3 & 1 & -1 & \text{(Dividend)} \\ & & 4 & 8 & 10 & 22 & \\ \hline & 2 & 4 & 5 & 11 & 21 & \end{array}$$

$$\underbrace{2\quad 4\quad 5\quad 11}_{\text{Quotient}}\quad \underbrace{21}_{\text{Remainder}}$$

The entries in the last row give, in order, the coefficients of the quotient and the remainder. The number at the far left in the first row is c if we are dividing by $x - c$. In the above case, the number is 2 because we are dividing by $x - 2$. The rest of the numbers in the first row are the coefficients, in order, of the dividend with 0's for missing terms. Thus,

$$2x^4 + (0)x^3 - 3x^2 + x - 1 = (x - 2)(2x^3 + 4x^2 + 5x + 11) + 21\,.$$

• EXAMPLE 4: Use synthetic division to find the quotient and remainder when $3x^4 + 7x^3 - 5x^2 + x - 11$ is divided by $x + 3$.

SOLUTION:

$$\begin{array}{r|rrrrr} -3 & 3 & 7 & -5 & 1 & -11 \\ & & -9 & 6 & -3 & 6 \\ \hline & 3 & -2 & 1 & -2 & -5 \end{array}$$

The numbers in the last row give the coefficients of the powers of x in the quotient and the remainder. Thus, the quotient is $3x^3 - 2x^2 + x - 2$ and the remainder is -5, or equivalently, $3x^4 + 7x^3 - 5x^2 + x - 11 = (x + 3)(3x^3 - 2x^2 + x - 2) - 5$. You can check this result using long division, or by multiplying $x + 3$ by $3x^3 - 2x^2 + x - 2$ and subtracting 5. •

• EXAMPLE 5: Use synthetic division to show that 3 is *not* a zero of $f(x) = 2x^3 + 5x^2 - 2x - 3$.

SOLUTION: If we show that the remainder when $f(x) = 2x^3 + 5x^2 - 2x - 3$ is divided by $x - 3$ is not zero, then 3 is not a zero of f:

$$\begin{array}{r|rrrr} 3 & 2 & 5 & -2 & -3 \\ & & 6 & 33 & 93 \\ \hline & 2 & 11 & 31 & 90 \end{array}$$

Thus, 3 is not a zero of f; in fact, $f(3) = 90$. •

If we divide a polynomial $f(x)$ by $x - c$, then the last number in the last row is the remainder that, by the Remainder Theorem, is $f(c)$. For this reason, synthetic division is sometimes called **synthetic substitution**. Synthetic division is one way to find the value of a polynomial $f(x)$ at a specific value of x.

• EXAMPLE 6: Find all rational zeros of $f(x) = x^3 - 3x - 2$.

SOLUTION: We divide each factor of the constant term -2 by each factor of the leading coefficient. Because the coefficient of x^3 is 1, all the possible rational number zeros are integers. The complete list of possible rational zeros is ± 1, ± 2. It is easy to check that $f(-1) = 0$ so that $x + 1$ is a factor. We use synthetic division to factor f:

$$\begin{array}{r|rrrr} -1 & 1 & 0 & -3 & -2 \\ & & -1 & 1 & 2 \\ \hline & 1 & -1 & -2 & 0 \end{array}$$

Therefore, $x^3 - 3x - 2 = (x + 1)(x^2 - x - 2)$. We can complete the factorization of f and obtain

$$x^3 - 3x - 2 = (x + 1)(x + 1)(x - 2) = (x + 1)^2(x - 2)\,.$$

Now, we can see that -1 and 2 are the rational zeros of f.

Alternatively, the list of possible rational zeros together with the graph of $f(x) = x^3 - 3x - 2$ (Figure 3.3.3) suggests that only -1 and 2 are possible rational zeros. Then, we

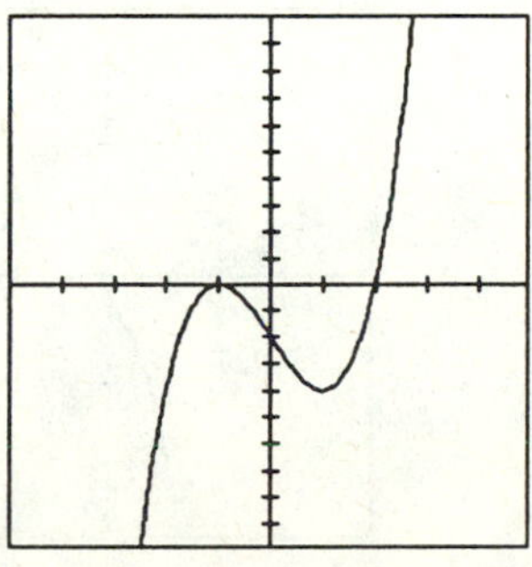

$f(x) = x^3 - 3x - 2$
$[-5, 5]$ by $[-10, 10]$
Figure 3.3.3

can compute $f(-1)$ and $f(2)$ to see that both -1 and 2 are zeros. Notice that the graph of f is tangent to the x-axis at the zero -1 of f. It doesn't cross the x-axis at $x = -1$ because the factor $x + 1$ is squared. (Why?) •

• EXAMPLE 7: Use synthetic division to show that $\frac{1}{2}$ is a rational zero of $f(x) = 6x^3 - 5x^2 + 3x - 1$. Then find all other real zeros of f.

SOLUTION: We want to show that $f(\frac{1}{2}) = 0$. The graph of f (Figure 3.3.4) suggests that this is true. We could substitute $\frac{1}{2}$ for x in the expression for f. The Remainder Theorem states that $f(c)$ is the remainder when $f(x)$ is divided by $x - c$. Thus, we need to show that the remainder when $f(x)$ is divided by $x - \frac{1}{2}$ is zero. We use synthetic division to divide f by $x - \frac{1}{2}$ because we also want to factor f:

$$\begin{array}{r|rrrr} \frac{1}{2} & 6 & -5 & 3 & -1 \\ & & 3 & -1 & 1 \\ \hline & 6 & -2 & 2 & 0 \end{array}$$

Thus, from the last line we see that $f(\frac{1}{2}) = 0$. We use the last line to factor f:

$$\begin{aligned} 6x^3 - 5x^2 + 3x - 1 &= (x - \tfrac{1}{2})(6x^2 - 2x + 2) \\ &= (x - \tfrac{1}{2})(2)(3x^2 - x + 1) \\ &= (2x - 1)(3x^2 - x + 1). \end{aligned}$$

Now, to find the other real zeros of f, we need to find the real zeros of $3x^2 - x + 1$. The quadratic formula can be used to find that the zeros of $3x^2 - x + 1$ are

$$\frac{1 \pm \sqrt{1 - 12}}{6} \quad \text{or} \quad \frac{1 \pm \sqrt{-11}}{6}.$$

These numbers are *not* real. Thus, $3x^2 - x + 1$ has no real zeros and f has only one real zero, namely $\frac{1}{2}$. This is certainly consistent with the graph in Figure 3.3.4. We will see in Section 3.4 that $\frac{1 \pm \sqrt{-11}}{6}$ are *complex* zeros of f. •

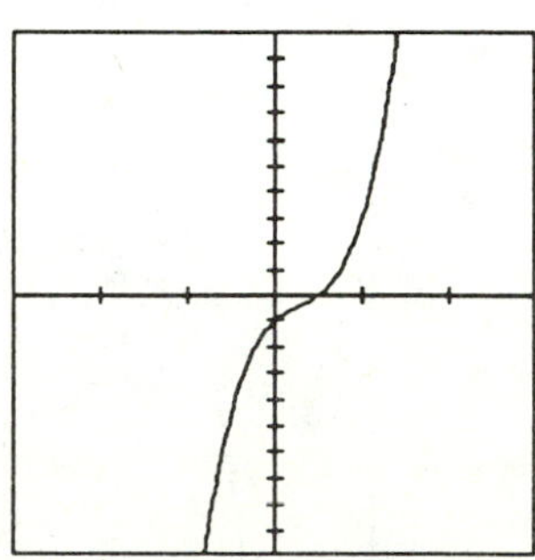

$f(x) = 6x^3 - 5x^2 + 3x - 1$
$[-3, 3]$ by $[-10, 10]$
Figure 3.3.4

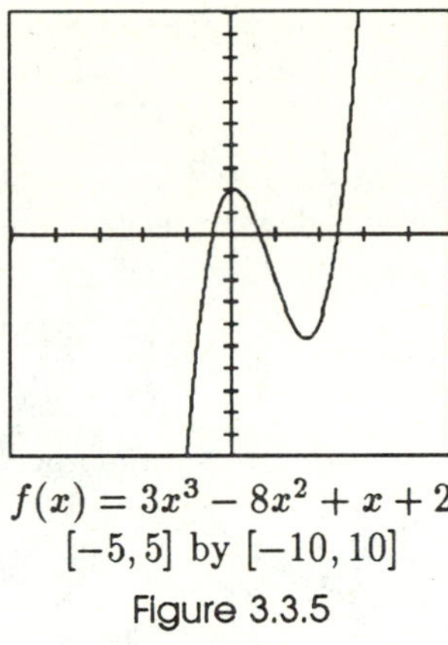

$f(x) = 3x^3 - 8x^2 + x + 2$
$[-5, 5]$ by $[-10, 10]$
Figure 3.3.5

• EXAMPLE 8: Find all real zeros of $f(x) = 3x^3 - 8x^2 + x + 2$. Classify them as integer, noninteger rational, or irrational.

SOLUTION: The possible rational roots are $\pm\frac{1}{3}, \pm\frac{2}{3}, \pm 1, \pm 2$. The graph of f in Figure 3.3.5 suggests that f has three real roots. The roots appear to be approximately equal to -0.4, 0.7, and 2.4. Thus, $\frac{2}{3}$ may be a rational root. We use synthetic division to check:

$$\begin{array}{r|rrrr} \frac{2}{3} & 3 & -8 & 1 & 2 \\ & & 2 & -4 & -2 \\ \hline & 3 & -6 & -3 & 0 \end{array}$$

Now, we can factor f:

$$\begin{aligned} 3x^3 - 8x^2 + x + 2 &= (x - \tfrac{2}{3})(3x^2 - 6x - 3) \\ &= (3x - 2)(x^2 - 2x - 1). \end{aligned}$$

Using the quadratic formula, we find that the zeros of $x^2 - 2x - 1$ are $\frac{2 \pm \sqrt{(-2)^2 - 4(-1)}}{2}$, or $1 \pm \sqrt{2}$. Thus, $\frac{2}{3}$ is a noninteger rational zero, and $1 + \sqrt{2}$ and $1 - \sqrt{2}$ are irrational. Remember that a graphing utility *cannot* distinguish between rational and irrational real zeros. •

• EXERCISES 3-3

Determine all real zeros of each function by algebraic methods. Classify them as integer, noninteger rational, or irrational. Check with a graphing utility.

1. $f(x) = x^2 - 3x + 4$
2. $f(x) = 2x^2 + 5x - 3$
3. $f(x) = x^2 - 3$
4. $g(x) = x^3 - 4x$
5. $g(x) = x^3 - 5x$
6. $f(x) = x^3 + x$

Show that the indicated number is a zero of $f(x)$ and then factor f.

7. $f(x) = 2x^3 + x^2 - 6x - 3;\ -\frac{1}{2}$

8. $f(x) = 3x^3 - 7x^2 + 10x - 8;\ \frac{4}{3}$

Use the Rational Zeros Theorem to list all possible rational zeros of each function. Draw a complete graph of each function in an appropriate viewing rectangle and indicate which of the possible rational zeros are actual zeros. Use this information to determine all real zeros (both rational and irrational).

9. $f(x) = x^3 - 2x^2 + 3x - 4$

10. $g(x) = 2x^3 - x + 1$

11. $f(x) = 2x^3 - 4x^2 - 5x + 10$

12. $g(x) = 4x^3 - x^2 + 2x - 6$

Use synthetic division to find the quotient and remainder when the first polynomial is divided by the second polynomial.

13. $x^3 + 2x^2 - 3x + 1;\ x + 2$

14. $-2x^3 - x^2 - 4;\ x - 1$

15. $x^4 - 2x^3 + x^2 - x + 2;\ x + 1$

16. $2x^4 - 3x^2 + x + 5;\ x + 3$

Draw a complete graph of each polynomial. Confirm that there is at least one rational root in each case. Use synthetic division to find the other quadratic factor, and then determine all real zeros. Classify each real zero as rational or irrational.

17. $f(x) = x^3 + 4x^2 - 4x - 1$

18. $f(x) = 6x^3 - 5x - 1$

19. $f(x) = 3x^3 - 7x^2 + 6x - 14$

20. $f(x) = 2x^3 - x^2 - 9x + 9$

Determine the real zeros of each polynomial. Classify them as rational or irrational. Carefully outline why you are sure of your results.

21. $f(x) = 2x^4 - 7x^3 - 2x^2 - 7x - 4$

22. $f(x) = 3x^4 - 2x^3 + 3x^2 + x - 2$

23. Does $f(x) = 3x^4 - 2x^2 + x - 1$ have any rational zeros? Any irrational zeros? If so, how many of each? Give both geometric and algebraic reasons for your answer.

24. Draw the graph of $f(x) = 900x^3 - 1200x^2 + 40x - 1$ in the viewing rectangle $[-2, 3]$ by $[-1000, 1000]$. What are the real zeros of f? Classify each zero as integer, noninteger rational, or irrational.

The framing for a building with rectangular sides and square ends is constructed from 480 feet of steel framing as shown in Figure 3.3.6. Assume that all of the steel is used in framing. Let x be the side length of each square end. Ignore the width and thickness of the steel.

25. Write an algebraic representation for the volume of the building as a function of x.

26. Draw a complete graph of the model in Exercise 25. What is the domain of the model? What values of x make sense in this problem situation?

27. What are the possible dimensions of the building if its volume is 3000 cubic feet?

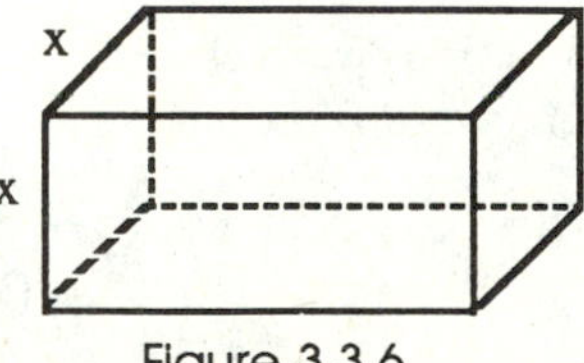

Figure 3.3.6

28. What are the dimensions of the building with maximum volume? What is the maximum volume?

29. Write an algebraic representation for the total surface area (area of the four sides plus the flat top) of the building as a function of x.

30. Draw a complete graph of the model in Exercise 29. What is the domain of the model? What values of x make sense in this problem situation?

31. In Exercise 27, what are the dimensions of the building with volume 3000 cubic feet that has the *least* surface area?

32. Are there values for x that make sense in the problem situation where the volume is equal in magnitude to the magnitude of the surface area? If so, what are the values? Give a graphical argument.

3.4 Complex Numbers as Zeros

In this section we begin the study of complex numbers and find complex number zeros of polynomials with real number coefficients. The Fundamental Theorem of Algebra is stated and used to determine the number of zeros of a polynomial. Historically, complex numbers were introduced to provide solutions to equations such as $x^2 + 1 = 0$. This equation has *no* real number solutions. Mathematicians have constructed a number system that contains both the real numbers and the solutions to equations like $x^2 + 1 = 0$. This number system is called the complex numbers.

Complex numbers are important in real world applications as well as in pure mathematics. For example, complex numbers are involved in the study of problems in mechanical vibrations of objects, in the theory of alternating electrical current, and in the flow of fluids.

The synthetic division process explained in Section 3.3 is valid for complex numbers. However, the quotient and remainder can be polynomials with complex number coefficients in this case. The Remainder Theorem and the Factor Theorem are valid for complex numbers. These theorems and the division algorithm are valid for polynomials with complex number coefficients.

Let i be a symbol with the property $i^2 + 1 = 0$. Formally, i is a solution to the equation $x^2 + 1 = 0$. Thus, $i^2 + 1 = 0$ or $i^2 = -1$. We also write $i = \sqrt{-1}$. Both i and $-i$ are solutions to the quadratic equation $x^2 + 1 = 0$.

• DEFINITION An expression of the form $a + bi$, where a and b are real numbers is called a **complex number**. The set of all such expressions is called the set of **complex numbers**, or simply the **complex numbers**. The real number a is called the **real part** of the complex number $a + bi$, and the real number b the **imaginary part** of the complex number. •

We agree that $a + 0i = a$. Then, the subset of complex numbers with $b = 0$ is just the set of real numbers. Before finding complex number zeros of polynomials we need to understand the arithmetic of complex numbers. Equality and the operations of addition, subtraction, multiplication, and division for complex numbers are defined as follows.

• DEFINITION Let $a + bi$ and $c + di$ be any two complex numbers.

1. Equality: $a + bi = c + di$ if and only if $a = b$ and $c = d$.
2. Addition: $(a + bi) + (c + di) = (a + c) + (b + d)i$
3. Subtraction: $(a + bi) - (c + di) = (a - c) + (b - d)i$
4. Multiplication: $(a + bi)(c + di) = (ac - bd) + (ad + bc)i$
5. Division: $\dfrac{a + bi}{c + di} = \dfrac{ac + bd}{c^2 + d^2} + \dfrac{bc - ad}{c^2 + d^2}i$, where $c^2 + d^2 \neq 0$. •

Thus, to add or subtract two complex numbers, we add or subtract their corresponding real and imaginary parts. To divide by $c + di$ we need $c^2 + d^2 \neq 0$. This means that not both c and d can be zero, that is, $c + di \neq 0 + 0i$. Addition and multiplication of complex numbers satisfies the commutative and associative properties. Furthermore, the distributive property holds for complex numbers.

• EXAMPLE 1: Write each of the following in the form $a + bi$, where a and b are real numbers.

(a) $(7 - 3i) + (4 + 5i)$ (b) $(7 - 3i) - (4 + 5i)$

(c) $(2 + 3i)(5 - i)$ (d) $\dfrac{4 - 3i}{2 + 5i}$

SOLUTION:

(a) We use the definition to determine the sum $(7 - 3i) + (4 + 5i)$:

$$\begin{aligned}(7 - 3i) + (4 + 5i) &= (7 + 4) + (-3 + 5)i \\ &= 11 + 2i.\end{aligned}$$

(b) We use the definition to determine the difference $(7 - 3i) - (4 + 5i)$:

$$\begin{aligned}(7 - 3i) - (4 + 5i) &= (7 - 4) + (-3 - 5)i \\ &= 3 - 8i.\end{aligned}$$

(c) First, we use the definition to find the product $(2 + 3i)(5 - i)$:

$$\begin{aligned}(2 + 3i)(5 - i) &= ((2)(5) - (3)(-1)) + ((2)(-1) + (3)(5))i \\ &= 13 + 13i.\end{aligned}$$

Alternatively, we can treat complex numbers as binomials and multiply in the expected way:

$$\begin{aligned}(2 + 3i)(5 - i) &= (2)(5) + 2(-i) + (3i)5 + (3i)(-i) \\ &= 10 - 2i + 15i - 3i^2 \\ &= 10 + 13i + 3 \\ &= 13 + 13i.\end{aligned}$$

Notice we replaced i^2 by -1 above.

(d) We use the definition to determine the quotient $\frac{4-3i}{2+5i}$:

$$\frac{4-3i}{2+5i} = \frac{(4)(2)+(-3)(5)}{2^2+5^2} + \frac{(-3)(2)-(4)(5)}{2^2+5^2}i$$
$$= -\frac{7}{29} - \frac{26}{29}i$$ •

In order to divide complex numbers without memorizing the definition, we need the following.

• DEFINITION The **complex conjugate** of $u = a + bi$ is $\overline{u} = a - bi$. •

• THEOREM 1 Let $u = a + bi$ be any complex number.

(a) $(a + bi) + (0 + 0i) = a + bi$ (b) $(a + bi)(1 + 0i) = a + bi$

(c) $u\bar{u} = (a + bi)(a - bi) = a^2 + b^2$

PROOF

(a) $(a + bi) + (0 + 0i) = (a + 0) + (b + 0)i = a + bi$

(b) $(a + bi)(1 + 0i) = ((a)(1) - (b)(0)) + ((a)(0) + (b)(1))i = a + bi$

(c) As in Example 1, we treat $a + bi$ and $a - bi$ as binomials and expand as follows:

$$u\bar{u} = (a + bi)(a - bi) = a^2 - abi + abi - b^2i^2$$
$$= a^2 + b^2.$$

With our agreement to write $0 + 0i$ as 0 and $1 + 0i$ as 1, Theorem 1 shows that 0 and 1 are the additive and multiplicative identities, respectively, for the complex numbers. The third part of Theorem 1 shows that the product of a complex number and its complex conjugate is a real number. We use this fact to divide complex numbers.

• EXAMPLE 2: Write each of the following in the form $a + bi$, where a and b are real numbers.

(a) $\frac{5+i}{2-3i}$ (b) $\frac{7-i}{2i}$

SOLUTION:

(a) We multiply the numerator and denominator of $\frac{5+i}{2-3i}$ by the complex conjugate of the denominator and simplify:

$$\frac{5+i}{2-3i} = \frac{(5+i)(2+3i)}{(2-3i)(2+3i)} = \frac{10+2i+15i-3}{4+9} = \frac{7}{13} + \frac{17}{13}i.$$

(b) We use the same procedure as in (a). Because the complex conjugate of $2i$ is $-2i$, we multiply the numerator and denominator of $\frac{7-i}{2i}$ by $-2i$:

$$\frac{7-i}{2i} = \frac{(7-i)(-2i)}{(2i)(-2i)} = \frac{-2-14i}{4} = -\frac{1}{2} - \frac{7}{2}i.$$

Alternatively, we can put $\frac{7-i}{2i}$ in the form $a+bi$ by multiplying numerator and denominator by i instead of the complex conjugate $-2i$ of the denominator:

$$\frac{7-i}{2i} = \frac{(7-i)(i)}{(2i)(i)} = \frac{1+7i}{-2} = -\frac{1}{2} - \frac{7}{2}i.$$ •

Whenever the denominator of a quotient of two complex numbers is of the form bi, the quotient can be put in the form $a+bi$ by multiplying numerator and denominator by i.

To find the value of a polynomial $f(x)$ when x is a complex number requires raising the complex number to various powers. In particular, we need to know the value of i^n for integer n. The values of i^n for some positive and negative values of n are given below:

$$i^1 = i \qquad i^5 = i^4 i = i \qquad i^{-1} = \frac{1}{i} = -i$$

$$i^2 = -1 \qquad i^6 = i^4 i^2 = -1 \qquad i^{-2} = \frac{1}{i^2} = -1$$

$$i^3 = i^2 i = -i \qquad i^7 = i^4 i^3 = -i \qquad i^{-3} = \frac{1}{i^3} = i$$

$$i^4 = i^2 i^2 = 1 \qquad i^8 = i^4 i^4 = 1 \qquad i^{-4} = \frac{1}{i^4} = 1$$

Notice that in the examples given every integer power of i is equal to one of the following: i, $-i$, -1, or 1. In fact, in the exercises we ask you to prove the following theorem.

• THEOREM 2 If n is an integer, then $i^n = i^r$ where r is the remainder when n is divided by 4.

For example, i^{271} and i^3 are equal because the remainder when 271 is divided by 4 is 3. Thus, $i^{271} = i^3 = -i$. Notice that $271 = 4(67) + 3$ so that

$$i^{271} = i^{4(67)+3} = (i^4)^{67} i^3 = (1)^{67} i^3 = i^3.$$

Every power of i is equal to one of i, $-i$, -1, or 1.

Direct substitution verifies that $\pm i$ are solutions to $x^2+1=0$. The quadratic formula can be used to find complex zeros of a general quadratic polynomial. First we need the following definition.

• DEFINITION If a is a positive number, the **principal square root of $-a$** is $i\sqrt{a}$, that is, $\sqrt{-a} = i\sqrt{a}$. In particular, $\sqrt{-1} = i$. •

• EXAMPLE 3: Find the zeros of x^2+x+1.

SOLUTION: We use the quadratic formula to solve the equation $x^2+x+1=0$:

$$x = \frac{-1 \pm \sqrt{1-4}}{2}.$$

Because $\sqrt{1-4} = \sqrt{-3} = i\sqrt{3}$, we can rewrite the solutions as $\frac{-1}{2} \pm \frac{\sqrt{3}}{2}i$. Thus, the two zeros are $-\frac{1}{2} + \frac{\sqrt{3}}{2}i$ and $-\frac{1}{2} - \frac{\sqrt{3}}{2}i$.

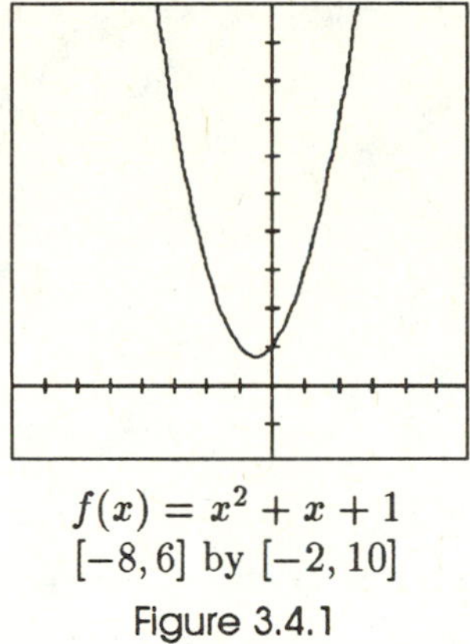

$f(x) = x^2 + x + 1$
$[-8, 6]$ by $[-2, 10]$
Figure 3.4.1

The graph of $f(x) = x^2 + x + 1$ is shown in Figure 3.4.1. Because f has no real zeros, its graph will not touch or cross the x-axis. •

Notice that the two zeros of $x^2 + x + 1$ found in Example 3 are complex conjugates. This is no accident. The following theorem is true about zeros of polynomials with real number coefficients.

• THEOREM 3 If $a + bi$ with $b \neq 0$ is a zero of a polynomial with real coefficients, then its complex conjugate $a - bi$ is *also* a zero of the polynomial.

In Section 3.2 we showed that a real number c was a zero of the polynomial f if and only if $x - c$ was a factor of f (Factor Theorem). In fact, this theorem is true even if c is a complex number. For example, $x^2 + 1 = (x - i)(x + i)$ where i and $-i$ are the complex zeros of $x^2 + 1$. Remember, by agreement, we use factors with noninteger coefficients only when we want to display zeros.

If a polynomial f with real number coefficients has a nonreal complex number zero, then we can use Theorem 3 to show that f has a quadratic factor with real number coefficients.

• THEOREM 4 If the complex number $a + bi$ with $b \neq 0$ is a zero of a polynomial f with real coefficients, then $x^2 - 2ax + a^2 + b^2$ is a factor of f.

PROOF Because $a + bi$ is a zero, Theorem 3 requires $a - bi$ to also be a zero. Then, both $x - (a + bi)$ and $x - (a - bi)$ are factors of f. Therefore, the product

$$(x - a - bi)(x - a + bi) = x^2 - 2ax + a^2 + b^2$$

is a factor of f.

• EXAMPLE 4: Show that $1 - 2i$ is a zero of $f(x) = 4x^4 + 17x^2 + 14x + 65$ and find all other zeros of f.

SOLUTION: To compute $f(1-2i)$ we need to compute $(1-2i)^2$ and $(1-2i)^4$:

$$\begin{aligned}(1-2i)^2 &= 1-4i+4i^2\\ &= -3-4i;\\ (1-2i)^4 &= ((1-2i)^2)^2\\ &= (-3-4i)^2\\ &= 9+24i+16i^2\\ &= -7+24i.\end{aligned}$$

Now,

$$\begin{aligned}f(1-2i) &= 4(1-2i)^4+17(1-2i)^2+14(1-2i)+65\\ &= 4(-7+24i)+17(-3-4i)+14(1-2i)+65\\ &= 0.\end{aligned}$$

Alternatively, we can use synthetic substitution to show that $f(1-2i)=0$:

$$\begin{array}{r|rrrrr} 1-2i & 4 & 0 & 17 & 14 & 65\\ & & 4-8i & -12-16i & -27-26i & -65\\ \hline & 4 & 4-8i & 5-16i & -13-26i & 0\end{array}$$

Thus, $1-2i$ is a zero of f. By Theorem 4,

$$x^2-(2)(1)x+1^2+(-2)^2 = x^2-2x+5$$

must be a factor of f. Using long division we can factor f:

$$f(x) = (x^2-2x+5)(4x^2+8x+13).$$

Finally, we use the quadratic formula to find the two zeros of $4x^2+8x+13$:

$$\begin{aligned}x &= \frac{-8\pm\sqrt{64-208}}{8}\\ &= \frac{-8\pm\sqrt{-144}}{8}\\ &= \frac{-8\pm 12i}{8}\\ &= \frac{-2\pm 3i}{2}\\ &= -1\pm\frac{3}{2}i.\end{aligned}$$

Thus, the four zeros of f are $1-2i$, $1+2i$, $-1+\frac{3}{2}i$, and $-1-\frac{3}{2}i$. •

The graph of f in the $[-10,10]$ by $[-500,500]$ viewing rectangle is a complete graph (Figure 3.4.2). The graph of f does not intersect the x-axis because f has *no* real zeros.

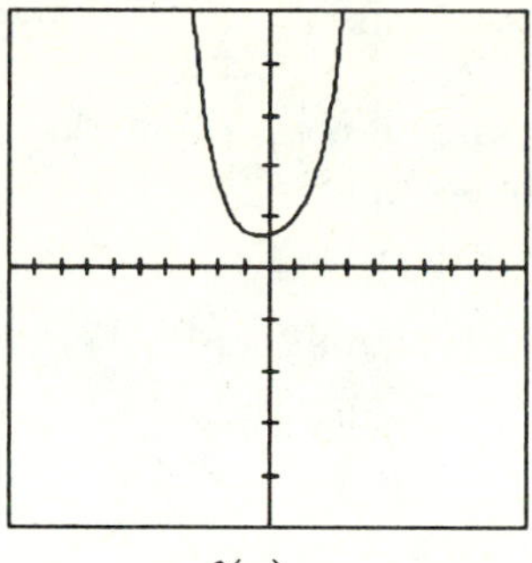

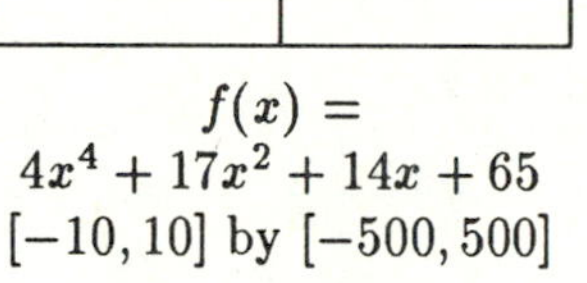

$f(x) =$
$4x^4 + 17x^2 + 14x + 65$
$[-10, 10]$ by $[-500, 500]$

Figure 3.4.2

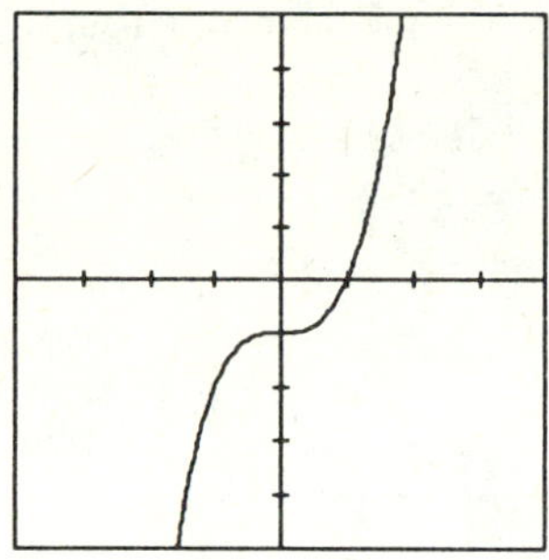

$f(x) = x^3 - 1$
$[-4, 4]$ by $[-5, 5]$

Figure 3.4.3

The graph of a function helps us find its real zeros but not its nonreal complex zeros. Finding the nonreal complex zeros can be more difficult than finding the real zeros and is often impossible.

• EXAMPLE 5: Find the zeros of $x^3 - 1$.

SOLUTION: A complete graph of $f(x) = x^3 - 1$ is given in Figure 3.4.3. This graph suggests that 1 is a zero of $x^3 - 1$ and that there are no other real zeros. We can confirm that 1 is a zero and factor f:

$$x^3 - 1 = (x - 1)(x^2 + x + 1).$$

In Example 3 we found that $-\frac{1}{2} + \frac{\sqrt{3}}{2}i$ and $-\frac{1}{2} - \frac{\sqrt{3}}{2}i$ were the two zeros of $x^2 + x + 1$. Thus, $x^3 - 1$ has three zeros: -1, $-\frac{1}{2} + \frac{\sqrt{3}}{2}i$, and $-\frac{1}{2} - \frac{\sqrt{3}}{2}i$. •

In Section 2.1 we gave the Cardan formulas for the zeros of a cubic equation. These formulas involved a complex number with third power equal to one. It can be shown that both

$$(-\frac{1}{2} + \frac{\sqrt{3}}{2}i)^3 = 1 \quad \text{and} \quad (-\frac{1}{2} - \frac{\sqrt{3}}{2}i)^3 = 1 .$$

Thus, either of the two complex numbers in Example 5 can be used for the complex number in the formula given in Section 2.1 for the exact zeros of a cubic equation.

The three solutions to $x^3 - 1 = 0$ found in Example 5 are called the **cube roots of unity**. Unity is just another way of saying *one*.

• DEFINITION The solutions to the equation $x^n - 1 = 0$ are called the **nth roots of unity**. •

It turns out that for any whole number n, there are n distinct nth roots of unity. Remember that we use the words *root* and *zero* interchangeably with polynomials. Thus, a is a root of a polynomial if and only if a is a zero of a polynomial.

• EXAMPLE 6: Find the four fourth roots of unity.

SOLUTION: We need to find the solutions to the equation $x^4 - 1 = 0$. We can look at the graph of $y = x^4 - 1$, or we can factor to see that $x^4 - 1$ has two real zeros:

$$x^4 - 1 = (x^2 - 1)(x^2 + 1)$$
$$x^4 - 1 = (x - 1)(x + 1)(x^2 + 1).$$

Thus, 1 and -1 are real zeros of $x^4 - 1$. The zeros of $x^2 + 1$ are i and $-i$. Therefore, the four fourth roots of unity are 1, -1, i, and $-i$. •

We state and use an important fact about polynomials. The proof of this is beyond the scope of this textbook.

• THEOREM 5 **(Fundamental Theorem of Algebra)** Every polynomial function of degree greater than 0 with real coefficients has at least one zero. This zero may be a real number or a nonreal complex number.

We use this theorem to write any polynomial of degree greater than 0 as a product of linear factors. Let $f(x)$ be a polynomial of degree n. By the Fundamental Theorem of Algebra, $f(x)$ has a zero, say c_1. Then, by the Factor Theorem, we can write $f(x) = (x - c_1)f_1(x)$. If c_1 is real, then the coefficients of f_1 are real. If c_1 is complex, then $\bar{c}_1$ is also a zero of f by Theorem 3. It follows from Theorem 4 that $f(x) = (x - c_1)(x - \bar{c}_1)f_1(x)$ where again the coefficients of f_1 are real. Now we apply the Fundamental Theorem of Algebra to $f_1(x)$. Let c_2 be a zero of the polynomial f_1, assured by the Fundamental Theorem of Algebra. Then, $f_1(x) = (x - c_2)f_2(x)$. Continuing in this way we can complete the factorization of f.

The above discussion can be summarized as follows.

• THEOREM 6 If f is a polynomial of degree n, then we can write $f(x) = a(x - c_1)(x - c_2) \cdots (x - c_n)$, where $c_1, c_2, \ldots, c_n$ are the zeros of f. The c_i are complex numbers and need not be distinct or real.

One consequence of Theorem 6 is that a polynomial of degree n cannot have *more* than n zeros. However, the c_i's in Theorem 6 need not be distinct. For example, we could have

$$f(x) = (x - 2)(x - 2)(x - 2)(x + 1)(x + 1)$$
$$= (x - 2)^3(x + 1)^2 .$$

In this case, we say that 2 is a zero of f of **multiplicity 3** and -1 is a zero of **multiplicity 2**. Counting multiplicities, we can conclude from Theorem 6 that a polynomial of degree n has n zeros.

The graph of $f(x) = (x - 2)^3(x + 1)^2$ in Figure 3.4.4 is typical of the behavior of functions near real zeros of even and odd multiplicity. Notice the graph of f is tangent to the x-axis at $x = -1$. This is what happens at real zeros of *even* multiplicity. The graph of f crosses the x-axis at $x = 2$ but is relatively flat near this zero. This is what happens at real zeros of *odd* multiplicity.

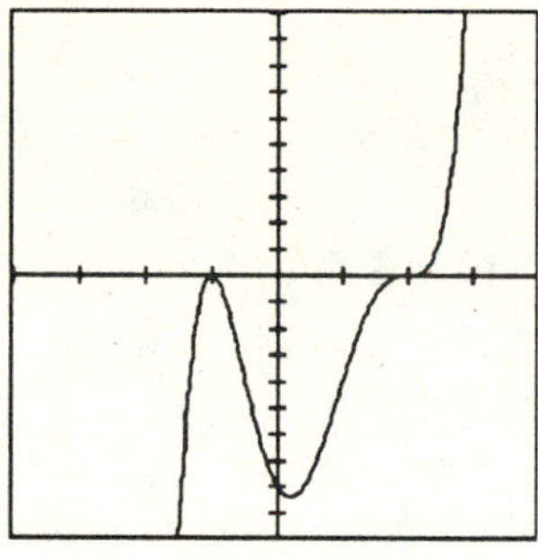

$f(x) = (x-2)^3(x+1)^2$
$[-4, 4]$ by $[-10, 10]$
Figure 3.4.4

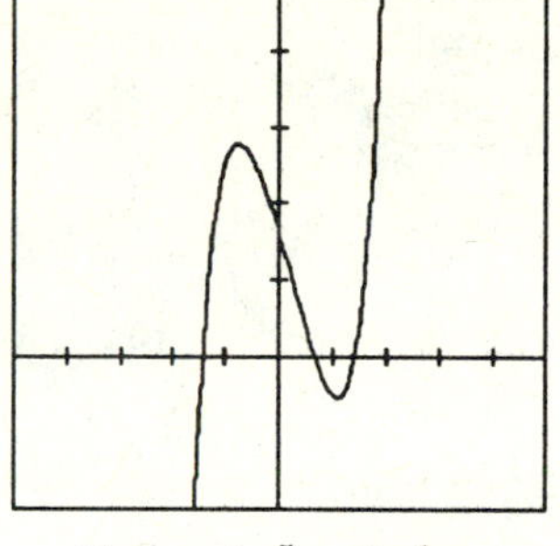

$f(x) = 3x^5 - 2x^4 +$
$6x^3 - 4x^2 - 24x + 16$
$[-5, 5]$ by $[-20, 50]$
Figure 3.4.5

• EXAMPLE 7: Write $f(x) = 3x^5 - 2x^4 + 6x^3 - 4x^2 - 24x + 16$ as a product of linear factors. Classify the zeros of f as rational, irrational, or nonreal complex.

SOLUTION: First we determine the list of possible rational zeros using the Rational Zeros Theorem:

$$\pm 1, \pm 2, \pm 4, \pm 8, \pm 16$$
$$\pm\frac{1}{3}, \pm\frac{2}{3}, \pm\frac{4}{3}, \pm\frac{8}{3}, \pm\frac{16}{3}$$

The graph of f in the $[-5, 5]$ by $[-20, 50]$ viewing rectangle (Figure 3.4.5) suggests that the list of possible rational zeros can be reduced to $-\frac{4}{3}$, $\frac{2}{3}$, $\frac{4}{3}$. It is important to note that it is not possible to determine whether a real zero is rational or irrational from the graph above. In fact, it turns out that only one of the three real zeros of f is rational.

We use synthetic division to show that $\frac{2}{3}$ is a zero of f and then factor f:

$\frac{2}{3}$	3	−2	6	−4	−24	16
		2	0	4	0	−16
	3	0	6	0	−24	0

Thus, $f(\frac{2}{3}) = 0$ and f factors as follows:

$$\begin{aligned} f(x) &= (x - \tfrac{2}{3})(3x^4 + 6x^2 - 24) \\ &= (x - \tfrac{2}{3})(3)(x^4 + 2x^2 - 8) \\ &= (3x-2)(x^4 + 2x^2 - 8) \\ &= (3x-2)(x^2-2)(x^2+4). \end{aligned}$$

The zeros of $x^2 - 2$ are $\pm\sqrt{2}$, and the zeros of $x^2 + 4$ are $\pm 2i$. Therefore, we can write

$$f(x) = (3x-2)(x-\sqrt{2})(x+\sqrt{2})(x-2i)(x+2i)\,.$$

Notice that the two real zeros $\sqrt{2}$ and $-\sqrt{2}$ are irrational. The zero $\frac{2}{3}$ is rational and the zeros $2i$ and $-2i$ are nonreal complex. •

If we want to find a polynomial with real number coefficients that has a given nonreal complex number as a zero, then, by Theorem 3, its complex conjugate must also be a zero.

• EXAMPLE 8: Find a polynomial of degree 3 with real number coefficients that has the following roots (zeros).

(a) $-\dfrac{1}{2}$, 2, and 3 (b) $-2+i$

SOLUTION:

(a) The zero $-\frac{1}{2}$ requires the factor $x+\frac{1}{2}$, or $2x+1$, if we want our polynomial to have integer coefficients. Thus, to have $-\frac{1}{2}$, 2, and 3 as zeros, our polynomial must have the product $(2x+1)(x-2)(x-3)$ as a factor. Because $(2x+1)(x-2)(x-3)$ is a polynomial of degree 3, any polynomial of degree 3 with real coefficients and $-\frac{1}{2}$, 2, and 3 as zeros must be of the form $a(2x+1)(x-2)(x-3)$ for some real number $a \neq 0$.

(b) Because $-2+i$ is to be a zero, we know that the complex conjugate $-2-i$ must also be a zero. Thus, the product $(x+2-i)(x+2+i)$ must be a factor of our polynomial. Because the product

$$(x+2-i)(x+2+i) = x^2+4x+5$$

is of degree 2, any polynomial of degree 3 with $-2+i$ and $-2-i$ as zeros must be of the form $(ax+b)(x^2+4x+5)$ for some real numbers a and b with $a \neq 0$. For example,

$$(x-1)(x^2+4x+5) = x^3-3x^2+x-5$$

is a polynomial of degree 3 with $-2+i$ as a zero. There are infinitely many other possibilities. •

• EXERCISES 3-4

Write each expression in the form $a+bi$ with a and b real numbers.

1. $2-3i+6$
2. $2-3i+6-4i$
3. $(2+3i)(2-i)$
4. $(2-i)(1+3i)$
5. $2(1+i)-1-i$
6. $3(6-i)-2(-1-3i)$
7. $3(2+i)^2-4i$
8. $(1-i)^3$

Determine the complex conjugate of each complex number.

9. $2-3i$
10. $-6i$
11. 2
12. $-1-\sqrt{2}i$

Write each expression in the form $a+bi$ with a and b real numbers.

13. $\dfrac{1}{2+i}$
14. $\dfrac{i}{2-i}$
15. $\dfrac{2+i}{2-i}$

16. $\dfrac{2+i}{3i}$ 17. $\dfrac{(2+i)^2(-i)}{1+i}$ 18. $\dfrac{(2-i)(1+2i)}{5+2i}$

Simplify each expression and write in the form $a + bi$.

19. i^{29} 20. i^{2946} 21. $(i^{102})^8$ 22. $i^{(-36)}$

Determine the number of real zeros and the number of nonreal complex zeros of each function.

23. $f(x) = x^2 - 2x + 7$
24. $f(x) = x^3 - 3x^2 + x + 1$
25. $f(x) = x^3 - x + 3$
26. $f(x) = x^4 - 2x^2 + 3x - 4$
27. $f(x) = x^4 - 5x^3 + x^2 - 3x + 6$
28. $f(x) = x^5 - 2x^2 - 3x + 6$

Find *all* the zeros of each polynomial. Classify each zero as integer, noninteger rational, irrational, or nonreal complex.

29. $f(x) = x^3 + 4x - 5$
30. $g(x) = x^3 - 10x^2 + 44x - 69$
31. $f(x) = x^4 + x^3 + 5x^2 - x - 6$
32. $g(x) = 3x^4 + 8x^3 + 6x^2 + 3x - 2$
33. Find the two square roots of unity; that is, solve the equation $x^2 = 1$.
34. Find the three cube roots of eight; that is, solve the equation $x^3 = 8$.

Find a polynomial with real coefficients satisfying the given conditions.

35. degree 2; zero $2 - 3i$
36. degree 3; zeros 1 and i
37. degree 3; zeros $1 - i$ and 3
38. degree 4; zeros $-2 + i$ and $1 - i$
39. (a) Can you find a polynomial of degree 3 with real coefficients that has -2 as its only real zero?
    (b) Can you find a polynomial of degree 3 with real coefficients that has $2i$ as its only complex nonreal zero?
40. (a) Find a polynomial $f(x)$ of degree 4 with real coefficients that has -3, $1 + i$, and $1 - i$ as its only zeros.
    (b) Find a polynomial $f(x)$ of degree 4 with real coefficients that has -3, $1 + i$, and $1 - i$ as its only zeros, and also satisfies $f(0) = 1$.
41. Show that $1 + i$ is a zero of $f(x) = 3x^3 - 7x^2 + 8x - 2$ and find all other zeros of f.
42. Show that $3 - 2i$ is a zero of $f(x) = x^4 - 6x^3 + 11x^2 + 12x - 26$ and find all the other zeros of f.

Write each polynomial as a product $f(x) = k(x - c_1)(x - c_2)\cdots(x - c_n)$ where n is the degree of the polynomial, k is a real number, and each c_i is a zero of f.

43. $f(x) = x^3 - x^2 + x - 1$
44. $f(x) = x^4 - 6x^2 + 5$
45. $f(x) = 2x^3 - x^2 + 3x - 4$
46. $f(x) = x^4 + 6x^3 + 7x^2 - 12x - 18$

Archimedes' law states that when a solid of density d_S is placed in a liquid of density d_L it will sink to a depth h that displaces an amount of liquid whose weight equals that of the solid. For a sphere of radius r, Archimedes' law becomes

$$\tfrac{\pi}{3}(3rh^2 - h^3)d_L = \tfrac{4}{3}\pi r^3 d_S.$$

47. If $r = 5$ feet, $d_L = 62.5$ lb/ft^3 (density of water), and $d_S = 20$ lb/ft^3, use zoom-in to determine h with error less than 0.01.
48. If $r = 5$ feet, $d_L = 62.5$ lb/ft^3 (density of water), and $d_S = 45$ lb/ft^3, use zoom-in to determine h with error less than 0.01.

49. If $r = 5$ feet, $d_L = 62.5$ lb/ft^3 (density of water), and $d_S = 70$ lb/ft^3, use zoom-in to determine h with error less than 0.01.

*50. A generalization of the *ideal gas law* $PV = nRT$ from physics and chemistry is given by the *Van der Waals* equation

$$(P + \frac{a}{V^2})(V - b) = nRT,$$

where R is the gas constant 0.08205 and n is the number of moles of gas. For isobutane it is known that $a = 12.87$ and $b = 0.1142$. Find a volume V of one mole of isobutane at a temperature T of 313 degrees and a pressure of two atmospheres by rewriting the above equation as a polynomial and using zoom-in.

51. Let $\bar{z} = a - bi$, the conjugate of $z = a + bi$. Prove that $z + \bar{z}$ is a real number for any complex number z.

52. Prove that $z \cdot \bar{z}$ is a real number for any complex number z.

53. Show that any polynomial with real coefficients can be written as a product of linear and quadratic factors each with real coefficients where the quadratic factor has no real zeros. *Hint*: Use Theorem 4.

Use the results of Exercise 53 to write each polynomial as a product of linear and quadratic factors with real coefficients where the quadratic factors have no real zeros.

54. $f(x) = 2x^3 + 7x^2 + 9x + 3$

55. $f(x) = (x^2 - 3)(x - 1 - i)(x - 1 + i)$

56. Verify that the complex number i is a zero of the polynomial $f(x) = x^3 - ix^2 + 2ix + 2$.

57. Verify that the complex conjugate $-i$ of i is *not* a zero of the polynomial in Exercise 56. Explain why this fact does not contradict Theorem 3 of this section.

58. Prove Theorem 2.

3.5 Rational Functions, Part 1

In this section we begin the study of rational functions. Specifically, we study rational functions that are quotients of two linear polynomials. Horizontal and vertical asymptotes are introduced. A problem situation that has a rational function for an algebraic representation is studied.

• DEFINITION A function f is called a **rational function** if it can be expressed in the form $f(x) = \frac{p(x)}{h(x)}$ where p and h are polynomials. •

The rational functions to be studied in this section are of the form $f(x) = \frac{ax+b}{cx+d}$. If the denominator is *not* a factor of the numerator, we will see that the graph of f can be obtained from the graph of the **reciprocal function**, $g(x) = \frac{1}{x}$ (Figure 3.5.1), using geometric transformations introduced earlier. First, we need to establish the important concept of asymptote of a function.

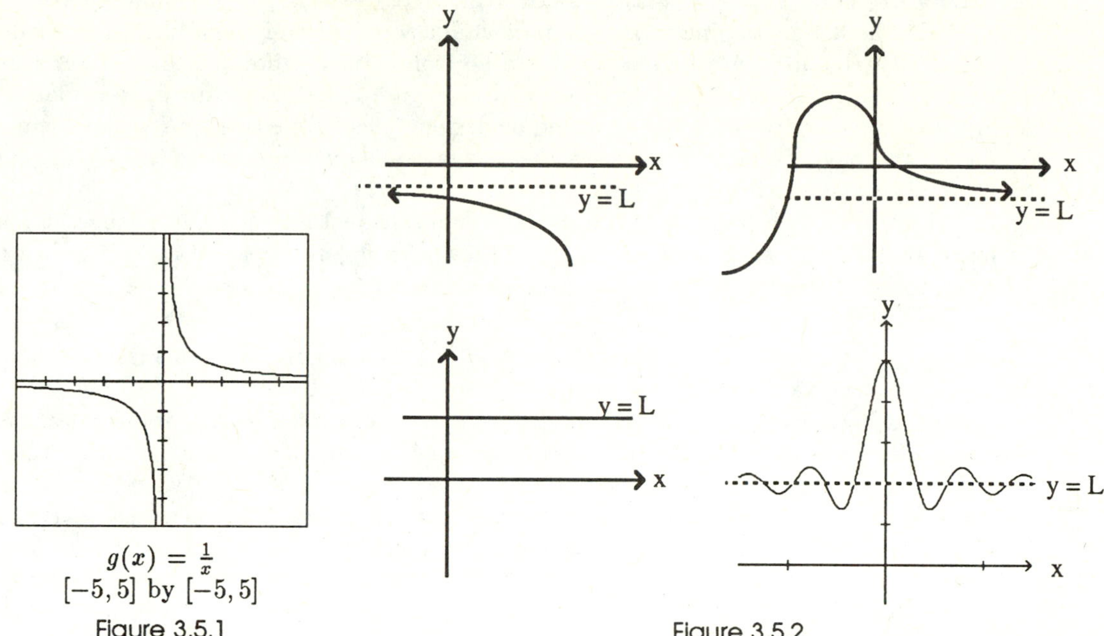

$g(x) = \frac{1}{x}$
$[-5, 5]$ by $[-5, 5]$
Figure 3.5.1

Figure 3.5.2

We can use a calculator to determine that the values of $g(x) = \frac{1}{x}$ get close to zero for values of x large in absolute value. We say that $g(x)$ approaches zero as $|x|$ approaches infinity and write $g(x) \to 0$ as $|x| \to \infty$. Thus, the end behavior of g is the same as the end behavior of the constant function $y = 0$. In Section 3.1 we called $t(x)$ an end behavior model for $s(x)$ if $\frac{s(x)}{t(x)} \to 1$ as $|x| \to \infty$. Although this definition is not satisfied, we will call the constant function $y = 0$ an end behavior model for $g(x) = \frac{1}{x}$ nonetheless. The horizontal line $y = 0$ (x-axis) is also called a *horizontal asymptote* of $g(x) = \frac{1}{x}$.

• DEFINITION The horizontal line $y = L$ is called a **horizontal asymptote** of f if $f(x) \to L$ as $x \to \infty$ or if $f(x) \to L$ as $x \to -\infty$. •

In the above definition, the notation $f(x) \to L$ as $x \to \infty$ means that as the values of x increase without bound, the corresponding values of $f(x)$ get close to L. Similarly, $f(x) \to L$ as $x \to -\infty$ means that for x negative with $|x|$ increasing without bound, the corresponding values of $f(x)$ get close to L. Figure 3.5.2 shows several graphical possibilities for the horizontal line $y = L$ to be a horizontal asymptote of a function. Notice in Figure 3.5.2 that a horizontal line is its own horizontal asymptote.

• EXAMPLE 1: Determine the horizontal asymptotes of $f(x) = \frac{1}{x} + 2$ and $g(x) = \frac{1}{x} - 3$. Determine the domain and range of each function.

SOLUTION: We know that $\frac{1}{x} \to 0$ as $|x| \to \infty$. Thus, $f(x) \to 2$ as $|x| \to \infty$ and $g(x) \to -3$ as $|x| \to \infty$. The horizontal line $y = 2$ is a horizontal asymptote and an end behavior model of f. Similarly, the horizontal line $y = -3$ is a horizontal asymptote and an end behavior model of g. The graphs of f (Figure 3.5.3) and g (Figure 3.5.4) confirm this information. The horizontal asymptotes have been added as dashed lines in the two figures and are not part of the graphs of f or g. Furthermore, these lines usually will not appear on the display screen of a graphing utility.

The domain of each function is the set of all nonzero real numbers. The range of f is the set of all real numbers not equal to 2, and the range of g is the set of all real numbers not equal to -3. (Why?) •

Notice that the value of the function $f(x) = \frac{1}{x} + 2$ in Figure 3.5.3 increases without bound for positive values of x that approach zero. We can obtain the same information numerically using a calculator. We say that $f(x)$ approaches infinity as x approaches zero through positive values and write $f(x) \to \infty$ as $x \to 0^+$. The plus sign in 0^+ means x takes on only values greater than zero. Notice also that $f(x) \to -\infty$ as $x \to 0^-$, where the minus sign in 0^- indicates that x takes on only values less than zero. The vertical line $x = 0$ (y-axis), is a *vertical asymptote* of $f(x) = \frac{1}{x} + 2$. Notice that $f(x) = \frac{1}{x} + 2$ is discontinuous at $x = 0$. (Why?)

In general, we use the notation $x \to h^+$ to indicate that x approaches h through values greater than h. Similarly, we use the notation $x \to h^-$ to indicate that x approaches h through values less than h.

• DEFINITION The vertical line $x = h$ is called a **vertical asymptote** of f if any *one* of the following four conditions holds:

(i) $f(x) \to \infty$ as $x \to h^+$. (ii) $f(x) \to -\infty$ as $x \to h^+$.

(iii) $f(x) \to \infty$ as $x \to h^-$. (iv) $f(x) \to -\infty$ as $x \to h^-$. •

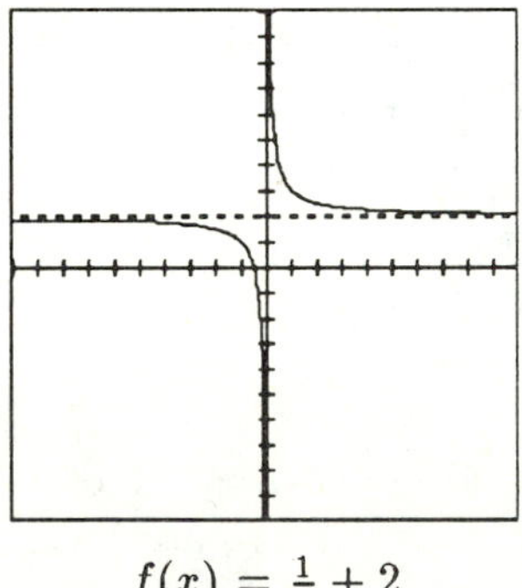

$f(x) = \frac{1}{x} + 2$
$[-10, 10]$ by $[-10, 10]$

Figure 3.5.3

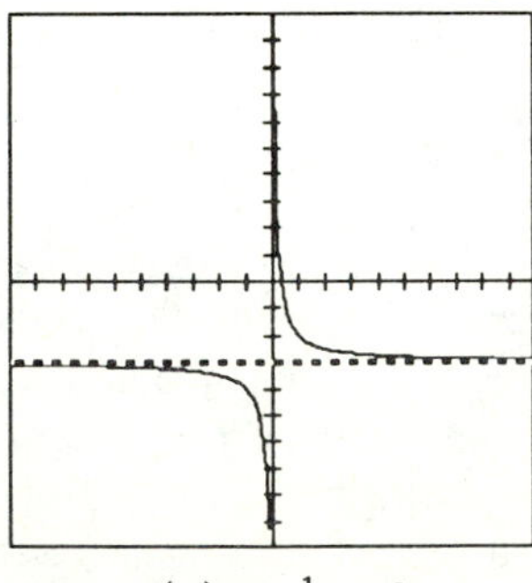

$g(x) = \frac{1}{x} - 3$
$[-10, 10]$ by $[-10, 10]$

Figure 3.5.4

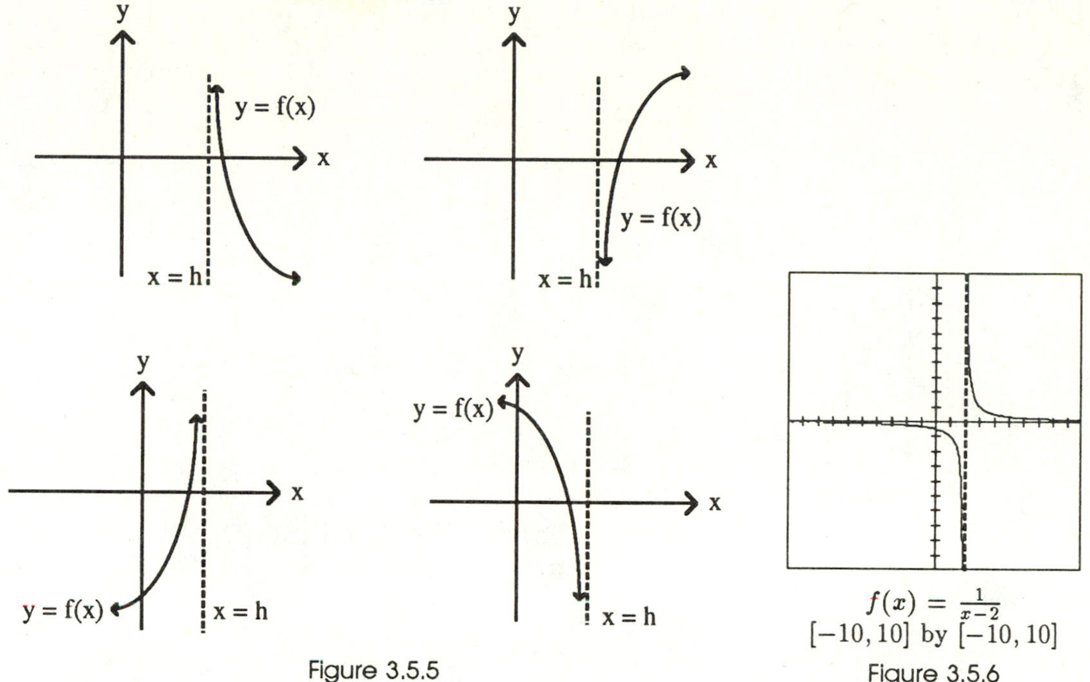

Figure 3.5.5

$f(x) = \frac{1}{x-2}$
$[-10, 10]$ by $[-10, 10]$

Figure 3.5.6

Figure 3.5.5 illustrates four of the many possibilities for the graph of f to have the vertical line $x = h$ as a vertical asymptote. Each corresponds to a condition in the above definition.

Notice the vertical asymptote and the end behavior model (horizontal asymptote) are crucial to the overall behavior of the function. For the rational functions in this section, the end behavior model is always a horizontal asymptote.

- EXAMPLE 2: Determine the horizontal and vertical asymptotes and draw a complete graph of each function.

(a) $f(x) = \dfrac{1}{x-2}$ (b) $f(x) = \dfrac{1}{x+3} - 1$

SOLUTION:

(a) A complete graph of $f(x) = \frac{1}{x-2}$ can be obtained by shifting the graph of $y = \frac{1}{x}$ horizontally to the right two units (Figure 3.5.6). The line $x = 2$ is a vertical asymptote of f, and the line $y = 0$ is a horizontal asymptote of f. (Why?) Notice that the line $x = 2$ is the y-axis shifted horizontally to the right two units. The vertical asymptote has been added as a dashed line in the figure and is not part of the graph of f. Notice how the vertical asymptote and the end behavior model account for all the important behavior of f.

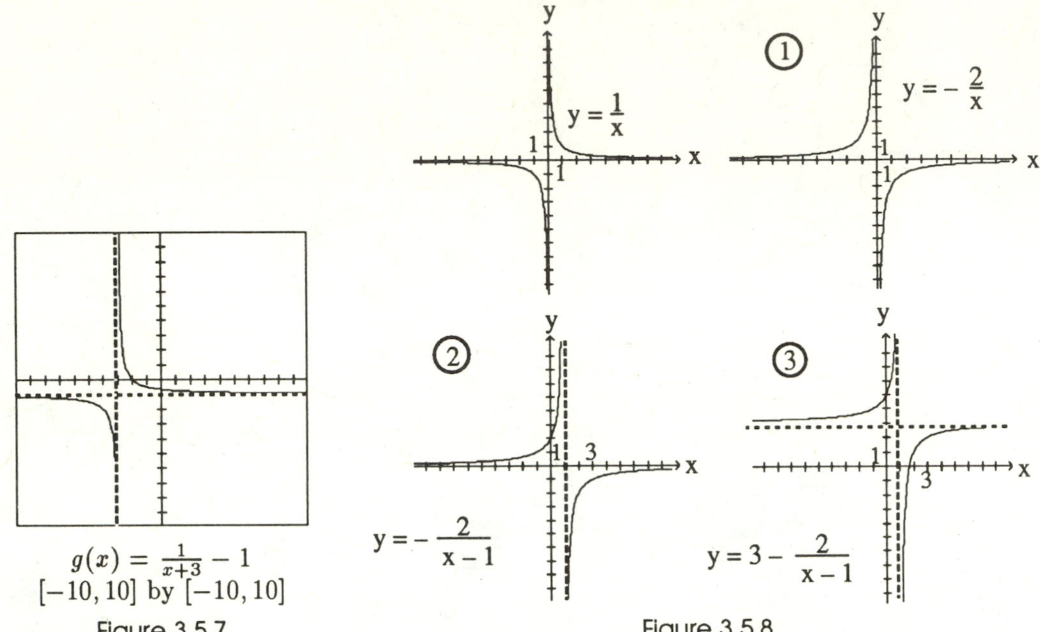

$g(x) = \frac{1}{x+3} - 1$
$[-10, 10]$ by $[-10, 10]$
Figure 3.5.7

Figure 3.5.8

(b) A complete graph of g can be obtained from the graph of $y = \frac{1}{x}$ by shifting horizontally left three units and then shifting vertically down one unit (Figure 3.5.7). Therefore, the line $x = -3$ is a vertical asymptote of g and the line $y = -1$ is a horizontal asymptote of g. These two lines have been added as dashed lines in the figure and are not part of the graph of g. Again notice how the vertical asymptote and end behavior model determine the complete graph. •

• EXAMPLE 3: Describe how the graph of $y = 3 - \frac{2}{x-1}$ is obtained from the graph of $y = \frac{1}{x}$.

SOLUTION: The following transformations performed in the specified order produce the graph of $y = 3 - \frac{2}{x-1}$.

1. Vertical stretch by a factor of 2 followed by a reflection through the x-axis to obtain $y = -\frac{2}{x}$.
2. Horizontal shift 1 unit right to obtain $y = -\frac{2}{x-1}$.
3. Vertical shift 3 units up to obtain $y = 3 - \frac{2}{x-1}$.

These transformations are illustrated in Figure 3.5.8. •

Examples 2 and 3 suggest how to obtain a complete graph of $f(x) = \frac{r}{x-h} + k$ from the graph of $y = \frac{1}{x}$ without using a graphing utility. More precisely, we have the following summary.

SUMMARY The graph of $f(x) = \frac{r}{x-h} + k$ is obtained from the graph of $y = \frac{1}{x}$ as follows:

1. Vertical stretch by a factor of $|r|$, followed by a reflection through the x-axis if $r < 0$ to obtain $y = \frac{r}{x}$.
2. Horizontal shift of $|h|$ units to obtain $y = \frac{r}{x-h}$. The shift is right if $h > 0$ and left if $h < 0$.
3. Vertical shift of $|k|$ units to obtain $f(s) = \frac{r}{x-h} + k$. The shift is up if $k > 0$ and down if $k < 0$.

The graph of f has the vertical line $x = h$ as a vertical asymptote and the horizontal line $y = k$ as a horizontal asymptote. The end behavior of f is $f(x) \to k$ as $|x| \to \infty$, and $y = k$ is an end behavior model for f.

The above summary can be used to obtain a complete graph of any rational function of the form $y = \frac{ax+b}{cx+d}$ *without* using a graphing utility. We must first use long division to rewrite y in a form in which we can identify the needed transformations.

• EXAMPLE 4: Determine the horizontal and vertical asymptotes and an end behavior model of $f(x) = \frac{2x-13}{x-5}$, and draw a complete graph. What are the domain and range of f?

SOLUTION: The quotient and remainder when $2x - 13$ is divided by $x - 5$ are 2 and -3, respectively. Thus, we can rewrite f as follows:

$$f(x) = 2 - \frac{3}{x-5}.$$

A graph of f is obtained from the graph of $y = \frac{1}{x}$ by first applying a vertical stretch by a factor of 3, followed by a reflection through the x-axis, followed by a horizontal shift right of 5 units, and finally a vertical shift up of 2 units. Thus, the line $x = 5$ is a vertical asymptote and the line $y = 2$ is a horizontal asymptote of f. The function $y = 2$ is an end behavior model for f.

Alternatively, we can use a calculator to show that $f(x) \to 2$ as $|x| \to \infty$, $f(x) \to -\infty$ as $x \to 5^+$, and $f(x) \to \infty$ as $x \to 5^-$. The graph of f given in Figure 3.5.9 confirms this information and is a complete graph.

We can see from the graph in Figure 3.5.9 that the domain of f is the set of all real numbers not equal to 5, and the range of f is the set of all real numbers not equal to 2. •

We can use composition of functions to describe how rational functions of the form $f(x) = \frac{ax+b}{cx+d}$ are obtained from $y = \frac{1}{x}$.

• EXAMPLE 5: Write $f(x) = \frac{2x-13}{x-5}$ as a composition of functions involving $g(x) = \frac{1}{x}$.

SOLUTION: In Example 4 we showed that $f(x) = \frac{2x-13}{x-5} = 2 - \frac{3}{x-5}$. Let $f_1(x) = x - 5$, $f_2(x) = -3x$ and $f_3(x) = x + 2$. It can be shown that $f = f_3 \circ f_2 \circ g \circ f_1$. •

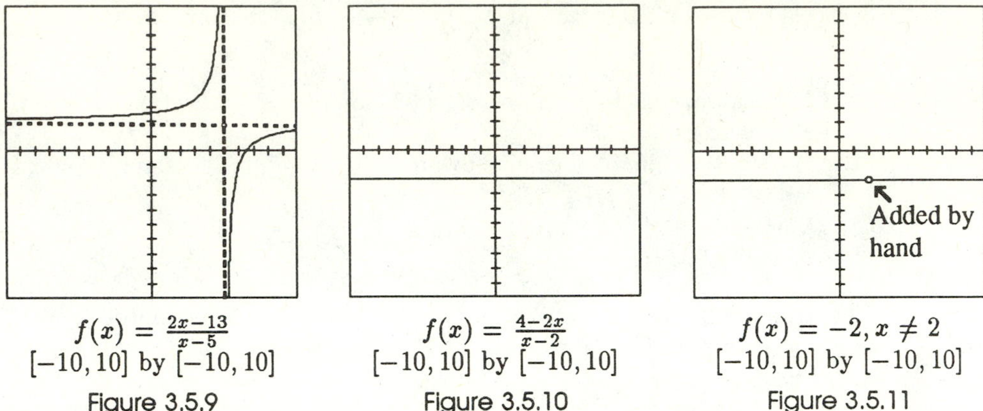

$f(x) = \frac{2x-13}{x-5}$
$[-10, 10]$ by $[-10, 10]$
Figure 3.5.9

$f(x) = \frac{4-2x}{x-2}$
$[-10, 10]$ by $[-10, 10]$
Figure 3.5.10

$f(x) = -2, x \neq 2$
$[-10, 10]$ by $[-10, 10]$
Figure 3.5.11

If the long division process used in Example 4 had produced zero as a remainder, then the graph would not have a vertical asymptote. The next example illustrates this point.

• EXAMPLE 6: Find the horizontal and vertical asymptotes and draw a complete graph of $f(x) = \frac{4-2x}{x-2}$.

SOLUTION: If we divide $4 - 2x$ by $x - 2$ the quotient is -2 and the remainder is 0. Thus, $x - 2$ is a *factor* of $4 - 2x$ and we can rewrite f as follows.

$$f(x) = \frac{4 - 2x}{x - 2} = \frac{-2(x - 2)}{x - 2} = -2, \quad \text{provided } x \neq 2 .$$

The computer graph of f given in Figure 3.5.10 also suggests that the graph of $f(x)$ is the horizontal line $y = -2$. Of course, f has *no* value when $x = 2$ because division by zero is not possible. A complete graph of f is the horizontal line $y = -2$ with the point $(2, -2)$ removed. We indicate that this point is missing by drawing an open circle at $(2, -2)$ (Figure 3.5.11). Therefore, f has no vertical asymptote and has $y = -2$ as a horizontal asymptote. •

Except for Example 6, the examples in this section suggest that we should expect a function to have a vertical asymptote if a variable is involved in the denominator of the function. Although this will often be the case, variables can occur in denominators without producing a vertical asymptote. Another illustration of this point is given in Example 7.

• EXAMPLE 7: Use a graph to find the horizontal and vertical asymptotes and describe the end behavior of $g(x) = \frac{x-2}{|x|+3}$. Then, confirm this information algebraically.

SOLUTION: A graph of g is given in Figure 3.5.12. Notice that g does not appear to have a vertical asymptote and seems to have *two* horizontal asymptotes. Because $|x| \geq 0$, the denominator of g is greater than or equal to 3 for all values of x. Thus, g has *no* vertical asymptote. To see why there are two horizontal asymptotes we need to consider separately the two cases $x \geq 0$ and $x \leq 0$.

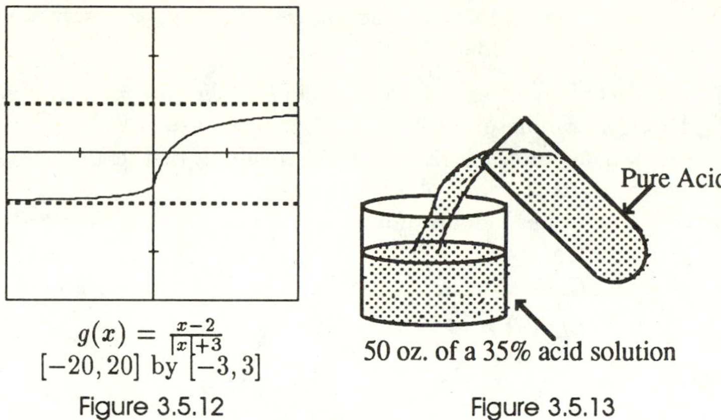

$g(x) = \frac{x-2}{|x|+3}$
$[-20, 20]$ by $[-3, 3]$

Figure 3.5.12

Figure 3.5.13

If $x \geq 0$, then $|x| = x$ and we can rewrite g as follows:

$$g(x) = \frac{x-2}{|x|+3} = \frac{x-2}{x+3} \quad (x \geq 0).$$

Now, we divide the numerator of g by the denominator of g and obtain the following form for g:

$$g(x) = \frac{x-2}{x+3} = 1 - \frac{5}{x+3} \quad (x \geq 0).$$

Notice that $y = 1$ is a horizontal asymptote of g because $g(x) = 1 - \frac{5}{x+3} \to 1$ as $x \to \infty$. The line $x = -3$ is *not* a vertical asymptote because this expression for g is not valid for $x < 0$.

For $x < 0$, we use $|x| = -x$ and division to obtain the following form for g:

$$g(x) = \frac{x-2}{|x|+3} = \frac{x-2}{-x+3} = -1 + \frac{1}{-x+3} \quad (x < 0).$$

Thus, $y = -1$ is a horizontal asymptote and the line $x = 3$ is *not* a vertical asymptote. (Why?) Therefore, g has no vertical asymptotes and has $y = 1$ and $y = -1$ as horizontal asymptotes. The end behavior of f is $f(x) \to 1$ as $x \to \infty$ and $f(x) \to -1$ as $x \to -\infty$. We can write g as a piecewise-defined function as follows:

$$g(x) = \begin{cases} -1 + \frac{1}{-x+3}, & \text{for } x < 0. \\ 1 - \frac{5}{x+3}, & \text{for } x \geq 0. \end{cases}$$

Notice that the two expressions for g agree when $x = 0$. •

Rational functions often occur as algebraic models of problem situations about mixing different types of solutions. For example, suppose we have 50 ounces of a 35% acid solution and want to increase the concentration (or percent) of acid by adding pure acid. If we add x ounces of pure acid, then the amount of pure acid in the new solution will be $x + 0.35(50)$, or $x + 17.5$, where the 17.5 represents the amount of pure acid in the 50 ounces of 35% acid solution (Figure 3.5.13).

• EXAMPLE 8: Pure acid is added to 50 ounces of a 35% acid solution. Let x be the number of ounces of pure acid added.

(a) Express the concentration C of acid of the new mixture as a function of x.
(b) Draw a complete graph of the function $y = C(x)$ indicating any horizontal or vertical asymptotes. Explain the meaning of the horizontal asymptote from the perspective of the problem situation.
(c) What are the domain and range of $y = C(x)$? What values of x make sense in the problem situation?
(d) How much pure acid should be added to the 35% acid solution to produce a mixture that is at least 75% acid?

SOLUTION:

(a) If we add x ounces of pure acid, the new mixture contains $x + 0.35(50)$ or $x + 17.5$ ounce of pure acid. Because the volume of the new solution is $x + 50$, the concentration of acid in the new solution is

$$C(x) = \frac{\text{pure acid}}{\text{total solution}} = \frac{x + 17.5}{x + 50}.$$

Actually $C(x)$ gives the decimal form of the percent of acid. For example, if $C(x) = 0.87$ for some value of x, then the concentration of acid would be 87%. We could multiply $\frac{x+17.5}{x+50}$ by 100, but this is not necessary as long as we remember to read the decimal form of $C(x)$ as a percent.

(b) The graph of $y = C(x)$ in the $[-100, 100]$ by $[-5, 5]$ viewing rectangle (Figure 3.5.14) suggests that $x = -50$ is a vertical asymptote and $y = 1$ is a horizontal asymptote. This information can be confirmed algebraically by using division to rewrite f in the form $f(x) = \frac{x+17.5}{x+50} = 1 - \frac{32.5}{x+50}$. Now, we can be sure that $x = -50$ is a vertical asymptote and $y = 1$ is a horizontal asymptote.
The concentration of acid can never exceed 100%. As a decimal 100% is equal to 1. Thus, the concentration should always be less than 1 but approach 1 for large x because increasing x means we are adding more pure acid. Therefore, the problem

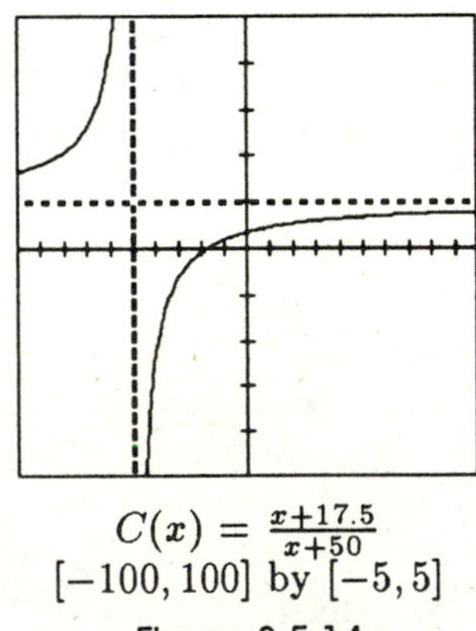

$C(x) = \frac{x+17.5}{x+50}$
$[-100, 100]$ by $[-5, 5]$
Figure 3.5.14

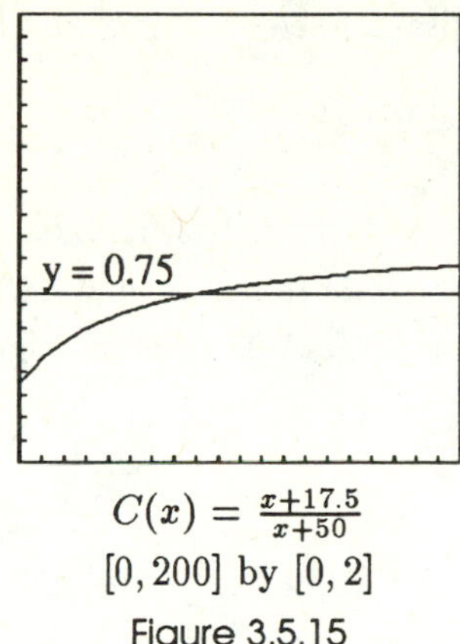

$C(x) = \frac{x+17.5}{x+50}$
[0, 200] by [0, 2]
Figure 3.5.15

situation explains why $C(x) \to 1^-$ as $x \to \infty$; that is, the values of $C(x)$ get close to 1 through values less than 1 as x approaches ∞.

(c) The function $y = \frac{x+17.5}{x+50}$ represents the problem situation only for $x \geq 0$. However, the domain of the function $y = C(x)$ is the set of all real numbers not equal to -50. The range of $y = C(x)$ is the set of all real numbers not equal to 1.

(d) A complete graph of the problem situation and the horizontal line $y = 0.75$ are given in Figure 3.5.15. We need to find the values of x for which $x \geq 0$ and $y = \frac{x+17.5}{x+50} \geq 0.75$. Thus, we need to find the values of x for which the graph of $y = \frac{x+17.5}{x+50}$ intersects or is *above* the graph of $y = 0.75$ in the first quadrant. (Why?) Using zoom-in we can find that these two graphs appear to intersect at $(80, 0.75)$. The concentration of acid will be at least 75% for $x \geq 80$. Thus, at least 80 ounces of pure acid must be added in order to produce a mixture that is at least 75% acid. •

Alternatively, we could solve the inequality $\frac{x+17.5}{x+50} \geq 0.75$ by first solving the equation $\frac{x+17.5}{x+50} = 0.75$ algebraically to find that the value of x that makes the concentration equal to 75% is exactly 80. Then, we observe that a value of x greater than 80 would make the concentration greater than 75%.

• EXERCISES 3-5

Draw a complete graph of each function without using a graphing utility. Identify and write the equations of all vertical and all horizontal asymptotes. Determine the end behavior and determine an end behavior model for each function. Check your answer with a graphing utility.

1. $y = \dfrac{2}{x+1}$
2. $y = -\dfrac{1}{x-2}$
3. $y = 4 + \dfrac{1}{x+1}$
4. $y = -2 + \dfrac{2}{x+3}$
5. $y = \dfrac{3x-1}{x+2}$
6. $y = \dfrac{8x+6}{2x-4}$
7. $y = \dfrac{2}{4-x}$
8. $y = \dfrac{x+2}{|x|+1}$

Write each function as a piecewise-defined function with no use of absolute value. Determine all points of discontinuity.

9. $y = \dfrac{|x| - 1}{x + 2}$

10. $y = \dfrac{2x - 3}{|x| - 1}$

11. Find the domain and range of the function in Exercise 2.

12. Find the domain and range of the function in Exercise 6.

A computer generated graph of $y = f(x)$ is given in Figure 3.5.16. Use it to complete Exercises 13 and 14.

13. As $x \to \infty$, $f(x) \to$?

14. As $x \to 3^-$, $f(x) \to$?

A computer generated graph of $y = g(x)$ is given in Figure 3.5.17. Use it to complete Exercises 15 and 16.

15. As $x \to -\infty$, $g(x) \to$?

16. As $x \to -2^+$, $g(x) \to$?

List the geometric transformations needed to produce a complete graph of the function f from a complete graph of $y = \frac{1}{x}$. Specify the order in which the transformations should be applied. In each case, draw a graph of the function without the aid of a graphing utility. Write the equations of any vertical or horizontal asymptotes. Check your answer with a graphing utility.

17. $f(x) = \dfrac{1}{x - 3}$

18. $f(x) = -1 - \dfrac{3}{x + 1}$

19. $f(x) = \dfrac{2}{x + 2}$

20. $f(x) = \dfrac{2x + 4}{x - 3}$

21–24. Express each function in Exercises 17–20 as a composition of functions of the form $g(x) = \frac{1}{x}$, $h(x) = ax$, and $k(x) = x + b$.

Draw a complete graph of each function. Give the domain and range and write equations of any asymptotes.

25. $y = \dfrac{x - 1}{2x - 2}$

26. $y = \dfrac{3x - 9}{6 - 2x}$

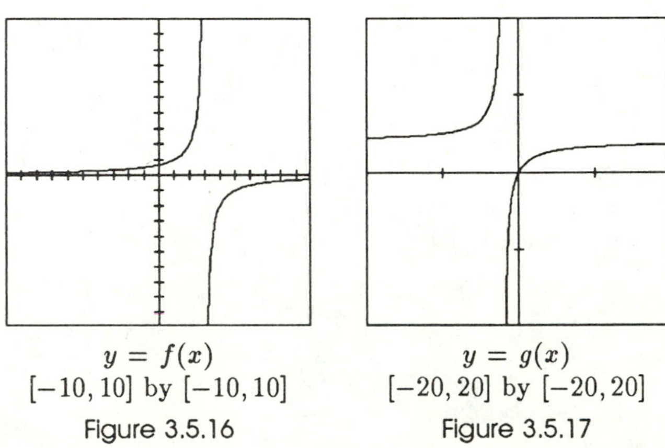

$y = f(x)$
$[-10, 10]$ by $[-10, 10]$
Figure 3.5.16

$y = g(x)$
$[-20, 20]$ by $[-20, 20]$
Figure 3.5.17

Use a calculator to complete each table for the function $f(x) = \frac{2x-13}{x-5}$ of Example 4.

27.

x	$f(x)$
1	
10	
100	
1,000	
10,000	
100,000	
1,000,000	

28.

x	$f(x)$
6	
5.1	
5.01	
5.001	
5.0001	
5.00001	
5.000001	

29. Use the table in Exercise 27 to determine numerically what happens to the values of the function $f(x)$ as $x \to \infty$.

30. Use the table in Exercise 28 to determine numerically what happens to the values of the function $f(x)$ as $x \to 5^+$.

31. Construct a table similar to the one in Exercise 28 to determine numerically what happens to the values of the function $f(x)$ as $x \to 5^-$.

32. Consider the function $f(x) = \frac{ax+b}{cx+d}$. List the transformations that will produce a complete graph of f from a complete graph of $y = \frac{1}{x}$, and specify an order in which they should be applied. *Hint:* Use long division.

33. The area of a rectangle is to be 300 square units.
    (a) Determine an algebraic representation that gives the length of this rectangle as a function of its width x.
    (b) Draw a complete graph of the function.
    (c) Write an equation for each asymptote.
    (d) What are the domain and range of this function?
    (e) What values of x make sense in this problem situation?
    (f) Use both an algebraic and a geometric method to find the width of the rectangle if the length is 2000 units.

34. Pure acid is added to 125 ounces of a 60% acid solution. Let x be the number of ounces of pure acid added.
    (a) Determine an algebraic representation of the acid concentration of the mixture.
    (b) Draw a complete graph of C. Write the equation of any asymptotes.
    (c) What values of x make sense in this problem situation? Draw a complete graph of the problem situation.
    (d) Use a computer drawn graph to determine how much pure acid should be added to the 60% solution to produce a new mixture that is at least 83% acid.
    (e) Solve (d) algebraically.

35. Boyle's law from physics states that for a certain gas $PV = 400$ where P is pressure and V is volume.
    (a) Draw a complete graph of volume as a function of pressure. Write an equation for each asymptote.
    (b) If $20 \leq V \leq 40$, what is the corresponding range for the values of P?
36. A chemistry student adds a certain amount of a 30% acid solution to 70 ounces of a 55% acid solution. Let x be the amount (in ounces) of 30% acid solution added.
    (a) Express the concentration C of acid of the mixture as a function of x.
    (b) Draw a complete graph of C. Write an equation for each asymptote.
    (c) What values of x make sense in this problem situation?
    (d) Use a computer drawn graph of C to determine how much of the 30% solution should be added to the 55% solution to produce a mixture of 40% acid.
37. Sally wishes to obtain 100 ounces of a 40% acid solution by combining a 60% acid solution and a 10% acid solution. How much of each solution should Sally use?
38. Find the domain and range of $f(x) = b + \frac{a}{x-h}$.
39. Find the domain and range of $f(x) = \frac{ax+b}{cx+d}$.
40. If possible, give an example of a function that has more than one vertical asymptote.

3.6 Rational Functions, Part 2

In this section we study rational functions $f(x) = \frac{p(x)}{h(x)}$ where p and h are polynomials and the degree of $p(x)$ is less than or equal to the degree of $h(x)$. In the previous section we studied rational functions where the degree of h was 1 and the degree of p was 0 or 1. The rational functions in this section can have several vertical asymptotes. However, the x-axis, or a line parallel to the x-axis, is an end behavior model and a horizontal asymptote of the function. In the next section we remove the restriction about the degree of $p(x)$ and see that an end behavior model of a general rational function is a polynomial.

The zeros of $h(x)$ are not in the domain of $f(x) = \frac{p(x)}{h(x)}$ because division by zero is not possible. Thus, the domain of $f(x) = \frac{p(x)}{h(x)}$ is the set of all real numbers for which $h(x) \neq 0$.

• EXAMPLE 1: Find the domain of $f(x) = \frac{x+3}{x^2-x}$.

SOLUTION: Notice that $x^2 - x = x(x-1)$. The denominator of f is zero for $x = 0$ or $x = 1$. Thus, the domain of f is the set of all real numbers different from 0 or 1, or $(-\infty, 0) \cup (0, 1) \cup (1, \infty)$. •

We want to assume that the polynomials $p(x)$ and $h(x)$ have no common factors to avoid certain trivial cases that can occur. The next example illustrates what can happen if we do not make this assumption.

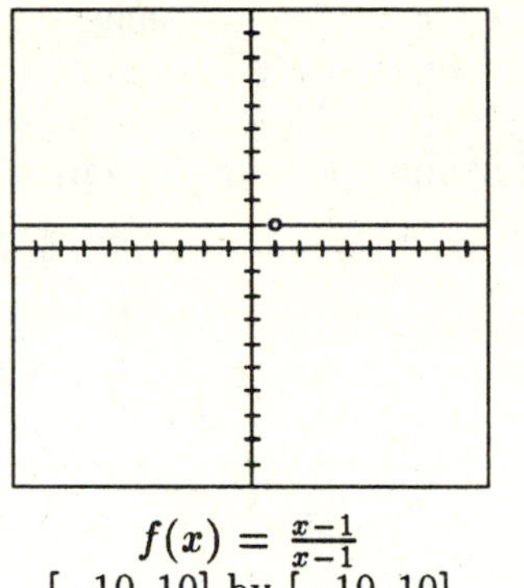

$f(x) = \frac{x-1}{x-1}$
$[-10, 10]$ by $[-10, 10]$

Figure 3.6.1

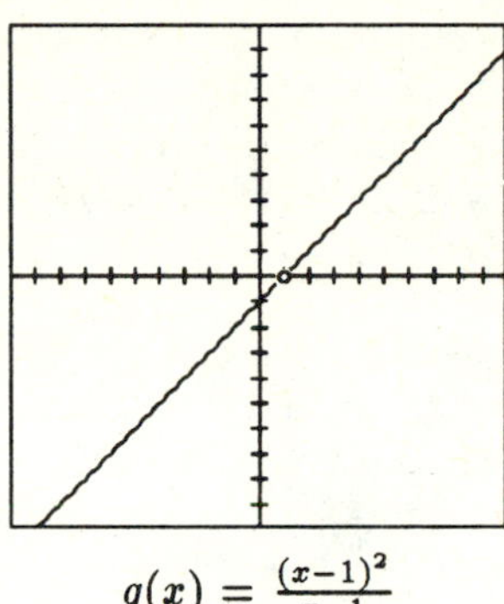

$g(x) = \frac{(x-1)^2}{x-1}$
$[-10, 10]$ by $[-10, 10]$

Figure 3.6.2

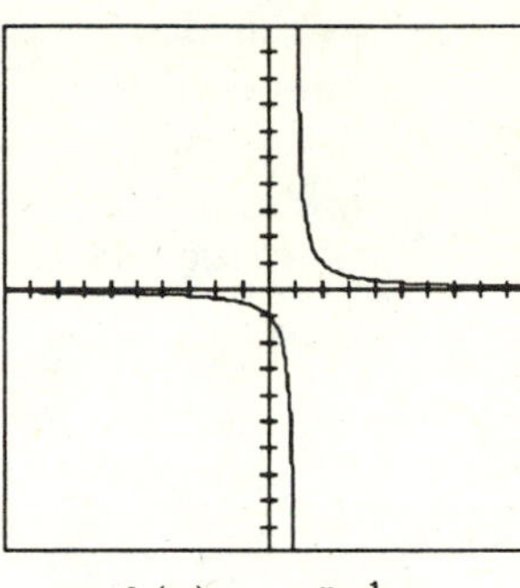

$h(x) = \frac{x-1}{(x-1)^2}$
$[-10, 10]$ by $[-10, 10]$

Figure 3.6.3

• EXAMPLE 2: Find the domain and vertical asymptotes of each function, and draw a complete graph.

(a) $f(x) = \dfrac{x-1}{x-1}$ (b) $g(x) = \dfrac{(x-1)^2}{x-1}$ (c) $h(x) = \dfrac{x-1}{(x-1)^2}$

SOLUTION: The domain of each function is the set of all real numbers different from 1. The graphs of f and g are given in Figures 3.6.1 and 3.6.2, respectively. Notice that the graph of f appears to be the horizontal line $y = 1$ and the graph of g the straight line $y = x - 1$. Neither graph has a vertical asymptote. We have added an open circle at $x = 1$ on both graphs to indicate that $x = 1$ is *not* in the domain of either function. The reason the graphs of f and g appear to be straight lines is that for all $x \neq 1$ we can rewrite f and g as follows:

$$f(x) = \frac{x-1}{x-1} = 1 \quad \text{and} \quad g(x) = \frac{(x-1)^2}{x-1} = x - 1\,, \quad \text{provided } x \neq 1.$$

The graph of h is given in Figure 3.6.3. Notice that the graph of h has a vertical asymptote at $x = 1$. We did not add the vertical asymptote as a dashed line in Figure 3.6.3 because it is reasonably clear from this graph that $x = 1$ is a vertical asymptote. We can rewrite h as follows:

$$h(x) = \frac{x-1}{(x-1)^2} = \frac{1}{x-1}.$$

From Section 3.5 we know that $y = \frac{1}{x-1}$ has a vertical asymptote at $x = 1$. •

Suppose that both p and h in the rational function $f(x) = \frac{p(x)}{h(x)}$ have $x - a$ as a common factor. Then a is not in the domain of f. Further, suppose that the exponent on $x - a$ in the numerator is greater than or equal to the exponent on $x - a$ in the denominator. The function f has a discontinuity at $x = a$ that is usually impossible to see on graphs produced by graphing utilities. Parts (a) and (b) of Example 2 demonstrate this situation. This type of discontinuity is called a *removable discontinuity*, because if we were to redefine the function appropriately at 1 there would be no discontinuity.

If the exponent on $x-a$ in the numerator of f is less than the exponent on $x-a$ in the denominator, then f has a vertical asymptote at $x=a$. Again, f has a point of discontinuity at a. We encountered this situation in (c) of Example 2. This type of discontinuity is an ***infinite discontinuity***. Notice that there is *no* way to redefine the function only at 1 to make it continuous at 1.

- **Vertical Asymptotes** Let $f(x) = \frac{p(x)}{h(x)} = \frac{(x-a)^r p_1(x)}{(x-a)^s h_1(x)}$, where $x-a$ is *not* a factor of either $p_1(x)$ or $h_1(x)$, r and s are integers, and $s \geq 1$. Then f has a discontinuity at $x = a$.

  (1) If $r \geq s$, then f has a removable discontinuity at a and does not have $x=a$ as a vertical asymptote.
  (2) If $r < s$ then f has an infinite discontinuity at a and has $x=a$ as a vertical asymptote.

For the remainder of this section except for the exercises we assume that the polynomials p and h in the rational function $f(x) = \frac{p(x)}{h(x)}$ have no common factors. The zeros of p are the zeros of f. (Why?) Furthermore, f has a vertical asymptote at each of the real zeros of h. We will not add dashed lines in computer drawn graphs for the asymptotes whenever their location is reasonably clear from the graphs.

- **EXAMPLE 3:** Draw a complete graph of $f(x) = \frac{x-1}{x^2-x-6}$. Determine the real zeros, the vertical and horizontal asymptotes, and an end behavior model for f.

SOLUTION: A graph of f is given in Figure 3.6.4. Notice that f is zero when x is 1 because the numerator of f is zero when x is 1. Further, f appears to have the lines $x=-2$ and $x=3$ as vertical asymptotes. We can confirm this observation by factoring the denominator:

$$x^2 - x - 6 = (x-3)(x+2).$$

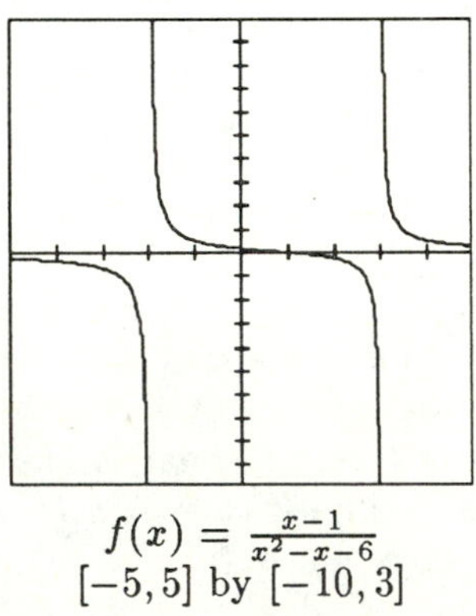

$f(x) = \frac{x-1}{x^2-x-6}$
$[-5,5]$ by $[-10,3]$
Figure 3.6.4

Now, we can see that the denominator is zero when x is -2 or 3. A calculator can be used to confirm the following information about the behavior of f when x is near -2 or 3:

$$f(x) \to -\infty \text{ as } x \to -2^-, \qquad f(x) \to -\infty \text{ as } x \to 3^-,$$
$$f(x) \to \infty \text{ as } x \to -2^+, \qquad f(x) \to \infty \text{ as } x \to 3^+.$$

Recall that $x \to -2^-$ means that x approaches -2 through values less than -2, and $x \to -2^+$ means that x approaches -2 through values greater than -2.

We can also use a calculator to check that the values of f approach zero as the absolute value of x approaches infinity:

$$f(x) \to 0 \text{ as } x \to \infty, \qquad f(x) \to 0 \text{ as } x \to -\infty.$$

Thus, the horizontal line $y = 0$ (x-axis) is a horizontal asymptote and an end behavior model of f. Now we can be sure that the graph of f in Figure 3.6.4 is a complete graph of f. Notice how the end behavior and the vertical asymptotes are critical to the overall behavior of f. •

There is another way to see that the function in Example 3 has $y = 0$ as a horizontal asymptote, or that $y = 0$ is an end behavior model for the function. Notice that x is an end behavior model for the numerator $x - 1$ of f, and x^2 is an end behavior model for the denominator $x^2 - x - 6$ of f. Thus, $\frac{x}{x^2} = \frac{1}{x}$ is an end behavior model for f, and $f(x) \to 0$ as $|x| \to \infty$.

In the next example we use the end behavior of the numerator and the denominator to determine the end behavior of the rational function.

• EXAMPLE 4: Determine an end behavior model for each function.

(a) $f(x) = \dfrac{2x^2 - 1}{3x^3 + 2x + 1}$ (b) $g(x) = \dfrac{2x^4 - 3x^2 + 1}{3x^4 - x^2 + x - 1}$

SOLUTION:

(a) Notice $2x^2$ and $3x^3$ are end behavior models for the numerator and denominator of f, respectively. Thus, $\frac{2x^2}{3x^3} = \frac{2}{3x}$ is an end behavior model for f, and $f(x) \to 0$ as $|x| \to \infty$. Now, we can see that $y = 0$ is a horizontal asymptote and a simpler end behavior model for f.

(b) The procedure used in (a) shows that $y = \frac{2x^4}{3x^4} = \frac{2}{3}$ is an end behavior model for g. Notice that $y = \frac{2}{3}$ is also a horizontal asymptote of g. •

The graphs of the functions $f(x) = \frac{2x^2-1}{3x^3+2x+1}$ (Figure 3.6.5) and $g(x) = \frac{2x^4-3x^2+1}{3x^4-x^2+x-1}$ (Figure 3.6.6) confirm the information found algebraically in Example 4. Notice that $y = 0$ is the horizontal asymptote of f and $y = \frac{2}{3}$ is the horizontal asymptote of g. The graph of g appears to have a vertical asymptote between $x = 0$ and $x = 1$. In the exercises you will investigate the behavior of g in more detail.

SUMMARY We can draw the following conclusions from Example 4. If $f(x) = \frac{p(x)}{h(x)}$ is a rational function with the degree of the polynomial p less than or equal to the degree

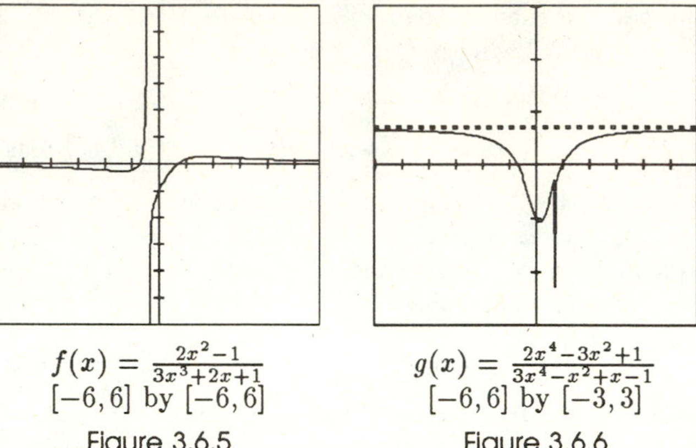

$f(x) = \frac{2x^2-1}{3x^3+2x+1}$
$[-6,6]$ by $[-6,6]$

Figure 3.6.5

$g(x) = \frac{2x^4-3x^2+1}{3x^4-x^2+x-1}$
$[-6,6]$ by $[-3,3]$

Figure 3.6.6

of the polynomial h, then an end behavior model of f is a horizontal line and the graph of f has this line as its *only* horizontal asymptote. Further, if the degree of p is strictly less than the degree of h, then an end behavior model of f is the horizontal line $y = 0$. Finally, if the degrees of p and h are equal, then an end behavior model of f is $y = \frac{a}{b}$ where a and b are the coefficients of the highest power of x in the numerator and denominator of f, respectively.

- EXAMPLE 5: Find all horizontal asymptotes and vertical asymptotes, and draw a rough sketch of $f(x) = \frac{3x^2+x-4}{2x^2-5x}$. Confirm the sketch with a graphing utility and determine the domain and range of f.

SOLUTION: We know from the above summary that $y = \frac{3}{2}$ is an end behavior model and a horizontal asymptote of f. Thus, $f(x) \to \frac{3}{2}$ as $|x| \to \infty$. Next, we factor the numerator and denominator of f:

$$f(x) = \frac{3x^2 + x - 4}{2x^2 - 5x} = \frac{(3x+4)(x-1)}{x(2x-5)}.$$

The denominator of f is zero when x is 0 or $\frac{5}{2}$. Thus, f has the lines $x = 0$ (y-axis) and $x = \frac{5}{2}$ as vertical asymptotes. The numerator of f is zero when x is $-\frac{4}{3}$ or 1. This means that the graph of f crosses the x-axis at $-\frac{4}{3}$ and 1. We know that $|f(x)| \to \infty$ as $x \to 0$ or $x \to \frac{5}{2}$, because f has vertical asymptotes at $x = 0$ and $x = \frac{5}{2}$. We can use a sign chart to describe the behavior of f near these two values of x.

Interval	$x < -\frac{4}{3}$	$-\frac{4}{3} < x < 0$	$0 < x < 1$	$1 < x < \frac{5}{2}$	$x > \frac{5}{2}$
Test value of x	-2	-1	0.5	2	3
Value of f	$\frac{1}{3}$	$-\frac{2}{7}$	1.375	-5	$\frac{26}{3}$
Sign of f	+	−	+	−	+

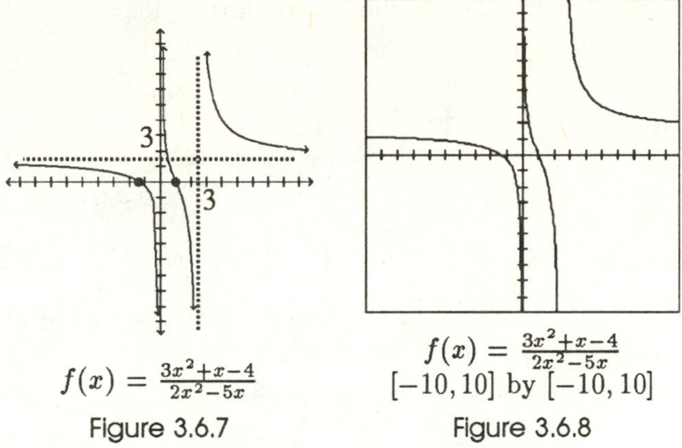

$f(x) = \frac{3x^2+x-4}{2x^2-5x}$

Figure 3.6.7

$f(x) = \frac{3x^2+x-4}{2x^2-5x}$
$[-10, 10]$ by $[-10, 10]$

Figure 3.6.8

We know that $|f(x)| \to \infty$ as $x \to \frac{5}{2}^+$ and $f(x) > 0$ for $x > \frac{5}{2}$. Thus, $f(x) \to \infty$ as $x \to \frac{5}{2}^+$. Similarly, we can use the above sign chart to determine the following information about the behavior of f near $x = 0$ and $x = \frac{5}{2}$, the vertical asymptotes of f:

$$f(x) \to -\infty \text{ as } x \to 0^-, \qquad f(x) \to -\infty \text{ as } x \to \frac{5}{2}^-,$$
$$f(x) \to \infty \text{ as } x \to 0^+, \qquad f(x) \to \infty \text{ as } x \to \frac{5}{2}^+.$$

We can obtain a rough sketch of a graph of f by using the above information. After drawing in the asymptotes that are not axes, we draw smooth curves satisfying all of the obtained information as shown in Figure 3.6.7. The computer drawn graph of f in Figure 3.6.8 confirms the rough sketch. Thus, the graph of f given in Figure 3.6.8 is a complete graph. We can see from this graph that the domain of f is the set of all real numbers different from 0 and $\frac{5}{2}$, and the range of f is the set of all real numbers. Notice how the vertical asymptotes and the end behavior are crucial to the overall behavior of f. •

• EXERCISES 3-6

Determine the domain by algebraic means. Check your answer with a graphing utility.

1. $f(x) = \dfrac{x^2+1}{x^2-1}$
2. $t(x) = \dfrac{2}{2x^2-x-3}$
3. $f(x) = \dfrac{x^2-4}{x^2-4x-1}$
4. $P(x) = \dfrac{4x-2}{x^2+5x+8}$
5. $f(x) = \dfrac{x^4-3x^2-5}{x^5-x}$
6. $f(x) = \dfrac{2x^2+5}{x^3-2x^2+x}$

Use algebraic means to sketch the graph of each rational function. Find all vertical and horizontal asymptotes. Determine an end behavior model and the end behavior of each function. Check your answer with a graphing utility.

7. $f(x) = \dfrac{2}{x-3}$
8. $g(x) = \dfrac{x-2}{x^2-2x-3}$
9. $f(x) = \dfrac{x-1}{x^2+3}$
10. $h(x) = \dfrac{3x^2-12}{4-x^2}$

Draw a complete graph of each function and determine the domain and range. Find all vertical and horizontal asymptotes and the end behavior.

11. $f(x) = \dfrac{x+1}{x^2}$

12. $f(x) = \dfrac{1}{x-1} + \dfrac{2}{x-3}$

13. $g(x) = \dfrac{x+1}{x^2-1}$

14. $f(x) = \dfrac{x^2+1}{x^4+1} + \dfrac{2}{x-3}$

15. $T(x) = \dfrac{x-1}{x^2+3x-1}$

16. $g(x) = \dfrac{2x}{(x-3)(x+2)}$

17. Let $y = \frac{2x^4-3x^2+1}{3x^4-x^2+x-1}$, the function of Example 4(b). A graph of y appears in Figure 3.6.6. Draw its graph in the $[-10, 10]$ by $[-10, 10]$ viewing rectangle.

 (a) How many vertical asymptotes are apparent?

 (b) Draw its graph in the $[-2, 2]$ by $[-2, 2]$ rectangle. Does your answer to part (a) change?

Determine the intervals on which each function is increasing and the intervals on which each function is decreasing.

18. $f(x) = \dfrac{x^2+1}{x}$

19. $g(x) = \dfrac{2x^3-3x+1}{x^2+4}$

Determine any local maximum and minimum values of the function.

20. $f(x) = \dfrac{x^3+1}{x}$

21. $g(x) = \dfrac{2x^4-2x^2+x+5}{x^2-3x-4}$

Let $f(x) = \frac{x-1}{x^2-x-6}$, the function of Example 3. Complete each table using a calculator.

22.

x	$f(x)$
−1	
−10	
−100	
−1,000	
−10,000	
−100,000	
−1,000,000	

23.

x	$f(x)$
3	
2.9	
2.99	
2.999	
2.9999	
2.99999	
2.999999	

24. Use the table in Exercise 22 to determine numerically what happens to the values of the function $f(x)$ as $x \to -\infty$.

25. Use the table in Exercise 23 to determine numerically what happens to the values of the function $f(x)$ as $x \to 3^-$.

• DEFINITION y is said to vary **inversely** with $f(x)$, or y is **inversely proportional** to $f(x)$, if $y = \frac{k}{f(x)}$ for some constant $k \neq 0$. •

26. Show that the length of a rectangle of fixed area varies inversely with its width. Use a graph to determine the width of a rectangle if it has area 337 square feet and length 26.25 inches. Use the graph to explain how such a rectangle can have width 0.1 inches, 0.01 inches and 0.0001 inches. How small can the width become?

27. The intensity of sound (loudness) varies inversely with the square of the distance from the sound. If the sound of a concert band measures 128 decibels at 82 feet, use a graph to determine the measure of the sound at 50 feet. Use the graph to explain what happens to the loudness as one gets very close to the band.

28. The area of a rectangle is to be 182 square units.

    (a) Draw a complete graph of an algebraic representation that gives the length of this rectangle as a function of its width x.

    (b) Write an equation for each asymptote.

    (c) What are the domain and range of this function?

    (d) What values of x make sense in this problem situation?

    (e) Use a computer drawn graph to find the width of the rectangle if the length is 2000 units.

29. Pure acid is added to 78 ounces of a 63% acid solution. Let x be the amount (in ounces) of pure acid added.

    (a) Express the concentration C of acid of the mixture as a function of x.

    (b) Draw a complete graph of C. Write the equation of any asymptotes.

    (c) What values of x make sense in this problem situation?

    (d) Use a computer drawn graph to determine how much pure acid should be added to the 63% solution to produce a new mixture that is at least 83% acid.

    (e) Solve (d) algebraically.

30. *Boyle's law* from physics states that for a certain gas $PV = 285$ where P is pressure and V is volume.

    (a) Draw a complete graph of volume as a function of pressure. Write an equation for each asymptote.

    (b) If $8 \leq V \leq 15$, what is the corresponding range for the values of P?

31. A chemistry student adds a certain amount of a 28% acid solution to 40 ounces of a 65% acid solution. Let x be the amount (in ounces) of 28% acid solution added.

    (a) Determine an algebraic representation of the acid concentration of the mixture.

    (b) Draw a complete graph of C. Write an equation for each asymptote.

    (c) What values of x make sense in this problem situation? Draw a complete graph of the problem situation.

    (d) Use a computer drawn graph of C to determine how much of the 28% solution should be added to the 65% solution to produce a mixture of 56% acid.

32. Sally wishes to obtain 250 ounces of a 55% acid solution by combining an 80% acid solution and a 25% acid solution. How much of each type of solution should Sally use?

*33. Suppose both p and h in the rational function $f(x) = \frac{p(x)}{h(x)}$ have a common factor $x - a$. Under what conditions is there a removable discontinuity at $x = a$?

*34. Suppose both p and h in the rational function $f(x) = \frac{p(x)}{h(x)}$ have a common factor $x - a$. Under what conditions is there an infinite discontinuity at $x = a$?

*35. Why do you think there is a distinction between removable and infinite discontinuities? Explain what you believe the difference to be.

3.7 Rational Functions, Part 3

In this section we complete the study of rational functions by focusing on rational functions $f(x) = \frac{p(x)}{h(x)}$ where the degree of $p(x)$ is greater than the degree of $h(x)$. In the previous two sections we have studied rational functions where the degree of the numerator was less than or equal to the degree of the denominator. We will see that the end behavior of f is the same as the end behavior of the quotient when $p(x)$ is divided by $h(x)$. A problem situation that has a rational function for an algebraic representation is studied.

In Section 3.1 we observed that the end behavior of a function can be determined by viewing its graph in a large viewing rectangle. Suppose we start with the graph of a function in a given viewing rectangle. Recall that the process of expanding the size of this viewing rectangle is called *zoom-out.* Usually, we have to zoom out several times to observe graphically the end behavior of a given function. It is possible to view the graph of a function in several viewing rectangles quickly because many graphing utilities have an automatic zoom-out feature.

• EXAMPLE 1: Draw a complete graph of $f(x) = \frac{2x^3-4x^2+3}{x-2}$. Find all vertical asymptotes and show that $y = 2x^2$ is an end behavior of model f.

SOLUTION: We can start with the graph of f in the standard viewing rectangle, $[-10, 10]$ by $[-10, 10]$, and zoom out to obtain the graph of f in the $[-10, 10]$ by $[-100, 100]$ viewing rectangle shown in Figure 3.7.1. The graph of f has a vertical asymptote at $x = 2$. We can zoom out again to obtain the graph of f shown in Figure 3.7.2. So few points are graphed for values of x near 2 that we can no longer see that f has a vertical asymptote at $x = 2$ in this figure.

However, we can now observe that the graph of f in this figure appears to be a parabola. If we divide $2x^3 - 4x^2 + 3$ by $x - 2$ we can see algebraically why the graph in Figure 3.7.2 appears to be a parabola:

$$\begin{array}{r|l} & 2x^2 \\ \hline x-2 & 2x^3 - 4x^2 + 3 \\ & 2x^3 - 4x^2 \\ \hline & \qquad\qquad\quad 3 \end{array}$$

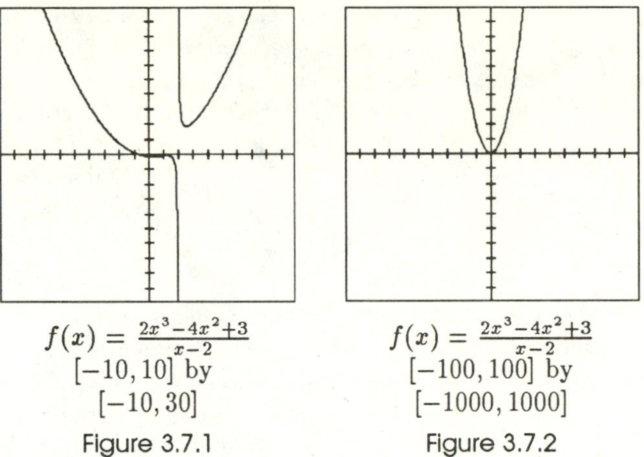

$f(x) = \frac{2x^3-4x^2+3}{x-2}$
$[-10, 10]$ by
$[-10, 30]$

Figure 3.7.1

$f(x) = \frac{2x^3-4x^2+3}{x-2}$
$[-100, 100]$ by
$[-1000, 1000]$

Figure 3.7.2

It follows that $2x^3 - 4x^2 + 3 = (x-2)(2x^2) + 3$. Thus, we can rewrite $f(x)$ as follows:

$$f(x) = \frac{2x^3 - 4x^2 + 3}{x-2} = \frac{(x-2)(2x^2)+3}{x-2} = 2x^2 + \frac{3}{x-2}.$$

The end behavior of f is the same as the end behavior of $2x^2$, because $\frac{3}{x-2} \to 0$ as $|x| \to \infty$. (Why?) It can be shown that $\frac{f(x)}{2x^2} \to 1$ as $|x| \to \infty$. Thus, $y = 2x^2$ is an end behavior model for f. Notice that $2x^2$ is the quotient when the numerator of f is divided by the denominator of f. If we were to overlay the graph of $y = 2x^2$ in Figure 3.7.2, it would be very hard to distinguish it from the graph of f. Now we can be sure that the graph of f in Figure 3.7.1 is a complete graph. •

The quotient when the numerator of a rational function is divided by the denominator is an important end behavior model. This particular end behavior model helps determine rough sketches of rational functions. For example, writing the function $f(x) = \frac{2x^3-4x^2+3}{x-2}$ in the form $f(x) = 2x^2 + \frac{3}{x-2}$ allows us to make important observations about the behavior of f that can be used to draw a rough sketch of the graph of f. For values of x near 2, the values of f are dominated by the values of $y = \frac{3}{x-2}$. More precisely we have the following.

$$\frac{3}{x-2} \to \infty \quad \text{as} \quad x \to 2^+ \qquad\qquad f(x) \to \infty \quad \text{as} \quad x \to 2^+$$

$$\frac{3}{x-2} \to -\infty \quad \text{as} \quad x \to 2^- \qquad\qquad f(x) \to -\infty \quad \text{as} \quad x \to 2^-$$

Thus, for values of x near 2, the values of f are approximately equal to the values of $y = 2(2)^2 + \frac{3}{x-2} = 8 + \frac{3}{x-2}$. For example, $f(2.1) = 38.82$ and the value of $y = 8 + \frac{3}{x-2}$ is 38 for $x = 2.1$.

For values of x far away from 2, the values of f are dominated by the values of $y = 2x^2$. For example, for $x = 5$ the value of $y = 2x^2$ is 50 and $f(5) = 51$. We can use the fact that $\frac{3}{x-2} \to 0$ as $|x| \to \infty$ to show that the values of f are approximately equal to the values of $y = 2x^2$ for values of x far away from 2.

Now we can obtain a rough sketch of the graph of f in the following way. Step 1: Draw a graph of $y = 2x^2$ and then erase a portion near $x = 2$ (Figure 3.7.3). Step 2: Complete this graph using the facts that $f(x) \to \infty$ as $x \to 2^+$ and $f(x) \to -\infty$ as $x \to 2^-$ (Figure 3.7.4).

Roughly speaking, the graph of f is the same as the graph of $y = 2x^2$ for values of x away from 2, and the graph of f is the same as the graph of $y = 8 + \frac{3}{x-2}$ for values of x near 2. Notice that 8 is the value at $x = 2$ of the end behavior model $y = 2x^2$ of f. Compare the rough sketch in Figure 3.7.4 with the computer drawn graph in Figure 3.7.1. In general, this important end behavior model for a rational function can be found as in Example 1.

Rational Function End Behavior The end behavior of the rational function $f(x) = \frac{p(x)}{h(x)}$ is the same as the quotient $q(x)$ when the polynomial $p(x)$ is divided by the polynomial $h(x)$. The function $y = q(x)$ is an end behavior model for f and is called the **end behavior asymptote** of f. If the degree of $q(x)$ is 1, the line $y = q(x)$ is called the **slant asymptote** of f. If the degree of $q(x)$ is 0, the line $y = q(x)$ is the *horizontal asymptote.*

• EXAMPLE 2: Determine the vertical asymptotes, the end behavior asymptote, and sketch a complete graph of $f(x) = \frac{2x^3+7x^2-4}{x^2+2x-3}$ without a graphing utility. Confirm the sketch with a graphing utility.

SOLUTION: First we determine the end behavior asymptote of f by long division:

$$\begin{array}{r|l} & 2x + 3 \\ \hline x^2 + 2x - 3 & 2x^3 + 7x^2 \qquad - 4 \\ & 2x^3 + 4x^2 - 6x \\ \hline & \qquad 3x^2 + 6x - 4 \\ & \qquad 3x^2 + 6x - 9 \\ \hline & \qquad\qquad\qquad 5 \end{array}$$

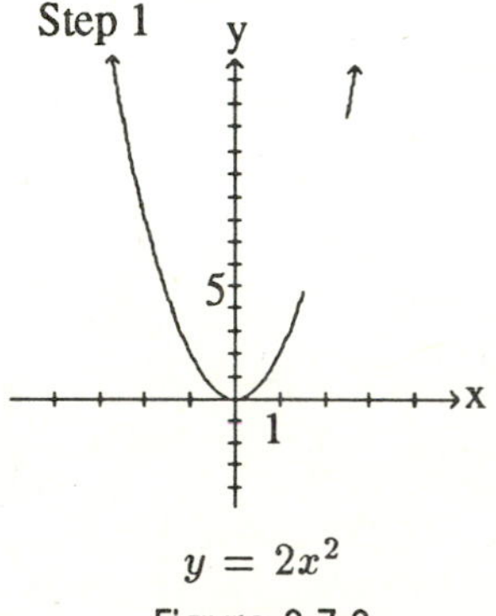

$y = 2x^2$

Figure 3.7.3

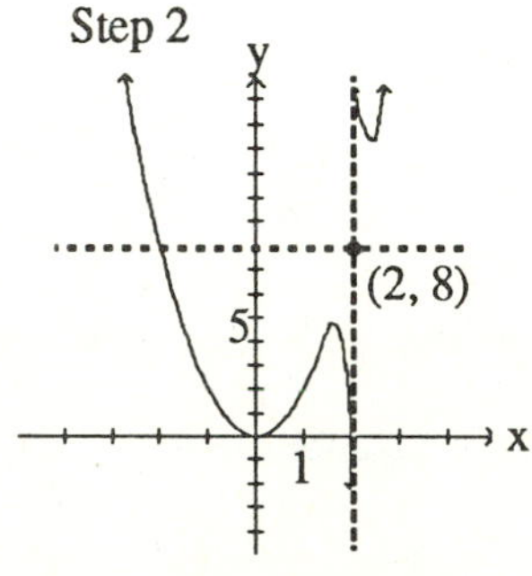

Figure 3.7.4

Thus, the end behavior asymptote of f is $y = 2x + 3$ and we can rewrite f as follows:

$$f(x) = 2x + 3 + \frac{5}{x^2 + 2x - 3}.$$

Because the end behavior asymptote has degree 1, the line $y = 2x + 3$ is a *slant* asymptote of the function.

Notice that the denominator of f can be factored:

$$x^2 + 2x - 3 = (x + 3)(x - 1).$$

Now, we can see that the vertical asymptotes of f are $x = -3$ and $x = 1$. A rough sketch of the graph of f can be obtained as follows. Step 1: Draw a graph of the end behavior asymptote $y = 2x+3$ of f and erase a portion near $x = -3$ and a portion near $x = 1$ (Figure 3.7.5). Near $x = 1$ the values of f are approximately equal to the values of $y = 5 + \frac{5}{4(x-1)}$, and near $x = -3$ the values of f are approximately equal to the values of $y = -3 + \frac{5}{(-4)(x+3)}$. Step 2: Use this information to complete a rough sketch of the graph of f (Figure 3.7.6). Alternatively, we can use a calculator to obtain the following information about the behavior of f near $x = -3$ and $x = 1$:

$$f(x) \to \infty \quad \text{as } x \to -3^-, \qquad f(x) \to -\infty \quad \text{as } x \to 1^-,$$

$$f(x) \to -\infty \quad \text{as } x \to -3^+, \qquad f(x) \to \infty \quad \text{as } x \to 1^+.$$

Adding this information to the graph in Figure 3.7.5 also gives the rough sketch in Figure 3.7.6.

The rough sketch in Figure 3.7.6 is a good approximation to the graph of f. Compare it with the complete graph of f given in Figure 3.7.7. It is the rough sketch in Figure 3.7.6 that allows us to determine that the computer graph in Figure 3.7.7 *is* a complete graph. •

In Section 3.1, we found that $y = a_n x^n$ is an end behavior model of the polynomial $y = a_n x^n + a_{n-1}x^{n-1} + \cdots + a_0$. In Example 2 above we found that $y = 2x + 3$ is the end behavior asymptote of $f(x) = \frac{2x^3+7x^2-4}{x^2+2x-3}$. The end behavior asymptote is also an end behavior model. It follows from Section 3.1 that $y = 2x$ is also an end behavior model of

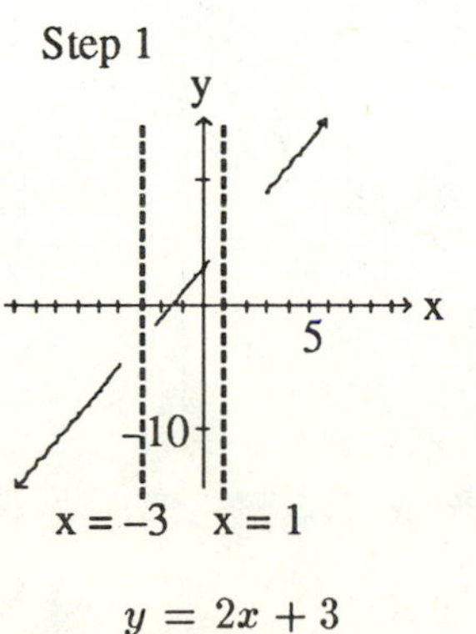

$y = 2x + 3$

Figure 3.7.5

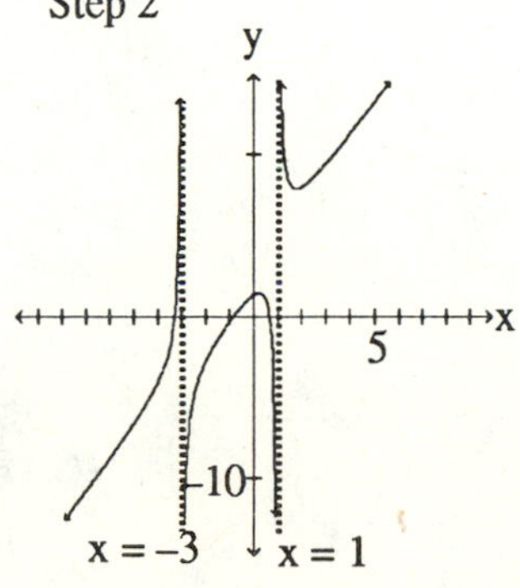

Figure 3.7.6

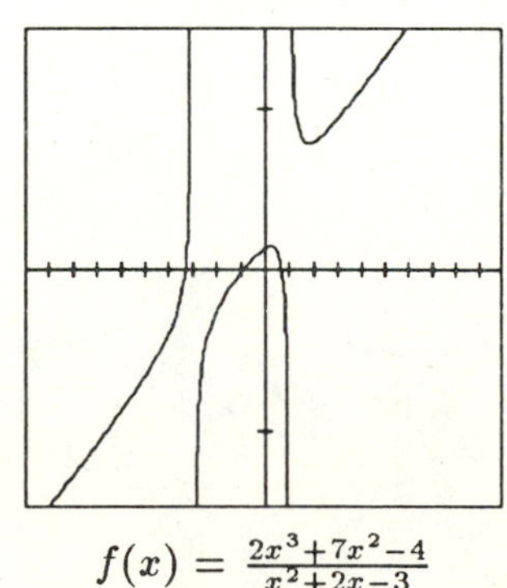

$f(x) = \frac{2x^3+7x^2-4}{x^2+2x-3}$

$[-10, 10]$ by $[-15, 15]$

Figure 3.7.7

f. For purposes of drawing a rough sketch of the graph of f the end behavior asymptote $y = 2x+3$ is the most useful of the end behavior models of f. The end behavior model $y = 2x$ is the most useful end behavior model to describe the end behavior of f. We summarize this discussion for any rational function.

SUMMARY Let $f(x) = \frac{p(x)}{h(x)}$ be a rational function and let $q(x) = a_n x^n + a_{n-1}x^{n-1} + \cdots + a_0$ be the quotient and $r(x)$ the remainder when $p(x)$ is divided by $h(x)$. Then,

1. the end behavior of f is the same as the end behavior model $y = a_n x^n$; and
2. $y = q(x)$ is the end behavior asymptote of f.

We can also use the technique illustrated in Example 2 to draw a rough sketch of a complete graph of $y = \frac{ax+b}{cx+d}$, the rational functions studied in Section 3.5. In the Exercises you will be asked to compare this method with the method of Section 3.5 of using geometric transformations to produce the graph of $y = \frac{ax+b}{cx+d}$ from the graph of $y = \frac{1}{x}$.

The graph in Figure 3.7.7 suggests that $f(x) = \frac{2x^3+7x^2-4}{x^2+2x-3}$ has a local maximum between $x = 0$ and $x = 1$ and a local minimum near $x = 2$. Graphing utilities are very useful if we need to find local maxima and minima of functions. However, we must be careful because sometimes there is behavior that is hidden from view on a computer screen. We need the techniques of calculus to be absolutely certain we have found all local extrema of a function.

• EXAMPLE 3: Determine the behavior of $f(x) = \frac{2x^4+7x^3+7x^2+2x}{x^3-x+50}$ in the interval $[-2, 1]$.

SOLUTION: The graph of f in the $[-20, 20]$ by $[-20, 20]$ viewing rectangle is shown in Figure 3.7.8. This graph leaves considerable doubt about the behavior of f between $x = -2$ and $x = 1$. If we zoom in a few times between $x = -2$ and $x = 1$ we get the graph shown in Figure 3.7.9.

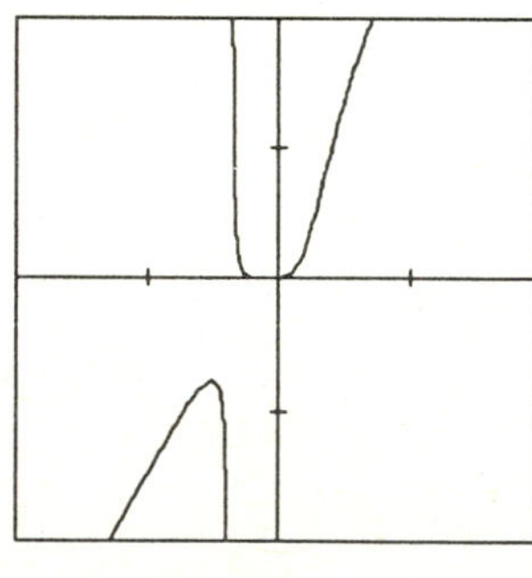

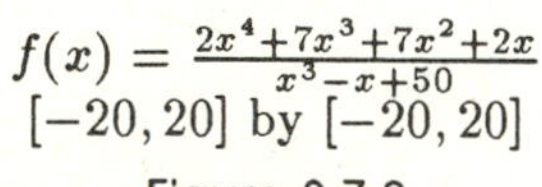

$f(x) = \frac{2x^4+7x^3+7x^2+2x}{x^3-x+50}$
$[-20, 20]$ by $[-20, 20]$

Figure 3.7.8

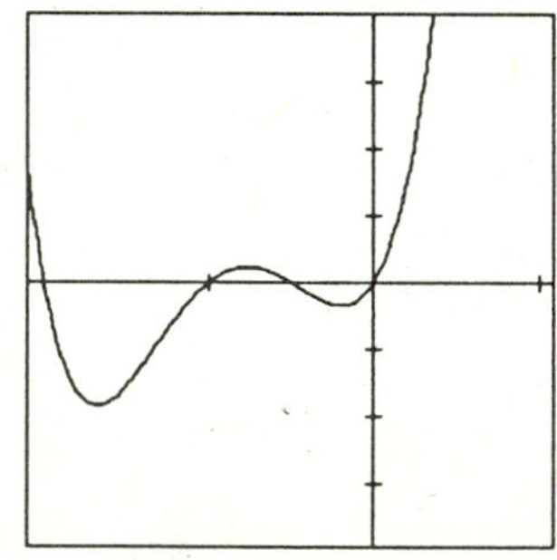

$f(x) = \frac{2x^4+7x^3+7x^2+2x}{x^3-x+50}$
$[-2.1, 1.1]$ by
$[-0.04, 0.04]$

Figure 3.7.9

Notice that f has two local minima and one local maximum between $x = -2$ and $x = 1$. These local extrema are very close to each other, which causes this behavior to be *hidden* unless we observe the graph in a small viewing rectangle. •

If we shift the graph of the function f in Example 3 horizontally or vertically, we can cause the *hidden behavior* to occur virtually anywhere in the coordinate plane. Although computer graphing is a very powerful and important tool, it does not remove the need for the study of calculus.

• EXAMPLE 4: Draw a complete graph of

$$f(x) = \frac{2x^4 + 7x^3 + 7x^2 + 2x}{x^3 - x + 50}.$$

SOLUTION: We can use long division to rewrite f in the following form:

$$f(x) = 2x + 7 + \frac{9x^2 - 91x - 350}{x^3 - x + 50}.$$

In this case, the end behavior asymptote is the slant asymptote $y = 2x + 7$. The graph of the denominator $y = x^3 - x + 50$ of f in the $[-10, 10]$ by $[-100, 100]$ viewing rectangle shows that the denominator of f has exactly one real zero (Figure 3.7.10). This zero is between -4 and -3 but closer to -4. Thus, f has one vertical asymptote. The graph in Figure 3.7.11 illustrates that the end behavior of f is the same as the end behavior model $y = 2x$. Thus, the graphs of f in Figures 3.7.8 and 3.7.9 give a complete graph of f. Notice that it takes two graphs to illustrate a complete graph of f. •

• EXAMPLE 5: Determine an end behavior model for

$$f(x) = \frac{3x^5 - 2x^4 + 3x^2 + 5x - 6}{x^2 - 3x + 6}.$$

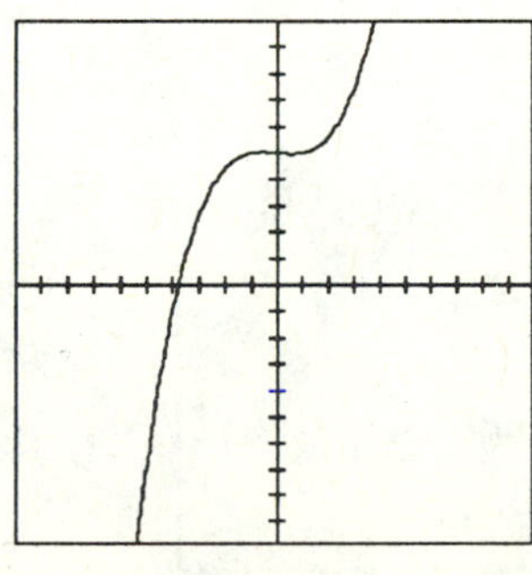

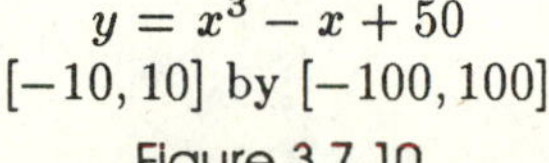

$y = x^3 - x + 50$
$[-10, 10]$ by $[-100, 100]$

Figure 3.7.10

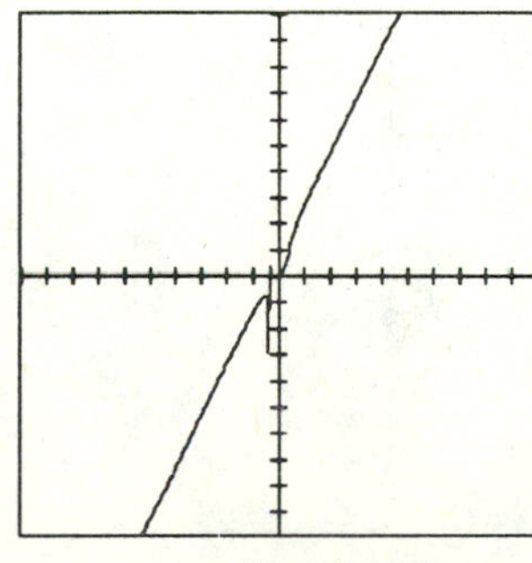

$f(x) = \frac{2x^4+7x^3+7x^2+2x}{x^3-x+50}$
$[-100, 100]$ by
$[-100, 100]$

Figure 3.7.11

SOLUTION: The graph of $y = 3x^3$ and f in the $[-50, 50]$ by $[-10{,}000, 10{,}000]$ viewing rectangle provides convincing geometric evidence that $y = 3x^3$ is an end behavior model for f (Figure 3.7.12). Notice that $3x^3$ is the ratio of the terms of highest degree in the numerator and denominator of f. In the Exercises we ask you to confirm the end behavior algebraically by finding the end behavior asymptote of f. •

We close this section with a problem situation that has a rational function for an algebraic model.

• EXAMPLE 6: Consider all rectangles with an area of 200 square feet. Let x be the width of one such rectangle.

(a) Determine an algebraic model for the perimeter P of the rectangle as a function of x.
(b) Determine the vertical asymptotes and the end behavior asymptote, and draw a complete graph of the function in (a).
(c) What portion of the graph in (b) represents the problem situation?
(d) Find the dimensions of a rectangle with an area of 200 square feet that has the least possible perimeter. What is this perimeter?

SOLUTION: If the width of a rectangle with an area of 200 is x, then its length must be $\frac{200}{x}$ (Figure 3.7.13).

(a) We can see from Figure 3.7.13 that the perimeter P is given by

$$P(x) = 2x + 2(\frac{200}{x}) = 2x + \frac{400}{x}.$$

(b) From the equation $P(x) = 2x + \frac{400}{x}$, we can see that the end behavior asymptote of P is $y = 2x$ and that $x = 0$ (y-axis) is a vertical asymptote. The graph of P in Figure 3.7.14 is a complete graph.

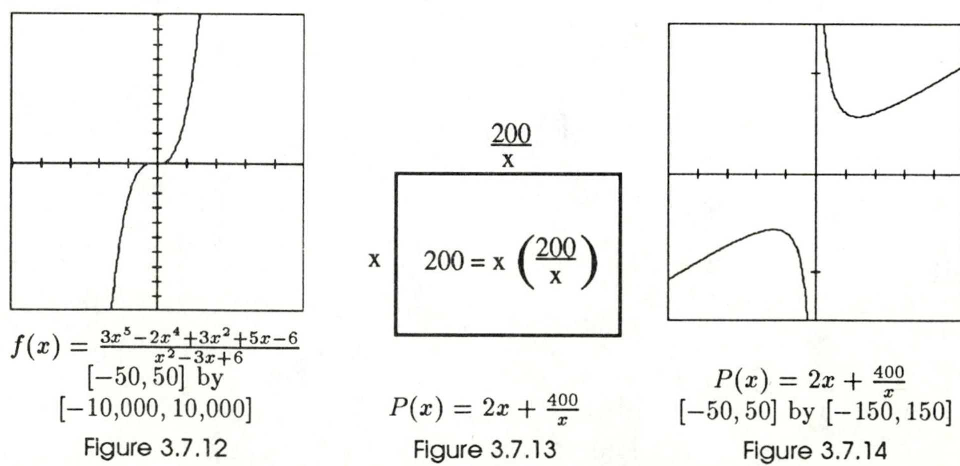

$f(x) = \frac{3x^5-2x^4+3x^2+5x-6}{x^2-3x+6}$
$[-50, 50]$ by $[-10{,}000, 10{,}000]$
Figure 3.7.12

$P(x) = 2x + \frac{400}{x}$
Figure 3.7.13

$P(x) = 2x + \frac{400}{x}$
$[-50, 50]$ by $[-150, 150]$
Figure 3.7.14

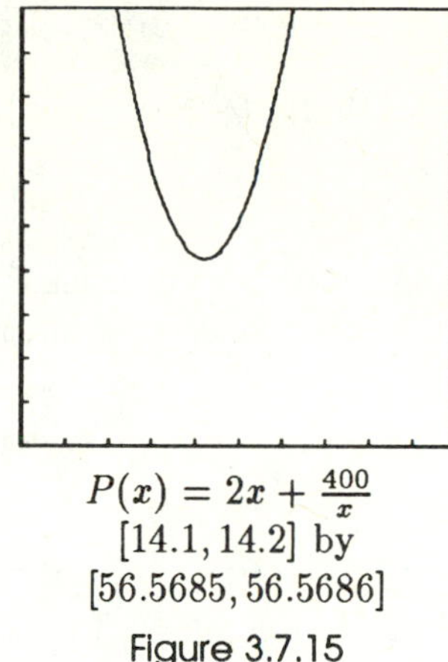

$P(x) = 2x + \frac{400}{x}$
[14.1, 14.2] by
[56.5685, 56.5686]

Figure 3.7.15

(c) The width x of the rectangle can be any positive real number. (Why?) Thus, the portion of the graph in Figure 3.7.14 to the right of the x-axis represents the problem situation.

(d) We can see from Figure 3.7.14 that P has a local minimum near $x = 15$. If we zoom in several times we obtain the graph of P given in Figure 3.7.15.

From this figure we can read that the value of x at the local minimum is 14.14 with error at most 0.01. The corresponding value of P is 56.56854 with error at most 0.00001. It was necessary to have a greater degree of accuracy on P in order to produce the value of x with error at most 0.01. The dimensions of the rectangle with an area of 200 square feet and the least possible perimeter are 14.14 feet by $\frac{200}{14.14}$ or 14.14 feet (a square) and the least perimeter is 56.56 feet. This result suggests that the exact solution is a $\sqrt{200} = 10\sqrt{2}$ foot by $\sqrt{200} = 10\sqrt{2}$ foot square with perimeter $40\sqrt{2}$ feet. In the study of calculus you will be able to verify the exact solution. •

• EXERCISES 3-7

Determine the domain by algebraic means. Check your answer with a graphing utility.

1. $f(x) = \dfrac{x^2+1}{x}$
2. $f(x) = \dfrac{x^2-1}{x}$
3. $f(x) = \dfrac{x^3+1}{x^2-1}$
4. $f(x) = \dfrac{x^3-1}{x^2+1}$
5. $f(x) = \dfrac{x^3-1}{x^2+4}$
6. $f(x) = \dfrac{x^3-1}{x^2-4}$

Solve each equation algebraically. Check your answer with a graphing utility.

7. $\dfrac{x-1}{x+2} = 3$
8. $\dfrac{1}{x} - \dfrac{2}{x-3} = 4$

Solve each inequality algebraically using the sign chart method. Check your answer with a graphing utility.

9. $\dfrac{1}{x-3} + 4 > 0$
10. $\dfrac{x-1}{x^2-4} < 0$

Consider the function f given in Example 1.

11. Graph f and $y = 2x^2$ in the $[-10, 10]$ by $[-100, 100]$ viewing rectangle.

12. Graph f and $y = 8 + \frac{3}{x-2}$ in the $[1, 3]$ by $[-50, 50]$ viewing rectangle.

Determine the end behavior asymptote of the function without using a graphing utility. Sketch a complete graph including all asymptotes. Check your answer with a graphing utility.

13. $f(x) = \dfrac{x^2 - 2x + 3}{x + 2}$

14. $g(x) = \dfrac{3x^2 - x + 5}{x^2 - 4}$

15. $f(x) = \dfrac{x^3 - 1}{x - 1}$

16. $h(x) = \dfrac{x^3 + 1}{x^2 + 1}$

17. $f(x) = \dfrac{x^2 - 3x - 7}{x + 3}$

18. $g(x) = \dfrac{2x^3 - 2x^2 - x + 5}{x - 2}$

19. $f(x) = \dfrac{3x^4 - 2x^2 + x - 1}{2x - 5}$

20. $g(x) = \dfrac{2x^5 - 3x^3 + 2x - 4}{x - 1}$

21. Sketch a complete graph of $f(x) = \frac{2x+4}{x-3}$ (Exercise 20, Section 3.5) using the method of Example 2. Compare this method with the method used in Section 3.5.

Determine the vertical asymptotes and the end behavior asymptote, and sketch a complete graph of the function. Determine the domain, the range, and all zeros of the function.

22. $f(x) = \dfrac{x^2 - 2x + 5}{x + 1}$

23. $g(x) = \dfrac{x^2 - 2x + 1}{x - 2}$

24. $f(x) = \dfrac{2x^3 - x^2 + 3x - 2}{x^3 + 3}$

25. $g(x) = \dfrac{x^3 - 2x + 1}{x - 2}$

26. $f(x) = \dfrac{x^4 - 2x^2 - x + 3}{x^2 + 4}$

27. $g(x) = \dfrac{x^4 - 2x^2 - x + 3}{x^2 - 4}$

Determine where the function is increasing and decreasing.

28. $f(x) = \dfrac{x^3 - 2x - 1}{3x + 5}$

29. $g(x) = \dfrac{2x^3 - x^2 + 1}{2 - x}$

Determine all local extrema, intervals when the function is increasing and decreasing, and all zeros of the function. Determine an end behavior model, and look for any hidden behavior.

30. $f(x) = \dfrac{x^3 - 2x^2 + x - 1}{2x - 1}$

31. $f(x) = \dfrac{2x^3 + x^2 - 24x - 12}{x^2 + x - 12}$

32. $f(x) = \dfrac{2x^4 + 3x^3 + x^2 + 2}{x^3 - x^2 + 20}$

Let $g(x) = \frac{3}{x-2}$ and $f(x) = 2x^2 + \frac{3}{x-2} = \frac{2x^3-4x^2+3}{x-2}$, the function of Example 1. Complete each table using a calculator.

33.

x	$g(x)$	$f(x)$
3		
2.1		
2.01		
2.001		
2.0001		
2.00001		
2.000001		

34.

x	$g(x)$	$f(x)$
1		
1.9		
1.99		
1.999		
1.9999		
1.99999		
1.999999		

35.

x	$f(x)$
1	
10	
100	
1,000	
10,000	
100,000	
1,000,000	

36.

x	$f(x)$
−1	
−10	
−100	
−1,000	
−10,000	
−100,000	
−1,000,000	

37. Use the table in Exercise 33 to determine numerically what happens to the values of the function $f(x)$ as $x \to 2^+$. Remember $x \to 2^+$ mean $x \to 2$ through values greater than 2.

38. Use the table in Exercise 34 to determine numerically what happens to the values of the function $f(x)$ as $x \to 2^-$. Remember $x \to 2^-$ mean $x \to 2$ through values less than 2.

39. Use the table in Exercise 35 to determine numerically what happens to the values of the function $f(x)$ as $x \to \infty$.

40. Use the table in Exercise 36 to determine numerically what happens to the values of the function $f(x)$ as $x \to -\infty$.

41. Let h and g be polynomial functions of degree greater than or equal to one. Under what conditions is the line $x = 2$ a vertical asymptote of $f(x) = \frac{g(x)}{(x-2)h(x)}$?

42. Consider all rectangles with an area of 375 square feet. Let x be the width of such a rectangle.
   (a) Write the perimeter P of the rectangle as a function of x.
   (b) Determine the vertical asymptotes and the end behavior asymptote, and draw a complete graph of the function in (a).
   (c) What portion of the graph in (b) represents the problem situation?
   (d) Find the dimensions of a rectangle with an area of 375 square feet that has the smallest possible perimeter. What is this perimeter?

43. A certain amount x of a 100% pure barium solution is added to 135 ounces of a 35% barium solution to obtain a 63% barium solution.
   (a) Determine the function P that gives the percent of barium in the new solution as a function of x, the amount of pure solution added.

(b) Write an algebraic representation involving P for the problem situation.

(c) Use computer graphing and zoom-in to determine x.

44. A certain amount x of an 87% barium solution is added to 135 ounces of a 35% barium solution to obtain a 63% barium solution.

(a) Determine the function P that gives the percent of barium in the new solution as a function of x, the amount of 87% solution added.

(b) Write an algebraic representation involving P for the problem situation.

(c) Use graphs and zoom-in to determine x.

45. Let $f(x) = \frac{1}{x}$. Show that $f \circ f(x) = x$ for all $x \neq 0$. Draw a complete graph of f. For what values of m is the line $y = mx$ a line of symmetry of f?

46. Determine the end behavior asymptote of the rational function of Example 5.

3.8 Radical Functions

In this section we investigate the graphs of radical functions, that is, functions that involve radicals such as functions of the form $y = \sqrt{x-2}$, $y = \sqrt[3]{x^2-1} + x$, and so forth. In particular, we see how to obtain a sketch of the graph of $y = a\sqrt[n]{bx+c} + d$ from what we know about the graph of $y = \sqrt[n]{x}$ and plotting two or three points of the graph of $y = a\sqrt[n]{bx+c} + d$. We solve equations and inequalities that involve radicals both algebraically and graphically. Problem situations with radical functions, equations, or inequalities as algebraic representations will be studied.

• DEFINITION Any expression of the form $\sqrt[n]{a}$, where n is an integer greater than 1, is called a **radical**. The integer n is called the **index** of the radical, a is called the **radicand**, and the symbol $\sqrt{}$ is called a **radical sign**. If *no* index appears in the radical, it is understood to be 2. •

Radical expressions can also be written using fractional exponents as suggested by the following examples:

$$\sqrt{x-2} = (x-2)^{1/2},$$

$$\sqrt[3]{x^2-1} = (x^2-1)^{1/3},$$

$$\sqrt[5]{x^4} = x^{4/5}.$$

If $a > 0$, the equation $x^2 = a$ has two real solutions; one is positive, and the other is negative. For example, the solutions to $x^2 = 4$ are 2 or -2. If $a < 0$, the equation $x^2 = a$ has *no* real solutions. The equation $x^2 = 0$ has only one real solution, namely 0.

• DEFINITION If $a > 0$, the positive solution to $x^2 = a$ is denoted by $\sqrt{a}$, and is called the **principal square root** of a. The negative solution is $-\sqrt{a}$. Also, $\sqrt{0} = 0$. •

For every real number a, the equation $x^3 = a$ has *exactly* one real solution. For example, the only real solution to $x^3 = -8$ is -2.

• DEFINITION The real solution to $x^3 = a$ is denoted by $\sqrt[3]{a}$ and is called the **principal cube root** of a. •

Let n be an integer greater than 1 and let a be a real number. If n is odd, the equation $x^n = a$ has *exactly* one real solution. If n is even, then the equation has real solutions if and only if $a \geq 0$.

• DEFINITION Let n be an integer greater than 1 and let a be a real number.

1. If n is odd, the real solution to $x^n = a$ is denoted by $\sqrt[n]{a}$ and is called the **principal nth root** of a.
2. If n is even and $a \geq 0$, the nonnegative real solution to $x^n = a$ is denoted by $\sqrt[n]{a}$ and is called the **principal nth root** of a. •

It follows from the above definition that the domain of the function $f(x) = \sqrt[n]{x}$ is the set of all real numbers if n is odd, and the set of all nonnegative real numbers if n is even. The graph of $y = \sqrt[n]{x}$ looks like the graph of $y = \sqrt{x}$ if n is even and looks like the graph of $y = \sqrt[3]{x}$ if n is odd.

• EXAMPLE 1: Draw and compare complete graphs of $y = \sqrt{x}$, $y = \sqrt[4]{x}$, and $y = \sqrt[6]{x}$. Determine the range of each function.

SOLUTION: A complete graph of each function is given in Figure 3.8.1. Notice that the graph of $y = \sqrt{x}$ is above the graph of $y = \sqrt[4]{x}$, which in turn is above the graph of $y = \sqrt[6]{x}$ in most of the interval $[0, \infty)$. The graphs of the three functions in Figure 3.8.2 suggest that $(0, 0)$ and $(1, 1)$ are common points. This can be confirmed by substitution. Now we can see that $\sqrt{x} > \sqrt[4]{x} > \sqrt[6]{x}$ in the interval $(1, \infty)$ and $\sqrt{x} < \sqrt[4]{x} < \sqrt[6]{x}$ in the interval $(0, 1)$. We have previously observed that the domain of each of these functions is $[0, \infty)$. The range of

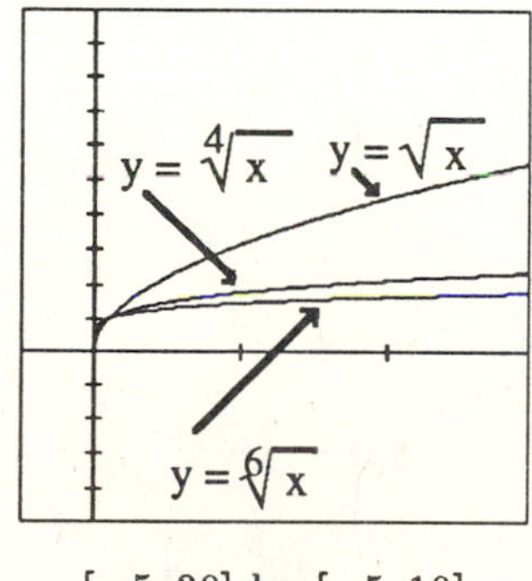

$[-5, 30]$ by $[-5, 10]$

Figure 3.8.1

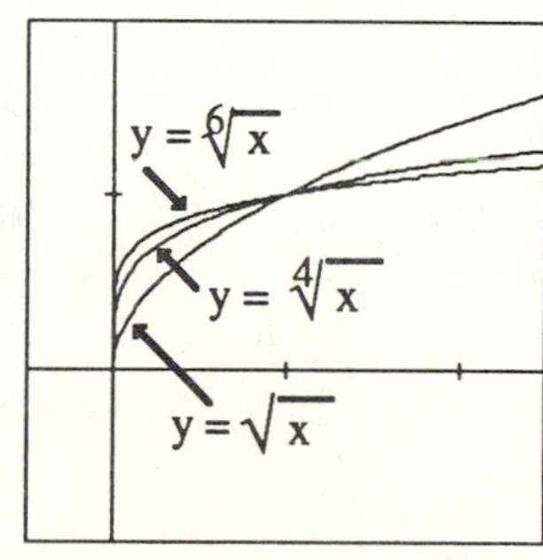

$[-0.5, 2.5]$ by $[-1, 2]$

Figure 3.8.2

each function is also $[0, \infty)$. On some graphing utilities the range may appear incorrectly to be $[a, \infty)$ for some $a > 0$ because the graph is very steep near the origin. •

The graphs of $y = \sqrt[n]{x}$ for n even are related as suggested by Example 1.

• EXAMPLE 2: Draw and compare complete graphs of $y = \sqrt[3]{x}$, $y = \sqrt[5]{x}$, and $y = \sqrt[7]{x}$. Determine the range of each function.

SOLUTION: The graphs of the three functions in Figure 3.8.3 and Figure 3.8.4 suggest that $(-1,-1)$, $(0,0)$, and $(1,1)$ are common points of intersection. Either figure gives a complete graph of each function. We can also see that $\sqrt[3]{x} > \sqrt[5]{x} > \sqrt[7]{x}$ in $(-1,0) \cup (1,\infty)$ and that $\sqrt[3]{x} < \sqrt[5]{x} < \sqrt[7]{x}$ in $(-\infty,-1) \cup (0,1)$. Notice that the domain and range of each function is $(-\infty, \infty)$. •

The graphs of $y = \sqrt[n]{x}$ for n odd are related as suggested by Example 2. Some graphing utilities produce only the portions of the graphs in Figure 3.8.4 that are to the right of the y-axis. If you are using such a graphing utility, Exercise 1 shows how to get around this difficulty and obtain a graph of $y = \sqrt[n]{x}$ for n odd that looks correct by graphing $y = \frac{|x|}{x} \sqrt[n]{|x|}$.

Reflection through the y-axis

• DEFINITION If we replace each point (x, y) of the graph of $y = f(x)$ by the point $(-x, y)$, then we say that the new graph has been obtained by **reflecting the graph of $y = f(x)$ through the y-axis.** •

Geometrically, the above definition means that each point (x, y) on the graph of $y = f(x)$ is replaced by its symmetric image with respect to the y-axis. Recall that a graph was symmetric with respect to the y-axis if we obtained the identical graph with this $(-x, y)$ for (x, y) replacement.

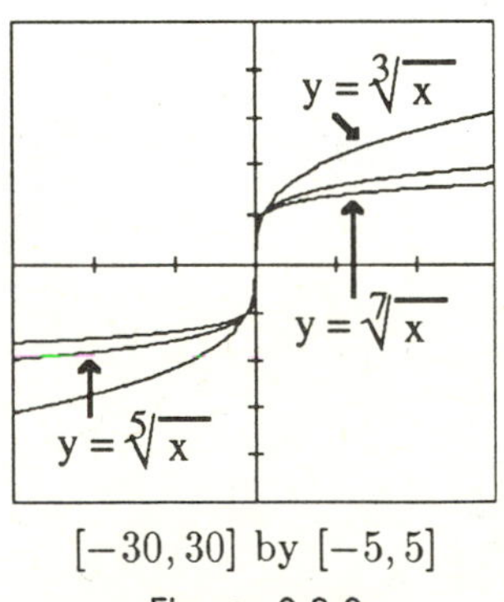

$[-30, 30]$ by $[-5, 5]$

Figure 3.8.3

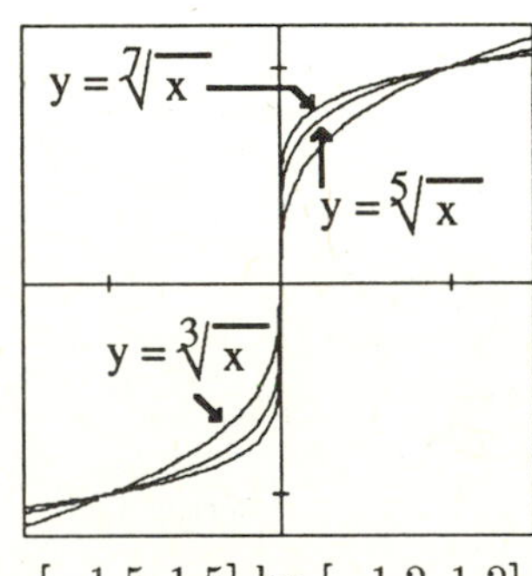

$[-1.5, 1.5]$ by $[-1.2, 1.2]$

Figure 3.8.4

• THEOREM 1 The graph of $y = f(-x)$ is the reflection through the y-axis of the graph of $y = f(x)$.

PROOF If (a, b) is on the graph of $y = f(x)$, then $b = f(a)$. We want to show that $(-a, b)$ is on the graph of $y = f(-x)$. We need to show that we get b when x is replace by $-a$ in $f(-x)$. Now, $f(-(-a)) = f(a) = b$. Therefore, $(-a, b)$ is on the graph of $y = f(-x)$. Similarly, if (r, s) is on the graph of $y = f(-x)$, then $(-r, s)$ is on the graph of $y = f(x)$. This means that the graph of $y = f(-x)$ is the reflection with respect to the y-axis of the graph of $y = f(x)$ (Figure 3.8.5).

Next, we will show that a sketch of a complete graph of $y = a\sqrt[n]{bx + c}+d$ can be obtained by plotting two or three points of its graph and using knowledge about the relationship of the graphs of $y = \sqrt[n]{x}$ and $y = a\sqrt[n]{bx + c} + d$.

• EXAMPLE 3: Use geometric transformation to draw a complete graph of each function.

(a) $y = \sqrt{x - 2}$ (b) $y = \sqrt{x + 3}$ (c) $y = \sqrt{3 - x}$

SOLUTION:

(a) A complete graph of $y = \sqrt{x - 2}$ can be obtained by shifting a complete graph of $y = \sqrt{x}$ horizontally to the right two units. Complete graphs of $y = \sqrt{x}$ and $y = \sqrt{x - 2}$ are given in Figure 3.8.6.

(b) A complete graph of $y = \sqrt{x + 3}$ can be obtained by shifting a complete graph of $y = \sqrt{x}$ horizontally to the left three units. Complete graphs of $y = \sqrt{x}$ and $y = \sqrt{x + 3}$ are given in Figure 3.8.7.

(c) The graph of $y = f(-x)$ can be obtained by reflecting the graph of $y = f(x)$ through the y-axis. We can rewrite y as follows:

$$y = \sqrt{3 - x} = \sqrt{-(x - 3)}.$$

Thus, a complete graph of $y = \sqrt{3 - x}$ can be obtained from a complete graph of $y = \sqrt{x}$ by applying, in order, the following transformations:

1. A reflection through the y-axis to get $y = \sqrt{-x}$.

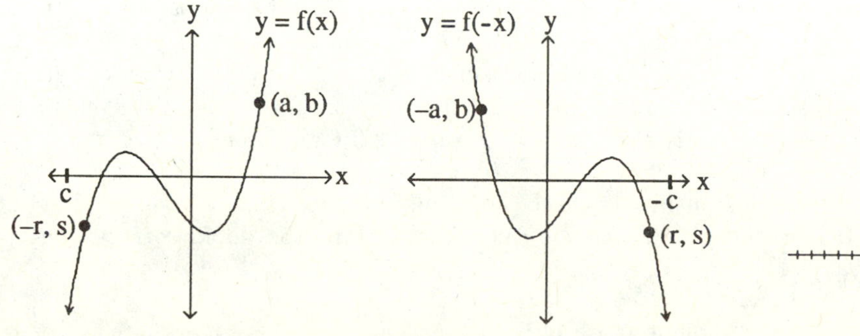

Figure 3.8.5

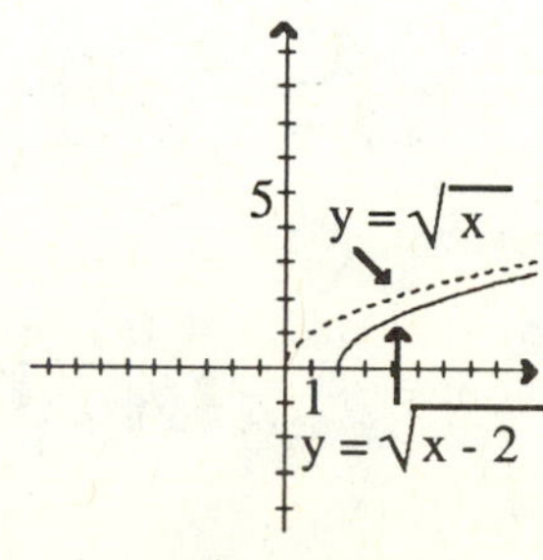

Figure 3.8.6

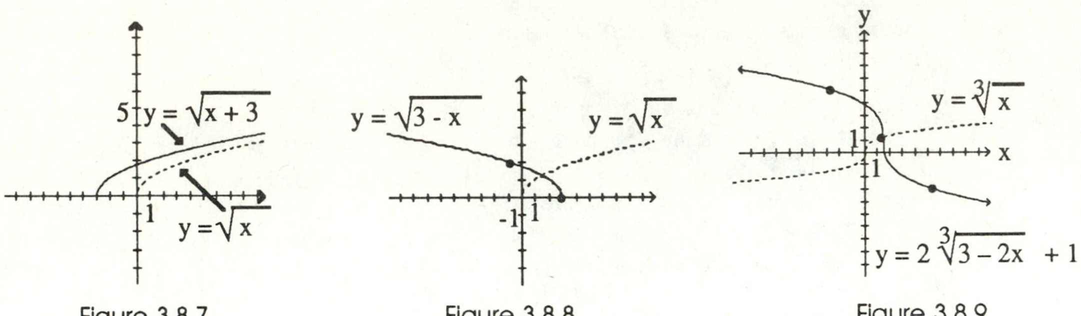

Figure 3.8.7 Figure 3.8.8 Figure 3.8.9

2. A horizontal shift right 3 units to get $y = \sqrt{-(x-3)}$.

   This means the graph of $y = \sqrt{3-x}$ looks like the graph of $y = \sqrt{x}$ except that it opens to the left with starting point $(3,0)$. Notice that the point $(-1,2)$ is also on the graph of $y = \sqrt{3-x}$ (Figure 3.8.8). Instead of using the geometric transformations above, we could use the two points $(3,0)$ and $(-1,2)$ to obtain a sketch of a complete graph of $y = \sqrt{3-x}$. •

In this way the actual transformations needed to obtain the graph of $y = \sqrt{3-x}$ from the graph of $y = \sqrt{x}$ are not used directly. What is important is that the existence of the transformations gives us enough information about the graph to produce a complete graph using only two points.

It can be shown that the graph of any radical function of the form $y = a\sqrt[n]{bx+c} + d$ can be obtained by applying geometric transformations to the graph of $y = \sqrt[n]{x}$. We can use this information to obtain a sketch of a complete graph by plotting two or three points, always being careful to include the point of $y = a\sqrt[n]{bx+c} + d$ that corresponds to the point $(0,0)$ of $y = \sqrt[n]{x}$. Examples 4 and 5 illustrate this.

• EXAMPLE 4: Sketch a complete graph of $y = 2\sqrt[3]{3-2x} + 1$.

SOLUTION: Notice that $\sqrt[3]{3-2x} = 0$ for $x = \frac{3}{2}$. Thus, $(\frac{3}{2}, 1)$ is a point of the graph of $y = 2\sqrt[3]{3-2x} + 1$. The point $(\frac{3}{2}, 1)$ corresponds to the point $(0,0)$ of $y = \sqrt[3]{x}$ under the transformations that carry $y = \sqrt[3]{x}$ to $y = 2\sqrt[3]{3-2x} + 1$. Now, we find a point to the left of $(\frac{3}{2}, 1)$ and a point to the right of $(\frac{3}{2}, 1)$ both of which lie on the graph of $y = 2\sqrt[3]{3-2x} + 1$. It is a good idea to find points that are not too close to $(\frac{3}{2}, 1)$. (Why?) Check that $(-\frac{5}{2}, 5)$ and $(\frac{11}{2}, -3)$ are such points. A complete graph of $y = 2\sqrt[3]{3-2x} + 1$ is given in Figure 3.8.9. The graph of $y = \sqrt[3]{x}$ is also shown in Figure 3.8.9 for reference. •

We can use a graphing utility to check the graph in Figure 3.8.9. If your graphing utility produces only half of the graph, then you can obtain a graph that looks correct by graphing $y = \frac{|3-2x|}{3-2x} 2\sqrt[3]{|3-2x|} + 1$. (Why?)

• EXAMPLE 5: Draw a complete graph of each function.

(a) $y = -2\sqrt{x-1} + 3$ (b) $y = 3\sqrt[4]{4-3x} - 2$

SOLUTION:

(a) The point $(1,3)$ on the graph of $y = -2\sqrt{x-1}+3$ corresponds to the point $(0,0)$ on the graph of $y = \sqrt{x}$. We can tell that the graph of $y = -2\sqrt{x-1}+3$ opens to the right of $(1,3)$ because the radicand $x-1$ must be positive. This means that $x \geq 1$. Thus, we look for a point on the graph of $y = -2\sqrt{x-1}+3$ to the right of $(1,3)$. Notice that $(5,-1)$ is such a point. A complete graph is given in Figure 3.8.10.

(b) The point $(\frac{4}{3},-2)$ on the graph of $y = 3\sqrt[4]{4-3x}-2$ corresponds to the point $(0,0)$ on the graph of $y = \sqrt[4]{x}$. We must have $4-3x \geq 0$ so that $x \leq \frac{4}{3}$. Thus, our graph opens to the left. Check that $(-4,4)$ is a point on the graph of $y = 3\sqrt[4]{4-3x}-2$ to the left of $(\frac{4}{3},-2)$. A complete graph is given in Figure 3.8.11. •

SUMMARY To obtain a sketch of the graph of $y = a\sqrt[n]{bx+c}+d$ we first graph that point $(-\frac{c}{b},d)$ that corresponds to $(0,0)$ of $y = \sqrt[n]{x}$. If n is even we plot one additional point; and if n is odd we plot two additional points, one on each side of $(-\frac{c}{b},d)$. Then, we draw a graph that has the basic shape like that of $y = \sqrt[n]{x}$ through the two or three points plotted.

Next, we will illustrate how to solve equations and inequalities involving radicals using both graphical and algebraic methods. We must be careful when we solve algebraically because we raise each side of an equation to a power. Performing such an operation leads to extraneous solutions; that is, solutions to equations cleared of radicals that are *not* solutions to the original equation.

• EXAMPLE 6: A real number plus its principle square root equals 1. Determine all such numbers.

SOLUTION: If x is such a number, then $x+\sqrt{x} = 1$. This equation can be solved graphically by finding the points of intersection of complete graphs of $y = x+\sqrt{x}$ and $y = 1$ (Figure 3.8.12). Because the graphs in Figure 3.8.12 are complete, we can see that this equation has

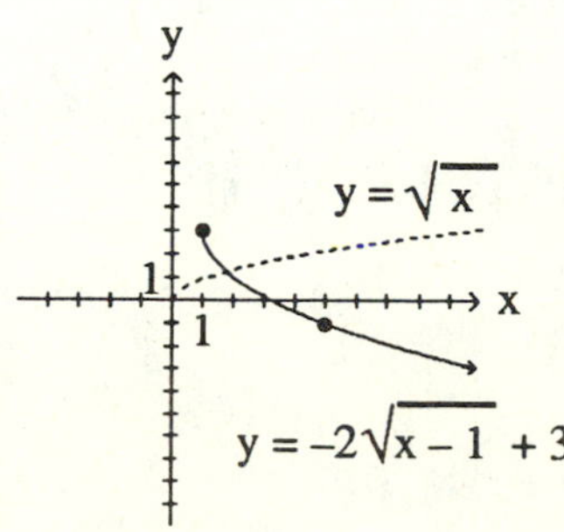

Figure 3.8.10

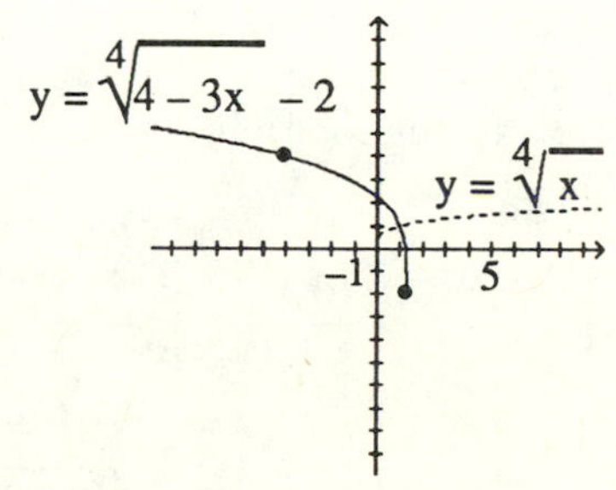

Figure 3.8.11

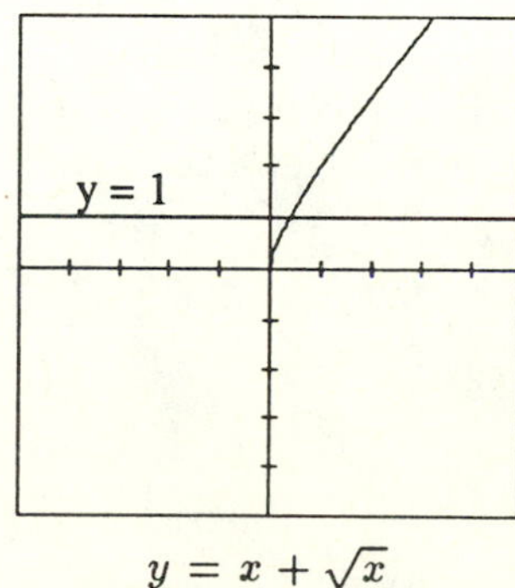

$y = x+\sqrt{x}$
$[-5,5]$ by $[-5,5]$

Figure 3.8.12

exactly one real solution and it is between 0 and 1. We can use zoom-in to determine that the solution to $x + \sqrt{x} = 1$ is 0.38 with error at most 0.01.

Alternatively, we can determine the solution algebraically. The procedure is to isolate the radical term on one side of the equation and then square each side of the equation:

$$\begin{aligned} x + \sqrt{x} &= 1 \\ \sqrt{x} &= 1 - x \\ x &= 1 - 2x + x^2 \\ 0 &= x^2 - 3x + 1. \end{aligned}$$

The new equation obtained by squaring is not necessarily equivalent to the original equation. However, every solution to the original equation is also a solution to the new equation. Using the quadratic formula we find that the possible solutions to the new equation are $\frac{3+\sqrt{5}}{2}$ and $\frac{3-\sqrt{5}}{2}$. Accurate to hundredths, these two numbers are 2.62 and 0.38. Checking in the original equation using 2.62 as an approximation to $\frac{3+\sqrt{5}}{2}$ will show that $\frac{3+\sqrt{5}}{2}$ is not a solution. Checking 0.38 in the original equation will convince you that $\frac{3-\sqrt{5}}{2}$ is a solution. Thus, $\frac{3-\sqrt{5}}{2}$ (approximately 0.38) is the only real solution to $x + \sqrt{x} = 1$, and $\frac{3+\sqrt{5}}{2}$ is an extraneous solution introduced by squaring. •

• EXAMPLE 7: Solve for x: $\sqrt{6x+12} - \sqrt{4x+9} = 1$.

SOLUTION: The radicand of each radical term must be greater than or equal to zero. This means that any solution to the equation must be a common solution of the following pair of inequalities:

$$\begin{aligned} 6x + 12 &\geq 0 \qquad & 4x + 9 &\geq 0 \\ 6x &\geq -12 & 4x &\geq -9 \\ x &\geq -2 & x &\geq -\frac{9}{4}. \end{aligned}$$

Thus, any solution to this equation must be greater than or equal to -2. (Why?)

If we rewrite the equation as $\sqrt{6x+12} - \sqrt{4x+9} - 1 = 0$, then we can solve graphically by finding the x-intercepts of $f(x) = \sqrt{6x+12} - \sqrt{4x+9} - 1$ (Figure 3.8.13). Notice that the domain of f is $[-2, \infty)$. Some graphing utilities may not show that the range is $[-2, \infty)$ because the graph is very steep near $x = -2$. It appears that 4 is the only real solution.

Alternatively, to solve the equation algebraically we isolate one radical term, square each side of the equation, and collect terms:

$$\begin{aligned} \sqrt{6x+12} - \sqrt{4x+9} &= 1 \\ \sqrt{6x+12} &= \sqrt{4x+9} + 1 \\ 6x + 12 &= 4x + 9 + 2\sqrt{4x+9} + 1 \\ 2x + 2 &= 2\sqrt{4x+9} \\ x + 1 &= \sqrt{4x+9}. \end{aligned}$$

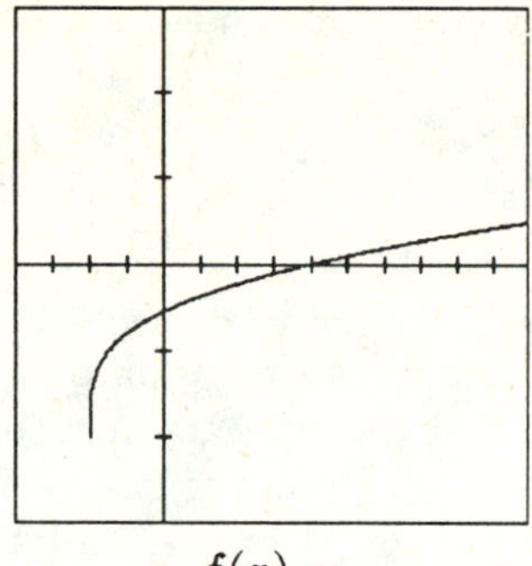

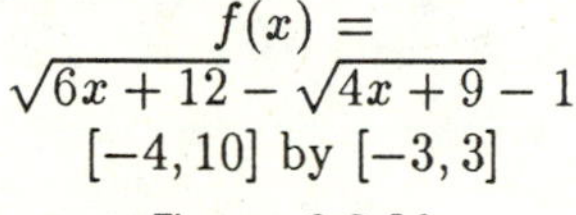

$f(x) =$
$\sqrt{6x+12} - \sqrt{4x+9} - 1$
$[-4, 10]$ by $[-3, 3]$

Figure 3.8.13

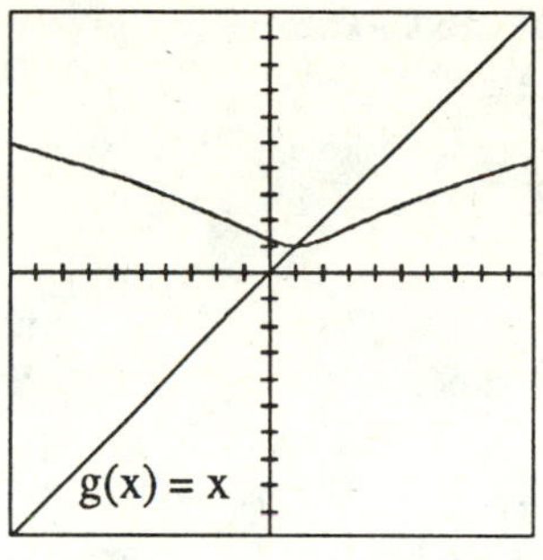

$f(x) = \sqrt[3]{x^2 - 2x + 2}$
$[-10, 10]$ by $[-10, 10]$

Figure 3.8.14

Squaring again, we obtain

$$\begin{aligned} x^2 + 2x + 1 &= 4x + 9 \\ x^2 - 2x - 8 &= 0 \\ (x-4)(x+2) &= 0. \end{aligned}$$

Thus, -2 and 4 are two possible real solutions. We can check in the original equation to see that 4 is a solution and -2 is not. •

If radicals other than square roots are involved, then algebraically we have to raise each side of the equation to higher powers.

• EXAMPLE 8: Solve for x: $\sqrt[3]{x^2 - 2x + 2} = x$.

SOLUTION: The complete graphs of $f(x) = \sqrt[3]{x^2 - 2x + 2}$ and $g(x) = x$ in Figure 3.8.14 suggest that the given equation has exactly one solution. We can use zoom-in to determine that the solution is 1.00 with error at most 0.01. This strongly suggests that 1 may be an exact solution.

Alternatively, we can find the solution algebraically by cubing each side of the given equation:

$$\begin{aligned} x &= \sqrt[3]{x^2 - 2x + 2} \\ x^3 &= x^2 - 2x + 2 \\ x^3 - x^2 + 2x - 2 &= 0 \\ x^2(x-1) + 2(x-1) &= 0 \\ (x-1)(x^2+2) &= 0. \end{aligned}$$

Now, we can see that 1 is the only possible real solution. Checking in the original equation shows that 1 is indeed the exact solution. •

In the next example we give only a graphical solution. It is not possible to solve this example algebraically.

• EXAMPLE 9: Solve for x: $\sqrt[5]{9-x^2} \geq x^2+1$.

SOLUTION: Let $f(x) = \sqrt[5]{9-x^2}$ and $g(x) = x^2+1$. We need to determine the values of x for which the graph of f lies above or intersects the graph of g. Figure 3.8.15 shows that the two graphs intersect in two points, and that the graph of f lies above the graph of g for all x between the x-coordinates of the two points of intersection. It can be shown that the graphs in Figure 3.8.15 are complete. Using zoom-in, we find the x-coordinates of the points of intersection to be -0.73 and 0.73 with error at most 0.01. The solution to the given inequality is $[-0.73, 0.73]$. •

Applications often give rise to functions involving radicals. In the next example we see how the intensity of illumination from a source of light depends on the distance from the source.

Suppose that a light source is situated on a pole x feet above the ground. The intensity of illumination I at a point P on the ground from the base of the pole is known to vary directly with $\sin\theta$ and inversely with the square of the distance d (Figure 3.8.16). Thus, $I = k\cdot\sin\theta\cdot\frac{1}{d^2} = k\cdot\frac{x}{d}\cdot\frac{1}{d^2} = \frac{kx}{d^3}$, where k is some positive real number. If we compute the value of I (usually in candle power) for a specific value of x and d, we can determine the value of the **constant of variation** k. We do not need the actual value of k for the next example.

• EXAMPLE 10: A pole is situated 24 feet from the front of a store. How high above the ground should a security light be placed in order to provide maximum illumination at the point on the ground at the base of the storefront closest to the base of the pole (Figure 3.8.16)?

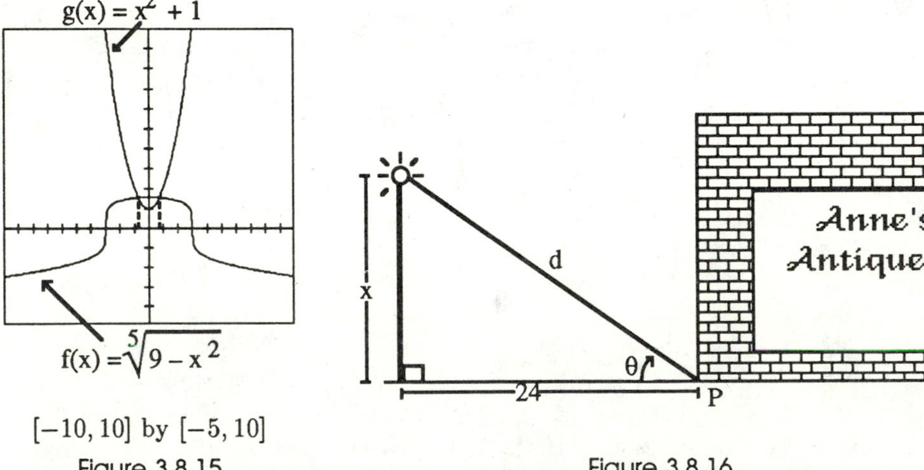

[−10, 10] by [−5, 10]

Figure 3.8.15

Figure 3.8.16

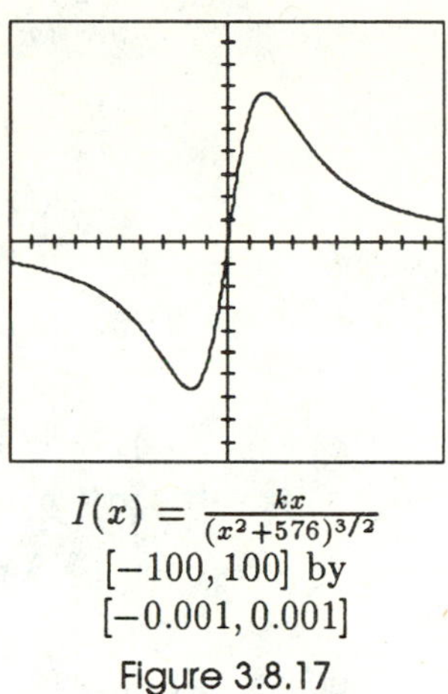

$I(x) = \frac{kx}{(x^2+576)^{3/2}}$
$[-100, 100]$ by
$[-0.001, 0.001]$

Figure 3.8.17

SOLUTION: We know that $I = \frac{kx}{d^3}$. Because $d = \sqrt{x^2 + 24^2} = \sqrt{x^2 + 576}$ we have $I(x) = \frac{kx}{(x^2+576)^{3/2}}$. We will draw a complete graph of I for $k = 1$. The graph of I for any other value of k is simply a vertical stretching or shrinking of $y = \frac{x}{(x^2+576)^{3/2}}$ by the factor k. Figure 3.8.17 gives the graph of $y = \frac{x}{(x^2+576)^{3/2}}$ in the $[-100, 100]$ by $[-0.001, 0.001]$ viewing rectangle. The value of x that gives the maximum value of I (k is positive) is the same no matter what the actual value of k. (Why?) Using zoom-in we can determine that $x = 16.97$ with error at most 0.01 gives the maximum value of I. Thus, the light should be placed on the pole 16.97 feet above the ground for maximum illumination. •

• EXERCISES 3-8

1[1]. (a) Use a computer graphing utility to plot a graph of $y = x^{1/3}$. (Enter as $x \wedge (1/3)$.)
(b) Use a computer graphing utility to plot a graph of $y = \frac{|x|}{x}(|\, x \,|)^{1/3}$.
(c) Which graph (part (a) or part (b)) is a correct complete graph of $y = x^{1/3}$?
(d) Explain why the answer to part (c) produces a correct graph of $y = x^{1/3}$.

2[1]. (a) Use a computer graphing utility to plot a graph of $y = x^{2/3}$.(Enter as $x \wedge (2/3)$.)
(b) Use a computer graphing utility to plot a graph of $y = (x^{1/3})^2$. (Enter as $(x \wedge (1/3)) \wedge 2$.)
(c) Use a computer graphing utility to plot a graph of $y = (x^2)^{1/3}$. (Enter as $(x \wedge 2) \wedge (1/3)$.)
(d) Which graph (part (a), (b), or (c)) is a correct graph of $y = x^{2/3}$?
(e) Explain why the answer to part (d) produces a correct graph of $y = x^{2/3}$.

3. Use a graphing utility to graph $y = \sqrt{x}$, $y = \sqrt[4]{x}$ and $y = \sqrt[6]{x}$ in the same viewing rectangle. For what values of x is $\sqrt[4]{x} < \sqrt{x}$? For what values of x is $\sqrt[6]{x} < \sqrt[4]{x}$?

Solve the equation algebraically. Check your answer with a graphing utility.

4. $\sqrt{x - 2} = 6$

5. $(2x - 1)^{1/2} = 2$

[1] Not applicable using most graphics calculators

6. $(2x-1)^{1/3}=2$

7. $2\sqrt{3-x}=-1$

8. $\sqrt[3]{x^2-1}=3$

9. $(x^2-1)^{1/3}=-\dfrac{1}{2}$

10. $x-\sqrt{x}=1$

11. $\sqrt{x-1}=\dfrac{x}{5}+1$

12. $\sqrt{x-3}-3\sqrt{x+12}=-11$

13. $\sqrt{x-5}-\sqrt{x+3}=-2$

Solve the inequality algebraically. Check your answer with a graphing utility.

14. $\sqrt{x+3}>6$

15. $\sqrt{x^2-4x-5}>x+2$

Solve the equation or inequality.

16. $\sqrt{x^3+2}=5$

17. $\sqrt{9-x^2}>x^2+1$

18. $\sqrt[3]{x^2-2x+1}=3x$

19. $2x+5<10+4\sqrt{3x-4}$

Sketch a complete graph of each function. Do not use a graphing utility. Check your answer with a graphing utility.

20. $y=-2\sqrt{x+3}$

21. $y=4+(x+2)^{1/2}$

22. $y=-3+\sqrt{x-5}$

23. $y=3-2(x-3)^{1/3}$

24. $y=2\sqrt[3]{3x-5}$

25. $y=-2+3(2-5x)^{1/3}$

26. $y=1-\sqrt[4]{2-4x}$

27. $y=2+3\sqrt[4]{2x+8}$

28–29. List the geometric transformations that can be applied to the graph of $y=\sqrt{x}$, $y=\sqrt[3]{x}$, or $y=\sqrt[4]{x}$ to obtain the given graph in Exercise 20 and in Exercise 23. Specify the order in which they should be applied.

Sketch a complete graph of the region satisfying the inequality without using a graphing utility. Check the boundary of your answer with a graphing utility.

30. $y<\sqrt{x-2}$

31. $y>4-\sqrt[3]{x+3}$

32. Draw a graph of $y=\frac{|x|}{x}$ without using a graphing utility. *Hint:* Consider two cases, $x\geq 0$ and $x<0$. What are the domain and range?

33. Show how information about complete graphs of $f(x)=x^3-1$ and $g(x)=\sqrt{x+1}$ can be used to determine the number of real solutions to $x^3-\sqrt{x+1}=1$. Do not use a graphing utility.

34. Show how to determine the number of real solutions to $\sqrt[3]{x-1}=4-x^2$. Do not use a graphing utility.

35. Use zoom-out to determine the end behavior of $f(x)=\sqrt{x^2-4x-5}$.

36. Determine the end behavior of $f(x)=\sqrt{ax^2+bx+c}$.

37. A number plus twice its square root equals 2. Find the numbers using an algebraic method. Check your solution with a graphing utility.

38. A number less twice its square root equals 2. Find the numbers using an algebraic method. Check your solution with a graphing utility.

39. A pole is situated 30 feet from the front of a store. How high above the ground should a security light bulb be placed in order to provide maximum illumination at the point on the base of the storefront closest to the base of the pole (see Figure 3.8.16)?

40. Determine the minimum *vertical* distance between the parabola $y=\frac{1}{10}(x-4)^2+28$ and the line $y=2x-11$. At what point on the parabola does it occur?

41. Penny is boating 20 miles offshore and wishes to reach a coastal city 60 miles further down the shore by steering the boat to a point P along the shore (Figure 3.8.18), and then driving the remaining distance. If her boat speed is 30 mph and her driving speed is 50 mph, where should she steer her boat to arrive at the city in the least amount of time?

    Let A denote the position of the boat, C the city, and AB the perpendicular to the shore line. Further, let x be the distance between B and P, and let T be the total time (in hours) for the trip.

    (a) Express T as a function of x.

    (b) Compute $T(0)$ and $T(60)$. Interpret these values in the problem situation.

    (c) Draw a complete graph of the function $y = T(x)$.

    (d) What values of x make sense in this problem situation? Draw a graph of the problem situation.

    (e) Where should she steer her boat in order to arrive at the city in the least amount of time?

    (f) What is the least amount of time?

42. The *surface* area S of a (right circular) cone excluding the base is given by $S = \pi r\sqrt{r^2 + h^2}$ where r is the radius and h is the height (Figure 3.8.19). The volume of the cone is $V = \frac{1}{3}\pi r^2 h$.

    (a) Let the height of a cone be fixed at 21 feet. Draw a complete graph of an algebraic representation of the surface area S as a function of the radius r.

    (b) What radius produces a surface area of 155 sq. ft.?

    (c) Let the volume be fixed at 380 cubic feet. Draw a complete graph of the model of the surface area as a function of the radius r. (*Hint:* Solve for h in the volume equation and substitute in the surface area equation.)

    (d) Find the dimensions of a cone with volume 380 cubic feet that has *minimal* surface area.

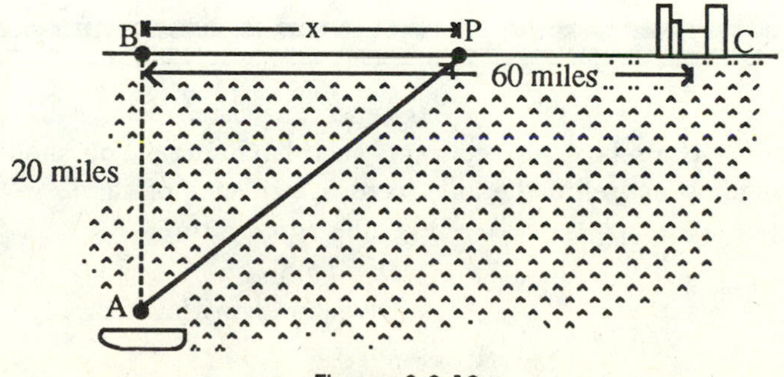

Figure 3.8.18

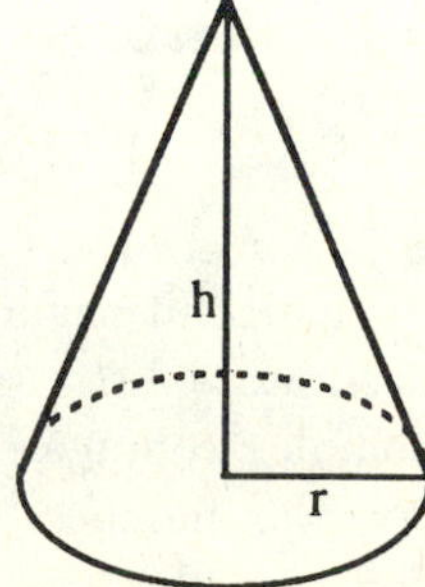

Figure 3.8.19

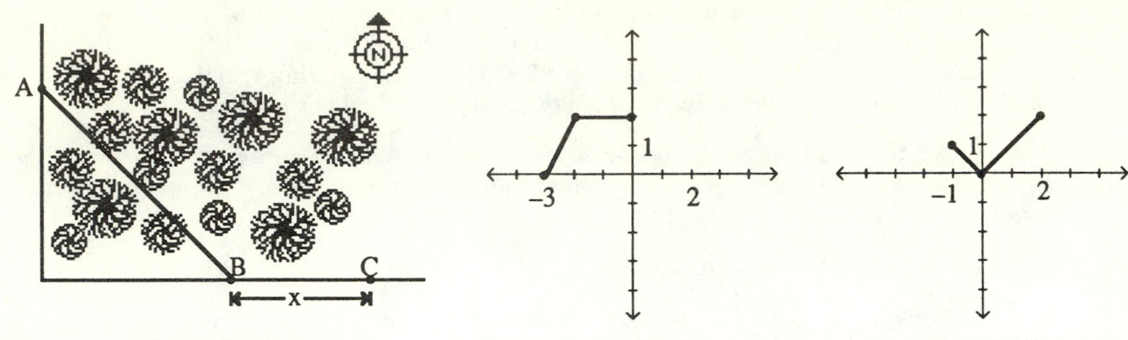

Figure 3.8.20 Figure 3.8.21 Figure 3.8.22

43. A person in a dense woods at point A can ride a trail bike due south along a road for 6 miles and then ride due east for 4 miles to arrive at point C. Or the rider could walk his bike directly through the woods to a point B along the east–west road and then ride along the road to point C (see Figure 3.8.20). Let x be the distance between points B and C. Assume the ground is level and the rider averages 10 mph while on the road and 2 mph while in the woods.
    (a) What path should the rider take to arrive at point C in the least amount of time?
    (b) Suppose the north–south road did not exist. Does this modification in the problem situation change the answer to (a)?
44. A complete graph of $y = f(x)$ is given in Figure 3.8.21. Draw a complete graph of each function and describe the geometric transformations needed to obtain its graph from the graph of $y = f(x)$.
    (a) $y = 1 + 2f(x)$ (b) $y = -f(x)$ (c) $y = -\frac{1}{2}f(x)$
    (d) $y = -1 + 2f(-x)$ (e) $y = -2f(x-2)$ (f) $y = 2 - f(x+1)$
45. A complete graph of $y = f(x)$ is given in Figure 3.8.22. Draw a complete graph of each function and describe the geometric transformations needed to obtain its graph from the graph of $y = f(x)$.
    (a) $y = f(-x)$ (b) $y = -f(x)$ (c) $y = -f(x-3)$
    (d) $y = -f(3-x)$ (e) $y = 2 - 3f(x+1)$ (f) $y = -1 + 2f(1-x)$

• KEY TERMS

Listed below are the key terms, vocabulary, and concepts in this chapter. You should be familiar with their definitions and meanings as well as be able to illustrate each of them.

Asymptote
Complex conjugate
Complex numbers
Constant of variation
Continuity
Cube roots of unity
Dividend
Division algorithm for polynomials
Divisor
End behavior

Extraneous solution
Factor Theorem
Fundamental Theorem of Algebra
Horizontal asymptote
Horizontal shifting
Imaginary part of a complex number
Index of a radical
Intermediate Value Property
Local extremum value
nth roots of unity
Point of discontinuity
Principal cube root
Principal n-th root
Principal square root
Principal square root of $-a$
Radical
Radical sign
Radicand
Rational function
Rational Zeros Theorem
Real part of a complex
Reciprocal function
Remainder
Remainder Theorem
Slant asymptote
Synthetic division
Vertical asymptote
Vertical stretching
Vertical shifting

The Review Exercises contain representative problems from each of the previous six sections. You should use these problems to test your understanding of the material covered in this chapter.

• REVIEW EXERCISES

In Exercises 1–2, determine the domain, range, and the values of x for which the function given by the graph is discontinuous. Assume the graph is complete.

1.

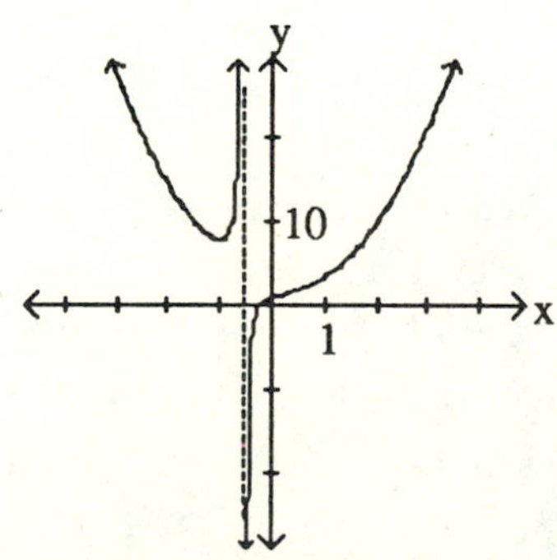

2.

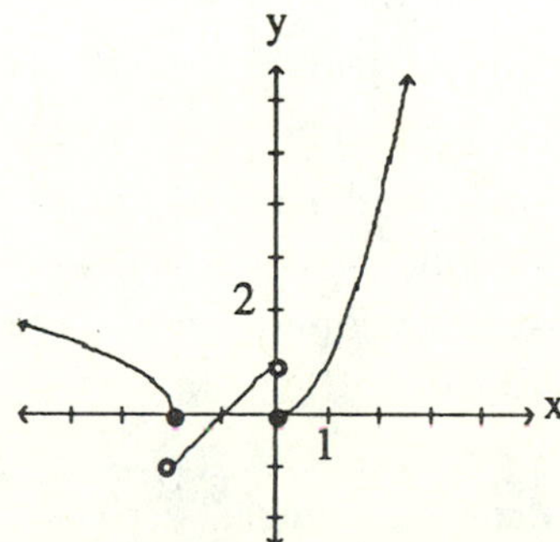

Determine the domain, range, and the points of discontinuity of each function.

3. $g(x) = \begin{cases} x^2 - 1, & x < 2 \\ \dfrac{3x}{2}, & x \geq 2 \end{cases}$

4. $V(x) = 5x^3 - 25x^2 + 30x + 9$

5. $y = \dfrac{1}{x+5}$

6. $f(x) = \text{INT}(x+2)$

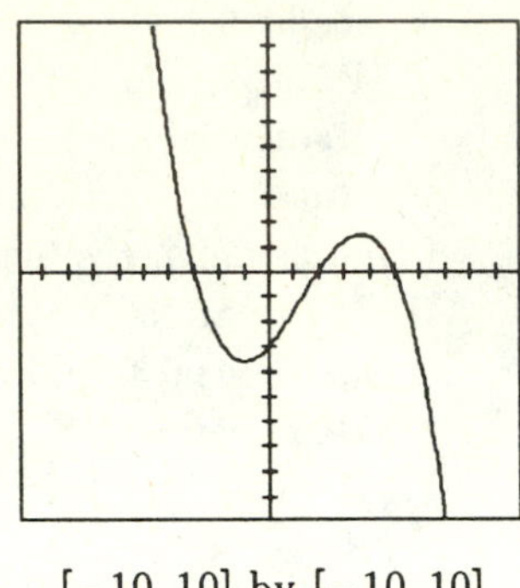

$[-10, 10]$ by $[-10, 10]$

Figure R.1

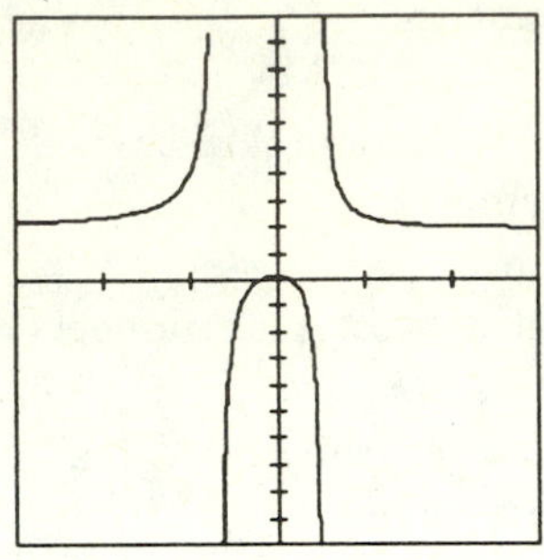

$[-30, 30]$ by $[-10, 10]$

Figure R.2

7. Graph the polynomial functions $f(x) = 5x^5 + 2x^2 - 4x + 7$ and $g(x) = 5x^5$ in a viewing rectangle that graphically illustrates that the end behavior of the two functions is the same. That is, a viewing rectangle in which the two graphs are nearly identical.

Assume the graph of $y = f(x)$ in Figure R.1 is complete.

8. As $x \to \infty$, $f(x) \to$?
9. As $x \to -\infty$, $f(x) \to$?

Assume the graph of $y = f(x)$ in Figure R.2 is complete.

10. As $x \to \infty$, $f(x) \to$?
11. As $x \to -\infty$, $f(x) \to$?

Find the quotient $q(x)$ and remainder $r(x)$ when $f(x)$ is divided by $h(x)$.

12. $f(x) = 3x^2 - 2x + 7$; $h(x) = x + 2$
13. $f(x) = 6x^3 - x^2 + 9x + 1$; $h(x) = 3x + 2$
14. Use the Remainder Theorem to determine the remainder when $f(x) = 3x^3 - 2x + 17$ is divided by $x - 1$. Check by using division.
15. Determine the remainder when $f(x) = 3x^{38} - 2x^{15} + 17x^2 - 3x + 12$ is divided by $x - 1$.
16. Use the Factor Theorem to determine if $x - 3$ is a factor of $x^3 - 2x^2 - 4x + 3$.
17. Determine the coordinates of a point (a, b) that must be added to the graph of $f(x) = \frac{x^2-4}{x+2}$ to make the function continuous for every real number.

Find all real zeros of each polynomial by factoring. Use a graphing utility to check your answer.

18. $f(x) = 8x - x^3$
19. $T(x) = x^4 - 13x^2 + 36$
20. Draw a complete graph of $f(x) = x^3 - x - 6$. Use the graph as an aid to factor the polynomial, and then determine all real zeros.
21. Determine *all* the zeros, real and nonreal complex, of the polynomial in Exercise 20.
22. Determine all the real zeros of $f(x) = x^5 - 5x^4 - x^3 + 5x^2 + 16x - 80$.
23. Draw the graph of $f(x) = 800x^3 + 780x^2 - 21x - 1$ in the $[-5, 5]$ by $[-150, 150]$ viewing rectangle.
    (a) How many real zeros are evident from this graph?
    (b) How many actual real zeros exist in this case?
    (c) Find all real zeros.

24. How many real zeros does $f(x) = 2x^4 + 2x^3 + 3x^2 + 2x + 1$ have? Why?

25. Determine a polynomial of degree 3 with real coefficients that has

    (a) 2 and -3 as zeros.

    (b) 2 and -3 as the *only* zeros.

    (c) 2 and $2 - i$ as zeros.

26. Is there a polynomial of degree 3 with real coefficients that has only nonreal complex zeros? Why?

27. If $f(x) = 3x^3 + 2kx^2 - kx + 2$, find a number k so that the graph of f contains $(1, 3)$.

Determine all real zeros of each function by algebraic methods. Classify them as integer, noninteger rational, or irrational. Check with a graphing utility.

28. $f(x) = x^3 + 3x^2 - 5x - 15$

29. $f(x) = x^4 - 8x^2 - 9$

Use the Rational Zeros Theorem to list all possible rational zeros of each function. Draw the graph of each function in an appropriate viewing rectangle and indicate which of the possible rational zeros are actual zeros. Determine all real zeros (both rational and irrational) of each function.

30. $f(x) = x^3 - 5x^2 + 2x - 10$

31. $f(x) = 3x^3 + 3x^2 - 4x - 2$

32. Use synthetic division to find the quotient and remainder when $2x^3 + x^2 - 4x + 5$ is divided by $x + 5$.

33. Draw a complete graph of $f(x) = x^3 + 7x^2 + 6x - 8$. Confirm that there is at least one *rational* root. Use synthetic division to find the other quadratic factor, and then determine all real zeros. Classify each real zero as rational or irrational.

34. Determine the real zeros of $f(x) = x^4 + 3x^3 - 3x^2 - 3x + 2$. Classify them as rational or irrational. Carefully outline why you are sure of your results.

35. A complete graph of $y = f(x)$ is given in Figure R.3. Draw a complete graph of $y = 1 + 2f(-x)$ and describe the geometric transformations needed to obtain its graph from the graph of $y = f(x)$.

36. Show that $x + 3$ is a factor of $g(x) = x^6 - 729$.

Write each expression in the form $a + bi$ with a and b real numbers.

37. $(5 - 2i)(1 - i)$

38. $3(2 + 3i) - 4 + 2i$

39. $\dfrac{3 + i}{5 + 4i}$

40. $(i^{93})^4$

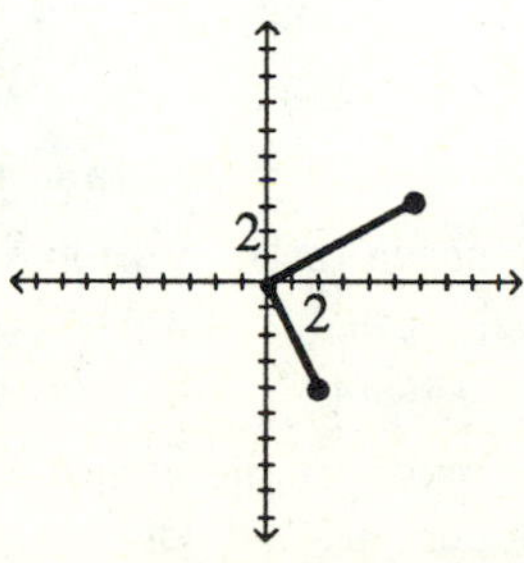

Figure R.3

Determine the number of real zeros and the number of nonreal complex zeros of each function.

41. $f(x) = 2x^3 + 3x^2 + 3x + 1$

42. $f(x) = x^3 - 2x^2 + 2x - 2$

43. Use computer graphing, the Rational Zeros Theorem, synthetic division, and the Fundamental Theorem of Algebra to find *all* the zeros of $f(x) = x^4 - 3x^3 - 12x - 16$. Classify each zero as integer, noninteger rational, irrational, or nonreal complex.

Find a polynomial with real coefficients satisfying the given condition.

44. degree 2; zero $1 - 2i$

45. degree 3; zeros $2 + 5i$ and 2

46. (a) Find a polynomial $f(x)$ of degree 4 with real coefficients that has 2, $3 + i$, and $3 - i$ as its *only* zeros.
    (b) Find a polynomial $f(x)$ of degree 4 with real coefficients that has 2, $3 + i$, and $3 - i$ as its only zeros, and also satisfies $f(0) = 1$.

47. Show that $2 - 3i$ is a zero of $f(x) = x^3 - 5x^2 + 17x - 13$ and find all other zeros of f.

Suppose $f(x)$ is a polynomial function of degree 4 with real coefficients.

48. What is the maximum possible number of real zeros?

49. What is the maximum possible number of nonreal complex zeros?

The function $f(x) = 0.04x(x - 11)(x - 18) + 30$ is an algebraic representation of the temperature (Fahrenheit) in Cleveland, Ohio, for a 24-hour period. Assume $x = 0$ is 6 A.M.

50. Without using a graphing utility, sketch a complete graph of the model. What is the domain of the model? What values of x make sense in this problem situation?

51. Use a graphing utility to draw a complete graph of the model and compare with your sketch in Exercise 50.

52. When will the temperature be 45° F?

53. What is the highest temperature and when is it achieved?

54. An object is shot straight up from the top of a 145-foot-tall tower with initial velocity 45 feet per second.
    (a) When will the object hit the ground?
    (b) When does the object reach its maximum height above the ground, and what is this maximum height?

55. The total daily revenue of Hank's Hamburger Haven is given by the equation $R = x \cdot p$, where x is the number of hamburgers sold and p is the price of one hamburger. Assume the price per hamburger is given by the supply equation $p = 0.3x + 0.05x^2 + 0.0007x^3$.
    (a) Determine for what positive values of x the supply function is increasing.
    (b) Draw a complete graph of the revenue function R.
    (c) What values of x make sense in the problem situation?
    (d) Determine the maximum possible daily revenue of Hank's Hamburger Haven and the number of hamburgers he needs to sell to achieve maximum revenue.

A liquid storage container on a truck is in the shape of a cylinder with hemispheres on each end as shown in Figure R.4. The cylinder and hemispheres have the same radius. The total length of the container is 140 feet.

56. Determine the volume V of the container as a function of the radius x.

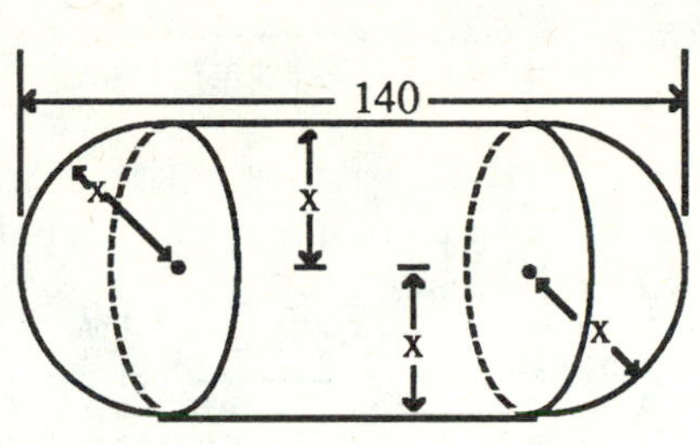

Figure R.4

[−10, 10] by [−10, 10]

Figure R.5

57. Draw a complete graph of the volume function $y = V(x)$.

58. What are the possible values of the radius determined by this problem situation?

59. What is the radius of the container with largest possible volume?

Biologists have determined that the polynomial function $P(t) = -0.00005t^3 + 0.003t^2 + 1.2t + 80$ approximates the population t days later of a certain group of wild pheasants left to reproduce on their own with no predators.

60. Draw a complete graph of $y = P(t)$.

61. Find the maximum pheasant population and when it occurs.

62. When will this pheasant population be extinct?

63. Create a scenario that could explain the "growth" exhibited by this pheasant population.

Sketch a complete graph of $y = f(x)$ without using a graphing utility. Check your answer with a graphing utility.

64. $y = 2(x-2)^2 + 3$

65. $y = 4 - 3\sqrt{x+1}$

Sketch a complete graph of each function without using a graphing utility. Identify and write an equation for each vertical and each horizontal asymptote. Check your answer with a graphing utility.

66. $y = \dfrac{5x}{x-3}$

67. $y = \dfrac{2+|x|}{x+1}$

68. (a) Find the domain and range of the function in Exercise 66.

(b) Write the function in Exercise 67 as a piecewise-defined function without using absolute value symbols.

A complete graph of $y = f(x)$ is given in Figure R.5. Use it to complete Exercises 69 and 70.

69. As $x \to \infty$, $f(x) \to$?

70. As $x \to -4^+$, $g(x) \to$?

71. List the geometric transformations needed to produce the graph of f from the graph of $y = \frac{1}{x}$. Specify the order in which the transformations should be applied. In each case, draw a complete graph of the function without the aid of a graphing utility. Write an equation for any vertical or horizontal asymptote. Check your answer with a graphing utility.

(a) $f(x) = \dfrac{5}{x-2}$

(b) $f(x) = 2 - \dfrac{3}{x+5}$

Find the domain of each function by algebraic means. Check your answer with a graphing utility.

72. $f(x) = \dfrac{3x-5}{x^2+x-6}$

73. $g(x) = \dfrac{5}{x^2-3x+1}$

74. $f(x) = \dfrac{x^3+1}{x^2+5}$

Sketch a complete graph of each function without using a graphing utility. Find all the vertical and horizontal asymptotes. Check your answer with a graphing utility.

75. $f(x) = \dfrac{7}{x+5}$

76. $g(x) = \dfrac{2x^2-6}{3-x^2}$

Draw a complete graph of each function and determine its domain and range. Find all the vertical asymptotes and the end behavior asymptote. Find all the zeros of the function.

77. $g(x) = \dfrac{3x^2}{(x-5)(x+4)}$

78. $f(x) = \dfrac{x^2-4}{x^2+4}$

79. $g(x) = \dfrac{2x^3+3x^2-6x-1}{x^3+2}$

Determine the intervals on which each function is increasing and the intervals on which each function is decreasing.

80. $f(x) = \dfrac{x^3+1}{x^2}$

81. $g(x) = \dfrac{-1}{x^3+x}$

Draw a complete graph of each function. Determine all local minima and maxima.

82. $f(x) = \dfrac{3x^2+10}{x}$

83. $f(x) = \dfrac{x^4+3x^3+2x^2-7}{x^2+2x-1}$

Sketch a complete graph. Determine an end behavior model. Find all vertical and horizontal asymptotes. Determine all zeros. Determine all extrema and the intervals where the function is increasing and decreasing.

84. $g(x) = \dfrac{x^3+1}{x^2-9}$

85. $f(x) = \dfrac{x^2-4x+13}{x+2}$

Determine a simple end behavior model and the end behavior asymptote of each function. Check your answer with a graphing utility.

86. $f(x) = \dfrac{x^3-5x^2-7x-1}{x+2}$

87. $g(x) = \dfrac{x^3-8}{x-2}$

88. $f(x) = \dfrac{x^2-x-2}{x^3+1}$

89. $f(x) = \dfrac{x^4-2x^3-8x^2+2x+3}{x+1}$

90. Let $g(x) = \frac{5}{x+1}$ and $f(x) = x^2 + \frac{5}{x+1} = \frac{x^3+x^2+5}{x+1}$. Complete the table using a calculator.

x	$g(x)$	$f(x)$
0		
−0.9		
−0.99		
−0.999		
−0.9999		
−0.99999		

91. Use the table in Exercise 90 to numerically determine what happens to the function values $f(x)$ as $x \to -1^+$.

92. Let $f(x) = \frac{x+5}{x^2-2x-8}$. Use a calculator to

(a) create a table to determine numerically what happens to the values of the function $f(x)$ as $x \to -\infty$.

(b) create a table to determine numerically what happens to the values of the function $f(x)$ as $x \to 4^-$.

Solve each equation algebraically. Check your answer with a graphing utility.

93. $\dfrac{2}{x} - \dfrac{3}{x+5} = 0$

94. $\dfrac{1}{x-3} + \dfrac{5}{x} = 2$

Solve each inequality algebraically using the sign chart method. Check your answer with a graphing utility.

95. $\dfrac{x-2}{x+5} < 0$

96. $\dfrac{x^2-4}{x+4} \geq 0$

97. Find the zeros, y-intercept, vertical asymptotes, and the end behavior asymptote of the function $f(x) = \frac{x^4-2x^3+3x^2+x-4}{x^2-5x-14}$. Draw a complete graph that shows all of this behavior. (More than one viewing rectangle may be needed.)

Solve each equation algebraically. Check your answer with a graphing utility.

98. $2(x-1)^{1/2} = 2$

99. $\sqrt[3]{x^2+3x+1} = 2$

100. $\sqrt{3x-5} = 2$

101. $(x^2+2x-3)^{1/3} = 5$

102. Use a graphical argument to show that if $t < 5$, then t is *not* a solution to the inequality $\sqrt{x+5} > 5 - \frac{x}{3}$. Solve the inequality.

Sketch a complete graph of each function without using a graphing utility. Check your answer with a graphing utility. List the geometric transformations that can be applied to the graph of $y = \sqrt{x}$ or $y = \sqrt[3]{x}$ to obtain the given graph. Specify the order in which they should be applied.

103. $y = \sqrt[3]{x-3}$

104. $y = -3 + 2\sqrt{5-x}$

105. Sketch a graph of the region satisfying the inequality $y > \sqrt{x-5}$ without using a graphing utility. Check the boundary of your answer with a graphing utility.

106. Judy is 5 feet 6 inches tall and walks at the rate of 4 feet per second away from a street light with a lamp 14.5 feet above level ground.

(a) Determine an algebraic representation of the length of Judy's shadow.

(b) At what rate is the length of Judy's shadow increasing?

(c) Express the distance D between the lamp and the tip of Judy's shadow as a function of time t.

(d) When will the distance D be 100 feet?

107. A balloon in the shape of a sphere is being inflated. Assume the radius r of the balloon is increasing at the rate of 3 inches per second and is zero when $t = 0$.

(a) Express the volume V of the balloon as a function of time t.

(b) Determine the volume of the balloon at $t = 5$ seconds.

(c) Suppose the balloon will burst when its volume is 15,000 cubic inches. When will the balloon burst?

108. The area of a triangle is to be 150 square units.
    (a) Express the height of this triangle as a function of the length of its base b and draw a complete graph of the function.
    (b) Write an equation for each asymptote.
    (c) What are the domain and range of this function?
    (d) What are the values of x that make sense in this problem situation?
    (e) Use a computer drawn graph to find the length of the base of the triangle if the height is 800 units.
109. Pure acid is added to 150 ounces of a 50% acid solution. Let x be the number of ounces of pure acid added.
    (a) Determine an algebraic representation of the acid concentration of the mixture.
    (b) Draw a complete graph of the algebraic representation. Write an equation for each asymptote.
    (c) What values of x make sense in this problem situation? Draw a complete graph of the problem situation.
    (d) Use a computer drawn graph to determine how much pure acid should be added to the 50% solution to produce a new mixture that is at least 78% acid.
    (e) Solve part (d) of this problem algebraically.
110. Lisa wishes to obtain 85 ounces of a 40% acid solution by combining a 72% acid solution with a 25% acid solution. How much of each solution should Lisa use?
111. Pure acid is added to 87 ounces of a 56% acid solution. Let x be the amount (in ounces) of pure acid added.
    (a) Express the concentration C of acid of the mixture as a function of x.
    (b) Draw a complete graph of C. Write an equation for each asymptote.
    (c) What values of x make sense in this problem situation?
    (d) Use a computer drawn graph to determine how much pure acid should be added to the 56% solution to produce a new mixture that is at least 78% acid.
    (e) Solve part (d) of this problem algebraically.
112. Boyle's law from physics states that for a certain gas $PV = 158$ where P is pressure and V is volume.
    (a) Draw a complete graph of volume as a function of pressure. Write an equation for each asymptote.
    (b) If $8 \leq V \leq 15$, what is the corresponding range for the values of P?
113. Consider all rectangles with an area of 500 square feet. Let x be the width of such a rectangle.
    (a) Write the perimeter P of the rectangle as a function of x.
    (b) Determine the vertical asymptotes and the end behavior, and draw a complete graph of the function in part (a).
    (c) What portion of the graph in part (b) represents the problem situation?
    (d) Find the dimensions of a rectangle with area 500 square feet that has the least possible perimeter. What is this perimeter?

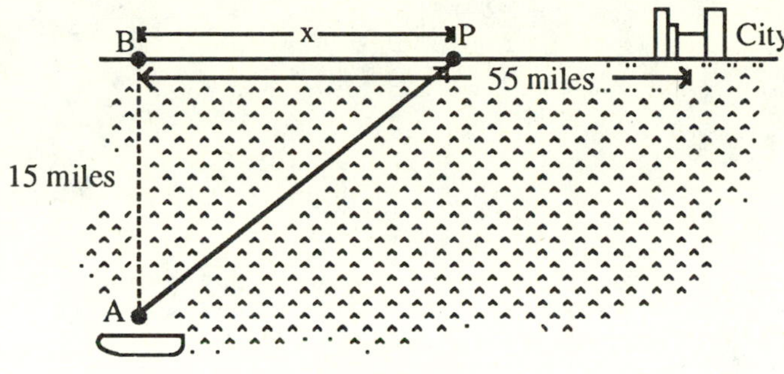

Figure R.6

114. Alan rode his bike 11 miles from his home to Columbus and then completed a 45 mile trip by car from Columbus to Marysville. Assume the average rate of the car was 41 miles per hour faster than the average rate of the bike.
   (a) Write the total time T required to complete the trip (bike and car) as a function of the rate x of the bike.
   (b) Draw a complete graph of the function in part (a) indicating any vertical and horizontal asymptotes and any zeros.
   (c) What are the values of x that make sense in this problem situation?
   (d) Use a geometric representation of the problem situation to find the rate of the bike if the total time of the trip was 55 minutes.

115. Jeri is boating 15 miles offshore and wishes to reach a coastal city 55 miles further down the shore by steering the boat to a point P along the shore (see Figure R.6), and then driving the remaining distance. Assume that her boat speed is 25 mph and her driving speed is 40 mph. Let A denote the position of the boat, C the city, and AB the perpendicular to the shore line. Further, let x be the distance between B and P, and let T be the total time (in hours) for the trip.
   (a) Determine an algebraic representation for the total time T of the trip.
   (b) Compute $T(0)$ and $T(55)$. Interpret these values in the problem situation.
   (c) Draw a complete graph of the function T.
   (d) What values of x make sense in this problem situation?
   (e) Where should she steer her boat in order to arrive at the city in the least amount of time?
   (f) What is the least amount of time?

4 Logarithmic and Exponential Functions

• Introduction

In this chapter we define the inverse relation of a relation. Inverses are extremely important in the study of mathematics. Exponential and logarithmic functions are introduced as inverses of each other. We see that the graphs of a relation and its inverse are symmetric with respect to the line $y = x$. Although the inverse of a function need not be a function, we show that inverses of one-to-one functions are functions.

Important applications about growth and decay are studied. For example, we study compound interest, annuities, population change, bacterial growth, and decay of radioactive substances. These problem situations have exponential functions as models, and algebraic solutions of such problems involve logarithmic functions. A special technique is needed to obtain a complete graph of most logarithmic functions with a graphing utility. Logarithm functions are used to help analyze data. Equations and inequalities that involve exponential and logarithmic functions are solved algebraically and graphically.

We assume that you can now quickly sketch a graph of $y = af(x + c) + d$ given the graph of $y = f(x)$. We make frequent use of this information to obtain graphs in this and subsequent chapters.

4.1 Inverse Relations

In this section we define the inverse relation of a given relation. Recall that a relation is any set of ordered pairs of real numbers. We show that the graph of an inverse relation can be obtained by reflecting the graph of the relation through the line $y = x$. If a relation is given by an algebraic expression, then we are able to determine the inverse relation algebraically. We show that the inverse of a one-to-one function is a function.

Roughly speaking, *inverse* means *undoing*. Suppose that we are given a certain process and know the effect of that process on an object. The inverse process, if there is one, is what has to be done to get back where we started. For example, *cubing* is a process that is *undone* by taking the principal *cube root*.

Cubing **Cube Root**

$$2 \xrightarrow{\text{(do)}} 8 \xrightarrow{\text{(undo)}} 2$$

Can you think of others?

We can think of a relation R as a *correspondence* between the domain of R and the range of R. For example $\{(-1,2),(0,4),(3,8)\}$ is a relation with domain $\{-1,0,3\}$, range $\{2,4,8\}$, and correspondence $-1 \to 2, 0 \to 4$, and $3 \to 8$. What should undoing mean for a relation R?

Domain of R **Range of R**

$$a \xrightarrow{R} b \xrightarrow{?} a$$

Whatever the inverse is, it should send b back to a. For our example, the correspondence for the inverse should be $2 \to -1, 4 \to 0$, and $8 \to 3$. This is the motivation for the following definition.

• DEFINITION The **inverse relation** of a relation R consists of those ordered pairs of real numbers obtained by interchanging the first and second entries of the ordered pairs in R. That is, (a,b) belongs to the inverse of R if and only if (b,a) belongs to R. The inverse relation is denoted by R^{-1}. The -1 in R^{-1} is not to be interpreted as an exponent. That is, $R^{-1} \neq \frac{1}{R}$ when R is a relation. •

• EXAMPLE 1: Show that the inverse of a relation that has a vertical line for its graph is a relation that has a horizontal line for its graph and vice versa.

SOLUTION: First we consider relations that have vertical lines for their graphs. We start with a particular case and then generalize our argument to all relations that have vertical lines for graphs. The relation $x = 3$ has a graph that is a vertical line which consists of all ordered pairs of the form $(3,b)$, where b is any real number. Thus, the inverse relation consists of the set of all ordered pairs of the form $(b,3)$, where b is any real number. The graph of the inverse relation is the horizontal line $y = 3$ (Figure 4.1.1). Similarly, we can

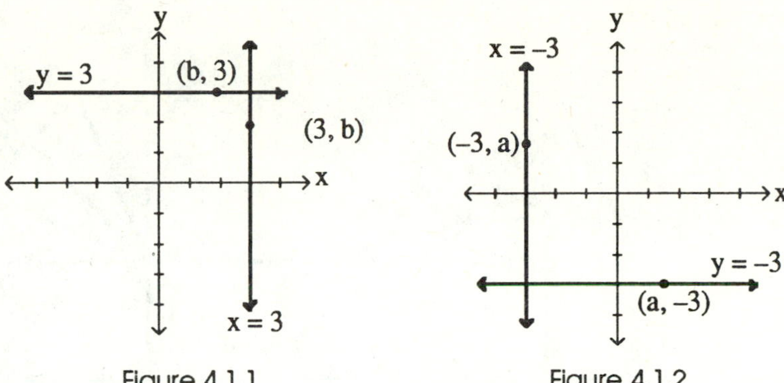

Figure 4.1.1

Figure 4.1.2

show that the graph of the inverse of the relation $x = a$ that has a vertical line for its graph is the relation $y = a$ that has a horizontal line for its graph.

Now we consider relations that have horizontal lines for their graphs. Again, we begin with a particular case and then generalize. The relation $y = -3$ has a graph that is a horizontal line which consists of all ordered pairs of the form $(a, -3)$, where a is any real number. The inverse relation consists of all ordered pairs of the form $(-3, a)$, where a is any real number. The graph of the inverse relation is the vertical line $x = -3$ (Figure 4.1.2). In general, the graph of the inverse relation of the relation $y = a$ that has a horizontal line for its graph is the relation $x = a$ that has a vertical line for its graph. •

Example 1 can be extended to lines that are not horizontal or vertical by using the concept of slope.

• EXAMPLE 2: Draw a complete graph of $y = 2x + 3$ and its inverse relation. Write an equation for the inverse relation.

SOLUTION: A complete graph of $y = 2x + 3$ is a straight line with slope 2. There is a one-to-one correspondence between the points on a complete graph of $y = 2x + 3$ and the points on a complete graph of the inverse of $y = 2x + 3$. Let (a, b) and (c, d) be any two distinct points on the graph of the inverse of $y = 2x + 3$. Then (b, a) and (d, c) are the corresponding distinct points on the graph of $y = 2x + 3$. We must have $\frac{c-a}{d-b} = 2$, because the slope of the line $y = 2x + 3$ is 2 and (b, a) and (d, c) are on this line. Now $a \neq c$ and $b \neq d$ because the line $y = 2x + 3$ is neither horizontal nor vertical. If we take the reciprocal of each side of $\frac{c-a}{d-b} = 2$ we find that $\frac{d-b}{c-a} = \frac{1}{2}$. This means that the points (a, b) and (c, d) are on a line with slope $\frac{1}{2}$. (Why?) Thus, a complete graph of the inverse of $y = 2x + 3$ is also a line.

The points $(3, 0)$ and $(7, 2)$ are on the graph of the inverse because $(0, 3)$ and $(2, 7)$ are on the graph of $y = 2x + 3$. We can use the points $(0, 3)$ and $(2, 7)$ to draw a complete graph of $y = 2x + 3$ and the points $(3, 0)$ and $(7, 2)$ to draw a complete graph of the inverse (Figure 4.1.3).

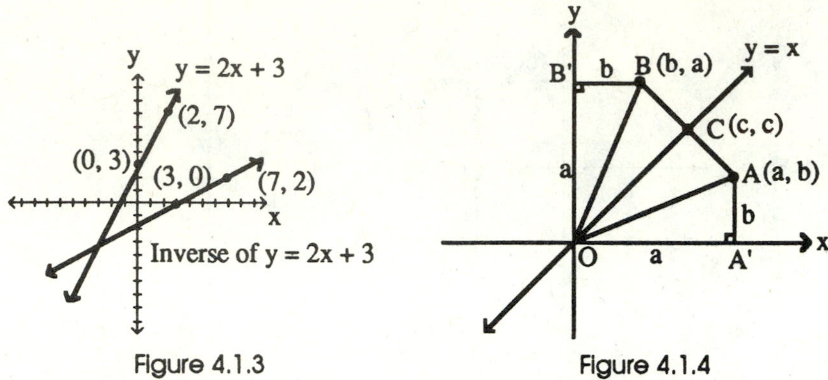

Figure 4.1.3 Figure 4.1.4

Next, we find an equation for the inverse of $y = 2x + 3$. First, we use the two points $(3, 0)$ and $(7, 2)$ of the inverse to compute the slope:

$$m = \frac{2-0}{7-3} = \frac{2}{4} = \frac{1}{2}.$$

Now, we use the point-slope formula to write an equation:

$$y - 0 = \frac{1}{2}(x-3)$$
$$2y = x - 3$$
$$x = 2y + 3.$$

Notice that the equation $x = 2y + 3$ for the inverse can be obtained by interchanging x and y in the equation $y = 2x + 3$. If we solve for y in terms of x we find that $y = \frac{1}{2}x - \frac{3}{2}$ is another form of the equation for the inverse of $y = 2x + 3$. In this form we can see that the inverse is a straight line with slope $\frac{1}{2}$ and y-intercept $-\frac{3}{2}$ as suggested by Figure 4.1.3. •

We give a generalization of Example 2 without proof.

• THEOREM 1 If R is the relation determined by a given equation or inequality in two variables, then R^{-1} is the relation determined by interchanging the two variables in the equation or inequality for the relation R.

The next theorem gives a geometric relationship between a point (a, b) on the graph of a relation and the corresponding point (b, a) on the graph of the inverse of the relation.

• THEOREM 2 The points (a, b) and (b, a) are symmetric with respect to the line $y = x$.

PROOF Without any loss of generality we can assume that $a > 0$, $b > 0$, and $a > b$. Let C be the point common to the line $y = x$ and the line through the points $A(a, b)$ and $B(b, a)$ (Figure 4.1.4). The slope of the line segment AB is $\frac{b-a}{a-b} = -1$. The slope of the line $y = x$ is 1. Therefore, AB is perpendicular to

OC. Let the coordinates of C be (c, c). Then, we determine the length of AC and BC as follows:

$$d(A, C) = \sqrt{(a-c)^2 + (b-c)^2},$$
$$d(B, C) = \sqrt{(b-c)^2 + (a-c)^2}.$$

Thus, the length of AC is equal to the length of BC. Therefore, the points $A(a, b)$ and $B(b, a)$ are symmetric with respect to the line $y = x$.

We can draw the following conclusion about the graph of the inverse of a relation from Theorem 2.

• **THEOREM 3** The graph of the inverse of a relation can be obtained by reflecting the graph of the relation through the line $y = x$.

• EXAMPLE 3: Draw a complete graph of the function $f(x) = x^2$ and its inverse relation f^{-1}. Write an equation for f^{-1}. Find the domain and range of f and f^{-1}. Is f^{-1} a function?

SOLUTION: A complete graph of $f(x) = x^2$ is given in Figure 4.1.5. The domain is $(-\infty, \infty)$, and the range is $[0, \infty)$. A complete graph of the inverse can be obtained in the following way. First draw a graph of $f(x) = x^2$ on a square piece of paper. Be sure to label the axes. Then, hold the line $y = x$ fixed and rotate the paper 180° about the line. The top of the paper will become the bottom of the paper and vice versa. Now, hold the paper up to a light and you will see the graph displayed in Figure 4.1.6. Notice that the axes have switched position.

We can check a few points to confirm that the graph in Figure 4.1.6 is the graph of the inverse of $f(x) = x^2$. Notice that $(-2, 4)$, $(0, 0)$, and $(2, 4)$ are three points on the graph of $f(x) = x^2$. The corresponding points $(4, -2)$, $(0, 0)$, and $(4, 2)$ are points on the graph of f^{-1}.

An equation can be obtained for f^{-1} by setting $y = f(x)$ and interchanging x and y in the equation $y = x^2$. Thus, $x = y^2$ is an equation for the inverse of $y = x^2$. The graph of $x = y^2$ is a parabola with the x-axis as its axis of symmetry. Parabolas will be studied in

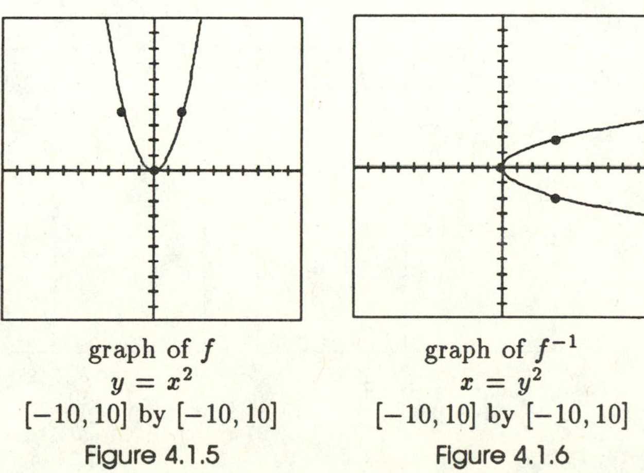

graph of f
$y = x^2$
$[-10, 10]$ by $[-10, 10]$
Figure 4.1.5

graph of f^{-1}
$x = y^2$
$[-10, 10]$ by $[-10, 10]$
Figure 4.1.6

more detail in Chapter 8. The domain of the inverse is $[0, \infty)$ and the range is $(-\infty, \infty)$. Notice that the domain and range of $y = x^2$ and its inverse $x = y^2$ are interchanged.

Finally, the vertical line test shows that f^{-1} is *not* a function. Thus, some functions have inverses that are *not* functions. •

The information found in Example 3 about the domain and range of a relation and its inverse is true in general.

• THEOREM 4 Let R be any relation. Then domain of R^{-1} = range of R and range of R^{-1} = domain of R.

Notice that the points $(4, -2)$ and $(4, 2)$ on the graph of the inverse of $y = x^2$ (Figure 4.1.6) correspond to the points $(-2, 4)$ and $(2, 4)$ of the graph of $y = x^2$ (Figure 4.1.5) under reflection with respect to the line $y = x$. The points $(-2, 4)$ and $(2, 4)$ lie on the horizontal line $y = 4$. The reason the inverse of $y = x^2$ is not a function is that there are horizontal lines that intersect the graph of $y = x^2$ in two points. This suggests the following theorem which we state without proof.

• THEOREM 5 f^{-1} is a function if and only if every horizontal line intersects the graph of f in at most one point.

• EXAMPLE 4: Let g be the function with domain $[0, \infty)$ defined by the rule $g(x) = x^2$. Show that g^{-1} is a function and find an algebraic rule for g^{-1}.

SOLUTION: A complete graph of g is given in Figure 4.1.7. Notice that the range of g is also $[0, \infty)$. It is important to observe that g, *by definition*, has no value when x is negative even though the rule for g makes sense for x negative. Every horizontal line intersects the graph of g in either one or zero points. Thus, g^{-1} is a function. The graph of g^{-1} is given in Figure 4.1.8.

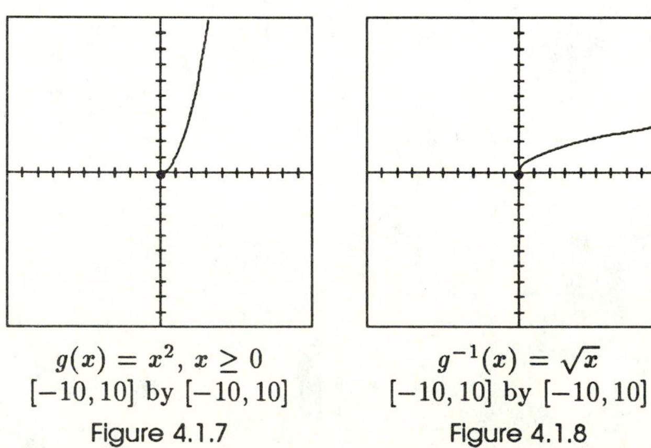

$g(x) = x^2,\ x \geq 0$
$[-10, 10]$ by $[-10, 10]$
Figure 4.1.7

$g^{-1}(x) = \sqrt{x}$
$[-10, 10]$ by $[-10, 10]$
Figure 4.1.8

The domain and range of g^{-1} are $[0, \infty)$. If we let $y = g(x)$, then we can interchange x and y in the equation $y = x^2$ and obtain the equation $x = y^2$ for g^{-1}. Because $x \geq 0$ and $y \geq 0$ we can write the equation for g^{-1} as $y = \sqrt{x}$. Thus, $g^{-1}(x) = \sqrt{x}$. •

There is another way to describe those functions with inverses that are also functions.

• **DEFINITION** A function f is said to be **one-to-one** if no two distinct ordered pairs of f have the same second entry. •

In Example 3 we found that $f(x) = x^2$ is not one-to-one because $(-2, 4)$ and $(2, 4)$ are two distinct ordered pairs of f with the same second entry. Saying that the function f is one-to-one is equivalent to saying that every horizontal line intersects the graph of f in at most one point.

Horizontal Line Test A function f is one-to-one if and only if every horizontal line intersects the graph of f in at most one point.

This fact together with Theorem 5 shows that the following is a theorem.

• **THEOREM 6** Let f be a function. f^{-1} is a function if and only if f is one-to-one.

In some textbooks the notation f^{-1} is used only for functions f whose inverses are also functions.

• **EXAMPLE 5:** Determine whether the following functions are one-to-one.

(a) $f(x) = -3x + 4$ (b) $g(x) = x^3 - 4x$

SOLUTION: A complete graph of $f(x) = -3x + 4$ is given in Figure 4.1.9. The graph is a straight line. Every horizontal line intersects this graph exactly once. Thus, f is a one-to-one function.

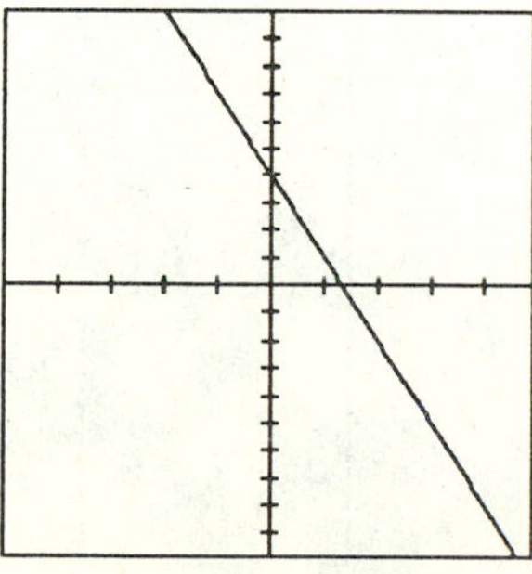

f is one-to-one
$f(x) = -3x + 4$
$[-5, 5]$ by $[-10, 10]$
Figure 4.1.9

A complete graph of $g(x) = x^3 - 4x$ is given in Figure 4.1.10. Each horizontal line between the local minimum of g and the local maximum of g intersects the graph of g exactly three times. Therefore, g is *not* one-to-one. •

The graph of any function of the form $f(x) = ax + b$, where $a \neq 0$, is a straight line that is neither horizontal nor vertical. Such functions are one-to-one, so their inverses are also functions. In the Exercises you will be asked to show that, as suggested by Example 2, the inverse of a linear function is also a linear function. The graph of a quadratic function $f(x) = ax^2 + bx + c$ is a parabola that opens upward or downward. Quadratic functions are never one-to-one. Therefore, the inverse relation of a quadratic function is *never* a function.

The function $y = x^3$ is increasing on $(-\infty, \infty)$, and the function $y = -x^3$ is decreasing on $(-\infty, \infty)$. Every horizontal line will intersect the graph of either function exactly once. Thus, $y = x^3$ and $y = -x^3$ are one-to-one and so have inverses. These two functions provide a special case of the following theorem we give without proof.

• THEOREM 7 Let f be a function with domain an interval I.

(a) If f is increasing on I, then it is a one-to-one function and its inverse is a function.
(b) If f is decreasing on I, then it is a one-to-one function and its inverse is a function.

Notice that the domain of the function g of Example 4 is the interval $[0, \infty)$, and g is increasing on this interval. By Theorem 7, g is a one-to-one function and g^{-1} is a function. Theorem 7 does not apply to the next example because the domain of that function is *not* an interval. However, the function is one-to-one so that its inverse is a function.

• EXAMPLE 6: Show that $f(x) = \frac{2x+7}{x+3}$ is one-to-one and find a rule for its inverse.

SOLUTION: We can use long division to write f in the form $f(x) = 2 + \frac{1}{x+3}$. A complete graph of the rational function f has a vertical asymptote at $x = -3$ and has $y = 2$ as

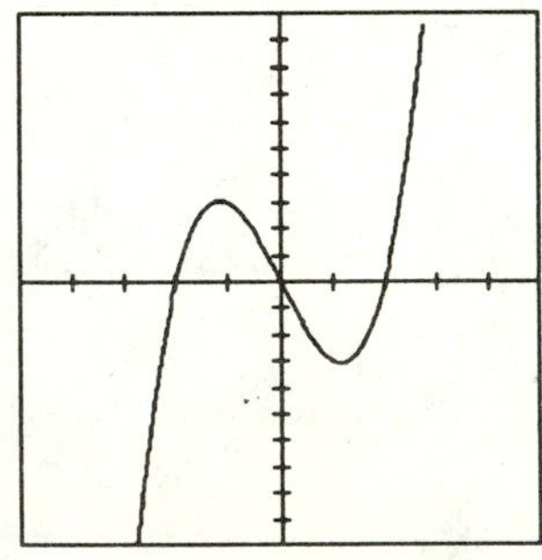

g is not one-to-one
$g(x) = x^3 - 4x$
$[-5, 5]$ by $[-10, 10]$
Figure 4.1.10

a horizontal asymptote (Figure 4.1.11). Notice that f is one-to-one and its inverse must be a function.

A complete graph of the inverse can be obtained by reflecting the graph of f through the line $y = x$ (Figure 4.1.12). Thus, the inverse will have $x = 2$ as a vertical asymptote and $y = -3$ as a horizontal asymptote. This follows because reflecting through the line $y = x$ sends horizontal lines to vertical lines and vice versa.

We can determine an equation for f^{-1} by setting $y = f(x)$, interchanging x and y, and then solving for y:

$$y = \frac{2x+7}{x+3},$$
$$x = \frac{2y+7}{y+3}.$$

If we solve this equation for y, we find that $y = \frac{-3x+7}{x-2}$. Thus, $f^{-1}(x) = \frac{-3x+7}{x-2}$. You can check that the computer generated graph of f^{-1} in the $[-10, 10]$ by $[-10, 10]$ viewing rectangle coincides with the graph in Figure 4.1.12. •

We must be very careful when we find rules for inverses that are functions. Suppose $y = f(x)$ and f^{-1} is a function. When we solve the equation $x = f(y)$ for y we may introduce extraneous solutions. This can lead to an incorrect rule for f^{-1}. Example 7 illustrates one such possibility.

• **EXAMPLE 7:** Let $f(x) = \sqrt{x+3}$. Show that f^{-1} is a function, find a rule for f^{-1}, and determine the domain and range of f^{-1}.

SOLUTION: The graph of f in Figure 4.1.13 shows that the domain of f is the interval $[-3, \infty)$, and f is increasing on $[-3, \infty)$ so by Theorem 7 f is one-to-one and f^{-1} is a function.

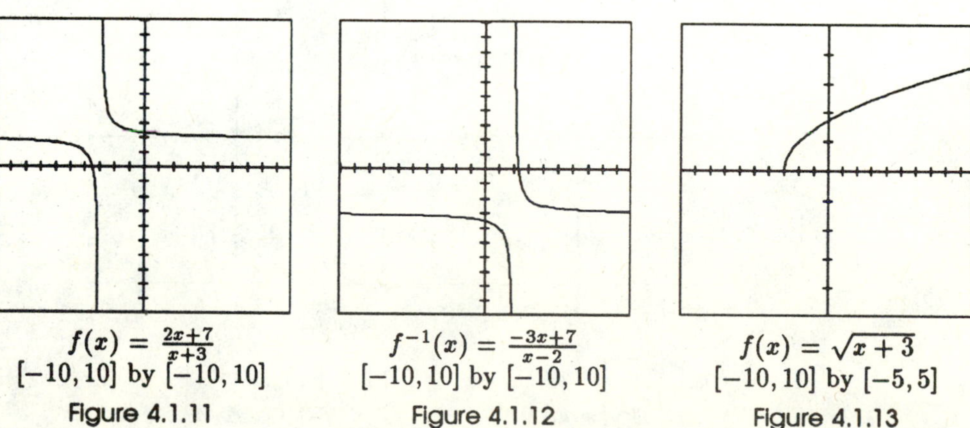

$f(x) = \frac{2x+7}{x+3}$
$[-10, 10]$ by $[-10, 10]$
Figure 4.1.11

$f^{-1}(x) = \frac{-3x+7}{x-2}$
$[-10, 10]$ by $[-10, 10]$
Figure 4.1.12

$f(x) = \sqrt{x+3}$
$[-10, 10]$ by $[-5, 5]$
Figure 4.1.13

To find a rule for f^{-1} we set $y = f(x)$, interchange x and y, and then solve for y:

$$y = \sqrt{x+3},$$
$$x = \sqrt{y+3},$$
$$x^2 = y+3,$$
$$y = x^2 - 3.$$

Thus, $f^{-1}(x) = x^2 - 3$ is a *rule* for f. However, because we squared each side of the equation $x = \sqrt{y+3}$, we may have introduced extraneous solutions and must be careful *not* to conclude that $f^{-1}(x) = x^2 - 3$ with no restriction on x without further investigation. Figure 4.1.13 shows that the domain of f^{-1} is $[0, \infty)$ because the domain of f^{-1} is the same as the range of f. Thus, $f^{-1}(x) = x^2 - 3$ *with the restriction* $x \geq 0$ correctly describes the inverse function. Simply stating that $f^{-1}(x) = x^2 - 3$ without restricting the domain is not correct. A complete graph of $f^{-1}(x) = x^2 - 3$, $x \geq 0$ coincides with the graph in Figure 4.1.14. •

The next example illustrates that a function can be one-to-one and have an inverse without satisfying the conditions of Theorem 7.

• EXAMPLE 8: Determine the domain, the range, where the function is increasing and decreasing, and show that it is one-to-one:

$$f(x) = \begin{cases} \frac{x}{x-1} & \text{for } x < 1 \\ x^2 + 1 & \text{for } x \geq 1 \end{cases}.$$

SOLUTION: A complete graph of f is given in Figure 4.1.15. The domain of f is $(-\infty, \infty)$ and the range of f is $(-\infty, 1) \cup [2, \infty)$. We can see that f is decreasing in $(-\infty, 1)$ and increasing in $[1, \infty)$. Notice that each horizontal line intersects the graph in Figure 4.1.15 in either zero or one point. The horizontal line test shows that f is one-to-one. •

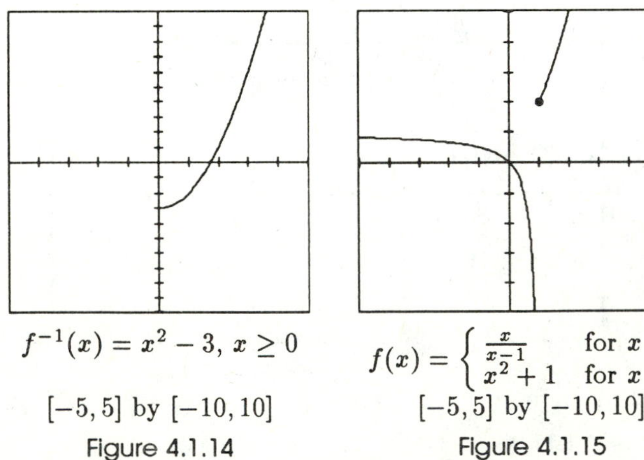

$f^{-1}(x) = x^2 - 3$, $x \geq 0$

$[-5, 5]$ by $[-10, 10]$

Figure 4.1.14

$f(x) = \begin{cases} \frac{x}{x-1} & \text{for } x < 1 \\ x^2 + 1 & \text{for } x \geq 1 \end{cases}$

$[-5, 5]$ by $[-10, 10]$

Figure 4.1.15

• **Rule for f^{-1}** Let f be a function with domain S, range T, and for which f^{-1} is also a function. If we have a rule for f, then a rule for f^{-1} can be determined in the following way:

1. Set $y = f(x)$.
2. Interchange x and y in the equation in 1.
3. Solve for y in 2.

The rule $y = f^{-1}(x)$ with y given by 3 is a rule for f^{-1}. The domain of f^{-1} is T and its range is S.

• EXERCISES 4-1

Let $(-2, 3)$ and $(4, 2)$ be two points on the graph of the inverse of the relation given by a line l.

1. Determine an equation for the inverse of the line l.
2. Determine an equation for the line l.

Consider the relation $y = 2x + 4$.

3. Is the point $(0, 4)$ on the graph of the inverse of the relation?
4. Is the point $(4, 0)$ on the graph of the inverse of the relation?
5. Find a point (a, b) that is on both the graph of the relation and the graph of its inverse.

Draw a complete graph of the inverse of each relation.

6.

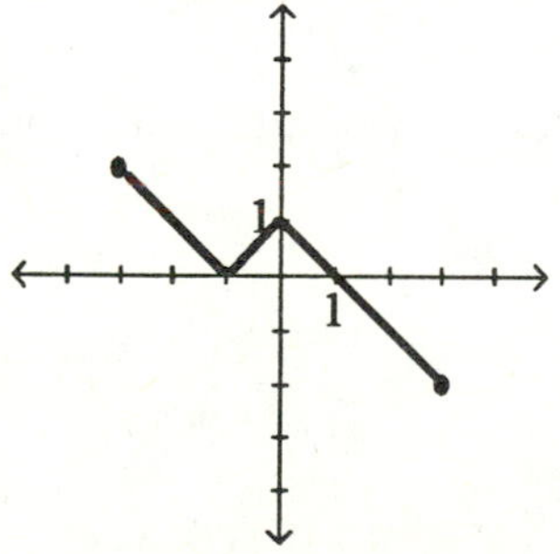

7.

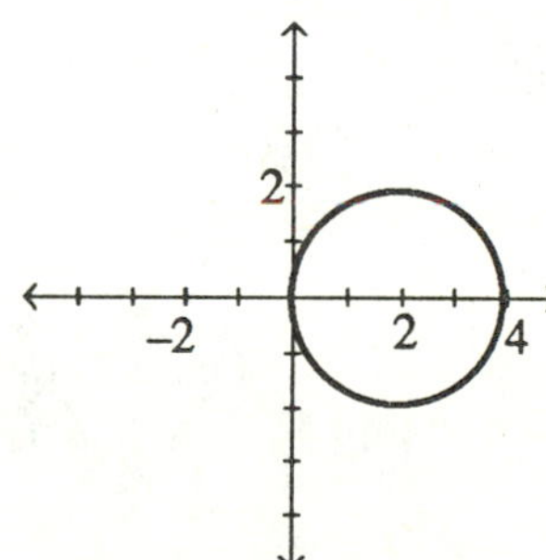

8. Let the domain of f be $(-\infty, 0]$ and let $f(x) = x^2$ when $x \leq 0$. Find a rule for the inverse of f. Is f^{-1} a function?
9. Let $f(x) = \sqrt{x-2}$. Find a rule for the inverse of f. Is f^{-1} a function? Determine the domain and range of f. Determine the domain and range of f^{-1}.
10. Explain geometrically the relationship between the domain and range of a function and the domain and range of its inverse.

Draw a complete graph of each relation and its inverse. Do not use a graphing utility. Determine an equation for the inverse. Specify which inverses are functions.

11. $y = 3 - 2x$
12. $y = \dfrac{4 + 7x}{3}$
13. $y = x^2 + 1$
14. $y^2 = x$
15. $x^2 + y^2 = 4$
16. $y = \sqrt{x - 3}$

Determine whether each function is one-to-one. Determine an equation for the inverse of each function. Is the inverse a function? For those functions with *inverses* that are functions, find an explicit rule for the inverse. Draw a graph of the function and its inverse relation without using a graphing utility. Check your answer with a graphing utility.

17. $f(x) = 3x - 6$
18. $g(x) = (x-2)^2$
19. $h(x) = \sqrt{x+2}$
20. $g(x) = x^3$
21. $f(x) = \dfrac{x+3}{x-2}$
22. $t(x) = \dfrac{2x-3}{x+1}$
23. Determine the domain and range of both the function and its inverse in Exercise 18.
24. Determine the domain and range of both the function and its inverse in Exercise 19.

Use a graphing utility to determine whether each function is one-to-one. Draw a complete graph of the function. Draw its inverse without the aid of a graphing utility. Identify the functions with inverses that are also functions.

25. $f(x) = x^3 - 8x$
26. $f(x) = x^4 - 2x + 3$
27. $f(x) = \dfrac{x^2 - 2x + 3}{x+2}$
28. $f(x) = x^3 - 2x - 6$
29. $f(x) = x^3 + 2x + 2$
30. $f(x) = \sin x$
31. Determine the domain and range of both the function and its inverse in Exercise 25.
32. Determine the domain and range of both the function and its inverse in Exercise 26.

*33. Show that if $f(x) = f(y)$ implies that $x = y$ for all x and y in the domain of f, then f is one-to-one.

*34. Show that if $x \neq y$, but $f(x) = f(y)$ for some x and y in the domain of f, then f is *not* one-to-one.

35. Prove that $f(x) = x^3 - 8x$ is *not* a one-to-one function.
36. Show, by example, that a relation that is not a function can have an inverse that is a function.
37. Find at least three functions such that $f^{-1} = f$.
38. If $f^{-1}(x) = f(x)$ for each x in the domain of f, then determine $f \circ f(x)$.
39. Let f be the function of Example 8. Determine the domain and range of f and f^{-1}, a rule for f^{-1}, and draw a complete graph of f^{-1}.

4.2 Exponential Functions

In this section we begin the study of exponential functions. We establish important properties of these functions as well as complete graphs. Problem situations about growth or decay of bacteria, radioactive substances, and population have exponential functions as models. Applications of this type are studied in this section.

In calculus you will be given a careful definition of an exponential function. In this textbook we assume that a^x is a real number for $a > 0$ and use a calculator or computer to compute the values of an exponential function. We further assume that in exponential functions $y = a^x$, the base is positive. The usual properties of exponents are given in the following theorem without proof.

• THEOREM 1 Let a, x, y be real numbers with $a > 0$.

(a) $a^0 = 1$ (b) $a^x a^y = a^{x+y}$ (c) $\dfrac{a^x}{a^y} = a^{x-y}$

(d) $a^{-x} = \dfrac{1}{a^x}$ (e)$(a^x)^y = a^{xy}$

• DEFINITION Let a be a positive real number. The function $f(x) = a^x$ with domain the set of all real numbers is called the **exponential function with base a.** •

• EXAMPLE 1: Draw and compare complete graphs of $f(x) = 2^x$ and $g(x) = 0.5^x$. Find the domain and range of each function.

SOLUTION: We can use a calculator or computer to determine the following end behavior of f:

$$f(x) \to \infty \quad \text{as} \quad x \to \infty,$$
$$f(x) \to 0 \quad \text{as} \quad x \to -\infty.$$

The y-intercept of the graph of f is 1 because $2^0 = 1$. A complete graph of f is given in Figure 4.2.1. The function f is increasing in $(-\infty, \infty)$.

We find the opposite end behavior for g:

$$g(x) \to 0 \quad \text{as} \quad x \to \infty,$$
$$g(x) \to \infty \quad \text{as} \quad x \to -\infty.$$

Notice that $g(x) = 0.5^x = \left(\frac{1}{2}\right)^x = 2^{-x}$ so that $g(x) = f(-x)$. Thus, a complete graph of g can be obtained by reflecting a complete graph of f through the y-axis. A complete graph of g is given in Figure 4.2.2. The function g is decreasing in $(-\infty, \infty)$.

The domain of each function is $(-\infty, \infty)$, the range of each function is $(0, \infty)$, and the x-axis is a horizontal asymptote for each function. •

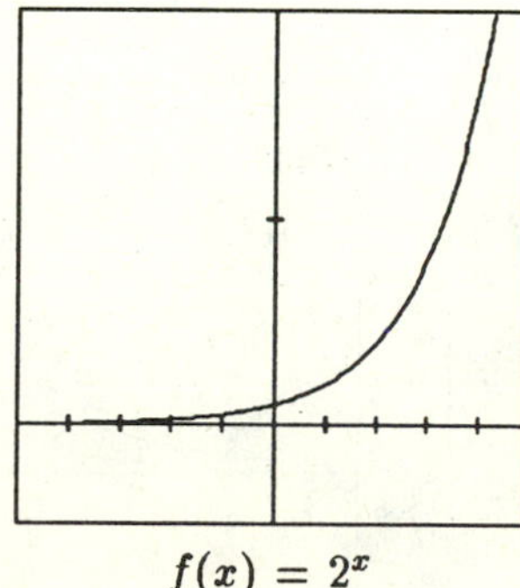

$f(x) = 2^x$
$[-5, 5]$ by $[-5, 20]$
Figure 4.2.1

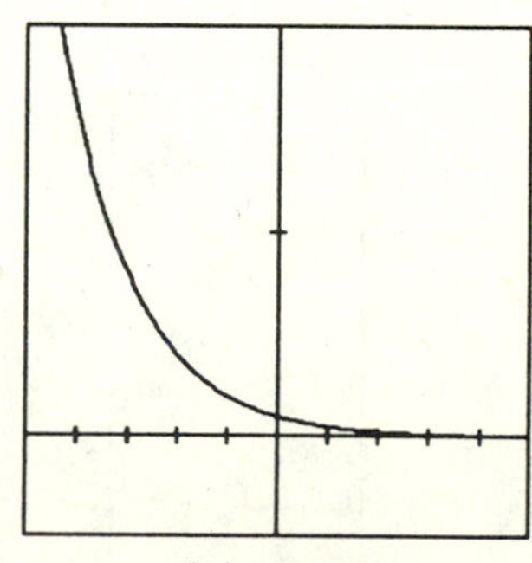

$g(x) = 0.5^x$
$[-5, 5]$ by $[-5, 20]$
Figure 4.2.2

If $a > 1$, the graph of $y = a^x$ looks essentially like the graph of $y = 2^x$. If $0 < a < 1$, the graph of $y = a^x$ looks essentially like the graph of $y = 0.5^x$.

• EXAMPLE 2: Draw and compare complete graphs of $y = 2^x$ and $y = 3^x$.

SOLUTION: A complete graph of each function is given in Figure 4.2.3. The end behavior of each function is the same, and each has y-intercept 1. The graphs in Figure 4.2.3 suggest that $2^x < 3^x$ in $(0, \infty)$ because the graph of $y = 2^x$ appears to be *below* the graph of $y = 3^x$. Indeed, this is true, but you may need to view the two graphs in several additional viewing rectangles to convince yourself of this fact. For $x < 0$, the graph of $y = 2^x$ is *above* the graph of $y = 3^x$. The graphs in Figure 4.2.4 illustrate this. Thus, $2^x > 3^x$ in $(-\infty, 0)$. •

If $a > b > 1$, the graphs of $y = a^x$ and $y = b^x$ are related as suggested in Example 2. In the Exercises you will compare these graphs for $0 < a < b < 1$.

A rough sketch of the graph of $y = a(b^{cx+d}) + k$ can be obtained by applying standard geometric transformations to the graph of $y = b^x$. First we show that c can be assumed to be equal to 1. Notice that we can use properties of exponents to rewrite the exponential expression 2^{3x+4} with base 2 as an exponential expression with base $2^3 = 8$.

$$2^{3x+4} = 2^{3(x+4/3)} = (2^3)^{x+4/3} = 8^{x+4/3}$$

Similarly, $y = a(b^{cx+d}) + k = a(b^c)^{x+d/c} + k$.

• EXAMPLE 3: Determine the domain, the range, and draw a complete graph of

(a) $y = 2(3^{x+1}) - 5$.　　(b) $y = 2^{3-2x}$.

SOLUTION:

(a) A complete graph of $y = 2(3^{x+1}) - 5$ can be obtained by applying, in order, the following transformation to the graph of $y = 3^x$.

1. Vertical stretch by a factor of 2 to obtain $y = 2(3^x)$.
2. Horizontal shift left 1 unit to obtain $y = 2(3^{x+1})$.
3. Vertical shift down 5 units to obtain $y = 2(3^{x+1}) - 5$.

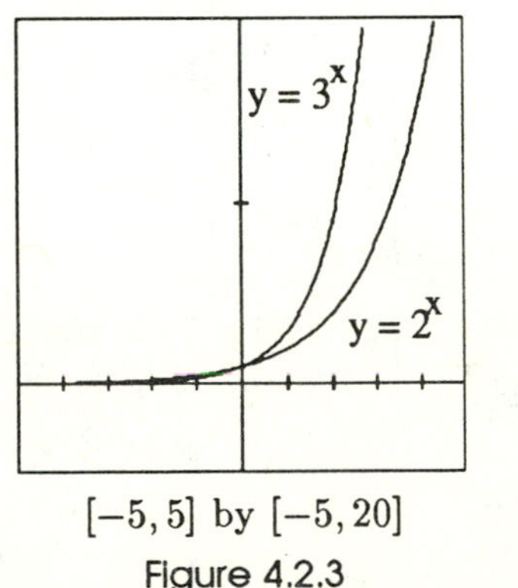

$[-5, 5]$ by $[-5, 20]$

Figure 4.2.3

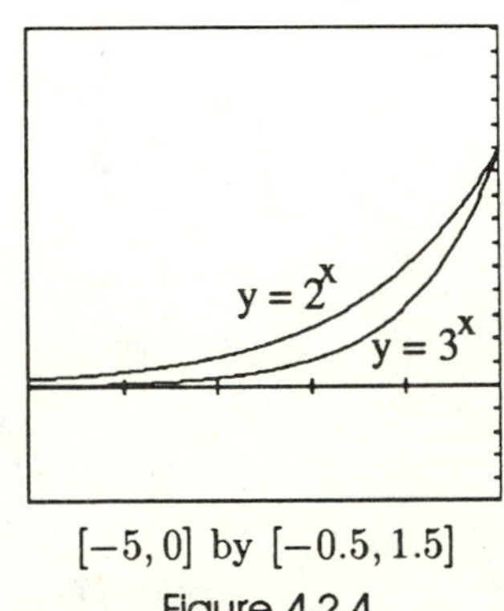

$[-5, 0]$ by $[-0.5, 1.5]$

Figure 4.2.4

The computer drawn graph in Figure 4.2.5 confirms this observation. The domain is $(-\infty, \infty)$, and the range is $(-5, \infty)$.

Alternatively, we can obtain a rough sketch in the following way. The function $y = 2(3^{x+1}) - 5$ must increase in $(-\infty, \infty)$ like $y = 3^x$ because the coefficient 2 of 3^{x+1} is positive. Notice that the function $y = -2(3^{x+1}) - 5$ is a decreasing function in $(-\infty, \infty)$. The -5 in $y = 2(3^{x+1}) - 5$ tells us that the line $y = -5$ is the horizontal asymptote of the function. We can see that the point $(0, 1)$ of $y = 3^x$ is moved to the point $(-1, -3)$ under the transformations that produce $y = 2(3^{x+1}) - 5$ from $y = 3^x$. This point, the horizontal asymptote, the y-intercept, and the increasing behavior allow us to produce the rough sketch in Figure 4.2.6.

(b) First we rewrite $y = 2^{3-2x}$ in the form $y = a^{x-h}$ for appropriate a and h:

$$2^{3-2x} = 2^{-2(x-\frac{3}{2})} = (2^{-2})^{x-1.5} = 0.25^{x-1.5}.$$

The graph of $y = 2^{3-2x} = 0.25^{x-1.5}$ must look like the graph of $y = 0.25^x$ except that it is shifted to the right 1.5 units. Thus, the function $y = 2^{3-2x}$ is a decreasing function in $(-\infty, \infty)$, and the x-axis is its horizontal asymptote. The point $(1.5, 1)$ on the graph of $y = 2^{3-2x}$ corresponds to the point $(0, 1)$ on the graph of $y = 0.25^x$, and $(0, 8)$ is the y-intercept of $y = 2^{3-2x}$. Figure 4.2.7 gives a rough sketch of $y = 2^{3-2x}$ using this information. •

Exponential functions of the form $y = a(b^{cx+d}) + k$ are either always increasing or always decreasing. Thus, such functions are one-to-one and their inverses are functions.

• EXAMPLE 4: Let $f(x) = 3^x$. Draw a complete graph of f^{-1}. Show that f^{-1} is a function.

SOLUTION: In Section 4.1 we showed that a complete graph of f^{-1} can be obtained by reflecting a complete graph of f through the line $y = x$. Complete graphs of f and f^{-1} are given in Figure 4.2.8.

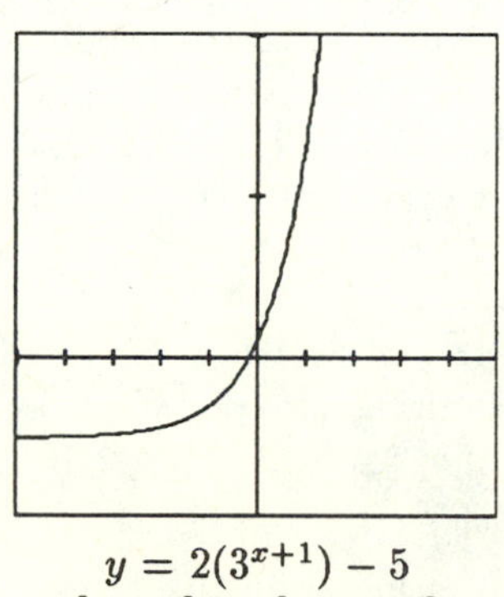

$y = 2(3^{x+1}) - 5$
$[-5, 5]$ by $[-10, 20]$
Figure 4.2.5

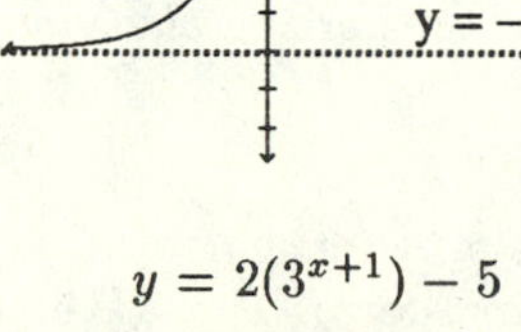

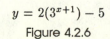

$y = 2(3^{x+1}) - 5$
Figure 4.2.6

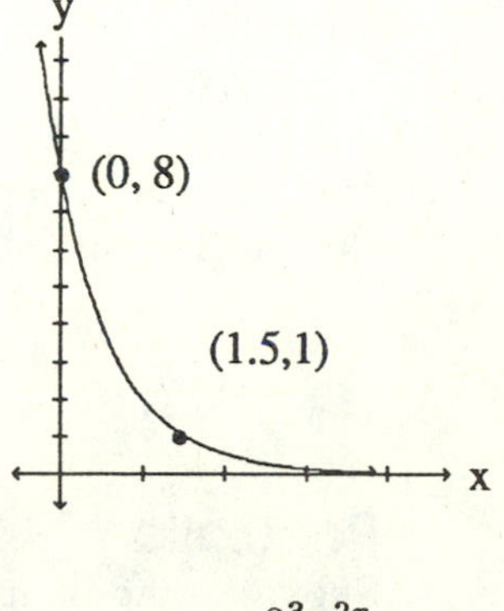

$y = 2^{3-2x}$
Figure 4.2.7

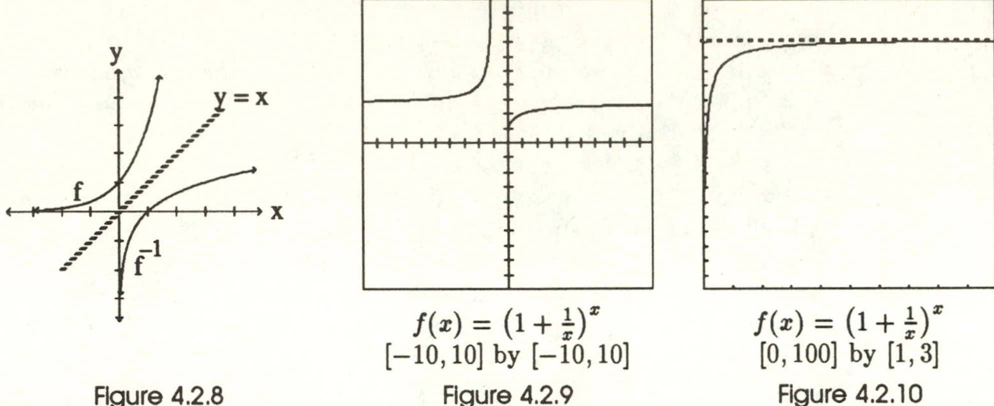

Figure 4.2.8

$f(x) = \left(1 + \frac{1}{x}\right)^x$
$[-10, 10]$ by $[-10, 10]$

Figure 4.2.9

$f(x) = \left(1 + \frac{1}{x}\right)^x$
$[0, 100]$ by $[1, 3]$

Figure 4.2.10

We can see from the graph in Figure 4.2.8 that f is one-to-one. Thus, f^{-1} is a function. Notice also that the graph of f^{-1} satisfies the vertical line test for a function. •

The horizontal asymptote of the function in the next example involves a number that is important in the study of mathematics and applications.

• EXAMPLE 5: Find the domain, the range, an end behavior model, and draw a complete graph of $f(x) = \left(1 + \frac{1}{x}\right)^x$.

SOLUTION: The graph of f in the $[-10, 10]$ by $[-10, 10]$ viewing rectangle is actually a complete graph (Figure 4.2.9). We must have $1 + \frac{1}{x} > 0$ because, by definition, the base of an exponential function must be positive. This means that $x > 0$ or $x < -1$. (Why?) Thus, the domain of f is $(-\infty, -1) \cup (0, \infty)$ as suggested by Figure 4.2.9. From this graph we can see that $f(x) \to \infty$ as $x \to -1^-$ and $f(x) \to 1$ as $x \to 0^+$. We can confirm this information using a calculator. The graph of f has a horizontal asymptote. The horizontal asymptote is $y = e$, where e is an irrational number that is equal to 2.718281828 accurate to 9 decimal places. The importance of the number e was first observed by Euler in the 1700's. Thus, $y = e$ is an end behavior model for f. You can approximate the y-coordinate of the horizontal asymptote graphically by a careful selection of viewing rectangles. The graph of f in $[0, 100]$ by $[1, 3]$ is a good starting point (Figure 4.2.10). From Figure 4.2.10, it appears that e is a bit larger than 2.7. It is possible to determine e graphically with accuracy up to the limits of machine precision. The range of f is $(1, e) \cup (e, \infty)$. •

In the Exercises we ask you to show that $y = e^r$ is an end behavior model of $f(x) = \left(1 + \frac{r}{x}\right)^x$. This means $y = e^r$ is the horizontal asymptote and thus $\left(1 + \frac{r}{x}\right)^x \to e^r$ as $x \to \infty$.

Population

Suppose that the population of a certain town is P and is increasing at constant rate r each year, where r is in decimal form. Then, the population of the town one year later is $P + Pr = P(1 + r)$. Two years later the population is $P(1 + r) + P(1 + r)r = P(1 + r)^2$. In general, the population t years later is $P(1+r)^t$. If the population of a town is decreasing at constant rate r each year, then the population one year later is $P - Pr = P(1-r)$. Two years later the population is $P(1-r) - P(1-r)r = P(1-r)^2$. In general, the population t years later is $P(1-r)^t$. These are examples of **exponential growth or decay.**

• EXAMPLE 6: The population of a town is 50,000 and is increasing at the rate of 2.5% each year.

(a) Write the population P as a function of time.
(b) Draw a complete graph of the function P in (a). What portion of the graph represents the problem situation?
(c) Determine when the population of the town will be 100,000.

SOLUTION:

(a) From the discussion preceding the example, we know that $P(t) = (50{,}000)(1+0.025)^t$, or $P(t) = (50{,}000)1.025^t$ where t represents the numbers of years after the population is 50,000.
(b) A complete graph of P is given in Figure 4.2.11. According to the statement of the example, only the portion of the graph in Figure 4.2.11 that is to the right of the y-axis represents the problem situation. However, if we also know that the population of the town had been increasing at the rate of 2.5% in prior years, then a portion of the graph to the left of the y-axis could also represent the problem situation. (Why?)
(c) We need to determine t so that $P(t) = (50{,}000)1.025^t = 100{,}000$. We can use the graph in Figure 4.2.11 to see that t is a little less than 30. Zoom-in can be used to determine that the value of t that makes $P(t) = 100{,}000$ is 28.07 with error at most 0.01. Thus, the population of the town will be 100,000 about 28.07 years later. •

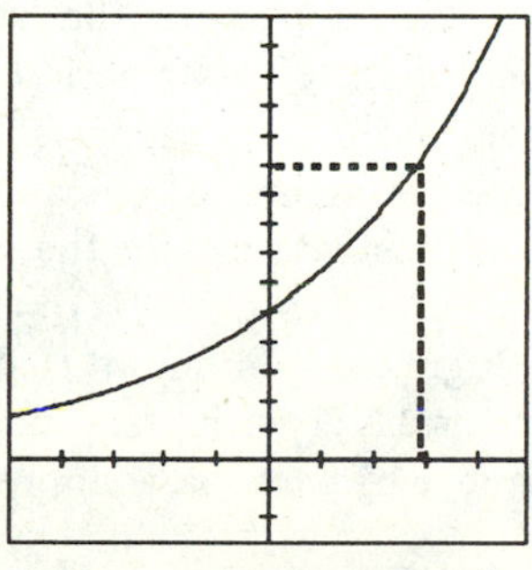

$P(t) = (50{,}000)1.025^t$
$[-50, 50]$ by
$[-30{,}000, 150{,}000]$

Figure 4.2.11

We can generalize the population example to general exponential growth and exponential decay problems. Suppose that there are P units of a substance present initially. If the substance P is growing at constant rate r, r in decimal form, then the amount of substance present t years later is $P(1+r)^t$. If the substance is decaying at constant rate r, then the amount of substance present t years later is $P(1-r)^t$. In general, $P \cdot a^t$ is a model for exponential growth if $a > 1$ and is a model for exponential decay if $0 < a < 1$. Finally, the unit of time need not be years. For example, t could be measured in months, days, or hours.

In the next example we see how to determine an exponential growth model when the doubling time for a substance is known.

Biological Growth

Assume that the number of bacteria in a certain bacterial culture doubles every hour and that there are 100 present initially. The number of bacteria after one hour is $200 = 2(100)$, after two hours is $400 = 2^2(100)$, and after three hours is $800 = 2^3(100)$. After t hours the number of bacteria present in the culture is given by $f(t) = (100)2^t$.

• EXAMPLE 7: Assume that the number of bacteria in a certain bacterial culture doubles every hour and that there are 100 present initially.

(a) How many bacteria are present after 7 hours and 15 minutes?
(b) Draw a graph that shows the number of bacteria present during the first 12 hours.
(c) Determine when the number of bacteria will be 350,000.

SOLUTION:

(a) From the discussion preceding this example, $f(t) = (100)2^t$ represents the number of bacteria present after t hours. Notice that $t = 7.25$ is equivalent to 7 hours and 15 minutes. Thus, the number of bacteria present after 7 hours and 15 minutes is $f(7.25) = (100)2^{7.25}$ which is about 15,222.
(b) The graph of f in the $[0, 13]$ by $[0, 450{,}000]$ viewing rectangle shows the number of bacteria present during the first 12-hour period (Figure 4.2.12). Notice how the values of f appear to be 0 during the first 5 hours. The number of bacteria is so large near the end of the 12-hour period that the number present during the first part of the 12-hour period, by comparison, seems to be 0. We need a second graph to better illustrate the values of f during the first six hours (Figure 4.2.13). Even this figure fails to illustrate clearly the values during the first hour.
(c) We need to determine t so that $f(t) = (100)2^t = 350{,}000$. The graph of f and $y = 350{,}000$ in Figure 4.2.14 shows that t is approximately 11.7. We can zoom in and find that t is 11.77 with error at most 0.01. Thus, after 11 hours and 46.2 minutes, the number of bacteria is approximately 350,000. •

Radioactive Decay

One example of exponential decay is that associated with radioactive substances. The rate of such decay is based on the half-life of the substance. The *half-life* of a radioactive substance

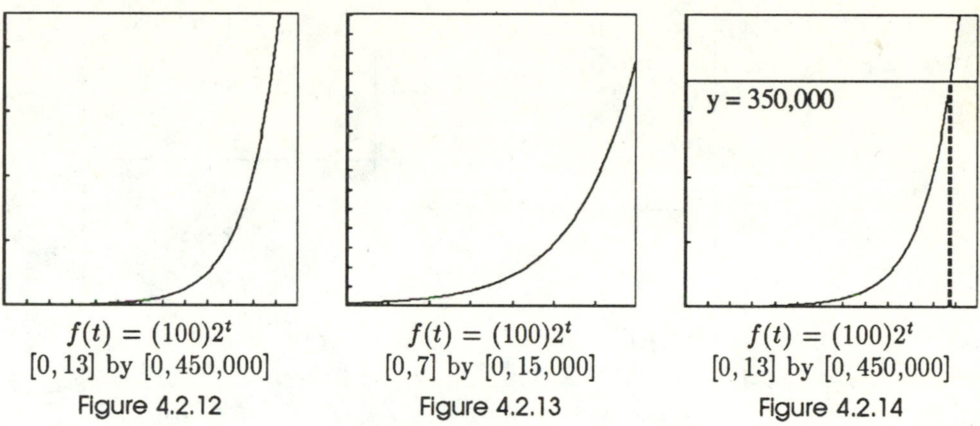

$f(t) = (100)2^t$
[0, 13] by [0, 450,000]
Figure 4.2.12

$f(t) = (100)2^t$
[0, 7] by [0, 15,000]
Figure 4.2.13

$f(t) = (100)2^t$
[0, 13] by [0, 450,000]
Figure 4.2.14

is the amount of time it takes for half of the substance to decay. Suppose that the half-life of a certain radioactive substance is 20 days, and there are 5 grams present initially. Then, after 20 days there are $5(\frac{1}{2})$ grams remaining. After 40 days there are $5(\frac{1}{2})^2$ grams remaining. In general, after t days there will be $5(\frac{1}{2})^{t/20}$ grams remaining. Notice the exponent on $\frac{1}{2}$ is 1 when t is 20, and 2 when t is 40.

• EXAMPLE 8: The half-life of a certain radioactive substance is 20 days, and there are 5 grams present initially.

(a) Write the amount A of substance remaining as a function of time.
(b) Draw a complete graph of the function in (a).
(c) When will there be less than 1 gram of the substance remaining?

SOLUTION:

(a) From the discussion preceding the example, we can write $A(t) = 5(0.5)^{t/20}$, where t is the number of days after the amount of radioactive substance is 5 grams.
(b) A complete graph of A is given in Figure 4.2.15.
(c) A is a decreasing function. If we find when $A(t) = 1$, then the solution we seek consists of all t greater than this number. We could graph A and $y = 1$ in the same viewing rectangle and use zoom-in to determine that the t coordinate of the point of intersection is 46.44. Thus, there will be less than 1 gram of the radioactive substance left after 46.44 days. •

In Section 4.4 we will learn how to solve problems like Example 8 algebraically using logarithms.

The exponential function $f(x) = a \cdot b^t$ is an extremely important model for compound growth or decay problems. The function is increasing and represents growth if $a > 0$ and $b > 1$. Similarly, the function is decreasing and represents decay if $a > 0$ and $0 < b < 1$. In Section 4.3, we will use this model in economic applications.

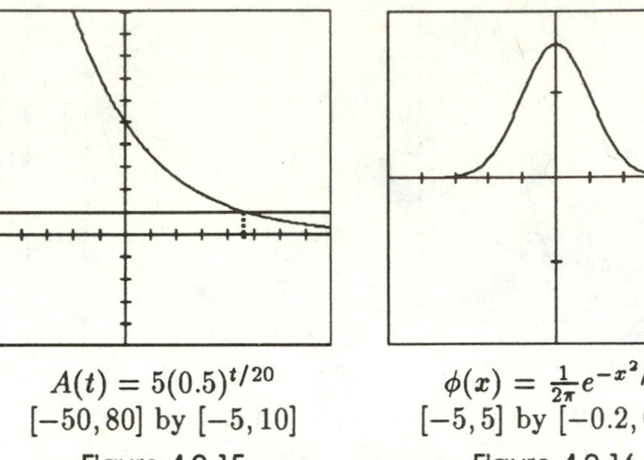

$A(t) = 5(0.5)^{t/20}$
$[-50, 80]$ by $[-5, 10]$
Figure 4.2.15

$\phi(x) = \frac{1}{2\pi}e^{-x^2/2}$
$[-5, 5]$ by $[-0.2, 0.2]$
Figure 4.2.16

We close this section by investigating an important graph that occurs in the study of statistics, the so called *bell-shaped curve* or *normal distribution curve*.

• EXAMPLE 9: Let $\phi(x) = \frac{1}{2\pi}e^{-x^2/2}$.

(a) Draw a complete graph of ϕ.
(b) Find any maximum and minimum values of ϕ.

SOLUTION: A complete graph of ϕ is given in Figure 4.2.16. Notice that $\phi(x) \to 0$ as $x \to \infty$ and as $x \to -\infty$. Thus, the x-axis is a horizontal asymptote of ϕ. The figure suggests the maximum value of ϕ occurs at $x = 0$. Thus, $\phi(0) = \frac{1}{2\pi}$ is the maximum value of ϕ. (Why?) We can use zoom-in to show that the maximum value of ϕ is 0.159 with error at most 0.01. Check that $\frac{1}{2\pi}$, accurate to thousandths, is 0.159. •

• EXERCISES 4-2

1. Graph $y = (\frac{1}{4})^x$ and $y = 4^{-x}$ in the same viewing rectangle.
2. Draw the graphs of $y = (\frac{1}{4})^x$, $y = (\frac{1}{3})^x$, $y = (\frac{1}{2})^x$ in the same $[-2, 4]$ by $[-1, 2]$ viewing rectangle. Solve each inequality.

   (a) $(\frac{1}{4})^x > (\frac{1}{3})^x$
   (b) $(\frac{1}{3})^x > (\frac{1}{2})^x$
   (c) $(\frac{1}{4})^x < (\frac{1}{3})^x$
   (d) $(\frac{1}{3})^x < (\frac{1}{2})^x$

Without using a graphing utility, draw a complete graph of each function. Check your answer with a graphing utility.

3. $f(x) = 1 + 2^x$
4. $f(x) = 1 - 3^x$
5. $f(x) = 2^{x-3}$
6. $y = 3 \cdot 2^{x+2}$
7. $g(x) = -2 - 2^{x+3}$
8. $D(x) = -2 + (\frac{1}{2})^{x+1}$
9. Determine the domain and range of the function in Exercise 5.

10. Determine the domain and range of the function in Exercise 7.

Use rules of exponents to solve each equation. Check with a graphing utility.

11. $2^x = 4^2$

12. $x^4 = 16$

13. $8^{x/2} = 4^{x+1}$

14. $(-8)^{5/3} = 2(4^{x/2})$

15. $(1-5x)^{1/2} = 20$

16. $x^{1/2} - (x-1.25)^{1/2} = 0.5$

*17. $e^{4x} - 2e^{2x} - 3 = 0$

*18. $3(2^{2x}) + 7(2^x) - 6 = 0$

Draw a complete graph of the function. Determine the domain of each function.

19. $f(x) = e^{2x-1}$

20. $g(x) = 2^{x^2-1}$

21. $f(x) = x3^x$

22. $f(x) = xe^{-x}$

Determine where each function is increasing and decreasing. Determine all local maximum and minimum values, and the range of each function.

23. $f(x) = e^{2x-1}$

24. $g(x) = 2^{x^2-1}$

25. $f(x) = x3^x$

26. $f(x) = xe^{-x}$

27. $f(x) = x2^{-x^2}$

28. $g(x) = -x10^{x^2/50}$

29. Let $f(x) = 2^x$. Show that f is one-to-one and sketch a graph of f and f^{-1} on the same coordinate system.

30. Let $f(x) = (\frac{1}{2})^x$. Show that f is one-to-one and sketch a graph of f and f^{-1} on the same coordinate system.

31. Investigate graphically the end behavior of $y = (1+2/x)^x$ by drawing the graph of $y = (1+2/x)^x$ and the line $y = e^2$ in the $[0, 100]$ by $[0, 10]$ viewing rectangle. Repeat in the $[100, 1000]$ by $[7, 8]$ viewing rectangle.

32. Investigate graphically the end behavior of $y = (1+3/x)^x$ by drawing the graph of $y = (1+3/x)^x$ and the line $y = e^3$ in the $[0, 100]$ by $[0, 25]$ viewing rectangle. Repeat in the $[100, 1000]$ by $[19, 21]$ viewing rectangle.

33. Investigate graphically the end behavior of $y = (1 + 0.1/x)^x$ by drawing the graph of $y = (1 + 0.1/x)^x$ and the line $y = e^{0.1}$ in the $[0, 0.1]$ by $[0.75, 1.25]$ viewing rectangle. Repeat in the $[1, 10]$ by $[1.08, 1.12]$ viewing rectangle.

34. Investigate graphically the behavior of $f(x) = (1+x)^{1/x}$ as $x \to 0$. What is the domain of f? Notice $f(0)$ is *not* defined, so f is *not* continuous at $x = 0$. How would you define the value of f at $x = 0$ in order to make f continuous in the interval $(-1, \infty)$? *Hint:* Use zoom-in. Does this suggest an alternate definition for e?

Draw a complete graph of each inequality.

35. $y < e^x - 1$

36. $y \geq 2e^{x-3}$

37. Find the simultaneous solutions to the system

$$\begin{cases} y = x^2 \\ y = 2^x \end{cases},$$

that is, solve $x^2 = 2^x$. For what values of x is $2^x > x^2$? For what values of x is $x^2 > 2^x$?

38. Find the simultaneous solution of the system

$$\begin{cases} y = x^3 \\ y = 3^x \end{cases},$$

that is, solve $x^3 = 3^x$. For what values of x is $3^x > x^3$? For what values of x is $x^3 > 3^x$?

39. Assume that the number of bacteria in a certain bacterial culture doubles every three hours and that there are 2500 present initially.
    (a) How many bacteria are present after 6 hours?
    (b) Draw a graph that shows the number of bacteria present during the first 24-hour period.
    (c) Determine when the number of bacteria will be 100,000.

40. Assume that the number of rabbits in a certain population doubles every month and that there are 20 rabbits present initially.
    (a) How many rabbits are present after one year? After 5 years?
    (b) Draw a graph that shows the number of rabbits present during the first year.
    (c) Determine when the number of rabbits will be 10,000.
    (d) Explain why this exponential growth model is not a good model for rabbit population growth over a long period of time. What factors influence population growth?

41. The population of a town is 475,000 and is increasing at the rate of 3.75% each year.
    (a) Write the population P as a function of time.
    (b) Draw a complete graph of the function P in (a). What portion of the graph represents the problem situation?
    (c) Determine when the population of the town will be one million.

42. The population of a town is 123,000 and is decreasing at the rate of 2.375% each year.
    (a) Write the population P as a function of time.
    (b) Draw a complete graph of the function P in (a). What portion of the graph represents the problem situation?
    (c) Determine when the population of the town will be 50,000.

43. The population of a small town in the year 1890 was 6250. Assume the population increased at the rate of 3.75% each year.
    (a) Write the population P as a function of time.
    (b) Draw a complete graph of the function P in (a). What portion of the graph represents the problem situation?
    (c) Determine the population of the town in 1915 and in 1940.

44. The half-life of a certain radioactive substance is 14 days, and there are 6.58 grams present initially.
    (a) Write the amount A of substance remaining as a function of time.
    (b) Draw a complete graph of the function in (a).
    (c) When will there be less than 1 gram of the substance remaining?

45. The half-life of a certain radioactive substance is 65 days, and there are 3.5 grams present initially.
    (a) Write the amount A of substance remaining as a function of time.
    (b) Draw a complete graph of the function in (a).
    (c) When will there be less than 1 gram of the substance remaining?

46. The half-life of a certain radioactive substance is 1.5 seconds. Let S be the amount of the substance initially (in grams).
    (a) Determine an algebraic representation for S.
    (b) Draw a complete graph of the algebraic representation in (a) if there are 2 grams of the substance initially.
    (c) Draw a graph of the problem situation in (b).
    (d) What is the initial amount of the substance needed if there is to be 1 gram left after 1 minute?
    *(e) Discuss the stability of this substance.

4.3 Economic Applications

In this section we continue the study of exponential functions. Simple interest, compound interest, and annuities are introduced. Economic applications are studied in this section.

Simple Interest

Suppose that \$200 is invested at 7% simple interest for one year. The *interest earned* at the end of one year is \$200 (0.07), or \$14. We say that \$200 is invested at *simple annual interest rate* 0.07 or at simple annual interest rate of 7%.

• DEFINITION Suppose that P dollars are invested at simple interest rate r, where r is a decimal. Then, P is called the **principal**, r the **rate of interest**, and Pr the **interest** received at the end of one interest period. •

Suppose that P dollars are invested at simple interest rate r. Then the total value of the investment at the end of one period is $P + Pr$, or $P(1 + r)$, that is, the principal *plus* the interest. At the end of the second period the value of the investment is $P + 2Pr$, or $P(1 + 2r)$. In general, the total value of the investment at the end of n periods using the simple interest rate model is given by the formula

$$\textbf{Simple Interest: } S = P(1 + nr).$$

• EXAMPLE 1: Five hundred dollars is invested at 7% simple annual interest. Determine the value of the investment 10 years later.

SOLUTION: The value of the investment is given by $S = P(1 + nr)$ where $P = 500$, $n = 10$, and $r = 0.07$. Thus, $S = 850$, and the value of the investment after 10 years is \$850. •

Compound Interest

Usually, financial institutions allow interest to compound; that is, they pay interest *on the interest.* In this case, the total value of the investment at the end of the first period is *also* $P + Pr = P(1 + r)$, the same as simple interest. This amount is considered the new principal for the second investment period. The interest on this investment for the second period is $P(1 + r)r$, and the value of the investment at the end of the second period is $P(1+r)+P(1+r)r$, or $P(1+r)^2$. At the end of the third period the value of the investment is $P(1+r)^3$. At the end of the nth period the value of the investment is $P(1+r)^n$. Interest accumulated in this way is called **compound interest**, and the value of the investment at the end of n periods is given by the formula

$$\textbf{Compound Interest: } S = P(1 + r)^n.$$

• EXAMPLE 2: Five hundred dollars is invested at 7% interest compounded annually. Determine the value of the investment 10 years later.

SOLUTION: The value of the investment is given by $S = P(1 + r)^n$ where $P = 500$, $n = 10$, and $r = 0.07$. Thus, $S = 983.58$ and the value of the investment after 10 years is \$983.58. •

Suppose that money is invested at 9% per year *compounded quarterly.* Usually we say that money is invested at 9% compounded quarterly. There are four interest periods per year, the time period is 3 months and the interest rate per quarter is $\frac{9}{4}$%, or 0.0225. For example, if \$200 is invested at 9% compounded quarterly, then at the end of n quarters ($3n$ months) the value of the investment is $200(1 + 0.0225)^n$, or $200(1.0225)^n$. We say that 9% is the *annual percentage rate* and $\frac{9}{4}$% is the *interest rate per quarter.*

• DEFINITION Suppose that a sum of money is invested at r% per year compounded k times a year. Then r% is called the **annual percentage rate** (APR) and $\frac{r}{k}$% the **interest rate per period.** •

Compound Interest Formula Suppose that P dollars are invested at annual interest rate r (r in decimal form), and that interest is compounded k times per year. The value of the investment at the end of n periods is given by the formula

$$S = P\left(1 + \frac{r}{k}\right)^n.$$

• EXAMPLE 3: Suppose that \$500 is invested at 9% APR compounded monthly.

(a) Find the value of the investment after 5 years.
(b) Draw a graph that shows the value of the investment during the first 30 years.
(c) Determine when the value of the investment is \$3000.

SOLUTION:

(a) The compound interest formula for this investment is $S = 500\left(1 + \frac{0.09}{12}\right)^n = 500(1.0075)^n$. We need to find the value S of the investment when $n = 60$. Now, $S = 500(1.0075)^{60} = 782.84$. The value of the investment after 5 years is \$782.84.

(b) We need a graph of $S = 500(1.0075)^n$ that shows the values of S for $0 \leq n \leq 360$. The graph of S in the $[0, 360]$ by $[0, 8000]$ viewing rectangle is given in Figure 4.3.1.

(c) We need to find a value of n for which $S = 500(1.0075)^n$ is 3000. In the next section we solve this equation algebraically. Now, we find a solution graphically by determining the x-coordinate of the point of intersection of the graph of $A = 500(1.0075)^x$ with the graph of $y = 3000$ (Figure 4.3.2). The answer appears to be about 240 months or 20 years. We can use zoom-in to find that x is 239.796 with error at most 0.01. Thus, the value of the investment will be \$3000 in about 239.796 months or 19.983 years. Depending on the way interest is paid, a better answer may be 20 years. •

• EXAMPLE 4: How much must be invested at 8% APR compounded daily so that the value of the investment 4 years later is \$5000?

SOLUTION: In the compound interest formula $S = P\left(1 + \frac{r}{k}\right)^n$ we have $S = 5000$ and $r = 0.08$. (Why?) The length of a compounding period is one day so that k is 365 and n is 5(365) or 1825. We ignore the one or two leap years that could occur in the 5-year period. Substituting these values in the compound interest formula we obtain

$$5000 = P\left(1 + \frac{0.08}{365}\right)^{1825}.$$

We can determine from this equation that $P = 3351.75$. Thus, if we invest \$3,351.75 at 8% APR compound daily, the value after 5 years will be \$5,000. •

• EXAMPLE 5: How long will it take to accumulate \$10,000 if \$1,000 is invested at 7.5% APR compounded quarterly?

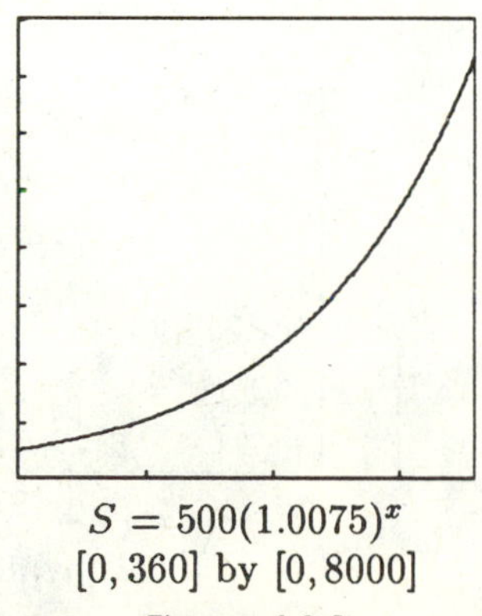

$S = 500(1.0075)^x$
$[0, 360]$ by $[0, 8000]$
Figure 4.3.1

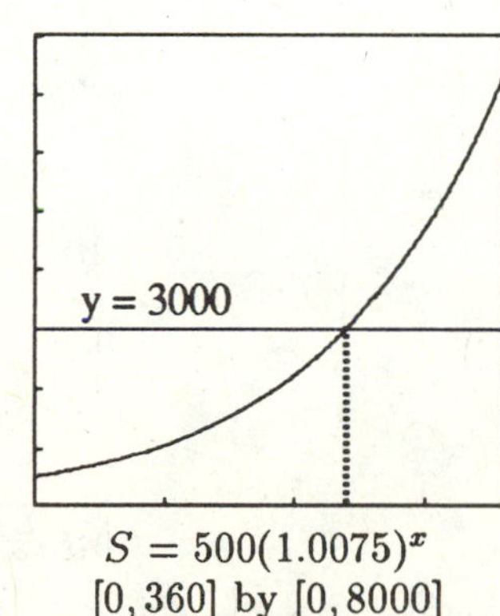

$S = 500(1.0075)^x$
$[0, 360]$ by $[0, 8000]$
Figure 4.3.2

SOLUTION: In the compound interest formula $S = P\left(1+\frac{r}{k}\right)^n$ we have $S = 10{,}000$, $P = 1{,}000$, $k = 4$, $r = 0.075$, and we need to determine n, the number of quarters required for the value of the investment to be \$10,000. With these substitutions we obtain

$$10{,}000 = 1000\left(1+\frac{0.075}{4}\right)^n.$$

We can rewrite the above equation in the following way:

$$10 = 1.01875^n.$$

We can determine n graphically by finding the x-coordinate of the point of intersection of the graphs of $y = 10$ and $y = 1.01875^x$ (Figure 4.3.3). We can estimate from this graph that $n(x)$ is about 125. If we use zoom-in, we will find that n is 123.95. Thus, after 123.95 quarters the value of the investment will be \$10,000. Depending on the way interest is paid, a better answer may be 124 quarters or 31 years. •

In the next section we will see how to solve equations like those in Examples 3 and 5 algebraically using logarithms.

• EXAMPLE 6: What APR rate compounded monthly is required for a \$2,000 investment to accumulate to \$4,000 in 7 years? Alternatively, we could solve $2 = (1+x)^{84}$ for x obtaining $x = 2^{1/84} - 1$, and use a calculator to determine x.

SOLUTION: The compound interest formula for this problem situation is

$$4000 = 2000\left(1+\frac{r}{12}\right)^{84}.$$

Let $x = \frac{r}{12}$. The above equation can be rewritten as

$$2 = (1+x)^{84}.$$

We can use the graphs of $y = 2$ and $y = (1+x)^{84}$ to determine x (Figure 4.3.4). Notice that we must choose a viewing rectangle with very small horizontal width because x represents

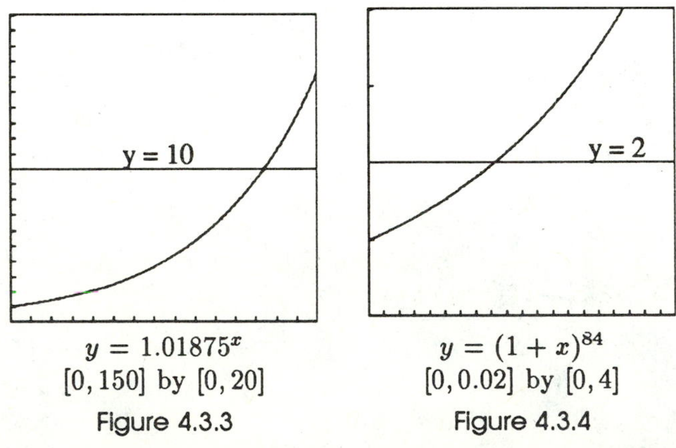

$y = 1.01875^x$
[0, 150] by [0, 20]
Figure 4.3.3

$y = (1+x)^{84}$
[0, 0.02] by [0, 4]
Figure 4.3.4

a monthly interest rate. We can use zoom-in to determine that x is 0.008285892. The APR is $12x$ or 0.0994296. Thus, investing \$2,000 at an 9.943% APR compounded monthly will produce \$4,000 in 7 years. Alternatively, we could solve $2 = (1+x)^{84}$ for x obtaining $x = 2^{1/84} - 1$, and use a calculator to determine x. •

• EXAMPLE 7: Suppose that \$1000 is invested at 8% APR compound interest. Determine the value of the investment after one year if the interest is compounded annually, quarterly, monthly, weekly, or daily.

SOLUTION: In the compound interest formula we set $P = 1000$ and $r = 0.08$. We use a calculator to complete Table 1. •

You might be surprised to see so little difference in the value of the investment in the table in Example 7. Even if we compound more often, the value of the investment will not increase very much. For example, compounding *hourly* gives

$$S = 1000\left(1 + \frac{0.08}{8760}\right)^{8760} = \$1083.29$$

where $8760 = (24)(365)$ is the number of hours in a 365-day year. It appears that S is approaching a value less than \$1084 as the number of compounding periods increases. From the Exercises of the previous section you can conclude that $\left(1 + \frac{0.08}{n}\right)^n \to e^{0.08}$ as $n \to \infty$. Thus, $S = 1000\left(1 + \frac{0.08}{n}\right)^n \to 1000e^{0.08} = 1083.29$ as $n \to \infty$. In general, if P dollars are invested at compound interest rate r for t years and the number of compounded periods approaches infinity, then the value of the investment approaches Pe^{rt}.

Compounding Period	Value of Investment After One Year
annually	$S = 1000(1 + 0.08) = \$1080.00$
quarterly	$S = 1000\left(1 + \frac{0.08}{4}\right)^4 = \$1082.43$
monthly	$S = 1000\left(1 + \frac{0.08}{12}\right)^{12} = \$1083.00$
weekly	$S = 1000\left(1 + \frac{0.08}{52}\right)^{52} = \$1083.22$
daily	$S = 1000\left(1 + \frac{0.08}{365}\right)^{365} = \$1083.28$

Table 1

• DEFINITION If P dollars are invested at APR r (in decimal form) **compounded continuously**, then the value of the investment after t years is given by $S = Pe^{rt}$, where e is approximately 2.718281828. •

• EXAMPLE 8: Suppose that \$1000 is invested at 8% compounded continuously. Find the value of the investment after 1 year and after 5 years.

SOLUTION: The value of the investment after 1 year at 8% compounded continuously is

$$S = 1000e^{0.08} = \$1083.29\,.$$

Most calculators have e built in so that this value can be computed easily with a calculator. We can also enter 2.718281828 for e and compute the value of S. Compare this value with the values found in Table 1 of this section.

The value of the investment after 5 years is

$$S = 1000e^{0.08(5)} = \$1491.82\,.$$ •

Annuities

• DEFINITION An **annuity** consists of a sequence of payments or deposits made at various times in the future. •

We will assume that the payments or deposits are equal and that the periods between payments or deposits are equal. We further assume that the payments or deposits are made at the *end* of each period. Such annuities are usually called **ordinary annuities** but we will simply refer to them as *annuities* in this textbook.

Suppose that n payments or deposits of R dollars each are made and that the interest rate per payment period is i. The following time line represents this problem situation:

Payment		R	R	R		R
Time	0	1	2	3	···	n

Notice that the first payment or deposit of R dollars occurs at the end of the first period. Usually the length of a period is a month. In Chapter 9 we will show that the value of this annuity after the nth payment or deposit is made is $S = R[(1+i)^n - 1]/i$.

• DEFINITION The **future value** S of an annuity consisting of n equal payments or deposits of R dollars each with interest rate i per period is given by

$$S = R\frac{(1+i)^n - 1}{i}.$$ •

Let r be the APR that gives interest rate i for the period length of the above annuity, and let k be the number of such periods in one year. Then, $i = \frac{r}{k}$ or $r = ki$. Let A be the amount that must be invested today at APR r compounded k times a year so that the value of the investment will be the same as the future value of the above annuity consisting of n equal payments or deposits of R dollars each. Then, we have the following:

$$A(1+i)^n = R\frac{(1+i)^n - 1}{i}$$

$$A = R\frac{1-(1+i)^{-n}}{i}$$

• DEFINITION The **present value** A of an annuity consisting of n equal payments or deposits of R dollars each with interest rate i per period is given by

$$A = R\frac{1-(1+i)^{-n}}{i}.$$

•

• EXAMPLE 9: What monthly payments are required for a 4-year \$9,000 car loan at 12.5% APR compounded monthly?

SOLUTION: This is an example of a 4-year annuity with present value \$9,000. (Why?) There are to be 60 equal monthly payments of R dollars each with monthly interest rate of $\frac{0.125}{12}$. We can determine R from the following equation:

$$9000 = R\frac{1-\left(1+\frac{0.125}{12}\right)^{-60}}{\frac{0.125}{12}}.$$

We can use a calculator to show that $R = 202.48$. Thus, 60 monthly payments of \$202.48 are required to pay for a \$9,000 car at 12.5% APR compound monthly. •

• EXAMPLE 10: Greg invests \$75 per month into an IRA annuity for 20 years at 11.8% APR. Determine the value of Greg's IRA investment after 20 years.

SOLUTION: By the assumptions of this section, the first \$75 deposit occurs at the end of the first month of the 20-year investment period. The monthly interest rate is $\frac{0.118}{12}$, and the number of deposits is 240. We use the future value formula to compute the value S of the investment after 20 years:

$$\begin{aligned} S &= R\frac{(1+i)^n - 1}{i} \\ &= 75\frac{\left(1+\frac{0.118}{12}\right)^{240} - 1}{\frac{0.118}{12}} \\ &= 72{,}225.47. \end{aligned}$$

Thus, the value of Greg's IRA investment after 20 years is \$72,225.47. •

Often annuities requiring *monthly* payments are given with APR rates which have no compounding frequency stated. In such cases, the APR rate is assumed to be compounded monthly.

• EXERCISES 4-3

1. Draw a complete graph of each function in the same viewing rectangle for $0 \le x \le 10$.
   (a) $y = 1000(1 + .05x)$ (b) $y = 1000(1.05)^x$
2. Draw a complete graph of each function in the same viewing rectangle for $0 \le x \le 10$.
   (a) $y = 1000(1 + .08x)$ (b) $y = 1000(1.08)^x$
3. Draw a complete graph of each function in the same viewing rectangle for $0 \le x \le 10$.
   (a) $y = 1000(1 + .06x)$ (b) $y = 1000(1 + .09x)$
4. Draw a complete graph of each function in the same viewing rectangle for $0 \le x \le 10$.
   (a) $y = 1000(1.06)^x$ (b) $y = 1000(1.09)^x$
5. Explain how the graphs in Exercises 1 and 2 are related to investing $1000 in different kinds of interest bearing savings accounts.
6. Explain how the graphs in Exercises 3 and 4 are related to investing $1000 in an interest bearing savings account of the same type.
7. Determine r if $2500 = 1000(1 + 12r)$.
8. Determine t if $2500 = 1000(1 + .07t)$.
9. Determine P if $2500 = P(1 + .08(16))$

Consider the equation $S = P(1 + rt)$. Solve the equation for the variable specified in terms of the other three variables.

10. r 11. t 12. P
13. Determine r if $2500 = 1000(1 + r)^{12}$.
14. Determine t if $2500 = 1000(1.07)^n$.
15. Determine P if $2500 = P(1.08)^{16}$.

Consider the equation $S = P(1 + r)^t$. Solve the equation for the variable specified in terms of the other three variables.

16. r 17. P
18. What is the value of an initial investment of $2575.00 at $8% compounded continuously for 6 years?
19. Determine when an investment doubles in value at 6% compounded continuously.
20. Determine r if $2300 = 1500\, e^{10r}$. Make up a problem situation for which the equation is a model.
21. Determine when an investment of $2300 accumulates to a value of $4150 if the investment earns interest at the rate of 9% compounded quarterly.
22. Determine when an investment of $1500 accumulates to a value of $3750 if the investment earns interest at the rate of 8% compounded monthly.

23. A \$1580 investment earns interest compounded annually. Determine the annual interest rate if the value of the investment is \$3000 after 8 years.

24. A \$22,000 investment earns interest compounded monthly. Determine the annual interest rate if the value of the investment is \$36,500 after 5 years.

25. Determine how much time is required for an investment to *double* in value if interest is earned at the rate of 5.75% compounded quarterly.

26. Determine how much time is required for an investment to *triple* in value if interest is earned at the rate of 6.25% compounded monthly.

27. Sally deposits \$500 in a bank that pays 6% annual interest. Assume that she makes no other deposits or withdrawals. How much has she accumulated after 5 years if the bank pays interest compounded

    (a) annually? (b) quarterly? (c) monthly?
    (d) daily? (e) continuously?

28. Pete deposits \$4500 in a bank that pays 7% annual interest. Assume that he makes no other deposits or withdrawals. How much has he accumulated after 8 years if the bank pays interest compounded

    (a) annually? (b) quarterly? (c) monthly?
    (d) daily? (e) continuously?

29. Draw a graph that shows the value during the first 10 years of a \$1000 investment that earns 8% simple interest. In the same viewing rectangle, draw the graph of the model for a \$1000 investment that earns 8% interest compounded monthly.

30. Compare the two graphs in Exercise 29 in the interval $[0,1]$ and in the interval $[1,10]$. Describe this comparison in terms of the problem situations given in Exercise 29.

Effective Annual Rate

The effective annual rate i_{eff} of APR r compounded k times per year is given by

$$i_{eff} = \left(1 + \frac{r}{k}\right)^k - 1.$$

Effective annual rates can be used to compare interest rates with compounding periods of different lengths as illustrated in Exercises 31 to 34. Effective annual rate is often called "effective yield" in business circles. It is not to be confused with annual percentage rate (APR).

31. What investment yields the greatest return: 6% compounded quarterly, or 5.75% compounded daily?

32. What investment yields the greatest return: 8.25% compounded monthly, or 8% compounded daily?

33. What investment yields the greatest return: 7% compounded quarterly, or 7.20% compounded daily?

34. What investment yields the greatest return: 8.5% compounded quarterly, or 8.40% compounded monthly?

35. Draw a complete graph of each function in the same viewing rectangle for $0 \leq x \leq 240$.

(a) $y = 100\dfrac{(1.005)^x - 1}{0.005}$ (b) $y = 100\dfrac{\left(1 + \frac{0.08}{12}\right)^x - 1}{\frac{0.08}{12}}$

36. Draw a complete graph of each function in the same viewing rectangle for $0 \leq x \leq 60$.

(a) $y = 200\dfrac{1 - \left(1 + \frac{0.08}{12}\right)^{-x}}{\frac{0.08}{12}}$ (b) $y = 200\dfrac{1 - (1.01)^{-x}}{0.01}$

37. Explain how the graphs in Exercises 35 are related to a retirement annuity requiring payments of \$100 per month.

38. Explain how the graphs in Exercises 36 are related to financing a car requiring monthly payments of \$200.

39. What are the monthly payments of a car loan of \$8,250 for 5 years at 13% annual interest? *Note:* Remember that the 13% is an APR rate, so the actual *monthly* interest rate is $\frac{0.13}{12}$.

40. What are the monthly payments of a home mortgage of \$81,500 for 30 years at 10.25% annual interest?

41. Amy contributes \$50 per month into an IRA annuity for 25 years. Assuming that the IRA earns 6.25% annual interest, what is the value of Amy's IRA account after 25 years?

42. Frank contributes \$50 per month into an IRA annuity for 15 years. Assuming that the IRA earns 5.5% annual interest, what is the value of Frank's IRA account after 15 years?

43. An \$86,000 mortgage for 30 years at 12% APR requires monthly payments of \$884.61. Suppose you decided to make monthly payments of 1050.00. When would the mortgage loan be completely paid?

44. Suppose you make payments of \$884.61 for the \$86,000 mortgage in Exercise 43 for 10 years and then make payments of \$1050 until the loan is paid. When will the mortgage loan be completely paid under these circumstances?

45. Consider an \$86,000 mortgage loan at 12% APR requiring monthly payments. Determine the required monthly payment if the loan has a term of

(a) 30 years. (b) 25 years. (c) 20 years. (d) 15 years.

46. Explain why the formula

$$B(n) = R\frac{1 - (1 + i)^{-(360-n)}}{i}$$

gives the outstanding loan balance of a 30-year mortgage loan requiring monthly payments of R dollars with APR $12i$ as a function of the number of payments (n) made.

47. (a) Draw a complete graph of

$$y = \$884.61\frac{1 - (1.01)^{-(360-x)}}{0.01}.$$

(b) Explain how this graph relates to the outstanding loan balance of an \$86,000 mortgage loan with a 30-year term requiring monthly payments of \$884.61.

48. Some mortgage loans are available that require payments every *two weeks* (26 times each year). Consider a mortgage loan of $80,000 for a 30-year term at 10% APR requiring monthly payments.

   (a) Determine the monthly payment.

   (b) Suppose 1/2 of the monthly payment was made every two weeks (26 times each year). When would the mortgage loan be completely paid? Assume the interest rate per two week interval is 1/2 of the monthly interest rate.

49. (a) Solve

$$86{,}000 = R\frac{1-(1.01)^{-x}}{0.01}$$

   for R.

   (b) Draw a complete graph of $R = f(x)$ for $0 \le x \le 360$.

   (c) Explain how the graph in (b) relates to a mortgage loan of $86,000 at 12% APR.

50. Solve for t using a graphing utility:

$$10{,}000 = 200\frac{1-(1.01)^{-t}}{0.01}.$$

   Explain how this problem relates to a car loan of $10,000 requiring monthly payments of $200.00.

51. Solve for i using a graphing utility:

$$10{,}000 = 238\frac{1-(1+i)^{-60}}{i}.$$

   Explain how this problem relates to a car loan of $10,000 requiring monthly payments of $238.00.

*52. Explain why $A(1+i)^n = S$ where S and A are the future value and present value, respectively, of the same annuity.

*53. Explain why $S = R + R(1+i) + R(1+i)^2 + R(1+i)^3 + \cdots + R(1+i)^{n-1}$ is the future value of an annuity of n payments of R dollars each made at equal intervals of time where the interest rate i is the interest rate per payment interval.

*54. Use the sum formula for a finite geometric series to show that

$$S = R\frac{(1+i)^n - 1}{i}.$$

*55. Use Exercises 53 and 54 to show that

$$A = R\frac{1-(1+i)^{-n}}{i}.$$

The Constant Percentage Depreciation Method

The following notation and definitions will be used in connection with depreciation and appreciation of an asset.

$C =$ original value (cost) of the asset

$n =$ length of term (useful life under consideration)

$S =$ value of the asset at end of term (in case of depreciation, S is called the *salvage value*)

$D = C - S$ is the *total depreciation* if $S < C$ and is the *total appreciation* if $S > C$

$B =$ *book value*, the value for accounting or tax purposes at any time t

$\dfrac{C-S}{n} =$ the annual depreciation or appreciation amount

A continuous model for *book value* at any time t using the *constant percentage* method is

$$B = C(1-r)^t \quad \text{where} \quad S = C(1-r)^n \quad \text{and} \quad 0 \le t \le n.$$

Consider a machine costing \$17,000 with a useful life of 6 years and that has a salvage value of \$1200.

56. Assume that the machine is depreciated using the constant percentage method. Use the fact that $S = C(1-r)^n$ to determine r.
57. Draw a complete graph of the *book value* model $y = B(t)$.
58. What values of t make sense in this problem situation?
59. What is the book value in 4 years, 3 months?
60. Draw the graph of *book value* of the machine using both the straight line method (see Exercise 45 from Section 1.5) and the constant percentage method in the same viewing rectangle for $0 \le t \le 6$.

Appreciation

The constant percentage depreciation method is closely related to a very common method of determining appreciation (inflation or growth).

61. Show that if $r < 0$, then $C(1-r)^n = S$ becomes $C(1+r)^n = S$ with $r > 0$. Further show that $S > C$, which means that the asset has *appreciated* over time.
62. Let $C =$ \$55,000 and $r = -0.08$. Draw a complete graph of $y = S(n) = c(1-r)^n$.

Inflation

The formula $S = C(1+r)^n$ is frequently used to model inflation. In such a case, r is the annual inflation rate, C is the value today, and S is the *inflated* value n years from now. The *purchasing power* of one dollar n years from now (assuming an annual inflation rate of r) is $C = 1/(1+r)^n$.

63. Assume an inflation rate of 8%. Use a graph to determine the value of a \$55,000 house in 7 years. (Assume no other factors affect the value of the house.)
64. What is the purchasing power of one dollar in 10 years if the annual inflation rate is 3%?
65. What is the purchasing power of one dollar in 10 years if the annual inflation rate is 8%?
66. What is the purchasing power of one dollar in 10 years if the annual inflation rate is 15%?

4.4 Logarithmic Functions

In this section we introduce logarithmic functions as inverses of exponential functions. In the past, logarithms were necessary for computation. Today calculators make computational use of logarithms obsolete. However, logarithmic functions are still very important in mathematics. For example, if x can be expressed as an exponential function of y, then we will see that y can be expressed as a logarithmic function of x. We give some of the important algebraic properties of logarithms and see how to use them to solve equations involving exponents. In the next section we see how to obtain their graphs with graphing utilities.

• THEOREM 1 Let $a > 0$ and $a \neq 1$. The inverse of the exponential function $y = a^x$ is a function.

PROOF In Section 4.2 we found that the function $y = a^x$ is increasing on $(-\infty, \infty)$ if $a > 1$ and is decreasing on $(-\infty, \infty)$ if $0 < a < 1$. Thus, the inverse of the one-to-one function $y = a^x$ is a function.

• DEFINITION The inverse of the exponential function $y = a^x$ is called the **logarithmic function with base a** and is denoted by $y = \log_a x$. The value of $\log_a x$ is called the **logarithm of x with base a**. The function $y = \log_{10} x$ is called the **common logarithmic function**, and the function $y = \log_e x$ is called the **natural logarithmic function**. •

Whenever the base of the logarithm is omitted, it is assumed to be 10. Thus, we write $\log_{10} x = \log x$. If the base is e, we write $\log_e x = \ln x$.

Most calculators have a key to evaluate the functions $y = \log x$ and $y = \ln x$. In the next section we see how to use a calculator to compute the values of $y = \log_a x$ for other values of a. Computers generally only have $y = \ln x$ built in. Unfortunately, $y = \ln x$ is often accessed on a computer by entering $y = \log x$. You should always check to see which logarithm function is built in on the computer you use. In the next section we use graphing utilities to graph $y = \log_a x$ for a different from 10 or e. In this section we graph logarithmic functions by using the fact that they are the inverses of exponential functions.

• EXAMPLE 1: Specify the domain, the range and draw a complete graph of each function.

(a) $y = \log_2 x$ (b) $y = \log x$

(c) $y = \ln x$ (d) $y = \log_{0.5} x$

SOLUTION: The domain of each of these functions is $(0, \infty)$, and the range is $(-\infty, \infty)$. This is because the corresponding exponential functions have domain $(-\infty, \infty)$ and range $(0, \infty)$. The function $y = \log_2 x$ is the inverse of $y = 2^x$, $y = \log x$ is the inverse of $y = 10^x$, $y = \ln x$ is the inverse of $y = e^x$, and $y = \log_{0.5} x$ is the inverse of $y = 0.5^x$. In Section 4.1, we showed that the graph of the inverse of a relation can be obtained by reflecting the graph of the relation through the line $y = x$. We use this property to draw the complete

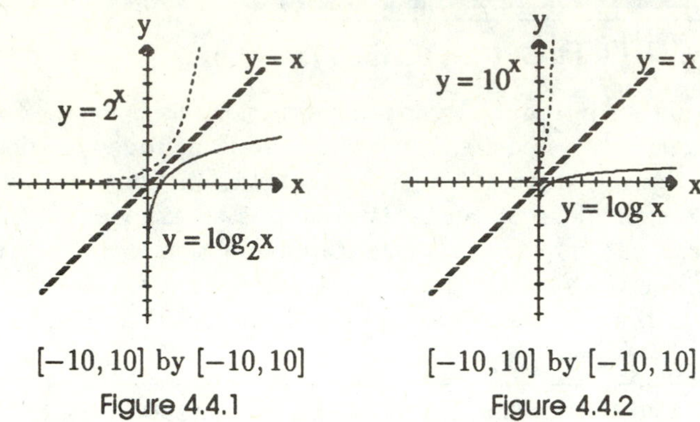

$[-10, 10]$ by $[-10, 10]$
Figure 4.4.1

$[-10, 10]$ by $[-10, 10]$
Figure 4.4.2

graphs of $y = \log_2 x$, $y = \log x$, $y = \ln x$, and $y = \log_{0.5} x$ that are given in Figures 4.4.1 through 4.4.4, respectively. •

We can compute the values of a logarithmic function by computing the values of the corresponding exponential function.

• EXAMPLE 2: Use the definition of logarithms to show that $\log_2 8 = 3$.

SOLUTION: To show that $\log_2 8 = 3$ we show that $(8, 3)$ is a solution to $y = \log_2 x$. Notice that $(3, 8)$ is a solution to $y = 2^x$, that is, $8 = 2^3$. Thus, $(8, 3)$ is a solution to $y = \log_2 x$ and $3 = \log_2 8$. Use Figure 4.4.1 to check this computation. •

We can generalize Example 2 in the following way.

SUMMARY If $a > 0$ and $a \neq 1$, then $y = \log_a x$ if and only if $x = a^y$.

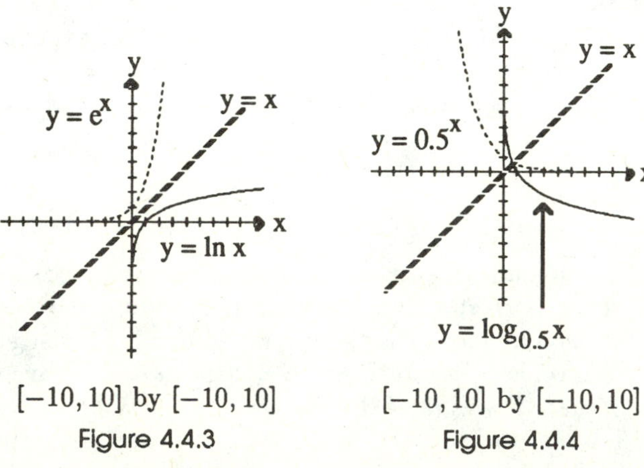

$[-10, 10]$ by $[-10, 10]$
Figure 4.4.3

$[-10, 10]$ by $[-10, 10]$
Figure 4.4.4

• EXAMPLE 3: Use exponents to compute $\log_5 \frac{1}{25}$.

SOLUTION: If we let $x = \log_5 \frac{1}{25}$, then $5^x = \frac{1}{25}$. Because $\frac{1}{25} = 5^{-2}$, we have $5^x = \frac{1}{25} = 5^{-2}$ so that $x = -2$. Therefore, $\log_5 \frac{1}{25} = -2$. •

• EXAMPLE 4: Use exponents to solve for x.

(a) $\log_5 x = 1.5$

(b) $\log_2(x-3) = 4$

(c) $\log_x 3 = 2$

(d) $\log_x \dfrac{1}{8} = -3$

SOLUTION:

(a) We can write $\log_5 x = 1.5$ in exponential form as $x = 5^{1.5}$. Thus, accurate to hundredths, the solution is 11.18.
(b) $\log_2(x-3) = 4$ is equivalent to $x - 3 = 2^4$. Thus, $x = 19$.
(c) $\log_x 3 = 2$ is equivalent to $x^2 = 3$. Thus, $x = \sqrt{3}$ because the base of a logarithm must be positive.
(d) $\log_x \frac{1}{8} = -3$ is equivalent to $x^{-3} = \frac{1}{8} = 2^{-3}$. Thus, $x = 2$. •

Properties of Logarithms

Some properties of logarithms are given in Theorems 2 and 3.

• THEOREM 2 Let $a > 0$ and $a \neq 1$. Then,

(a) $a^{\log_a x} = x$ for every positive real number x.

(b) $\log_a a = 1$.

(c) $\log_a 1 = 0$.

PROOF

(a) We know that $y = \log_a x$ is equivalent to $x = a^y$. Thus, $x = a^y = a^{\log_a x}$.
(b) The logarithmic form of $a^1 = a$ is $\log_a a = 1$.
(c) The logarithmic form of $a^0 = 1$ is $0 = \log_a 1$.

• THEOREM 3 Let a, r, and s be positive real numbers with $a \neq 1$. Then,

(a) $\log_a rs = \log_a r + \log_a s$.

(b) $\log_a \dfrac{r}{s} = \log_a r - \log_a s$.

(c) $\log_a r^c = c \log_a r$ for every real number c.

(d) $\log_a a^x = x$ for every real number x.

PROOF Let $\log_a r = u$ and $\log_a s = v$. Then, in exponential form, we have $a^u = r$ and $a^v = s$.

(a) $rs = a^u a^v = a^{u+v}$. The logarithmic form of $rs = a^{u+v}$ is $\log_a rs = u + v$. Substituting for u and v we find

$$\log_a rs = u + v = \log_a r + \log_a s\,.$$

(b) $\frac{r}{s} = \frac{a^u}{a^v} = a^{u-v}$. The exponential form of $\frac{r}{s} = a^{u-v}$ is $\log_a \frac{r}{s} = u - v$. Substituting for u and v we find

$$\log_a \frac{r}{s} = \log_a r - \log_a s\,.$$

(c) $r^c = (a^u)^c = a^{uc}$. Thus, $\log_a r^c = uc$. Substituting for u we find

$$\log_a r^c = c\log_a r\,.$$

(d) Let $r = a$ and $c = x$ in (c) and we obtain $\log_a a^x = x\log_a a$. By Theorem 2(b), we know that $\log_a a = 1$. Thus, $\log_a a^x = x$.

If we let $f(x) = \log_a x$, then the inverse of f is the function $f^{-1}(x) = a^x$. It follows from Theorem 2(a) that $f^{-1} \circ f(x) = x$ for every positive real number x. Similarly, Theorem 3(d) gives $f \circ f^{-1}(x) = x$ for every real number x. Notice that the domain of $f^{-1} \circ f$ is $(0, \infty)$ and the domain of $f \circ f^{-1}$ is $(-\infty, \infty)$.

Before the easy availability of calculators, the formulas in Theorem 3 were used to compute the value of complicated arithmetic expressions. This was possible because a logarithmic function is one-to-one. This means that if the number is known, its logarithm is also known and vice versa. This was accomplished by first using Theorem 3 to express the logarithm of the complicated expression in terms of the logarithms of its simpler components. Then, logarithm tables were used to finish the computation. The basic idea is illustrated in the next example.

• EXAMPLE 5: Express $\log_a(x^3y^{3/2})/\sqrt{z}$ in terms of $\log_a x$, $\log_a y$, and $\log_a z$.

SOLUTION: We use Theorem 3 to obtain the following:

$$\begin{aligned}\log_a \frac{x^3y^{3/2}}{\sqrt{z}} &= \log_a(x^3y^{3/2}) - \log_a \sqrt{z}\\ &= \log_a x^3 + \log_a y^{3/2} - \log_a z^{1/2}\\ &= 3\log_a x + \frac{3}{2}\log_a y - \frac{1}{2}\log_a z\end{aligned}$$ •

Notice that the logarithm of an expression that involves products, quotients and exponents can be expressed as sums and differences of products of logarithms and numbers. On the other hand, Example 6 shows that an exponential function with an exponent that involves sums and differences can be expressed as a product and quotient of exponential expressions.

• EXAMPLE 6: Express $a^{2x+3y-z}$ in terms of a^x, a^y, and a^z.

SOLUTION: We use properties of exponents to obtain the following:

$$\begin{aligned}a^{2x+3y-z} &= a^{2x} \cdot a^{3y} \cdot a^{-z}\\ &= \frac{(a^x)^2 \cdot (a^y)^3}{a^z}.\end{aligned}$$ •

In the previous two sections we used graphs to solve equations involving exponents. Now we can obtain an algebraic solution using logarithms.

• EXAMPLE 7: The population of a town is 50,000 and is increasing at the rate of 2.5% per year. Determine when the population of the town is 100,000.

SOLUTION: This was (c) of Example 6 of Section 4.2. In that section we answered this question using a graph and zoom-in. Now we use logarithms. The population of the town at any time t is $P(t) = 50{,}000(1.025)^t$. Thus, we need to solve the following equation for t:

$$50{,}000(1.025)^t = 100{,}000.$$

If we divide each side of the above equation by 50,000 we obtain the following equivalent equation:

$$1.025^t = 2.$$

At this point we use the important *well-defined property* of a function f that states that if $a = b$, then $f(a) = f(b)$. In our case, this means we can take the logarithm of each side of an equation to obtain a new equation. Because $1.025^t = 2$, we can use the well-defined property of the logarithm base 10 function to continue the algebraic procedure in the following way:

$$\begin{aligned} 1.025^t &= 2 \\ \log 1.025^t &= \log 2 \\ t \log 1.025 &= \log 2 \text{ (Why?)} \\ t &= \frac{\log 2}{\log 1.025}. \end{aligned}$$

Thus, accurate to hundredths, t is 28.07. Check that this is the solution obtained in Example 6 of Section 4.2. •

Data Analysis

Scientists often try to determine the relationship between variables empirically. They collect enough experimental data to establish that a given rule is plausible. Once a rule is conjectured, additional evidence can be used to verify or support the rule. We illustrate this idea using logarithms.

Suppose $y = ax^b$ is a power function of x. We further assume a, x, and y are positive. The following equations are equivalent:

$$\begin{aligned} y &= ax^b \\ \ln y &= \ln ax^b \\ \ln y &= \ln a + \ln x^b \\ \ln y &= \ln a + b \ln x. \end{aligned}$$

Notice that $\ln y$ is a *linear* function of $\ln x$.

• **THEOREM 4** Let a, x, and y be positive. Then $y = ax^m$ if and only if $\ln y$ is a linear function of $\ln x$.

PROOF We have already showed that if $y = ax^m$, then $\ln y = \ln a + m \ln x$. Thus, $\ln y$ is a linear function of $\ln x$.

Conversely, suppose $\ln y$ is a linear function of $\ln x$. Then there are real numbers m and c so that $\ln y = m \ln x + c$. The natural logarithm function is one-to-one, and has range $(-\infty, \infty)$. Thus, there is a unique positive real number a such that $\ln a = c$. We substitute this form for c in the expression for $\ln y$:

$$\begin{aligned} \ln y &= m \ln x + c \\ \ln y &= m \ln x + \ln a \\ \ln y &= \ln x^m + \ln a \\ \ln y &= \ln ax^m \\ y &= ax^m. \end{aligned}$$

• **EXAMPLE 8:** Use the information in Table 1 and logarithms to determine a power rule model for y in terms of x.

SOLUTION: According to Theorem 4 we can show that y is a power function of x by showing that $\ln y$ is a linear function of $\ln x$. Figure 4.4.5 shows the four points $(\ln x, \ln y)$ for x and y as given by the rows of Table 1. Notice that the four points appear to be on a straight line. We could draw a line through the four points and estimate the slope m and y-intercept b to write $\ln y = m \ln x + b$. In fact, even if the points did not lie on a line we could draw a "best" fitting line through the data. This line would give rise to a "best" fitting power function through the data points (x, y).

x	y
1	5
3	31.2
5	73.1
8	160

Table 1

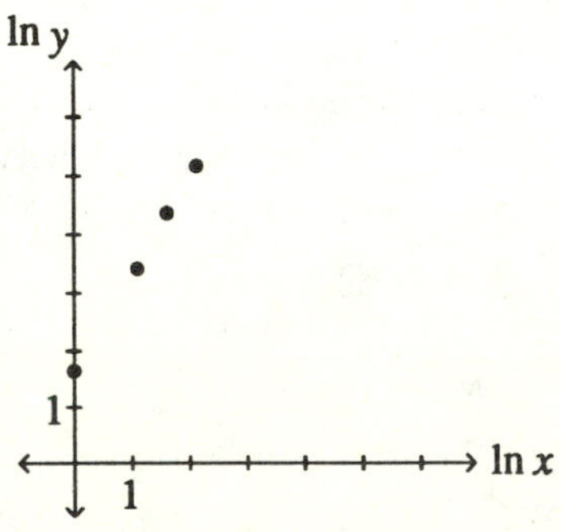

Figure 4.4.5

Alternatively, we could use two of the four pairs of values $(\ln x, \ln y)$ to determine m and b in $\ln y = m \ln x + b$. First we use $(\ln 1, \ln 5)$:

$$\begin{aligned}\ln 5 &= m \ln 1 + b\\ \ln 5 &= m \cdot 0 + b\\ \ln 5 &= b.\end{aligned}$$

Thus, $b = \ln 5$. Next, we use $(\ln 8, \ln 160)$:

$$\ln 160 = m \ln 8 + \ln 5.$$

Solving for m we find $m = 1.666666667$, which suggests that $m = \frac{5}{3}$. In fact, $m = \frac{5}{3}$ is an exact solution to $\ln 160 = m \ln 8 + \ln 5$:

$$\begin{aligned}\ln 160 &= m \ln 8 + \ln 5\\ \ln 160 &= \ln(8^m)(5)\\ 160 &= 5(8^m)\\ 32 &= 8^m\\ 2^5 &= 2^{3m}\\ m &= \frac{5}{3}.\end{aligned}$$

Generally, we would simply approximate the value of m.

We can use the values of m and b to write y as a power function of x:

$$\begin{aligned}\ln y &= m \ln x + b\\ \ln y &= \frac{5}{3} \ln x + \ln 5\\ \ln y &= \ln(5x^{5/3})\\ y &= 5x^{5/3}.\end{aligned}$$

•

The power function model determined in Example 8 can be used to predict the value of y given a new value of x.

• EXERCISES 4-4

Compute without using a calculator.

1. $\log_4 16$
2. $\log_{10} 10$
3. $\log_{1/2} 16$
4. $\ln 1$
5. $\log_2(-3)$
6. $\log_4(-2)^4$

Solve each equation. Do not use a calculator.

7. $\log_9 x = 2$
8. $\log_2 x = 5$
9. $\log_x \frac{1}{125} = -3$
10. $\log_3(x+1) = 2$
11. $\log_x 81 = 9$
12. $\log_2 x^2 = -2$
13. $\log_3 |x| = 1$
14. $\log_6(x^2 - 2x + 1) = 0$

Sketch a complete graph. Use the inverse property. Do not use a graphing utility.

15. $f(x) = \log_3 x$
16. $g(x) = \log_5(x)$
17. $y = \log_{1/4} x$
18. $y = \log_2(-x)$

Express in logarithmic form.

19. $x + y = 2^8$
20. $(1+r)^n = P$

Express in exponential form.

21. $\log_3 \frac{x}{y} = -2$
22. $\frac{\ln P}{\ln(1+r)} = n$

Use properties of logarithms to write the expression as a sum, difference, or product of simple logarithms (logarithms of expressions that do not involve sums, products, quotients, or exponents).

23. $\log[5000x(1+r)^{360}]$
24. $\log(\sqrt[5]{216z^3})$
25. Explain why the functions $f(x) = \log_2 x^2$ and $g(x) = 2\log_2 x$ are different even though Theorem 3(c) implies that $\log_2 x^2 = 2\log_2 x$.

Use a calculator to investigate numerically the end behavior of the function. That is, find what $f(x)$ approaches as $|x| \to \infty$.

26. $f(x) = a^{1/x}$
27. $f(x) = \frac{\ln x}{x}$
28. $f(x) = x^{1/x}$
29. $f(x) = \frac{x}{\ln x}$

In Exercises 30–34, use logarithms to solve the problem algebraically.

30. The population of a town is 475,000 and is increasing at the rate of 3.75% each year. Determine when the population of the town will be one million.
31. The population of a town is 123,000 and is decreasing at the rate of 2.375% each year. Determine when the population of the town will be 50,000.
32. The population of a small town in the year 1890 was 6250. Assume the population increased at the rate of 3.75% each year. Determine the population of the town in 1915 and in 1940.
33. The half-life of a certain radioactive substance is 14 days, and there are 6.58 grams present initially. When will there be less than 1 gram of the substance remaining?
34. The half-life of a certain radioactive substance is 65 days, and there are 3.5 grams present initially. When will there be less than 1 gram of the substance remaining?

35. Consider the data in Tables 2, 3, and 4. Which data have a power rule for an algebraic representation? Why?

36. For Tables 2, 3, and 4 with a power rule algebraic representation, determine the explicit power rule.

It is established in the Exercises in Section 4.5 that if the points $(x, \ln y)$ are on the same line with slope m and y-intercept c then an algebraic representation of y in terms of x is the *exponential* function $y = ab^x$ where $\ln a = c$ and $\ln b = m$.

37. Which data in Tables 2, 3, and 4 have an exponential algebraic representation?

38. For the tables in Exercise 37 that have an exponential model, determine the explicit exponential rule.

Use an algebraic method to solve the equation for t. Check with a graphing utility.

39. $2500 = 1000(1.08)^t$

40. $6000 = 4600(1.05)^t$

41. $10{,}000 = 3500(1.07)^t$

42. Explain how Exercises 39–41 relate to interest bearing saving accounts.

43. Solve for t: $S = P(1 + r)^t$

44. Use an algebraic method to determine how long it will take for an initial deposit of \$1250 to double in value at 7% compounded monthly.

45. Use an algebraic method to determine how long it will take for an initial deposit of \$1250 to triple in value at 7% compounded monthly.

Use an algebraic method to solve the equation for t. Check with a graphing utility.

46. $10{,}000 = 225\dfrac{1 - (1.01)^{-t}}{0.01}$

47. $10{,}000 = 300\dfrac{1 - (1.01)^{-t}}{0.01}$

48. Explain how Exercises 46 and 47 relate to a car loan of \$10,000.

49. Use a graphing utility to solve $10{,}000 = 250[1 - (1 + i)^{-60}]/i$ for i. *Note:* Think about solving this equation algebraically. There is no known method of expicitly solving this equation for i!

50. Explain how Exercise 49 relates to a car loan of \$10,000 requiring monthly payments of \$250.

x	y
2	7.48
3	7.14
7.5	1.94
7.7	0.84

Table 2

x	y
4	2 816
6.5	31,908
8.5	122,019
10	275,000

Table 3

x	y
8	23.84
12	58.2
15	113.69
40	30,092.66

Table 4

Planet	Period (days)	Length of Semi-Major Axes (miles)
Earth	365	92,600,000
Mercury	88	36,000,000
Venus	225	67,100,000
Mars	687	141,700,000
Jupiter	4330	483,400,000
Saturn	10,750	886,100,000

Table 5

Planetary Motion

Table 5 gives the period of one complete revolution about the sun for the listed planets along with the length of the semi-major orbit axes.

51. Let P be the orbit period and x be the length of the major axis. Plot $\ln P$ against $\ln x$. That is, plot the pairs $(\ln x, \ln P)$. Verify the relationship is linear and estimate the slope m and y-intercept.
52. Use Exercise 51 to show that $P = ax^m$ for constants a and m. This result was discovered by Kepler in the early 1700's and is known as *Kepler's third law of planetary motion.*
53. Predict the period of Pluto's orbit if its semi-major orbit axis length is 2,200,000,000 miles.

Blood Pressure

The *at rest blood pressure* P and weight x of various primates were measured as shown in Table 6.

54. Plot $\ln P$ against $\ln x$. That is, plot the pairs $(\ln x, \ln P)$. Verify the relationship is linear. Determine the slope m and y-intercept.
55. Use Exercise 54 to find a formula expressing the blood pressure as a function of weight.

Weight x (lbs.)	Blood Pressure P
20	106
50	133
80	150
110	162
125	167
140	172

Table 6

4.5 More on Logarithms

In this section we obtain a formula that gives the logarithms of numbers in one base in terms of the logarithms of numbers in another base. This formula, called the change of base formula, can be used to compute logarithms with any base on a calculator and to obtain a complete graph of any logarithmic function with a graphing utility.

• THEOREM 1 **(Change of base formula)** Let a and b be positive real numbers. Then,

(a) $\log_b x = \dfrac{\log_a x}{\log_a b}$, (b) $\log_b a = \dfrac{1}{\log_a b}$.

PROOF

(a) The exponential form of $y = \log_b x$ is $x = b^y$. Next, we apply the logarithm base a to both sides of the equation $x = b^y$ and solve for y:

$$x = b^y$$
$$\log_a x = \log_a b^y$$
$$\log_a x = y \log_a b$$
$$y = \frac{\log_a x}{\log_a b}.$$

Thus, $y = \log_b x = \frac{\log_a x}{\log_a b}$.

(b) Let $x = a$ in the formula in (a). Then we have

$$\log_b x = \frac{\log_a x}{\log_a b}$$
$$\log_b a = \frac{\log_a a}{\log_a b}.$$

Because $\log_a a = 1$, we have

$$\log_b a = \frac{1}{\log_a b}.$$

• EXAMPLE 1: Use a calculator to compute each value.

(a) $\log_5 4$ (b) $\log_2 9$

SOLUTION: Most calculators have both logarithms base 10, [log], and logarithms base e, [ln], as built-in functions.

(a) We can use logarithm base 10 and enter $\log_5 4$ as $\frac{\log 4}{\log 5}$. Thus, $\log_5 4 = 0.86$ accurate to hundredths. Because $5^{\log_5 4} = 4$, the calculator answer for $\log_5 4$ can be checked by raising 5 to that power to see if you get approximately 4. That is, we can use a calculator to check that $5^{0.86}$ is 3.9912984, which is approximately 4.

(b) We can use logarithm base e and enter $\log_2 9$ as $\log_2 9 = \frac{\ln 9}{\ln 2}$ to find that $\log_2 9$ is 3.17 accurate to hundredths. •

Theorem 1(a) is the relationship we need to graph logarithmic functions with a graphing utility. Most calculators have both $\log x$ and $\ln x$ as built-in functions. Generally computers have $\ln x$ as a built-in function that is accessed by entering $\log x$. As long as a graphing utility has one logarithmic function built in, we can use this logarithmic function to graph *any* other logarithmic function.

• EXAMPLE 2: Express $\log_3 x$ in terms of $\log x$ and $\ln x$. Then, draw a complete graph of $y = \log_3 x$.

SOLUTION: From Theorem 1(a) we can write

$$\log_3 x = \frac{\log x}{\log 3} \quad \text{or} \quad \log_3 x = \frac{\ln x}{\ln 3}.$$

Now, depending on which logarithmic function is built in on your graphing utility, a complete graph of $y = \log_3 x$ can be obtained by entering $y = \frac{\log x}{\log 3}$ or $y = \frac{\ln x}{\ln 3}$ (Figure 4.5.1). •

A rough sketch of the graph of $y = a\log_b(cx + d) + k$ can be obtained by applying standard geometric transformations to the graph of $y = \log_b x$.

• EXAMPLE 3: Determine the domain, the range, and draw a complete graph of $y = \log_3(-x)$.

SOLUTION: A complete graph of $y = \log_3(-x)$ can be obtained by reflecting the graph of $y = \log_3 x$ through the y-axis (Figure 4.5.2). We can also obtain a complete graph of $y = \log_3(-x)$ with a graphing utility by entering $y = \frac{\ln(-x)}{\ln 3}$. The domain of $y = \log_3(-x)$ is $(-\infty, 0)$, and the range is $(-\infty, \infty)$. •

• EXAMPLE 4: Determine the domain, the range, the asymptotes, and draw a complete graph of each function.

(a) $y = \log_5(x + 2)$ (b) $y = \log_5(3 - x)$

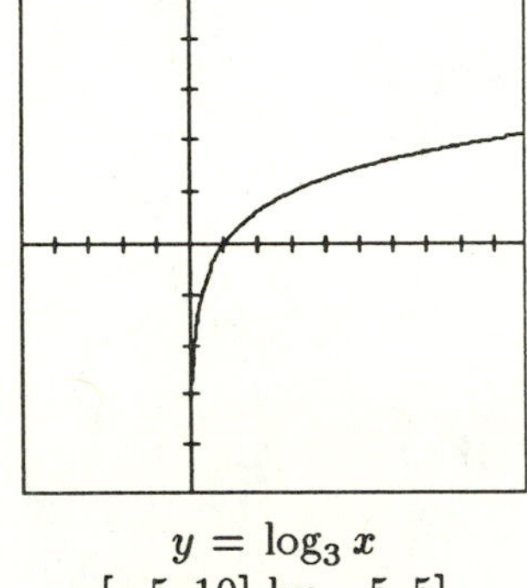

$y = \log_3 x$
$[-5, 10]$ by $-5, 5]$
Figure 4.5.1

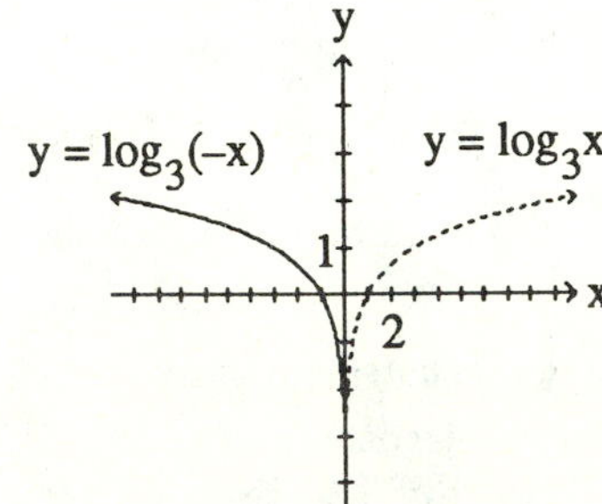

Figure 4.5.2

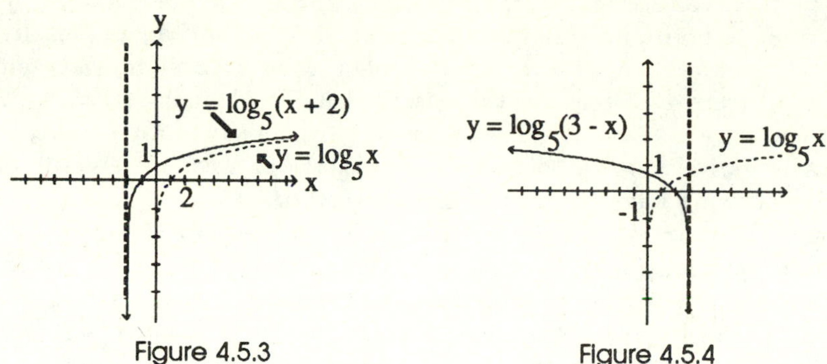

Figure 4.5.3

Figure 4.5.4

SOLUTION:

(a) A complete graph of $y = \log_5(x+2)$ can be obtained by shifting a complete graph of $y = \log_5 x$ horizontally to the left 2 units (Figure 4.5.3). We can also obtain a complete graph with a graphing utility by writing $y = \log_5(x+2) = \frac{\log(x+2)}{\log 5}$. The domain of $y = \log_5(x+2)$ is $(-2, \infty)$ and the range is $(-\infty, \infty)$. The line $x = -2$ is a vertical asymptote of $y = \log_5(x+2)$ because the y-axis ($x = 0$) is a vertical asymptote of $y = \log_5 x$.

(b) Notice that $y = \log_5(3-x) = \log_5((-1)(x-3))$. Thus, a complete graph of $y = \log_5(3-x)$ can be obtained by reflecting the graph of $y = \log_5 x$ through the y-axis and then shifting the resulting graph 3 units to the right (Figure 4.5.4). You can check this result with a graphing utility. The domain of $y = \log_5(3-x)$ is $(-\infty, 3)$ and the range is $(-\infty, \infty)$. The line $x = 3$ is a vertical asymptote of $y = \log_5(3-x)$. (Why?) •

In the next example, we determine the effect of the coefficient a in $y = \log_b(ax+c)$.

• EXAMPLE 5: Determine the domain, the range, the asymptotes, and draw a complete graph of $y = \log_2(3x+5)$.

SOLUTION: We rewrite y as follows:

$$\begin{aligned} y &= \log_2(3x+5) \\ &= \log_2\left((3)\left(x+\frac{5}{3}\right)\right) \\ &= \log_2(3) + \log_2\left(x+\frac{5}{3}\right). \end{aligned}$$

Now, we can see that a graph of $y = \log_2(3x+5)$ can be obtained by shifting the graph of $y = \log_2 x$ horizontally to the left $\frac{5}{3}$ units, and then shifting the resulting graph vertically $\log_2 3$ units which is approximately 1.58 units. The computer drawn graph of $y = \log_2(3x+5)$ given in Figure 4.5.5 confirms this observation. The domain of $y = \log_2(3x+5)$ is $(-\frac{5}{3}, \infty)$, and the range is $(-\infty, \infty)$. The line $x = -\frac{5}{3}$ is a vertical asymptote.

Alternatively, we can obtain a rough sketch of a complete graph of $y = \log_2(3x+5)$ in the following way. We can see that the domain of $y = \log_2(3x+5)$ is $(-\frac{5}{3}, \infty)$ because we must have $3x + 5 > 0$. Therefore, the line $x = -\frac{5}{3}$ must be a vertical asymptote. The graph of $y = \log_2(3x+5)$ must be like $y = \log_2 x$. Thus, we can find two points of the graph of $y = \log_2(3x+5)$ and use them to obtain a rough sketch of the graph. The x- and y-intercepts are a natural pair of points to find. If $\log_2(3x+5) = 0$, then $3x + 5 = 2^0$ so that $x = -\frac{4}{3}$. The x-intercept is the point $(-\frac{4}{3}, 0)$. The y-intercept is $\log_2 5$, which is approximately 2.3. Now, we can use the points $(-\frac{4}{3}, 0)$ and $(0, \log_2 5)$ together with the fact that the graph of $y = \log_2(3x+5)$ is like the graph of $y = \log_2 x$ to draw a rough sketch of the graph of $y = \log_2(3x+5)$. •

• **EXAMPLE 6:** Determine the domain, the range, and sketch a complete graph of $y = -3\log_4(2x-5) + 1$.

SOLUTION: To obtain a computer drawn graph we enter y in the form $y = -3\frac{\log(2x-5)}{\log 4} + 1$. A complete graph is given in Figure 4.5.6. We can see from this graph that the domain is $(\frac{5}{2}, \infty)$ and the range is $(-\infty, \infty)$. The domain is $(\frac{5}{2}, \infty)$ because we must have $2x - 5 > 0$. Thus, $x = \frac{5}{2}$ is a vertical asymptote.

To obtain a rough sketch of $y = -3\log_4(2x-5) + 1$ we use any convenient pair of points of the graph. If $x = 3$, then $y = -3\log_4(1) + 1 = 1$ because $\log_4 1 = 0$. If $x = \frac{9}{2}$, then $y = -3\log_4(9-5) + 1 = -2$ because $\log_4 4 = 1$. Because we know the general shape of $y = -3\log_4(2x-5) + 1$, we can use the points $(3, 1)$ and $(\frac{9}{2}, -2)$ to draw a rough sketch (Figure 4.5.6). •

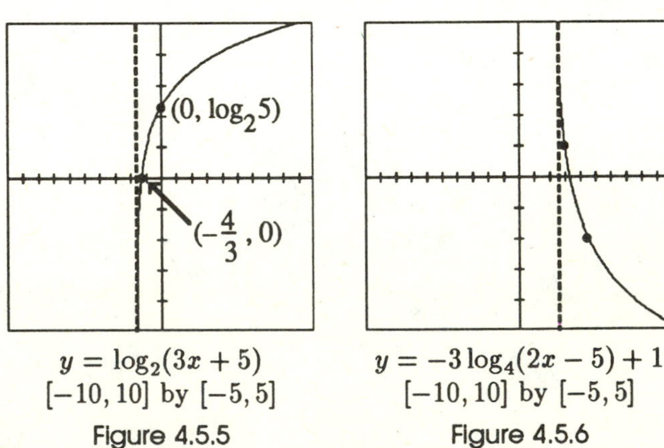

$y = \log_2(3x+5)$
$[-10, 10]$ by $[-5, 5]$
Figure 4.5.5

$y = -3\log_4(2x-5) + 1$
$[-10, 10]$ by $[-5, 5]$
Figure 4.5.6

In the next example we look at graphs of more complicated functions involving logarithmic functions.

• EXAMPLE 7: Determine the domain, the range, the asymptotes, and draw a complete graph of each function.

(a) $y = \log|x+2|$ (b) $y = \log_3 \sqrt{x-1}$ (c) $y = \log \dfrac{2x+1}{x-3}$

SOLUTION:

(a) The graph of $y = \log|x+2|$ in the $[-10, 10]$ by $[-5, 5]$ viewing rectangle is given in Figure 4.5.7. Notice that the graph is symmetric with respect to the vertical line $x = -2$ and has this line for a vertical asymptote. Because $|x+2| \geq 0$ for all x, the domain is the set of all real numbers different from -2, and the range is $(-\infty, \infty)$. The right branch of $y = \log|x+2|$ is the graph of $y = \log x$ shifted horizontally left 2 units. The left branch is the reflection of the right branch through the line $x = -2$. We can write y as a piecewise-defined function:

$$y = \begin{cases} \log -(x+2) & \text{for } x < -2 \\ \log(x+2) & \text{for } x \geq -2 \end{cases}$$

(b) The graph of $y = \log_3 \sqrt{x-1}$ in Figure 4.5.8 is a complete graph. This graph looks very much like the graph of $y = \log_3(x-1)$. We rewrite y to see why this is true:

$$\begin{aligned} y &= \log_3 \sqrt{x-1} \\ &= \log_3(x-1)^{1/2} \\ &= \frac{1}{2}\log_3(x-1). \end{aligned}$$

The domain is $(1, \infty)$, and the range is $(-\infty, \infty)$. The line $x = 1$ is a vertical asymptote.

(c) The domain of this function consists of those values of x for which $\frac{2x+1}{x-3} > 0$. The solution to this inequality is $(-\infty, -\frac{1}{2}) \cup (3, \infty)$. (Why?) Thus, the domain

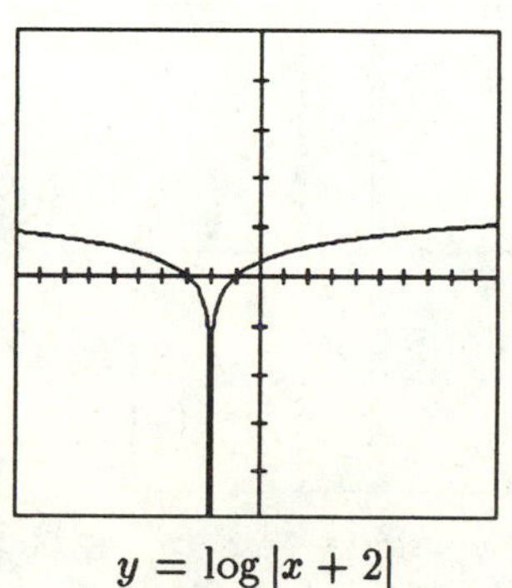

$y = \log|x+2|$
$[-10, 10]$ by $[-5, 5]$
Figure 4.5.7

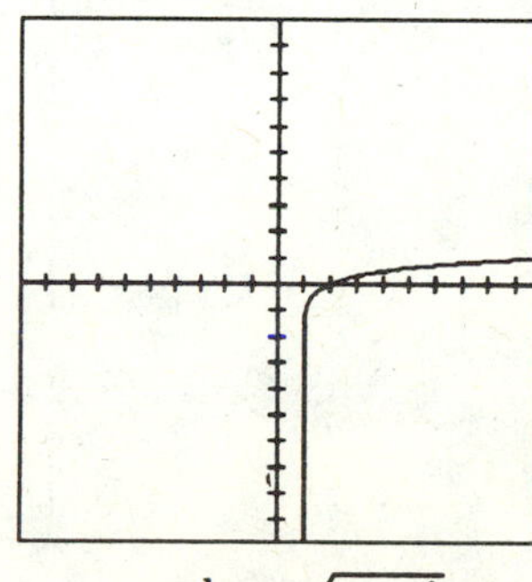

$y = \log_3 \sqrt{x-1}$
$[-10, 10]$ by $[-5, 5]$
Figure 4.5.8

is $(-\infty, -\frac{1}{2}) \cup (3, \infty)$. A complete graph is given in Figure 4.5.9. The range is $(-\infty, \infty)$. The lines $x = -\frac{1}{2}$ and $x = 3$ are vertical asymptotes. This graph appears to have a horizontal asymptote. If we write $\frac{2x+1}{x-3} = 2 + \frac{7}{x-3}$ we can see why there is a horizontal asymptote. Notice that $2 + \frac{7}{x-3} \to 2$ as $x \to \infty$ or as $x \to -\infty$. Therefore,

$$\log \frac{2x+1}{x-3} = \log\left(2 + \frac{7}{x-3}\right) \to \log 2$$

as $x \to \infty$ or as $x \to -\infty$. The line $y = \log 2 = 0.30$ is a horizontal asymptote. •

We close this section with an example that examines the effect of the monthly payment on the term (length) of a loan. Suppose that an \$86,000 loan at APR 12% compounded monthly is required to purchase a certain home. The interest payment required the first month is (0.01)(86,000) = \$860. Thus, monthly payments must exceed \$860 in order to pay off the loan. We assume the interest rate and payment are fixed.

• EXAMPLE 8: Consider an \$86,000 loan at APR 12% compounded monthly.

(a) Draw a graph that shows how the term of the loan depends on the amount of the monthly payment.
(b) What monthly payments are required to pay off the loan in 25 years?
(c) If the amount in (b) is increased by \$50, how long will it take to pay off the loan?

SOLUTION:

(a) This loan is an example of an annuity with present value \$86,000 and monthly interest rate 0.01. Let x (in dollars) be the amount of the monthly payment and t (in months) be the number of months required to pay off the loan. The present value formula of Section 4.3 gives the following algebraic representation:

$$86{,}000 = x\frac{1 - 1.01^{-t}}{0.01}.$$

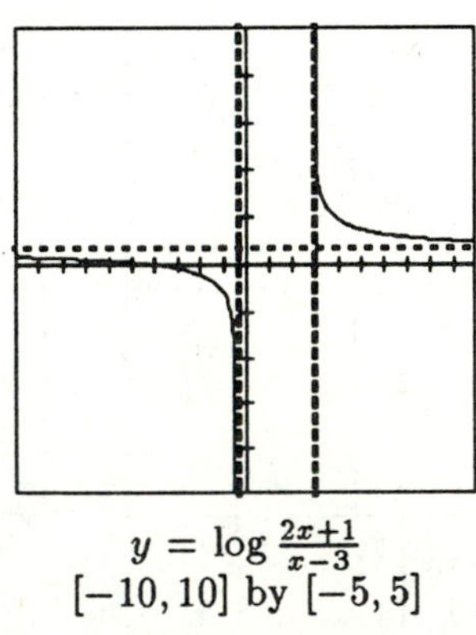

$y = \log \frac{2x+1}{x-3}$
$[-10, 10]$ by $[-5, 5]$
Figure 4.5.9

Next, we solve the above equation for t:

$$\begin{aligned}
86{,}000 &= x\frac{1-1.01^{-t}}{0.01}\\
860 &= x(1-1.01^{-t})\\
\frac{860}{x} &= 1-1.01^{-t}\\
1.01^{-t} &= 1-\frac{860}{x}\\
1.01^{-t} &= \frac{x-860}{x}\\
1.01^{t} &= \frac{x}{x-860}\\
\ln(1.01^{t}) &= \ln\left(\frac{x}{x-860}\right)\\
t &= \frac{1}{\ln 1.01}\ln\left(\frac{x}{x-860}\right).
\end{aligned}$$

Figure 4.5.10 gives the graph of t as a function of x in the $[0, 2000]$ by $[0, 360]$ viewing rectangle. Notice that t is a decreasing function in the interval $(0, 2000)$. This is reasonable because as the monthly payment x increases we would expect the term t (length) of the loan to decrease. However, notice how sharply the graph decreases. This means that small increases in the monthly payments dramatically reduce the term of the loan.

(b) Twenty five years is the same as 300 months. We can use zoom-in to determine that the x-coordinate of the point of intersection of the graph in Figure 4.5.10 with the horizontal line $t = 300$ is 905.77. This means that monthly payments of \$905.77 are required to pay off the loan in 25 years.

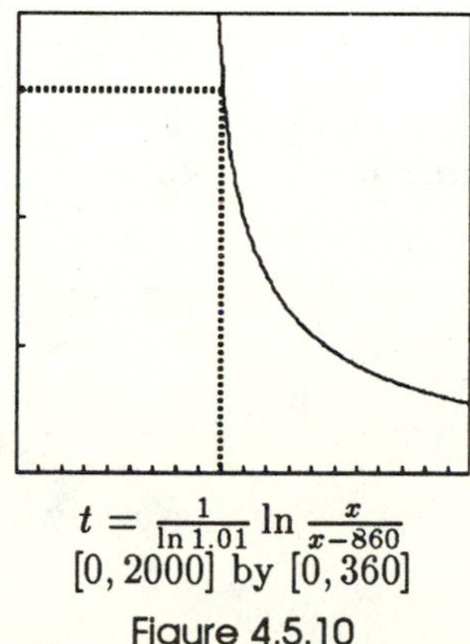

$t = \frac{1}{\ln 1.01}\ln\frac{x}{x-860}$
$[0, 2000]$ by $[0, 360]$

Figure 4.5.10

(c) We can use the equation

$$t = \frac{1}{\ln 1.01} \ln\left(\frac{x}{x-860}\right)$$

to determine that the number of months required to pay off the loan with monthly payments of \$955.77 is 231.21. This is equivalent to 19.27 years. Thus, the term of the loan is decreased by nearly $5\frac{3}{4}$ years (about 23%) if monthly payments are increased by \$50 (about 6%). •

• EXERCISES 4-5

1. Use a graphing utility to draw complete graphs of $y = \log_2 x$, $y = \ln x$, $y = \log_3 x$, $y = \log_5 x$, and $y = \log x$ in the same viewing rectangle.

Solve each inequality graphically.

2. $\log_2 x < \log_3 x$
3. $\log_2 x > \log_3 x$
4. $\ln x < \log_3 x$
5. $\ln x > \log_3 x$

Without using a graphing utility, sketch a complete graph of each function. Check your answer with a graphing utility.

6. $y = 2\ln(x+3)$
7. $y = -1 - \ln(x-1)$
8. $y = 1 + \log_3(x-2)$
9. $y = 2 - \log_2(2x+6)$
10. Determine the domain and range of the function in Exercise 7.
11. Determine the domain and range of the function in Exercise 8.

Draw a complete graph and determine the domain of each function.

12. $y = \log_5 \sqrt{x-3}$
13. $y = \log_2(x+3)^2$
14. $y = \ln(5-2x)$
15. $y = \dfrac{\ln\sqrt{x}}{x-1}$
16. $y = \dfrac{1}{2}\ln\left(\dfrac{1+x}{1-x}\right)$
17. $y = \ln(x + \sqrt{x^2+1})$

Determine the end behavior of each function. That is, determine what $f(x)$ approaches as $|x| \to \infty$.

18. $y = x \ln x$
19. $y = x^2 \ln x$
20. $y = x^2 \ln|x|$
21. $y = \dfrac{\ln x}{x}$

Draw a complete graph of each function. Determine where the function is increasing and decreasing. Determine all local maximum and minimum values.

22. $f(x) = x \ln x$
23. $g(x) = x^2 \ln x$
24. $f(x) = x^2 \ln|x|$
25. $g(x) = \dfrac{\ln x}{x}$

Sketch the graph of each inequality. Do not use a graphing utility.

26. $y < \ln(3-x)$
27. $y \geq 2 - \ln(x+2)$

Solve each system of equations using a graphing utility.

28. $x + y = 6$
    $y = \ln(x-2)$
29. $y = x^2 - 2$
    $y - 1 = \ln|x+5|$

30. Show that $y = \frac{1}{2}(e^x - e^{-x})$ and $y = \ln(x + \sqrt{x^2+1})$ are inverses of each other. Give a graphical argument.

For each pair of functions f and g, find the solutions to $f(x) > g(x)$, $f(x) < g(x)$, and $f(x) = g(x)$ graphically.

31. $f(x) = 3^x$; $g(x) = x^3$

32. $f(x) = e^x$; $g(x) = x^e$

33. $f(x) = 10^x$; $g(x) = x^x$

34. $f(x) = e^x$; $g(x) = (\ln x)^x$

35. $f(x) = \ln x$; $g(x) = \ln(\ln x)$

36. $f(x) = \ln x$; $g(x) = x^{1/3}$

37. How would you graph $y = \log_x 4$ using a graphing utility?

Use the results of Exercise 37 to graph the equation.

38. $y = \log_x 4$

39. $y = \log_x(x-1)$

40. Determine the domain and range of the function in Exercise 38.

41. Determine the domain and range of the function in Exercise 39.

Earthquake Intensity

Earthquake intensity is reported using the Richter scale,

$$R = \log\left(\frac{a}{T}\right) + B,$$

where R is the magnitude (of intensity), a is the amplitude in microns of the vertical ground motion at the receiving station, T is the period of the seismic wave (in seconds), and B is a factor that accounts for the weakening of the seismic wave with increasing distance from the epicenter of the earthquake.

42. What is the magnitude on the Richter scale of an earthquake if $a = 250$, $T = 2$, and $B = 4.250$?

43. Explain why an earthquake of magnitude 6 on the Richter scale is 100 times more intense than one with the same epicenter of magnitude 4 on the Richter scale. Assume T and B are constant. The change in the vertical ground motion (amplitude) is directly related to the *intensity* of the earthquake.

44. Draw a complete graph of the Richter scale model (magnitude R as a function of amplitude a). Assume $T = 2$ and $B = 4.250$. What are the values of a that make sense in this problem situation?

Light Absorption

The *Beer-Lambert law of light absorption* is given by

$$\log\left(\frac{I}{I_0}\right) = Kx,$$

where I_0 and I denote the intensity of light of a particular type before and after passing through a body of material, respectively, and x denotes the length of the path followed by the beam of light passing through the material. K is a constant.

45. Let $I_0 = 12$ lumens and assume $K = -0.00235$. Write I as a function of x.

46. Draw a complete graph of $I = f(x)$.

Suppose I_0 represents the intensity of light measured at the surface of a lake. The Beer-Lambert law can be used to determine the intensity I of the light measured at a depth of x feet from the surface.

47. Assume the constants given in Exercise 45 for Lake Erie. What is the intensity of the light at a depth of 30 feet?

48. Suppose for Lake Superior, $K = -0.0125$ and the surface intensity of the light is 12 lumens. What is the intensity at a depth of 30 feet?

RL circuit

It is known from the theory of electricity that in an RL circuit the current I in the circuit is given by

$$I = \frac{V}{R}(1 - e^{-(Rt)/L})$$

as a function of time t. Here V, R, and L are constants for voltage, resistance, and self-inductance, respectively.

49. Assume $V = 10$ volts, $R = 3$ ohms, and $L = 0.03$ Henries. Draw a complete graph of $I = f(t)$.

50. Show that $I \to V/R = 10/3$ as $t \to \infty$. Give a graphical argument.

Rule of 72

51. Graph $y = r$ and $y = \ln(1 + r)$ for $0 \leq r \leq 1$ in the same viewing rectangle. When is $\ln(1 + r)$ approximately equal to r?

52. Show how using 0.72 as an approximate value for $\ln 2$ you can derive an easy to remember "rule" that says money invested at annual rate r compounded annually will double in value after $72/100r$ years. For example, at 6% interest compounded annually, you would expect money to double in value after $72/[100(0.06)] = 72/6 = 12$ years. Is this a good "rule of thumb?"

Semilogarithmic Graphing

Exponential relations appear to be linear when plotted on *semilogarithmic* graph paper. Before calculators and computers were readily available, techniques for data analysis frequently involved semilogarithmic graph paper. The same analysis can be carried out using *ordinary* graph paper and a scientific or graphing calculator.

53. Plot the pairs $(x, \log y)$ on regular graph paper using the values x and of y from Table 1. Verify that the points are collinear (on the same line).

54. Compute the slope m and the y-intercept y_0 of the line in Exercise 53. Verify that the data can be represented by the *exponential* relationship $y = ab^x$ where $m = \log b$ and $\log a = y_0$.

55. Prove that if f is the exponential function $f(x) = ab^x$ then $\log f(x)$ varies *linearly* with x. That is, show that $\log f(x) = mx + y_0$ for some constants m and y_0.

56. Determine the actual exponential function which provides a model of the data in Exercise 53. *Hint:* Use Exercise 55.

x	y
2.3	43.8
4	283.5
5.5	1473.1
7	7654.5
9	68,890.5

Table 1

57. Let $f(x) = ab^x$. Show that

$$\log b = \frac{\log f(x_2) - \log f(x_1)}{x_2 - x_1}$$

for any pair of real numbers x_1, x_2 with $x_1 \neq x_2$. Assume $a > 0$ and $b > 1$.

58. Consider a \$75,000 mortgage loan at 10.50% APR compounded monthly.
    (a) Draw a graph that shows how the term of the loan depends on the amount of the monthly payment.
    (b) What monthly payments are required to pay off the loan in 25 years?
    (c) If the amount in (b) is increased by \$50, how long will it take to pay off the loan?

59. Consider a mortgage loan of \$110,000 at 9% APR compounded monthly.
    (a) Draw a graph that shows how the term of the loan depends on the amount of the monthly payment.
    (b) What monthly payments are required to pay off the loan in 25 years?
    (c) If the amount in (b) is increased by \$50, how long will it take to pay off the loan?

4.6 Equations, Inequalities, and Extreme-Value Problems

In this section we solve equations and inequalities that involve exponential or logarithmic functions both algebraically and graphically. We find local maximum and minimum values of functions that involve exponential and logarithmic functions. You will find that solving these equations and inequalities algebraically involves moving from the exponential form to the logarithmic form and vice versa, and applying the properties of logarithms.

• EXAMPLE 1: Solve $\log x = -2.5$.

SOLUTION: One way to solve this equation is to write it in the following equivalent exponential form:

$$x = 10^{-2.5}.$$

Now, the exponentiation key on a calculator can be used to show that x is approximately 0.0031623. •

When we solve equations and inequalities involving logarithms algebraically we often apply the properties of logarithms to algebraic expressions. However, we must be careful because using these properties need not lead to equivalent equations. For example, suppose we replace $\log \frac{x}{x-2}$ by $\log x - \log(x-2)$ or vice versa. The graphs of $y = \log \frac{x}{x-2}$ and $y = \log x - \log(x-2)$ in Figure 4.6.1 show that $\log \frac{x}{x-2} = -1$ has a solution but $\log x - \log(x-2) = -1$ does not. In fact, the domain $(2, \infty)$ of $y = \log x - \log(x-2)$ is a subset of the domain $(-\infty, 0) \cup (2, \infty)$ of $y = \log \frac{x}{x-2}$, but *not* vice versa. Thus, the domains are different.

We may introduce extraneous solutions, but we do not lose solutions if we replace $\log_a f(x) - \log_a g(x)$ by $\log_a(f(x)/g(x))$ because the domain of $y = \log_a f(x) - \log_a g(x)$ is contained in the domain of $y = \log_a(f(x)/g(x))$. Replacement the other way may lose solutions because the domain of $y = \log_a(f(x)/g(x))$ generally contains numbers *not* in the domain of $y = \log_a f(x) - \log_a g(x)$. Similarly, we replace $\log_a f(x) + \log_a g(x)$ by $\log_a f(x)g(x)$ and not vice versa when solving algebraically for precisely the same reason. If $r > 0$, then both $\log_a r^s$ and $s \log_a r$ are defined and equal for all real numbers s. With these relationships in mind, we are ready to solve algebraically. To be safe, always check to see if a solution is extraneous.

• **EXAMPLE 2:** Solve $\frac{1}{2}\log_5(x+6) - \log_5 x = 0$.

SOLUTION: Our strategy is to rewrite the equation in the form $\log_5 f(x) = c$ and use the exponential form $f(x) = 5^c$ to solve for x. Remember, we replace $\log_a g(x) - \log_a h(x)$ by $\log_a(g(x)/h(x))$ and $\log_a g(x) + \log_a h(x)$ by $\log_a g(x)h(x)$, but not the other way, to avoid

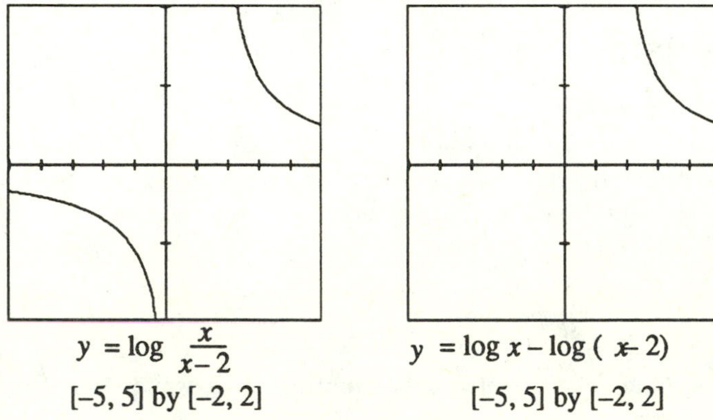

Figure 4.6.1

losing solutions. We proceed as follows:

$$\begin{aligned}
\frac{1}{2}\log_5(x+6) - \log_5 x &= 0 \\
\log_5(x+6)^{1/2} - \log_5 x &= 0 \\
\log_5 \sqrt{x+6} - \log_5 x &= 0 \\
\log_5 \frac{\sqrt{x+6}}{x} &= 0 \\
\frac{\sqrt{x+6}}{x} &= 5^0 \\
\frac{\sqrt{x+6}}{x} &= 1.
\end{aligned}$$

Next, we solve $\sqrt{x+6}/x = 1$ in the usual way:

$$\begin{aligned}
\frac{\sqrt{x+6}}{x} &= 1 \\
\sqrt{x+6} &= x \\
x+6 &= x^2 \\
0 &= x^2 - x - 6 \\
0 &= (x-3)(x+2).
\end{aligned}$$

We must check that -2 and 3 are solutions to the original equation. (Why?) A calculator can be used to show that $\frac{1}{2}\log_5(3+6) - \log_5 3 = 0$. (Remember that $\log_5 a = \log a/\log 5$.) Thus, 3 is a solution to the original equation. Notice -2 is an extraneous solution to the original equation because $\log_5 x$ is not defined at $x = -2$. Therefore, only 3 is a solution to the original equation. The graph of $y = \frac{1}{2}\log_5(x+6) - \log_5 x$ in Figure 4.6.2 suggests that the domain of this function is $(0, \infty)$ and that 3 is a solution to the original equation.

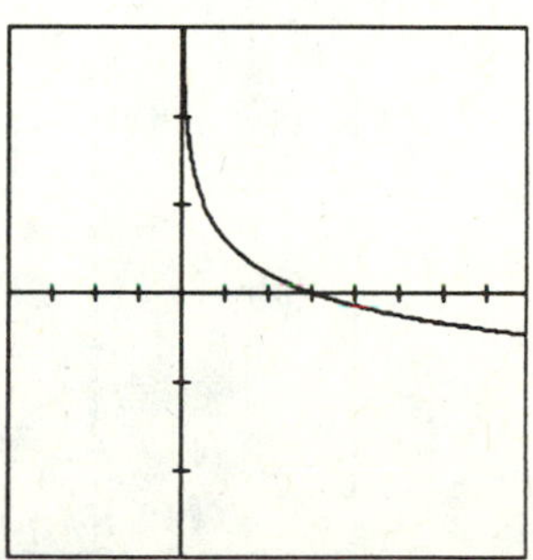

$y = \frac{1}{2}\log_5(x+6) - \log_5 x$
$[-4, 8]$ by $[-3, 3]$

Figure 4.6.2

Because this graph is complete there can only be one solution to the original equation. (Why?) •

• EXAMPLE 3: Solve $\log_2(2x+1) - \log_2(x-3) = 2\log_2 3$.

SOLUTION: The following equations need not be equivalent. However, all solutions to the original equation are also solutions to each new equation. (Why?)

$$\begin{aligned} \log_2(2x+1) - \log_2(x-3) &= 2\log_2 3 \\ \log_2 \frac{2x+1}{x-3} &= \log_2 9 \\ \frac{2x+1}{x-3} &= 9 \\ 9x - 27 &= 2x + 1 \\ 7x &= 28 \\ x &= 4. \end{aligned}$$

To check that 4 is a solution to the original equation we need to show that $\log_2 9 - \log_2 1 = 2\log_2 3$. Now, $\log_2 9 = 2\log_2 3$ and $\log_2 1 = 0$, so this equation is true, and 4 is the solution to the original equation.

The graph of $y = \log_2(2x+1) - \log_2(x-3) - 2\log_2 3$ in Figure 4.6.3 is complete and suggests that the domain of this function is $(3, \infty)$ and that 4 is the only solution to the original equation. (Why?) •

The next two examples involve equations containing exponential expressions.

• EXAMPLE 4: Solve $1.08^x = 3$.

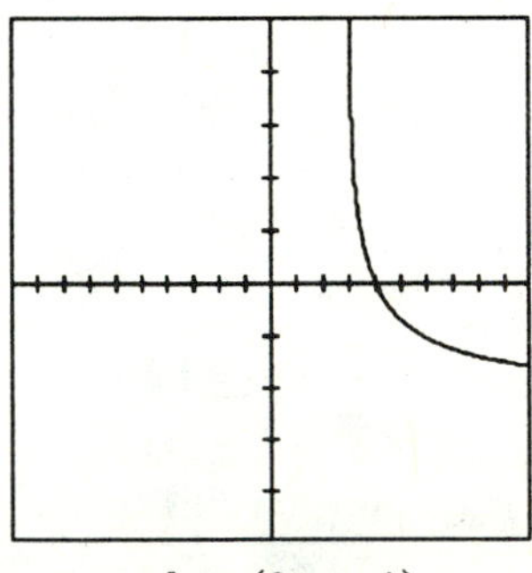

$y = \log_2(2x+1) - \log_2(x-3) - 2\log_2 3$
$[-10, 10]$ by $[-5, 5]$
Figure 4.6.3

SOLUTION: We take the logarithm base 10 of each side of the equation:

$$\begin{aligned} 1.08^x &= 3 \\ \log 1.08^x &= \log 3 \\ x \log 1.08 &= \log 3 \\ x &= \frac{\log 3}{\log 1.08}. \end{aligned}$$

We can check that this value of x gives a solution to the original equation. This solution is approximately 14.274915. •

• EXAMPLE 5: Solve $\frac{3^x - 3^{-x}}{2} = 5$.

SOLUTION: It can be shown that the graphs of $y = \frac{3^x - 3^{-x}}{2}$ and $y = 5$ in Figure 4.6.4 are complete graphs. This figure suggests that there is only one real solution to this equation, and its value is approximately 2. We could use zoom-in to determine this solution more accurately.

Alternatively, we can find this solution algebraically. The following equations are equivalent:

$$\begin{aligned} \frac{3^x - 3^{-x}}{2} &= 5 \\ 3^x - 3^{-x} &= 10 \\ (3^x)^2 - 1 &= (10)3^x \quad \text{(Multiply each side by } 3^x\text{.)} \\ (3^x)^2 - (10)3^x - 1 &= 0. \end{aligned}$$

This equation is quadratic in 3^x. The quadratic formula gives

$$3^x = \frac{10 \pm \sqrt{104}}{2} = 5 \pm \sqrt{26}.$$

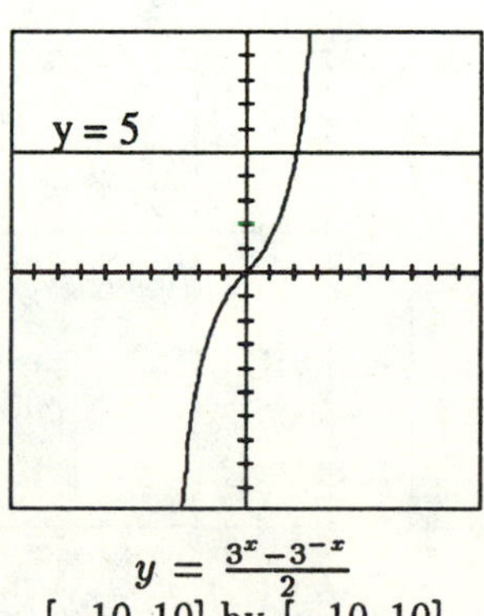

$y = \frac{3^x - 3^{-x}}{2}$
[−10, 10] by [−10, 10]
Figure 4.6.4

The equation $3^x = 5 - \sqrt{26}$ has no real solution because $5 - \sqrt{26}$ is negative. We can use logarithms to solve the equation $3^x = 5 + \sqrt{26}$ as we did in Example 4. You will find that x is approximately 2.1048721. •

The algebraic technique needed to solve an inequality involving logarithms is similar to the technique used to solve equations. The resulting inequalities are not necessarily equivalent, so solutions must be checked in the original inequality.

• EXAMPLE 6: Solve $3\log_4 x - 1 < 0$.

SOLUTION: Notice that we must have $x > 0$. (Why?) We can rewrite the inequality as $\log_4 x < \frac{1}{3}$. Figure 4.6.5 gives the graph of $y = \log_4 x$ and $y = \frac{1}{3}$. We can use zoom-in to show that the solution is $0 < x < 1.59$. •

In the next two examples we find local maximum and minimum values of functions involving exponential and logarithmic functions.

• EXAMPLE 7: Find the domain, the range, the local maximum and minimum values, and draw a complete graph of $y = \ln x(3 - x)$.

SOLUTION: The domain of this function consists of those values of x for which $x(3-x) > 0$. It can be shown that the solution to $x(3 - x) > 0$ is $(0, 3)$. The graph of $y = \ln x(3 - x)$ in Figure 4.6.6 is a complete graph.

We can use zoom-in to determine that the local maximum value of this function is 0.81 with error at most 0.01. There are *no* local minimum values, and the range is $(-\infty, 0.81]$. •

• EXAMPLE 8: Find the domain, the range, the local maximum and minimum values, and draw a complete graph of $y = 10(e^{(x-100)/50} + e^{(100-x)/50})$.

SOLUTION: The domain of this function is the set of all real numbers. The graph in Figure 4.6.7 is a complete graph.

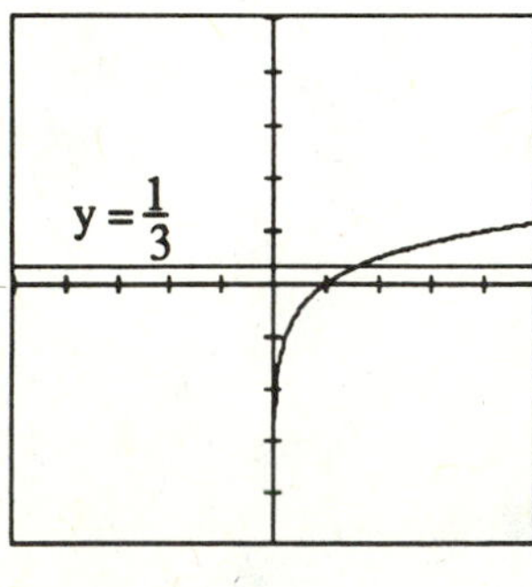

$y = \log_4 x$
$[-5, 5]$ by $[-5, 5]$
Figure 4.6.5

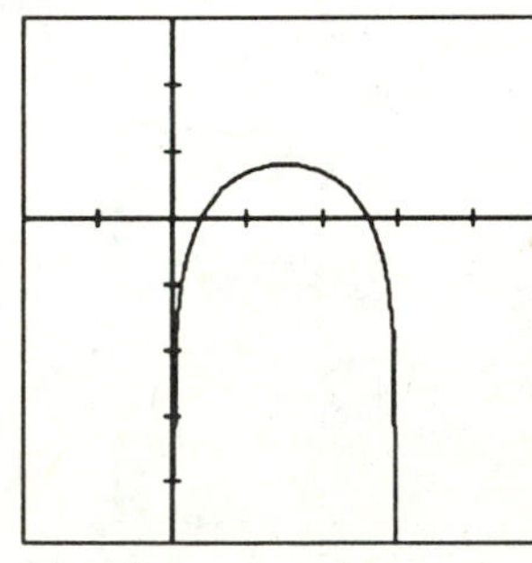

$y = \ln x(3 - x)$
$[-2, 5]$ by $[-5, 3]$
Figure 4.6.6

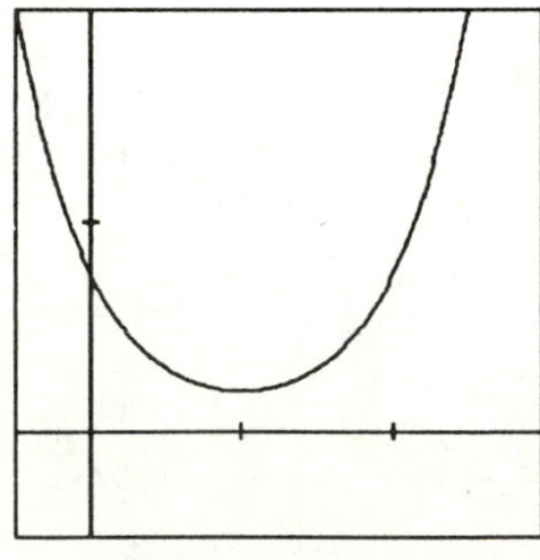

$y = 10(e^{(x-100)/50} + e^{(100-x)/50})$
$[-50, 300]$ by $[-50, 200]$
Figure 4.6.7

We can use zoom-in to determine that the local minimum value is 20 with error at most 0.01. There is *no* local maximum value and the range is $[20, \infty)$. It can be shown that the coordinates of the lowest point are exactly $(100, 20)$. •

In the next example we introduce another application that has an exponential function for a model. Scientists have established that standard atmospheric pressure of 14.7 pounds per square inch is reduced by half for each 3.6 miles of vertical ascent. This rule for atmospheric pressure holds for altitudes up to 50 miles.

• EXAMPLE 9: Express the atmospheric pressure P as a function of altitude and determine when the atmospheric pressure will be 4 pounds per square inch. What is the pressure at an altitude of 50 miles?

SOLUTION: The information above gives $P = 14.7(0.5)^{h/3.6}$ where h is the altitude in miles. To determine when atmospheric pressure is 4 pounds per square inch, we must solve the following equation:

$$14.7(0.5)^{h/3.6} = 4.$$

We can solve this equation as follows:

$$\begin{aligned} 14.7(0.5)^{h/3.6} &= 4 \\ (0.5)^{h/3.6} &= \frac{4}{14.7} \\ \frac{h}{3.6}\log(0.5) &= \log\left(\frac{4}{14.7}\right) \\ h &= \frac{3.6\log(4/14.7)}{\log(0.5)}. \end{aligned}$$

Using a calculator we find that h is approximately 6.7598793. Thus, the atmospheric pressure will be 4 pounds per square inch at an altitude of approximately 6.7598793 miles.

To find the pressure at an altitude of 50 miles we need to find $P(50)$. Using a calculator we find that $P(50)$ is approximately 0.000969. Thus, the atmospheric pressure at an altitude of 50 miles is about 0.000969 pounds per square inch. •

• EXERCISES 4-6

Solve each equation algebraically. Check your answer with a graphing utility.

1. $\log x = 4$
2. $\ln x = -1$
3. $\log_4(x-5) = -1$
4. $\log_x(1-x) = 1$
5. $\log_x(12-x) = 2$
6. $3\log_2 x = 4$
7. $\ln x + \ln 2 = 3$
8. $\frac{1}{2}\log_3(x+1) = 2$
9. $\log_2 x + \log_2(x+3) = 2$
10. $\log_4(x+1) - \log_4 x = 1$

Compare the domain and range of each pair of functions.

11. $y = 2\log x$; $y = \log x^2$
12. $y = \log x + \log(x+1)$; $y = \log(x(x+1))$

13. $y = \log_5 \sqrt{x+6} - \log_5 x;\ y = \log_5 \dfrac{\sqrt{x+6}}{x}$

14. $y = \log_2 x - \log_2(x+1);\ y = \log_2 \left(\dfrac{x}{x+1}\right)$

Find the *exact* solutions to each equation.

15. $(1.06)^x = 4.1$

16. $(0.98)^x = 1.6$

17. $\dfrac{2^x - 2^{-x}}{3} = 4$

18. $\dfrac{2^x + 2^{-x}}{2} = 3$

19–22. Use a calculator to find a decimal approximation to each solution in Exercises 15–18. Then use a graphing utility to check the solution.

Draw a complete graph of each function. Determine the domain and the range, and find where the function is increasing or decreasing. Find all local maximum and minimum values.

23. $y = xe^x$

24. $y = x^2e^{-x}$

25. $y = \dfrac{3^x - 3^{-x}}{3}$

26. $y = \dfrac{3^x + 3^{-x}}{3}$

27. $y = \ln(x^2 + 2x)$

28. $y = \ln\left(\dfrac{x}{2+x}\right)$

Solve each inequality.

29. $2\log_2 x - 4\log_2 3 > 0$

30. $2\log_3(x+1) - 2\log_3 6 < 0$

Solve each equation or inequality.

31. $e^x + x = 5$

32. $\ln|x| - e^{2x} \geq 3$

33. $e^x < 5 + \ln x$

34. $e^{2x} - 8x + 1 = 0$

Solve each equation.

35. $\dfrac{\ln x}{x} = \dfrac{1}{10}\ln 2$

36. $2^x = x^{10}$

37. Explain why Exercises 35 and 36 have the same *positive* solutions.

38. Show that $f(x) = (e^x + e^{-x})/2$ is an even function and that $g(x) = (e^x - e^{-x})/2$ is an odd function.

Probability Density Function

The function $f(x) = (25/(r\sqrt{2\pi}))e^{-(x-u)^2/(2r^2)}$ is an adjusted normal probability density function with mean u and variance r^2. Let $u = 11.6$ and $r^2 = 2.35$.

39. Draw a complete graph of f.

40. Determine the domain of f.

41. Show that the graph of f is symmetric with respect to the line $x = u$.

42. Determine all x where $f(x) = 2$.

Approximations for e^x

Let

$$f_1(x) = x + 1$$

$$f_2(x) = \frac{x^2}{2!} + x + 1$$

$$f_3(x) = \frac{x^3}{3!} + \frac{x^2}{2!} + x + 1$$

$$f_4(x) = \frac{x^4}{4!} + \frac{x^3}{3!} + \frac{x^2}{2!} + x + 1$$

$$\vdots$$

$$f_n(x) = \frac{x^n}{n!} + \frac{x^{n-1}}{(n-1)!} + \cdots + x + 1$$

$$\vdots$$

Note: $n! = n(n-1)(n-2)\cdots 3 \cdot 2 \cdot 1$.

43. Show that $f_n(x) = \dfrac{x^n}{n!} + f_{n-1}(x)$.
44. Draw the graphs of $y = f_1(x)$, $y = f_2(x)$, $y = f_3(x)$, $y = f_4(x)$, $y = f_5(x)$ and $y = e^x$ in the $[-5, 5]$ by $[-25, 25]$ viewing rectangle.
45. For what values of x does $f_3(x)$ approximate e^x with an error less than 0.1?
46. For what values of x does $f_5(x)$ approximate e^x with an error less than 0.1?

Potential Energy

In a certain molecular structure, the total potential energy E between two ions is given by

$$E = \frac{-5.6}{r} + 10e^{-r/3},$$

where r is the distance separating the nuclei.

47. Draw a complete graph of E as a function of r. Determine the domain of E.
48. What values of r make sense in the problem situation?
49. Find any relative maximum and minimum values of the total potential energy on $(0, \infty)$ and the number r for which they occur.

The Sinking Fund Depreciation Method

The following notation and definitions will be used in connection with depreciation of an asset:

C = original value (cost) of the asset.

n = length of term (useful life under consideration).

S = value of the asset at the end of term; S is called the ***salvage value.***

$D = C - S$ is the ***total depreciation.***

B = ***book value***, the value for accounting and/or tax purposes at any time t.

A continuous model for ***book value*** at any time t using the ***sinking fund*** method is

$$B = C - \left(\frac{C - S}{((1+i)^n - 1)/i}\right)\left(\frac{(1+i)^t - 1}{i}\right)$$

where $0 \le t \le n$ and i is an annual interest rate. In this case, the annual depreciation charge is assumed to earn interest at rate i per year, and the annual depreciation charges accumulate to $D = C - S$, the total depreciation.

50. A machine costs $17,000, has a useful life of 6 years, and has salvage value of $1200. Draw a complete graph of the "book value" model $y = B(t)$. What values of t make sense in the problem situation? Assume $i = 5\%$.
51. Consider the machine in Exercise 50. What is the book value of the machine in 4 years, 3 months?
52. Consider the machine in Exercise 50. Draw three graphs of the book value of the machine using
    (a) the straight line method (see Exercise 45, Section 1.5).
    (b) the constant percentage method (see Exercise 56, Section 4.3).
    (c) the sinking fund method.
    For each graph use the same viewing rectangle for $0 \le t \le 6$.
53. Use the graphs in Exercise 52 to explain which method allows the greatest depreciation in the first 2 years of the useful life of the machine.

Mortgage Interest Rates

There is no known method to determine an exact, explicit solution for the following two problems. Use a graphing zoom-in method.

54. Amy and Steve can afford to make monthly mortgage payments of $800. They want to purchase a home requiring a 30-year, $92,000 mortgage. What APR interest rate will be required?
55. Jill and Benny can afford to make monthly mortgage payments of $1200. They want to purchase a home requiring a 30-year $130,000 mortgage. What APR interest rate will be required?

• KEY TERMS

Listed below are the key terms, vocabulary, and concepts in this chapter. You should be familiar with their definitions and meanings as well as be able to illustrate each of them.

Annuity
Common logarithmic function
Compound interest
Compounded continuously
Data analysis
e
Exponential function
Exponential growth

Future value
Half-life
Horizontal line test
Interest
Inverse relation
Logarithmic function
Natural logarithmic function
One-to-one function
Present value
Principal
Radioactive decay
Rate of interest
Well-defined property of functions

• REVIEW EXERCISES

This section contains representative problems from each of the previous six sections. You should use these problems to test your understanding of the material covered in this chapter.

1. Draw a complete graph of the inverse of the relation given by the graph of Figure R.1.

Draw a complete graph of each relation and its inverse. Do not use a graphing utility. Determine an equation for the inverse. Specify which inverses are functions.

2. $y = 5 - 3x$
3. $y = x^2 - 2$

Determine if each function is one-to-one. Determine an equation for the inverse of each function. Is the inverse a function? For those functions whose inverses are *functions* find an explicit rule $y = f^{-1}(x)$ for the inverse. Draw a graph of the function and its inverse relation without using a graphing utility. Check your answer with a graphing utility.

4. $f(x) = (x + 2)^2$
5. $h(x) = 2\sqrt{x - 4}$
6. Determine the domain and range of both the function and its inverse in Exercise 4.

Use a graphing utility to determine if each function is one-to-one. Sketch a complete graph of the inverse of each function without the aid of a graphing utility. Identify the functions whose inverses are also functions.

7. $f(x) = x^3 - 8x$
8. $f(x) = \dfrac{x^2 - 2x + 3}{x + 2}$

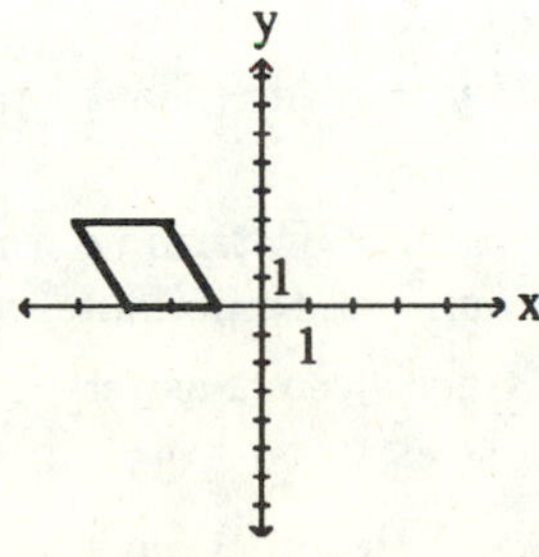

Figure R.1

9. Determine the domain and range of both the function and its inverse in Exercise 8.

Without using a graphing utility, draw a complete graph of each function. Determine the domain and range. Check your answer with a graphing utility.

10. $f(x) = 2 + 3^x$

11. $f(x) = 3 - 2^{x+2}$

Without using a graphing utility, sketch a complete graph of each function. Check your answer with a graphing utility.

12. $y = 2 + \ln(x - 3)$

13. $y = \ln(x - 2) - 1$

14. $y = 3 \ln x$

15. $y = 2 \ln(x + 4)$

16. Use rules of exponents to solve $2^{x+1} = 4^3$.

17. Use a graphing utility to draw a complete graph of $f(x) = (8x)2^{-x}$. Determine the domain and range.

Determine where each function is increasing and decreasing. Determine all local maximum and minimum values.

18. $f(x) = e^{-2x}$

19. $g(x) = (8x)2^{-x}$

Compute without using a calculator.

20. $\log_5 125$

21. $\log_3(-9)$

Solve each equation. Do not use a calculator.

22. $\log_x \frac{1}{9} = -2$

23. $\log_4 x = 3$

24. Use properties of logarithms to write the expression $\log[(1 + r)^{12}/r]$ as a sum, difference, or product of simple logarithms.

25. Use properties of logarithms to solve $z = (23 + x)^M$ for M.

26. Use a calculator to investigate numerically the end behavior of the function $f(x) = x^2/\ln x$. That is, find what $f(x)$ approaches as $x \to \infty$.

27. Use a calculator to investigate the behavior of $g(x) = x/\ln(x - 2)$ as $x \to 2^+$.

28. Use a calculator to investigate the behavior of $f(x) = x/\ln x^2$ as $x \to 1^-$.

29. Solve $\ln x > \log_2 x$ graphically.

Draw a complete graph. Determine the domain and range.

30. $y = \dfrac{\ln \sqrt{x + 2}}{x - 2}$

31. $y = \log_3(x - 5)^2$

32. Determine the end behavior of $y = x^3 \ln x$. That is, determine what $f(x)$ approaches as $|x| \to \infty$.

33. Draw a complete graph of $y = x^3 \ln x$. Determine where the function is increasing and decreasing. Determine all local maximum and minimum values.

Solve each equation algebraically. Check your answer with a graphing utility.

34. $5 \log_3 x = 2$

35. $\log_5 x + \log_5(x - 4) = 1$

36. Compare the domain and range of the functions $y = \log_3 x - \log_3(x+2)$ and $y = \log_3[x/(x+2)]$.

37. Find the *exact* solution to the equation $(1.5)^x = 0.90$.

38. Use a calculator to find a decimal approximation to the solution in Exercise 37. Then use a graphing utility to check the solution.

Draw a complete graph of each function. Determine the domain and the range, and find where the function is increasing or decreasing. Find all local maximum and minimum values.

39. $y = \ln\left(\dfrac{x}{x-3}\right)$

40. $y = \dfrac{2^x + 2^{-x}}{3}$

Solve algebraically. Check with a graphing utility.

41. $3\log_5(x-1) - 2\log_5 4 > 0$

42. $\log_2(x-1) + \log_2(x+2) = 2$

43. $\log_3(x+5) + 2\log_3 x > 1$

Use a graphing utility to solve each equation or inequality.

44. $e^x + \ln x = 5$

45. $e^{2x} + 3e^x \leq 10$

46. Linda deposits \$500 in a bank that pays 7% annual interest. Assume that she makes no other deposits or withdrawals. How much has she accumulated after 5 years if the bank pays interest compounded

(a) annually? (b) quarterly? (c) monthly?

(d) daily? (e) continuously?

47. Determine when an investment of \$1500 accumulates to a value of \$2280 if the investment earns interest at the rate of 7% compounded monthly.

48. Determine how much time is required for an investment to *double* in value if interest is earned at the rate of 6.25% compounded quarterly.

49. Draw a graph that shows the value at any time t of a \$500 investment that earns 9% simple interest. In the same viewing rectangle, draw the graph that shows the value at any time t of a \$500 investment that earns 9% interest, compounded monthly.

50. The population of a town is 625,000 and is increasing at the rate of 4.05% each year.

(a) Write the population P as a function of time.

(b) Draw a complete graph of the function P in (a). What portion of the graph represents the problem situation?

(c) Determine when the population of the town is one million.

51. The half-life of a certain radioactive substance is 21 days, and there are 4.62 grams present initially.

(a) Write the amount A of substance remaining as a function of time.

(b) Draw a complete graph of the function in (a).

(c) When will there be less than one gram of the substance remaining?

Earthquake intensity is reported using the Richter scale,

$$R = \log\left(\frac{a}{T}\right) + B,$$

where R is the magnitude, a is the amplitude in microns of the vertical ground motion at the receiving station, T is the period of the seismic wave (in seconds), and B is a factor that accounts for the weakening of the seismic wave with increasing distance from the epicenter of the earthquake.

52. What is the magnitude on the Richter scale of an earthquake if $a = 275$, $T = 2.5$, and $B = 4.250$?

53. Draw a complete graph of the Richter scale model (magnitude R as a function of amplitude a). Assume $T = 2.5$ and $B = 4.250$. What are the values of a that make sense in this problem situation?

5

Trigonometric Functions

• Introduction

In this chapter we begin the study of trigonometry. More than 2000 years ago the Greeks started the development of this branch of mathematics to make precise measurements of the angles and sides of triangles. This theory is still extremely important today and is extensively used in numerous applications including space science.

Two units of measure for angles are used in this textbook. One is called *degrees* and the other *radians*. Radian measure is introduced in Section 5.5. We will see that an angle with measure $360°$ also has measure 2π radians. In this textbook, we almost always give degree measurements in decimal form.

We start with a brief discussion of angles. Then, right triangle trigonometry is studied in the second section of this chapter and applications that involve right triangles are investigated. The six basic trigonometric functions of acute angles are defined. Then, these definitions are extended to arbitrary angles and complete graphs of the basic trigonometric functions are obtained. Calculators are used to compute values of trigonometric functions, and graphing utilities are used to draw complete graphs of functions of the form $y = af(bx + c) + d$ where f is one of the six basic trigonometric functions.

There are important relationships among the trigonometric functions. These relationships are called identities, and their study begins in this chapter. We solve equations, find local extrema, zeros, asymptotes, and determine the increasing and decreasing behavior of trigonometric functions. Applications with models that involve trigonometric functions are studied. In particular, we begin the study of sinusoidal functions, which are important in applications.

5.1 Angle

In this section we define angle and specify the measure of an angle in degrees. We say what is meant by acute angle, right angle, central angle, and an angle in standard position. The terms complementary angles, supplementary angles, quadrantal angle, and coterminal angles are defined.

In geometry, an angle is defined as the union of two rays with a common endpoint. In trigonometry, an angle is defined in terms of a rotation.

• DEFINITION Let ℓ_1 be a ray with initial point O. Let ℓ_1 be rotated counterclockwise about 0 to a new position given by the ray ℓ_2 in Figure 5.1.1. This determines the **angle** with **initial side** ℓ_1, **terminal side** ℓ_2, and the rotation specified by Figure 5.1.1. The amount of rotation is the **measure** of the angle. The point O is called the **vertex** of the angle. •

In trigonometry the measure of an angle can be positive or negative. If the rotation of ℓ_1 to ℓ_2 is counterclockwise, then the measure of the angle is positive (Figure 5.1.2). If the rotation of ℓ_1 to ℓ_2 is clockwise, then the measure of the angle is negative. The measure of an angle determined by one complete counterclockwise rotation is defined to be 360 degrees. We use the notation 360° to stand for 360 degrees. Notice that the initial side and terminal side of a 360° angle coincide (Figure 5.1.3). The measure of an angle given by one-half of a complete counterclockwise rotation is 180° (Figure 5.1.3). (Why?) An angle that is formed by $\frac{1}{360}$ of a complete rotation counterclockwise has measure one **degree. A minute** is $\frac{1}{60}$ of a degree, and a **second** is $\frac{1}{60}$ of a minute. Because of the accuracy of calculators, angles are now commonly measured in degrees in decimal form rather than the historical use of degrees, minutes and seconds.

• DEFINITION An angle is said to be **acute** if its measure is between 0° and 90°. An angle with measure 90° is called a **right angle**. •

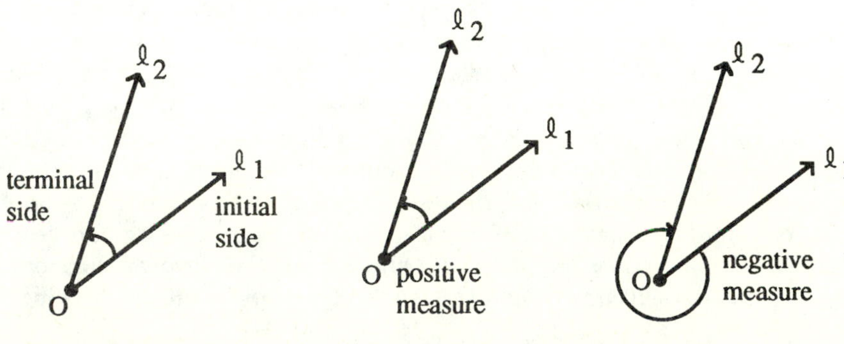

Figure 5.1.1 Figure 5.1.2

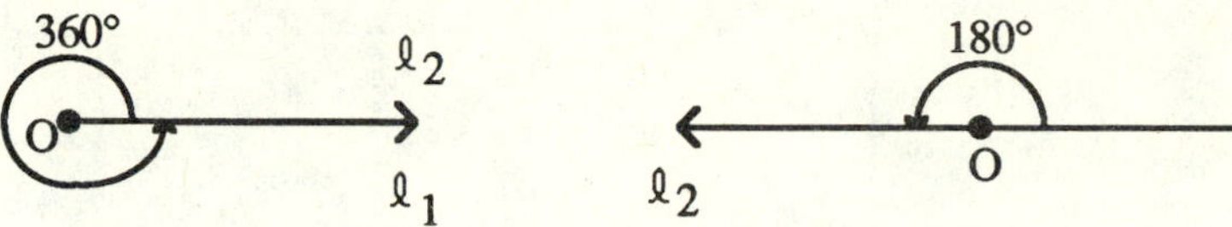

Figure 5.1.3

• DEFINITION An angle in the coordinate plane is said to be in **standard position** if the vertex is at the origin and the initial side is along the positive x-axis (Figure 5.1.4). •

• EXAMPLE 1: Determine the measure and draw the angle specified by each rotation in standard position.

(a) $\frac{3}{4}$ rotation, counterclockwise (b) $\frac{1}{2}$ rotation, clockwise

(c) 1.5 rotation, counterclockwise (d) $\frac{2}{3}$ rotation, clockwise

SOLUTION: The measure of the angles in (a)–(d) are 270°, −180°, 540°, and −240°, respectively. The angles are drawn in Figure 5.1.5. •

Notice that the terminal side of the angles in (a)–(c) of Example 1 lie along a coordinate axis. An angle in standard position with terminal side along a coordinate axis is called a **quadrantal angle**. An angle in standard position is said to be in a given quadrant if its terminal side lies in that quadrant.

• EXAMPLE 2: Determine the quadrant of the following angles in standard position with specified measure.

(a) 2268° (b) −1008° (c) 630°

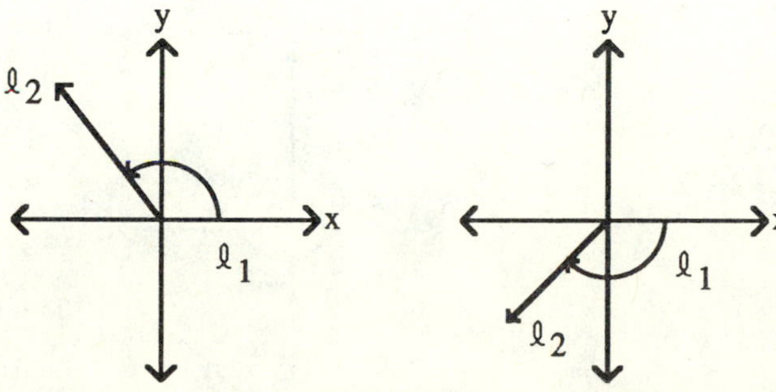

Figure 5.1.4

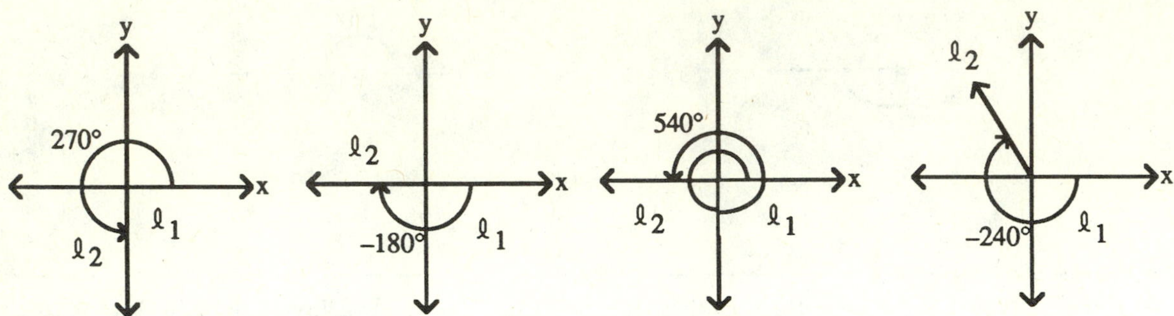

Figure 5.1.5

SOLUTION:

(a) One complete rotation is 360°. Now, $\frac{2268}{360} = 6.3$. Thus, 2268° is 6.3 complete counterclockwise rotations. The positive y-axis is 0.25 of a counterclockwise rotation, and the negative x-axis is 0.5 of a counterclockwise rotation. It follows that 0.3 of a rotation must be in the second quadrant. Therefore, 2268° is in the second quadrant.

(b) Now, $\frac{-1008}{360} = -2.8$. Thus, −1008° is 2.8 complete clockwise rotations. In a clockwise direction, 0.8 of a rotation is in the first quadrant. (Why?) Thus, −1008° is in the first quadrant.

(c) $\frac{630}{360} = 1.75$. Thus, the terminal side of 630° is along the negative y-axis. Therefore, 630° is a quadrantal angle and does *not* lie in any quadrant. •

Specifying the initial and terminal side of an angle does *not* determine a unique angle. For example, angles in standard position with measure 60° or 420° have the same terminal side (Figure 5.1.6). (Why?)

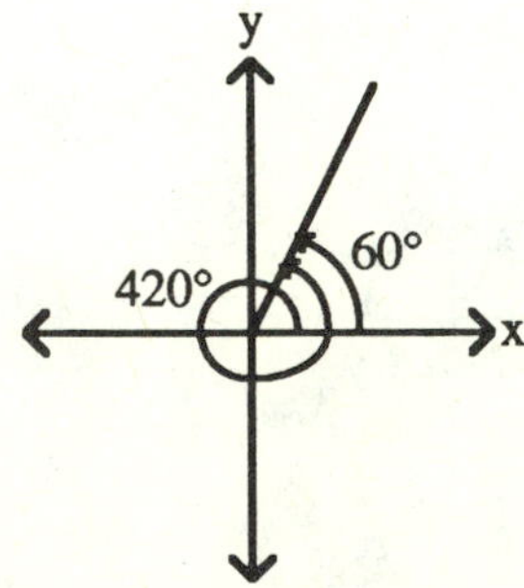

Figure 5.1.6

• DEFINITION Two angles in standard position with the same terminal sides are said to be **coterminal**. •

Coterminal angles differ by an integer multiple of $360°$. For example, $180°$ and $-180°$ are coterminal because $180° - (-180°) = 360°$ (Figure 5.1.7). Similarly, $1125°$ and $45°$ are coterminal because $1125° - 45° = 1080° = 3(360)°$ (Figure 5.1.7). Notice that $-180° - 180° = -360°$ and $45° - 1125° = -1080° = (-3)(360)°$. Thus, it does not matter in which order the subtraction is performed to test for coterminal angles. We can summarize in the following way.

• THEOREM 1 Let the measure of an angle be θ. Then, the measure of any angle coterminal with θ is of the form $\theta + n \cdot 360°$ where n is an integer.

An immediate consequence of Theorem 1 is given by the following theorem.

• THEOREM 2 If θ is the measure of an angle, then there is an integer n and a measure θ' such that $\theta = \theta' + n \cdot 360°$, where $0 \leq \theta' < 360°$.

• EXAMPLE 3: Give the measures of three angles coterminal with an angle of measure $-60°$.

SOLUTION: Any angle coterminal with an angle of measure $-60°$ must have measure $-60° + n \cdot 360°$. The angles of measure $300°$, $1020°$, and $-780°$ where we have chosen n to be 1, 3, and -2, respectively, are coterminal with an angle of measure $-60°$. Of course, other integer values for n could have been selected. •

Central Angles

If the vertex of an angle is the center of a circle, then the angle is called a **central angle**. If the measure of a central angle is known, then the length of the portion of the circumference of a circle intercepted by the angle can be determined.

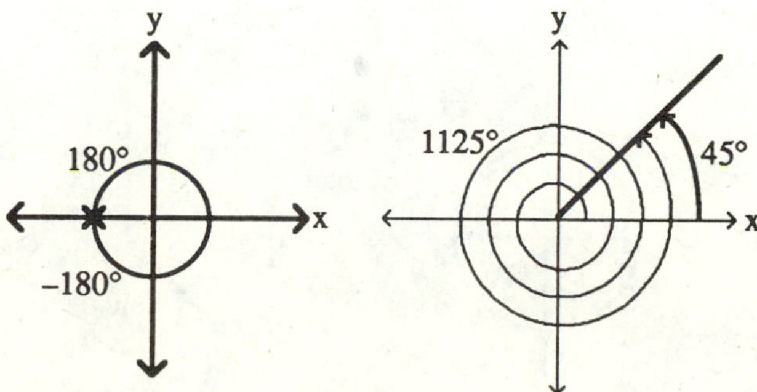

Figure 5.1.7

• **EXAMPLE 4:** Determine the length of the arc of a circle of radius 2 intercepted (subtended) by a central angle of measure

(a) 45°. (b) 112°.

SOLUTION:

(a) The circumference of a circle with radius 2 is 4π because the circumference of a circle of radius r is $2\pi r$. A central angle of 45° subtends $\frac{45}{360} = \frac{1}{8}$ of the circumference (Figure 5.1.8). Thus, the length of the arc is $\frac{1}{8}(4\pi) = \frac{\pi}{2}$ units.

(b) The length of the arc in this case is $\frac{112}{360}(4\pi) = \frac{56\pi}{45} = 3.091$ units. •

• **THEOREM 3** Let S be the length of the arc of a circle of radius r subtended by a central angle of measure θ°. Then, $S = \frac{\theta}{360}(2\pi r)$.

Notice that the lengths of the arcs in Example 4 depend on the radius of the circle. If we divide the length of the arc by the length of the radius we get a new measure of the angle that is called *radian measure*. We study radian measure in Section 5.5.

Navigation

In navigation, the **course**, or **path**, of an object is sometimes given as the measure of the *clockwise* angle θ that the **line of travel** makes with *due north* (Figure 5.1.9). By convention, the angle measure θ is positive even though the angle rotation is clockwise.

We also say that the "line of travel" has **bearing** θ.

• **EXAMPLE 5:** The captain of a boat steers a course of 35° from home port. Use a drawing to illustrate the path of the boat.

SOLUTION: We draw a ray from home port in the due north direction. The boat travels in the direction given by the ray obtained by rotating the north ray 35° clockwise (Figure 5.1.10). •

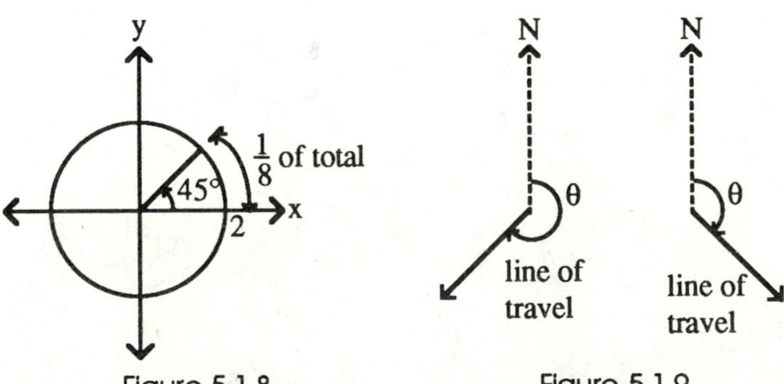

Figure 5.1.8

Figure 5.1.9

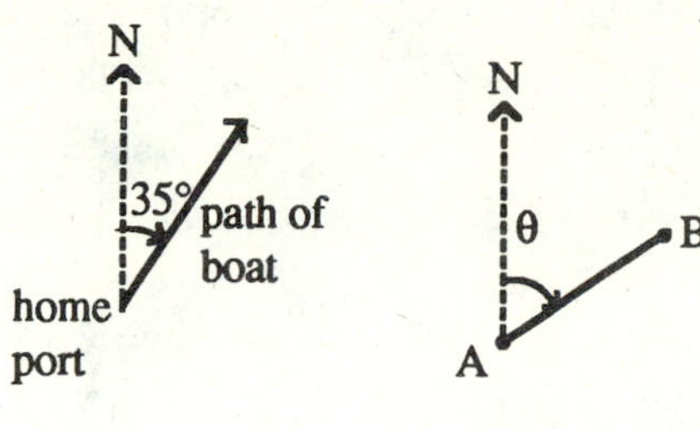

Figure 5.1.10 Figure 5.1.11

If an object at B is sighted from the point A, the **bearing** from A to B is the same as the *line of travel* from A to B. (Figure 5.1.11).

• EXERCISES 5-1

Determine the measure and draw the angle in standard position.

1. $\frac{1}{2}$ counterclockwise rotation
2. $\frac{1}{3}$ clockwise rotation
3. $\frac{4}{3}$ counterclockwise rotation
4. $\frac{5}{3}$ clockwise rotation
5. 2.5 counterclockwise rotations
6. 3.5 clockwise rotations

Determine the quadrant of the terminal side of the angle in standard position.

7. $-160°$
8. $280°$
9. $452°$
10. $-827°$
11. $1150°$
12. $-455°$

Determine the measure of an angle θ coterminal with the given angle satisfying the specified condition.

13. $48°$, $360° \leq \theta \leq 720°$
14. $110°$, $0° \leq \theta \leq -360°$
15. $-15°$, $180° \leq \theta \leq 540°$
16. $-250°$, $360° \leq \theta \leq 720°$

Determine four *different* angles, two with positive measures and two with negative measures, coterminal with the angle.

17. $55°$
18. $-22°$
19. $410°$
20. $-150°$

Assume the given point is on the terminal side of an angle θ in standard position where $0° \leq \theta \leq 360°$. Determine θ.

21. $(-1, 0)$
22. $(0, 5)$
23. $(3, 3)$
24. $(-2, 2)$
25. $(5, -5)$
26. $(10, 0)$

Assume the given point is on the terminal side of an angle θ in standard position. Determine the distance r from the point to the origin.

27. $(2, 2)$
28. $(5, -2)$
29. $(-2, 3)$
30. $(4, 5)$
31. $(-2, -8)$
32. $(-4, 6)$

Determine a formula for the measure of *all* angles coterminal with the angle of specified measure.

33. 37°
34. −125°
35. −72°
36. 458°

Determine the length of the arc of a circle with the specified radius subtended by a central angle of the given measure.

37. $r = 2$, $\theta = 30°$
38. $r = 3.75$, $\theta = 122°$
39. $r = 5.76$, $\theta = 155°$
40. $r = 20.55$, $\theta = 72°$

Consider a family of circles with center at the origin each with the specified radius r (Figure 5.1.12). Let θ be a central angle with initial side on the positive x-axis. Compute the ratio s/r of the arc length s to the radius r for each circle.

41. $\theta = 50°$, $r_1 = 1$, $r_2 = 2$, $r_3 = 2.5$ and $r_4 = 5$
42. $\theta = 38°$, $r_1 = 1$, $r_2 = 1.75$, $r_3 = 5$ and $r_4 = 7.2$

Nautical Miles

Let A and B be two points on the earth's surface. Suppose further that A and B lie on a circle with center O located at the center of the earth. We assume the earth is a sphere. If $\angle AOB$ has measure one minute ($\frac{1}{60}$ of a degree), then the distance between A and B on the circumference of the circle is *one* **nautical mile** (Figure 5.1.13). Assume the diameter of the earth is 7,952 statute miles. (A **statute mile** is a standard mile of 5280 feet.)

43. How many feet are in one nautical mile?
44. Points A and B are 257 nautical miles apart. How far apart are A and B in *statute* miles?
45. Draw a diagram to illustrate a navigational course of 35°, 128°, and 310°, respectively.
46. At a certain time an airplane is between two signal towers that lie on an east–west line. The bearing from the plane to each tower is 340° and 37°, respectively. Use a drawing to show the exact location of the plane.

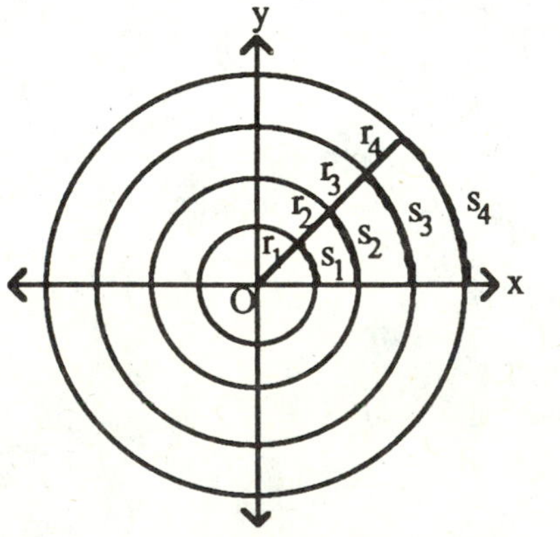

Figure 5.1.12

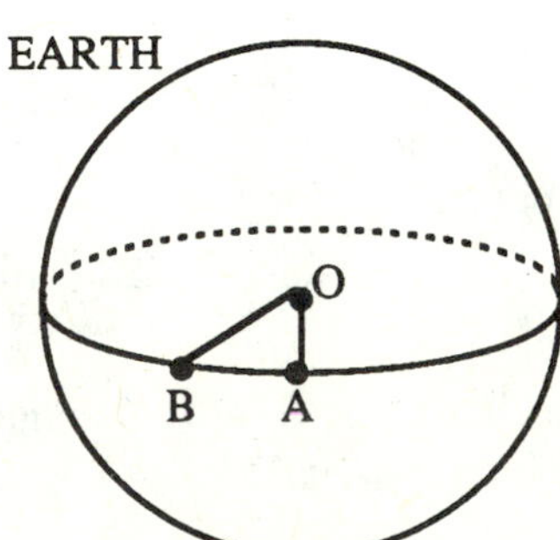

Figure 5.1.13

47. Let A be the origin of a coordinate plane. Draw a diagram showing the location of the point B if the bearing from A to B is 22°, 185°, and 292°, respectively.

Two rational numbers, $\frac{a}{b}$ and $\frac{c}{d}$ ($b \neq 0$, $d \neq 0$), are equal when $ad = bc$. That is, if $b \neq 0$ and $d \neq 0$, $\frac{a}{b} = \frac{c}{d}$ if and only if $ad = bc$. Technically the equal sign between $\frac{a}{b}$ and $\frac{c}{d}$ is different from the equal sign between ad and bc because they are defined relative to different sets of objects (in this case, different sets of numbers).

48. Devise a definition of equality of angles that involves coterminality (e.g., we want $35° = -325° = 395° = \ldots$).

49. Explain how reducing a rational number to lowest terms (e.g., $\frac{6}{8} = \frac{3}{4}$) is similar to finding an angle θ' equal (in the sense of Exercise 48) to θ with $0° \leq \theta' < 360°$.

5.2 Right Triangle Trigonometry

In this section we define the six trigonometric functions of an acute angle. These functions can be evaluated by using the lengths of the sides of a *right triangle* containing the acute angle or by using a calculator. A triangle is called a **right triangle** if one of its angles has measure 90°. The sum of the measures of the three angles of *any* triangle is 180°. This means that *each* of the other two angles of a right triangle is acute and the sum of their measures is 90° (Figure 5.2.1). We use the notation $\angle A$, $\angle B$, and $\angle C$ to represent the name of the angle of the triangle at the vertices A, B, and C, respectively. We will study applications that can be modeled with right triangles in Section 5.3. In this section we find the measure of the remaining parts of a right triangle, given the measures of two sides or the measures of one side and one of the angles.

Let θ be any acute angle. We can construct a right triangle containing θ by choosing a point B on the terminal side of θ and drawing a line segment from B that is *perpendicular* to the initial side of θ (Figure 5.2.2). The triangle ABC is a right triangle with $\angle A = \theta$. Let the side opposite A have length a, the side opposite B have length b, and the side opposite C have length c. We call BC the **side opposite** to θ, AC the **side adjacent** to θ, and AB the **hypotenuse** of the right triangle in Figure 5.2.2. Notice that BC is the side adjacent to $\angle B$, AC is the side opposite $\angle B$, and the measure of $\angle B$ is 90° minus the measure of θ.

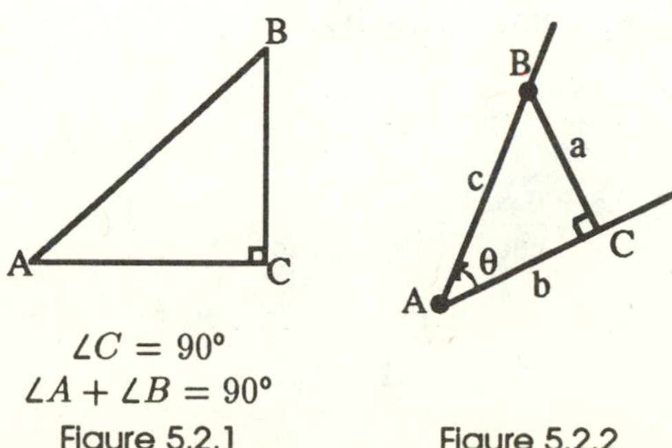

$\angle C = 90°$
$\angle A + \angle B = 90°$
Figure 5.2.1

Figure 5.2.2

• **Agreement** We use the symbol θ to represent both the name of the angle and the measure of the angle. Thus, the measure of $\angle B$ is $90° - \theta$. We often label the side opposite an angle with the lower-case letter corresponding to the angle name. For example, the side BC of the triangle in Figure 5.2.2 will often be called the *side* a.

• **DEFINITION** The six trigonometric functions of the acute angle θ in Figure 5.2.2 are defined as follows:

$$\textbf{sine } \theta = \frac{\text{length of opposite side}}{\text{length of hypotenuse}} = \frac{a}{c},$$

$$\textbf{cosecant } \theta = \frac{\text{length of hypotenuse}}{\text{length of opposite side}} = \frac{c}{a},$$

$$\textbf{cosine } \theta = \frac{\text{length of adjacent side}}{\text{length of hypotenuse}} = \frac{b}{c},$$

$$\textbf{secant } \theta = \frac{\text{length of hypotenuse}}{\text{length of adjacent side}} = \frac{c}{b},$$

$$\textbf{tangent } \theta = \frac{\text{length of opposite side}}{\text{length of adjacent side}} = \frac{a}{b},$$

$$\textbf{cotangent } \theta = \frac{\text{length of adjacent side}}{\text{length of opposite side}} = \frac{b}{a}.$$ •

We often use the abbreviation *trig* for *trigonometry* or *trigonometric*. In addition, we use the following abbreviations for the trig functions:

$$\textbf{sin } \theta = \text{sine } \theta \qquad \textbf{csc } \theta = \text{cosecant } \theta$$

$$\textbf{cos } \theta = \text{cosine } \theta \qquad \textbf{sec } \theta = \text{secant } \theta$$

$$\textbf{tan } \theta = \text{tangent } \theta \qquad \textbf{cot } \theta = \text{cotangent } \theta$$

Next, we give some basic properties of the trig functions of acute angles.

• **THEOREM 1** If θ is an acute angle, then

(a) $\sin\theta = \dfrac{1}{\csc\theta}$, $\csc\theta = \dfrac{1}{\sin\theta}$ (b) $\cos\theta = \dfrac{1}{\sec\theta}$, $\sec\theta = \dfrac{1}{\cos\theta}$

(c) $\tan\theta = \dfrac{1}{\cot\theta}$, $\cot\theta = \dfrac{1}{\tan\theta}$ (d) $\tan\theta = \dfrac{\sin\theta}{\cos\theta}$, $\cot\theta = \dfrac{\cos\theta}{\sin\theta}$

PROOF From the above definition we have the following:

$$\frac{1}{\csc\theta} = \frac{1}{\frac{c}{a}} = \frac{a}{c} = \sin\theta.$$

Taking reciprocals of each side of the above equation gives the following:

$$\frac{1}{\csc\theta} = \sin\theta,$$
$$\frac{1}{\frac{1}{\csc\theta}} = \frac{1}{\sin\theta},$$
$$\csc\theta = \frac{1}{\sin\theta}.$$

Thus, (a) is true. The remaining proofs are left as exercises.

In Section 5.4 we will see that Theorem 1 is true even if θ is not acute, provided that the denominators of the fractions are *not* zero.

• THEOREM 2 If θ is an acute angle, then

(a) $0 < \sin\theta < 1$, (b) $0 < \cos\theta < 1$,

(c) $\csc\theta > 1$, (d) $\sec\theta > 1$.

PROOF Refer to Figure 5.2.2. We know that $c = \sqrt{a^2 + b^2}$. It follows that $0 < a < c$ and $0 < b < c$. (Why?) Thus, $0 < \frac{a}{c} < 1$ and $0 < \frac{b}{c} < 1$. That is, $0 < \sin\theta < 1$ and $0 < \cos\theta < 1$. Next (c) follows from (a) and $\csc\theta = \frac{1}{\sin\theta}$. Similarly, (d) follows from (b) and $\sec\theta = \frac{1}{\cos\theta}$.

• DEFINITION Two angles are said to be **complementary** if the sum of their measures is 90°. •

The two acute angles of a right triangle are complementary. Notice in Figure 5.2.2 that the side adjacent to $\angle B$ is the side opposite $\angle A$, and the side opposite $\angle B$ is the side adjacent to $\angle A$. This observation together with the above definition of the trig functions give the following *identities* about complementary angles. We will give a definition of identity in Section 5.4.

Identities for Complementary Angles The following equations are true for any acute angle θ:

$$\sin(90^\circ - \theta) = \cos\theta, \qquad \csc(90^\circ - \theta) = \sec\theta,$$
$$\cos(90^\circ - \theta) = \sin\theta, \qquad \sec(90^\circ - \theta) = \csc\theta,$$
$$\tan(90^\circ - \theta) = \cot\theta, \qquad \cot(90^\circ - \theta) = \tan\theta.$$

The sine of the complement of θ is equal to the cosine of θ. Similar statements can be made about the tangent and secant of the complement of θ. (Why?) If we agree that sine, tangent, and secant are the co-functions of cosine, cotangent, and cosecant, respectively, then the above identities can be summarized in the following way:

Complementary Angles The trig function of the complement of θ is equal to the co-function of θ.

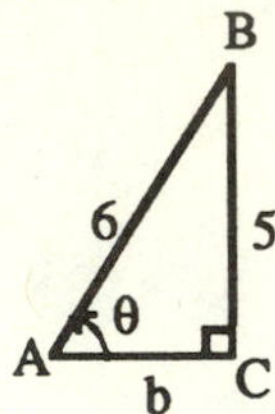

Figure 5.2.3

• EXAMPLE 1: Let θ be an acute angle with $\sin\theta = \frac{5}{6}$. Find the values of all trig functions at θ.

SOLUTION: Figure 5.2.3 gives a right triangle containing an acute angle θ with opposite side equal to 5 and hypotenuse equal to 6. The Pythagorean theorem gives $b^2 + 5^2 = 6^2$ so that $b = \sqrt{11}$. The definitions of the trig functions give the following for their exact values at θ:

$$\sin\theta = \frac{5}{6} \qquad \csc\theta = \frac{6}{5}$$

$$\cos\theta = \frac{\sqrt{11}}{6} \qquad \sec\theta = \frac{6}{\sqrt{11}}$$

$$\tan\theta = \frac{5}{\sqrt{11}} \qquad \cot\theta = \frac{\sqrt{11}}{5}$$

•

The complementary angle identities can be used to compute the values of the trig functions at $90° - \theta$ if the values at θ are known. For example, using Figure 5.2.3 and Example 1 we have

$$\sin(90° - \theta) = \cos\theta = \frac{\sqrt{11}}{6} \quad \text{and} \quad \cos(90° - \theta) = \sin\theta = \frac{5}{6}.$$

Notice that using the definition of the trig functions gives the same values.

Using a Calculator

In Example 1 we described an angle θ by specifying two of the three sides of a right triangle that contained the angle θ. If the measure of an angle is known, then a scientific calculator can be used to compute the values of the trig functions at that angle. Calculators generally only have keys reserved for $\sin\theta$, $\cos\theta$, and $\tan\theta$. To compute $\cot\theta$, $\sec\theta$, and $\csc\theta$, we use the identities $\cot\theta = \frac{1}{\tan\theta}$, $\sec\theta = \frac{1}{\cos\theta}$, $\csc\theta = \frac{1}{\sin\theta}$ given in Theorem 1.

• EXAMPLE 2: Find the value of the six trig functions at $42°$.

SOLUTION: A scientific calculator will compute directly the sine, cosine, or tangent of any angle θ by using the appropriate trig key. Some calculators require that θ be entered before a trig key is pressed. Other calculators require that a trig key be pressed before an angle is entered. Check the owner's manual of your calculator to see how the trig keys on your calculator work. Be sure that your calculator is set in *degree* mode rather than radian mode. Then, use your calculator and the reciprocal relationships of Theorem 1 to compute the trig values at 42°. The following approximations were obtained using a *Casio* graphing calculator:

$$\sin 42^\circ = 0.6691306064, \qquad \csc 42^\circ = \frac{1}{\sin 42^\circ} = 1.49447655,$$

$$\cos 42^\circ = 0.7431448255, \qquad \sec 42^\circ = \frac{1}{\cos 42^\circ} = 1.34563273,$$

$$\tan 42^\circ = 0.9004040443, \qquad \cot 42^\circ = \frac{1}{\tan 42^\circ} = 1.110612515.$$

•

In Example 1 we found the value of all trig functions at θ where $\sin\theta = \frac{5}{6}$. In Example 3 we see how to approximate the value of θ with a calculator. In Section 6.3 we will provide detail about why this calculator procedure works.

• EXAMPLE 3: Use a calculator to approximate the acute angle θ if $\sin\theta = \frac{5}{6}$. Then, compute the values of the other trig functions at θ. Compare with the values obtained in Example 1.

SOLUTION: Scientific calculators will compute the *acute* angle θ given its sine value in decimal form using two keys. You may need to enter the sine value, $5 \div 6$, before or after pressing the two keys. Usually, the two keys are [INV] [SIN], [2nd F] [SIN], or [SHIFT] [SIN]. Again, you need to check your calculator owner's manual to determine which keys to use and whether to enter the sine value before or after pressing the two keys. Be sure that your calculator is set in *degree* mode rather than radian mode. We will discuss radian measure in Section 5.5. With a calculator set in degree mode you will find that the solution θ to $\sin\theta = \frac{5}{6}$ is approximately 56.44269024°. Next, we store 56.44269024 in the memory of the calculator and compute the values of the other trig functions at θ:

$$\sin 56.44269^\circ = 0.8333333, \qquad \csc 56.44269024^\circ = 1.2,$$

$$\cos 56.44269024^\circ = 0.5527707984, \qquad \sec 56.44269024^\circ = 1.809068067,$$

$$\tan 56.44269024^\circ = 1.507556723, \qquad \cot 56.44269024^\circ = 0.663324958.$$

•

If we approximate the exact values for the trig functions at θ given in Example 1 with a calculator, we will find that they agree with the above values. For example, $\tan\theta = \frac{5}{\sqrt{11}} = 1.507556723$ and $\cos\theta = \frac{\sqrt{11}}{6} = 0.5527707984$.

• EXAMPLE 4: Use a calculator to determine the value of θ if $\sec\theta = 2.3$. Then, compute the values of the other trig functions at θ.

SOLUTION: The secant function is *not* built in on most calculators. In order to find θ we first compute the value of $\cos\theta$. Now, $\cos\theta = \frac{1}{\sec\theta}$ so that $\cos\theta = 0.4347826087$. We can use the calculator keys [INV] [COS], [2nd F] [COS], or [SHIFT] [COS] to find that $\theta = 64.22853826°$. After storing this number the other trig values at θ can be computed with a calculator:

$$\sin 64.22853826^\circ = 0.9005354425, \qquad \csc 64.22853826^\circ = 1.110450464,$$

$$\cos 64.22853826^\circ = 0.4347826087, \qquad \sec 64.22853826^\circ = 2.3,$$

$$\tan 64.22853826^\circ = 2.071231518, \qquad \cot 64.22853826^\circ = 0.4828045495.$$

•

If we did not want the measure of θ in Example 4, then the other trig values at θ could be computed using the procedure of Example 1. However, we often need to determine the measure of angles. Thus, we need to know how to use a calculator to determine the measure of θ if a trig value of θ is known.

Before the easy availability of scientific calculators, tables were used to compute the values of the trig functions. In addition, knowing the values of the trig functions for the special angles of 30°, 45°, and 60° was very important. We give the next example for historical reasons. It is still a good idea to commit to memory the relationships of the two triangles given in the following example.

• EXAMPLE 5: Use triangles to determine the values of the trig functions at 30°, 45°, and 60°.

SOLUTION: Triangle ABC in Figure 5.2.4 is equilateral with the length of each side equal to 2. The measure of each angle is 60°. The line from B perpendicular to AC bisects AC and the angle at B. Thus, the length of AD or DC is 1, and the measure of angle ABD or DBC is 30°. The length of BD is $\sqrt{3}$ because ABD is a right triangle. Triangle ABD can be used to find the values of the trig functions at 30° and 60° (Table 1). Some of the values are left for the exercises.

$\sin 30^\circ = \frac{1}{2}$	$\csc 30^\circ = 2$
$\cos 30^\circ = \frac{\sqrt{3}}{2}$	$\sec 30^\circ = ?$
$\tan 30^\circ = ?$	$\cot 30^\circ = \sqrt{3}$
$\sin 60^\circ = \frac{\sqrt{3}}{2}$	$\csc 60^\circ = ?$
$\cos 60^\circ = ?$	$\sec 60^\circ = 2$
$\tan 60^\circ = \sqrt{3}$	$\cot 60^\circ = ?$

Table 1

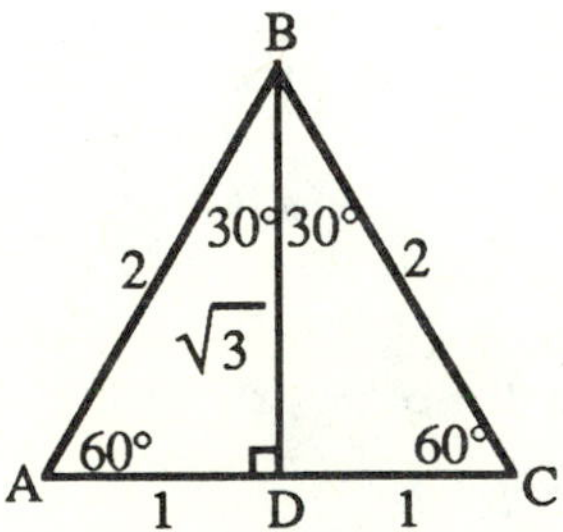

Figure 5.2.4

The triangle in Figure 5.2.5 is an isoceles right triangle. Each of the non-right angles is 45°, and the length of the hypotenuse is $\sqrt{2}$. Again, some of the values are left for the exercises (Table 2).

$\sin 45° = \frac{1}{\sqrt{2}}$	$\csc 45° = ?$
$\cos 45° = ?$	$\sec 45° = \sqrt{2}$
$\tan 45° = ?$	$\cot 45° = 1$

Table 2 •

Solving Right Triangles

If the measure of two sides of a triangle or the measure of one side and one angle are known, then the measure of the other parts (sides or angles) of the right triangle can be determined. Determining the measures of the missing parts of a right triangle is often referred to as *solving a right triangle.* Example 6 illustrates this.

• EXAMPLE 6: One angle of a right triangle has measure 37° and the hypotenuse has length 8. Find the measure of the remaining parts of the right triangle.

SOLUTION: The right triangle in Figure 5.2.6 has the angle at A with measure 37° and the hypotenuse with length 8. We need to determine a, b, and $\angle B$. Notice $\angle C$ is the 90° angle so that the measure of $\angle B$ is 53°. (Why?) We can use the following equations to solve for a and b:

$$\sin 37° = \frac{a}{8} \qquad \cos 37° = \frac{b}{8}$$

Thus, $a = 8\sin 37°$ and $b = 8\cos 37°$. Accurate to hundredths, a is 4.81 and b is 6.39. Suppose we want to check our estimates for a and b. We know that $\sqrt{a^2+b^2}$ should be 8. (Why?) The calculator we use gives 7.998012253 when we compute $\sqrt{(4.81)^2+(6.39)^2}$. This number is close enough to 8 for us to be confident that 4.81 and 6.39 are correct. •

There is another way to check the computations in Example 6. If the full calculator approximations are used for $a = 8\sin 37°$ and $b = 8\cos 37°$, then most calculators will give 8 as the value of $\sqrt{(8\sin 37°)^2+(8\cos 37°)^2}$. This happens because the exact value of $\sqrt{(8\sin 37°)^2+(8\cos 37°)^2}$ is 8 and most calculators carry extra digits (not shown) so that machine round-off produces 8 when this expression is evaluated.

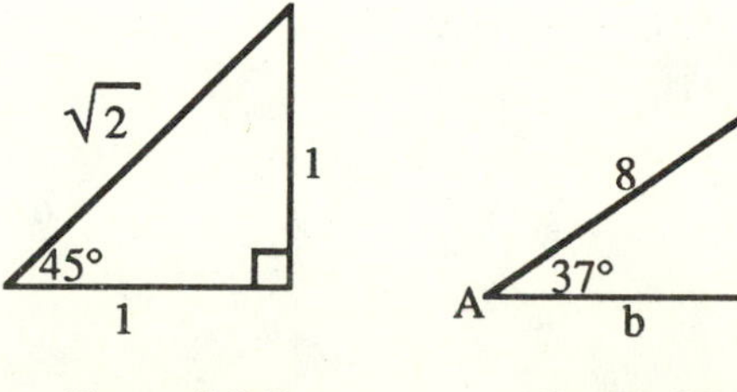

Figure 5.2.5 Figure 5.2.6

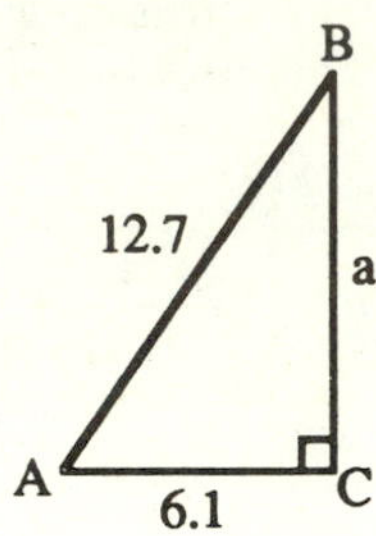

Figure 5.2.7

In the next example the measures of two sides of a right triangle are given, and the measures of the other parts are determined. It does not matter which two sides are given.

• EXAMPLE 7: The hypotenuse and one other leg of a right triangle have measures 12.7 and 6.1, respectively. Find the measures of the remaining parts of the right triangle.

SOLUTION: We must determine a and the measures of $\angle A$ and $\angle B$ of Figure 5.2.7. Now, $\cos A = \frac{6.1}{12.7}$. We can use the calculator keys [INV] [COS], [2nd F] [COS], or [SHIFT] [COS] to determine that the measure of A is 61.29°, accurate to hundredths. We can use the Pythagorean theorem to determine that $a = \sqrt{12.7^2 - 6.1^2}$ is 11.14, accurate to hundredths. Notice that $\sin A = \frac{a}{12.7}$, so that the equation $a = 12.7 \sin A$ could also be used to find the value of a. •

• **Note Concerning Accuracy** In the above example we assumed that the given lengths 12.7 and 6.1 of the sides were exact to produce answers with error at most 0.01. Some textbooks assume that numbers like 12.7 and 6.1 are accurate only to tenths. In this textbook, assume given information is exact unless otherwise specified, and find answers with error at most 0.01.

• EXERCISES 5-2

In this exercise set, all angles are acute angles or right angles.

Find the value of the six trig functions at each angle.

1. 23°
2. 59°
3. 38°
4. 83°
5. Complete Table 1.
6. Complete Table 2.
7. Approximate the values in the solution to Example 1 and compare with the values of the six trig functions at $\theta = 56.44269024°$ in the solution to Example 3.

Consider the triangle ABC in Figure 5.2.8.

8. Find all six trig function values at $\angle A$.

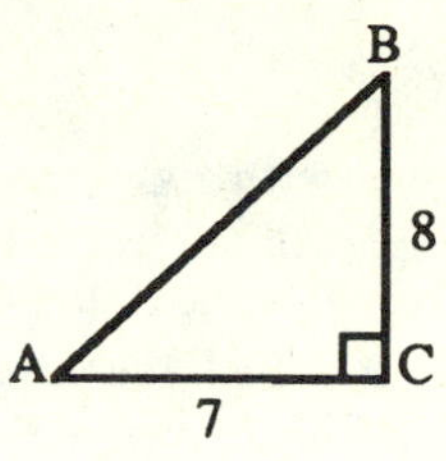

Figure 5.2.8

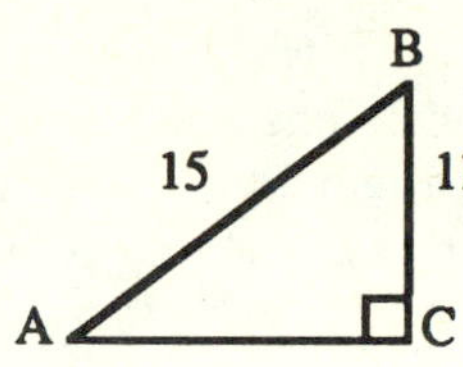

Figure 5.2.9

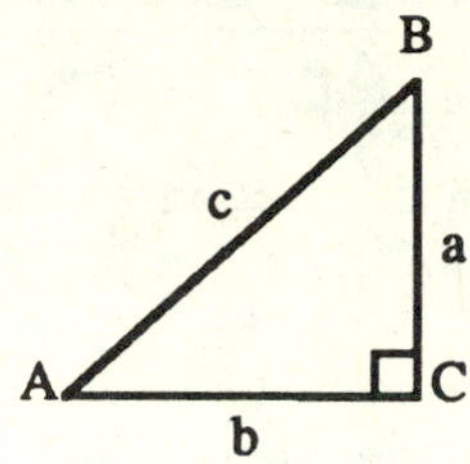

Figure 5.2.10

9. Find all six trig function values at $\angle B$.

Consider the triangle ABC in Figure 5.2.9.

10. Find all six trig function values at $\angle A$.
11. Find all six trig function values at $\angle B$.

Use a right triangle to determine the values of all trig functions at θ.

12. $\sin\theta = \frac{3}{5}$
13. $\tan\theta = \frac{1}{3}$

Use a calculator to determine θ. Make sure your calculator is in degree mode.

14. $\sin\theta = \frac{1}{2}$
15. $\sin\theta = 0.8245$
16. $\cos\theta = \frac{4}{5}$
17. $\cos\theta = 0.125$
18. $\tan\theta = 1$
19. $\tan\theta = 3$
20. $\tan\theta = 0.423$
21. $\tan\theta = 2.80$
22. Prove (b) of Theorem 1.
23. Prove (c) of Theorem 1.
24. Prove (d) of Theorem 1.
25. Use a right triangle to explain how $\tan\theta$ could be very large (greater than 1000) or very small (less than 0.001).
26. Determine θ if $\csc\theta = 2$.
27. Determine θ if $\sec\theta = 3.81$.
28. Determine θ if $\cot\theta = \frac{3}{5}$.
29. Determine θ if $\cot\theta = 1.875$.

Solve the right triangle in Figure 5.2.10.

30. $\angle A = 20°$, $a = 12.3$
31. $a = 3$, $b = 4$
32. $\angle A = 41°$, $c = 10$
33. $\angle A = 55°$, $b = 15.58$
34. $b = 5$, $c = 7$
35. $a = 20.2$, $c = 50.75$
36. $a = 2$, $b = 9.25$
37. $a = 5$, $\angle B = 59°$
38. $c = 12.89$, $\angle B = 12.55°$
39. $\angle A = 10.2°$, $c = 14.5$
40. $\angle A = 21.82°$, $b = 13.65$
41. Let $\theta = 10°$, $20°$, $25°$, $48°$, $55°$, and $69°$. Compute $\sin^2\theta + \cos^2\theta$ for each value of θ.

5.3 Applications

In this section we study applications that can be modeled with right triangles. Solving the applications requires finding the measure of sides or angles of the right triangles.

Suppose that an observer sights the top of an object, say a pole (Figure 5.3.1). The angle that the line of sight makes with the horizontal is called the **angle of elevation.** If the object is below the observer, then the angle that the line of sight makes with the horizontal is called the **angle of depression** (Figure 5.3.2).

- Accuracy Agreement Unless we specify otherwise all answers will be given accurate to hundredths. Furthermore, we will carry full calculator approximations until all computations are completed. Then, we round all answers to hundredths.

• EXAMPLE 1: The angle of elevation of the top of a building from a point 100 feet away from the building on level ground is 65° (Figure 5.3.3). Determine the height h of the building.

SOLUTION: Notice that $\tan 65° = \frac{h}{100}$ so that $h = 100 \tan 65°$. Accurate to hundredths, h is 214.45 feet. •

• EXAMPLE 2: The angle of depression of a buoy from a point on a lighthouse 130 feet above the surface of the water is 6° (Figure 5.3.4). Find the distance x from the base of the lighthouse to the buoy.

SOLUTION: The angle θ in Figure 5.3.4 must be 6°, because the angle of depression is 6°. (Why?) Now, $\tan \theta = \frac{130}{x}$ so that $\tan 6° = \frac{130}{x}$. Solving for x we have $x = \frac{130}{\tan 6°}$. Thus, x is 1236.87 feet. •

• EXAMPLE 3: A guy wire 75 feet long runs from an antenna to a point on level ground 10 feet from the base of the antenna (Figure 5.3.5). Determine the angle α the guy wire makes with the horizontal, the angle β the guy wire makes with the antenna, and the distance h above ground of the point B at which the guy wire is attached to the antenna.

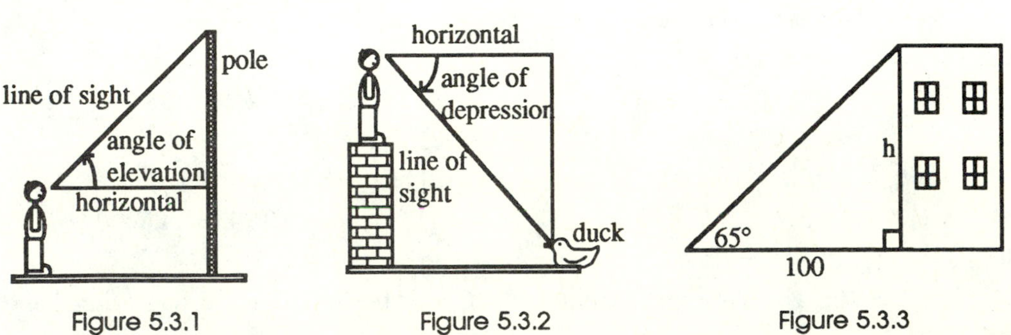

Figure 5.3.1 Figure 5.3.2 Figure 5.3.3

SOLUTION: We can use the calculator keys [INV] [COS], [2nd F] [COS], or [SHIFT] [COS] to determine that $\alpha = 82.34°$ from the equation $\cos\alpha = \frac{10}{75}$. Thus, β is 7.66°. (Why?) We can determine h from the equation $75^2 = 10^2 + h^2$, or from the equation $\tan\alpha = \frac{h}{10}$. We find that h is 74.33 feet if we use the equation $75^2 = 10^2 + h^2$. However, if we use $\tan\alpha = \frac{h}{10}$ with $\alpha = 82.34°$ we would find that h is 74.35. This discrepancy in the second decimal place occurs because α was rounded to hundredths *before* the value of h was computed. If we use the full calculator approximation of 82.33774434 for α we would find that h is 74.33 when rounded. It is a good practice to finish *all* computations before rounding. Thus, the guy wire makes an 82.34° angle with the horizontal, a 7.66° angle with the antenna, and the guy wire is attached to a point 74.33 feet above ground. •

• EXAMPLE 4: From the top of a 100-foot building a man observes a car moving toward him. If the angle of depression of the car changes from 22° to 46° during the period of observation, how far does the car travel?

SOLUTION: Let A and B be the positions of the car that correspond to the 22° and 46° angles of depression. Let the man be at point D at the top of the building and let x be the distance traveled by the car (Figure 5.3.6). The angles α and β are equal to 22° and 46°, respectively. (Why?) Let d be the length of BC. From triangle ADC we have $\tan\alpha = \frac{100}{x+d}$, or $\tan 22° = \frac{100}{x+d}$. From triangle BDC we have $\tan\beta = \frac{100}{d}$, or $\tan 46° = \frac{100}{d}$. We can rewrite these two equations as follows:

$$x + d = \frac{100}{\tan 22°}, \qquad d = \frac{100}{\tan 46°}.$$

Next, we solve these equations for x by substituting the value of d from the second equation into the first equation:

$$\begin{aligned} x &= \frac{100}{\tan 22°} - d \\ &= \frac{100}{\tan 22°} - \frac{100}{\tan 46°} \\ &= 150.94. \end{aligned}$$

Thus, the car travels 150.94 feet during the period of observation. •

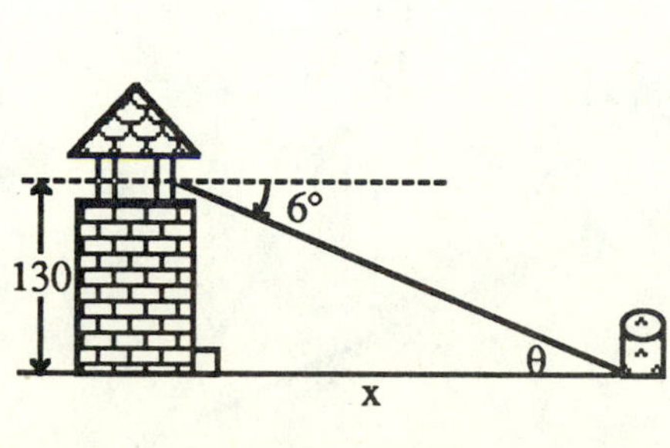

Figure 5.3.4

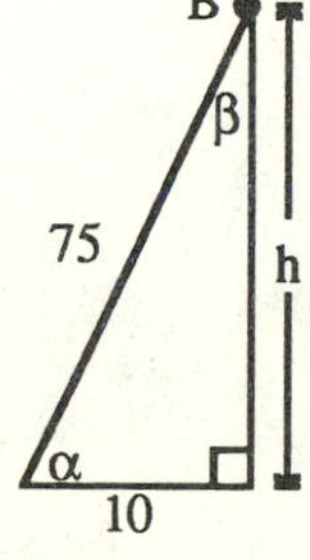

Figure 5.3.5

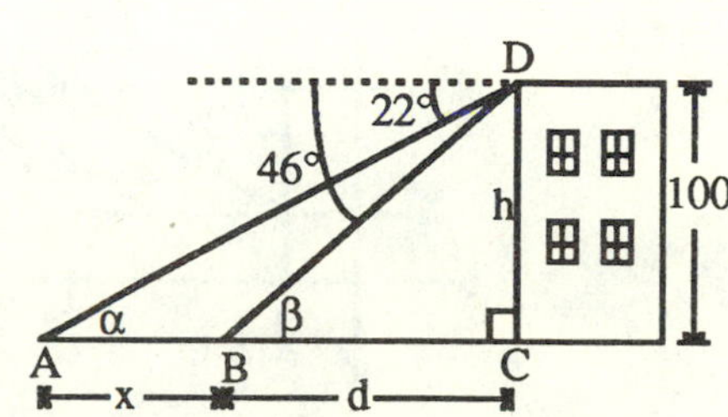

Figure 5.3.6

Distance Formula

We determine a formula for the distance between any two points in the coordinate plane using the Pythagorean relationship.

To find the distance between the points $P(x_1, y_1)$ and $Q(x_2, y_2)$, we form a right triangle as illustrated in Figure 5.3.7. The coordinates of the point R are (x_2, y_1). The distance from P to R is $|x_2 - x_1|$ and the distance from R to Q is $|y_2 - y_1|$. We need to use absolute value because, in general, Q need not be higher than P or to the right of P. Now, using the Pythagorean theorem, we can find the distance $D(P, Q)$ between P and Q.

$$\begin{aligned} D(P,Q) &= \sqrt{|x_2 - x_1|^2 + |y_2 - y_1|^2} \\ &= \sqrt{(x_2 - x_1)^2 + (y_2 - y_1)^2} \end{aligned}$$

The last step follows because $|x_2 - x_1|^2 = (x_2 - x_1)^2$ and $|y_2 - y_1|^2 = (y_2 - y_1)^2$. In fact, $|a|^2 = a^2$ for any real number a. (Why?)

- **Distance Formula.** The distance D between the points (x_1, y_1) and (x_2, y_2) is given by

$$D = \sqrt{(x_2 - x_1)^2 + (y_2 - y_1)^2}\,.$$

- **EXAMPLE 5:** Find the distance between the points $(-1, -3)$ and $(2, 5)$.

SOLUTION: We use the distance formula:

$$\begin{aligned} D &= \sqrt{(2-(-1))^2 + (5-(-3))^2} \\ &= \sqrt{3^2 + 8^2} \\ &= \sqrt{73} \end{aligned}$$

Thus, the distance between $(-1, -3)$ and $(2, 5)$ is $\sqrt{73}$, or approximately 8.54 units. •

- **EXAMPLE 6:** A naval boat travels at a speed of 35 mph from its home port on a course of 53° for two hours and then changes to a course of 143° for three hours. Determine the

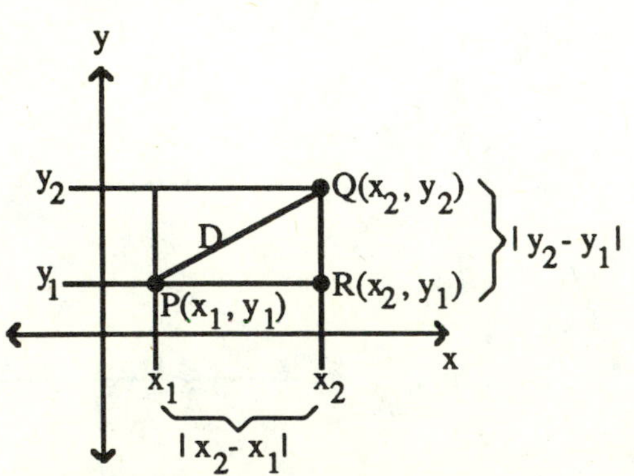

Figure 5.3.7

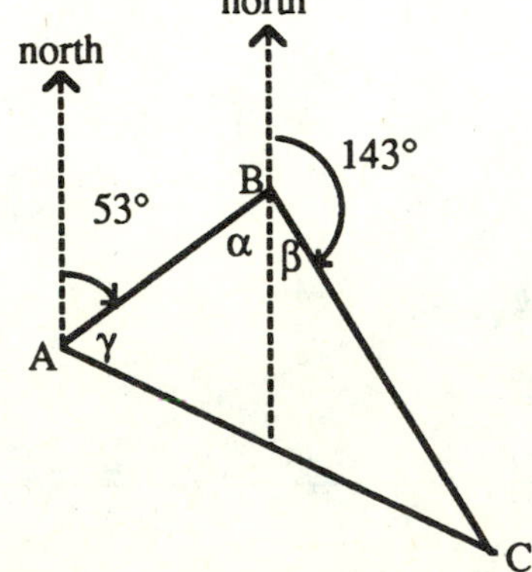

Figure 5.3.8

distance from the boat to its home port (Figure 5.3.8). What is the bearing from the boat's home port to its location at the end of five hours?

SOLUTION: In Figure 5.3.8 we let A represent the boat's home port, B the position of the boat when it changes course, and C the position of the boat at the end of the trip. Then the angles marked α and β in Figure 5.3.8 have measure 53° and 37°, respectively. (Why?) Thus, $\alpha + \beta = 90°$. The triangle ABC is a right triangle with hypotenuse AC. $d(A, B) = 70$ miles because the boat travels for two hours on a course of 53° at 35 mph. Similarly, $d(B, C) = 105$ miles. The distance from the boat to A is $d(A, C)$. Now,

$$d(A, C) = \sqrt{70^2 + 105^2} = 126.19\,.$$

Thus, the boat is 126.19 miles from A.

The bearing from A to C is $53° + \gamma$ where $\sin\gamma = \frac{d(B,C)}{d(A,C)} = \frac{105}{126.19} = 0.83$. Using the inverse calculator keys we find $\gamma = 56.31°$. Thus, the bearing is $53° + 56.31° = 109.31°$. •

• EXERCISES 5-3

Find the distance between the points in the plane with the given coordinates.

1. $(0, 0)$, $(3, 4)$
2. $(-1, 2)$, $(2, 3)$
3. $(-3, -2)$, $(6, -2)$
4. (a, b), $(2, 3)$

Let $f(x) = 2 - x^2$ and $g(x) = \sqrt{6 - x}$. Find the distance between the points.

5. $(2, f(2))$, $(-3, g(-3))$
6. $(-3, f(-3))$, $(4, g(4))$
7. The angle of elevation of the top of a building from a point 100 feet away from the building on level ground is 45°. Determine the height of the building.
8. The angle of elevation of the top of a building from a point 80 feet away from the building on level ground is 70°. Determine the height of the building.
9. The angle of depression of a buoy from a point on a lighthouse 120 feet above the surface of the water is 10°. Find the distance from the lighthouse to the buoy.
10. The angle of depression of a buoy from a point on a lighthouse 100 feet above the surface of the water is 3°. Find the distance from the lighthouse to the buoy.
11. The angle of elevation of the top of a building from a point 250 feet away from the building on level ground is 23°. Determine the height of the building.
12. A guy wire 30 meters long runs from an antenna to a point on level ground 5 meters from the base of the antenna. Determine the angle the guy wire makes with the horizontal, the angle the guy wire makes with the antenna, and how far above ground the guy wire is attached to the antenna.
13. A building casts a shadow 130 feet long when the angle of elevation of the sun (measured from the horizon) is 38°. How tall is the building?
14. An observer on the ground is one mile from a building 1200 feet tall. What is the angle of elevation from the observer to the top of the building?

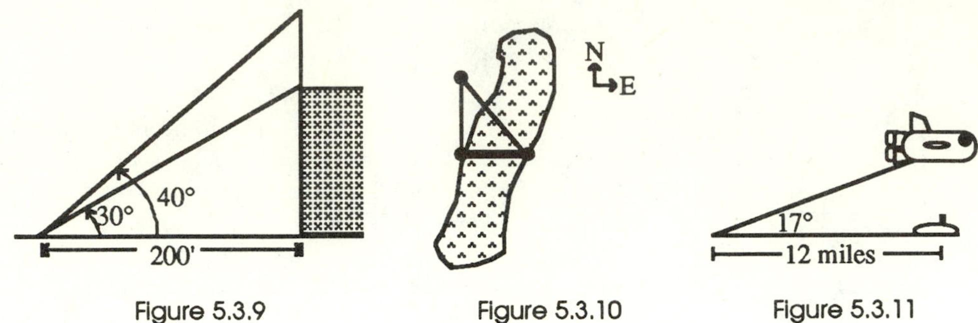

Figure 5.3.9 Figure 5.3.10 Figure 5.3.11

15. The angle of elevation from an observer to the top of a building located 200 feet from the observer is 30° (Figure 5.3.9). The angle of elevation from the observer to the top of a microwave tower (on top of the building) is 40°. What is the height of the tower?

16. From the top of a 100-foot building a man observes a car moving toward him. If the angle of depression of the car changes from 15° to 33° during the period of observation, how far does the car travel?

17. A boat travels at a speed of 30 mph from its home port on a course of 95° for two hours and then changes to a course of 185° for two hours. Determine the distance from the boat to its home port and the bearing from the home port to the boat.

18. A boat travels at a speed of 40 mph from its home base on a course of 65° for two hours and then changes to course of 155° for four hours. Determine the distance from the boat to its home port and the bearing from the home port to the boat.

19. A point on the north rim of the Grand Canyon is 7256 feet above sea level. A point on the south rim directly across from the other point is 6159 feet above sea level. The canyon is 3180 feet wide (horizontal distance) between the two points. What is the angle of depression from the north rim point to the south rim point?

20. A ranger spots a fire from a 73-foot tower at Yellowstone National Park. She measures the angle of depression to be 1°20′. How far is the fire from the tower? (*Note:* 20′ = 20 minutes, and there are 60 minutes in one degree.)

21. A foot bridge is to be constructed along an east–west line across a river gorge (Figure 5.3.10). The bearing of a line of sight from a point 325 feet due north of the west end of the bridge to the east end of the bridge is 117°. How long is the bridge?

22. The angle of elevation of a space shuttle launched from Cape Canaveral is measured to be 17° relative to the point of launch when it is directly over a ship 12 miles down range (Figure 5.3.11). What is the altitude of the shuttle when it is directly over the ship?

23. A truss for a barn roof is constructed as shown in Figure 5.3.12. What is the height of the vertical center span?

24. A hot air balloon over Park City, Utah is 760 feet above the ground. The angle of depression from the balloon to a small lake is 5.25°. How far is the lake from a point on the ground directly under the balloon?

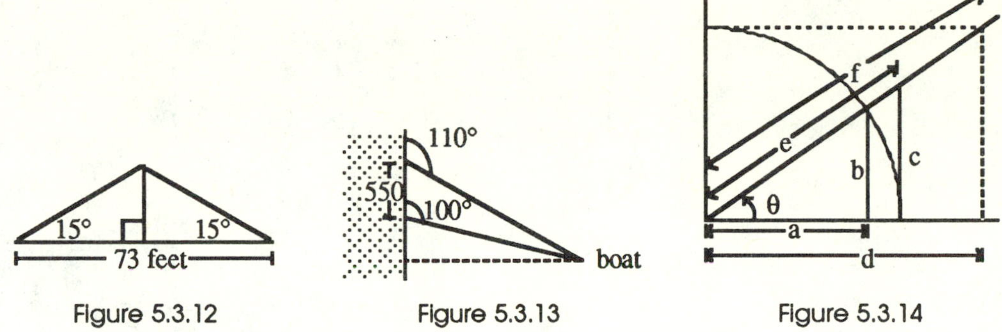

Figure 5.3.12 Figure 5.3.13 Figure 5.3.14

25. The bearings of lines of sight from two points on the shore to a boat are 110° and 100°, respectively (Figure 5.3.13). Assume the two points are 550 feet apart. How far is the boat from the nearest point on shore?

26. Boats A and B leave from ports on opposite sides of a large lake. The ports are on a due east–west line. Boat A steers a course of 105°, and boat B steers a course of 195°. Boat A averages 23 mph and collides with boat B (it was a foggy night). What was boat B's average speed?

27. Figure 5.3.14 shows an acute central angle θ of a circle with radius 1, and six line segments with lengths a through f. Identify each of these lengths with the six trig function values at θ.

28. Explain how a right triangle can be used to determine the exact value of $\cos\theta$ and $\tan\theta$ if $\sin\theta = a$.

29. Explain how a right triangle can be used to determine the exact value of $\sin\theta$ and $\tan\theta$ if $\cos\theta = a$.

30. Determine $\cos\theta$ if $\cos(90° - \theta) = a$.

31. Determine $\sin\theta$ if $\sin(90° - \theta) = a$.

5.4 Trigonometric Functions of Any Angle

In this section the trig functions for any angle are defined using the coordinates of a point on the terminal side of the angle. If the measure of an angle is known, a calculator is used to find the values of the trig functions of the angle. We define the reference triangle and reference angle for any nonquadrantal angle and use a calculator to find the measure of the reference angle. As an application of reference angles, we determine the measure of the angle a line makes with the x-axis. Fundamental identities about trig functions are established.

Let θ be a nonquadrantal angle in standard position, and $P(x, y)$ a point other than the origin on the terminal side of θ. Figure 5.4.1 gives one possibility for θ. There are infinitely many possibilities for a rotation that produces the terminal side shown in Figure 5.4.1. Moreover, the point $P(x, y)$ could have been in any of the four quadrants. Let r be the distance from the origin to P. Then, $r = \sqrt{x^2 + y^2} \neq 0$ because $(x, y) \neq (0, 0)$.

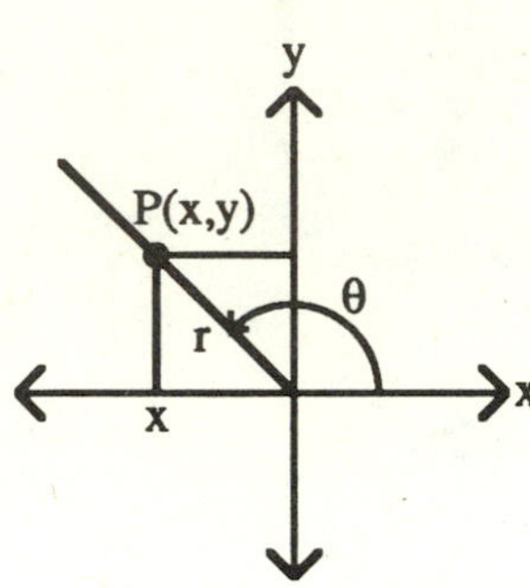

Figure 5.4.1

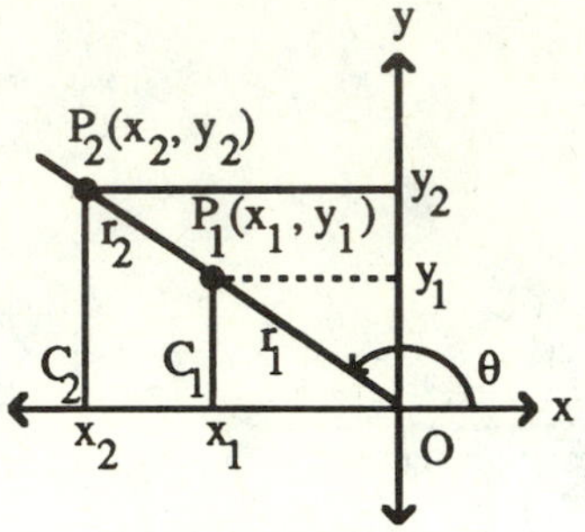

Figure 5.4.2

• **DEFINITION** Let θ be an angle in standard position, $P(x,y)$ a point other than the origin on the terminal side of P, and $r = \sqrt{x^2+y^2}$. The six trig function values of θ are defined as follows:

$$\sin\theta = \frac{y}{r}, \qquad \csc\theta = \frac{r}{y} \ (\text{if } y \neq 0),$$
$$\cos\theta = \frac{x}{r}, \qquad \sec\theta = \frac{r}{x} \ (\text{if } x \neq 0),$$
$$\tan\theta = \frac{y}{x} \ (\text{if } x \neq 0), \qquad \cot\theta = \frac{x}{y} \ (\text{if } y \neq 0).$$ •

First, we show that the above definition of the six trig functions is meaningful.

• **THEOREM 1** Let θ be an angle in standard position. The six trig function values of θ are independent of the particular point P chosen on the terminal side of θ to make the definitions.

PROOF We give the proof for nonquadrantal angles. It can be shown that the theorem is also true for quadrantal angles. Let $P_1(x_1,y_1)$ and $P_2(x_2,y_2)$ be any two points different from $(0,0)$ on the terminal side of θ (Figure 5.4.2). The triangles P_1OC_1 and P_2OC_2 are *similar*. Let $r_1 = d(P_1,O)$ and $r_2 = d(P_2,O)$. It follows that

$$\frac{y_1}{r_1} = \frac{y_2}{r_2},$$

where $r_1 = \sqrt{x_1^2+y_1^2}$ and $r_2 = \sqrt{x_2^2+y_2^2}$. Thus, we obtain the same value for $\sin\theta$ no matter which point on the terminal side of θ is used to compute $\sin\theta$. Similarly,

$$\frac{x_1}{r_1} = \frac{x_2}{r_2}, \quad \frac{y_1}{x_1} = \frac{y_2}{x_2}, \quad \frac{r_1}{y_1} = \frac{r_2}{y_2}, \quad \frac{r_1}{x_1} = \frac{r_2}{x_2}, \quad \text{and} \quad \frac{x_1}{y_1} = \frac{x_2}{y_2}.$$

Therefore, the other trig function values of θ do not depend on the particular point P chosen on the terminal side of θ to make the definitions.

If θ is acute, it can be shown that the new definitions of the six trig functions agree with the definitions given in Section 5.2.

It follows from the definition of the trig functions that the trig function values of any angle *coterminal* with θ are the same as the trig function values of θ. For example, the trig function values of $400° = 40° + 360°$ and $-320° = 40° - 360°$ are the same as the trig function values of $40°$. The trig functions have the following important property.

- **Circular Property of Trig Functions** Let θ be any angle and k any integer. The six trig function values at θ are equal to the values at $\theta + k(360°)$.

If the coordinates of a point on the terminal side of an angle are given, then the trig function values of the angle can be computed.

- **EXAMPLE 1:** Determine the values of the trig functions at an angle θ in standard position with the following point on its terminal side:

(a) $P_1(2,3)$ (b) $P_2(-3,4)$

SOLUTION:

(a) Let θ be any angle with the point $P_1(2,3)$ on its terminal side (Figure 5.4.3). The distance r from the origin to P_1 is $\sqrt{2^2+3^2} = \sqrt{13}$. We use the definitions to determine the values of the six trig functions at θ:

$$\sin\theta = \frac{3}{\sqrt{13}}, \qquad \csc\theta = \frac{\sqrt{13}}{3},$$

$$\cos\theta = \frac{2}{\sqrt{13}}, \qquad \sec\theta = \frac{\sqrt{13}}{2},$$

$$\tan\theta = \frac{3}{2}, \qquad \cot\theta = \frac{2}{3}.$$

A calculator could be used to determine decimal approximations to the above values. Notice that all trig function values are positive if the terminal side lies in the first quadrant. (Why?)

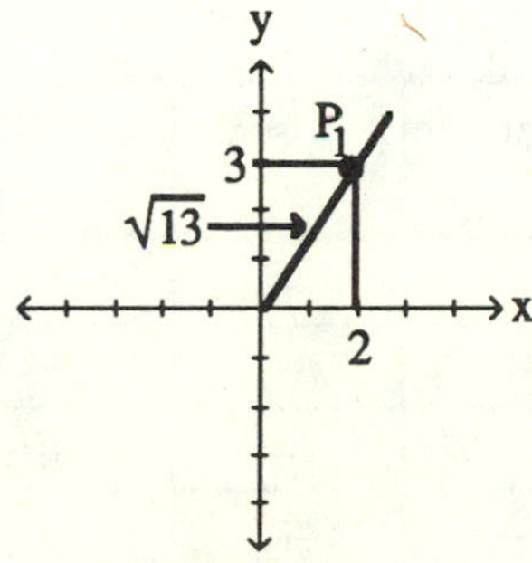

Figure 5.4.3

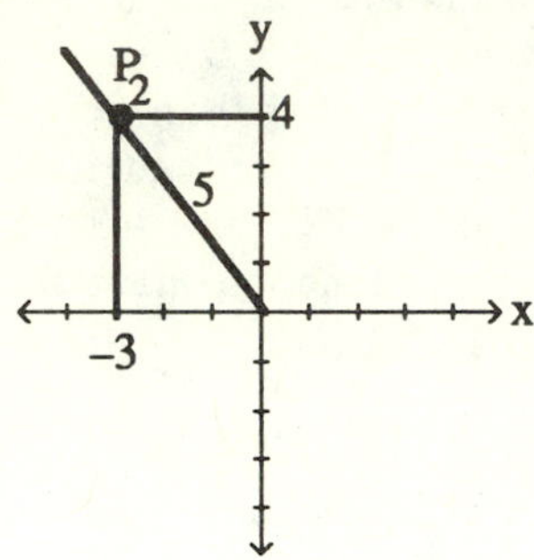

Figure 5.4.4

(b) Let θ be any angle with the point $(-3,4)$ on its terminal side (Figure 5.4.4). The distance r from the origin to P_2 is $\sqrt{(-3)^2+4^2}=5$. Again, we use the definitions to determine the values of the six trig functions at θ:

$$\sin\theta=\frac{4}{5}, \qquad \csc\theta=\frac{5}{4},$$
$$\cos\theta=\frac{-3}{5}=-\frac{3}{5}, \qquad \sec\theta=\frac{5}{-3}=-\frac{5}{3},$$
$$\tan\theta=\frac{4}{-3}=-\frac{4}{3}, \qquad \cot\theta=\frac{-3}{4}=-\frac{3}{4}.$$

Notice that $\cos\theta$, $\sec\theta$, $\tan\theta$, and $\cot\theta$ are negative, and only $\sin\theta$ and $\csc\theta$ are positive. This is true for any angle in standard position with terminal side in the second quadrant. (Why?) •

Let θ be any angle in standard position and $P(x,y)$ a point other than the origin on the terminal side of θ. The distance $r=\sqrt{x^2+y^2}$ of P from the origin is positive. The signs of x and y depend on the quadrant in which P lies. The sign of a trig function value can be positive or negative and depends on the signs of the x- and y-coordinates of P. Table 1 gives the signs of the values of the trig functions at θ for those angles where the terminal side of θ does *not* lie on the x- or y-axis.

Quadrant of θ	$\sin\theta$	$\cos\theta$	$\tan\theta$	$\csc\theta$	$\sec\theta$	$\cot\theta$
I	+	+	+	+	+	+
II	+	−	−	+	−	−
III	−	−	+	−	−	+
IV	−	+	−	−	+	−

Table 1

Signs of the Trig Functions

• EXAMPLE 2: Find the quadrant of the angle θ if $\tan\theta < 0$ and $\sin\theta < 0$.

SOLUTION: If $\tan\theta < 0$, then θ is in Quadrant II or IV. If $\sin\theta < 0$, then θ is in Quadrant III or IV. Thus, θ must be in Quadrant IV. •

Quadrantal Angles

• EXAMPLE 3: Determine the six trig function values of θ if the terminal side of θ lies on

(a) the positive x-axis.

(b) the positive y-axis.

(c) the negative x-axis.

(d) the negative y-axis.

SOLUTION: We leave (b), (c), and (d) for the exercises.

(a) The point $(1,0)$ is on the terminal side of θ. Thus, $r = \sqrt{1^2+0^2} = 1$, and

$$\sin\theta = \frac{y}{r} = \frac{0}{1} = 0, \qquad \csc\theta \text{ (no value)},$$
$$\cos\theta = \frac{x}{r} = \frac{1}{1} = 1, \qquad \sec\theta = 1,$$
$$\tan\theta = \frac{y}{x} = \frac{0}{1} = 0, \qquad \cot\theta \text{ (no value)}.$$
•

Example 3 can be summarized as shown in Table 2. (Why?)

θ coterminal with	$\sin\theta$	$\cos\theta$	$\tan\theta$	$\csc\theta$	$\sec\theta$	$\cot\theta$
0°	0	1	0	no value	1	no value
90°	1	0	no value	1	no value	0
180°	0	−1	0	no value	−1	no value
270°	−1	0	no value	−1	no value	0

Table 2

From now on we will assume that angles are in standard position unless specified otherwise. Let $P(x, y)$ be a point on the terminal side of a nonquadrantal angle θ, and draw a perpendicular from P to the x-axis to form a right triangle. Figure 5.4.5 shows four possibilities for the right triangle corresponding to the quadrant containing the terminal side of θ. The right triangle is called the *reference triangle* determined by θ, and the *acute angle* θ' made by the terminal side of θ and the positive or negative x-axis is called the *reference angle* determined by θ. Any angle coterminal with θ produces the same reference angle.

- EXAMPLE 4: Sketch the reference triangle and reference angle θ' determined by each angle θ of Example 1. Compare the trig function values of θ and θ'. Determine the measures of θ and θ'.

SOLUTION:

(a) The reference triangle and reference angle θ' of an angle θ determined by $P_1(2, 3)$ are shown in Figure 5.4.6. Note θ and θ' are coterminal. The side opposite to θ' has length 3, the side adjacent to θ' has length 2, and the hypotenuse has length $\sqrt{13}$. The right triangle definitions of the trig functions give the following:

$$\sin\theta' = \frac{3}{\sqrt{13}}, \qquad \csc\theta' = \frac{\sqrt{13}}{3},$$
$$\cos\theta' = \frac{2}{\sqrt{13}}, \qquad \sec\theta' = \frac{\sqrt{13}}{2},$$
$$\tan\theta' = \frac{3}{2}, \qquad \cot\theta' = \frac{2}{3}.$$

In Example 1(a) we found that

$$\sin\theta = \frac{3}{\sqrt{13}}, \qquad \csc\theta = \frac{\sqrt{13}}{3},$$
$$\cos\theta = \frac{2}{\sqrt{13}}, \qquad \sec\theta = \frac{\sqrt{13}}{2},$$
$$\tan\theta = \frac{3}{2}, \qquad \cot\theta = \frac{2}{3}.$$

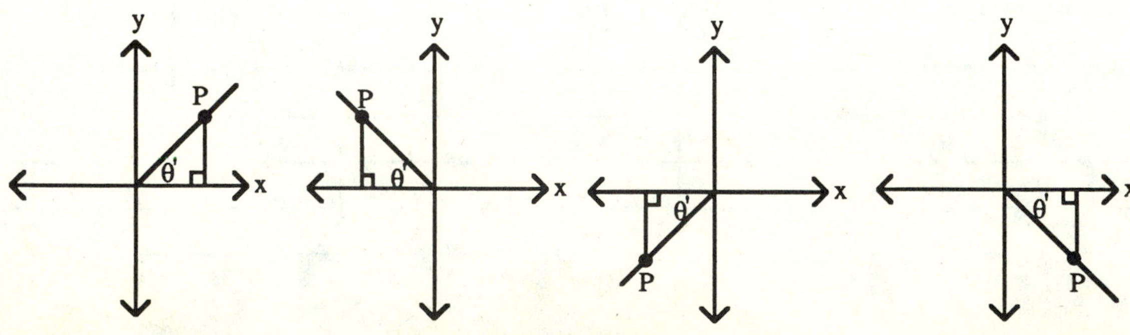

Figure 5.4.5

Therefore, the trig function values of θ and θ' are identical in this case. We can use the calculator keys [INV] [TAN], [2nd F] [TAN], or [SHIFT] [TAN] and the equation $\tan\theta' = \frac{3}{2}$ to show that θ' is 56.31°. Thus, θ has measure $56.31° + k(360°)$ for some integer k. Of course, we could have used other trig functions of θ' to determine θ'.

(b) The reference triangle and reference angle θ' of an angle determined by $P_2(-3,4)$ are shown in Figure 5.4.7. The side opposite to θ' has length 4, the side adjacent to θ' has length 3, and the hypotenuse has length 5. The right triangle definitions of the trig functions give the following:

$$\sin\theta' = \frac{4}{5}, \qquad \csc\theta' = \frac{5}{4},$$
$$\cos\theta' = \frac{3}{5}, \qquad \sec\theta' = \frac{5}{3},$$
$$\tan\theta' = \frac{4}{3}, \qquad \cot\theta' = \frac{3}{4}.$$

In Example 1(b) we found that

$$\sin\theta = \frac{4}{5}, \qquad \csc\theta = \frac{5}{4},$$
$$\cos\theta = -\frac{3}{5}, \qquad \sec\theta = -\frac{5}{3},$$
$$\tan\theta = -\frac{4}{3}, \qquad \cot\theta = -\frac{3}{4}.$$

Notice that $\sin\theta = \sin\theta'$, $\csc\theta = \csc\theta'$, but $\cos\theta = -\cos\theta'$, $\sec\theta = -\sec\theta'$, $\tan\theta = -\tan\theta'$, and $\cot\theta = -\cot\theta'$. We can use a calculator and the equation $\sin\theta' = \frac{4}{5}$ to show that $\theta' = 53.13°$. Then, θ can be any angle coterminal with $180° - \theta' = 126.87°$. Thus, θ has measure $126.87° + k(360°)$ for some integer k. •

Example 4 can be extended as follows.

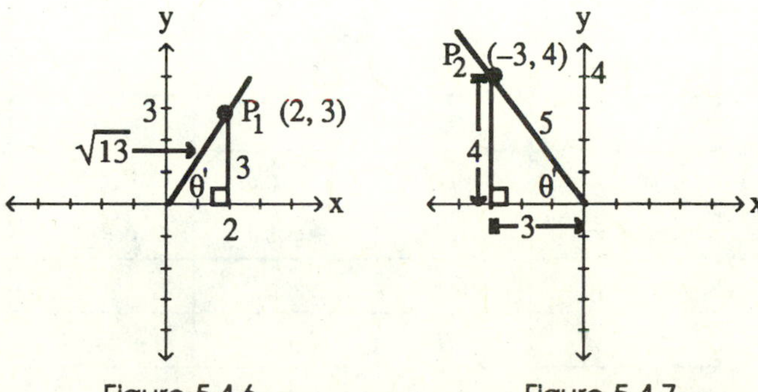

Figure 5.4.6

Figure 5.4.7

• **THEOREM 2** Except for sign, the trig function values of θ and its reference angle θ' are the same. They are the same if θ is acute. Table 3 gives a complete summary.

• EXAMPLE 5: The terminal side of an angle θ is in the fourth quadrant and lies on the line $y = -2.5x$. Determine all trig function values of θ, the measure of the reference angle θ', and the measure of θ if $0° \leq \theta \leq 360°$.

SOLUTION: The point $(2, -5)$ lies on the line $y = -2.5x$ and is in the fourth quadrant. Thus, $(2, -5)$ is on the terminal side of θ (Figure 5.4.8). The distance from the origin to $(2, -5)$ is $\sqrt{29}$. Using the reference angle and adjusting for the algebraic sign we obtain the following:

$$\sin\theta = -\frac{5}{\sqrt{29}}, \qquad \csc\theta = -\frac{\sqrt{29}}{5},$$
$$\cos\theta = \frac{2}{\sqrt{29}}, \qquad \sec\theta = \frac{\sqrt{29}}{2},$$
$$\tan\theta = -\frac{5}{2}, \qquad \cot\theta = -\frac{2}{5}.$$

We can see from Figure 5.4.8 that $\tan\theta' = \frac{5}{2}$. We can use a calculator to determine that the measure of the acute angle θ' is $68.2°$. If $0° \leq \theta < 360°$, then $\theta = 360° - 68.2° = 291.8°$. •

The terminal side of the angle θ of Example 5 lies in the fourth quadrant. Only $\cos\theta$ and $\sec\theta$ are positive in this case. This is consistent with Table 1. You should check that the signs of the trig function values of the angles in Example 1 also agree with Table 1.

Let ℓ_1 and ℓ_2 be two intersecting lines, and α and β the two angles determined by ℓ_1 and ℓ_2 as illustrated in Figure 5.4.9. Notice that $\alpha + \beta = 180°$. Thus, either α or β is acute, or $\alpha = \beta = 90°$.

Quadrant containing θ

	I	II	III	IV
$\sin\theta =$	$\sin\theta'$	$\sin\theta'$	$-\sin\theta'$	$-\sin\theta'$
$\cos\theta =$	$\cos\theta'$	$-\cos\theta'$	$-\cos\theta'$	$\cos\theta'$
$\tan\theta =$	$\tan\theta'$	$-\tan\theta'$	$\tan\theta'$	$-\tan\theta'$
$\cot\theta =$	$\cot\theta'$	$-\cot\theta'$	$\cot\theta'$	$-\cot\theta'$
$\sec\theta =$	$\sec\theta'$	$-\sec\theta'$	$-\sec\theta'$	$\sec\theta'$
$\csc\theta =$	$\csc\theta'$	$\csc\theta'$	$-\csc\theta'$	$-\csc\theta'$

Table 3

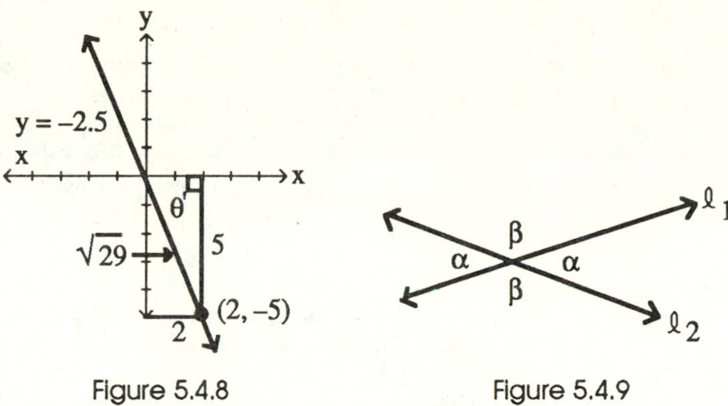

Figure 5.4.8 Figure 5.4.9

• DEFINITION The smaller, acute angle of Figure 5.4.9, or either angle if $\alpha = \beta = 90°$, is called the **angle between ℓ_1 and ℓ_2**. If $\alpha = \beta = 90°$, then ℓ_1 and ℓ_2 are perpendicular, and are also said to be **orthogonal** or **normal**. •

By convention the angle between two parallel lines is 0°.

• EXAMPLE 6: Determine the angle between the x-axis and the line $y = -2x + 3$.

SOLUTION: The angle between the x-axis and the line $y = -2x + 3$ is the same as the angle between the x-axis and the line $y = -2x$. (Why?) Let θ be the angle $y = -2x$ makes with the positive x-axis and θ' its reference angle (Figure 5.4.10). The point $(-1, 2)$ is on the terminal side of θ. Thus, $\tan\theta = -2$ and $\tan\theta' = 2$. The angle between the x-axis and the line $y = -2x$ is the acute angle θ', the reference angle of θ. We can use a calculator to determine that θ' is 66.43°. Thus, the angle between the x-axis and the line $y = -2x + 3$ is 63.43°. •

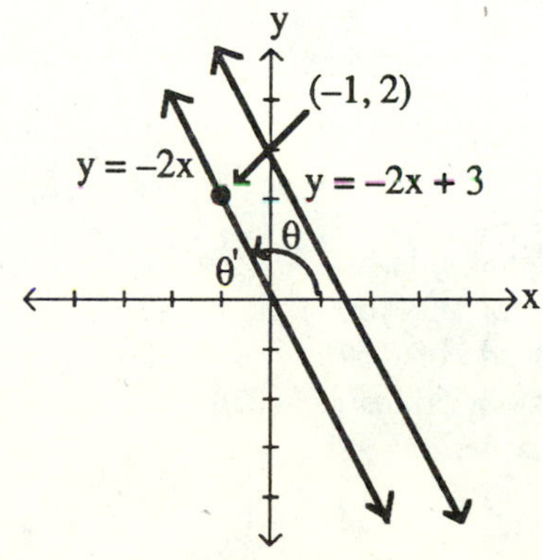

Figure 5.4.10

If the measure of an angle is known, then a calculator can be used to compute directly the trig function values of the angle. Because only three of the trig functions appear on a calculator, we need to use the following extension of Theorem 1 of Section 5.2. First we give the following definition.

• DEFINITION An equation is called an **identity** if and only if the equation is true for all values of the variables common to the domains of the expressions on each side of the equation. •

• THEOREM 3 (Reciprocal Identities) The following equations are identities.

(a) $\sin\theta = \dfrac{1}{\csc\theta}$, $\csc\theta = \dfrac{1}{\sin\theta}$ (b) $\cos\theta = \dfrac{1}{\sec\theta}$, $\sec\theta = \dfrac{1}{\cos\theta}$

(c) $\tan\theta = \dfrac{1}{\cot\theta} = \dfrac{\sin\theta}{\cos\theta}$, $\cot\theta = \dfrac{1}{\tan\theta} = \dfrac{\cos\theta}{\sin\theta}$

PROOF The proof of this theorem is a direct application of the definitions of the trig functions. We prove the second part of (b) and leave the rest of the proofs for the exercises. Let $P(x, y)$ be a point other than the origin on the terminal side of θ and $r = \sqrt{x^2 + y^2}$. Then,

$$\sec\theta = \frac{r}{x} = \frac{1}{\frac{x}{r}} = \frac{1}{\cos\theta}.$$

We use $\csc\theta = \frac{1}{\sin\theta}$, $\sec\theta = \frac{1}{\cos\theta}$, and $\cot\theta = \frac{1}{\tan\theta}$ to compute the values of $\csc\theta$, $\sec\theta$, and $\cot\theta$, respectively, with a calculator.

• EXAMPLE 7: Determine all trig function values of each angle.

(a) 470° (b) −670°

SOLUTION:

(a) We use the SIN, COS, and TAN keys together with Theorem 2 to compute the six trig function values of 470°:

$\sin 470° = 0.9396926208$, $\csc 470° = 1.064177772$,

$\cos 470° = -0.3420201433$, $\sec 470° = -2.9238044$,

$\tan 470° = -2.747477419$, $\cot 470° = -0.3639702343$. •

The terminal side of 470° lies in the second quadrant (Figure 5.4.11). Notice that the signs of the above values are consistent with Table 1.

(b) The terminal side of −670° lies in the first quadrant (Figure 5.4.12). By Table 1, the trig function values of −670° should all be positive. We compute the values with a calculator:

$\sin -670° = 0.7660444431$, $\csc -670° = 1.305407289$,

$\cos -670° = 0.6427876097$, $\sec -670° = 1.555723827$,

$\tan -670° = 1.191753593$, $\cot -670° = 0.8390996309$. •

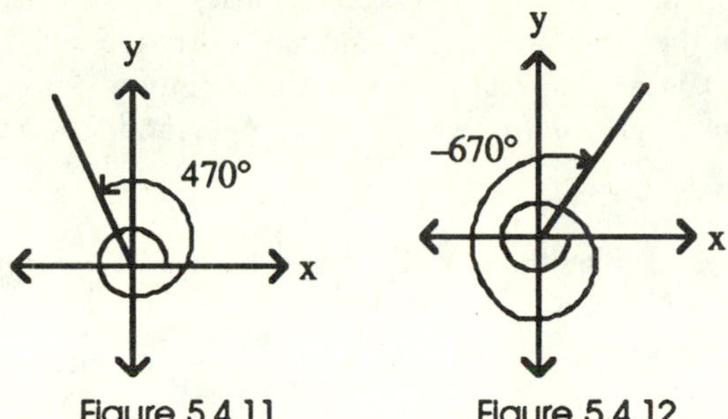

Figure 5.4.11 Figure 5.4.12

Trigonometric Identities

There are many important relationships among the trig functions. The fundamental reciprocal identities were given in Theorem 3. Theorem 4 establishes three additional very basic and fundamental identities that should be committed to memory.

• **THEOREM 4** The following equations are true for any angle θ in the domain of the corresponding trig functions:

(a) $\sin^2\theta + \cos^2\theta = 1$, (b) $1 + \tan^2\theta = \sec^2\theta$,

(c) $1 + \cot^2\theta = \csc^2\theta$.

PROOF Let $P(x, y)$ be a point other than the origin on the terminal side of θ. Then

$$\sin\theta = \frac{y}{r}, \qquad \csc\theta = \frac{r}{y},$$

$$\cos\theta = \frac{x}{r}, \qquad \sec\theta = \frac{r}{x},$$

$$\tan\theta = \frac{y}{x}, \qquad \cot\theta = \frac{x}{y},$$

where $r^2 = x^2 + y^2$. If we divide each side of the equation $x^2 + y^2 = r^2$ by r^2 we obtain

$$x^2 + y^2 = r^2$$

$$\frac{x^2}{r^2} + \frac{y^2}{r^2} = \frac{r^2}{r^2}$$

$$\left(\frac{x}{r}\right)^2 + \left(\frac{y}{r}\right)^2 = 1$$

$$\cos^2\theta + \sin^2\theta = 1.$$

Dividing each side of $x^2 + y^2 = r^2$ by x^2 ($x \neq 0$) establishes the equation in (b). Similarly, dividing each side of the same equation by y^2 ($y \neq 0$) gives the equation in (c).

The equation in Theorem 4(a) is an identity because the common domain of $\sin^2\theta+\cos^2\theta$ and of 1 is the set of all angles, and $\sin^2\theta+\cos^2\theta=1$ for every angle θ. The equation in Theorem 4(b) is an identity because the common domain of $1+\tan^2\theta$ and of $\sec^2\theta$ consists of those angles θ for which $\cos\theta\neq 0$, and $1+\tan^2\theta = \sec^2$ for those values of θ.

The identities of Theorems 3 and 4 can be used to establish other identities.

• EXAMPLE 8: Show algebraically that $\dfrac{1+\sec\theta}{\sin\theta+\tan\theta}=\csc\theta$ is an identity.

SOLUTION: You may be surprised that the complicated expression on the left-hand side of the equation is just another way of writing $\csc\theta$. The two graphs in Figure 5.4.13 support the validity of this identity by suggesting that the graphs of

$$y=\frac{1+\sec\theta}{\sin\theta+\tan\theta}$$

and $y = \csc\theta$ are identical.

A reasonable start towards verifying the identity algebraically is to rewrite the most complicated side of the equation by converting all functions to sines and cosines using identities and then simplify:

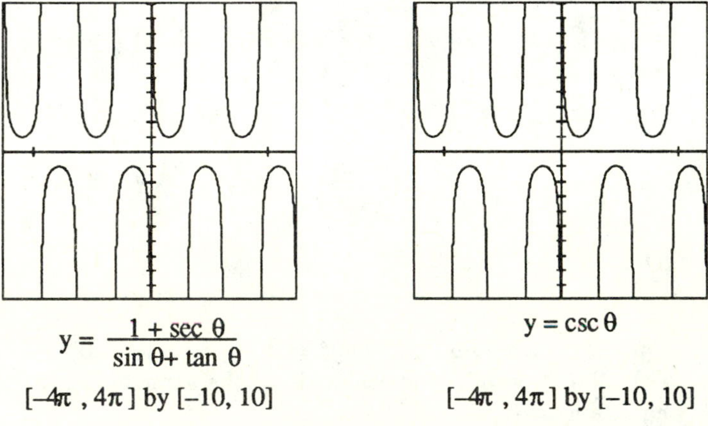

Figure 5.4.13

$$\frac{1+\sec\theta}{\sin\theta+\tan\theta} = \frac{1+\frac{1}{\cos\theta}}{\sin\theta+\frac{\sin\theta}{\cos\theta}} = \frac{\frac{1+\cos\theta}{\cos\theta}}{\frac{\sin\theta\cos\theta+\sin\theta}{\cos\theta}}$$

$$= \frac{1+\cos\theta}{\sin\theta\cos\theta+\sin\theta} = \frac{1+\cos\theta}{(\sin\theta)(1+\cos\theta)}$$

$$= \frac{1}{\sin\theta} = \csc\theta.$$

Thus, $\frac{1+\sec\theta}{\sin\theta+\tan\theta} = \csc\theta$. •

There is a subtle issue in the above example that should be mentioned. As functions, the left-hand side $f(\theta) = \frac{1+\sec\theta}{\sin\theta+\tan\theta}$ and the right-hand side $g(\theta) = \csc\theta$ are not the same because their *domains* are different. Notice that $\frac{\pi}{2}$ is in the domain of g but *not* in the domain of f. The graph of f actually has removable discontinuities at odd multiples of $\frac{\pi}{2}$ that are not visible in Figure 5.4.13.

Section 6.2 will be devoted to verifying identities.

• EXERCISES 5-4

Use a calculator unless indicated otherwise. Angles are measured in degrees.

Determine the values of the six trig functions at each angle.

1. 83°
2. −108°
3. −400°
4. 305°
5. Complete (b) and (c) of Example 3.
6. Complete (d) of Example 3.

Explain how the signs of the trigonometric functions are determined in Table 1 for each quadrant.

7. Quadrant I
8. Quadrant II
9. Quadrant III
10. Quadrant IV

Let θ be an angle in standard position with the point P on the terminal side of θ. Sketch the reference triangle and the reference angle θ' determined by each angle θ. Find the values of the six trig functions at θ and θ'. Determine the measures of θ' and θ.

11. $P = (-3, 0)$
12. $P = (0, -5)$
13. $P = (4, 3)$
14. $P = (-1, 2)$
15. $P = (-4, -6)$
16. $P = (5, -2)$
17. $P = (22, -22)$
18. $P = (-8, 1)$

Sketch the reference triangle and the reference angle θ' determined by each angle θ.

19. $156°$
20. $-305°$
21. $614°$
22. $213°$

23–26. Determine the trig function values of θ and θ' for each angle θ in Exercises 19–22.

Let θ be an angle in standard position with terminal side given by the equation. Find the values of the six trig functions at θ.

27. $y = x$ in the third quadrant.
28. $y = 2x$ in the first quadrant.
29. $y = \frac{3}{4}x$ in the first quadrant.
30. $y = -x$ in the second quadrant.
31. $y = \frac{2}{3}x$ in the third quadrant.
32. $y = -\frac{3}{5}x$ in the fourth quadrant.

Determine the measure of the reference angle θ' for the given angle θ.

33. $106°$
34. $-584°$
35. $-157°$
36. $287°$
37. $214°$
38. $-200°$

Determine the angle between the x-axis and each line.

39. $y = 2x$
40. $2x - 3y = 4$
41. $y = -3x + 1$
42. $3x - 5y = -2$

θ is an angle in standard position. Find the quadrant containing the terminal side of θ.

43. $\sin\theta < 0$ and $\tan\theta > 0$
44. $\cos\theta > 0$ and $\tan\theta < 0$
45. $\tan\theta > 0$ and $\sec\theta < 0$
46. $\sin\theta > 0$ and $\cos\theta < 0$

Verify the identity algebraically and confirm graphically.

47. $\cot\theta\tan\theta = 1$
48. $\sin\theta\sec\theta = \tan\theta$
49. $\sec^2\theta - \sin^2\theta = \cos^2\theta + \tan^2\theta$
50. $\dfrac{1}{\sin^2\theta} + \dfrac{1}{\cos^2\theta} = \dfrac{1}{\sin^2\theta\cos^2\theta}$

51. A wire stretches from the top of a vertical pole to a point on level ground 16 feet from the base of the pole. If the wire makes an angle of 62° with the ground, determine the height of the pole and the length of the wire.

52. A lighthouse at L stands three miles from the closest point P along a straight shore. Determine the distance from P to a point Q along the shore if the angle PQL measures 35°.

53. Using a sextant, a surveyor determines that the angle of elevation of a mountain peak is 35°. Moving 1000 feet further away from the mountain the surveyor determines the angle of elevation to be 30°. Determine the height of the mountain.

	θ	$\sin\theta$	$\cos\theta$	$\tan\theta$	$\cot\theta$	$\sec\theta$	$\csc\theta$
(a)	0						
(b)	30°						
(c)	45°						
(d)	60°						
(e)	90°						
(f)	180°						
(g)	270°						
(h)	360°						

Table 4

54. Prove (a) and (c) of Theorem 4.
55. Use properties of a 30°–60° right triangle and a 45° right triangle (see Figures 5.2.4 and 5.2.5) to determine the *exact* value of each trig function in Table 4.
56. Use properties of a 30°–60° right triangle and a 45° right triangle (see Figures 5.2.4 and 5.2.5) to determine the *exact* value of each trig function in Table 5.

	θ	$\sin\theta$	$\cos\theta$	$\tan\theta$	$\cot\theta$	$\sec\theta$	$\csc\theta$
(a)	240°						
(b)	315°						
(c)	−30°						
(d)	480°						
(e)	−540°						
(f)	−135°						
(g)	13230°						
(h)	36000°						

Table 5

5.5 Radian Measure and Graphs of $\sin x$ and $\cos x$

In this section we introduce radian measure and convert degree measure to radian measure and vice versa. Using radians, we define the trig functions for real numbers. The graphs of $\sin x$ and $\cos x$ will be drawn and used to solve equations.

Radian Measure

Recall from Section 5.1 that an angle θ is called a central angle if its vertex is the center of a circle. Any angle can be made a central angle by placing its vertex at the center of some circle with initial and terminal sides along radii of the circle.

• DEFINITION A central angle with positive measure in a circle of radius r that subtends an arc of length r is said to have measure **1 radian** (Figure 5.5.1). •

Because the length of the circumference of a circle of radius r is $2\pi r$, the circumference consists of 2π segments of length r. Thus, the radian measure of the angle of one complete rotation, or revolution counterclockwise, is 2π. An angle has *negative* measure if the rotation that produces the angle is clockwise. We use this convention for angles measured in degrees or radians. The radian measure of the angle of one-half revolution clockwise is $-\pi$. These are special cases of the following more general result.

• THEOREM 1 Let the central angle θ in a circle of radius r subtend an arc of length s (Figure 5.5.2). The radian measure of θ is s/r if θ has positive measure, or $-s/r$ if θ has negative measure.

We can obtain the following relationships from Theorem 1.

• Degree-Radian Equivalent Measures

$$360^\circ = 2\pi \text{ radians}, \qquad 180^\circ = \pi \text{ radians},$$

$$1^\circ = \frac{\pi}{180} \text{ radians}, \qquad 1 \text{ radian} = \left(\frac{180}{\pi}\right)^\circ.$$

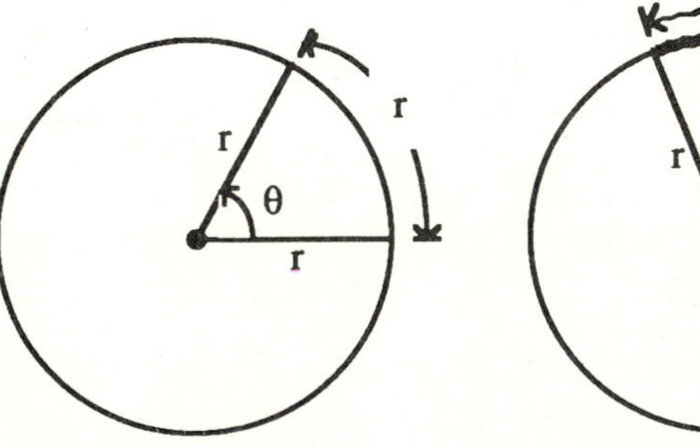

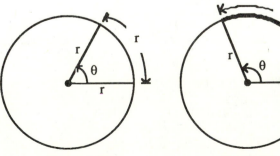

Figure 5.5.1 Figure 5.5.2

We can use a calculator to see that 1° is approximately 0.01745329252 radians and 1 radian is approximately 57.59577951°. To convert radian measure to degree measure we multiply the radian measure by $\frac{180}{\pi}$, and to convert degree measure to radian measure we multiply the degree measure by $\frac{\pi}{180}$.

• **EXAMPLE 1:** Find the radian measure of an angle whose measure is

(a) 120°. (b) 500°. (c) -200°.

SOLUTION: We know that $1^\circ = \frac{\pi}{180}$ radians. Thus, each degree measure must be multiplied by $\frac{\pi}{180}$.

(a) $\frac{\pi}{180}(120) = \frac{2\pi}{3}$. Thus, $120^\circ = \frac{2\pi}{3}$ radians. We can use a calculator to show that $\frac{2\pi}{3}$ radians is approximately 2.094395102 radians.
(b) $500^\circ = \frac{\pi}{180}(500)$ radians, which is approximately 8.72664626 radians.
(c) $-200^\circ = \frac{\pi}{180}(-200)$ radians, which is approximately -3.490658504 radians. •

• **EXAMPLE 2:** Find the degree measure of an angle whose measure is

(a) $\frac{7\pi}{6}$ radians. (b) -8 radians.

SOLUTION: We know that 1 radian $= \left(\frac{180}{\pi}\right)^\circ$. Thus, each radian measure must be multiplied by $\frac{180}{\pi}$.

(a) $\frac{7\pi}{6}$ radians $= \frac{180}{\pi}\left(\frac{7\pi}{6}\right) = 210^\circ$.
(b) -8 radians $= \frac{180}{\pi}(-8)$, which is approximately -458.3662361°. •

Arc Length If we use the radian measure θ of an angle as the name of the angle, then we can rewrite Theorem 1 in the form $|\theta| = \frac{s}{r}$, or $s = r|\theta|$. This equation is not valid if θ is in degrees. The variable s is the length of the arc subtended by the angle of measure θ radians.

• **Radian Measure Convention** From now on we will *not* usually write the word *radian* when giving the measure of an angle in radians. For example, if the measure of an angle is 2 radians we usually say that the measure of the angle is 2. We continue to use the established abbreviation for degree measure. That is, 2° stands for an angle whose measure is 2 degrees.

• **EXAMPLE 3:** In a circle of radius 3 inches, find the length of the arc subtended by an angle of measure

(a) 2. (b) -60°.

SOLUTION:

(a) Now, $r = 3$ and $\theta = 2$ so that $s = r|\theta| = 6$ (Figure 5.5.3). Thus, the length of the arc subtended by an angle of measure 2 radians in a circle of radius 3 inches is 6 inches.

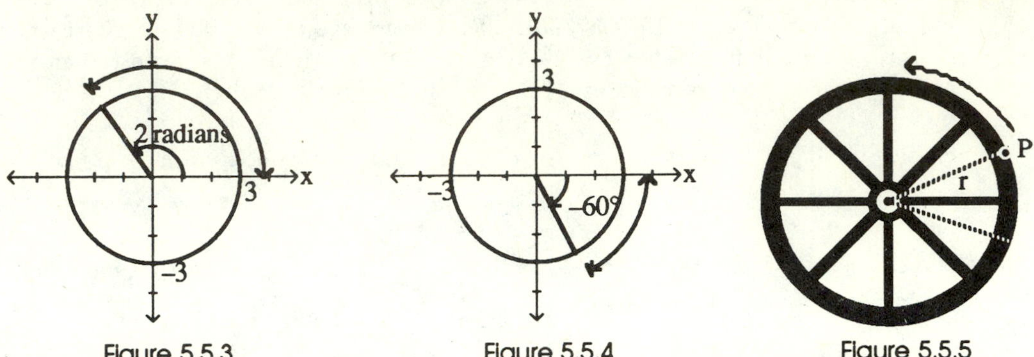

Figure 5.5.3 Figure 5.5.4 Figure 5.5.5

(b) First, we convert $-60°$ to $-\frac{\pi}{3}$ radians. Then, $r = 3$ and $\theta = -\frac{\pi}{3}$ so that $s = r|\theta| = \pi$ (Figure 5.5.4). Thus, the length of the arc subtended by an angle of measure $-60°$ in a circle of radius 3 inches is π inches. •

Angular and Linear Speed

A wheel of radius r is rotating at a constant rate (Figure 5.5.5). Let P be a point on the circumference of the wheel.

• DEFINITION The **angular speed** of the wheel is the angle formed or swept out in one unit of time by the line segment from the center of the wheel to the point P on the circumference of the wheel. The **linear speed** of the point P is the distance P travels in one unit of time. •

• EXAMPLE 4: A wheel of radius 18 inches is rotating at 850 revolutions per minute (rpm). Determine the

(a) angular speed of the wheel.
(b) linear speed of a point on the circumference of the wheel.

SOLUTION:

(a) The number of revolutions per minute is 850, and each revolution produces an angle of 2π radians. Thus, the angle formed by the wheel in one minute is $850(2\pi) = 1700\pi$. The angular speed of the wheel is 1700π radians per minute.
(b) The distance traveled by a point on the wheel in one minute is given by $s = r\theta$ where r is 1.5 feet and θ is 1700π radians. Thus, the linear speed of a point on the wheel is 8011.06 feet per minute. •

Circular Functions

A circle with center at the origin and radius 1 is called a *unit circle* (Figure 5.5.6). We construct a vertical number line tangent to the unit circle at the point $(1, 0)$, that is, a

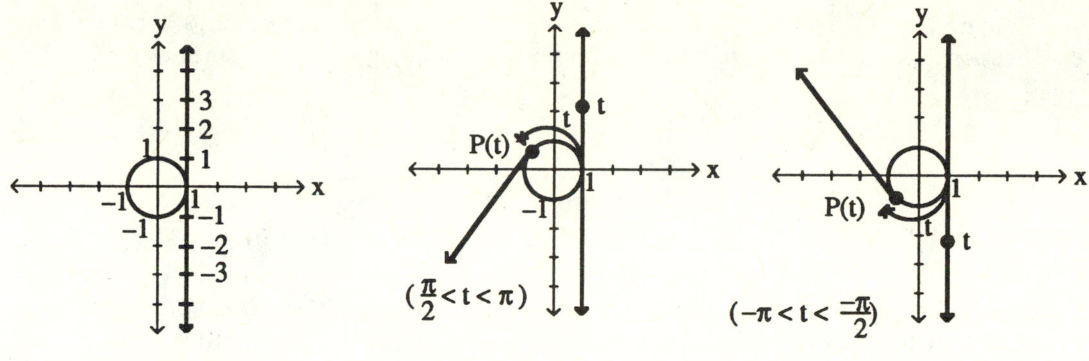

Figure 5.5.6

Figure 5.5.7

line parallel to the y-axis through the point $(1,0)$. Every real number t on the number line corresponds to a point $P(t)$ on the unit circle in the following way. If $t > 0$, we wrap the positive portion of the vertical line in Figure 5.5.6 counterclockwise around the unit circle until the point with coordinate t on the number line meets the unit circle (Figure 5.5.7). We name this point $P(t)$. If $t < 0$, the negative portion of the vertical line is wrapped clockwise around the unit circle.

For example, the points $\frac{\pi}{2}$, π, $\frac{3\pi}{2}$, and $-\frac{\pi}{2}$ on the vertical number line correspond to the points $(0,1)$, $(-1,0)$, $(0,-1)$, and $(0,-1)$, respectively, on the unit circle (Figure 5.5.8).

We know from Theorem 1 that $s = |\theta|$ because the radius of the unit circle is 1. It follows that the radian measure of the central angle θ determined by the arc of the circle from $(1,0)$ to $P(t)$ is t (Figure 5.5.9). (Why?) There are infinitely many points on the vertical number line that correspond to the same point on the unit circle. For example, the points $\frac{\pi}{2}$, $\frac{\pi}{2}+2\pi$, $\frac{\pi}{2}+4\pi$, and so forth, on the vertical number line correspond to the point $(0,1)$ on the unit circle. However, there is a one-to-one correspondence between the points of $[0,2\pi)$ of the vertical number line and the points of the unit circle. This follows because the circumference of a unit circle is 2π.

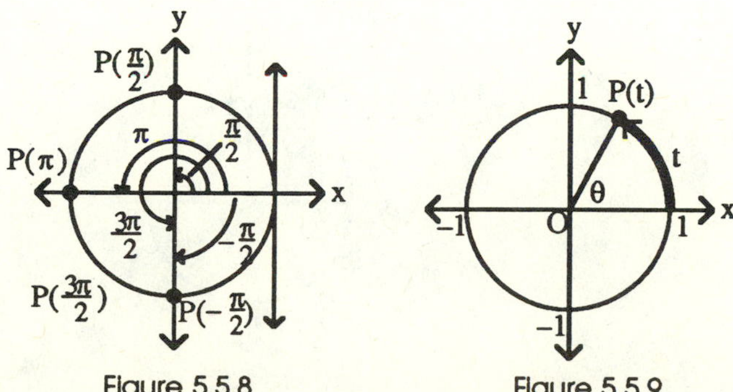

Figure 5.5.8

Figure 5.5.9

Let $P(t) = (x, y)$ be the point on the unit circle corresponding to the real number t (Figure 5.5.10). Let θ be any angle in standard position with $P(t)$ on its terminal side. In Section 5.4, we gave the following definitions for the trig function values at θ:

$$\sin\theta = \frac{y}{r}, \qquad \csc\theta = \frac{r}{y}\ (y \neq 0),$$

$$\cos\theta = \frac{x}{r}, \qquad \sec\theta = \frac{r}{x}\ (x \neq 0),$$

$$\tan\theta = \frac{y}{x}\ (x \neq 0), \qquad \cot\theta = \frac{x}{y}\ (y \neq 0).$$

Now, $r = \sqrt{x^2 + y^2} = 1$ because (x, y) is on a unit circle. Thus, $y = \sin\theta$ and $x = \cos\theta$. The coordinates of $P(t)$ are the sine and cosine of an angle in standard position with $P(t)$ on its terminal side.

• **DEFINITION** Let t be any real number, and let $P(t) = (x, y)$ be the point on the unit circle corresponding to t. We define the six trig function values at t as follows:

$$\sin t = y, \qquad \csc t = \frac{1}{y}\ (y \neq 0),$$

$$\cos t = x, \qquad \sec t = \frac{1}{x}\ (x \neq 0),$$

$$\tan t = \frac{y}{x}\ (x \neq 0), \qquad \cot t = \frac{x}{y}\ (y \neq 0).$$ •

Because of the previous discussion, we know that these definitions of the trig functions are consistent with the definitions in Section 5.4. Furthermore, it can be shown that the reciprocal identities of Theorem 2 of Section 5.4 hold for the trig functions defined as above. The trig functions are often called **circular functions** because of the above definition that involves the unit circle. Because of the wrapping property illustrated in Figure 5.5.7, the circular property of the trig functions given in Section 5.4 can be extended to the trig functions with angles measured in radians.

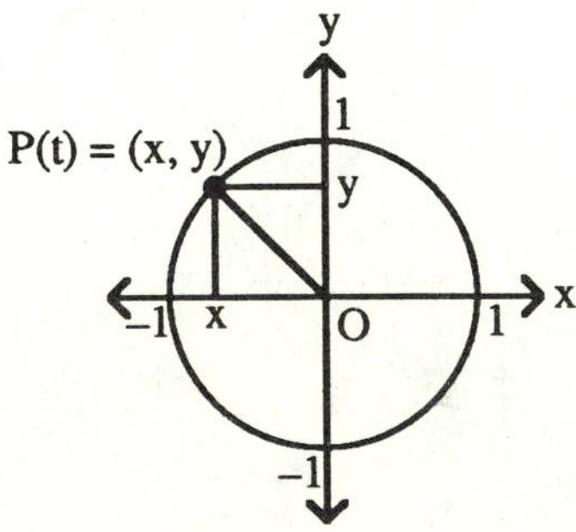

Figure 5.5.10

• **Circular Property of Trig Functions** Let k be any integer and x any real number. The six trig function values at x are equal to the values at $x + 2k\pi$.

• EXAMPLE 5: Determine the sign of the six trig function values at 4.5.

SOLUTION: An angle of measure 4.5 radians is in Quadrant III. (Why?) Thus, only $\tan 4.5$ and $\cot 4.5$ are positive. The other values are negative. •

A calculator can be used to compute the trig function values at a real number t. In this case, t is also the radian measure of the angle determined by the arc from $(1,0)$ to $P(t)$ on the unit circle. To compute the trig function value at a real number t with a calculator, we must put the calculator in *radian mode.*

• EXAMPLE 6: Use a calculator to find the values of the trig functions at

(a) 4. (b) −3.5.

SOLUTION: Put your calculator in radian mode and then use the trig keys to make the following computations. Remember we now give answers accurate to hundredths unless otherwise specified.

(a) $\sin 4 = -0.76$ $\csc 4 = -1.32$

$\cos 4 = -0.65$ $\sec 4 = -1.53$

$\tan 4 = 1.16$ $\cot 4 = 0.86$

(b) $\sin(-3.5) = 0.35$ $\csc(-3.5) = 2.85$

$\cos(-3.5) = -0.94$ $\sec(-3.5) = -1.07$

$\tan(-3.5) = -0.37$ $\cot(-3.5) = -2.67$ •

Graphs of sin x and cos x

• EXAMPLE 7: Determine the domain, the range, and the zeros of $f(x) = \sin x$ and draw a complete graph. Show that $f(x + 2k\pi) = f(x)$ for any integer k.

SOLUTION: Let x be any real number and $P(x)$ the corresponding point on the unit circle (Figure 5.5.11). The coordinates of $P(x)$ are $(\cos x, \sin x)$, and the central angle determined by the arc of the circle from $(1,0)$ to $P(x)$ has radian measure x. We have drawn Figure 5.5.11 with $0 < x < \frac{\pi}{2}$. If $x = 0$, then $\sin x = 0$. As x varies from 0 to $\frac{\pi}{2}$, $\sin x$ varies from 0 to 1. Similarly, the following values of x produce the corresponding values of $\sin x$

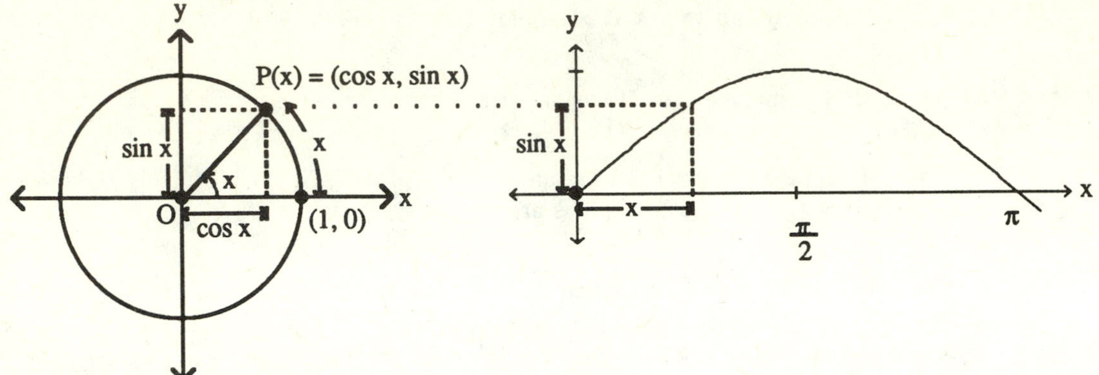

Figure 5.5.11

in the specified order:

$$
\begin{aligned}
0 \le x \le \frac{\pi}{2}, &\qquad 0 \le \sin x \le 1, \\
\frac{\pi}{2} \le x \le \pi, &\qquad 1 \ge \sin x \ge 0, \\
\pi \le x \le \frac{3\pi}{2}, &\qquad 0 \ge \sin x \ge -1, \\
\frac{3\pi}{2} \le x \le 2\pi, &\qquad -1 \le \sin x \le 0.
\end{aligned}
$$

Notice that $\sin x = 0$ in $[0, 2\pi]$ when $x = 0, \pi$, or 2π.

Because of the circular property of the sine function, as x varies from 2π to 4π, $\sin x$ varies through the same values in the same order as above:

$$
\begin{aligned}
2\pi \le x \le \frac{5\pi}{2}, &\qquad 0 \le \sin x \le 1, \\
\frac{5\pi}{2} \le x \le 3\pi, &\qquad 1 \ge \sin x \ge 0, \\
3\pi \le x \le \frac{7\pi}{2}, &\qquad 0 \ge \sin x \ge -1, \\
\frac{7\pi}{2} \le x \le 2\pi, &\qquad -1 \le \sin x \le 0.
\end{aligned}
$$

In fact, as x varies from $2n\pi$ to $(2n + 2)\pi$, where n is a positive integer, $\sin x$ varies through the same values in the same order as when x varies from 0 to 2π. If x is negative, then we mark off an arc of length $|x|$ on the unit circle starting at $(1, 0)$ and move clockwise (Figure 5.5.12).

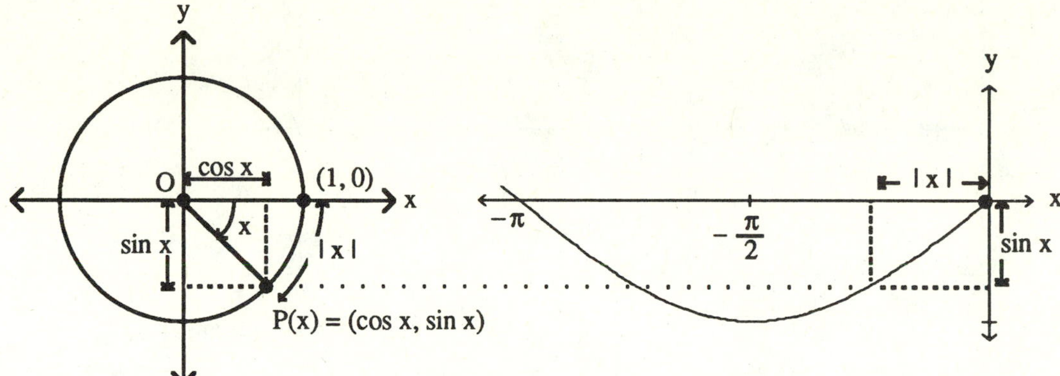

Figure 5.5.12

Notice that the following values of x produce the corresponding values of $\sin x$ in the specified order:

$$\begin{aligned} 0 \geq x &\geq -\frac{\pi}{2}, & 0 &\geq \sin x \geq -1, \\ -\frac{\pi}{2} \geq x &\geq -\pi, & -1 &\leq \sin x \leq 0, \\ -\pi \geq x &\geq -\frac{3\pi}{2}, & 0 &\leq \sin x \leq 1, \\ -\frac{3\pi}{2} \geq x &\geq -2\pi, & 1 &\geq \sin x \geq 0. \end{aligned}$$

The zeros of f in $[-2\pi, 0]$ occur when $x = -2\pi, -\pi$, or 0. Similarly, as x varies from $(-2n-2)\pi$ to $-2n\pi$ where n is a positive integer, $\sin x$ varies through the same values in the same order as when x varies from -2π to 0. Now, we can see that the range of $f(x) = \sin x$ is $[-1, 1]$ and the domain of f is the set of all real numbers. Moreover, $f(x + 2k\pi) = f(x)$ for any integer k. The computer drawn graph of $f(x) = \sin x$ given in Figure 5.5.13 is a complete graph. The zeros of f occur at integer multiples of π, that is, when $x = k\pi, k$ any integer. •

• DEFINITION A function f is said to be **periodic** if there is a positive real number h such that $f(x + h) = f(x)$ for every x in the domain of f. The smallest such positive number h is called the **period** of f. •

It follows from Example 7 that 2π is a candidate for the period of the function $f(x) = \sin x$. It is clear from the graph in Figure 5.5.13 that no smaller number can be the period. Thus, 2π is the period of $\sin x$. The graph of $f(x) = \sin x$ is completely determined by its graph in the interval $[0, 2\pi]$, because the length of this interval is 2π. Therefore, the graph of f in Figure 5.5.14 is also a complete graph if we indicate that f is periodic with period 2π.

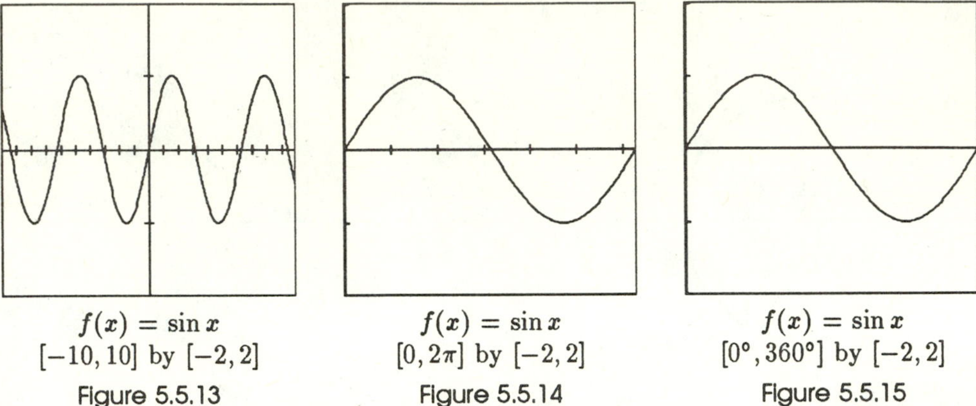

$f(x) = \sin x$
$[-10, 10]$ by $[-2, 2]$
Figure 5.5.13

$f(x) = \sin x$
$[0, 2\pi]$ by $[-2, 2]$
Figure 5.5.14

$f(x) = \sin x$
$[0^\circ, 360^\circ]$ by $[-2, 2]$
Figure 5.5.15

• EXAMPLE 8: Find the coordinates of the points at which $f(x) = \sin x$ has a local maximum value or a local minimum value.

SOLUTION: The range of f is $[-1, 1]$ so that $f\left(\frac{\pi}{2}\right) = \sin\frac{\pi}{2} = 1$ is a local maximum value of f. Similarly, $f\left(-\frac{\pi}{2}\right) = \sin\left(-\frac{\pi}{2}\right) = -1$ is a local minimum value of f. Now, for any integer k, the circular property of the trig functions gives $f\left(\frac{\pi}{2} + 2k\pi\right) = f\left(\frac{\pi}{2}\right) = 1$ and $f\left(-\frac{\pi}{2} + 2k\pi\right) = f\left(-\frac{\pi}{2}\right) = -1$. Thus, f has local maximum values at $\frac{\pi}{2} + 2k\pi$ and local minimum values at $-\frac{\pi}{2} + 2k\pi$. •

The domain of $f(x) = \sin x$ in Example 7 is the set of all real numbers. We can interpret radian measures of angles as real numbers. Thus, we can view x in $\sin x$ as an angle. In this case the domain of f consists of all angles with measure in radians. Moreover, if θ is the equivalent degree measure of x radians, the domain of $f(\theta) = \sin\theta$ consists of all angles with measures in degrees. The graph of $f(\theta) = \sin\theta$ in the $[0^\circ, 360^\circ]$ by $[-2, 2]$ viewing rectangle represents one complete period of f with angle measure in degrees (Figure 5.5.15). Compare this graph with the one in Figure 5.5.14. From now on, we usually display computer drawn trig graphs with real numbers (radians) as viewing rectangle parameters.

Notice that the graph of $f(x) = \sin x$ in Figure 5.5.13 is *symmetric with respect to the origin*. Thus $\sin x$ is an *odd* function and $\sin(-x) = -\sin x$. Alternatively, we can use the definition of the sine function to show that $\sin(-x) = -\sin x$. Let $P(x)$ and $P(-x)$ be the points on the unit circle corresponding to x and $-x$ (Figure 5.5.16). We have drawn Figure 5.5.16 assuming $x > 0$ and $P(x)$ is in the first quadrant. The following argument is valid without these restrictions. If the coordinates of $P(x)$ are (a, b), then the coordinates of $P(-x)$ are $(a, -b)$. We also know that $b = \sin x$ and $-b = \sin(-x)$ by the definition of the trig functions. Replacing b by $\sin x$ in $-b = \sin(-x)$ gives $-\sin x = \sin(-x)$.

• EXAMPLE 9: Determine all solutions of $\sin x = 0.4$ in the interval $0 \le x \le 2\pi$.

SOLUTION: Figure 5.5.17 gives the graphs of $f(x) = \sin x$ and $g(x) = 0.4$ in the $[0, 2\pi]$ by $[-2, 2]$ viewing rectangle. There are two values of x satisfying $\sin x = 0.4$; one is in the first

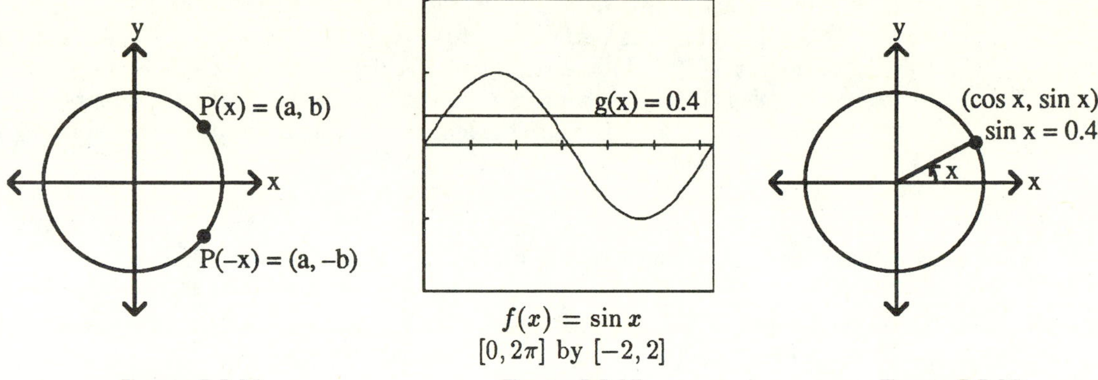

Figure 5.5.16

$f(x) = \sin x$
$[0, 2\pi]$ by $[-2, 2]$

Figure 5.5.17

Figure 5.5.18

quadrant, and the other is in the second quadrant. (Why?) One way to determine the two values of x is to use zoom-in. Alternatively, we can use a calculator to find the two values of x. Suppose x is in the first quadrant (Figure 5.5.18). We can put a calculator in radian mode (Why?) and use the INV SIN, 2nd F SIN, or SHIFT SIN keys to compute directly that $x = 0.41$. Now suppose that x is in the second quadrant (Figure 5.5.19). In this case the reference angle x' is the value found previously using a calculator. Thus, $x' = 0.41$ and $x = \pi - 0.41 = 2.73$ is the second solution of $\sin x = 0.4$ in $0 \leq x \leq 2\pi$. If you obtained $x = 23.58$, check the mode of your calculator. •

We will obtain a complete graph of $\cos x$ by first establishing that $\cos x = \sin(x + \frac{\pi}{2})$ for all x. Then, the graph of $y = \cos x$ is simply the graph of $g(x) = \sin x$ shifted horizontally left $\frac{\pi}{2}$ units.

• THEOREM 2 $\cos t = \sin(t + \frac{\pi}{2})$ for all t.

PROOF We give the proof with t in the first quadrant. The technique used in this case is valid regardless of the quadrant of the angle t. Figure 5.5.20 shows

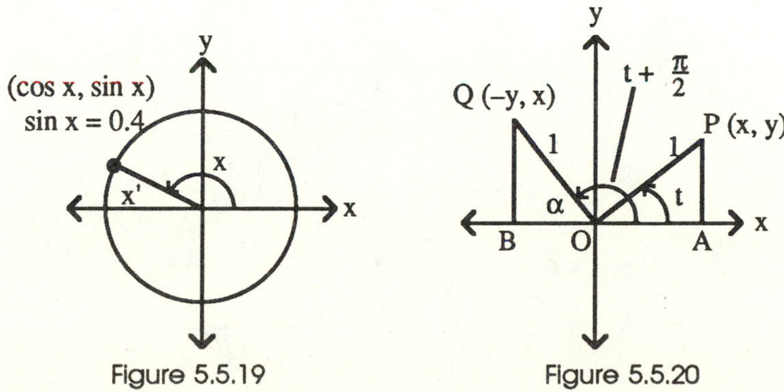

Figure 5.5.19

Figure 5.5.20

the angles t and $t + \frac{\pi}{2}$. We construct the triangles so that the length of OP and OQ are 1, and OQ is perpendicular to OP. The measure of the angle α of triangle OQB is $\pi - (t + \frac{\pi}{2}) = \frac{\pi}{2} - t$. Thus, the right triangles OQB and OPA are congruent. (Why?) Let the coordinates of P be (x, y). Then, the coordinates of Q are $(-y, x)$. Now, $x = \cos t$ because the length of OP is 1. Similarly, $x = \sin(t + \frac{\pi}{2})$ because the length of OQ is 1. Therefore, $\cos t = x = \sin(t + \frac{\pi}{2})$.

• **EXAMPLE 10:** Determine the domain, the range, the period, and the zeros of $f(x) = \cos x$ and draw a complete graph. Show that $f(x + 2k\pi) = f(x)$ for any integer k.

SOLUTION: From Theorem 2 we know that $f(x) = \cos x = \sin(x + \frac{\pi}{2})$. Thus, the graph of f is the graph of $y = \sin x$ shifted horizontally left $\frac{\pi}{2}$ units. The computer drawn graph of f (solid graph) in Figure 5.5.21 confirms this observation and is a complete graph. The period of f must be the same as $\sin x$. Thus, the period of $\cos x$ is 2π. The domain of $f(x) = \cos x$ is the set of all real numbers; the range is $[-1, 1]$. For any integer k, $f(x + 2k\pi) = f(x)$ by the circular property of the cosine function. This fact also follows from the equation $f(x) = \sin(x + \frac{\pi}{2})$ and Example 7 as follows:

$$\begin{aligned} f(x + 2k\pi) &= \sin(x + 2k\pi + \frac{\pi}{2}) \\ &= \sin((x + \frac{\pi}{2}) + 2k\pi) \\ &= \sin(x + \frac{\pi}{2}) \\ &= f(x). \end{aligned}$$

The zeros of $f(x) = \cos x$ occur when $x = k\pi - \frac{\pi}{2}$. (Why?) We can rewrite $k\pi - \frac{\pi}{2}$ in the form $(2k - 1)\frac{\pi}{2}$. Thus, the zeros of $\cos x$ occur at odd multiples of $\frac{\pi}{2}$. •

Notice that the graph of $f(x) = \cos x$ in Figure 5.5.21 is *symmetric with respect to the y-axis* so that it is an even function. In the exercises, you will be asked to show that $f(x) = \cos x$ is an even function using the wrapping property of the trig functions.

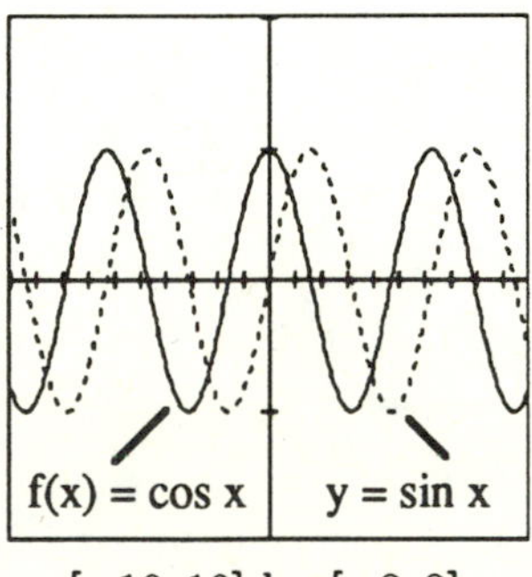

$[-10, 10]$ by $[-2, 2]$

Figure 5.5.21

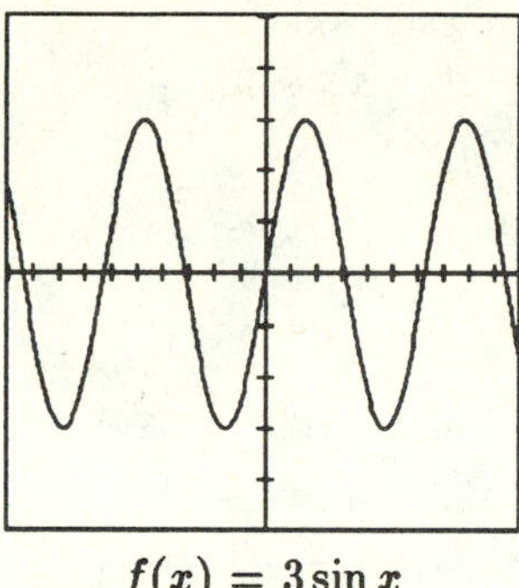

$f(x) = 3\sin x$
$[-10, 10]$ by $[-5, 5]$
Figure 5.5.22

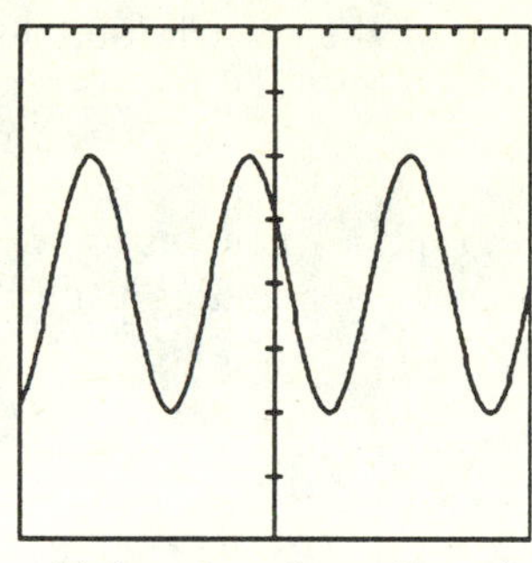

$f(x) = 2\cos(x+1) - 4$
$[-10, 10]$ by $[-8, 0]$
Figure 5.5.23

• EXAMPLE 11: Determine the domain, the range, and the period of each function, and draw a complete graph.

(a) $f(x) = 3\sin x$ (b) $g(x) = 2\cos(x+1) - 4$

SOLUTION:

(a) The graph of $f(x) = 3\sin x$ can be obtained by applying a vertical stretch factor of 3 to $y = \sin x$. The computer drawn graph of f in Figure 5.5.22 confirms this observation and is a complete graph. The domain of f is the set of all real numbers, the range is $[-3, 3]$, and the period of f is 2π. (Why?)

(b) The graph of $g(x) = 2\cos(x+1)-4$ can be obtained by applying a vertical stretch factor of 2 to $y = \cos x$, followed by a horizontal shift left one unit and then a vertical shift down 4 units. The graph of f in Figure 5.5.23 is complete. The domain of f is $(-\infty, \infty)$, the range is $[-6, -2]$, and the period is 2π. (Why?) •

The order in which the transformations were applied in Example 11(b) is important. You will see in the Exercises that changing the order will not necessarily lead to the same graph.

• EXERCISES 5-5

Use a calculator unless stated otherwise. Angles are measured in degrees or radians.

Find the equivalent radian measure.

1. $1°$
2. $55°$
3. $30°$
4. $72°$
5. $-120°$
6. $-380°$

Find the equivalent degree measure of each radian measure.

7. 1
8. 4
9. $\dfrac{3\pi}{4}$
10. $\dfrac{2\pi}{3}$
11. -2.3
12. $\dfrac{15\pi}{6}$

Determine the values of the six trig functions at each angle.

13. $200°$
14. 3.8 radians
15. -2 radians
16. $-115°$

Sketch a complete graph without using a graphing utility. State the domain, range, and period. Check your answer with a graphing utility.

17. $y = \sin 3x$
18. $g(x) = 2 + \sin x$
19. $y = 2 + \sin(x - 45°)$
20. $y = -1 - \cos(x - \pi)$
21. $y = -1 + 3\cos(x - 2)$
22. $y = 2 - 2\cos(x + 3)$
23. $y = -1 + 3\sin(x - 2)$
24. $y = 2 - 2\sin(x + 180°)$

List, in order, the geometric transformations that can be used to obtain each graph from the graph of $y = \sin x$ or $y = \tan x$.

25. $y = -2\sin(x - 180°)$
26. $y = 2 + \tan\left(x - \frac{\pi}{4}\right)$

Use the circular properties of the trig functions to verify the identity.

27. $\cos(-x) = \cos x$ (cosine is an *even* function)
28. $\tan(-x) = -\tan x$ (tangent is an *odd* function)

Use the circular properties of the trig functions to verify each statement. Do not use a calculator. *Hint:* Draw a unit circle and approximate arcs of length 2 and 3 on the circumference from the point $(1, 0)$.

29. $\sin 2 > \sin 3$
30. $\cos 2 > \cos 3$
31. $\sin 10 < 0$
32. $\cos(\sin 3) > 0$

Solve for x if $0 \leq x \leq 2\pi$.

33. $\sin x = 0.5$
34. $\sin x = -0.6$
35. $\sin x = 0.8$
36. $\sin x = \dfrac{\sqrt{3}}{2}$
37. $\cos x = 0.6$
38. $\cos x = -0.4$

Solve the inequality if $0° \leq \theta \leq 360°$.

39. $\sin\theta < 0.6$
40. $3\cos\theta < 4$
41. Which of the following statements are true for every α between $0°$ and $90°$ or every x satisfying $0 < x < \frac{\pi}{2}$? Justify your conclusion with graphs.
    (a) $\sin(90° - \alpha) = \cos\alpha$
    (b) $\cos\left(\frac{\pi}{2} - x\right) = \sin x$
    (c) $\sin(\pi + x) = -\cos x$
    (d) $\tan(90° - \alpha) = \dfrac{1}{\tan\alpha}$
42. Show that no number less than 2π can be the period of $y = \sin x$.

Let θ be a central angle of a circle of radius r and s the length of the subtended arc. Find the arc length s.

43. $\theta = 22°$, $r = 15$ inches
44. $\theta = 3$ radians, $r = 5.6$ feet

A simple winch that is used to lift heavy objects is positioned 10 feet above ground level. Assume that the radius of the winch is r feet as shown in Figure 5.5.24. For the given radius r and winch rotation θ, determine the distance that the object is lifted above ground.

45. $r = 4$ inches, $\theta = 720°$

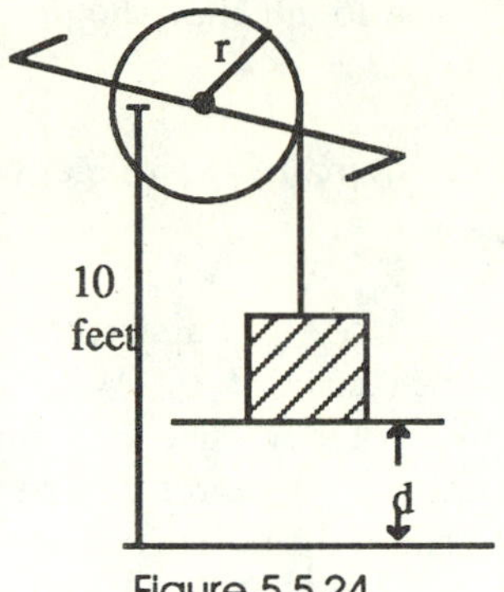

Figure 5.5.24

46. $r = 1$ foot, $\theta = 720°$

Consider a wheel of radius 5 feet rotating at 1200 rpm (revolutions per minute).

47. Determine the angular speed of the wheel.
48. Determine the linear speed of a point on the circumference of the wheel.

Consider a wheel of radius 2.8 feet rotating at 600 rpm.

49. Determine the angular speed of the wheel.
50. Determine the linear speed of a point on the circumference of the wheel.
51. Jeffrey can obtain a speed of 42 mph on his exercise bike in high gear. The bike wheels are 30 inches in diameter, the pedal sprocket is 16 inches in diameter, and the wheel sprocket is 5 inches (in high gear). Find the angular speed of the wheel and of both sprockets. *Note:* The linear speed of a point on the circumference of the wheel is also 42 mph.
52. A regular pentagon is inscribed in a circle of radius 10 inches. What is the perimeter of the pentagon?
53. A regular pentagon is inscribed in a circle. The length of each side of the pentagon is 22 inches. Determine the radius of the circle.
54. Show that applying a vertical stretch factor of 2 to the graph of $y = \sin x$ followed by applying a vertical shift up 3 units is not the same as applying a vertical shift up 3 units to the graph of $y = \sin x$ followed by applying a vertical stretch factor of 2.
55. Determine the two equations of the resulting graphs from Exercise 54. Confirm that they are different graphs.

5.6 Graphs of tan x, cot x, sec x, and csc x

In this section we determine the graphs of the other four basic trig functions and we investigate the role of the parameter a in the graph of $y = \sin ax$ and other trig functions. Equations involving trig functions are solved.

In the previous section we found that the period of $\sin x$ or $\cos x$ was 2π. Now, we see that the period of $\tan x$ or $\cot x$ is π.

- EXAMPLE 1: Determine the domain, the range, the zeros, and the asymptotes of $f(x) = \tan x$, and draw a complete graph.

SOLUTION: Recall that $\tan x = \frac{\sin x}{\cos x}$. This means that the domain of $f(x) = \tan x$ consists of all real numbers for which $\cos x \neq 0$. (Why?) In the previous section we found that $\cos x = 0$ at all odd multiples of $\frac{\pi}{2}$. Thus, the domain of f consists of all real numbers *not* equal to $(2n+1)\frac{\pi}{2}$, n an integer. The zeros of f occur at the values of x for which $\sin x = 0$. (Why?) In Section 5.5 we showed that $\sin x = 0$ when $x = n\pi$, n an integer.

We know that both $\sin x$ and $\cos x$ are periodic with period 2π. Thus, $\tan x = \frac{\sin x}{\cos x}$ must be periodic. It is natural to conjecture that the period of $\tan x$ is also 2π. However, the graph of f in Figure 5.6.1 suggests that the period of f is π. This will be verified in the next example. Thus, the graph of f can be obtained by piecing together, end-to-end, copies of the graph of f in $\left[-\frac{\pi}{2}, \frac{\pi}{2}\right]$. Therefore, the graph of f in Figure 5.6.2 is complete and shows three periods of f. The graph of f has vertical asymptotes at the odd multiples of $\frac{\pi}{2}$, the zeros of the denominator ($\cos x$) of f. This means that $f(x) \to \infty$ or $f(x) \to -\infty$ as $x \to (2n+1)\frac{\pi}{2}$ from the left or right, respectively. •

We can see from Figure 5.6.2 that the graph of $f(x) = \tan x$ is symmetric with respect to the origin. Thus, $f(x) = \tan x$ is an odd function. We can also see that the function $f(x) = \tan x$ is increasing in each interval $\left(-\frac{\pi}{2} + k\pi, \frac{\pi}{2} + k\pi\right)$, k an integer. In the next example we confirm that the period of f is π.

- EXAMPLE 2: Show that $\tan(x+\pi) = \tan x$ for all $x \neq (2k-1)\frac{\pi}{2}$, k an integer.

SOLUTION: Let $P_1(a,b)$ be a point on the terminal side of the angle x (Figure 5.6.3). This figure is drawn with x in the first quadrant. The actual details that follow are valid independent of the quadrant in which x lies. The point $P_2(-a,-b)$ is on the terminal side

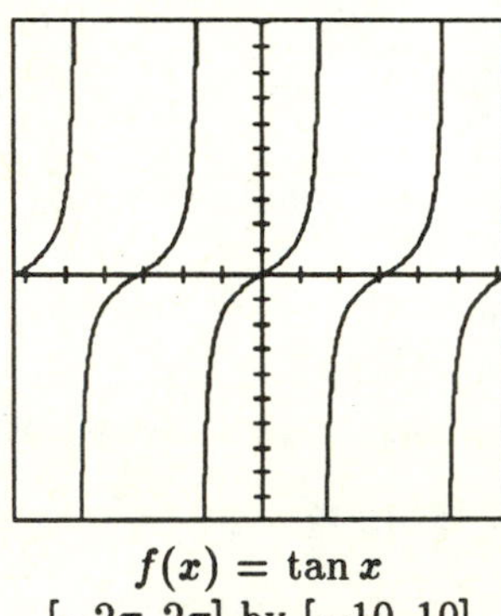

$f(x) = \tan x$
$[-2\pi, 2\pi]$ by $[-10, 10]$
Figure 5.6.1

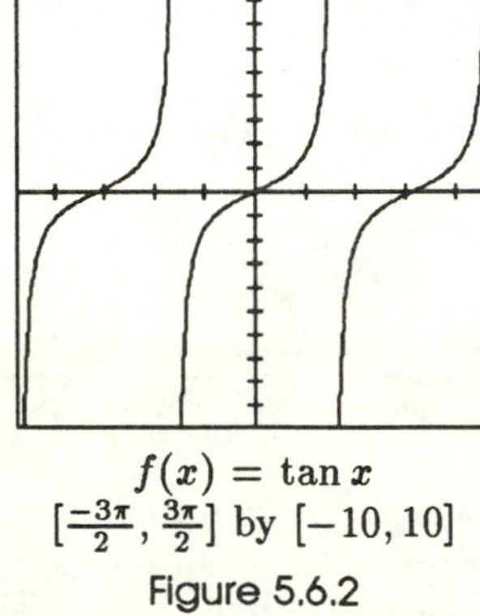

$f(x) = \tan x$
$[\frac{-3\pi}{2}, \frac{3\pi}{2}]$ by $[-10, 10]$
Figure 5.6.2

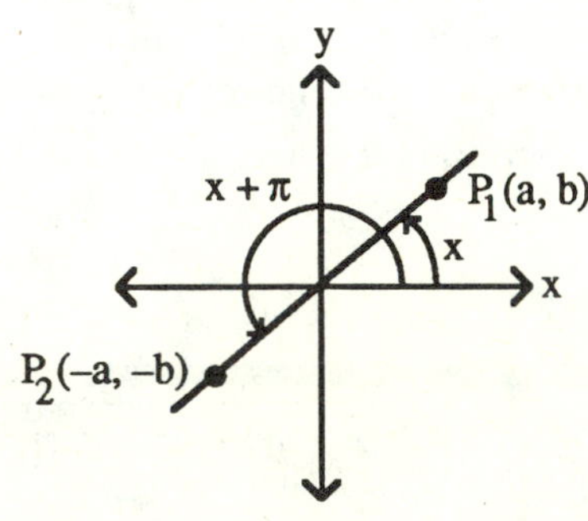

Figure 5.6.3

of $x + \pi$. (Why?) Using the coordinates of P_1, we have $\tan x = \frac{b}{a}$. Using the coordinates of P_2, we have $\tan(x+\pi) = \frac{-b}{-a} = \frac{b}{a}$. Thus, $\tan(x+\pi) = \tan x$. •

It can be shown that no number smaller than π can be a period for $\tan x$. Thus, the period of $\tan x$ is π.

• EXAMPLE 3: Determine the domain, the range, the asymptotes, the zeros, and the period of $f(x) = \cot x$, and draw a complete graph.

SOLUTION: Investigating the graph of $\cot x$ is very similar to that of $\tan x$. We can use the identity

$$\cot x = \frac{1}{\tan x}$$

to show that period of $f(x) = \cot x$ must be π:

$$\begin{aligned}\cot(x+\pi) &= \frac{1}{\tan(x+\pi)}\\ &= \frac{1}{\tan x}\\ &= \cot x.\end{aligned}$$

Furthermore, the range of

$$f(x) = \cot x = \frac{1}{\tan x}$$

is the set of all real numbers. (Why?) We can see that

$$f(x) = \cot x = \frac{\cos x}{\sin x}$$

has a vertical asymptote at any value of x for which $\sin x = 0$. In the previous section we showed that $\sin x = 0$ at any value of x of the form $k\pi$, where k is an integer. Therefore, the domain of f consists of all real numbers different from $k\pi$, k any integer, and the vertical asymptotes are $x = k\pi$. The computer drawn graph of $f(x) = \cot x$ in Figure 5.6.4 is complete and shows four complete periods of the function. •

Now, $\cot x = 0$ if and only if $\cos x = 0$. (Why?) In the previous section we showed that $\cos x = 0$ at any value of x of the form $(2k-1)\frac{\pi}{2}$, k an integer. Thus, $\cot x = 0$ at the odd multiples of $\frac{\pi}{2}$.

• EXAMPLE 4: Determine the domain, the range, the asymptotes, the zeros, and the period of $f(x) = \csc x$ and $g(x) = \sec x$, and draw complete graphs.

SOLUTION: We can use the identity

$$\csc x = \frac{1}{\sin x}$$

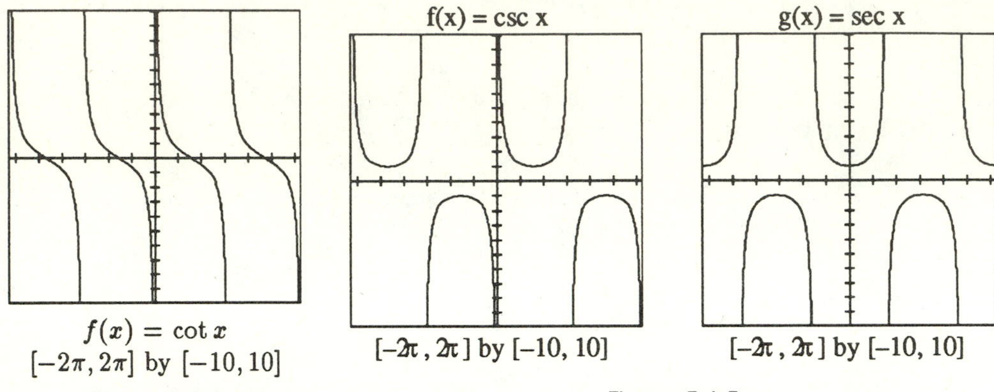

$f(x) = \cot x$
$[-2\pi, 2\pi]$ by $[-10, 10]$

Figure 5.6.4

$[-2\pi, 2\pi]$ by $[-10, 10]$ $[-2\pi, 2\pi]$ by $[-10, 10]$

Figure 5.6.5

and Example 3 to see that $f(x) = \csc x$ has a vertical asymptote at $k\pi$, k any integer. Thus, the domain of f consists of all real numbers different from $k\pi$, k any integer. Similarly, we can use the identity

$$\sec x = \frac{1}{\cos x}$$

and Example 1 to see that $g(x) = \sec x$ has vertical asymptotes $x = (2k+1)\frac{\pi}{2}$, k any integer. Thus, the domain of g consists of all real numbers different from $(2k+1)\frac{\pi}{2}$, k any integer. As in Example 2, we can use the above two identities to show that the periods of $f(x) = \csc x = \frac{1}{\sin x}$ and $g(x) = \sec x = \frac{1}{\cos x}$ are 2π. The graphs of f and g in Figure 5.6.5 are complete and show two periods of each function. We can see from these graphs that the ranges of f and g are $(-\infty, -1] \cup [1, \infty)$. Alternatively, we can use the fact that the ranges of $y = \sin x$ and $y = \cos x$ are $[-1, 1]$, and we can use the identities

$$\csc x = \frac{1}{\sin x}, \qquad \sec x = \frac{1}{\cos x}$$

to see that the ranges of f and g are $(-\infty, -1] \cup [1, \infty)$. Notice that neither f nor g has any real zeros. •

Solving Equations

Let f be any one of the six basic trig functions. The equation $f(x) = a$, where a is a real number, need not have a solution. For example, if f is the sine or cosine function, we can use the graphs in Figure 5.6.6 to see that the equation $f(x) = a$ has solutions if $|a| \leq 1$ and *no* solutions if $|a| > 1$. Similarly, Figure 5.6.5 shows that the equations $\sec x = 0.5$ and $\csc x = -0.5$ have no solutions. However, if $f(x) = a$ has a solution, then the circular properties of the trig functions show that the equation has *infinitely many* solutions.

- **Solving Trig Equations** A good strategy to find all solutions is first to solve the equation in an interval of length equal to the period of the trig function. Then add $2k\pi$ ($k\pi$ for $\tan x$ and $\cot x$), where k is any integer, to *each* solution found in the interval of length equal to the period to determine all the solutions.

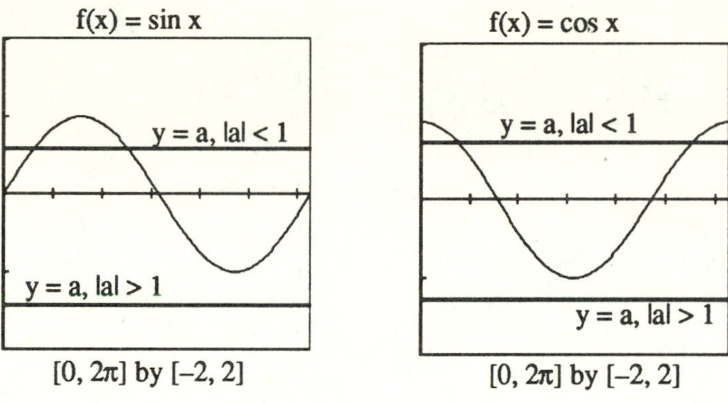

Figure 5.6.6

We can solve the equation $f(x) = a$ with a graphing utility using zoom-in or with a calculator. However, we must be careful when we use a calculator to solve such equations. Figures 5.6.2 and 5.6.4 show that the equations $\tan x = a$ and $\cot x = a$ have exactly one solution in any interval of length π, the period of the tangent or cotangent. Suppose that $f(x)$ is $\sin x$, $\cos x$, $\sec x$, or $\csc x$. The graphs in Figures 5.6.5 and 5.6.6 show that the equations $f(x) = 1$ and $f(x) = -1$ have exactly one solution in any interval of length 2π. However, the equation $f(x) = a$ with $|a| \neq 1$ has either no solutions or exactly two solutions in any interval of length 2π. Calculators give one solution only to $f(x) = a$ when $f(x)$ is $\sin x$, $\cos x$, or $\tan x$ using the 2nd F, SHIFT, or INV key together with the trig key. Furthermore, calculators produce an x in the interval $[-\frac{\pi}{2}, \frac{\pi}{2}]$ for $\sin x = a$ or $\tan x = a$ (Figure 5.6.7). In this case, $|x|$ is the reference angle for any solution to $\sin x = a$ or $\tan x = a$. Calculators produce an x in the interval $[0, \pi]$ for $\cos x = a$ (Figure 5.6.8). In this case, x is the reference angle if $0 \leq x \leq \frac{\pi}{2}$ and $\pi - x$ is the reference angle if $\frac{\pi}{2} < x \leq \pi$ for any solution to $\cos x = a$.

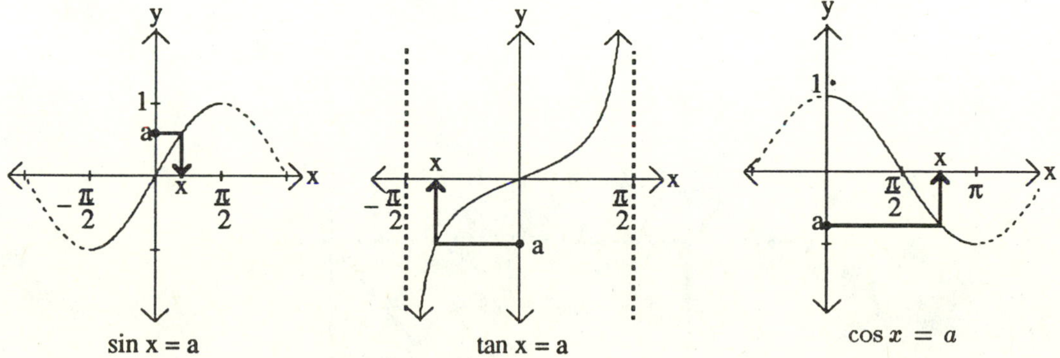

Figure 5.6.7

Figure 5.6.8

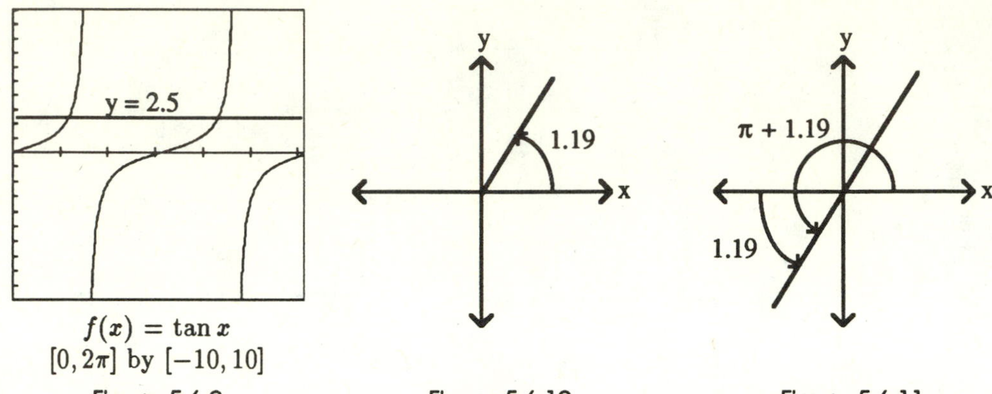

$f(x) = \tan x$
$[0, 2\pi]$ by $[-10, 10]$
Figure 5.6.9

Figure 5.6.10

Figure 5.6.11

• EXAMPLE 5: Determine all solutions to the following equations in the interval $[0, 2\pi)$.

(a) $\tan x = 2.5$ (b) $\cos x = -0.7$ (c) $\csc x = -1.6$

SOLUTION:

(a) We can use Figure 5.6.9 to see that the equation $\tan x = 2.5$ has two solutions in the interval $[0, 2\pi)$. One solution is in the first quadrant, and the other is in the third quadrant. We can *estimate* from Figure 5.6.9 that the two solutions to $\tan x = 2.5$ in $[0, 2\pi]$ are 1.2 and 4.3. A calculator gives only the first quadrant solution 1.19 (Figure 5.6.10). Now, 1.19 is the reference angle for the other solution in the third quadrant (Figure 5.6.11). Thus, the third quadrant solution is $\pi + 1.19$, or 4.33.

(b) We can use Figure 5.6.6 to see that the equation $\cos x = -0.7$ has two solutions in the interval $[0, 2\pi)$. One is in the second quadrant, and the other is in the third quadrant. A calculator gives only the second quadrant solution 2.35 (Figure 5.6.12). The angle $\pi - 2.35 = 0.80$ is the reference angle for the other solution in the third quadrant (Figure 5.6.13). Thus, the third quadrant solution is $\pi + 0.80$, or 3.94.

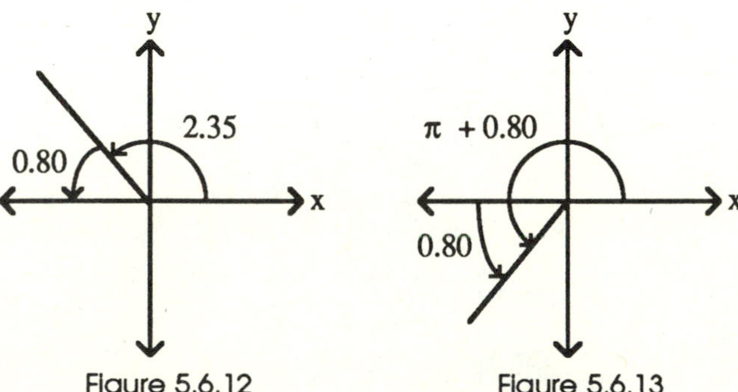

Figure 5.6.12

Figure 5.6.13

(c) Figure 5.6.5 shows that the equation $\csc x = -1.6$ has two solutions in the interval $[0, 2\pi)$. One is in the third quadrant, and the other is in the fourth quadrant. To use a calculator to find the solutions we must first observe that $\sin x = \frac{1}{\csc x} = -0.625$. A calculator gives -0.68 as the value of x. Now, $|-0.68| = 0.68$ is the reference angle for the two solutions we seek. The third quadrant solution is $\pi + 0.68 = 3.82$, and the fourth quadrant solution is $2\pi - 0.68 = 5.61$ (Figure 5.6.14). •

• EXAMPLE 6: Determine all solutions to each of the following equations in degrees.

(a) $\sin x = 0.7$ (b) $\sec x = 2.5$ (c) $\cot x = -5$

SOLUTION:

(a) $\sin x$ is periodic with period $2\pi = 360°$. There are two solutions to this equation in the interval $[0°, 360°)$ (Figure 5.6.15). The two solutions are in the first and second quadrants. A calculator gives $44.43°$ as the solution to $\sin x = 0.7$. Now, $44.43°$ is the reference angle for the second quadrant solution. Thus, the second quadrant solution is $180° - 44.43° = 135.57°$. Any other solution is of the form $44.42° + k360°$ or $135.57° + k360°$, k any integer.

(b) First, we convert $\sec x = 2.5$ to the equivalent equation $\cos x = 0.4$. (How?) Now, $\cos x$ is periodic with period $360°$, and the equation $\cos x = 0.4$ has two solutions in the interval $[0, 360°)$ (Figure 5.6.16). One solution is in the first quadrant, and the other is in the fourth quadrant. With a calculator we find that $66.42°$ is a solution. This is the first quadrant solution. The solution in the fourth quadrant has $66.42°$ as a reference angle. Thus, the fourth quadrant solution is $360° - 66.42° = 293.58°$. Any other solution is of the form $66.42° + k360°$ or $293.58° + k360°$, k any integer.

(c) First, we must convert $\cot x = -5$ to the equivalent equation $\tan x = -0.2$. (How?) Now, $\tan x$ is periodic with period $\pi = 180°$. Notice that $\tan x = a$ has exactly one solution in any interval of length π (Figure 5.6.2). Because calculators give values in the interval $[-\frac{\pi}{2}, \frac{\pi}{2}]$ as solutions to $\tan x = a$ and the length of the interval $[-\frac{\pi}{2}, \frac{\pi}{2}]$ is π, we can use the calculator values directly to solve any equation of the

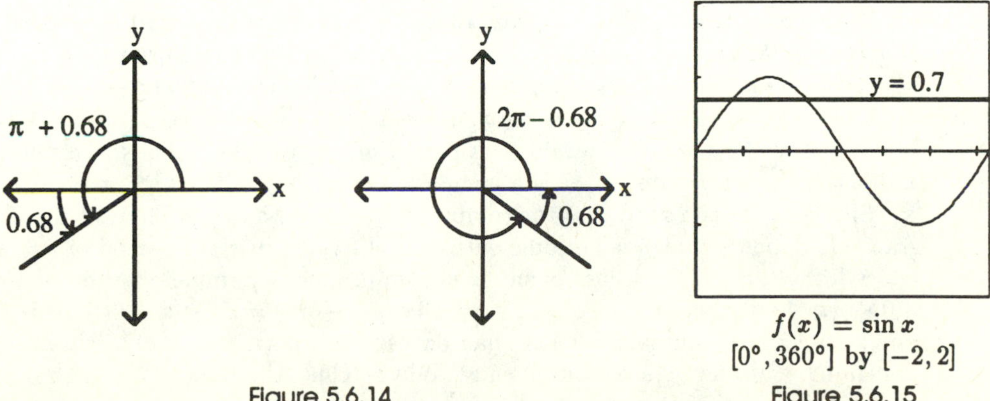

$f(x) = \sin x$
$[0°, 360°]$ by $[-2, 2]$

Figure 5.6.14

Figure 5.6.15

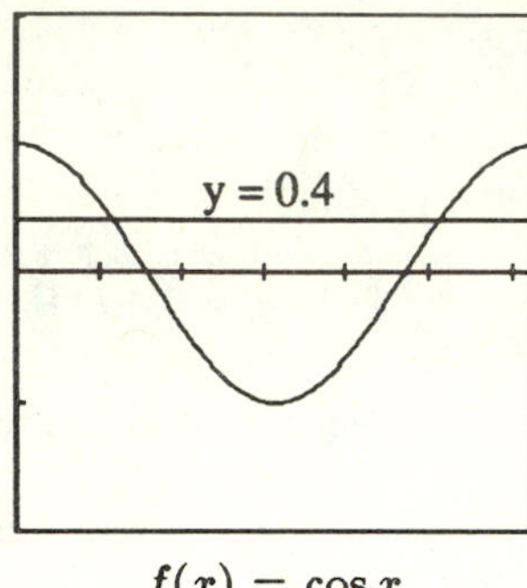

$f(x) = \cos x$
$[0°, 360°]$ by $[-2, 2]$
Figure 5.6.16

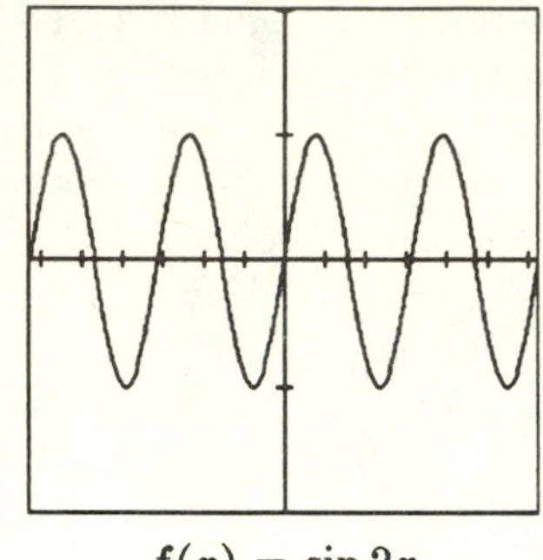

$f(x) = \sin 2x$
$[-2\pi, 2\pi]$ by $[-2, 2]$
Figure 5.6.17

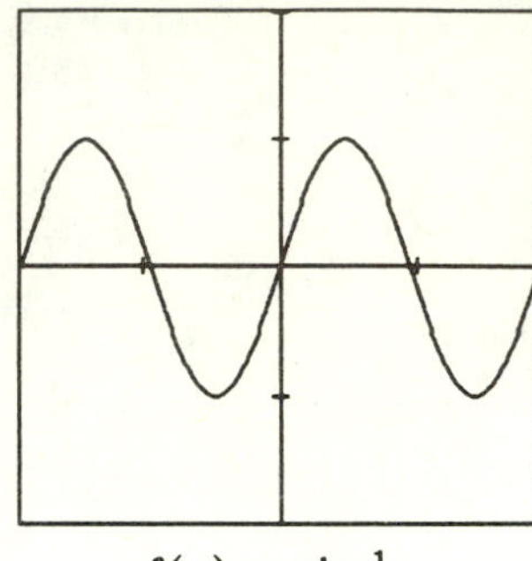

$f(x) = \sin \frac{1}{3}x$
$[-6\pi, 6\pi]$ by $[-2, 2]$
Figure 5.6.18

form $\tan x = a$. A calculator gives $-11.31°$ as the solution to $\tan x = -0.2$. Any other solution is of the form $-11.31° + k180°$, k any integer. •

Horizontal Stretch or Shrink

Consider trig functions of the form $\sin ax$, $\cos ax$, and so forth. If $a > 0$, we will show that the effect of the parameter a is to stretch or shrink horizontally the graph of the corresponding basic trig function. If $a < 0$, the effect of the parameter a is to stretch or shrink horizontally the graph of the corresponding basic trig function and then reflect the resulting graph through the y-axis.

• EXAMPLE 7: Draw a complete graph of $f(x) = \sin 2x$ and $g(x) = \sin \frac{1}{3}x$. Show that each function is periodic and determine its period.

SOLUTION: The periodicity of the sine and cosine functions makes it reasonable to expect that f and g are periodic. We will show that the computer drawn graph of f in Figure 5.6.17 is a complete graph. It appears that f is periodic with period π, one-half the period of $y = \sin x$. As $2x$ varies from 0 to 2π, the values of $\sin 2x$ should produce a complete period of $f(x) = \sin 2x$. If $0 \le 2x \le 2\pi$, then $0 \le x \le \pi$. Notice that $f(x + \pi) = \sin 2(x + \pi) = \sin(2x + 2\pi) = \sin 2x = f(x)$. No number smaller than π can be the period of $f(x) = \sin 2x$ because twice that number would be a period of $\sin x$. (Why?) Thus, the period of f is π, and the graph in Figure 5.6.17 is a complete graph showing four periods.

Similarly, we can show that the computer drawn graph of g in Figure 5.6.18 is a complete graph. It appears that y is periodic with period 6π, 3 times the period of $y = \sin x$. As $\frac{1}{3}x$ varies from 0 to 2π, the values of $\sin \frac{1}{3}x$ should produce a complete period of $g(x) = \sin \frac{1}{3}x$. If $0 \le \frac{1}{3}x \le 2\pi$, then $0 \le x \le 6\pi$. Notice that $g(x + 6\pi) = \sin \frac{1}{3}(x + 6\pi) = \sin\left(\frac{1}{3}x + 2\pi\right) = \sin \frac{1}{3}x = g(x)$. No number smaller than 6π can be a period of $g(x) = \sin \frac{1}{3}x$ because $\frac{1}{3}$ of that number would be a period of $\sin x$. (Why?) Thus, the period of g is 6π, and the graph of g in Figure 5.6.18 is complete showing two periods. •

• EXAMPLE 8: Draw a complete graph of $f(x) = \sin(-2x)$ and $g(x) = \cos(-2x)$. Show that each function is periodic and determine its period.

SOLUTION: The graph of $f(x) = \sin(-2x)$ can be obtained by reflecting the graph of $y = \sin 2x$ through the y-axis. The computer drawn graph of f in Figure 5.6.19 confirms this observation. Similarly, the graph of $g(x) = \cos(-2x)$ can be obtained by reflecting the graph of $y = \cos 2x$ through the y-axis. However, the graph of $y = \cos 2x$ is symmetric with respect to the y-axis because the cosine function is an even function. Thus, the graphs of g and $y = \cos 2x$ should be the same. The graph of g is given in Figure 5.6.20. You should overlay the graph of $y = \cos 2x$ to convince yourself that the two graphs are identical. •

The effect of the coefficient 2 of x in $f(x) = \sin 2x$ is to shrink horizontally or compress the graph of $y = \sin x$ by a factor of $\frac{1}{2}$. That is, the portion of the graph of $y = \sin x$ for $0 \leq x \leq 2\pi$ becomes the portion of the graph of $f(x) = \sin 2x$ for $0 \leq x \leq \pi$. This is a *horizontal shrinking* of $y = \sin x$ by a factor of $\frac{1}{2}$. Similarly, the effect of the coefficient $\frac{1}{3}$ of x in $g(x) = \sin \frac{1}{3}x$ is to stretch horizontally or expand the graph of $y = \sin x$ by a factor of 3. That is, the portion of the graph of $y = \sin x$ for $0 \leq x \leq 2\pi$ becomes the portion of the graph of $g(x) = \sin \frac{1}{3}x$ for $0 \leq x \leq 6\pi$. This is a *horizontal stretching* of $y = \sin x$ by a factor of 3.

Example 7 is a special case of the following theorem which we state without proof.

• THEOREM 1 Let a be any real number. The function $f(x) = \sin ax$ is periodic with period $\frac{2\pi}{|a|}$.

• EXAMPLE 9: Draw a complete graph of $f(x) = \tan 3x$ and $g(x) = \tan \frac{1}{4}x$. Show that each function is periodic and determine its period.

SOLUTION: The period of $\tan x$ is π. As $3x$ varies from 0 to π, the values of $\tan 3x$ should produce a complete period of $f(x) = \tan 3x$. If $0 \leq 3x \leq \pi$, then $0 \leq x \leq \frac{\pi}{3}$. Thus, the period of $f(x) = \tan 3x$ is $\frac{\pi}{3}$, one-third the period of $y = \tan x$. The graph of $f(x) = \tan 3x$ in Figure 5.6.21 is complete and shows six periods.

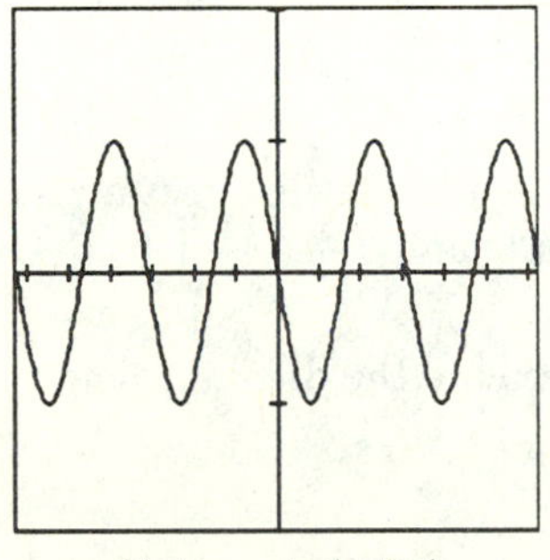

$f(x) = \sin(-2x)$
$[-2\pi, 2\pi]$ by $[-2, 2]$
Figure 5.6.19

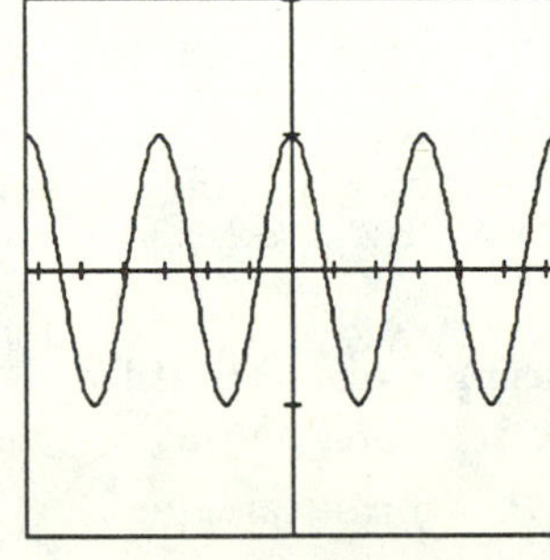

$f(x) = \cos(-2x)$
$[-2\pi, 2\pi]$ by $[-2, 2]$
Figure 5.6.20

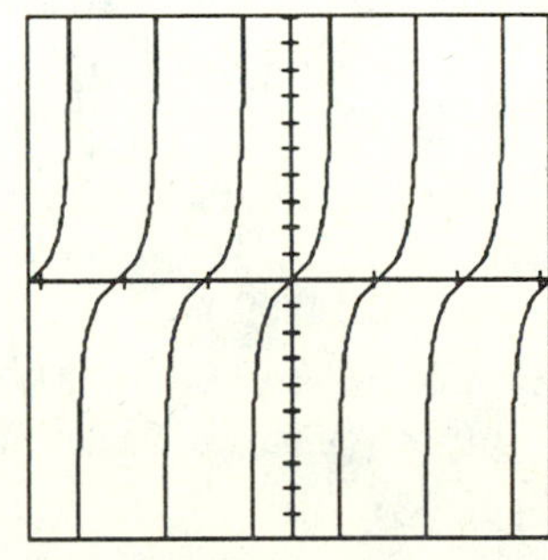

$f(x) = \tan 3x$
$[-\pi, \pi]$ by $[-10, 10]$
Figure 5.6.21

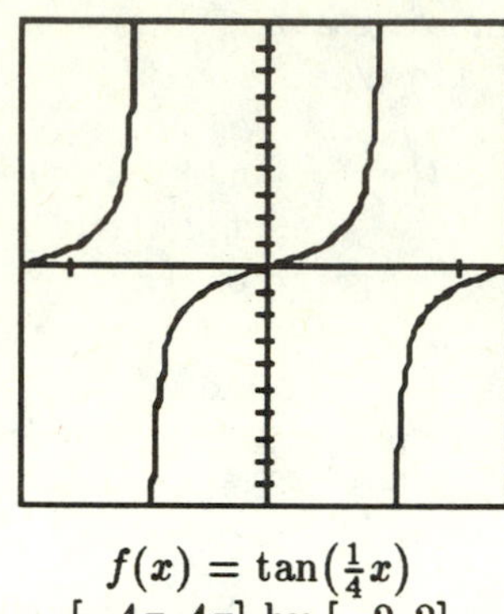

$f(x) = \tan(\frac{1}{4}x)$
$[-4\pi, 4\pi]$ by $[-2, 2]$
Figure 5.6.22

As $\frac{1}{4}x$ varies from 0 to π, the values of $\tan\frac{1}{4}x$ should produce a complete period of $g(x) = \tan\frac{1}{4}x$. If $0 \le \frac{1}{4}x \le \pi$, then $0 \le x \le 4\pi$. Thus, the period of $g(x) = \tan\frac{1}{4}x$ is 4π, four times the period of $y = \tan x$. The graph of $g(x) = \tan\frac{1}{4}x$ in Figure 5.6.22 is complete and shows two periods. •

Example 9 and Theorem 1 are special cases of the following more general theorem which we give without proof.

• **THEOREM 2** If $b \ne 0$, the graph of $y = f(bx)$ can be obtained by horizontally stretching or shrinking the graph of $y = f(x)$ by the factor $\frac{1}{|b|}$, followed by a reflection in the y-axis if $b < 0$. If, in addition, f is periodic with period p, then $y = f(bx)$ is periodic with period $\frac{p}{|b|}$.

• EXERCISES 5-6

Use a calculator unless stated otherwise. Angles are measured in degrees or radians.

1. Use a graphing utility to draw the graph of $y = \sin x$ in the viewing rectangles $[0, 2\pi]$ by $[-2, 2]$, $[-2\pi, 0]$ by $[-2, 2]$, $[-8\pi, -6\pi]$ by $[-2, 2]$, and $[10\pi, 12\pi]$ by $[-2, 2]$. (Use $\pi = 3.14159$.)
2. Use a graphing utility to draw the graph of $y = \tan x$ in the viewing rectangles $[-\frac{\pi}{2}, \frac{\pi}{2}]$ by $[-10, 10]$, $[\frac{7\pi}{2}, \frac{9\pi}{2}]$ by $[-10, 10]$, $[-\frac{11\pi}{2}, -\frac{9\pi}{2}]$ by $[-10, 10]$, and $[\frac{71\pi}{2}, \frac{73\pi}{2}]$ by $[-10, 10]$. (Use $\pi = 3.14159$.)

Draw a complete graph without using a graphing utility. Determine the domain, range, period, and asymptotes (if any). Check using a graphing utility.

3. $y = 3\tan x$
4. $y = -\tan x$
5. $y = \frac{1}{2}\sec x$
6. $y = \sec(-x)$
7. $y = 3\csc x$
8. $y = 2\tan x$
9. $y = 3\sin x$
10. $y = -4\sin 2x$
11. $y = -3\tan\frac{1}{2}x$
12. $y = 2\cot\frac{1}{2}x$
13. $y = 2\csc x$
14. $y = -2\sec\frac{1}{2}x$

15. $y = \sin 2x$
16. $y = \sin(-4x)$
17. $y = \sec 2x$
18. $y = \sec\left(-\frac{1}{2}x\right)$
19. $y = \sin \frac{1}{2}x$
20. $y = \tan 2x$
21. $y = 2\sin 3x$
22. $y = \frac{1}{2}\sin \frac{1}{2}x$
23. $y = 3\cos\left(-\frac{1}{2}x\right)$
24. $y = 2\tan \frac{1}{2}x$

Draw a complete graph. Determine the domain, range, period, and asymptotes (if any) of each function.

25. $y = 2\sin x$
26. $y = -3 + \sin \frac{\pi}{2}x$
27. $y = 2\tan \pi x$
28. $y = -\tan \frac{\pi}{2}x$
29. $y = 3\sec 2x$
30. $y = 4\csc \frac{1}{3}x$

List, in proper order, the geometric transformations that can be used to obtain each graph from the graph of $y = \sin x$, $y = \cos x$, or $y = \tan x$.

31. $y = 2\sin 3x$
32. $g(x) = \tan \frac{1}{2}x$
33. $y = 3\cos \pi x$
34. $y = -3\sin\left(-\frac{\pi}{2}x\right)$

Solve the equation for $-2\pi \leq x \leq 2\pi$.

35. $\sin x = 0.75$
36. $\cos x = 0.3$
37. $\tan x = 3.25$
38. $\cot x = -5.6$

Solve the inequality for $-2\pi \leq x \leq 2\pi$.

39. $2\sin x > \frac{3}{2}\cos 2x$
40. $-\sin x < 3\cos \frac{1}{2}x$
41. $2\tan x < 5$
42. $3\sec \frac{1}{2}x > \cos x$

Prove the identity.

43. $\cot x = \cot(x + \pi)$
44. $\sec x = \sec(x + 2\pi)$
45. $\csc x = \csc(x + 2\pi)$
46. $\cot x = \cot(x - \pi)$

We know all real solutions to $\sin x = 0.75$ are given by $x + 2k\pi$, $k = 0, \pm 1, \pm 2, \ldots$ where x is 0.85 or 2.29. Use zoom-in to confirm.

47. For $k = 0$ (i.e., for $x = 0.85$ and $x = 2.29$)
48. For $k = 2$
49. For $k = -3$
50. For $k = 5$

Find all real solutions.

51. $\sin x = 0.25$
52. $\tan x = 4$
53. $\cos x = 0.42$
54. $\sec x = 3$
55. $\sin x < 0.15$
56. $\tan x > 2$
57. $\sin 3x = 0.55$
58. $\cos 2x = 0.85$
59. $\csc \frac{1}{2}x \leq 4$
60. $3\sin \frac{1}{2}x > 2\cos x$

5.7 Trigonometric Graphs

In this section we determine the graph of any function of the form $y = af(bx + c) + d$ where f is one of the six basic trig functions. We also find the local extrema and determine the increasing and decreasing behavior of such functions. An application that involves finding a local extremum of a trigonometric graph will be investigated.

We know that the graph of $y = af(x)$ can be obtained by vertically stretching or shrinking the graph of $y = f(x)$ by the factor $|a|$, followed by a reflection in the x-axis if $a < 0$. The graph of $y = f(x + c)$ can be obtained by shifting the graph of $y = f(x)$ horizontally $|c|$ units left if $c > 0$ or right if $c < 0$. Similarly, the graph of $y = f(x) + d$ can be obtained by shifting the graph of $y = f(x)$ vertically $|d|$ units up if $d > 0$ or down if $d < 0$. In the previous section we showed that if f is one of the six basic trig functions, then the function $y = f(bx)$ is periodic with period equal to the period of f divided by $|b|$. In the last section we also found that the graph of $y = f(bx)$ could be obtained by horizontally stretching or shrinking the graph of $y = f(x)$ by the factor $\frac{1}{|b|}$, followed by a reflection in the y-axis if $b < 0$. Now, we combine all of this information to determine complete graphs of functions of the form $y = af(bx + c) + d$ where f is one of the six basic trig functions.

If f is $\sin x$ or $\cos x$, then f has a largest and a smallest value in the sense of the following definition.

• DEFINITION The function value $f(a)$ is called an **absolute maximum** of f if $f(x) \leq f(a)$ for all x in the domain of f. The function value $f(a)$ is called an **absolute minimum** of f if $f(x) \geq f(a)$ for all x in the domain of f. •

Notice $f(x) = x^2$ has zero as an absolute minimum, and $g(x) = x^3 - 4x$ has neither an absolute minimum nor an absolute maximum.

• EXAMPLE 1: Determine the period, the local maximum and minimum values, the absolute maximum and minimum values, and the increasing and decreasing behavior of each function and draw a complete graph.

(a) $f(x) = 3\sin 2x$ (b) $g(x) = 4\tan 3x$

SOLUTION:

(a) A complete graph of f can be obtained by vertically stretching the graph of $y = \sin 2x$ by a factor of 3. From the previous section, we know that the period of $\sin 2x$ is $\frac{2\pi}{2} = \pi$. Thus, the graph of f in Figure 5.7.1 is complete and shows four periods. Because the period of f is π its complete behavior can be determined from any interval of length π, say $[0, \pi]$. In this interval, f has a local maximum of 3 at $x = \frac{\pi}{4}$ and a local minimum of -3 at $x = \frac{3\pi}{4}$. (Why?) Because f is periodic, there are infinitely many points at which f has a local maximum value or a local minimum value. In fact, f has a local maximum value of 3 at $\frac{\pi}{4} + k\pi$ and a local minimum value of -3 at $\frac{3\pi}{4} + k\pi$ for each integer k because f has period π. We can also see that 3 is an absolute maximum and -3 an absolute minimum of this function. This function alternates increasing and decreasing between consecutive extremum points. That is, if x_1, x_2, and x_3 are consecutive values of x at which f has extremum values, then f increases in $[x_1, x_2]$ and decreases in $[x_2, x_3]$, or vice versa. In fact, f decreases in $[\frac{\pi}{4}+k\pi, \frac{3\pi}{4}+k\pi]$ and increases in $[\frac{3\pi}{4}+k\pi, \frac{\pi}{4}+(k+1)\pi]$ for each integer k. (Why?)

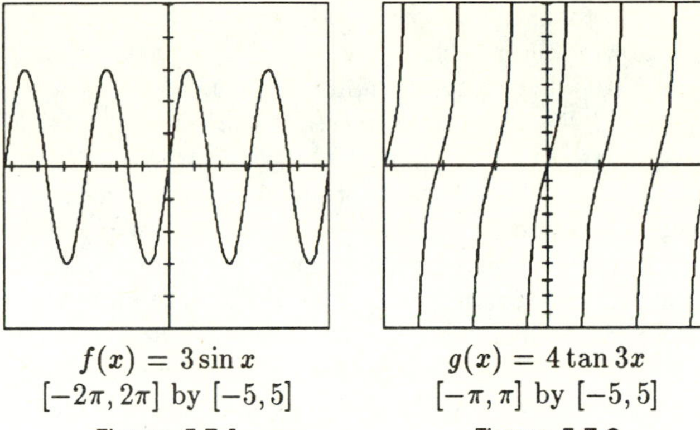

$f(x) = 3\sin x$
$[-2\pi, 2\pi]$ by $[-5, 5]$
Figure 5.7.1

$g(x) = 4\tan 3x$
$[-\pi, \pi]$ by $[-5, 5]$
Figure 5.7.2

(b) The function $g(x) = 4\tan 3x$ is periodic with period $\frac{\pi}{3}$. (Why?) A complete graph of g can be obtained by vertically stretching the graph of $y = \tan 3x$ by a factor of 4. Thus, the graph of g in Figure 5.7.2 is complete. We can see that g has no local extremum values. Furthermore, g is increasing in each interval $\left(-\frac{\pi}{6} + k\pi, \frac{\pi}{6} + k\pi\right)$, k an integer. •

The function $f(x) = 3\sin 2x$ of Example 1(a) has 3 for an absolute maximum value and -3 for an absolute minimum value, because $y = \sin 2x$ has 1 for an absolute maximum value and -1 for an absolute minimum value.

• DEFINITION Let f be a periodic function with an absolute maximum value of M and an absolute minimum value of m. Then, $\frac{1}{2}(M - m)$ is called the **amplitude** of f. •

The function $f(x) = 3\sin 2x$ of Example 1(a) has an amplitude of 3. The function $g(x) = 4\tan 3x$ has *no* amplitude because g has neither an absolute maximum value nor an absolute minimum value. The function $y = af(bx + c) + d$ has *no* amplitude if f is the tangent, cotangent, secant, or cosecant.

• EXAMPLE 2: Determine the period, the amplitude, the local maximum and minimum values and where they occur, and the increasing and decreasing behavior of each function, and draw a complete graph.

(a) $f(x) = 2\cos\left(x - \frac{\pi}{2}\right)$ (b) $g(x) = 3\sec(x + 1)$

SOLUTION:

(a) A complete graph of f can be obtained by vertically stretching the graph of $y = \cos x$ by a factor of 2 and then shifting the resulting graph horizontally right $\frac{\pi}{2}$ units. Thus, f is periodic with the same period as $y = \cos x$, and the graph of f in Figure 5.7.3 is complete and shows two periods. The amplitude of f is 2, and f is periodic with period 2π. The complete behavior of f can be determined from any interval of length 2π, say $[0, 2\pi]$. Now, f has a local maximum value of 2 at $\frac{\pi}{2}$ and a local minimum value of -2 at $x = \frac{3\pi}{2}$. Thus, f has a local maximum value of 2 at $x = \frac{\pi}{2} + 2k\pi$, a local minimum value of -2 at $x = \frac{3\pi}{2} + 2k\pi$, is decreasing in $[\frac{\pi}{2}+2k\pi, \frac{3\pi}{2}+2k\pi]$, and is increasing in $[\frac{3\pi}{2}+2k\pi, \frac{\pi}{2}+2(k+1)\pi]$ for each integer k.

(b) A complete graph of g can be obtained by vertically stretching the graph of $y = \sec x$ by a factor of 3 and then shifting the resulting graph horizontally left 1 unit. Thus, g is periodic with period 2π, and the graph of g in Figure 5.7.4 is complete and shows two periods. The function g has *no* amplitude because g has neither an absolute maximum value nor an absolute minimum value. The local extremum values of $y = \sec x$ in $[0, 2\pi]$ are 1 and -1, and they occur when $x = 0$ and $x = \pi$, respectively. Thus, using the vertical stretch and horizontal shift mentioned above, the local extremum values of $g(x) = 3\sec(x+1)$ are 3 and -3, and these values occur when $x = 0 - 1 = -1$ and $x = \pi - 1$ (about 2.14), respectively. You can confirm these values using zoom-in. Thus, g has a local minimum value of 3 at $x = -1 + 2k\pi$, a local maximum value of -3 at $x = \pi - 1 + 2k\pi$, is increasing in $[-1+2k\pi, \frac{\pi}{2} - 1 + 2k\pi)$ and $(\frac{\pi}{2} - 1 + 2k\pi, \pi - 1 + 2k\pi]$, and is decreasing in $[\pi - 1 + 2k\pi, \frac{3\pi}{2} - 1 + 2k\pi)$ and $(\frac{3\pi}{2} - 1 + 2k\pi, 2\pi - 1 + 2k\pi]$ for each integer k. •

We sometimes say that the graph of $f(x) = 2\cos(x - \frac{\pi}{2})$ is obtained from the graph of $y = 2\cos x$ by a ***phase shift*** of $\frac{\pi}{2}$ units, or a ***phase shift right*** $\frac{\pi}{2}$ units. Similarly, the graph of $g(x) = 3\sec(x+1)$ is said to be obtained from the graph of $y = 3\sec x$ by a ***phase shift*** of -1 units, or a ***phase shift left*** 1 unit. The complete definitions will be given after the next example.

• EXAMPLE 3: Determine the domain, range, period, and amplitude of $f(x) = -4\cot(2x+\frac{\pi}{3})$, and then draw a complete graph.

SOLUTION: First we rewrite the expression for f as follows:

$$f(x) = -4\cot 2(x + \frac{\pi}{6}).$$

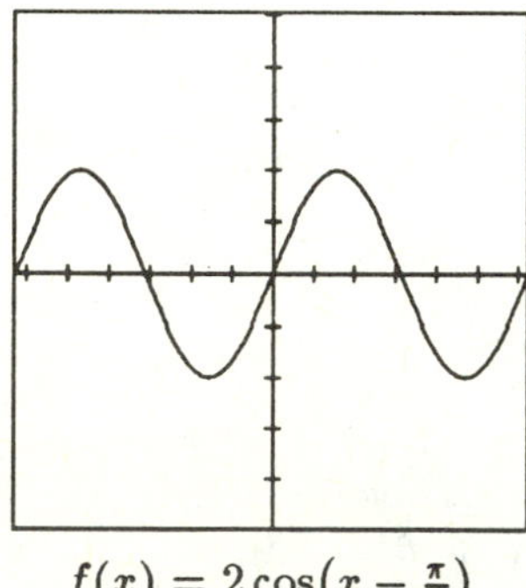

$f(x) = 2\cos(x - \frac{\pi}{2})$
$[-2\pi, 2\pi]$ by $[-5, 5]$
Figure 5.7.3

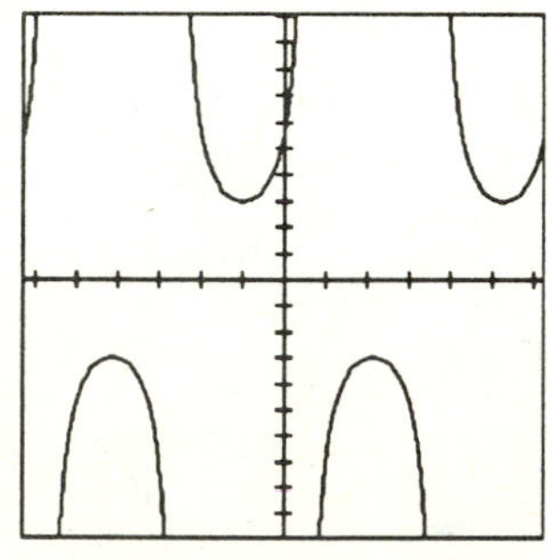

$g(x) = 3\sec(x+1)$
$[-2\pi, 2\pi]$ by $[-10, 10]$
Figure 5.7.4

Now we can see that the graph of f can be obtained by applying, in order, the following transformations to the graph of $y = \cot x$:

1. A vertical stretch by a factor of 4 to obtain $y = 4\cot x$.
2. A reflection in the x-axis to obtain $y = -4\cot x$.
3. A horizontal shrink by a factor of $\frac{1}{2}$ to obtain $y = -4\cot 2x$.
4. A horizontal shift left $\frac{\pi}{6}$ units to obtain $f(x) = -4\cot 2(x + \frac{\pi}{6})$.

Further, f has no amplitude because f has neither an absolute maximum nor an absolute minimum. The range of f is the set of all real numbers. The period of f is $\frac{\pi}{2}$, because the period of $\cot x$ is π. The vertical asymptotes of f are at $x = -\frac{\pi}{6} + k\frac{\pi}{2}$ and $x = \frac{\pi}{2} - \frac{\pi}{6} + k\frac{\pi}{2} = \frac{\pi}{3} + k\frac{\pi}{2}$, k an integer. (Why?) Thus, the domain of f consists of all real numbers different from $-\frac{\pi}{6} + k\frac{\pi}{2}$ or $\frac{\pi}{3} + k\frac{\pi}{2}$, where k is any integer. Thus, the graph of f in Figure 5.7.5 is complete.

The horizontal shift is $\frac{\pi}{6}$ and not $\frac{\pi}{3}$ because $f(x) = -4\cot(2x + \frac{\pi}{3}) = -4\cot 2(x + \frac{\pi}{6})$. •

• DEFINITION Let f be a periodic function. The number $-\frac{c}{b}$ is called the **phase shift** of $y = f(bx + c)$ and is the horizontal shift required in obtaining the graph of $y = af(bx + c) + d$ from the graph of f. The phase shift of the function of Example 3 is $-\frac{\pi}{6}$. •

• EXAMPLE 4: Find the domain, range, period, amplitude, and phase shift of $f(x) = 2\sin(3x + 4) - 1$, and draw a complete graph.

SOLUTION: We know that this graph should look very much like the graph of $y = \sin x$. We can express f in the following form.

$$f(x) = 2\sin\left[3\left(x + \frac{4}{3}\right)\right] - 1$$

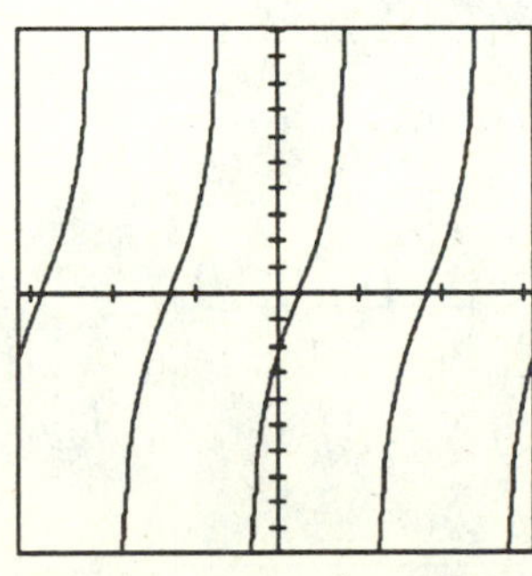

$f(x) = -4\cot(2x + \frac{\pi}{3})$
$[-\pi, \pi]$ by $[-10, 10]$
Figure 5.7.5

It follows from Theorem 1 of Section 5.6 that f is periodic with period $\frac{2\pi}{3}$. We can also say that the effect of the coefficient 3 of x is to shrink the graph of $y = \sin x$ horizontally by a factor of $\frac{1}{3}$. That is, the portion of the graph of $y = \sin x$ for $0 \leq x \leq 2\pi$ becomes the portion of the graph of $y = \sin 3x$ for $0 \leq x \leq \frac{2\pi}{3}$. This means that there are three periods of the graph of f in the interval $[0, 2\pi]$. Next, the graph of $y = \sin 3x$ is shifted horizontally to the left $\frac{4}{3}$ units to obtain the graph of $y = \sin 3\left(x + \frac{4}{3}\right)$. Thus, the phase shift is $-\frac{4}{3}$. Then, the resulting graph is vertically stretched by a factor of 2 to obtain the graph of $y = 2\sin 3\left(x + \frac{4}{3}\right)$. Therefore, the amplitude is 2. Finally, the last graph is shifted vertically down 1 unit to obtain the graph of $f(x) = 2\sin(3x+4) - 1$. The computer drawn graph of f in Figure 5.7.6 confirms these observations.

The domain of f is all real numbers. Figure 5.7.6 suggests the range is $[-3, 1]$. In general, it will not be necessary to go through the geometric transformations used above to produce the graph of f. We can see from the rule for f that the sine wave is centered vertically about the line $y = -1$ and rises 2 units above this line at its highest points and falls 2 units below at its lowest points. (Why?) Then, we draw one complete sine wave with amplitude 2 starting at the point $\left(-\frac{4}{3}, -1\right) = (1.33, -1)$ and ending at the point $\left(-\frac{4}{3} + \frac{2\pi}{3}, -1\right) = (0.76, -1)$ (Figure 5.7.7). Notice that the points $\left(-\frac{4}{3}, -1\right)$ and $\left(-\frac{4}{3} + \frac{2\pi}{3}, -1\right)$ on the graph of f correspond to the points $(0, 0)$ and $(2\pi, 0)$, respectively, on the graph of $y = \sin x$. •

We can generalize Example 4 as follows.

SUMMARY Let $f(x) = a\sin(bx + c) + d$. The graph of f is a sine wave centered vertically about the line $y = d$ with one period starting at the point $(-\frac{c}{b}, d)$ and ending at the point $(-\frac{c}{b} + \frac{2\pi}{|b|}, d)$. The number $|a|$ is called the *amplitude* of the sine wave. The graph rises $|a|$ units above $y = d$ at its highest points and falls $|a|$ units below $y = d$ at its lowest points. f is periodic with period $\frac{2\pi}{|b|}$. The number $-\frac{c}{b}$ is the ***phase shift*** and gives the horizontal shift necessary to produce the graph of f from the graph of $y = \sin x$. If $a > 0$, the graph starts out increasing from $\left(-\frac{c}{b}, d\right)$; if $a < 0$, the graph starts out decreasing from $\left(-\frac{c}{b}, d\right)$ (Figure 5.7.8).

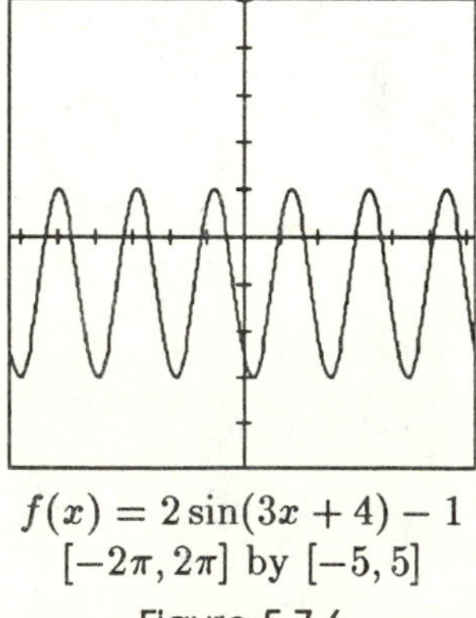

$f(x) = 2\sin(3x + 4) - 1$
$[-2\pi, 2\pi]$ by $[-5, 5]$

Figure 5.7.6

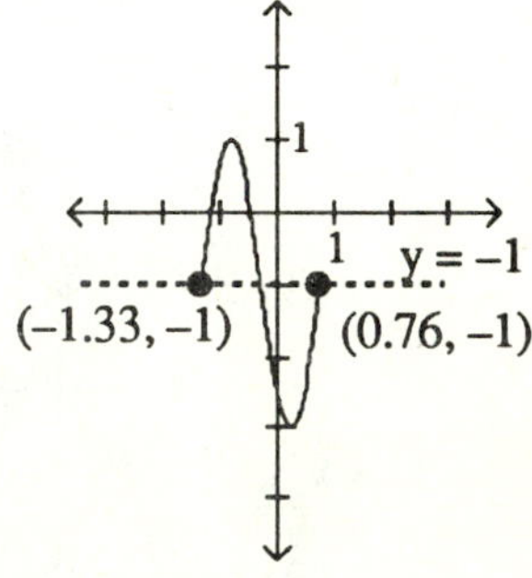

Figure 5.7.7

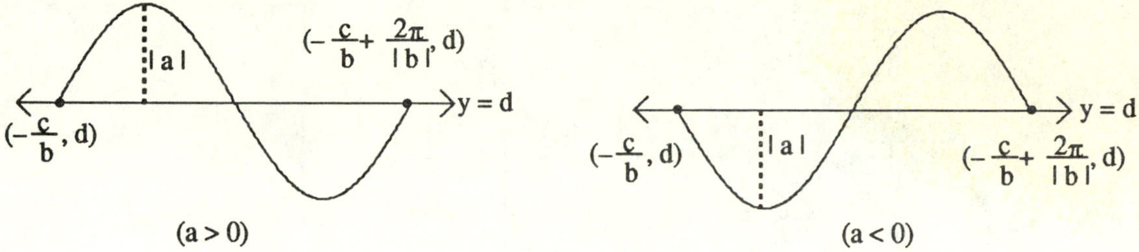

Figure 5.7.8

The graph of $g(x) = a\cos(bx+c)+d$ is related to the graph of $y = \cos x$ in exactly the same way that the graph of $h(x) = a\sin(bx+c)+d$ is related to the graph of $y = \sin x$. This is also true for the trig functions $\sec x$ and $\csc x$, except that there is no amplitude for these two functions.

The graph of $f(x) = a\tan(bx+c)+d$ is related to the graph of $y = \tan x$ in the same way that the graph of $g(x) = a\sin(bx+c)+d$ is related to the graph of $y = \sin x$ with two exceptions. The period of $f(x) = a\tan(bx+c)+d$ is $\frac{\pi}{|b|}$, where π is the period of $y = \tan x$, and f has no amplitude.

Sinusoids

• **DEFINITION** Any function of the form $f(x) = a\sin(bx+c)+d$, where a, b, c, and d are real numbers, is called a **sinusoid**. •

For example, the function f of Example 4 is a sinusoid. In Theorem 2 of Section 5.5 we showed that $\cos x = \sin\left(x + \frac{\pi}{2}\right)$. From this it follows that any function of the form $f(x) = a\cos(bx+c)+d$ is also a sinusoid. There are other more complicated expressions that turn out to be sinusoidal. Sinusoids will be studied in more detail in Chapter 6. It turns out that the function of the next example is a sinusoid.

• EXAMPLE 5: A cross section of a tunnel is a semicircle of radius 20 feet. The interior surfaces of the tunnel form a rectangular opening as illustrated in Figure 5.7.9. Determine the dimensions of a tunnel opening with maximum cross-sectional area, and determine the value of the maximum area.

SOLUTION: Let $P(x, y)$ be the upper right-hand corner of the rectangle and θ the angle OP makes with the positive x-axis (Figure 5.7.9). The area A of the rectangle is $2xy$. Notice that $\cos\theta = \frac{x}{20}$ and $\sin\theta = \frac{y}{20}$. Thus, $A = 2xy = 800\sin\theta\cos\theta$. Figure 5.7.10 gives a complete graph of $A = 800\sin\theta\cos\theta$. The values of θ that make sense in this problem situation are $0 < \theta < \frac{\pi}{2}$ (Figure 5.7.11). Notice that A is 0 when $x = \frac{\pi}{2}$. We can use zoom-in to estimate that the coordinates of the highest point of A in $[0, \frac{\pi}{2}]$ are $(0.785, 400)$ with error at most 0.01. Thus, $x = 20\cos(0.785) = 14.15$ and $y = 20\sin(0.785) = 14.15$. It turns out that the exact coordinates of the highest point of A between $\theta = 0$ and $\theta = \frac{\pi}{2}$ are $(\frac{\pi}{4}, 400)$. Thus,

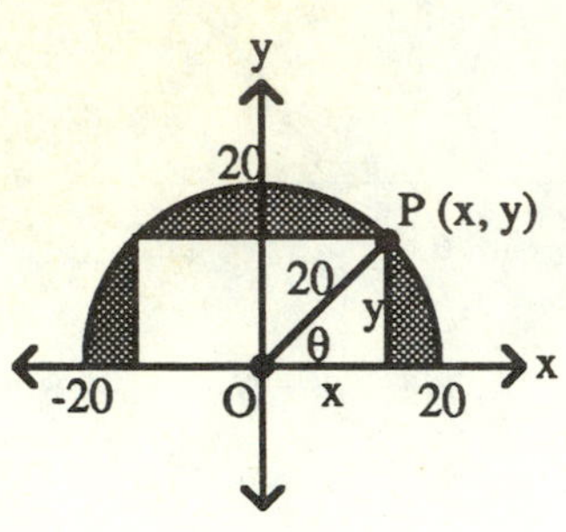

Figure 5.7.9

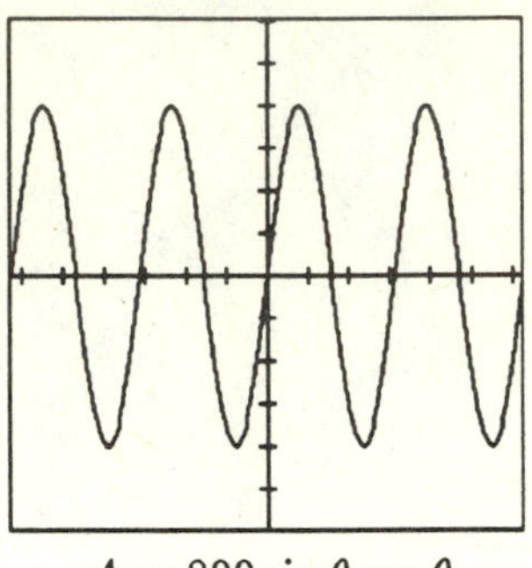

$A = 800 \sin\theta \cos\theta$
$[-2\pi, 2\pi]$ by $[-600, 200]$

Figure 5.7.10

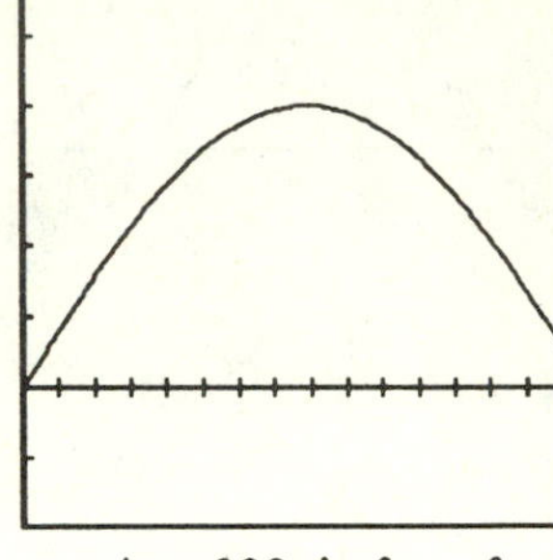

$A = 800 \sin\theta \cos\theta$
$[0, \frac{\pi}{2}]$ by $[-600, 600]$

Figure 5.7.11

$x = 20\cos\frac{\pi}{4}$ and $y = 20\sin\frac{\pi}{4}$ for the rectangle with maximum cross-sectional area. The exact value of $\cos\frac{\pi}{4}$ or $\sin\frac{\pi}{4}$ is $\frac{1}{\sqrt{2}}$, or $\frac{\sqrt{2}}{2}$. Therefore, $x = y = 10\sqrt{2}$. The dimensions of the rectangle with maximum area are $20\sqrt{2}$ feet by $10\sqrt{2}$ feet, and the maximum area is 400 square feet. •

In Chapter 6 we will show that $\sin\theta\cos\theta = \frac{1}{2}\sin 2\theta$. This means that the algebraic representation of Example 5 is a sinusoid.

• EXERCISES 5-7

Sketch a complete graph without using a graphing utility. Check your answer with a graphing utility.

1. $y = 2 + \sin\left(x - \frac{\pi}{4}\right)$
2. $y = -1 - \cos(x - \pi)$
3. $y = -1 + 3\sin(x - 2)$
4. $y = 2 - 2\sin(x + 3)$
5. $f(x) = 2 - 3\sin(2x)$
6. $g(x) = -2 + \tan\left(\frac{1}{2}x\right)$
7. $y = 2 + 3\cos\left(\dfrac{x-1}{2}\right)$
8. $y = -2 - 4\sin(2x + 6)$
9. $T(x) = -3 + 2\sin(2x - \pi)$
10. $y = 2 - 3\cos(4x - 2\pi)$

Determine the domain, range, period, and asymptotes (if any) of each function in the specified Exercise.

11. Exercise 3
12. Exercise 5
13. Exercise 6
14. Exercise 8

Determine where the function is increasing and decreasing and all local extrema for x in $[0, 2\pi]$.

15. $y = 3\sin 2x$
16. $y = 1 + 2\cos(x - \pi)$
17. $y = -\cot(x - \pi)$
18. $y = 2\sec\left(x + \frac{\pi}{2}\right)$
19. $y = 2 - 3\sin(4x - \pi)$
20. $y = -3 + 2\cos\left(\frac{1}{3}x - \frac{\pi}{6}\right)$

Find all real solutions.

21. $2 = 4\sin 3x$

22. $3 = 2\sin 3x$

23. $\cos \frac{1}{2}x = 0.24$

24. $4\sin 2x < x$

Find all real solutions for x in $[0, 5]$.

25. $5\sin\left(\frac{1}{4}x - \pi\right) = 3\cos\left(x + \frac{\pi}{2}\right)$

26. $\tan(x - \pi) = 3\sin 2x$

27. $3\sin x > 2\cos(x - 1)$

28. $3\sin^2 x > 2.65$ [*Note*: $\sin^2 x = (\sin x)^2$]

Write an equation of a sinusoid with given amplitude A, phase shift B, and period P, and draw a complete graph.

29. $A = 3,\ B = \dfrac{\pi}{2},\ P = \pi$

30. $A = \dfrac{1}{3},\ B = 2,\ P = 4$

31. $A = 2,\ B = -\dfrac{\pi}{4},\ P = 4\pi$

32. $A = 5,\ B = -1,\ P = 1$

List the geometric transformations which if applied in order to $y = \sin x$ or $y = \tan x$ will produce the indicated graph.

33. $y = 3 - 4\sin(2x - \pi)$

34. $2 + 3\sin\left(\frac{1}{2}x + \frac{\pi}{2}\right)$

35. $y = 2 - \tan(x - \pi)$

36. $1 + 2\tan(2x - \pi)$

Frequency

The *frequency* of a periodic function is the reciprocal of its period. One *cycle* of a periodic function is the graph in one period. It follows that the frequency represents the number of cycles of a periodic function in an interval of length *one unit*. Determine the period and frequency.

37. $y = 2\sin 3x$

38. $y = -4\cos \frac{1}{2}x$

39. $y = 3\sin \frac{\pi}{2}x$

40. $y = 2\cos 2\pi x$

Harmonic Motion

It can be shown that $y = A\cos Bt$ is a sinusoidal model for the bouncing motion of an object hung from the end of a spring, or *simple harmonic motion*. Assume a coordinate system has been introduced so the vertical axis of the spring coincides with the y-axis and the rest position of the object is at the origin. Further assume that at $t = 0$ the object has been stretched to a position $(0, A)$ when $A < 0$ and released (when $t > 0$). Let y be the vertical position of the object as a function of time t. We assume there is no friction to stop the motion once it is started.

An object on a spring oscillates between $y = -15$ and $y = 15$, and the frequency of oscillation is $\frac{1}{4}$ of a cycle per second.

41. Determine the time interval of one complete cycle (i.e., the period) and then determine A and B of the algebraic representation.

42. Draw a complete graph of the algebraic representation.

43. What part of the graph represents the problem situation?

44. At what times is the object 5 units from the rest position?

An object on a spring oscillates between $y = -28$ and $y = 28$, and the *frequency* of oscillation is 3 cycles per second.

45. Determine the time interval of one complete cycle (i.e., the period) and then determine A and B of the algebraic representation.

46. Draw a complete graph of the algebraic representation.

47. What part of the graph represents the problem situation?

48. At what times is the object 5 units from the rest position?

Damped Motion

Actual oscillatory motion of an object on a spring is affected by friction so eventually the object returns to an at rest position. Such motion is called damped.

49. Draw a complete graph of $y = (20 - x)\cos(x - 3)$.

50. Explain how a portion of the graph in Exercise 49 could be a model for the motion of an object hung from a stretched spring. (This motion is called *dampened* oscillatory motion.)

51. Determine the length of a belt running around two wheels of radii 22 inches and 15 inches, respectively. Assume the wheels are 50 inches apart. (*Hint:* See Figure 5.7.12. Compute d, then compute $\angle\alpha$.)

52. Determine the length of a belt running around two wheels of radii 18 inches and 11 inches, respectively. Assume the wheels are 36 inches apart. (*Hint:* See Figure 5.7.12. Compute d, then compute $\angle\alpha$.)

A playground merry-go-round 16 feet in diameter rotates at 20 rpm (revolutions per minute).

53. Determine the angular velocity in radians per second.

54. Determine the speed (in mph) of a boy located at a point 2 feet from the center.

55. Determine the speed (in mph) of a girl located at the edge of the merry-go-round.

56. Solve Example 5 using a non-trig algebraic representation for the cross-sectional area. (*Hint:* $x^2 + y^2 = 400$.)

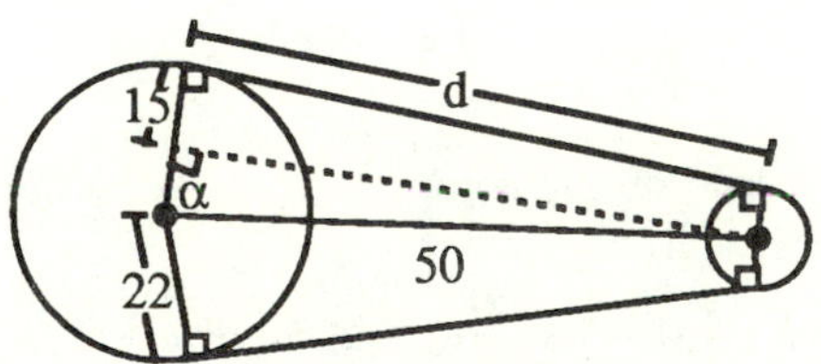

Figure 5.7.12

• KEY TERMS

Listed below are the key terms, vocabulary, and concepts in this chapter. You should be familiar with their definitions and meanings as well as be able to illustrate each of them.

Absolute maximum
Absolute mimimum
Acute angle
Amplitude
Angle
Angular speed
Arc length
Bearing
Central angle
Circular property of trig functions
Complementary angles
Coterminal angles
Degrees
Distance formula
Horizontal shifting
Horizontal shrinking
Hypotenuse
Identities
Initial side
Linear speed
Minute
Opposite side
Period
Periodic functions
Perpendicular (orthogonal)
Phase shift
Positive and negative measure
Quadrantal angle
Radians
Reciprocal identities
Right angle
Right triangle
Second
Sinusoids
Solving right triangles
Standard position
Supplementary angle relationships
Supplementary angles
Terminal side
Trigonometric identities
Trigonometric functions
Vertex of an angle
Zeros of periodic functions

• REVIEW EXERCISES

This section contains representative problems from each of the previous seven sections. You should use these problems to test your understanding of the material covered in this chapter.

Find the value of the six trig functions at

1. $32°$
2. 2 radians
3. Complete Table 1. Find *exact* values.

Use properties of a $30°$–$60°$ right triangle and a $45°$ right triangle to determine the *exact* values of each trig function in Table 2.

10. Consider the triangle ABC in Figure R.1. Find all six trig functions of $\angle A$.

$\sin 30° = ?$	$\csc \frac{\pi}{2} = ?$
$\cos 270° = ?$	$\sec 270° = ?$
$\tan 135° = ?$	$\cot \frac{\pi}{4} = ?$
$\sin \frac{5\pi}{6} = ?$	$\csc 210° = ?$
$\cos \frac{2\pi}{3} = ?$	$\sec 330° = ?$
$\cot 135° = ?$	$\cot \frac{5\pi}{4} = ?$

Table 1

11. Use a right triangle to determine the values of all trig functions at θ with $\cos\theta = \frac{5}{7}$.

12. Use the [INV] [COS], [2nd F] [COS] or [SHIFT] [COS] keys on your calculator to determine $0° \leq \theta \leq 90°$ with $\cos\theta = \frac{5}{13}$. Make sure your calculator is in degree mode.

13. Show that $\cot\theta = \dfrac{\cos\theta}{\sin\theta}$.

14. Determine θ if $\cot\theta = \dfrac{12}{5}$.

Solve the right triangle in Figure R.2.

15. $\angle A = 35°$, $c = 15$

16. $b = 8$, $c = 10$

17. Find the equivalent radian measure of 60°.

18. Find the equivalent degree measure of $\frac{3\pi}{4}$ radians.

19. Determine the measure and draw the angle in standard position.
    (a) $\frac{3}{4}$ counterclockwise rotation
    (b) $2\frac{1}{2}$ counterclockwise rotations

	θ	$\sin\theta$	$\cos\theta$	$\tan\theta$	$\cot\theta$	$\sec\theta$	$\csc\theta$
4.	0						
5.	$-\frac{\pi}{6}$						
6.	$\frac{3\pi}{4}$						
7.	60°						
8.	−135°						
9.	300°						

Table 2

20. Let θ be an angle in standard position with the point P on the terminal side of θ. Find the values of the six trig functions at θ if $P = (-3, -2)$.
21. Determine θ in Exercise 14 if $0° \leq \theta \leq 360°$.
22. Determine *all* values of θ that satisfy the conditions of Exercise 20.
23. Let θ be an angle in standard position with terminal side given by the equation $y = \frac{4}{3}x$ in the first quadrant. Find the values of the six trig functions at θ.

Verify the identity.

24. $1 + \cot^2\theta = \dfrac{1}{\sin^2\theta}$
25. $\dfrac{\sin^2\theta}{1 - \cos\theta} = 1 + \cos\theta$
26. Use the circular properties of the trig functions to verify $\cos 1 > \cos 4$. Do not use a calculator.
27. Determine the angle between the x-axis and the line $2x - 7y = 5$.

θ is an angle in standard position. Find the quadrant containing θ if $0° \leq \theta \leq 360°$.

28. $\sin\theta < 0$ and $\tan\theta < 0$
29. $\cos\theta < 0$ and $\csc\theta > 0$
30. Find an acute angle α satisfying the condition $\cos(90° - \alpha) = \cos\alpha$.
31. Determine the measure of an angle θ coterminal with the given angle satisfying the specified condition.

    (a) $34°$; $360° \leq \theta \leq 720°$
    (b) $\frac{-3\pi}{4}$; $0 \leq \theta \leq 2\pi$
32. A central angle θ subtends an arc of length s on a circle of diameter 10 units. Determine s for the following central angles.

    (a) $\theta = \frac{\pi}{4}$
    (b) $\theta = 100°$

Find *all* real solutions to each equation.

33. $\cos x = \dfrac{\sqrt{3}}{2}$
34. $\tan x = 0$

Sketch a complete graph without using a graphing utility. Check your answer with a graphing utility.

35. $y = \sin(x + 45°)$
36. $y = 3 + 2\cos x$
37. $y = -\sin(x + \frac{\pi}{2})$
38. $y = -2 - 3\sin(x - \pi)$
39. Determine a complete graph, domain, range, and period of the function $y = 3\sin 2x$.
40. Sketch a complete graph of $y = 1 + 2\cos 3x$ without using a graphing utility. Check your answer with a graphing utility. Determine the domain, range, and period.

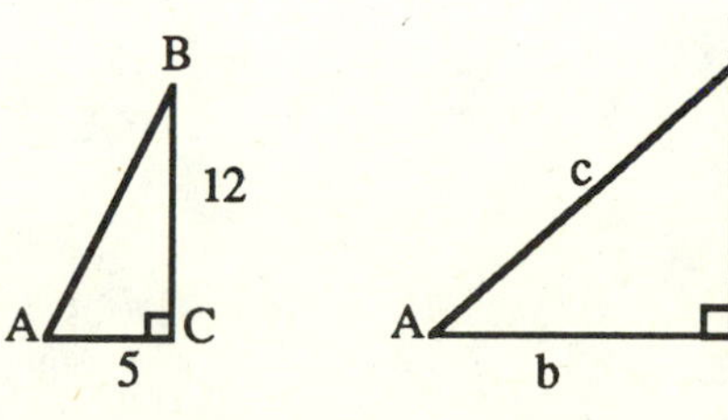

Figure R.1 Figure R.2

41. Determine the domain, range, period, and asymptotes (if any) of the function $y = 1 + 2\sec 3x$.

List, in proper order, the geometric transformations that can be used to obtain each graph from the graph of $y = \sin x$ or $y = \tan x$.

42. $y = 4 + \sin(2x - 1)$

43. $g(x) = -5 + \tan\left(\frac{1}{2}x\right)$

Solve the equation given that $0 \leq x \leq 2\pi$.

44. $\sin x = 0.45$

45. $1 - x = \cos x$

46. Solve the inequality $3\cos x < x + 2$.

True or false? Give an argument based on graphs.

47. $\sin(x + \pi) = \sin x$

48. $\cos\left(x + \frac{\pi}{2}\right) = -\cos x$

49. If $\cos\theta < 0$ and $\tan\theta < 0$, then θ is in which quadrant?

50. The angle of elevation of the top of a building from a point 100 meters away from the building on level ground is 78°. Determine the height of the building.

51. A tree casts a shadow 51 feet long when the angle of elevation of the sun (measured from the horizon) is 25°. How tall is the tree?

52. From the top of a 150 foot building a man observes a car moving toward him. If the angle of depression of the car changes from 18° to 42° during the observation, how far does the car travel?

53. A lighthouse L stands 4 miles from the closest point P along a straight shore. Find the distance from P to a point Q along the shore if the angle PLQ measures 22°.

54. An airplane at a certain time is between two signal towers that are on line with bearing 0° (along a north–south line). The bearing between the plane and each tower is 23° and 128° respectively. Use a drawing to show the exact location of the plane.

55. The bearings of lines of sight from a boat to two points on the shore are 115° and 123°, respectively. Assume the two points are 855 feet apart. How far is the boat from the nearest point on shore if the shore is a straight line?

Sketch the reference triangle and the reference angle θ' determined by each angle. Determine the reference angle and the values of the six trig functions at θ' and θ.

56. 128°

57. −204°

6 Analytic Trigonometry

• Introduction

In this chapter we continue the study of trig functions, find their local extreme values and zeros, and determine their end behavior. We see how to write certain trig functions as sinusoidal functions, that is, as functions of the form $a \sin(bx + c) + d$. Applications involving sinusoidal functions will be investigated.

Trigonometric identities involving double-angles and half-angles will be established. The inverse trig functions are defined and their complete graphs determined. Equations and inequalities involving trig functions are solved both algebraically and graphically. The law of sines and the law of cosines are established and used in applications. Two formulas for the area of a triangle that use the measure of its angles and sides will be obtained. A formula for the angle between two lines is determined.

6.1 More on Trigonometric Graphs

The investigation of graphs of trig functions continues in this section. We see how to rewrite certain sums of trig functions as sinusoidal functions, and we determine the end behavior and local extrema of functions that involve trig functions. We investigate an application that has a model involving a trig function.

In Section 5.7 we indicated that a sinusoid is any function that can be written in the form $f(x) = a\sin(bx + c) + d$, where a, b, c, and d are real numbers. The graph of a sinusoid is shaped like the graph of the sine function. (Why?) We showed in Theorem 2 of Section 5.5 that $\cos x = \sin(x + \frac{\pi}{2})$. Thus, $f(x) = \cos x$ is a sinusoid. We investigate other sinusoidal functions in Examples 1 and 2.

In the Exercises you will show that the following theorem is a special case of Theorem 2 that appears later in this section.

• **THEOREM 1** For all real numbers a and b, there exist real numbers A and α such that

$$a \sin x + b \cos x = A \sin(x + \alpha).$$

More precisely, we have the following:

(1) If $a = 0$, then $\alpha = \frac{\pi}{2}$ and $A = b$.
(2) If $a \neq 0$, then $-\frac{\pi}{2} \leq \alpha \leq \frac{\pi}{2}$ with $\tan \alpha = \frac{b}{a}$. If $a > 0$, then $A = \sqrt{a^2 + b^2}$, and if $a < 0$, then $A = -\sqrt{a^2 + b^2}$.

• EXAMPLE 1: Determine the domain, range and period of $f(x) = 2 \sin x + 5 \cos x$, and draw a complete graph.

SOLUTION: The graph in Figure 6.1.1 appears to be the graph of a sinusoid. We can estimate from this figure that the amplitude is 5.4, the period 2π, and the phase shift 1.2 radians left. This information can be confirmed geometrically by overlaying the graph of $y = 5.4 \sin(x + 1.2)$ in Figure 6.1.1. Theorem 1 confirms algebraically that f is a sinusoid and that the graph in Figure 6.1.1 is complete. •

• EXAMPLE 2: Determine A and α so that $A \sin(x + \alpha) = 2 \sin x + 5 \cos x$.

SOLUTION: From Theorem 1 we have $A = \sqrt{2^2 + 5^2} = \sqrt{29}$ and $\tan \alpha = \frac{5}{2}$. Thus, $\alpha = 1.19$. Accurate to tenths, $A = 5.4$, and $\alpha = 1.2$. •

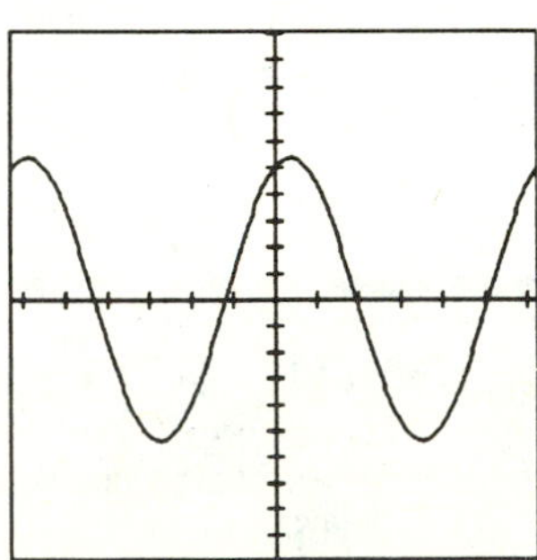

$f(x) = 2 \sin x + 5 \cos x$
$[-2\pi, 2\pi]$ by $[-10, 10]$
Figure 6.1.1

Notice that the values of A and α found in Example 2 are consistent with the approximations found in Example 1. We can now be sure that the function of Example 1 is a sinusoid. That is,

$$f(x) = 2\sin x + 5\cos x = \sqrt{29}\sin(x+\alpha)$$

where $-\frac{\pi}{2} \leq \alpha \leq \frac{\pi}{2}$ and $\tan\alpha = 2.5$.

• EXAMPLE 3: Determine A and α so that $A\sin(x+\alpha) = -3\sin x + 2\cos x$ and confirm geometrically.

SOLUTION: From Theorem 1 we have $A = -\sqrt{(-3)^2 + 2^2} = -\sqrt{13}$ and $\tan\alpha = -\frac{2}{3}$. Thus, $\alpha = -0.59$. The graph of $y = -3\sin x + 2\cos x$ in Figure 6.1.2 appears to be a sinusoid with period 2π, amplitude 3.6, and phase shift 0.6 radians right. We can overlay the graph of $y = -\sqrt{13}\sin(x - 0.59)$ in Figure 6.1.2 to confirm geometrically that $-3\sin x + 2\cos x = -\sqrt{13}\sin(x-0.59)$. •

• EXAMPLE 4: Show that $f(x) = 3\sin(2x-1) + 4\cos(2x+3)$ is a sinusoid.

SOLUTION: The graph in Figure 6.1.3 suggests that f is a sinusoid with period π, amplitude 6.6, and phase shift 0.7 radians right. This information can be confirmed geometrically by overlaying the graph of $y = 6.6\sin 2(x-0.7)$ in Figure 6.1.3. An algebraic confirmation is provided by Theorem 2. •

Example 4 is a special case of the following theorem which will be proved in the next section.

• THEOREM 2 For all real numbers a, b, d, h, and k, there exist real numbers A and α such that

$$a\sin(bx+h) + d\cos(bx+k) = A\sin(bx+\alpha).$$

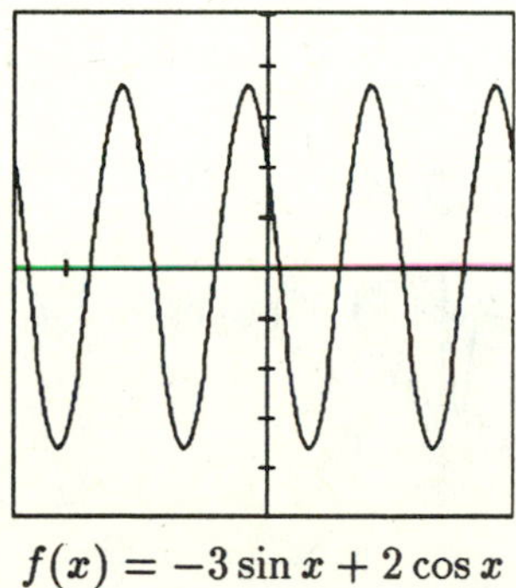

$f(x) = -3\sin x + 2\cos x$

$[-4\pi, 4\pi]$ by $[-5, 5]$

Figure 6.1.2

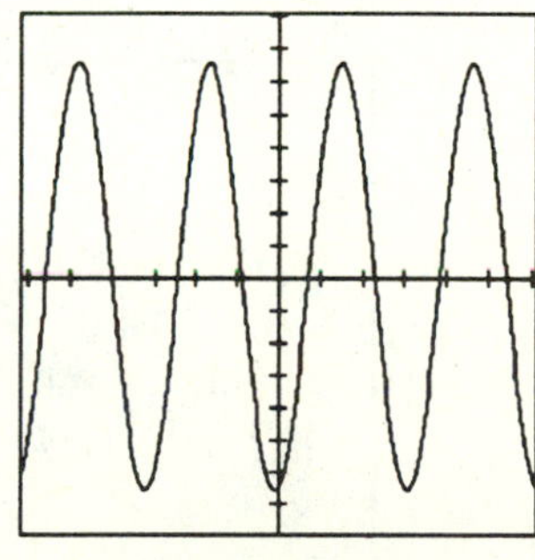

$f(x) =$
$3\sin(2x-1) + 4\cos(2x+3)$

$[-2\pi, 2\pi]$ by $[-8, 8]$

Figure 6.1.3

More precisely, we have the following:

(1) If $a\cos h - d\sin k = 0$ then $\alpha = \frac{\pi}{2}$ and $A = a\sin h + d\cos k$.
(2) If $a\cos h - d\sin k \neq 0$, then $-\frac{\pi}{2} \leq \alpha \leq \frac{\pi}{2}$ with $\tan\alpha = \frac{a\sin h + d\cos k}{a\cos h - d\sin k}$ and $A = \pm\sqrt{(a\sin h + d\cos k)^2 + (a\cos h - d\sin k)^2}$. A and $a\cos h - d\sin k$ have the same sign.

We can use Theorem 2 to determine A and α so that $3\sin(2x-1) + 4\cos(2x+3) = A\sin(2x+\alpha)$. This confirms that the function of Example 4 is a sinusoid. In the Exercises, you will be asked to use the formulas of Theorem 2 to determine the values of A and α.

The functions in the next two examples are not sinusoids.

- EXAMPLE 5: Determine the domain, the range, the period, the local maximum and minimum values in one period, and draw a complete graph of $f(x) = \sin 2x + \cos 3x$.

SOLUTION: We show that the graph in Figure 6.1.4 is a complete graph. The domain of f is the set of all real numbers. (Why?) The function f is periodic because both $\sin 2x$ and $\cos 3x$ are periodic. Certainly 2π is a candidate for the period of f. Notice that $y = \sin 2x$ is periodic with period π and $y = \cos 3x$ is periodic with period $\frac{2\pi}{3}$. Thus, the period of f must be a positive integer multiple of both π and $\frac{2\pi}{3}$. Their least common multiple is 2π, and so f has period 2π.

The graph of f in the $[0, 2\pi]$ by $[-3, 3]$ viewing rectangle displays one complete period of f (Figure 6.1.5). We need to find the smallest and largest value of f in this figure to determine the range of f. Notice that f has six extreme values in Figure 6.1.5. We do not include the endpoints in our consideration about extreme values of f because, when the entire graph is considered, f does *not* have extreme values at the endpoints.

The absolute maximum of f in $[0, 2\pi]$ is 1.91 and occurs when x is 4.11. The other two local maximum values are 1.22 and 0.22 and occur when x is 0.22 and 1.91, respectively. Similarly, the absolute minimum of f in $[0, 2\pi]$ is -1.91 and occurs when x is 5.32. The other two local minimum values are -0.22 and -1.22 and occur when x is 1.23 and 2.92,

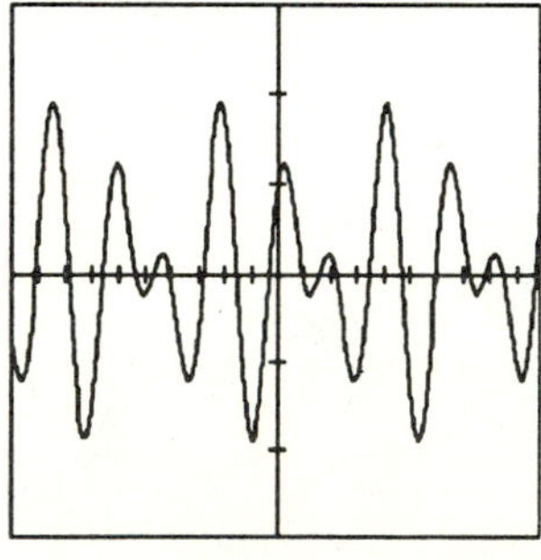

$f(x) = \sin 2x + \cos 3x$
$[-10, 10]$ by $[-3, 3]$
Figure 6.1.4

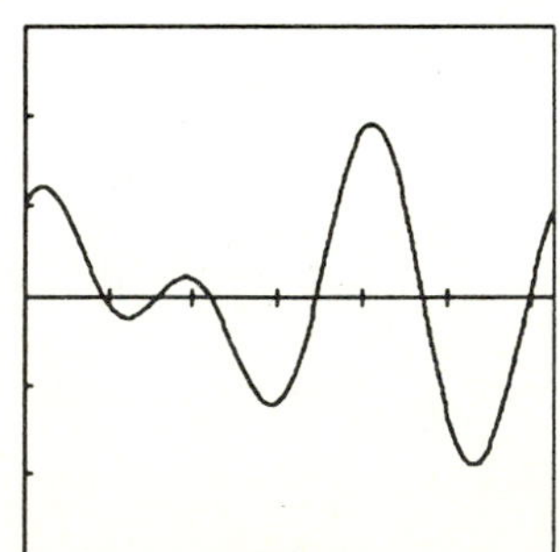

$f(x) = \sin 2x + \cos 3x$
$[0, 2\pi]$ by $[-3, 3]$
Figure 6.1.5

respectively. Thus, the range of f is $[-1.91, 1.91]$ with error at most 0.01 in the endpoints of the interval. •

Figure 6.1.5 shows that $f(x) = \sin 2x + \cos 3x$ has six zeros in the interval $[0, 2\pi]$. Zoom-in could be used to find these zeros. They can also be determined algebraically, but several trig identities are needed to get started. In Section 6.2 we establish the necessary identities and find the zeros algebraically.

• EXAMPLE 6: Determine the domain, the range and the period of $f(x) = \sin \frac{x}{2} + \sin \frac{x}{3}$, and draw a complete graph.

SOLUTION: The domain of f is the set of all real numbers. Notice that $\sin \frac{x}{2}$ is periodic with period 4π and $\sin \frac{x}{3}$ is periodic with period 6π. (Why?) The period of f is the least common multiple of 4π and 6π. (Why?) Thus, the period of f is 12π. The graph of f in Figure 6.1.6 is a complete graph displaying two complete periods and confirms these observations.

The absolute maximum of f in $[0, 12\pi]$ is 1.91 and occurs when x is 3.62. The absolute minimum of f in $[0, 2\pi]$ is -1.91 and occurs when x is 34.08. Thus, the range of f is $[-1.91, 1.91]$. •

The function in the next example has interesting end behavior. Its graph turns out to be symmetric with respect to the y-axis. Recall that a graph of an equation is symmetric with respect to the y-axis if the equation is unchanged when x is replaced by $-x$. We have also seen that a graph of a function f is symmetric with respect to the y-axis if $f(-x) = f(x)$, that is, if f is an even function.

• EXAMPLE 7: Determine the end behavior and draw a complete graph of $y = x \sin x$. Show that the graph is symmetric with respect to the y-axis.

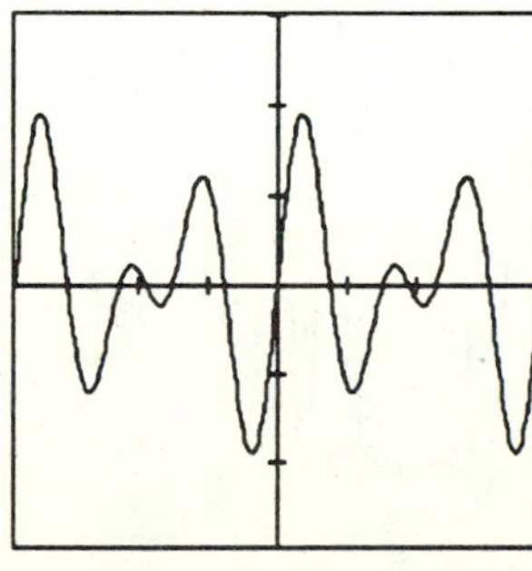

$f(x) = \sin \frac{x}{2} + \sin \frac{x}{3}$
$[-12\pi, 12\pi]$ by $[-3, 3]$
Figure 6.1.6

SOLUTION: Notice that the equation for this function is unchanged when x is replaced by $-x$. We use the fact that $\sin x$ is an odd function, that is, $\sin(-x) = -\sin x$:

$$\begin{aligned} y &= x\sin x \\ &= (-x)\sin(-x) \\ &= (-x)(-\sin x) \\ &= x\sin x \end{aligned}$$

Thus, f is an even function, and its graph is symmetric with respect to the y-axis. You may need to experiment with different viewing rectangles to convince yourself that the graph in Figure 6.1.7 is a complete graph. The amplitude of the sine-like wave gets larger and larger as $|x| \to \infty$. In fact, the graph bounces back and forth between the lines $y = x$ and $y = -x$ as illustrated in Figure 6.1.8.

Thus, the end behavior of $y = x\sin x$ is a sine-like wave with amplitude that approaches ∞ as $|x| \to \infty$. •

The graph of the function f in the next example is symmetric with respect to the origin because it is an odd function. That is, $f(-x) = -f(x)$. Moreover, this function and the function of Example 9 have very interesting behavior near $x = 0$.

• EXAMPLE 8: Determine the end behavior and the behavior near $x = 0$ of $f(x) = \sin\frac{1}{x}$, and draw a complete graph.

SOLUTION: The graph of f in Figure 6.1.9 suggests that the end behavior of f is the horizontal line $y = 0$. We can use zoom-out to confirm this conjecture. That is, f has the x-axis as a horizontal asymptote. This figure also shows that the behavior of f near $x = 0$ is peculiar. The graph of f for $-0.1 \le x \le 0.1$ in Figure 6.1.10 gives a closer look at the behavior of f near $x = 0$.

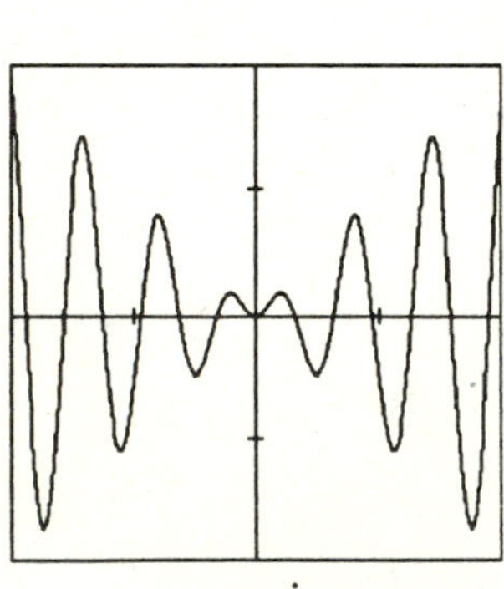

$y = x\sin x$
$[-20, 20]$ by $[-20, 20]$
Figure 6.1.7

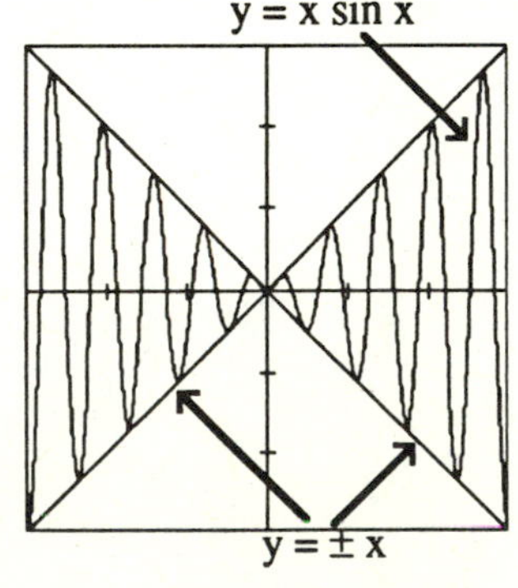

$[-30, 30]$ by $[-30, 30]$
Figure 6.1.8

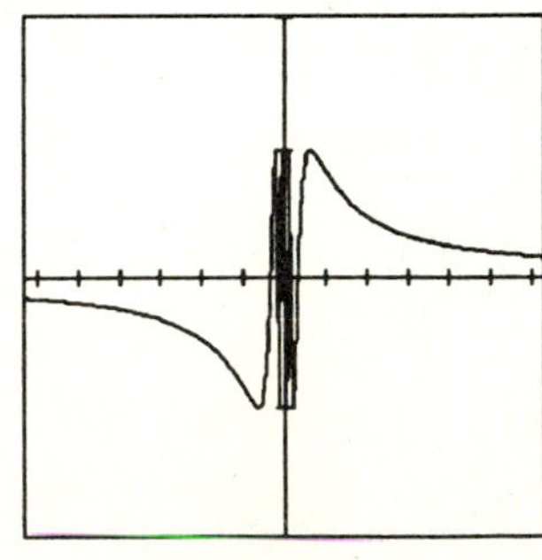

$f(x) = \sin\frac{1}{x}$
$[-2\pi, 2\pi]$ by $[-2, 2]$
Figure 6.1.9

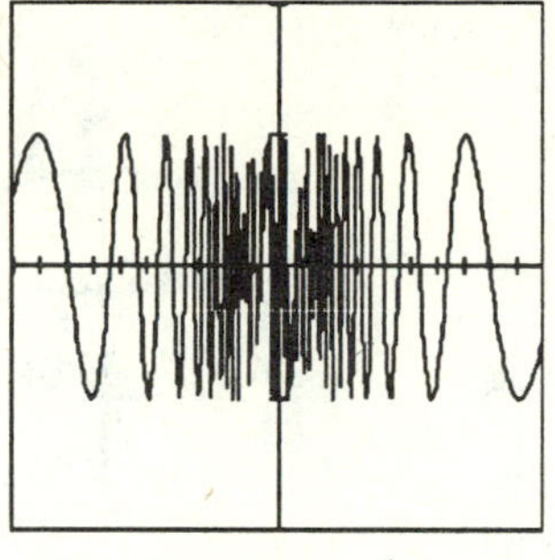

$f(x) = \sin \frac{1}{x}$
$[-0.1, 0.1]$ by $[-2, 2]$
Figure 6.1.10

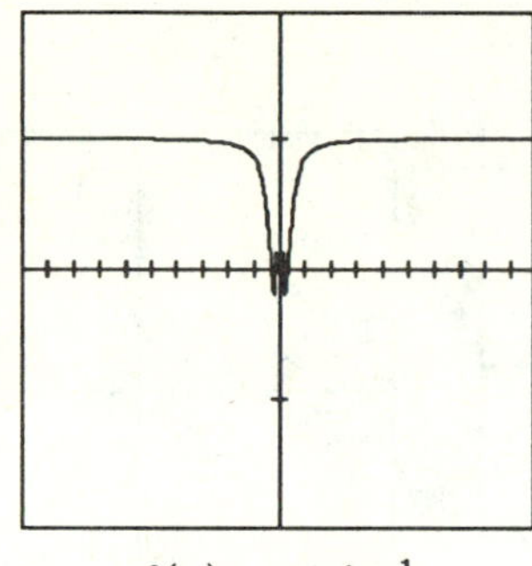

$f(x) = x \sin \frac{1}{x}$
$[-10, 10]$ by $[-2, 2]$
Figure 6.1.11

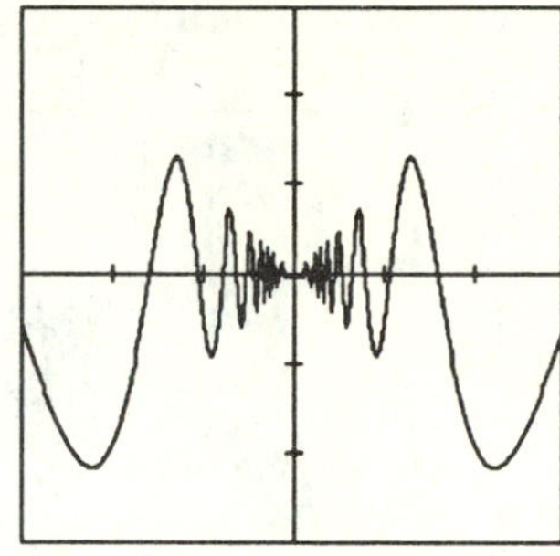

$f(x) = x \sin \frac{1}{x}$
$[-0.3, 0.3]$ by $[-0.3, 0.3]$
Figure 6.1.12

The function $f(x) = \sin \frac{1}{x}$ takes on all values between -1 and 1 in smaller and smaller intervals near $x = 0$ and $f(0)$ does not exist. For example, $f(x) = \sin \frac{1}{x}$ takes on all values between -1 and 1 in the interval $\left[\frac{1}{12\pi}, \frac{1}{10\pi}\right]$. We can see this as follows. If $\frac{1}{12\pi} \le x \le \frac{1}{10\pi}$, then $12\pi \ge \frac{1}{x} \ge 10\pi$. Thus, $\sin \frac{1}{x}$ takes on the same values in the interval $\frac{1}{12\pi} \le x \le \frac{1}{10\pi}$ that $\sin x$ does in the interval $10\pi \le x \le 12\pi$. Notice that $\frac{1}{12\pi}$ is approximately 0.0265 and $\frac{1}{10\pi}$ is approximately 0.0318. As x approaches 0 from either side, the graph of f consists of linking together end-to-end complete sine-like waves that have smaller and smaller periods. The graphs of f in Figures 6.1.9 and 6.1.10 constitute a complete graph of f. •

The function in the next example also has a horizontal asymptote. Its graph is symmetric with respect to the y-axis because it is an even function.

• EXAMPLE 9: Determine the end behavior, the behavior near $x = 0$, and draw a complete graph of $f(x) = x \sin \frac{1}{x}$.

SOLUTION: The graph of f in Figure 6.1.11 shows that $f(x) \to 1$ as $|x| \to \infty$. That is, f has the line $y = 1$ as a horizontal asymptote. As in the previous example, the behavior near $x = 0$ is peculiar. The graph of f in Figure 6.1.12 gives a closer look near $x = 0$.

As we saw in Example 8, $\sin \frac{1}{x}$ takes on all values between -1 and 1 in smaller and smaller intervals near $x = 0$. The effect of multiplying $\sin \frac{1}{x}$ by x is to cause the graph of f to bounce back and forth between $y = x$ and $y = -x$ near $x = 0$ (Figure 6.1.13). Therefore, the graphs of f in Figures 6.1.11 and 6.1.12 constitute a complete graph of f. •

• EXAMPLE 10: Determine the domain and the range of $f(x) = 2^{\sin x}$, and draw a complete graph.

SOLUTION: The graph of f in Figure 6.1.14 suggests that the domain of f is the set of all real numbers. The range of f appears to be about $\left[\frac{1}{2}, 2\right]$. The function f is periodic with period 2π because $\sin x$ is periodic with period 2π. (Why?) Because $\sin x$ takes on all values between -1 and 1 and 2^x is an increasing function, f takes on all values between 2^{-1}

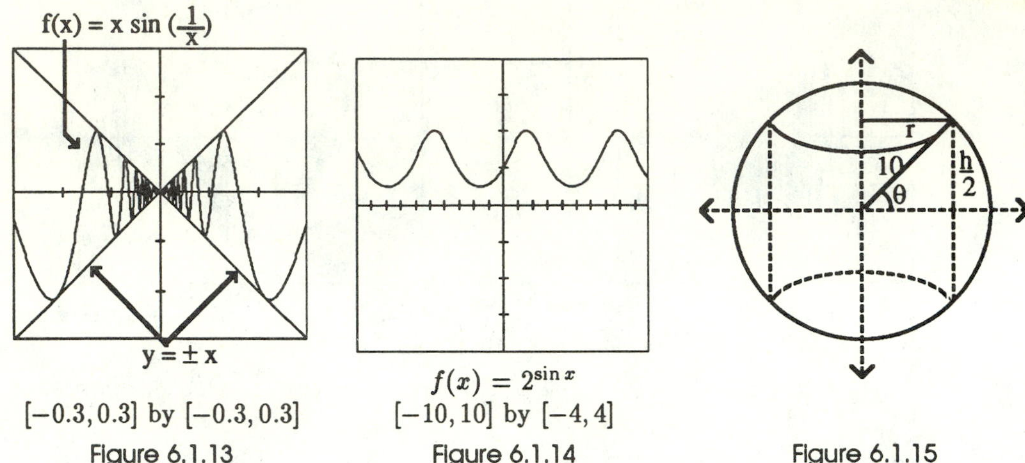

$[-0.3, 0.3]$ by $[-0.3, 0.3]$
Figure 6.1.13

$f(x) = 2^{\sin x}$
$[-10, 10]$ by $[-4, 4]$
Figure 6.1.14

Figure 6.1.15

and 2^1. Thus, the domain of f is the set of all real numbers, and the range of f is $\left[\frac{1}{2}, 2\right]$. Therefore, the graph of f in Figure 6.1.14 is a complete graph. •

A hole is drilled through the center of a sphere of radius 10 (Figure 6.1.15). The portion of the sphere drilled out consists of a right circular cylinder and two spherical caps. Let r and h be the radius and height, respectively, of the cylinder drilled out. We want to determine the volume of the hole in the form of a right circular cylinder that is created by the drilling. Draw a line (radius of sphere) from the center of the sphere to the top of the cylinder as illustrated in Figure 6.1.15. Let θ be the angle that this radius makes with the horizontal.

• EXAMPLE 11: A hole is drilled through the center of a sphere as illustrated in Figure 6.1.15.

(a) Express the volume V of the right circular cylinder as a function of the angle θ defined above.
(b) Draw a complete graph of V. What portion of the graph represents the problem situation?
(c) Determine the dimensions of a cylinder of the largest possible such hole that can be created in the sphere. What is the maximum volume?

SOLUTION:

(a) The volume of the cylinder is $V = \pi r^2 h$. Refer to Figure 6.1.15. Notice that $\sin\theta = (h/2)/10$ and $\cos\theta = \frac{r}{10}$. Thus, $h = 20\sin\theta$, $r = 10\cos\theta$ and $V = 2000\pi\sin\theta\cos^2\theta$.
(b) The graph of V in the $[-10, 10]$ by $[-4000, 4000]$ viewing rectangle is shown in Figure 6.1.16. It can be shown that V is periodic with period 2π. Thus, this graph is a complete graph. We can see from Figure 6.1.15 that $0 < \theta < \frac{\pi}{2}$. Therefore, only the portion of Figure 6.1.16 between 0 and $\frac{\pi}{2}$ represents the problem situation (Figure 6.1.17).

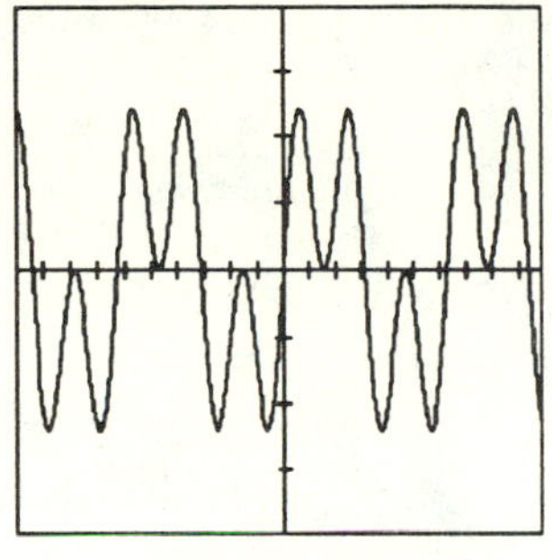

$V = 2000\pi \sin\theta \cos^2\theta$
$[-10, 10]$ by
$[-4000, 4000]$
Figure 6.1.16

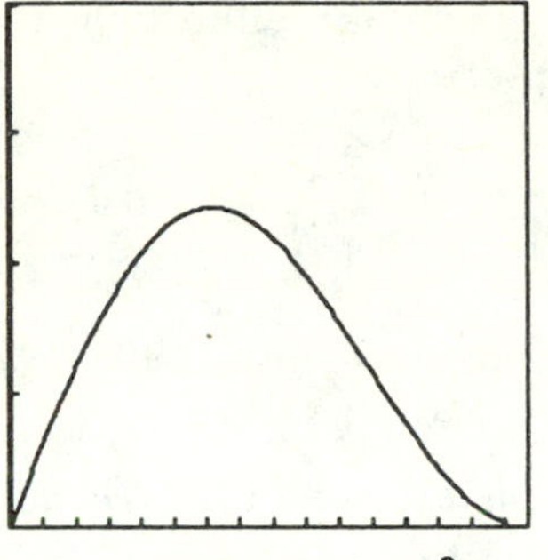

$V = 2000\pi \sin\theta \cos^2\theta$
$[0, \frac{\pi}{2}]$ by
$[0, 4000]$
Figure 6.1.17

(c) We can use zoom-in to determine that coordinates of the highest point of the graph of V in $0 \leq \theta \leq \frac{\pi}{2}$ are $(0.615, 2418.399)$ with error of at most 0.01. Thus, the maximum volume is 2418.399 cubic units and occurs when θ is 0.615 radians. We can use $h = 20\sin\theta$ and $r = 10\cos\theta$ to determine that h is 11.54 and r is 8.17. •

• EXERCISES 6-1

Draw a complete graph and show that it is a sinusoid $y = A\sin(bx + \alpha)$ by estimating A, b, and α and overlaying the graph of $y = A\sin(bx + \alpha)$ to check your estimated values.

1. $y = 3\sin x + 2\cos x$
2. $y = -5\sin x + 3\cos x$
3. $y = -\sin(x + \pi) + \cos x$
4. $y = 4\sin 2x - 3\cos 2x$

Use the formulas in Theorem 2 to verify the values of amplitude and phase shift observed in Example 4.

5. Determine the amplitude factor.
6. Determine the phase shift factor.

Determine the domain, range and period of the function, and draw a complete graph. Determine all local and absolute maximum and minimum values in the interval $[0, 2\pi]$ of each function.

7. $f(x) = \sin x + \cos 2x$
8. $g(x) = 2\sin x + 3\cos 2x$
9. $f(x) = \sin 3x + \cos x$
10. $t(x) = 3\sin 2x - \cos x$

Draw a complete graph and determine the end behavior of the function.

11. $y = \dfrac{\sin x}{x}$
12. $y = x^2 \sin x$
13. $y = (-x)\sin\dfrac{1}{x}$
14. $y = x^2\sin\dfrac{1}{x}$
15. $y = \ln|\sin x|$
16. $y = e^{\sin x}$

Draw a complete graph of the function. Examine the graph for symmetry. Indicate any horizontal or vertical asymptotes.

17. $y = |\sin x|$

18. $y = \dfrac{x}{\sin x}$

19. $y = \left|x \sin \dfrac{1}{x}\right|$

20. $y = \dfrac{\sin x}{x^2 - 4}$

A right circular cylinder is inscribed in a sphere of radius 20.

21. Express the volume V of the cylinder as a function of the angle θ.

22. Draw a complete graph of V. Which portion of the graph represents the problem situation?

23. Determine the dimensions of a cylinder of maximum possible volume that can be inscribed in the sphere. What is the maximum volume?

24. Find *all* maximum and minimum values of $g(x) = 3\cos x + 3\sin x$ and the associated values of x where they occur.

25. Determine an algebraic (non-trigonometric) representation of the problem situation of Example 11. Draw its complete graph and the graph of the problem situation. Find the maximum volume of the right circular cylinder drilled out and compare with the solution given in Example 11.

The Sinusodial Equation $T = A + B\sin(\frac{\pi}{6}\{[360 - (360 - t)]/30\} + C)$ is a model for the mean daily temperature (T) as a function of time t measured in days. (Adapted from "Seasonal Temperature Patterns of Selected Cities in and Around Ohio." *Ohio Journal of Science* (1) pp. 8–10, 1986). A is the mean yearly temperature, B is $\frac{1}{2}$ of the total mean temperature variation, and C is a phase shift factor for the beginning of a temperature cycle. In Columbus, Ohio, the mean yearly temperature is 58° F and the total mean temperature variation is 56°. Assume the temperature on April 1 is the mean yearly temperature, so $C = 0$, and t is the number of days past April 1.

26. Determine a trigonometric representation of the problem situation.

27. Draw a complete graph of the trigonometric representation.

28. What portion of the graph represents the problem situation?

29. About what month and day is it the hottest and coldest during the year in Columbus?

Investigate graphically what happens to the values of $f(x)$ as $x \to 0$.

30. $f(x) = \dfrac{\sin x}{x}$

31. $f(x) = \dfrac{1 - \cos x}{x^2}$

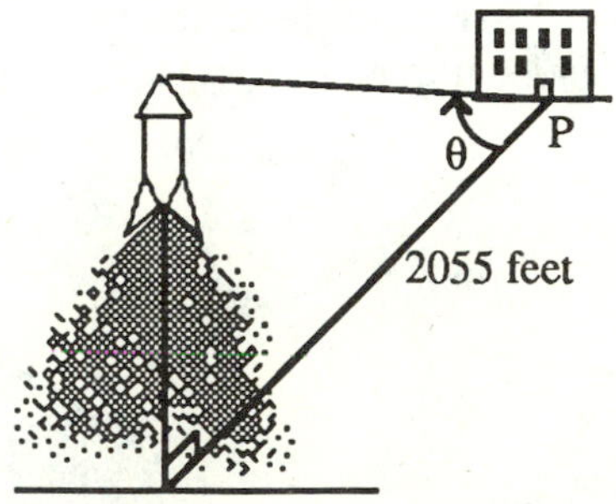

Figure 6.1.18

32. $f(x) = \dfrac{1 - \cos x}{\sin x}$

33. $f(x) = x \cot x$

34. $f(x) = \dfrac{1 - \cos x}{x^2 + x}$

A rocket is launched straight up from ground level at 200 feet per second. A tracking device is located at a point P that is 2055 feet from the launch point (Figure 6.1.18).

35. Write $\angle P = \theta$ as a function of time t.

36. Determine the angle θ 15 seconds after launch.

6.2 Trigonometric Identities

In this section we establish important trig identities about sums and differences of angles. These formulas will be used to obtain the *double-angle* and *half-angle* formulas. We use these formulas to help solve an equation algebraically and to show that the slopes of perpendicular lines are negative reciprocals.

In Section 5.4 we established the following basic identities.

- Basic Identities

$$\tan\theta = \frac{\sin\theta}{\cos\theta} \qquad \sec\theta = \frac{1}{\cos\theta} \qquad \sin^2\theta + \cos^2\theta = 1$$

$$1 + \tan^2\theta = \sec^2\theta$$

$$\cot\theta = \frac{\cos\theta}{\sin\theta} \qquad \csc\theta = \frac{1}{\sin\theta} \qquad 1 + \cot^2\theta = \csc^2\theta$$

The Addition Formulas

Let α and β be two angles in standard position. Let A and B be the points on the unit circle on the terminal side of α and β, respectively (Figure 6.2.1). By definition of the trig functions, the coordinates of A are $(\cos\alpha, \sin\alpha)$ and the coordinates of B are $(\cos\beta, \sin\beta)$.

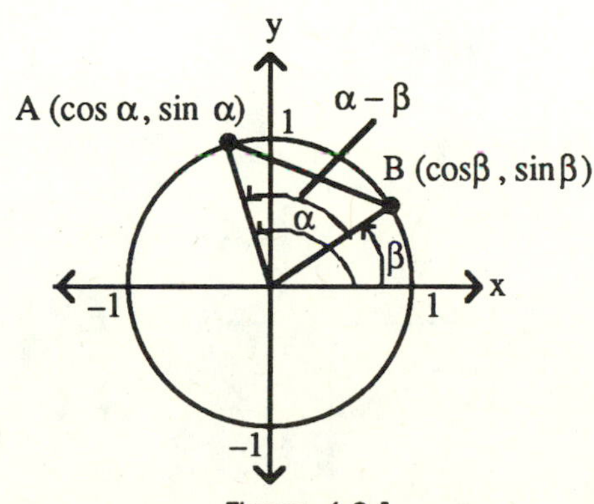

Figure 6.2.1

If we place the angle $\alpha - \beta$ in standard position, the point P with coordinates $(\cos(\alpha - \beta), \sin(\alpha - \beta))$ is on the terminal side of $\alpha - \beta$ and also on the unit circle (Figure 6.2.2).

• THEOREM 1 For all angles α and β, $\cos(\alpha - \beta) = \cos\alpha\cos\beta + \sin\alpha\sin\beta$.

PROOF Refer to Figures 6.2.1 and 6.2.2. The lengths of the line segments AB and CP are equal. We obtain a formula for $\cos(\alpha - \beta)$ by setting the square of the lengths of the line segments AB and CP equal to each other and simplifying the resulting equation. First we compute the square of $d(A, B)$:

$$\begin{aligned} d^2(A,B) &= (\cos\alpha - \cos\beta)^2 + (\sin\alpha - \sin\beta)^2 \\ &= \cos^2\alpha - 2\cos\alpha\cos\beta + \cos^2\beta + \sin^2\alpha - 2\sin\alpha\sin\beta + \sin^2\beta \\ &= (\sin^2\alpha + \cos^2\alpha) + (\sin^2\beta + \cos^2\beta) - 2\cos\alpha\cos\beta - 2\sin\alpha\sin\beta \\ &= 2 - 2\cos\alpha\cos\beta - 2\sin\alpha\sin\beta. \end{aligned}$$

The last step in the above computation is valid because $\sin^2\alpha + \cos^2\alpha = 1$ and $\sin^2\beta + \cos^2\beta = 1$. Next, we compute the square of $d(C, P)$:

$$\begin{aligned} d^2(C,P) &= (1 - \cos(\alpha - \beta))^2 + (0 - \sin(\alpha - \beta))^2 \\ &= 1 - 2\cos(\alpha - \beta) + \cos^2(\alpha - \beta) + \sin^2(\alpha - \beta) \\ &= 2 - 2\cos(\alpha - \beta). \end{aligned}$$

Finally, we simplify the equation $d^2(C,P) = d^2(A,B)$.

$$\begin{aligned} d^2(C,P) &= d^2(A,B) \\ 2 - 2\cos(\alpha - \beta) &= 2 - 2\cos\alpha\cos\beta - 2\sin\alpha\sin\beta \\ -2\cos(\alpha - \beta) &= -2(\cos\alpha\cos\beta + \sin\alpha\sin\beta) \\ \cos(\alpha - \beta) &= \cos\alpha\cos\beta + \sin\alpha\sin\beta \end{aligned}$$

In Section 5.2 we showed that $\sin(90° - \theta) = \cos\theta$ and $\cos(90° - \theta) = \sin\theta$ for any *acute* angle θ. In Example 1 we see that these formulas are also true for angles that are

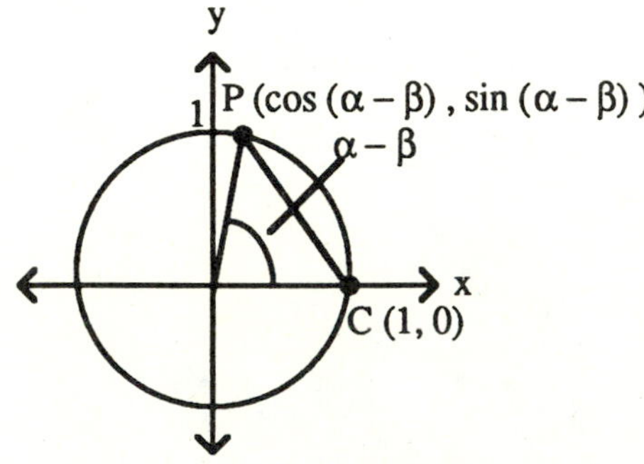

Figure 6.2.2

not acute. We will also establish a relationship for **supplementary angles**; that is, two angles with sum π, or 180°.

• EXAMPLE 1: Let θ be any angle. Show that

(a) $\cos(\frac{\pi}{2} - \theta) = \sin\theta$.

(b) $\sin(\frac{\pi}{2} - \theta) = \cos\theta$.

(c) $\cos(\pi - \theta) = -\cos\theta$.

(d) $\sin(\pi - \theta) = \sin\theta$.

SOLUTION:

(a) Notice that the graph of $y = \cos(\frac{\pi}{2} - \theta)$ in Figure 6.2.3 appears to be identical to the graph of $y = \sin\theta$. If we replace α by $\frac{\pi}{2}$ and β by θ in $\cos(\alpha - \beta) = \cos\alpha\cos\beta + \sin\alpha\sin\beta$, then we obtain the following formula:

$$\begin{aligned}\cos\left(\frac{\pi}{2} - \theta\right) &= \cos\frac{\pi}{2}\cos\theta + \sin\frac{\pi}{2}\sin\theta \\ &= (0)\cos\theta + (1)\sin\theta \\ &= \sin\theta.\end{aligned}$$

(b) In the formula $\cos\left(\frac{\pi}{2} - \theta\right) = \sin\theta$ from (a), we replace θ by $\frac{\pi}{2} - \theta$ to obtain the following:

$$\begin{aligned}\cos\left(\frac{\pi}{2} - \theta\right) &= \sin\theta \\ \cos\left(\frac{\pi}{2} - \left(\frac{\pi}{2} - \theta\right)\right) &= \sin\left(\frac{\pi}{2} - \theta\right) \\ \cos\theta &= \sin\left(\frac{\pi}{2} - \theta\right).\end{aligned}$$

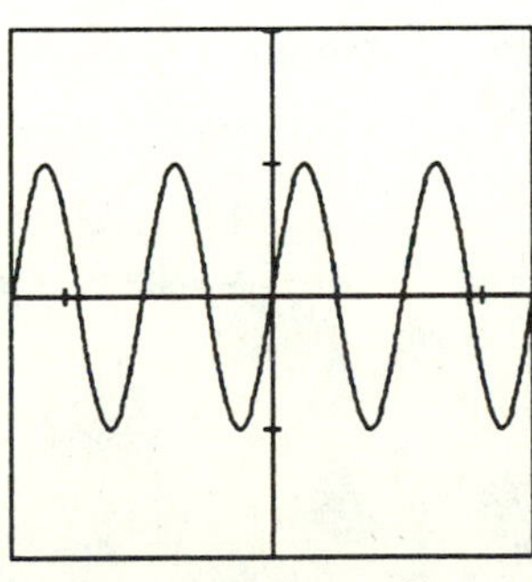

$y = \cos(\frac{\pi}{2} - \theta)$
$[-4\pi, 4\pi]$ by $[-2, 2]$

Figure 6.2.3

(c) Replace α by π and β by θ in the formula for $\cos(\alpha - \beta)$:

$$\begin{aligned}\cos(\alpha - \beta) &= \cos\alpha\cos\beta + \sin\alpha\sin\beta \\ \cos(\pi - \theta) &= \cos\pi\cos\theta + \sin\pi\sin\theta \\ &= (-1)\cos\theta + (0)\sin\theta \\ &= -\cos\theta.\end{aligned}$$

(d) In the formula $\cos\left(\frac{\pi}{2} - \theta\right) = \sin\theta$ from (a), we replace θ by $\pi - \theta$ to obtain the following:

$$\begin{aligned}\cos\left(\frac{\pi}{2} - \theta\right) &= \sin\theta \\ \cos\left(\frac{\pi}{2} - (\pi - \theta)\right) &= \sin(\pi - \theta) \\ \cos - \left(\frac{\pi}{2} - \theta\right) &= \sin(\pi - \theta) \\ \cos\left(\frac{\pi}{2} - \theta\right) &= \sin(\pi - \theta) \qquad (\cos(-x) = \cos x) \\ \sin\theta &= \sin(\pi - \theta).\end{aligned}$$

•

We can use Theorem 1 and Example 1 to obtain formulas for $\cos(\alpha + \beta)$, $\sin(\alpha + \beta)$, and $\sin(\alpha - \beta)$.

• THEOREM 2 Let α and β be any two angles. Then,

(a) $\cos(\alpha + \beta) = \cos\alpha\cos\beta - \sin\alpha\sin\beta$.
(b) $\sin(\alpha + \beta) = \sin\alpha\cos\beta + \cos\alpha\sin\beta$.
(c) $\sin(\alpha - \beta) = \sin\alpha\cos\beta - \cos\alpha\sin\beta$.

PROOF

(a) Replacing β by $-\beta$ in Theorem 1, we obtain

$$\begin{aligned}\cos(\alpha - \beta) &\equiv \cos\alpha\cos\beta + \sin\alpha\sin\beta \\ \cos(\alpha - (-\beta)) &= \cos\alpha\cos(-\beta) + \sin\alpha\sin(-\beta) \\ \cos(\alpha + \beta) &= \cos\alpha\cos\beta - \sin\alpha\sin\beta.\end{aligned}$$

The last step follows because $\cos(-\beta) = \cos\beta$ and $\sin(-\beta) = -\sin\beta$.

(b) From Example 1(a) we have $\sin(\alpha + \beta) = \cos\left(\frac{\pi}{2} - (\alpha + \beta)\right)$. Next, we apply Theorem 1:

$$\begin{aligned}\sin(\alpha + \beta) &= \cos\left(\frac{\pi}{2} - (\alpha + \beta)\right) \\ &= \cos\left(\left(\frac{\pi}{2} - \alpha\right) - \beta\right) \\ &= \cos\left(\frac{\pi}{2} - \alpha\right)\cos\beta + \sin\left(\frac{\pi}{2} - \alpha\right)\sin\beta \\ &= \sin\alpha\cos\beta + \cos\alpha\sin\beta.\end{aligned}$$

The last step follows from Example 1(a) and 1(b).

(c) Replace β by $-\beta$ in the formula from (b).

$$\begin{aligned}\sin(\alpha+\beta) &= \sin\alpha\cos\beta+\cos\alpha\sin\beta\\ \sin(\alpha+(-\beta)) &= \sin\alpha\cos(-\beta)+\cos\alpha\sin(-\beta)\\ \sin(\alpha-\beta) &= \sin\alpha\cos\beta-\cos\alpha\sin\beta\,.\end{aligned}$$

The addition formulas for sine and cosine can be used to establish addition formulas for the tangent function.

• **THEOREM 3** Let α and β be any two angles. Then,

(a) $\tan(\alpha+\beta) = \dfrac{\tan\alpha+\tan\beta}{1-\tan\alpha\tan\beta}$, provided $\tan\alpha\tan\beta \neq 1$.

(b) $\tan(\alpha-\beta) = \dfrac{\tan\alpha-\tan\beta}{1+\tan\alpha\tan\beta}$, provided $\tan\alpha\tan\beta \neq -1$.

PROOF Part (b) follows from (a) by replacing β by $-\beta$ and using $\tan(-\beta) = -\tan\beta$. We establish (a) using $\tan\theta = \frac{\sin\theta}{\cos\theta}$ and the addition formulas for sine and cosine:

$$\begin{aligned}\tan(\alpha+\beta) &= \frac{\sin(\alpha+\beta)}{\cos(\alpha+\beta)}\\ &= \frac{\sin\alpha\cos\beta+\cos\alpha\sin\beta}{\cos\alpha\cos\beta-\sin\alpha\sin\beta}.\end{aligned}$$

Next, we divide the numerator and denominator of the right-hand side of the above equation by $\cos\alpha\cos\beta$:

$$\begin{aligned}\tan(\alpha+\beta) &= \frac{\sin\alpha\cos\beta+\cos\alpha\sin\beta}{\cos\alpha\cos\beta-\sin\alpha\sin\beta}\\ &= \frac{\dfrac{\sin\alpha\cos\beta}{\cos\alpha\cos\beta}+\dfrac{\cos\alpha\sin\beta}{\cos\alpha\cos\beta}}{\dfrac{\cos\alpha\cos\beta}{\cos\alpha\cos\beta}-\dfrac{\sin\alpha\sin\beta}{\cos\alpha\cos\beta}}\\ &= \frac{\tan\alpha+\tan\beta}{1-\tan\alpha\tan\beta}\end{aligned}$$

We summarize as follows:

• The Addition Formulas

$$\sin(\alpha+\beta) = \sin\alpha\cos\beta+\cos\alpha\sin\beta, \qquad \cos(\alpha+\beta) = \cos\alpha\cos\beta-\sin\alpha\sin\beta,$$

$$\sin(\alpha-\beta) = \sin\alpha\cos\beta-\cos\alpha\sin\beta, \qquad \cos(\alpha-\beta) = \cos\alpha\cos\beta+\sin\alpha\sin\beta,$$

$$\tan(\alpha+\beta) = \frac{\tan\alpha+\tan\beta}{1-\tan\alpha\tan\beta} \quad (\tan\alpha\tan\beta\neq 1),$$

$$\tan(\alpha-\beta) = \frac{\tan\alpha-\tan\beta}{1+\tan\alpha\tan\beta} \quad (\tan\alpha\tan\beta\neq -1).$$

Prior to the widespread availability of calculators, the addition formulas were often used to compute exact values of trig functions. We illustrate this past use of the addition formulas in Example 2 for historical reasons.

- EXAMPLE 2: Compute the *exact* trig function values of 15°.

SOLUTION: Notice that $15° = 60° - 45°$. In Section 5.2, we determined the exact values of the trig functions at 60° and 45° (Figure 6.2.4).

Now, we apply the addition formulas for sine and cosine and use the exact values given in Figure 6.2.4:

$$\begin{aligned}
\sin 15° &= \sin(60° - 45°) \\
&= \sin 60° \cos 45° - \cos 60° \sin 45° \\
&= \left(\frac{\sqrt{3}}{2}\right)\left(\frac{1}{\sqrt{2}}\right) - \left(\frac{1}{2}\right)\left(\frac{1}{\sqrt{2}}\right) \\
&= \frac{\sqrt{3}-1}{2\sqrt{2}}; \\
\cos 15° &= \cos(60° - 45°) \\
&= \cos 60° \cos 45° + \sin 60° \sin 45° \\
&= \left(\frac{1}{2}\right)\left(\frac{1}{\sqrt{2}}\right) + \left(\frac{\sqrt{3}}{2}\right)\left(\frac{1}{\sqrt{2}}\right) \\
&= \frac{\sqrt{3}+1}{2\sqrt{2}}; \\
\tan 15° &= \frac{\sin 15°}{\cos 15°} = \frac{\frac{\sqrt{3}-1}{2\sqrt{2}}}{\frac{\sqrt{3}+1}{2\sqrt{2}}} = \frac{\sqrt{3}-1}{\sqrt{3}+1}.
\end{aligned}$$

The cosecant, secant, and cotangent of 15° are the reciprocals of the sine, cosine, and tangent of 15°, respectively. •

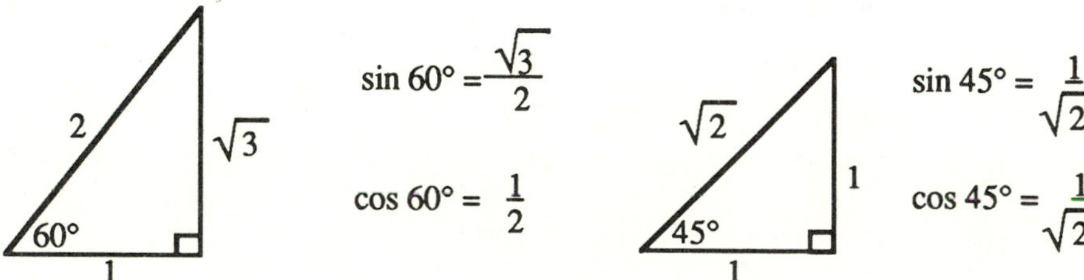

Figure 6.2.4

With calculators we simply approximate the values of the trig functions of 15° directly. Even with the exact values in Example 2, we would likely approximate these numbers with a calculator if they were to be used in other calculations. The addition formulas are still important today, as the next example illustrates.

• EXAMPLE 3: Let ℓ_1 be a line that is neither horizontal nor vertical. Show that the slope of a line ℓ_2 that is perpendicular to ℓ_1 is the negative reciprocal of the slope of ℓ_1.

SOLUTION: Without any loss of generality we may assume that ℓ_1 and ℓ_2 intersect at the origin as illustrated in Figure 6.2.5.

If θ is the angle that ℓ_1 makes with the positive x-axis, then $90+\theta$ is the angle ℓ_2 makes with the positive x-axis. The tangent of an angle is the ratio of the length of the opposite side to the length of the adjacent side. Thus, $\tan\theta = \frac{b}{a}$. However, $\frac{b}{a}$ is also the slope of the line ℓ_1. We need to show that $\tan(90° + \theta)$ is the negative reciprocal of $\tan\theta$. (Why?) You might be tempted to use the formula for $\tan(\alpha + \beta)$, but the tangent function is not defined at 90°. We use $\tan(90° + \theta) = \frac{\sin(90°+\theta)}{\cos(90°+\theta)}$:

$$\begin{aligned}
\sin(90° + \theta) &= \sin 90° \cos\theta + \cos 90° \sin\theta \\
&= (1)\cos\theta + (0)\sin\theta \\
&= \cos\theta; \\
\cos(90° + \theta) &= \cos 90° \cos\theta - \sin 90° \sin\theta \\
&= (0)\cos\theta - (1)\sin\theta \\
&= -\sin\theta.
\end{aligned}$$

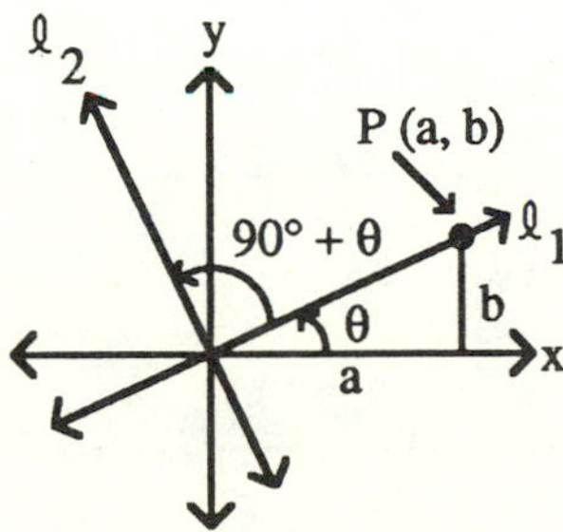

Figure 6.2.5

Next, we compute $\tan(90° + \theta)$:

$$\begin{aligned}\tan(90° + \theta) &= \frac{\sin(90° + \theta)}{\cos(90° + \theta)} \\ &= \frac{\cos\theta}{-\sin\theta} \\ &= -\frac{1}{\frac{\sin\theta}{\cos\theta}} \\ &= -\frac{1}{\tan\theta}.\end{aligned}$$

Therefore, the slope of ℓ_2 is the negative reciprocal of the slope of ℓ_1. •

Double-Angle Formulas

The double-angle formulas are important special cases of the addition formulas that permit us to write $\sin 2\theta$, $\cos 2\theta$, and $\tan 2\theta$ in terms of $\sin\theta$, $\cos\theta$, and $\tan\theta$.

• **THEOREM 4** Let θ be any angle. Then, the following **double-angle** formulas are true:

(a) $\sin 2\theta = 2\sin\theta\cos\theta$

(b) $$\begin{aligned}\cos 2\theta &= \cos^2\theta - \sin^2\theta \\ &= 1 - 2\sin^2\theta \\ &= 2\cos^2\theta - 1\end{aligned}$$

(c) $\tan 2\theta = \dfrac{2\tan\theta}{1 - \tan^2\theta}$

PROOF

(a) Notice that the graphs of $y = \sin 2\theta$ and $y = 2\sin\theta\cos\theta$ in Figure 6.2.6 appear to be identical. This can be confirmed by graphing both functions in the same viewing rectangle. Let $\alpha = \beta = \theta$ in the formula for $\sin(\alpha + \beta)$

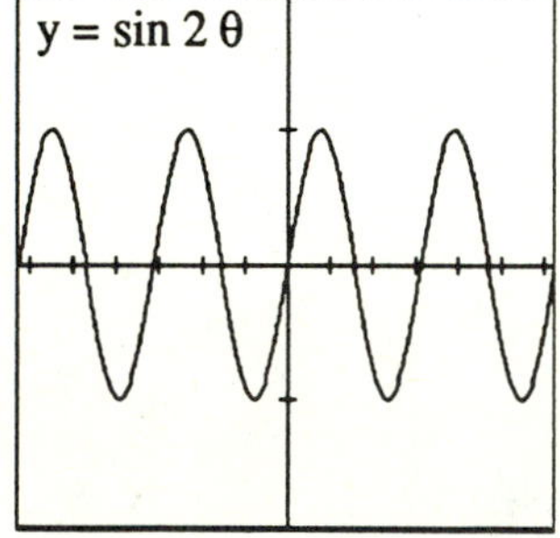

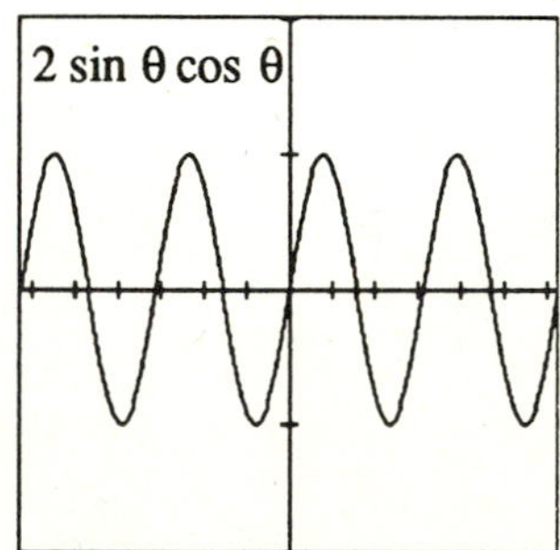

$[-2\pi, 2\pi]$ by $[-2, 2]$

Figure 6.2.6

and we obtain the following:

$$\begin{aligned}\sin(\alpha+\beta) &= \sin\alpha\cos\beta + \cos\alpha\sin\beta\\ \sin(\theta+\theta) &= \sin\theta\cos\theta + \cos\theta\sin\theta\\ \sin 2\theta &= 2\sin\theta\cos\theta.\end{aligned}$$

(b) Let $\alpha = \beta = \theta$ in the formula for $\cos(\alpha+\beta)$:

$$\begin{aligned}\cos(\alpha+\beta) &= \cos\alpha\cos\beta - \sin\alpha\sin\beta\\ \cos(\theta+\theta) &= \cos\theta\cos\theta - \sin\theta\sin\theta\\ \cos 2\theta &= \cos^2\theta - \sin^2\theta.\end{aligned}$$

We can use the identity $\sin^2\theta+\cos^2\theta = 1$ to replace $\cos^2\theta$ by $1-\sin^2\theta$ in the above equation:

$$\begin{aligned}\cos 2\theta &= \cos^2\theta - \sin^2\theta\\ &= (1-\sin^2\theta) - \sin^2\theta\\ &= 1 - 2\sin^2\theta.\end{aligned}$$

The last part of (b) as well as (c) are left for the Exercises.

Half-Angle Formulas

The double-angle formulas can be used to obtain formulas that permit us to write $\sin\frac{1}{2}\theta$, $\cos\frac{1}{2}\theta$, $\tan\frac{1}{2}\theta$ in terms of $\sin\theta$, $\cos\theta$, and $\tan\theta$.

• THEOREM 5 Let θ be any angle, then the following **half-angle formulas** are true:

(a) $\sin\dfrac{1}{2}\theta = \pm\sqrt{\dfrac{1-\cos\theta}{2}}$ (b) $\cos\dfrac{1}{2}\theta = \pm\sqrt{\dfrac{1+\cos\theta}{2}}$

(c) $\tan\dfrac{1}{2}\theta = \dfrac{\sin\theta}{1+\cos\theta}$

The signs in (a) and (b) are chosen according to the quadrant of $\frac{1}{2}\theta$.

PROOF We use the following two identities:

$$\begin{aligned}\sin^2\theta + \cos^2\theta &= 1,\\ \cos^2\theta - \sin^2\theta &= \cos 2\theta.\end{aligned}$$

(a) Subtracting, we obtain

$$\begin{aligned}2\sin^2\theta &= 1 - \cos 2\theta\\ \sin^2\theta &= \frac{1-\cos 2\theta}{2}\\ \sin\theta &= \pm\sqrt{\frac{1-\cos 2\theta}{2}}.\end{aligned}$$

Now, replacing θ by $\frac{1}{2}\theta$ gives the desired formula.

(b) Adding, we obtain the following:

$$2\cos^2\theta = 1 + \cos 2\theta$$
$$\cos^2\theta = \frac{1+\cos 2\theta}{2}$$
$$\cos\theta = \pm\sqrt{\frac{1+\cos 2\theta}{2}}.$$

Again, replacing θ by $\frac{1}{2}\theta$ gives the desired formula.

(c) First, we multiply the numerator and denominator of the right-hand side of $\tan\frac{1}{2}\theta = \sin\frac{1}{2}\theta/\cos\frac{1}{2}\theta$ by $2\cos\frac{1}{2}\theta$:

$$\tan\tfrac{1}{2}\theta = \frac{\sin\frac{1}{2}\theta}{\cos\frac{1}{2}\theta} = \frac{(\sin\frac{1}{2}\theta)(2\cos\frac{1}{2}\theta)}{(\cos\frac{1}{2}\theta)(2\cos\frac{1}{2}\theta)} = \frac{2\sin\frac{1}{2}\theta\cos\frac{1}{2}\theta}{2\cos^2\frac{1}{2}\theta}.$$

The numerator can be simplified as follows:

$$2\sin\tfrac{1}{2}\theta\cos\tfrac{1}{2}\theta = \sin 2(\tfrac{1}{2}\theta) = \sin\theta.$$

The denominator can be simplified by replacing θ by $\frac{1}{2}\theta$ in Theorem 4(b):

$$\cos 2\theta = 2\cos^2\theta - 1$$
$$\cos\theta = 2\cos^2\frac{1}{2}\theta - 1$$
$$1 + \cos\theta = 2\cos^2\frac{1}{2}\theta.$$

Thus,

$$\tan\frac{1}{2}\theta = \frac{\sin\theta}{1+\cos\theta}.$$

The proof of Theorem 2 in Section 6.1 is an application of the formulas for sine and cosine.

• EXAMPLE 4: Prove Theorem 2 of Section 6.1.

SOLUTION: We need to show that for all real numbers a, b, d, h, and k, there exist real numbers A and α such that

$$a\sin(bx+h) + d\cos(bx+k) = A\sin(bx+\alpha). \tag{1}$$

We use the addition formulas for sine and cosine to expand and simplify each side of equation (1). We can rewrite the right-hand side of equation (1) as follows:

$$A\sin(bx+\alpha) = A(\sin bx\cos\alpha + \cos bx\sin\alpha) = (A\cos\alpha)\sin bx + (A\sin\alpha)\cos bx. \tag{2}$$

Similarly, we can rewrite the left-hand side of equation (1) as follows:

$$a\sin(bx+h)+d\cos(bx+k) = a(\sin bx\cos h+\cos bx\sin h)+d(\cos bx\cos k-\sin bx\sin k)$$
$$= (a\cos h - d\sin k)\sin bx + (a\sin h + d\cos k)\cos bx. \tag{3}$$

Comparing equations (2) and (3), we obtain the following equations:

$$A\cos\alpha = a\cos h - d\sin k, \tag{4}$$
$$A\sin\alpha = a\sin h + d\cos k. \tag{5}$$

If the right-hand side of equation (4) is zero, we choose $\alpha = \frac{\pi}{2}$. Then, equation (4) is satisfied because $\cos\frac{\pi}{2} = 0$. Equation (5) will be satisfied if we choose $A = a\sin h + d\cos k$ because $\sin\frac{\pi}{2} = 1$. If the right-hand side of equation (4) is not zero, then we divide each side of equation (5) by the corresponding side of equation (4) to obtain

$$\tan\alpha = \frac{A\sin\alpha}{A\cos\alpha} = \frac{a\sin h + d\cos k}{a\cos h - d\sin k}. \tag{6}$$

Now, we choose α so that equation (6) is satisfied and $-\frac{\pi}{2} < \alpha < \frac{\pi}{2}$. Next, we square each side of equations (4) and (5) and add to obtain the following result:

$$A^2\cos^2\alpha + A^2\sin^2\alpha = (a\cos h - d\sin k)^2 + (a\sin h + d\cos k)^2$$
$$A^2(\cos^2\alpha + \sin^2\alpha) = (a\cos h - d\sin k)^2 + (a\sin h + d\cos k)^2$$
$$A^2 = (a\cos h - d\sin k)^2 + (a\sin h + d\cos k)^2$$
$$A = \pm\sqrt{(a\cos h - d\sin k)^2 + (a\sin h + d\cos k)^2}. \tag{7}$$

Now, $\cos\alpha > 0$ because $-\frac{\pi}{2} < \alpha < \frac{\pi}{2}$. Thus, it follows from equation (4) that A and $a\cos h - d\sin k$ have the same sign. Therefore, if $a\cos h - d\sin k > 0$, we use the + sign in equation (7), and if $a\cos h - d\sin k < 0$, we use the − sign in equation (7). •

In Section 6.1 we observed that the function $f(x) = \sin 2x + \cos 3x$ was periodic of period 2π. Thus, the graph of f in Figure 6.2.7 is complete. We use the identities of this section to help us find the zeros of $f(x) = \sin 2x + \cos 3x$ algebraically.

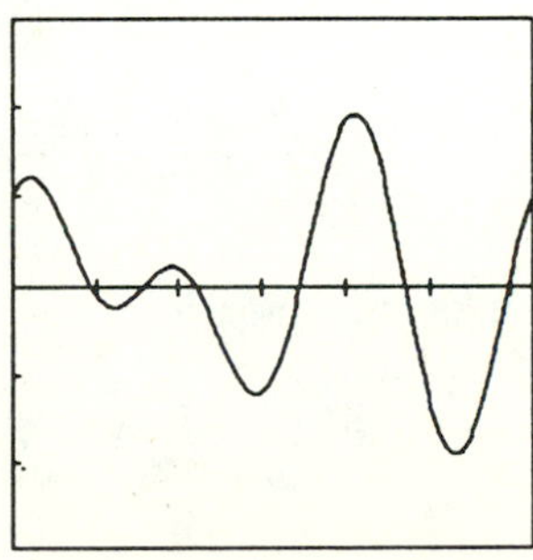

$f(x) = \sin 2x + \cos 3x$
$[0, 2\pi]$ by $[-3, 3]$
Figure 6.2.7

• EXAMPLE 5: Find all the solutions of $\sin 2x + \cos 3x = 0$ that are in the interval $[0, 2\pi]$.

SOLUTION: First we write $\cos 3x$ in terms of $\sin x$ and $\cos x$:

$$\begin{aligned}\cos 3x &= \cos(2x + x) \\ &= \cos 2x \cos x - \sin 2x \sin x \\ &= (1 - 2\sin^2 x)\cos x - 2\sin x \cos x \sin x \\ &= \cos x - 2\sin^2 x \cos x - 2\sin^2 x \cos x \\ &= \cos x - 4\sin^2 x \cos x.\end{aligned}$$

Now, we simplify the original equation as follows:

$$\begin{aligned}\sin 2x + \cos 3x &= 0 \\ 2\sin x \cos x + (\cos x - 4\sin^2 x \cos x) &= 0 \\ (\cos x)(2\sin x + 1 - 4\sin^2 x) &= 0.\end{aligned}$$

Therefore, either $\cos x = 0$ or $2\sin x + 1 - 4\sin^2 x = 0$. The solutions of $\cos x = 0$ in $[0, 2\pi]$ are $\frac{\pi}{2}$ and $\frac{3\pi}{2}$. The other equation can be rewritten

$$4\sin^2 x - 2\sin x - 1 = 0.$$

Notice that this equation is quadratic in $\sin x$. The quadratic formula gives the following:

$$\begin{aligned}\sin x &= \frac{2 \pm \sqrt{4 + 16}}{8} \\ &= \frac{2 \pm 2\sqrt{5}}{8} \\ &= \frac{1 \pm \sqrt{5}}{4}.\end{aligned}$$

The solutions to $\sin x = (1 \pm \sqrt{5})/4$ in $[0, 2\pi]$ accurate to hundredths are 0.94 and 2.20. The solutions to $\sin x = (1 - \sqrt{5})/4$ in $[0, 2\pi]$ are 3.46 and 5.97. Therefore, the *six* solutions to $\sin 2x + \cos 3x = 0$ in $[0, 2\pi]$ accurate to hundredths are 0.94,1.57 $\left(\frac{\pi}{2}\right)$, 2.20, 3.46, 4.71 $\left(\frac{3\pi}{2}\right)$, and 5.97. This information is confirmed by the graph in Figure 6.2.7. •

• EXERCISES 6-2

Simplify (express in terms of a single trig function).

1. $\sin(\theta + 90^\circ)$
2. $\tan(180^\circ + x)$
3. $\cos\left(\dfrac{\pi}{2} + x\right)$
4. $\cot(\theta + 2\pi)$
5. $\dfrac{1 + \tan x}{1 + \cot x}$
6. $\dfrac{\sec x + \tan x}{\sec x + \tan x - \cos x}$

Use algebraic means to verify the following identities.

7. $\tan x + \cot x = \sec x \cdot \csc x$
8. $\tan x + \cot x = 2\csc 2x$

9. $\tan x \cdot \sin x + \cos x = \sec x$

10. $\cos^4 x - \sin^4 x = \cos 2x$

11. $\dfrac{1+\cos 2x}{\sin 2x} = \dfrac{\sin 2x}{1-\cos 2x}$

12. $\cot x - \tan x = 2\cot 2x$

13. $\dfrac{\tan x + 1}{\cot x + 1} = \tan x$

14. $\csc x - \cot x = \tan \dfrac{x}{2}$

Use graphs to investigate whether the equation is an identity. If not, determine a *counterexample.*

15. $\dfrac{\sin t - \cos t}{\cos t} + 1 = \tan t$

16. $\csc(x+\pi) = -\sec x$

17. $\cos(3\pi + x) = -\cos x$

18. $\dfrac{1}{\tan t} + \dfrac{\sin t}{\cos t - 1} = -\csc t$

Verify the following identities by graphing each side of the equation and comparing the graphs. Then verify the identity algebraically.

19. $1 + \cot^2\theta = \csc^2\theta$

20. $1 - 2\sin^2 x = 2\cos^2 x - 1$

21. $\cos^2\theta + 1 = 2\cos^2\theta + \sin^2\theta$

22. $\sin\theta + \cos\theta\cot\theta = \csc\theta$

Determine each *exact* value. Check your answers with a calculator.

23. $\cos\dfrac{\pi}{12}$, $\sin\dfrac{5\pi}{12}$

24. $\tan\dfrac{9\pi}{8}$, $\cos 22.5°$

If $\sin x = \frac{2}{3}$ and $\frac{\pi}{2} < x < \pi$, then determine the trig value.

25. $\cos x$, $\sin x$

26. $\tan 2x$, $\cos\frac{x}{2}$

If $\cos x = -\frac{1}{2}$ and $\pi < x < \frac{3\pi}{2}$, then determine the trig value.

27. $\sec x$

28. $\sin x$

29. $\tan 2x$

30. $\cos\frac{x}{2}$

Find the domain and range of the function, and draw a complete graph.

31. $f(x) = \sec x^2$

32. $g(x) = x^2 \sin\left(x - \frac{\pi}{2}\right)$

33. $g(x) = 3\sin^2 x$

34. $f(x) = \dfrac{\tan x}{x^2}$

Write each as an expression involving only $\sin\theta$ and $\cos\theta$.

35. $\sin 2\theta + \cos\theta$

36. $\sin 2\theta + \cos 3\theta$

37. $\sin 3\theta + \cos 2\theta$

38. $\sin 4\theta + \cos 3\theta$

Solve the equation for x in the interval $[0, 2\pi]$. Find exact answers when possible. Check your answer using a graphing utility.

39. $\sin x = \dfrac{1}{2}$

40. $\cos x = 0$

41. $\sin x = \dfrac{\sqrt{3}}{2}$

42. $\tan x = 1$

43. $\sin^2 x - 1 = 0$

44. $2\sin^2 x + \sin x - 1 = 0$

45. $\sin^2 x - 2\sin x = 0$

46. $\cos 2x + \sin 3x = 0$

Determine the quadrant of the angle θ satisfying the specified conditions.

47. $\cos t > 0$ and $\sin 2t < 0$

48. $\cos t < 0$ and $\sin 2t > 0$

Draw the graph of the equation. If it is a sinusoid, give the period, amplitude and phase shift.

49. $y = 3\sin x + 5\sin(x + 2)$

50. $y = 3\sin(2x - 1) + 5\sin(2x + 3)$

51. Use the addition formula for sine to prove Theorem 1 of Section 6.1 directly.

6.3 Inverse Trigonometric Functions

In Section 4.1 we defined the inverse relation of a relation. In particular, the inverse relation of a given function is a function if and only if the given function is one-to-one. The inverse relations of the six basic trig functions are not functions. However, in this section we see how to restrict the domains of the basic trig functions so that the inverses of the new functions are functions. These inverses are referred to as the **inverse trig functions**. We determine complete graphs of the inverse trig functions.

Figure 6.3.1 contains a complete graph of $y = \sin x$, and Figure 6.3.2 contains a complete graph of the inverse of this function. The inverse is *not* a function because it fails the vertical line test. Each vertical line should intersect the graph of the inverse of $y = \sin x$ in Figure 6.3.2 at most once if the inverse is to be a function. However, vertical lines $x = c$ with $-1 \leq c \leq 1$ intersect the graph of the inverse of $y = \sin x$ infinitely many times.

In Section 4.1 we used the horizontal line test to decide if the inverse of a function is itself a function. This test states that the inverse of a function f is also a function if and only if every horizontal line intersects the graph of f at most once. Notice that the horizontal line $y = c$ with $-1 \leq c \leq 1$ intersects the graph of $y = \sin x$ infinitely many times (Figure 6.3.1). A function is said to be one-to-one if and only if every horizontal line intersects the graph of f in at most one point. Thus, the inverse of a function f is also a function if and only if f is one-to-one.

Notice in Figure 6.3.1 that $y = \sin x$ takes on every value in $[-1, 1]$ *exactly* once if the x-values are restricted to the interval $\left[-\frac{\pi}{2}, \frac{\pi}{2}\right]$. Thus, the function $f(x) = \sin x$ with domain $\left[-\frac{\pi}{2}, \frac{\pi}{2}\right]$ is a one-to-one function. Notice that every horizontal line intersects the graph of f in at most one point (Figure 6.3.3). Therefore, the inverse of f is also a function for this specified domain (Figure 6.3.4).

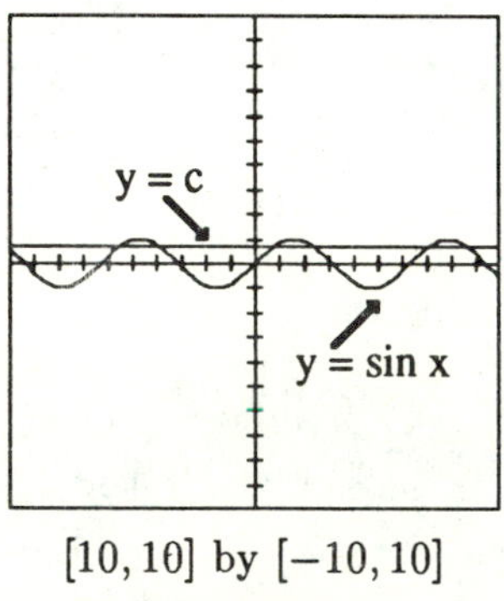

[10, 10] by [−10, 10]

Figure 6.3.1

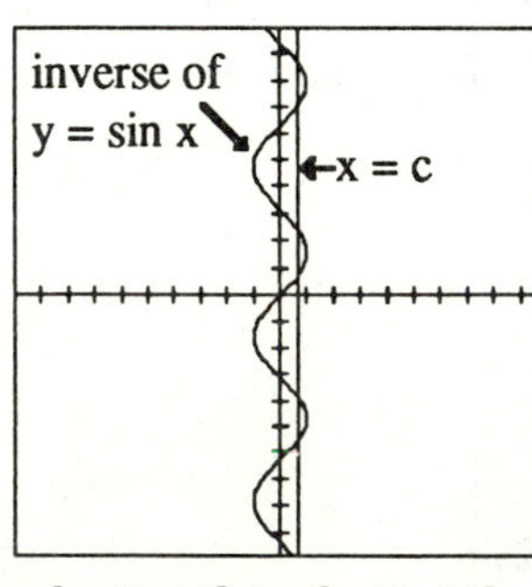

[−10, 10] by [−10, 10]

Figure 6.3.2

Traditionally, the inverse notation $y = \sin^{-1} x$ has been reserved for the *function* with the graph in Figure 6.3.4, that is, the inverse of $f(x) = \sin x$ with domain $-\frac{\pi}{2} \leq x \leq \frac{\pi}{2}$. We will hold to this custom. Thus, even though the relation with the graph in Figure 6.3.2 is the inverse of the sine function, the notation $y = \sin^{-1} x$ will be reserved for the function with the graph in Figure 6.3.4. That is, $y = \sin^{-1} x$ is, by definition, the inverse of the function with the graph in Figure 6.3.3. More precisely, we have the following definition.

• DEFINITION The **inverse sine function**, denoted by $y = \sin^{-1} x$, is the function with domain $-1 \leq x \leq 1$ and range $-\frac{\pi}{2} \leq y \leq \frac{\pi}{2}$ that satisfies $\sin y = x$. •

It is also customary to use the notation $y = \textbf{arcsin}\, x$ interchangeably with $y = \sin^{-1} x$. We give the following theorem without proof.

• THEOREM 1 Let $f(x) = \sin x$ with $-\frac{\pi}{2} \leq x \leq \frac{\pi}{2}$. Then $f^{-1}(x) = \sin^{-1} x$ and

(a) $f^{-1} \circ f(x) = x$ for all x in $\left[-\frac{\pi}{2}, \frac{\pi}{2}\right]$.
(b) $f \circ f^{-1}(x) = x$ for all x in $[-1, 1]$.

The domain and range of the composite function $f^{-1} \circ f$ of Theorem 1 is $\left[-\frac{\pi}{2}, \frac{\pi}{2}\right]$, and the domain and range of the composite function $f \circ f^{-1}$ is $[-1, 1]$. Therefore, $f^{-1} \circ f \neq f \circ f^{-1}$ because the domains of the two functions are different. Many times in the previous chapter we used the calculator key INV, 2nd F, or SHIFT together with the SIN key to solve an equation of the form $\sin x = 0.21$. The unique number produced by a calculator in this case is $\sin^{-1}(0.21)$. Thus, we have implicitly used the idea of inverse functions when we solved trig equations.

• EXAMPLE 1: Find each of the following.

(a) $\sin^{-1}(0.5)$ (b) $\sin^{-1}(-0.7)$ (c) $\sin^{-1}(1.2)$

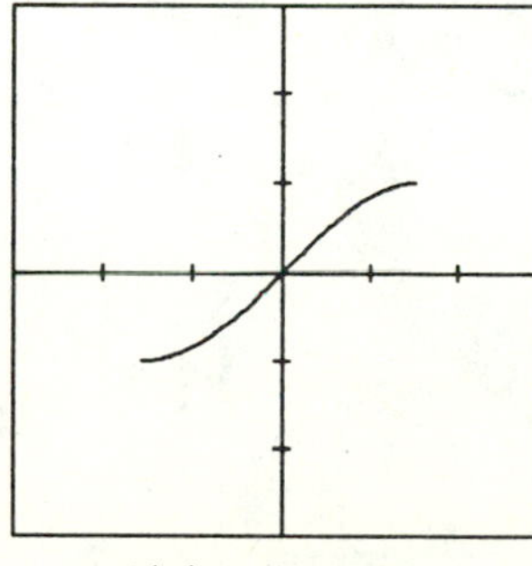

$f(x) = \sin x$ for
$-\frac{\pi}{2} \leq x \leq \frac{\pi}{2}$
$[-3, 3]$ by $[-3, 3]$
Figure 6.3.3

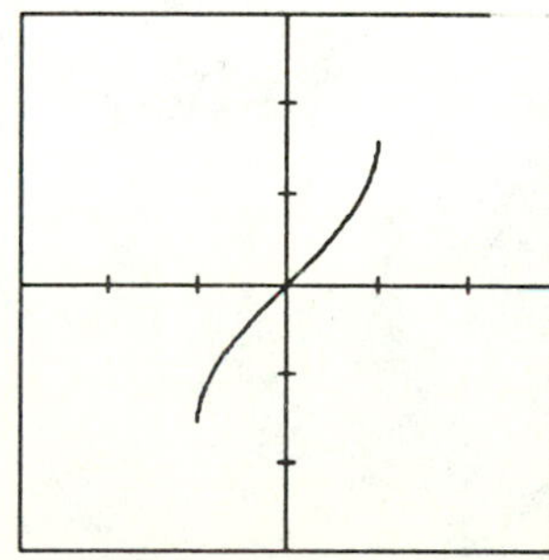

Inverse of f
$[-3, 3]$ by $[-3, 3]$
Figure 6.3.4

SOLUTION:

(a) By definition, $y = \sin^{-1}(0.5)$ if and only if $\sin y = 0.5$ and $-\frac{\pi}{2} \le y \le \frac{\pi}{2}$. If you happen to remember that $\sin\frac{\pi}{6} = \sin 30° = 0.5$, then $\sin^{-1}(0.5) = \frac{\pi}{6}$. Otherwise, we can use a calculator to determine $\sin^{-1}(0.5)$. With a calculator set in radian mode you will find that $\sin^{-1}(0.5)$ is 0.5235987756. This is a decimal approximation to $\frac{\pi}{6}$. If you put your calculator in degree mode you will find $\sin^{-1}(0.5) = 30°$.

(b) You can use a calculator to show that $\sin^{-1}(-0.7) = -0.7753975$ radians or $\sin^{-1}(-0.7) = -44.427004°$.

(c) If you use a calculator you get an error message for $\sin^{-1}(1.2)$. This happens because there is *no* angle with sine equal to 1.2. The sine of any angle or real number is a number in the interval $[-1, 1]$. •

In Example 1(a), we wrote $y = \sin^{-1}(0.5)$ in the form $\sin y = 0.5$. There is an important subtle distinction between solving the equation $\sin y = 0.5$ and finding $\sin^{-1}(0.5)$. The equation $\sin y = 0.5$ has infinitely many solutions. That is, there are infinitely many real numbers whose sine equals 0.5. However, the notation $\sin^{-1}(0.5)$ stands for that *single* number in $\left[-\frac{\pi}{2}, \frac{\pi}{2}\right]$ with sine equal to 0.5.

• EXAMPLE 2: Find $\cos(\sin^{-1}\mu)$ and $\tan(\sin^{-1}\mu)$.

SOLUTION: $\sin^{-1}\mu$ stands for an angle between $-\frac{\pi}{2}$ and $\frac{\pi}{2}$ with sine equal to μ. If we call this angle θ, we can put θ in standard form as illustrated in Figure 6.3.5. If $\mu > 0$, θ is in the first quadrant; and if $\mu < 0$, θ is in the fourth quadrant. We can use the Pythagorean theorem to find that the length of the side adjacent to θ is $\sqrt{1-\mu^2}$. Thus,

$$\cos(\sin^{-1}\mu) = \sqrt{1-\mu^2}.$$

Notice that the cosine of the angle θ is positive whether μ is positive or negative.

Next,

$$\tan(\sin^{-1}\mu) = \frac{\mu}{\sqrt{1-\mu^2}}.$$

If $\mu < 0$, then $\tan\theta < 0$; and if $\mu > 0$, then $\tan\theta > 0$. This agrees with the fact that if $\mu < 0$, then θ is in quadrant IV; and if $\mu > 0$, then θ is in quadrant I. •

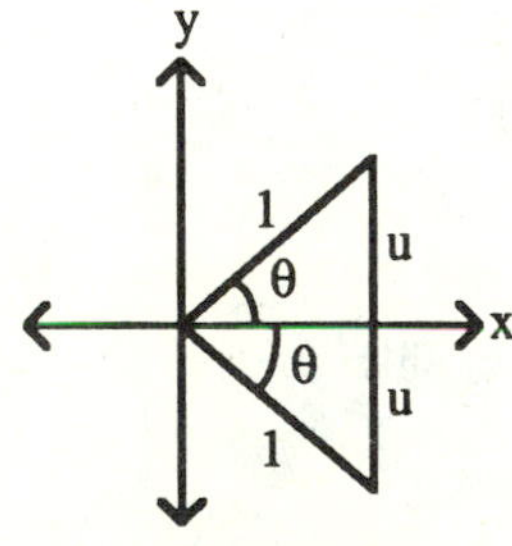

Figure 6.3.5

• EXAMPLE 3: Determine the domain and the range of each function, and draw a complete graph.

(a) $f(x) = \sin^{-1}(2x)$ (b) $g(x) = \sin^{-1}(\frac{1}{3}x)$

SOLUTION:

(a) In Section 5.6, we found that the effect of the coefficient 2 of x in $y = h(2x)$ is to shrink horizontally the graph of $y = h(x)$ by a factor of $\frac{1}{2}$. Thus, the graph of $f(x) = \sin^{-1}(2x)$ can be obtained by applying a horizontal shrink of $\frac{1}{2}$ to the graph of $y = \sin^{-1} x$. This is confirmed by the computer drawn complete graph of $f(x) = \sin^{-1}(2x)$ in Figure 6.3.6. The domain of $f(x) = \sin^{-1}(2x)$ is $\left[-\frac{1}{2}, \frac{1}{2}\right]$ and its range is $\left[-\frac{\pi}{2}, \frac{\pi}{2}\right]$. Alternatively, $-1 \leq 2x \leq 1$ implies $-\frac{1}{2} \leq x \leq \frac{1}{2}$.

(b) The graph of $g(x) = \sin^{-1}\left(\frac{1}{3}x\right)$ can be obtained by applying a horizontal stretch of 3 to the graph of $y = \sin^{-1} x$ (Figure 6.3.7). The domain of $g(x) = \sin^{-1}\left(\frac{1}{3}x\right)$ is $[-3, 3]$, and its range is $\left[-\frac{\pi}{2}, \frac{\pi}{2}\right]$. •

• EXAMPLE 4: Find the domain and the range of $f(x) = 3\sin^{-1}\left(\frac{1}{4}x + 2\right) - 5$, and draw a complete graph.

SOLUTION: We rewrite the expression for f as follows:

$$f(x) = 3\sin^{-1}\left(\frac{1}{4}(x+8)\right) - 5.$$

The graph of f can be obtained from the graph of $y = \sin^{-1} x$ by applying, in order, the following transformations.

1. A horizontal stretch by a factor of 4.
2. A horizontal shift of -8 units.
3. A vertical stretch by a factor of 3.
4. A vertical shift of -5 units.

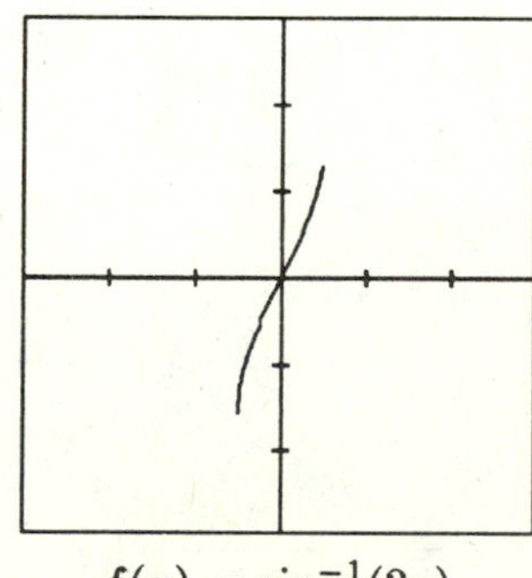

$f(x) = \sin^{-1}(2x)$
$[-3, 3]$ by $[-3, 3]$
Figure 6.3.6

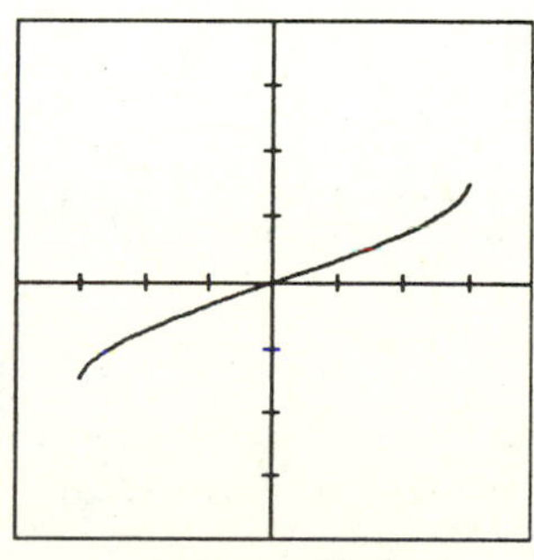

$g(x) = \sin^{-1}(\frac{1}{3}x)$
$[-4, 4]$ by $[-4, 4]$
Figure 6.3.7

This is confirmed by the computer drawn complete graph of f in Figure 6.3.8. The domain of f is $[-12, -4]$, and its range is $\left[-\frac{3\pi}{2} - 5, \frac{3\pi}{2} - 5\right]$, or $[-9.71, -0.29]$ accurate to hundredths. •

Complete graphs of $y = \cos x$ and $y = \tan x$ are given in Figures 6.3.9 and 6.3.10, respectively. Every possible value of each function is obtained for some value of x in the domain represented by the x values in the unshaded portions of the two graphs.

Restricting the domain of $y = \cos x$ to $[0, \pi]$ and the domain of $y = \tan x$ to $-\frac{\pi}{2}, \frac{\pi}{2}$ produces one-to-one functions. We use these restrictions to define the inverse cosine and inverse tangent functions.

• DEFINITION The **inverse cosine function**, denoted by $y = \cos^{-1} x$ or $y = \arccos x$, is the function with domain $[-1, 1]$, range $[0, \pi]$ that satisfies $\cos y = x$. •

• DEFINITION The **inverse tangent function**, denoted by $y = \tan^{-1} x$ or $y = \arctan x$, is the function with domain $(-\infty, \infty)$, range $(-\frac{\pi}{2}, \frac{\pi}{2})$ that satisfies $\tan y = x$. •

The graphs of $y = \cos^{-1} x = \arccos x$ and $y = \tan^{-1} x = \arctan x$ are given in Figures 6.3.11 and 6.3.12, respectively. The domain and range of $y = \cos^{-1} x$ are $[-1, 1]$ and $[0, \pi]$, respectively. The domain and range of $y = \tan^{-1} x$ are the set of all real numbers and $\left(-\frac{\pi}{2}, \frac{\pi}{2}\right)$, respectively. Notice that $y = -\frac{\pi}{2}$ and $y = \frac{\pi}{2}$ are horizontal asymptotes of $y = \tan^{-1} x$.

The inverse cotangent, secant, and cosecant functions are so rarely used that we will not give definitions of these three functions.

• EXAMPLE 5: Without using a calculator, find the value of each expression.

(a) $\sin^{-1}\left(\tan \frac{3\pi}{4}\right)$ (b) $\cos\left(\tan^{-1} \frac{1}{2}\right)$

(c) $\sin(\arccos(-\frac{2}{3}) - \arctan \frac{3}{5})$

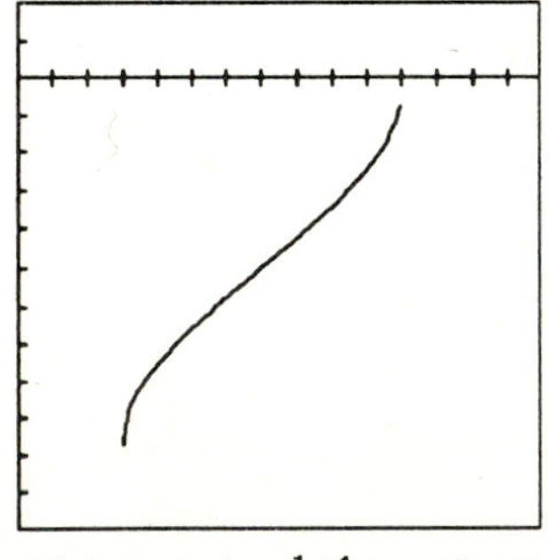

$f(x) = 3\sin^{-1}(\frac{1}{4}x+2)-5$
$[-15, 0]$ by $[-12, 2]$
Figure 6.3.8

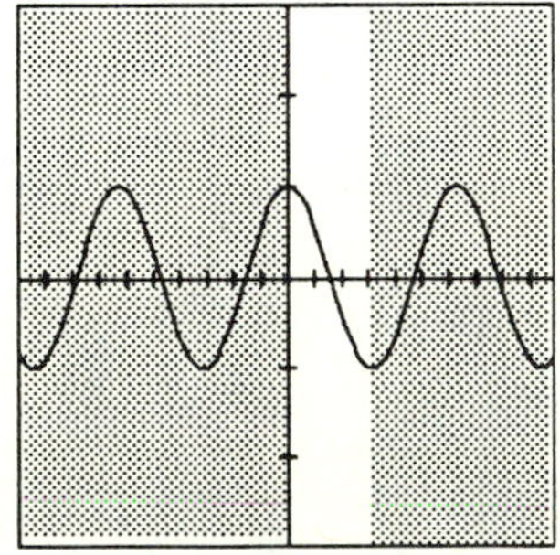

$y = \cos x$
$[-10, 10]$ by $[-3, 3]$
Figure 6.3.9

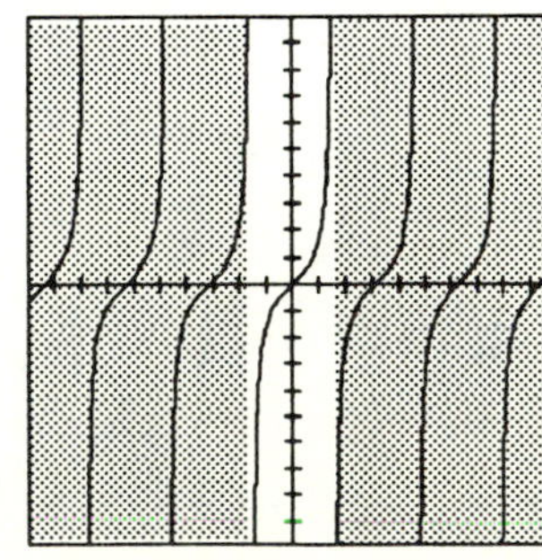

$y = \tan x$
$[-10, 10]$ by $[-10, 10]$
Figure 6.3.10

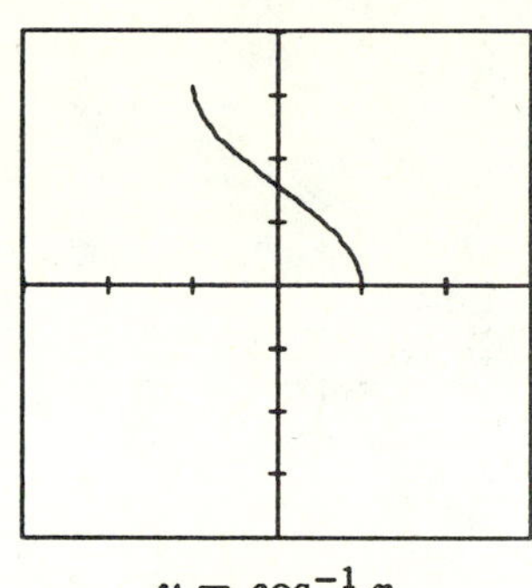

$y = \cos^{-1} x$
$[-3, 3]$ by $[-4, 4]$

Figure 6.3.11

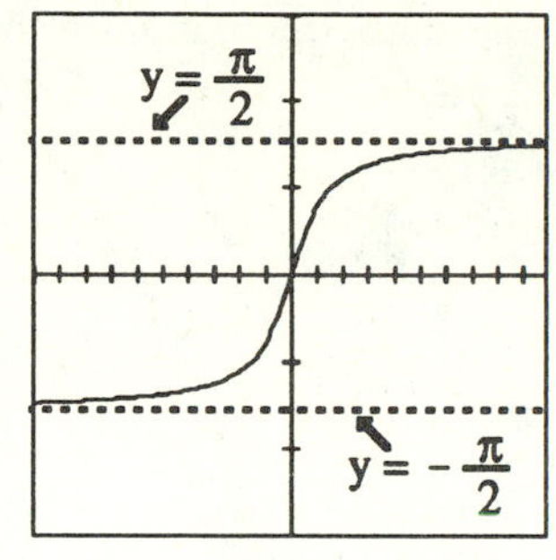

$y = \tan^{-1} x$
$[-10, 10]$ by $[-3, 3]$

Figure 6.3.12

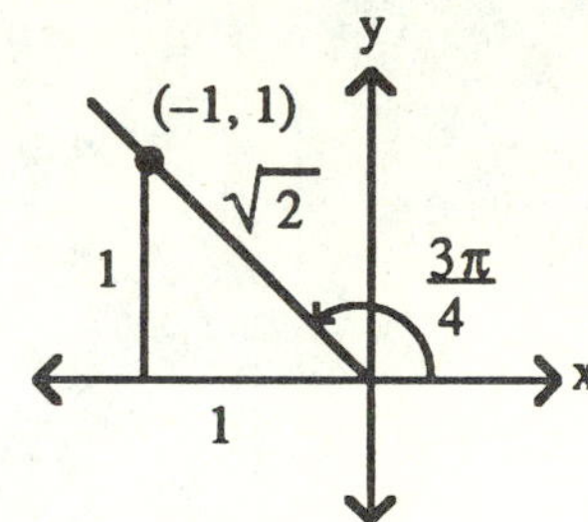

Figure 6.3.13

SOLUTION:

(a) We know from Section 5.1 that the point $(-1, 1)$ is on the terminal side of $\frac{3\pi}{4}$ (Figure 6.3.13). Thus, $\tan \frac{3\pi}{4} = -1$ and $\sin^{-1}\left(\tan \frac{3\pi}{4}\right) = \sin^{-1}(-1)$. Because $-\frac{\pi}{2} \le \sin^{-1}(-1) \le \frac{\pi}{2}$ and $\sin\left(-\frac{\pi}{2}\right) = -1$, we have $\sin^{-1}(-1) = -\frac{\pi}{2}$. Therefore, $\sin^{-1}\left(\tan \frac{3\pi}{4}\right) = -\frac{\pi}{2}$.

(b) Let $\theta = \tan^{-1} \frac{1}{2}$. Then, θ is an acute angle because $-\frac{\pi}{2} \le \tan^{-1} \frac{1}{2} \le \frac{\pi}{2}$ and $\tan \theta = \frac{1}{2} > 0$. Figure 6.3.14 shows θ in standard position. Thus, $\cos\left(\tan^{-1} \frac{1}{2}\right) = \cos \theta = 2/\sqrt{5}$.

(c) Let $\alpha = \arccos\left(-\frac{2}{3}\right)$ and $\beta = \arctan \frac{3}{5}$. Figures 6.3.15 and 6.3.16 give the standard positions of the angles α and β.

Because $0 \le \alpha \le \pi$ and $\cos \alpha = -\frac{2}{3} < 0$, α must be in the second quadrant. β must be in the first quadrant. (Why?) We can use the addition formulas from

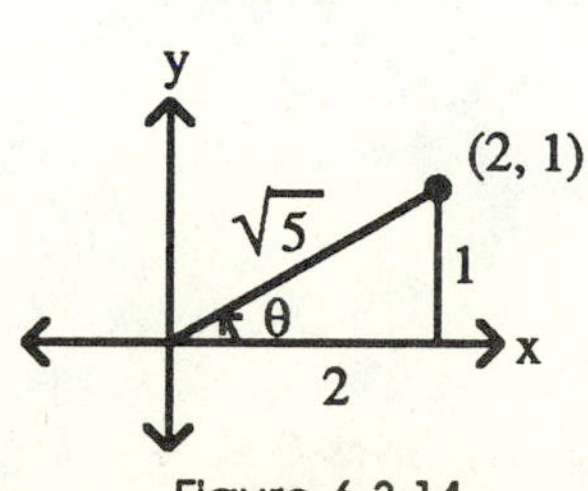

Figure 6.3.14

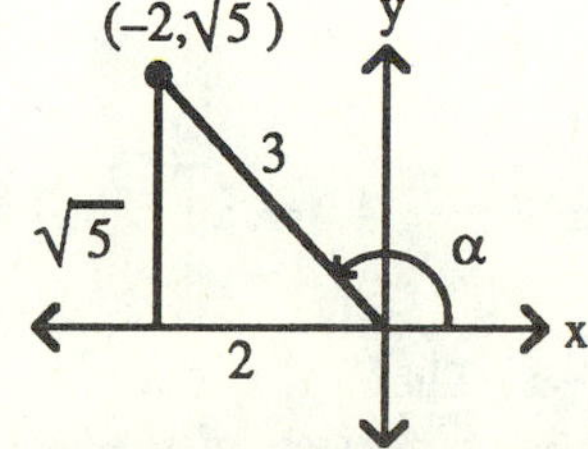

Figure 6.3.15

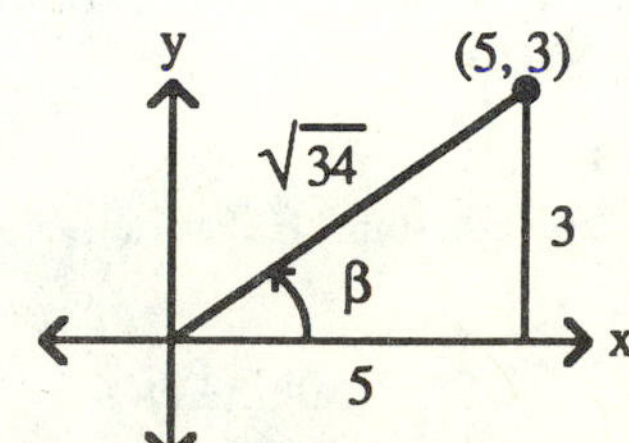

Figure 6.3.16

Section 6.2 to help make the computation:

$$\begin{aligned}\sin\left(\arccos\left(-\frac{2}{3}\right) - \arctan\frac{3}{5}\right) &= \sin(\alpha - \beta)\\ &= \sin\alpha\cos\beta - \cos\alpha\sin\beta\\ &= \left(\frac{\sqrt{5}}{3}\right)\left(\frac{5}{\sqrt{34}}\right) - \left(-\frac{2}{3}\right)\left(\frac{3}{\sqrt{34}}\right)\\ &= \frac{5\sqrt{5}+6}{3\sqrt{34}}.\end{aligned}$$

•

Calculators can be used to check our solutions to the above example. For example, we can use a calculator to compute directly the number in (b) and then compare this value with a decimal approximation to $2/\sqrt{5}$.

• EXAMPLE 6: Solve for x.

(a) $\sin x = 0.6$ (b) $\cot x = 2.5$

SOLUTION:

(a) We use a calculator to determine that $\sin^{-1} 0.6 = 0.6435011$. Now x must be an angle in the first or second quadrant because $\sin x > 0$. Therefore, x can be any angle whose terminal side is the same as the terminal side of 0.6435011 in the first quadrant, or whose terminal side is the same as $\pi - 0.6435011 = 2.4980915$ in the second quadrant. Recall that angles with the same terminal side are said to be coterminal. Thus, the solutions to $\sin x = 0.6$ are $0.6435011 + 2k\pi$ or $2.4980915 + 2k\pi$, where k is any integer.

(b) The period of $\cot x$ is π. Because $\cot x > 0$, x is in the first quadrant. Calculators do not have $\cot^{-1} x$ as a built-in function. However, if $\cot x = 2.5$, then $\tan x = 0.4$. Now, we use a calculator to find that $\tan^{-1} 0.4 = 0.3805064$. Thus, x is $0.3805064 + k\pi$ where k is any integer. •

In the last example of this section we establish an identity involving inverse trig functions.

• EXAMPLE 7: $\sin^{-1} x + \cos^{-1} x = \frac{\pi}{2}$ for all x in $[-1, 1]$.

SOLUTION: Let $\theta = \sin^{-1} x$. Then, $x = \sin\theta$ and $-\frac{\pi}{2} \le \theta \le \frac{\pi}{2}$ so that $-\frac{\pi}{2} \le -\theta \le \frac{\pi}{2}$ is also true. Adding $\frac{\pi}{2}$ to each portion of the last inequality gives $0 \le \frac{\pi}{2} - \theta \le \pi$. Now $\cos\left(\frac{\pi}{2} - \theta\right) = \sin\theta = x$. Because $0 \le \frac{\pi}{2} - \theta \le \pi$, it follows that $\frac{\pi}{2} - \theta = \cos^{-1} x$. Replacing

θ by $\sin^{-1} x$ gives

$$\frac{\pi}{2} - \theta = \cos^{-1} x$$
$$\frac{\pi}{2} - \sin^{-1} x = \cos^{-1} x$$
$$\frac{\pi}{2} = \sin^{-1} x + \cos^{-1} x.$$

EXERCISES 6-3

Use a calculator to evaluate. Express your answer in degrees.

1. $\sin^{-1}(0.362)$
2. $\arcsin(-1.67)$
3. $\tan^{-1}(0.125)$
4. $\tan^{-1}(-2.8)$

Use a calculator to evaluate. Express your answer as a real number.

5. $\sin^{-1}(0.46)$
6. $\cos^{-1}(-0.853)$
7. $\tan^{-1}(2.37)$
8. $\tan^{-1}(-22.8)$

Compute the exact value without using a calculator. Express your answer as a real number.

9. $\sin^{-1} 1$
10. $\tan^{-1} \sqrt{3}$
11. $\sin^{-1} \dfrac{\sqrt{2}}{2}$
12. $\cos^{-1} \left(-\dfrac{\sqrt{3}}{2}\right)$
13. $\tan^{-1}(-\sqrt{3})$
14. $\sin^{-1}(-1)$

Use a calculator to evaluate. Express your answer as a real number.

15. $\sin(\sin^{-1}(0.36))$
16. $\sin(\arccos(0.568))$
17. $\sin^{-1}(\cos 20)$
18. $\cos(\sin^{-1}(-0.125))$
19. $\sin(\sin^{-1} 1.2)$
20. $\sin^{-1}(\sin 1.2)$
21. $\sin(\sin^{-1} 2)$
22. $\sin^{-1}(\sin 2)$
23. $\tan^{-1}(\sin 2)$
24. $\tan(\arctan 3)$
25. $\sin^{-1}(\sin(\cos^{-1} 0.23))$
26. $\cos(\tan^{-1} 3.5 - \sin^{-1} 0.35)$

Compute the exact value without using a calculator.

27. $\cos\left(\sin^{-1} \dfrac{1}{2}\right)$
28. $\sin(\tan^{-1} 1)$
29. $\cos\left(2 \sin^{-1} \dfrac{1}{2}\right)$
30. $\cos(\tan^{-1} \sqrt{3}) - \sin(\tan^{-1} 0)$
31. $\cos(\sin^{-1} 0.6)$
32. $\sin(\tan^{-1} 2)$

Draw a complete graph. Determine the domain and range.

33. $y = 1 - \arcsin x$
34. $y = 3 + \cos^{-1}(x - 2)$
35. $y = -0.25 \tan^{-1}(x - \pi)$
36. $g(x) = 3 \arccos(\frac{1}{2}x - \pi)$
37. $y = \sin(\sin^{-1} x)$
38. $y = \sin^{-1}(\sin x)$

Solve for x. Find the exact solution(s).

39. $\sin(\sin^{-1} x) = x$

40. $\cos^{-1}(\cos x) = x$

41. $2\sin^{-1} x = 1$

42. $\tan^{-1} x = -1$

Find an equivalent algebraic expression not involving trig functions.

43. $\sin(\tan^{-1} x)$

44. $\cos(\tan^{-1} x)$

45. $\tan(\sin^{-1} x)$

46. $\cot(\cos^{-1} x)$

Solve for x. Find *all* solutions.

47. $\tan x = 2.3$

48. $\sin x = -0.75$

49. $\sec x = 3$

50. $\cot x = -5$

Describe what happens to the values of $f(x)$.

51. $f(x) = \dfrac{\sin^{-1} x}{x}$ as $x \to 0$

52. $f(x) = 2\tan^{-1} x$ as $|x| \to \infty$

Solve the inequality in $[-2\pi, 2\pi]$.

53. $(\sin x)(\tan^{-1} x) \geq 0$

54. $\dfrac{\sin^{-1} 2x}{\sin x} \geq 2$

Verify the following identities.

55. $\sin^{-1}(-x) = -\sin^{-1} x$ for $|x| \leq 1$

56. $\sin^{-1} x = \tan^{-1} \dfrac{x}{\sqrt{1-x^2}}$ for $|x| < 1$

57. $\arccos\theta + \arcsin\theta = 90°$

58. $\cos(\sin^{-1} x) = \sqrt{1-x^2}$ for $|x| \leq 1$

59. The length L of the shadow cast by a tower 50 feet tall depends on θ, the angle of elevation of the sun (measured from the horizontal). Express θ as a function of L. Draw a complete graph of the function. Which portion of the graph represents the problem situation?

60. A revolving light beacon L stands 3 miles from the closest point P along a straight shore. Let θ be the angle PLQ (Figure 6.3.17). Express this angle as a function of the distance from

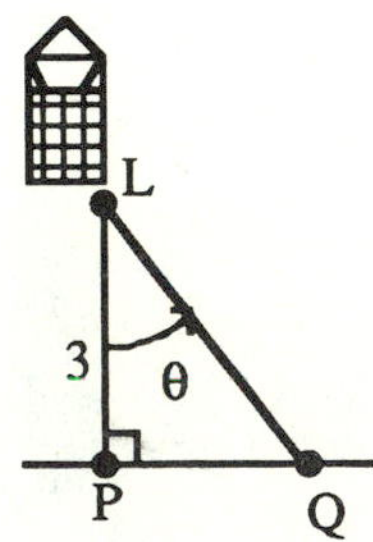

Figure 6.3.17

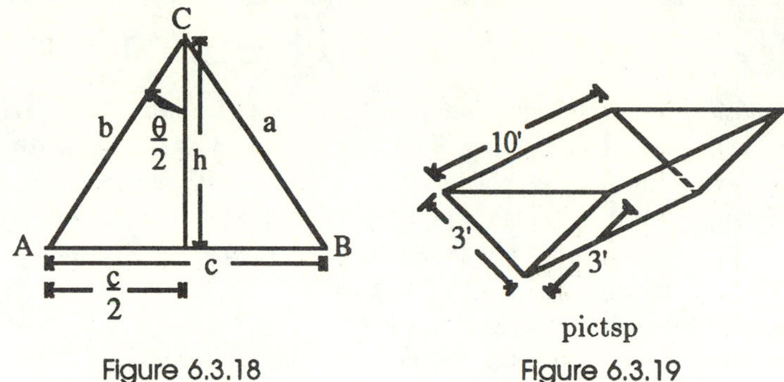

Figure 6.3.18

Figure 6.3.19

P to Q. Draw a complete graph of the function. Which portion of the graph represents the problem situation?

*61. Let triangle ABC be isosceles with $a = b$, and let $\angle C = \theta$ (Figure 6.3.18). Show that the area of the triangle is $\frac{1}{2}a^2 \sin\theta$. *Hint:* Let h be the altitude of the triangle. Notice that the area is $\frac{1}{2}ch$. Find $\sin\frac{\theta}{2}$ and $\cos\frac{\theta}{2}$ and use the half-angle formulas.

The ends of a feeding trough 10 feet long are isosceles triangles with 3-foot sides, as shown in Figure 6.3.19. Let θ be the angle between the two equal sides.

62. Use the area result of Exercise 61 to write a *function* that gives the angle θ as a function of the *volume* V of the trough.

63. Draw a complete graph of the function in Exercise 62. Explain why the graph represents only a portion of the problem situation.

64. What values of θ make sense in the problem situation?

65. Write an *equation* that relates angle θ and V for all values of θ and V that make sense in the problem situation.

66. Is there a maximum volume? What is the associated angle θ?

67. For what values of x is $\sin^{-1}\left(\dfrac{2x}{x^2+1}\right) = 2\tan^{-1}x$?

68. How would you define $\csc^{-1}x$?

69. How could you compute $\sin^{-1}x$ for all $|x| < 1$ on a calculator that had only a $\tan^{-1}$ key?

70. Show that for $A > 0$, $A\cos(Bx - C) = a\cos Bx + b\sin Bx$ where $A = \sqrt{a^2+b^2}$ and $C = \tan^{-1}\left(\frac{b}{a}\right)$.

71. Use Exercise 70 to write $\sin x + \cos x$ in the form $A\cos(Bx - C)$. Check your answer with a graphing utility.

72. Use Exercise 70 to write $20\sin(10\pi t) + 30\cos(10\pi t)$ in the form $A\cos(Bt - C)$. Check your answer with a graphing utility.

73. Use Exercise 70 to write $3\sin 2x + 4\cos 2x$ in the form $A\cos(Bx - C)$. Check your answer with a graphing utility.

74. Generalize Exercise 70 to write $4\sin x - 2\cos x$ in the form $A\cos(Bx - C)$. Check your answer with a graphing utility.

6.4 Solving Equations and Inequalities

In this section we solve equations and inequalities that involve trig functions both algebraically and graphically. You will again see that a graphing method is more general and powerful. The identities established in previous sections are used to transform equations into equivalent forms to which algebraic techniques can be applied. Periodicity of functions is used to facilitate finding solutions regardless of the method used. We also determine local maximum and minimum values of trig functions.

The circular property of the trig functions gives $\sin(x + 2k\pi) = \sin x$ for any integer k. This property is used to solve the equation in Example 1.

• EXAMPLE 1: Solve $\sin x = 0.8$.

SOLUTION: Figure 6.4.1 shows that the equation $\sin x = 0.8$ has infinitely many solutions. The circular property of $\sin x$ suggests that we first solve the equation $\sin x = 0.8$ in $[0, 2\pi]$. There are two solutions in this interval. We can estimate from Figure 6.4.1 that the two solutions in $[0, 2\pi]$ are about 0.9 and 2.2. The solution near 0.9 represents an angle in the first quadrant and the solution near 2.2 an angle in the second quadrant. This is not to be confused with the first quadrant portion of the graphs in Figure 6.4.1. The first quadrant solution is $\sin^{-1} 0.8 = 0.927$, and the second quadrant solution is $\pi - \sin^{-1} 0.8 = 2.214$. (Why?) The complete solution of $\sin x = 0.8$ consists of all real numbers of the form $0.927 + 2k\pi$ or $2.214 + 2k\pi$, where k is any integer. •

We must be careful when we use a complete period of periodic functions to solve equations, inequalities, and so forth. For example, Figure 6.4.1 suggests that the equation $\sin x = 0$ has three solutions in $[0, 2\pi]$. However, we can ignore the zero at 2π, because it is obtained when we add the period 2π to the zero at 0. In some cases, we may need to look at the interval $[0, 2\pi)$ instead of $[0, 2\pi]$ to simplify observations.

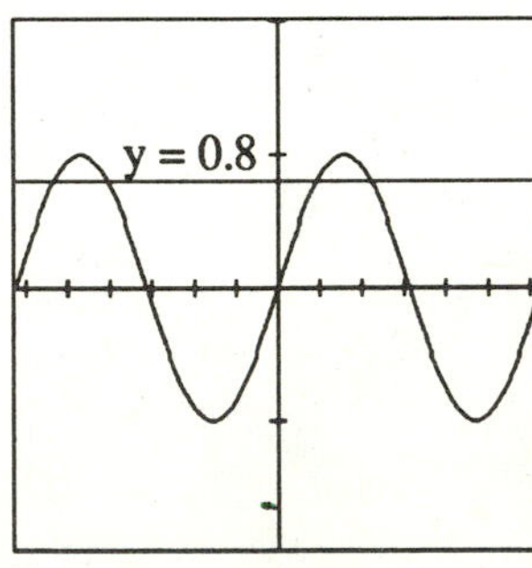

$y = \sin x$
$[-2\pi, 2\pi]$ by $[-2, 2]$
Figure 6.4.1

Any function, no matter how complicated, that involves only the six basic trig functions is periodic. Periodicity is useful when solving equations.

• EXAMPLE 2: Solve $2\cos^2 t + \sin t + 1 = 0$.

SOLUTION: It can be shown that the left-hand side of this equation is a periodic function with period 2π (Figure 6.4.2). If we find the solutions to the equation that are in the interval $[0, 2\pi]$, then we can easily find all the rest of the solutions of the equation. There appears to be one solution in $[0, 2\pi]$ at about 4.7. The identity $\sin^2 t + \cos^2 t = 1$ can be used to replace $\cos^2 t$ by $1 - \sin^2 t$. In this way we obtain an equivalent equation that involves only one trig function:

$$\begin{aligned} 2\cos^2 t + \sin t + 1 &= 0 \\ 2(1 - \sin^2 t) + \sin t + 1 &= 0 \\ 2 - 2\sin^2 t + \sin t + 1 &= 0 \\ -2\sin^2 t + \sin t + 3 &= 0 \\ 2\sin^2 t - \sin t - 3 &= 0. \end{aligned}$$

We have transformed the original equation into an equivalent equation that is quadratic in $\sin t$. Notice that this quadratic equation can be factored. We could, of course, use the quadratic formula:

$$\begin{aligned} 2\sin^2 t - \sin t - 3 &= 0 \\ (2\sin t - 3)(\sin t + 1) &= 0. \end{aligned}$$

Now, t is a solution to the equation $2\cos^2 t + \sin t + 1 = 0$ if and only if t is a solution to one of the equations

$$2\sin t - 3 = 0 \quad \text{or} \quad \sin t + 1 = 0.$$

The first of these equations can be rewritten in the form $\sin t = \frac{3}{2}$ to see that it has *no* solutions because $|\sin t| \leq 1$ for all t. If we rewrite $\sin t + 1 = 0$ in the form $\sin t = -1$,

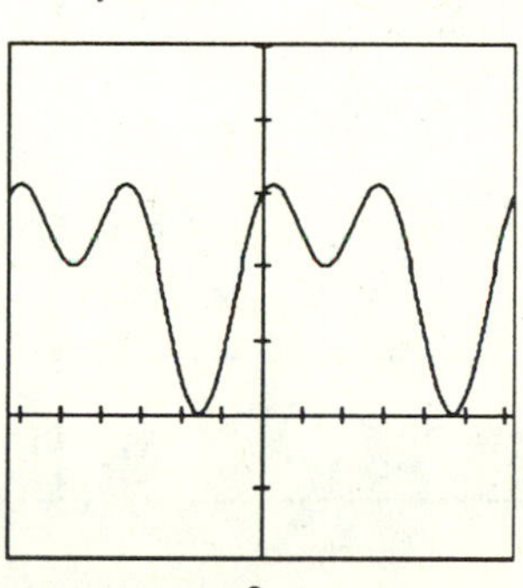

$f(t) = 2\cos^2 t + \sin t + 1$
$[-2\pi, 2\pi]$ by $[-2, 5]$
Figure 6.4.2

we can see that $\frac{3\pi}{2}$ (about 4.71) is the only solution to this equation in the interval $[0, 2\pi]$. Using the circular property of the trig functions we can conclude that the solutions to the original equation consist of the numbers of the form $\frac{3\pi}{2} + 2k\pi$, where k is any integer. These are precisely the angles that are coterminal with $\frac{3\pi}{2}$. The graph of $f(t) = 2\cos^2 t + \sin t + 1$ in $[0, 2\pi]$ by $[-5, 5]$ confirms that there is only one solution of $2\cos^2 t + \sin t + 1 = 0$ in the interval $[0, 2\pi]$ (Figure 6.4.3). You may need to zoom-in around $t = \frac{3\pi}{2}$ to convince yourself that there is only one x-intercept. •

Once we observed in Example 2 that the function $f(t) = 2\cos^2 t + \sin t + 1$ was periodic of period 2π we were able to focus our attention on the interval $[0, 2\pi]$. This considerably simplifies both the algebraic and graphical methods of solution.

From now on, answers given will either be exact or accurate to hundredths unless otherwise stated. When you work the example yourself you will be able to tell whether the answers we give are exact or approximations.

• EXAMPLE 3: Solve $\tan x = 3\cos x$.

SOLUTION: Again, we need to find the solutions to this equation only in the interval $[0, 2\pi]$ because the function $f(x) = \tan x - 3\cos x$ is periodic with period 2π. The graph of f in Figure 6.4.4 shows that the equation $\tan x = 3\cos x$ has two solutions in $[0, 2\pi]$. Notice that f has $x = \frac{\pi}{2}$ and $x = \frac{3\pi}{2}$ as vertical asymptotes. (Why?) Zoom-in could be used to determine the solutions in the interval $[0, 2\pi]$. We give an algebraic solution to this equation.

Often it helps to rewrite trig equations so that the only trig functions involved are the *sine* and *cosine* functions. We are able to do this because $\tan x = \frac{\sin x}{\cos x}$, $\cot x = \frac{\cos x}{\sin x}$, $\sec x = \frac{1}{\cos x}$, and $\csc x = \frac{1}{\sin x}$. If we replace $\tan x$ by $\frac{\sin x}{\cos x}$ in $\tan x = 3\cos x$ we obtain $\frac{\sin x}{\cos x} = 3\cos x$. It seems reasonable to multiply each side of the last equation by $\cos x$. However, this process can introduce extraneous solutions. This means we must check all

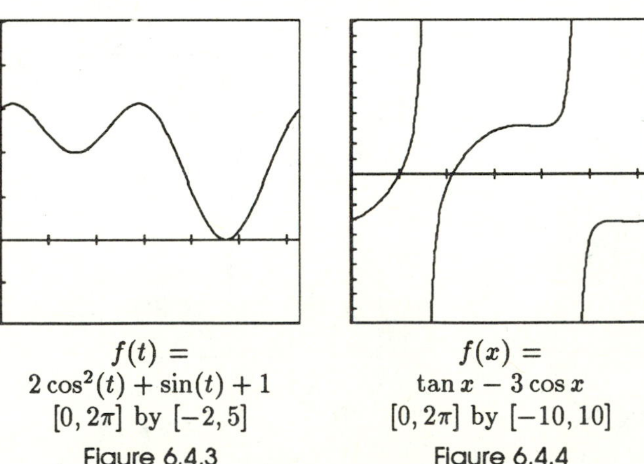

$f(t) =$
$2\cos^2(t) + \sin(t) + 1$
$[0, 2\pi]$ by $[-2, 5]$
Figure 6.4.3

$f(x) =$
$\tan x - 3\cos x$
$[0, 2\pi]$ by $[-10, 10]$
Figure 6.4.4

solutions in the original equation. We can proceed as follows:

$$\tan x = 3\cos x$$
$$\frac{\sin x}{\cos x} = 3\cos x$$
$$\sin x = 3\cos^2 x$$
$$\sin x = 3(1 - \sin^2 x)$$
$$3\sin^2 x + \sin x - 3 = 0.$$

The quadratic formula gives $\sin x = (-1 \pm \sqrt{37})/6$. (Why?) We can use a calculator to determine that the number $(-1 - \sqrt{37})/6$ is strictly less than -1. Thus, the equation $\sin x = (-1 - \sqrt{37})/6$ has no solutions. However, the equation $\sin x = (-1 + \sqrt{37})/6$ has two solutions in the interval $[0, 2\pi]$. One is in the first quadrant and the other is in the second quadrant because $(-1 + \sqrt{37})/6 > 0$.

We can use a calculator to determine that $x = \sin^{-1}[(-1 + \sqrt{37})/6] = 1.01$ is the solution in the first quadrant. The solution in the second quadrant is $\pi - 1.01 = 2.13$. You can check that the values are also solutions to the original equation. Thus, the solutions to the original equation are $1.01 + 2k\pi$ or $2.13 + 2k\pi$, where k is any integer. •

From now on we will use the following convention.

• DEFINITION If f is a periodic function with period h, then we will say that the solutions of the equation $f(x) = 0$ are **periodic** with period h. •

• Zeros of Periodic Functions Let f be a periodic function with period h, and let $x_1, x_2, \ldots, x_n$ be all the solutions to $f(x) = 0$ in an interval of length the period h. Then, $x_i + kh$ for $i = 1, 2, \ldots, n$ and k any integer are all the solutions to $f(x) = 0$.

If one of the zeros occurs at the left-hand endpoint of an interval of length h, then there is also a zero at the right-hand endpoint. You should use only one of these two zeros in the above description of the zeros of a periodic function.

• EXAMPLE 4: Solve $3\tan^4 x = 1 + \sec^2 x$.

SOLUTION: First, we show that the solutions to this equation are periodic with period π by showing that the function $f(x) = 3\tan^4 x - 1 - \sec^2 x$ is periodic with period π (Figure 6.4.5). We have shown in Section 5.6 that $\tan x$ is periodic with period π. Thus, we need to show only that $\sec^2 x$ is periodic with period π. (Why?) Notice that

$$\sec^2(x + \pi) = (-\sec x)^2 = \sec^2 x\,.$$

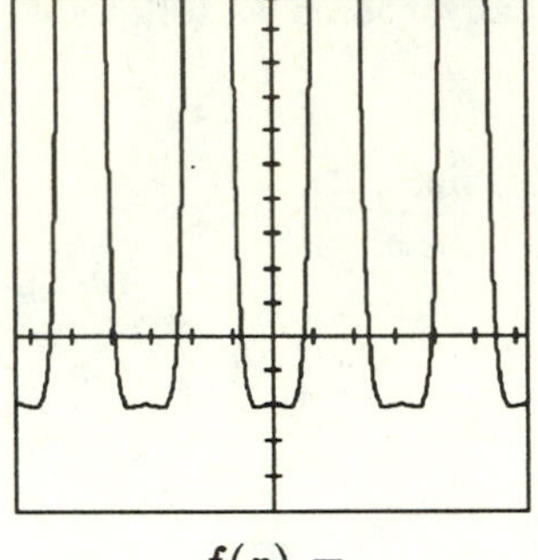

$f(x) =$
$3\tan^4 x - 1 - \sec^2 x$
$[-2\pi, 2\pi]$ by $[-5, 10]$
Figure 6.4.5

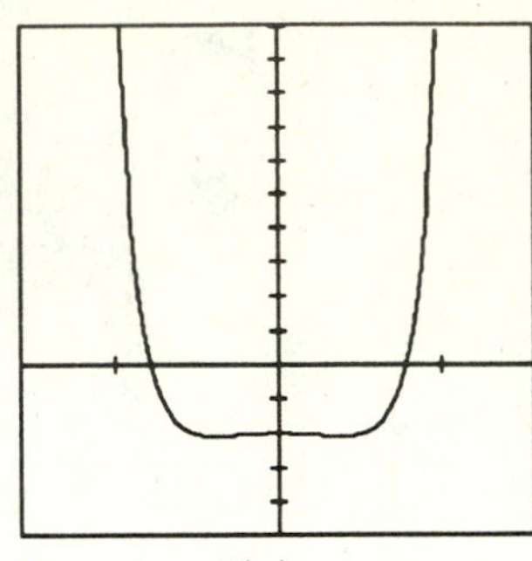

$f(x) =$
$3\tan^4 x - 1 - \sec^2 x$
$[-\frac{\pi}{2}, \frac{\pi}{2}]$ by $[-5, 10]$
Figure 6.4.6

Now, we can be sure that f has period π. The graph of f in Figure 6.4.6 suggests that this equation has two solutions in the interval $[-\frac{\pi}{2}, \frac{\pi}{2}]$. The graph of f has $x = -\frac{\pi}{2}$ and $x = \frac{\pi}{2}$ as vertical asymptotes. (Why?)

We find the solutions algebraically by using the identity $1 + \tan^2 x = \sec^2 x$ to rewrite the equation:

$$\begin{aligned} 3\tan^4 x &= 1 + \sec^2 x \\ 3\tan^4 x &= 1 + 1 + \tan^2 x \\ 3\tan^4 x - \tan^2 x - 2 &= 0 \\ (3\tan^2 x + 2)(\tan^2 x - 1) &= 0. \end{aligned}$$

Thus, we need to solve the following equations:

$$\begin{aligned} 3\tan^2 x + 2 &= 0 \\ \tan^2 x &= -\frac{2}{3} \end{aligned} \qquad \begin{aligned} \tan^2 x - 1 &= 0 \\ \tan^2 x &= 1 \\ \tan x &= \pm 1 \end{aligned}$$

The equation $\tan^2 x = -\frac{2}{3}$ has no solutions. One solution to $\tan x = 1$ is $\tan^{-1} 1$. Similarly, one solution to $\tan x = -1$ is $\tan^{-1}(-1)$. Now, $\tan^{-1} 1 = \frac{\pi}{4}$ and $\tan^{-1} 1 = -\frac{\pi}{4}$. Thus, the two solutions to $f(x) = 0$ in $\left[-\frac{\pi}{2}, \frac{\pi}{2}\right]$ are $\frac{\pi}{4}$ or $-\frac{\pi}{4}$. The periodicity of the solutions implies that the solutions to the original equation are $-\frac{\pi}{4} + k\pi$ or $\frac{\pi}{4} + k\pi$, where k is any integer. These solutions can also be rewritten in the form $\frac{\pi}{4} + k\frac{\pi}{2}$, where k is any integer. (Why?) •

In Example 5 we factor the left side of the equation by grouping. We may not always include a graph with each example, but it is a good idea for you to start by viewing a graph to obtain preliminary information about the functions involved in an equation or inequality.

• EXAMPLE 5: Solve $2\cot x \cos x - 3\cos x + 6\cot x - 9 = 0$.

SOLUTION: The function $f(x) = 2\cot x\cos x - 3\cos x + 6\cot x - 9$ can be shown to be periodic with period 2π and to have two zeros in any interval of length 2π. We can group terms and factor as follows:

$$\begin{aligned}(2\cot x\cos x - 3\cos x) + (6\cot x - 9) &= 0\\ (\cos x)(2\cot x - 3) + 3(2\cot x - 3) &= 0\\ (\cos x + 3)(2\cot x - 3) &= 0.\end{aligned}$$

The factor $\cos x + 3$ can never be zero. (Why?) To solve the equation $\cot x = \frac{3}{2}$ with a calculator you will need to solve the equivalent equation $\tan x = \frac{2}{3}$. (We used $\tan x = \frac{1}{\cot x}$.) The solutions to the original equation are periodic with period π because they are also solutions to the equation $\tan x = \frac{2}{3}$. The equation $\tan x = \frac{2}{3}$ has exactly one solution in any interval of length π and $\tan^{-1}\left(\frac{2}{3}\right) = 0.59$. Thus, the solutions to the original equation are $0.59 + k\pi$, where k is any integer. •

The same observations about periodicity for equations can be used to help solve inequalities involving trig functions.

• EXAMPLE 6: Solve $\sin x < \cot x$.

SOLUTION: $\sin x$ is periodic with period 2π and $\cot x$ is periodic with period π. Thus, we can solve this inequality by first finding the solutions in the interval $[0, 2\pi]$. We need to determine where the graph of $f(x) = \sin x - \cot x$ is *below* the x-axis (Figure 6.4.7). Notice that f has $x = 0$, $x = \pi$, and $x = 2\pi$ as vertical asymptotes. We can see from this figure that f has two zeros in the interval $[0, 2\pi]$. These zeros can be found graphically using zoom-in or algebraically.

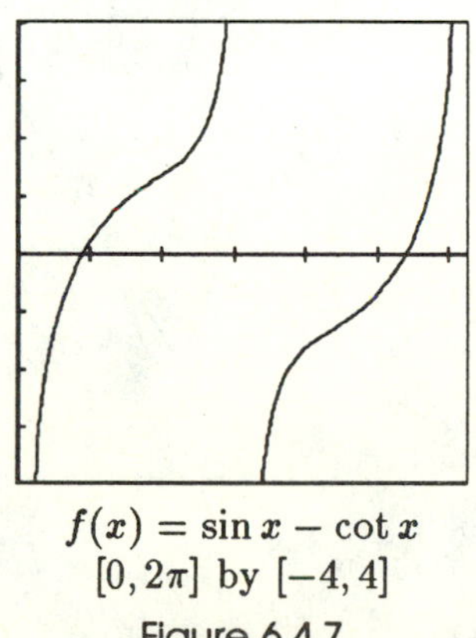

$f(x) = \sin x - \cot x$
$[0, 2\pi]$ by $[-4, 4]$

Figure 6.4.7

We will determine the zeros algebraically:

$$\sin x - \cot x = 0$$
$$\sin x - \frac{\cos x}{\sin x} = 0$$
$$\sin^2 x - \cos x = 0$$
$$1 - \cos^2 x - \cos x = 0$$
$$\cos^2 x + \cos x - 1 = 0.$$

Using the quadratic formula we find that $\cos x = (-1 \pm \sqrt{5})/2$. Now, $\cos x = (-1-\sqrt{5})/2$ has no real solutions. (Why?) The solutions to $\cos x = (-1+\sqrt{5})/2$ in $[0, 2\pi]$ are 0.90 and 5.38. The solution to the original inequality in $[0, 2\pi]$ is $(0, 0.90) \cup (\pi, 5.38)$. Furthermore, the solutions to the original inequality are periodic with period 2π. •

We were able to use algebraic techniques in all of the previous examples to obtain solutions. Generally, it is difficult or even impossible to find zeros of trig functions algebraically. The periodicity of the functions in the previous examples helped simplify finding solutions. However, we should not expect all equations and inequalities to involve only periodic functions. The next example involves a function that is not periodic.

• EXAMPLE 7: Solve $\sin 2x \leq x$.

SOLUTION: Figure 6.4.8 gives a complete graph of $f(x) = \sin 2x$ and $g(x) = x$. Notice that neither g nor $f - g$ is periodic. We need to determine those values of x for which the graph of $f(x) = \sin 2x$ is *below* or intersects the graph of $g(x) = x$. We can see from this figure that the graphs of f and g have three points in common. One of these is $(0, 0)$. The graphs of both f and g are symmetric with respect to the origin because $f(-x) = -f(x)$ and $g(-x) = -g(x)$. We can use zoom-in to determine that the point of intersection in the first quadrant is $(0.948, 0.948)$ with error at most 0.01. By symmetry, $(-0.948, -0.948)$ is

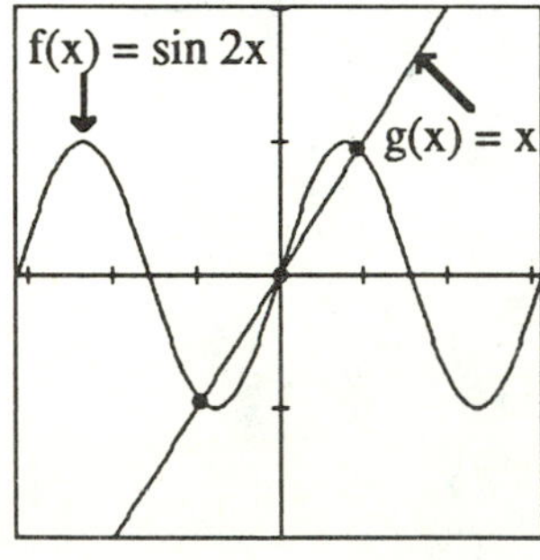

$[-\pi, \pi]$ by $[-2, 2]$

Figure 6.4.8

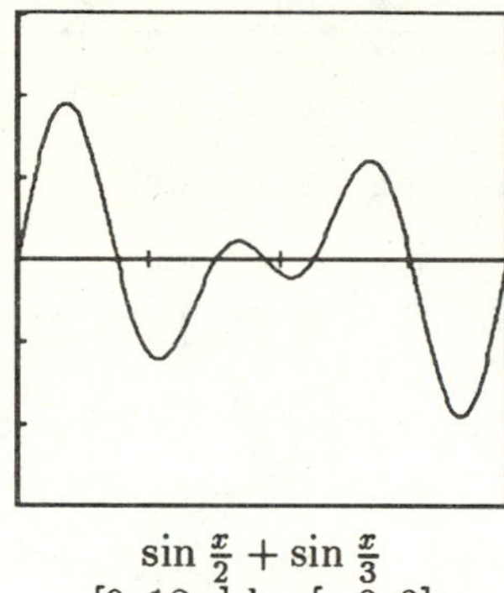

$\sin\frac{x}{2} + \sin\frac{x}{3}$
$[0, 12\pi]$ by $[-3, 3]$
Figure 6.4.9

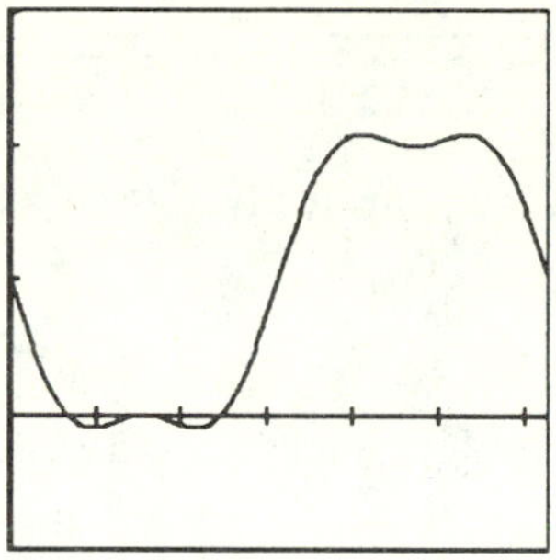

$f(x) = \sin^3 x - 2\sin x + 1$
$[0, 2\pi]$ by $[-1, 3]$
Figure 6.4.10

the point of intersection in the third quadrant. Thus, the graph of f is below the graph of g in $[-0.948, 0] \cup [0.948, \infty)$. Therefore, $\sin 2x \leq x$ in $[-0.948, 0] \cup [0.948, \infty)$. •

In Section 6.1 we showed that the function of the next example is periodic with period 12π.

• EXAMPLE 8: Determine all the zeros of $f(x) = \sin\frac{x}{2} + \sin\frac{x}{3}$ and the local maximum and minimum values of f in $[0, 12\pi]$.

SOLUTION: Because f is periodic with period 12π, we can determine the zeros of f in $[0, 12\pi]$ and then add $12k\pi$, k any integer, to each of them. The graph of f in Figure 6.4.9 is a complete graph of f and shows that f has seven zeros in $[0, 12\pi]$. However, the zero at 12π is obtained by adding 12π to the zero at 0. Thus, we need to determine only the other six zeros of f in $[0, 12\pi]$.

Two of the zeros are 0 and 6π because both $\sin\frac{x}{2}$ and $\sin\frac{x}{3}$ are zero for these values. We can use zoom-in to determine that the solutions, excluding 12π, to the equation $\sin\frac{x}{2} + \sin\frac{x}{3} = 0$ in the interval $[0, 12\pi]$ are 0, 7.54, 15.08, 18.85, 22.62, or 30.16 with error at most 0.01. Now, adding $12k\pi$, k any integer, to these six values gives all solutions to the equation.

In Example 6 of Section 6.1 we found that f had a local maximum value of 1.91 at 3.62. The other two local maximum values in $[0, 2\pi]$ are 0.22 and 1.22 and occur at 16.79 and 26.97, respectively. The three local minimum values of f in $[0, 12\pi]$ are -1.22, -0.22, and -1.91, and they occur at 10.72, 20.91, and 34.08, respectively. •

• EXAMPLE 9: Find all solutions to $\sin^3 x - 2\sin x + 1 = 0$ in the interval $[0, 2\pi]$.

SOLUTION: The function $f(x) = \sin^3 x - 2\sin x + 1$ is periodic with period 2π. The graph of f in Figure 6.4.10 is a complete graph and shows that f has three zeros in $[0, 2\pi]$. We can use zoom-in to determine that these zeros are 0.666, 1.571 or 2.475 with error at most 0.01. •

• EXERCISES 6-4

Solve the equation algebraically. Check with a graphing utility.

1. $\sin x = 0.7$
2. $\cos x = 0.9$
3. $\sin x = 1.3$
4. $\tan x = 2.75$

Solve the equation algebraically in the interval $[0, 2\pi]$. Check with a graphing utility.

5. $\sin 2x = 1$
6. $\sin 3t = 1$
7. $2\sin x = 1$
8. $2\sin 3x = 1$
9. $3\cos 2x = 1$
10. $\sin 2t = \sin t$
11. $\sin^2 x = 0$
12. $\sin x \tan x + \sin x = 0$
13. $\cos x(\sin x - 1) = 0$
14. $\sin^2 x - 1 = 0$

Solve the equation algebraically. Check with a graphing utility.

15. $\sin^2 \theta - 2\sin\theta = 0$
16. $2\cos 2t = 1$
17. $2\sin^2 x + 3\sin x - 2 = 0$
18. $\cos(\sin x) = 1$
19. $3\sin t = 2\cos^2 t$
20. $2\cos^2 x + \cos x - 1 = 0$
21. $\cos 2x + \cos x = 0$
22. $\sin 2t = \sin t$
23. $1 = \csc x - \cot x$
24. $\tan^2 x \cos x + 5\cos x = 0$

Solve the inequality algebraically. Check with a graphing utility.

25. $\sin x < \frac{1}{2}$ for $0 \le x \le 2\pi$
26. $|\sin x| < \frac{1}{2}$ for $0 \le x \le 2\pi$
27. $\sin 2x < \cos 2x$
28. $\tan x > 0$

Solve the equation or inequality.

29. $3\sin x = x$
30. $\tan x = x$ for $-\pi \le x \le \pi$
31. $3\sin 2x - x^2 = 0$
32. $\sin x < \cos(x - \pi)$
33. $5\sin 2x < 3\cos x$
34. $x \sin x \ge 1$ for $-10 \le x \le 10$
35. $\sin 3x + \cos x = 0$
36. $3\sin 2x = \cos x$

Draw the graph in the indicated interval. Determine all local maximum and minimum values in the interval.

37. $g(x) = 2 - 3\sin(\frac{1}{2}x - \frac{\pi}{2})$ for $0 \le x \le 4\pi$
38. $f(x) = x^2 \sin x$ for $-\pi \le x \le \pi$
39. $g(x) = -1 + 2\cos(\pi x - 2)$ for $0 \le x \le 3$
40. $f(x) = e^{-x/2}\sin 2x$ for $0 \le x \le 8$
41. Determine all extrema for the function in Example 8.

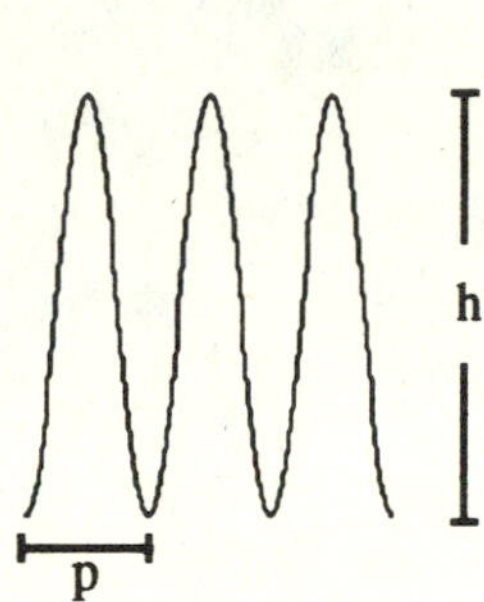

Figure 6.4.11

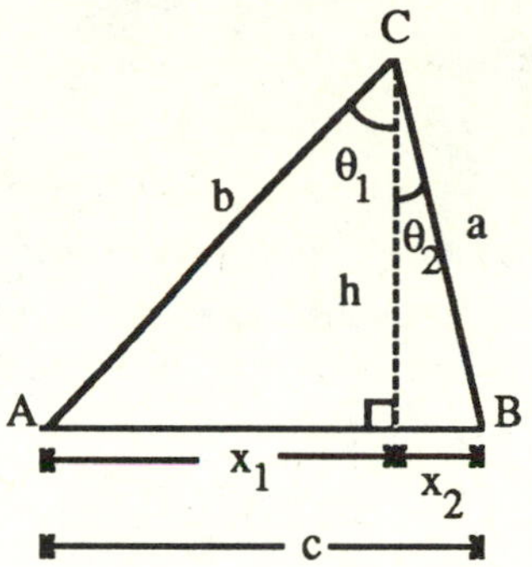

Figure 6.4.12

42. Determine all extrema for the function in Example 9.

Applications Involving Sinusoids

43. Show that $g(x) = 2 - 3\cos(2x - \pi)$ is a sinusoid by finding A, B, C, and D so that $A + B\sin(Cx + D) = g(x)$.

The height of a tidal wave has a sinusoidal representation as a function of time t with period p minutes and *maximum* height h feet. (See Figure 6.4.11.) Assume the height is 0 at time $t = 0$.

44. Show that $y = \frac{h}{2} + \frac{h}{2}\sin(\frac{2\pi}{p}(t - \frac{p}{4}))$ is a sinusoid representation of the tidal wave.

45. Draw a complete graph of the sinusoid representation if the period is 40 minutes and the *maximum* height is 27 feet.

The Cannon Problem

A cannon shell is fired with initial velocity V feet per second. If the cannon barrel makes an angle of θ with the ground, then the horizontal distance the shell travels before it hits the ground is given by $d = \frac{V^2}{16}\sin\theta\cos\theta$.

46. Let $V = 500$ feet per second. Draw a complete graph of d as a function of θ. What portion of the graph represents the problem situation?

47. At what angle should the cannon be aimed to hit a target 2350 feet away if the initial velocity is 500 feet per second?

48. What is the maximum horizontal distance the shell travels if the initial velocity is 500 feet per second?

49. What is the maximum horizontal distance the shell travels if the initial velocity is 1200 feet per second?

Area of a Triangle

The area $\mathcal{A}$ of a triangle ABC is given by $\mathcal{A} = \frac{1}{2}ab\sin C$. In Figure 6.4.12, $x_1 + x_2 = c$ and h is the altitude of the triangle.

50. Derive the area formula. Hints:
 i) Note that $\mathcal{A} = \frac{1}{2}x_1h + \frac{1}{2}x_2h$,
 ii) $\frac{1}{2}ab\sin C = \frac{1}{2}ab\sin(\theta_1 + \theta_2)$
 $= \frac{1}{2}ab(\sin\theta_1\cos\theta_2 + \sin\theta_2\cos\theta_1)$,
 iii) Compute $\sin\theta_1$, $\cos\theta_2$, $\sin\theta_2$ and $\cos\theta_1$.

51. A single cell in a beehive is a regular hexagonal prism open at the front with a trihedral top at the back. It can be shown that the surface area of a cell is given by

$$S(\theta) = 6ab + \frac{3}{2}b^2\left(-\cot\theta + \frac{\sqrt{3}}{\sin\theta}\right)$$

where θ is the trihedral angle, a is the depth of the cell, and $2b$ is the length of the line segment through the center connecting opposite vertices of the hexagonal front. Assume $a = 1.75$ (inches) and $b = 0.65$ (inches).

(a) Draw a complete graph of the trigonometric representation for the surface area.
(b) What values of θ make sense in the problem situation?
(c) What value of θ gives the minimum surface area? (*Remark:* This answer is remarkably close to the observed angle in nature.)
(d) What is the minimum surface area?

52. The range, D, of a projectile shot at an angle of elevation, θ, and with an initial velocity, v, is given by

$$D(\theta) = \left(\frac{v^2}{g}\right)\sin 2\theta$$

where g is the acceleration due to gravity (32 feet per second per second). Assume $v = 85$ feet per second. (See Figure 6.4.13.)

(a) Draw a complete graph of the trigonometric representation.
(b) What values of θ make sense in this problem situation?
(c) What value of θ gives the maximum range?
(d) A target is 215 feet down range. At what angle should the projectile be aimed in order to hit the target?
(e) Derive a formula for the trigonometric representation of the problem situation.
(f) Compare with the cannon problem (46–49).

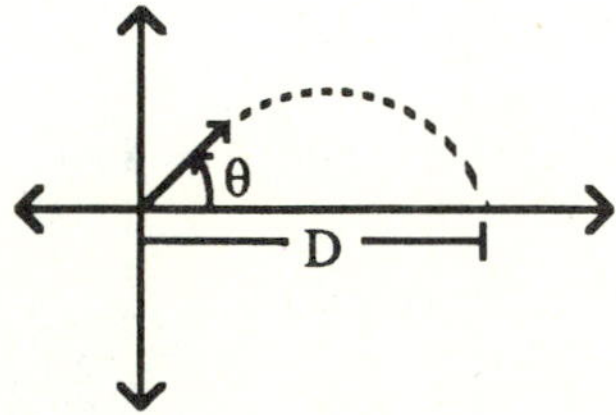

Figure 6.4.13

53. Determine the exact solutions to Example 9.

54. For what values of x is $\log(\tan x) = \log(\sin x) - \log(\cos x)$?

55. For what values of x is $\log(\cot x) = \log(\cos x) - \log(\sin x)$?

6.5 The Law of Sines

In Section 5.2 we solved right triangles. Recall that to solve a triangle means to find the remaining parts of a triangle when some of its angles and sides are specified. In this and the next section we solve triangles that do not necessarily contain a right angle. A triangle that does not contain a right angle is called an **oblique triangle**. In this section we solve such triangles using a formula called the *law of sines* that is valid for any triangle. We also investigate applications whose models involve a triangle. Solving the application will require the use of the law of sines to find the missing parts of the triangle.

We use A, B, and C to denote the vertices of a triangle. The angles at A, B, and C are denoted by α, β, and γ, respectively. The measures of the sides opposite A, B, and C are denoted by a, b, and c, respectively (Figure 6.5.1). Again, α, β, and γ stand for the name of the angle as well as the measure of the angle. At most one angle in an oblique triangle can have measure greater than 90°. (Why?)

If three of the six parts α, β, γ, a, b, or c of a triangle are specified, then often a *unique* triangle is determined. In some cases, no triangle is determined or two triangles are determined. If three angles are specified and the sum of their measures is 180°, then infinitely many triangles are possible. Recall that two triangles are *similar* if they have corresponding angles equal. There are infinitely many triangles similar to a given triangle. Figure 6.5.2 suggests one way to find additional triangles similar to a given triangle ABC. Any pair of sides of triangle ABC can be extended or shrunk as suggested by Figure 6.5.2 to form additional triangles similar to ABC.

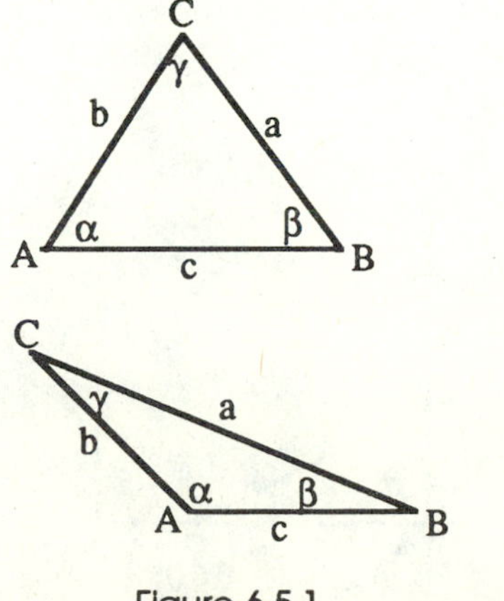

Figure 6.5.1

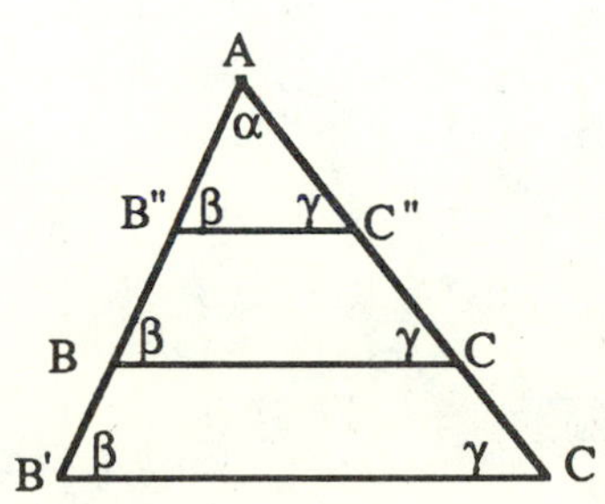

Figure 6.5.2

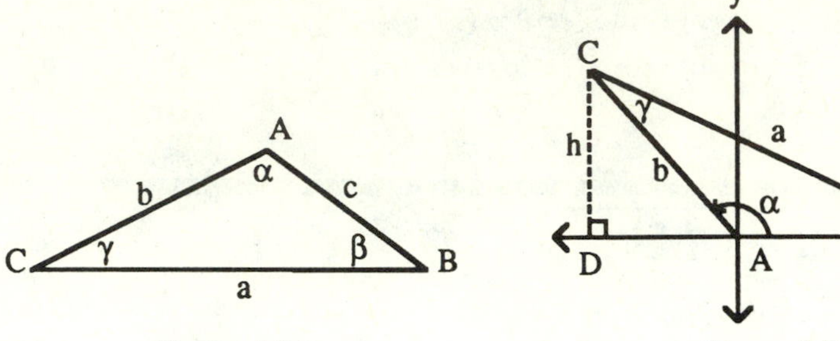

Figure 6.5.3

Figure 6.5.4

In this section we determine the triangle(s) when two angles and a side are specified, or, in some instances, when two sides and an angle are specified. The remaining possibilities are covered in the next section. First, we need the following important theorem.

• THEOREM 1 **(The Law of Sines)** In any triangle ABC (Figure 6.5.3) we have

$$\frac{\sin\alpha}{a} = \frac{\sin\beta}{b} = \frac{\sin\gamma}{c}.$$

PROOF Without any loss of generality we can put the vertex A at the origin and the vertex B on the positive x-axis at $(c, 0)$ (Figure 6.5.4). We have drawn the triangle ABC to suggest α is greater than 90°. The same proof is valid if α is acute.

We choose D on the x-axis so that CD is perpendicular to the x-axis. Then, the distance h from C to the x-axis is a height, or altitude, of the triangle ABC. Because α is in standard position we have $\sin\alpha = \frac{h}{b}$. From the right triangle BCD we have $\sin\beta = \frac{h}{a}$. Thus, $h = b\sin\alpha$ and $h = a\sin\beta$. It follows that $b\sin\alpha = a\sin\beta$, or $\frac{\sin\alpha}{a} = \frac{\sin\beta}{b}$. If we put the vertex C on the positive x-axis we obtain Figure 6.5.5. Then, the same argument as above gives $\frac{\sin\alpha}{a} = \frac{\sin\gamma}{c}$. Therefore, $\frac{\sin\alpha}{a} = \frac{\sin\beta}{b} = \frac{\sin\gamma}{c}$.

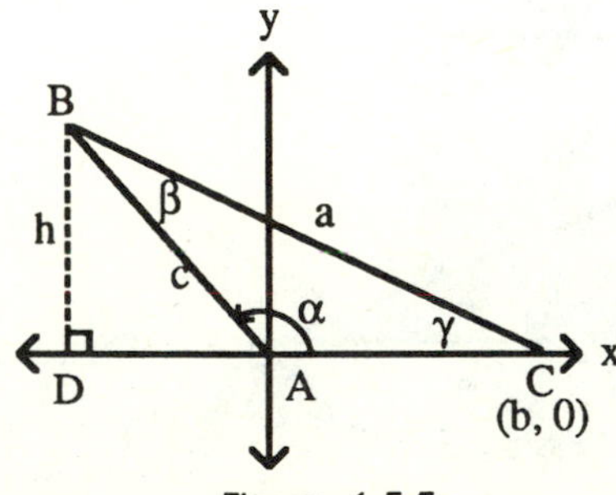

Figure 6.5.5

Two Angles and a Side Specified

If the measures of any two angles of a triangle are given, then the measure of the third angle can be determined. (Why?) If the measure of one side is also given, the law of sines can be used to find the measures of the remaining two sides as illustrated in Example 1.

• EXAMPLE 1: Solve triangle ABC if $\alpha = 36°$, $\beta = 48°$, and $a = 8$.

SOLUTION: First, $\gamma = 180° - (36° + 48°) = 96°$. Next, we use the law of sines:

$$\frac{\sin\alpha}{a} = \frac{\sin\beta}{b} \qquad \frac{\sin\alpha}{a} = \frac{\sin\gamma}{c}$$

$$\frac{\sin 36°}{8} = \frac{\sin 48°}{b} \qquad \frac{\sin 36°}{8} = \frac{\sin 96°}{c}$$

$$b = \frac{8\sin 48°}{\sin 36°} \qquad c = \frac{8\sin 96°}{\sin 36°}$$

$$b = 10.11 \qquad c = 13.54$$

•

Notice that the procedure of Example 1 works no matter which pair of angles is given or which side is given. Thus, if the measure of two angles and a side are given, then a unique triangle is determined, provided we choose a pair of angles whose sum is less than 180°.

Two Sides and an Angle Specified

We use the law of cosines in Section 6.6 to determine the triangle(s) if two sides and the *included* angle are given (Figure 6.5.6). This situation can be handled directly using the definitions of the trig functions. However, the method of Section 6.6 is considerably simpler. In this section, we determine the triangle(s) if two sides and an angle *opposite* one of the sides are given. This is sometimes called the ambiguous case because there may be *no* triangle, exactly *one* triangle, or *two* triangles in this situation.

Without any loss of generality we may assume that a, b, and β are given. We first assume that β is acute and put β in standard position with C in the first quadrant. Then, A *must* be on the positive x-axis (Figure 6.5.7).

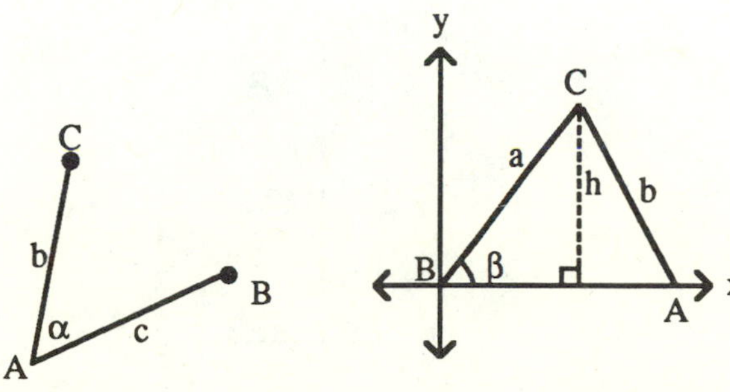

Figure 6.5.6 Figure 6.5.7

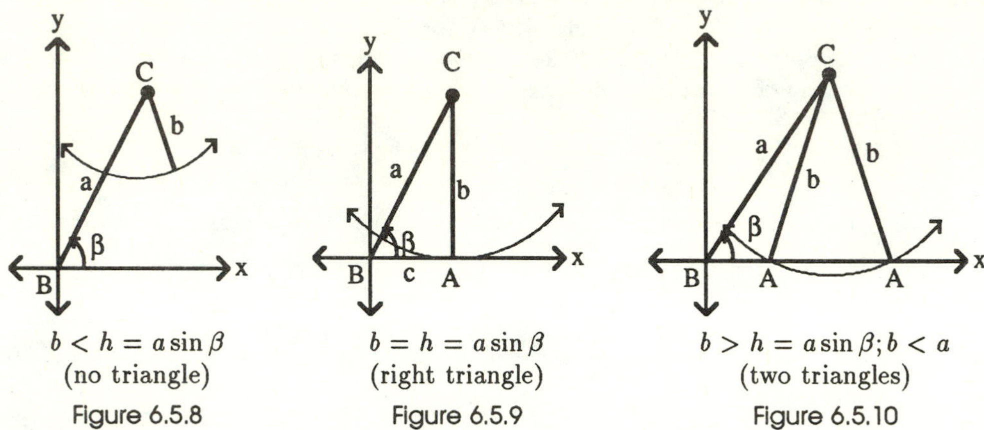

$b < h = a \sin \beta$
(no triangle)
Figure 6.5.8

$b = h = a \sin \beta$
(right triangle)
Figure 6.5.9

$b > h = a \sin \beta ; b < a$
(two triangles)
Figure 6.5.10

If h is the distance from C to the x-axis, then $h = a \sin \beta$ and gives the altitude of any such triangle. (Why?) We can think of b as swinging like a pendulum. If $b < h = a \sin \beta$, then *no* triangle is determined (Figure 6.5.8). If $b = h = a \sin \beta$, then a *right* triangle, as shown in Figure 6.5.9, is determined. If $b > h = a \sin \beta$ and $b < a$, then *two* triangles are determined (Figure 6.5.10). If $b > h = a \sin \beta$ and $b \geq a$, then *one* triangle is determined (Figure 6.5.11).

It may seem that we should get two triangles in this last case; the second triangle formed using the origin, C, and the point where the arc in Figure 6.5.10 intersects the negative x-axis. This cannot happen because we have assumed that β is acute and A is in the first quadrant. (Why?) We illustrate the four situations above with β acute in Example 2.

• EXAMPLE 2: Solve triangle ABC if $\beta = 30°$, $a = 6$, and

(a) $b = 2$. (b) $b = 3$. (c) $b = 5$. (d) $b = 7$.

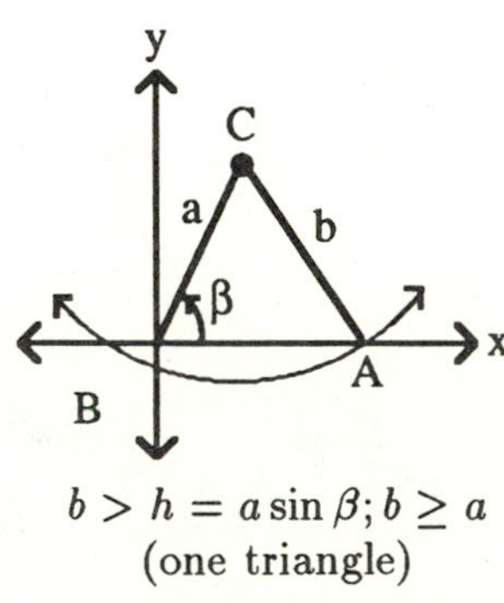

$b > h = a \sin \beta ; b \geq a$
(one triangle)
Figure 6.5.11

SOLUTION: Any such triangle must have height $h = a\sin\beta = 6\sin 30° = 3$.

(a) Because $b = 2 < a\sin\beta = 3$ there should be no solution. We use the law of sines:

$$\frac{\sin\alpha}{a} = \frac{\sin\beta}{b}$$
$$\frac{\sin\alpha}{6} = \frac{\sin 30°}{2}$$
$$\sin\alpha = 1.5.$$

There is no solution in this case because $|\sin\alpha| \leq 1$ for any α.

(b) There should be exactly one solution in this case because $b = 3 = a\sin\beta$. We use the law of sines:

$$\frac{\sin\alpha}{a} = \frac{\sin\beta}{b}$$
$$\frac{\sin\alpha}{6} = \frac{\sin 30°}{3}$$
$$\sin\alpha = 1.$$

The only solution to $\sin\alpha = 1$ with $0° < \alpha < 180°$ is $\alpha = 90°$. Thus, $\gamma = 180° - (30° + 90°) = 60°$. Next, we solve for c:

$$\frac{\sin\alpha}{a} = \frac{\sin\gamma}{c}$$
$$\frac{\sin 90°}{6} = \frac{\sin 60°}{c}$$
$$c = 6\sin 60°$$
$$c = 3\sqrt{3}.$$

We have used the fact that the exact value of $\sin 60°$ is $\frac{\sqrt{3}}{2}$. If a calculator is used, we would find $c = 5.20$ accurate to hundredths. In this case we get the triangle in Figure 6.5.12.

(c) There are two solutions in this case. First, we find α:

$$\frac{\sin\alpha}{a} = \frac{\sin\beta}{b}$$
$$\frac{\sin\alpha}{6} = \frac{\sin 30°}{5}$$
$$\sin\alpha = \frac{6\sin 30°}{5}$$
$$\sin\alpha = \frac{3}{5}.$$

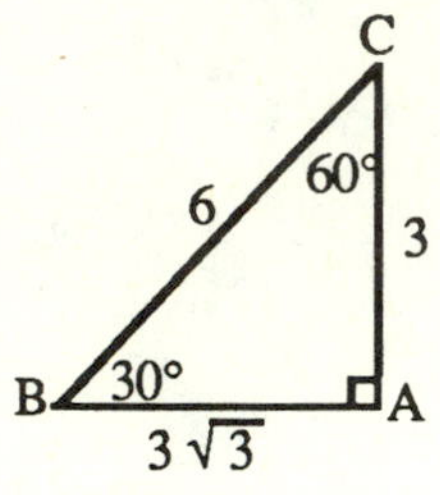

Figure 6.5.12

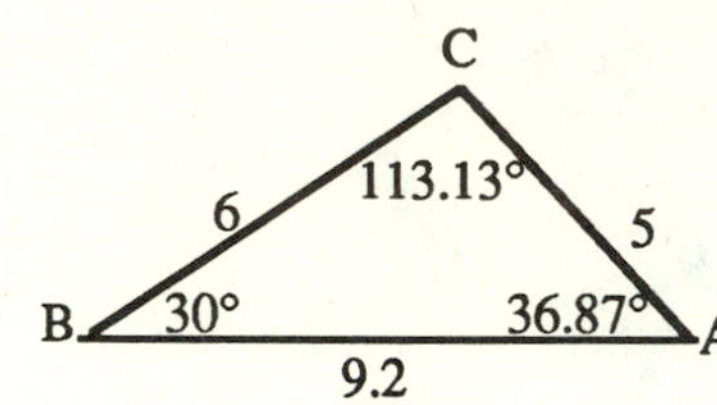

Figure 6.5.13

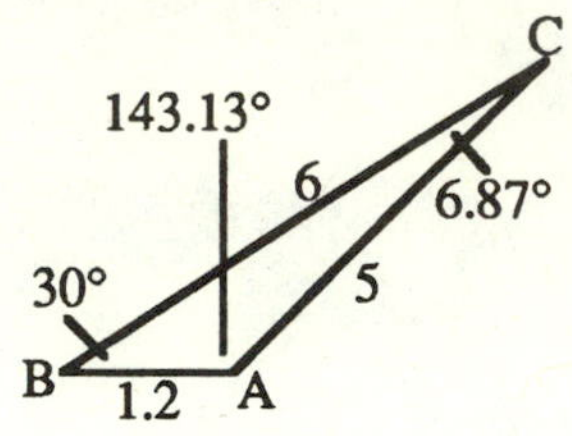

Figure 6.5.14

There are two solutions to $\sin\alpha = \frac{3}{5}$ with $0° < \alpha < 180°$, namely 36.87° and 143.13°. If $\alpha = 36.87°$, then $\gamma = 113.13°$ and the law of sines determines c:

$$\frac{\sin\beta}{b} = \frac{\sin\gamma}{c}$$
$$\frac{\sin 30°}{5} = \frac{\sin 113.13°}{c}$$
$$c = 9.20.$$

Thus, $\alpha = 36.87°$, $\gamma = 113.13°$, and $c = 9.20$ (Figure 6.5.13). If $\alpha = 143.13°$, then $\gamma = 6.87°$ and the law of sines determines c:

$$\frac{\sin\beta}{b} = \frac{\sin\gamma}{c}$$
$$\frac{\sin 30°}{5} = \frac{\sin 6.87°}{c}$$
$$c = 1.20.$$

Thus, $\alpha = 143.13°$, $\gamma = 6.87°$, and $c = 1.20$ (Figure 6.5.14).

(d) There is one solution in this case. We use the law of sines to find α:

$$\frac{\sin\alpha}{a} = \frac{\sin\beta}{b}$$
$$\frac{\sin\alpha}{6} = \frac{\sin 30°}{7}$$
$$\sin\alpha = \frac{6\sin 30°}{7}$$
$$\sin\alpha = \frac{3}{7}.$$

There are two solutions to $\sin\alpha = \frac{3}{7}$ with $0 < \alpha < 180°$, namely 25.38° and 154.62°. However, if α is 154.62°, then $\alpha + \beta$ is 184.62°, which is not possible.

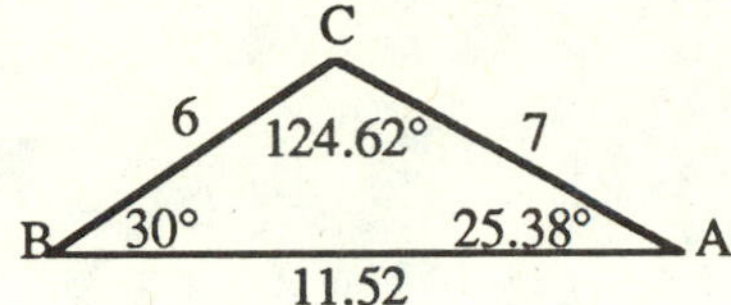

Figure 6.5.15

Therefore, $\alpha = 25.38°$ and $\gamma = 124.62°$:

$$\frac{\sin \beta}{b} = \frac{\sin \gamma}{c}$$

$$\frac{\sin 30°}{7} = \frac{\sin 124.62°}{c}$$

$$c = 11.52.$$

We get the triangle in Figure 6.5.15. •

Figure 6.5.16 illustrates the corresponding four cases if β is assumed to be obtuse. Notice that there is either exactly one solution or no solution if β is obtuse because A must be on the positive x-axis by the initial placement of the proposed triangle.

Next, we consider applications that involve the use of the law of sines.

• EXAMPLE 3: A forest ranger at an observation point A along a straight road sights a fire in the direction 32° east of north. Another ranger at a second observation point B on the road, 10 miles due east of A, sights the same fire at 48° west of north. Find the distance from each observation point to the fire, and find the shortest distance from the road to the fire.

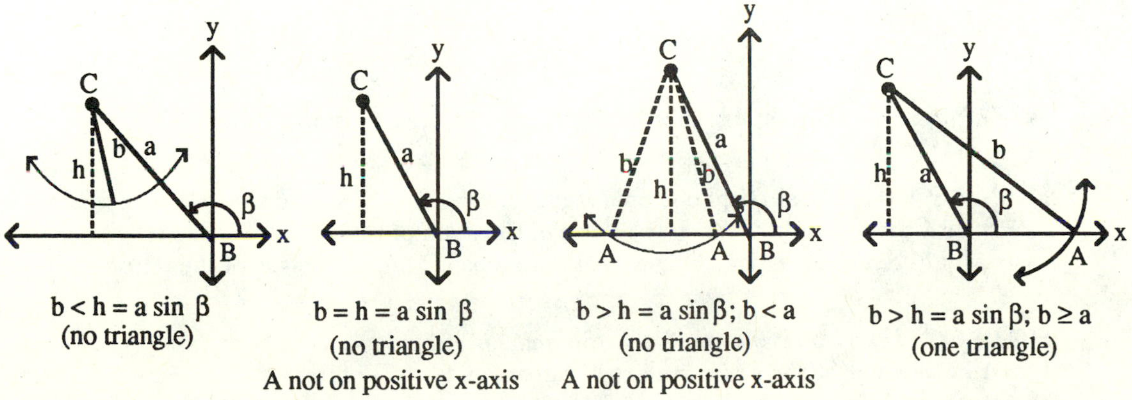

Figure 6.5.16

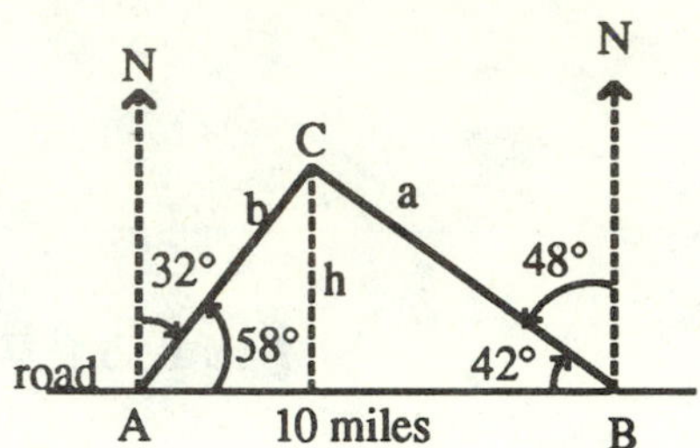

Figure 6.5.17

SOLUTION: Assume the fire is located at point C (Figure 6.5.17). Because the fire is 32° east of north, α must be $90° - 32°$ or 58°. Similarly, β must be 42°. Let h be the length of the line segment from C perpendicular to the road. Then, h is the shortest distance from the road to the fire. We must determine a, b, and h in Figure 6.5.17.

In this triangle we know that $\alpha = 58°$, $\beta = 42°$, $\gamma = 80°$, and $c = 10$. We use the law of sines:

$$\frac{\sin\alpha}{a} = \frac{\sin\gamma}{c} \qquad \frac{\sin\beta}{b} = \frac{\sin\gamma}{c}$$

$$\frac{\sin 58°}{a} = \frac{\sin 80°}{10} \qquad \frac{\sin 42°}{b} = \frac{\sin 80°}{10}$$

$$a = \frac{10\sin 58°}{\sin 80°} \qquad b = \frac{10\sin 42°}{\sin 80°}$$

$$a = 8.61 \qquad b = 6.79$$

Thus, the fire is 6.79 miles from A and 8.61 miles from B. Further, the perpendicular distance h from the fire to the road can be found from either

$$\sin 58° = \frac{h}{b} \qquad \text{or} \qquad \sin 42° = \frac{h}{a}.$$

We find that the shortest distance from the road to the fire is 5.76 miles. •

• EXAMPLE 4: A vertical telephone pole stands by the side of a road that slopes at an angle of 10° with the horizontal. When the angle of elevation of the sun is 62°, the telephone pole casts a 14.5-foot shadow downhill parallel to the road. Find the height of the telephone pole.

SOLUTION: Let C and B denote the top and bottom of the telephone pole, respectively. Let A be the end of the shadow along the road. Refer to Figure 8.5.18. Because the angle of elevation of the sun is 62°, γ must be 28°. Because the road makes a 10° angle with the horizontal, β must be 100°. Thus, α is 52°. The length of the shadow is $c = 14.5$, and the

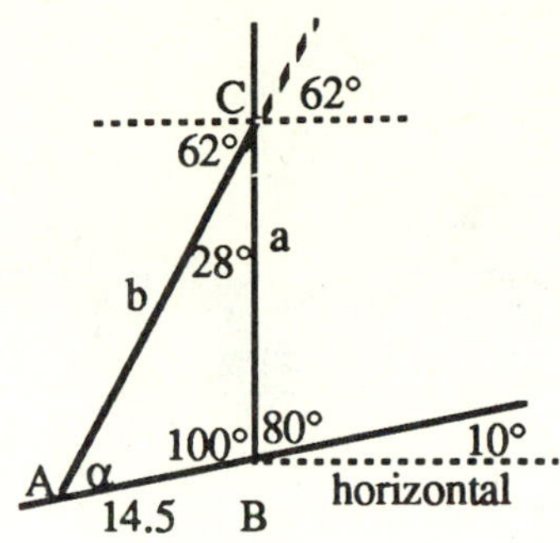

Figure 6.5.18

height of the pole is a. We use the law of sines to determine a.

$$\frac{\sin\alpha}{a} = \frac{\sin\gamma}{c}$$
$$\frac{\sin 52°}{a} = \frac{\sin 28°}{14.5}$$
$$a = \frac{14.5\sin 52°}{\sin 28°}$$
$$a = 24.34.$$

Thus, the height above the ground of the telephone pole is 24.34 feet. •

• EXERCISES 6-5

1. Give the side lengths a, b, and c for *two different* triangles ABC whose interior angles are $\alpha = 32°$, $\beta = 75°$, and $\gamma = 73°$.
2. Give the side lengths a, b, and c of *two different* triangles ABC whose interior angles are $\alpha = 54°$, and $\gamma = 16°$.

Solve the triangle.

3. $\alpha = 40°$, $\beta = 30°$, $b = 10$
4. $\alpha = 60°$, $a = 3$, $b = 4$
5. $\beta = 30°$, $a = 12$, $b = 6$
6. $\alpha = 50°$, $\beta = 62°$, $a = 4$
7. $\alpha = 33°$, $\beta = 79°$, $b = 7$
8. $\beta = 85°$, $a = 4$, $b = 6$
9. $\alpha = 50°$, $a = 4$, $b = 5$
10. $\beta = 38°$, $a = 16$, $b = 20$
11. $\beta = 38°$, $a = 16$, $b = 12$
12. $\beta = 116°$, $a = 11$, $b = 13$
13. $\beta = 116°$, $a = 11$, $b = 10$
14. $\beta = 116°$, $a = 11$, $b = 8$
15. $\beta = 152°$, $a = 8$, $b = 10$
16. $\gamma = 103°$, $\beta = 16°$, $c = 12$
17. If $a = 10$ and $\beta = 42°$, determine the values of b for which α has

    (a) two values. (b) one value. (c) no value.

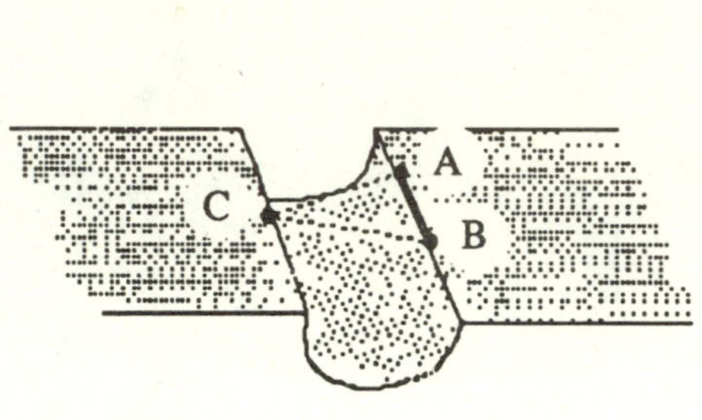

Figure 6.5.19

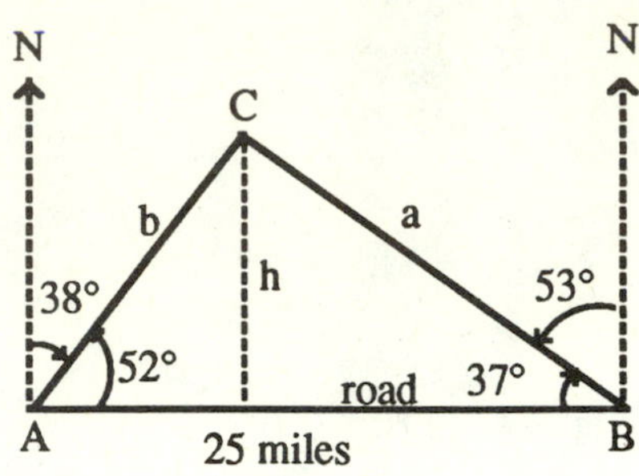

Figure 6.5.20

18. Two markers A and B on the same side of a canyon rim are 56 feet apart (Figure 6.5.19). A hiker is located across the rim at point C. A surveyor determines that $\angle BAC = 72°$ and $\angle ABC = 53°$.

    (a) What is the distance between the hiker and point A?

    (b) What is the distance between the two canyon rims? (Assume they are parallel.)

19. A forest ranger at an observation point A along a straight road sights a fire in the direction 38° east of north. Another ranger at an observation point B on the road, 25 miles due east of A, sights the same fire at 53° west of north (Figure 6.5.20). Find the distance from each observation point to the fire, and find the shortest distance from the road to the fire.

20. A vertical telephone pole stands by the side of a road that slopes at an angle of 15° with the horizontal. When the angle of elevation of the sun is 62°, the telephone pole casts a 16-foot shadow downhill parallel to the road (Figure 6.5.21). Find the height of the telephone pole.

21. A hot air balloon is seen over Park City, Utah, simultaneously by two observers at points A and B. Assume the balloon and the observers are in the same vertical plane and the angles of elevation and the distances are as shown in Figure 6.5.22. How high above ground is the balloon?

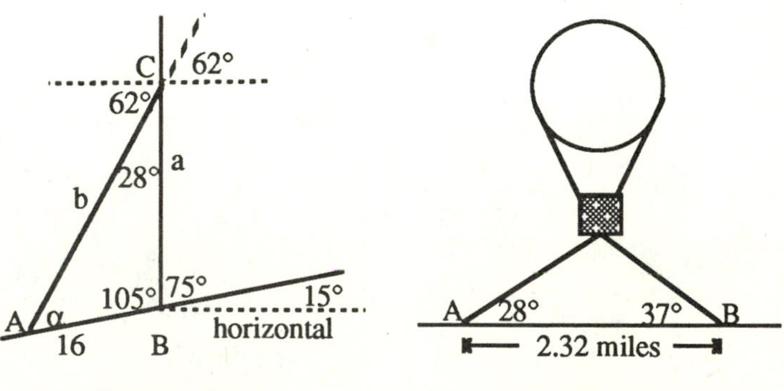

Figure 6.5.21

Figure 6.5.22

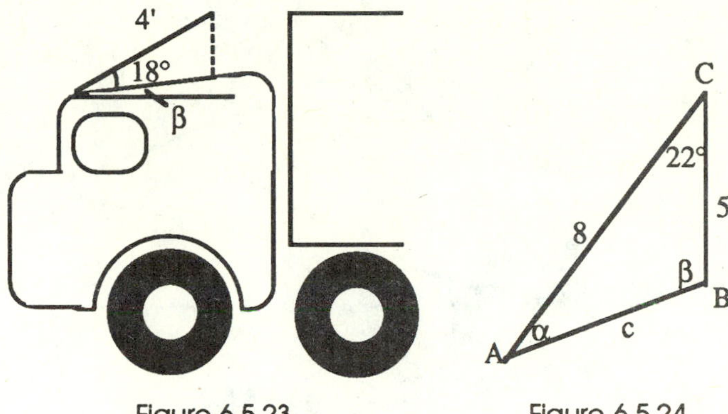

Figure 6.5.23 Figure 6.5.24

22. A 4-foot airfoil is attached to the cab of a truck as shown in Figure 6.5.23. The angle between the airfoil and the cab top is 18°. What is the length of a *vertical* brace positioned as shown in Figure 6.5.23 if angle β is 10°?

23. Solve the triangle ABC in Figure 6.5.24, given that $a = 5$, $b = 8$, and $\gamma = 22°$. *Hint*: Erect a perpendicular from A to the line through B and C.

King of the Mountains

"Just a few feet, it seems, can make a difference. Last March, George Wallerstein, an astronomer at the University of Washington, stunned mountaineers and geologists by declaring that the Himalayan mountain known as K-2 might be 36 ft. taller than Mount Everest, long thought to be the world's highest peak. This month, however, an eight-man Italian expedition, led by Geologist Ardito Desio, 90, refuted that claim. Using satellite signals and surveying techniques, they found that Everest towers 29,108 ft. above sea level—80 ft. taller than previously believed and 840 ft. higher than K-2.

To accomplish their lofty task, the Italians carried computerized radio receivers to stations on each mountain. The instruments used timed signals from U.S. Navstar satellites to calculate the exact longitude, latitude and altitude of each receiver. Armed with these coordinates, the researchers then measured the angles formed by the peaks and the receiver stations with a surveyor's theodolite, as well as the distance between the stations. Since the length of one side of each triangle and two angles were known, the peaks' heights could be accurately determined by simple geometry.

Wallerstein readily concedes that his measurement, taken while climbing K-2 may have been inaccurate. 'Mine was done with a secondhand receiver, while theirs was made with first-rate equipment,' he says. 'I guess it proves that in this business there's no place for amateurs.' " (From *Time* magazine, November 2, 1987.)

24. Using the (oversimplified) diagram in Figure 6.5.25 and "simple geometry" determine the height of the mountain peak above one of the radio receiver stations. The following information is known:

$$h_2 - h_1 = 895 \quad \text{feet}, \quad d = 2.63 \quad \text{miles}, \quad \angle z_1 = 41°, \quad \text{and} \quad \angle z_2 = 63° \,.$$

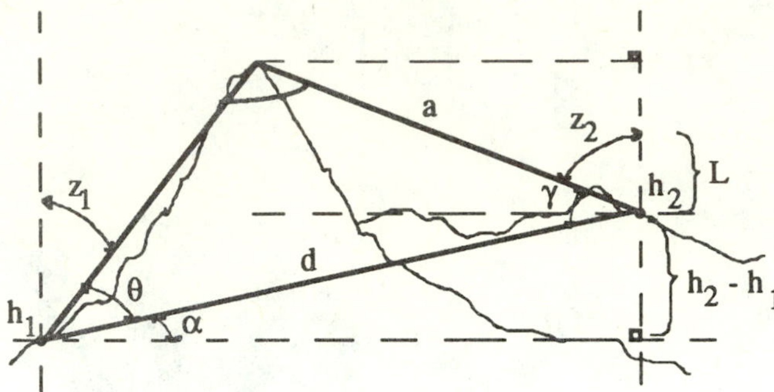

Figure 6.5.25

6.6 The Law of Cosines and Area of Triangles

In this section we obtain a formula called the *law of cosines* that is valid for any triangle. Applications whose solutions require finding missing parts of a triangle using the law of cosines will be investigated. We obtain formulas that give the area of a triangle in terms of the measures of its sides and angles. We use the law of cosines to obtain a formula for the angle between two lines.

• THEOREM 1 **(The Law of Cosines)** Let ABC be a triangle labeled in the usual way (Figure 6.6.1). Then,

$$\begin{aligned} a^2 &= b^2 + c^2 - 2bc\cos\alpha\,, \\ b^2 &= a^2 + c^2 - 2ac\cos\beta, \quad \text{and} \\ c^2 &= a^2 + b^2 - 2ab\cos\gamma\,. \end{aligned}$$

PROOF Without any loss of generality we can put the vertex A at the origin and the vertex B on the positive x-axis at $(c, 0)$ (Figure 6.6.2). We have drawn the triangle ABC as if angle α is obtuse. The proof will be valid even if α is acute. We choose D on the x-axis so that CD is perpendicular to the x-axis. Then, the distance h from C to the x-axis is an altitude of the triangle ABC. Let the coordinates of D be $(d, 0)$. Because α is in standard position we have

$$\cos\alpha = \frac{d}{b} \quad \text{and} \quad \sin\alpha = \frac{h}{b}\,.$$

Thus, $d = b\cos\alpha$ and $h = b\sin\alpha$. Notice that $d(B,C) = a$. Using the distance formula we have

$$\begin{aligned} (d(B,C))^2 &= (c-d)^2 + (0-h)^2 \\ a^2 &= (c-d)^2 + h^2. \end{aligned}$$

Substituting the above values of d and h we obtain

$$\begin{aligned} a^2 &= (c-d)^2 + h^2 \\ &= (c - b\cos\alpha)^2 + (b\sin\alpha)^2 \\ &= c^2 - 2bc\cos\alpha + b^2\cos^2\alpha + b^2\sin^2\alpha \\ &= c^2 - 2bc\cos\alpha + b^2(\cos^2\alpha + \sin^2\alpha) \\ &= b^2 + c^2 - 2bc\cos\alpha. \end{aligned}$$

This is the first of the three formulas to be obtained. Now, put the angle β in standard position with A on the positive x-axis (Figure 6.6.3). If we proceed in a manner similar to the above derivation, we will obtain $b^2 = a^2 + c^2 - 2ac\cos\beta$. This is the second of the three formulas. If we put the angle γ in standard position, we will obtain the last of the three formulas.

There are two situations remaining to complete the investigation of the possible triangles determined when three of its six parts are specified.

Three Sides Specified

Suppose that the measure of the three sides of a triangle are specified. In order that there actually be such a triangle, the sum of the measures of any two sides must exceed the measure of the third side. (Why?)

- EXAMPLE 1: Solve triangle ABC if $a = 9$, $b = 7$, and $c = 5$.

SOLUTION: We use the law of cosines to find angle α (Figure 6.6.4):

$$\begin{aligned} a^2 &= b^2 + c^2 - 2bc\cos\alpha \\ 9^2 &= 7^2 + 5^2 - 2(7)(5)\cos\alpha \\ \cos\alpha &= \frac{7^2 + 5^2 - 9^2}{(2)(7)(5)} \\ \cos\alpha &= -\frac{1}{10}. \end{aligned}$$

The only solution to $\cos\alpha = -\frac{1}{10}$ with $0° < \alpha < 180°$ is $95.74°$.

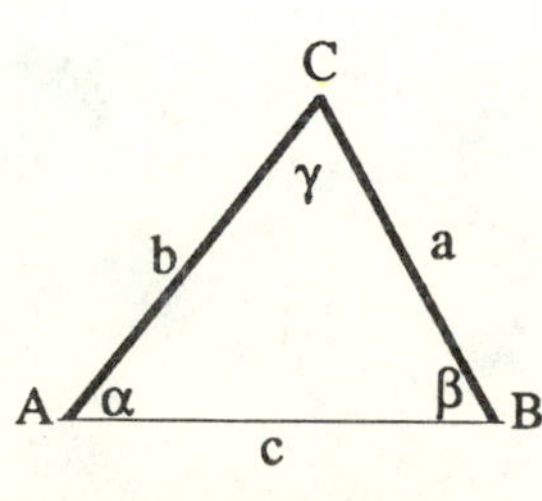

Figure 6.6.1

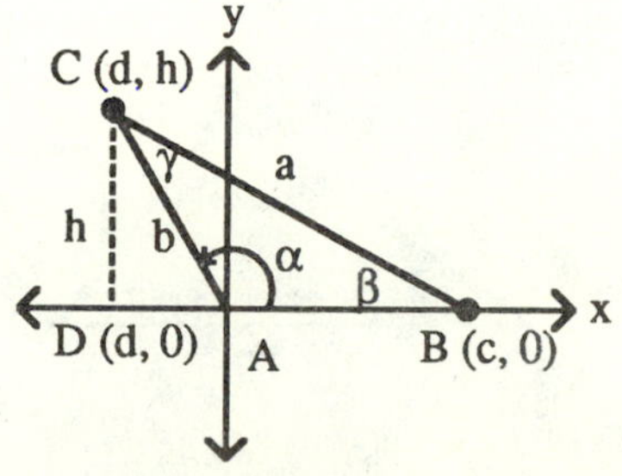

Figure 6.6.2

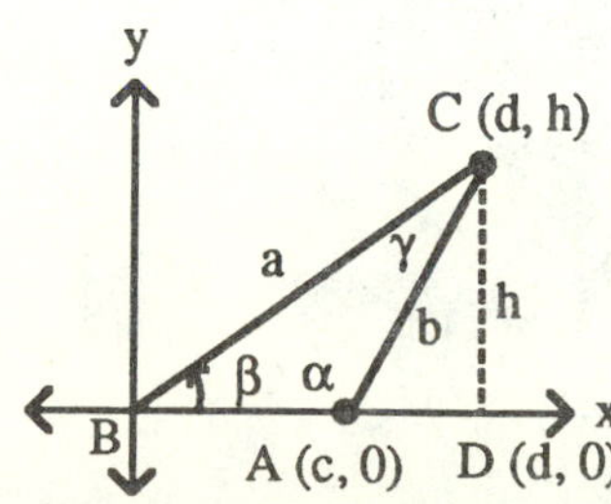

Figure 6.6.3

We use the law of cosines to find angle β:

$$b^2 = a^2 + c^2 - 2ac\cos\beta$$
$$7^2 = 9^2 + 5^2 - 2(9)(5)\cos\beta$$
$$\cos\beta = \frac{9^2 + 5^2 - 7^2}{(2)(9)(5)}$$
$$\cos\beta = \frac{19}{30}.$$

Thus, $\beta = 50.70°$. It follows that $\gamma = 33.56°$. •

Two Sides and the Included Angle Specified

If two sides and an angle opposite one of those sides are given, we found in Section 6.1 that zero, one, or two triangles were possible. If two sides and an included angle of measure between 0° and 180° are specified, then precisely one triangle is determined as illustrated in Example 2.

• EXAMPLE 2: Solve triangle ABC if $a = 4$, $b = 7$, and $\gamma = 42°$.

SOLUTION: We use the law of cosines to find the third side:

$$c^2 = a^2 + b^2 - 2ab\cos\gamma$$
$$c^2 = 4^2 + 7^2 - 2(4)(7)\cos 42°$$
$$c^2 = 23.383890$$
$$c = 4.8356892.$$

We did not round c to hundredths because we need to use the value of c again. Next, we use the law of sines to determine α:

$$\frac{\sin\alpha}{a} = \frac{\sin\gamma}{c}$$
$$\frac{\sin\alpha}{4} = \frac{\sin 42°}{4.8356892}$$
$$\sin\alpha = 0.5534935$$
$$\alpha = 33.61°.$$

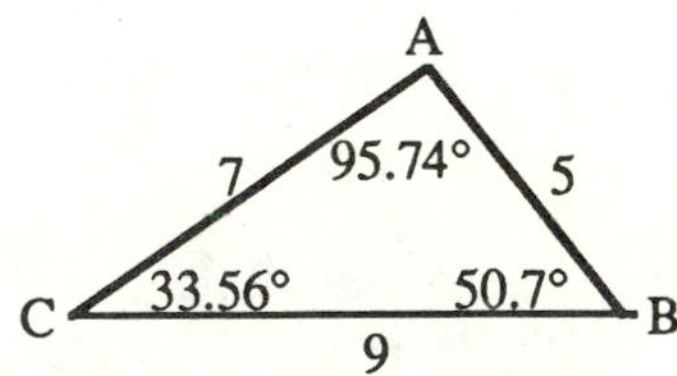

Figure 6.6.4

Thus, $\alpha = 33.61°$, $\beta = 104.39°$, $c = 4.84$, and we have determined the triangle in Figure 6.6.5. •

Now we have covered all possibilities if three of the six parts of a triangle are specified. We have the following summary.

SUMMARY Let three of the six parts of a triangle be given.

Parts Given	Number of Possible Triangles
Three angles (sum equals 180°)	Infinitely many, they are all similar
Two angles (sum less than 180°), one side	One
One angle, two sides	Zero, one, or two
Three sides (sum of any two greater than the third)	One

Area of Triangles

• THEOREM 2 Let ABC be a triangle labeled in the usual way (Figure 6.6.6). Then, the area $\mathcal{A}$ of the triangle is given by

$$\mathcal{A} = \frac{1}{2}bc\sin\alpha = \frac{1}{2}ac\sin\beta = \frac{1}{2}ab\sin\gamma.$$

PROOF First, we put the angle α of triangle ABC in standard position (Figure 6.6.6). We have drawn angle α as if it is acute. The same proof works if α is obtuse. Let h be an altitude of the triangle ABC. Then, the area of the triangle is given by $\mathcal{A} = \frac{1}{2}ch$. Notice that $\sin\alpha = \frac{h}{b}$ so that $h = b\sin\alpha$. Substituting this

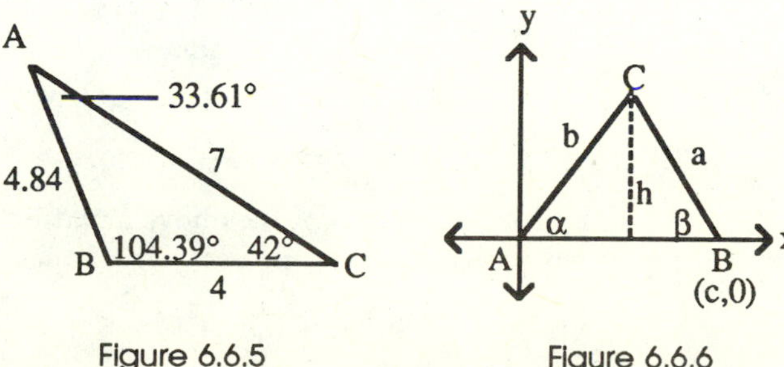

Figure 6.6.5 Figure 6.6.6

value for h into the area formula gives

$$\begin{aligned} \mathcal{A} &= \frac{1}{2}ch \\ &= \frac{1}{2}cb\sin\alpha \\ &= \frac{1}{2}bc\sin\alpha. \end{aligned}$$

If we interchange b with a and β with α we obtain the second form for A. The third form follows in a similar way.

• EXAMPLE 3: Find the area of the triangle ABC if $a = 8$, $b = 5$, and $\gamma = 52°$.

SOLUTION: From Theorem 2 we know that the area is given by the following formula:

$$\begin{aligned} \mathcal{A} &= \frac{1}{2}ab\sin\gamma \\ &= \frac{1}{2}(8)(5)\sin 52° \\ &= 15.76. \end{aligned}$$

Thus, the area is 15.76 square units. •

The next theorem gives the area of a triangle in terms of the measure of its three sides.

• THEOREM 3 **(Heron's Formula.)** Let ABC be a triangle labeled in the usual way. Then, the area $\mathcal{A}$ of the triangle is given by

$$\mathcal{A} = \sqrt{s(s-a)(s-b)(s-c)},$$

where $s = \frac{1}{2}(a+b+c)$ is one-half of the perimeter.

PROOF From Theorem 2 we have $\mathcal{A} = \frac{1}{2}ab\sin\gamma$. We proceed as follows:

$$\begin{aligned} \mathcal{A} &= \frac{1}{2}ab\sin\gamma \\ 2\mathcal{A} &= ab\sin\gamma \\ 4\mathcal{A}^2 &= a^2b^2\sin^2\gamma \\ &= a^2b^2(1-\cos^2\gamma) \\ &= a^2b^2 - a^2b^2\cos^2\gamma. \end{aligned}$$

Next, we multiply each side of the above equation by 4 and then use the law of cosines to replace $2ab\cos\gamma$ by $a^2+b^2-c^2$:

$$\begin{aligned} 4\mathcal{A}^2 &= a^2b^2 - a^2b^2\cos^2\gamma \\ 16\mathcal{A}^2 &= 4a^2b^2 - (2ab\cos\gamma)^2 \\ &= 4a^2b^2 - (a^2+b^2-c^2)^2. \end{aligned}$$

It can be verified that the right-hand side of the above equation is the same as $(a+b+c)(a+b-c)(b+c-a)(c+a-b)$. Thus,

$$16A^2 = (a+b+c)(a+b-c)(b+c-a)(c+a-b)\,.$$

Notice that $a+b+c = 2s$, $a+b-c = 2s-2c$, $b+c-a = 2s-2a$, and $a+c-b = 2s-2b$. Substituting these values we obtain

$$16A^2 = 2s(2s-2c)(2s-2a)(2s-2b)$$
$$16A^2 = 16s(s-c)(s-a)(s-b)$$
$$A = \sqrt{s(s-a)(s-b)(s-c)}.$$

• EXAMPLE 4: Find the area of the triangle ABC if $a = 9$, $b = 7$, and $c = 5$.

SOLUTION: First, $s = \frac{1}{2}(a+b+c) = 10.5$. Then, we compute the area using Heron's formula:

$$\begin{aligned} a &= \sqrt{s(s-a)(s-b)(s-c)} \\ &= \sqrt{(10.5)(1.5)(3.5)(5.5)} \\ &= 17.41. \end{aligned}$$

Thus, the area of the triangle is 17.41 square units. •

Next, we consider applications that involve the use of the law of cosines.

• EXAMPLE 5: In order to determine the distance between two points A and B on opposite sides of a building, a surveyor chooses a point C that is 110 feet from A and 160 feet from B (Figure 6.6.7). If the measure of the angle at C is 54°, find the distance between A and B.

SOLUTION: The distance from A to B in the triangle ABC is c. Now, we use the law of cosines:

$$c^2 = a^2 + b^2 - 2ab\cos\gamma$$
$$c^2 = 160^2 + 110^2 - 2(160)(110)\cos 54°$$
$$c = 130.42.$$

Thus, the distance between A and B is 130.42 feet. •

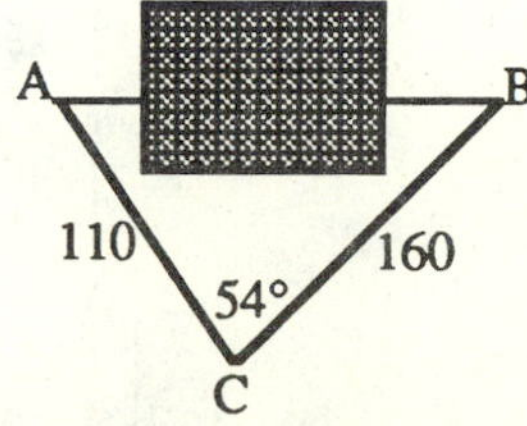

Figure 6.6.7

• EXAMPLE 6: In major league baseball, the four bases form a square with sides of length 90 feet. The front edge of the pitching rubber is 60.5 feet from home plate. Find the distance from the front edge of the pitching rubber to first base.

SOLUTION: Let P be the front edge of the pitching rubber and H be home plate (Figure 6.6.8). Let c denote the distance from the front edge of the pitching rubber to first base. The measure of angle γ is 45° because the angle at each base is 90° and a diagonal of a square bisects the angle at a vertex.

The law of cosines gives the following for the distance c:

$$c^2 = (60.5)^2 + (90)^2 - 2(60.5)(90)\cos 45^\circ$$
$$c = 63.72.$$

Thus, the distance from the front edge of the pitching rubber to first base is 63.72 feet. •

Angle Between Two Lines

In Section 5.4 we defined the angle between two lines to be the acute angle determined by the lines, or 90° if the two lines are perpendicular. In the next example we use the law of cosines to determine the angle between two lines.

• EXAMPLE 7: Let ℓ_1 and ℓ_2 be two lines with slopes m_1 and m_2, respectively. The angle between ℓ_1 and ℓ_2 is either

$$\alpha = \cos^{-1}\left(\frac{1 + m_1 m_2}{\sqrt{(1+m_1^2)(1+m_2^2)}}\right) \quad \text{or} \quad 180^\circ - \alpha \quad \text{if} \quad \alpha > 90^\circ.$$

SOLUTION: Without loss of generality, we can assume that the line ℓ_2 makes the larger counterclockwise angle with the positive x-axis. With this agreement, Figure 6.6.9 illustrates the three possibilities for ℓ_1 and ℓ_2 corresponding to m_1 and m_2 positive, m_2 negative and m_1 positive, or m_1 and m_2 negative. In each case, let P and Q be the intersection of the vertical line $x = 1$ with ℓ_1 and ℓ_2, respectively. Notice that the coordinates of P are $(1, m_1)$ and the coordinates of Q are $(1, m_2)$. (Why?)

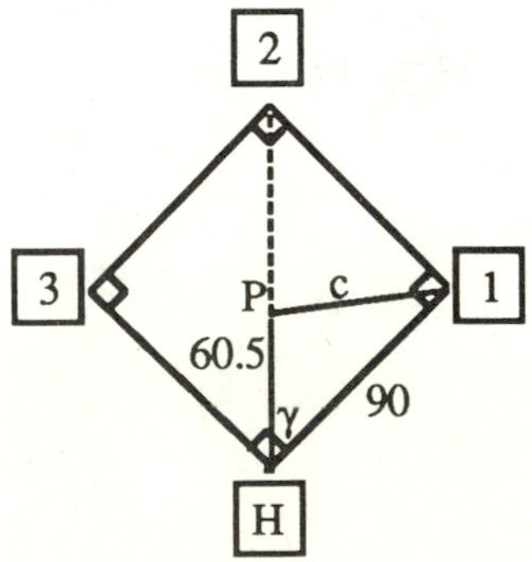

Figure 6.6.8

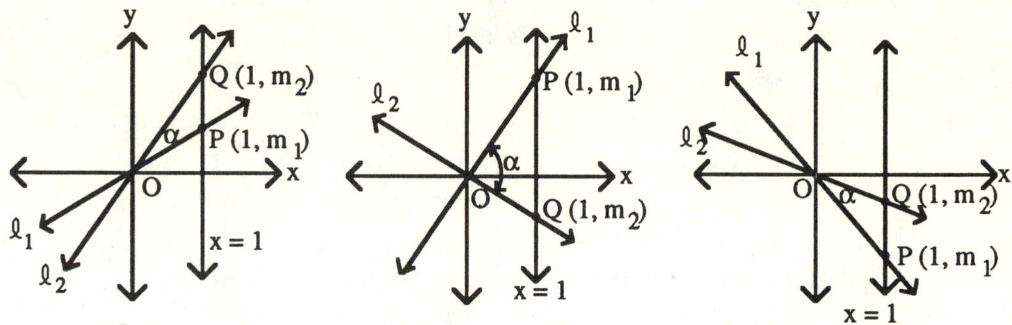

Figure 6.6.9

Let α be the angle of the triangle OPQ between OP and OQ. Either α or $180° - \alpha$ is acute. Thus, either α or $180° - \alpha$ is the angle between ℓ_1 and ℓ_2.

Now, $d(P,Q) = |m_1 - m_2|$, $d(O,P) = \sqrt{1+m_1^2}$, and $d(O,Q) = \sqrt{1+m_2^2}$. (Why?) We use the law of cosines to determine α:

$$d^2(P,Q) = d^2(O,P) + d^2(O,Q) - 2d(O,P)d(O,Q)\cos\alpha$$

$$|m_1 - m_2|^2 = (\sqrt{1+m_1^2})^2 + (\sqrt{1+m_2^2})^2 - 2(\sqrt{1+m_1^2})(\sqrt{1+m_2^2})\cos\alpha$$

$$m_1^2 - 2m_1m_2 + m_2^2 = 1 + m_1^2 + 1 + m_2^2 - 2(\sqrt{(1+m_1^2)(1+m_2^2)})\cos\alpha$$

$$\cos\alpha = \frac{1+m_1m_2}{\sqrt{(1+m_1^2)(1+m_2^2)}}$$

$$\alpha = \cos^{-1}\left(\frac{1+m_1m_2}{\sqrt{(1+m_1^2)(1+m_2^2)}}\right).$$

If $\cos\alpha > 0$, then α is acute. If $\cos\alpha < 0$, then $90° < \alpha < 180°$ and $180° - \alpha$ is acute. If $\cos\alpha = 0$, then $\alpha = 90°$ and ℓ_1 is perpendicular to ℓ_2. •

• EXERCISES 6-6

Solve the triangle ABC.

1. $a = 1, b = 5, c = 4$
2. $a = 1, b = 5, c = 8$
3. $a = 3.2, b = 7.6, c = 6.4$
4. $\alpha = 21°, \beta = 17°, c = 15$
5. $\alpha = 55°, b = 12, c = 7$
6. $\beta = 125°, a = 25, c = 41$
7. $\beta = 103°, b = 13, a = 18$
8. $\alpha = 36°, b = 17, a = 14$
9. $\beta = 110°, b = 13, c = 15$
10. $a = 5, b = 7, c = 6$
11. If $a = 8$ and $\beta = 58°$, determine the values of b for which α has

    (a) two values. (b) one value. (c) no value.

12. If $a = 12$ and $\beta = 32°$, determine the values of b for which α has
    (a) two values. (b) one value. (c) no value.

Find the area of the triangle ABC.

13. If $b = 6$, $c = 8$, and $\alpha = 47°$
14. If $a = 17$, $c = 14$, and $\beta = 103°$
15. If $\alpha = 15°$, $\beta = 65°$, and $a = 8$
16. If $\alpha = 10°$, $\gamma = 110°$, and $c = 12.3$
17. If $a = 2$, $b = 6$, and $c = 7$
18. If $a = 20$, $b = 36$, and $c = 50$
19. In order to determine the distance between two points A and B on opposite sides of a lake, a surveyor chooses a point C that is 860 feet from A and 175 feet from B (Figure 6.6.10). If the measure of the angle at C is 78°, find the distance between A and B.
20. Consider the ball diamond problem of Example 6 (Figure 6.6.8). Determine
    (a) the distance from the front edge of the pitching rubber to third base.
    (b) the distance between first and third bases.
    (c) the angle between home plate, the front edge of the pitching rubber, and first base.
    (d) the angle between the front edge of the pitching rubber, first base and home plate.
21. In women's softball, the four bases form a square with sides of length 60 feet. The front edge of the pitching rubber is 40 feet from home plate. Find the distance from the front edge of the pitching rubber to first base.
22. Find the radian measure of the largest angle of the triangle with sides of length 4, 5, and 6.
23. The sides of a parallelogram are of lengths 18 feet and 26 feet, and one angle is 39°. Find the length of the longer diagonal.
24. Two observers are 600 feet apart on opposite sides of a flag pole. The angles of elevation from the observers to the top of the pole are 19° and 21°. Find the height of the flag pole.
25. A blimp is sighted simultaneously by two observers: A at the top of a 650 foot tower and B at the base of the tower. Find the distance of the blimp from observer A if the angle of elevation (from the horizontal) as viewed by A is 32° and the angle of elevation as viewed by B is 56°. How high is the blimp?
26. In a parallelogram, two adjacent sides meet at an angle of 35° and are 3 feet and 8 feet in length. What is the length of the shorter diagonal of the parallelogram?

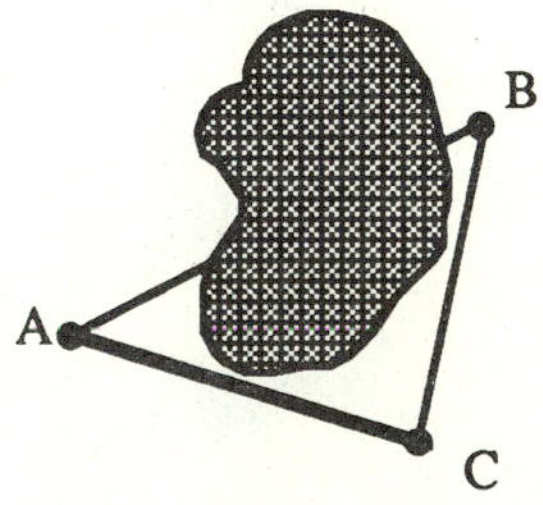

Figure 6.6.10

27. Two observers are 400 feet apart on opposite sides of a tree. If the angles of elevation from the observers to the top of the tree are $15°$ and $20°$, how tall is the tree?

• KEY TERMS

Listed below are the key terms, vocabulary, and concepts in this chapter. You should be familiar with their definitions and meanings as well as be able to illustrate each of them.

- Addition formulas
- Amplitude
- Arccos
- Arcsin
- Arctan
- Area of triangles
- Circular properties of trig functions
- Coterminal
- Double-angle formulas
- End behavior
- Half-angle formulas
- Heron's formula
- Horizontal shifting
- Horizontal shrinking
- Identities for inverse trig functions
- Initial side
- Inverse trigonometric functions
- Law of cosines
- Law of sines
- Oblique triangle
- Periodic functions
- Phase shift
- Reference angle
- Sinusoid
- Solving oblique triangles
- Standard position
- Supplementary angle relationships
- Terminal sides
- Trigonometric equations
- Trigonometric functions
- Trigonometric identities
- Trigonometric inequalities
- Unit circle
- Zeros of periodic functions

• REVIEW EXERCISES

This section contains representative problems from each of the previous six sections. You should use these problems to test your understanding of the material covered in this chapter.

Solve the triangle in Figure R.1.

1. $\alpha = 79°$, $\beta = 33°$, and $a = 7$

2. $a = 5$, $b = 8$, and $\beta = 110°$

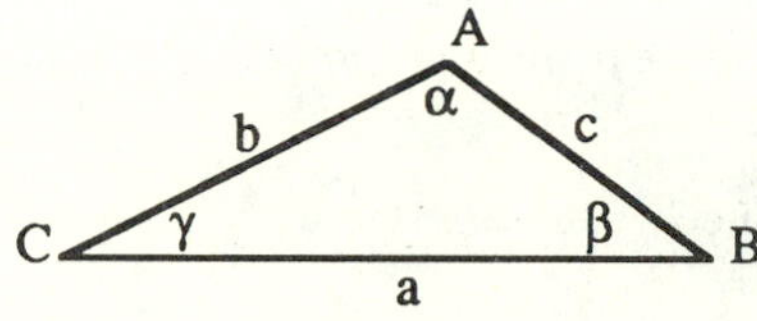

Figure R.1

3. $a = 14.7$, $\alpha = 29.3°$, and $\gamma = 37.2°$
4. $a = 8$, $b = 3$, and $\beta = 30°$
5. $\alpha = 34°$, $\beta = 74°$, and $c = 5$
6. $c = 41$, $\alpha = 22.9°$, and $\gamma = 55.1°$
7. $a = 5$, $b = 7$, and $c = 6$
8. $\alpha = 85°$, $a = 6$, and $b = 4$
9. If $a = 12$ and $\beta = 28°$, determine the values of b for which α has (a) two values, (b) one value, and (c) no value.

Find the area of the triangle ABC.

10. $a = 3$, $b = 5$, and $c = 6$
11. $a = 10$, $b = 6$, and $\gamma = 50°$

Write as a sinusoid $y = A\sin(x + \alpha)$ and confirm by graphing.

12. $y = 2\sin x - 5\cos x$
13. $y = -4\sin x + 3\cos x$

Verify the following identities by graphing each side of the equation and comparing the graphs. Then verify the identity algebraically.

14. $\tan^2 x - \sin^2 x = \sin^2 x \tan^2 x$
15. $2\sin\theta\cos^3\theta + 2\sin^3\theta\cos\theta = \sin 2\theta$

Draw a complete graph. Determine the domain and range.

16. $y = 3x\cos x + \sin 2x$
17. $y = 2\sin 3x - 5\cos 2x$
18. $y = \dfrac{2x - 1}{\tan x}$
19. $f(x) = x^2 \sin x$

Draw a complete graph and determine the end behavior of the function.

20. $y = \dfrac{1}{x}\sin(\dfrac{1}{x})$
21. $y = x^3 \cos x$

Investigate graphically what happens to the values of $f(x)$ as $x \to 0$.

22. $f(x) = \dfrac{1}{x}\sin(\dfrac{1}{x})$
23. $f(x) = \dfrac{\sin x}{x^2}$
24. Determine the domain, range, period, and draw a complete graph of the function $h(x) = 2\sin x - \cos 2x$.
25. Determine the zeros in the interval $[0, 2\pi]$ of the function in Exercise 24.
26. Determine all local and absolute maximum and minimum values in the interval $[0, 2\pi]$ of the function in Exercise 24.
27. Use an algebraic method to determine all real solutions of $0.25 - \cos^2 x = 0$ in the interval $[0, \pi]$. Check your answer with a graphing utility.

Use algebraic means to verify the following identities. Confirm your analysis by graphing.

28. $\csc x - \cos x \cot x = \sin x$
29. $\dfrac{\tan\theta + \sin\theta}{2\tan\theta} = \cos^2\dfrac{\theta}{2}$
30. $\dfrac{1 + \tan\theta}{1 - \tan\theta} + \dfrac{1 + \cot\theta}{1 - \cot\theta} = 0$
31. $\sin 3\theta = 3\cos^2\theta\sin\theta - \sin^3\theta$

Use a graphing utility to investigate if the equation is an identity. If not, determine a *counterexample*.

32. $\sec x - \sin x \tan x = \cos x$
33. $(\sin^2\alpha - \cos^2\alpha)(\tan^2\alpha + 1) = \tan^2\alpha - 1$

Determine the *exact* value. Check your answer with a calculator.

34. $\cos\dfrac{\pi}{8}$
35. $\sin\dfrac{7\pi}{12}$

If $\sin x = \frac{2}{3}$ and $\frac{\pi}{2} < x < \pi$, then determine the trig value.

36. $\sin \dfrac{x}{2}$

37. $\tan x$

38. Find the domain and range of the function $g(x) = 5\cos^2 2x$, and draw a complete graph.

39. Write $\sin 3\theta + \cos 3\theta$ as an expression involving only $\sin\theta$ and $\cos\theta$.

Solve the equation for x in the interval $[0, 2\pi]$. Find exact answers when possible. Check your answer using a graphing utility.

40. $2\cos^2 x + \cos x - 1 = 0$

41. $\sin x = \dfrac{\sqrt{2}}{2}$

42. Use a calculator to evaluate $\sin^{-1}(0.766)$. Express your answer in degrees.

43. Use a calculator to evaluate $\sin^{-1}(0.479)$. Express your answer as a real number.

Compute the exact value without using a calculator.

44. $\sin^{-1}(\dfrac{\sqrt{3}}{2})$

45. $\tan^{-1} 1$

Use a calculator to evaluate. Express your answer as a real number.

46. $\sin(\sin^{-1}(0.25))$

47. $\cos^{-1}(\tan(0.2))$

48. $\tan(\sin^{-1} 2)$

49. $\sin^{-1}(\sin(\sqrt{3}\cos 3))$

50. Compute the exact value of $\tan(\sin^{-1} 0.5)$ without using a calculator.

Draw a complete graph. Do not use a graphing utility. Determine the domain and range. Check with a graphing utility.

51. $y = \arccos x - 1$

52. $f(x) = (\sin^{-1} x)^2$

53. Solve $2\sin^{-1} x = \sqrt{2}$ for x. Find the exact solution(s).

54. Find an equivalent algebraic expression *not involving trig functions* for $\sin(\cos^{-1} x)$.

55. Solve $\cos x = -0.707$ for x. Find *all* solutions.

56. Solve the inequality $\sin 2x/[\cos(x/2)] \geq 1$ in $[-2\pi, 2\pi]$.

Solve the equation algebraically in the interval $[0, 2\pi]$. Check with a graphing utility.

57. $2\sin 2x = 1$

58. $2\cos x = 1$

Solve the equation algebraically. Find *all* real solutions. Check with a graphing utility.

59. $\sin 3x = \sin x$

60. $\sin^2 x - 2\sin x - 3 = 0$

61. $\cos 2t = \cos t$

62. $\sin(\cos x) = 1$

Solve the inequality algebraically. Check with a graphing utility.

63. $2\cos 2x > 1$ for $0 \leq x \leq 2\pi$

64. $\sin 2x > 2\cos x$ for $0 \leq x \leq 2\pi$

65. Suppose $\cos x = 0.2$ with $0 \leq x \leq \frac{\pi}{2}$. Compute $\sin(2x)$ exactly.

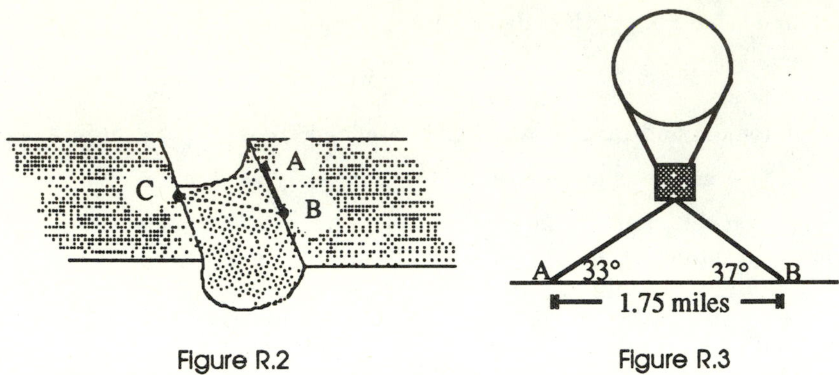

Figure R.2 Figure R.3

66. Draw the graph of $g(x) = 5 + 2x\sin(2x + \pi)$ for $-2\pi \leq x \leq 2\pi$. Determine all local maximum and minimum values in the interval.

67. Two markers A and B on the same side of a canyon rim are 80 feet apart (Figure R.2). A hiker is located across the rim at point C. A surveyor determines that $\angle BAC = 70°$ and $\angle ABC = 65°$.

    (a) What is the distance between the hiker and point A?

    (b) What is the distance between the two canyon rims? (Assume they are parallel.)

68. A hot air balloon is seen over Hanover, Indiana simultaneously by two observers at points A and B. Assume the balloon and the observers are in the same vertical plane and the angles of elevation and the distances are as shown in Figure R.3. How high above ground is the balloon?

69. In order to determine the distance between two points A and B on opposite sides of a lake, a surveyor chooses a point C that is 900 feet from A and 225 feet from B (Figure R.4). If the measure of the angle at C is 70°, find the distance between A and B.

70. Find the radian measure of the largest angle of the triangle with sides of length 9, 8, and 10.

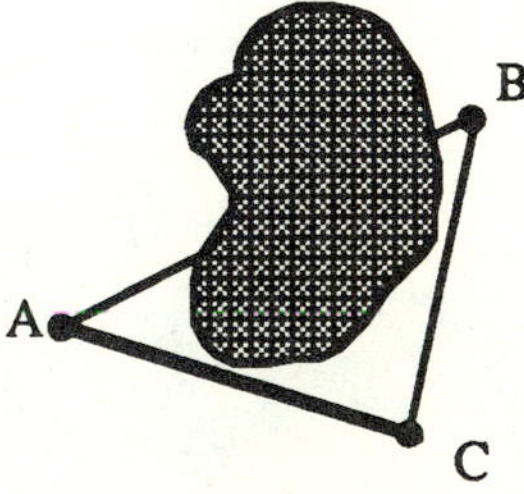

Figure R.4

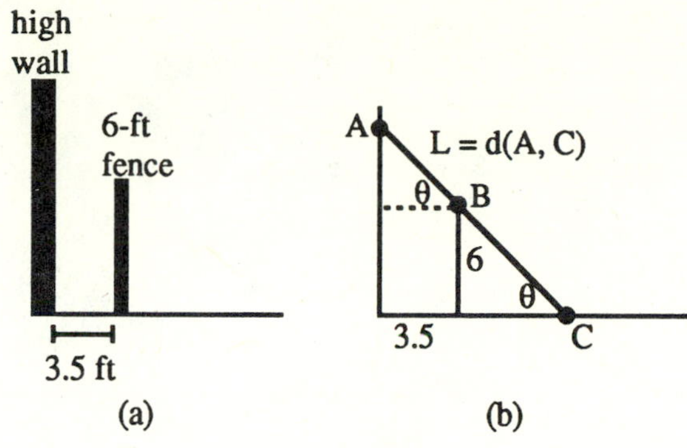

Figure R.5

71. A 6-foot fence stands 3.5 feet from a high wall (Figure R.5). Let L be the length of a ladder extending from some point C on the ground to some point A on the high wall. The shortest such ladder must just touch the fence. A *shorter* ladder would reach from C to a point slightly below A on the wall. Let θ be the angle of inclination shown in Figure R.5.
    (a) Determine a trigonometric representation of the length of such a ladder (as a function of θ).
    (b) Draw a complete graph of the trigonometric representation.
    (c) What values of θ make sense in the problem situation?
    (d) Find the angle of inclination of the shortest such ladder.

72. Sarah is in a rowboat at a point P 1.5 miles from a straight shore. Her home is on the shore at a point C that is 6 miles from the point on the shore (A) closest to Sarah. Assume that she can walk twice as fast as she can row and that she walks 3 miles per hour. Assume she rows to a point B and then walks to her home at C. Let θ be the angle shown in Figure R.6.
    (a) Determine a trigonometric representation of the time it takes Sarah to reach home.
    (b) Draw a complete graph of the trigonometric representation.
    (c) What values of θ make sense in this problem situation?
    (d) What is her quickest way home?

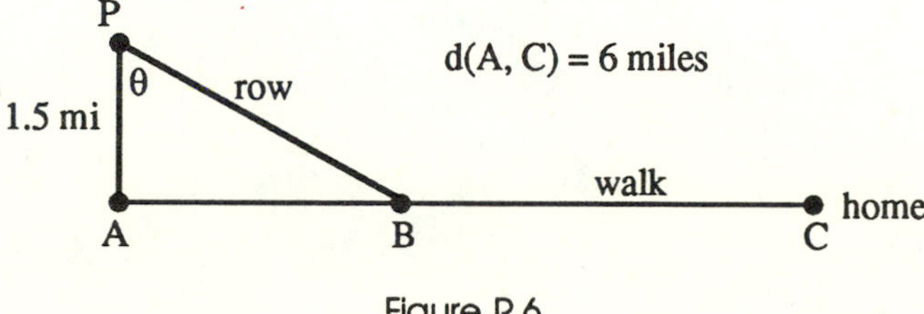

Figure R.6

7

Trigonometry of Complex Numbers, Vectors, Polar Equations, and Parametric Equations

• Introduction

In this chapter we represent complex numbers geometrically and write them in trigonometric form. This form is used to do the arithmetic of complex numbers. We state De Moivre's theorem for complex numbers and use it to determine integer powers and roots of complex numbers.

Polar coordinates are introduced and their relationship with rectangular coordinates investigated. Complete graphs of equations involving the polar coordinates r and θ are obtained.

The study of vectors is started in this chapter. We use both geometric methods and components of vectors to do the arithmetic of vectors. Applications that involve vectors are studied. Vectors are used to describe the path of objects moving in a plane.

We define curves in the plane using parametric equations. Parametric equations are used to obtain graphs of relations and to simulate motion. We use both a polar grapher and a parametric grapher to obtain graphs in this chapter. Many graphing utilities are capable of drawing graphs of polar or parametric equations. There are relatively simple programs that allow graphing calculators to obtain graphs of polar and parametric equations. See the *Graphing Calculator and Computer Graphing Laboratory Manual* for details.

7.1 Trigonometry and the Geometry of Complex Numbers

In this section we represent complex numbers geometrically and obtain trigonometric representations for complex numbers. Trigonometric representations are used to do the arithmetic of complex numbers.

In Section 3.4 we introduced the *complex numbers*, which consist of all numbers of the form $a + bi$, where a and b are real numbers. We gave algebraic rules to perform the arithmetic of complex numbers. Now we give a geometric representation for a complex number and investigate another way to do the arithmetic of complex numbers.

Geometric Representation of Complex Numbers

There is a one-to-one correspondence between the set of all complex numbers and the points in the coordinate plane. The complex number $a+bi$ corresponds to the point $P(a, b)$ of the coordinate plane (Figure 7.1.1). When the coordinate plane is viewed as a representation of the complex numbers, it is called the **complex plane**, or **Gaussian plane**, after the famous mathematician K. F. Gauss. The x-axis of the complex plane is called the **real axis,** and the y-axis is called the **imaginary axis**. Figure 7.1.2 illustrates the correspondence for the complex numbers $2 + 3i$, $2 - 3i$, and $-3 - 2i$. We also say that Figure 7.1.2 gives the graph of the complex numbers $2 + 3i$, $2 - 3i$, and $-3 - 2i$.

The distance between the complex number $a + bi$ and the origin O is $\sqrt{a^2 + b^2}$ (Figure 7.1.1). We call $\sqrt{a^2 + b^2}$ the **absolute value** or **modulus** of the complex number $a + bi$. In general, the absolute value of a complex number z is denoted by $|z|$. If $z = a + bi$ is real ($b = 0$), then $|z| = \sqrt{a^2} = |a|$.

In Section 3.4, we gave the following definition for addition of complex numbers:

$$(a + bi) + (c + di) = (a + c) + (b + d)i.$$

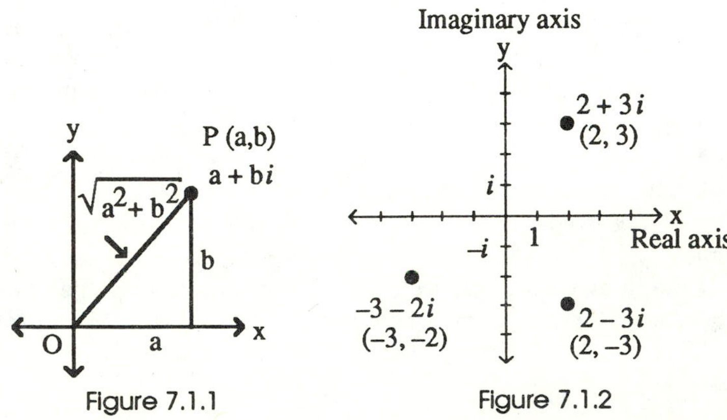

Figure 7.1.1

Figure 7.1.2

We give a geometric interpretation of addition of complex numbers. Let $P = (a, b)$, $Q = (c, d)$, and $R = (a + c,\ b + d)$, the points corresponding to the complex numbers $a + bi$, $c + di$, and $(a + c) + (b + d)i$, respectively. We represent the complex numbers $a + bi$ and $c + di$ by the line segments OP and OQ, respectively (Figure 7.1.3). It can be shown that the line segment OR is a diagonal of the parallelogram determined by the line segments OP and OQ.

Notice in Figure 7.1.3 that PR is parallel to OQ and has the same length as OQ. Thus, the point R can be determined by drawing a line segment from P to R that is parallel to OQ and has the same length as OQ as shown in Figure 7.1.4.

• EXAMPLE 1: Use the line segment representation of complex numbers to determine the following complex numbers.

(a) $(2 + i) + (1 + 3i)$ (b) $(3 + 5i) - (2 + i)$

SOLUTION:

(a) Let $P = (2, 1)$ and $Q = (1, 3)$ be the points corresponding to the complex numbers $2 + i$ and $1 + 3i$, respectively. Draw a line segment parallel to OQ with initial point P and terminal point R (Figure 7.1.5). Notice that the coordinates of R are $(3, 4)$ and that by simple addition $(2 + i) + (1 + 3i) = 3 + 4i$.

(b) Let $R = (3, 5)$, and $P = (2, 1)$ be the points corresponding to the complex numbers $3 + 5i$ and $2 + i$, respectively. Let $z = (3 + 5i) - (2 + i)$, and let Q be the point in the plane corresponding to z. Then OR must be a diagonal of a parallelogram determined by OP and OQ because $3 + 5i = (2 + i) + z$ (Figure 7.1.6). Furthermore, the line segment PR must be parallel to OQ and must have the same length as OQ. Thus, the coordinates of Q must be $(1, 4)$. Thus, z is the complex number $1 + 4i$ determined by the point $Q = (1, 4)$. Notice that by subtraction $(3 + 5i) - (2 + i) = 1 + 4i$. •

Trigonometric Form for Complex Numbers

Let $P(a, b)$ be the point in the complex plane that corresponds to the complex number $z = a + bi$. Let θ be any angle in standard position with the point P on its terminal side. Let r be the length of the segment OP. It follows that $r = \sqrt{a^2 + b^2}$, $\cos\theta = \frac{a}{r}$, and

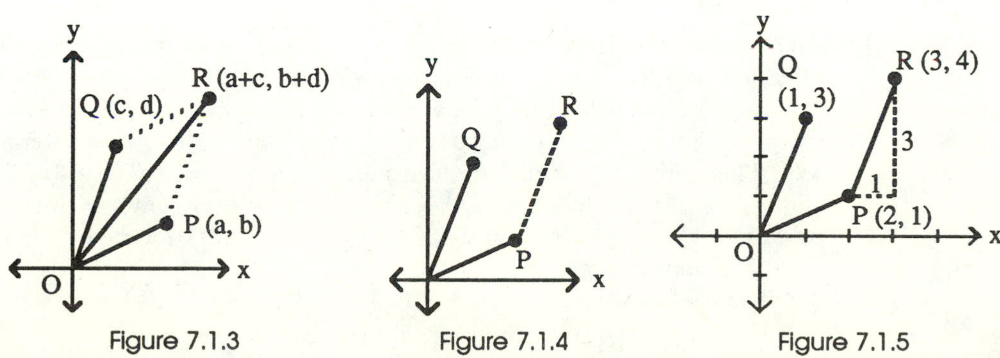

Figure 7.1.3 Figure 7.1.4 Figure 7.1.5

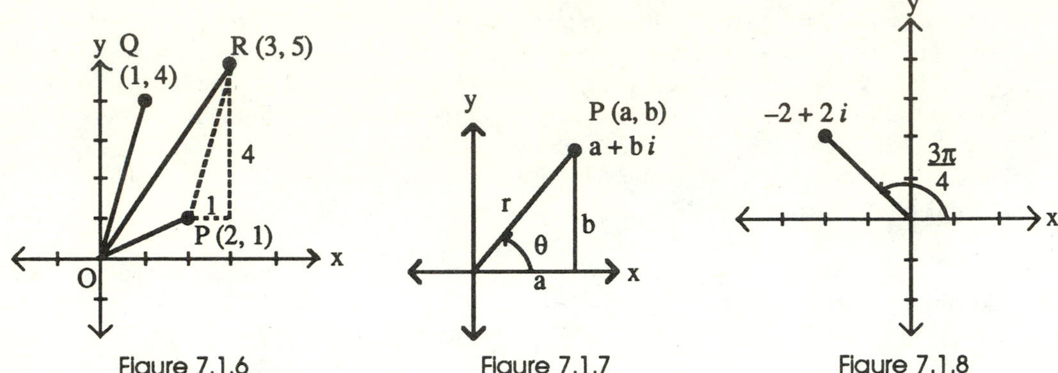

Figure 7.1.6 Figure 7.1.7 Figure 7.1.8

$\sin\theta = \frac{b}{r}$ (Figure 7.1.7). Thus, $a = r\cos\theta$ and $b = r\sin\theta$. We can write $z = a + bi = (r\cos\theta) + (r\sin\theta)i = r(\cos\theta + i\sin\theta)$. Often this form for z is abbreviated to $z = r\,\text{cis}\,\theta$. The angle θ can be given in degrees or radians.

• **DEFINITION** Let $z = a + bi \neq 0$. A trigonometric form for z is given by

$$z = r(\cos\theta + i\sin\theta) = r\,\text{cis}\,\theta$$

where $r = |z| = \sqrt{a^2 + b^2}$, $\cos\theta = \frac{a}{r}$, and $\sin\theta = \frac{b}{r}$. •

The trig form for $z = a + bi$ is *not* unique because there are infinitely many choices for the angle θ. In fact, we can write $z = r(\cos(\theta + 2k\pi) + i\sin(\theta + 2k\pi))$ for any integer k if θ is measured in radians. If θ is measured in degrees then $z = r(\cos(\theta + k\cdot 360°) + i\sin(\theta + k\cdot 360°))$ for any integer k. For the complex number 0, we take $r = 0$ and θ can be any angle.

• **EXAMPLE 2:** Determine the trig form for the following complex numbers with $0 \leq \theta < 2\pi$. Then, give all possible trig forms. Also give the trig form in degrees for (a) and (b).

(a) $-2 + 2i$ (b) $3 - 3\sqrt{3}\,i$ (c) $-3 - 4i$

SOLUTION:

(a) The distance from the origin to $-2 + 2i$ is $\sqrt{(-2)^2 + 2^2} = \sqrt{8} = 2\sqrt{2}$. The angle between 0 and 2π that $-2 + 2i$ makes with the positive x-axis is $3\pi/4$ (Figure 7.1.8). (Why?) Thus, we can write $z = -2+2i$ as $z = 2\sqrt{2}(\cos(3\pi/4) + i\sin(3\pi/4))$. Another trig form for z can be obtained by using any other angle coterminal with $3\pi/4$. Therefore, all possible trig forms are given by $z = 2\sqrt{2}(\cos(3\pi/4 + 2k\pi) + i\sin(3\pi/4 + 2k\pi)) = 2\sqrt{2}\,\text{cis}(3\pi/4 + 2k\pi)$, where k is any integer. In degrees, the trig form for z is given by $z = 2\sqrt{2}(\cos(135° + k\cdot 360°) + i\sin(135° + k\cdot 360°))$, where k is any integer.

(b) The distance from the origin to $3 - 3\sqrt{3}\,i$ is $\sqrt{(3)^2 + (-3\sqrt{3})^2} = \sqrt{36} = 6$. Thus, $3 - 3\sqrt{3}\,i = 6(\frac{1}{2} - i\frac{\sqrt{3}}{2})$. (Why?) Recall that $\cos\frac{\pi}{3} = \frac{1}{2}$ and $\sin\frac{\pi}{3} = \frac{\sqrt{3}}{2}$. Therefore, the angle that $3-3\sqrt{3}\,i$ makes with the positive x-axis is $\frac{5\pi}{3}$ (Figure 7.1.9). Thus, we can write $z = 3-3\sqrt{3}\,i$ as $z = 6\left(\cos\frac{5\pi}{3} + i\sin\frac{5\pi}{3}\right)$. All possible trig forms for z can be obtained by using any other angle coterminal with $\frac{5\pi}{3}$. Therefore, all possible trig forms are given by $z = 6\,\text{cis}\left(\frac{5\pi}{3} + 2k\pi\right)$, where k is any integer. In degrees, the trig form of z is given by $z = 6\,\text{cis}(300° + k \cdot 360°)$, where k is any integer.

(c) The distance from the origin to $-3-4i$ is $\sqrt{(-3)^2+(-4)^2} = \sqrt{25} = 5$. Thus, $-3-4i = 5\left(-\frac{3}{5} - \frac{4}{5}i\right)$. Let θ be the angle $-3-4i$ makes with the positive x-axis (Figure 7.1.10). Then, θ is in the third quadrant and $\tan\theta = \frac{-4}{-3} = \frac{4}{3}$. Thus, $\theta' = \tan^{-1}(\frac{4}{3})$ is the reference angle of θ. We can use a calculator to show that θ' is about 0.93 radians. It follows that $\theta = \pi + 0.93 = 4.07$ radians. (Why?) Thus, $z = 5(\cos 4.07 + i\sin 4.07)$. All trig forms for z are given by $z = 5\,\text{cis}(4.07 + 2k\pi)$, where k is any integer. •

The trig forms for complex numbers can be used to multiply or divide complex numbers.

• **THEOREM 1** Let $z_1 = r_1(\cos\theta_1 + i\sin\theta_1)$ and $z_2 = r_2(\cos\theta_2 + i\sin\theta_2)$. Then,

(a) $z_1z_2 = r_1r_2(\cos(\theta_1+\theta_2) + i\sin(\theta_1+\theta_2)) = r_1r_2\,\text{cis}(\theta_1+\theta_2)$.

(b) $\frac{z_1}{z_2} = \frac{r_1}{r_2}(\cos(\theta_1-\theta_2) + i\sin(\theta_1-\theta_2)) = \frac{r_1}{r_2}\,\text{cis}(\theta_1-\theta_2)$, if $z_2 \neq 0$.

PROOF

(a) We multiply z_1 by z_2 in the usual way:

$$\begin{aligned} z_1z_2 &= [r_1(\cos\theta_1 + i\sin\theta_1)][r_2(\cos\theta_2 + i\sin\theta_2)] \\ &= r_1r_2[(\cos\theta_1\cos\theta_2 - \sin\theta_1\sin\theta_2) + i(\sin\theta_1\cos\theta_2 + \cos\theta_1\sin\theta_2)] \\ &= r_1r_2(\cos(\theta_1+\theta_2) + i\sin(\theta_1+\theta_2)) \\ &= r_1r_2\,\text{cis}(\theta_1+\theta_2). \end{aligned}$$

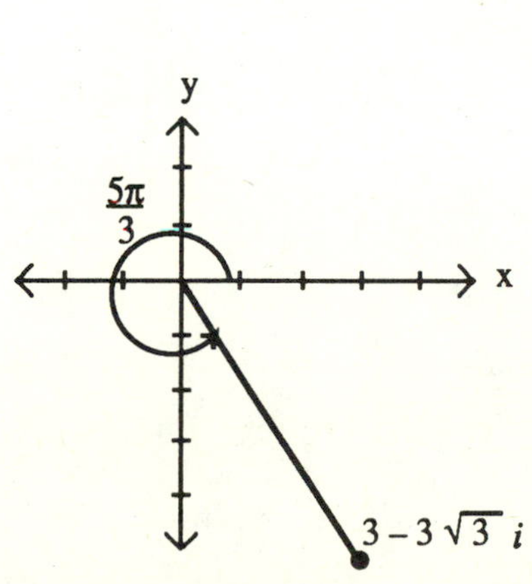

Figure 7.1.9

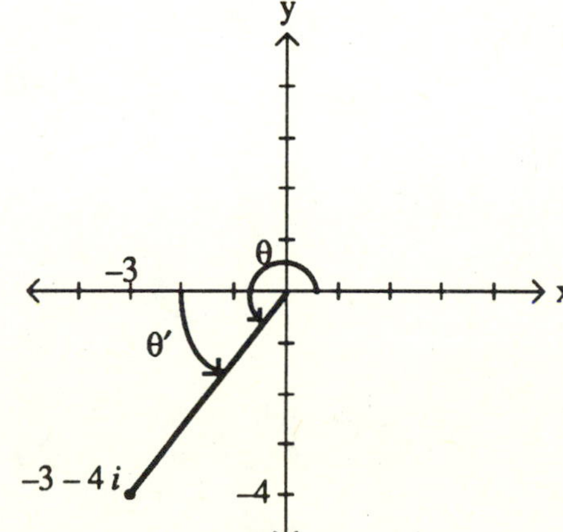

Figure 7.1.10

The last step follows because

$$\cos(\theta_1+\theta_2) = \cos\theta_1\cos\theta_2 - \sin\theta_1\sin\theta_2\,, \quad \text{and}$$
$$\sin(\theta_1+\theta_2) = \sin\theta_1\cos\theta_2 + \cos\theta_1\sin\theta_2.$$

(b) We divide z_1 by z_2 by first multiplying numerator and denominator by the complex conjugate of the portion of the denominator between the parentheses:

$$\begin{aligned}
\frac{z_1}{z_2} &= \frac{r_1(\cos\theta_1 + i\sin\theta_1)}{r_2(\cos\theta_2 + i\sin\theta_2)} \\
&= \frac{r_1}{r_2}\frac{(\cos\theta_1 + i\sin\theta_1)}{(\cos\theta_2 + i\sin\theta_2)}\cdot\frac{(\cos\theta_2 - i\sin\theta_2)}{(\cos\theta_2 - i\sin\theta_2)} \\
&= \frac{r_1}{r_2}\frac{(\cos\theta_1\cos\theta_2 + \sin\theta_1\sin\theta_2) + i(\sin\theta_1\cos\theta_2 - \cos\theta_1\sin\theta_2)}{(\cos^2\theta_2 + \sin^2\theta_2)}. \\
&= \frac{r_1}{r_2}(\cos(\theta_1-\theta_2) + i\sin(\theta_1-\theta_2)) \\
&= \frac{r_1}{r_2}\operatorname{cis}(\theta_1-\theta_2)
\end{aligned}$$

• **EXAMPLE 3:** Use the trig forms to determine z_1z_2 and $\frac{z_1}{z_2}$ if $z_1 = -2+2i$ and $z_2 = 3-i3\sqrt{3}$.

SOLUTION: In Example 2 we found that $-2+2i = 2\sqrt{2}(\cos\frac{3\pi}{4} + i\sin\frac{3\pi}{4})$ and $3-i3\sqrt{3} = 6(\cos\frac{5\pi}{3} + i\sin\frac{5\pi}{3})$.

We apply Theorem 1 to compute z_1z_2:

$$\begin{aligned}
z_1z_2 &= (-2+2i)(3-i3\sqrt{3}) \\
&= \left[2\sqrt{2}\left(\cos\frac{3\pi}{4} + i\sin\frac{3\pi}{4}\right)\right]\left[6\left(\cos\frac{5\pi}{3} + i\sin\frac{5\pi}{3}\right)\right] \\
&= 12\sqrt{2}\left[\cos\left(\frac{3\pi}{4}+\frac{5\pi}{3}\right) + i\sin\left(\frac{3\pi}{4}+\frac{5\pi}{3}\right)\right] \\
&= 12\sqrt{2}\left(\cos\frac{29\pi}{12} + i\sin\frac{29\pi}{12}\right) \\
&= 12\sqrt{2}\left(\cos\frac{5\pi}{12} + i\sin\frac{5\pi}{12}\right).
\end{aligned}$$

The last step follows because $\frac{29\pi}{12} = \frac{5\pi}{12} + 2\pi$. If we multiply z_1 and z_2 directly we obtain

$$(-2+2i)(3-i3\sqrt{3}) = (-6+6\sqrt{3}) + (6+6\sqrt{3})i.$$

A calculator can be used to verify that $-6+6\sqrt{3} = 12\sqrt{2}\cos\frac{5\pi}{12}$ and $6+6\sqrt{3} = 12\sqrt{2}\sin\frac{5\pi}{12}$.

We apply Theorem 1 to compute $\frac{z_1}{z_2}$:

$$\begin{aligned}\frac{z_1}{z_2} &= \frac{2\sqrt{2}\left(\cos\frac{3\pi}{4} + i\sin\frac{3\pi}{4}\right)}{6\left(\cos\frac{5\pi}{3} + i\sin\frac{5\pi}{3}\right)} \\ &= \frac{\sqrt{2}}{3}\left[\cos\left(\frac{3\pi}{4} - \frac{5\pi}{3}\right) + i\sin\left(\frac{3\pi}{4} - \frac{5\pi}{3}\right)\right] \\ &= \frac{\sqrt{2}}{3}\left[\cos\left(-\frac{11\pi}{12}\right) + i\sin\left(-\frac{11\pi}{12}\right)\right] \\ &= \frac{\sqrt{2}}{3}\left(\cos\frac{11\pi}{12} - i\sin\frac{11\pi}{12}\right).\end{aligned}$$

The last step follows because cosine is an even function and sine is an odd function. If we divide z_1 by z_2 directly we obtain

$$\begin{aligned}\frac{-2+2i}{3-i3\sqrt{3}} &= \frac{-2+2i}{3-i3\sqrt{3}} \cdot \frac{3+i3\sqrt{3}}{3+i3\sqrt{3}} \\ &= \frac{(-6-6\sqrt{3}) + i(6-6\sqrt{3})}{9+27} \\ &= -\frac{1+\sqrt{3}}{6} + i\frac{1-\sqrt{3}}{6}.\end{aligned}$$

Again a calculator can be used to show that $-\frac{1+\sqrt{3}}{6} = \frac{\sqrt{2}}{3}\cos\frac{11\pi}{12}$ and $\frac{1-\sqrt{3}}{6} = -\frac{\sqrt{2}}{3}\sin\frac{11\pi}{12}$. •

• EXERCISES 7-1

Compute. Express the answer in the form $a + bi$ where a and b are real numbers.

1. $(-1+2i)+(3+5i)$
2. $3(2-3i)$
3. $(2-4i)(3+2i)$
4. $(-1+2i)^2$
5. $\dfrac{1+i}{1-i}$
6. $\dfrac{2-3i}{4+5i}$

Graph in the complex plane.

7. 3
8. $3i$
9. $4-4i$
10. $-4+3i$
11. $2(\cos 30° + i\sin 30°)$
12. $3(\cos\frac{\pi}{4} + i\sin\frac{\pi}{4})$

Consider complex numbers z_1 and z_2 as represented in Figure 7.1.11.

13. If $z_1 = a + bi$, determine a, b, and $|z_1|$.
14. If $z_2 = a + bi$, determine a, b, and $|z_2|$.
15. Determine the trig form of z_1.
16. Determine the trig form of z_2.

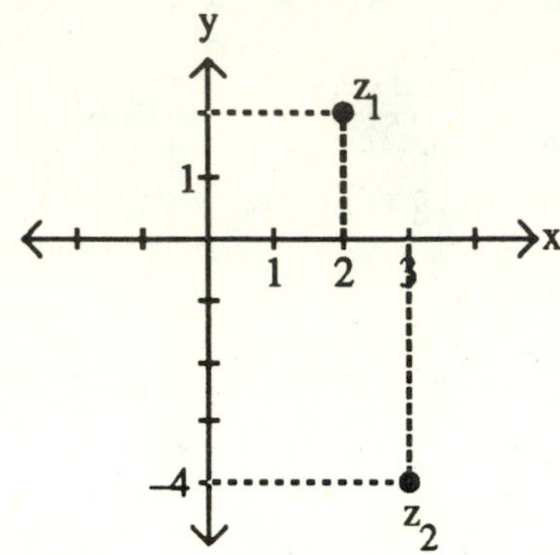

Figure 7.1.11

Let $z_1 = -2 + 3i$, $z_2 = 3 + 4i$, and $z_3 = 2 - 5i$. Perform the indicated computation and write in the form $a + bi$.

17. $z_1 + z_2$

18. $4z_3$

19. $z_1 z_3$

20. $|z_1 - z_2|$

21. $|z_1 z_2|$

22. $\dfrac{z_2}{z_3}$

23. Determine $(1 + 3i) + (4 + 4i)$ using a geometric representation.

24. Determine $(-3 + i) - (4 + 2i)$ using a geometric representation.

Write the complex number in the form $a + bi$.

25. $2(\cos 60° + i \sin 60°)$

26. $5(\cos \frac{3\pi}{4} + i \sin \frac{3\pi}{4})$

27. $3.4(\cos 5 + i \sin 5)$

28. $10(\cos 6\pi + i \sin 6\pi)$

Write in trig form with $0 \le \theta < 2\pi$, then give all possible trig forms.

29. $2 + 3i$

30. $-4 + 4i$

31. $\sqrt{2} - 8i$

32. $-1 - 2i$

33. Prove that $\bar{u} = r \operatorname{cis}(-\theta)$ if $u = r \operatorname{cis} \theta$. (Recall $\bar{u} = a - bi$ if $u = a + bi$ and $\operatorname{cis} \theta = \cos \theta + i \sin \theta$.)

34. Show that $|u \cdot v| = |u||v|$ for complex numbers u and v.

35. Prove that for complex numbers u and v, $u \cdot v = 0$ if and only if $u = 0$ or $v = 0$.

Use Theorem 1 to compute. (Note: $\operatorname{cis} \theta = \cos \theta + i \sin \theta$.)

36. $(2 \operatorname{cis} 30°)(3 \operatorname{cis} 60°)$

37. $\dfrac{2 \operatorname{cis} 30°}{3 \operatorname{cis} 60°}$

38. $(\sqrt{2} \operatorname{cis} 3)(\sqrt{3} \operatorname{cis} 5)$

39. $(2 - 3i)(-3 + 4i)$

40-43. Write the two complex numbers in Exercises 36–39 in the form $a + bi$, perform the operations, and verify the results are the same as that obtained using the trig form and Theorem 1.

44. Use a calculator to confirm that $-6 + 6\sqrt{3} = 12\sqrt{2} \cos \frac{5\pi}{12}$ and $6 + 6\sqrt{3} = 12\sqrt{2} \sin \frac{5\pi}{12}$ (see Example 3).

7.2 De Moivre's Theorem, Roots of Complex Numbers, and Binomial Theorem

In this section we state the Binomial Theorem and a famous theorem of De Moivre. De Moivre's theorem is used to determine roots of complex numbers. We investigate the geometric relationship among the roots of a complex number.

Binomial Theorem

De Moivre's theorem is about raising a complex number to an integer power. Raising a binomial to a power is usually tedious. For example, we can use multiplication to establish the following formulas:

$$(a+b)^2 = a^2 + 2ab + b^2,$$
$$(a+b)^3 = a^3 + 3a^2b + 3ab^2 + b^3,$$
$$(a+b)^4 = a^4 + 4a^3b + 6a^2b^2 + 4ab^3 + b^4.$$

In general, the following theorem is true.

• THEOREM 1 **(The Binomial Theorem).** Let n be a positive integer. Then,

$$(a+b)^n = a^n + na^{n-1}b + \frac{n(n-1)}{1\cdot 2}a^{n-2}b^2 + \frac{n(n-1)(n-2)}{1\cdot 2\cdot 3}a^{n-3}b^3 + \cdots + \frac{n(n-1)\cdots(n-k+1)}{1\cdot 2\cdots k}a^{n-k}b^k + \cdots + nab^{n-1} + b^n.$$

If we use the pattern suggested by the coefficient of general term

$$\frac{n(n-1)\cdots(n-k+1)}{1\cdot 2\cdots k}a^{n-k}b^k$$

of the expansion of $(a+b)^n$ to compute the coefficient of ab^{n-1}, we obtain

$$\frac{n(n-1)\cdots 2}{1\cdot 2\cdots(n-1)}ab^{n-1},$$

which simplifies to nab^{n-1}. (Why?)

• EXAMPLE 1: Use the Binomial Theorem to expand $(a+b)^5$.

SOLUTION: Applying the Binomial Theorem we obtain

$$(a+b)^5 = a^5 + 5a^4b + \frac{5\cdot 4}{1\cdot 2}a^3b^2 + \frac{5\cdot 4\cdot 3}{1\cdot 2\cdot 3}a^2b^3 + \frac{5\cdot 4\cdot 3\cdot 2}{1\cdot 2\cdot 3\cdot 4}ab^4 + b^5.$$

We can simplify the coefficients and obtain the following form:

$$(a+b)^5 = a^5 + 5a^4b + 10a^3b^2 + 10a^2b^3 + 5ab^4 + b^5.$$ •

The binomial coefficients form an interesting triangular array of numbers called **Pascal's triangle**. The first row of the triangle consists of the single coefficient 1 of the expansion of $(a+b)^0$. The second row consists of the two coefficients 1 and 1 of the expansion of $(a+b)^1$. The third row consists of the coefficients 1, 2, and 1 of the expansion of $(a+b)^2 = a^2 + 2ab + b^2$. Continuing in this way we obtain the following triangular array.

$$\begin{array}{ccccccccccccc}
 & & & & & & 1 & & & & & & \\
 & & & & & 1 & & 1 & & & & & \\
 & & & & 1 & & 2 & & 1 & & & & \\
 & & & 1 & & 3 & & 3 & & 1 & & & \\
 & & 1 & & 4 & & 6 & & 4 & & 1 & & \\
 & 1 & & 5 & & 10 & & 10 & & 5 & & 1 & \\
 & \cdot & \cdot & & \cdot & & \cdot & & \cdot & & \cdot & \cdot & \\
\cdot & \cdot & & \cdot & & \cdot & & \cdot & & \cdot & & \cdot & \cdot
\end{array}$$

Notice how each entry in a row, other than the first and the last, is the sum of the two closest entries of the previous row. For example, the entries in the seventh row above are 1, $6 = 1+5$, $15 = 5+10$, $20 = 10+10$, $15 = 10+5$, $6 = 5+1$, and 1.

The Binomial Theorem can be used to expand more complicated expressions.

• EXAMPLE 2: Use the Binomial Theorem to expand $(2x - 3y^2)^4$.

SOLUTION: We use the Binomial Theorem with $a = 2x$, $b = -3y^2$, and $n = 4$:

$$\begin{aligned}
(2x-3y^2)^4 &= (2x)^4 + 4(2x)^3(-3y^2) + \frac{4\cdot 3}{1\cdot 2}(2x)^2(-3y^2)^2 \\
&\quad + \frac{4\cdot 3\cdot 2}{1\cdot 2\cdot 3}(2x)(-3y^2)^3 + (-3y^2)^4 \\
&= 16x^4 - 96x^3y^2 + 216x^2y^4 - 216xy^6 + 81y^8.
\end{aligned}$$

•

De Moivre's Theorem

The Binomial Theorem is fairly complicated. However, the formula for raising a complex number to an integer power is considerably simpler.

• THEOREM 2 **(De Moivre's Theorem).** Let n be any integer. Then,

$$[r(\cos\theta + i\sin\theta)]^n = r^n(\cos n\theta + i\sin n\theta),$$
$$(r\operatorname{cis}\theta)^n = r^n \operatorname{cis}(n\theta).$$

Let a be a complex number and n a positive integer. If z is a complex number satisfying $z^n = a$, then z is called an **nth root of a**. Every complex number a has n different nth roots. De Moivre's theorem can be used to find all the nth roots of any complex number. If z is a complex number satisfying $z^n = 1$, then z is called an *nth root of unity*. We begin by investigating the geometry of roots of unity.

• EXAMPLE 3: Determine the two square roots of unity, give their trig forms in degrees, and graph them.

SOLUTION: The square roots of unity are the two solutions of $z^2 = 1$ (Figure 7.2.1). They are 1 and -1. The trig forms, in degrees, of the two square roots of unity are

$$1 = \cos 0° + i \sin 0°,$$
$$-1 = \cos 180° + i \sin 180°.$$

Notice that 1 and -1 are 180° apart on the unit circle, and that $180° = \frac{360°}{2}$. •

• **EXAMPLE 4:** Determine the four fourth roots of unity, give their trig forms in radians, and graph them.

SOLUTION: The fourth roots of unity are the four solutions of $z^4 = 1$. We can solve this equation algebraically:

$$z^4 = 1$$
$$z^4 - 1 = 0$$
$$(z^2 - 1)(z^2 + 1) = 0$$
$$(z - 1)(z + 1)(z^2 + 1) = 0.$$

Now we can see that the four fourth roots of unity are 1, -1, $-i$, and i (Figure 7.2.2). Next, we give their trig forms in radians:

$$1 = \cos 0 + i \sin 0,$$
$$i = \cos \frac{\pi}{2} + i \sin \frac{\pi}{2},$$
$$-1 = \cos \pi + i \sin \pi,$$
$$-i = \cos \frac{3\pi}{2} + i \sin \frac{3\pi}{2}.$$

Notice that the fourth roots of unity are spaced 90° apart on the unit circle and that $90° = \frac{360°}{4}$. •

Examples 3 and 4 are special cases of the following theorem, which we give without proof.

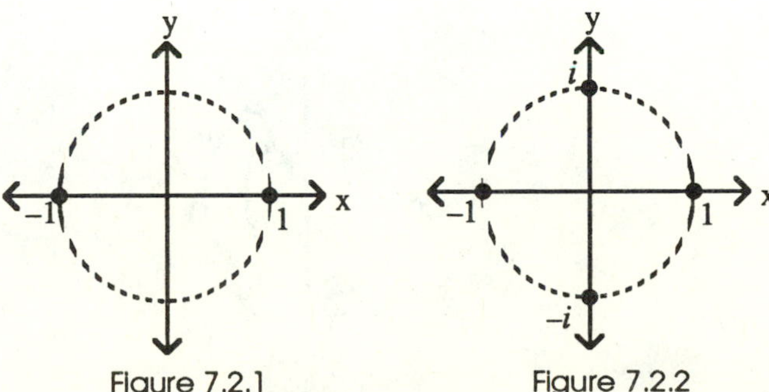

Figure 7.2.1 Figure 7.2.2

• **THEOREM 3** The nth roots of a complex number z are spaced $\left(\frac{360}{n}\right)^\circ$ apart on a circle of radius $\sqrt[n]{|z|}$.

• **EXAMPLE 5:** Determine the cube roots of -1.

SOLUTION: The cube roots of -1 are the three solutions of $z^3 = -1$. Notice that -1 is a cube root of -1 because $(-1)^3 = -1$. If the cube roots of -1 are indeed spaced 120° apart, then we can see from Figure 7.2.3 that the other two cube roots of -1 should be $\cos 60^\circ + i \sin 60^\circ$ and $\cos(-60^\circ) + i\sin(-60^\circ)$. We use De Moivre's theorem to show that these complex numbers are cube roots of -1:

$$\begin{aligned}(\cos 60^\circ + i\sin 60^\circ)^3 &= \cos 180^\circ + i \sin 180^\circ \\ &= -1 + i(0) \\ &= -1, \\ (\cos(-60^\circ) + i\sin(-60^\circ))^3 &= \cos(-180^\circ) + i\sin(-180^\circ) \\ &= -1 + i(0) \\ &= -1.\end{aligned}$$

Alternatively, we can write

$$\begin{aligned}\cos 60^\circ + i \sin 60^\circ &= \frac{1}{2} + i\frac{\sqrt{3}}{2}, \\ \cos(-60^\circ) + i\sin(-60^\circ) &= \frac{1}{2} - i\frac{\sqrt{3}}{2},\end{aligned}$$

and verify directly that $(\frac{1}{2} + i\frac{\sqrt{3}}{2})^3 = -1$ and $(\frac{1}{2} - i\frac{\sqrt{3}}{2})^3 = -1$.

Notice that $z^3 = -1$ is equivalent to $z^3 + 1 = 0$ and that $z^3 + 1 = (z+1)(z^2 - z + 1)$. The complex numbers $\frac{1}{2} + i\frac{\sqrt{3}}{2}$ and $\frac{1}{2} - i\frac{\sqrt{3}}{2}$ are the two solutions of $z^2 - z + 1 = 0$. •

• **EXAMPLE 6:** Use De Moivre's theorem to determine the value of $(-1+i)^{12}$.

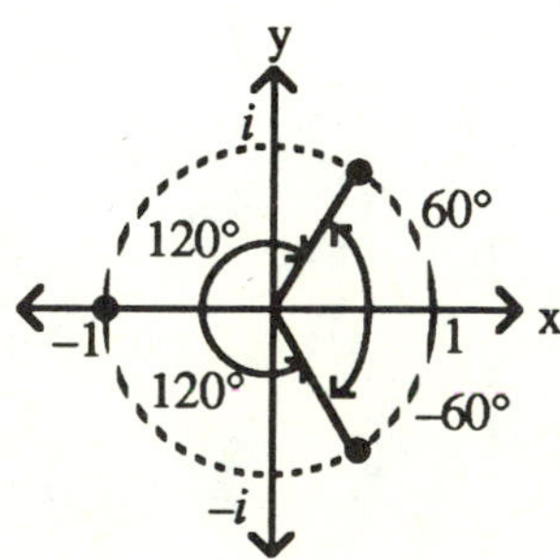

Figure 7.2.3

SOLUTION: In trig form, $-1+i=\sqrt{2}(\cos\frac{3\pi}{4}+i\sin\frac{3\pi}{4})$. Thus, we can compute as follows:

$$
\begin{aligned}
(-1+i)^{12} &= \left[\sqrt{2}\left(\cos\frac{3\pi}{4}+i\sin\frac{3\pi}{4}\right)\right]^{12} \\
&= 2^6\left(\cos 9\pi + i\sin 9\pi\right) \\
&= 64\left(\cos\pi + i\sin\pi\right) \\
&= -64.
\end{aligned}
$$

•

We can conclude from Example 6 that $-1+i$ is a 12th root of -64. In the next example we illustrate a systematic procedure for determining the nth roots of a complex number.

• EXAMPLE 7: Determine the cube roots of -8.

SOLUTION: First, we put -8 in trig form:

$$-8 = 8(\cos\pi + i\sin\pi).$$

If $z=r(\cos\theta+i\sin\theta)$ is a cube root of -8, then $z^3=-8$. We proceed as follows:

$$
\begin{aligned}
z^3 &= -8 \\
[r(\cos\theta+i\sin\theta)]^3 &= 8(\cos\pi+i\sin\pi) \\
r^3(\cos 3\theta + i\sin 3\theta) &= 8(\cos\pi+i\sin\pi).
\end{aligned}
$$

Now, $r^3=8$ and r is real, so $r=2$. Next, 3θ can be any angle coterminal with π. Thus,

$$3\theta = \pi + 2k\pi$$

where k is any integer. It follows that

$$\theta = \frac{\pi}{3} + \frac{2k\pi}{3}.$$

Notice that

$$\theta = \begin{cases} \frac{\pi}{3} & \text{if } k=0 \\ \pi & \text{if } k=1 \\ \frac{5\pi}{3} & \text{if } k=2 \end{cases}.$$

Any other integer value of k produces an angle coterminal with one of $\frac{\pi}{3}$, π, or $\frac{5\pi}{3}$. The three distinct complex numbers $2(\cos\frac{\pi}{3}+i\sin\frac{\pi}{3})$, $2(\cos\pi+i\sin\pi)$, and $2(\cos\frac{5\pi}{3}+i\sin\frac{5\pi}{3})$ are the three cube roots of -8. Notice that

$$2\left(\cos\frac{\pi}{3}+i\sin\frac{\pi}{3}\right) = 2\left(\frac{1}{2}+i\frac{\sqrt{3}}{2}\right) = 1+i\sqrt{3},$$

$$2\left(\cos\pi+i\sin\pi\right) = -2, \quad \text{and}$$

$$2\left(\cos\frac{5\pi}{3}+i\sin\frac{5\pi}{3}\right) = 2\left(\frac{1}{2}-i\frac{\sqrt{3}}{2}\right) = 1-i\sqrt{3}.$$

You should check that the cube of each of these numbers is -8. •

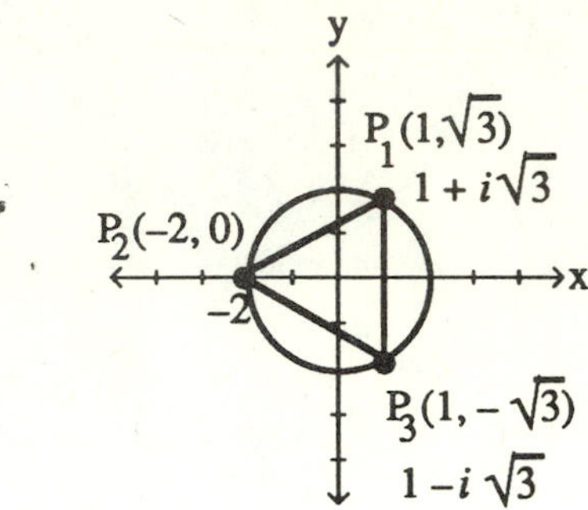

Figure 7.2.4

• EXAMPLE 8: Verify directly that the three cube roots of -8 lie on a circle of radius 2 centered at the origin and that they determine an equilateral triangle.

SOLUTION: We know from the trig form of the three complex numbers given in Example 7 that each has distance 2 from the origin. Thus, $1+i\sqrt{3}$, -2, and $1-i\sqrt{3}$ lie on the circle of radius 2 centered at the origin (Figure 7.2.4). To verify that the three cube roots of -8 form an equilateral triangle, we need to show that the distances between the points $P_1(1,\sqrt{3})$, $P_2(-2,0)$, and $P_3(1,-\sqrt{3})$ are equal.

$$d(P_1,P_2)=\sqrt{3^2+(\sqrt{3})^2}=\sqrt{12}$$

$$d(P_2,P_3)=\sqrt{(-3)^2+(\sqrt{3})^2}=\sqrt{12}$$

$$d(P_3,P_1)=\sqrt{0^2+(-2\sqrt{3})^2}=\sqrt{12}$$

Thus, the points P_1, P_2, and P_3 form an equilateral triangle. •

Examples 7 and 8 can be generalized in the following way.

• THEOREM 4 Let $c=r(\cos\theta+i\sin\theta)$ be any complex number with $r\neq 0$ and n any positive integer. Then, c has n distinct nth roots z_k given by

$$z_k=\sqrt[n]{r}\left(\cos\left(\frac{\theta+2k\pi}{n}\right)+i\left(\sin\frac{\theta+2k\pi}{n}\right)\right),$$

for $k=0,1,2,\ldots,n-1$. The n distinct nth roots lie on the circle of radius $\sqrt[n]{r}$ centered at the origin and determine a *regular* polygon with n equal sides.

If the measure of θ is given in degrees, then we must replace $2k\pi$ by $k\cdot 360°$ in Theorem 4.

• DEFINITION A polygon with all sides equal is said to be **regular**. A polygon with n equal sides and n equal angles is called a **regular n-gon**. •

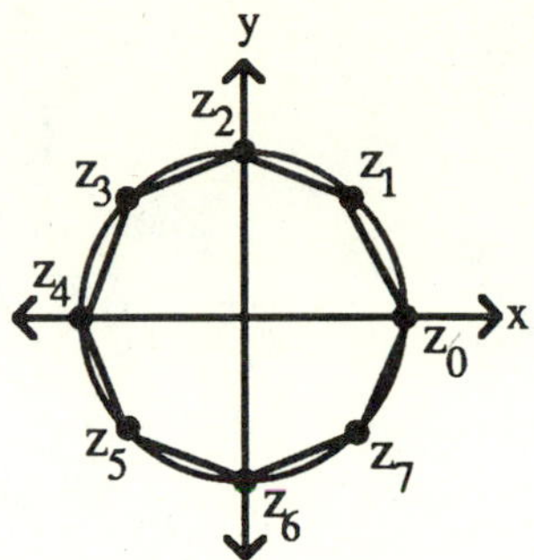

Figure 7.2.5

• EXAMPLE 9: Find the eight 8th roots of unity and draw the regular octagon (8-gon) determined by the roots.

SOLUTION: We want to find the eight 8th roots of the complex number 1. In trig form, we have $1 = \cos 0 + i \sin 0$. By Theorem 4, the eight 8th roots of 1 are given by

$$z_k = \cos \frac{0 + 2k\pi}{8} + i \sin \frac{0 + 2k\pi}{8} = \cos \frac{k\pi}{4} + i \sin \frac{k\pi}{4},$$

for $k = 0, 1, \ldots, 7$. We can write these eight complex numbers in the form $a + bi$ as follows:

$$z_0 = \cos 0 + i \sin 0 = 1,$$

$$z_1 = \cos \frac{\pi}{4} + i \sin \frac{\pi}{4} = \frac{1}{\sqrt{2}} + \frac{1}{\sqrt{2}} i,$$

$$z_2 = \cos \frac{\pi}{2} + i \sin \frac{\pi}{2} = i,$$

$$z_3 = \cos \frac{3\pi}{4} + i \sin \frac{3\pi}{4} = -\frac{1}{\sqrt{2}} + \frac{1}{\sqrt{2}} i,$$

$$z_4 = \cos \pi + i \sin \pi = -1,$$

$$z_5 = \cos \frac{5\pi}{4} + i \sin \frac{5\pi}{4} = -\frac{1}{\sqrt{2}} - \frac{1}{\sqrt{2}} i,$$

$$z_6 = \cos \frac{3\pi}{2} + i \sin \frac{3\pi}{2} = -i,$$

$$z_7 = \cos \frac{7\pi}{4} + i \sin \frac{7\pi}{4} = \frac{1}{\sqrt{2}} - \frac{1}{\sqrt{2}} i.$$

The above eight complex numbers all lie on the circle of radius 1 centered at the origin and determine the regular 8-gon illustrated in Figure 7.2.5. •

• EXERCISES 7-2

Write in trig form with $0 \le \theta < 2\pi$.

1. $3-4i$
2. $-2+2i$
3. $5+3i$
4. $8-5i$
5. $-2i$
6. -10

Use Theorem 1 of Section 7.1 to compute. (Note: $\operatorname{cis}\theta = \cos\theta + i\sin\theta$.)

7. $(2\operatorname{cis}30°)(3\operatorname{cis}60°)$
8. $\dfrac{2\operatorname{cis}45°}{3\operatorname{cis}30°}$
9. Write the two complex numbers in Exercise 7 in the form $a+bi$, perform the multiplication, and verify that the result is the same as that obtained using the trig form and Theorem 1 of Section 7.1.
10. Write the two complex numbers in Exercise 8 in the form $a+bi$, perform the division, and verify that the result is the same as that obtained using the trig form and Theorem 1 of Section 7.1.

Compute.

11. $\left(\operatorname{cis}\dfrac{\pi}{4}\right)^5$
12. $(1+i)^5$
13. $(2\operatorname{cis}\pi)^6$
14. $(3+4i)^{20}$

Find the nth roots of each complex number for the specified value of n. Graph each nth root in the complex plane.

15. $n=4,\quad 1$
16. $n=6,\quad 1$
17. $n=4,\quad 2+2i$
18. $n=4,\quad -2+2i$
19. $n=6,\quad -2+2i$
20. $n=5,\quad 32$
21. Determine z and the three cube roots of z if one cube root of z is $1+\sqrt{3}i$. Graph the roots.
22. Determine z and the four fourth roots of z if one fourth root of z is $-2-2i$. Graph the roots.
23. Solve $z^4 = 5-5i$ and graph the solutions.
24. Solve $z^6 = 2i$ and graph the solutions.
25. In Example 6 we found that $-1+i$ was a 12th root of -64. Determine the other eleven 12th roots of -64 and graph each in the complex plane.
26. Show geometrically that we can multiply two complex numbers by multiplying their absolute values and adding their arguments (Figure 7.2.6). That is, if $z_1 = r_1\operatorname{cis}\alpha$ and $z_2 = r_2\operatorname{cis}\beta$, then $z_1z_2 = r_1r_2\operatorname{cis}(\alpha+\beta)$. Graph z_1. Connect z_1 to the point $P=(1,0)$. Graph z_2. Construct $\angle\alpha$ on OQ. Construct $\angle\gamma$ on OQ with vertex at Q. Triangles Oz_2S and OPz_1 are similar. Thus, $\dfrac{d(0,S)}{d(0,z_1)} = \dfrac{d(0,z_2)}{d(0,P)} = \dfrac{d(0,z_2)}{1}$ so $d(0,S) = d(0,z_1)\cdot d(0,z_2) = r_1\cdot r_2$.
27. Construct a regular pentagon (5-gon) of radius 4; that is, each of the five vertices are at a distance 4 from the origin.
28. Construct a regular 10-gon of radius 6.

Expand.

29. $(x+y)^6$
30. $(a-2b^2)^4$

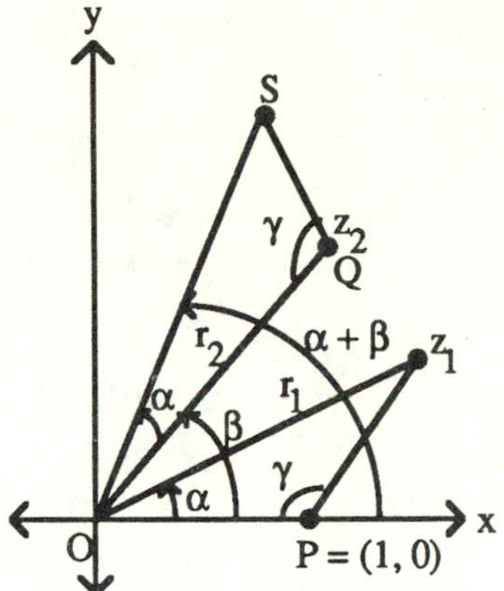

Figure 7.2.6

31. $(x - y)^6$

32. $(1 + 0.08)^5$

33. What is the coefficient of the x^5y^3 term in $(x + y)^8$?

34. What is the coefficient of the x^3y^5 term in $(x + y)^8$?

Factorial and the binomial coefficient. We use $n!$ to denote the product $n(n-1)(n-2)\cdots 2\cdot 1$ for a positive integer n. For example, $3! = 3\cdot 2\cdot 1 = 6$, $5! = 5\cdot 4\cdot 3\cdot 2\cdot 1 = 120$.

35. Show that the binomial coefficient of the general term $a^{n-k}b^k$ in the expansion of $(a + b)^n$ is $\dfrac{n!}{(n-k)!k!}$

36. Let $\binom{n}{k} = \dfrac{n!}{(n-k)!k!}$ for positive integers $k < n$. Compute

(a) $\binom{5}{2}$ (b) $\binom{8}{5}$ (c) $\binom{8}{3}$ (d) $\binom{10}{1}$

37. Show that (a) $\binom{n}{k} = \binom{n}{n-k}$ and (b) $\binom{n}{1} = n$.

38. What would be sensible definitions for $0!$, $\binom{n}{0}$, and $\binom{n}{n}$?

7.3 Polar Coordinates and Graphs

In this section we see how to represent points in the coordinate plane with polar coordinates. This process is very closely related to finding the trig form for a complex number. We also graph polar equations, that is, equations in the two polar variables r and θ.

Let P be a point in the coordinate plane with rectangular coordinates (x, y) (Figure 7.3.1). The location of P can be specified by giving its distance r from the origin O and an

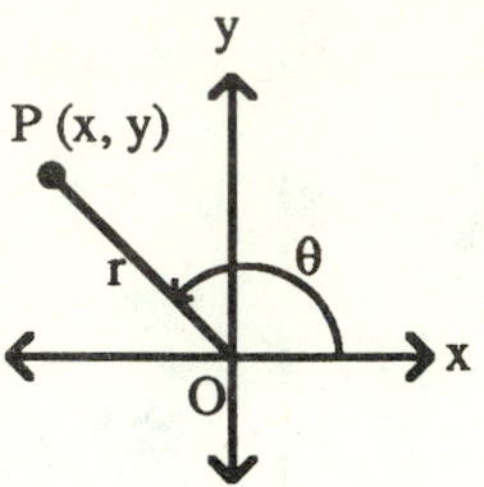

Figure 7.3.1

angle θ that the line through O and P makes with the positive x-axis. The measure of θ can be given in degrees or radians. We say that (r, θ) are **polar coordinates** of P. Notice that $r^2 = x^2 + y^2$ and $\tan\theta = \frac{y}{x}$ if $x \neq 0$. Polar coordinates of the point (x, y) are ***not unique*** because $(r, \theta + 2n\pi)$, n any integer, are also polar coordinates for the point P. For example, $(2, \frac{\pi}{3})$, $(2, \frac{7\pi}{3})$, and $(2, -\frac{5\pi}{3})$ are all polar coordinates for the same point P (Figure 7.3.2).

We permit the variable r to take on negative values with the following meaning. Let $r > 0$. Figure 7.3.3 illustrates how to graph the polar coordinates $(-r, \theta)$. The distance of the point $(-r, \theta)$ from the origin is r but in the direction opposite to the direction given by the terminal side of θ in (r, θ). Notice that $(-r, \theta)$ also has polar coordinates $(r, \theta + \pi)$. For example, $(-2, \frac{\pi}{3})$ and $(2, \frac{4\pi}{3})$ are polar coordinates for the same point (Figure 7.3.4). Notice that $\frac{4\pi}{3} = \frac{\pi}{3} + \pi$. In fact, $(-2, \frac{\pi}{3} + 2n\pi)$ are also polar coordinates for the point $(-2, \frac{\pi}{3})$.

• THEOREM 1 Let (r, θ) be polar coordinates for a point P. Then, P also has the following for polar coordinates:

$$(r, \theta + 2n\pi), \quad n \text{ any integer}$$

$$(-r, \theta + \pi + 2n\pi) = (-r, \theta + (2n+1)\pi), \quad n \text{ any integer}$$

The origin O is assumed to have polar coordinates $(0, \theta)$, where θ can be any angle.

Theorem 2 gives the relationship between rectangular and polar coordinates of a point.

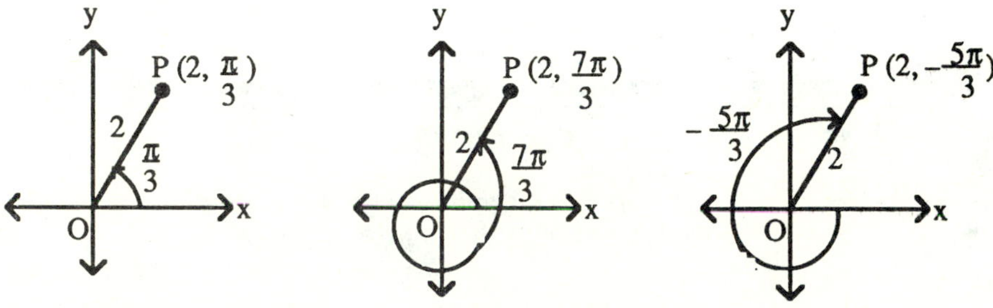

Figure 7.3.2

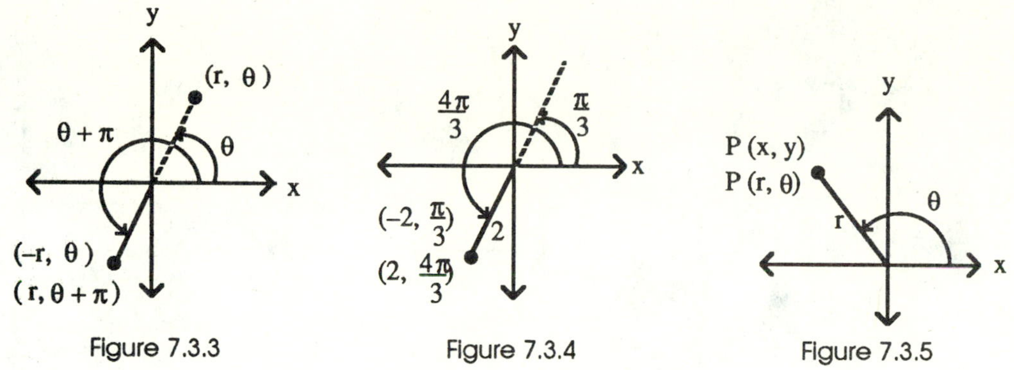

Figure 7.3.3 Figure 7.3.4 Figure 7.3.5

• **THEOREM 2** Let a point P have rectangular coordinates (x, y) and polar coordinates (r, θ). Then, $x = r\cos\theta$, $y = r\sin\theta$, $r^2 = x^2 + y^2$, and $\tan\theta = \frac{y}{x}$ $(x \neq 0)$.

PROOF Let P be a point with rectangular coordinates (x, y) and polar coordinates (r, θ). Figure 7.3.5 gives one possibility for the point P with $r > 0$ and θ in the second quadrant. We know that $r = \sqrt{x^2 + y^2}$ so that $r^2 = x^2 + y^2$. From the definitions of the trig functions at θ, we have $\sin\theta = \frac{y}{r}$, $\cos\theta = \frac{x}{r}$, and $\tan\theta = \frac{y}{x}$. Thus, $y = r\sin\theta$ and $x = r\cos\theta$. This proof is valid if θ is in any of the other three quadrants.

Figure 7.3.6 displays the relationship if $r < 0$ and θ in the second quadrant. In this case P is in Quadrant IV, so $x > 0$ and $y < 0$. Again, this proof does not depend on the quadrant that contains θ. Let Q be the corresponding point in the second quadrant with polar coordinates $(-r, \theta)$. Then Q has rectangular coordinates $(-x, -y)$. (Why?) Because $r < 0$ we have $|r| = -r$. We use the point Q to determine the trig functions at θ:

$$\sin\theta = \frac{-y}{|r|} = \frac{-y}{-r} = \frac{y}{r},$$

$$\cos\theta = \frac{-x}{|r|} = \frac{-x}{-r} = \frac{x}{r},$$

$$\tan\theta = \frac{-y}{-x} = \frac{y}{x}.$$

Thus, $y = r\sin\theta$, $x = r\cos\theta$, and $\tan\theta = \frac{y}{x}$. Also $r^2 = x^2 + y^2$ because $|r| = \sqrt{(-x)^2 + (-y)^2}$.

• **EXAMPLE 1**: Find the rectangular coordinates of the points with the following polar coordinates.

(a) $(3, \frac{5\pi}{6})$ (b) $(2, -200°)$

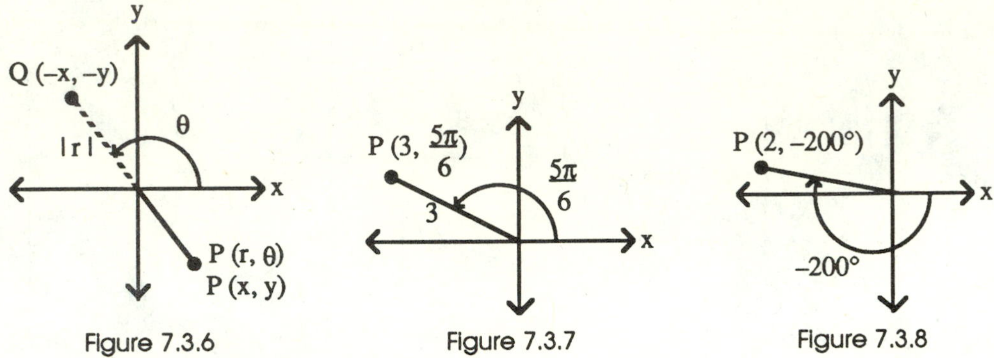

Figure 7.3.6 Figure 7.3.7 Figure 7.3.8

SOLUTION:

(a) Refer to Figure 7.3.7. Applying Theorem 2, we find that the rectangular coordinates of the point with polar coordinates $(3, \frac{5\pi}{6})$ are $(3\cos\frac{5\pi}{6}, 3\sin\frac{5\pi}{6}) = (-\frac{3\sqrt{3}}{2}, \frac{3}{2})$.

(b) Refer to Figure 7.3.8. Applying Theorem 2, we find that the rectangular coordinates of the point are $(2\cos(-200°), 2\sin(-200°)) = (-1.88, 0.68)$. •

Suppose that we are given the rectangular coordinates (x, y) of a point P and want to find polar coordinates (r, θ) for the point P. We can choose $r = \sqrt{x^2+y^2}$. However, we must be careful if we use the inverse tangent keys on a calculator to determine θ from $\tan\theta = \frac{y}{x}$. If P is in Quadrant I or IV, then $\tan^{-1}\frac{y}{x}$ is a correct angle to pair with $r = \sqrt{x^2+y^2}$ to obtain polar coordinates for P. This is not true if θ is in Quadrant II or III, as the next example illustrates.

• EXAMPLE 2: Find polar coordinates for the points with the following rectangular coordinates.

(a) $(3, -5)$ (b) $(-1, 1)$ (c) $(-2, -3)$

SOLUTION:

(a) The point P with rectangular coordinates $(3, -5)$ is in the fourth quadrant (Figure 7.3.9). If (r, θ) are polar coordinates of P with $r > 0$, then $r = \sqrt{3^2+(-5)^2} = \sqrt{34}$. Also, $\tan\theta = -\frac{5}{3}$. Now, $\tan^{-1}(-\frac{5}{3}) = -1.03$ is in the fourth quadrant, so we can choose $\theta = \tan^{-1}(-\frac{5}{3}) = -1.03$. Thus, $(\sqrt{34}, -1.03)$ or $(5.83, -1.03)$ give polar coordinates of P. This can be confusing because $(5.83, -1.03)$ look like rectangular coordinates. We have to depend on the context to determine the type of coordinates being specified.

(b) The point P with rectangular coordinates $(-1, 1)$ is in the second quadrant (Figure 7.3.10). Again, we can choose $r = \sqrt{(-1)^2+1^2} = \sqrt{2}$. Now, $\tan^{-1}(-1) = -\frac{\pi}{4}$ is not in the second quadrant. However, $\pi + \tan^{-1}(-1) = \frac{3\pi}{4}$ is in the second quadrant

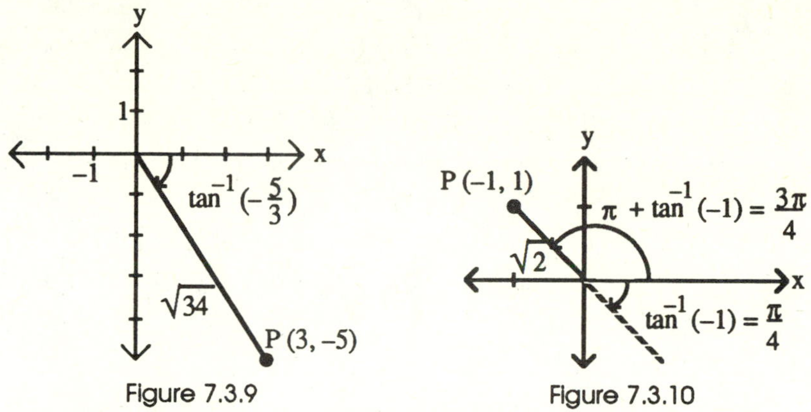

Figure 7.3.9

Figure 7.3.10

and can be paired with $r = \sqrt{2}$ to give polar coordinates for P. Thus, $(\sqrt{2}, \frac{3\pi}{4})$ are polar coordinates for P.

(c) The point P with rectangular coordinates $(-2, -3)$ is in the third quadrant (Figure 7.3.11). Again, we can choose $r = \sqrt{(-2)^2 + (-3)^2} = \sqrt{13}$. Now, $\tan^{-1}(\frac{-3}{-2}) = 0.98$ is in the first quadrant. However, $\pi + \tan^{-1}(\frac{-3}{-2}) = 4.12$ is in the third quadrant. Thus, $(\sqrt{13}, 4.12)$, or $(3.61, 4.12)$ are polar coordinates for P. •

Example 2 can be generalized as follows.

• **THEOREM 3** Let P have rectangular coordinates (x, y). Then, the following are polar coordinates for P where $r = \sqrt{x^2 + y^2}$.

(a) $(r, \tan^{-1} \frac{y}{x})$, if (x, y) is in the first or fourth quadrant.
(b) $(r, \pi + \tan^{-1} \frac{y}{x})$ if (x, y) is in the second or third quadrant.

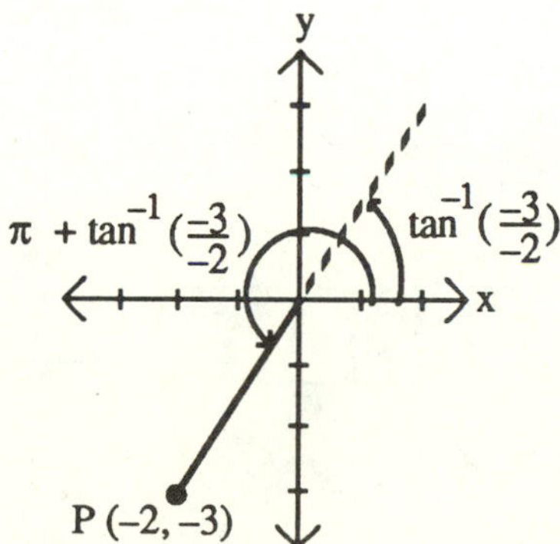

Figure 7.3.11

Polar Graphs

• DEFINITION The **graph of an equation** in the two polar variables r and θ is the set of all points (r, θ) in the coordinate plane where (r, θ) is a solution to the equation. •

Distinct solutions may produce the same point in the plane. For example, if (r, θ) and $(r, \theta + 2\pi)$ are solutions to a given equation, they correspond to the same point in the coordinate plane.

• EXAMPLE 3: Draw a complete graph of each polar equation.

(a) $r = 2$ (b) $\theta = 2$

SOLUTION:

(a) The solutions to the polar equation $r = 2$ consist of all points $(2, \theta)$ when θ is any angle. The set of all such points is the *circle* with center at the origin and radius 2 (Figure 7.3.12).

(b) The solutions to the equation $\theta = 2$ consist of all points $(r, 2)$ where r is any real number. The set of all such points is a *straight line* through the origin (Figure 7.3.13). •

If r is a periodic function of θ, then we can use the periodicity to help draw a complete graph. Example 4 illustrates this situation.

• EXAMPLE 4: Draw a complete graph of the polar equation $r = 3 + 3\sin\theta$.

SOLUTION: The function r is periodic with period 2π. (Why?) Thus, we will obtain a complete graph by restricting θ to the interval $[0, 2\pi]$. In fact, if we draw the graph of $r = 3+3\sin\theta$ for $2\pi \leq \theta \leq 4\pi$ we will completely retrace the graph of r for θ in the interval $0 \leq \theta \leq 2\pi$. For $0 \leq \theta \leq \frac{\pi}{2}$, we have $0 \leq \sin\theta \leq 1$ and $3 \leq 3 + 3\sin\theta \leq 6$. Thus, as θ increases from 0 to $\frac{\pi}{2}$, $\sin\theta$ increases from 0 to 1, and r increases from 3 to 6. Similarly, we can determine the remaining information given in Table 1.

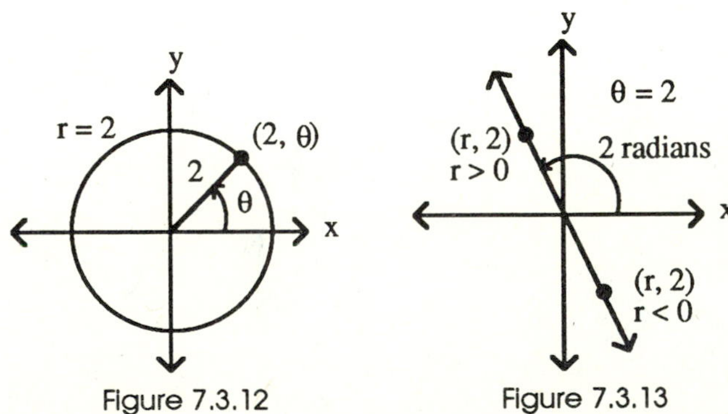

Figure 7.3.12 Figure 7.3.13

θ	$\sin\theta$	$r = 3 + 3\sin\theta$
$0 \le \theta \le \frac{\pi}{2}$	increases from 0 to 1	increases from 3 to 6
$\frac{\pi}{2} \le \theta \le \pi$	decreases from 1 to 0	decreases from 6 to 3
$\pi \le \theta \le \frac{3\pi}{2}$	decreases from 0 to -1	decreases from 3 to 0
$\frac{3\pi}{2} \le \theta \le 2\pi$	increases from -1 to 0	increases from 0 to 3

Table 1

You may need to compute several specific values to convince yourself that the graph in Figure 7.3.14 is complete. Notice that the graph is traced in a counterclockwise fashion starting with $(3, 0)$ and ending with $(3, 2\pi)$ as θ varies from 0 to 2π. •

The graph in Figure 7.3.14 is called a **cardioid**, which refers to its heartlike shape. In fact, the graph of any function of the form $r = a \pm a\cos\theta$ or $r = a \pm a\sin\theta$ is called a cardioid.

In addition to specifying the viewing rectangle $L \le x \le R$, $B \le y \le T$ we usually need to specify the range of values of θ used to obtain a graph. For example, we could have indicated that the graph in Figure 7.3.14 was drawn for θ in $[0, 2\pi]$. However, because of the periodicity of the function r this was not necessary. That is, the graph in Figure 7.3.14 is also the graph of $r = 3 + 3\sin\theta$ for θ in $(-\infty, \infty)$ or for θ in $[a, a + 2n\pi]$, n any integer.

The polar equation $r = 3 + 3\sin\theta$ of Example 4 defines r as a function of θ. That is, each value of θ determines a unique value of r. However, the graph in Figure 7.3.14 fails the vertical line test for a function. This is not a contradiction because the vertical line test is used to determine if y is a function of x in an equation involving x and y where x and y are the usual rectangular coordinates. If we convert the polar equation to rectangular form, we

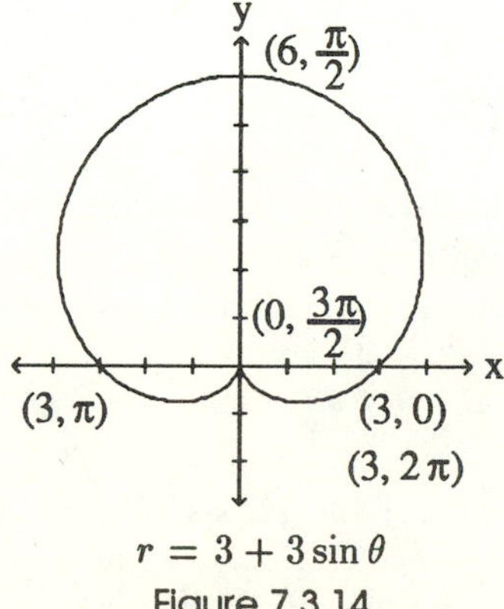

$r = 3 + 3\sin\theta$

Figure 7.3.14

can see directly that the equation does *not* define y as a function of x:

$$r = 3 + 3\sin\theta$$
$$r = 3 + 3\frac{y}{r}$$
$$r^2 = 3r + 3y$$
$$x^2 + y^2 = 3\sqrt{x^2+y^2} + 3y.$$

Now, we can see that y is *not* a function of x because each of the rectangular pairs $(0,0)$ and $(0,6)$ is a solution to the equation.

Most graphing utilities are able to obtain graphs of polar equations. If you are using a programmable graphing calculator, you probably will be able to write or obtain a short program that will draw graphs of polar equations. The *Graphing Calculator and Computer Graphing Laboratory Manual* accompanying this textbook contains a simple program that permits graphs of curves defined by polar equations to be obtained with a graphing calculator. The graphing utility *Master Grapher* used to draw graphs in this textbook has a polar equation option.

If a polar graphing utility is used, then the information given in Table 1 can be determined graphically by observing the drawing of the graph with θ restricted to the four intervals in the table. That is, setting the minimum value of θ to 0 and the maximum value of θ to $\frac{\pi}{2}$ will produce the first quadrant portion of the graph in Figure 7.3.14 drawn in counterclockwise fashion starting with $(3,0)$ and ending with $(6,\frac{\pi}{2})$.

• EXAMPLE 5: Draw a complete graph of the polar function $r = 5\cos 2\theta$.

SOLUTION: r is a periodic function of θ with period π. (Why?) Even though the values of r for $\pi \le \theta \le 2\pi$ are the same as the values of r for $0 \le \theta \le \pi$, we obtain distinct points on the graph because the angles are different. For example, $r = 4.70$ for $\theta = 0.175$ or 3.316, but the polar coordinates $(4.7, 0.175)$ and $(4.7, 3.316)$ name different points. However, we will get a complete graph if we let θ vary from 0 to 2π. Again, the graph of $r = 5\cos 2\theta$ for $2\pi \le \theta \le 4\pi$ retraces the graph of r for $0 \le \theta \le 2\pi$. We can determine the following information about r.

Figure 7.3.15 shows the graph of $r = 5\cos 2\theta$ for $0 \le \theta \le \frac{\pi}{2}$, $\frac{\pi}{2} \le \theta \le \pi$, $\pi \le \theta \le \frac{3\pi}{2}$, $\frac{3\pi}{2} \le \theta \le 2\pi$, and $0 \le \theta \le 2\pi$. As θ varies from 0 to $\frac{\pi}{4}$, the first quadrant portion of the graph in $0 \le \theta \le \frac{\pi}{2}$ is produced in counterclockwise fashion. The third quadrant portion of the graph in $0 \le \theta \le \frac{\pi}{2}$ is produced counterclockwise as θ varies from $\frac{\pi}{4}$ to $\frac{\pi}{2}$. The graphs for the ranges $\frac{\pi}{2} \le \theta \le \pi$, $\pi \le \theta \le \frac{3\pi}{2}$, and $\frac{3\pi}{2} \le \theta \le 2\pi$ can be similarly analyzed. The graph in $0 \le \theta \le 2\pi$ is complete and is called a **four-leafed rose**. To obtain a cumulative effect of the graph of $r = 5\cos 2\theta$ in the above θ-intervals, try drawing the graph for $0 \le \theta \le \frac{\pi}{2}$, $0 \le \theta \le \pi$, $0 \le \theta \le \frac{3\pi}{2}$, and $0 \le \theta \le 2\pi$. •

Notice that the length of the petals of the graph in $0 \le \theta \le 2\pi$ of Figure 7.3.15 is 5 units. The graph of $r = \cos 2\theta$ is also a four-leafed rose but the length of each petal is only one unit. If $a > 1$, the effect of a in $r = af(\theta)$ is to **stretch radially** the graph of

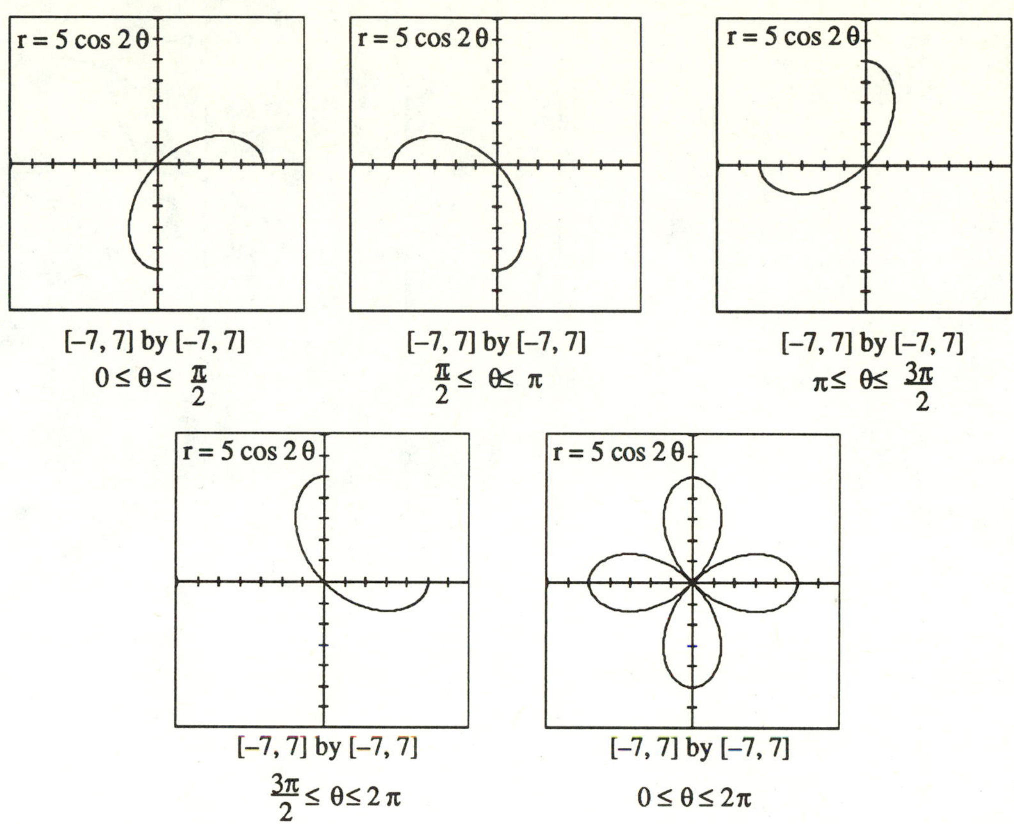

Figure 7.3.15

$r = f(\theta)$ by the factor a. If $0 < a < 1$, the effect of a in $r = af(\theta)$ is to **shrink radially** the graph of $r = f(\theta)$ by the factor a.

In the Exercises, you will investigate the graphs of $r = a \sin n\theta$ and $r = a \cos n\theta$ for a any real number and n any integer. If n is an odd integer greater than one, the graphs are n-leafed roses. If n is an even positive integer, then the graphs are $2n$-leafed roses. The above statements are also true if n is negative, except we must say $|n|$-leafed or $2|n|$-leafed rose in this case.

The graph in the next example is called a **spiral of Archimedes**.

• EXAMPLE 6: Draw the graph of $r = \theta$ for $0 \le \theta \le 2\pi$, $0 \le \theta \le 4\pi$, $0 \le \theta \le 8\pi$, $-4\pi \le \theta \le 0$, $-8\pi \le \theta \le 0$ and $-8\pi \le \theta \le 8\pi$.

SOLUTION: The six computer drawn graphs are given in Figure 7.3.16. The graph of $r = \theta$ goes around and around getting further from the origin as $|\theta|$ increases. •

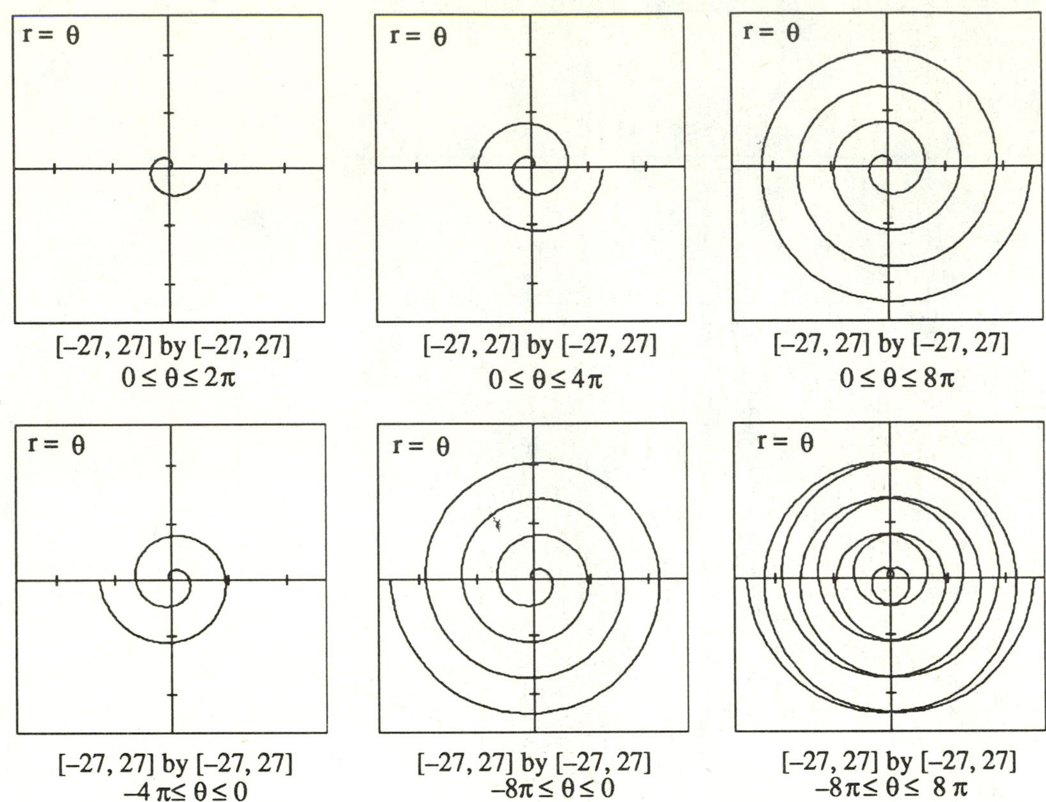

Figure 7.3.16

• EXERCISES 7-3

Compute the polar coordinates (r, θ) with $0 \leq \theta < 360°$ for the point with given rectangular coordinates.

1. $(1, 1)$
2. $(-10, 0)$
3. $(-3, 4)$
4. $(-2, 5)$

Compute *all* polar coordinate pairs for the point with given rectangular coordinates.

5. $(2, 2)$
6. $(-2, 5)$

Compute the rectangular coordinates for the point with given polar coordinates.

7. $(0, \pi)$
8. $(2, 30°)$
9. $(-3, 135°)$
10. $(-2, -120°)$

Compute the distance between the two points.

11. $(2.5, 30°)$, $(4, 90°)$
12. $(-3, 120°)$, $(2, 60°)$

Sketch a complete graph of each equation without using a polar graphing utility. If possible, check your answer with a polar graphing utility.

13. $r = 3$
14. $\theta = \frac{\pi}{4}$
15. $\theta = -\frac{\pi}{2}$
16. $r = -1$
17. $r = 3\cos\theta$
18. $r = 2\sin\theta$
19. $r = 2 + 2\cos\theta$
20. $r = 4 + 4\sin\theta$
21. $r = -1 + 2\cos\theta$
22. $r = \sin 2\theta$

Use a polar graphing utility to draw the polar graph for θ in the given intervals.

23. $r = 3\theta$; $0 \le \theta \le \frac{\pi}{2}$, $0 \le \theta \le \pi$, $0 \le \theta \le 2\pi$, $-\pi \le \theta \le \pi$, and $0 \le \theta \le 4\pi$.
24. $r = 5\sin\theta$, $0 \le \theta \le \frac{\pi}{2}$, $0 \le \theta \le \pi$, $0 \le \theta \le 2\pi$, $-\pi \le \theta \le \pi$, and $0 \le \theta \le 4\pi$.
25. $r = 5\cos 3\theta$, $0 \le \theta \le \frac{\pi}{2}$, $0 \le \theta \le \pi$, $0 \le \theta \le 2\pi$, $-\pi \le \theta \le \pi$, and $0 \le \theta \le 4\pi$.
26. $r = 5\sin 2\theta$, $0 \le \theta \le \frac{\pi}{2}$, $0 \le \theta \le \pi$, $0 \le \theta \le 2\pi$, $-\pi \le \theta \le \pi$, and $0 \le \theta \le 4\pi$.

Use a polar graphing utility to draw a complete graph of the polar equation. Specify an interval $a \le \theta \le b$ of smallest length that gives a complete graph.

27. $r = 2\theta$
28. $r = 0.275\theta$
29. $r = 2 - 3\cos\theta$
30. $r = 2\sin 3\theta$
31. $r = 3\sin^2\theta$
32. $r = 2 - 2\sin\theta$
33. Use a polar graphing utility to draw a complete graph of $r = 5\sin n\theta$ for $n = 1, 2, 3, 4, 5$, and 6. Explain how the number of rose "petals" is related to n.
34. Use a polar graphing utility to draw a complete graph of $r = 5\cos n\theta$ for $n = 1, 2, 3, 4, 5$, and 6. Explain how the number of rose "petals" is related to n.
35. Use a polar graphing utility to draw a complete graph of $r = a\sin 2\theta$ for $a = 1, 2, 3, 4, 5$, and 10. What is the effect of the parameter a on the graph?
36. Use a polar graphing utility to draw a complete graph of $r = a\cos 3\theta$ for $a = 1, 2, 3, 4, 5$, and 10. What is the effect of the parameter a on the graph?

Determine the length of *one* rose "petal."

37. $r = 5\sin 3\theta$
38. $r = 5\sin 2\theta$
39. $r = 8\sin 5\theta$
40. $r = 8\sin 4\theta$

Draw the graph of $r = 5\sin 2\theta$ and $r = 5\sin 3\theta$ in the standard viewing rectangle.

41. Predict what the graph of $r = 5\sin(2.5\theta)$ looks like, then draw a complete graph of $r = 5\sin(2.5\theta)$ with a polar graphing utility.
42. Give an algebraic argument for the appearance of the graph in Exercise 41.
43. Predict the graph of $r = 5\sin(3.5\theta)$. Confirm your graph by using a polar graphing utility.
44. Predict the graph of $r = a\sin\left(\frac{m}{n}\theta\right)$ for m and n relatively prime positive integers (m and n have no common factors).

Polar Coordinate System

Let ℓ be the half line consisting of the positive x-axis. In the *polar coordinate system*, this half line is called the *polar axis* and the end O (origin) is called the *pole* (Figure 7.3.17). If $r > 0$, the point P with *polar coordinates* (r, θ) is the point r units from the pole and on the terminal side of angle

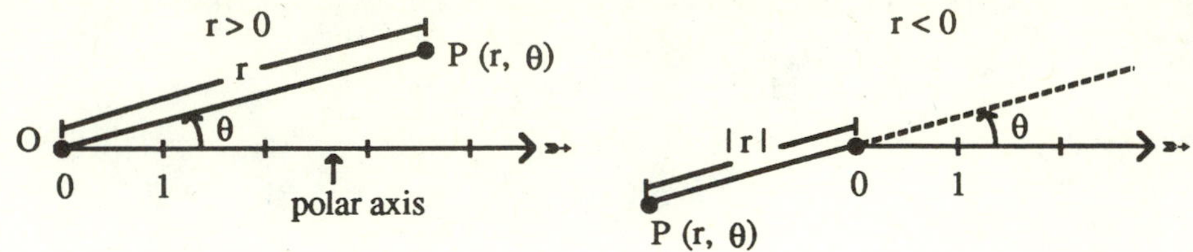

Figure 7.3.17

θ. If $r < 0$, the point P is the point $|r|$ units from the pole and on the half-line which is obtained by extending the terminal side of angle θ through the pole.

45. Find the polar coordinates of the vertices of a square with side length a whose center is at the pole and with two sides parallel to the polar axis.

46. Find the polar coordinates of the vertices of a regular pentagon if the center is at the pole, one vertex is on the polar axis, and the distance from the center to a vertex is a.

Before computers, polar graphs were often sketched using polar graph paper as shown in Figure 7.3.18.

47. Use polar graph paper (you can make it yourself using ruler, protractor, and compass) and sketch a complete graph of $r = 3\sin 2\theta$ for $0 \leq \theta \leq 360°$. Plot points for $\theta = 0 + k \cdot 30°$, $k = 0, 1, 2, \ldots, 12$.

Solve the polar system by using a polar graphing utility.

48. $r = 5\sin\theta$
$r = 5\cos\theta$

49. $r = 5\sin\theta$
$r = 5\sin 3\theta$

50. $r = \dfrac{\theta}{\pi}$
$r = 3\sin\theta$

51. $r = 2\cos\theta - 1$
$r = 3\sin 2\theta$

52. Show that if P_1 and P_2 have polar coordinates (r_1, θ_1) and (r_2, θ_2) respectively, then the distance between P_1 and P_2 is $d(P_1, P_2) = \sqrt{r_1^2 + r_2^2 - 2r_1r_2\cos(\theta_2 - \theta_1)}$.

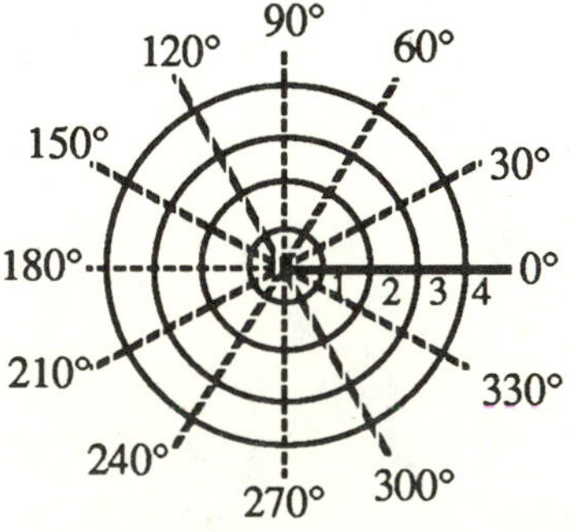

Figure 7.3.18

7.4 Vectors

In this section we begin the study of vectors. We investigate the arithmetic of vectors and use them with applications about motion. Many physical concepts can be represented by vectors. For example, the initial idea to use vectors grew out of the study of forces in physics. A **force** is a quantity that has both magnitude and direction. Forces are vectors that are often represented by arrows. The arrow points in the direction that the force is applied, and the length of the arrow represents the magnitude of the force.

A **vector** is a **directed line segment**. For example, $\overrightarrow{AB}$ denotes the vector determined by the line segment AB with direction assigned from A to B (Figure 7.4.1). A is called the **initial point** and B the **terminal point** of the vector $\overrightarrow{AB}$. $\overrightarrow{BA}$ is the vector determined by the line segment AB with direction assigned from B to A. B is the initial point of $\overrightarrow{BA}$ and A the terminal point. The **magnitude** of the vector $\overrightarrow{AB}$ is the length of the line segment AB. We denote the magnitude of $\overrightarrow{AB}$ by $|\overrightarrow{AB}|$. Two **vectors are equal** if and only if they have the same magnitude and direction. There are infinitely many vectors equal to a given vector $\overrightarrow{AB}$. The collection of vectors equal to $\overrightarrow{AB}$ consists of the vectors obtained by applying horizontal and/or vertical shifts to $\overrightarrow{AB}$; Figure 7.4.2 gives some of the possibilities. If a vector $\overrightarrow{CD}$ has been obtained by applying a combination of horizontal and vertical shifts to $\overrightarrow{AB}$, then $\overrightarrow{CD}$ is called a **parallel shift** of $\overrightarrow{AB}$.

In a coordinate system, we can be more precise about the meaning of the vector $\overrightarrow{AB}$. Consider the two points $A = (1, 2)$ and $B = (5, 4)$ (Figure 7.4.3). The magnitude of $\overrightarrow{AB}$ is $\sqrt{(5-1)^2 + (4-2)^2} = \sqrt{20}$, the distance from A to B. The direction of $\overrightarrow{AB}$ is the angle θ that $\overrightarrow{AB}$ makes with the positive x-axis. The magnitude of $\overrightarrow{BA}$ is also $\sqrt{20}$, but its direction with the positive x-axis is $\pi + \theta$ (Figure 7.4.4).

In a coordinate plane, we have the following theorem about equality of vectors.

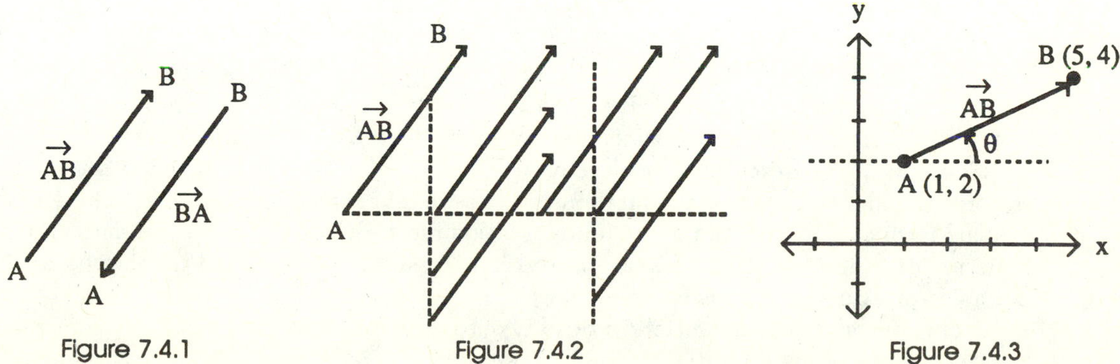

Figure 7.4.1 Figure 7.4.2 Figure 7.4.3

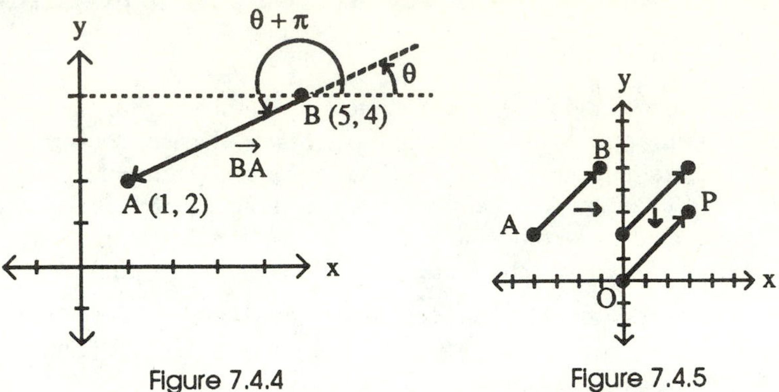

Figure 7.4.4

Figure 7.4.5

• THEOREM 1 The **vectors $\overrightarrow{AB}$ and $\overrightarrow{CD}$ are equal** if and only if $\overrightarrow{AB}$ can be obtained from $\overrightarrow{CD}$ by a *parallel shift*. In other words, $\overrightarrow{AB}$ can be obtained from $\overrightarrow{CD}$ by applying appropriate horizontal or vertical shifts.

It follows from Theorem 1 that if $\overrightarrow{AB}$ and $\overrightarrow{CD}$ are equal, then $d(A, B) = d(C, D)$. Furthermore, vectors are not changed under parallel shifting.

• EXAMPLE 1: Let $A = (-4, 2)$, $B = (-1, 6)$, $O = (0, 0)$, and $P = (3, 4)$. Show that $\overrightarrow{AB} = \overrightarrow{OP}$.

SOLUTION: The length of each vector is $\sqrt{3^2 + 4^2} = 5$. If we shift $\overrightarrow{AB}$ horizontally to the right 4 units, and then shift the resulting vector vertically down 2 units we obtain the vector $\overrightarrow{OP}$ (Figure 7.4.5). •

Notice that the coordinates of A in $\overrightarrow{AB} = \overrightarrow{OP}$ give the shift factors needed to slide $\overrightarrow{OP}$ to $\overrightarrow{AB}$. In Example 1, shifting $\overrightarrow{OP}$ horizontally left 4 units and up 2 units produces $\overrightarrow{AB}$. The opposites of the coordinates of A give the shift factors for obtaining $\overrightarrow{OP}$ from $\overrightarrow{AB}$

Theorem 1 and Example 1 can be generalized as follows. Let $\overrightarrow{AB}$ be any vector. There is a *unique* point P such that $\overrightarrow{AB} = \overrightarrow{OP}$, where $O = (0, 0)$ is the origin. Thus, every vector can be represented by a directed line segment with initial point the origin of a coordinate system. In this case, the vector is completely determined by its terminal point.

Choosing the representation of a vector with initial point $(0, 0)$ from among a set of equal vectors is similar to choosing a reduced form of a set of equal fractions. For example, $\frac{1}{2}$ is the reduced form of $\frac{2}{4}, \frac{3}{6}, \ldots$, and is the representation most often used. In future mathematics courses, you will learn that equality of vectors and equality of rational numbers are each examples of **equivalence relations**.

From now on we will use letters in **bold** type to represent vectors. If (x, y) is the terminal point of the vector $\boldsymbol{v}$ with initial point $(0, 0)$, we will write $\boldsymbol{v} = (x, y)$ because $\boldsymbol{v}$ is completely determined by its terminal point. We call x the **x-component** and y the **y-component** of the vector $\boldsymbol{v}$. The magnitude of $\boldsymbol{v}$ is the length $\sqrt{x^2+y^2}$ of $\boldsymbol{v}$ and is denoted by $|\boldsymbol{v}|$.

• EXAMPLE 2: Let $A = (2,5), B = (4,6), C = (-3,4)$, and $D = (-5,-2)$ be four points. Determine the vector $\boldsymbol{v} = (x,y)$ that is equal to the vector

(a) $\overrightarrow{AB}$. (b) $\overrightarrow{CD}$.

SOLUTION:

(a) Let $P = (x,y)$ be the terminal point of $\boldsymbol{v}$ (Figure 7.4.6). Then $d(O,P) = d(A,B)$ and OP and AB have the same slope. (Why?) The slope of AB is $\frac{1}{2}$. (Why?) Thus, the slope $\frac{y}{x}$ is also equal to $\frac{1}{2}$. It follows that $\frac{y}{x} = \frac{1}{2}$ or $2y = x$. Now $d(A,B) = \sqrt{2^2+1^2} = \sqrt{5}$ so that $d(O,P) = \sqrt{x^2+y^2} = \sqrt{5}$, or $x^2+y^2 = 5$. Substituting $2y$ for x in $x^2+y^2 = 5$ we obtain

$$x^2+y^2 = 5$$
$$4y^2+y^2 = 5$$
$$y^2 = 1.$$

It follows that $y = 1$ because P is in the first quadrant. (Why?) Thus, $x = 2$ and $\boldsymbol{v} = (2,1)$.

Alternatively, the terminal point B of $\overrightarrow{AB}$ is 2 units right and 1 unit up from the initial point A. Therefore, the terminal point P of $\overrightarrow{OP}$ must be 2 units right and 1 unit up from the initial point $(0,0)$. Thus, $P = (2,1)$ and $\boldsymbol{v} = (2,1)$.

(b) Let $P = (x,y)$ be the terminal point of $\boldsymbol{v}$ (Figure 7.4.7). Both x and y must be negative. (Why?) Notice P must be 2 units left and 6 units down from the origin so that $P = (-2,-6)$. We can use the fact that OP and CD have the same slope to obtain $y = 3x$, and $d(O,P) = d(C,D) = \sqrt{40}$ to obtain $x^2+y^2 = 40$. Again $P = (-2,-6)$ so that $\boldsymbol{v} = (-2,-6)$. •

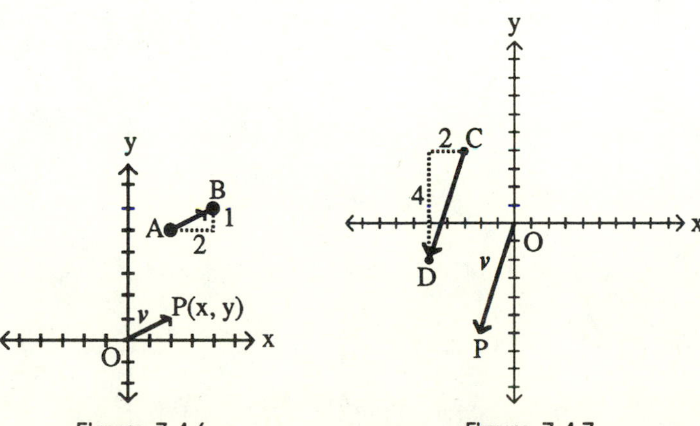

Figure 7.4.6 Figure 7.4.7

Notice in Example 2(a) that the coordinates of the vector $\boldsymbol{v} = (2,1)$ can be obtained by subtracting the coordinates of the point $A = (2,5)$ from the coordinates of the point $B = (4,6)$. That is, $\boldsymbol{v} = (2,1) = (4-2, 6-5)$. Similarly, for the vector of Example 2(b) we have $\boldsymbol{v} = (-2,-6) = (-5-(-3), -2-4)$. These are special cases of the following theorem.

• THEOREM 2 Let $A = (a_1, a_2)$ and $B = (b_1, b_2)$ be two points, and let $\boldsymbol{v} = (x, y)$ be the vector with initial point the origin that is equal to $\overrightarrow{AB}$. Then $x = b_1 - a_1$ and $y = b_2 - a_2$.

Just as we say $\frac{2}{4} = \frac{1}{2}$ for rational numbers, we will say that $\overrightarrow{AB} = (b_1 - a_1, b_2 - a_2)$ for the vector $\overrightarrow{AB}$ in Theorem 2.

Addition of Vectors

Let $\boldsymbol{u}$ and $\boldsymbol{v}$ be any two vectors. We define the **sum** $\boldsymbol{u} + \boldsymbol{v}$ as follows. Draw a vector equal to $\boldsymbol{v}$ with initial point the terminal point of $\boldsymbol{u}$ and terminal point P (Figure 7.4.8). We may use $\boldsymbol{v}$ for the name of this vector because it is equal to $\boldsymbol{v}$. The vector $\boldsymbol{u} + \boldsymbol{v}$ is the vector with initial point O and terminal point P.

If we close the parallelogram in Figure 7.4.8 with a vector equal to $\boldsymbol{u}$, then the diagonal would represent $\boldsymbol{v} + \boldsymbol{u}$ (Figure 7.4.9). Thus, $\boldsymbol{u} + \boldsymbol{v} = \boldsymbol{v} + \boldsymbol{u}$ so that addition of vectors is *commutative*.

Notice that the addition of vectors is essentially the same as the geometric addition of complex numbers described in Section 7.1. If (a, b) is the terminal point of $\boldsymbol{u}$ and (c, d) is the terminal point of $\boldsymbol{v}$, then the following can be shown to be valid:

$$\begin{aligned} \boldsymbol{u} + \boldsymbol{v} &= (a, b) + (c, d) \\ &= (a + c, b + d). \end{aligned}$$

In other words, if we know the *components* of $\boldsymbol{u}$ and $\boldsymbol{v}$, then the *components* of the sum $\boldsymbol{u}+\boldsymbol{v}$ are the sum of the *components* of $\boldsymbol{u}$ and of $\boldsymbol{v}$. In fact, complex numbers can be represented by vectors. For example, the vector $\boldsymbol{u}$ with initial point $(0,0)$ and terminal point $(2,3)$ is a vector representation of the complex number $2 + 3i$ (Figure 7.4.10). It can be shown that addition of complex numbers is preserved by this representation. That is, if $a + bi$ corresponds to $\boldsymbol{v}$ and $c + di$ corresponds to $\boldsymbol{u}$, then $(a + c) + (b + d)i$ corresponds to $\boldsymbol{v} + \boldsymbol{u}$.

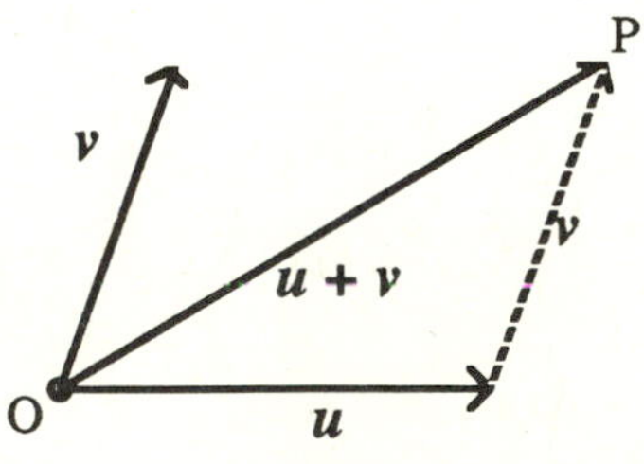

Figure 7.4.8

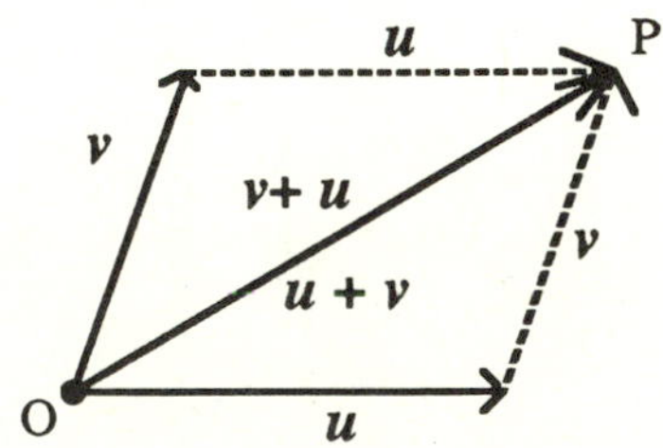

Figure 7.4.9

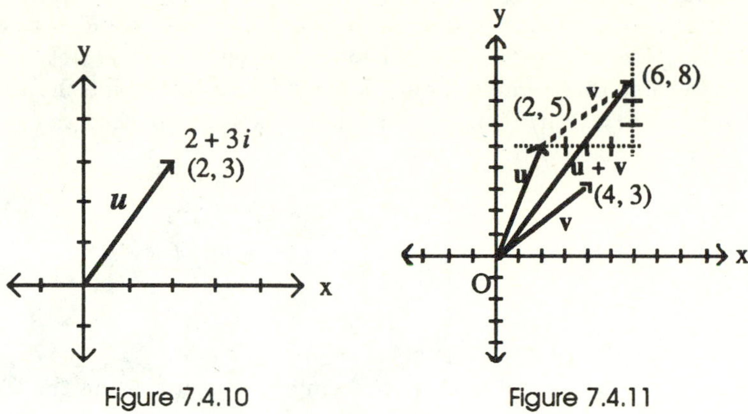

Figure 7.4.10 Figure 7.4.11

• EXAMPLE 3: Use the geometric definition of addition of vectors to find $\boldsymbol{u}+\boldsymbol{v}$, where $\boldsymbol{u} = (2,5)$ and $\boldsymbol{v} = (4,3)$. Compare with componentwise addition.

SOLUTION: The vector equal to $\boldsymbol{v} = (4,3)$ with initial point $(2,5)$ has terminal point $(6,8)$ (Figure 7.4.11). (Why?) Thus, $\boldsymbol{u}+\boldsymbol{v} = (6,8)$. Componentwise addition gives

$$\begin{aligned} \boldsymbol{u}+\boldsymbol{v} &= (2,5)+(4,3) \\ &= (2+4, 5+3) \\ &= (6,8). \end{aligned}$$

•

• DEFINITION The vector with $(0,0)$ as both the initial and terminal point is denoted by $\mathbf{0}$ and is called the **zero vector**. •

Notice that $\boldsymbol{u}+\mathbf{0} = \mathbf{0}+\boldsymbol{u} = \boldsymbol{u}$ for any vector $\boldsymbol{u}$.

• DEFINITION Let $\boldsymbol{u}$ be any vector. The vector with the same magnitude as $\boldsymbol{u}$ but with opposite direction is denoted by $-\boldsymbol{u}$ and is called the **additive inverse** or **negative** of $\boldsymbol{u}$. •

Notice that $\boldsymbol{u}+(-\boldsymbol{u}) = (-\boldsymbol{u})+\boldsymbol{u} = \mathbf{0}$ for any vector $\boldsymbol{u}$. Some of the important properties about vectors are summarized in Theorem 3, which we give without proof.

• THEOREM 3 Let $\boldsymbol{u}$, $\boldsymbol{v}$, and $\boldsymbol{w}$ be arbitrary vectors. Then,

(1) $\boldsymbol{u}+\boldsymbol{v} = \boldsymbol{v}+\boldsymbol{u}$ (Commutative Property)
(2) $(\boldsymbol{u}+\boldsymbol{v})+\boldsymbol{w} = \boldsymbol{u}+(\boldsymbol{v}+\boldsymbol{w})$ (Associative Property)
(3) $\boldsymbol{u}+\mathbf{0} = \boldsymbol{u}$ (Additive Identity Property)
(4) $\boldsymbol{u}+(-\boldsymbol{u}) = \mathbf{0}$ (Additive Inverse Property)

Subtraction of Vectors

The difference of two vectors $\boldsymbol{u}$ and $\boldsymbol{v}$ is denoted by $\boldsymbol{u}-\boldsymbol{v}$ and is that vector which, when added to $\boldsymbol{v}$ gives $\boldsymbol{u}$ (Figure 7.4.12). That is, $\boldsymbol{v}+(\boldsymbol{u}-\boldsymbol{v})=\boldsymbol{u}$. Notice that the vector $\boldsymbol{u}-\boldsymbol{v}$ has initial point the terminal point of $\boldsymbol{v}$, and terminal point the terminal point of $\boldsymbol{u}$. If $\boldsymbol{u}=(a,b)$ and $\boldsymbol{v}=(c,d)$ then subtraction can be accomplished *componentwise* because the following can be shown to be true:

$$\begin{aligned}\boldsymbol{u}-\boldsymbol{v}&=(a,b)-(c,d)\\&=(a-c,b-d).\end{aligned}$$

Multiplication by a Scalar

We use a vector and a real number to form a new vector using the following definition.

• DEFINITION Let a be any real number and $\boldsymbol{u}$ any vector. The **scalar multiple** of $\boldsymbol{u}$ by a is defined as follows.

(1) If $a=0$, then $a\boldsymbol{u}=\boldsymbol{0}$.
(2) If $a>0$, then the vector $a\boldsymbol{u}$ has the same direction as $\boldsymbol{u}$ and has magnitude $a|\boldsymbol{u}|$.
(3) If $a<0$, then the vector $a\boldsymbol{u}$ has direction opposite to $\boldsymbol{u}$ and has magnitude $|a||\boldsymbol{u}|$. •

Figure 7.4.13 gives some examples of scalar multiples of a vector $\boldsymbol{u}$. Notice that a scalar multiple of a vector $\boldsymbol{u}$ is a stretch or shrink of $\boldsymbol{u}$ in the same or opposite direction of $\boldsymbol{u}$.

Some of the important properties about scalar multiplication are summarized in Theorem 4, which we give without proof.

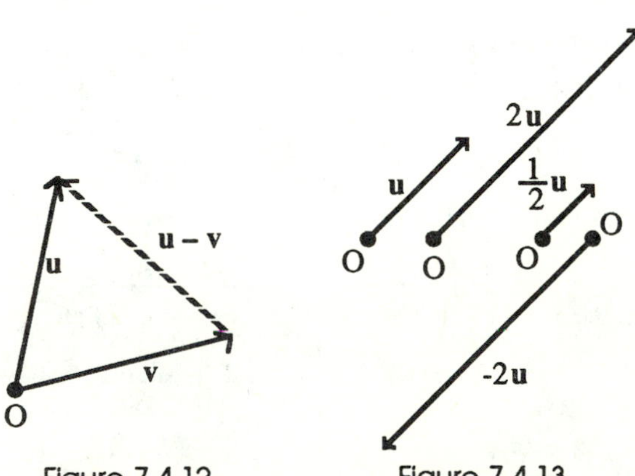

Figure 7.4.12 Figure 7.4.13

• THEOREM 4 Let $\boldsymbol{u}$ and $\boldsymbol{v}$ be any two vectors and a and b any two scalars. Then,

(1) $a(\boldsymbol{u}+\boldsymbol{v}) = a\boldsymbol{u} + a\boldsymbol{v}$ (2) $(a+b)\boldsymbol{u} = a\boldsymbol{u} + b\boldsymbol{u}$

(3) $(ab)\boldsymbol{u} = a(b\boldsymbol{u})$ (4) $(1)\boldsymbol{u} = \boldsymbol{u}$

(5) $(-1)\boldsymbol{u} = -\boldsymbol{u}$ (6) $a\boldsymbol{0} = \boldsymbol{0}$, $0\boldsymbol{u} = \boldsymbol{0}$

If $\boldsymbol{u} = (a, b)$, then scalar multiplication can be accomplished *componentwise* because the following can be shown to be true:

$$\begin{aligned} c\boldsymbol{u} &= c(a, b) \\ &= (ca, cb). \end{aligned}$$

Linear Combinations of Vectors

• DEFINITION The vector $\boldsymbol{w}$ is said to be a **linear combination** of the vectors $\boldsymbol{u}$ and $\boldsymbol{v}$ if there are scalars a and b such that $\boldsymbol{w} = a\boldsymbol{u} + b\boldsymbol{v}$. •

Let $\boldsymbol{i} = (1, 0)$ and $\boldsymbol{j} = (0, 1)$ (Figure 7.4.14). Every vector can be expressed as a linear combination of $\boldsymbol{i}$ and $\boldsymbol{j}$. This can be determined as follows. Let $\boldsymbol{u} = (a, b)$ be any vector:

$$\begin{aligned} \boldsymbol{u} &= (a, b) \\ &= (a, 0) + (0, b) \\ &= a(1, 0) + b(0, 1) \\ &= a\boldsymbol{i} + b\boldsymbol{j}. \end{aligned}$$

Notice that the scalars in this case are the x- and y-components of the vector $\boldsymbol{u} = (a, b)$ (Figure 7.4.15). The magnitude of each of the vectors $\boldsymbol{i}$ and $\boldsymbol{j}$ is 1. Vectors with magnitude 1 are called **unit vectors**. Hence $\boldsymbol{i}$ and $\boldsymbol{j}$ are unit vectors.

• EXAMPLE 4: Let $A = (-3, 2)$ and $B = (2, 6)$. Express the vector $\overrightarrow{AB}$ as a linear combination of $\boldsymbol{i}$ and $\boldsymbol{j}$.

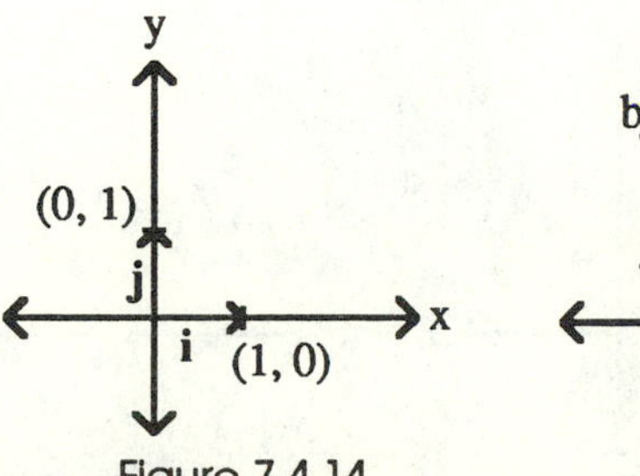

Figure 7.4.14

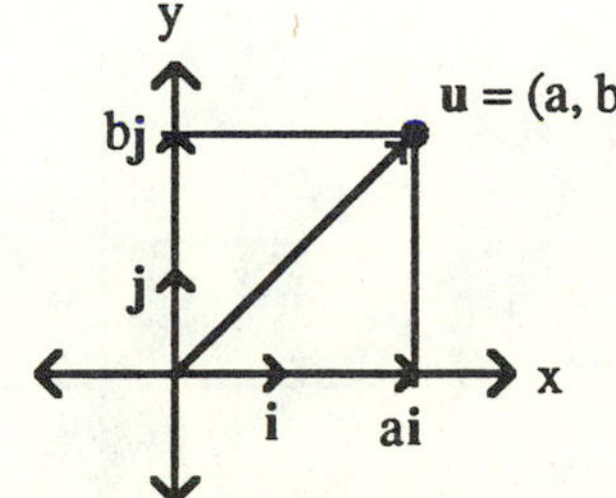

Figure 7.4.15

SOLUTION: We use Theorem 2 and compute as follows:

$$\begin{aligned}\overrightarrow{AB} &= (2-(-3), 6-2)\\ &= (5,4)\\ &= 5(1,0)+4(0,1)\\ &= 5\boldsymbol{i}+4\boldsymbol{j}.\end{aligned}$$

•

If we know an angle that gives the direction of a vector, then we can determine its x- and y-components. This is very similar to the procedure used in Section 7.1 to find the trig form of a complex number.

• THEOREM 5 Let $\boldsymbol{u} = (a,b)$ be any vector and P the point with coordinates (a,b) (Figure 7.4.16). If θ is an angle in standard position with terminal side OP, then $a = |\boldsymbol{u}|\cos\theta$ and $b = |\boldsymbol{u}|\sin\theta$.

PROOF We know that $r = \sqrt{a^2+b^2} = |\boldsymbol{u}|$. From the definition of the trig functions at θ we have $\cos\theta = \frac{a}{r}$ and $\sin\theta = \frac{b}{r}$. Thus, $a = r\cos\theta = |\boldsymbol{u}|\cos\theta$ and $b = r\sin\theta = |\boldsymbol{u}|\sin\theta$.

• EXAMPLE 5: Determine the x- and y-components of the vector $\boldsymbol{u}$ that makes an angle of $115°$ with the positive x-axis and has length 6 (Figure 7.4.17).

SOLUTION: By Theorem 5, the x-coordinate of $\boldsymbol{u}$ is $6\cos 115° = -2.54$ and the y-coordinate is $6\sin 115° = 5.44$. Thus, $\boldsymbol{u} = (-2.54, 5.44)$. •

• EXAMPLE 6: Let $\boldsymbol{u} = (3,2)$ and $\boldsymbol{v} = (-2,5)$. Determine

(a) $2\boldsymbol{u} - 3\boldsymbol{v}$.

(b) $|\boldsymbol{u}|$ and $|\boldsymbol{v}|$.

(c) the angle that each vector makes with the positive x-axis.

(d) the angle with initial side $\boldsymbol{u}$ and terminal side $\boldsymbol{v}$.

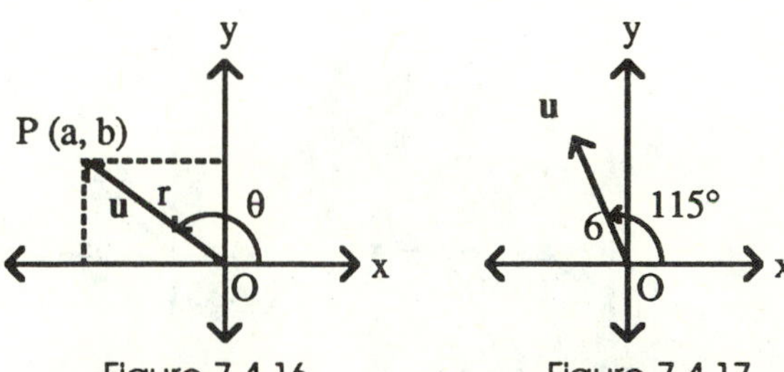

Figure 7.4.16 Figure 7.4.17

SOLUTION:

(a) $$\begin{aligned} 2\boldsymbol{u} - 3\boldsymbol{v} &= 2(3,2) - 3(-2,5) \\ &= (6,4) - (-6,15) \\ &= (12,-11) \end{aligned}$$

(b) $$\begin{aligned} |\boldsymbol{u}| &= \sqrt{3^2 + 2^2} = \sqrt{13} \\ |\boldsymbol{v}| &= \sqrt{(-2)^2 + 5^2} = \sqrt{29} \end{aligned}$$

(c) The angle α that $\boldsymbol{u}$ makes with the positive x-axis is acute (Figure 7.4.18). Thus, $\alpha = \tan^{-1}(\frac{2}{3}) = 33.69°$. The angle β that $\boldsymbol{v}$ makes with the positive x-axis is in the *second* quadrant. Now, $\tan^{-1}(-\frac{5}{2}) = -68.20°$ is *not* in the second quadrant. Thus, $\beta = 180° + \tan^{-1}(-\frac{5}{2}) = 111.80°$.

(d) We can see from Figure 7.4.18 that the angle θ with initial side $\boldsymbol{u}$ and terminal side $\boldsymbol{v}$ is $\beta - \alpha$, or $78.11°$. •

Example 6(d) can be solved directly without finding the angles that $\boldsymbol{u}$ and $\boldsymbol{v}$ make with the positive x-axis. Let $A = (3,2)$ and $B = (-2,5)$ be the terminal points of the vector $\boldsymbol{u}$ and $\boldsymbol{v}$ in Figure 7.4.18. Then from the Law of Cosines we have

$$\begin{aligned} d^2(A,B) &= |\boldsymbol{u}|^2 + |\boldsymbol{v}|^2 - 2|\boldsymbol{u}||\boldsymbol{v}|\cos(\beta - \alpha) \\ 34 &= 12 + 29 - 2\sqrt{13}\sqrt{29}\cos(\beta - \alpha) \\ \beta - \alpha &= \cos^{-1}\left(\frac{13 + 29 - 34}{2\sqrt{13}\sqrt{29}}\right) \\ \beta - \alpha &= 78.11° \end{aligned}$$

We can generalize this approach in the following way.

• **Angle between Vectors** Let $\boldsymbol{u}$ and $\boldsymbol{v}$ be vectors (Figure 7.4.19). The angle β between $\boldsymbol{u}$ and $\boldsymbol{v}$ is given by

$$\beta = \cos^{-1}\left(\frac{|\boldsymbol{u}|^2 + |\boldsymbol{v}|^2 - d^2}{2|\boldsymbol{u}||\boldsymbol{v}|}\right)$$

where d is the distance between the terminal points of $\boldsymbol{u}$ and $\boldsymbol{v}$.

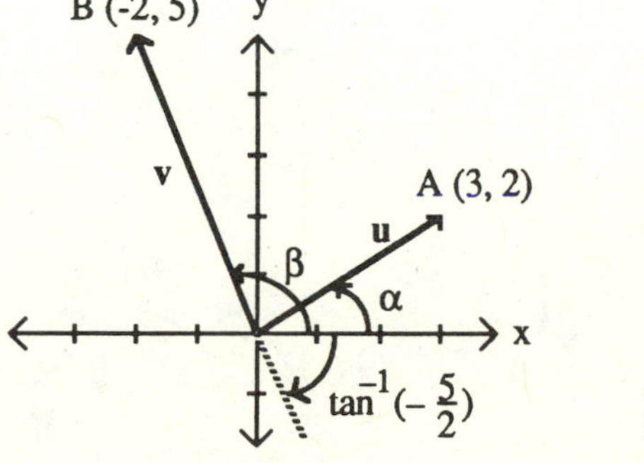

Figure 7.4.18

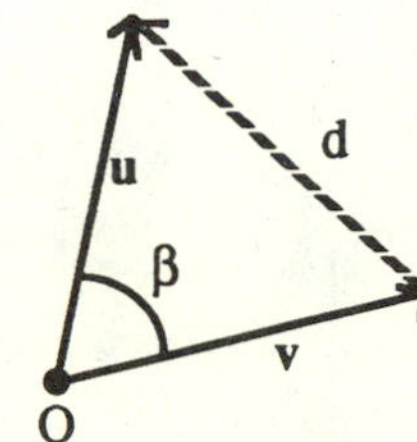

Figure 7.4.19

The **velocity** of a moving object is a vector because it has both magnitude and direction. Usually, the magnitude of velocity is called **speed**.

• EXAMPLE 7: A plane is flying on a bearing of 65° east of north at 500 mph. Express the velocity of the plane as a vector.

SOLUTION: Let $\boldsymbol{v}$ be the velocity vector (Figure 7.4.20). The angle that $\boldsymbol{v}$ makes with the positive x-axis is 25°. The magnitude of $\boldsymbol{v}$ is 500 because the plane is flying at 500 mph. It follows from Theorem 5 that

$$\begin{aligned}\boldsymbol{v} &= (500\cos 25°, 500\sin 25°)\\ &= (453.15, 211.31).\end{aligned}$$ •

• EXAMPLE 8: A plane is flying on a bearing of 65° east of north at 500 mph. A tail wind that adds to the plane's velocity is blowing in the direction 35° east of north at 80 mph. Express the actual velocity of the plane as a vector. Determine the actual speed and direction of the plane.

SOLUTION: In Example 7 we found that the velocity of the plane in still air was $\boldsymbol{v} = (453.15, 211.31)$. The velocity $\boldsymbol{u}$ of the wind can be expressed as a vector as follows (Figure 7.4.21). The angle $\boldsymbol{u}$ makes with the positive x-axis is 55° because $\boldsymbol{u}$ is 35° east of north. Thus, $\boldsymbol{u} = (80\cos 55°, 80\sin 55°)$, or $\boldsymbol{u} = (45.89, 65.53)$. The actual speed and direction of the plane is given by the vector $\boldsymbol{u}+\boldsymbol{v}$:

$$\begin{aligned}\boldsymbol{u}+\boldsymbol{v} &= (45.89, 65.53) + (453.15, 211.31)\\ &= (499.04, 276.84).\end{aligned}$$

The actual speed of the plane is $|\boldsymbol{u}+\boldsymbol{v}| = \sqrt{(499.04)^2 + (276.84)^2} = 570.68$ mph, in the direction $\tan^{-1}(\frac{276.84}{499.04}) = 29.02°$ north of east. We could also say that the direction of the plane is 60.98° east of north. •

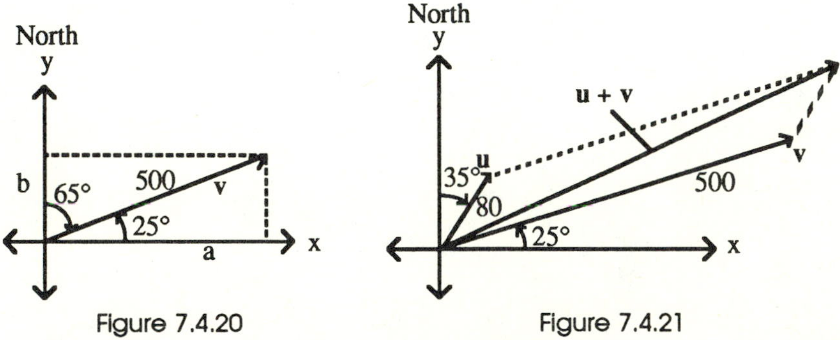

Figure 7.4.20

Figure 7.4.21

Usually pilots face a slightly different problem. The actual speed and direction ($\boldsymbol{u}+\boldsymbol{v}$) are known as well as the tail or head wind vector ($\boldsymbol{u}$). The flight bearing must be established. This is illustrated in Exercises 38 and 39.

• EXERCISES 7-4

Let $\boldsymbol{u} = (-1,3)$, $\boldsymbol{v} = (2,4)$, and $\boldsymbol{w} = (2,-5)$ be vectors. Determine the following.

1. $\boldsymbol{u}+\boldsymbol{v}$
2. $\boldsymbol{u}-\boldsymbol{v}$
3. $\boldsymbol{u}-\boldsymbol{w}$
4. $3\boldsymbol{v}$
5. $2\boldsymbol{u}+3\boldsymbol{w}$
6. $2\boldsymbol{u}-4\boldsymbol{v}$
7. $|\boldsymbol{u}+\boldsymbol{v}|$
8. $|\boldsymbol{u}-\boldsymbol{v}|$

Use the geometric definitions of addition and subtraction of vectors to find the sum $\boldsymbol{u}+\boldsymbol{v}$, the difference $\boldsymbol{u}-\boldsymbol{v}$, and then compare with the results obtained using the respective componentwise definitions.

9. $\boldsymbol{u} = (1,3)$, $\boldsymbol{v} = (3,6)$
10. $\boldsymbol{u} = (-1,2)$, $\boldsymbol{v} = (4,-2)$

Let $A = (-1,2)$, $B = (3,4)$, $C = (-2,5)$, and $D = (2,-8)$. Write each as a vector with *initial point at the origin.*

11. $\overrightarrow{AB}$
12. $3\overrightarrow{AB}$
13. $2\overrightarrow{AB}-\overrightarrow{BA}$
14. $\overrightarrow{AB}+\overrightarrow{CD}$
15. $\overrightarrow{AB}-\overrightarrow{CD}$
16. $2\overrightarrow{AB}+3\overrightarrow{CD}$

Compute.

17. $|\overrightarrow{AB}|$
18. $|2\overrightarrow{AB}|$
19. $|\overrightarrow{AB}+\overrightarrow{BA}|$

Show geometrically that $\overrightarrow{AB} = \overrightarrow{OB} - \overrightarrow{OA}$.

20. $A = (2,3)$, $B = (5,2)$
21. $A = (-2,4)$, $B = (2,6)$

Express the vector $\overrightarrow{AB}$ as linear combination of $\boldsymbol{i} = (1,0)$ and $\boldsymbol{j} = (0,1)$.

22. $A = (3,0)$, $B = (0,6)$
23. $A = (2,1)$, $B = (5,0)$
24. $A = (3,-2)$, $B = (2,6)$
25. $A = (-2,1)$, $B = (-3,-5)$

Find the angle each vector makes with the positive x-axis. Then determine the angle between the vectors.

26. $\boldsymbol{u} = (3,4)$, $\boldsymbol{v} = (1,0)$
27. $\boldsymbol{u} = (-1,2)$, $\boldsymbol{v} = (3,2)$
28. $\boldsymbol{u} = (-1,2)$, $\boldsymbol{v} = (3,-4)$
29. $\boldsymbol{u} = (2,-3)$, $\boldsymbol{v} = (-3,-5)$
30. Let $\boldsymbol{r} = (1,2)$ and $\boldsymbol{s} = (2,-1)$. Show that the vector $\boldsymbol{v} = (5,7)$ can be expressed as a linear combination of $\boldsymbol{r}$ and $\boldsymbol{s}$. Explain what this means geometrically.
31. Let $\boldsymbol{r} = (1,2)$ and $\boldsymbol{s} = (2,-1)$. Show that *any* vector $\boldsymbol{v} = (x,y)$ can be expressed as a linear combination of $\boldsymbol{r}$ and $\boldsymbol{s}$.
32. Let $\boldsymbol{r} = (a,b)$ and $\boldsymbol{s} = (c,d)$. Show that any vector $\boldsymbol{v} = (x,y)$ can be expressed as a linear combination of $\boldsymbol{r}$ and $\boldsymbol{s}$ for almost all values of a, b, c, and d. For what values of a, b, c, and d is it *impossible* to express $\boldsymbol{v}$ as a linear combination of $\boldsymbol{r}$ and $\boldsymbol{s}$?
33. A plane is flying on a bearing of 25° west of north at 530 mph. Express the velocity of the plane as a vector.

34. A plane is flying on a bearing of 10° east of south at 460 mph. Express the velocity of the plane as a vector.

35. A plane is flying on a bearing of 20° west of north at 325 mph. A tail wind is blowing in the direction 40° west of north at 40 mph. Express the actual velocity of the plane as a vector. Determine the actual speed and direction of the plane.

36. A plane is flying on a bearing of 10° east of south at 460 mph. A tail wind is blowing in the direction 20° west of south at 80 mph. Express the actual velocity of the plane as a vector. Determine the actual speed and direction of the plane.

37. A sailboat under auxiliary power is proceeding on a bearing of 25° north of west at 6.25 mph in still water. Then a tail wind blowing 15 mph in the direction 35° south of west alters the course of the sailboat. Express the actual velocity of the sailboat as a vector. Determine the actual speed and direction of the boat.

38. A pilot must actually fly due west at a constant speed of 382 mph against a head wind of 55 mph blowing in the direction of 22° south of east. What direction and speed must the pilot maintain to keep on course (due west)?

39. A jet fighter pilot must actually fly due north at a constant speed of 680 mph against a head wind of 80 mph blowing in the direction of 10° west of south. What direction and speed must the pilot maintain to keep on course (due north)?

Motion Problems

40. A ball is tossed up in the air at an angle of 70° with the horizontal and with initial velocity 36 feet per second (Figure 7.4.22). Neglecting air resistance *and* gravity, specify the position of the ball as a vector 1 second, 2.5 seconds, and 4 seconds after the ball is released. Express the position of the ball at time t, where t is the number of seconds after the ball is released. Explain why this is not a realistic model. (When does the ball return to the ground?)

Consider an object dropped with initial velocity *zero* (a so-called "freely falling body") from a tower at point A (Figure 7.4.23). Let point B be the position of the object t seconds after it is released (neglecting air resistance and wind effects). It is a fact from elementary physics that on Earth the

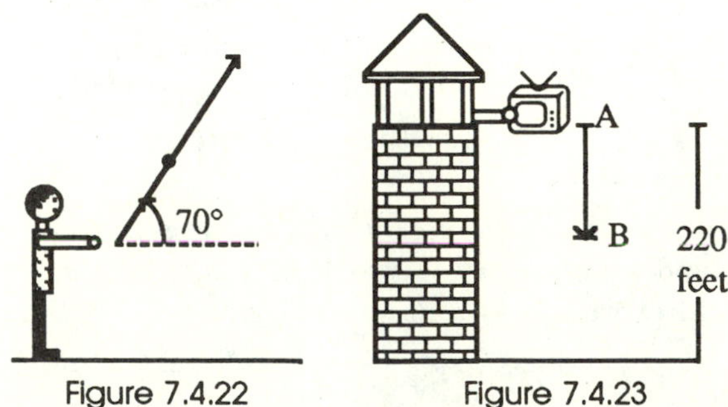

Figure 7.4.22 Figure 7.4.23

vector $\overrightarrow{AB}$ is equal to the "effect of gravity" vector $\boldsymbol{g} = (0, -16t^2)$. Assume point A is 220 feet above the ground.

41. Determine a vector that describes the speed and position of the object after 2 seconds, 3 seconds, and 4 seconds. Specify the coordinates of point B.
42. When will the object hit the ground? At what speed?
43. Rework Exercise 40 accounting for the **effect of gravity**. That is, the ball is tossed up and it eventually falls back to the ground (due to gravity). *Hint: Add* the two vectors $\boldsymbol{u}$ and $\boldsymbol{g}$ where $\boldsymbol{u}$ is the vector in Exercise 40 (neglecting the effect of gravity) and $\boldsymbol{g}$ is the "effect of gravity" vector.
44. When will the ball of Exercise 40 hit the ground? At what speed?
45. How high above the ground will the ball of Exercise 40 reach? How far will the ball travel in the horizontal direction?
46. Let ℓ_1 and ℓ_2 be perpendicular lines intersecting at the origin with ℓ_1 in the first quadrant and third quadrant. Let the slope of ℓ_1 be given by b/a with $a \neq 0$ and $b \neq 0$. Show that the slope of the lines bisecting the angle between ℓ_1 and ℓ_2 is $\frac{a+b}{a-b}$. (*Note:* this problem was suggested by Richard Liu, a student in Mrs. Pam Giles's 1988-89 C^2PC class in Sandy, Utah.)

7.5 Parametric Equations

In this section we introduce parametric equations for a curve. We show that the graph of a polar function $r = f(\theta)$ can be obtained parametrically. The graphs of many important relations can often be obtained as graphs of parametric equations for the curve defined by the relation. Most graphing utilities can easily produce the graph of curves defined by parametric equations.

• DEFINITION Let f and g be two functions defined on an interval I. The relation consisting of all ordered pairs $(f(t), g(t))$ for t in I is called a **plane curve** C, or more simply a **curve** C. The equations $x = f(t)$ and $y = g(t)$ for t in I are called **parametric equations** for C. The variable t is called a **parameter**. The parametric equations $x = f(t)$ and $y = g(t)$, with t in I, are said to give a **parametrization** of C in terms of t. •

Figure 7.5.1 shows several possibilities for plane curves.

If I is the interval $[a, b]$, then the points $(x(a), y(a))$ and $(x(b), y(b))$ are called **end points** of the curve C. It is possible that a curve described parametrically intersects itself. This means that more than one value of t produces the same point in the coordinate plane. Usually a curve is not the graph of a function. However, any function $y = F(x)$ can be defined parametrically by the equations $x = f(t)$, $y = g(t)$ where $f(t) = t$ and $g(t) = F(t)$.

• EXAMPLE 1: Describe the plane curve C defined by $x = 1 - 2t$ and $y = 2 - t$ for t in $[0, 1]$.

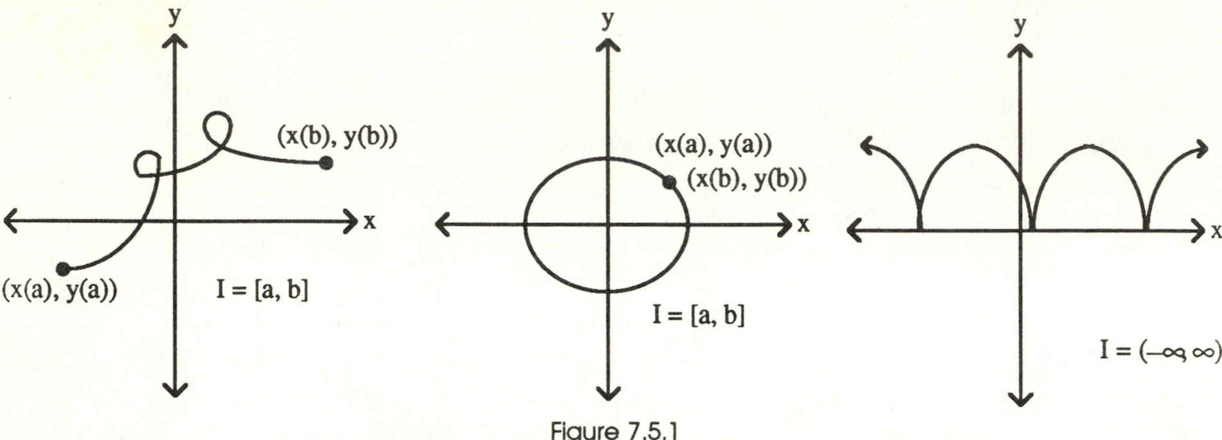

Figure 7.5.1

SOLUTION: The point corresponding to $t = 0$ is $(1, 2)$ and to $t = 1$ is $(-1, 1)$. Thus, the points $(1, 2)$ and $(-1, 1)$ are endpoints of C. Notice that x varies from 1 to -1 and y varies from 2 to 1 as t varies from 0 to 1. In other words, as t increases from 0 to 1, x decreases from 1 to -1 and y decreases from 2 to 1. We solve $y = 2 - t$ for t ($t = 2 - y$) and substitute into $x = 1 - 2t$ obtaining $x = 1 - 2(2 - y)$ or $2y - x = 3$.

The graph of $2y - x = 3$ is a straight line. Thus, C is the portion of the straight line between $(-1, 1)$ and $(1, 2)$ (Figure 7.5.2). •

In Example 1, we were able to remove the parameter t and obtain an equation involving only the variables x and y. Usually it is impossible to remove the parameter. Thus, in such cases a graph will need to be determined directly from the parametric equations.

The *Graphing Calculator and Computer Graphing Laboratory Manual* accompanying this textbook contains a simple program that permits graphs of curves defined by parametric equations to be obtained with a graphing calculator. The graphing utility *Master Grapher* used to draw graphs in this textbook has a parametric option. For example, the graph in Example 1 can be obtained by entering $x(t) = 1 - 2t$, $y(t) = 2 - t$, and choosing the range

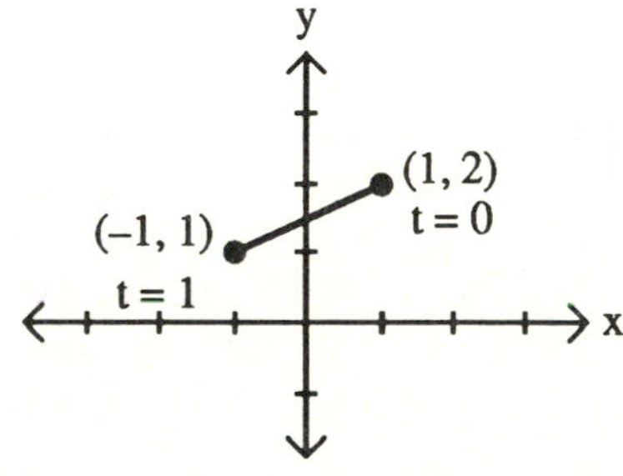

Figure 7.5.2

on t to be $0 \le t \le 1$. We use the notation **t min** and **t max** for the smallest and largest values of t entered into the graphing utility. The graph in Figure 7.5.2 can be confirmed using a parametric graphing utility.

• EXAMPLE 2: Find parametric equations for the circle with center at the origin and radius 2.

SOLUTION: Let $P(x, y)$ be a point on the circle (Figure 7.5.3). Let θ be the angle OP makes with the positive x-axis. We use θ as the parameter to obtain parametric equations for this circle. From the definition of the trig functions we have $x = 2\cos\theta$ and $y = 2\sin\theta$. Notice that we obtain a complete graph of the circle by letting θ vary through *any* interval of length 2π. (Why?) In particular, $x = 2\cos\theta$, $y = 2\sin\theta$, for $t\min = 0$ and $t\max = 2\pi$ gives a parametrization of the circle centered at the origin with radius 2. •

The domain of the relation graphed in Figure 7.5.3 is $-2 \le x \le 2$ and the range is $-2 \le y \le 2$. In general, the domain and range of a curve are not the same as the range of values of the parameter that produce the curve. For example, $0 \le \theta \le 2\pi$ will produce the graph in Figure 7.5.3.

• EXAMPLE 3: Draw the graph of the curve defined by $x = t\cos t$, $y = t\sin t$ for $0 \le t \le 4\pi$.

SOLUTION: The graph of the curve is shown in Figure 7.5.4. •

The graph in Figure 7.5.4 is exactly the same as one of the *polar* graphs of $r = \theta$ in Figure 7.3.16. This suggests that polar graphs and parametric equations are related. Indeed, this is true and the following theorem gives the connection.

• THEOREM 1 The polar graph $r = f(\theta)$ is the curve defined parametrically by $x(t) = f(t)\cos t$ and $y(t) = f(t)\sin t$.

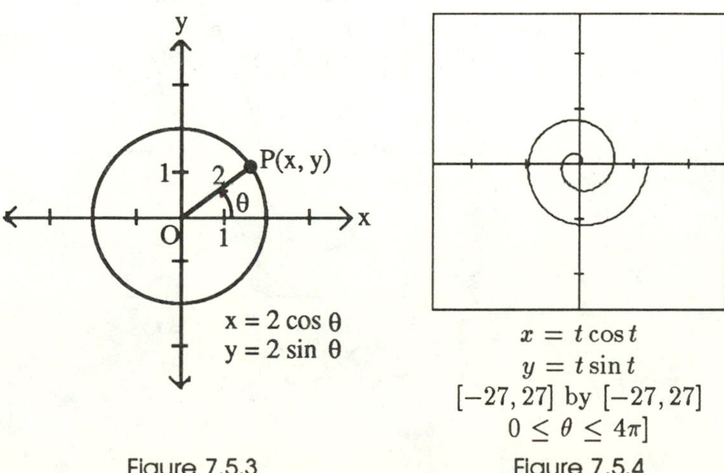

$x = t\cos t$
$y = t\sin t$
$[-27, 27]$ by $[-27, 27]$
$0 \le \theta \le 4\pi]$

Figure 7.5.3

Figure 7.5.4

PROOF Let (x, y) be the rectangular coordinates of a point P on the polar graph $r = f(t)$ (Figure 7.5.5). By definition of the trig functions we have

$$\sin t = \frac{y}{r}, \qquad \cos t = \frac{x}{r}.$$

Thus, $x = r\cos t = f(t)\cos t$ and $y = r\sin t = f(t)\sin t$. Notice that we have replaced θ by t in $r = f(\theta)$.

Theorem 1 allows us to draw a graph of any polar equation using a parametric graphing utility. In fact, this theorem gives the basis for using a parametric graphing program to obtain polar graphs.

• EXAMPLE 4: Determine parametric equations for the four-leafed rose $r = 7\sin 2\theta$. Use a parametric graphing utility to draw a complete graph of the curve.

SOLUTION: Applying Theorem 1 we have $x(t) = 7\sin 2t\cos t$ and $y(t) = 7\sin 2t\sin t$ where we have replaced θ by t. A complete graph is produced for $0 \le t \le 2\pi$ as shown in Figure 7.5.6. •

• EXAMPLE 5: Draw the graph of the curve defined by $x(t) = 5\sin t$ and $y(t) = 5t\cos t$ for $0 \le t \le 20$.

SOLUTION: The curve is shown in Figure 7.5.7. The x coordinates of the points are between -5 and 5 regardless of the value of t. (Why?) The y coordinates are between -100 and 100 for all values of t in $0 \le t \le 20$. (Why?) •

• EXAMPLE 6: Draw a complete graph of the curve defined by the parametric equations $x = 5t - 5\sin t$ and $y = 5 - 5\cos t$.

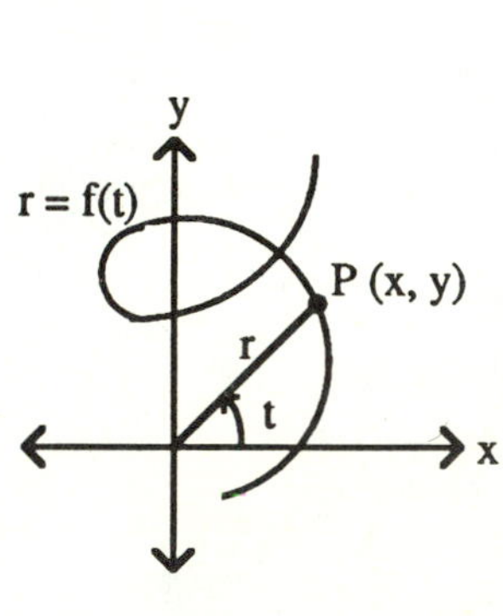

Figure 7.5.5

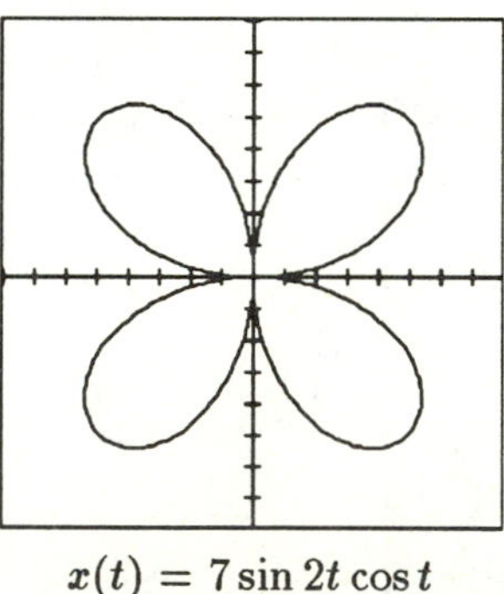

$x(t) = 7\sin 2t\cos t$
$y(t) = 7\sin 2t\sin t$
$0 \le t \le 2\pi$
$[-8, 8]$ by $[-8, 8]$

Figure 7.5.6

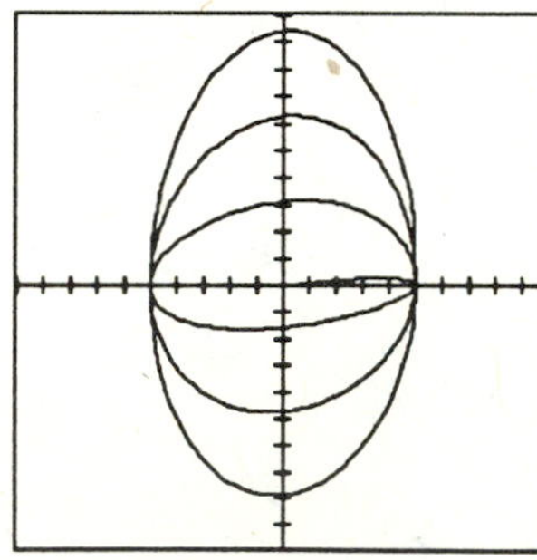

$x(t) = 5\sin t$
$y(t) = 5t\cos t$
$0 \le t \le 20$
$[-10, 10]$ by $[-100, 100]$

Figure 7.5.7

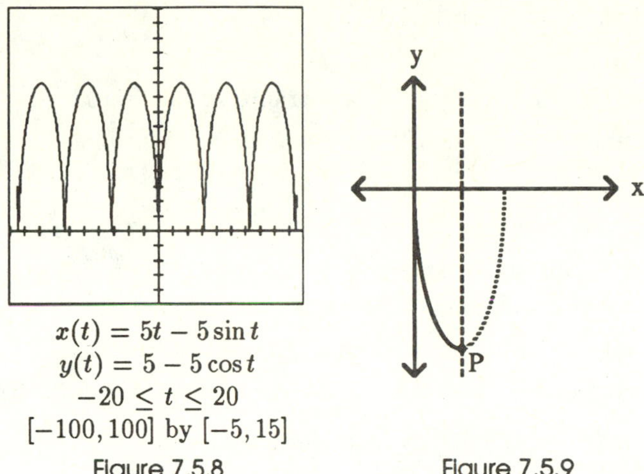

$x(t) = 5t - 5\sin t$
$y(t) = 5 - 5\cos t$
$-20 \le t \le 20$
$[-100, 100]$ by $[-5, 15]$

Figure 7.5.8

Figure 7.5.9

SOLUTION: The values of $\cos t$ vary between -1 and 1. Thus, we can see from the equation for y that the values of y vary between 0 and 10. (Why?) Furthermore, y is a periodic function of t with period 2π. The values of $\sin t$ vary between -1 and 1. Thus, $-5\sin t$ varies from -5 to 5. The values of x vary from $-\infty$ to ∞ as the values of t vary from $-\infty$ to ∞. (Why?) The graph in Figure 7.5.8 is a complete graph. You may need to view the graph in additional viewing rectangles to convince yourself that this is a complete graph of the curve. •

The graph in Figure 7.5.8 is called a cycloid. In fact, the graph of the parametric equations $x = a(t - \sin t)$ and $y = a(1 - \cos t)$ for any real number a and all real t is a **cycloid**. Graphs of cycloids are similar to the graph in Figure 7.5.8.

Cycloids are important in physics. They are sometimes called the *curve of quickest descent*. Suppose a cycloid is reflected through the x-axis. Then, out of all possible frictionless curves connecting the origin to the first local minimum (point P) to the right of $x = 0$, a cycloid is the one that an object will slide down in the *least* amount of time (Figure 7.5.9).

• EXERCISES 7-5

Describe the curve defined parametrically. Sketch the graph without using a graphing utility. Check using a parametric graphing utility.

1. $x(t) = 1 + t$ and $y(t) = t$ with t in $[0, 5]$
2. $x(t) = 1 + t$ and $y(t) = t$ with t in $[-3, 2]$
3. $x(t) = 2 - 3t$ and $y(t) = 5 + t$ with t in $[-1, 3]$
4. $x(t) = 2 - 3t$ and $y(t) = 5 + t$ with t in $(-\infty, \infty)$
5. $x(t) = t$ and $y(t) = \frac{2}{t}$ with t in $(0, 5]$
6. $x(t) = t + 2$ and $y(t) = \frac{2}{t}$ with t in $[-3, 0) \cup (0, 3]$

7. $x(t) = 3\cos t$ and $y(t) = 3\sin t$ with t in $[0, 4\pi]$
8. $x(t) = 4\sin t$ and $y(t) = 4\cos t$ with t in $\left[-\frac{\pi}{2}, \frac{\pi}{2}\right]$
9. Which of the curves in Exercises 1–8 are graphs of functions?
10. Which of the curves in Exercises 1–8 are graphs of one-to-one functions?
11. What are the domain and range of the curve in Exercise 2?
12. What are the domain and range of the curve in Exercise 3?
13. What are the domain and range of the curve in Exercise 5?
14. What are the domain and range of the curve in Exercise 7?

Find parametric equations and draw the graph of each circle.

15. centered at the origin with radius 5
16. centered at $(10, 0)$ with radius 4
17. centered at $(0, 10)$ with radius 6
18. centered at (a, b) with radius r

Draw a complete graph.

19. $x(t) = 4t - 2$ and $y = 8t^2$ with t in $(-\infty, \infty)$
20. $x(t) = 4\cos t$ and $y = 8\sin t$ with t in $(-2\pi, 2\pi)$
21. $x(t) = 1 + \frac{1}{t}$ and $y = t - \frac{1}{t}$ with t in $(0, 20]$
22. $x(t) = 2t - 2\sin t$ and $y(t) = 2 - 2\cos t$ with t in $[0, 40]$
23. $x(t) = 4t - 4\sin t$ and $y(t) = 4 - 4\cos t$ with t in $[0, 40]$
24. $x(t) = 6\cos t - 4\cos(\frac{3}{2}t)$ and $y(t) = 6\sin t - 4\sin(\frac{3}{2}t)$ with t in $[0, 2\pi]$. (This is an *epicycloid*, the path of a point on a circle rolling on another circle.)
25. $x(t) = 5\cos^3 t$ and $y(t) = 5\sin^3 t$ with t in $[0, 2\pi]$. (This is a *hypocycloid* of four cusps.)
26. $x(t) = 4(\cos t + t\sin t)$ and $y(t) = 4(\sin t - t\cos t)$ with t in $[0, 50]$. (This is the *involute* of a circle.) *Hint:* Use a large viewing rectangle.
27. Find the domain and range of the curve in Exercise 25.
28. Find the domain and range of the curve in Exercise 26.
29. Let f be a function. Show that the curve defined parametrically by $x(t) = t$ and $y(t) = f(t)$ with t in the interval I is the *function* f with domain equal to the interval I.
30. Let $f(x) = 3 - x^2$. Show that the curve defined parametrically by $x(t) = at + b$ and $y(t) = f(t)$ with t in the interval I is a *function* $y = g(x)$. Find a rule for $g(x)$. What is the domain of g?
31. Find the maximum value (if any) of $y(t)$ for the curve defined parametrically by $x(t) = 2t - 1$ and $y(t) = 4 - t^2$.

Write the polar equation in parametric form. Graph *both* the parametric and polar form with an appropriate graphing utility.

32. $r = 3\theta$ for θ in $[0, 10\pi]$
33. $r = 5\sin\theta$ for θ in $[-\pi, \pi]$
34. $r = 5\sin 2\theta$ for θ in $[0, \pi]$
35. $r = 5\sin 3\theta$ for θ in $[0, 2\pi]$
36. Draw a complete graph of $x(t) = a(t - \sin t)$ and $y(t) = a(1 - \cos t)$ for $a = 1, 2, 3, 4$, and 5. Use zoom-in and determine the period of the curve and the maximum y-coordinate with least positive x-coordinate.

37. Based on the results of Exercise 36, make a conjecture for the *period* and the maximum y-value with least positive x-coordinate of the cycloid $x(t) = a(t - \sin t)$ and $y(t) = a(1 - \cos t)$ for an arbitrary value of a.

7.6 Motion Problems and Parametric Equations

In this section we continue the study of parametric equations. We see that the path of objects acted upon by certain forces can be described by parametric equations. We asume all motion occurs in a two dimensional plane.

In Section 7.4 we found that velocity can be described with a vector because it has both magnitude and direction. We know from Section 1.4 that if we assume that the only force acting on an object is due to gravity, then the distance above ground of an object thrown straight up with initial velocity v_0 from a point s_0 above ground is given by $s = -16t^2 + v_0t + s_0$. Actually, the coefficient of t^2 is $-\frac{g}{2}$, where g is an acceleration constant due to gravity. A usual approximation for g on Earth at sea level is 32 ft/sec^2.

In Example 1 we consider an object that is propelled with an initial velocity that acts in a direction that makes an angle different from 90° with the horizontal. The velocity vector will have both a horizontal and vertical component in this case. We also neglect air resistance when we find parametric equations describing the path of the object.

• EXAMPLE 1: An object is launched with an initial velocity of 50 ft/sec at an angle of 60° with the positive x-axis. Assume that the only force acting on the object is due to gravity.

(a) Find parametric equations that model the motion of the object.
(b) When will the object hit the ground?
(c) How far does the object travel in the horizontal direction?
(d) Find the maximum height attained by the object, and when it reaches that height.
(e) Show that the position at any time t of the object written as a vector can be expressed as the sum of two vectors; one of the vectors depends only on the initial velocity and the other only on gravity.

SOLUTION: Let t be the time (in seconds) that the object is in flight. Assume the object is launched from the origin.

(a) Let $\boldsymbol{v}_0$ be the initial velocity vector. The horizontal component of the initial velocity vector is $50 \cos 60° = 25$ and the vertical component is $50 \sin 60° = 25\sqrt{3}$ (Figure 7.6.1). Thus, $\boldsymbol{v}_0 = (25, 25\sqrt{3})$. Let $P(t) = P(x, y)$ be the position of the object at time t. Then, both x and y are functions of time. By assumption, the only force acting in the horizontal direction on the object is the horizontal component 25 of $\boldsymbol{v}_0$. Using the standard distance formula $d = rt$ we have $x(t) = 25t$. In the vertical direction, gravity and the vertical component of $\boldsymbol{v}_0$ are the only forces assumed to be acting on the object. We can use the formula of Section 1.4 to write $y(t) = -16at^2 + 25\sqrt{3}t$.

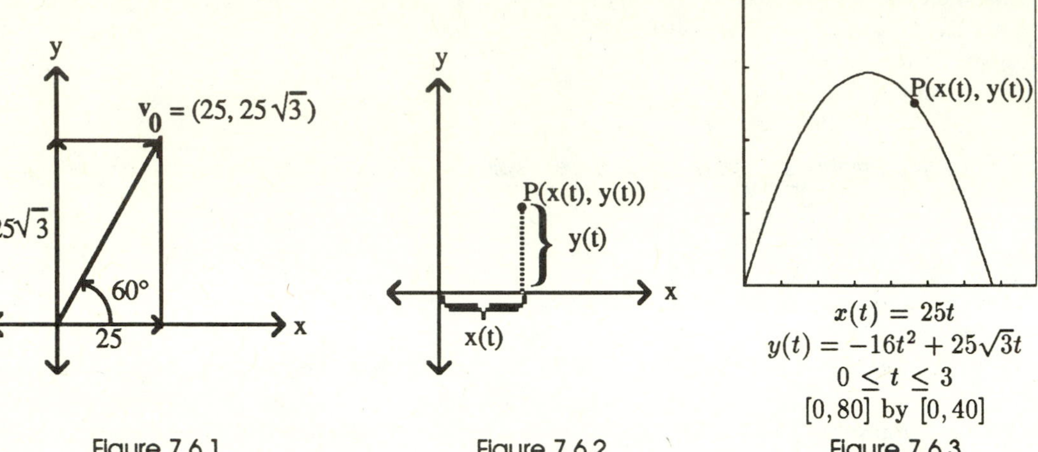

Figure 7.6.1

Figure 7.6.2

$x(t) = 25t$
$y(t) = -16t^2 + 25\sqrt{3}t$
$0 \leq t \leq 3$
$[0, 80]$ by $[0, 40]$

Figure 7.6.3

The position of the object at any time t is given by $P(x(t), y(t))$ where $x(t) = 25t$ and $y(t) = -16t^2 + 25\sqrt{3}t$ (Figure 7.6.2).

(b) The object will hit the ground when $y(t) = 0$:

$$y(t) = -16t^2 + 25\sqrt{3}t = 0$$
$$t(25\sqrt{3} - 16t) = 0.$$

Now, $y(t) = 0$ when $t = 0$ or $t = \frac{25\sqrt{3}}{16}$. Thus, the object will hit the ground when $t = \frac{25\sqrt{3}}{16} = 2.7063294$, or approximately 2.71 seconds after launch.

(c) We need to determine the value of $x(t) = 25t$ when the object hits the ground. Using the value of t from (b) we will find that $x(t) = 67.66$ when $t = 2.7063294$. Thus, the object travels 67.66 feet in the horizontal direction.

(d) The function $y(t) = -16t^2 + 25\sqrt{3}t$ is a parabola that opens downward and has zeros of 0 and $\frac{25\sqrt{3}}{16}$. Thus, the largest value of y is attained when t is half way between its two zeros, or when $t = \frac{25\sqrt{3}}{32}$. The value of t is 1.3531647 and the corresponding values of x and y are 33.83 and 29.30, respectively. The object reaches its maximum height of 29.3 feet 1.35 seconds after launch. The graph of the parametric equations in Figure 7.6.3 can be used to confirm this information.

(e) The position at any time t of the object in vector form is

$$p = (25t, 25\sqrt{3}t - 16t^2).$$

We can rewrite p as follows:

$$\begin{aligned} p &= (25t, 25\sqrt{3}t - 16t^2) \\ &= (25t, 25\sqrt{3}t) + (0, -16t^2) \\ &= t(25, 25\sqrt{3}) + (0, -16t^2). \end{aligned}$$

Notice that the vector $\boldsymbol{v}_1 = t(25, 25\sqrt{3}) = t\boldsymbol{v}_0$ depends only on the initial velocity vector, and the vector $\boldsymbol{v}_2 = (0, -16t^2)$ depends only on gravity (Figure 7.6.4). Thus, $\boldsymbol{p} = t\boldsymbol{v}_0 + \boldsymbol{v}_2$. •

A parametric graphing utility actually simulates the path (motion) of an object. That is, by observing a computer or graphing calculator draw the graph of $P(x(t), y(t))$ for $a \leq t \leq b$, we can see the path of the object from time $t = a$ to time $t = b$ as it actually evolves.

• EXAMPLE 2: A baseball is hit when it is 3 feet above the ground and leaves the bat with initial velocity of 150 feet per second and at an angle of elevation of 20°. A 6 mph wind is blowing in the horizontal direction against the batter. A 20 foot high fence is 400 feet from home plate. Will the hit be a home run?

SOLUTION: First we determine parametric equations for the motion of the ball. Let t be the time in seconds after the ball is hit and $P(t) = (x(t), y(t))$ be the coordinates of the position of the ball. Now $P(0) = (0, 3)$. We need to determine $x(t)$ and $y(t)$. The initial velocity that acts on the ball due to the bat is $\boldsymbol{v}_0 = (150 \cos 20°, 150 \sin 20°)$. Thus, the effect of the bat on the ball at any time t is given by the vector $\boldsymbol{v}_1 = (150t \cos 20°, 150t \sin 20°)$. The effect on the ball due to gravity at any time t is $\boldsymbol{v}_2 = (0, 3 - 16t^2)$. It can be shown that 6 mph is the same as 8.8 feet per second. The effect on the ball of the wind is given by the vector $\boldsymbol{v}_3 = (-8.8t, 0)$ because the wind is blowing in the horizontal direction against the batter (Figure 7.6.5). Therefore, the path of the ball is given by

$$\begin{aligned} P(t) &= \boldsymbol{v}_1 + \boldsymbol{v}_2 + \boldsymbol{v}_3 \\ &= (150t \cos 20°, 150t \sin 20°) + (0, 3 - 16t^2) + (-8.8t, 0) \\ &= (150t \cos 20° - 8.8t, 150t \sin 20° + 3 - 16t^2). \end{aligned}$$

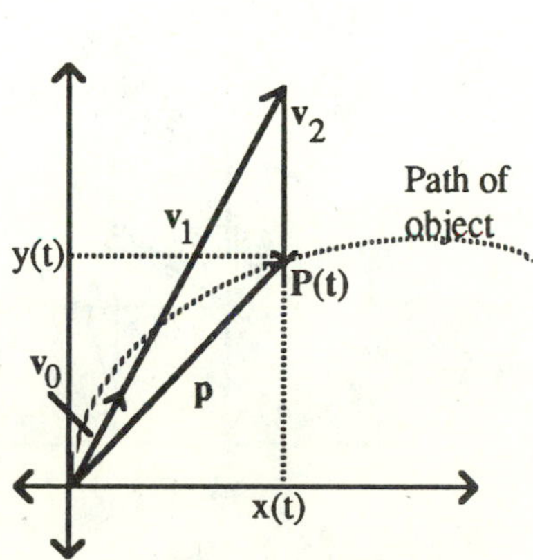

Figure 7.6.4

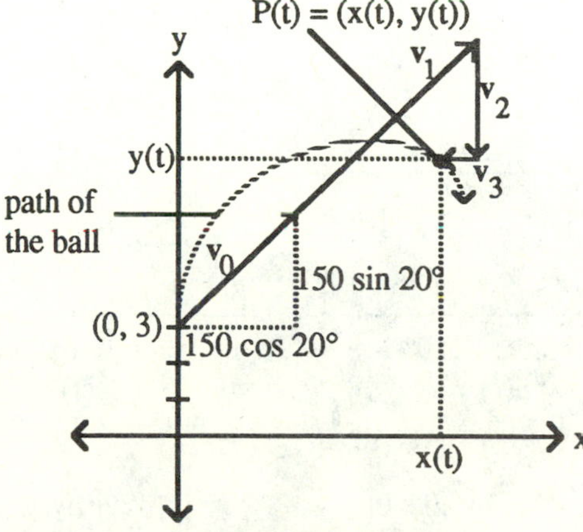

Figure 7.6.5

It now follows that

$$x(t) = 150t\cos 20^\circ - 8.8t\,, \quad \text{and}$$
$$y(t) = 150t\sin 20^\circ + 3 - 16t^2\,.$$

The ball will reach the fence when $x(t) = 400$. This occurs when t is 3.03 seconds. At this value of t we have $y(t) = 11.7$. Thus, the ball does not clear the fence. We need not do the above computations to determine if the hit is a home run. The graph of the curve defined by the above parametric equations together with the graph of the horizontal line $y = 20$ and the vertical line $x = 400$ in Figure 7.6.6 clearly shows that the hit is *not* a home run. Again, by watching a computer or graphing calculator draw a graph of the parametric equations we can see a simulation of the actual path of the ball. •

• EXAMPLE 3: Suppose the ball of Example 2 leaves the bat at an angle of elevation of 30° but all other conditions of the example remain the same. Will this hit be a home run?

SOLUTION: The parametric equations for the path of the ball in this case are the following:

$$x(t) = 150t\cos 30^\circ - 8.8t$$
$$y(t) = 150t\sin 30^\circ + 3 - 16t^2.$$

Notice only the angle value changes from the previous example. The graph in Figure 7.6.7 shows that this hit is a home run; and a big one at that! In the Exercises you will show that the ball reaches the fence in 3.30 seconds and the height of the ball at this time is 76.17 feet. •

• EXAMPLE 4: Let $C(0, 20)$ be the center of a circle with radius 20 and let $P(x, y)$ be any point on the circle. Find parametric equations for the circle in terms of the parameter θ, where θ is the angle that PC makes with the positive x-axis (Figure 7.6.8).

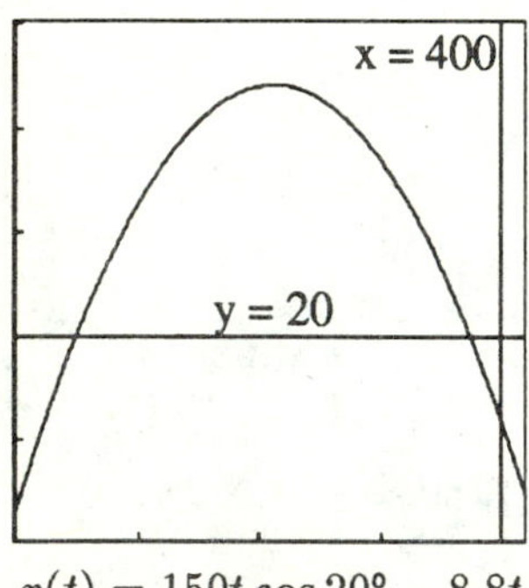

$x(t) = 150t\cos 20^\circ - 8.8t$
$y(t) =$
$150t\sin 20^\circ + 3 - 16t^2$
$0 \le t \le 5$
$[0, 420]$ by $[0, 50]$

Figure 7.6.6

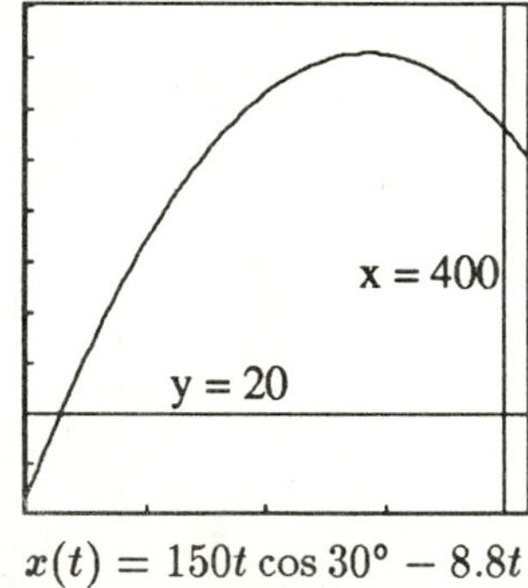

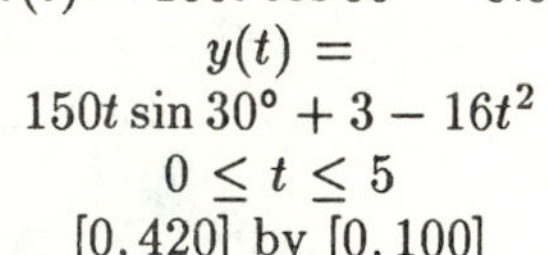

$x(t) = 150t\cos 30^\circ - 8.8t$
$y(t) =$
$150t\sin 30^\circ + 3 - 16t^2$
$0 \le t \le 5$
$[0, 420]$ by $[0, 100]$

Figure 7.6.7

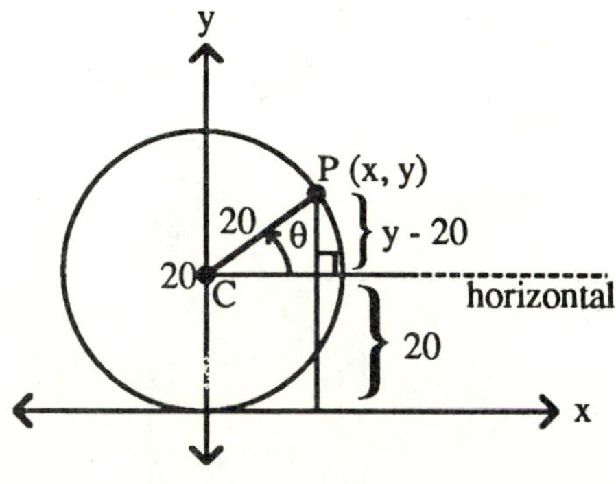

Figure 7.6.8

SOLUTION: We can think of θ as a central angle of the circle. Notice that $\cos\theta = \frac{x}{20}$ and $\sin\theta = \frac{y-20}{20}$. Thus, $x = 20\cos\theta$ and $y = 20 + 20\sin\theta$. A range on θ that produces the graph of the circle exactly once is given by any interval of length 2π. The graph in Figure 7.6.8 can be confirmed by using a parametric graphing utility to graph $x(\theta) = 20\cos\theta$, $y(\theta) = 20 + 20\sin\theta$ with $\theta\,\mathrm{min} = 0$ and $\theta\,\mathrm{max} = 2\pi$. •

Now we use the information gained in Examples 1 and 4 to help solve the following problem posed by Neal Koblitz in the March, 1988 issue of *The American Mathematical Monthly*.

• EXAMPLE 5: You are standing on the ground at point D (Figure 7.6.9), a distance of 75 feet from the bottom of a ferris wheel that is 20 feet in radius. Your arm is at the same level as the bottom of the ferris wheel. Your friend is on the ferris wheel, which makes one revolution (counterclockwise) every 12 seconds. At the instant when she is at point A you throw a ball to her at 60 ft/sec at an angle of 60° above the horizontal. Take $g = 32$ ft/sec^2, and neglect air resistance. How close does the ball get to your friend?

SOLUTION: We assume that the ferris wheel is in a rectangular coordinate system with a diameter along the y-axis and the bottom at the origin. Then, the ferris wheel is a circle with center $C(0, 20)$ and radius 20 (Figure 7.6.10). We know from Example 4 that parametric equations for the position of the point $P = (x_A(\theta), y_A(\theta))$ on the ferris wheel are given by $x_A(\theta) = 20\cos\theta$ and $y_A(\theta) = 20 + 20\sin\theta$, where θ is the angle that the radius CP makes with the positive x-axis.

Because the ferris wheel makes one complete revolution every 12 seconds we can write θ as a function of time t (in seconds) as follows: $\theta = \frac{2\pi}{12}t = \frac{\pi}{6}t$. Thus, in terms of the parameter t (time), we have the following parametric equations for the position P of the

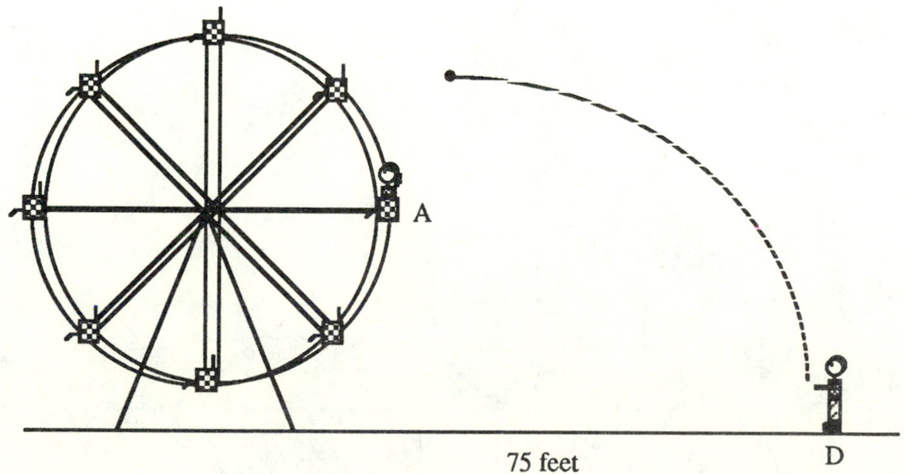

Figure 7.6.9

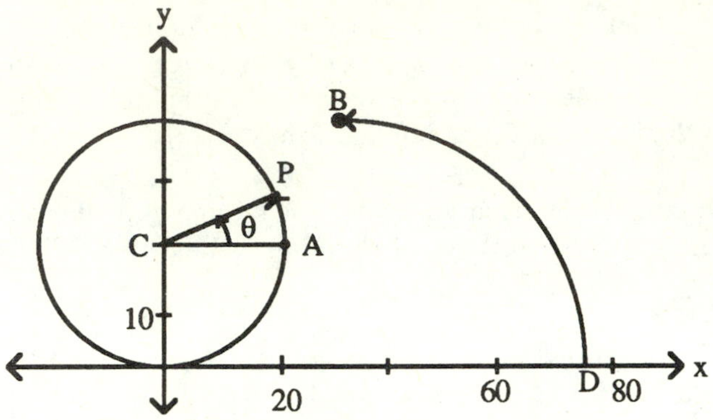

Figure 7.6.10

friend on the ferris wheel:

$$x_A(t) = 20\cos\left(\frac{\pi}{6}t\right)$$
$$y_A(t) = 20 + 20\sin\left(\frac{\pi}{6}t\right).$$

Next, we determine parametric equations for the path $B = (x_B(t), y_B(t))$ of the ball. The initial velocity at which the ball is thrown has horizontal and vertical components $60\cos 120°$ and $60\sin 120°$, respectively (Figure 7.6.11). Notice $\cos 120°$ is negative so that the horizontal component acts to the left.

Thus, we have the following parametric equations for the path of the ball:

$$x_B(t) = 75 + (60\cos 120°)t = 75 - 30t,$$
$$y_B(t) = -16t^2 + (60\sin 120°)t = -16t^2 + 30\sqrt{3}\,t.$$

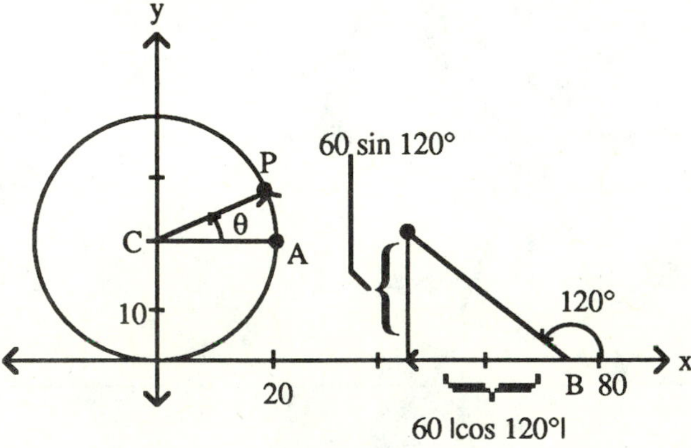

Figure 7.6.11

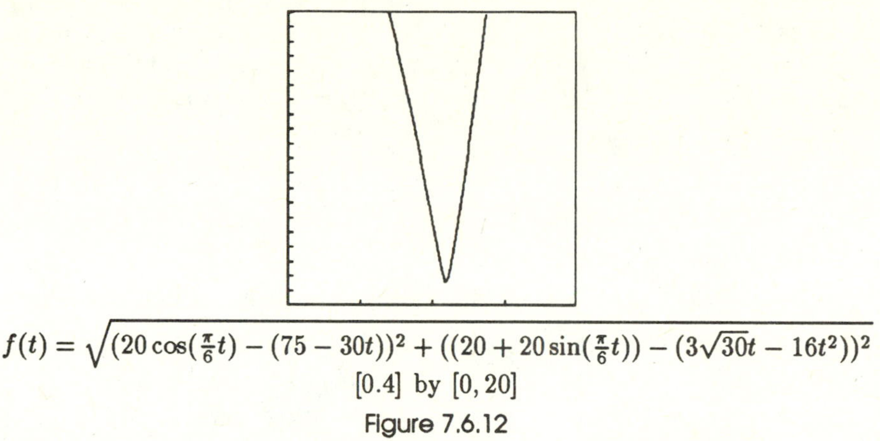

$$f(t) = \sqrt{(20\cos(\tfrac{\pi}{6}t) - (75 - 30t))^2 + ((20 + 20\sin(\tfrac{\pi}{6}t)) - (3\sqrt{30}t - 16t^2))^2}$$

[0.4] by [0, 20]

Figure 7.6.12

The parametric graphing utility that we use and graphing calculators permit us to draw the two curves simultaneously. That is, we can see the points being plotted on each curve for the same value of t. This simultaneous feature of the graphing utility permits us to actually simulate the real world problem situation. It can be shown that the ball comes close to the friend on the ferris wheel, but they do not meet. In fact, the distance between the ball and the friend steadily decreases to a minimum that is less than 2 and then increases again. The minimum occurs for t between 2.1 and 2.3 seconds. •

• EXAMPLE 6: Find the minimum distance between the ball and the friend on the ferris wheel and the time at which the minimum occurs (Figure 7.6.9).

SOLUTION: We have shown that $x_A(t) = 20\cos(\frac{\pi}{6}t)$ and $y_A(t) = 20 + 20\sin(\frac{\pi}{6}t)$ are parametric equations for the path of the friend on the ferris wheel, and that $x_B(t) = 75 - 30t$ and $y_B(t) = 30\sqrt{3}\,t - 16t^2$ are parametric equations for the path of the ball. For each t, the distance between the friend and the ball is given by $D(t) = \sqrt{(x_A(t) - x_B(t))^2 + (y_A(t) - y_B(t))^2}$. We can use $y_B(t)$ to show that the ball will hit the ground in a little more than 3 seconds. Thus, the graph of D in the [0, 4] by [0, 20] viewing rectangle given in Figure 7.6.12 shows a complete graph of the problem situation. We can use zoom-in to show that D has an absolute minimum of 1.59 and this value of D occurs at $t = 2.19$. Therefore, the minimum distance between the ball and the friend on the ferris wheel is 1.59 feet and occurs 2.19 seconds after the ball is thrown. Furthermore, this answer has error at most 0.01. •

• EXERCISES 7-6

Draw a complete graph.

1. $x(t) = 2t - 6\sin t$ and $y(t) = 2 - 6\cos t$
2. $x(t) = \cos^2 t$ and $y(t) = (1 - \cos t)\sin t$
3. $x(t) = 2 - 5\cos t$ and $y(t) = -2 + 4\cos t$

4. $x(t) = \frac{4}{1+\sin t}\cos t$ and $y(t) = \frac{4}{1+\sin t}\sin t$

A dart is thrown upward with an initial velocity of 58 feet per second at an angle of elevation 41°. Consider the position of the dart at any time t ($t = 0$ when the dart is thrown). Neglect air resistance.

5. Find parametric equations that model the problem situation.
6. Draw a complete graph of the model.
7. What portion of the graph represents the problem situation?
8. When will the dart hit the ground?
9. Find the maximum height of the dart. At what time will it reach maximum height?
10. How far does the dart travel in the horizontal direction? Neglect air resistance.
11. An arrow is shot with an initial velocity of 205 feet per second and at an angle of elevation of 48°. Find when and where the arrow will strike the ground.
12. A golfer hits a ball with an initial velocity of 133 feet per second and at an angle of 36° from the horizontal. Find when and where the ball will hit the ground.
13. Will the ball in Exercise 12 clear a fence 9 feet high that is at a distance of 275 feet from the golfer?
14. With what initial velocity must a ball be thrown from the ground at an angle of 35° from the horizontal if it is to travel a horizontal distance of 255 feet?
15. A batter hits a ball with an initial velocity of 92 feet per second and at an angle of 55° from the horizontal. Find the maximum height attained and the total horizontal distance traveled by the ball.

Chris and Linda are standing 78 feet apart. At the same time they each throw a softball toward each other. Linda throws her ball with an initial velocity of 45 feet per second with an angle of inclination of 44°. Chris throws her ball with an initial velocity of 41 feet per second with an angle of inclination of 39°.

16. Find two sets of parametric equations that represent a model of the problem situation.
17. Draw complete graphs of both sets of parametric equations in the same viewing rectangle.
18. What values of t make sense in this problem situation?
19. Find the maximum height of each ball. How far does each ball travel in the horizontal direction? When does each ball hit the ground? Whose ball hits first?
20. By choosing t max carefully (guess and check), estimate how close the two balls get, and the time when they are closest (minimum distance).
21. Use the distance formula and zoom-in to find the minimum distance (and the time at which it occurs) with error less than 0.01.
22. A river boat's paddle wheel has diameter of 26 feet and makes 1 revolution clockwise in 2 seconds at full speed. Write parametric equations describing the position of a point A on the paddle. Assume at $t = 0$, A is at the very top of the wheel as shown in Figure 7.6.13.
23. In Exercise 22, how far has point A, which is fixed on the wheel, moved in one minute?
24. Solve the ferris wheel and ball problem (Example 5) if the radius of the ferris wheel is 26 feet and the ball is thrown with initial velocity 76 feet per second at an angle of 52° with the horizontal

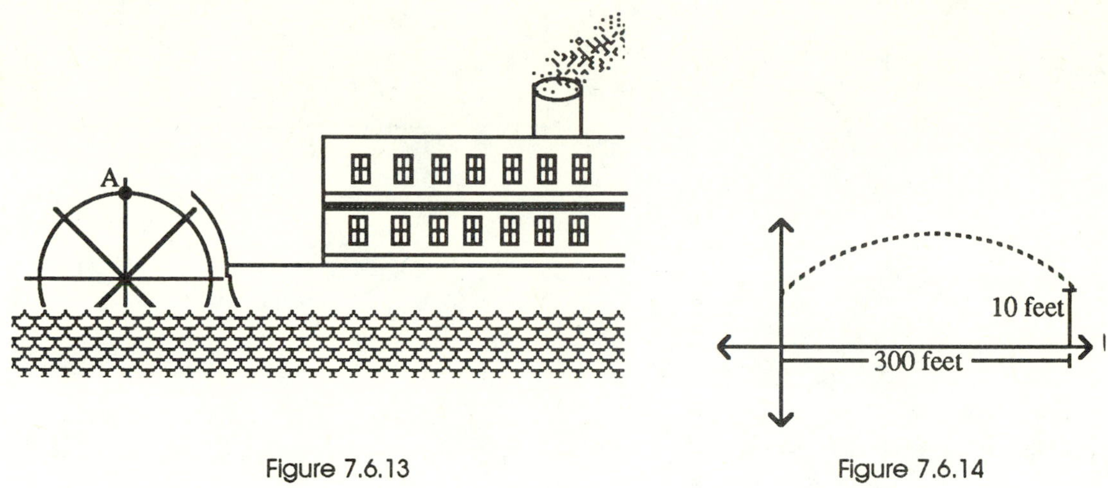

Figure 7.6.13

Figure 7.6.14

from a distance of 62 feet from the bottom of the ferris wheel. Use parametric simulation and estimate your answer. (Assume the same number of seconds per ferris wheel revolution.)

25. Solve Exercise 24 using the distance formula and zoom-in.

26. Use the parametric simulation and vary the position of the ball thrower (from 62 feet) and see how close you can get the ball to the friend on the ferris wheel. Use a guess and check method. (Keep all other values the same.)

27. Use the parametric simulation and vary the angle of elevation (from 52°) and see how close you can get the ball and the friend on the ferris wheel. Use a guess and check method. (Keep all other values the same.)

An NFL punter at the 15 yard line kicks a football downfield with initial velocity 85 feet per second at an angle of elevation of 56°.

28. Draw a complete graph of the problem situation.

29. How far downfield will the football first hit the field?

30. Determine the maximum height of the ball above the field.

31. What is the "hang time?" (That is, the total time the football is in the air.)

A Major League baseball player hits a fast ball with initial velocity 103 feet per second on a path 300 feet (measured horizontally) from the home run fence. The fence is 10 feet high. Draw a complete graph of the problem situation and determine if the hit is a home run (clears the fence). We assume (unless stated otherwise) that gravity is the only force affecting the path of the ball (Figure 7.6.14).

32. The hit is at an angle of elevation of 35°.

33. The hit is at an angle of elevation of 43°.

34. The hit is at an angle of elevation of 49°.

35. The hit is at an angle of elevation of 41° and there is a wind of 22 feet per second blowing in the same direction as the horizontal path of the ball.

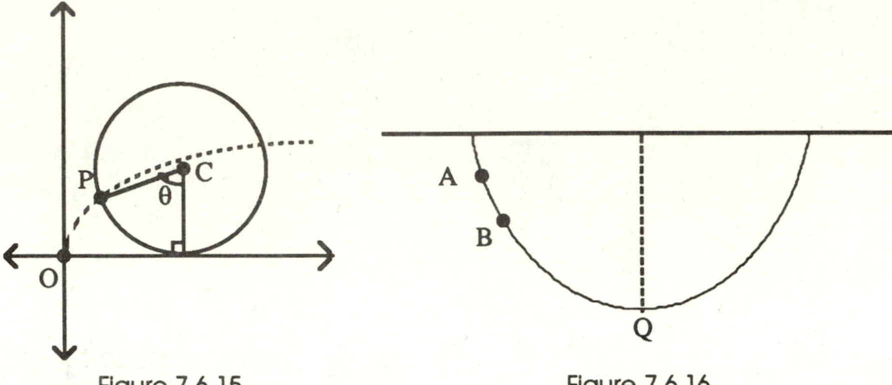

Figure 7.6.15

Figure 7.6.16

36. The hit is at an angle of elevation of 41° and there is a wind of 22 feet per second blowing in the opposite direction as the horizontal path of the ball.
37. The hit is at an angle of elevation of 41° and there is a wind of 22 feet per second blowing in the opposite direction as the horizontal path of the ball; the wind is blowing with angle of depression 12°.
38. Show algebraically that in Example 3, the ball reaches the fence in 3.30 seconds and is 76.17 feet above the ground at that time.

The Cycloid Revisited

*39. Show that the path of a fixed point P on a circle, with radius a and center C, that rolls on a straight line is the **cycloid**

$$x(\theta) = a(\theta - \sin\theta),$$
$$y(\theta) = a(1 - \cos\theta),$$

where θ represents the angle the radius CP has turned through. Assume the starting position is at the origin and θ is a positive angle, even though the rotation is clockwise (Figure 7.6.15).

*40. Design an experiment in physics to show the following remarkable property of cycloids. If Q is the lowest point of an arch of an inverted cycloid, then the time it takes for a frictionless bead to slide down the curve to point Q is *independent* of the starting point. That is, if A and B are dropped at the same time they will land at Q at the same time (Figure 7.6.16).

• KEY TERMS

Listed below are the key terms, vocabulary, and concepts in this chapter. You should be familiar with their definitions and meanings as well as be able to illustrate each of them.

Acceleration
Addition of vectors
Cardioid
Cis notation
Complex numbers
Complex plane

Cycloid
DeMoivre's theorem
Directed line segment
Gaussean plane
Imaginary axis
Initial point of vectors
Initial velocity
Linear combination
Magnitude of vectors
Modulus
Motion simulation
Parameters
Parametric equations
Parametrization
Plane curve
Polar coordinates
Polar graphs
Radial shrinking
Radial stretching
Real axis
Rectangular coordinates
Regular n-gon
Roots of unity
Scalar multiplication of vectors
Scalars
Speed
Spiral of Archimedes
Terminal point of vectors
Trig form of complex numbers
Vector
Zero vector

• REVIEW EXERCISES

This section contains representative problems from each of the previous six sections. You should use these problems to test your understanding of the material covered in this chapter.

Compute. Express the answers in the form $a + bi$ where a and b are real numbers.

1. $(4+3i)-(2-i)$
2. $(1-i)(3+2i)$
3. $\dfrac{-2+4i}{1-i}$

Graph in the complex plane.

4. $-5+2i$
5. $2(\cos\frac{\pi}{3} + i\sin\frac{\pi}{3})$

Consider the complex number z_1 as represented in Figure R.1.

6. If $z_1 = a + bi$, determine a, b, and $|z_1|$.

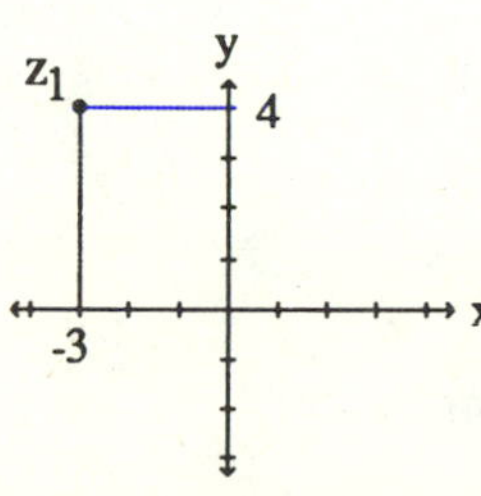

Figure R.1

7. Determine the trig form of z_1.

Let $z_1 = -1 + 4i$, $z_2 = 3 + 4i$, and $z_3 = 5 - 2i$. Perform the indicated computation and write in the form $a + bi$.

8. $z_1 - z_2$

9. $z_2 z_3$

10. $|z_1 z_3|$

Write the complex number in the form $a + bi$.

11. $6(\cos 30^\circ + i \sin 30^\circ)$

12. $2.5(\cos 4 + i \sin 4)$

Write in trig form with $0 \leq \theta \leq 2\pi$, then give all possible trig forms.

13. $3 - 3i$

14. $-1 + \sqrt{2}i$

15. $3 - 5i$

16. $-2 - 2i$

Use Theorem 1 of Section 7.1 to compute. (*Note:* $\operatorname{cis}\theta = \cos\theta + i\sin\theta$.)

17. $(3 \operatorname{cis} 30^\circ)(4 \operatorname{cis} 60^\circ)$

18. $\dfrac{3 \operatorname{cis} 30^\circ}{4 \operatorname{cis} 60^\circ}$

19. Write the two complex numbers in Exercise 17 in the form $a + bi$, perform the multiplication, and verify that the result is the same as that obtained using the trig form and Theorem 1 of Section 7.1.

20. Write the two complex numbers in Exercise 18 in the form $a+bi$, perform the division, and verify that the result is the same as that obtained using the trig form and Theorem 1 of Section 7.1.

Compute.

21. $(1 - i)^8$

22. $(3 \operatorname{cis} \frac{\pi}{4})^5$

Find the nth roots of each complex number for the specified value of n. Graph each nth root in the complex plane.

23. $n = 4$, $3 + 3i$

24. $n = 3$, 8

25. Determine z and the three cube roots of z if one cube root of z is $2 + 2i$.

26. Solve $z^4 = 4 - 4i$ and graph the solutions.

Compute a polar coordinate pair for the point with given rectangular coordinates.

27. $(-4, 5)$

28. $(8, -1)$

Compute *all* polar coordinate pairs for the point with given rectangular coordinates.

29. $(-2, 2)$

30. $(3, -1)$

Compute the rectangular coordinates for the point with given polar coordinates.

31. $(-4, 225^\circ)$

32. $(3, \frac{4\pi}{3})$

33. Compute the distance between the points $(-2, 120^\circ)$ and $(3, 60^\circ)$.

Sketch a complete graph of each equation without using a graphing utility. Check your answer with a graphing utility.

34. $r = 5$

35. $r = 2\sin\theta$

36. $r = 3 + 3\cos\theta$

Use a graphing utility to draw the polar graph for θ in the given interval.

37. $r = 4\theta$; $0 \leq \theta \leq \frac{\pi}{2}$, $0 \leq \theta \leq \pi$, $0 \leq \theta \leq 2\pi$, $-\pi \leq \theta \leq \pi$, and $0 \leq \theta \leq 4\pi$.

38. $r = 5\sin 3\theta$; $0 \leq \theta \leq \frac{\pi}{2}$, $0 \leq \theta \leq \pi$, $0 \leq \theta \leq 2\pi$, $-\pi \leq \theta \leq \pi$, and $0 \leq \theta \leq 4\pi$.

Use a graphing utility to draw a complete graph of the polar equation. Specify an interval $a \leq \theta \leq b$ of smallest length that gives a complete graph.

39. $r = 2\cos 4\theta$

40. $r = 3\theta$

41. Determine the length of *one* rose "petal" for $r = 6\sin 2\theta$.

Solve the polar system by using a graphing utility.

42. $r = 3\cos\theta$
$r = 3\sin\theta$

43. $r = 2 + \sin\theta$
$r = 1 - \cos\theta$

Let $\boldsymbol{u} = (2, -1)$, $\boldsymbol{v} = (4, 2)$, and $\boldsymbol{w} = (1, -3)$ be vectors. Determine the following.

44. $\boldsymbol{u} - \boldsymbol{v}$

45. $2\boldsymbol{u} - 3\boldsymbol{w}$

46. $|\boldsymbol{u} + \boldsymbol{v}|$

47. Use the geometric definitions of addition and subtraction of vectors to find the sum $\boldsymbol{u} + \boldsymbol{v}$, the difference $\boldsymbol{u} - \boldsymbol{v}$, and then compare with the results obtained using the respective componentwise definitions for $\boldsymbol{u} = (2, 4)$ and $\boldsymbol{v} = (-1, 3)$.

Let $A = (2, -1)$, $B = (3, 1)$, $C = (-4, 2)$, and $D = (1, -5)$. Write each as a vector with *initial point at the origin.*

48. $3\overrightarrow{AB}$

49. $\overrightarrow{AB} + \overrightarrow{CD}$

50. Compute $|\overrightarrow{AB} + \overrightarrow{CD}|$.

Express the vector $\overrightarrow{AB}$ as a linear combination of $\boldsymbol{i} = (1, 0)$ and $\boldsymbol{j} = (0, 1)$.

51. $A = (4, 0)$, $B = (2, 1)$

52. $A = (3, 1)$, $B = (5, 1)$

Find the angle each vector makes with the positive x-axis. Then determine the angle between the vectors.

53. $\boldsymbol{u} = (4, 3)$, $\boldsymbol{v} = (2, 5)$

54. $\boldsymbol{u} = (-2, 4)$, $\boldsymbol{v} = (6, 4)$

55. Let $\boldsymbol{r} = (1, 2)$ and $\boldsymbol{s} = (2, -1)$. Show that *any* vector $\boldsymbol{v} = (x, y)$ can be expressed as a linear combination of $\boldsymbol{r}$ and $\boldsymbol{s}$.

Describe the curve defined parametrically. Sketch the graph without using a graphing utility. Check using a parametric graphing utility.

56. $x(t) = 2 + 3t$ and $y(t) = t$ with t in $[0, 5]$

57. $x(t) = 2t$ and $y(t) = \frac{3}{t}$ with t in $[1, 4]$

58. $x(t) = 4\cos\theta$ and $y(t) = 4\sin\theta$ with θ in $[-2\pi, 2\pi]$

59. Which of the curves in Exercises 56–58 are graphs of functions?

60–62. What are the domain and range of the curves in Exercises 56–58?

63. Which of the curves in Exercises 56–58 are graphs of one-to-one functions?

64. Find parametric equations and draw the graph of the circle centered at $(5, 0)$ with radius 4.

Draw a complete graph.

65. $x(t) = 3t + 2$ and $y = 6t^2$ with t in $(-\infty, \infty)$

66. $x(t) = 2 - \frac{1}{t}$ and $y = t + \frac{1}{t}$ with t in $(0, 20]$

67. $x(t) = \sin^2 t$ and $y(t) = (1 - \sin t \cos t)$ with t in $[0, 2\pi]$

68. $x(t) = t + \sin t$ and $y(t) = 2 + \cos t$ with t in $[-10, 10]$

69. $x(t) = 12\cos t - 8\cos(\frac{3}{2}t)$ and $y(t) = 12\sin t - 8\sin(\frac{3}{2}t)$ with t in $[0, 2\pi]$ (This is an *epicycloid*, the path of a point on a circle rolling on another circle.)

70. Find the domain and range of the curve in Exercise 69.

71. Write the polar equation $r = 2\theta$ for θ in $[0, 10\pi]$ in parametric form. Graph *both* the parametric and polar form with an appropriate graphing utility.
72. A plane is flying on a bearing of 10° east of south at 460 mph. A tail wind is blowing in the direction 20° west of south at 80 mph. Express the actual velocity of the plane as a vector. Determine the actual speed and direction of the plane.
73. A plane is flying on a bearing of 10° east of south at 460 mph. A 30 mph head wind is blowing in the direction of 20° east of north. Express the actual velocity of the plane as a vector. Determine the actual speed and direction of the plane.
74. A sailboat under auxiliary power is proceeding on a bearing of 25° north of west at 6.25 mph in still water. Then a tail wind blowing 15 mph in the direction 35° south of west alters the course of the sailboat. Express the actual velocity of the sailboat as a vector. Determine the actual speed and direction of the boat.

A dart is thrown upward with an initial velocity of 60 feet per second at an angle of elevation 35°. Consider the position of the dart at any time t ($t = 0$ when the dart is thrown). Neglect air resistance.

75. Find parametric equations that model the problem situation.
76. Draw a complete graph of the model.
77. What portion of the graph represents the problem situation?
78. When will the dart hit the ground?
79. Find the maximum height of the dart. At what time will it reach maximum height?
80. A golfer hits a ball with an initial velocity of 125 feet per second and at an angle of 33° from the horizontal. Find when and where the ball will hit the ground.
81. A batter hits a ball with an initial velocity of 100 feet per second and at an angle of 50° from the horizontal. Find the maximum height attained and the distance traveled by the ball.

An NFL punter at the 20-yard line kicks a football downfield with initial velocity 80 feet per second at an angle of elevation of 55°.

82. Draw a complete graph of the problem situation.
83. How far downfield will the football first hit the field?
84. Determine the maximum height above the field of the ball.
85. What is the "hang time?" (That is, the total time the football is in the air.)

8 Conics

• Introduction

In this chapter we study conic sections. A **conic section**, or **conic**, is any curve obtained by intersecting a double-napped right circular cone with a plane. The three basic conic sections are the **parabola**, the **ellipse**, and the **hyperbola**. A cone is the set of points that result when a line is rotated around another line, the axis of symmetry, that it intersects. A nappe is similar to an ice cream cone and is the portion of the cone formed by the rotation of one of the half lines. Figure 8.0.1 illustrates how the three conics are obtained as intersections of double-napped right circular cones and planes. Notice that the planes intersect one nappe to form a parabola or an ellipse and both nappes to form a hyperbola.

The position of the plane relative to the cone can sometimes result in **degenerate conics**. For example, if the plane contains the axis of symmetry of the cone, then the curve of intersection is a pair of intersecting lines. A pair of intersecting lines is a *degenerate hyperbola*.

Conics are important in applications. In this chapter we see that the path of any planet around the sun is an ellipse. The path of a comet can be hyperbolic. Conics have important *reflective properties*. Certain microphones and automotive headlamps are constructed using the reflective properties of parabolas. The reflective properties of parabolas and hyperbolas are used in optics. Ellipses are used in medicine. For example, the lithotripter is a device designed to break up kidney stones and remove the need for surgery. It uses shock waves and the reflective properties of an ellipse. Hyperbolas are used in navigation.

We will see that any conic is a graph of a quadratic equation in two variables of the form

$$Ax^2 + Bxy + Cy^2 + Dx + Ey + F = 0$$

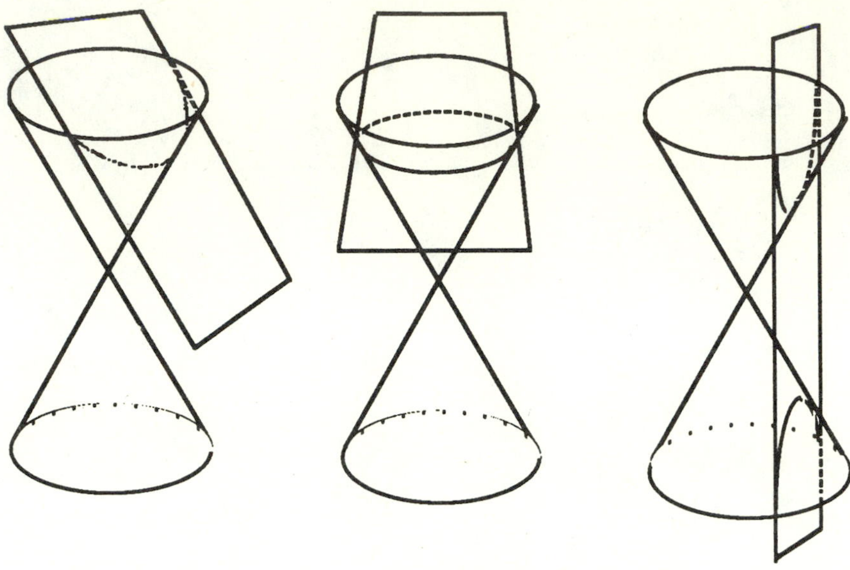

Figure 8.0.1

where at least one of A, B, or C must be different from zero. When $B = 0$, the above equation can be rewritten in a form to help make a quick sketch of its graph. This can also be done when $B \neq 0$, but the algebraic procedure is very tedious. We treat the case $B \neq 0$ using the quadratic formula and a function graphing utility.

Systems of equations and inequalities that involve equations quadratic in two variables are solved. Polar equations for conics are obtained in this chapter using the focus-directrix definition of a conic.

8.1 Parabolas

In this section we give the focus-directrix definition of a parabola and obtain the standard form for the equation of a parabola with line of symmetry parallel to a coordinate axis. In Section 8.4 we consider parabolas with line of symmetry *not* parallel to a coordinate axis. An application that uses the reflective properties of a parabola is investigated.

The path of motion of an object thrown through the air is approximately a parabola. If air resistance is neglected, or if the object is thrown in a vacuum, then the path traveled by the object is a parabola. Parabolas have important reflective properties that are discussed in this section.

In Sections 1.6 and 1.7 we showed that the graph of any quadratic function $y = ax^2 + bx + c$ can be obtained from the graph of $y = x^2$ by using vertical stretching or shrinking, reflection through the x-axis, vertical shifting, and horizontal shifting. The graph of $y = ax^2 + bx + c$ is a parabola that opens up if $a > 0$ and opens down if $a < 0$. The line of symmetry of this parabola is a vertical line. If we rewrite

$$y = ax^2 + bx + c = a(x - h)^2 + k,$$

then the line of symmetry is the vertical line $x = h$ and the vertex is (h, k) (Figure 8.1.1).

• DEFINITION Let L be any line and $F(x_0, y_0)$ any point not on L. The set of all points $P(x, y)$ in the plane determined by L and F that are equidistant from L and F is called a **parabola** (Figure 8.1.2). The point F is called the **focus** of the parabola, and the line L is called the **directrix** of the parabola. •

Notice that the midpoint of the line segment from F perpendicular to L is on the graph of the parabola. This point V is the **vertex** of the parabola. The vertex and focus lie on the same side of the directrix. The point P' that is symmetric to any point P on the parabola with respect to the line through F and V is a point of the parabola. (Why?) The line through F and V is a line of symmetry of the parabola.

We use the word *focus* in the definition of a parabola because of the *reflective properties* of a parabola. For example, light will be reflected by a parabola parallel to its line of symmetry if the source of the light is placed at the focus. Headlamps of motor vehicles use this principle. Similarly, sound waves traveling into the opening of a parabola parallel to its line of symmetry will be reflected through the focus of the parabola. Parabolic microphones use this principle.

In the first example we determine an equation of a parabola with given vertex, focus, and directrix.

• EXAMPLE 1: Determine an equation for the parabola with focus $(0, a)$ and directrix $y = -a$ where $a > 0$ (Figure 8.1.3). Draw a complete graph of the parabola and give the coordinates of the vertex.

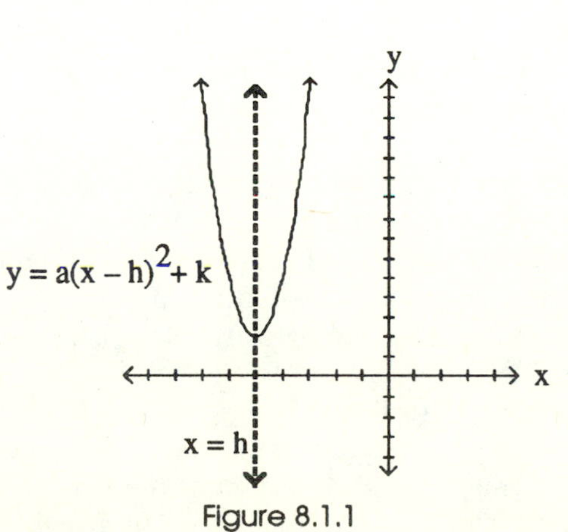

Figure 8.1.1

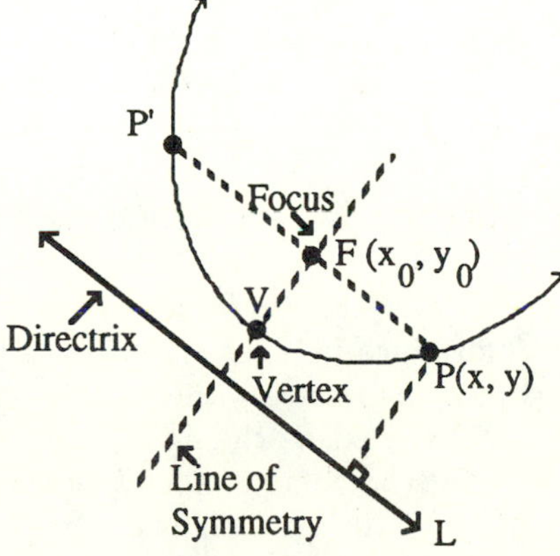

Figure 8.1.2

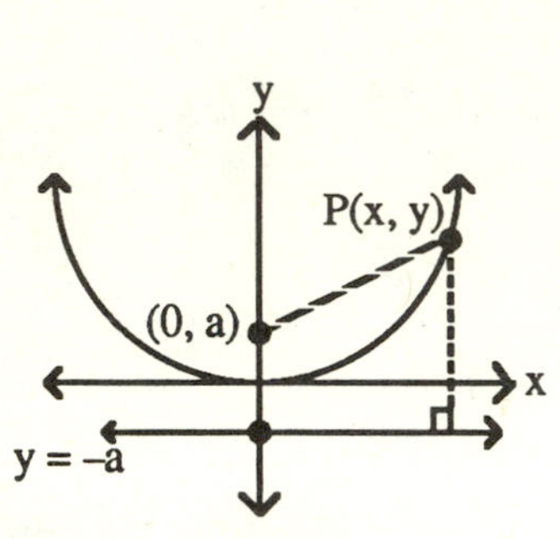

Figure 8.1.3

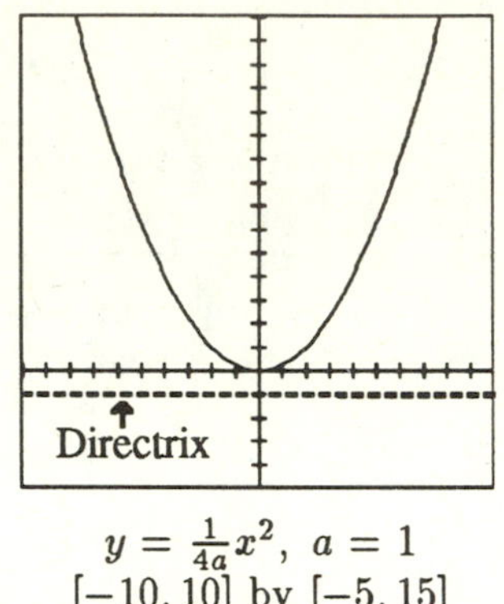

$y = \frac{1}{4a}x^2$, $a = 1$
$[-10, 10]$ by $[-5, 15]$

Figure 8.1.4

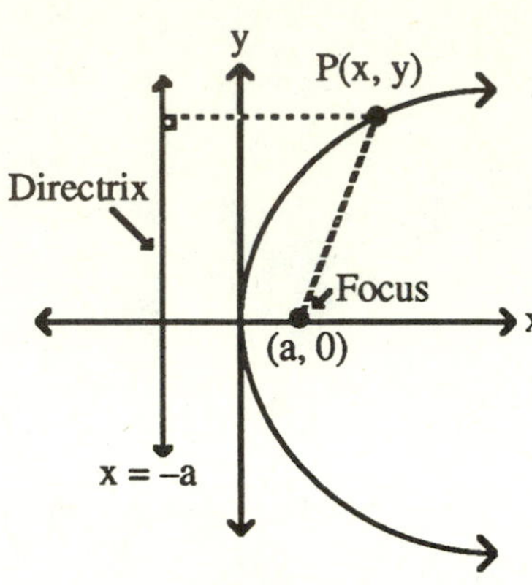

Figure 8.1.5

SOLUTION: The distance from $P(x, y)$ to the line $y = -a$ is $|y + a|$, and the distance from $P(x, y)$ to $(0, a)$ is $\sqrt{(x-0)^2 + (y-a)^2}$. (Why?) Applying the definition of a parabola, we obtain the following equation:

$$\sqrt{x^2 + (y-a)^2} = |y + a|. \tag{1}$$

Next, we square each side of equation (1) and simplify:

$$\begin{aligned} \sqrt{x^2 + (y-a)^2} &= |y+a| \\ x^2 + (y-a)^2 &= (y+a)^2 \\ x^2 + y^2 - 2ay + a^2 &= y^2 + 2ay + a^2 \\ x^2 &= 4ay. \end{aligned} \tag{2}$$

Thus, every point on the parabola is a solution to equation (2). Because the steps used to derive equation (2) from equation (1) are reversible, every solution to equation (2) corresponds to a point on the parabola.

We can rewrite equation (2) in the form $y = \frac{1}{4a}x^2$ to see that the graph can be obtained by starting with $y = x^2$ and applying a vertical stretch or shrink by the factor $\frac{1}{4a}$. The coordinates of the vertex are $(0, 0)$. The computer drawn graph for $a = 1$ in Figure 8.1.4 is a complete graph and suggests what the graphs for $a > 0$ look like. •

An equation for the parabola with focus $(0, a)$ and directrix $y = -a$ with $a < 0$ is also $x^2 = 4ay$. However, the graph of this equation is the reflection with respect to the x-axis of the graph in Example 1 because $a < 0$ in this case. Furthermore, $|a|$ is equal to half the distance between the focus and the directrix in each case. If the parabola opens up (focus above directrix), then $a > 0$; if the parabola opens down (focus below directrix), then $a < 0$. The y-axis is the line of symmetry of this parabola and is also the line through the focus perpendicular to the directrix. The vertex is $(0, 0)$.

If the directrix of a parabola is the line $x = -a$ and the focus is the point $(a, 0)$ (Figure 8.1.5), then, as in Example 1, we would find that an equation of the parabola is $y^2 = 4ax$. That is, we only need to interchange x and y in the equation found in Example 1. The

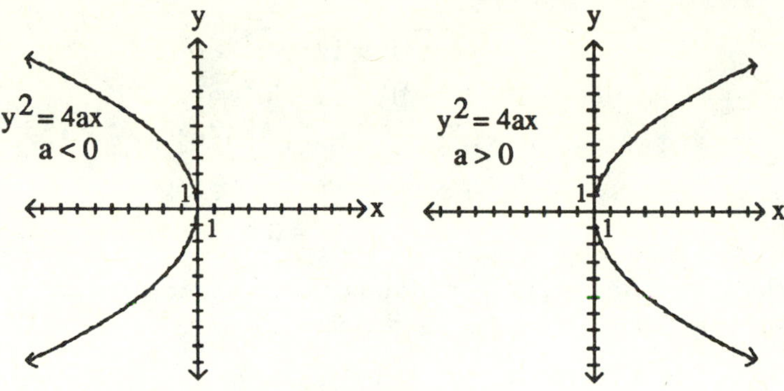

Figure 8.1.6

graph of $y^2 = 4ax$ is a parabola that opens to the right if $a > 0$, and a parabola that opens to the left if $a < 0$. In either case the line of symmetry of $y^2 = 4ax$ is the x-axis and the vertex is $(0,0)$ (Figure 8.1.6).

The next example illustrates that any parabola with directrix parallel to the x-axis can be obtained from $y = x^2$ by applying appropriate geometric transformations.

• EXAMPLE 2: Determine an equation for the parabola with focus $(3,6)$ and directrix $y = -2$. Give the coordinates of the vertex and the line of symmetry, and sketch a complete graph. Give a sequence of geometric transformations that will produce the graph of this parabola from the graph of $y = x^2$.

SOLUTION: The distance from $P(x,y)$ to $(3,6)$ is $\sqrt{(x-3)^2 + (y-6)^2}$, and the distance from $P(x,y)$ to $y = -2$ is $|y+2|$ (Figure 8.1.7). We apply the definition of a parabola and

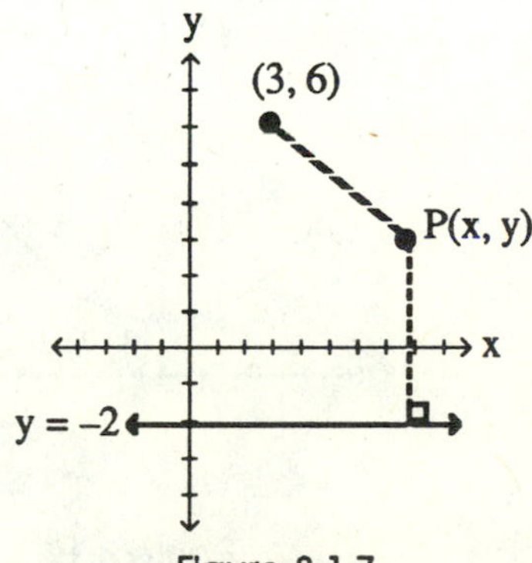

Figure 8.1.7

then simplify the resulting equation:

$$\begin{aligned}
\sqrt{(x-3)^2+(y-6)^2} &= |y+2| \\
(x-3)^2+(y-6)^2 &= (y+2)^2 \\
(x-3)^2+y^2-12y+36 &= y^2+4y+4 \\
(x-3)^2 &= 16y-32 \\
(x-3)^2 &= 16(y-2). \qquad (3)
\end{aligned}$$

Notice that the vertex is $(3,2)$ and the line of symmetry is $x = 3$.

We can rewrite equation (3) in the following way:

$$y = \frac{1}{16}(x-3)^2 + 2.$$

Now, we can see that a complete graph can be obtained from $y = x^2$ by applying a vertical shrink with factor $\frac{1}{16}$, followed by a horizontal shift right 3 units, and then a vertical shift up 2 units (Figure 8.1.8). •

Equation (3), $(x-3)^2 = 16(y-2)$, is often called the *standard form* for the equation of the parabola of Example 2. It follows from equation (2) that the coefficient 16 of $y-2$ is four times the distance between the focus and vertex, or, equivalently, four times the perpendicular distance between the vertex and directrix. (Why?) We can use this information to simplify writing equations for parabolas.

• EXAMPLE 3: Write an equation, determine the line of symmetry, the focus, and sketch a complete graph of the parabola with

(i) vertex $(-2,1)$ and directrix $y = -2$.
(ii) vertex $(3,-1)$ and directrix $y = 1$.

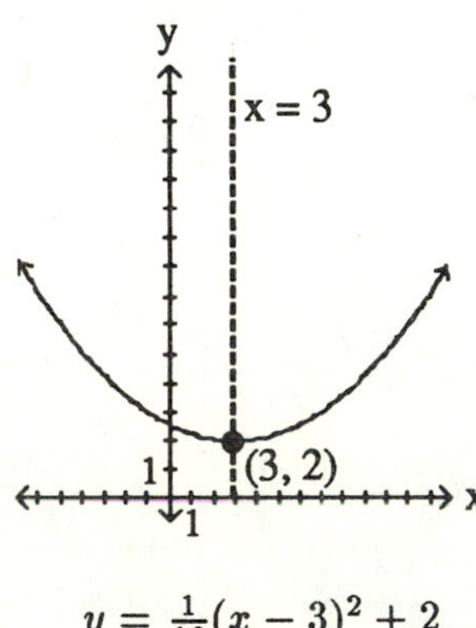

$y = \frac{1}{16}(x-3)^2 + 2$

Figure 8.1.8

SOLUTION:

(i) This parabola opens upward because the vertex is above the directrix. The line of symmetry is $x = -2$, and an equation for the parabola is $4a(y-1) = (x+2)^2$ where $a > 0$. The perpendicular distance between the vertex and directrix is 3. Thus, $a = 3$ and an equation for the parabola is $12(y-1) = (x+2)^2$. The focus is 3 units above the vertex and is on the line $x = -2$; its coordinates are $(-2, 4)$. A complete graph of $12(y-1) = (x+2)^2$ or $y = \frac{1}{12}(x+2)^2 + 1$ can be obtained from the graph of $y = x^2$ by applying a vertical shrink with factor $\frac{1}{12}$, followed by a horizontal shift left 2 units, and then a vertical shift up 1 unit. (Why?) Alternatively, we can use the fact that the graph of $12(y-1) = (x+2)^2$ is a parabola that opens upward with vertex $(-2, 1)$ and compute a few more points to produce a quick sketch. The x-intercepts and y-intercepts are good points to include. However, you may only get one additional point this way. In this case, there are no x-intercepts and $(0, \frac{4}{3})$ is the y-intercept. Symmetry can be used to produce $(-4, \frac{4}{3})$ as a second point. (Why?) A complete graph is given in Figure 8.1.9. A function graphing utility could be used to confirm this graph.

(ii) This parabola opens downward because the vertex is below the directrix. The line of symmetry is $x = 3$, and an equation for the parabola is $4a(y+1) = (x-3)^2$ where $a < 0$. The perpendicular distance between the vertex and directrix is 2. Thus, $a = -2$, and an equation for the parabola is $-8(y+1) = (x-3)^2$. The focus is 2 units below the vertex and is on the line $x = 3$; its coordinates are $(3, -3)$. Again we can use geometric transformations or plot three points to sketch a complete graph. There are no x-intercepts and $(0, -\frac{17}{8})$ is the y-intercept. Thus, $(6, -\frac{17}{8})$ is another point on the graph. A complete graph is given in Figure 8.1.10. •

We can extend Example 3 to any parabola with directrix parallel to the y-axis. We use the fact that an equation of such a parabola has the form $(y-k)^2 = 4a(x-h)$ where (h, k) is the vertex of the parabola.

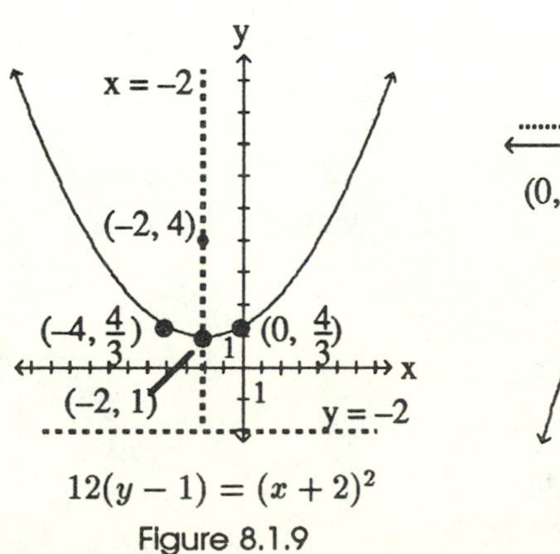

$12(y-1) = (x+2)^2$

Figure 8.1.9

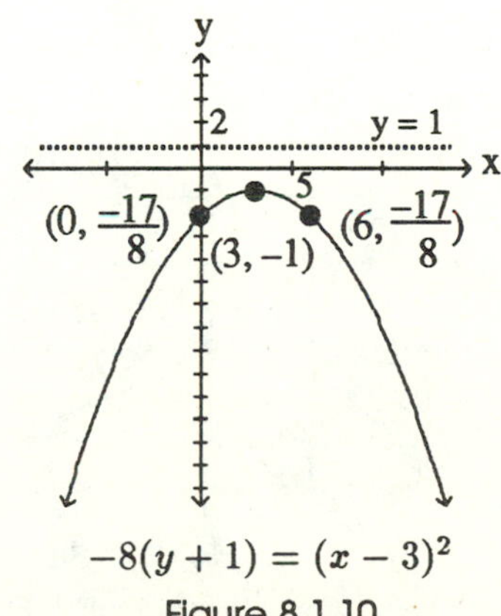

$-8(y+1) = (x-3)^2$

Figure 8.1.10

• EXAMPLE 4: Write an equation, determine the line of symmetry, the focus, and sketch a complete graph of the parabola with

(i) vertex $(-2, -3)$ and directrix $x = -\frac{7}{2}$.
(ii) vertex $(4, 1)$ and directrix $x = 6$.

Check each graph with a graphing utility.

SOLUTION:

(i) This parabola opens to the right because the vertex is to the right of the directrix. The line of symmetry is $y = -3$, and an equation for the parabola is $(y+3)^2 = 4a(x+2)$ where $a > 0$. The perpendicular distance between the vertex and directrix is $\frac{3}{2}$. Thus, $a = \frac{3}{2}$, and an equation for the parabola is $(y+3)^2 = 6(x+2)$. The focus is $\frac{3}{2}$ units to the right of the vertex and is on the line $y = -3$; its coordinates are $\left(-\frac{1}{2}, -3\right)$. We can determine two points in addition to the vertex to sketch the complete graph given in Figure 8.1.11. The points $(0, 0.46), (0, -6.46)$ are the two y-intercepts.

If we solve for y in terms of x we obtain $y + 3 = \pm\sqrt{6(x+2)}$ or $y = -3 \pm \sqrt{6(x+2)}$. A complete graph of this parabola can be obtained using a function graphing utility to overlay the graphs of the two functions $y = -3 + \sqrt{6(x+2)}$ and $y = -3 - \sqrt{6(x+2)}$ in the same viewing rectangle (Figure 8.1.12). The graph of $y = -3 + \sqrt{6(x+2)}$ is the upper half of the parabola, and the graph of $y = -3 - \sqrt{6(x+2)}$ is the lower half of the parabola. Notice the gap between the two graphs near the vertex. This gap occurs because the graph of this parabola is very steep near the vertex. Gaps often occur in graphs of conics obtained with function graphing utilities, and even with special conic graphing utilities. These gaps can often be narrowed if high resolution graphs are drawn. Most graphs in this textbook will have the gaps removed or narrowed as much as possible.

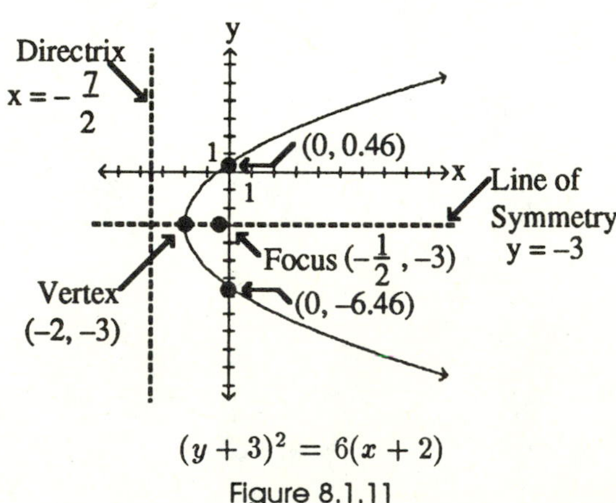

$(y+3)^2 = 6(x+2)$

Figure 8.1.11

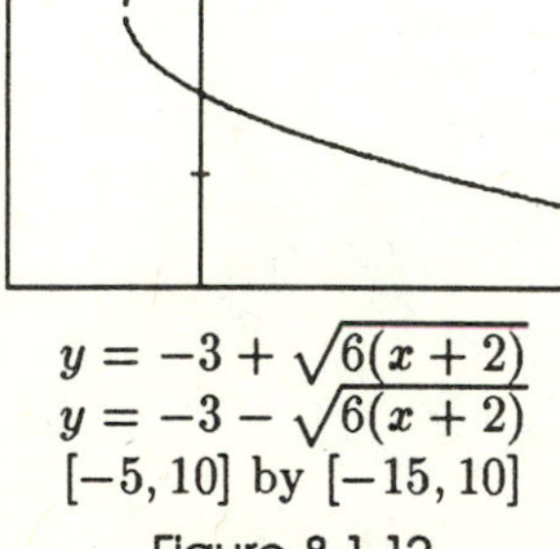

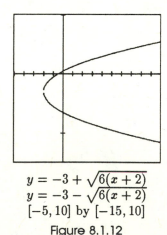

$y = -3 + \sqrt{6(x+2)}$
$y = -3 - \sqrt{6(x+2)}$
$[-5, 10]$ by $[-15, 10]$

Figure 8.1.12

(ii) This parabola opens to the left because the vertex is to the left of the directrix. The line of symmetry is $y = 1$, and an equation for the parabola is $(y-1)^2 = 4a(x-4)$ where $a < 0$. The perpendicular distance between the vertex and directrix is 2. Thus, $a = -2$, and an equation for the parabola is $(y-1)^2 = -8(x-4)$. The focus is 2 units to the left of the vertex and is on the line $y = 1$; its coordinates are $(2, 1)$. A complete graph is sketched in Figure 8.1.13. In the exercises you will be asked to confirm this graph with a function graphing utility. •

Geometric transformations can be used to produce the complete graphs in Example 4. However, usually it is much simpler to determine the vertex and two additional points to obtain a compete graph of any parabola. The equations found in Examples 3 and 4 are called the *standard form* for the equation of a parabola.

Equations of Parabolas in Standard Form

If a parabola has a line of symmetry parallel to or coincident with a coordinate axis, then it is the graph of an equation of the type given in Theorem 1.

• THEOREM 1

(i) The graph of the equation $(x-h)^2 = 4a(y-k)$ is a parabola with vertex (h, k) and directrix $y = c$ where $c = k - a$. The line of symmetry is $x = h$, and the focus is $(h, k + a)$.

(ii) The graph of the equation $(y-k)^2 = 4a(x-h)$ is a parabola with vertex (h, k) and directrix $x = c$ where $c = h - a$. The line of symmetry is $y = k$, and the focus is $(h + a, k)$.

• DEFINITION The equations of Theorem 1 are called **standard forms for the equation of a parabola.** •

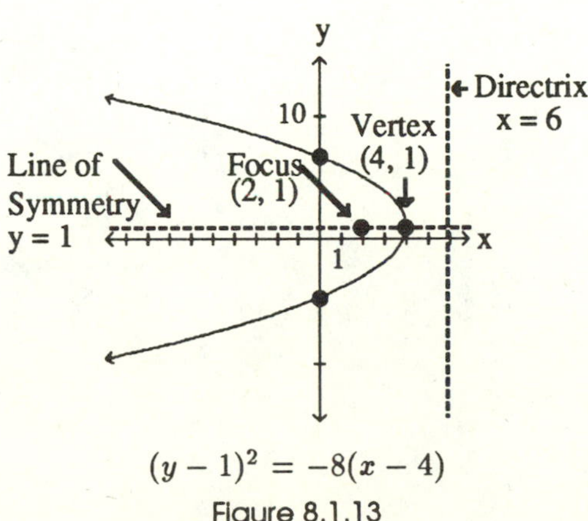

$(y-1)^2 = -8(x-4)$

Figure 8.1.13

The graphs of the equations of Theorem 1 can be obtained by applying horizontal and vertical shifts to the graphs of the equations $x^2 = 4ay$ and $y^2 = 4ax$.

It is not necessary to memorize the formulas in Theorem 1 for the directrix or the focus. In practice, you will use completing the square to put an equation of a parabola in standard form. Then, the values of a, h, and k can be read and their meaning used to calculate the focus and directrix. We illustrate this use of the information about the standard form of the equation of a parabola in the next example.

• EXAMPLE 5: Find the vertex, line of symmetry, focus, and directrix of each parabola, and draw a complete graph.

(a) $(x+1)^2 = -12(y-3)$
(b) $(y-2)^2 = -4(x+3)$

SOLUTION:

(a) The graph of $(x+1)^2 = -12(y-3)$ is of the form $(x-h)^2 = 4a(y-k)$ with $h=-1$ and $k=3$. This graph is a parabola that opens down with vertex $(-1,3)$, line of symmetry $x=-1$, and directrix parallel to the x-axis. Because $4a=-12$, $a=-3$. The focus is $(-1,0)$ because it is on the line $x=-1$ and 3 units below the vertex $(-1,3)$. The directrix is $y=6$ because it is a horizontal line 3 units above the vertex. In this case, $(0, \frac{35}{12})$ is the y-intercept. We find the x-intercepts by setting $y=0$. Now, $(x+1)^2 = 36$ so that $x+1=\pm 6$. Thus, $(-7,0)$ and $(5,0)$ are the two x-intercepts. A complete graph is sketched in Figure 8.1.14. Solving for y in terms of x we obtain $y = -\frac{1}{12}(x+1)^2 + 3$. Now we can use a function graphing utility to confirm the graph in Figure 8.1.14.

(b) The graph of $(y-2)^2 = -4(x+3)$ is of the form $(y-k)^2 = 4a(x-h)$ with $k=2$ and $h=-3$. This graph is a parabola that opens to the left with vertex $(-3,2)$, line of symmetry $y=2$, and directrix parallel to the y-axis. Because $4a=-4$, we see that $a=-1$. The focus is $(-4,2)$ because it is on the line $y=2$ and 1 unit to the left of the vertex $(-3,2)$. The directrix is $x=-2$ because it is a vertical line 1 unit to the right of the vertex. There is no y-intercept and $(-4,0)$ is the x-intercept.

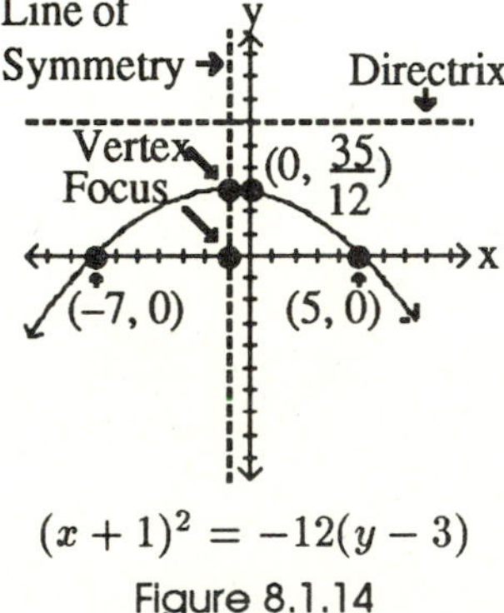

$(x+1)^2 = -12(y-3)$

Figure 8.1.14

Thus, $(-4, 4)$ is also a point on the graph. (Why?) A complete graph is sketched in Figure 8.1.15. Again solving for y in terms of x we obtain $y = 2 \pm \sqrt{-4(x+3)}$. Check that overlaying the graphs of $y = 2 + \sqrt{-4(x+3)}$ and $y = 2 - \sqrt{-4(x+3)}$ in the appropriate viewing rectangle agrees with the graph in Figure 8.1.15. The graph of $y = 2 + \sqrt{-4(x+3)}$ gives the upper half of the parabola, and the graph of $y = 2 - \sqrt{-4(x+3)}$ gives the lower half of the parabola. Your graph may have a gap near the vertex. •

From the standard form we can see that the equation of any parabola with line of symmetry parallel to a coordinate axis is quadratic in one variable and linear in the other. Thus, parabolas with line of symmetry parallel to or coincident with a coordinate axis are graphs of quadratic equations in two variables. We can use completing the square to find the standard form of such a parabola.

• EXAMPLE 6: Find the standard form, vertex, focus, directrix, and line of symmetry of each parabola, and sketch a complete graph. Check using a graphing utility.

(a) $x^2 + 6x - 2y + 11 = 0$ (b) $y^2 - 6x + 2y + 13 = 0$

SOLUTION:

(a) First, we put the equation in standard form by completing the square on the terms involving the variable x. Note the equation is quadratic in the variable x:

$$x^2 + 6x - 2y + 11 = 0$$
$$x^2 + 6x = 2y - 11$$
$$x^2 + 6x + 9 = 2y - 11 + 9$$
$$(x+3)^2 = 2y - 2$$
$$(x+3)^2 = 2(y-1).$$

From the standard form of the equation, we can see that this parabola opens up with vertex $(-3, 1)$ and has line of symmetry $x = -3$. Now, $4a = 2$ so that $a = 0.5$.

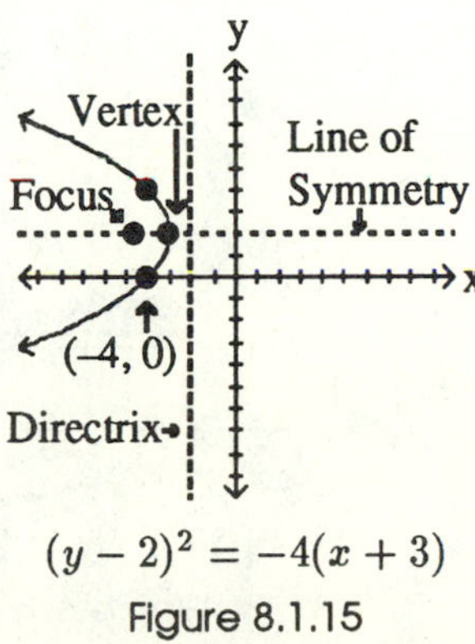

$(y-2)^2 = -4(x+3)$

Figure 8.1.15

Thus, the focus is $(-3, 1.5)$, and the directrix is $y = 0.5$. The y-intercept is $(0, \frac{11}{2})$. We can use this information to draw a rough sketch of the graph. However, we will use a graphing utility to determine the graph. Solving for y in terms of x we obtain $y = 0.5x^2 + 3x + 5.5$. The computer drawn graph given in Figure 8.1.16 is complete and confirms our algebraic analysis.

(b) This time we complete the square on the terms involving the variable y because the equation is quadratic in y:

$$\begin{aligned} y^2 - 6x + 2y + 13 &= 0 \\ y^2 + 2y &= 6x - 13 \\ y^2 + 2y + 1 &= 6x - 13 + 1 \\ (y+1)^2 &= 6x - 12 \\ (y+1)^2 &= 6(x-2). \end{aligned}$$

This parabola opens to the right, has vertex $(2, -1)$, and line of symmetry $y = -1$. Now $4a = 6$, so $a = 1.5$. The focus is $(3.5, -1)$ and the directrix is $x = 0.5$. We could, of course, use the above standard form to solve for y in terms of x. Alternatively, we can rewrite the original equation in the form $y^2 + 2y + (13 - 6x) = 0$ and use the quadratic formula to solve for y:

$$\begin{aligned} y^2 + 2y + (13 - 6x) &= 0 \\ y &= \frac{-2 \pm \sqrt{4 - 4(13 - 6x)}}{2} \\ &= -1 \pm \sqrt{1 - (13 - 6x)} \quad \text{(Why?)} \\ &= -1 \pm \sqrt{6x - 12}. \end{aligned}$$

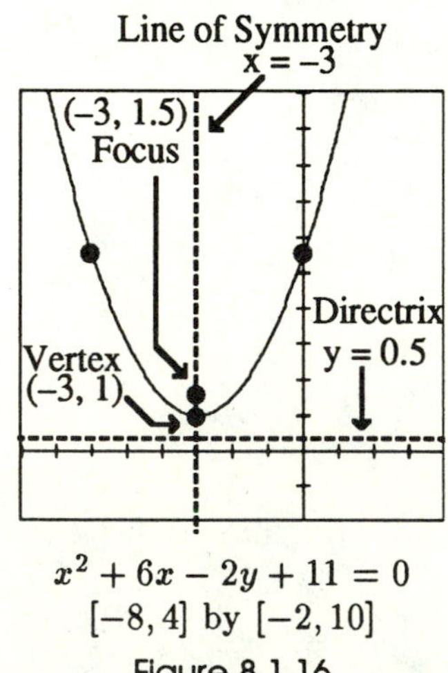

$x^2 + 6x - 2y + 11 = 0$
$[-8, 4]$ by $[-2, 10]$
Figure 8.1.16

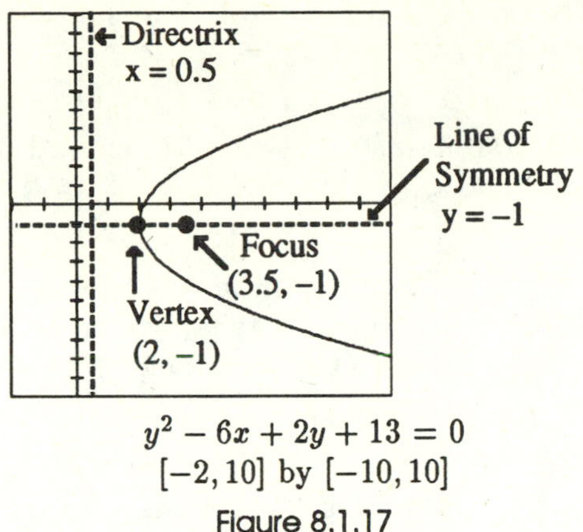

$y^2 - 6x + 2y + 13 = 0$
$[-2, 10]$ by $[-10, 10]$

Figure 8.1.17

Thus, $y = -1 + \sqrt{6x - 12}$ and $y = -1 - \sqrt{6x - 12}$. The computer drawn graphs of these two functions given in Figure 8.1.17 give a complete graph of the parabola and confirm our algebraic analysis. •

• EXAMPLE 7: A parabolic microphone is formed by revolving the portion of the parabola $15y = x^2$ between $x = -9$ and $x = 9$ about its line of symmetry. Where should the sound receiver be placed?

SOLUTION: The receiver should be placed at the focus of the parabola $15y = x^2$. Now, $4a = 15$ so that $a = 3.75$. The focus is at $(0, 3.75)$. Thus, the receiver should be placed 3.75 units from the vertex along its line of symmetry. •

• EXERCISES 8-1

Determine an equation, the vertex, and the line of symmetry of the parabola, and sketch a complete graph.

1. focus $(2, -1)$ and directrix $y = 3$
2. focus $(2, -1)$ and directrix $y = -3$
3. focus $(2, -1)$ and directrix $x = 3$
4. focus $(2, -1)$ and directrix $x = -3$

Determine an equation, the line of symmetry, and the focus of the parabola, and sketch a complete graph.

5. vertex $(3, 2)$ and directrix $x = 5$
6. vertex $(3, 2)$ and directrix $x = -5$
7. vertex $(3, 2)$ and directrix $y = 5$
8. vertex $(3, 2)$ and directrix $y = -5$

Determine an equation, the line of symmetry, and the directrix of the parabola, and sketch a complete graph.

9. focus $(-2, 0)$ and vertex $(-2, -6)$
10. focus $(-2, 0)$ and vertex $(-2, 4)$
11. focus $(3, 2)$ and vertex $(-1, 2)$
12. focus $(2, 4)$ and vertex $(0, 4)$

List a sequence of geometric transformations that when applied to the graph of $y = x^2$ will result in the graph of the given parabola.

13. $12(y+1) = (x-3)^2$

14. $y = 2x^2 - 6x + 9$

List a sequence of geometric transformations that when applied to the graph of $y^2 = x$ will result in the graph of the given parabola.

15. $(y+1)^2 = 3(x-2)$

16. $y^2 - 2y - 6 = 2x$

17. Consider the graph of $y = 4x^2$. Explain how a horizontal shrink of factor $\frac{1}{2}$ applied to the graph of $y = x^2$ produces the same graph as applying a vertical stretch of factor 4 to $y = x^2$. (*Hint*: What happens to the point $(1, 1)$ on $y = x^2$?)

18. Consider the graph of $x = 4y^2$. Explain how a vertical shrink of factor $\frac{1}{2}$ applied to $x = y^2$ produces the same graph as applying a horizontal stretch of factor 4 to the graph of $x = y^2$.

19. Given that $y = Ax^2 + Bx + C$ is the equation of a parabola, show that the point $\left(-\frac{B}{2A}, \frac{4AC-B^2+1}{4A}\right)$ is the focus.

20. Draw a complete graph of the parabola $y^2 - 2y + 4 = x$ using only a *function* graphing utility.

Find the standard form, vertex, focus, directrix, and line of symmetry of each parabola, and sketch a complete graph without using a graphing utility. Check your graph with a graphing utility.

21. $x^2 + 2x - y + 3 = 0$

22. $3x^2 - 6x - 6y + 10 = 0$

23. $y^2 - 2y + 4x - 12 = 0$

24. $9y^2 - 9x - 6y - 5 = 0$

25. A parabolic microphone is formed by revolving the portion of the parabola $10y = x^2$ between $x = -7$ and $x = 7$ about its line of symmetry. Where should the sound receiver be placed for best reception?

26. A parabolic microphone is formed by revolving the portion of the parabola $18y = x^2$ between $x = -12$ and $x = 12$ about its line of symmetry. Where should the sound receiver be placed for best reception?

27. A parabolic headlight is formed by revolving the portion of the parabola $y^2 = 12x$ between the lines $y = -4$ and $y = 4$ about its line of symmetry. Where should the headlight bulb be placed for maximum illumination?

28. A parabolic headlight is formed by revolving the portion of the parabola $y^2 = 15x$ between the lines $y = -3$ and $y = 3$ about its line of symmetry. Where should the headlight bulb be placed for maximum illumination?

29. A large parabolic search light formed by revolving a parabola about its line of symmetry has its light source placed for maximum illumination at the focus of the parabola. The focus is 10 inches from the vertex of the search light. Determine an equation for the parabola. Explain what assumptions you make.

30. A large parabolic search light formed by revolving a parabola about its line of symmetry has its light source placed for maximum illumination at the focus of a parabola. The focus is 8 inches from the vertex of the search light. Determine an equation for the parabola. Explain what assumptions you make.

Find the points of intersection of the line and the parabola (if any).

31. $y = 2x^2 - 6x + 7$, $2x + 3y - 6 = 0$

32. $x = 3y^2 - 2y + 6$, $x - 4y - 10 = 0$

33. A parabola of the form $y = ax^2 + b$ passes through the points $(-1, 4)$ and $(2, 8)$. Determine an equation of the parabola.

34. A parabola of the form $y = ax^2 + b + c$ passes through the points $(0, 2)$, $(2, 5)$ and $(3, 8)$. Determine an equation of the parabola.

35. Find all points on the parabola $y^2 = -8x$ that are closest to the point $(-10, 0)$.

A paraboloid of revolution is determined by rotating the portion of the parabola $y = 6 - x^2$ above the x-axis about its line of symmetry. A cylinder with line of symmetry the y-axis and base on the x-axis is inscribed in the paraboloid of revolution. Let (x, y) be a point on the parabola where x represents the radius of the cylinder.

36. Determine an algebraic representation of the volume of the cylinder as a function of x.

37. Draw a complete graph of the algebraic representation.

38. Draw a complete graph of the problem situation. What values of x make sense in the problem situation?

39. Determine the dimensions of the cylinder of maximum volume and the maximum volume.

40. Devise a real world problem scenario that fits Exercise 36.

41. The line $y = 3x - \frac{9}{8}$ is tangent to the curve $y = 2x^2$.
    (a) Find the point of tangency, P.
    (b) Find an equation of the line ℓ connecting P and the focus of $y = 2x^2$.
    (c) Determine the angle between the line ℓ and the tangent line.
    (d) What is the angle between the tangent line and a line through P that is parallel to the line of symmetry of $y = 2x^2$.
    (e) Do the angle sizes found in (c) and (d) help explain the reflective property of the parabola $y = 2x^2$?

42. Use a function graphing utility to confirm the graph in Example 4(ii).

8.2 Ellipses

In this section we give the two-foci definition of an ellipse and obtain the standard form for the equation of an ellipse with lines of symmetry parallel to the coordinate axes. We also consider applications that have an ellipse for an algebraic model.

• DEFINITION Let F_1 and F_2 be any two points in the coordinate plane. An **ellipse** is the set of all points P in the plane with the property that the sum of the distance from P to F_1 and the distance from P to F_2 is a constant (Figure 8.2.1). Each of the points F_1 and F_2 is called a **focus** of the ellipse. •

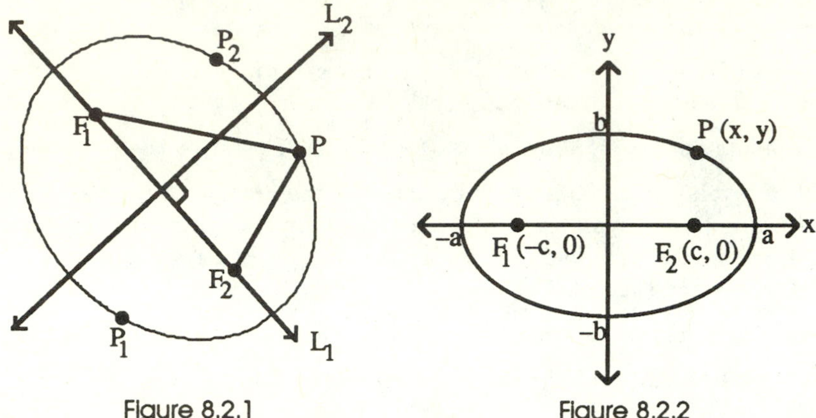

Figure 8.2.1

Figure 8.2.2

There are *no* points P satisfying the condition that the sum of the distance from P to F_1 and the distance from P to F_2 is a constant unless that constant is greater than or equal to the distance between F_1 and F_2. (Why?) Let L_1 be the line determined by F_1 and F_2, and L_2 the perpendicular bisector of the line segment joining F_1 and F_2. The points P_1 and P_2 that are symmetric to any point P on the ellipse with respect to L_1 and L_2, respectively, are also on the ellipse. (Why?) Thus, the ellipse is symmetric with respect to both L_1 and L_2. The point of intersection of L_1 and L_2 is called the **center** of the ellipse. An ellipse is symmetric with respect to its center. (Why?)

According to the laws of Newtonian mechanics the paths of objects due to forces such as gravitational attraction are conic sections. For example, if forces acting on the earth due to celestial bodies other than the sun are neglected, then the path of the earth around the sun is an ellipse with the sun as one focus. Orbits of the other planets and many comets around the sun are also ellipses with the sun as one focus.

Ellipses also have important reflective properties. If a source of light or sound is placed at one focus of an ellipse it will be reflected through the other focus. This principle was used to design a machine that is able to break up kidney stones and remove the need for surgery.

In this section we consider ellipses with lines of symmetry parallel to the coordinate axes. In Section 8.4 we remove this restriction. We begin by determining an equation for an ellipse with given foci. In this example the constant is $2a$.

• EXAMPLE 1: Let $F_1 = (-c, 0)$, $F_2 = (c, 0)$ be the two foci of the ellipse consisting of the points P where $d(P, F_1) + d(P, F_2) = 2a$ with $a > c > 0$ (Figure 8.2.2). Determine an equation for the ellipse, the x-intercepts, the y-intercepts, and the center of the ellipse.

SOLUTION: We can use the distance formula to obtain an equation for the ellipse:

$$d(P, F_1) + d(P, F_2) = 2a$$

$$\sqrt{(x+c)^2 + y^2} + \sqrt{(x-c)^2 + y^2} = 2a.$$

Next, we obtain an alternate form of this equation that does not involve radicals:

$$\begin{aligned}
\sqrt{(x+c)^2+y^2}+\sqrt{(x-c)^2+y^2} &= 2a\\
\sqrt{(x+c)^2+y^2} &= 2a-\sqrt{(x-c)^2+y^2}\\
(x+c)^2+y^2 &= 4a^2-4a\sqrt{(x-c)^2+y^2}+(x-c)^2+y^2\\
x^2+2cx+c^2+y^2 &= 4a^2-4a\sqrt{(x-c)^2+y^2)}+x^2-2cx+c^2+y^2\\
2cx &= 4a^2-4a\sqrt{(x-c)^2+y^2}-2cx\\
4a\sqrt{(x-c)^2+y^2} &= 4a^2-4cx\\
a\sqrt{(x-c)^2+y^2} &= a^2-cx.
\end{aligned}$$

We continue by squaring each side of the above equation.

$$\begin{aligned}
a\sqrt{(x-c)^2+y^2} &= a^2-cx\\
a^2(x-c)^2+a^2y^2 &= a^4-2a^2cx+c^2x^2\\
a^2x^2-2a^2cx+a^2c^2+a^2y^2 &= a^4-2a^2cx+c^2x^2\\
a^2x^2-c^2x^2+a^2y^2 &= a^4-a^2c^2\\
(a^2-c^2)x^2+a^2y^2 &= a^2(a^2-c^2).
\end{aligned}$$

We divide each side of the above equation by $a^2(a^2-c^2)$ and obtain the following:

$$\frac{x^2}{a^2}+\frac{y^2}{a^2-c^2}=1.$$

Because $a>c$ we can define a positive number b by $b^2=a^2-c^2$ and rewrite the above equation in the following form:

$$\frac{x^2}{a^2}+\frac{y^2}{b^2}=1. \tag{1}$$

Thus, every point on the ellipse is a solution to equation (1). It can also be shown that every solution of (1) corresponds to a point on the ellipse. This form is called the *standard form* for the equation of an ellipse. The x-intercepts are $\pm a$, and the y-intercepts are $\pm b$. (Why?) The center of the ellipse is $(0,0)$. •

In Figure 8.2.2, the line segment from $(-a,0)$ to $(a,0)$ is called the **major axis** or **major diameter** of the ellipse, and the line segment from $(0,-b)$ to $(0,b)$ is the **minor axis** or **minor diameter**. We call a the **semimajor axis length** and b the **semiminor axis length**. Notice that the equation $\frac{x^2}{a^2}+\frac{y^2}{b^2}=1$ is not changed if x is replaced by $-x$ or if y is replaced by $-y$. This confirms algebraically that the graph in Figure 8.2.2 is symmetric with respect to both the x-axis and the y-axis. It follows that the graph is also symmetric with respect to the origin, the center of the ellipse.

It can be shown that the standard form for the equation of the ellipse with foci $F_1(0,-c)$ and $F_2(0,c)$ on the y-axis that satisfies $d(P,F_1)+d(P,F_2)=2a$ with $a>c>0$ is

$$\frac{x^2}{b^2}+\frac{y^2}{a^2}=1, \tag{2}$$

where $b^2 = a^2 - c^2$. This equation is the same as equation (1) except that the number under x^2 is smaller than the number under y^2. A complete graph of this ellipse is given in Figure 8.2.3. The *major axis* is the line segment from $(0, -a)$ to $(0, a)$, and the *minor axis* is the line segment from $(-b, 0)$ to $(b, 0)$. The ellipse in Figure 8.2.3 is a complete graph of equation (2).

• EXAMPLE 2: Determine the endpoints of the major and minor axes, the standard form of the equation, and sketch a complete graph of each ellipse.

(a) The sum of the distances from the foci is 10; the foci are $(-3, 0)$ and $(3, 0)$.
(b) The sum of the distances from the foci is 10; the foci are $(0, -3)$ and $(0, 3)$.

SOLUTION:

(a) The sum of the distances from any point on the ellipse to the two foci is $2a$. Thus, $2a = 10$ and $a = 5$. The foci are $(0, -3)$ and $(0, 3)$, so $c = 3$. Now, $b^2 = a^2 - c^2 = 5^2 - 3^2 = 16$. The endpoints of the major axis are $(-5, 0)$ and $(5, 0)$, and the endpoints of the minor axis are $(0, -4)$ and $(0, 4)$. The standard form for the equation is

$$\frac{x^2}{25} + \frac{y^2}{16} = 1.$$

A complete graph is sketched in Figure 8.2.4.

(b) Again, we find $a = 5$, $c = 3$, and $b = 4$. However, the standard form for the equation is

$$\frac{x^2}{16} + \frac{y^2}{25} = 1.$$

The endpoints of the major axis are $(0, -5)$ and $(0, 5)$, and the endpoints of the minor axis are $(-4, 0)$ and $(4, 0)$. A complete graph is sketched in Figure 8.2.5. •

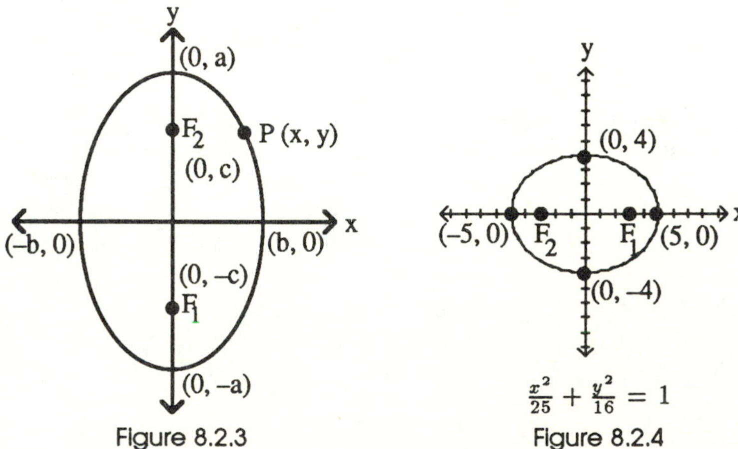

Figure 8.2.3

Figure 8.2.4

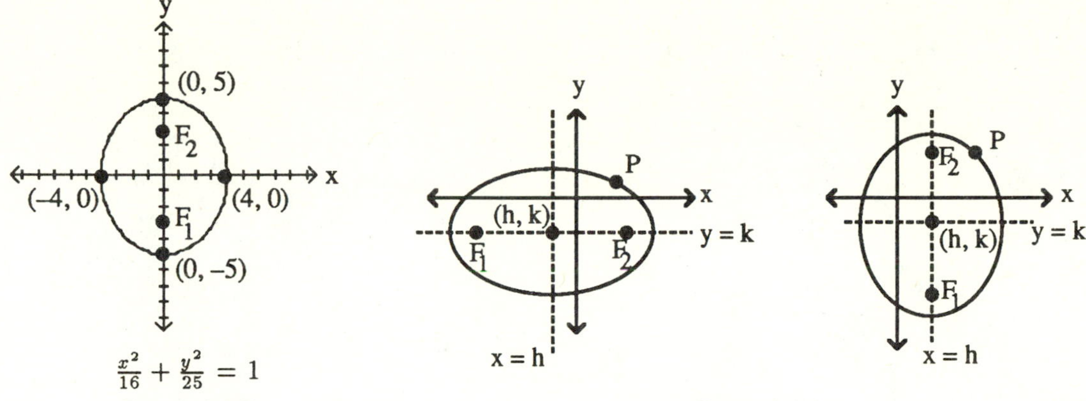

Figure 8.2.5

Figure 8.2.6

The graphs in Example 2 can be confirmed with a function graphing utility by solving for y in terms of x. Note that two functions will have to be entered to graph each ellipse.

Equations of Ellipses in Standard Form

If an ellipse has lines of symmetry parallel to or coincident with the coordinate axes, then it is the graph of an equation of the type given in Theorem 1.

• THEOREM 1 Let $a > b$. The graph of the equation $\frac{(x-h)^2}{a^2} + \frac{(y-k)^2}{b^2} = 1$ or the equation $\frac{(x-h)^2}{b^2} + \frac{(y-k)^2}{a^2} = 1$ is an ellipse with center at (h, k), and lines of symmetry the lines $x = h$ and $y = k$. The foci of the first equation are $(h \pm c, k)$ where $c = \sqrt{a^2 - b^2}$ and lie on the line $y = k$. The foci of the second equation are $(h, k \pm c)$ where $c = \sqrt{a^2 - b^2}$ and lie on the line $x = h$ (Figure 8.2.6).

• DEFINITION The equations of Theorem 1 are called **standard forms for the equation of an ellipse.** •

The graphs of the equations of Theorem 1 can be obtained by applying horizontal and vertical shifts to the graphs of equations (1) and (2).

It is not necessary to memorize the formulas in Theorem 1 for the foci. In practice, you will use completing the square to put an equation of an ellipse in standard form. Then, the values of a, b, h and k can be read and used to compute c, the foci, and the endpoints of the major and minor axes.

• EXAMPLE 3: Give the endpoints of the major and minor axes, the coordinates of the center and foci, the lines of symmetry, and sketch a complete graph of the following ellipses.

(a) $\frac{(x+2)^2}{25} + \frac{(y-3)^2}{9} = 1$ (b) $\frac{(x-4)^2}{25} + \frac{(y+1)^2}{169} = 1$

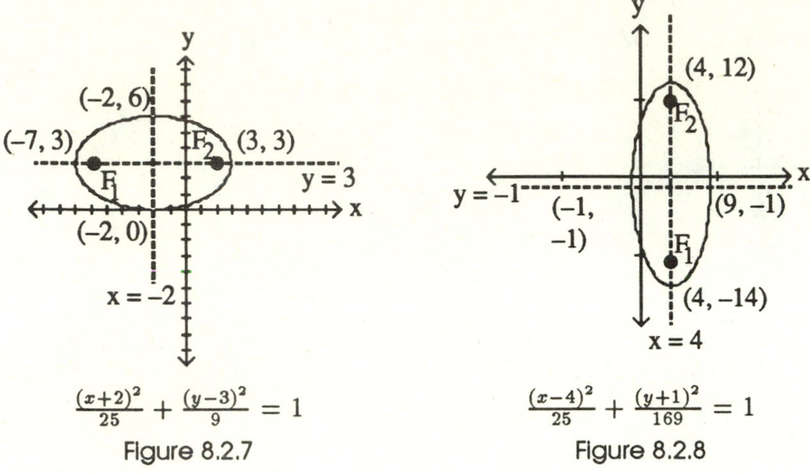

$\frac{(x+2)^2}{25} + \frac{(y-3)^2}{9} = 1$

Figure 8.2.7

$\frac{(x-4)^2}{25} + \frac{(y+1)^2}{169} = 1$

Figure 8.2.8

SOLUTION:

(a) If we start with the graph of

$$\frac{x^2}{25} + \frac{y^2}{9} = 1$$

and shift horizontally left 2 units and vertically up 3 units we get the desired graph (Figure 8.2.7). Thus, the center of the ellipse is $(-2, 3)$, and the lines of symmetry are the lines $x = -2$ and $y = 3$. Now $a = 5$ and $b = 3$ so that $c^2 = a^2 - b^2 = 25 - 9 = 16$. The foci of $\frac{x^2}{25} + \frac{y^2}{9} = 1$ are $(-4, 0)$ and $(4, 0)$. The foci of $\frac{(x+2)^2}{25} + \frac{(y-3)^2}{9} = 1$ are $(-6, 3)$ and $(2, 3)$ because they are obtained by shifting the foci of $\frac{x^2}{25} + \frac{y^2}{9} = 1$ left 2 units and up 3 units. The endpoints of the major axis are $(-7, 3)$ and $(3, 3)$, and the endpoints of the minor axis are $(-2, 0)$ and $(-2, 6)$.

(b) Shifting the graph of

$$\frac{x^2}{25} + \frac{y^2}{169} = 1$$

horizontally right 4 units and vertically down 1 unit produces the desired graph (Figure 8.2.8). The center of the ellipse is $(4, -1)$. The lines of symmetry are $x = 4$ and $y = -1$. Now $a = 13$, $b = 5$, so that $c = 12$. The foci are $(4, -13)$ and $(4, 11)$. The endpoints of the major axis are $(4, -14)$ and $(4, 12)$, and the endpoints of the minor axis are $(-1, -1)$ and $(9, -1)$. •

If $a^2 = b^2$ then the equations in Theorem 1 can be rewritten as follows.

$$\frac{(x-h)^2}{a^2} + \frac{(y-k)^2}{a^2} = 1$$

$$(x-h)^2 + (y-k)^2 = a^2$$

The graph of this equation is a circle with center (h, k) and radius a. Thus, a circle is a degenerate case of an ellipse. In a circle, there is just one focus, and it is the center of the circle.

If the equations of Theorem 1 are cleared of fractions and parentheses are expanded, then the new equations are of the form $Ax^2 + Bxy + Cy^2 + Dx + Ey + F = 0$ where $B = 0$. Thus, ellipses with lines of symmetry parallel to or coincident with the coordinate axes are graphs of quadratic equations in two variables. The complete graphs obtained in Example 3 can be checked with a function graphing utility by first solving for y in terms of x.

If we start with an equation in expanded form with $B = 0$, then we can use completing the square to write the equation in standard form and sketch its graph.

• EXAMPLE 4: Use completing the square to write $9x^2 + 5y^2 - 54x + 40y + 116 = 0$ in standard form. Give the endpoints of the major and minor axes, the coordinates of the center and foci, the lines of symmetry, and sketch a complete graph.

SOLUTION: We group the terms involving x and the terms involving y and then complete the square:

$$\begin{aligned}
9x^2 + 5y^2 - 54x + 40y + 116 &= 0 \\
9(x^2 - 6x) + 5(y^2 + 8y) &= -116 \\
9(x^2 - 6x + 9) + 5(y^2 + 8y + 16) &= -116 + 9(9) + 5(16) \\
9(x-3)^2 + 5(y+4)^2 &= 45 \\
\frac{9(x-3)^2}{45} + \frac{5(y+4)^2}{45} &= 1 \\
\frac{(x-3)^2}{5} + \frac{(y+4)^2}{9} &= 1.
\end{aligned}$$

The graph is an ellipse with center $(3, -4)$ and lines of symmetry $x = 3$ and $y = -4$. The endpoints of the major axis are $(3, -7)$ and $(3, -1)$, and the endpoints of the minor axis are $(3 - \sqrt{5}, -4) = (0.76, -4)$ and $(3 + \sqrt{5}, -4) = (5.24, -4)$. Notice that $a^2 = 9$, $b^2 = 5$ so that $c^2 = 4$. Thus, the foci are $(3, -6)$ and $(3, -2)$. Solving the above equation for y in terms of x we obtain

$$y = -4 \pm 3\sqrt{1 - \frac{(x-3)^2}{5}}.$$

Now, we can use a function graphing utility and enter

$$y = -4 + 3\sqrt{1 - \frac{(x-3)^2}{5}}$$

and

$$y = -4 - 3\sqrt{1 - \frac{(x-3)^2}{5}}$$

to obtain the complete graph shown in Figure 8.2.9. This graph confirms our algebraic computations. •

• EXAMPLE 5: Use the quadratic formula and a function graphing utility to obtain a complete graph of the ellipse $5x^2 + 3y^2 - 20x + 6y = -8$.

SOLUTION: First we rewrite the equation:

$$5x^2 + 3y^2 - 20x + 6y = -8$$
$$3y^2 + 6y + (5x^2 - 20x + 8) = 0.$$

Now, we use the quadratic formula to solve for y:

$$y = \frac{-6 \pm \sqrt{36 - 12(5x^2 - 20x + 8)}}{6}.$$

It is *not* necessary to simplify this equation before using a graphing utility. The graph of

$$y = \frac{-6 + \sqrt{36 - 12(5x^2 - 20x + 8)}}{6}$$

produces the upper half of the complete graph of the ellipse in Figure 8.2.10, and the graph of

$$y = \frac{-6 - \sqrt{36 - 12(5x^2 - 20x + 8)}}{6}$$

produces the lower half. •

The next example is a real-world application that takes advantage of the reflective properties of an ellipse. Light or sound emitted at one focus of an ellipse is reflected by the ellipse through the other focus (Figure 8.2.11). A lithotripter is a special device that is used to break up kidney stones with shock waves through water. Roughly speaking a lithotripter is constructed as follows. An ellipse is rotated about its major axis and cut in half to form half of an *ellipsoid* (Figure 8.2.12). Careful measurements are made to put the patient's kidney stones at one focus and a source that produces ultra high frequency shock waves at the other focus. The shock waves are reflected by the ellipsoid to the other focus to break up the kidney stones.

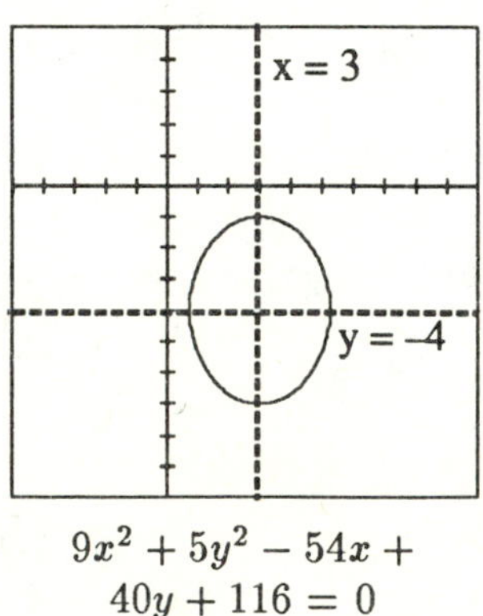

$9x^2 + 5y^2 - 54x + 40y + 116 = 0$
$[-5, 10]$ by $[-10, 5]$
Figure 8.2.9

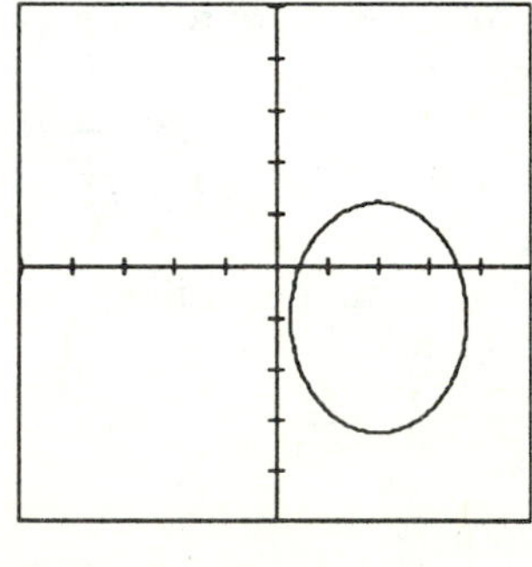
$5x^2 + 3y^2 - 20x + 6y = -8$
$[-5, 5]$ by $[-5, 5]$
Figure 8.2.10

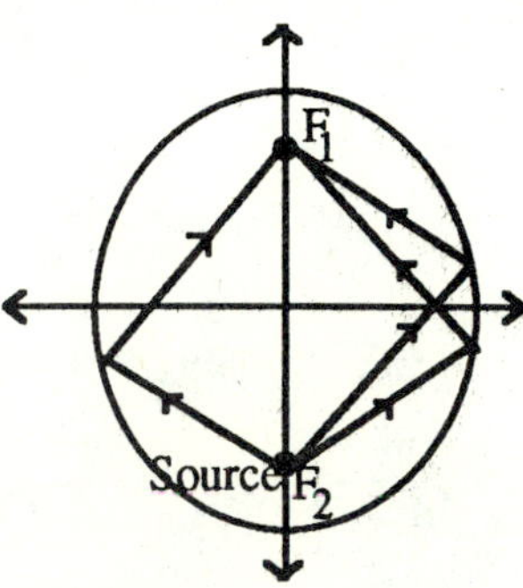

Figure 8.2.11

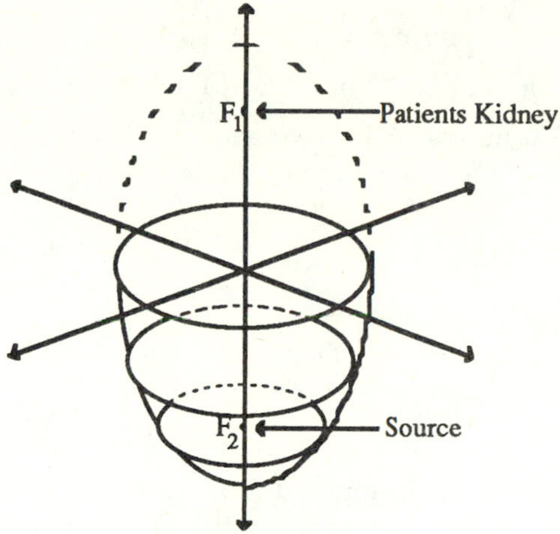

Figure 8.2.12

Assume that the center of the ellipse that is rotated to construct a lithotripter is $(0,0)$, the two ends of the major axis are $(-6,0)$ and $(6,0)$, and one end of the minor axis is $(0,-2.5)$. These are measurements in inches of an actual lithotripter.

• **EXAMPLE 6:** Determine the coordinates of the foci of the lithotripter described above.

SOLUTION: An equation for the ellipse that is rotated is $\frac{x^2}{6^2} + \frac{y^2}{2.5^2} = 1$. Thus, $a = 6$ and $b = 2.5$. Now, $c^2 = a^2 - b^2$ so that $c^2 = 29.75$ and $c = 5.45$. Therefore, the foci are at $(-5.45, 0)$ and $(5.45, 0)$. The shock source (an underwater spark discharge) is placed at the negative focus and the patient's kidney at the positive focus. •

• EXERCISES 8-2

Determine the foci and standard form of the equation of the ellipse, and draw a complete graph. Do not use a graphing utility.

1. major axis of length 12 along the x-axis, minor axis of length 9 along the y-axis
2. major axis of length 6 along the y-axis, minor axis of length 5 along the x-axis

Determine the endpoints of the major and minor axes and standard form of the equation of the ellipse, and sketch a complete graph. Do not use a graphing utility.

3. foci at $(3,0)$ and $(-3,0)$, minor axis of length 14
4. foci at $(0,4)$ and $(0,-4)$, minor axis of length 9

Determine the endpoints of the major and minor axes and the standard form of the equation of the ellipse, and sketch a complete graph. Do not use a graphing utility.

5. The sum of the distances from the foci is 9; the foci are $(-4,0)$ and $(4,0)$.

6. The sum of the distances from the foci is 11; the foci are $(-0.5, 0)$ and $(0.5, 0)$.
7. foci at $(1, 3)$ and $(1, 9)$ and major axis of length 12
8. center at $(1, -4)$, foci at $(1, -2)$ and $(1, -6)$, and minor axis of length 10

Give the endpoints of the major and minor axes, the coordinates of the center and foci, and the lines of symmetry of the ellipse, and sketch a complete graph. Do not use a graphing utility.

9. $\dfrac{(x-1)^2}{2} + \dfrac{(y+3)^2}{4} = 1$
10. $(x+3)^2 + 4(y-1)^2 = 16$
11. $\dfrac{(x+2)^2}{5} + 2(y-1)^2 = 1$
12. $\dfrac{(x-4)^2}{16} + 16(y+4)^2 = 8$

Draw a complete graph using only a function graphing utility.

13. $2x^2 - 4x + y^2 - 6 = 0$
14. $3x^2 - 6x + 2y^2 + 8y + 5 = 0$

Write the equation of the ellipse in standard form, give the endpoints of the major and minor axes, the coordinates of the center and foci, and the lines of symmetry of the ellipse, and draw a complete graph. Check your answer using a graphing utility.

15. $9x^2 + 4y^2 - 18x + 8y - 23 = 0$
16. $3x^2 + 5y^2 - 12x + 30y + 42 = 0$
17. $4x^2 + y^2 - 32x + 16y + 124 = 0$
18. $2x^2 + 3y^2 + 12x - 24y + 60 = 0$
19. Explain why the graph of $5x^2 + y^2 + 5 = 0$ has no points.
20. A lithotripter is made by rotating the portion of an ellipse below its minor axis about its major axis. The major diameter (length of the major axis) is 26 inches, and the maximum depth from the major axis is 10 inches. Where should the shock source and patient be placed for maximum effect? Give the appropriate measurements.
21. There are real elliptical pool tables that have been constructed with a single pocket at one of the foci. Suppose such a table has major diameter 6 feet and minor diameter 4 feet.
    (a) Explain how a pool shark who knows conic geometry is at a great advantage over a "mark" who knows no conic geometry.
    (b) How should the ball be hit so it bounces off the cushion directly into the pocket? Give specific measurements.

The United States Capitol dome is in the shape of an ellipsoid of revolution. Assume the dome is formed by revolving the portion of an ellipse above its major axis about its minor axis. Further assume the diameter of the dome (major axis of the ellipse) is 400 feet and its maximum height is 60 feet (semiminor axis, one-half the minor axis length).

22. In 1849, a U.S. senator standing along the major axis of the ellipsoid under the capitol dome had his very private whispered conversation with a lady overheard by another senator standing quite a distance away (true story). How could this have happened?
23. Give the likely exact position(s) of the two senators in Exercise 22.

The **eccentricity** e of an ellipse in standard form is defined by $e = c/a = \sqrt{a^2 - b^2}/a$. Compute the eccentricity of the ellipse.

24. $4x^2 + \dfrac{1}{8}y^2 = 16$
25. $2x^2 + 3y^2 + 6y - 4x - 1 = 0$

Planet	e	a (miles)
Mercury	0.20	3.6×10^7
Earth	0.02	9.3×10^7
Mars	0.09	1.4×10^8
Saturn	0.50	8.9×10^8
Neptune	0.01	2.8×10^9
Pluto	0.25	3.7×10^9

Table 1

26. Show that $b = \sqrt{a^2(1-e^2)}$ for an ellipse in standard form, where e is its eccentricity.

Determine an equation and sketch a complete the graph of the ellipse. Assume that the center is at the origin.

27. eccentricity 0.1 and major axis length 8
28. eccentricity 0.5 and major axis length 6
29. Show that $0 < e < 1$ for every ellipse.
30. What is the change in the nature of the ellipse as its eccentricity approaches 0?
31. What is the change in the nature of the ellipse as its eccentricity approaches 1?
32. Explain how eccentricity is a measure of the circularity of an ellipse.
33. What should the eccentricity of a circle be?

A *Hohmann* ellipse is an ideal trajectory for space flight between two planets that requires a minimum of energy for the journey. Assume two planets 12,800,000 miles apart are located at the foci of a Hohmann ellipse. Determine the equation in standard form of the ellipse with the following eccentricity.

34. 0.2
35. 0.4

Table 1 gives the eccentricity e of the elliptical orbits of selected planets and the length of their semimajor axes (one-half the length of the major axes). Determine an equation and draw the graph of each orbit in the same viewing rectangle. The sun is one of the foci. Assume the center of each orbit is at the origin, and each major axis is along the x-axis.

36. Earth and Mercury
37. Earth, Mars, and Saturn
38. Neptune and Pluto

8.3 Hyperbolas

In this section we give the two-foci definition of a hyperbola and obtain the standard form for the equation of a hyperbola with lines of symmetry parallel to the coordinate axes. The standard form is used to obtain a rough sketch of its graph. We use completing the square to put equations of hyperbolas into standard form. An application about lenses involving reflective properties of a hyperbola will be investigated.

In Chapter 1, we drew the graph of $xy = 1$, or $y = \frac{1}{x}$, which is an example of a *rectangular hyperbola*. A **rectangular hyperbola** is any hyperbola whose asymptotes are parallel to or coincident with the coordinate axes. Hyperbolas of this type will be investigated in the exercises of the next section.

The force of attraction between objects can cause the path of an object to be part of a hyperbola. For example, some comets travel through our solar system on hyperbolic paths. Applications involving navigation or locating sources of sound sometimes involve hyperbolas. We will study one such application in Section 8.5.

• DEFINITION Let F_1 and F_2 be any two points in the coordinate plane. A **hyperbola** is the set of all points P in the plane such that the absolute value of the difference of the distances from P to F_1 and from P to F_2 is a constant (Figure 8.3.1). Each of the points F_1 and F_2 is called a **focus** of the hyperbola. •

There are *no* points P satisfying the condition that the absolute value of the difference of the distances from P to F_1 and from P to F_2 is a constant unless that constant is less than the distance between the two foci. (Why?) Let L_1 be the line determined by F_1 and F_2, and L_2 the perpendicular bisector of the line segment joining F_1 and F_2. The points P_1 and P_2 that are symmetric to any point P on the hyperbola with respect to L_1 and L_2, respectively, are on the hyperbola. (Why?) Thus, the hyperbola is symmetric with respect to both L_1 and L_2. The point of intersection of L_1 and L_2 is called the **center** of the hyperbola. A hyperbola is symmetric with respect to its center. The two points of the hyperbola that lie on L_1 are called **vertices** of the hyperbola. The line segment between the vertices is called the **transverse axis**. The line L_1 is called the **principal axis** of the hyperbola and the line L_2 the **conjugate axis**.

In this section we will consider hyperbolas with lines of symmetry parallel to the coordinate axes. This restriction will be removed in Section 8.4. We begin by determining an equation for a hyperbola with given foci.

It is convenient to let $2a$ be the difference of the distances from P to F_1 and F_2.

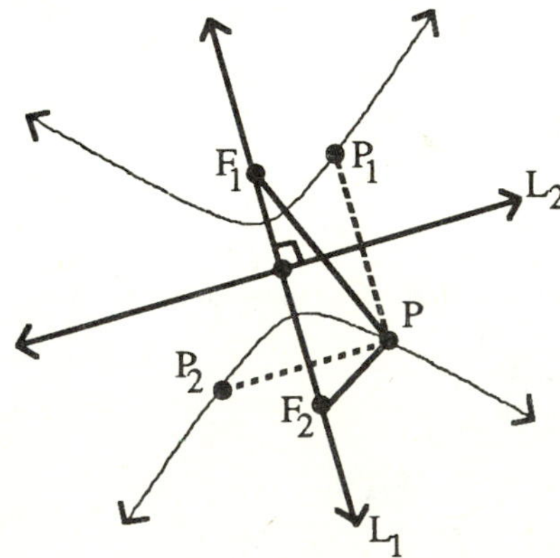

Figure 8.3.1

• EXAMPLE 1: Let $F_1 = (-c, 0)$ and $F_2 = (c, 0)$ be the two foci of the hyperbola consisting of the points P where $|d(P, F_1) - d(P, F_2)| = 2a$ with $0 < a < c$ (Figure 8.3.2). Determine an equation for the hyperbola, the x-intercepts, the y-intercepts, and the center of the hyperbola.

SOLUTION: We can use the distance formula to obtain an equation for the hyperbola:

$$|d(P, F_1) - d(P, F_2)| = 2a$$
$$|\sqrt{(x+c)^2 + y^2} - \sqrt{(x-c)^2 + y^2}| = 2a$$
$$\sqrt{(x+c)^2 + y^2} - \sqrt{(x-c)^2 + y^2} = \pm 2a$$
$$\sqrt{(x+c)^2 + y^2} = \sqrt{(x-c)^2 + y^2} \pm 2a.$$

As in the previous section we can obtain the following alternate form of this equation that does not involve radicals:

$$\frac{x^2}{a^2} - \frac{y^2}{c^2 - a^2} = 1.$$

Because $a < c$, we can define a positive number b by $b^2 = c^2 - a^2$ and rewrite the above equation in the following form:

$$\frac{x^2}{a^2} - \frac{y^2}{b^2} = 1. \tag{1}$$

Thus, every point on the hyperbola is a solution to equation (1). It can also be shown that every solution to (1) corresponds to a point on the hyperbola. This form is called a *standard form* for the equation of a hyperbola. Notice that the x-intercepts are $\pm a$ and are also the *vertices* of the hyperbola. The distance between the center $(0, 0)$ and each vertex is one-half of the absolute value of the difference of the distances from P to F_1 and from P to F_2. There are *no* y-intercepts. •

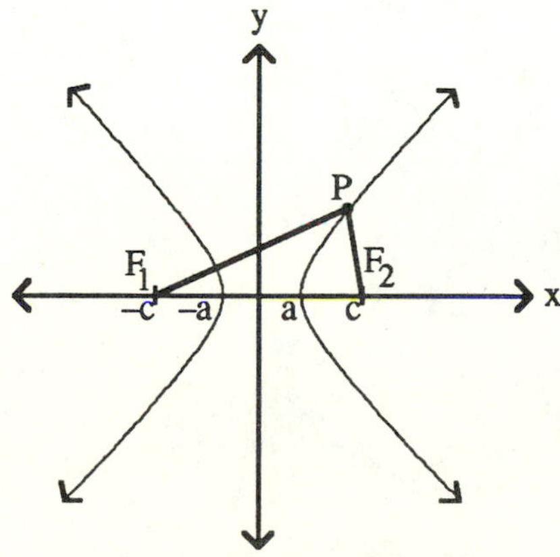

Figure 8.3.2

Notice that the equation $\frac{x^2}{a^2} - \frac{y^2}{b^2} = 1$ of Example 1 is not changed if x is replaced by $-x$ or if y is replaced by $-y$. This confirms algebraically that the graph in Figure 8.3.2 is symmetric with respect to the x-axis, y-axis, and the origin.

It can be shown that a standard form for the equation of the hyperbola with foci $F_1(0, -c)$ and $F_2(0, c)$ on the y-axis that satisfies $|d(P, F_1) - d(P, F_2)| = 2a$ with $0 < a < c$ is

$$\frac{y^2}{a^2} - \frac{x^2}{b^2} = 1 \tag{2}$$

where $b^2 = c^2 - a^2$ (Figure 8.3.3). Notice that this is the equation of Example 1 with x and y interchanged. The vertices $\pm a$ are the y-intercepts and there are *no* x-intercepts. The center is $(0, 0)$.

Asymptotes

Equation (1) of the hyperbola in Example 1 can be rewritten as follows:

$$\frac{x^2}{a^2} - \frac{y^2}{b^2} = 1$$

$$\frac{x^2}{a^2} - 1 = \frac{y^2}{b^2}$$

$$y^2 = \frac{b^2x^2}{a^2} - b^2$$

$$y^2 = \frac{b^2x^2}{a^2}\left(1 - \frac{a^2}{x^2}\right). \tag{3}$$

We can see from equation (3) that as the absolute value of x gets very large the equation of the hyperbola is approximately $y^2 = \frac{b^2x^2}{a^2}$. Thus, an end behavior model of the hyperbola of Example 1 is $y^2 = \frac{b^2x^2}{a^2}$. This will be illustrated in Example 3. Notice that the graph of $y^2 = \frac{b^2x^2}{a^2}$ is the pair of intersecting lines $y = \frac{b}{a}x$ and $y = -\frac{b}{a}x$. These two lines are called

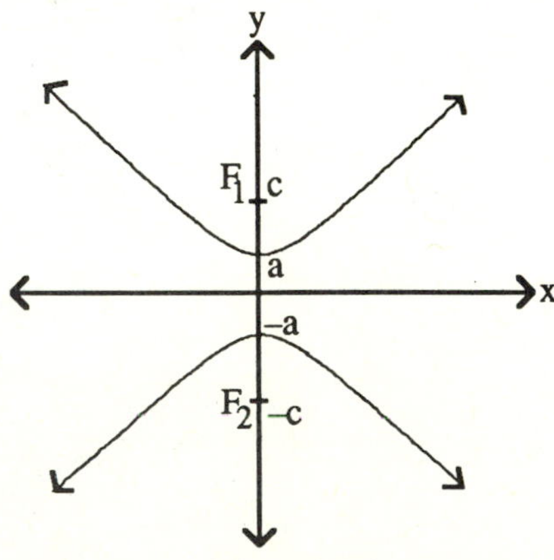

Figure 8.3.3

the **asymptotes** of the hyperbola $\frac{x^2}{a^2} - \frac{y^2}{b^2} = 1$ and can be used to help sketch the graph. The asymptotes $y = \pm\frac{b}{a}x$ are the diagonals of the dashed rectangle of width 2a and length 2b in Figure 8.3.4. In the exercises you will show that the asymptotes of $\frac{y^2}{a^2} - \frac{x^2}{b^2} = 1$ are $y = \pm\frac{a}{b}x$.

• EXAMPLE 2: Determine the vertices, center, a standard form of the equation, and asymptotes of each hyperbola, and sketch a complete graph.

(a) The absolute value of the difference of the distance from the foci is 8; the foci are $(-5,0)$ and $(5,0)$.
(b) The absolute value of the difference of the distance from the foci is 8; the foci are $(0,-5)$ and $(0,5)$.

SOLUTION:

(a) The absolute value of the difference of the distances from any point on the hyperbola to the two foci is $2a$. Thus, $2a = 8$ and $a = 4$. The foci are $(-5,0)$ and $(5,0)$, so $c = 5$. Now, $b^2 = c^2 - a^2 = 25 - 16 = 9$. The standard form for the equation is $\frac{x^2}{16} - \frac{y^2}{9} = 1$. The vertices are $(\pm 4, 0)$, and the center is $(0,0)$. The asymptotes are $y = \pm\frac{b}{a}x = \pm\frac{3}{4}x$. A complete graph of the hyperbola is sketched in Figure 8.3.5.
(b) Again, we find $a = 4$, $c = 5$, and $b = 3$. However, the standard form for the equation is $\frac{y^2}{16} - \frac{x^2}{9} = 1$. Notice this equation is the equation of (a) with x and y interchanged. The vertices are $(0, \pm 4)$, the center is $(0,0)$, and the asymptotes are $y = \pm\frac{a}{b}x = \pm\frac{4}{3}x$. A complete graph of the hyperbola is sketched in Figure 8.3.6. •

• EXAMPLE 3: Show that the asymptotes $y = \pm\frac{3}{4}x$ of Example 2(a) correctly describe the end behavior of the hyperbola $\frac{x^2}{16} - \frac{y^2}{9} = 1$.

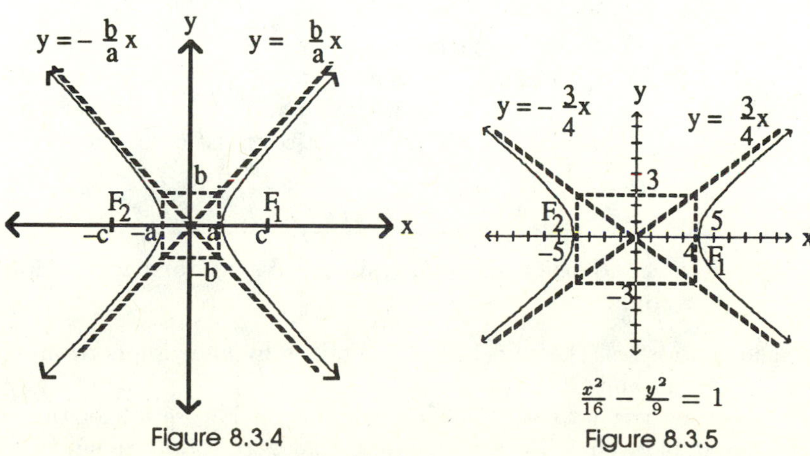

Figure 8.3.4

Figure 8.3.5

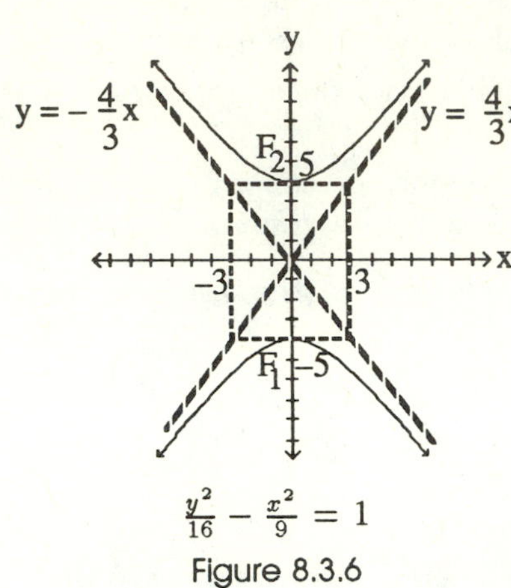

$\frac{y^2}{16} - \frac{x^2}{9} = 1$

Figure 8.3.6

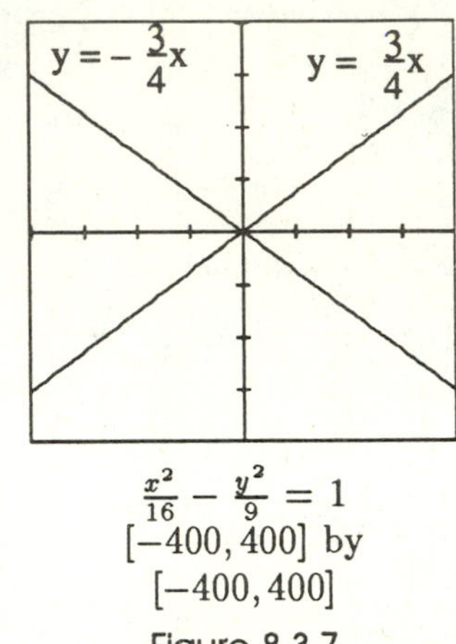

$\frac{x^2}{16} - \frac{y^2}{9} = 1$
$[-400, 400]$ by
$[-400, 400]$

Figure 8.3.7

SOLUTION: If we use a sufficiently large viewing rectangle, the graphs of the hyperbola $\frac{x^2}{16} - \frac{y^2}{9} = 1$ and the pair of lines $y = \pm\frac{3}{4}x$ appear to be identical (Figure 8.3.7). This means that an end behavior model of the hyperbola is the pair of lines $y = \pm\frac{3}{4}x$. •

Equation of Hyperbolas in Standard Form

If a hyperbola has lines of symmetry parallel to or coincident with the coordinate axes, then it is the graph of an equation of the type given in Theorem 1.

• THEOREM 1

(i) The graph of the equation $\frac{(x-h)^2}{a^2} - \frac{(y-k)^2}{b^2} = 1$ is a hyperbola with center (h, k) and lines of symmetry $x = h$ and $y = k$. The vertices are $(h \pm a, k)$, the foci are $(h \pm c, k)$ where $c = \sqrt{a^2 + b^2}$, and the asymptotes are $y - k = \pm\frac{b}{a}(x - h)$ (Figure 8.3.8). Notice that the branches of the hyperbola open left and right.

(ii) The graph of the equation $\frac{(y-k)^2}{a^2} - \frac{(x-h)^2}{b^2} = 1$ is a hyperbola with center (h, k) and lines of symmetry $x = h$ and $y = k$. The vertices are $(h, k \pm a)$, the foci are $(h, k \pm c)$ where $c = \sqrt{a^2 + b^2}$, and the asymptotes are $y - k = \pm\frac{a}{b}(x - h)$ (Figure 8.3.9). Notice that the branches of the hyperbola open up and down.

• DEFINITION The equations of Theorem 1 are called **standard forms for the equation of a hyperbola.** •

The equations of Theorem 1 can be obtained by applying horizontal and vertical shifts to equations (1) and (2).

It is not necessary to memorize the formulas in Theorem 1 for the foci and the asymptotes. In practice, you will use completing the square to put an equation of a hyperbola in

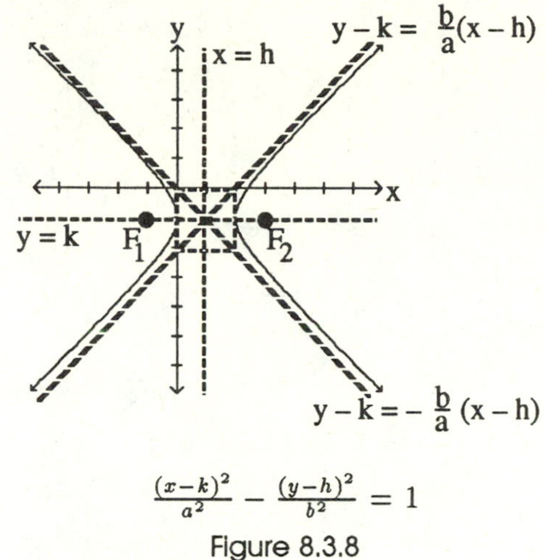

$\frac{(x-k)^2}{a^2} - \frac{(y-h)^2}{b^2} = 1$

Figure 8.3.8

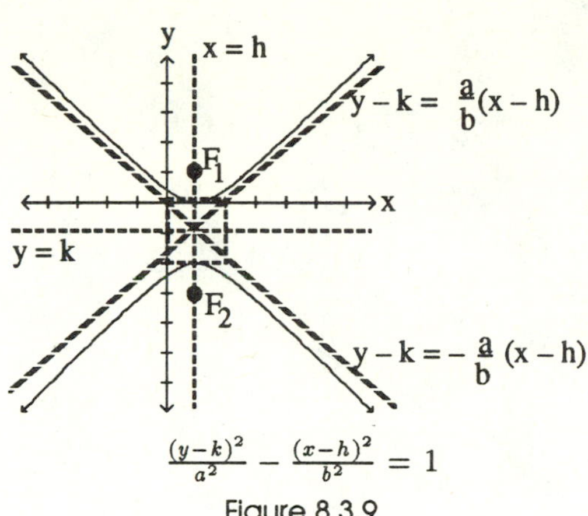

$\frac{(y-k)^2}{a^2} - \frac{(x-h)^2}{b^2} = 1$

Figure 8.3.9

standard form. Then, the values of a, b, h and k can be read and used to compute c, the foci, and the asymptotes.

• EXAMPLE 4: Determine the coordinates of the center, foci, vertices, and the asymptotes of each hyperbola, and sketch a complete graph.

(a) $\dfrac{(x+2)^2}{9} - \dfrac{(y-1)^2}{16} = 1$ (b) $\dfrac{(y+3)^2}{4} - \dfrac{(x-2)^2}{25} = 1$

SOLUTION:

(a) The center is $(-2, 1)$. Because $a = 3$ and $b = 4$, $c^2 = a^2 + b^2 = 9 + 16 = 25$. Thus, $c = 5$. Because the branches of the hyperbola open left and right, the vertices are on the line $y = 1$ and are 3 units on either side of the center. Thus, the vertices are $(-5, 1)$ and $(1, 1)$. The foci are also on the line $y = 1$, but are 5 units on either side of the center. (Why?) The foci are $(-7, 1)$ and $(3, 1)$. The asymptotes are $y - 1 = \pm\frac{4}{3}(x + 2)$, and the graph sketched in Figure 8.3.10 is complete.

(b) The center is $(2, -3)$. Because $a = 2$ and $b = 5$, $c^2 = a^2 + b^2 = 4 + 25 = 29$. Thus, $c = \sqrt{29}$. The branches of this hyperbola open up and down, so the vertices are on the line $x = 2$ and are 2 units on either side of the center. Thus, the vertices are $(2, -5)$ and $(2, -1)$. The foci are also on the line $x = 2$, but are $\sqrt{29}$ units on either side of the center. The foci are $(2, -3 - \sqrt{29}) = (2, -8.39)$ and $(2, -3 + \sqrt{29}) = (2, 2.39)$. The asymptotes are $y + 3 = \pm\frac{2}{5}(x - 2)$, and the graph sketched in Figure 8.3.11 is complete. •

If the equations in Theorem 1 are cleared of fractions and if parentheses are expanded, then the resulting equations are of the form $Ax^2 + Bxy + Cy^2 + Dx + Ey + F = 0$ where $B = 0$. Thus, hyperbolas with lines of symmetry parallel to or coincident with the coordinate axes are graphs of quadratic equations in two variables. The coefficients A and C will have unlike

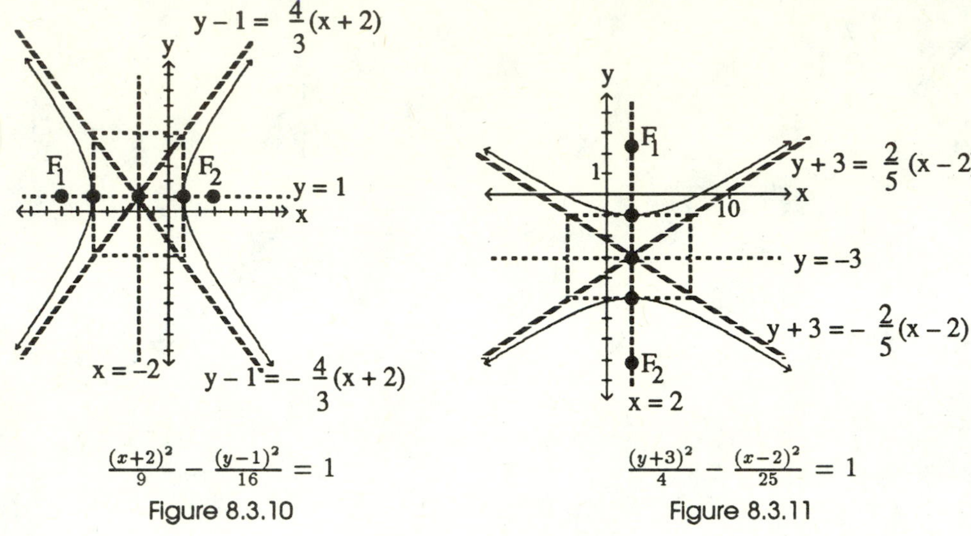

$\frac{(x+2)^2}{9} - \frac{(y-1)^2}{16} = 1$

Figure 8.3.10

$\frac{(y+3)^2}{4} - \frac{(x-2)^2}{25} = 1$

Figure 8.3.11

signs in this case. However, when the equations of Theorem 1 of Section 8.2 are cleared of fractions, the coefficients A and C have like signs. The complete graphs obtained in Example 4 can be checked with a function graphing utility after first solving for y in terms of x.

As in the previous section, if we have an equation in expanded form with $B = 0$, then we can use completing the square to write the equation in standard form and sketch a complete graph, or we can use the quadratic formula and a function graphing utility to draw a complete graph.

• EXAMPLE 5: Write each of the following equations in standard form. Determine the center, foci, vertices, lines of symmetry, asymptotes, and sketch a complete graph. Use the quadratic formula and a function graphing utility to check the graphs.

(a) $x^2 - 4y^2 + 2x - 24y = 39$ (b) $9y^2 - 5x^2 + 30x - 36y = 54$

SOLUTION:

(a) We group the terms involving x and the terms involving y and then complete the square:

$$\begin{aligned}
x^2 - 4y^2 + 2x - 24y &= 39\\
(x^2 + 2x) - 4(y^2 + 6y) &= 39\\
(x^2 + 2x + 1) - 4(y^2 + 6y + 9) &= 39 + 1 - 4(9)\\
(x + 1)^2 - 4(y + 3)^2 &= 4\\
\frac{(x + 1)^2}{4} - \frac{(y + 3)^2}{1} &= 1.
\end{aligned}$$

Now, we can see that the graph of this equation is a hyperbola with center $(-1, -3)$, and lines of symmetry $x = -1$ and $y = -3$. Notice that $a = 2$ and $b = 1$ so that $c^2 = 5$. The vertices are $(-3, -3)$ and $(1, -3)$, and the foci are $(-1 - \sqrt{5}, -3) = (-3.24, -3)$ and $(-1 + \sqrt{5}, -3) = (1.24, -3)$. The asymptotes are $y + 3 = \pm\frac{1}{2}(x + 1)$. A complete graph is sketched in Figure 8.3.12.

To check this graph with a function graphing utility we solve the original equation for y using the quadratic formula:

$$\begin{aligned}
&x^2 - 4y^2 + 2x - 24y = 39 \\
&-4y^2 - 24y + (x^2 + 2x - 39) = 0 \\
&4y^2 + 24y - (x^2 + 2x - 39) = 0 \\
&y = \frac{-24 \pm \sqrt{576 + 16(x^2 + 2x - 39)}}{8}.
\end{aligned}$$

The graph of

$$y = \frac{-24 + \sqrt{576 + 16(x^2 + 2x - 39)}}{8}$$

produces the portion of the hyperbola in Figure 8.3.12 that is above the line $y = -3$, and the graph of

$$y = \frac{-24 - \sqrt{576 + 16(x^2 + 2x - 39)}}{8}$$

produces the portion below the line $y = -3$ confirming the above algebraic computations.

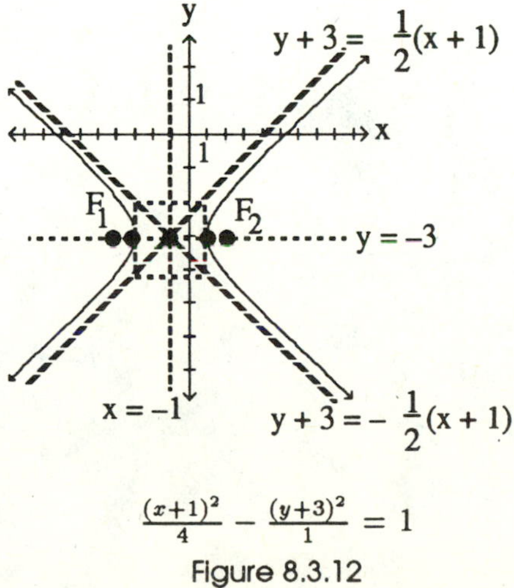

$\frac{(x+1)^2}{4} - \frac{(y+3)^2}{1} = 1$

Figure 8.3.12

(b) Again, we group terms and complete the square:

$$9y^2 - 5x^2 + 30x - 36y = 54$$
$$9(y^2 - 4y) - 5(x^2 - 6x) = 54$$
$$9(y^2 - 4y + 4) - 5(x^2 - 6x + 9) = 54 + 9(4) - 5(9)$$
$$9(y-2)^2 - 5(x-3)^2 = 45$$
$$\frac{(y-2)^2}{5} - \frac{(x-3)^2}{9} = 1.$$

The graph of this equation is a hyperbola with center at $(3,2)$ and lines of symmetry $x = 3$ and $y = 2$. Notice $a = \sqrt{5}$ and $b = 3$ so that $c^2 = 14$. The vertices are $(3, 2-\sqrt{5}) = (3, -0.24)$ and $(3, 2+\sqrt{5}) = (3, 4.24)$, and the foci are $(3, 2-\sqrt{14}) = (3, -1.74)$ and $(3, 2+\sqrt{14}) = (3, 5.74)$. The asymptotes are $y - 2 = \pm\frac{\sqrt{5}}{3}(x-3)$. The complete graph in Figure 8.3.13 can be confirmed with a graphing utility as in (a). •

Reflective Property of Hyperbolas

It can be proved that hyperbolas have a reflective property that is important in the construction of reflecting telescopes. The right branch of the hyperbola with foci F_1 and F_2 in Figure 8.3.14 is coated with a substance that causes it to reflect light. A light ray aimed toward the focus at F_1 is reflected by the surface through the focus at F_2 as illustrated in Figure 8.3.14. This fact is used to make a telescope using a parabolic lens and a hyperbolic lens. The main lens is parabolic with focus F_1 and vertex F_2, and the secondary lens is hyperbolic with foci F_1 and F_2 (Figure 8.3.15). The eye is positioned at the point F_2. The key to this lens arrangement is that one focus of the hyperbola *is* the focus of the parabola.

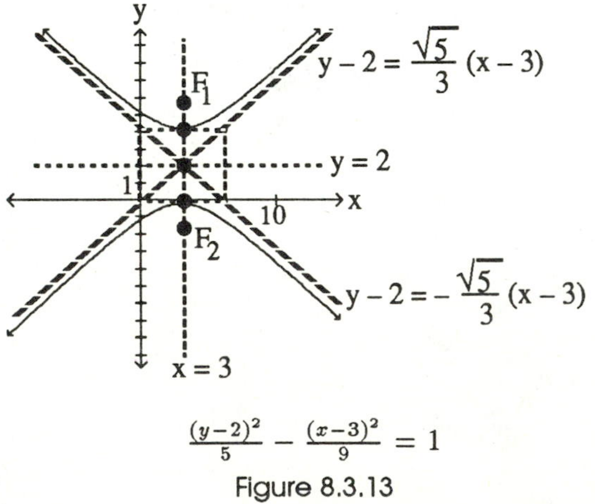

$\frac{(y-2)^2}{5} - \frac{(x-3)^2}{9} = 1$

Figure 8.3.13

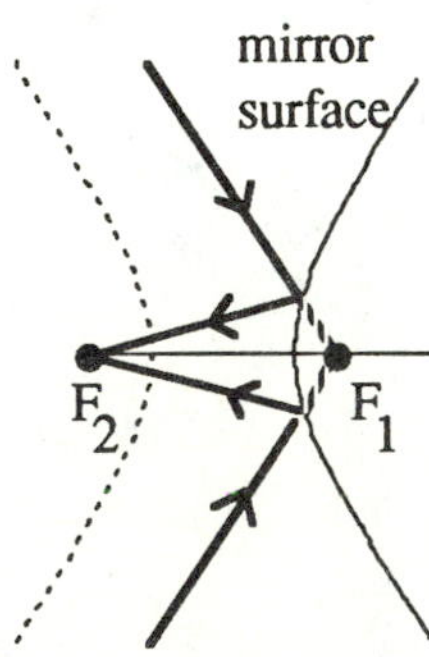

Figure 8.3.14

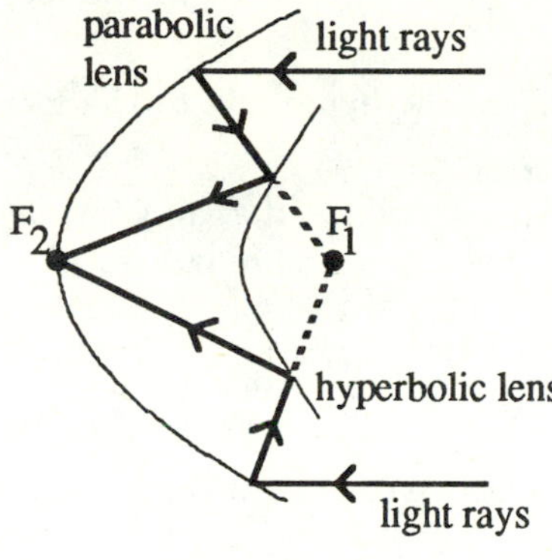

Figure 8.3.15

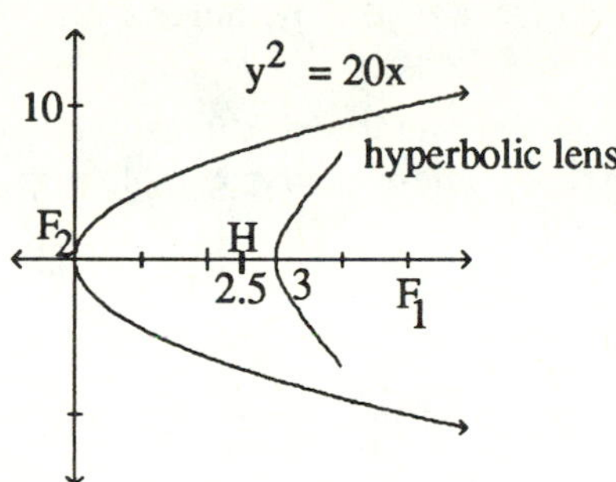

Figure 8.3.16

• EXAMPLE 6: The parabolic lens in Figure 8.3.15 is determined by a portion of the parabola $y^2 = 20x$ with $F_2 = (0,0)$. The vertex of the right branch of the hyperbola is at $(3,0)$. Determine

(a) the focus of the parabolic lens.
(b) the foci of the hyperbolic lens.
(c) an equation for the hyperbola.

SOLUTION:

(a) The focus F_1 of the parabolic lens is on the positive x-axis a units to the right of the origin, where $4a = 20$ (Figure 8.3.16). Thus, the focus of the parabola is $F_1 = (5,0)$.
(b) The foci of the hyperbolic lens are the vertex and focus of the parabolic lens. Thus, $(0,0)$ and $(5,0)$ are the foci of the hyperbolic lens.
(c) The foci of the hyperbola are $(0,0)$ and $(5,0)$. Thus, the center H of the hyperbola is $(2.5,0)$ and $a = 3 - 2.5 = 0.5$. Because $c = 5 - 2.5 = 2.5$, $b^2 = c^2 - a^2 = (2.5)^2 - (0.5)^2 = 6$. The equation of the hyperbola in standard form is

$$\frac{(x-2.5)^2}{0.25} - \frac{y^2}{6} = 1 .$$

•

• EXERCISES 8-3

Determine the standard form of the equation, and sketch a complete graph of the hyperbola. Do not use a graphing utility.

1. The difference of the distances from any point on the hyperbola to the foci is 4; the foci are $(-3,0)$ and $(3,0)$.
2. The difference of the distances from any point on the hyperbola to the foci is 8; the foci are $(0,-8)$ and $(0,8)$.

3. The difference of the distances from any point on the hyperbola to the foci is 3; the foci are $(-3, 2)$ and $(3, 2)$.
4. The difference of the distances from any point on the hyperbola to the foci is 6; the foci are $(1, -2)$ and $(9, -2)$.

Determine the coordinates of the center, foci, vertices, the lines of symmetry, and the asymptotes of the hyperbola, and sketch a complete graph. Do not use a graphing utility.

5. $\dfrac{x^2}{4} - \dfrac{(y-3)^2}{5} = 1$
6. $\dfrac{(y-3)^2}{9} - \dfrac{(x+2)^2}{4} = 1$
7. $4(y-1)^2 - 9(x-3)^2 = 36$
8. $4(x-2)^2 - 9(y+4)^2 = 1$

Use the quadratic formula and a function graphing utility to draw a complete graph.

9. $2x^2 - y^2 + 4x + 6 = 0$
10. $3y^2 - 5x^2 + 2x - 6y - 9 = 0$

Identify the conic. Write an equation of the conic in standard form; give the foci, vertices, center, the lines of symmetry, and asymptotes (if any), and sketch a complete graph. Check your answer using a graphing utility.

11. $9x^2 - 4y^2 - 36x + 8y - 4 = 0$
12. $y^2 - 4y - 8x + 20 = 0$
13. $25y^2 - 9x^2 - 50y - 54x - 281 = 0$
14. $9x^2 + 16y^2 + 54x - 32y - 47 = 0$
15. $4y^2 - 9x^2 - 18x - 8y - 41 = 0$
16. $16x^2 - y^2 - 32x - 6y - 57 = 0$
17. $2x^2 - 6x + 5y - 13 = 0$
18. $9x^2 + 4y^2 - 18x + 16y - 11 = 0$

The *eccentricity* e of a hyperbola is defined by $c/a = \sqrt{a^2 + b^2}/a$. Note a and b are both positive so eccentricity is positive. Compute the eccentricity of the hyperbola.

19. $x^2 - 4y^2 + 4 = 0$
20. $4x^2 - 9y^2 - 24x - 18y - 9 = 0$

Write an equation for the conic in standard form and sketch a complete graph.

21. Eccentricity 2 and transverse axis length 8.
22. Eccentricity 5 and transverse axis length 8.
23. Eccentricity 100 and transverse axis length 8.
24. Show that the eccentricity of a hyperbola is greater than one.
25. What does the eccentricity measure in a hyperbola?
26. What is the change in the nature of the hyperbola as the eccentricity gets very large?

A comet with hyperbolic path will pass between the sun and the earth. One focus of the hyperbola is the sun. Assume that the line through the earth and sun is a line of symmetry of the hyperbola and that the distance between the earth and the sun is 93,000,000 miles. Determine the equation in standard form of the hyperbola representing the path of the comet satisfying the following condition.

27. Eccentricity 20
28. Eccentricity 400

Consider a parabolic and hyperbolic lens arrangement as illustrated in Figure 8.3.17. Assume that the generating equation of the parabolic lens is $y^2 = 24x$ and that the vertex of the right branch of the generating hyperbola is $(4.5, 0)$.

29. What are the coordinates of the foci of the hyperbola?
30. Determine the standard form of the equation of the generating hyperbola.

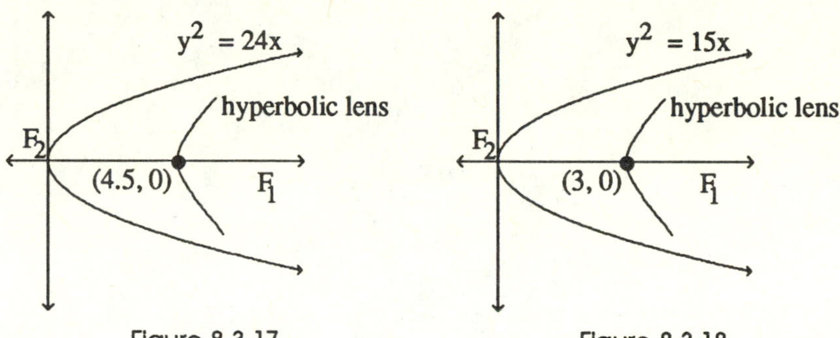

Figure 8.3.17

Figure 8.3.18

31. Draw complete graphs of both conics in the same viewing rectangle. Explain how this lens arrangement works.

Consider a parabolic and hyperbolic lens arrangement as illustrated in Figure 8.3.18. Assume that the generating equation of the parabolic lens is $y^2 = 15x$ and that the vertex of the right branch of the generating hyperbola is $(3, 0)$.

32. What are the coordinates of the foci of the hyperbola?
33. Determine the standard form of the equation of the generating hyperbola.
34. Draw complete graphs of both conics in the same viewing rectangle. Explain how this lens arrangement works.

A certain comet's path is known to be hyperbolic with the sun as one focus. Assume a space station is located at the center of the hyperbola. Further assume it has been determined that the distance between the comet and the space station is 5,000,000 miles at the closest point and that the sun and the space station are 12,000,000 miles apart.

35. Determine the standard form of the equation of the hyperbola representing the comet's path.
36. Draw a complete graph of the algebraic representation of the path of the comet. What portion of the graph represents the problem situation?
37. Describe the set of all points P satisfying the condition that $d(P, F_1) - d(P, F_2) = k$ when F_1 and F_2 are distinct fixed points and k is a positive real number.

8.4 Quadratic Forms and Conics

In this section we use a function graphing utility to determine complete graphs of equations of the form

$$Ax^2 + Bxy + Cy^2 + Dx + Ey + F = 0 \qquad (1)$$

where not all of A, B, and C are zero. Such an expression $Ax^2 + Bxy + Cy^2 + Dx + Ey + F$ is called a **quadratic form** in the variables x and y. A complete graph of equation (1) is either an ordinary conic section or a degenerate conic. For example, degenerate forms are circles, parallel lines, intersecting lines, one line, a single point, or no graph.

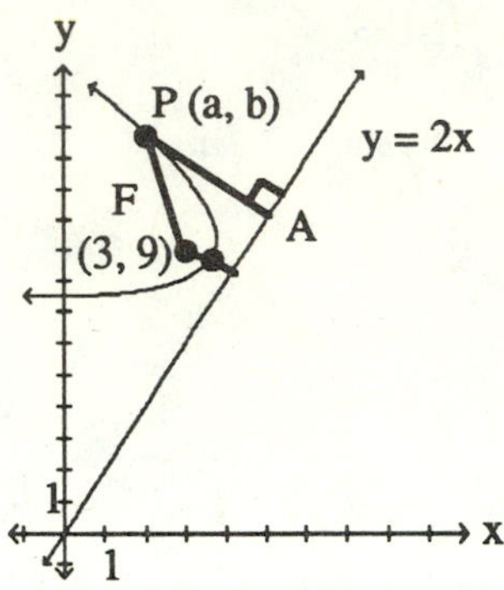

Figure 8.4.1

In the previous sections of this chapter we found that conic sections with lines of symmetry parallel to the coordinate axes are the graphs of equations of the type given in (1) with $B = 0$. If the lines of symmetry are not parallel to the coordinate axes then we obtain an equation with $B \neq 0$. Example 1 illustrates this situation.

• EXAMPLE 1: Determine an equation for the parabola with focus $F(3,9)$ and directrix $y = 2x$, and determine the coordinates of the vertex.

SOLUTION: Let $P(a,b)$ be any point on the parabola (Figure 8.4.1). Then choose the point A on $y = 2x$ so that PA is perpendicular to the line $y = 2x$. The parabola consists of all points (a,b) satisfying $d(P,F) = d(P,A)$. Now, $d(P,F) = \sqrt{(a-3)^2 + (b-9)^2}$. To determine $d(P,A)$ we must find the coordinates of the point A.

The slope of any line perpendicular to $y = 2x$ is $-\frac{1}{2}$. (Why?) An equation of the line through P perpendicular to $y = 2x$ is $y - b = -\frac{1}{2}(x-a)$. A is the point of intersection of the lines $y = 2x$ and $y - b = -\frac{1}{2}(x-a)$. The x-coordinate of the point of intersection can be found by solving each equation for y and setting these two expressions equal to each other:

$$\begin{aligned} 2x &= -\frac{1}{2}(x-a) + b \\ \frac{5x}{2} &= \frac{a}{2} + b \\ x &= \frac{a+2b}{5}. \end{aligned}$$

The y-coordinate of A is $2(\frac{a+2b}{5})$. Thus, A has coordinates $(\frac{a+2b}{5}, \frac{2a+4b}{5})$. Next we find $d(P, A)$:

$$\begin{aligned} d(P, A) &= \sqrt{(a - \frac{a+2b}{5})^2 + (b - \frac{2a+4b}{5})^2} \\ &= \sqrt{(\frac{4a-2b}{5})^2 + (\frac{-2a+b}{5})^2} \\ &= \sqrt{(\frac{2}{5})^2(2a-b)^2 + (-\frac{1}{5})^2(2a-b)^2} \\ &= \sqrt{\frac{1}{5}(2a-b)^2}\,. \end{aligned}$$

We use the definition of a parabola to determine an equation:

$$\begin{aligned} d(P, F) &= d(P, A) \\ \sqrt{(a-3)^2 + (b-9)^2} &= \sqrt{\frac{1}{5}(2a-b)^2} \\ (a-3)^2 + (b-9)^2 &= \frac{1}{5}(2a-b)^2 \\ 5a^2 - 30a + 45 + 5b^2 - 90b + 405 &= 4a^2 - 4ab + b^2 \\ a^2 + 4ab + 4b^2 - 30a - 90b + 450 &= 0. \end{aligned}$$

The above equation holds for every point (a, b) on the parabola. Thus, we have the following equation for the parabola:

$$x^2 + 4xy + 4y^2 - 30x - 90y + 450 = 0.$$

Next, we find the coordinates of the vertex. An equation for the line through F perpendicular to $y = 2x$ is $y - 9 = -\frac{1}{2}(x - 3)$. It can be shown that the lines $y = 2x$ and $y - 9 = -\frac{1}{2}(x - 3)$ intersect at $(\frac{21}{5}, \frac{42}{5})$. The vertex V of the parabola is the midpoint of the line segment joining $F(3, 9)$ and $(\frac{21}{5}, \frac{42}{5})$. Thus, $V = (\frac{18}{5}, \frac{87}{10}) = (3.6, 8.7)$. •

Graphing Conics with a Function Grapher

If $C \neq 0$, then equation (1) is quadratic in y and can be rewritten and the quadratic formula used to solve for y:

$$\begin{aligned} &Ax^2 + Bxy + Cy^2 + Dx + Ey + F = 0 \\ &Cy^2 + (Bx + E)y + (Ax^2 + Dx + F) = 0\,. \\ &y = \frac{-(Bx + E) \pm \sqrt{(Bx + E)^2 - 4C(Ax^2 + Dx + F)}}{2C}\,. \end{aligned}$$

A complete graph of equation (1) can be obtained by drawing the graphs of

$$f(x) = \frac{-(Bx+E) + \sqrt{(Bx+E)^2 - 4C(Ax^2+Dx+F)}}{2C}$$

and

$$g(x) = \frac{-(Bx+E) - \sqrt{(Bx+E)^2 - 4C(Ax^2+Dx+F)}}{2C}$$

in an appropriate viewing rectangle.

• EXAMPLE 2: Use the quadratic formula and a function graphing utility to determine a complete graph of the parabola $x^2 + 4xy + 4y^2 - 30x - 90y + 450 = 0$ of Example 1.

SOLUTION: The equation in Example 1 can be rewritten and the quadratic formula used to solve for y in terms of x:

$$x^2 + 4xy + 4y^2 - 30x - 90y + 450 = 0$$

$$4y^2 + (4x-90)y + (x^2 - 30x + 450) = 0$$

$$y = \frac{-(4x-90) \pm \sqrt{(4x-90)^2 - 16(x^2 - 30x + 450)}}{8}.$$

It is not necessary to simplify the above expression before using a function graphing utility. However, simplifying sometimes leads to functions that are easier to enter into a graphing utility. It can be shown that the above expression for y simplifies to

$$y = \frac{45 - 2x \pm \sqrt{2225 - 60x}}{4}.$$

The graph of

$$f(x) = \frac{45 - 2x + \sqrt{225 - 60x}}{4} \quad \text{and} \quad g(x) = \frac{45 - 2x - \sqrt{225 - 60x}}{4}$$

in the $[-5,5]$ by $[-5,20]$ viewing rectangle gives a complete graph of the parabola (Figure 8.4.2). The graph of f gives the top portion of the parabola and the graph of g the lower portion. Depending on the resolution of your graph, you may see a gap in your graph. This gap occurs where the graph of f and graph of g meet. This is *not* at the vertex. (Why?) Generally, we remove or reduce gaps as much as possible in this textbook. •

Next we consider the case $C = 0$ in equation (1). If $C = 0$, then equation (1) becomes

$$(Bx + E)y + (Ax^2 + Dx + F) = 0. \qquad (2)$$

This equation is linear in y. We can solve for y unless the coefficient of y is 0. This happens if $B = E = 0$. If $B = E = 0$, then the graph of equation (2) consists of a pair of vertical lines, one vertical line, or no graph depending on whether $Ax^2 + Dx + F$ has two real zeros, one real zero, or no real zeros. (Why?) In this case we cannot use a function graphing utility to obtain a graph of equation (2).

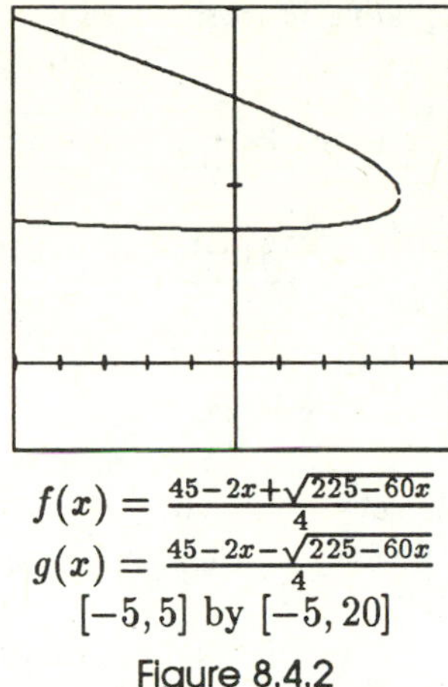

$f(x) = \frac{45-2x+\sqrt{225-60x}}{4}$
$g(x) = \frac{45-2x-\sqrt{225-60x}}{4}$
$[-5, 5]$ by $[-5, 20]$
Figure 8.4.2

Now we determine the graph if one or both of B and E are different from zero. If $B = 0$ and $E \neq 0$, then we can solve equation (2) for y as follows:

$$y = -\frac{Ax^2 + Dx + F}{E}.$$

The graph is a parabola in this case because $A \neq 0$. This is true because not all of A, B, and C can be zero and we are assuming that $B = C = 0$. Notice that a function graphing utility can be used to graph equation (2). Finally, suppose that $B \neq 0$. We can have $E = 0$ or $E \neq 0$. If $-\frac{E}{B}$ is *not* a zero of $Ax^2 + Dx + F$, then $-\frac{E}{B}$ is not in the domain of the function defined by $(Bx + E)y + (Ax^2 + Dx + F) = 0$ so that

$$y = \frac{Ax^2 + Dx + F}{Bx + E}.$$

A function graphing utility can be used to graph equation (2) in this case. If $-\frac{E}{B}$ is a zero of $Ax^2 + Dx + F$, then $Bx + E$ is a factor of $Ax^2 + Dx + F$. Thus $Ax^2 + Dx + F = (Bx + E)(Gx + H)$ for some G and H. Then, we can rewrite equation (2) as follows:

$$(Bx + E)y + (Ax^2 + Dx + F) = 0$$
$$(Bx + E)y + (Bx + E)(Gx + H) = 0$$
$$(Bx + E)(y + (Gx + H)) = 0.$$

Thus, in this case the graph of equation (2) is a pair of intersecting lines given by $Bx + E = 0$ and $y + Gx + H = 0$. Only one of these graph can be obtained with a function graphing utility. Therefore, except for the case $Bx + E = 0$ and the case $C = B = E = 0$, we can obtain the graph of equation (1) with a function graphing utility.

• EXAMPLE 3: Identify and draw a complete graph of $3x^2 + 4xy + 3y^2 - 12x + 2y + 7 = 0$.

SOLUTION: We use the quadratic formula to solve for y in terms of x and determine the following pair of functions:

$$f(x) = \frac{-(4x+2) + \sqrt{(4x+2)^2 - 12(3x^2 - 12x + 7)}}{6},$$
$$g(x) = \frac{-(4x+2) - \sqrt{(4x+2)^2 - 12(3x^2 - 12x + 7)}}{6}.$$

The graphs of f and g in Figure 8.4.3 give a complete graph of the conic. From this figure we can identify this conic as an ellipse. •

• EXAMPLE 4: Estimate the coordinates of the center, the endpoints of the major and minor axes, and the lines of symmetry of the ellipse of Example 3.

SOLUTION: We magnify the view in Figure 8.4.3 to obtain the graph in Figure 8.4.4, from which we can *estimate* that the endpoints of the major axis have coordinates $(0.7, 0)$ and $(7.3, -6)$. The center is the midpoint of the major axis, and its coordinates are $(\frac{0.7+7.3}{2}, \frac{0+(-6)}{2}) = (4, -3)$. It turns out that the coordinates of the center are exactly $(4, -3)$. The slope of the major axis is $\frac{-6-0}{7.3-0.7} = -0.9$. Thus, the line of symmetry containing the major axis has equation $y = -0.9(x - 0.7)$. The line through the center $(4, -3)$ perpendicular to the major axis is the other line of symmetry. This line has equation $y+3 = 1.1(x-4)$. (Why?) We can estimate the endpoints of the minor axis to be $(2.7, -4.5)$ and $(5.3, -1.5)$. •

We can use the distance formula and the endpoints found in Example 3 to show that the lengths of the major and minor axes are 8.9 and 4.0, respectively. Thus, if the ellipse of Example 3 were in standard form with center at the origin and major axis on the x-axis, we would have $a = 4.5$ and $b = 2.0$. We can use $c^2 = a^2 - b^2$ to determine that $c = 4.0$. Therefore, the two foci are 4 units on either side of the center $(4, -3)$ on the line

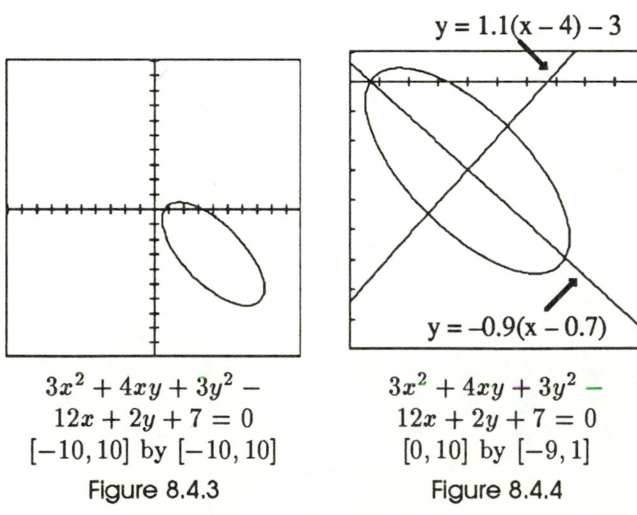

$3x^2 + 4xy + 3y^2 - 12x + 2y + 7 = 0$
$[-10, 10]$ by $[-10, 10]$
Figure 8.4.3

$3x^2 + 4xy + 3y^2 - 12x + 2y + 7 = 0$
$[0, 10]$ by $[-9, 1]$
Figure 8.4.4

$y = -0.9(x - 0.7)$. The line $y = -0.9(x - 0.7)$ makes an angle of $-42.0°$ with the positive x-axis because $\tan^{-1}(-0.9) = -42.0°$. Thus, it appears that the ellipse in standard form with center at $(4, -3)$, $a = 4.5$, and $b = 2.0$ has been rotated $42.0°$ clockwise to produce the ellipse in Figure 8.4.4. It turns out that the exact value of the angle is $45°$.

We can use the above information to obtain the graph of the ellipse $3x^2 + 4xy + 3y^2 - 12x + 2y + 7 = 0$ in the following way from the graph of

$$\frac{x^2}{(4.5)^2} + \frac{y^2}{(2.0)^2} = 1.$$

We apply, in order, the following transformations:

1. A horizontal shift 4 units right to obtain $\frac{(x-4)^2}{(4.5)^2} + \frac{y^2}{(2.0)^2} = 1$.
2. A vertical shift 3 units down to obtain $\frac{(x-4)^2}{(4.5)^2} + \frac{(y+3)^2}{(2.0)^2} = 1$.
3. A $42.0°$ rotation clockwise of the conic of 2 about the point $(4, -3)$ to obtain $3x^2 + 4xy + 3y^2 - 12x + 2y + 7 = 0$.

The **translation** (horizontal and vertical shift) factors and the angle of rotation can be computed exactly, but the computation is tedious. It is much faster to approximate these values graphically. It is easy to check these approximations with some graphing utilities, such as *Master Grapher*, which include a special conic grapher.

It turns out that every conic in the plane can be obtained by applying geometric transformations similar to the ones above to an appropriate conic in standard form with center at the origin. The angle of rotation needed can be assumed to be of measure between $-45°$ and $45°$. If the conic is rotated through such an angle then its new equation contains an xy term with nonzero coefficient.

Alternatively, if we start with a given conic in general form $Ax^2 + Bxy + Cy^2 + Dx + Ey + F = 0$, then it is possible to position the coordinate axes x' and y' so that an equation of the conic is one of the following (Figure 8.4.5):

Parabola	**Ellipse**	**Hyperbola**
$(y')^2 = 4ax'$	$\frac{(x')^2}{a^2} + \frac{(y')^2}{b^2} = 1$	$\frac{(x')^2}{a^2} - \frac{(y')^2}{b^2} = 1$
$(x')^2 = 4ay'$	$\frac{(x')^2}{b^2} + \frac{(y')^2}{a^2} = 1$	$\frac{(y')^2}{b^2} - \frac{(x')^2}{a^2} = 1$

The availability of graphing utilities reduces the need for determining the geometric transformations needed to obtain the graph of a given general conic from the graph of a conic in standard form centered at the origin. Now that we can quickly obtain a complete graph of a conic we rarely need to know the precise rotation or shifting factors. Historically, one major use of these factors was to obtain a quick sketch of a graph.

• **Understanding** From now on in this textbook we will assume that either a special conic graphing utility or the quadratic formula and a function graphing utility are used to draw the graphs of conics.

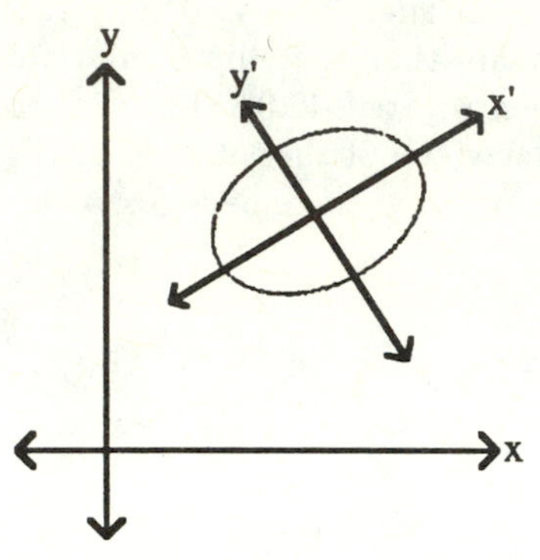

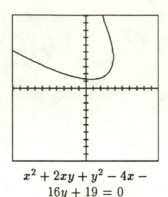

Figure 8.4.5

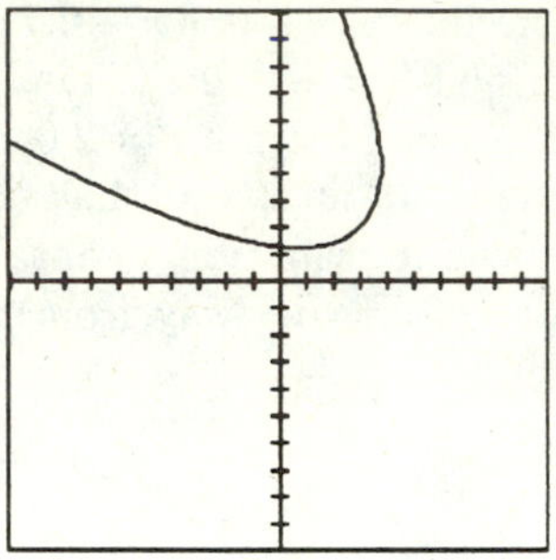

$x^2 + 2xy + y^2 - 4x -$
$16y + 19 = 0$
$[-10, 10]$ by $[-10, 10]$

Figure 8.4.6

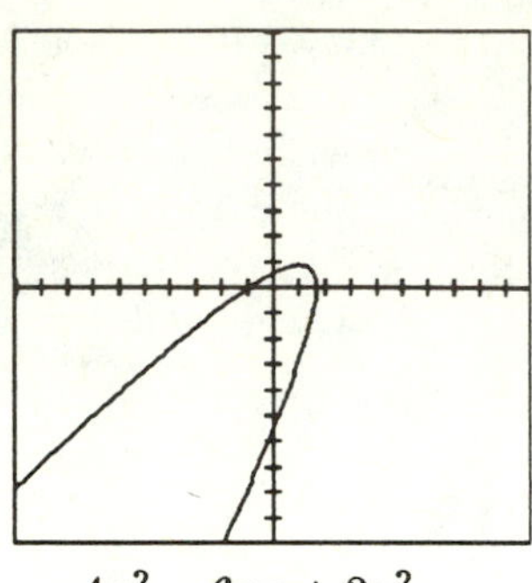

$4x^2 - 6xy + 2y^2 -$
$3x + 10y - 6 = 0$
$[-10, 10]$ by $[-10, 10]$

Figure 8.4.7

• EXAMPLE 5: Identify and draw a complete graph of each conic.

(a) $x^2 + 2xy + y^2 - 4x - 16y + 19 = 0$
(b) $4x^2 - 6xy + 2y^2 - 3x + 10y - 6 = 0$

SOLUTION:

(a) The graph in Figure 8.4.6 is a complete graph. Thus, the conic in (a) is a parabola. However, we must be careful to zoom out in order to be sure that the conic is a parabola. A given viewing rectangle may only display one branch of a hyperbola or one end of an ellipse. Such a figure may suggest that the graph is a parabola when it is not. This point is illustrated in (b).

(b) The graph of this equation in the standard viewing rectangle suggests this conic is a parabola (Figure 8.4.7). Notice that the graph in Figure 8.4.8 clearly shows that this conic is a hyperbola. •

We give the following theorem without proof to be used as you work the exercises.

• THEOREM 1 Consider the equation $Ax^2 + Bxy + Cy^2 + Dx + Ey + F = 0$.

(a) The graph of this equation is (possibly degenerate)
1. a parabola if $B^2 - 4AC = 0$,
2. an ellipse if $B^2 - 4AC < 0$, or
3. a hyperbola if $B^2 - 4AC > 0$.

(b) The angle θ at which a conic in standard form with lines of symmetry parallel to the coordinate axes needs to be rotated to coincide (after appropriate horizontal and vertical shifts) with the given conic is $\frac{1}{2}\tan^{-1}(\frac{B}{A-C})$ if $A \neq C$, and 45° if $A = C$.

The angle of rotation in Theorem 1 could be clockwise (negative) or counterclockwise (positive) depending on the particular standard form we start with. Notice that $A = C = 3$

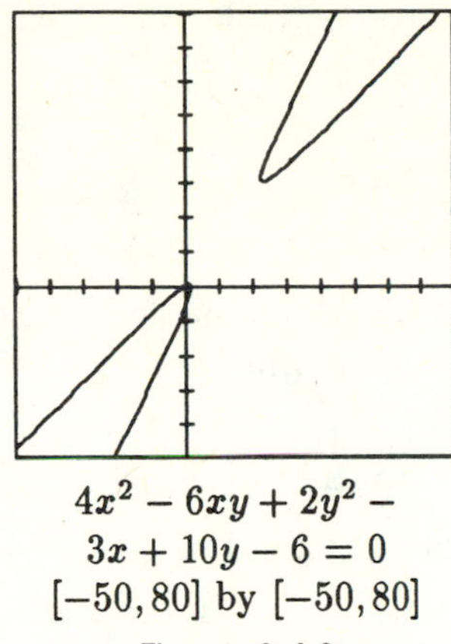

$4x^2 - 6xy + 2y^2 -$
$3x + 10y - 6 = 0$
$[-50, 80]$ by $[-50, 80]$

Figure 8.4.8

for the conic of Example 4. Thus, a 45° rotation clockwise of $\frac{(x-4)^2}{(4.5)^2} + \frac{(y+3)^2}{(2.0)^2} = 1$ produces $3x^2 + 4xy + 3y^2 - 12x + 2y + 7 = 0$. It can be shown that a 45° rotation counterclockwise of $\frac{(x-4)^2}{(2.0)^2} + \frac{(y+3)^2}{(4.5)^2} = 1$ produces the same ellipse.

- EXAMPLE 6: Identify the conic $2x^2 - 2xy + y^2 - 4x - 21 = 0$ and determine an angle θ that a conic in standard form needs to be rotated so that its lines of symmetry coincide with the lines of symmetry of the given conic.

SOLUTION: Now $A = 2$, $B = -2$, and $C = 1$, so $B^2 - 4AC = -4 < 0$ and the conic is an ellipse. Next $\theta = \frac{1}{2}\tan^{-1}\left(\frac{B}{A-C}\right) = \frac{1}{2}\tan^{-1}(-2) = -31.7°$. Check that a rotation of the given ellipse clockwise 31.7° about its center makes the lines of symmetry of the ellipse parallel to the coordinate axis, that is, makes the lines of symmetry horizontal and vertical. •

Some conic graphing utilities have features that allow a given conic to be rotated, translated (shifted), or graphs of functions to be overlayed with graphs of conics. *Master Grapher* is such a utility. The type of conic graphing utility you are using will influence how you do the exercises in this section.

• EXERCISES 8-4

Draw a complete graph and identify the conic.

1. $2x^2 - xy + 3y^2 - 3x + 4y - 6 = 0$
2. $-x^2 + 3xy + 4y^2 - 5x - 10y - 20 = 0$
3. $2x^2 - 4xy + 8y^2 - 10x + 4y - 13 = 0$
4. $2x^2 - 4xy + 2y^2 - 5x + 6y - 15 = 0$
5. $10x^2 - 8xy + 6y^2 + x - 5y + 20 = 0$
6. $10x^2 - 8xy + 6y^2 + x - 5y - 30 = 0$
7. $3x^2 - 2xy - 5x + 6y - 10 = 0$
8. $5xy - 6y^2 + 10x - 17y + 20 = 0$
9. $-3x^2 + 7xy - 2y^2 - x + 20y - 15 = 0$
10. $-3x^2 + 7xy - 2y^2 - 2x + 3y - 10 = 0$

Estimate the coordinates of the center, the endpoints of the major and minor axes, and the equations of the lines of symmetry of the ellipse.

11. $2x^2 - xy + 3y^2 + 2x + 6y - 10 = 0$
12. $2x^2 - 2xy + y^2 + 2x - 3y - 2 = 0$

Estimate the coordinates of the center and the equations of the asymptotes of the hyperbola.

13. $2x^2 - 4xy + y^2 + x + y - 2 = 0$
14. $2x^2 - 3xy + y^2 + 2x - 3y - 2 = 0$

Determine a pair of equations that model the end behavior of the hyperbola. Draw the two equations and the hyperbola in the same large viewing rectangle to check your answer.

15. $2x^2 + 5xy - y^2 + 2x - 8y + 10 = 0$
16. $-3x^2 + 8xy + 4y^2 - 10x + 5y - 30 = 0$
17. $20x^2 - 120xy - 15y^2 + 10x - 80y + 30 = 0$
18. $-5x^2 + 25xy + 6y^2 - 100x + 200y + 36 = 0$

Notice that $4x^2 + 9y^2 = 36$ can be expressed as $\left(\frac{x}{3}\right)^2 + \left(\frac{y}{2}\right)^2 = 1$.

19. Graph $4x^2 + 9y^2 = 36$ and $x^2 + y^2 = 1$ in the same viewing rectangle.

Give a sequence of transformations with order specified that when applied to a conic in standard form centered at the origin results in the given conic.

20. $4x^2 + 9y^2 = 36$
21. $9x^2 + 4y^2 - 18x + 16y - 11 = 0$
22. $9x^2 + 4y^2 + 36x - 18y + 4 = 0$
23. $xy - 2 = 0$
24. $xy + 2y - 1 = 0$
25. Determine an equation for the ellipse with foci $(-1, 1)$ and $(1, 2)$ such that the sum of the distances of its points from the foci is the constant 3.
26. Determine an equation for the hyperbola with foci $(-1, 1)$ and $(1, 2)$ such that the absolute value of the difference of the distances of its points from the foci is the constant 1.
27. Construct a new coordinate system for the conic in Exercise 25 such that the origin is at its center and the x-axis is along the major axis. Determine an equation in standard form of the conic relative to this new coordinate system.
28. Construct a new coordinate system for the conic in Exercise 26 such that the origin is at its center and the x-axis is along the major axis. Determine an equation in standard form of the conic relative to this new coordinate system.

Use the formula for the rotation angle θ given in Theorem 1 to determine the angle θ at which a conic in standard form needs to be rotated so the lines of symmetry coincide with the lines of symmetry of the given conic.

29. $2x^2 - 3xy + 2y^2 - 6x + 4y - 10 = 0$
30. the conic of Exercise 1

Carefully approximate the center and the lengths of the major and minor axes of the ellipse. Then list the transformations with order specified that when applied to an ellipse in standard form centered at the origin produces the given ellipse.

31. the ellipse of Exercise 3
32. the ellipse of Exercise 6

Consider 74 oz of a 56% acid solution. Let x oz of pure acid be added to the solution. How much pure acid must be added to obtain a mixture that is 81% acid?

33. Determine an algebraic representation of the final acidity percentage as a function of the amount of pure acid added.
34. Draw a complete graph and identify the type of graph.

35. What are the domain and range of the algebraic representation?

36. What values of x make sense in this problem situation?

37. Show how the intersection of two graphs can be used to solve the problem. Find the solution.

Consider 74 oz of a 56% acid solution. Let x oz of pure water be added to the solution. How much pure water must be added to obtain a mixture that is 35% acid?

38. Determine an algebraic representation of the final acidity percentage as a function of the amount of pure water added.

39. Draw a complete graph and identify the type of graph.

40. What are the domain and range of the algebraic representation?

41. What values of x make sense in this problem situation?

42. Show how the intersection of two graphs can be used to solve the problem. Find the solution.

43. **A rectangular hyperbola** has asymptotes coincident with or parallel to the coordinate axes. What conditions on A, B, C, D, E, and F in the general form of a conic will result in a rectangular hyperbola? What angle of rotation applied to a rectangular hyperbola will result in a conic of the form $Ax^2 + Cy^2 + Dx + Ey + F = 0$?

44. Graph the following degenerate conics.

(a) $x^2 = -1$ (b) $x^2 = 0$ (c) $x^2 = 1$

(d) $x^2 - xy = 0$ (e) $x^2 + y^2 = 0$ (f) $x^2 - y^2 = 0$

(g) $x^2 + y^2 + 1 = 0$

8.5 Nonlinear Systems of Equations and Inequalities

Some of the systems of equations solved in Section 2.2 were nonlinear. However, except for a few special exercises, each equation of the systems represented a function. In this section we solve systems of equations and inequalities where an equation or inequality need not represent a function. For example, equations of conics are involved in some of the systems. Two applications, one with a nonlinear system of equations for its model, and the other with a nonlinear system of inequalities, are investigated.

Three observers are strategically located with synchronized watches. They record the precise time that they hear sound being emitted from a certain source. For example, the source of sound may be a gun being fired. Let d_1, d_2, and d_3 be the distances of the observers A, B, and C, respectively, from the source of sound. We assume sound travels at the rate of 1100 feet per second. Suppose observer B hears the sound two seconds before observer A does. Then the sound source must be 2200 feet further from observer A than from observer B, that is, $d_1 - d_2 = 2200$. We can consider the positions of observers A and B to be foci of a hyperbola with $2a = 2200$. It follows that the sound source is a point of the hyperbola. Recall that a hyperbola is defined as the set of all points such that the absolute value of the *difference* of the distances from two fixed points (foci) is a constant denoted by $2a$. We must also have $a < c$, where c is half the distance between the foci. In this case we must be sure that the distance between observers A and B is greater than 2200 feet. Suppose observer C hears the sound four seconds before observer A does. Then the sound source must be 4400 feet further from observer A than from observer C, that is, $d_1 - d_3 = 4400$. We can consider the positions of observers A and C to be the foci of a hyperbola with $2a = 4400$. The location of the gun is at one point of intersection of the two hyperbolas. Thus, the location of the gun is a simultaneous solution to a system of equations consisting of the equations of the two hyperbolas.

- EXAMPLE 1: Observer A is located at the origin, observer B is located at $(0, 4000)$, and observer C is located at $(7000, 0)$. Observer B hears the sound of a gun 2 seconds before observer A, and observer C hears the sound 4 seconds before observer A. Determine a model whose solution locates the position of the gun when it was fired.

SOLUTION: Let P be the position of the gun (Figure 8.5.1). The hyperbola determined by A and B is defined by $|d(P, A) - d(P, B)| = 2200$, and the hyperbola determined by A and C is defined by $|d(P, A) - d(P, C)| = 4400$. Let P have coordinates (x, y). Next we determine an equation for each hyperbola. The center of the hyperbola determined by A, C, and the gun is $(3500, 0)$. Also $2a = 4400$, $c = 3500$, so that $b^2 = c^2 - a^2 = 7{,}410{,}000$. Thus, an equation is

$$\frac{(x-3500)^2}{(2200)^2} - \frac{y^2}{7{,}410{,}000} = 1.$$

Similarly, an equation determined by A, B, and the gun is

$$\frac{(y-2000)^2}{(1100)^2} - \frac{x^2}{2{,}790{,}000} = 1.$$

The location of the gun is at one point of intersection of the two hyperbolas above. In other words, the location of the gun is a simultaneous solution to the following system of equations:

$$\begin{cases} \dfrac{(x-3500)^2}{(2200)^2} - \dfrac{y^2}{7{,}410{,}000} = 1 \\[2ex] \dfrac{(y-2000)^2}{(1100)^2} - \dfrac{x^2}{2{,}790{,}000} = 1. \end{cases}$$

•

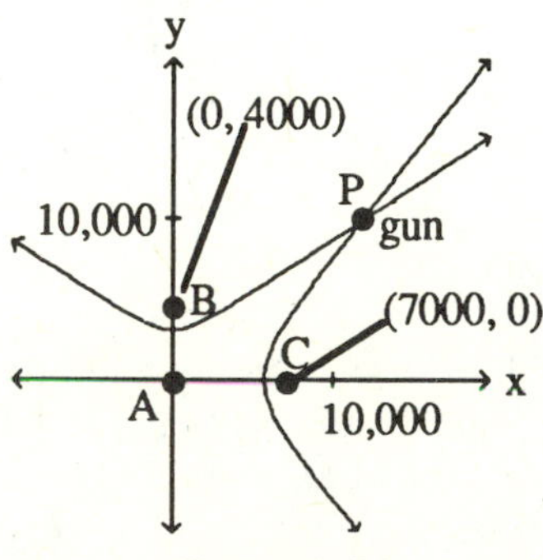

Figure 8.5.1

• EXAMPLE 2: Find the simultaneous solutions to

$$\begin{cases} \dfrac{(x-3500)^2}{(2200)^2} - \dfrac{y^2}{7{,}410{,}000} = 1 \\ \dfrac{(y-2000)^2}{(1100)^2} - \dfrac{x^2}{2{,}790{,}000} = 1. \end{cases}$$

SOLUTION: Solving the first equation for y we obtain

$$y = \pm\sqrt{7{,}410{,}000\left[\frac{(x-3500)^2}{(2200)^2} - 1\right]},$$

and solving the second equation for y we obtain

$$y = 2000 \pm 1100\sqrt{\frac{x^2}{2{,}790{,}000} + 1}.$$

The graph of the resulting four functions in the $[-10{,}000, 20{,}000]$ by $[-10{,}000, 20{,}000]$ viewing rectangle given in Figure 8.5.2 gives a complete graph of the system of equations and shows that there are four simultaneous solutions. We can use zoom-in to determine that the four solutions are $(162.63, 3105.20)$, $(1241.99, 629.24)$, $(6414.37, -2365.08)$, and $(11{,}714.31, 9792.52)$. •

Now we are in a position to locate the position of the gun in Example 1. We see in the next example that only one of the four possible positions found in Example 2 is the actual location of the gun.

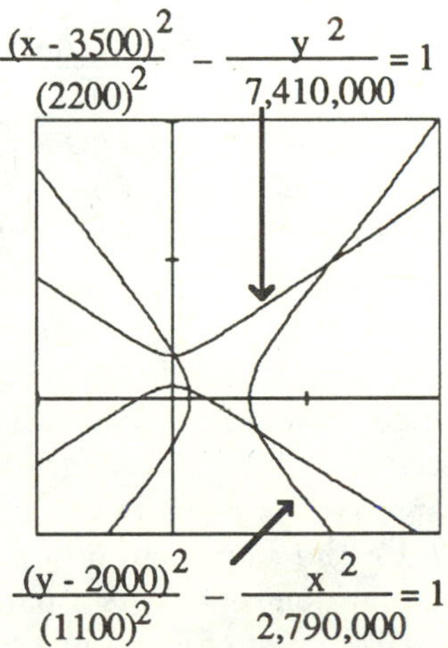

(x - 3500)^2 / (2200)^2 - y^2 / 7,410,000 = 1
(y - 2000)^2 / (1100)^2 - x^2 / 2,790,000 = 1

$[-10{,}000, 20{,}000]$ by
$[-10{,}000, 20{,}000]$

Figure 8.5.2

• EXAMPLE 3: Determine the position of the gun of Example 1.

SOLUTION: Let P be the position of the gun. From Example 2 we know there are four possibilities for P. From the statement of the problem we know that $d(P,A)-d(P,B) = 2200$ because P must be further from A than B. This means P is on the upper half of the hyperbola determined by A and B. (Why?) Similarly, $d(P,A) - d(P,C) = 4400$ because P must be further from A than C. This means that P is on the right branch of the hyperbola determined by A and C. (Why?) Therefore, the gun must be located at $(11{,}714.31, 9792.52)$. Alternatively, P cannot be $(162.63, 3105.20)$ or $(1241.99, 629.24)$ because $d(P,A)-d(P,C) = -4400$ in both of these cases. Similarly, P cannot be $(6414.37, -2365.08)$ because $d(P,A) - d(P,B) = -2200$ in this case. It can be shown directly that $d(P,A) - d(P,C) = 4400$ and $d(P,A) - d(P,B) = 2200$ if $P = (11{,}714.31, 9792.52)$. •

The same principle used to locate the gun is used in navigation; radio waves are used in this case instead of sound.

• EXAMPLE 4: Find the simultaneous solutions to

$$\begin{cases} x^2 - 2xy + y^2 - 8x - 8y + 48 = 0 \\ 5x^2 + xy + 6y^2 - 79x - 73y + 196 = 0 \end{cases}$$

SOLUTION: We can use Theorem 1 of the previous section to see that the first conic is a parabola and the second an ellipse. Solving the two equations for y gives the following:

$$y = \frac{2x + 8 \pm \sqrt{(2x+8)^2 - 4(x^2 - 8x + 48)}}{2}$$

$$y = \frac{73 - x \pm \sqrt{(73 - x)^2 - 24(5x^2 - 79x + 196)}}{12}.$$

Now we can use a function grapher to obtain a complete graph of each equation in the system. Figure 8.5.3 gives a complete graph of each equation and shows that there are two points of intersection. We can use zoom-in to determine that the solutions are $(3.26, 11.74)$ and $(15.05, 4.60)$. •

It is interesting to try to solve the system of equations in Example 4 algebraically. For example, the two expressions for y can be equated and an algebraic process used to eliminate the radicals. This will lead to a polynomial equation of degree 4 and is an extremely tedious procedure. A graphing approach is much easier and faster.

Suppose that we replace the equal sign in $Ax^2 + Bxy + Cy^2 + Dx + Ey + F = 0$ by an inequality sign. If the equation represents a parabola or an ellipse, then the graph of a corresponding inequality will consist of the points on one side or the other of the conic. The conic itself will be included in the graph if $\geq$ or $\leq$ are involved. If the conic is a hyperbola, then the graph of a corresponding inequality consists of the points between the two branches or the points outside of the two branches. Again, the hyperbola may be part of the graph of an inequality. The definitions of the conics can be used to help establish these facts.

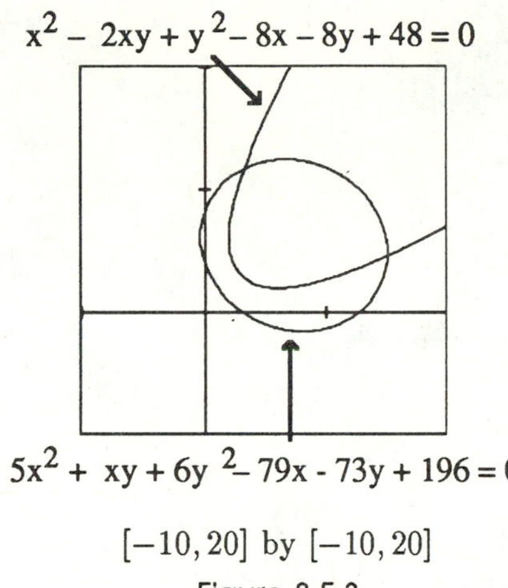

[−10, 20] by [−10, 20]

Figure 8.5.3

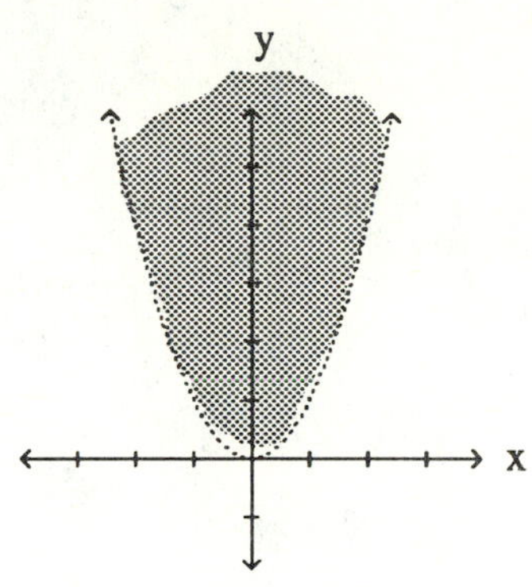

Figure 8.5.4

• EXAMPLE 5: Solve

$$\begin{cases} y > x^2, \\ 2x + 3y < 4. \end{cases}$$

SOLUTION: We know that the graph of $y > x^2$ consists of the portion on one side or the other of $y = x^2$. The parabola $y = x^2$ is the boundary of the region. A point can be checked to determine which portion is the solution of the inequality. The point $(0, 2)$ is *inside* of the parabola and is a solution to the inequality because $2 > 0$. Thus, the graph of $y > x^2$ consists of the portion *inside* of the parabola $y = x^2$. This is the shaded portion in Figure 8.5.4. The graph of the parabola $y = x^2$ is called the **boundary** of the shaded region. Notice that the parabola is dashed to indicate it is not part of the solution. In Section 1.5 we found that the solution to $2x + 3y < 4$ consists of the points on one side or the other of the line $2x + 3y = 4$. Again, checking a single point will determine which side of the line is the solution to the inequality. You can verify that the solution is the portion below the line (Figure 8.5.5). Therefore, the solution to the system of inequalities consists of the intersection of the shaded portions of Figures 8.5.4 and 8.5.5. The solution to the system is the shaded portion of Figure 8.5.6. The parabola and line are dashed to indicate they are *not* part of the solution. •

• EXAMPLE 6: Solve

$$\begin{cases} x^2 - 2xy + y^2 - 8x - 8y + 48 \le 0, \\ 5x^2 + xy + 6y^2 - 79x - 73y + 196 > 0. \end{cases}$$

SOLUTION: Complete graphs of $x^2 - 2xy + y^2 - 8x - 8y + 48 = 0$ and $5x^2 + xy + 6y^2 - 79x - 73y + 196 = 0$ are given in Figure 8.5.3 of Example 4. We can use Figure 8.5.3 to convince ourselves that $(4, 4)$ is a point *interior* to both the parabola and the ellipse. Notice that

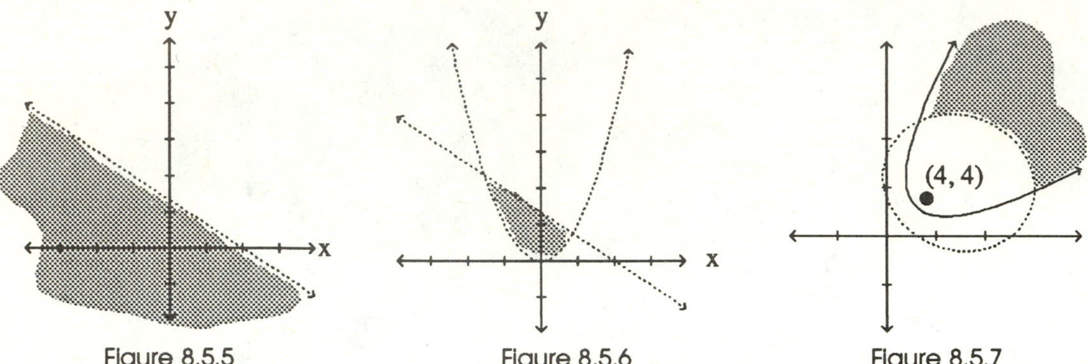

Figure 8.5.5 Figure 8.5.6 Figure 8.5.7

$4^2 - 2(4)(4) + 4^2 - 8(4) - 8(4) + 48 = -16$ *is* less than 0 and $5(4^2) + (4)(4) + 6(4^2) - 79(4) - 73(4) + 196 = -220$ ***is not*** greater than 0. Thus, $(4, 4)$ is a solution to the first inequality but ***not*** to the second inequality. Therefore, the solution to this system of inequalities consists of the points *inside* the parabola and *outside* of the ellipse (Figure 8.5.7). The portion of the ellipse inside the parabola is not a solution to the second inequality. Thus, this portion of the boundary of the solution region is dashed to indicate it is not part of the solution to the system. Similarly, the portion of the parabola outside of the ellipse is a solution to both inequalities. Therefore, this portion of the boundary of the solution region is drawn with a solid curve to indicate it is part of the solution to the system. •

Let x and y represent the number of cc of red and white paint, respectively, that are mixed together to produce a new color. The new color depends on the amounts of each paint used.

• EXAMPLE 7: Red paint at \$1.00 per cc is mixed with white paint at \$0.20 per cc in a container with capacity 1050 cc. Color mixologists have determined that the product of x and y must be greater than 100,000 to produce *eye pleasing* colors. What are the possible mixtures that produce eye pleasing colors with total cost at most \$500?

(a) Determine an algebraic representation for the problem situation.
(b) Draw a complete graph of the algebraic representation in (a).
(c) Which portions of the graph in (b) represent the solutions to the problem situation?

SOLUTION: Let x be the amount (in cc) of red paint used, and y the amount (in cc) of white paint used. There are many possible solutions to this problem. For example, we can show that $x = 400$ and $y = 400$ is a solution. First, the mixing container will hold 800 cc of paint. Second, the cost of this mixture is $400(0.2) + 400(1) = \$480$, and \$480 is less

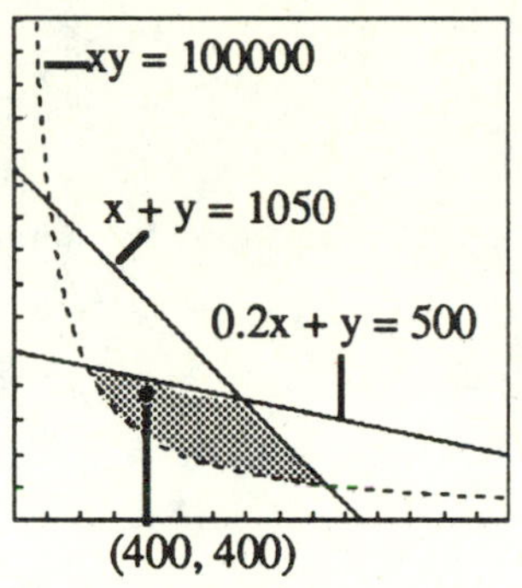

[0, 1500] by [0, 1500]

Figure 8.5.8

than \$500. Finally, (400)(400) = 160,000 is greater than 100,000 so the resulting color will be *eye pleasing.*

(a) We must have $x + y \leq 1050$ because the container holds at most 1050 cc. Next, $0.2x + y \leq 500$ because the total cost of the mixture is to be at most \$500. Finally, $xy > 100{,}000$ because the color of the mixture is to be eye pleasing. An algebraic representation of the problem situation is given by the following system of inequalities:

$$\begin{aligned} x + y &\leq 1050, \\ 0.2x + y &\leq 500, \\ xy &> 100{,}000. \end{aligned}$$

Of course, we must also have $x \geq 0$ and $y \geq 0$. (Why?)

(b) Figure 8.5.8 shows the graphs of $x + y = 1050$, $0.2x + y = 500$, and $xy = 100{,}000$ in $[0, 1500]$ by $[0, 1500]$. The complete graph of the algebraic representation in (a) consists of the region in the first quadrant that is below each of the lines $x + y = 1050$ and $0.2x + y = 500$, and above the branch of the hyperbola $xy = 100{,}000$. This is the shaded region in Figure 8.5.8. The straight line portions of the boundary are included, but the curved portion of the boundary given by the hyperbola is *not.*

(c) All pairs (x, y) given by the shaded portion of Figure 8.5.8, including the straight line boundary segments and excluding the curved boundary segment, represent solutions to the problem situation. Notice that $(400, 400)$ is a point in the shaded region that corresponds to the solution found earlier. •

• EXERCISES 8-5

Find the simultaneous solutions to each system.

1. $4x^2 + 9y^2 = 36$, $x + 2y = 2$
2. $4x^2 + 9y^2 = 36$, $x - 2y = 2$
3. $9x^2 - 4y^2 = 36$, $x + 2y = 4$
4. $9x^2 - 4y^2 = 36$, $x - 2y = 4$

5. $9x^2 - 4y^2 = 36,\ 4x^2 + 9y^2 = 36$

6. $9x^2 - 4y^2 = 36,\ 9x^2 + 4y^2 = 36$

7. $x^2 + 4y^2 = 4,\ y = 2x^2 - 3$

8. $x^2 - 4y^2 = 4,\ y = 2x^2 - 3$

9. $x^2 - 4y^2 = 4,\ x = 2y^2 - 3$

10. $x^2 + 4y^2 = 4,\ x = 2y^2 - 3$

Draw a complete graph of each equation and find the simultaneous solutions to each system.

11. $x^2 + xy + y^2 + x + y - 6 = 0,\ 2x^2 - 3xy + 2y^2 + x + y - 8 = 0.$

12. $5x^2 - 40xy + 20y^2 - 17x + 25y + 50 = 0,\ xy - 3x = 0.$

13. $2x^2 - 3xy + 2y^2 + x + y - 8 = 0,\ 2x^2 - 8xy + 3y^2 + x + y - 10 = 0.$

14. $3x^2 - 5xy + 6y^2 - 7x + 5y - 9 = 0,\ x^2 + y^2 - 2x - 6 = 0.$

Draw a graph representing the solutions to the system of inequalities.

15. $y \geq x^2,\ x^2 + y^2 \leq 4$

16. $y \leq x^2 + 4,\ x^2 + y^2 \geq 4$

17. $x^2 - y^2 < 4,\ x^2 + y^2 < 4$

18. $x^2 + 4y^2 - 2x + 4y - 6 > 0,\ y < \frac{1}{2}x + 1$

19. $x^2 + 2xy - 5y^2 + 2x + 4y - 10 > 0,\ 4x^2 + xy + 4y^2 - 5x + 5y - 15 < 0$

20. $x^2 + 2xy + 5y^2 + 2x + 4y - 10 > 0,\ 5y^2 + 2x + 3y - 6 < 0$

Consider an acid solution of 84 oz that is 58% pure acid. How many ounces of pure acid (x) must be added to obtain a solution that is 70% to 80% acid? Let y be the concentration of acid in the final solution.

21. Determine a model consisting of a pair of inequalities and an equation whose simultaneous solution can be used to determine the solution to the problem situation.

22. Draw complete graphs of each inequality and the equation of the model of Exercise 21 and a graph of the solution to the model.

23. What values of x and y make sense in this problem situation?

24. Solve the problem.

Suppose three observers A, B, and C, listening for illegal dynamite explosions are positioned so the angle of the line between A and B and the line between B and C is 90°. Further suppose A and B are 6000 feet apart and B and C are 2000 feet apart. A hears the explosion 4.06 seconds before B, and C hears the explosion 0.63 seconds before B.

25. Determine a model whose solution locates the position of the explosion.

26. Solve the simultaneous system of equations in the model of Exercise 25.

27. Determine the position of the explosion.

"Hush Dog" shoes are made from two types of leather, A and B. Type A costs \$0.25 per square inch, and type B cost \$0.65 per square inch. The "Hush Dog" Company must buy at least 3000 square inches of A per day and 8600 square inches of B per day (or the supplier will not do business). However, the total square inches of material provided by the supplier per day must be less than 30,000 square inches. Experience has shown the president of the company that the total cost of the material must remain less than \$10,000 per day or no one will buy "Hush Dogs" (because they would need to be priced too high).

28. Determine a system of inequalities whose solution provides a model for all possible combinations of the amounts of leather A and leather B that could be used in this problem situation.

29. Graph the region that represents all possible combinations of the amounts of leather A and leather B that could be used in this problem situation.

8.6 Polar Equations of Conics

In Section 8.1 we defined a parabola as the locus of all points P such that the perpendicular distance from P to a fixed line (directrix) equals the distance from P to a fixed point (focus) not on the directrix. In this section we give focus-directrix definitions for nondegenerate hyperbolas and ellipses. A circle is a degenerate ellipse. Then, we obtain polar equations for nondegenerate conics using their focus-directrix definitions.

First, we will extend the notion of eccentricity discussed in earlier exercises.

Eccentricity

• DEFINITION The **eccentricity e of an ellipse or a hyperbola** is defined by $e = \frac{c}{a}$, where $2c$ is the distance between foci and $2a$ is the length of the major axis of an ellipse or the transverse axis of a hyperbola. The eccentricity of a parabola is defined to be 1. The eccentricity of a circle is defined to be 0. •

For an ellipse we have $0 < c < a$ so $e = \frac{c}{a} < 1$. In the exercises you will see that the closer e is to 0, the more circular the ellipse is. Similarly, for a hyperbola $c > a > 0$ so $e = \frac{c}{a} > 1$. As $e \to \infty$, the graph of the hyperbola becomes more like a pair of parallel lines.

• EXAMPLE 1: Determine the eccentricity of each conic.

(a) $\dfrac{x^2}{16} + \dfrac{y^2}{9} = 1$ (b) $\dfrac{x^2}{4} + \dfrac{y^2}{25} = 1$ (c) $\dfrac{x^2}{9} - \dfrac{y^2}{16} = 1$ (d) $\dfrac{y^2}{16} - \dfrac{x^2}{9} = 1$

SOLUTION:

(a) Here $a = 4$ and $b = 3$ because we must have $a > b$. Now, $c^2 = a^2 - b^2 = 16 - 9 = 7$. Thus, $e = \frac{c}{a} = \frac{\sqrt{7}}{4}$. Notice $e < 1$.

(b) Here $a = 5$ and $b = 2$. Now, $c^2 = a^2 - b^2 = 25 - 4 = 21$. Thus, $e = \frac{c}{a} = \frac{\sqrt{21}}{5}$ and is again less than 1.

(c) Here $a^2 = 9$, $b^2 = 16$, and $c^2 = a^2 + b^2 = 25$. Thus, $e = \frac{c}{a} = \frac{5}{3}$. Notice $e > 1$.

(d) Here $a^2 = 16$, $b^2 = 9$, and $c^2 = a^2 + b^2 = 25$. Thus, $e = \frac{c}{a} = \frac{5}{4}$, and again $e > 1$. •

• THEOREM 1 Let $\frac{x^2}{a^2} + \frac{y^2}{b^2} = 1$ be an ellipse with foci on the x-axis and let l be the vertical line $x = \frac{a}{e}$ where e is the eccentricity of the ellipse. (Figure 8.6.1). Let P be any point on the ellipse and Q the point on l such that PQ is perpendicular to l. Then $d(P,F)/d(P,Q) = e$ where $F = (c,0)$ and $c > 0$.

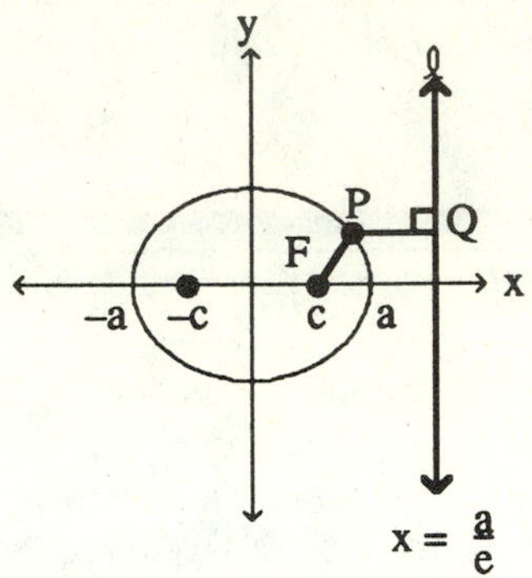

Figure 8.6.1

PROOF If we solve $\frac{x^2}{a^2}+\frac{y^2}{b^2}=1$ for y^2 we obtain $y^2=b^2-\frac{b^2x^2}{a^2}$. Now we compute $d(P,F)$ where $P=(x,y)$ is on the ellipse:

$$
\begin{aligned}
d(P,F) &= \sqrt{(x-c)^2+y^2} \\
&= \sqrt{(x-c)^2+b^2-\frac{b^2x^2}{a^2}} \\
&= \sqrt{x^2-2cx+c^2+b^2-\frac{b^2x^2}{a^2}} \\
&= \sqrt{x^2-2cx+a^2-b^2+b^2-\frac{b^2x^2}{a^2}} \\
&= \sqrt{x^2\left(1-\frac{b^2}{a^2}\right)-2cx+a^2} \\
&= \sqrt{x^2\left(\frac{a^2-b^2}{a^2}\right)-2cx+a^2} \\
&= \sqrt{\frac{c^2}{a^2}x^2-2cx+a^2} \\
&= \sqrt{\frac{c^2}{a^2}\left(x^2-\frac{2a^2}{c}x+\frac{a^4}{c^2}\right)} \\
&= \sqrt{\frac{c^2}{a^2}\left(x-\frac{a^2}{c}\right)^2} \\
&= \sqrt{\frac{c^2}{a^2}\left(x-\frac{a}{e}\right)^2} \\
&= \frac{c}{a}\left|x-\frac{a}{e}\right|
\end{aligned}
$$

$$= e\left|x - \frac{a}{e}\right|$$
$$= e \cdot d(P, Q).$$

Theorem 1 means that the vertical line $x = \frac{a}{e}$ can be taken as the directrix and the point $(c, 0)$ as the focus for an ellipse in a focus-directrix definition. Later we will use the fact that the distance between the directrix and focus is $\frac{a}{e} - c$. It can be shown that the ellipse $\frac{x^2}{a^2} + \frac{y^2}{b^2} = 1$ consists of precisely those points P for which $d(P, F)/d(P, Q) = e$. The vertical line $x = -\frac{a}{e}$ and the point $(-c, 0)$ can also be chosen as a focus and directrix of the ellipse of Theorem 1. In fact, Theorem 1 is also true for this pair. Again, the distance between the directrix and focus is $\frac{a}{e} - c$.

Consider an ellipse $\frac{x^2}{b^2} + \frac{y^2}{a^2} = 1$ with foci $(0, -c)$ and $(0, c)$ on the y-axis. Again, $e = \frac{c}{a}$, and we can choose the horizontal line $y = \frac{a}{e}$ as the directrix to pair with the focus $(0, c)$, or the horizontal line $y = -\frac{a}{e}$ as the directrix to pair with the focus $(0, -c)$, where $c > 0$. Again, the distance between the focus and directrix is $\frac{a}{e} - c$.

We state a theorem without proof for hyperbolas similar to Theorem 1.

• **THEOREM 2** Let $\frac{x^2}{a^2} - \frac{y^2}{b^2} = 1$ be a hyperbola with foci on the x-axis and let l be the vertical line $x = \frac{a}{e}$ where e is the eccentricity of the hyperbola (Figure 8.6.2). Let P be any point on the hyperbola and Q the point on l such that PQ is perpendicular to l. Then $d(P, F)/d(P, Q) = e$ where $F = (c, 0)$ and $c > 0$.

Notice that $0 < \frac{a}{e} < a$ because $e > 1$. We can also use the vertical line $x = -\frac{a}{e}$ and the point $(-c, 0)$ as directrix and focus for the hyperbola. For the hyperbola $\frac{y^2}{a^2} - \frac{x^2}{b^2} = 1$ the horizontal line $y = \frac{a}{e}$ and the point $(0, c)$ with $c^2 = a^2 + b^2$ can be chosen as the directrix and focus, or the horizontal line $y = -\frac{a}{e}$ and the point $(0, -c)$ can be chosen as the directrix and focus. In all cases, the distance between the focus and directrix is $c - \frac{a}{e}$, the *opposite* of the distance for an ellipse.

Now we can give the following alternative definition of a conic.

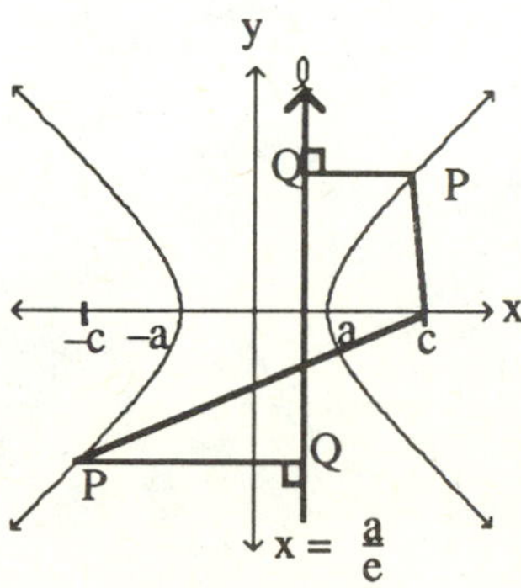

Figure 8.6.2

• DEFINITION Let F be a fixed point and l a line *not* containing F. A **conic** is the locus of all points P with $d(P, F)/d(P, Q) = e$ where e is a positive constant called the eccentricity and Q is the point on l with PQ perpendicular to l. F is called a **focus** and l a **directrix** for the conic. •

It turns out that the conic in the above definition is an ellipse if $e < 1$, a parabola if $e = 1$, and a hyperbola if $e > 1$.

Next we obtain polar equations for nondengenerate conics. We choose our coordinate system so that a focus of the conic is at the *origin* and a directrix is parallel to the x- or y-axis.

Polar Equations

We consider a new representation of a nondegenerate conic. Let l be the line $x = -h$ with $h > 0$. Let $F = (0, 0)$ be a focus and l a directrix of the conic. Let $P(r, \theta)$ be any point on the conic in polar form and Q the point on l with PQ perpendicular to l (Figure 8.6.3). First we assume that P is on the same side of l as F and that we have chosen the polar coordinates of P so that $r > 0$. Then, $d(P, F) = r$ and $D(P, Q) = r\cos\theta - (-h) = h + r\cos\theta$. The following equation is true for P:

$$
\begin{aligned}
d(P, F) &= e \cdot d(P, Q) \\
r &= e(h + r\cos\theta) \\
r - er\cos\theta &= eh \\
r &= \frac{eh}{1 - e\cos\theta}.
\end{aligned}
$$

Now, suppose that P is on the opposite side of l (Figure 8.6.4). Notice if there is such a point, then we must have $e > 1$. We show that the alternate coordinates of P given by (r, θ), where $r < 0$ as suggested by Figure 8.6.4 satisfy the same equation obtained above.

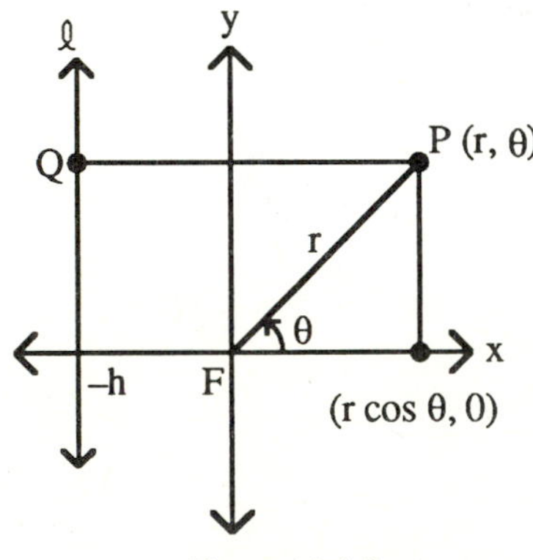

Figure 8.6.3

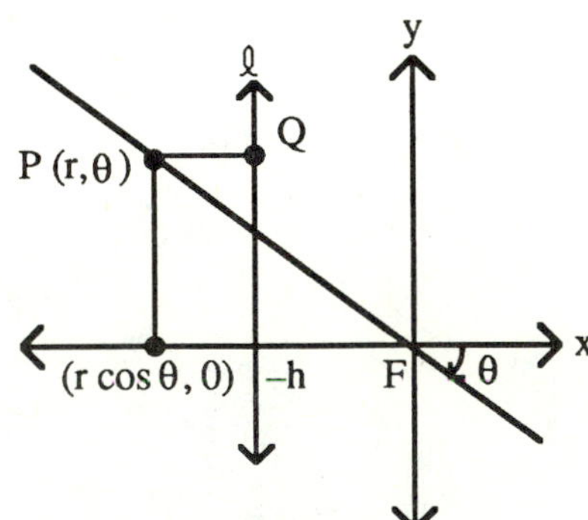

Figure 8.6.4

Notice that $\cos\theta > 0$ so that $r\cos\theta < \theta$:

$$\begin{aligned} d(P,F) &= e \cdot d(P,Q) \\ -r &= e(-h - r\cos\theta) \\ r &= eh + er\cos\theta \\ r &= \frac{eh}{1 - e\cos\theta}. \end{aligned}$$

The *Graphing Calculator and Computer Graphing Laboratory Manual* accompanying this textbook contains a simple program that permits graphs of curves defined by polar equations to be obtained with a graphing calculator. The graphing utility *Master Grapher*, used to draw graphs in this textbook, also has a polar equation option.

- EXAMPLE 2: Draw complete graphs of $r = \frac{2e}{1-e\cos\theta}$ for e equal to 0.25, 0.5, 1.5, and 4.

SOLUTION: Notice that $x = -2$ is a directrix for each of these polar equations, and the origin is a focus for each equation. Complete graphs are given in Figures 8.6.5–8.6.8. Notice that the first two graphs are ellipses with the first one more circular. The last two graphs are hyperbolas with the branches of the last one appearing least parabolic. •

The next situation we consider is where the directrix is to the right of the y-axis (Figure 8.6.9). Let the directrix be the line l defined by $x = h$, $h > 0$. First, we assume that P is on the same side of l as F and (r, θ) is chosen so that $r > 0$:

$$\begin{aligned} d(P,F) &= e \cdot d(P,Q) \\ r &= e(h - r\cos\theta) \\ r &= \frac{eh}{1 + e\cos\theta}. \end{aligned}$$

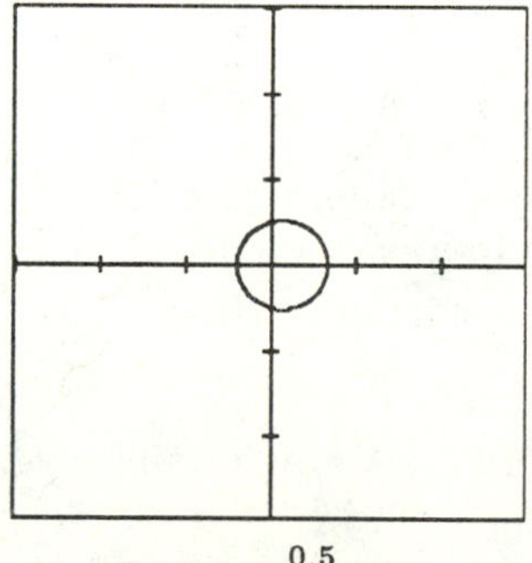

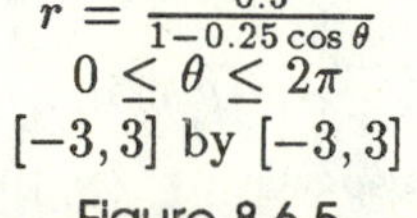

$r = \frac{0.5}{1-0.25\cos\theta}$
$0 \le \theta \le 2\pi$
$[-3,3]$ by $[-3,3]$
Figure 8.6.5

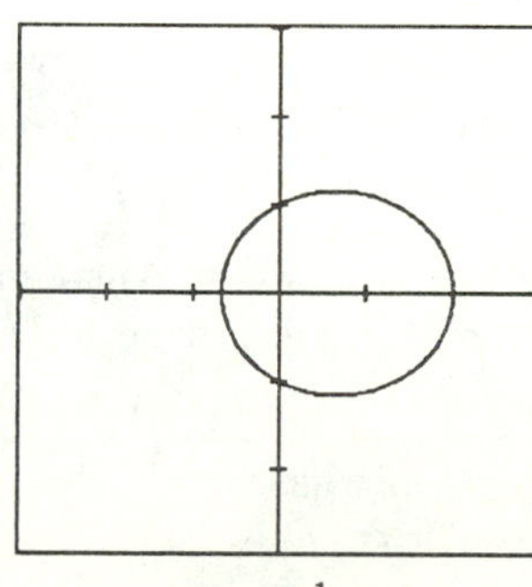

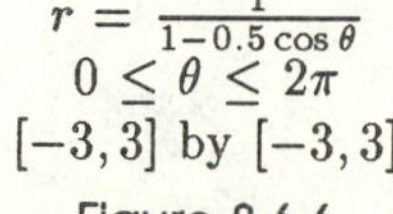

$r = \frac{1}{1-0.5\cos\theta}$
$0 \le \theta \le 2\pi$
$[-3,3]$ by $[-3,3]$
Figure 8.6.6

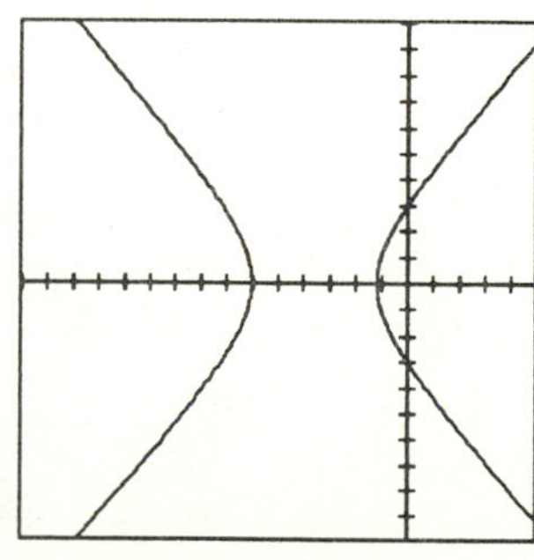

$r = \frac{3}{1-1.5\cos\theta}$
$0 \le \theta \le 2\pi$
$[-15,5]$ by $[-10,10]$
Figure 8.6.7

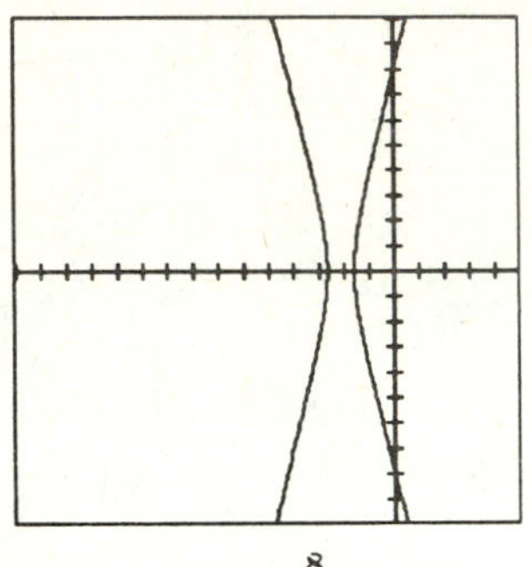

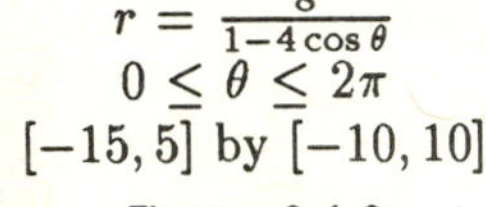

$r = \frac{8}{1-4\cos\theta}$
$0 \le \theta \le 2\pi$
$[-15, 5]$ by $[-10, 10]$

Figure 8.6.8

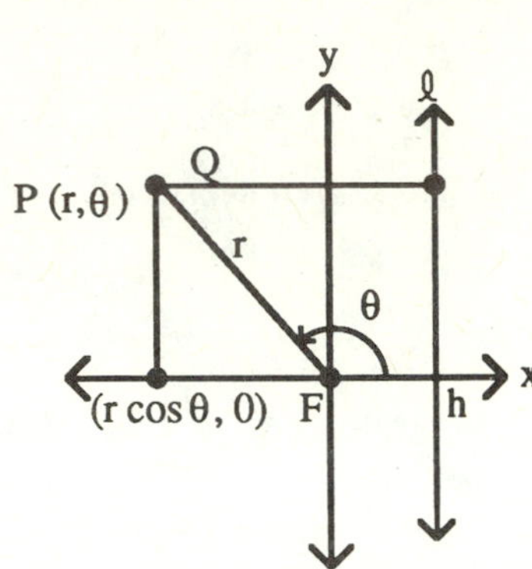

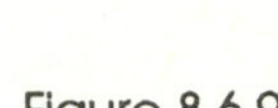

Figure 8.6.9

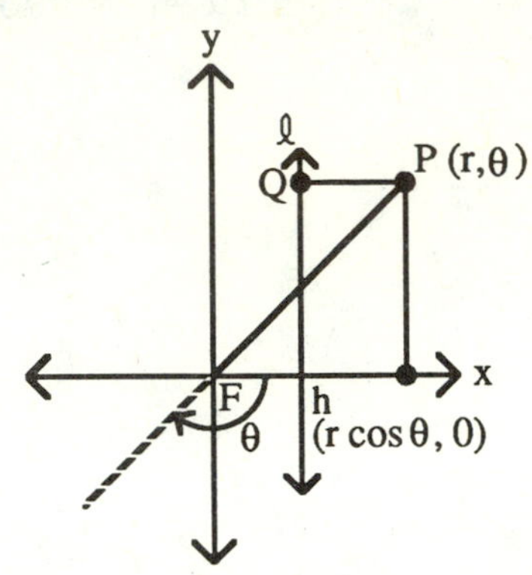

Figure 8.6.10

Now, suppose P is on the other side of l (Figure 8.6.10). We show that the coordinates (r, θ) where $r < 0$ as suggested by Figure 8.6.10 satisfy the same equation obtained above. Notice that $\cos\theta < 0$ so that $r\cos\theta > 0$:

$$\begin{aligned} d(P, F) &= e \cdot d(P, Q) \\ -r &= e(r\cos\theta - h) \\ r &= \frac{eh}{1 + e\cos\theta}. \end{aligned}$$

Notice h is the distance between a focus and directrix. For an ellipse, $h = \frac{a}{e} - c$, and for a hyperbola, $h = c - \frac{a}{e}$. Thus, eh is $a - ce$ for an ellipse and $ce - a$ for a hyperbola. It can be shown that $h = 2a$ for a parabola. The numerators of the above equation can be replaced by $a - ce$, $ce - a$, or $2ae$ to give alternative polar forms for conics.

It can be shown that if the focus is $(0, 0)$ then an equation of a conic is also given by

1. $r = \frac{eh}{1+e\sin\theta}$, if the directrix is $y = h$, with $h > 0$.
2. $r = \frac{eh}{1-e\sin\theta}$, if the directrix is $y = -h$, with $h > 0$.

Again, h can be replaced by $\frac{a}{e} - c$, $c - \frac{a}{e}$, or $2a$ to give alternative forms.

We summarize the above discussion in Theorem 3. However, we do not need to restrict h to be positive because the theorem is true for $h < 0$.

• THEOREM 3 A nondegenerate conic has one of the following for a polar equation where e is the eccentricity of the conic:

$$r = \frac{he}{1 \pm e\cos\theta}, \qquad r = \frac{he}{1 \pm e\sin\theta}.$$

For the first equation, $x = \pm h$ are directrices, respectively; for the second equaton, $y = \pm h$ are directrices, respectively. Furthermore, he can be replaced by

$\pm(ce - a)$ or $\pm 2ae$. Moreover, the graph of any such equation is a conic. The conic is an ellipse if $0 < e < 1$, a parabola if $e = 1$, and a hyperbola if $e > 1$.

• EXAMPLE 3: Identify and draw a complete graph of $r = \frac{6}{4-3\cos\theta}$. Specify a directrix and a range for θ that produces a complete graph. Determine the standard form for the equation of the conic.

SOLUTION: We divide numerator and denominator of this equation by 4 to put it into the form given in Theorem 3. (Why?)

$$r = \frac{6}{4-3\cos\theta},$$
$$r = \frac{1.5}{1-0.75\cos\theta}.$$

Now, we can see that $e = 0.75$ (why?) and that $he = 1.5$, or $h = \frac{1.5}{e} = 2$. Thus, this conic is an ellipse with directrix $x = -2$. (Why?) A complete graph is given in Figure 8.6.11. We obtain a complete graph by letting θ vary from 0 to 2π. In fact, any interval of length 2π for θ gives a complete graph. (Why?)

One focus of the ellipse is the origin. The ends of the major axis occur when $\theta = 0$ and $\theta = \pi$. (Why?) Thus, the endpoints of the major axis are $(-0.86, 0)$ and $(6, 0)$ in rectangular form. The center of the ellipse is halfway between the endpoints of the major axis. Thus, the coordinates of the center are $(2.57, 0)$. This also means $c = 2.57$ because c is the distance from the center to a focus. Now, the length of the major axis is $2a = 6 - (-0.86) = 6.86$, so $a = 3.43$. We use the equation $b^2 = a^2 - c^2$ to determine that $b = 2.27$. The standard form for the equation of the conic is

$$\frac{(x-2.57)^2}{(3.43)^2} + \frac{y^2}{(2.27)^2} = 1.$$ •

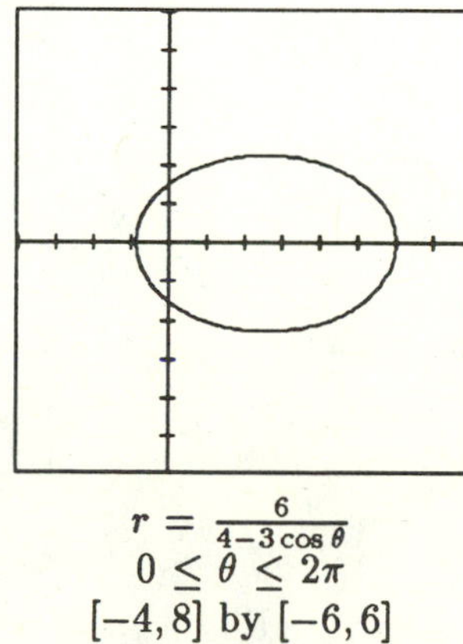

$r = \frac{6}{4-3\cos\theta}$
$0 \le \theta \le 2\pi$
$[-4, 8]$ by $[-6, 6]$

Figure 8.6.11

We must be careful before concluding that a given graph is an ellipse. We mentioned earlier in the section that the shape of an ellipse is more circular as e gets closer to 0. However, to distinguish visually between circles and ellipses it is necessary to have the physical length representing one unit on each axis be the same. (Why?)

• EXAMPLE 4: Identify and draw a complete graph of $r = \frac{3}{1+\cos\theta}$. Specify a directrix and a range for θ that produces a complete graph.

SOLUTION: We can see that $e = 1$, $h = 3$, and that this conic is a parabola with directrix $x = 3$. A complete graph is given in Figure 8.6.12 and can be obtained by letting θ vary from 0 to 2π. •

• EXAMPLE 5: Identify and draw a complete graph of $r = \frac{4}{3-4\sin\theta}$. Specify a directrix and a range for θ that produces a complete graph.

SOLUTION: One focus of this conic is at the origin. First we divide numerator and denominator by 3.

$$r = \frac{\frac{4}{3}}{1 - \frac{4}{3}\sin\theta}$$

Thus, $e = \frac{4}{3}$, and this conic is a hyperbola. Also $h = \frac{4}{3e} = 1$ so that $y = -1$ is a directrix. We obtain a complete graph by allowing θ to vary from 0 to 2π (Figure 8.6.13). •

It can be shown that any interval of length 2π produces a complete graph of a conic with equation given by Theorem 3. (Why?) In the Exercises you will see that the graphs of the equations given in Theorem 3 are conics even if the numerators are negative. The graphs are also conics if the denominators are replaced by $-1 \pm e\cos\theta$ and $-1 \pm e\sin\theta$.

We assume that the orbit of the Earth around the sun is an ellipse with the sun as focus. The length of the semi-major axis is 9.3×10^7 miles, and the eccentricity of the ellipse is

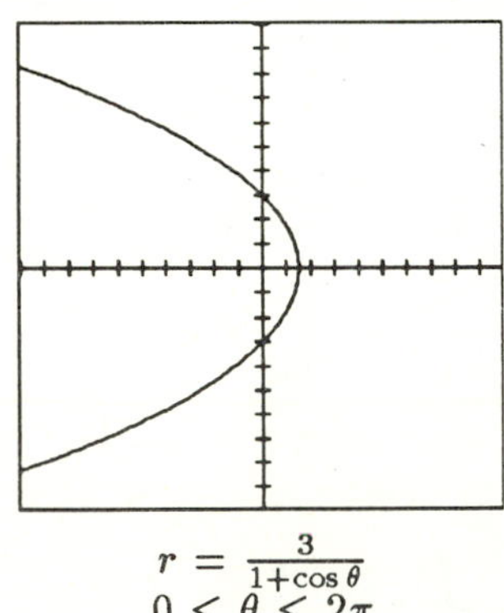

$r = \frac{3}{1+\cos\theta}$
$0 \le \theta \le 2\pi$
$[-10, 10]$ by $[-10, 10]$
Figure 8.6.12

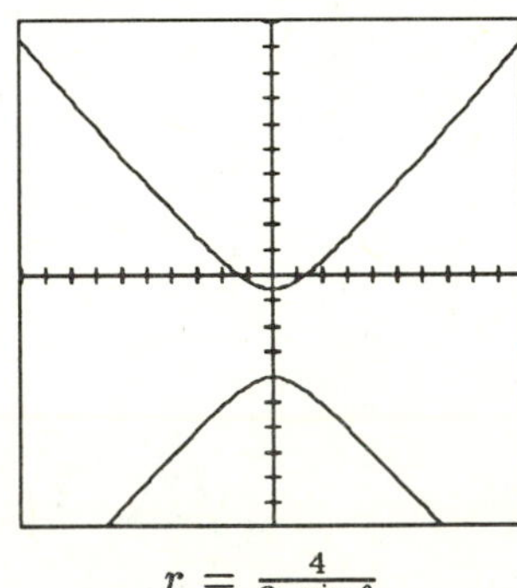

$r = \frac{4}{3-\sin\theta}$
$0 \le \theta \le 2\pi$
$[-10, 10]$ by $[-10, 10]$
Figure 8.6.13

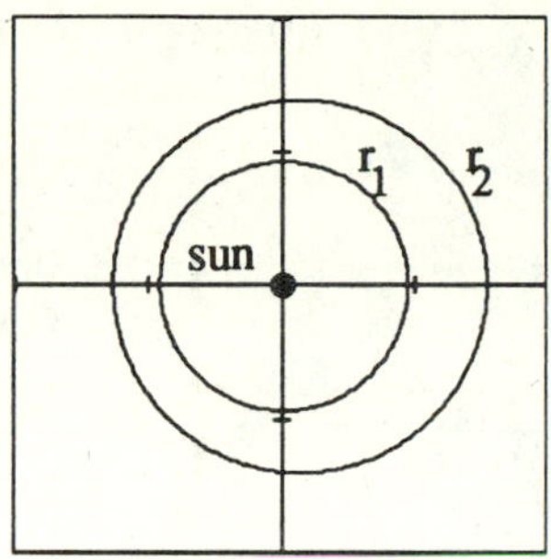

$0 \le \theta \le 2\pi$
$[-200{,}000{,}000, 200{,}000{,}000]$
by
$[-200{,}000{,}000, 200{,}000{,}000]$

Figure 8.6.14

0.02. Similarly, the orbit of the planet Mars is an ellipse with the sun as a focus, semi-major axis length of 1.4×10^8 miles, and eccentricity 0.1. We assume the orbits are in the same plane and have major axes along the x-axis.

- EXAMPLE 6: Draw complete graphs of the elliptical orbits of the Earth and Mars in the same viewing rectangle.

SOLUTION: We use the polar equation $r = \frac{a-ce}{1-e\cos\theta}$ of Theorem 3 where e is the eccentricity, a the length of the semi-major axis, and c half the distance between the foci. We also know that $e = \frac{c}{a}$ so that $c = ea$. Thus, we can write the polar equation in the form $r = \frac{a-e^2a}{1-e\cos\theta}$.

An equation for the orbit of the Earth is

$$r_1 = \frac{93{,}000{,}000 - (0.02)^2(93{,}000{,}000)}{1 - 0.02\cos\theta} = \frac{9.3 \times 10^7 - (0.02)^2(9.3 \times 10^7)}{1 - 0.02\cos\theta},$$

and for the orbit of Mars is

$$r_2 = \frac{140{,}000{,}000 - (0.1)^2(140{,}000{,}000)}{1 - 0.1\cos\theta} = \frac{1.40 \times 10^8 - (0.1)^2(1.40 \times 10^8)}{1 - 0.1\cos\theta}.$$

Figure 8.6.14 gives a complete graph of each orbit. •

• EXERCISES 8-6

Determine the eccentricity of the conic and draw a complete graph. Check using a graphing utility.

1. $4x^2 + y^2 = 4$
2. $4x^2 + 25y^2 = 100$
3. $9x^2 - 16y^2 = 144$
4. $4x^2 - y^2 = 4$
5. $2x^2 - 3y^2 - 4x + 6y - 10 = 0$
6. $-6x^2 + 8y^2 - 6x + 9y + 10 = 0$

Determine the center, length of the transverse axis, and the equations of the asymptotes.

7. the conic of Exercise 5
8. the conic of Exercise 6

Estimate the eccentricity of the ellipse. (*Hint*: First estimate a and b. Then compute c and e.) Draw a complete graph.

9. $2x^2 - 3xy + 6y^2 - 8x + 10y - 5 = 0$

10. $10x^2 + 52xy + 300y^2 - 100x + 250y - 500 = 0$

11. Draw a complete graph of $r = \frac{2e}{1-e\cos\theta}$ for $e = 0.8, 0.9, 1, 1.1, 1.2, 1.3$ in the same viewing rectangle.

12. Draw a complete graph of $r = \frac{2e}{1+e\cos\theta}$ for $e = 0.8, 0.9, 1, 1.1, 1.2$ in the same viewing rectangle.

13. Draw a complete graph of $r = \frac{2e}{1-e\sin\theta}$ for $e = 0.8, 0.9, 1, 1.1, 1.2$ in the same viewing rectangle.

14. Draw a complete graph of $r = \frac{2e}{1+e\sin\theta}$ for $e = 0.8, 0.9, 1, 1.1, 1.2$ in the same viewing rectangle.

Identify and draw a complete graph of the conic in polar form. Specify a directrix and a range for θ that produces a complete graph.

15. $r = \dfrac{5}{1 - \cos\theta}$

16. $r = \dfrac{4}{2 - 4\sin\theta}$

17. $r = \dfrac{3}{4 - 2\cos\theta}$

18. $r = \dfrac{4}{3 + 4\sin\theta}$

19. $r = \dfrac{-8}{6 + 3\cos\theta}$

20. $r = \dfrac{-4}{4 - 2\sin\theta}$

Draw a complete graph. Identify the conic. Determine the standard form of the conic.

21. $r = \dfrac{16}{12 + 9\cos\theta}$

22. $r = \dfrac{-4}{2 + 6\sin\theta}$

23. Consider an ellipse with major axis length $2a$ equal to 8 and length c of a focus from the center equal to 2. Compute the eccentricity and draw the graph of r_1 and r_2 in the same viewing rectangle. What conclusions can you draw?

$$r_1 = \frac{a - ce}{1 + e\cos\theta} \qquad r_2 = \frac{ce - a}{1 - e\cos\theta}$$

24. Consider an ellipse with major axis length $2a$ equal to 12 and length c of a focus from the center equal to 3.5. Compute the eccentricity and draw the graph of r_1 and r_2 in the same viewing rectangle. What conclusions can you draw?

$$r_1 = \frac{a - ce}{1 + e\cos\theta} \qquad r_2 = \frac{ce - a}{1 - e\cos\theta}$$

25. Consider the graph $r = \frac{eh}{-1-e\cos\theta}$ for $e = 7$ and $h = 2$. Is the graph a conic? What conic? Why?

26. Show that if an ellipse has major axis length $2a$ and minor axis length $2b$ then a polar equation of the ellipse is

$$r = \frac{b^2}{a + \sqrt{a^2 - b^2}\cos\theta}$$

where a focus is at the origin and the directrix is $x = h > 0$.

Let $a = 4$ and $b = \sqrt{12}$ in an ellipse. Compute the eccentricity and draw the polar graphs of r_1, r_2, r_3, and r_4 in the same viewing rectangle. What conclusions can you draw?

27. $r_1 = \dfrac{b^2}{a + \sqrt{a^2 - b^2}\cos\theta}$

28. $r_2 = \dfrac{-b^2}{a + \sqrt{a^2 - b^2}\cos\theta}$

29. $r_3 = \dfrac{-b^2}{a - \sqrt{a^2 - b^2}\cos\theta}$

30. $r_4 = \dfrac{b^2}{a - \sqrt{a^2 - b^2}\cos\theta}$

Draw the graph of the elliptical orbits of the indicated planets in the same viewing rectangle. Table 1 gives the eccentricity and the length of the semi-major axis. Assume the major axes are along the x-axis.

31. Mercury and Earth

32. Earth and Saturn

33. Earth, Saturn, and Pluto

34. All of the planets listed in the table.

35. Sketch the elliptical orbit of the Earth and the orbit of the asteroid Icarus in the same viewing rectangle. The eccentricity of the asteroid Icarus' orbit about the sun is 0.80 and its semi-major axis length is about 100,000,000 miles.

36. The elliptical orbit of Halley's comet has eccentricity about 0.97. Its major axis length is 3,348,000,000. What is its minor axis length? Draw a complete graph of its orbit.

37. How are $r_1 = \frac{h}{2-\cos t}$, $h > 0$ and $r_2 = \frac{h}{2-\cos t}$, $h < 0$ related?

38. (a) In the same viewing rectangle, draw complete graphs of

$$r = \frac{\frac{1}{2}h}{1 - \frac{1}{2}\cos t}$$

for $h = \frac{1}{4}, \frac{1}{2}, 1, 2, 5, 10$, and 20.

Planet	e	a (miles)
Mercury	0.2	3.6×10^7
Earth	0.02	9.3×10^7
Mars	0.09	1.4×10^8
Saturn	0.5	8.9×10^8
Neptune	0.01	2.8×10^9
Pluto	0.25	3.7×10^9

Table 1

(b) Notice the eccentricity in (a) is constant. Describe the effects of changing the value of h. What is the line $y = h$?

(c) Repeat (a) for $h = -\frac{1}{4}, -\frac{1}{2}, -1, -2, -5, -10$, and 20.

39. What is the graph of $r = \frac{eh}{-1+e\cos\theta}$? Why?

40. What is the graph of $r = \frac{eh}{-1+e\sin\theta}$? Why?

• KEY TERMS

Listed below are the key terms, vocabulary, and concepts in this chapter. You should be familiar with their definitions and meanings as well as be able to illustrate each of them.

Asymptotes
Center
Circle
Conic sections
Degenerate conics
Directrix
Eccentricity
Ellipse
Ellipsoid
End behavior
Endpoints
Expanded form of conic equations
Focus
Hyperbola
Hyperboloid
Lines of symmetry
Major axis
Major diameter
Minor axis
Minor diameter
Parabola
Paraboloid
Polar form of conic equations
Quadratic form
Reflective properties of conics
Semimajor axis
Semiminor axis
Standard form of conic equations
Transverse axis
Vertex

• REVIEW EXERCISES

This section contains representative problems from each of the previous six sections. You should use these problems to test your understanding of the material covered in this chapter.

Determine an equation for the parabola. Give the coordinates of the vertex and the line of symmetry, and draw a complete graph.

1. focus $(-2, 1)$ and directrix $y = -3$
2. focus $(-2, 1)$ and directrix $x = -3$

3. Determine an equation, the line of symmetry, and the focus of the parabola with vertex $(1, 4)$ and directrix $x = 7$, and draw a complete graph.

4. Determine an equation, the line of symmetry, and the directrix of the parabola with focus $(0,2)$ and vertex $(0,-4)$, and draw a complete graph.

List a sequence of geometric transformations that when applied to the graph of $y = x^2$ will result in the graph of the given parabola.

5. $6(y-1) = (x-3)^2$

6. $x^2 - 2x - 6 = 2y$

7. The parabola $y^2 - 4y + 2 = x$ does not define y as a function of x. How can it be graphed using only a function graphing utility?

Find the standard form, vertex, focus, directrix, line of symmetry, and draw a complete graph of each parabola without using a graphing utility. Check your graph with a graphing utility.

8. $x^2 + 4x - y + 6 = 0$

9. $6y^2 - 12y - 12x + 20 = 0$

10. Find the points of intersection of the line and the parabola (if any): $y = x^2 - 4x + 5$, $2x + y - 7 = 0$.

11. A parabola of the form $y = ax^2 + b$ passes through the points $(2,4)$ and $(5,8)$. Determine an equation of the parabola.

12. Determine the foci, standard form of the equation, and draw a complete graph of the ellipse without using a graphing utility: Major axis along the x-axis of length 10, minor axis along the y-axis of length 4.

Determine the endpoints of the major and minor axes, the standard form of the equation, and draw a complete graph of the ellipse. Do not use a graphing utility.

13. The sum of the distances from the foci is 16; the foci are $(-5,0)$ and $(5,0)$.

14. foci at $(2,0)$ and $(-2,0)$; minor axis of length 10

15. foci at $(2,3)$ and $(2,9)$; major axis of length 12

Give the endpoints of the major and minor axes, the coordinates of the center and foci, the lines of symmetry, and draw a complete graph of the ellipse. Do not use a graphing utility.

16. $\dfrac{(x+3)^2}{4} + \dfrac{(y-1)^2}{6} = 1$

17. $(x-1)^2 + 8(y+2)^2 = 16$

Write the equation of the ellipse in standard form, give the endpoints of the major and minor axes, the coordinates of the center and foci, the lines of symmetry, and draw a complete graph. Check your answer using a graphing utility.

18. $4x^2 + 9y^2 + 8x - 36y + 4 = 0$

19. $4x^2 + 5y^2 - 24x + 10y + 21 = 0$

Determine the standard form of the equation, and draw a complete graph of the hyperbola. Do not use a graphing utility.

20. The absolute value of the difference of the distances from any point on the hyperbola to the foci is 2; the foci are $(-2,0)$ and $(2,0)$.

21. The absolute value of the difference of the distances from any point on the hyperbola to the foci is 10; the foci are $(0,-2)$ and $(0,10)$.

Determine the coordinates of the center, foci, vertices, the lines of symmetry, and the asymptotes of the hyperbola, and draw a complete graph. Do not use a graphing utility.

22. $\dfrac{x^2}{9} - \dfrac{(y+2)^2}{6} = 1$

23. $3(y-3)^2 - 5(x+1)^2 = 15$

Identify the conic. Write the equation of the conic in standard form, give the foci, vertices, center, the axes of symmetry, asymptotes (if any) and draw a complete graph. Check your answer using a graphing utility.

24. $25x^2 - 9y^2 - 50x - 36y - 236 = 0$
25. $y^2 - x + 6y + 9 = 0$
26. $5x^2 + 4y^2 + 30x - 8y + 29 = 0$
27. $4y^2 - 9x^2 + 24y + 35 = 0$
28. Explain how a conic of the form $Ax^2 + By^2 + Cx + Dy + E = 0$ can be graphed using a *function* graphing utility, not a special conic graphing utility.

Draw a complete graph and identify the conic.

29. $4x^2 + 3xy - y^2 - 10x - 5y - 20 = 0$
30. $2x^2 - 4xy + 2y^2 + 6x - 5y - 15 = 0$
31. $6x^2 - 8xy + 10y^2 - 5x + y - 30 = 0$
32. $-2x^2 + 7xy - 3y^2 + 3x - 2y - 10 = 0$
33. Estimate the coordinates of the center and the equations of the asymptotes of the hyperbola $2x^2 - 3xy + y^2 + 2x - 3y - 2 = 0$.
34. Estimate the coordinates of the center, the endpoints of the major and minor axes, and the equations of the lines of symmetry of the ellipse $3x^2 - xy + 2y^2 + 6x + 2y - 10 = 0$.
35. Determine a pair of equations that give an end behavior model of the hyperbola $4x^2 + 8xy - 3y^2 + 5x - 10y - 30 = 0$. Draw the two equations and the hyperbola in the same large viewing rectangle to check your answer.
36. Explain what transformations applied first to a circle in standard form will result in the ellipse $9x^2 + 16y^2 = 144$.
37. What transformations applied first to a conic in standard form will result in the conic $4x^2 + 9y^2 + 16x - 18y - 11 = 0$?

Find the simultaneous solutions to each system.

38. $9x^2 + 25y^2 = 225$, $x + 2y = 2$
39. $9x^2 - 25y^2 = 225$, $x + 2y = 4$
40. $9x^2 - 25y^2 = 225$, $9x^2 + 25y^2 = 225$
41. $x^2 + 9y^2 = 9$, $y = 2x^2 + 3$

Find the number of simultaneous solutions to each system.

42. $2x^2 - 3xy + 2y^2 + x + y - 8 = 0$, $3x^2 - 8xy + 2y^2 + x + y - 10 = 0$
43. $3x^2 - 5xy + 6y^2 - 7x + 5y - 9 = 0$, $x^2 + y^2 - 2y - 6 = 0$
44. Find the simultaneous solutions to the system in Exercise 42.
45. Find the simultaneous solutions to the system in Exercise 43.

Draw a graph representing the solution to the system of inequalities.

46. $x^2 + y^2 < 9$, $y > x^2$
47. $x^2 - y^2 > 9$, $x^2 + y^2 < 9$

Determine the eccentricity and draw a complete graph of the conic. Check using a graphing utility.

48. $x^2 - 9y^2 = 9$
49. $4x^2 + 9y^2 + 54y + 45 = 0$
50. Determine the center, length of the transverse axis, and the equations of the asymptotes for the conic in Exercise 48.
51. Draw a complete graph of $r = \frac{2e}{1 - e\cos\theta}$ for $e = 0.75$, 1, and 1.25 in the same viewing rectangle.

52. Draw a complete graph of $r = \frac{2e}{1-e\sin\theta}$ for $e = 0.75$, 1, and 1.25 in the same viewing rectangle.

Identify and draw a complete graph of the conic in polar form. Specify a directrix and a range for θ that produces a complete graph.

53. $r = \dfrac{4}{1-2\cos\theta}$ 54. $r = \dfrac{2}{4+\sin\theta}$ 55. $r = \dfrac{8}{8-4\sin\theta}$

56. A parabolic microphone is formed by revolving the portion of the parabola $12y = x^2$ between $x = -6$ and $x = 6$ about its axis of symmetry. Where should the sound receiver be placed for best reception?

57. A parabolic headlight is formed by revolving the portion of the parabola $y^2 = 10x$ between the lines $y = -3$ and $y = 3$ about its axis of symmetry. Where should the headlight bulb be placed for maximum lumination?

58. There are real elliptical pool tables that have been constructed with a single pocket at one of the foci. Suppose such a table has major diameter 8 feet and minor diameter 6 feet. How should the ball be hit so it bounces off the cushion directly into the pocket? Give specific measurements.

A certain comet's path is known to be hyperbolic with one focus at the sun. Assume a space station is located at the center of the hyperbola. Further assume it has been determined that the distance between the comet and the space station is 6,000,000 miles at the closest point, and that the sun and the space station are 12,000,000 miles apart.

59. Determine the equation in standard form of the hyperbola representing the comet's path.

60. Draw a complete graph of the algebraic representation of the path of the comet. What portion of the graph represents the problem situation?

Consider an acid solution of 80 oz that is 65% acid. Let x oz of pure acid be added to the solution. How much pure acid must be added to obtain a mixture that is 81% acid?

61. Determine an algebraic representation of the final acidity percentage as a function of the amount of pure acid added.

62. Draw a complete graph and identify the type of graph.

63. What is the domain and range?

64. What values of x make sense in this problem situation?

65. Show how the intersection of two graphs can be used to solve the problem. Find the solution.

Suppose three observers A, B, and C, listening for illegal dynamite explosions, are positioned so the angle of the line between A and B and the line between B and C is 90°. Further, suppose A and B are 4000 feet apart and B and C are 1000 feet apart. A hears the explosion 1 second before B, and C hears the explosion 2 seconds before B.

66. Determine a model whose solution will locate the position of the explosion.

67. Solve the simultaneous system of equations in the model of Exercise 66.

68. Determine the position of the explosion.

"Hush Dog" shoes are made from two types of leather, A and B. Type A costs \$0.25 per square inch while type B costs \$0.65 per square inch. The "Hush Dog" Company must buy at least 4000 square inches of A per day and 8000 square inches of B per day (or the supplier will not do business).

However, the total square inches of material provided by the supplier per day must be less than 30,000 square inches. Experience has shown the president that the total cost of the material must remain less than $10,000 per day or no one will buy "Hush Dogs" (because they would need to be priced very high).

69. Determine a system of inequalities whose solution provides a model for all possible combinations of the amounts of leather A and leather B that could be used in this problem situation.

70. Graph the region that represents all possible combinations of the amounts of leather A and leather B that could be used in this problem situation.

The ellipse near the Washington Monument in Washington, D.C. has eccentricity of about 0.52. Its major axis length is 458 meters.

71. What is its minor axis length?

72. Determine an equation for the boundary of the ellipse, assuming its center is at the origin.

9 Sequences, Series, Matrices, Three-Dimensional Geometry

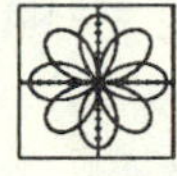

• Introduction

In this chapter we introduce sequences and series and see how to obtain the sum of certain infinite series using the principle of mathematical induction. The binomial theorem is used in applications involving probability and error analysis. We approximate certain functions with polynomials and graphically investigate the error in such approximations.

Matrices are introduced in this chapter and used to solve systems of linear equations in several variables. We write equations for lines and planes in three-dimensional space and give an introduction to three-dimensional analytic geometry. We study surfaces in three dimensions, including surfaces that are graphs of functions of two variables. Maximum and minimum values of functions of two variables are determined graphically. Applications that involve determining a local extremum are investigated.

We represent the complex zeros of a polynomial as the zeros of a related function of two variables. Then, a three-dimensional graphing utility is used to approximate these zeros.

9.1 Sequences and Mathematical Induction

In this section sequences are introduced. We investigate sequences where the nth term is given by an explicit function of n or by a recursive formula. Mathematical induction is used to verify a formula for the amount of money accumulated in an account that pays compound interest. Arithmetic and geometric progressions are introduced and used to investigate arithmetic and geometric means.

The domains of most of the functions in this textbook have been intervals or unions of intervals. Now we study functions with domains that are subsets of the positive integers.

• DEFINITION A **sequence** is a function whose domain is either the set of all positive integers or an initial subset of positive integers $\{1, 2, \ldots, N\}$ for some fixed positive integer N. A sequence with domain the set of all positive integers is called an **infinite sequence**, and a sequence with domain $\{1, 2, \ldots, N\}$ for some fixed positive integer N is called a **finite sequence**. •

Let f be a sequence and n a positive integer. The number $a_n = f(n)$ in the range of the sequence f is called the **nth term of the sequence.** If the domain of f is $\{1, 2, \ldots, N\}$ then a_n is defined only for $n \leq N$. We often use the notation $\{a_n\}$ to denote a sequence with nth term a_n. Sometimes the sequence is expressed in list form as $a_1, a_2, \ldots, a_n, \ldots$.

• EXAMPLE 1: List the first three terms and the fifteenth term of each sequence with the given nth term. Graph each sequence.

(a) $a_n = \dfrac{n^2 - 1}{n}$ (b) $a_n = (-1)^n 2^n$

SOLUTION:

(a) To find the first three terms of the sequence we substitute 1, 2, and 3 for n in the nth term formula $a_n = \frac{n^2-1}{n}$. Thus, $a_1 = 0$, $a_2 = \frac{3}{2}$, and $a_3 = \frac{8}{3}$ are the first three terms; $a_{15} = \frac{224}{15}$ is the fifteenth term. The graph of this sequence consists of the infinitely many points on the graph of $f(x) = \frac{x^2-1}{x}$ where the x-coordinate is a positive integer. Figure 9.1.1 gives a complete graph of the rational function f. Five points of the graph of the sequence $\{a_n\}$ are marked with solid dots.

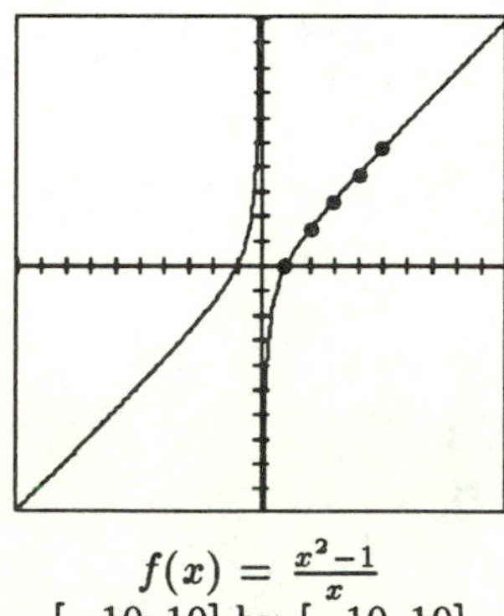

$f(x) = \frac{x^2-1}{x}$
$[-10, 10]$ by $[-10, 10]$
Figure 9.1.1

(b) Similarly, $a_1 = -2, a_2 = 4$, and $a_3 = -8$ are the first three terms of this sequence; $a_{15} = -2^{15} = -32{,}768$ is the fifteenth term. The graph of the sequence consists of infinitely many points, some of which are shown as solid dots in Figure 9.1.2. Notice that the points above the x-axis are the points on the graph of $y = 2^x$ with positive even integer x-coordinates, and the points below the x-axis are the points on the graph of $y = -2^x$ with positive odd integer x-coordinates. •

Some sequences are defined by giving the first few terms and a rule for obtaining the nth term from preceding terms. We call such a rule a **recursive formula**, and we say that the sequence is defined **recursively**.

• EXAMPLE 2: Determine the first four terms, the eighth term, and a formula for the nth term of the sequence $\{a_n\}$ defined by $a_1 = 4$, and $a_n = 3 + a_{n-1}$ for $n \geq 2$.

SOLUTION: Notice that $a_2 = 3 + a_1 = 7, a_3 = 3 + a_2 = 10$, and $a_4 = 3 + a_3 = 13$. If we know a given term of this recursively defined sequence, we can always determine the next term. We can also work backwards to determine a given term of the sequence:

$$\begin{aligned} a_8 &= 3 + a_7 \\ &= 3 + (3 + a_6) \\ &= 6 + (3 + a_5) \\ &= 9 + (3 + a_4) \\ &= 25. \end{aligned}$$

One way to determine the nth term of this recursively defined sequence is to compute terms until we are able to conjecture a formula for a_n in terms of n and a_1. Notice the

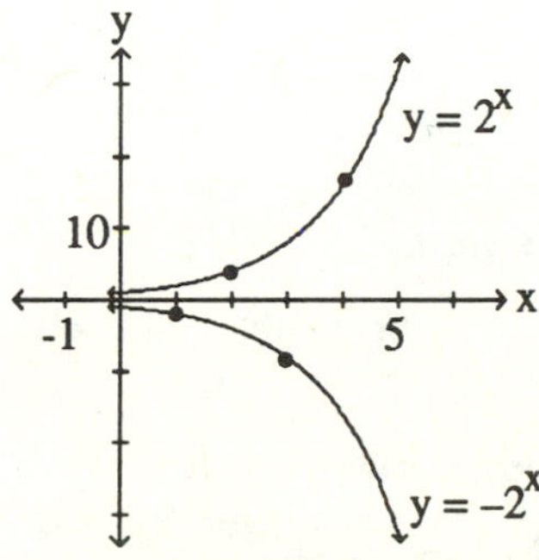

Figure 9.1.2

following:

$$a_2 = 3 + a_1$$
$$a_3 = 3 + a_2 = 2(3) + a_1$$
$$a_4 = 3 + a_3 = 3(3) + a_1$$
$$a_5 = 3 + a_4 = 4(3) + a_1.$$

It appears that

$$a_n = 3(n-1) + a_1 = 3(n-1) + 4.$$

We can write $a_n = 3n + 1$. (Why?) •

Later in the section we will use mathematical induction to verify the formula we discovered for the nth term of the sequence of Example 2. It can be difficult to determine the nth term of a sequence defined recursively as an explicit function of n. Try determining the nth term of the sequence of the next example as an explicit function of n.

• EXAMPLE 3: Determine the first six terms of the sequence $\{a_n\}$ defined by $a_1 = a_2 = 1$, and $a_n = a_{n-1} + a_{n-2}$ for $n \geq 3$.

SOLUTION: We must determine a_n for $n = 3, 4, 5$, and 6:

$$a_3 = a_2 + a_1 = 2,$$
$$a_4 = a_3 + a_2 = 3,$$
$$a_5 = a_4 + a_3 = 5,$$
$$a_6 = a_5 + a_4 = 8.$$ •

The sequence of Example 3 is called a *Fibonacci sequence*. Any sequence $\{a_n\}$ that satisfies the recursive definition $a_n = a_{n-1} + a_{n-2}$ for $n \geq 3$ is called a **Fibonacci sequence**. Consecutive terms of a Fibonacci sequence occur frequently in nature. For example, botanists have known for a long time that many natural spirals, like the growth of leaves around a plant stem, or the spirals on the surface of a pine cone, are associated with the terms of a Fibonacci sequence.

Mathematical Induction

Proofs by *mathematical induction* are based on the following fundamental axiom about subsets of the set of positive integers.

• Axiom of Induction Let S be a subset of the set of positive integers with the following two properties:

(i) S contains the integer 1.
(ii) S contains the integer $k+1$ whenever S contains the integer k.

Then, S is the entire set of positive integers.

Suppose S is a subset of positive integers satisfying property (i) above; that is, 1 is an element of S. If S also satisfies property (ii), then S contains the integer 2. (Why?) Now that S contains the integer 2, it must contain the integer 3 by another application of property (ii). Continuing in this way, it seems reasonable that S must contain all positive integers.

Proofs by mathematical induction are given by applying the following theorem, which is a direct application of the Axiom of Induction.

• THEOREM 1 **(Principle of Mathematical Induction).** Let P_n be a statement that is defined for each positive integer n. All the statements P_n are true provided the following two conditions are satisfied.

(i) P_1 is true.

(ii) P_{k+1} is true whenever P_k is true.

Theorem 1 can be explained in the following way. Think of each statement P_k as a domino. If the statement P_k is true, then it is tipped over. Statement (ii) means that if P_k is tipped over it strikes and knocks P_{k+1} over. Now, statement (i) tells us that P_1 is tipped over (Figure 9.1.3). By statement (ii), P_2 is tipped over by P_1. Another application of (ii) tells us that P_3 is tipped over by P_2. Continuing in this way every domino is tipped over.

Theorem 1 is applied to give a proof by mathematical induction in the following way. Let P_n be a statement we wish to show is true for all positive integers n. First, we show that the statement P_1 is true. Next, we assume that the statement P_k is true and show that the statement P_{k+1} is true. Then, it follows from Theorem 1 that P_n is true for all positive integers n.

• EXAMPLE 4: Use mathematical induction to prove that $a_n = 3n + 1$ for the sequence of Example 2.

SOLUTION: This sequence was defined by $a_1 = 4$, and $a_n = 3 + a_{n-1}$ for $n \geq 2$. Let P_n be the statement that $a_n = 3n+1$. We show that P_n satisfies properties (i) and (ii) of Theorem 1. Now, $a_1 = 4$ and $3(1) + 1 = 4$, so the formula $a_n = 3n + 1$ is valid for $n = 1$. Thus, P_1 is true. Next we show that if P_k is true, then P_{k+1} is true. This means that the formula $a_n = 3n + 1$ is assumed valid for $n = k$, so $a_k = 3k + 1$. Now we have to show that P_{k+1} is true; that is, the formula $a_n = 3n + 1$ is valid for $n = k + 1$. In other words, we must show that $a_{k+1} = 3(k + 1) + 1$. By definition, $a_{k+1} = 3 + a_k$. By assumption, $a_k = 3k + 1$.

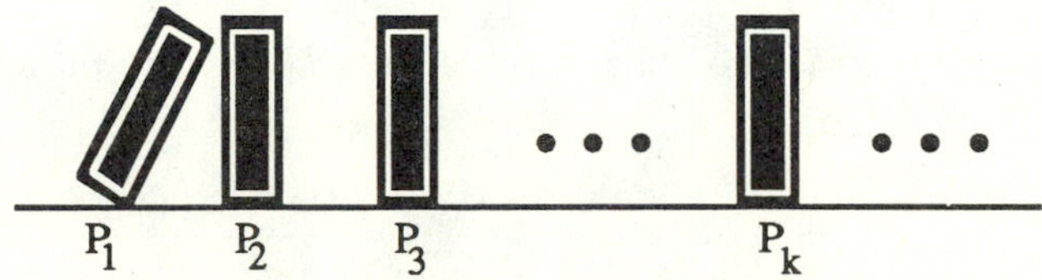

Figure 9.1.3

Making this substitution for a_k in the equation $a_{k+1} = 3 + a_k$ gives:

$$\begin{aligned} a_{k+1} &= 3 + a_k \\ &= 3 + 3k + 1 \\ &= 3(k+1) + 1. \end{aligned}$$

This means that the formula $a_n = 3n+1$ is valid for $n = k+1$. Thus, P_{k+1} is true whenever P_k is true. It follows from Theorem 1 that $a_n = 3n + 1$ for every positive integer n. •

• EXAMPLE 5: Use mathematical induction to prove that

$$1^2 + 3^2 + 5^2 + \cdots + (2n-1)^2 = \frac{n(2n-1)(2n+1)}{3}$$

is true for every positive integer n.

SOLUTION: Let P_n be the statement

$$1^2 + 3^2 + 5^2 + \cdots + (2n-1)^2 = \frac{n(2n-1)(2n+1)}{3}.$$

For $n = 1$, the left-hand side of the above equation is $1^2 = 1$, and the right-hand side is $\frac{1(2-1)(2+1)}{3} = 1$. Thus, P_1 is true.

Next, we must show that if P_k is true, then P_{k+1} is true. Thus, we assume that P_k is true. This means

$$1^2 + 3^2 + 5^2 + \cdots + (2k-1)^2 = \frac{k(2k-1)(2k+1)}{3}$$

is true. Now we need to show that P_{k+1} is true. That is, we must show the following is true:

$$\begin{aligned} 1^2 + 3^2 + 5^2 + \cdots + (2(k+1)-1)^2 &= \frac{(k+1)(2(k+1)-1)(2(k+1)+1)}{3}, \\ 1^2 + 3^2 + 5^2 + \cdots + (2k+1)^2 &= \frac{(k+1)(2k+1)(2k+3)}{3}. \end{aligned}$$

The second to last term on the left-hand side of the above equation is $(2k-1)^2$. (Why?) We use this fact to rewrite the left-hand side of the equation:

$$\begin{aligned} 1^2+3^2+5^2+\cdots+(2k+1)^2 &= 1^2+3^2+5^2+\cdots+(2k-1)^2+(2k+1)^2 \\ &= \frac{k(2k-1)(2k+1)}{3}+(2k+1)^2 \\ &= (2k+1)\left[\frac{k(2k-1)}{3}+2k+1\right] \\ &= (2k+1)\left(\frac{2k^2-k+6k+3}{3}\right) \\ &= \frac{(2k+1)(2k^2+5k+3)}{3} \\ &= \frac{(2k+1)(k+1)(2k+3)}{3} \\ &= \frac{(k+1)(2k+1)(2k+3)}{3}. \end{aligned}$$

Thus,

$$1^2+3^2+5^2+\cdots+(2k+1)^2 = \frac{(k+1)(2k+1)(2k+3)}{3},$$

so that P_{k+1} is true. By Theorem 1,

$$1^2+3^2+5^2+\cdots+(2n-1)^2 = \frac{n(2n-1)(2n+1)}{3}$$

for all positive integers n. •

Suppose that A dollars are invested in an account that compounds interest. If i is the interest rate paid per compounding period, then as we showed in Section 4.3, the amount in the account after n periods is $A(1+i)^n$. For example, if \$200 is invested in an account that pays 9% interest compounded monthly, then $i = \frac{0.09}{12}$ and the amount in the account after n months is $200\left(1+\frac{0.09}{12}\right)^n$.

• **EXAMPLE 6:** Suppose A dollars are invested in a compound interest bearing account that pays interest rate i per compounding period. Use mathematical induction to prove that the amount in the account at the end of the nth period is $A(1+i)^n$.

SOLUTION: Let P_n be the statement that the amount in the account at the end of the nth period is $A(1+i)^n$. First, we show that P_1 is true. At the end of the first period the amount in the account is the initial investment of A dollars plus the interest Ai earned during the first period. Thus, the amount in the account at the end of the first period is

$$A+Ai = A(1+i).$$

Next, we show that if P_k is true, then P_{k+1} is true. Thus, we assume that P_k is true. This means that the amount in the account at the end of the kth period is $A(1+i)^k$. We must show that P_{k+1} is true. That is, the amount in the account at the end of the $(k+1)$st period is $A(1+i)^{k+1}$. The interest earned during the $(k+1)$st period on this investment is $A(1+i)^k i$. Thus, the amount in the account at the end of the $(k+1)$st period is

$$A(1+i)^k + A(1+i)^k i = A(1+i)^k(1+i) = A(1+i)^{k+1}.$$

Therefore, P_{k+1} is true and, by Theorem 1, P_n is true for all positive integers n. This means that the amount in the account at the end of n periods is $A(1+i)^n$ for every positive integer n. •

There are many problem situations involving growth of quantities that are similar to the growth of money invested in an account that compounds interest. The next example illustrates the underlying feature common to such problems.

• EXAMPLE 7: A certain quantity is growing at a constant rate between successive equal intervals of time. Determine a recursive formula for the accumulated amount of the quantity at the end of any period.

SOLUTION: Let i be the constant growth rate and let S_k be the accumulated amount of the quantity after k growth periods. Then, by assumption,

$$S_k = S_{k-1} + iS_{k-1} = S_{k-1}(1+i)$$

for $k \geq 2$. (Why?)

Let A be the amount of the quantity present initially; that is, at the beginning of the first period. Then, $S_1 = A + iA = A(1+i)$. Thus,

$$S_1 = A(1+i)$$
$$S_n = S_{n-1}(1+i), \quad n \geq 2$$

gives a recursive definition for the sequence $\{S_n\}$ of numbers consisting of the amounts of the quantity accumulated at the end of n periods. •

In the Exercises we will ask you to show that the recursive definition of Example 7 is equivalent to $S_n = A(1+i)^n$, the expression for compound or exponential growth verified in Example 6, that was introduced in Section 4.3.

Arithmetic Progression

• DEFINITION A sequence $\{a_n\}$ is called an **arithmetic progression** or **arithmetic sequence** if successive terms differ by a fixed real number. That is, there is a real number d such that $a_k = a_{k-1} + d$ for all integers $k \geq 2$. The number d is called the **common difference** of the arithmetic progression. •

Let a_1 be the first term of an arithmetic progression with common difference d. The second term is $a_2 = a_1 + d$. The third term is $a_3 = a_2 + d = a_1 + 2d$. The fourth term is $a_4 = a_3 + d = a_1 + 3d$. Mathematical induction can be used to establish the following formula.

- **Arithmetic Progression** The nth term of an arithmetic progression with common difference d is given by $a_n = a_1 + (n-1)d$.

Notice that the sequence of Examples 2 and 4 is an arithmetic progression with common difference 3.

• EXAMPLE 8: The first two terms of an arithmetic progression are -8 and -2, respectively. Find the tenth term and a formula for the nth term.

SOLUTION: We know that $a_1 = -8$, $a_2 = -2$, and $a_2 = a_1 + d$. Thus, $d = 6$. The tenth term is $a_{10} = a_1 + 9d = -8 + 9(6) = 46$. The nth term is

$$a_n = a_1 + (n-1)d = -8 + 6(n-1) = 6n - 14.$$ •

• EXAMPLE 9: The third and eighth terms of an arithmetic progression are 13 and 3, respectively. Determine the first term and a formula for the nth term.

SOLUTION: We use $a_n = a_1 + (n-1)d$ to establish the following pair of equations:

$$a_3 = a_1 + 2d = 13,$$
$$a_8 = a_1 + 7d = 3.$$

Subtracting, we obtain $5d = -10$, so $d = -2$. Substituting $d = -2$ in either of the above equations gives $a_1 = 17$. Thus, $a_n = a_1 + (n-1)d = 17 - 2(n-1)$, or $a_n = 19 - 2n$. •

We must be careful not to be misled by Examples 8 and 9. Suppose that 2 and 4 are the first two terms of a sequence. There is only one possible arithmetic sequence with 2 and 4 as the first two terms, namely, the sequence with nth term $a_n = 2n$. (Why?) However, if we are not given that the sequence is arithmetic, then there are infinitely many possibilities for the sequence. Here is one way to illustrate this. Let g be any function whose domain includes the positive integers. The sequence $\{b_n\}$ where $b_n = 2n + (n-1)(n-2)g(n)$ has 2 and 4 as its first two terms. Notice that $b_3 = 6 + 2g(3)$ can be any real number by making an appropriate choice for the function g. We were able to completely determine the sequences of Examples 8 and 9 from two of its terms because we also knew the sequences were arithmetic.

Geometric Progression

• DEFINITION A sequence $\{a_k\}$ is called a **geometric progression** or **geometric sequence** if the ratio of successive terms is a fixed nonzero number. That is, there is a number $r \neq 0$ such that $a_{k+1}/a_k = r$ for every positive integer k. The number r is called the **common ratio** of the sequence. If such a sequence is finite, then it is called a **finite geometric progression**; if the sequence is infinite, it is called an **infinite geometric progression**. •

• THEOREM 2 If $\{a_k\}$ is a geometric progression with common ratio r, then $a_n = a_1 r^{n-1}$ for every positive integer n.

PROOF Notice $a_2/a_1 = r$, so $a_2 = a_1 r$. The formula is valid for $n = 1$ because $a_1 = a_1 r^0$. (Why?) We use mathematical induction to prove that $a_n = a_1 r^{n-1}$. Let P_n be the statement $a_n = a_1 r^{n-1}$. We know that P_1 is true. Assume that P_k is true. This means that $a_k = a_1 r^{k-1}$. By definition, $a_{k+1}/a_k = r$. Thus, $a_{k+1} = a_k r = (a_1 r^{k-1})r = a_1 r^k$. Therefore, P_{k+1} is true whenever P_k is true. It follows that P_n is true for every positive integer n. That is, $a_n = a_1 r^{n-1}$ for every positive integer n.

• EXAMPLE 10: Two hundred dollars are deposited in an account that pays 9% APR compounded monthly. Let P_n be the amount in the account at the end of the nth month. Show that $P_1, P_2, \ldots, P_n, \ldots$ is a geometric progression. Determine P_1 and the common ratio.

SOLUTION: The monthly interest rate is $\frac{0.09}{12} = 0.0075$, so $P_n = 200(1.0075)^n$. Thus, $P_1 = 200(1.0075)$, $P_2 = 200(1.0075)^2$, $P_3 = 200(1.0075)^3$, and so forth. Notice that $P_{k+1}/P_k = 1.0075$. Thus, the sequence $P_1, P_2, \ldots, P_n, \ldots$ is a geometric progression with common ratio 1.0075, and $P_1 = 200(1.0075) = 201.50$. •

• DEFINITION The **arithmetic mean**, or **average** of a and b is defined to be $\frac{a+b}{2}$. The numbers $c_1, c_2, \ldots, c_k$ are called the **k arithmetic means of a and b** if the sequence $a, c_1, c_2, \ldots, c_k, b$ is an arithmetic progression. •

• EXAMPLE 11: Insert four arithmetic means between 3 and 11.

SOLUTION: We must determine c_1, c_2, c_3, and c_4 so that $3, c_1, c_2, c_3, c_4, 11$ is an arithmetic progression. This means that $a_1 = 3$ and $a_6 = 11$. We also know that $a_6 = a_1 + 5d = 11$. Thus, $5d = 8$ or $d = 1.6$. The four arithmetic means are $a_2 = a_1 + d = 4.6$, $a_3 = a_2 + d = 6.2$, $a_4 = a_3 + d = 7.8$, and $a_5 = a_4 + d = 9.4$. •

• EXERCISES 9-1

Determine the first four terms and the tenth term of the sequence. Draw a graph of the sequence.

1. $a_n = (-1)^{n+1}$
2. $a_n = 2n - 1$
3. $a_n = \dfrac{n+1}{n}$
4. $a_n = (-1)^n 2n$
5. $a_n = \dfrac{n^3 - 1}{n+1}$
6. $a_n = \dfrac{4}{n+2}$

Determine the first four terms of the sequence and a rule for the nth term. That is, determine a_n as an explicit function of n. Use induction to verify the rule for the nth term.

7. $a_1 = 3$; $a_n = a_{n-1} + 1$, $n = 2, 3, \ldots$

8. $a_1 = -2$; $a_n = a_{n-1} + 2$, $n = 2, 3, \ldots$

9. $a_1 = 2$; $a_n = 2a_{n-1}$, $n = 2, 3, \ldots$

10. $a_1 = \frac{3}{2}$; $a_n = \frac{1}{2}a_{n-1}$, $n = 2, 3, \ldots$

Determine the common difference, the tenth term, and a formula for the nth term of the arithmetic progression.

11. $6, 10, 14, 18, \ldots$

12. $-4, 1, 6, 11, \ldots$

13. The fourth and seventh terms of an arithmetic progression are -8 and 4, respectively. Determine the first term and a formula for the nth term.

14. The fifth and ninth terms of an arithmetic progression are -5 and -17, respectively. Determine the first term and a formula for the nth term.

Which sequences could be arithmetic or geometric? For arithmetic or geometric sequences, state the appropriate common difference or ratio and determine a formula for the nth term.

15. $5, 10, 20, 40, \ldots$

16. $-0.25, 1, -4, 16, -64, \ldots$

17. $-16, -9, -2, 5, \ldots$

18. $10.1, 10.201, 10.30301, 10.4060401, \ldots$

19. $-2, 1, -\frac{1}{2}, \frac{1}{4}, \ldots$

20. $1, 5, 7, 11, 17, \ldots$

The half-life of a certain unstable radioactive substance is one week. Suppose 1000 grams of the substance exists today. Let n represent the number of weeks the substance exists.

21. Determine an infinite geometric sequence that is a model of the amount of the substance at week n; $n = 1, 2, 3, \ldots$. List the first 10 terms of the sequence. What is the common ratio?

22. When will there be only 0.05 grams of the substance remaining?

23. Will the substance ever be reduced to nothing?

The height of a certain fast growing plant increases at the rate of 2.5% per month. Assume the plant is 15 inches in height today. Let n represent the number of months the plant grows and assume the plant dies in 10 months.

24. Determine a finite geometric sequence that is a model of the height of the plant after n months. Write out all the terms of the sequence. What is the common ratio?

25. How long would the plant need to live to double in height?

Prove by mathematical induction.

26. The nth term of an arithmetic progression is $a_n = a_1 + (n-1)d$ where a_1 is the first term and d is the common difference.

27. $1 + 2 + \cdots + n = \dfrac{n(n+1)}{2}$

28. $1 + 3 + 5 + \cdots + (2n-1) = n^2$

29. $1 + 5 + 9 + \cdots + (4n-3) = n(2n-1)$

30. $1 + 2 + 2^2 + \cdots + 2^{n-1} = 2^n - 1$

31. $\frac{1}{1 \cdot 2} + \frac{1}{2 \cdot 3} + \frac{1}{3 \cdot 4} + \cdots + \frac{1}{n(n+1)} = \frac{n}{n+1}$

Determine a simple rule for the nth term and give the next two terms of the sequence. Many rules are possible.

32. $2, 5, 8, 11, \ldots$

33. $-3, 0, 3, 6, 9, \ldots$

34. $0, 3, 8, 15, 24, \ldots$

35. $1, 2, 6, 24, 120, \ldots$

36. Sue had \$1250 in a savings account three years ago. What will the value of her account be two years from now assuming no deposits or withdrawals are made and the account earns 6.5% interest compounded annually?

37. Frank has \$12,876 in a savings account today. He made no deposits or withdrawals during the past six years. What was the value of his account six years ago? Assume the account earned interest at 5.75% compounded monthly.

38. Insert two arithmetic means between 3 and 20.

39. Insert three arithmetic means between 3 and 20.

40. Insert four arithmetic means between 3 and 20.

41. Insert five arithmetic means between 3 and 20.

Rabbits and the Fibonacci Sequence

Assume rabbits become fertile one month after birth and each male–female pair of fertile rabbits gives birth to one new male–female pair of rabbits each month. Further assume the rabbit colony begins with one newborn male–female pair of rabbits and no rabbits die for 12 months. Let a_n represent the number of *pairs* of rabbits in the colony after $n - 1$ months.

42. Explain why $a_1 = 1$, $a_2 = 1$, and $a_3 = 2$.

43. Determine a_4, a_5, a_6, a_7, and a_8.

44. Explain why the sequence in Exercise 43 is a model for the size of the rabbit colony.

45. Investigate the value of $a_n = 10n \sin\left(\frac{\pi}{n}\right)$ for $n = 10$; 100; 1000; and 10,000. Compute the circumference of a circle of radius 5. What conclusions can you draw?

46. Draw a graph of a_n for $1 \leq n \leq 100$ where a_n is defined in Exercise 45. How does the graph relate to Exercise 45?

47. Compute the first nine terms of

$$a_n = \frac{1}{\sqrt{5}}\left(\frac{1+\sqrt{5}}{2}\right)^n - \frac{1}{\sqrt{5}}\left(\frac{1-\sqrt{5}}{2}\right)^n.$$

Do you recognize this sequence?

48. Consider the sequence $a_1 = 1$ and $a_n = na_{n-1}$. Compute the first six terms of the sequence and determine a rule for a_n as an explicit function of n.

Convergent Sequences

Let a_n be a sequence. It can be established that a sequence a_n *converges* to a real number K, denoted by $a_n \to K$ as $n \to \infty$ if the line $y = K$ is a horizontal asymptote (an end behavior model) of the graph of a_n. Show that each sequence converges to a number K. Determine K and draw a complete graph of a_n and the line $y = K$ in the same viewing rectangle.

49. $a_n = \dfrac{2n}{n+1}$

50. $a_n = \left(1 + \dfrac{.05}{n}\right)^n$

51. $a_n = 3 + \frac{(-1)^n}{n}$

52. $a_n = n \cdot \sin\left(\frac{\pi}{2n}\right)$

Fixed Point Iteration

The following special sequence can be used to solve some equations of the form $x = f(x)$. Let f be a continuous function on an interval $[a, b]$. Define a sequence a_n as follows:

$$a_1 = G \quad \text{for some constant} \quad G \quad \text{in } (a, b),$$
$$a_n = f(a_{n-1}), \quad n \geq 2.$$

In some cases, $a_n \to S$ where S is a solution to the equation $x = f(x)$. Let G be any first approximation (a first guess) to the solution.

53. Compute the first eight terms of the sequence a_n for the function $f(x) = \left(\frac{1}{2}\right)^x$ and first approximation $G = 1$.
54. Draw complete graphs of $y = x$ and $f(x) = \left(\frac{1}{2}\right)^x$ in the same viewing rectangle.
55. Solve the equation $x = \left(\frac{1}{2}\right)^x$ and compare the solution with the eighth term of the sequence in Exercise 53. What conclusions can you draw?
56. Compute the first eight terms of the sequence a_n for the function $f(x) = 3 + \sqrt{x}$ and first approximation $G = 4$.
57. Draw complete graphs of $y = x$ and $f(x) = 3 + \sqrt{x}$ in the same viewing rectangle.
58. Solve the equation $x = 3 + \sqrt{x}$ and compare the solution with the eighth term in Exercise 56. What conclusions can you draw?
59. Show that the recursive definition of S_n of Example 7 is equivalent to the compound growth model $S_n = A(1 + i)^n$.

Geometric Means

The positive numbers $c_1, c_2, \ldots, c_k$ are called the k geometric means of the positive numbers a and b if $a, c_1, c_2, \ldots, c_k, b$ is a finite geometric progression.

60. Insert three geometric means between 2 and 81.
61. Insert four geometric means between 6 and 102.
62. Suppose that a positive number b is a geometric mean of the positive numbers a and c. Prove that $b = \sqrt{ac}$.

*63. Let $\{a_n\}$ be a Fibonacci sequence and let $b_n = a_{n+1}/a_n$.
 (a) Compute the first 10 terms of the sequence $\{b_n\}$.
 (b) Make a conjecture about $b_n \to ?$ as $n \to \infty$.

9.2 Series and the Binomial Theorem

We introduce finite and infinite series in this section. Formulas are obtained for the sum of some finite series, and the meaning of the sum of an infinite series is explored. A formula for the partial sums of a geometric progression is given and used to establish the future value of an annuity. We continue the study of the binomial theorem and use it in applications involving probability and error analysis.

Summation Notation

There are times when we want to determine the sum of the first n terms of a sequence. Consider the sequence $\{a_k\}$. Let

$$S_1 = a_1,$$
$$S_2 = a_1 + a_2,$$
$$S_3 = a_1 + a_2 + a_3,$$

and, in general, let

$$S_n = a_1 + a_2 + \cdots + a_n.$$

Each S_n is an example of a *finite* series. The number S_n is called the **nth partial sum** of the sequence $\{a_k\}$. We also use the symbol $\sum_{k=1}^{n} a_k$, read "the sum from $k = 1$ to n of a_k," to represent the nth partial sum. Thus,

$$S_n = \sum_{k=1}^{n} a_k = a_1 + a_2 + \cdots + a_n.$$

The Greek capital letter Σ (sigma) stands for sum and the symbol a_k for the kth term of the sequence $\{a_k\}$. The expression $\sum_{k=1}^{n} a_k = a_1 + a_2 + \cdots + a_n$ is called a *finite series.* The variable k is called the **index of summation** or the **summation variable.**

In the previous section we used mathematical induction to show that

$$1^2 + 3^2 + 5^2 + \cdots + (2n-1)^2 = \frac{n(2n-1)(2n+1)}{3}.$$

Notice that the left-hand side of the above equation is the nth partial sum of the sequence $\{(2k-1)^2\}$. We can rewrite the above equation in the following way:

$$S_n = \sum_{k=1}^{n} (2k-1)^2 = \frac{n(2n-1)(2n+1)}{3}.$$

• EXAMPLE 1: Determine a formula for the partial sums of the sequence $\{k\}$.

SOLUTION: First $S_n = \sum_{k=1}^{n} k = 1+2+\cdots+n$ is the sum of the first n positive integers. We write this nth partial sum twice, reversing the order of the terms in the second expression for S_n:

$$S_n = 1 + 2 + \cdots + n,$$
$$S_n = n + (n-1) + \cdots + 1.$$

Adding, we obtain

$$\begin{aligned} 2S_n &= (n+1) + (n+1) + \cdots + (n+1) \\ &= n(n+1). \end{aligned}$$

Thus, $S_n = \frac{n(n+1)}{2}$. In the Exercises of the previous section, you established this formula using mathematical induction. •

Several properties about the summation notation are stated in the following theorem, which we give without proof.

• THEOREM 1 Let $\{a_k\}$ and $\{b_k\}$ be two sequences, and let n be a positive integer. Then,
(i) $\sum_{k=1}^{n}(a_k + b_k) = \sum_{k=1}^{n} a_k + \sum_{k=1}^{n} b_k$.
(ii) $\sum_{k=1}^{n}(a_k - b_k) = \sum_{k=1}^{n} a_k - \sum_{k=1}^{n} b_k$.
(iii) $\sum_{k=1}^{n} ca_k = c\sum_{k=1}^{n} a_k$ for every number c.

Geometric Series

We will develop a formula for the nth partial sum S_n of the geometric progression $a_1, a_1 r, \ldots, a_1 r^{n-1}, \ldots$. We start with expressions for S_n and rS_n:

$$S_n = \sum_{k=1}^{n} a_1 r^{k-1} = a_1 + a_1 r + \cdots + a_1 r^{n-1},$$

$$rS_n = \sum_{k=1}^{n} a_1 r^{k} = a_1 r + a_1 r^2 + \cdots + a_1 r^{n}.$$

Subtracting and simplifying we obtain

$$\begin{aligned} S_n - rS_n &= a_1 - a_1 r^n \\ (1-r)S_n &= a_1 - a_1 r^n \\ S_n &= \frac{a_1 - a_1 r^n}{1-r} \\ S_n &= a_1 \frac{1-r^n}{1-r}. \end{aligned}$$

If the terms of a finite series form a geometric progression, then the finite series is called a **finite geometric series**. We have proved the following theorem.

• THEOREM 2 The sum of the *finite* geometric series $\sum_{k=1}^{n} a_1 r^{k-1} = a_1 + a_1 r + a_2 r^2 + \cdots + a_1 r^{n-1}$, or the nth partial sum of a geometric progression with first term a_1 and common ratio r is given by

$$S_n = a_1 \frac{1-r^n}{1-r}.$$

• EXAMPLE 2: Determine the fourth partial sum of the geometric progression $2, \frac{2}{3}, \frac{2}{9}, \ldots$.

SOLUTION: The first term is $a_1 = 2$. The common ratio is

$$r = \frac{\frac{2}{3}}{2} = \frac{1}{3}.$$

As a check, we can verify that $(\frac{2}{9})/(\frac{2}{3})$ is also $\frac{1}{3}$. We replace n by $4, a_1$ by 2, and r by $\frac{1}{3}$ in the formula of Theorem 2:

$$\begin{aligned} S_n &= a_1 \frac{1-r^n}{1-r}, \\ S_4 &= 2\frac{1-\left(\frac{1}{3}\right)^4}{1-\frac{1}{3}} \\ &= \frac{2\left(1-\frac{1}{81}\right)}{\frac{2}{3}} \\ &= \frac{240}{81} \\ &= \frac{80}{27}. \end{aligned}$$

Of course, we can check this result by direct computation:

$$S_4 = 2 + \frac{2}{3} + \frac{2}{9} + \frac{2}{27} = \frac{54+18+6+2}{27} = \frac{80}{27}.$$ •

Infinite Series

In your previous mathematics courses you have probably dealt only with *finite* sums, that is, with finite series $\sum_{k=1}^{n} a_k = a_1 + a_2 + a_3 + \cdots + a_n$. Now, under appropriate conditions, we are able to assign a meaning to an *infinite* sum of numbers.

• DEFINITION The expression

$$\sum_{k=1}^{\infty} a_k = a_1 + a_2 + \cdots + a_n + \cdots$$

is called an **infinite series** or **series**. •

For the above infinite series we set

$$\begin{aligned} S_1 &= a_1, \\ S_2 &= a_1 + a_2, \\ S_3 &= a_1 + a_2 + a_3, \end{aligned}$$

and, in general,

$$S_n = a_1 + a_2 + a_3 + \cdots + a_n.$$

The number S_n is called the **nth partial sum** of the infinite series $\sum_{k=1}^{\infty} a_k$. Notice that the S_n are simply the partial sums of the sequence $\{a_k\}$.

If the line $y = a$ is an end behavior model of the graph of the sequence $\{a_n\}$, then we write $a_n \to a$ as $n \to \infty$ and say that the sequence $\{a_n\}$ **converges** to a as n approaches infinity. This idea was illustrated in Exercises 49–52 of the previous section.

• DEFINITION Let $S_n = \sum_{k=1}^{n} a_k$. If $S_n \to S$ as $n \to \infty$, then S is said to be the *sum* of the *infinite* series $\sum_{k=1}^{\infty} a_k$. We also say that the series **converges** to S. •

We investigate infinite series of the form

$$\sum_{k=1}^{\infty} a_1 r^{k-1} = a_1 + a_1 r + a_1 r^2 + \cdots + a_n r^{n-1} + \cdots,$$

where the terms a_k of the series are a geometric progression. This series is called a **geometric series**. Let

$$f(x) = a_1 \frac{1 - r^x}{1 - r}.$$

Notice that for positive integers n, $f(n)$ is equal to the nth partial sum of the series $\sum_{k=1}^{\infty} a_1 r^{k-1}$. We investigate the behavior of the sequence of partial sums S_n as $n \to \infty$ by investigating the end behavior of f because the graph of $\{S_n\}$ consists of the points on the graph of f with positive integer x-coordinates. Notice that f can be rewritten in the form

$$f(x) = \frac{a_1}{1 - r} - \frac{a_1}{1 - r} r^x.$$

The claim that an *infinite series* has a finite sum may go against your intuition. For example, you might think that the infinite series $\frac{1}{2} + \frac{1}{4} + \frac{1}{8} + \cdots$ has no sum because it appears that adding positive numbers will *never* lead to a real number sum. In the next example we show that this series indeed has a finite sum.

• EXAMPLE 3: Determine an end behavior model for $\sum_{k=1}^{\infty} 1/2^k$ and use it to find the sum of the infinite series.

SOLUTION: Let S_n be the nth partial sum of the infinite series $\sum_{k=1}^{\infty} 1/2^k$. Then, by Theorem 2,

$$S_n = \frac{1}{2} \frac{1 - \left(\frac{1}{2}\right)^n}{1 - \frac{1}{2}} = 1 - \left(\frac{1}{2}\right)^n$$

because $a_1 = \frac{1}{2}$ and $r = \frac{1}{2}$. Let $f(x) = 1 - \left(\frac{1}{2}\right)^x$. The graph of $\{S_n\}$ consists of the points on the graph of f with positive integer x-coordinates. A graph of f and the line $y = 1$ is given in Figure 9.2.1. We can see from this figure that $y = 1$ is an end behavior model for f. Thus, $f(x) \to 1$ as $x \to \infty$. Now, $y = 1$ is also an end behavior model for the graph of $\{S_n\}$ because the graph of $\{S_n\}$ consists of the points on the graph of f with positive integer x-coordinates. It follows that $S_n \to 1$ as $n \to \infty$, so the sum of the infinite series $\sum_{k=1}^{\infty} 1/2^k$ is 1. •

The previous example is famous in the history of mathematics and is sometimes referred to as the *Zeno paradox* (see Exercise 44).

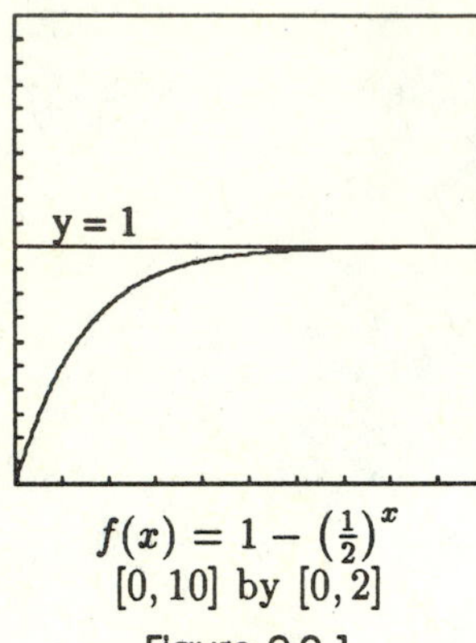

$f(x) = 1 - \left(\frac{1}{2}\right)^x$
[0, 10] by [0, 2]
Figure 9.2.1

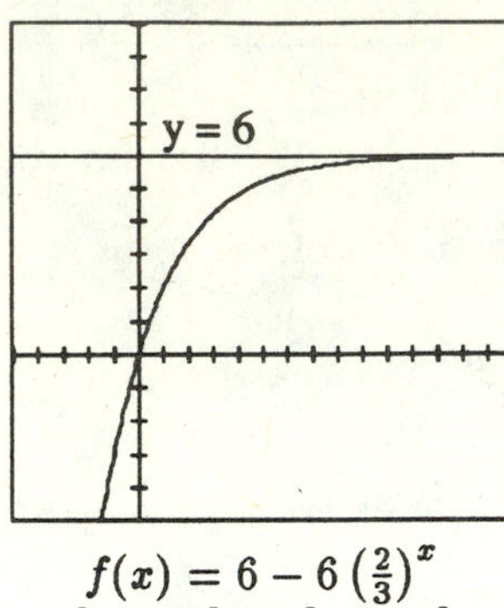

$f(x) = 6 - 6\left(\frac{2}{3}\right)^x$
[−5, 15] by [−5, 10]
Figure 9.2.2

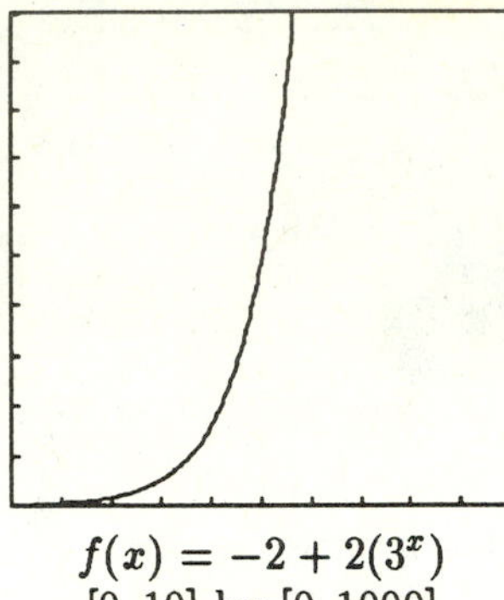

$f(x) = -2 + 2(3^x)$
[0, 10] by [0, 1000]
Figure 9.2.3

- EXAMPLE 4: Determine an end behavior model of $f(x) = a_1 \frac{1-r^x}{1-r}$ as $x \to \infty$ if

(a) $r = \dfrac{2}{3}$. (b) $r = 3$.

SOLUTION:

(a) Now,

$$f(x) = a_1 \frac{1-r^x}{1-r} = \frac{a_1}{1-r} - \frac{a_1 r^x}{1-r}.$$

Thus, for $r = \frac{2}{3}$, $f(x) = 3a_1 - 3a_1\left(\frac{2}{3}\right)^x$. If $a_1 > 0$, the graph of f can be obtained from the graph of $y = \left(\frac{2}{3}\right)^x$ by applying a reflection through the x-axis, followed by a vertical stretch with factor $3a_1$, and then a vertical shift up of $3a_1$ units. If $a_1 < 0$, the reflection is not needed but the vertical shift is down. Figure 9.2.2 shows the graph of f with $a_1 = 2$. Notice that an end behavior model of f for x positive is the horizontal asymptote $y = 6$. Thus, $f(x) \to 6$ as $x \to \infty$. In general, we can use the method of Chapter 4 to show that for $0 < r < 1$ the function

$$f(x) = a_1 \frac{1-r^x}{1-r} \to \frac{a_1}{1-r}$$

as $x \to \infty$.

(b) For $r = 3$, $f(x) = -\frac{1}{2}a_1 + \frac{1}{2}a_1(3^x)$. Figure 9.2.3 displays the behavior of f as $x \to \infty$ for $a_1 = 4$. We can see that $f(x) \to \infty$ as $x \to \infty$. In general, we can use the method of Chapter 4 to show that $f(x) \to \infty$ as $x \to \infty$ for any $a_1 \neq 0$. •

Sum of a Geometric Series

Let

$$S_n = a_1 \frac{1-r^n}{1-r}.$$

It can be shown that

$$S_n \to \frac{a_1}{1-r}$$

as $n \to \infty$ if $|r| < 1$. This means the following theorem is true.

• THEOREM 3 If $|r| < 1$, then the infinite geometric series $\sum_{k=1}^{\infty} a_1 r^{k-1} = a_1 + a_1 r + \cdots + a_1 r^{n-1} + \cdots$ has sum $a_1/(1-r)$.

• EXAMPLE 5: Find the sum of the infinite series

$$\sum_{k=1}^{\infty} \frac{2}{3^{k-1}} = 2 + \frac{2}{3} + \frac{2}{3^2} + \cdots.$$

SOLUTION: This series is geometric of the form $\sum_{k=1}^{\infty} a_1 r^{k-1}$ with $a_1 = 2$ and $r = \frac{1}{3}$. By Theorem 3, this series converges to the sum 3. Alternatively, let S_n be the nth partial sum of this series. By Theorem 2,

$$S_n = 2\frac{1-\left(\frac{1}{3}\right)^n}{1-\frac{1}{3}} = 3 - 3\left(\frac{1}{3}\right)^n.$$

The sum 3 is confirmed by the graphs of $f(x) = 3 - 3\left(\frac{1}{3}\right)^x$ and $y = 3$ in Figure 9.2.4. •

Annuities

In Section 4.3 we stated that if R dollars are deposited in an account at the end of each period (all the same length) and the interest rate per period is i, then the amount in the account at the end of n periods is $S_n = R\frac{(1+i)^n-1}{i}$. The sequence of deposits is an example of an *annuity*, and S_n is called the *future value* of the annuity. We will now show why this future value formula is correct.

The deposit of R dollars at the end of the first period earns interest for $n-1$ periods. The value of this deposit at the end of the nth period is $R(1+i)^{n-1}$. (Why?) Similarly, the

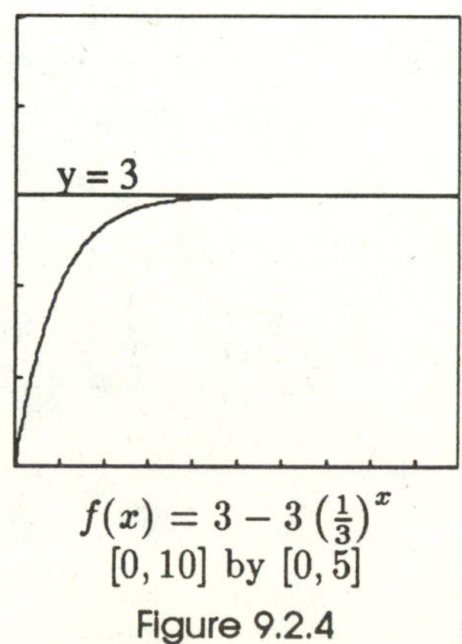

$f(x) = 3 - 3\left(\frac{1}{3}\right)^x$
[0, 10] by [0, 5]
Figure 9.2.4

deposit of R dollars at the end of the second period earns interest for $n-2$ periods. The value of this deposit at the end of the nth period is $R(1+i)^{n-2}$ (Figure 9.2.5). The total value of the n deposits at the end of the nth period is

$$S_n = R(1+i)^{n-1} + R(1+i)^{n-2} + \cdots + R(1+i)^{n-k} + \cdots + R(1+i) + R.$$

The finite sequence R, $R(1+i)$, ..., $R(1+i)^{n-1}$ is a geometric progression with $a_1 = R$ and common ratio $(1+i)$. Notice that S_n is the nth partial sum of this sequence. (Why?) By Theorem 2,

$$S_n = a_1\frac{1-r^n}{1-r} = R\frac{1-(1+i)^n}{1-(1+i)} = R\frac{(1+i)^n - 1}{i}.$$

This is the future value annuity formula given in Section 4.3.

• EXAMPLE 6: Bob deposits \$75 at the end of each month in an IRA account that pays 9.5% APR compounded monthly. Find the value of Bob's IRA account at the end of 20 years.

SOLUTION: This is an example of an annuity with monthly interest rate $i = \frac{0.095}{12}$. The value of this annuity in 20 years is given by

$$S_n = R\frac{(1+i)^n - 1}{i},$$

where $n = 240$, $i = \frac{0.095}{12}$, and $R = 75$. We can use a calculator to show that $S_{240} =$ \$101,449.11. •

Binomial Theorem

In Section 7.2 we gave the following form for the binomial theorem:

$$(a+b)^n = a^n + na^{n-1}b + \frac{n(n-1)}{1\cdot 2}a^{n-2}b^2 + \frac{n(n-1)(n-2)}{1\cdot 2\cdot 3}a^{n-3}b^3 + \cdots + \frac{n(n-1)\cdots(n-k+1)}{1\cdot 2\cdots k}a^{n-k}b^k + \cdots + b^n.$$

Here n can be any positive integer. This is another example of a finite series. This formula can be established using mathematical induction. We will use this formula to give an application involving probability. We introduce notation to simplify this formula for $(a+b)^n$.

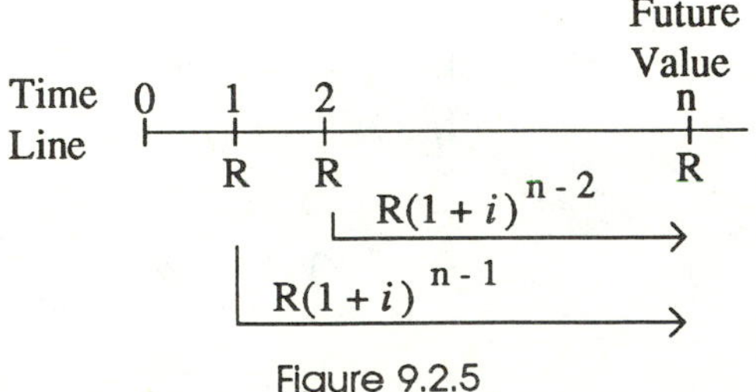

Figure 9.2.5

For any positive integer n we define $\mathbf{n!}$, read **n factorial**, by

$$n! = 1 \cdot 2 \cdots n.$$

Notice that $n!$ is the product of all positive integers starting with 1 and ending with n. For example, $3! = 1 \cdot 2 \cdot 3 = 6$ and $5! = 1 \cdot 2 \cdot 3 \cdot 4 \cdot 5 = 120$. We define $0! = 1$. We can write the coefficient of $a^{n-k}b^k$ in the expansion of $(a+b)^n$ in the following way:

$$\begin{aligned}\frac{n(n-1)\cdots(n-k+1)}{1 \cdot 2 \cdots k} &= \frac{n(n-1)\cdots(n-k+1)(n-k)(n-k-1)\cdots 1}{(1 \cdot 2 \cdots k)(n-k)(n-k-1)\cdots 1} \\ &= \frac{n!}{k!(n-k)!}\end{aligned}$$

These numbers are called **binomial coefficients** and are denoted by

$$\binom{n}{k} = \frac{n!}{k!(n-k)!}.$$

Thus, the coefficient of $a^{n-k}b^k$ in $(a+b)^n$ is the binomial coefficient $\binom{n}{k}$. This is true for $k = 0, 1, \ldots, n$. Notice that $\binom{n}{0} = \frac{n!}{0!(n-0)!} = 1$, $\binom{n}{n} = \frac{n!}{n!(n-n)!} = 1$, $\binom{n}{1} = \frac{n!}{1!(n-1)!} = n$, and $\binom{n}{n-1} = \frac{n!}{(n-1)!1!} = n$. Using this notation we can rewrite the binomial theorem in the following way:

$$\begin{aligned}(a+b)^n = \binom{n}{0} a^n + \binom{n}{1} a^{n-1}b + \binom{n}{2} a^{n-2}b^2 + \cdots + \binom{n}{k} a^{n-k}b^k \\ + \cdots + \binom{n}{n-1} ab^{n-1} + \binom{n}{n} b^n\end{aligned}$$

We usually omit writing $\binom{n}{0}$ and $\binom{n}{n}$ because they are equal to 1.

Mathematical Probability

Consider the process of *tossing four fair coins.* Any possible result of tossing the four coins is called an **outcome**. For example, the first coin could be a head and the other three tails. We denote this outcome by $HTTT$. A collection of one or more *outcomes* is called an **event**. For example, the four *outcomes* $HTTT$, $THTT$, $TTHT$, and $TTTH$ make up the *event* of obtaining exactly one head when tossing four coins. The outcome $TTHT$ means that a head occurs on the third coin and tails on the other three. These outcomes are *equally likely,* as is obtaining a head or a tail on a given coin. The **mathematical probability** of an event E, denoted by $p(E)$, is the ratio of the number of possible outcomes of the event E to the total number of possible outcomes:

$$p(E) = \frac{\text{number of outcomes of event } E}{\text{total number of outcomes}}.$$

There are 16 total possible outcomes when tossing four coins. In the exercises we ask you to list all of them. We saw above that there are four outcomes with exactly one head. We denote this event of obtaining exactly one head when four coins are tossed by HT^3. This expression records one head and three tails on a toss of four coins. Thus, the probability of the event HT^3 is

$$p(HT^3) = \frac{4}{16} = 0.25.$$

Notice the probability of *not* obtaining exactly one head on a toss of four coins is $\frac{12}{16} = 0.75 = 1 - 0.25$. (Why?)

The binomial expansion provides a way to organize the probabilities of the events of obtaining exactly i tails on a toss of four fair coins for $i = 0, 1, 2, 3, 4$. Consider the binomial expansion of $(H + T)^4$:

$$(H+T)^4 = \binom{4}{0} H^4 + \binom{4}{1} H^3T + \binom{4}{2} H^2T^2 + \binom{4}{3} HT^3 + \binom{4}{4} T^4.$$

Each coefficient $\binom{4}{i}$ gives the number of ways exactly i tails can occur on a toss of four coins. We can think of $(H + T)^4$ as representing all possible outcomes arranged by the number of tails that occur for the toss of four coins. If we let $H = T = \frac{1}{2}$, the probability of a head or a tail on one toss of a single coin, then we can rewrite the above expression for $(H + T)^4$ in the following way:

$$\begin{aligned} 1 = \left(\frac{1}{2} + \frac{1}{2}\right)^4 &= \frac{\binom{4}{0}}{16} + \frac{\binom{4}{1}}{16} + \frac{\binom{4}{2}}{16} + \frac{\binom{4}{3}}{16} + \frac{\binom{4}{4}}{16} \\ &= \frac{1}{16} + \frac{4}{16} + \frac{6}{16} + \frac{4}{16} + \frac{1}{16}. \end{aligned}$$

Each term $\binom{4}{i}/16$ represents the probability of the event consisting of exactly i tails occurring on a toss of four coins. This generalizes to tosses of any number of fair coins as illustrated in the next example. We note that a toss of n fair coins and n tosses of one fair coin are equivalent regarding events about the number of heads or tails that occur.

• EXAMPLE 7: Determine the probability that exactly three tails occur on seven tosses of a fair coin.

SOLUTION: By the discussion preceding this example, the desired probability is given by substituting $H = T = \frac{1}{2}$ in $\binom{7}{3} H^4T^3$. Thus, the probability of exactly three tails occurring on seven tosses of a fair coin is

$$\binom{7}{3}\left(\frac{1}{2}\right)^4\left(\frac{1}{2}\right)^3 = \frac{7!}{4!3!}\frac{1}{2^7} = \frac{35}{128}. \quad \bullet$$

The above ideas can be further generalized. Notice that $p(H) = p(T) = \frac{1}{2}$ and $p(H) + p(T) = 1$ for the coin toss problem. In general, suppose that a given process is repeated n times and the process has one of two possible outcomes. For example, a coin is tossed n times and each coin results in a head or a tail; or n light bulbs are selected and each bulb is either good or defective. Let H and T represent the two possible outcomes. If $p(H) = k$ then $k \leq 1$ and $p(T) = 1 - k$. We give the following theorem without proof.

• **THEOREM 4** Suppose a certain process has possible outcome H or T, and the process is repeated n times. Let $p(H) = k \leq 1$ and $p(T) = 1 - k$. Let E be the event consisting of those outcomes where the number of T's occurring is exactly r. Then,

$$p(E) = \binom{n}{r} p(H)^{n-r} p(T)^r$$

$$= \frac{n!}{r!(n-r)!} k^{n-r}(1-k)^r.$$

Notice $p(H) + p(T) = 1$. In the Exercises, you will apply this theorem to defective bats and light bulbs.

• **EXAMPLE 8:** Suppose it is known that one out of 1000 of a certain brand of computer chip is defective. Five computer chips are selected at random. What is the probability that a lot of five will contain one defective chip?

SOLUTION: Let H represent the outcome that a selected computer chip is good, and T that it is defective. Then, $p(H) = \frac{999}{1000} = 0.999$ and $p(T) = \frac{1}{1000} = 0.001$. The event E with exactly one defective computer chip can be represented by $\binom{5}{1} H^4 T$. The probability of this event is

$$\binom{5}{1} p(H)^4 p(T) = (5)(0.99)^4(0.001)$$

$$= 0.00498002998.$$

Thus, the probability that one of the five selected computer chips is defective is about 0.00498. •

We can draw the following conclusion from Example 8. About one lot of five out of 200 lots of five will contain one defective chip. (Why?)

Experimental Probability

Consider an experiment of tossing four coins 50 times. Suppose an observer records the number of times exactly one head occurs. We expect 12 or 13 of the observations to be HT^3 because $p(HT^3) = \frac{1}{4}$. (Why?) It may happen that no heads occur. However, this is most unlikely. How unlikely? Such issues are important in the study of statistics and probability.

If we know the mathematical probability of an event, then we expect the experimental evidence to be consistent with the mathematical probability. In this sense, the mathematical probability is a model of the real world experiment. For example, the mathematical probability of obtaining a head on a single toss of a fair coin is $\frac{1}{2}$. If a given coin is tossed four times, heads need not occur twice. In fact, the mathematical probability that exactly two heads occur is $\binom{4}{2} p(H)^2 p(T)^2 = \frac{6}{16}$. Suppose a given fair coin is tossed n times and H_n is the number of times a head occurs. Because the mathematical probability of obtaining a head on a given toss is $\frac{1}{2}$, we should expect that $H_n/n \to \frac{1}{2}$ as $n \to \infty$. In other words, the observed probability of a given event should be a good approximation to the mathematical probability if a large number of experiments are performed.

• EXERCISES 9-2

Use summation notation to write the sum. Assume that the suggested patterns continue.

1. $2+5+8+11+\cdots+29$
2. $-1+2+7+14+23+\cdots+62$
3. $\dfrac{1}{4}+\dfrac{1}{16}+\dfrac{1}{64}+\cdots$
4. $-2+2-2+2-2+2-\cdots$

Use summation notation to write the nth partial sum of the sequence. Assume that the suggested patterns continue.

5. $-8, -6, -4, \ldots$
6. $2, 5, 10, 17, 26, \ldots$
7. $3, -6, 9, -12, \ldots$
8. $\dfrac{5}{2}, \dfrac{5}{3}, \dfrac{5}{4}, \ldots$

Determine a formula for the nth term of the sequence of partial sums. Which sequences could be arithmetic? Geometric?

9. $1, 1+\frac{1}{2}, 1+\frac{1}{2}+\frac{1}{4}, 1+\frac{1}{2}+\frac{1}{4}+\frac{1}{8}, \ldots$
10. $1, 1+2, 1+2+3, 1+2+3+4, \ldots$
11. $-2, -2+1, -2+1-\frac{1}{2}, -2+1-\frac{1}{2}+\frac{1}{4}, \ldots$
12. $2, 2+4, 2+4+6, 2+4+6+8, \ldots$

Determine a real number sum, if it exists, of the infinite series.

13. $\dfrac{1}{8}+\dfrac{1}{16}+\dfrac{1}{32}+\dfrac{1}{64}+\cdots$
14. $-23-20-17-14-\cdots$
15. $1+3+5+\cdots$
16. $-\dfrac{1}{4}-\dfrac{1}{8}-\dfrac{1}{16}-\dfrac{1}{32}-\cdots$
17. $\dfrac{3}{10}+\dfrac{3}{100}+\dfrac{3}{1000}+\cdots$
18. $\dfrac{1}{2}+\dfrac{1}{8}+\dfrac{1}{32}+\dfrac{1}{128}+\cdots$
19. Explain why it makes sense to claim $0.9999\ldots = 1$. Use an infinite geometric series for your argument.
20. Explain why it makes sense to claim $0.333\ldots = \frac{1}{3}$. Use an infinite geometric series for your argument.
21. Determine an end behavior model for $f(x) = 3\frac{1-(2/3)^x}{1-2/3}$. Explain how the answer relates to the infinite series $3 + 2 + 4/3 + 8/9 + \cdots$.
22. Determine an end behavior model for

$$f(x) = 2\frac{1-1.05^x}{1-1.05}.$$

Explain how the answer relates to the infinite series $2 + 2.1 + 2.205 + 2.31525 + \cdots$.

23. Determine an end behavior model for

$$f(x) = -5\frac{1-(-1/3)^x}{1-(-1/3)}.$$

Explain how the answer relates to the infinite series $-5 + \frac{5}{3} - \frac{5}{9} + \frac{5}{27} - \cdots$.

24. Determine an end behavior model for

$$f(x) = 2\frac{1-0.95^x}{1-0.95}.$$

Explain how the answer relates to the infinite series $2 + 1.9 + 1.805 + 1.71475 + \cdots$.

All annuity deposits or payments are assumed to be made at the end of the period.

25. Determine the first four terms, the common ratio, and a formula for the nth partial sum of a finite geometric series representing the future value of an annuity of \$125 per month. Assume an APR of 8% compounded monthly. Let S_n represent the future value of the annuity at the end of the nth month. Determine the future value of the annuity in 15 years.

26. Determine the first four terms, the common ratio, and a formula for the nth partial sum of a finite geometric series representing the future value of an annuity of \$250 per month. Assume an APR of 9% compounded monthly. Let S_n represent the future value of the annuity at the end of the nth month. Determine the future value of the annuity in 20 years.

27. Determine the first four terms, the common ratio, and a formula for the nth partial sum of a finite geometric series representing a car loan of \$225 per month. Assume an APR interest rate of 12% compounded monthly. Let A_n represent the present value of a car loan requiring n monthly payments. Determine the car loan amount if 48 payments are made.

28. Determine the first four terms, the common ratio, and a formula for the nth partial sum of a finite geometric series representing a car loan of \$350 per month. Assume an APR of 10% compounded monthly. Let A_n represent the present value of a car loan requiring n monthly payments. Determine the car loan amount if 36 payments are made.

The Harmonic Series

It can be established that the infinite series $1 + \frac{1}{2} + \frac{1}{3} + \cdots$ does *not* converge. That is, the sequence of partial sums does not converge. A graphing calculator program

$0 \to A : 1 \to N$

Label 1

$N\ \Delta$

$1 \div N + A \to A\ \Delta$

$N + 1 \to N$

Goto 1

can be used to display the sequence of partial sums S_n of the harmonic series.

29. Use the above program to generate the first 15 partial sums of the harmonic series. (Note: the symbol Δ is the *pause* command.) When the program is run the first value of N will be

displayed. Press [EXE] to get the associated partial sum. Press [EXE] repeatedly to obtain N and the associated Nth partial sum.

30. Remove the two pause commands (△). Run the program for 5 seconds, 10 seconds and 15 seconds. (Press [AC] to stop the program.) Record the value of N and A. (Press [ALPHA] [N] [EXE] to recall N, etc.)

31. Let the modified program of Exercise 30 run for 2 minutes, 10 minutes, 30 minutes, 1 hour, 2 hours, and overnight. Record the results. What conclusions can you draw?

32. Explain how the harmonic series program works.

33. Modify the harmonic series program to show how the infinite geometric series $\frac{1}{2} + \frac{1}{4} + \frac{1}{8} + \cdots$ converges (has a finite sum). That is, show how the sequence of partial sums converges to the number 1.

34. List all the possible outcomes when four coins are tossed.

Determine the mathematical probabilities of the following events. Assume all the coins are fair and have two distinct sides.

35. Obtaining exactly two heads in a toss of five coins.

36. Obtaining exactly three heads in a toss of five coins.

37. Obtaining exactly four tails in a toss of nine coins.

38. Obtaining exactly two heads in a toss of nine coins.

39. Suppose a fair coin with two distinct sides is tossed 50 times and *each* time it shows a head. What is the probability that the 51st toss results in a head? How likely is obtaining no tails (all heads) on a toss of 50 coins? Be specific.

40. Suppose the probability of producing a defective bat is 0.02. Four bats are selected at random. What is the probability that the lot of four bats contains

    (a) no defective bats?

    (b) one defective bat?

41. Suppose the probability of producing a defective light bulb is 0.0004. Ten light bulbs are selected at random. What is the probability that the lot of 10 contains

    (a) no defective light bulbs?

    (b) two defective light bulbs?

42. Can a sequence be both arithmetic and geometric? If so, give an example.

The Precocity of Gauss

In the nineteenth century a young boy, Karl Frederick Gauss (later to become a famous mathematician), amazed his teacher by deducing the sum of the first 100 positive integers by considering the sum of $1 + 100; 2 + 99; 3 + 98; \cdots; 50 + 51$ and then concluding the sum to be $50(101)$ or 5050 in his head.

43. Generalize Gauss's method to produce an inductive argument that $1+2+3+\cdots+n=\frac{n(n+1)}{2}$. (See Example 1)

Zeno's Paradox

Frank shoots an arrow at a target 100 feet away. He hits the target in the bullseye. Zeno argues that the arrow never hits the target. He says in order for the arrow to reach the target it must travel one half of the total distance, then one half of the remaining distance, then one half of the remaining distance, and so forth. He argues that the process of "halving" the distances lasts *forever*, so the arrow never reaches the target.

44. Explain how Zeno's mathematics needs a "tune-up."

45. Conduct an experiment of tossing a quarter 50 times, 100 times, 200 times, 300 times and 400 times. Record the number of heads. Compute H_n/n where H_n is the number of observed heads. Does H_n/n approximate $\frac{1}{2}$ for n large? Should it?

9.3 Polynomial Approximations to Functions

In this section we use polynomials to approximate functions. Graphs are used to determine a range of values of the independent variable for which the approximation is reasonable. Graphs are also used to estimate the error in using a value of a polynomial to approximate the value of a function.

In advanced mathematics courses you will see that certain functions can be represented by infinite series of the form $\sum_{k=0}^{\infty} a_k x^k$ for a range of values of x.

• DEFINITION A **power series** is any series of the form

$$\sum_{k=0}^{\infty} a_k x^k = a_0 + a_1 x + a_2 x^2 + \cdots + a_n x^n + \cdots. \quad \bullet$$

The partial sums of a power series are *polynomials*:

$$S_1 = a_0,$$

$$S_2 = a_0 + a_1 x,$$

$$S_3 = a_0 + a_1 x + a_2 x^2,$$

$$\vdots$$

$$S_n = a_0 + a_1 x + \cdots + a_n x^{n-1} = \sum_{k=0}^{n-1} a_k x^k.$$

Let f be a function with domain and range that are subsets of the real numbers. We say that the power series $\sum_{k=0}^{\infty} a_k x^k$ represents f at $x = b$ if $\sum_{k=0}^{\infty} a_k b^k$ converges to $f(b)$; that is, the sequence of partial sums $S_n(b) = \sum_{k=0}^{n-1} a_k b^k \to f(b)$ as $n \to \infty$. We write $f(x) =$

$\sum_{k=0}^{\infty} a_k x^k$ for those values of x for which the series converges to $f(x)$. It can happen that an infinite power series converges for only one value of x. Let $f_n(x) = S_{n+1}(x) = \sum_{k=0}^{n} a_k x^k$. Notice $f_n(x)$ is a polynomial of degree n if $a_n \neq 0$:

$$\begin{aligned}
f_0(x) &= a_0, \\
f_1(x) &= a_0 + a_1 x, \\
f_2(x) &= a_0 + a_1 x + a_2 x^2, \\
&\vdots \\
f_n(x) &= a_0 + a_1 x + \cdots + a_n x^n.
\end{aligned}$$

It may happen that the nth degree polynomial $f_n(x)$ is a good approximation to f over some range of values of x. We illustrate this with several examples in this section.

We begin with a somewhat simpler example that illustrates the main ideas of the section.

- EXAMPLE 1: Use the first six terms of the binomial expansion of $(1+0.06)^{10}$ to approximate the value of $(1.06)^{10}$.

SOLUTION: We use the binomial theorem to expand $(1+0.06)^{10}$:

$$\begin{aligned}
(1+0.06)^{10} = 1 &+ \binom{10}{1}(0.06) + \binom{10}{2}(0.06)^2 + \binom{10}{3}(0.06)^3 \\
&+ \binom{10}{4}(0.06)^4 + \binom{10}{5}(0.06)^5 + \cdots + (0.06)^{10}.
\end{aligned}$$

Next, we use a calculator to compute the value c of the first six terms:

$$\begin{aligned}
c &= 1 + \binom{10}{1}(0.06) + \binom{10}{2}(0.06)^2 + \binom{10}{3}(0.06)^3 \\
&\quad + \binom{10}{4}(0.06)^4 + \binom{10}{5}(0.06)^5 \\
&= 1 + 10(0.06) + 45(0.06)^2 + 120(0.06)^3 \\
&\quad + 210(0.06)^4 + 252(0.06)^5 \\
&= 1.790837555.
\end{aligned}$$

It is interesting to note that the exact value of c is 1.7908375552. If we use a calculator we find $(1.06)^{10} = 1.790847697$. Notice that the two values agree through the fourth decimal place. •

The calculator value of 1.790847697 for $(1.06)^{10}$ is in error of at most 10^{-9} because the calculator we use computes with 13 digits and displays only 10. This means the first 9 digits are accurate and the last digit could be off by one. Your calculator may be different. We will assume all calculator computations are correct to the number of digits displayed.

Example 1 would be of no value if the only point was to compute $(1.06)^{10}$, because calculators do this very well. A more interesting question is to determine a range of values of x for which the fifth degree polynomial $f(x)$ determined by the first six terms of the binomial expansion of $(1+x)^{10}$ is a good approximation to $g(x) = (1+x)^{10}$. We can conclude from the computation of Example 1 that

$$f(x) = 1 + 10x + 45x^2 + 120x^3 + 210x^4 + 252x^5. \quad \text{(Why?)}$$

• EXAMPLE 2: Determine a range of values for x for which $f(x) = 1 + 10x + 45x^2 + 120x^3 + 210x^4 + 252x^5$ is a good approximation to $g(x) = (1+x)^{10}$. Estimate the error in approximating $g(0.06) = (1.06)^{10}$ by $f(0.06)$.

SOLUTION: Figure 9.3.1 shows the graphs of f and g in the $[-3, 1]$ by $[-1, 10]$ viewing rectangle. This figure suggests that f is a good approximation to g for x in $-0.2 \le x \le 0.2$. It may be true that f is a good approximation to g beyond $x = 0.2$. However, we cannot determine this from Figure 9.3.1. The range of values of x for which f appears to be a good approximation depends on the viewing rectangle used.

If we zoom in several times around $x = 0.06$ we obtain the graphs in Figure 9.3.2. We can see from this figure that $f(0.06)$ is an underestimate of $g(0.06)$ of about 10^{-5}. Thus, the value $f(0.06) = 1.790837555$ found in Example 1 approximates $g(0.06) = (1.06)^{10}$ with error of about 10^{-5}. •

Power Series

It is known that

$$\sin x = x - \frac{x^3}{3!} + \frac{x^5}{5!} - \frac{x^7}{7!} + \cdots = \sum_{k=0}^{\infty} (-1)^k \frac{x^{2k+1}}{(2k+1)!}.$$

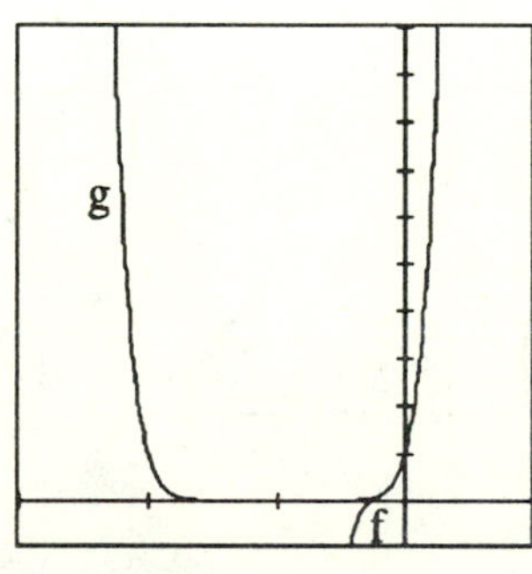

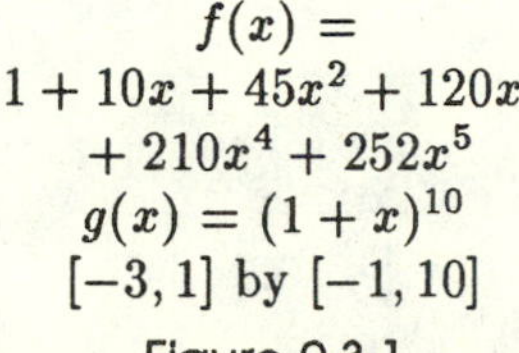
$f(x) =$
$1 + 10x + 45x^2 + 120x^3$
$+ 210x^4 + 252x^5$
$g(x) = (1+x)^{10}$
$[-3, 1]$ by $[-1, 10]$

Figure 9.3.1

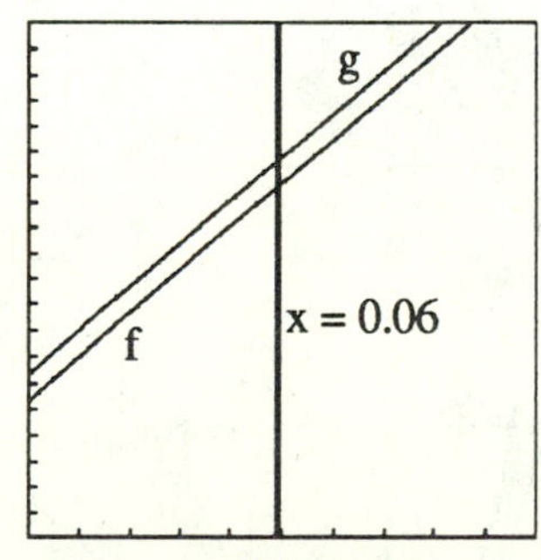

$f(x) =$
$1 + 10x + 45x^2 + 120x^3$
$+ 210x^4 + 252x^5$
$g(x) = (1+x)^{10}$
$[0.059995, 0.060005]$ by
$[1.7907, 1.7909]$

Figure 9.3.2

Let

$$f_n(x) = \sum_{k=0}^{n} \frac{(-1)^k x^{2k+1}}{(2k+1)!} = x - \frac{x^3}{3!} + \frac{x^5}{5!} - \cdots + (-1)^n \frac{x^{2n+1}}{(2n+1)!}.$$

Notice that $f_n(x)$ is polynomial of degree $2n+1$. This representation of $\sin x$ by a power series is valid for all real numbers. This means that for any real number a there is a positive integer n such that the polynomial value $f_n(a)$ is a good approximation to $\sin a$. We may need to choose n very large!

Every power series converges for some interval of values of x. The interval can be a single point. We call this interval the **interval of convergence**.

• EXAMPLE 3: Determine a range of values of x for which

$$f_n(x) = \sum_{k=0}^{n} \frac{(-1)^k x^{2k+1}}{(2k+1)!}$$

is a good approximation to $f(x) = \sin x$ when $n = 1, 2$, or 3.

SOLUTION: Notice $f_1(x) = x - \frac{x^3}{3!} = x - \frac{1}{6}x^3$. Figure 9.3.3 shows the graphs of f and f_1 in the $[-5,5]$ by $[-1.5,1.5]$ viewing rectangle. It appears that $f_1(x)$ is a good approximation to $f(x) = \sin x$ for x in $-1 \le x \le 1$. Similarly, Figure 9.3.4 shows that $f_2(x) = x - \frac{x^3}{6} + \frac{x^5}{120}$ is a good approximation to $f(x) = \sin x$ for x in $-2 \le x \le 2$. Finally, Figure 9.3.5 shows that $f_3(x) = x - \frac{x^3}{6} + \frac{x^5}{120} - \frac{x^7}{5040}$ is a good approximation to $f(x) = \sin x$ for x in $-3 \le x \le 3$. •

It can be shown that the range of values of x for which

$$f_n(x) = \sum_{k=0}^{n} \frac{(-1)^k x^{2k+1}}{(2k+1)!}$$

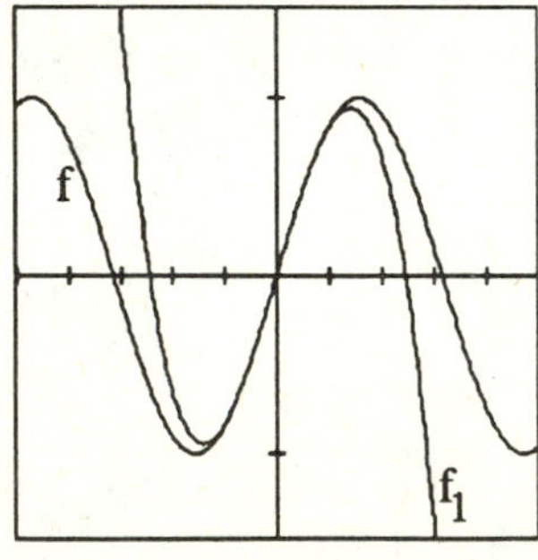

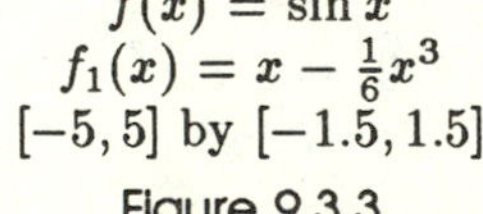
$f(x) = \sin x$
$f_1(x) = x - \frac{1}{6}x^3$
$[-5,5]$ by $[-1.5,1.5]$
Figure 9.3.3

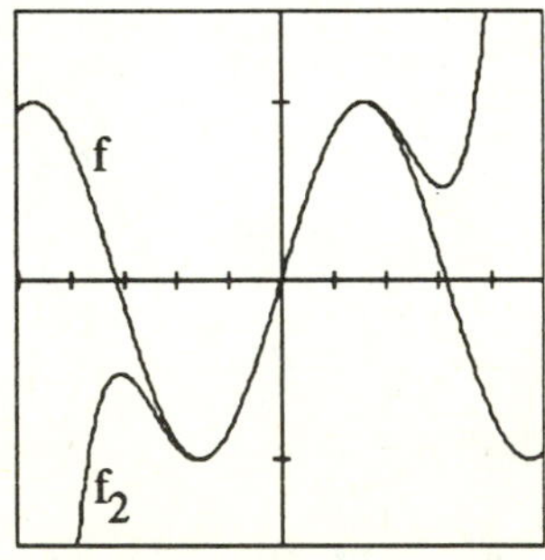

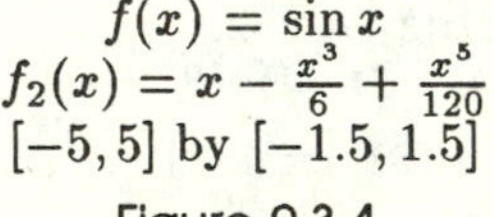
$f(x) = \sin x$
$f_2(x) = x - \frac{x^3}{6} + \frac{x^5}{120}$
$[-5,5]$ by $[-1.5,1.5]$
Figure 9.3.4

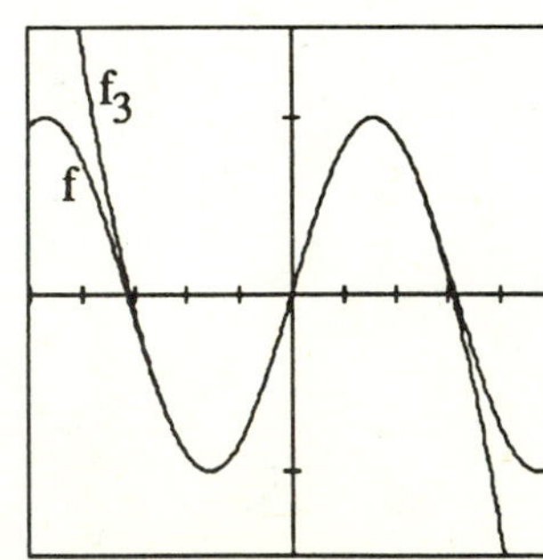

$f(x) = \sin x$
$f_3(x) = x - \frac{x^3}{6} + \frac{x^5}{120} - \frac{x^7}{5040}$
$[-5,5]$ by $[-1.5,1.5]$
Figure 9.3.5

is a good approximation to f increases as n increases. This is not true for the function of the next example.

It is known that the power series expansion $\frac{1}{1-x} = \sum_{k=0}^{\infty} x^k$ is valid for $-1 < x < 1$. This is the familiar geometric series introduced in Section 9.2. Notice that the common ratio is x.

• EXAMPLE 4: Determine a range of values of x for which $f_5(x) = \sum_{k=0}^{5} x^k = 1 + x + x^2 + x^3 + x^4 + x^5$ is a good approximation to $f(x) = \frac{1}{1-x}$. Estimate the error in approximating $f(0.2)$ and $f(0.3)$ by $f_5(0.2)$ and $f_5(0.3)$, respectively.

SOLUTION: The graphs in Figure 9.3.6 show that $f_5(x)$ is a good approximation to $f(x) = \frac{1}{1-x}$ in $-0.7 \leq x \leq 0.7$. It can be shown that, for any $n, f_n(x) = \sum_{k=0}^{n} x^k$ is a good approximation of $\frac{1}{1-x}$ in only a portion of the interval $(-1, 1)$.

Now, $f(0.2) = 1.25$. (Why?) The graphs of f and f_5 in Figure 9.3.7 show that $f_5(0.2)$ is an underestimate of $f(0.2)$ by about 0.0001. Similarly, the graphs of f and f_5 in Figure 9.3.8 show that $f_5(0.3)$ is an underestimate of $f(0.3)$ by about 0.001. •

The above example illustrates that the approximation to $f(x) = \frac{1}{1-x} = \sum_{k=i}^{\infty} a_k x^k$ by the polynomial function $f_n(x) = \sum_{k=0}^{n} a_k x^k = a_0 + a_1 x + \cdots + a_n x^n$ is less accurate near the end of the interval of convergence.

Generalized Binomial Expansion

In Section 7.2 we showed that

$$(a+b)^n = a^n + \binom{n}{1} a^{n-1} b + \cdots + \binom{n}{k} a^{n-k} b^k + \cdots + b^n$$

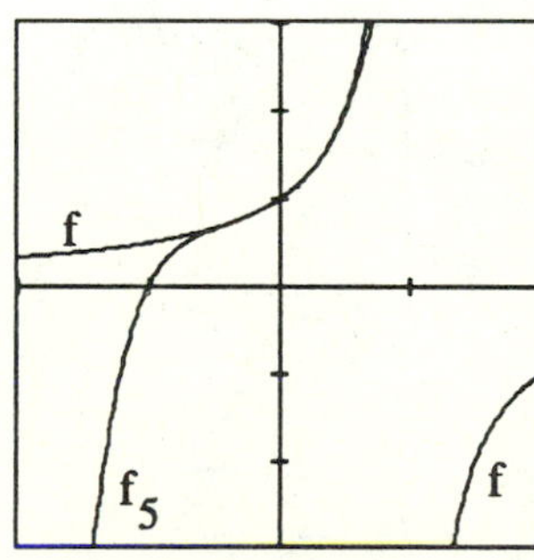

$f(x) = \frac{1}{1-x}$
$f_5(x) =$
$1 + x + x^2 + x^3 + x^4 + x^5$
$[-2, 2]$ by $[-3, 3]$

Figure 9.3.6

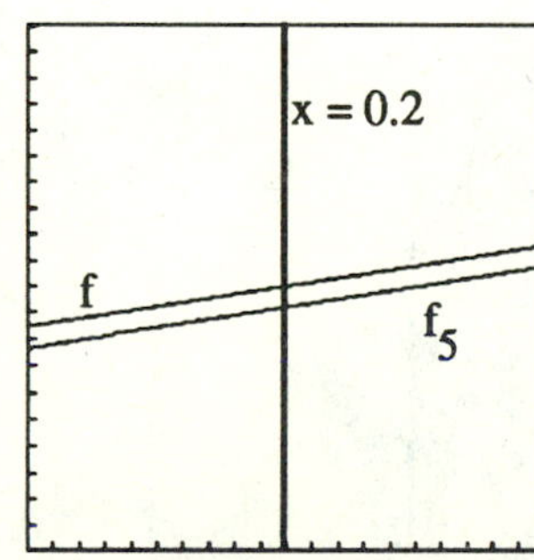

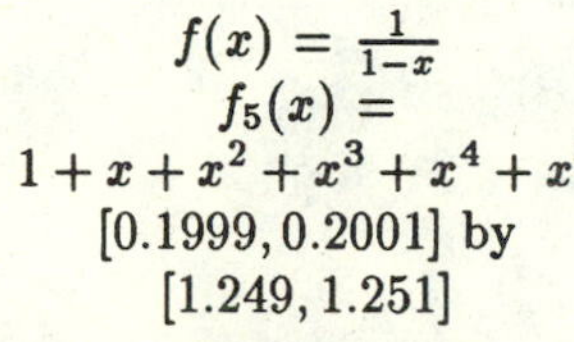

$f(x) = \frac{1}{1-x}$
$f_5(x) =$
$1 + x + x^2 + x^3 + x^4 + x^5$
$[0.1999, 0.2001]$ by
$[1.249, 1.251]$

Figure 9.3.7

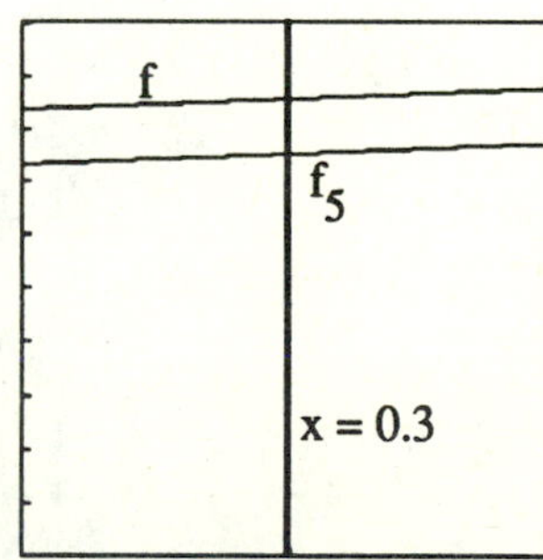

$f(x) = \frac{1}{1-x}$
$f_5(x) =$
$1 + x + x^2 + x^3 + x^4 + x^5$
$[0.299, 0.301]$
by $[1.42, 1.43]$

Figure 9.3.8

where

$$\binom{n}{k} = \frac{n!}{k!(n-k)!} = \frac{n(n-1)\cdots(n-k+1)}{k!}.$$

Thus,

$$(1+x)^n = 1 + \binom{n}{1}x + \binom{n}{2}x^2 + \cdots + x^n.$$

It turns out that this formula is valid for $n = \frac{1}{2}$ and for other noninteger numbers. In the case $n = \frac{1}{2}$ we get the following *infinite series* representation:

$$\sqrt{1+x} = (1+x)^{1/2} = 1 + \binom{0.5}{1}x + \binom{0.5}{2}x^2 + \cdots + \binom{0.5}{k}x^k + \cdots$$

where

$$\binom{0.5}{k} = \frac{0.5(0.5-1)\cdots(0.5-k+1)}{k!}$$

using the formula

$$\binom{n}{r} = \frac{n!}{r!(n-r)!} = \frac{n(n-1)\cdots(n-r+1)}{r!}$$

with n replaced by $\frac{1}{2}$. Notice that $\binom{0.5}{1} = 0.5$, $\binom{0.5}{2} = \frac{0.5(0.5-1)}{2!} = -\frac{1}{8}$, $\binom{0.5}{3} = \frac{0.5(0.5-1)(0.5-2)}{3!} = \frac{1}{16}$, and so forth. The above expansion of $\sqrt{1+x} = (1+x)^{1/2}$ is called **a generalized binomial expansion.**

• EXAMPLE 5: Let $f(x) = \sqrt{1+x}$, $f_3(x) = 1 + \frac{1}{2}x - \frac{1}{8}x^2 + \frac{1}{16}x^3$, and $f_4(x) = 1 + \frac{1}{2}x - \frac{1}{8}x^2 + \frac{1}{16}x^3 - \frac{5}{128}x^4$. Estimate the error in approximating $\sqrt{1.5} = f(0.5)$ by $f_3(0.5)$ and by $f_4(0.5)$.

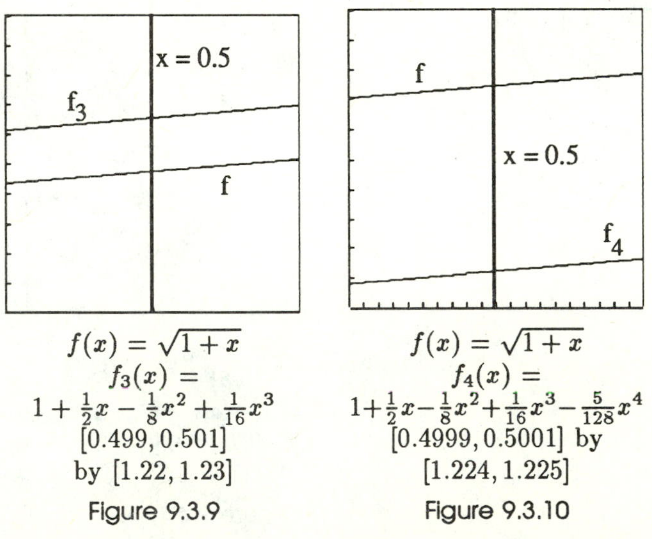

$f(x) = \sqrt{1+x}$
$f_3(x) =$
$1 + \frac{1}{2}x - \frac{1}{8}x^2 + \frac{1}{16}x^3$
[0.499, 0.501]
by [1.22, 1.23]

Figure 9.3.9

$f(x) = \sqrt{1+x}$
$f_4(x) =$
$1+\frac{1}{2}x-\frac{1}{8}x^2+\frac{1}{16}x^3-\frac{5}{128}x^4$
[0.4999, 0.5001] by
[1.224, 1.225]

Figure 9.3.10

SOLUTION: It can be shown that the above power series expansion represents f for $|x| \leq 1$. The graphs of f and f_3 in Figure 9.3.9 show that $f_3(0.5)$ is an overestimate of $f(0.5)$ by about 0.002. Similarly, the graphs of f and f_4 in Figure 9.3.10 show that $f_4(0.5)$ is an underestimate of $f(0.5)$ by about 0.0006. These results can be confirmed by direct calculator computation. •

We can make two observations from the above example. Let $f(x) = \sum_{k=0}^{\infty} a_k x^k$, $f_n(x) = \sum_{k=0}^{n} a_k x^k$, and let c be a value of x in the interval of convergence of f. Then, as n increases, $f_n(c)$ becomes a better approximation to $f(c)$. Additionally, if the a_k alternate in sign, the $f_n(c)$ will alternate between underestimates and overestimates of $f(c)$.

• EXERCISES 9-3

Consider the infinite series $1 + \frac{1}{1!} + \frac{1}{2!} + \frac{1}{3!} + \cdots$. Assume the suggested pattern continues.

1. Compute the first 10 partial sums of the series.
2. It can be shown the series converges to a number ℓ. Make a conjecture about the value of ℓ.

Let $f_1(x) = 1-x$, $f_2(x) = 1-x+x^2$, $f_3(x) = 1-x+x^2-x^3, \ldots, f_n(x) = 1-x+x^2-x^3+\cdots+(-1)^n x^n$.

3. Graph $f_1, f_2, f_3, f_4, f_5, f_6, f_7$, and f_8 in the same viewing rectangle, $[-2, 2]$ by $[-5, 5]$.
4. Draw a complete graph of $f(x) = \frac{1}{x+1}$. Compare this graph with the graphs in Exercise 3 to conclude that the power series $1 - x + x^2 - x^3 + \cdots + (-1)^n x^n + \cdots$ represents $f(x) = \frac{1}{x+1}$ for some values of x.
5. Based on Exercises 3 and 4, make a conjecture about the interval of convergence of the power series expansion

$$\frac{1}{x+1} = 1 - x + x^2 - x^3 + \cdots + (-1)^n x^n + \cdots.$$

6. Relate the power series expansion of $\frac{1}{x+1}$ to a geometric series.

Let $f(x) = (1+x)^6$ and

$$\begin{aligned}
f_1(x) &= 1 + 6x, \\
f_2(x) &= 1 + 6x + 15x^2, \\
f_3(x) &= 1 + 6x + 15x^2 + 20x^3, \\
f_4(x) &= 1 + 6x + 15x^2 + 20x^3 + 15x^4, \\
f_5(x) &= 1 + 6x + 15x^2 + 20x^3 + 15x^4 + 6x^5, \\
f_6(x) &= 1 + 6x + 15x^2 + 20x^3 + 15x^4 + 6x^5 + x^6.
\end{aligned}$$

7. Verfiy that $f(x) = (1+x)^6 = f_6(x)$.
8. Draw complete graphs of $f_1, f_2, \ldots, f_6$ in the same viewing rectangle.
9. Use a graphing argument to determine the error in using f_4 to approximate the value of $f(x) = (1+x)^6$ when $x = 1$.
10. Compute $|f(1) - f_4(1)|$ and compare with the error estimate determined in Exercise 9.

11. Is $f_4(1)$ an underestimate or an overestimate of $f(1)$? Why?

12–14. Repeat Exercises 9–11 using f_5 to approximate $f(x) = (1+x)^6$ when $x = 1$.

Let

$$f_1(x) = 1 + x,$$

$$f_2(x) = 1 + x + \frac{x^2}{2!},$$

$$f_3(x) = 1 + x + \frac{x^2}{2!} + \frac{x^3}{3!},$$

$$\vdots$$

$$f_n(x) = 1 + x + \frac{x^2}{2!} + \frac{x^3}{3!} + \cdots + \frac{x^n}{n!}.$$

15. Draw complete graphs of the six polynomial functions $f_1, f_2, \ldots, f_6$ in the same viewing rectangle.

16. Draw a complete graph of $f(x) = e^x$ and compare with the graphs in Exercise 15. Draw complete graphs of $f(x) = e^x$ and f_6 in the same viewing rectangle.
    (a) Is $f_6(-8)$ a good estimate for e^{-8}? Why?
    (b) Is $f_6(3)$ a good estimate for e^3? Why?
    (c) On what interval is $f_6(x)$ a good approximation for e^x?

17. Use a graphing argument to determine the error in using $f_3(3)$ to approximate e^3. Repeat using $f_4(3), f_5(3)$, and $f_6(3)$. Are they underestimates or overestimates?

Let

$$f_1(x) = 1 - \frac{x^2}{2!},$$

$$f_2(x) = 1 - \frac{x^2}{2!} + \frac{x^4}{4!},$$

$$f_3(x) = 1 - \frac{x^2}{2!} + \frac{x^4}{4!} - \frac{x^6}{6!},$$

$$\vdots$$

$$f_n(x) = 1 - \frac{x^2}{2!} + \frac{x^4}{4!} - \frac{x^6}{6!} + \cdots + \frac{(-1)^n x^{2n}}{(2n)!}.$$

18. Draw complete graphs of the six polynomial functions $f_1, f_2, \ldots, f_6$ in the same viewing rectangle.

19. Draw a complete graph of $f(x) = \cos x$ and compare with the graphs in Exercise 18. Draw complete graphs of $f(x) = \cos x$ and f_6 in the same viewing rectangle.
    (a) Is $f_6(-8)$ a good estimate for $\cos(-8)$? Why?
    (b) Is $f_6(3)$ a good estimate for $\cos(3)$? Why?

20. Explain why $f_{10}(3)$ is a very good estimate for cos(3). What is the error in the estimate? Is $f_{10}(3)$ an underestimate or an overestimate?

Let

$$f_n(x) = 1 + \frac{1}{3}x + \frac{\frac{1}{3}\left(-\frac{2}{3}\right)}{2!}x^2 + \frac{\frac{1}{3}\left(-\frac{2}{3}\right)\left(-\frac{5}{3}\right)}{3!}x^3 + \cdots + \frac{\frac{1}{3}\left(-\frac{2}{3}\right)\left(-\frac{5}{3}\right)\cdots\left(\frac{1}{3}-n+1\right)}{n!}x^n.$$

21. Determine $f_1, f_2, \ldots, f_6$.

22. Draw complete graphs of the polynomial function $f_1, f_2, \ldots, f_6$ in the same viewing rectangle and compare with the graph of $f(x) = \sqrt[3]{1+x}$.

23. On what interval do you think $f_n(x)$, for n large, closely approximates $f(x) = \sqrt[3]{1+x}$?

24. Use a graphing argument to estimate the error in using $f_3(-0.25)$ as an estimate for $\sqrt[3]{0.75}$.

25. Use a graphing argument to estimate the error in using $f_6(-0.25)$ as an estimate for $\sqrt[3]{0.75}$.

26. Give a graphing argument that

$$\tan^{-1} x = x - \frac{x^3}{3} + \frac{x^5}{5} - \frac{x^7}{7} + \cdots + \frac{(-1)^{n+1}x^{2n-1}}{2n-1} + \cdots$$

for $-1 < x < 1$ by graphing $f(x) = \tan^{-1} x$ and the polynomial function consisting of the first eight terms of the power series $x - \frac{x^3}{3} + \frac{x^5}{5} - \cdots$ in the same viewing rectangle.

27. Give a graphing argument that

$$\ln(1+x) = x - \frac{x^2}{2} + \frac{x^3}{3} + \cdots + (-1)^{n+1}\frac{x^n}{n} + \cdots$$

for $0 < x < 1$ by graphing $g(x) = \ln(1+x)$ and the polynomial function consisting of the first eight terms of the power series $x - \frac{x^2}{2} + \frac{x^3}{3} - \frac{x^4}{4} + \cdots$ in the same viewing rectangle.

28. Give a graphing argument that

$$\frac{e^x - e^{-x}}{2} = x + \frac{x^3}{3!} + \frac{x^5}{5!} + \cdots + \frac{x^{2n-1}}{(2n-1)!} + \cdots$$

for $-2 < x < 2$ by graphing $g(x) = \frac{e^x - e^{-x}}{2}$ and the polynomial function equal to the first eight terms of the power series $x + \frac{x^3}{3!} + \frac{x^5}{5!} + \cdots + \frac{x^{2n-1}}{(2n-1)!} + \cdots$ in the same viewing rectangle.

The ***hyperbolic cosine*** (cosh) function is defined by

$$\cosh x = \frac{e^x + e^{-x}}{2}.$$

29. Use the fact that $e^x = 1 + x + \frac{x^2}{2!} + \frac{x^3}{3!} + \cdots$ for all real x (see Exercises 15–17) to show that $\cosh x = 1 + \frac{x^2}{2!} + \frac{x^4}{4!} + \frac{x^6}{6!} + \cdots$ for all real x. Assume that you can add infinite series term by term.

30. Graph $g(x) = \cosh x$ and the polynomial function f consisting of the first seven terms of the power series $1 + \frac{x^2}{2!} + \frac{x^4}{4!} + \cdots$ in the same viewing rectangle. What conclusions can you draw? For what values of x is f a good approximation of $g(x) = \cosh x$?

It is known that $\sin x = x - \frac{x^3}{3!} + \frac{x^5}{5!} - \frac{x^7}{7!} + \cdots$ for all real x.

31. How many terms of the power series $x - \frac{x^3}{3!} + \frac{x^5}{5!} - \cdots$ must be used to obtain a good polynomial approximation to the sine function in $[-\frac{\pi}{2}, \frac{\pi}{2}]$? Use a graphing argument.

32. How many terms of the power series $x - \frac{x^3}{3!} + \frac{x^5}{5!} - \cdots$ must be used to obtain a good polynomial approximation to the sine function in $[-\pi, \pi]$? Use a graphing argument.

33. How many terms of the power series $x - \frac{x^3}{3!} + \frac{x^5}{5!} - \cdots$ must be used to obtain a good polynomial approximation to the sine function in $[-2\pi, 2\pi]$? Use a graphing argument.

Euler's Formula and the Five Most Important Constants in Algebra

34. It has been established in the previous Exercises that $\sin x = x - \frac{x^3}{3!} + \frac{x^5}{5!} - \cdots$, $\cos x = 1 - \frac{x^2}{2!} + \frac{x^4}{4!} - \cdots$, and $e^x = 1 + x + \frac{x^2}{2!} + \frac{x^3}{3!} + \cdots$. Show that $\cos x + i \sin x = e^{ix}$ (*Euler's Formula*) by expanding e^u where $u = ix$.

35. Use Exercise 34 to show that $e^{\pi i} + 1 = 0$. This equation links the five most important constants in algebra: $0, 1, e, \pi$, and i!

9.4 Matrices and Systems of Equations

In Section 2.2 we solved systems of equations algebraically and graphically. In this section we use matrices to solve systems of linear equations.

• DEFINITION Let $x_1, x_2, \ldots, x_n$ be variables, b a real number, and $a_1, a_2, \ldots, a_n$ real numbers not all of which are zero. Any equation of the form

$$a_1x_1 + a_2x_2 + \cdots + a_nx_n = b$$

is called a **linear equation in the variables $x_1, x_2, \ldots, x_n$**. •

The use of matrices arises out of streamlining the elimination process for solving systems of linear equations much like synthetic division results from streamlining long division. We illustrate the main idea of this simplification process with an example.

• EXAMPLE 1: Solve $\begin{cases} x - 2y + z = -1 \\ 2x + 3y - 2z = -3 \\ x + 3y - 2z = -2 \end{cases}$

SOLUTION: We multiply the first equation by -2 and add the second equation to obtain $7y - 4z = -1$. It turns out that the following system of equations is equivalent to the original system of equations. Notice that the original second equation $2x + 3y - 2 = -3$ has been replaced by the new equation $7y - 4z = -1$:

$$\begin{cases} x - 2y + z = -1 \\ 7y - 4z = -1 \\ x + 3y - 2z = -2 \end{cases}$$

Next, we multiply the first equation by -1 and add the third equation to obtain $5y - 3z = -1$. The following system of equations is also equivalent to the original system of equations:

$$\begin{cases} x - 2y + z = -1 \\ 7y - 4z = -1 \\ 5y - 3z = -1 \end{cases}$$

Now, we multiply the second equation by -5, the third equation by 7, and add to obtain $-z = -2$. We obtain the following equivalent system of equations:

$$\begin{cases} x - 2y + z = -1 \\ 7y - 4z = -1 \\ -z = -2 \end{cases}$$

Finally, we multiply the second equation by $\frac{1}{7}$ and the third equation by -1 to obtain the following system of equations that is equivalent to the original and all of the above systems of equations:

$$\begin{cases} x - 2y + z = -1 \\ y - \frac{4}{7}z = -\frac{1}{7} \\ z = 2 \end{cases}$$

We can now read the solution to the original system of equations from the above system of equations. The third equation gives $z = 2$. Substituting this value for z into the second equation gives $y = 1$. Then, substituting the values of y and z into the first equation gives $x = -1$. Thus, the solution to the original system of equations is the ordered triple $(-1, 1, 2)$. •

Notice the form of the final system of equations in the above example. The second equation does not contain the variable x, and the third equation does not contain the variable x or y. The coefficient of the first variable in each equation is 1.

We use rectangular arrays of numbers to record the five systems of equations that occur in the above example. A rectangular array of numbers is called a *matrix*. The numbers are the coefficients of x, y, z and the constants that occur on the right-hand sides of the above equations:

$$\begin{pmatrix} 1 & -2 & 1 & -1 \\ 2 & 3 & -2 & -3 \\ 1 & 3 & -2 & -2 \end{pmatrix} \xrightarrow{-2r_1+r_2} \begin{pmatrix} 1 & -2 & 1 & -1 \\ 0 & 7 & -4 & -1 \\ 1 & 3 & -2 & -2 \end{pmatrix} \xrightarrow{-r_1+r_3}$$

$$\begin{pmatrix} 1 & -2 & 1 & -1 \\ 0 & 7 & -4 & -1 \\ 0 & 5 & -3 & -1 \end{pmatrix} \xrightarrow{-5r_2+7r_3} \begin{pmatrix} 1 & -2 & 1 & -1 \\ 0 & 7 & -4 & -1 \\ 0 & 0 & -1 & -2 \end{pmatrix} \xrightarrow[-r_3]{\frac{1}{7}r_2}$$

$$\begin{pmatrix} 1 & -2 & 1 & -1 \\ 0 & 1 & -\frac{4}{7} & -\frac{1}{7} \\ 0 & 0 & 1 & 2 \end{pmatrix}$$

The second matrix has been obtained from the first matrix by replacing the second row by -2 times the first row added to the second row. We use the expression $-2r_1 + r_2$ above the first arrow to record this change. The third matrix has been obtained from the second matrix by replacing the third row by -1 times the first row added to the third row. The expression $-r_1+r_3$ above the second arrow records this change. The next change is recorded by $-5r_2 + 7r_3$, and the last change by $\frac{1}{7}r_2$ and $-r_3$. (Why?) Notice that the elimination process used to solve the system of Example 1 corresponds to operations on the rows of an associated matrix.

• DEFINITION Let m and n be positive integers. An $m \times n$ **matrix** A is an m by n rectangular array of numbers:

$$A = \begin{pmatrix} a_{11} & a_{12} & a_{13} & \cdots & a_{1n} \\ a_{21} & a_{22} & a_{23} & \cdots & a_{2n} \\ \vdots & \vdots & \vdots & & \vdots \\ a_{m1} & a_{m2} & a_{m3} & \cdots & a_{mn} \end{pmatrix}.$$

The numbers a_{ij} are called **elements** of the matrix. The first subscript i of a_{ij} records the *row* containing the element a_{ij}, and the second subscript j the *column* containing the element a_{ij}. If $m = n$, the matrix is called a **square matrix**. Capital letters are generally used to denote matrices. •

Notice that the element a_{32} of the above matrix occurs in the third row and second column. The matrices we used above to represent the equivalent systems of Example 1 are all 3×4 matrices. We sometimes say they are all of *size* 3×4.

Consider the following system of m linear equations in the n variables $x_1, x_2, \ldots, x_n$:

$$\begin{cases} a_{11}x_1 + a_{12}x_2 + \cdots + a_{1n}x_n = b_1 \\ a_{21}x_1 + a_{22}x_2 + \cdots + a_{2n}x_n = b_2 \\ \vdots \qquad\qquad \vdots \qquad\qquad\qquad \vdots \qquad \vdots \\ a_{m1}x_1 + a_{m2}x_2 + \cdots + a_{mn}x_n = b_m \end{cases}$$

The $m \times (n+1)$ matrix

$$\begin{pmatrix} a_{11} & a_{12} & \cdots & a_{1n} & b_1 \\ a_{21} & a_{22} & \cdots & a_{2n} & b_2 \\ \vdots & \vdots & & & \vdots \\ a_{m1} & a_{m2} & \cdots & a_{mn} & b_m \end{pmatrix}$$

is a *matrix model* of the system. We call the unique matrix determined by the system the **matrix of the system**.

For the system of equations of Example 1, a matrix model of the system is the 3×4 matrix A where A is given by

$$A = \begin{pmatrix} 1 & -2 & 1 & -1 \\ 2 & 3 & -2 & -3 \\ 1 & 3 & -2 & -2 \end{pmatrix}.$$

Matrix Row Operations

The matrix method used to solve Example 1 can be extended to any system of m linear equations in n variables.

• THEOREM 1 Applying any one the following row operations to a matrix of a system of linear equations results in a matrix that corresponds to an equivalent system of equations.

(a) Interchanging any two rows. We use $r_{i,j}$ to indicate the process of interchanging rows i and j.

(b) Multiplying all elements of a row by the nonzero number k. We use kr_i to indicate that all elements of the ith row have been multiplied by k.

(c) Adding k times the elements in one row to the corresponding elements in another row. We use $kr_i + r_j$ to indicate that the elements in the ith row have been multiplied by k and added to the corresponding elements of the jth row.

• DEFINITION The row operations of Theorem 1 are called **elementary row operations**. •

Any matrix of a system of equations can be brought to a form from which the solutions can be read as we did in Example 1.

• DEFINITION A matrix is said to be in **row echelon form** if the following conditions are satisfied.

(a) The first nonzero entry in each row is a 1.

(b) The index, j, of the column in which the first nonzero entry of a row occurs is less than the column index of the first nonzero entry of the next row.

(c) Any rows consisting entirely of zeros occur at the bottom of the matrix. •

The solutions to a system of linear equations can be read from the row echelon form of the matrix of the system. We illustrate this in the next example.

• EXAMPLE 2: Solve
$$\begin{cases} x - 2y + z = 7 \\ 2x + 3y - w = 0 \\ y + 2z - 3w = 2 \\ -x - y + 3z - w = 7 \end{cases}$$

SOLUTION: We begin with the matrix of the system and use row operations to bring it to row echelon form. We say that we are solving the system by reducing the matrix of the system to row echelon form.

$$\begin{pmatrix} 1 & -2 & 1 & 0 & 7 \\ 2 & 3 & 0 & -1 & 0 \\ 0 & 1 & 2 & -3 & 2 \\ -1 & -1 & 3 & -1 & 7 \end{pmatrix} \xrightarrow{-2r_1+r_2} \begin{pmatrix} 1 & -2 & 1 & 0 & 7 \\ 0 & 7 & -2 & -1 & -14 \\ 0 & 1 & 2 & -3 & 2 \\ -1 & -1 & 3 & -1 & 7 \end{pmatrix} \xrightarrow{r_1+r_4}$$

$$\begin{pmatrix} 1 & -2 & 1 & 0 & 7 \\ 0 & 7 & -2 & -1 & -14 \\ 0 & 1 & 2 & -3 & 2 \\ 0 & -3 & 4 & -1 & 14 \end{pmatrix} \xrightarrow{r_{2,3}} \begin{pmatrix} 1 & -2 & 1 & 0 & 7 \\ 0 & 1 & 2 & -3 & 2 \\ 0 & 7 & -2 & -1 & -14 \\ 0 & -3 & 4 & -1 & 14 \end{pmatrix} \xrightarrow{-7r_2+r_3}$$

$$\begin{pmatrix} 1 & -2 & 1 & 0 & 7 \\ 0 & 1 & 2 & -3 & 2 \\ 0 & 0 & -16 & 20 & -28 \\ 0 & -3 & 4 & -1 & 14 \end{pmatrix} \xrightarrow{3r_2+r_4} \begin{pmatrix} 1 & -2 & 1 & 0 & 7 \\ 0 & 1 & 2 & -3 & 2 \\ 0 & 0 & -16 & 20 & -28 \\ 0 & 0 & 10 & -10 & 20 \end{pmatrix} \xrightarrow{\frac{1}{10}r_4}$$

$$\begin{pmatrix} 1 & -2 & 1 & 0 & 7 \\ 0 & 1 & 2 & -3 & 2 \\ 0 & 0 & -16 & 20 & -28 \\ 0 & 0 & 1 & -1 & 2 \end{pmatrix} \xrightarrow{r_{3,4}} \begin{pmatrix} 1 & -2 & 1 & 0 & 7 \\ 0 & 1 & 2 & -3 & 2 \\ 0 & 0 & 1 & -1 & 2 \\ 0 & 0 & -16 & 20 & -28 \end{pmatrix} \xrightarrow{16r_3+r_4}$$

$$\begin{pmatrix} 1 & -2 & 1 & 0 & 7 \\ 0 & 1 & 2 & -3 & 2 \\ 0 & 0 & 1 & -1 & 2 \\ 0 & 0 & 0 & 4 & 4 \end{pmatrix} \xrightarrow{\frac{1}{4}r_4} \begin{pmatrix} 1 & -2 & 1 & 0 & 7 \\ 0 & 1 & 2 & -3 & 2 \\ 0 & 0 & 1 & -1 & 2 \\ 0 & 0 & 0 & 1 & 1 \end{pmatrix}$$

The last matrix corresponds to the following system of equations that is equivalent to the original system.

$$\begin{cases} x - 2y + z = 7 \\ y + 2z - 3w = 2 \\ z - w = 2 \\ w = 1 \end{cases}$$

From the last equation we can read $w = 1$. Substituting $w = 1$ in the third equation gives $z = 3$. Continuing, we find $y = -1$ and $x = 2$. You can check that the ordered 4-tuple $(2, -1, 3, 1)$ is a solution to each of the four original equations. •

After a little practice you should be able to perform several row operations on a matrix at one time. We will often do this to save space.

The first two examples have unique solutions. Systems of linear equations can have infinitely many solutions or no solutions. The next two examples illustrate this.

• EXAMPLE 3: Solve $\begin{cases} x + y - 2z = -2 \\ 2x - 3y + z = 1 \\ 2x + y - 3z = -2 \end{cases}$

SOLUTION: We use row operations to obtain the row echelon form of the matrix of the system. That is, we reduce the matrix of the system to row echelon form. The first step and the last step involve two row operations:

$$\begin{pmatrix} 1 & 1 & -2 & -2 \\ 2 & -3 & 1 & 1 \\ 2 & 1 & -3 & -2 \end{pmatrix} \xrightarrow[-2r_1+r_3]{-2r_1+r_2} \begin{pmatrix} 1 & 1 & -2 & -2 \\ 0 & -5 & 5 & 5 \\ 0 & -1 & 1 & 2 \end{pmatrix} \xrightarrow{r_{2,3}}$$

$$\begin{pmatrix} 1 & 1 & -2 & -2 \\ 0 & -1 & 1 & 2 \\ 0 & -5 & 5 & 5 \end{pmatrix} \xrightarrow{-5r_2+r_3} \begin{pmatrix} 1 & 1 & -2 & -2 \\ 0 & -1 & 1 & 2 \\ 0 & 0 & 0 & -5 \end{pmatrix} \xrightarrow[-\frac{1}{5}r_3]{-r_2}$$

$$\begin{pmatrix} 1 & 1 & -2 & -2 \\ 0 & 1 & -1 & -2 \\ 0 & 0 & 0 & 1 \end{pmatrix}$$

The third row of the last matrix corresponds to the false equation $0 = 1$. This means the system has no solutions. Such a system is called an *inconsistent* system. •

In the next example we solve a system of two linear equations in three variables. In the next section we will see that a complete graph of either linear equation is a plane in three-dimensional space. The system of equations has no solutions if the two planes are parallel, and infinitely many solutions if the two planes intersect in a line or a plane. In the latter case, the two planes are really the same plane.

• EXAMPLE 4: Solve $\begin{cases} x + y + z = 3 \\ 2x + y + 4z = 8 \end{cases}$

SOLUTION: We reduce the matrix of the system to row echelon form:

$$\begin{pmatrix} 1 & 1 & 1 & 3 \\ 2 & 1 & 4 & 8 \end{pmatrix} \xrightarrow{-2r_1+r_2} \begin{pmatrix} 1 & 1 & 1 & 3 \\ 0 & -1 & 2 & 2 \end{pmatrix} \xrightarrow{-r_2} \begin{pmatrix} 1 & 1 & 1 & 3 \\ 0 & 1 & -2 & -2 \end{pmatrix}.$$

If we use elementary row operations to make all other entries zero in a column containing the leading nonzero entry one of a row, the new form is called the **reduced row echelon form** of the matrix. It turns out that it is particularly easy to read the solutions of a system of equations from the reduced row echelon form when there are infinitely many solutions. If we add -1 times the second row of the above matrix to the first row we obtain the reduced row echelon form:

$$\begin{pmatrix} 1 & 0 & 3 & 5 \\ 0 & 1 & -2 & -2 \end{pmatrix}.$$

The following system of equations corresponds to the above matrix:

$$x + 3z = 5$$
$$y - 2z = -2$$

We see that $x = -3z + 5$ and $y = 2z - 2$. Thus, the solutions to the system consist of all ordered triples $(-3z+5,\ 2z-2,\ z)$ where z can be any real number. In the next section you will see why the solutions considered as points in three-dimensional space form a *line*. •

We can use the reduced row echelon form of the matrix of a system to read the solutions to any system. We illustrate this with Example 2 by continuing the reduction process from the previously obtained row echelon form of the system of equations:

$$\begin{pmatrix} 1 & -2 & 1 & 0 & 7 \\ 0 & 1 & 2 & -3 & 2 \\ 0 & 0 & 1 & -1 & 2 \\ 0 & 0 & 0 & 1 & 1 \end{pmatrix} \xrightarrow{2r_2+r_1} \begin{pmatrix} 1 & 0 & 5 & -6 & 11 \\ 0 & 1 & 2 & -3 & 2 \\ 0 & 0 & 1 & -1 & 2 \\ 0 & 0 & 0 & 1 & 1 \end{pmatrix} \underset{-2r_3+r_2}{\xrightarrow{-5r_3+r_1}}$$

$$\begin{pmatrix} 1 & 0 & 0 & -1 & 1 \\ 0 & 1 & 0 & -1 & -2 \\ 0 & 0 & 1 & -1 & 2 \\ 0 & 0 & 0 & 1 & 1 \end{pmatrix} \underset{r_4+r_3}{\xrightarrow{\substack{r_4+r_1 \\ r_4+r_2}}} \begin{pmatrix} 1 & 0 & 0 & 0 & 2 \\ 0 & 1 & 0 & 0 & -1 \\ 0 & 0 & 1 & 0 & 3 \\ 0 & 0 & 0 & 1 & 1 \end{pmatrix}$$

The final step gives the reduced row echelon form of the matrix of the system of Example 2. From this matrix we can read $x = 2$, $y = -1$, $z = 3$, and $w = 1$ which agrees with our earlier result. However, it is usually not worth the extra time needed to obtain the reduced row echelon form when solving systems of equations that have unique solutions.

The set $\{(-3z + 5,\ 2z - 2,\ z) : z \text{ any real number}\}$ of solutions to the system of equations of Example 4 is sometimes called a one-parameter family of solutions. We call this a one-parameter family because the variable z can take on any real number. In the next example the set of solutions forms a two-parameter family of solutions.

• EXAMPLE 5: Solve $\begin{cases} x + 2y - 3z = -1 \\ 2x + 3y - 4z + w = -1 \\ 3x + 5y - 7z + w = -2 \end{cases}$

SOLUTION: We bring the matrix of the system to reduced row echelon form:

$$\begin{pmatrix} 1 & 2 & -3 & 0 & -1 \\ 2 & 3 & -4 & 1 & -1 \\ 3 & 5 & -7 & 1 & -2 \end{pmatrix} \underset{-3r_1+r_3}{\xrightarrow{-2r_1+r_2}} \begin{pmatrix} 1 & 2 & -3 & 0 & -1 \\ 0 & -1 & 2 & 1 & 1 \\ 0 & -1 & 2 & 1 & 1 \end{pmatrix}$$

$$\underset{-r_2+r_3}{\xrightarrow{2r_2+r_1}} \begin{pmatrix} 1 & 0 & 1 & 2 & 1 \\ 0 & -1 & 2 & 1 & 1 \\ 0 & 0 & 0 & 0 & 0 \end{pmatrix} \xrightarrow{-r_2} \begin{pmatrix} 1 & 0 & 1 & 2 & 1 \\ 0 & 1 & -2 & -1 & -1 \\ 0 & 0 & 0 & 0 & 0 \end{pmatrix}$$

The following system of equations corresponds to the above matrix:

$$x + z + 2w = 1$$
$$y - 2z - w = -1$$

We see that $x = -z-2w+1$ and $y = 2z+w-1$. Thus, the solutions to the system consist of all ordered 4-tuples $(-z-2w+1,\ 2z+w-1,\ z,\ w)$ where z and w can be any real numbers. The set $\{(-z-2w+1, 2z+w-1, z, w) : z \text{ and } w \text{ any real numbers}\}$ of solutions is called a two-parameter family of solutions because the two variables z and w can independently take on any real number. •

In the next example, we investigate a real world problem situation that has a system of linear equations as a model.

• EXAMPLE 6: A chemist wants to prepare a 60 liter mixture that is 40% acid using three concentrations of acid on hand. The first is 15% acid, the second is 35% acid, and the third is 55% acid. Because of the amounts of acid solution on hand, the chemist wants to use twice as much of the 35% solution as the 55% solution. How much of each solution should he use?

SOLUTION: Let x, y, and z be the amounts (in liters) of the 15%, 35%, and 55% acid solutions, respectively, needed to make the 60 liter 40% acid solution. Then, $x+y+z=60$ because there are to be 60 liters of the solution. Notice $0.15x + 0.35y + 0.55z = 0.4(60)$ because the mixture is to be 40% acid. Finally, $y = 2z$. (Why?) We must find the simultaneous solution to the following system of equations:

$$\begin{aligned} x+y+z&=60,\\ 0.15x+0.35y+0.55z&=24,\\ y-2z&=0. \end{aligned}$$

We reduce the matrix of the system to row echelon form:

$$\begin{pmatrix} 1 & 1 & 1 & 60\\ 0.15 & 0.35 & 0.55 & 24\\ 0 & 1 & -2 & 0 \end{pmatrix} \xrightarrow{-0.15r_1+r_2} \begin{pmatrix} 1 & 1 & 1 & 60\\ 0 & 0.2 & 0.4 & 15\\ 0 & 1 & -2 & 0 \end{pmatrix}$$

$$\xrightarrow{r_{2,3}} \begin{pmatrix} 1 & 1 & 1 & 60\\ 0 & 1 & -2 & 0\\ 0 & 0.2 & 0.4 & 15 \end{pmatrix} \xrightarrow{-0.2r_2+r_3} \begin{pmatrix} 1 & 1 & 1 & 60\\ 0 & 1 & -2 & 0\\ 0 & 0 & 0.8 & 15 \end{pmatrix}$$

$$\xrightarrow{\frac{1}{0.8}r_3} \begin{pmatrix} 1 & 1 & 1 & 60\\ 0 & 1 & -2 & 0\\ 0 & 0 & 1 & 18.75 \end{pmatrix}$$

The last matrix is equivalent to the following system of equations:

$$\begin{aligned} x+y+z&=60,\\ y-2z&=0,\\ z&=18.75. \end{aligned}$$

We can read from this system that $z = 18.75$. Thus, $y = 37.50$ and $x = 3.75$. Consequently, 3.75 liters of 15% acid, 37.50 liters of 35%, and 18.75 liters of 55% acid are needed to make a 60 liter 40% acid solution. Notice the amount of 35% solution is twice the amount of 55% solution. •

• EXERCISES 9-4

Consider the matrix $A = \begin{pmatrix} -1 & 2 & 3 & -1 \\ 4 & 0 & 1 & 4 \\ 2 & 6 & -2 & 3 \end{pmatrix}$.

1. What is the size of the matrix? How many rows are there? How many columns are there?
2. Determine $a_{21}, a_{33}, a_{23}, a_{14}$ and a_{24}.
3. Determine the result of applying the following elementary row operations to A.
   (a) $r_1 + r_2$ (b) $2r_1 + r_3$ (c) $4r_1 + r_2$
4. Determine the equivalent row echelon form of A.
5. Determine the equivalent reduced row echelon form of A.

Consider the matrix $A = \begin{pmatrix} 3 & 0 & -9 & 2 & 1 \\ 1 & 1 & -1 & 0 & 3 \\ -6 & 0 & 18 & -4 & -2 \end{pmatrix}$.

6. What is the size of the matrix?
7. Determine $a_{13}, a_{32}, a_{23}, a_{25}$ and a_{54}.
8. Determine the result of applying the following elementary row operations to A.
   (a) $\frac{1}{3}r_1$ (b) $2r_1 + r_3$ (c) $-\frac{1}{3}r_1 + r_2$
9. Determine the equivalent row echelon form of A.
10. Determine the equivalent reduced row echelon form of A.
11. Determine the reduced row echelon form of the matrix of the system of Example 1.
12. Determine the reduced row echelon form of the matrix of the system of Example 3.

Write a matrix model for each system of equations. Solve the system by reducing the matrix of the system to row echelon form or reduced row echelon form.

13. $x - 3y = 6$
    $2x + y = 19$
14. $1.3x - 2y = 3.3$
    $5x + 1.6y = 3.4$
15. $x + y = 3$
    $x - y = 5$
    $2x + y = 4$
16. $x - y + z = 6$
    $x + y + 2z = -2$
17. $x - y + z = 0$
    $2x - 3z = -1$
    $-x - y + 2z = -1$
18. $2x - y = 0$
    $x + 3y - z = -3$
    $3y + z = 8$
19. $x + y - 2z = 2$
    $3x - y + z = 4$
    $-2x - 2y + 4z = 6$
20. $2x - y = 10$
    $x - z = -1$
    $y + z = -9$
21. $x + y - 2z = 2$
    $3x - y + z = 1$
    $-2x - 2y + 4z = -4$
22. $1.25x + z = 2$
    $y - 5.5z = -2.75$
    $3x - 1.5y = -6$

23. $x + y - z = 4$
$y + w = 4$
$x - y = 1$
$x + z + w = 4$

24. $\frac{1}{2}x - y + z - w = 1$
$-x + y + z + 2w = -3$
$x - z = 2$
$y + w = 0$

25. $2x + y + z + 2w = -3.5$
$x + y + z + w = -1.5$

26. $2x + y + 4w = 6$
$x + y + z + w = 5$

27. At a zoo in Pittsburgh, Pennsylvania, children ride a train for 25 cents but adults must pay \$1.00. On a given day, 1088 passengers paid a total of \$545 for the rides. How many passengers were children? Adults?

28. One silver alloy is 42% silver and another silver alloy is 30% silver. How many grams of each are required to produce 50 grams of a new alloy that is 34% silver?

29. Amy inherits \$20,000. She is advised to divide the money into three amounts and then make three different investments. The first earns 6% APR, the second 8% APR and the third 10% APR. How much is invested at 6%, 8%, and 10% if the amount of the first investment is twice that of the second investment and the total annual interest received is \$1640.00?

30. A scientist observes that data derived from an experiment seem to be parabolic when plotted on ordinary graph paper. Three of the observed data points are $(1, 59), (5, 75)$ and $(10, 50)$. Determine an algebraic representation of the parabola that is a model for this problem situation.

31. A scientist observes that data derived from an experiment seem to be parabolic when plotted on ordinary graph paper. Three of the observed data points are $(1, 10), (2, 8)$ and $(3, 4)$. Determine an algebraic representation of the parabola that is a model for this problem situation.

32. Joe has three employees of different abilities. They are assigned to work together on a task. Bo can do the task by himself in 6 hours, Jim and Bill working together can do the task in 1.2 hours and Bo and Bill working together can do the task in 1.5 hours. How long will it take Bo, Bill, and Jim working together to complete the task?

9.5 Three-Dimensional Geometry

In this section we introduce the Cartesian coordinates, the cylindrical coordinates, and the spherical coordinates of points in three-dimensional space. Graphs of lines, planes, and quadric surfaces are drawn. We obtain two forms for an equation of a line. Formulas for the distance between two points and the midpoint of a line segment are determined. We investigate systems of linear equations in three variables whose solutions are lines.

Cartesian Coordinates

A **Cartesian coordinate system** can be introduced in three-dimensional space by placing three mutually perpendicular number lines (coordinate lines) in 3-space so that they intersect at the zero on each line (Figure 9.5.1). The point O of intersection of the three lines is called the **origin**. The three lines are called **coordinate axes**. The **x-axis** appears to go

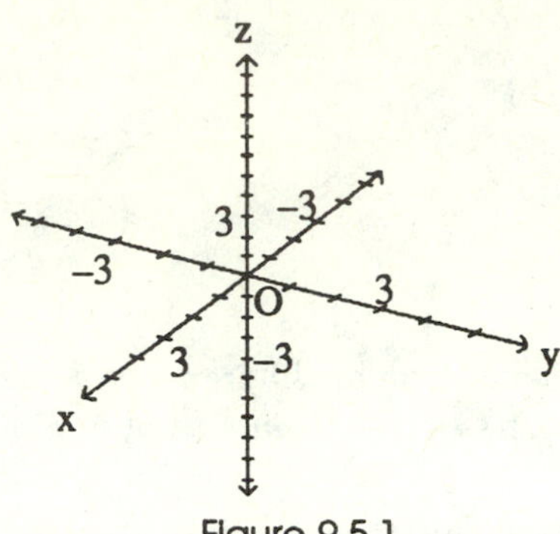

Figure 9.5.1

from front to back in Figure 9.5.1, the **y-axis** from left to right, and the **z-axis** from top to bottom. The positive directions are from O forward on the x-axis, from O to the right on the y-axis, and from O upward on the z-axis. Each pair of coordinate axes determines a coordinate plane. The x- and y-axes determine the xy-plane, the x- and z-axes determine the xz-plane, and the y- and z-axes determine the yz-plane.

The system is called **right-handed** because a counterclockwise rotation of the positive x-axis in the xy-plane will cause a right-handed screw to advance along the positive z-axis. A right-handed screw tightens under a clockwise turn and loosens under a counterclockwise turn. The three coordinate planes divide 3-space into eight *octants*. The octants *above* the *first*, *second*, *third*, and *fourth* quadrants of the xy-plane are called the **first**, **second**, **third**, and **fourth octants**, respectively. The octants *below* the *first*, *second*, *third*, and *fourth* quadrants of the xy-plane are called the **fifth**, **sixth**, **seventh**, and **eighth octants**, respectively. Often, we show only the first octant when we draw a coordinate system in 3-space.

There is a one-to-one correspondence between the points P in 3-space and ordered triples (x, y, z) of real numbers. The plane through $P(a, b, c)$ and parallel to the xy-plane intersects the z-axis at c, the plane through $P(a, b, c)$ and parallel to the xz-plane intersects the y-axis at b, and the plane through $P(a, b, c)$ and parallel to the yz-plane intersects the x-axis at a. We call a the **x-coordinate** of $P(a, b, c)$, b the **y-coordinate** of $P(a, b, c)$, and c the **z-coordinate** of $P(a, b, c)$. The coordinates of the origin are $(0, 0, 0)$. Points on the x-axis have coordinates $(x, 0, 0)$, points on the y-axis have coordinates $(0, y, 0)$ and points on the z-axis have coordinates $(0, 0, z)$. The points $(x, 0, 0), (0, y, 0)$, and $(0, 0, z)$ are sometimes called the **projection of the point (x, y, z) onto the x-axis, y-axis, and z-axis**, respectively.

Points in the xy-plane have coordinates $(x, y, 0)$. Plotting ordered triples in the xy-plane is exactly like plotting points in two dimensions that we have had considerable experience with. Points in the xz-plane have coordinates $(x, 0, z)$, and points in the yz-plane have coordinates $(0, y, z)$. Plotting points in these coordinate planes is also like plotting points in two dimensions. The points $(x, y, 0)$, $(x, 0, z)$, and $(0, x, y)$ are sometimes called the **projection of the point (x, y, z) onto the xy-plane, xz-plane, and the yz-plane**, respectively.

• EXAMPLE 1: Plot the points with coordinates $P(2,3,-4)$ and $Q(-2,-3,3)$. Name the octant containing each point.

SOLUTION: One way to plot $(2,3,-4)$ is to first plot the point $(2,3,0)$ in the xy-plane in the usual way. Then, we move down 4 units from the point $(2,3,0)$ along a line parallel to the z-axis to locate $(2,3,-4)$ (Figure 9.5.2). To plot $(-2,-3,3)$ we first locate $(-2,-3,0)$ in the xy-plane and then move up 3 units from this point along a line parallel to the z-axis. The point $(2,3,-4)$ is in Octant V, and the point $(-2,-3,3)$ is in Octant III. •

Notice that $x = 2$ is an equation satisfied by every point $(2,y,z)$ lying in the plane that is perpendicular to the x-axis at $(2,0,0)$ and parallel to the yz-plane. Similarly, $y = 3$ is an equation satisfied by every point $(x,3,z)$ lying in the plane that is perpendicular to the y-axis at $(0,3,0)$ and parallel to the xz-plane. Also, $z = -4$ is an equation satisfied by every point $(x,y,-4)$ lying in the plane that is perpendicular to the z-axis at $(0,0,-4)$ and parallel to the xy-plane. The point $(2,3,-4)$ of Example 1 is the point of intersection of the three planes $x = 2, y = 3$, and $z = -4$. (Why?)

Distance Formula

Let $P_1(x_1,y_1,z_1)$ and $P_2(x_2,y_2,z_2)$ be any two points in 3-space. The points $P_1(x_1,y_1,z_1)$ and $Q(x_2,y_2,z_1)$ both lie in the plane $z = z_1$ (Figure 9.5.3). The points $P_2(x_2,y_2,z_2)$ and $Q(x_2,y_2,z_1)$ lie on a line parallel to the z-axis that is perpendicular to the plane $z = z_1$ at Q. The points P_1, P_2, and Q form a right triangle. The line segment P_1Q lies in the plane $z = z_1$ and has length $\sqrt{(x_2-x_1)^2+(y_2-y_1)^2}$. (Why?) The line segment P_2Q is parallel to the z-axis and has length $|z_2 - z_1|$. (Why?). Applying the Pythagorean theorem, we obtain

$$
\begin{aligned}
d(P_1,P_2) &= \sqrt{d^2(P_1,Q) + d^2(P_2,Q)} \\
&= \sqrt{\left(\sqrt{(x_2-x_1)^2+(y_2-y_1)^2}\right)^2 + (|z_2-z_1|)^2} \\
&= \sqrt{(x_2-x_1)^2+(y_2-y_1)^2+(z_2-z_1)^2}.
\end{aligned}
$$

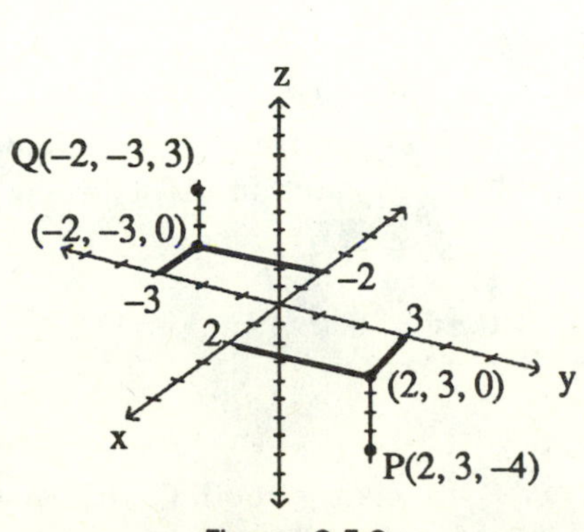

Figure 9.5.2

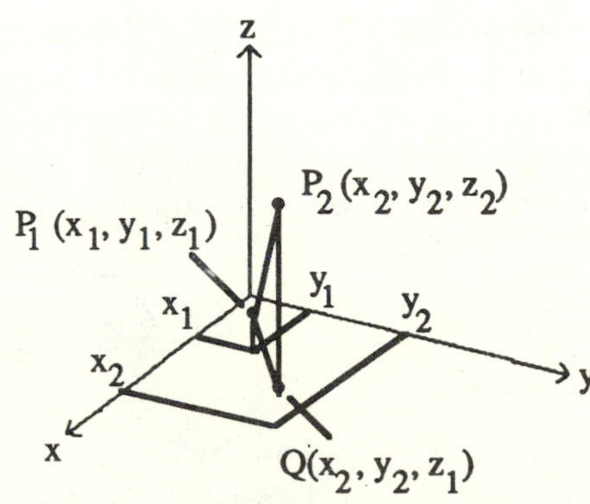

Figure 9.5.3

- **Distance Formula** The distance between any two points

$$P_1(x_1, y_1, z_1) \quad \text{and} \quad P_2(x_2, y_2, z_2)$$

in 3-space is given by

$$d(P_1, P_2) = \sqrt{(x_2 - x_1)^2 + (y_2 - y_1)^2 + (z_2 - z_1)^2}.$$

The techniques used in two dimensions to obtain the coordinates of the midpoint of a line segment can be extended to 3-space to verify the following formula.

- **Midpoint Formula** The coordinates of the midpoint of the line segment determined by the points (x_1, y_1, z_1) and (x_2, y_2, z_2) is

$$\left(\frac{x_1 + x_2}{2}, \frac{y_1 + y_2}{2}, \frac{z_1 + z_2}{2}\right).$$

• EXAMPLE 2: Find the distance between the points $P(-2, 3, -1)$ and $Q(4, -1, 5)$ and the coordinates of the midpoint of the line segment PQ.

SOLUTION: The distance formula gives

$$\begin{aligned} d(P, Q) &= \sqrt{(4+2)^2 + (-1-3)^2 + (5+1)^2} \\ &= \sqrt{88} \\ &= 9.38. \end{aligned}$$

The coordinates of the midpoint of the line segment PQ are

$$\left(\frac{-2+4}{2}, \frac{3-1}{2}, \frac{-1+5}{2}\right) = (1, 1, 2).$$

•

There are two other coordinate systems frequently used to locate points in 3-space: cylindrical coordinates and spherical coordinates. Cylindrical coordinates are often used when a three-dimensional graph has a line of symmetry, and spherical coordinates are used when the graph has a point of symmetry.

Cylindrical Coordinates

Essentially, these are the polar coordinates (r, θ) in the xy-plane $(z = 0)$ together with the z coordinate. Thus, every point in 3-space has coordinates (r, θ, z) (Figure 9.5.4). If (x, y, z) and (r, θ, z) are Cartesian and cylindrical coordinates, respectively, for a point, then the third coordinates agree and the other coordinates are related in the following way:

$$x = r\cos\theta, \qquad r^2 = x^2 + y^2,$$

$$y = r\sin\theta, \qquad \tan\theta = \frac{y}{x}.$$

Spherical Coordinates

The graphing utility we use to draw graphs in 3-space involves both Cartesian and spherical coordinates. The spherical coordinates (ρ, ϕ, θ) of a point P are illustrated in Figure 9.5.5.

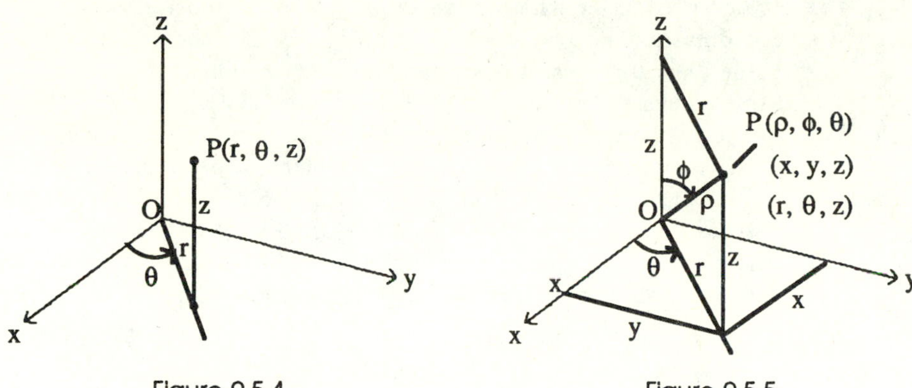

Figure 9.5.4

Figure 9.5.5

The Greek letter ρ (rho) is the distance of the point P from the origin O. The Greek letter θ has the same meaning as in polar coordinates of a plane or cylindrical coordinates in 3-space. The Greek letter ϕ (phi) is the measure of the angle from the positive z-axis to the line segment OP.

• EXAMPLE 3: Sketch a complete graph of each equation.

(a) $y = x$ (b) $\theta = 45^\circ$ (c) $r = 2$ (d) $\rho = 2$

SOLUTION:

(a) In the xy-plane ($z = 0$), the graph of $y = x$ is a straight line. In any plane $z = c$, the graph of $y = x$ is the straight line through $(0, 0, c)$ and parallel to the line $y = x$ in the xy-plane. Thus, the graph of $y = x$ is the plane perpendicular to the xy-plane that contains the line $y = x$ (Figure 9.5.6).

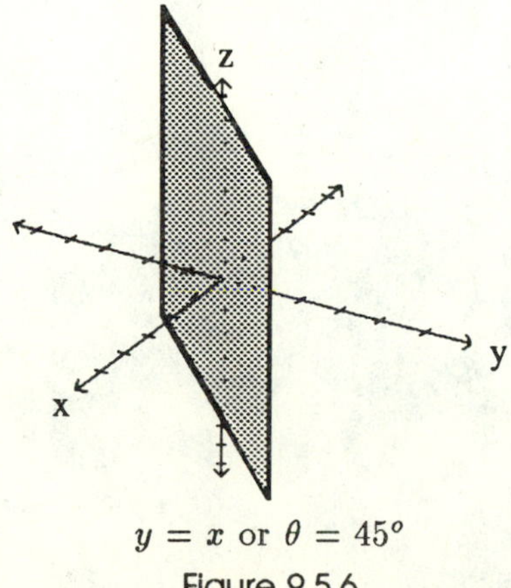

$y = x$ or $\theta = 45^\circ$

Figure 9.5.6

(b) It doesn't matter whether we think of θ in cylindrical coordinates or spherical coordinates. The graph will be the same. Notice the graph of $\theta = 45°$ in the xy-plane ($z = 0$) is also the straight line $y = x$. Thus, the graph of $\theta = 45°$ is the same as the graph of $y = x$ in (a) (Figure 9.5.6).

(c) In the xy-plane, the graph of $r = 2$ is a circle with radius 2 centered at the origin. In each plane $z = c$ parallel to the xy-plane, the graph of $r = 2$ is a circle with radius 2 centered at $(0, 0, c)$. Thus, the graph of $r = 2$ is a right circular cylinder with axis of symmetry the z-axis (Figure 9.5.7).

(d) The graph of $\rho = 2$ consists of all points in 3-space at a distance 2 from the origin. This is a sphere of radius 2 with center at the origin (Figure 9.5.8). •

Quadric Surfaces

The equation $r = 2$ of Example 3(c) can also be written in the form $x^2 + y^2 = 4$. (Why?) Similarly, the equation $\rho = 2$ of Example 3(d) can be written in the form $x^2 + y^2 + z^2 = 4$. (Why?) These equations are quadratic in the variables involved. Their graphs are called *quadric surfaces*.

• DEFINITION A surface that is a graph of an equation quadratic in the variables x, y, and z is called a **quadric surface**. •

In Chapter 8, we studied the graphs of equations quadratic in two variables x and y. The extension of this study to 3-space is the study of quadric surfaces. You will find the terminology very similar.

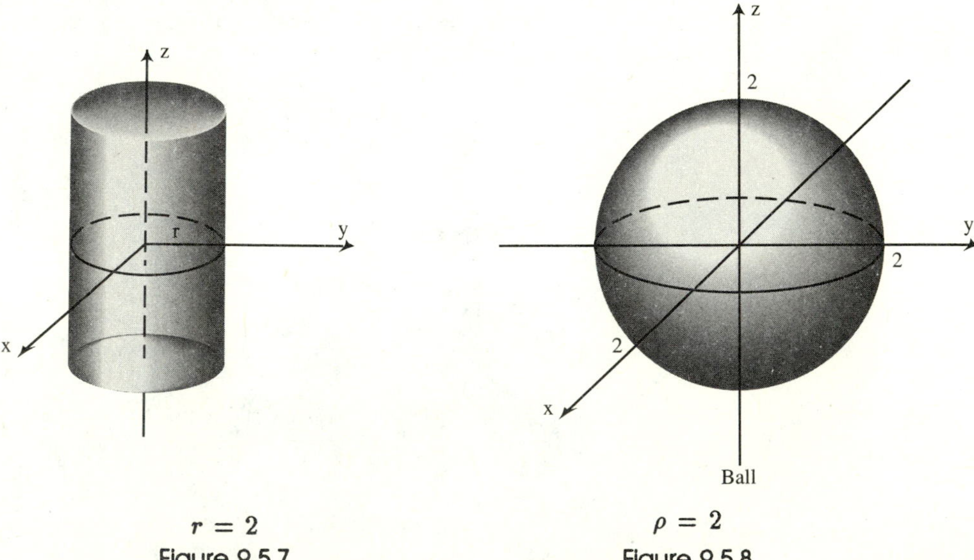

$r = 2$
Figure 9.5.7

$\rho = 2$
Figure 9.5.8

• EXAMPLE 4: Sketch a complete graph of $\dfrac{x^2}{4} + \dfrac{y^2}{9} = 1$.

SOLUTION: In the xy-plane the graph of $\frac{x^2}{4} + \frac{y^2}{9} = 1$ is an ellipse with center at the origin, major axis of length 6, and minor axis of length 4. The graph in any plane $z = c$ is an ellipse with center $(0, 0, c)$, with major axis of length 6, and minor axis of length 4. The endpoints of the major axis are $(0, \pm 3, c)$ and the endpoints of the minor axis are $(\pm 2, 0, c)$. The complete graph is the cylinder shown in Figure 9.5.9. •

• DEFINITION The graph of $\dfrac{x^2}{a^2} + \dfrac{y^2}{b^2} = 1$ is called an **elliptic cylinder**. •

Notice if $a = b$ in $\frac{x^2}{a^2} + \frac{y^2}{b^2} = 1$, then the surface is a right circular cylinder similar to the graph obtained in Example 3(c).

• EXAMPLE 5: Sketch a complete graph of $\dfrac{x^2}{a^2} + \dfrac{y^2}{b^2} + \dfrac{z^2}{c^2} = 1$.

SOLUTION: Notice that the x-intercepts are $(\pm a, 0, 0)$, the y-intercepts are $(0, \pm b, 0)$ and the z-intercepts are $(0, 0, \pm c)$. We can assume that $a > 0, b > 0$, and $c > 0$. If $|x| > a$, then $\frac{x^2}{a^2} > 1$ and there are no solutions to the equation

$$\frac{x^2}{a^2} + \frac{y^2}{b^2} + \frac{z^2}{c^2} = 1.$$

(Why?) Similarly, if $|y| > b$ or $|z| > c$, there are no solutions. Thus, we must have $|x| \leq a$, $|y| \leq b$, and $|z| \leq c$. The graph of this quadric surface is contained in the box determined by the planes $x = \pm a, y = \pm b$, and $z = \pm c$. If we set $z = 0$ in

$$\frac{x^2}{a^2} + \frac{y^2}{b^2} + \frac{z^2}{c^2} = 1$$

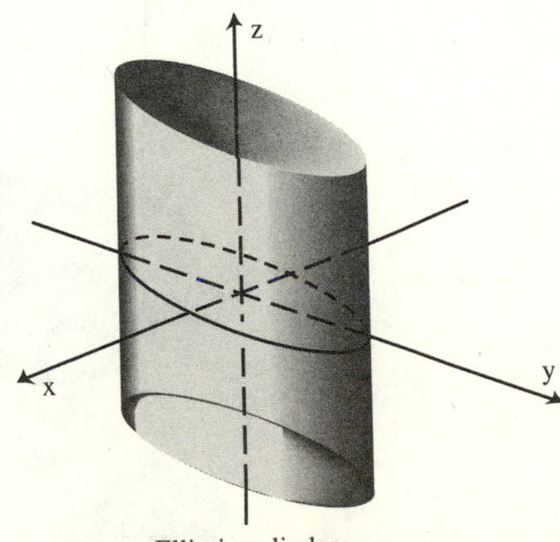

Elliptic cylinder

Figure 9.5.9

we obtain the graph of the equation in the xy-plane. The graph of $\frac{x^2}{a^2}+\frac{y^2}{b^2}=1$ is an ellipse and is called the *trace* of the graph of the original equation in the plane $z=0$. The traces of

$$\frac{x^2}{a^2}+\frac{y^2}{b^2}+\frac{z^2}{c^2}=1$$

in the xz- or yz-plane are also ellipses. Let $z=d$ with $-c<d<c$. The trace of

$$\frac{x^2}{a^2}+\frac{y^2}{b^2}+\frac{z^2}{c^2}=1$$

in the plane $z=d$ is also an ellipse. The lengths of the major and minor axes of

$$\frac{x^2}{a^2}+\frac{y^2}{b^2}=1-\frac{d^2}{c^2}, \quad \text{or} \quad \frac{x^2}{\frac{a^2(c^2-d^2)}{c^2}}+\frac{y^2}{\frac{b^2(c^2-d^2)}{c^2}}=1$$

are less than the lengths of the major and minor axes of

$$\frac{x^2}{a^2}+\frac{y^2}{b^2}=1,$$

because $1-\frac{d^2}{c^2}=(c^2-d^2)/c^2$ is a number between 0 and 1. The trace in the plane $z=d$ is the same as the trace in the plane $z=-d$. Thus, this graph is symmetric with respect to the xy-plane. If necessary, we can analyze the equation further to see that the graph in Figure 9.5.10 is complete. In the next section, we will see how to confirm this graph with a graphing utility. •

• **DEFINITION** The graph of $\frac{x^2}{a^2}+\frac{y^2}{b^2}+\frac{z^2}{c^2}=1$ is called an **ellipsoid**. •

If $a=b=c=r$ in

$$\frac{x^2}{a^2}+\frac{y^2}{b^2}+\frac{z^2}{c^2}=1,$$

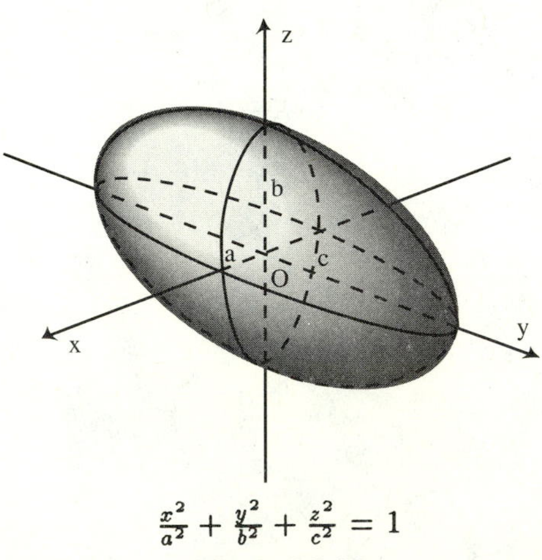

$\frac{x^2}{a^2}+\frac{y^2}{b^2}+\frac{z^2}{c^2}=1$

Figure 9.5.10

then the surface is a sphere of radius r.

• DEFINITION The **trace** of a surface in a plane is the intersection of the surface with the plane. •

The trace of a surface in the xy-, xz-, or yz-plane can be obtained by setting $z = 0, y = 0$, and $x = 0$, respectively, in the equation of the surface.

• EXAMPLE 6: Sketch a complete graph of

(a) $\frac{z}{c} = \frac{x^2}{a^2} + \frac{y^2}{b^2}$. (b) $\frac{z}{c} = \frac{y^2}{b^2} - \frac{x^2}{a^2}$.

SOLUTION:

(a) Except for the origin $(0,0,0)$, the graph of

$$\frac{z}{c} = \frac{x^2}{a^2} + \frac{y^2}{b^2}$$

lies entirely above the xy-plane if $c > 0$, and it lies entirely below the xy-plane if $c < 0$. (Why?) Without loss of generality we can assume $c > 0$. The trace of

$$\frac{z}{c} = \frac{x^2}{a^2} + \frac{y^2}{b^2}$$

in the plane $z = d$ with $d > 0$ is the ellipse $\frac{x^2}{a^2} + \frac{y^2}{b^2} = \frac{d}{c}$, or

$$\frac{x^2}{\frac{a^2 d}{c}} + \frac{y^2}{\frac{b^2 d}{c}} = 1.$$

The endpoints of the major and minor axes are

$$(\pm\sqrt{\frac{a^2 d}{c}}, 0, d) \quad \text{and} \quad (0, \pm\sqrt{\frac{b^2 d}{c}}, d).$$

The trace of

$$\frac{z}{c} = \frac{x^2}{a^2} + \frac{y^2}{b^2}$$

in the xz-plane is the parabola $z/c = x^2/a^2$, and the trace in the yz-plane is the parabola $z/c = y^2/b^2$. These parabolas lie above the x-axis and y-axis, respectively, because $c > 0$ by assumption. If c was negative, then the parabolas would lie below the x- and y-axes. The graph in Figure 9.5.11 is complete and can be confirmed with the graphing utility introduced in the next section.

(b) The trace of

$$\frac{z}{c} = \frac{y^2}{b^2} - \frac{x^2}{a^2}$$

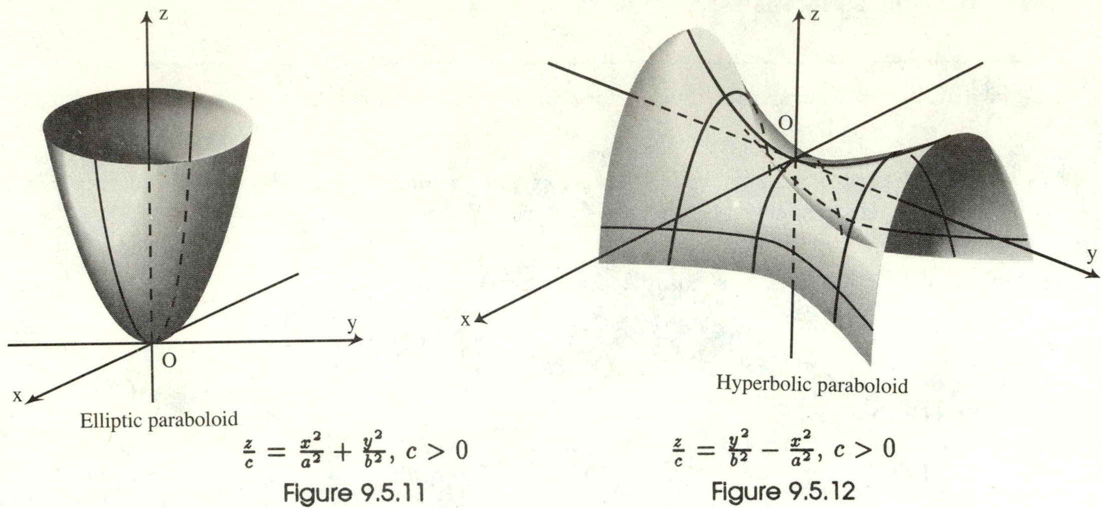

$\frac{z}{c} = \frac{x^2}{a^2} + \frac{y^2}{b^2}$, $c > 0$

Figure 9.5.11

$\frac{z}{c} = \frac{y^2}{b^2} - \frac{x^2}{a^2}$, $c > 0$

Figure 9.5.12

in the xy-plane ($z = 0$) is the pair of intersecting lines $y = \pm\frac{b}{a}x$. The trace of

$$\frac{z}{c} = \frac{y^2}{b^2} - \frac{x^2}{a^2}$$

in the plane $z = d$ is the hyperbola

$$\frac{d}{c} = \frac{y^2}{b^2} - \frac{x^2}{a^2}, \quad \text{or} \quad 1 = \frac{y^2}{\frac{db^2}{c}} - \frac{x^2}{\frac{da^2}{c}}.$$

The asymptotes are the lines $y = \pm\frac{b}{a}x$ in the plane $z = d$ parallel to the xy-plane. (Why?) If $\frac{d}{c} > 0$, the vertices are

$$(0, \pm\sqrt{\frac{db^2}{c}}, d),$$

and if $\frac{d}{c} < 0$, then the vertices are

$$(\pm\sqrt{\frac{-da^2}{c}}, 0, d).$$

The graph in Figure 9.5.12 with $c > 0$ is complete and appears to resemble a *saddle* near the origin. •

• DEFINITION The graph of

$$\frac{z}{c} = \frac{x^2}{a^2} + \frac{y^2}{b^2}$$

(Figure 9.5.11) is called an **elliptic paraboloid**. The graph of

$$\frac{z}{c} = \frac{y^2}{b^2} - \frac{x^2}{a^2}$$

(Figure 9.5.12) is called a **hyperbolic paraboloid**. •

Vectors in 3-Space

The study in Chapter 7 of vectors in two dimensions can be easily extended to vectors in 3-space. Two vectors are said to be **equal** if either vector can be obtained from the other by one or more of the following transformations: a shift parallel to the x-axis, a shift parallel to the y-axis, a shift parallel to the z-axis. Using these transformations, every vector in 3-space is equal to a unique vector of the form $\boldsymbol{v} = (x, y, z)$ with initial point the origin $(0,0,0)$ and terminal point the point (x, y, z). We call this representation the **standard form for** the vector. Similar to Chapter 7, we use 3-tuples to represent points and vectors, with the context clarifying a particular use of the given representation.

Addition, subtraction, and scalar multiplication of vectors in 3-space can be accomplished componentwise:

$$(x_1, y_1, z_1) \pm (x_2, y_2, z_2) = (x_1 \pm x_2, y_1 \pm y_2, z_1 \pm z_2),$$
$$c(x, y, z) = (cx, cy, cz),$$

The **length**, or **norm** of the vector $\boldsymbol{v} = (x, y, z)$ is defined by

$$|\boldsymbol{v}| = \sqrt{x^2 + y^2 + z^2}.$$

Two vectors in 3-space are **parallel** if and only if, in *standard form*, one vector is a scalar multiple of the other.

• EXAMPLE 7: Find the standard form and length of the vector with initial point $A = (-1, 2, 3)$ and terminal point $B = (2, -3, 4)$.

SOLUTION: The procedure used in Chapter 7 also works in 3-space. Let $\boldsymbol{v}$ be the standard form for the vector $\overrightarrow{AB}$, and let $\overrightarrow{OA}$ and $\overrightarrow{OB}$ be the vectors with initial point O and terminal point A and B, respectively. Then

$$\begin{aligned} \boldsymbol{v} &= \overrightarrow{OB} - \overrightarrow{OA} \\ &= (2, -3, 4) - (-1, 2, 3) \\ &= (3, -5, 1) \end{aligned}$$

•

Equation of a Line

We determine two forms for an equation of the line ℓ through $P_0(x_0, y_0, z_0)$ and parallel to the vector $\boldsymbol{v} = (A, B, C)$. Let $P(x, y, z)$ be any point on the line ℓ (Figure 9.5.13). The vector

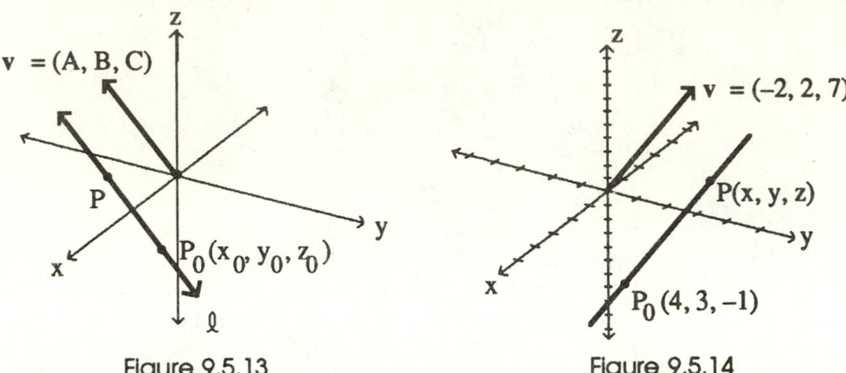

Figure 9.5.13

Figure 9.5.14

$\overrightarrow{OP} - \overrightarrow{OP_0}$ must be a scalar multiple of $\boldsymbol{v}$. (Why?) Thus, there is a real number t such that

$$\overrightarrow{OP} - \overrightarrow{OP_0} = t\boldsymbol{v},$$

$$(x - x_0, y - y_0, z - z_0) = t(A, B, C).$$

Notice that $x - x_0 = tA$, $y - y_0 = tB$, and $z - z_0 = tC$.

• DEFINITION Let $P_0 = (x_0, y_0, z_0)$ be a point in 3-space and $\boldsymbol{v} = (A, B, C)$ any vector. A **parametric form** for the equation of the line through P_0 and parallel to $\boldsymbol{v}$ is given by

$$x - x_0 = tA, \quad y - y_0 = tB, \quad \text{and} \quad z - z_0 = tC$$

where the parameter t can be any real number. •

If we solve for t in each of the equations of the parametric form for a line we obtain a *Cartesian equation* of the line ℓ.

• DEFINITION Let $P_0 = (x_0, y_0, z_0)$ be a point in 3-space and $\boldsymbol{v} = (A, B, C)$ any vector. A **Cartesian form** for the equation of the line through P_0 and parallel to $\boldsymbol{v}$ is given by

$$\frac{x - x_0}{A} = \frac{y - y_0}{B} = \frac{z - z_0}{C}.$$ •

• EXAMPLE 8: Determine a parametric form and a Cartesian form for an equation of the line through the point $P_0(4, 3, -1)$ and parallel to the vector $\boldsymbol{v} = (-2, 2, 7)$.

SOLUTION: Let $P(x, y, z)$ be any point on the line and t any real number (Figure 9.5.14). A parametric equation for the line is given by

$$x - 4 = -2t, \quad y - 3 = 2t, \quad z + 1 = 7t \quad \text{or}$$

$$x = 4 - 2t, \quad y = 3 + 2t, \quad \text{and} \quad z = -1 + 7t.$$

If we solve each equation for t we obtain

$$\frac{x-4}{-2} = t, \quad \frac{y-3}{2} = t, \quad \text{and} \quad \frac{z+1}{7} = t.$$

Thus, a Cartesian equation of the line is given by

$$\frac{x-4}{-2} = \frac{y-3}{2} = \frac{z+1}{7}. \quad \text{(Why?)}$$

•

• EXAMPLE 9: Find Cartesian and parametric forms for an equation of the line determined by the points $A = (3, 0, -2)$ and $B = (-1, 2, -5)$.

SOLUTION: The line is parallel to the vector $\boldsymbol{v} = \overrightarrow{OB} - \overrightarrow{OA} = (-4, 2, -3)$. We can use the point A and $\boldsymbol{v}$ to obtain a Cartesian form:

$$\frac{x-3}{-4} = \frac{y}{2} = \frac{z+2}{-3}.$$

Notice, that $x = 3 - 4t, y = 2t$, and $z = -2 - 3t$ gives a parametric form for the equation. If we use $-\boldsymbol{v}$ instead of $\boldsymbol{v}$ we obtain the following Cartesian form:

$$\frac{x-3}{4} = \frac{y}{-2} = \frac{z+2}{3}.$$

If we use $\boldsymbol{v}$ and the point B we obtain the following equivalent form:

$$\frac{x+1}{-4} = \frac{y-2}{2} = \frac{z+5}{-3}.$$

•

Planes

We have observed in this section that graphs of the equations $x = a$, $y = b$, $z = c$, and $y = x$ are planes. It is known that the graph of any equation of the form $ax + by + cz = d$, where not all of a, b, and c are zero, is a plane. Moreover, every plane in 3-space is the graph of such an equation.

• EXAMPLE 10: Draw a graph of the planes $x + y + z = 3$ and $2x + y + 4z = 8$.

SOLUTION: Notice that 3 is the x-intercept, the y-intercept, and the z-intercept of the plane $x+y+z = 3$. Figure 9.5.15 shows the first octant portion of the plane $x+y+z = 3$ as well as the trace of the plane in each coordinate plane. Of course, the complete graph extends outside of the first octant.

The x-intercept of $2x + y + 4z = 8$ is 4, the y-intercept is 8, and the z-intercept is 2. (Why?) Figure 9.5.16 gives the first octant portion of the graph of $2x + y + 4z = 8$ together with the trace in each of the coordinate planes. •

In Example 4 of the previous section we used matrices to show that the simultaneous solution to

$$x + y + z = 3,$$
$$2x + y + 4z = 8,$$

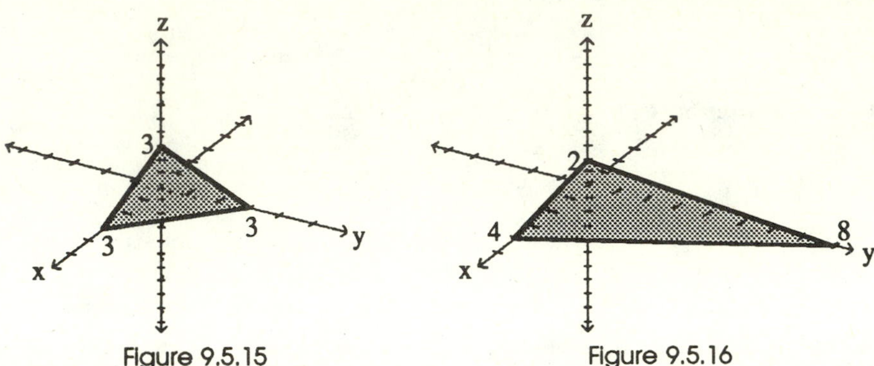

Figure 9.5.15

Figure 9.5.16

was the set of all points of the form $(-3z + 5,\ 2z - 2,\ z)$, where z is any real number. The graph of each equation is a plane. The two planes in Figures 9.5.15 and 9.5.16 provide a geometric representation of this system of equations. A geometric representation of the solutions to the system consists of the intersection of the two planes. It turns out that these two planes intersect in a line as shown in the next example.

• EXAMPLE 11: Show that the simultaneous solution to

$$\begin{cases} x + y + z = 3 \\ 2x + y + 4z = 8 \end{cases}$$

is a line. Give a parameterization of the line and draw its graph.

SOLUTION: The graph of each equation is a plane. Two planes can be parallel, the same, or intersect in a line. The solution to this system is $(-3z + 5, 2z - 2, z)$, where z is any real number. We can see that this collection of points forms a line and that $(-3z + 5, 2z - 2, z)$ gives a parameterization of the line. More explicitly, if we let $z = t$, then every solution to the above system of equations is of the form

$$x = 5 - 3t, \quad y = -2 + 2t, \quad \text{and} \quad z = t,$$

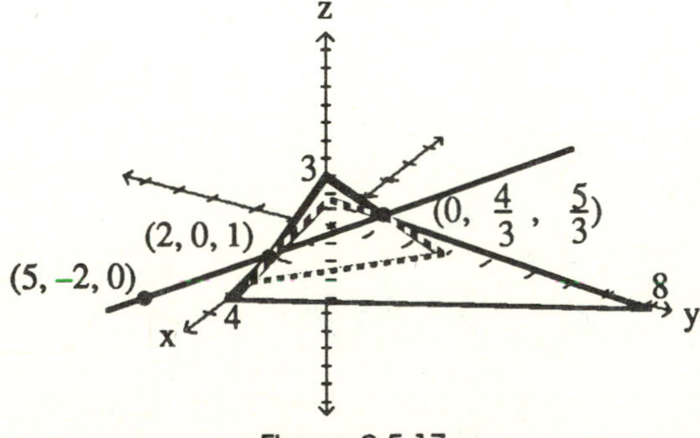

Figure 9.5.17

where t is any real number. This is a parametric form for the line through the point $(5,-2,0)$ and parallel to the vector $\boldsymbol{v}=(-3,2,1)$. The line intersects the xz-plane in the point $(2,0,1)$, and the yz-plane in the point $(0,\frac{4}{3},\frac{5}{3})$. (Why?) The graph of the line of solutions together with the first octant portion of each plane is given in Figure 9.5.17. •

• EXERCISES 9-5

Plot the points in 3-space.

1. $(3,4,2)$
2. $(2,-3,6)$
3. $(1,-2,-4)$
4. $(-2,3,-5)$

Name the octant containing the point in

5. Exercise 1
6. Exercise 2
7. Exercise 3
8. Exercise 4

Sketch a graph of the plane in the first octant.

9. $x+y+3z=9$
10. $x+y-2z=8$
11. $x+z=3$
12. $2y+z=6$
13. $x-3y=6$
14. $x=3$

Consider the system $\begin{cases} x+y+z=6 \\ x+y+2z=8 \end{cases}$.

15. (a) Sketch a graph in the first octant of the planes given by $x+y+z=6$ and $x+y+2z=8$.
    (b) Describe the intersection of the two planes. Sketch and determine an equation in parametric form.
    (c) Plot four points in the first octant satisfying the equation found in (b).

Consider the system $\begin{cases} x+2y=6 \\ x+2y+z=6 \end{cases}$.

16. (a) Sketch a graph in the first octant of the planes given by $x+2y=6$ and $x+2y+z=6$.
    (b) What is the intersection of the two planes? Sketch and determine an equation in parametric form.
    (c) Plot four points in the first octant satisfying the equation found in (b).
17. Compute the distance between the points $(-1,2,5)$ and $(3,-4,6)$.
18. Compute the distance between the points $(2,-1,-8)$ and $(6,-3,4)$.
19. Determine the midpoint of the line segment between the points $(-1,2,5)$ and $(3,-4,6)$.
20. Determine the midpoint of the line segment between the points $(2,-1,-8)$ and $(6,-3,4)$.
21. Let $v=(2,4,3)$ and $P=(4,6,1)$. Determine an equation for the line ℓ containing P parallel to $\boldsymbol{v}$.
22. Let $v=(1,5,3)$ and $P=(2,3,6)$. Determine an equation for the line ℓ containing P parallel to $\boldsymbol{v}$.
23. Sketch a graph of the vector v and line ℓ in Exercise 21.
24. Sketch a graph of the vector v and line ℓ in Exercise 22.

Equation of a Line Through Two Points in 3-Space

Let $A=(x_1,y_1,z_1)$ and $B=(x_2,y_2,z_2)$ be two distinct points in 3-space.

25. Show that the equation of the line through AB in *Cartesian* equation form is

$$\frac{x - x_1}{x_2 - x_1} = \frac{y - y_1}{y_2 - y_1} = \frac{z - z_1}{z_2 - z_1}.$$

26. Write the equation of the line through AB in parametric form.

Let $A = (-1, 2, 4), B = (0, 6, -3)$, and $C = (2, -4, 1)$. Use the result of Exercise 25 in Exercises 27–30.

27. Write an equation for the line through points A and B in
    (a) Cartesian form.
    (b) parametric form.
28. Write an equation for the line through points A and C in
    (a) Cartesian form.
    (b) parametric form.
29. Write an equation for the line through points B and C in
    (a) Cartesian form.
    (b) parametric form.
30. Write an equation for the line through points M and A where M is the midpoint of line segment BC in
    (a) Cartesian form.
    (b) parametric form.

Describe and sketch the quadric surface.

31. $4x^2 + 9y^2 = 36$
32. $x^2 + 9z^2 = 9$
33. $144x^2 + 64y^2 + 36z^2 = 576$
34. $x^2 + y^2 + z^2 = 64$
35. $9x^2 + 4y^2 = 36z$
36. $16x^2 + y^2 = 16z$
37. $9y^2 - 4x^2 = 36z$
38. $\sqrt{64 - x^2 - y^2} = z$
39. $x^2 + z^2 = 4$
40. $9y^2 + z^2 = 9$

Compute the cylindrical coordinates of the points with Cartesian coordinates

41. $(4, 0, 0)$
42. $(0, 0, 5)$
43. $(4, 4, 3)$
44. $(2, 2, -3)$

Compute the spherical coordinates of the points with Cartesian coordinates.

45. $(1, 0, 0)$
46. $(0, 0, -1)$
47. $(1, 1, \sqrt{2})$
48. $(1, -1, -\sqrt{2})$
49. Given that (ρ, ϕ, θ) are the spherical coordinates of a point with Cartesian coordinates (x, y, z), show that $\rho^2 = x^2 + y^2 + z^2$.

Describe a complete graph in 3-space.

50. $y = 2x$
51. $\theta = 30°$
52. $r = 3$
53. $\rho = 4$

9.6 Graphs of Functions of Two Variables

In this section we define a function of two variables and use a graphing utility to obtain complete graphs of such functions. We investigate contour curves, level curves, and the domain and range of functions of two variables. We explain what is meant by a local maximum value or a local minimum value of a function of two variables and use a graphing utility to approximate such values. Applications whose solutions involve finding a local maximum value or a local minimum value of a function of two variables are investigated.

A function consists of a domain and a rule that assigns a unique element of the range to each element of the domain. If the domain D of a function f is a subset of the xy-plane, and the range R is a subset of the real numbers, then f is called a *function of two variables.* We can think of f as a mapping from D to R (Figure 9.6.1). If (x, y) is an ordered pair in D, then we let the real number $z = f(x, y)$ be the corresponding element in the range of f. Here, x and y are called independent variables and z the dependent variable.

In the previous section we determined that the graph of the equation $x^2 + y^2 + z^2 = 64$ is a sphere centered at the origin with radius 8. This equation does not define z as a function of x and y. Notice that both $\pm\sqrt{62}$ are paired with the domain element $(1, 1)$. (Why?) The upper half of the sphere is the graph of $z = \sqrt{64 - x^2 - y^2}$, and the lower half of the sphere is the graph of $z = -\sqrt{64 - x^2 - y^2}$. Each of these equations defines z as a function of x and y. We will generally not take the time to verify that a given equation in three variables defines one variable as a function of the other two.

• DEFINITION A **function f of two variables** defined on a *domain* D in the xy-plane is a rule that assigns to each point (x, y) in D a unique real number $f(x, y)$. •

• EXAMPLE 1: Determine the domain and the range of the function $f(x, y) = x^2 + y^2$.

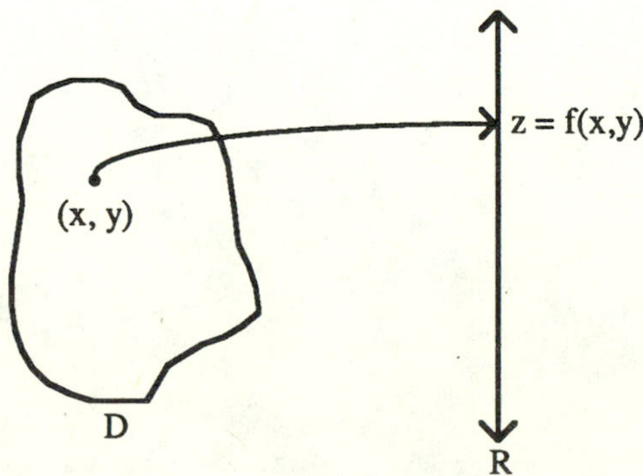

Figure 9.6.1

SOLUTION: Notice that x^2+y^2 is defined and real for all pairs of real numbers (x, y). Thus, the domain of f is the entire xy-plane. For each pair (x, y) of real numbers, $x^2+y^2 \geq 0$. The range of f is $[0, \infty)$. In the previous section we determined complete graphs of *elliptic paraboloids* $\frac{z}{c} = \frac{x^2}{a^2} + \frac{y^2}{b^2}$. If we set $a = b = c = 1$ in this equation we obtain $z = x^2 + y^2$. Figure 9.6.2 gives a complete graph of f. This graph confirms the domain and range of f. •

We use the expression R^2 to denote the xy-plane. Thus, we can say that the domain of the function $f(x, y) = x^2 + y^2$ of Example 1 is R^2. Actually, any plane in three-dimensional space is a copy of the xy-plane and is often referred to as R^2. The symbol R^3 is often used to denote three-dimensional space.

• EXAMPLE 2: Determine the domain of the function $f(x, y) = \frac{x+y}{\sqrt{4-x^2-y^2}}$.

SOLUTION: We must have $4 - x^2 - y^2 > 0$ in order for $f(x, y)$ to be real. (Why?) Thus, $x^2 + y^2 < 4$. The domain of f is the *interior* of the circle with center $(0, 0)$ and radius 2. •

In the previous section we sketched graphs of some quadric surfaces. We now say what we mean by a graph of any function of two variables.

• DEFINITION Let $z = f(x, y)$ be a function of two variables with domain D. The **graph of the function f** is the set of all points (x, y, z) in 3-space where (x, y) is in D and $z = f(x, y)$. A graph of such a function is called a **surface**. •

Three-Dimensional Graphing Utilities

We use the graph of $z = f(x, y) = x^2 + y^2$ in Figure 9.6.2 to explain how some surface or 3-D graphing utilities work. The plane $x = 2$ intersects the graph of $z = x^2 + y^2$ in the

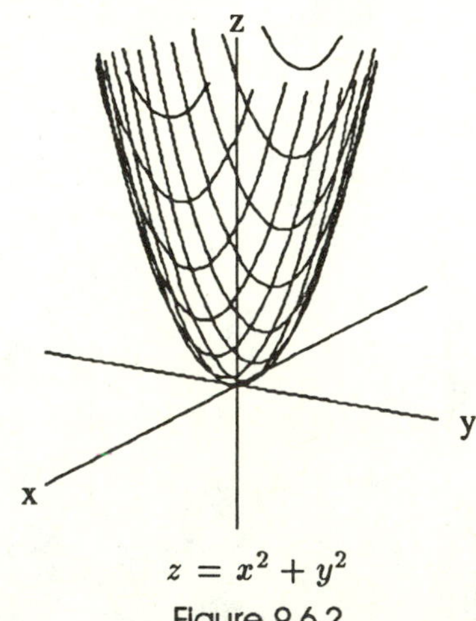

$z = x^2 + y^2$
Figure 9.6.2

parabola $z = 4 + y^2$. In the previous section we called this graph the *trace* of $z = x^2 + y^2$ in the plane $x = 2$. The trace $z = 4 + y^2$ is a function of y. Figure 9.6.3 shows the trace of $z = x^2 + y^2$ in the planes $x = 6, x = 4, x = 2$, and $x = 0$. The trace $z = a^2 + y^2$ is a function of y for each fixed value a of x.

Notice how the traces in Figure 9.6.3 suggest the complete graph given in Figure 9.6.2. The traces of the graph of $z = f(x, y)$ in the planes $x = a$ are sometimes called **x-sections** or **x-cuts** of the graph of f. For $x = a, z = f(a, y)$ is a function of y. Similarly, the traces of the graph of $z = f(x, y)$ in the planes $y = a$ are called **y-sections** or **y-cuts**. For $y = a, z = f(x, a)$ is a function of x. Notice how the y-sections $y = 6$, $y = 4$, $y = 2$, and $y = 0$ of $z = x^2 + y^2$ in Figure 9.6.4 also suggest the complete graph given in Figure 9.6.2. The y-section of $z = x^2 + y^2$ for $y = a$ is the parabola $z = x^2 + a^2$. The equation of this trace defines z as a function of x. Figure 9.6.2 is a combination of both x-cuts and y-cuts.

Along an x-section or a y-section an ordinary function grapher can be used to graph the trace. This is one of the ideas used to develop the 3-D version of *Master Grapher* used to obtain the graphs in this and the next section. Consult the *Graphing Calculator and Computer Graphing Laboratory Manual* that accompanies this textbook for more detail about 3-D graphing using *Master Grapher*.

You might be tempted to also consider **z-cuts** or **z-sections** $z = a$ when graphing $z = f(x, y)$. However, the equation $a = f(x, y)$ of this trace need not define y as a function of x or x as a function of y. In general, we would not be able to use a function grapher to graph z-cuts for a function $z = f(x, y)$. For this reason, the 3-D grapher we use in this textbook does not draw z-cuts.

The z-cuts are sometimes called the **contour curves** of a surface. Projections of the contour curves onto the xy-plane are called **level curves** and are used to construct topographical maps that describe the physical features of areas such as mountains or rivers. See Exercises 49 and 50 for more detail.

The rectangular parallelepiped determined by $A \le x \le B, C \le y \le D$, and $E \le z \le F$ is called the **viewing box $[A, B]$ by $[C, D]$ by $[E, F]$**. The 3-D grapher we use allows

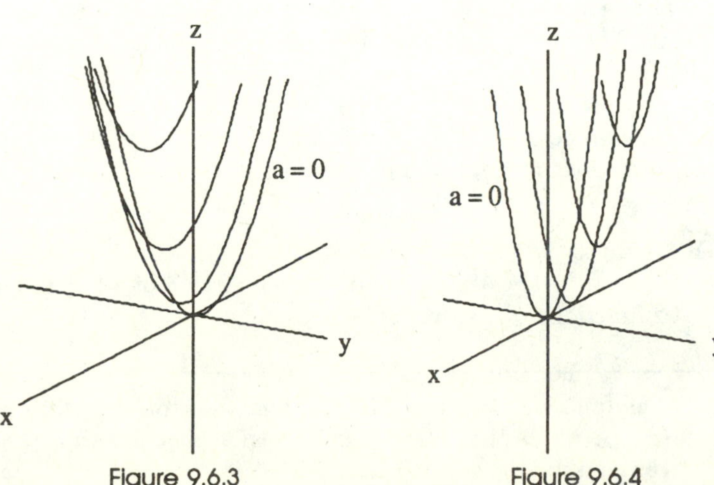

Figure 9.6.3 Figure 9.6.4

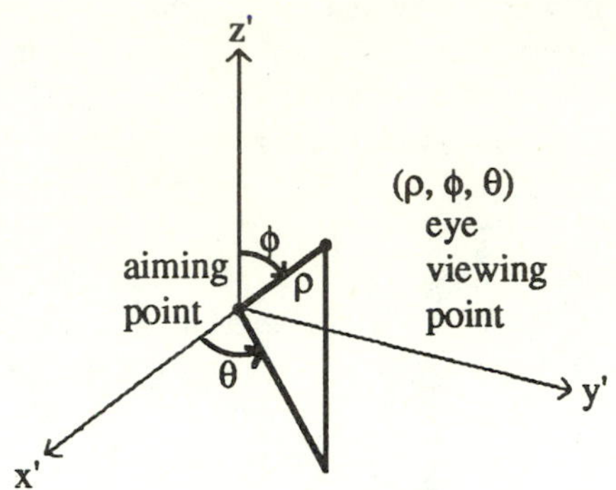

Figure 9.6.5

us to choose the viewing box, and then to choose x-sections, or y-sections, or both x- and y-sections to obtain 3-D graphs. After the traces are drawn a perspective projection is used to obtain the image that appears on a computer screen. The hidden line option can be selected. This option usually "hides" (doesn't plot) points that should be hidden from view by portions of the surface. We can specify the number of x-sections and y-sections to be used in $[A, B]$ and $[C, D]$, respectively. We can also determine the resolution of the traces used to produce the final graph. The spherical coordinates (ρ, ϕ, θ) of a viewing point at which we place our "eye," and the "aiming" point at which we direct our vision can be specified to view a graph (Figure 9.6.5). We will call the point at which we place our "eye" the **viewing point** and the point to which we direct our view the **aiming point**. The values of ρ, ϕ, and θ give the spherical coordinates of the viewing point at which we place our eye relative to the aiming point. That is, we place a three-dimensional coordinate system $x'y'z'$ at the aiming point and specify the spherical coordinates of the viewing point relative to the $x'y'z'$-coordinate system. The points of the surface are computed with respect to the xyz-coordinate system. By changing the values of ρ, ϕ, and θ we can simulate moving around in 3-space and viewing the surface from several different vantage points with respect to the aiming point. The viewing point and the aiming point can be chosen independently, but the viewing point depends on the aiming point.

The computer drawn graphs in this section will specify the viewing box, the aiming point, the viewing point, and the function used to produce the graph. In practice, we view x-sections, y-sections or both of a function until we determine a complete graph.

Maximum and Minimum Values

The notions about local extrema and absolute extrema of a function of one variable extend naturally to functions of two variables.

• DEFINITION The function $z = f(x, y)$ has a **local maximum value** or **local maximum** at (x_0, y_0) if there is a rectangular region $a \leq x \leq b, c \leq y \leq d$ in the xy-plane containing (x_0, y_0) such that $f(x, y) \leq f(x_0, y_0)$ for all (x, y) in the rectangular region that are in the domain of f. The local maxium value is $z_0 = f(x_0, y_0)$. If $f(x,y) \leq f(x_0, y_0)$ for all (x, y) in the domain of f, then f has $z_0 = f(x_0, y_0)$ as an **absolute maximum value.** •

• DEFINITION The function $z = f(x, y)$ has a **local minimum value** or **local minimum at** (x_0, y_0) if there is a rectangular region $a \leq x \leq b, c \leq y \leq d$ in the xy-plane containing (x_0, y_0) such that $f(x,y) \geq f(x_0, y_0)$ for all (x, y) in the rectangular region that are in the domain of f. The local minimum value is $z_0 = f(x_0, y_0)$. If $f(x,y) \geq f(x_0, y_0)$ for all (x, y) in the domain of f, then f has $z_0 = f(x_0, y_0)$ as an **absolute minimum** value. •

Notice that the function $z = x^2 + y^2$ of Figure 9.6.2 has a local minimum value of 0 at $(0,0)$. This local minimum is also an absolute minimum of the function. This function has neither a local nor absolute maximum.

• EXAMPLE 3: Determine the domain, the range, any local extrema, and draw a complete graph of $z = \sqrt{64 - x^2 - y^2}$.

SOLUTION: You may need to view the graph of $z = \sqrt{64 - x^2 - y^2}$ in several viewing boxes before you are sure about its complete graph. The x-section $x = a$ is the graph of the equation $z = \sqrt{64 - a^2 - y^2}$, which is the upper half of the circle with center $(a, 0, 0)$ and radius $\sqrt{64 - a^2}$. Similarly, the y-section $y = b$ is the graph of the equation $z = \sqrt{64 - x^2 - b^2}$ which is the upper half of the circle with center $(0, b, 0)$ and radius $\sqrt{64 - b^2}$. Figure 9.6.6 gives a complete graph of $z = \sqrt{64 - x^2 - y^2}$, which is the upper half of the

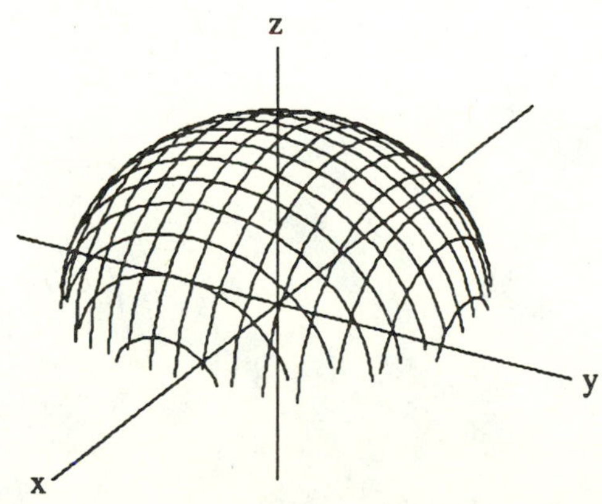

$z = \sqrt{64 - x^2 - y^2}$
$(\rho, \phi, \theta) = (100, 60°, 30°)$,
aiming point $(0,0,0)$
$[-20, 20]$ by $[-20, 20]$ by $[-20, 20]$

Figure 9.6.6

sphere with center at the origin and radius 8. Notice that the value of z is 8 at $(0,0)$ and that 8 is both a local maximum and an absolute maximum of z. Similarly, 0 is both a local minimum and an absolute minimum of z. The value 0 occurs at any point on the circle $x^2 + y^2 = 8$ in the xy-plane. The domain is the disk $x^2 + y^2 \leq 8$ which consists of all the interior points and the boundary of the circle with center at $(0,0)$ and radius 8. The range is $[0,8]$. •

The graph in Figure 9.6.6 would be enhanced if we added the trace in the xy-plane. However, the 3-D grapher we are using does not draw z-cuts.

• EXAMPLE 4: Determine the domain, the range, any local extrema, and draw a complete graph of $z = y^2 - x^2$.

SOLUTION: The domain of the function $z = y^2 - x^2$ is R^2, and the range is $(-\infty, \infty)$. (Why?) The x-sections $x = a$ are the graphs of the equations $z = y^2 - a^2$, which are parabolas with vertices $(a, 0, -a^2)$ that open upward (Figure 9.6.7). These vertices lie on the parabola $z = -x^2$ which is the trace of the function $z = y^2 - x^2$ in the xz-plane. This trace gives the illusion that the surface has a local maximum at the origin, which is not true.

The y-sections $y = b$ are the graphs of the equations $z = b^2 - x^2$, which are parabolas with vertices $(0, b, b^2)$ that open downward (Figure 9.6.8). Each of these vertices lies on the parabola $z = y^2$ which is the trace of the function $z = y^2 - x^2$ in the yz-plane. This trace gives the illusion that the surface has a local minimum at the origin, which is not true.

We say that the function $z = y^2 - x^2$ has a "saddle point" at the origin, because the graph looks like a saddle near the origin. This function has *no* local extrema or absolute extrema. A complete graph using both x- and y-sections and the hidden line option is given in Figure 9.6.9. •

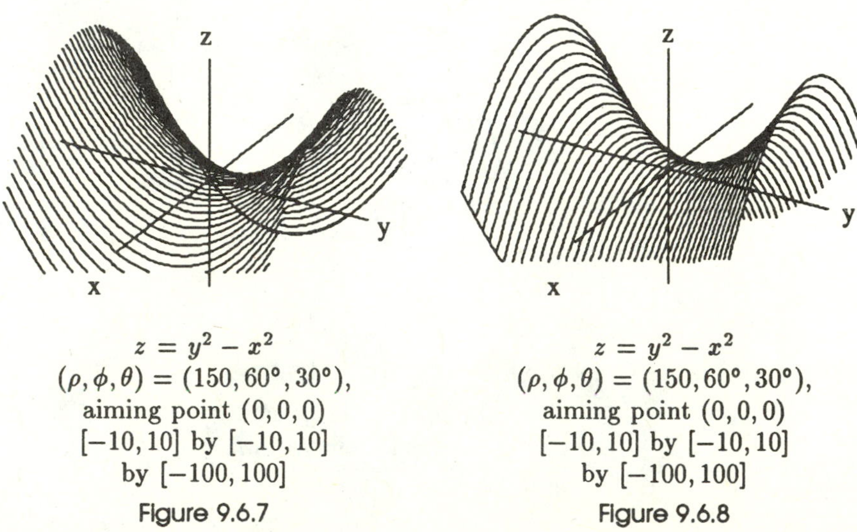

$z = y^2 - x^2$
$(\rho, \phi, \theta) = (150, 60°, 30°)$,
aiming point $(0,0,0)$
$[-10,10]$ by $[-10,10]$
by $[-100,100]$

Figure 9.6.7

$z = y^2 - x^2$
$(\rho, \phi, \theta) = (150, 60°, 30°)$,
aiming point $(0,0,0)$
$[-10,10]$ by $[-10,10]$
by $[-100,100]$

Figure 9.6.8

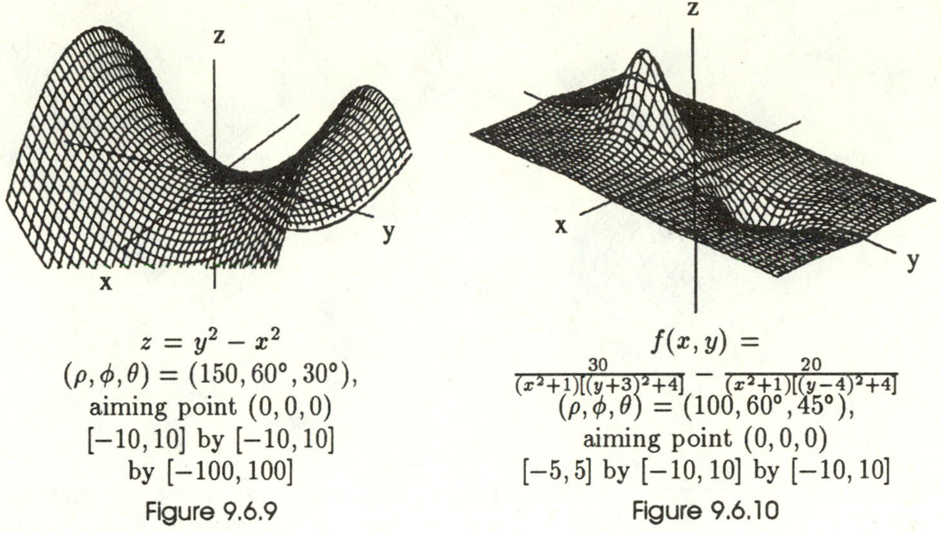

$z = y^2 - x^2$
$(\rho, \phi, \theta) = (150, 60°, 30°)$,
aiming point $(0, 0, 0)$
$[-10, 10]$ by $[-10, 10]$
by $[-100, 100]$

Figure 9.6.9

$f(x, y) =$
$\frac{30}{(x^2+1)[(y+3)^2+4]} - \frac{20}{(x^2+1)[(y-4)^2+4]}$
$(\rho, \phi, \theta) = (100, 60°, 45°)$,
aiming point $(0, 0, 0)$
$[-5, 5]$ by $[-10, 10]$ by $[-10, 10]$

Figure 9.6.10

• EXAMPLE 5: Determine the domain, the range, any local extrema, and draw a complete graph of

$$f(x, y) = \frac{30}{(x^2 + 1)[(y + 3)^2 + 4]} - \frac{20}{(x^2 + 1)[(y - 4)^2 + 4]}.$$

SOLUTION: The denominators of the fractions in the rule for the function f are different from zero for any element in R^2. Thus, the domain of f is R^2. We will determine the range of f from a complete graph. After experimenting with several viewing boxes we can show that the graph of f in Figure 9.6.10 is complete except there is one other extremum point. Actually f has a relative maximum value of about 0.0012 when x is 0 and y is 52.3 in addition to the local high point above the xy-plane, and the local low point below the xy-plane in Figure 9.6.10. Notice how the values of f get close to zero as the distance of the domain elements (x, y) from $(0, 0)$ increases. This means an end behavior model of f is the xy-plane. There are no other extrema. Thus, the range of f consists of those real numbers between and including the local minimum and maximum values of f shown in Figure 9.6.10.

The graph of f in Figure 9.6.11 gives a closer view of this local maximum of f. We can continue to zoom in and estimate that the coordinates of the high point are about $(0, -3.4, 6.9)$. Thus, f has a local maximum value of about 6.9 when $x = 0$ and $y = -3.4$.

The graph of f in Figure 9.6.12 gives a view of this local minimum of f from *under* the surface. We can continue to zoom in and estimate that the coordinates of the low point are $(0, 4.8, -3.8)$. Thus, f has a local minimum value of about -3.8 when $x = 0$ and $y = 4.8$. Notice that the local maximum of f at $(0, -3.4, 6.9)$ is actually an absolute maximum, and the local minimum of f is actually an absolute minimum. Thus, the range of f is about $[-3.8, 6.9]$. •

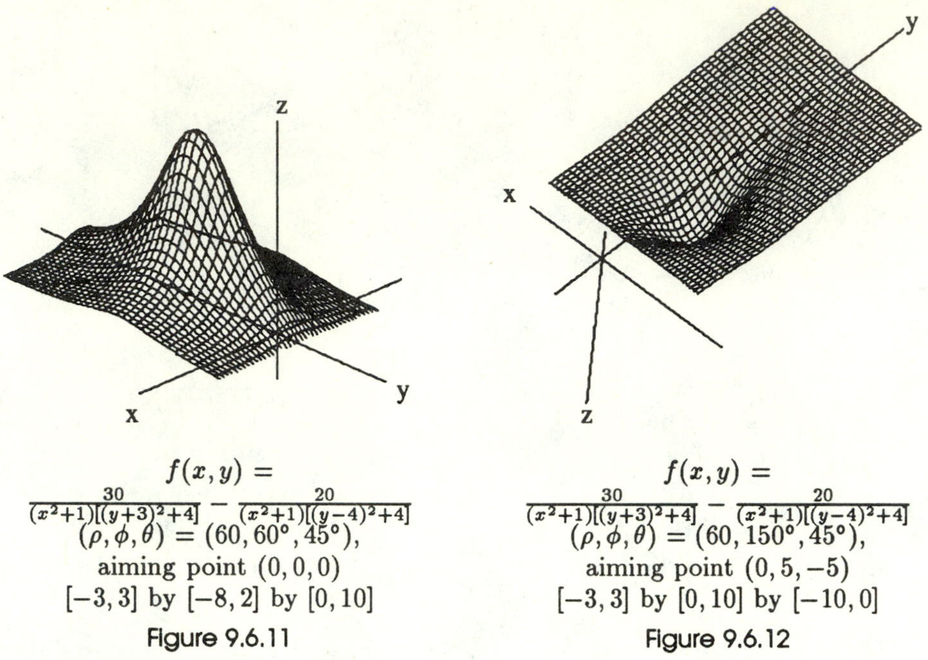

$f(x,y) = \frac{30}{(x^2+1)[(y+3)^2+4]} - \frac{20}{(x^2+1)[(y-4)^2+4]}$
$(\rho, \phi, \theta) = (60, 60°, 45°)$,
aiming point $(0,0,0)$
$[-3,3]$ by $[-8,2]$ by $[0,10]$

Figure 9.6.11

$f(x,y) = \frac{30}{(x^2+1)[(y+3)^2+4]} - \frac{20}{(x^2+1)[(y-4)^2+4]}$
$(\rho, \phi, \theta) = (60, 150°, 45°)$,
aiming point $(0,5,-5)$
$[-3,3]$ by $[0,10]$ by $[-10,0]$

Figure 9.6.12

Suppose we want to construct a box with a top that has capacity 300 cubic units. For economic reasons we would like to use the least amount of material to construct the box. This means we would like to find the minimum possible value of the surface area of the box.

Let the dimensions of the base of the box be x by y, and let h be the height of the box (Figure 9.6.13). The surface area S of the box is $2xy + 2xh + 2yh$. (Why?) We have $xyh = 300$ because the box is to have volume 300. If we solve $xyh = 300$ for h and substitute into the expression for the surface area of the box we obtain

$$\begin{aligned} S &= 2xy + 2xh + 2yh \\ &= 2xy + 2x\left(\frac{300}{xy}\right) + 2y\left(\frac{300}{xy}\right) \\ &= 2xy + \frac{600}{y} + \frac{600}{x}. \end{aligned}$$

Thus, the surface area S of such a box is a function of two variables x and y, the dimensions of the base of the box.

- EXAMPLE 6: Determine the dimensions of a box with top that has capacity 300 cubic units and has minimum possible surface area.

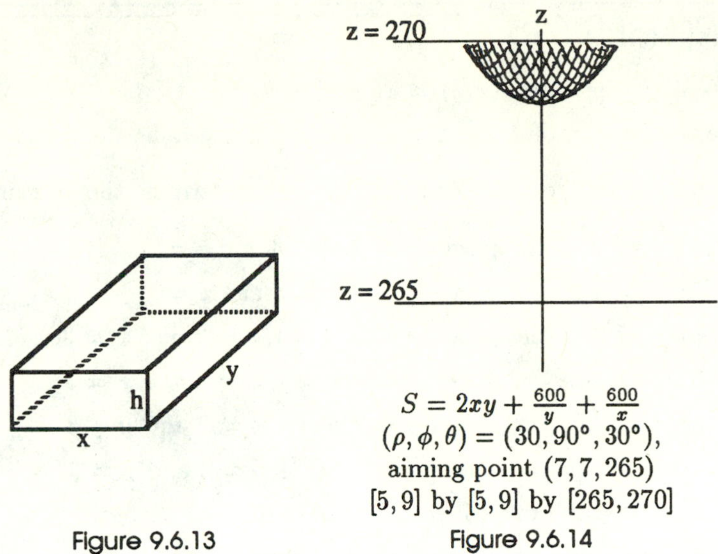

Figure 9.6.13

$S = 2xy + \frac{600}{y} + \frac{600}{x}$
$(\rho, \phi, \theta) = (30, 90°, 30°)$,
aiming point $(7, 7, 265)$
$[5, 9]$ by $[5, 9]$ by $[265, 270]$

Figure 9.6.14

SOLUTION: Let S be the surface area of the box. By the discussion preceding the example we know that

$$S = 2xy + \frac{600}{y} + \frac{600}{x},$$

where x and y are the dimensions of the base of the box. Both x and y must be positive. We need to determine the minimum value of S in the first octant. By experimenting with several viewing boxes, you will be convinced that S has a unique local minimum value in the first octant, and that this value of S is also an absolute minimum of S in the first octant.

The graph of S in Figure 9.6.14 gives a good view of this local minimum of S. Notice that our viewing "eye" is looking through the plane $z = 265$ because $\phi = 90°$ and the aiming point is $(7, 7, 265)$. We can continue to zoom in and estimate that the coordinates of this low point of S are about $(6.7, 6.7, 269)$. Because $x = y = 6.7$, it follows that $h = \frac{300}{xy} = 6.7$. This means that the dimensions of a box with a top and volume 300 cubic units that has minimum surface area is a *cube* of side length about 6.7 units. (Why?) The corresponding value of the surface area is about 269 square units.

The techniques of calculus can be used to show that the minimum value of S in the first octant is 268.88 with error at most 0.01 and the value of S occurs when x and y are exactly equal to $\sqrt[3]{300} = 6.69$.

It is very tedious to use a surface graphing utility and zoom-in to obtain accurate solutions of extrema problems. We will be content to estimate the values. Your estimates may differ from ours by a sizable margin. •

• EXERCISES 9-6

Let $f(x, y) = 9 - x^2 - y^2$ and $g(x, y) = x^2 + y^2 - 9$.

1. Determine the domain of f and of g.
2. Determine the domain of $z = \sqrt{f(x, y)}$.
3. Determine the domain of $z = \sqrt{g(x, y)}$.
4. Determine the domain of $z = \dfrac{1}{f(x, y)}$.
5. Draw the trace of f in the planes $z = 0, z = 3$ and $z = 9$.
6. Draw the trace of g in the planes $z = 0, z = 3$ and $z = 9$.
7. Draw the trace of f in the planes $x = 0, x = 1, x = 2$ and $x = 3$.
8. Draw the trace of g in the planes $x = 0, x = 1, x = 2$ and $x = 3$.
9. Draw the trace of f in the planes $y = 0, y = 1, y = 2$ and $y = 3$.
10. Draw the trace of g in the planes $y = 0, y = 1, y = 2$ and $y = 3$.
11. Draw a complete graph of f. Identify all extrema.
12. Draw a complete graph of g. Identify all extrema.

Let $f(x, y) = 9 - (x - 2)^2 - (y - 3)^2$ and $g(x, y) = \sqrt{9 - (x - 2)^2 - (y - 3)^2}$.

13. Determine the domain of f.
14. Determine the domain of g. *Hint*: When will $(x - 2)^2 + (y - 3)^2 \leq 9$?
15. Draw a complete graph of f and identify all extrema.
16. Draw a complete graph of g and identify all extrema.

Determine the domain and draw a complete graph of the function. Estimate the coordinates (a, b, c) of any extrema. Identify any *saddle* points.

17. $z = 25 - x^2 - y^2$
18. $z = \sqrt{25 - x^2 - y^2}$
19. $z = \sqrt{25 - x^2 + 4x - y^2 + 9y + 13}$
20. $z = 5\sin(\frac{1}{2}x)$
21. $z = \dfrac{1}{\sqrt{9 - x^2 - y^2}}$
22. $z = \dfrac{1}{9 - x^2 - y^2}$
23. $z = y^2 - 8y - x^2 + 16$
24. $z = \ln(\cos x) - \ln(\cos y)$
25. $z = 24y^2 + 1.6y^3 - y^4 - 32x^2$
26. $z = 20xe^{-x^2-y^2}$
27. $z = 0.7y^2 + 0.05y^3 - 0.03y^4 - x^2 + 5$
28. $z = 1 + \dfrac{1}{x} + \dfrac{1}{y}$

Consider the function $z = 4xy + \frac{300}{x} + \frac{100}{y}$.

29. Let $x = 2, 4, 6$, and 8. For each x, compute z for $y = 1, 3, 7$, and 10.
30. What happens to the values of z as $x \to \infty$ and $y \to \infty$?
31. Graph z in the region $0 \leq x \leq 10$, $0 \leq y \leq 10$, and $100 \leq z \leq 200$. Use $(0, 0, 150)$ as your aiming point.

32. Estimate the local minimum value and where it occurs.

The front and back of a box is constructed from material that costs \$1 per square inch. The material for the top and bottom of the box costs \$2 per square inch and the material for the two sides costs \$3 per square inch. Let the box have length x, width y, and height h.

33. Determine an algebraic representation in terms of x and y which gives the total cost C of the materials used in construction of a box with volume 50 cubic inches .

34. What values of x and y make sense in the problem situation of Exercise 33?

35. Draw a complete graph of $z = C(x, y)$ in the first octant.

36. Estimate the minimum cost of a box with volume 50 cubic inches and the dimensions of such a box of minimum cost.

A box with a top is constructed that must have volume 500 cubic inches. Let the box have length x, width y, and height h.

37. Determine an algebraic representation of the total surface area S of the box in terms of x and y.

38. What values of x and y make sense in the problem situation?

39. Draw a complete graph of $z = S(x, y)$ in the first octant.

40. What are the dimensions of the box with minimum surface area? What is the minimum surface area?

A box with *no* top is constructed that must have volume 500 cubic inches. Let the box have length x, width y, and height h.

41. Determine an algebraic representation of the total surface area S of the box in terms of x and y.

42. What values of x and y make sense in the problem situation?

43. Draw a complete graph of $z = S(x, y)$ in the first octant.

44. What are the dimensions of the box with minimum surface area? What is the minimum surface area?

Determine an end behavior model for the function $z = f(x, y)$. Use a graphical argument.

45. $f(x, y) = xe^{-|x+y|}$

46. $f(x, y) = \dfrac{1}{2 + x^2 + y^2}$

47. $f(x, y) = 2x \sin \dfrac{1}{x}$

48. $f(x, y) = 3 + \dfrac{2}{x} + \dfrac{4}{y}$

Level curves on a topographical map are used to identify terrain at the same elevation. The intersection of a plane $z = h$ with the surface $z = f(x, y)$ is a *contour curve* of height h. That is, the trace of the surface z in the plane $z = h$ is a contour curve. The *projections* of contour curves of a surface onto the xy-plane are called *level curves*.

49. Draw contour curves of the surface $z = 32 - x^2 - y^2$ of height $h = 28, h = 23, h = 16$, $h = 7$, and $h = 0$.

50. Draw the level curves of the surface $z = 32 - x^2 - y^2$ for the heights in Exercise 49. Determine an equation of each level curve.

9.7 Complex Zeros Graphically, The Modulus Surface

In this section we introduce the modulus surface of a function f of a single variable and use the surface to determine the zeros of f graphically. We have seen that the real zeros of a function f with independent variable x correspond to the x-intercepts of the graph of f. Prior to this section we have not determined graphically the nonreal complex zeros of a function of one variable.

Consider the polynomial function $f(x) = x^2 + 1$. The zeros of f are the complex numbers i and $-i$. The graph in Figure 9.7.1 gives no clue about the nonreal complex zeros of $f(x) = x^2 + 1$. Of course, we can determine from this graph that f has no real zeros. Any function of one variable can be used to define a surface in 3-space that provides an important and useful *geometric representation* of the zeros of the function.

Recall that the *absolute value* or *modulus* $|u|$ of the complex number $u = a + bi$ is defined by $|u| = \sqrt{a^2 + b^2}$ and is a *real* number.

• DEFINITION Let f be any function of one variable. The **modulus surface** of f is the surface defined by $z = g(x, y) = |f(u)|$, where u is the complex number $x + yi$. The function g is called the **modulus function** of f. •

The value of z in $z = |f(u)|$ is a nonnegative *real number* for every complex number $x + yi$. If we interpret the xy-plane as the complex plane, then we can think of $z = |f(u)|$ as a function of the two variables x and y where $u = x + yi$ (Figure 9.7.2). The point $(x, y, 0)$ corresponds to the complex number $x + yi$. The horizontal axis of the xy-plane is the real axis and the vertical axis of the xy-plane is the imaginary axis.

The trace of $z = |f(u)|$ in the xz-plane ($y = 0$) is the graph of $z = |f(x + (0)i)| = |f(x)|$. Let $f(x) = x^3 - 4x$ (Figure 9.7.3). The trace of $z = |f(u)|$ in the xz-plane is the graph of

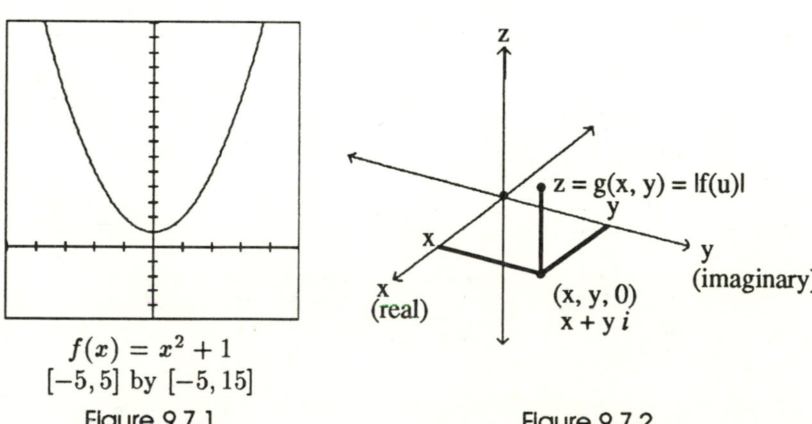

$f(x) = x^2 + 1$
$[-5, 5]$ by $[-5, 15]$
Figure 9.7.1

Figure 9.7.2

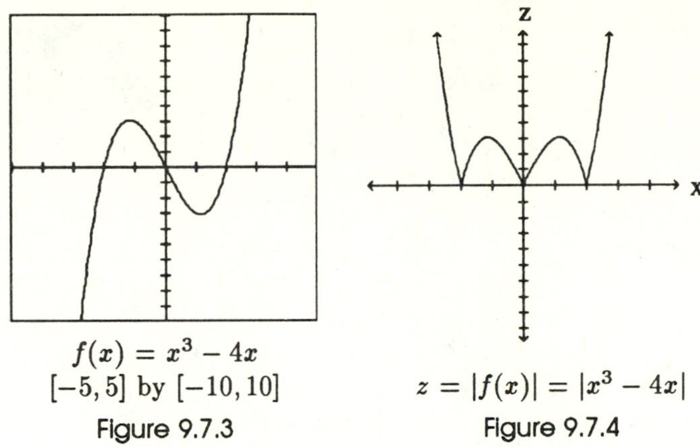

$f(x) = x^3 - 4x$
$[-5, 5]$ by $[-10, 10]$
Figure 9.7.3

$z = |f(x)| = |x^3 - 4x|$
Figure 9.7.4

$z = |x^3 - 4x|$ (Figure 9.7.4). The real zeros of f show up as the x-intercepts of the trace of the modulus function $z = |f(u)|$ in the xz-plane. These are points on the x-axis where the modulus function $z = |f(u)|$ of $f(x) = x^3 - 4x$ touches the xy-plane. This observation is a special case of the more general result.

• THEOREM 1 Let $u = x + yi$ and let f be a function with domain and range that are subsets of the complex numbers. The complex number $a + bi$ is a zero of f if and only if the modulus surface $z = g(x, y) = |f(u)|$ touches the complex plane at $a + bi$. That is, $a + bi$ is a zero of f if and only if $(a, b, 0)$ is a point on the graph of $z = g(x, y) = |f(u)|$.

PROOF Assume $a + bi$ is a zero of f. Then $f(a + bi) = 0$. It follows that $z = g(a, b) = |f(a + bi)| = 0$. Thus, the modulus surface $z = g(x, y) = |f(u)|$ touches the xy-plane at $(a, b, 0)$, or at $a + bi$ when the xy-plane is considered the complex plane. Notice that $(a, b, 0)$ is a point on the graph of $z = g(x, y) = |f(u)|$ in this case. Moreover, the modulus surface never goes through the xy-plane (complex plane) because $z = |f(u)| \geq 0$ for all x and y.

Now, assume the modulus surface $z = g(x, y) = |f(u)|$ touches the complex plane at $a + bi$. That is, assume that $(a, b, 0)$ is a point of the graph of $z = |f(u)|$. This means that $z = 0$ at $u = a + bi$. Thus, $|f(a + bi)| = 0$. It follows that $f(a + bi) = 0$ because the modulus of a complex number is zero if and only if the complex number is zero.

• EXAMPLE 1: Determine a geometric representation of the zeros of $f(x) = x^2 + 1$. That is, draw a graph that shows all points where the modulus surface of f touches the xy-plane.

SOLUTION: The modulus surface of f is the graph of $z = g(x, y) = |f(u)|$ where $u = x + yi$:

$$\begin{aligned} z &= |f(u)| \\ &= |f(x+yi)| \\ &= |(x+yi)^2 + 1| \\ &= |x^2 + 2xyi + y^2i^2 + 1| \\ &= |x^2 - y^2 + 1 + 2xyi| \\ &= \sqrt{(x^2-y^2+1)^2 + (2xy)^2}. \quad \text{(Why?)} \end{aligned}$$

A graph of the modulus surface of f is given in Figure 9.7.5. The surface appears to touch the xy-plane at $(0, 1, 0)$ and $(0, -1, 0)$. We have added the coordinate axes and the lines $y = 1$ and $y = -1$ to the graph in Figure 9.7.5 to help locate the coordinates of these points. *Master Grapher* will overlay lines and the coordinate axes on a graph. We know from Chapter 4 that $f(x) = x^2 + 1$ has exactly two zeros, namely $\pm i$. Thus, Figure 9.7.5 provides a **geometric representation** of the zeros of $f(x) = x^2 + 1$! •

The graphs of $x = a$ and $y = b$ in 3-space are planes. However, in this section, when we refer to the lines $x = a$ and $y = b$ we mean the lines $x = a$ and $y = b$ in the xy-plane.

In the Exercises, we will ask you to give a geometric representation of the zeros of $f(x) = x^2 - 1$. You will find that the modulus surface of this function is the surface of Example 1 rotated 90° about the z-axis. The two minimum points where the modulus surface of $f(x) = x^2 - 1$ touches the xy-plane are on the x-axis. (Why?) This means the zeros of f are real.

• EXAMPLE 2: Determine a geometric representation of the zeros of $f(x) = x^2 + x + 1$.

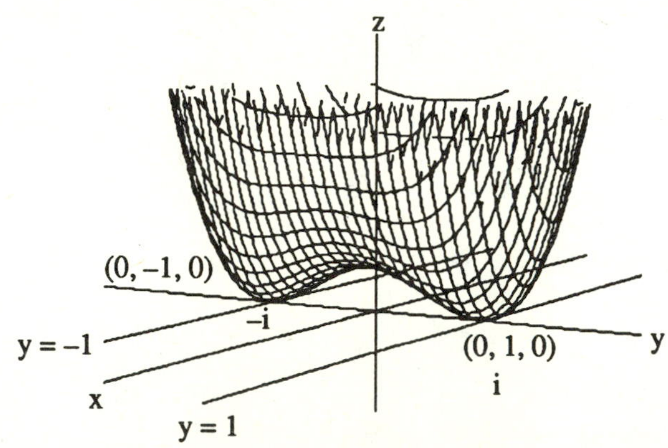

$z = \sqrt{(x^2 - y^2 + 1)^2 + 4x^2y^2}$
$(\rho, \phi, \theta) = (20, 80°, 30°)$, aiming point $(0, 0, 0)$
$[-2, 2]$ by $[-2, 2]$ by $[-5, 5]$
Figure 9.7.5

SOLUTION: We can use the quadratic formula to show that the two zeros of f are $\frac{-1\pm\sqrt{3}i}{2}$ or $-0.5 \pm 0.87i$. First, we determine the modulus function $z = g(x, y) = |f(u)|$ of f:

$$\begin{aligned} z &= |f(u)| \\ &= |f(x+yi)| \\ &= |(x+yi)^2 + (x+yi) + 1| \\ &= |(x^2 - y^2 + x + 1) + (2xy + y)i| \\ &= \sqrt{(x^2 - y^2 + x + 1)^2 + (2xy + y)^2}. \end{aligned}$$

A geometric representation of the two complex zeros of f is given by the graph of $z = \sqrt{(x^2 - y^2 + x + 1)^2 + (2xy + y)^2}$, the modulus surface of f (Figure 9.7.6). We have added the coordinate axes and the lines $x = -0.5$, $y = 0.87$ and $y = -0.87$ to help locate the zeros from the graph. This graph confirms the zeros we determined algebraically using the quadratic formula. •

• EXAMPLE 3: Determine a geometric representation of the zeros of $f(x) = x^3 - x^2 + x - 1$. Use the graph to estimate the zeros of f.

SOLUTION: Any polynomial of odd degree always has at least one real zero. (Why?) It turns out that f has two nonreal complex zeros. First, we determine the modulus function of f:

$$\begin{aligned} z &= |f(x+yi)| \\ &= |(x+yi)^3 - (x+yi)^2 + (x+yi) - 1| \\ &= |(x^3 + 3x^2yi - 3xy^2 - y^3i) - (x^2 + 2xyi - y^2) + x + yi - 1| \\ &= |(x^3 - 3xy^2 - x^2 + y^2 + x - 1) + (3x^2y - y^3 - 2xy + y)i| \\ &= \sqrt{(x^3 - 3xy^2 - x^2 + y^2 + x - 1)^2 + (3x^2y - y^3 - 2xy + y)^2}. \end{aligned}$$

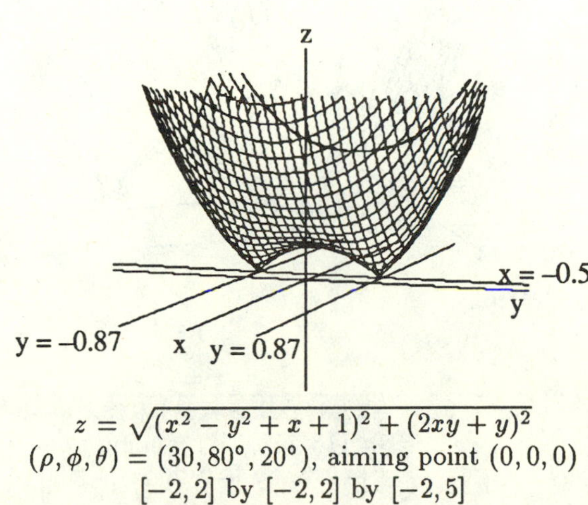

$z = \sqrt{(x^2 - y^2 + x + 1)^2 + (2xy + y)^2}$
$(\rho, \phi, \theta) = (30, 80°, 20°)$, aiming point $(0, 0, 0)$
$[-2, 2]$ by $[-2, 2]$ by $[-2, 5]$

Figure 9.7.6

We used the binomial theorem to expand $(x + yi)^3$ in the above computation. A geometric representation of the zeros of f is given by the graph of

$$z = \sqrt{(x^3 - 3xy^2 - x^2 + y^2 + x - 1)^2 + (3x^2y - y^3 - 2xy + y)^2},$$

the modulus surface of f (Figure 9.7.7). It appears that the modulus surface touches the xy-plane at two points on the y-axis and one point on the x-axis. By using zoom-in we can estimate these points to be $(1,0,0)$, $(0,1,0)$, and $(0,-1,0)$. We have added the coordinate axes and the lines $x = 1, y = 1$, and $y = -1$ to help establish these estimates. These points correspond to the complex numbers $1, i$, and $-i$ which are good estimates for the zeros of $f(x) = x^3 - x^2 + x - 1$. In this case, we can show that $1, i$, and $-i$ are *exact* zeros of f. (How?) •

The function $f(x) = 4x^4 + 17x^2 + 14x + 65$ of the next example has no real zeros. In the Exercises we ask you to confirm this fact.

• EXAMPLE 4: Determine a geometric representation of the zeros of $f(x) = 4x^4 + 17x^2 + 14x + 65$. Use the graph to estimate the zeros of f.

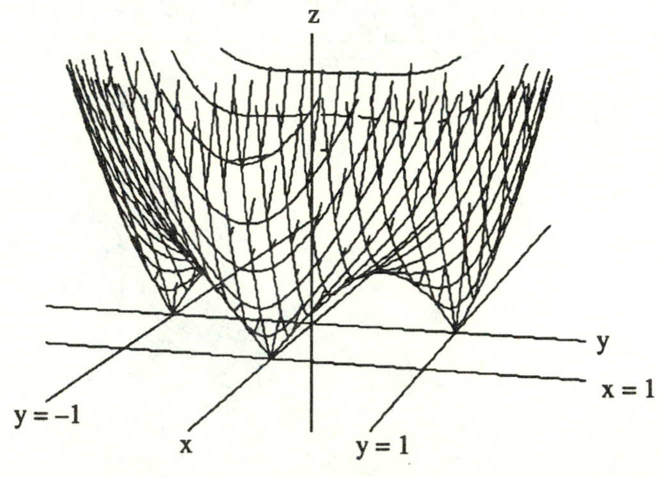

$z = \sqrt{(x^3 - 3xy^2 - x^2 + y^2 + x - 1)^2 + (3x^2y - y^3 - 2xy + y)^2}$
$(\rho, \phi, \theta) = (15, 75°, 15°)$, aiming point $(0,0,0)$
$[-2,2]$ by $[-2,2]$ by $[-3,3]$

Figure 9.7.7

SOLUTION: We need to expand $(x+yi)^4$ to obtain the modulus function of f. The binomial theorem gives

$$\begin{aligned}(x+yi)^4 &= x^4 + \binom{4}{1} x^3(yi) + \binom{4}{2} x^2(yi)^2 + \binom{4}{3} x(yi)^3 + (yi)^4 \\ &= x^4 + 4x^3yi + 6x^2y^2i^2 + 4xy^3i^3 + y^4i^4 \\ &= x^4 + 4x^3yi - 6x^2y^2 - 4xy^3i + y^4.\end{aligned}$$

Next, we determine the modulus function $z = |f(x+yi)|$:

$$\begin{aligned}z &= |f(x+yi)| \\ &= |4(x+yi)^4 + 17(x+yi)^2 + 14(x+yi) + 65| \\ &= |4(x^4 + 4x^3yi - 6x^2y^2 - 4xy^3i + y^4) + 17(x^2 + 2xyi - y^2) + 14x + 14yi + 65| \\ &= |(4x^4 - 24x^2y^2 + 4y^4 + 17x^2 - 17y^2 + 14x + 65) + (16x^3y - 16xy^3 + 34xy + 14y)i| \\ &= \sqrt{(4x^4 - 24x^2y^2 + 4y^2 + 17x^2 - 17y^2 + 14x + 65)^2 + (16x^3y - 16xy^3 + 34xy + 14y)^2}\end{aligned}$$

The modulus surface of f is given in Figure 9.7.8. The coordinate axes and the lines $x = \pm 1, y = \pm 1.5$, and $y = \pm 2$ have been added to this figure. This figure suggests that $1 \pm 2i$ and $-1 \pm 1.5i$ are the four zeros of f. In general, considerable experimentation is needed to determine the coordinates of the points where a modulus surface touches the xy-plane without prior knowledge of the zeros. Overlaying lines of the form $x = a$ and $y = b$ helps this determination. In Example 4 of Section 3.4 we found that $1 \pm 2i$ and $-1 \pm 1.5i$ were the *exact* zeros of f. •

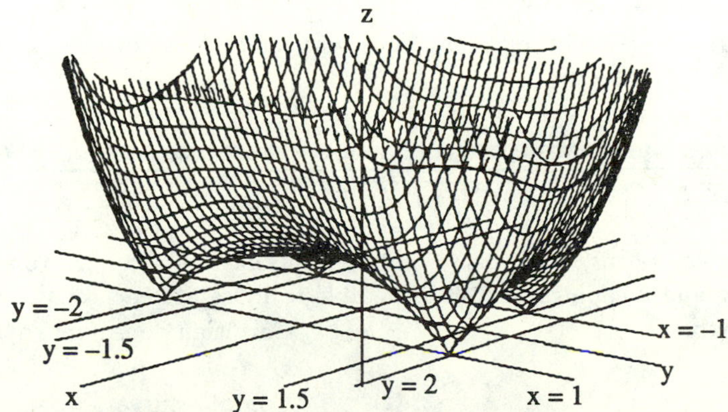

$$z = \sqrt{(4x^4 - 24x^2y^2 + 4y^2 + 17x^2 - 17y^2 + 14x + 65)^2 + (16x^3y - 16xy^3 + 34xy + 14y)^2}$$
$(\rho, \phi, \theta) = (25, 75°, 40°)$, aiming point $(0, 0, 0)$
$[-3, 3]$ by $[-3, 3]$ by $[-300, 300]$

Figure 9.7.8

The algebraic manipulations in this section are a little tedious. However, a graphical method to determine the real and nonreal complex zeros of a function is powerful because there are no formulas for the exact zeros of polynomials of degree 5 or higher.

• EXERCISES 9-7

Write the complex number in the form $a + bi$.

1. $(1 - 2i)^3$
2. $(1 + i)^5$
3. $-2(x + yi)^3$
4. $3(x + yi)^5$

Determine the modulus function of f.

5. $f(x) = x^2 - 1$
6. $f(x) = x^2 + 4$
7. $f(x) = x^2 - 2x + 5$
8. $f(x) = x^3 - 4x$
9. $f(x) = x^3 - 8$
10. $f(x) = x^4 - 1$

11–16. Determine the zeros of each function f in Exercise 5–10. Identify them as real or nonreal complex.

17–22. Draw a complete graph of the modulus surface determined in Exercises 5–10. Compare the points where the modulus surface touches the xy or complex plane with the zeros determined in Exercises 11–16.

23. Confirm that the function of Example 4 has no real zeros.
24. Use a geometric representation to estimate all zeros of $f(x) = x^4 + x^3 + x^2 + x + 1$.
25. Use a geometric representation to estimate all zeros of $f(x) = x^4 + 2x^3 + 3x^2 + 2x + 2$.
26. Use a geometric representation to estimate all zeros of $f(x) = x^5 - 1$.
27. Confirm that the modulus surface of $f(x) = x^4 + 2x^2 + 1$ has only two extrema and that f has four zeros. Why doesn't this modulus surface have four extrema?
28. Explain how a level curve plot of the modulus function of f can be used to accurately estimate all the zeros of f.

• KEY TERMS

Listed below are the key terms, vocabulary, and concepts in this chapter. You should be familiar with their definitions and meanings as well as be able to illustrate each of them.

Absolute maximum
Absolute minimum
Aiming point
Arithmetic mean
Arithmetic progression
Arithmetic sequence
Binomial series
Binomial theorem
Cartesian coordinate system
Cartesian equation of line
Contour curve
Cylindrical coordinates
Element of a matrix
Elementary row operations
Ellipsoid
Elliptic cylinder

Elliptic paraboloid
Equation of line in parametric form
Experimental probability
Fibonacci sequence
Finite sequence
Finite series
Function of two variables
Generalized binomial expansion
Geometric mean
Geometric progression
Geometric sequence
Geometric series
Harmonic series
Hyperbolic paraboloid
Infinite sequence
Infinite series
Interval of convergence
Level curve
Linear equation
Local maximum value
Local minimum value
Mathematical induction
Mathematical probability
$m \times n$ matrix
Matrix
Matrix of the system
Modulus
Modulus function
Modulus surface
nth term of a sequence partial sum
Outcome
Plane
Polynomial approximation
Quadric surface
Recursive formula
Reduced row echelon form
Right-handed system
Row and column of a matrix
Row echelon form
Sequence
Series
Spherical coordinates
Square matrix
Sum of an infinite series
Surface
Surface area
3-space
x-cuts
y-cuts
z-cuts
Zeno's paradox
Zeros of a function

• REVIEW EXERCISES

Determine the first four terms and the tenth term of the sequence. Draw a graph of the sequence.

1. $a_n = (-1)^{n+1}(n-1)$
2. $a_n = 2n^2 - 1$
3. The fourth and seventh terms of an arithmetic progression are -8 and 16, respectively. Determine the first term and a formula for the nth term.
4. The fifth and ninth terms of an arithmetic progression are -5 and 13, respectively. Determine the first term and a formula for the nth term.

Use summation notation to write the sum. Assume that the suggested patterns continue.

5. $-2 - 5 - 8 - 11 - \cdots - 29$
6. $4 + 16 + 64 + 256 + \cdots$

Use summation notation to write the nth partial sum of the sequence. Assume that the suggested patterns continue.

7. $8, 6, 4, \ldots$

8. $-3, 6, -9, 12, \ldots$

Determine a real number sum, if it exists, of the infinite series.

9. $\frac{1}{16} + \frac{1}{32} + \frac{1}{64} + \cdots$

10. $-1 - 3 - 5 - \cdots$

11. Determine an end behavior model for $f(x) = 5\frac{1-(2/3)^x}{1-2/3}$. Explain how the answer relates to the infinite series $5 + 10/3 + 20/9 + 40/27 + \cdots$.

Let

$$f_1(x) = 1 - x$$

$$f_2(x) = 1 - x + \frac{x^2}{2!},$$

$$f_3(x) = 1 - x + \frac{x^2}{2!} - \frac{x^3}{3!}$$

$$\vdots$$

$$f_n(x) = 1 - x + \frac{x^2}{2!} - \frac{x^3}{3!} + \cdots + (-1)^n \frac{x^n}{n!}$$

12. Draw complete graphs of the six polynomial functions $f_1, f_2, \ldots, f_6$ in the same viewing rectangle.

13. Draw a complete graph of $f(x) = e^{-x}$ and compare with the graphs in Exercise 12. Draw complete graphs of $f(x) = e^{-x}$ and f_6 in the same viewing rectangle.
   (a) Is $f_6(-2)$ a good estimate for e^2? Why?
   (b) Is $f_6(4)$ a good estimate for e^{-4}? Why?
   (c) On what interval is $f_6(x)$ a good approximation for e^{-x}?

14. Use a graphing argument to determine the error in using $f_3(-2)$ to approximate e^2. Repeat using $f_4(-2), f_5(-2)$, and $f_6(-2)$. Are they underestimates or overestimates?

Write a matrix model for each system of equations. Solve the system by reducing the matrix of the system to row echelon form or reduced row echelon form.

15. $x + y = 3$
$x - y = 1$
$2x + y = 5$

16. $x - y + z = 6$
$x + y + 2z = -4$

17. $x - y + z = 0$
$2x - 3z = 4$
$-x - y + 2z = -4$

18. $4x - 2y = 0$
$x + 3y - z = -3$
$6y + 2z = 16$

19. $x + y - 2z = -4$
$3x - y + z = -7$
$-2x - 2y + 4z = 8$

20. $2x - 4y = 10$
$x - z = -1$
$2y - x = -5$

Plot the points in 3-space.

21. $(3, 1, 5)$

22. $(-2, 3, -5)$

Name the octant containing the point in

23. Exercise 21.

24. Exercise 22.

Sketch a graph of the plane in the first octant.

25. $2x + y + 3z = 12$

26. $x + z = 5$

27. Compute the distance between the points $(1, 2, -5)$ and $(-3, 4, 6)$.

28. Determine the midpoint of the line segment between the points $(-1, 2, 5)$ and $(5, -2, 7)$.

29. Write an equation for the line through points $(2, 1, 5)$ and $(6, 4, 8)$ in
    (a) Cartesian form.
    (b) parametric form.

30. Write an equation for the line through points $(3, -4, 6)$ and $(2, 8, -2)$ in
    (a) Cartesian form.
    (b) parametric form.

Describe and sketch the quadric surface.

31. $9x^2 + y = 9$

32. $4x^2 + 9z^2 = 36$

33. $16x^2 + 64y^2 + z^2 = 64$

34. $x^2 + y^2 + z^2 = 81$

Let $f(x, y) = 16 - x^2 - y^2$ and $g(x, y) = x^2 + y^2 - 16$.

35. Determine the domain of f and of g.

36. Determine the domain of $z = \sqrt{f(x, y)}$.

37. Determine the domain of $z = \sqrt{g(x, y)}$.

38. Determine the domain of $z = \dfrac{1}{f(x, y)}$.

39. Draw the trace of f in the planes $x = 0, x = 1, x = 2, x = 3$ and $x = 4$.

40. Draw the trace of g in the planes $y = 0, y = 1, y = 2, y = 3$ and $y = 4$.

Determine the domain and draw a complete graph of the function. Estimate the coordinates (a, b, c) of any extrema. Identify any *saddle* points.

41. $z = 9 - x^2 - y^2$

42. $z = \sqrt{9 - x^2 - y^2}$

43. $z = \sqrt{5 - x^2 + y^2 + 2x - 6y}$

44. $z = 0.25(x - 3)y$

Determine the modulus function of f.

45. $f(x) = x^2 - 4$

46. $f(x) = x^2 + 5$

47. $f(x) = x^2 - 2x + 6$

48. $f(x) = x^3 - 4x - 3$

49–52. Determine the zeros of the functions f in Exercises 45–48. Identify them as real or nonreal complex.

53–56. Draw a complete graph of the modulus surfaces determined by the modulus functions in Exercises 45–48. Compare the points where the modulus surface touches the xy or complex plane with the zeros determined in Exercises 49–52.

The half-life of a certain unstable radioactive substance is one week. Suppose 2000 grams of the substance exists today. Let n represent the number of weeks the substance exists.

57. Determine an infinite geometric sequence that is a model of amount of the substance at week n; $n = 1, 2, 3, \ldots$. List the first 10 terms of the sequence. What is the common ratio?

58. When will there be only one gram of the substance remaining?

59. Janice had \$1550 in a savings account three years ago. What will the value of her account be four years from now assuming no deposits or withdrawals are made and the account earns 8% interest compounded annually?

60. Determine the first five terms, the common ratio, and a formula for the nth partial sum of a finite geometric series representing a car loan of \$280 per month. Assume an APR interest rate of 12% compounded monthly. Let A_n represent the present value of a car loan requiring n monthly payments. Determine the car loan amount if 60 payments are made.

61. At a zoo in Columbus, Ohio, children ride a train for 50 cents but adults must pay \$1.50. On a given day, 95 passengers paid a total of \$75.50 for the rides. How many passengers were children? Adults?

62. One silver alloy is 50% silver and another silver alloy is 30% silver. How many grams of each are required to produce 100 grams of a new alloy that is 40% silver?

The front and back of a box is constructed from material that costs \$1 per square inch. The material for the top and bottom of the box costs \$3 per square inch and the material for the two sides costs \$5 per square inch. Let the box have length x, width y, and height h.

63. Determine an algebraic representation in terms of x and y which gives the total cost C of the materials used in construction of a box with volume 400 cubic inches.

64. What values of x and y make sense in the problem situation of Exercise 63?

65. Draw a complete graph of $z = C(x, y)$ in the first octant.

66. Estimate the minimum cost of a box with volume 400 cubic inches and the dimensions of such a box of minimum cost.

A box with a top is constructed that must have volume 750 cubic inches. Let the box have length x, width y, and height h.

67. Determine an algebraic representation of the total surface area S of the box in terms of x and y.

68. What values of x and y make sense in the problem situation?

69. Draw a complete graph of $z = S(x, y)$ in the first octant.

70. What are the dimensions of the box with minimum surface area? What is the minimum surface area?

Answers

• CHAPTER 1

Exercises 1.1

1. A: $(1,3)$

   B: $(4,2)$

   C: $(1,-2)$

   D: $(-4,-3)$

   E: $(-2,-1)$

   F: $(-4,0)$

   G: $(-3,1)$

3.

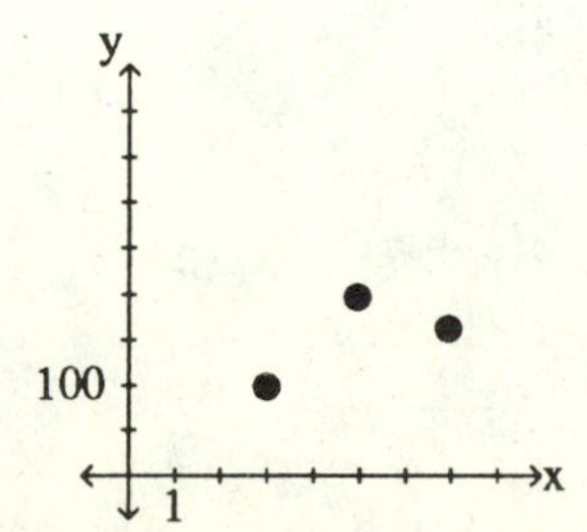

5. (a) Triangle

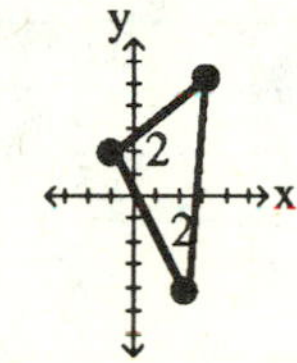

   (b) Trapezoid

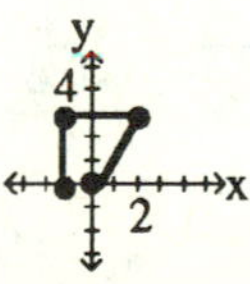

   (c) Parallelogram

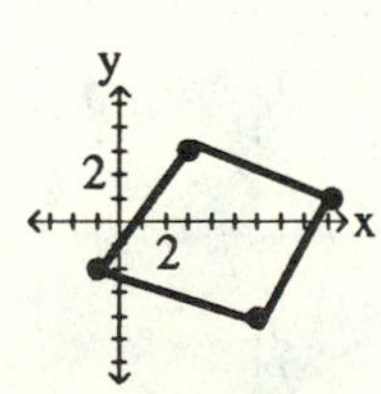

   (d) Pentagon

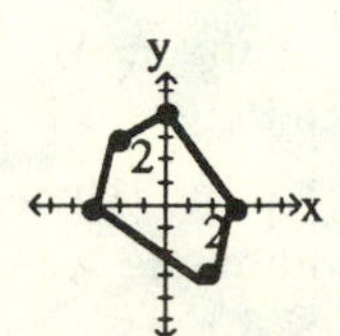

7. $T = -13$

9. Answers will vary

11. (a)

Height (ft)	Weight (lbs)
2	140
1	100
4	160
6	200

   (b) Domain: All nonnegative real numbers less than or equal to 6.

   Range: All nonnegative real numbers less than or equal to 200.

13.

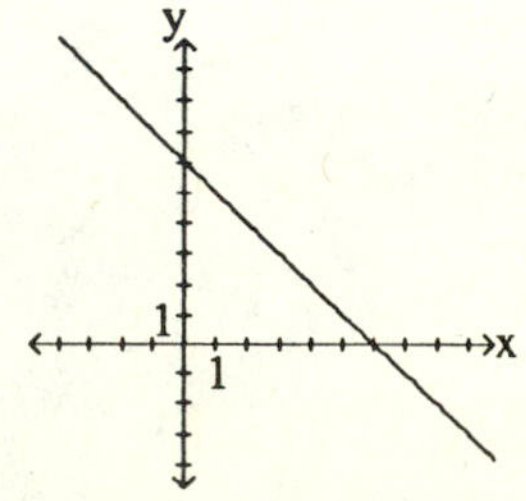

15.

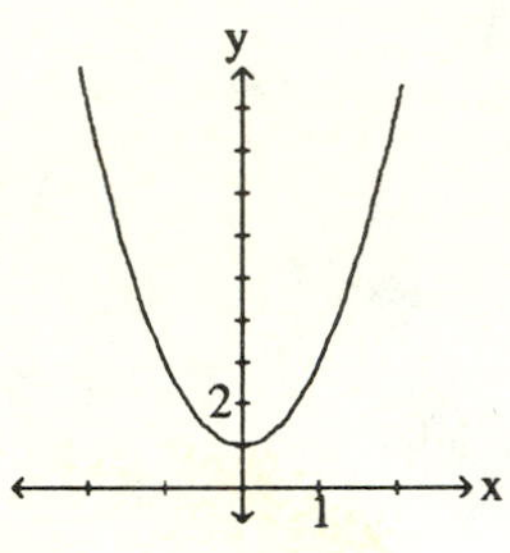

17.

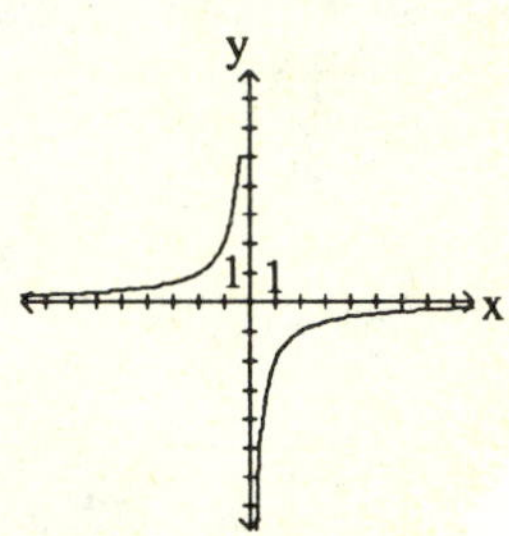

19. (a)

Input No.	Output No.
	(square root)
-3	-
-2	-
0	0
1	1
2	1.41
3	1.73
4	2
5	2.24
6	2.45

(b) and (c)

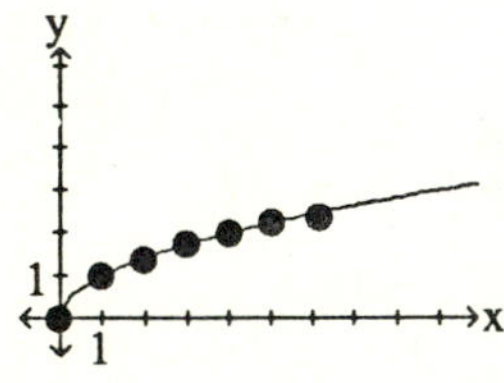

(d) Domain and Range: All nonnegative real numbers

21. (a)

Input	Output (sin)
0	0
0.52	0.50
1.05	0.87
1.57	1.00
2.09	0.87
2.62	0.50
3.14	0
3.67	-0.50
4.19	-0.87
4.71	-1.00
5.24	-0.87
5.76	-0.50
6.28	0
6.81	0.50
7.33	0.87
7.85	1.00
8.38	0.87
8.90	0.50
9.42	0
9.95	-0.50

(b) and (c)

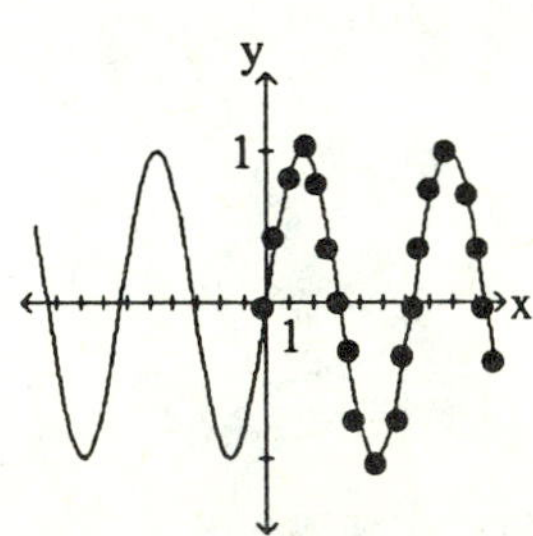

(d) Domain: All real numbers.

Range: All real numbers ≥ -1 and ≤ 1.

23. $ab = 100$

25. $w = 0.25z - 35.75$

*27. (a) $y \leq x$ where x represents the first number & y the second.

(b) See graph.

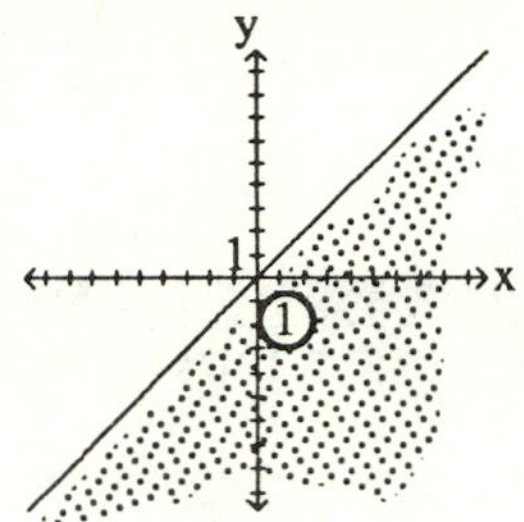

(c) Domain and Range: All Real Numbers

Exercises 1.2

1. (a) and (c).
3. yes
5. yes
7. no
9. yes
11. $g(0) = 1$

$g(1) = \frac{1}{2}$

$g(3) = \frac{1}{4}$

$g(-5) = -\frac{1}{4}$

$g(t) = \frac{1}{t+1}$

$g(-t) = \frac{1}{-t+1}$

$g(\frac{2}{t}) = \frac{t}{2+t}$

$g(a+h) = \frac{1}{a+h+1}$

$\frac{g(a+h)-g(a)}{h} = \frac{-1}{(a+h+1)(a+1)}$

13. (a) $V(0) = 0$; $V(-5) = -10,000$;

$V(4) = 300;$
$V(11) = 0;$

(b) $x = 0, 6.5$ or 11

(c) $x = 0.5, 6$ or 11.5

(d) $x = 13.5$

(e) $x = -0.75, 7$ or 10.5

15.

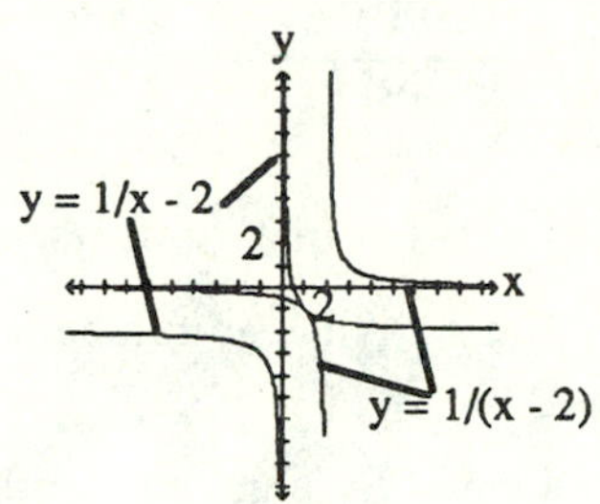

17. b, d, e

19. (d)

21. A function.

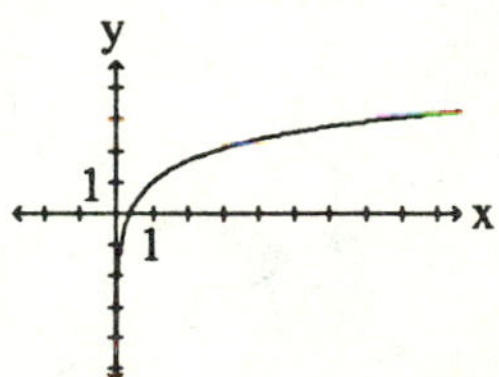

23. A function.

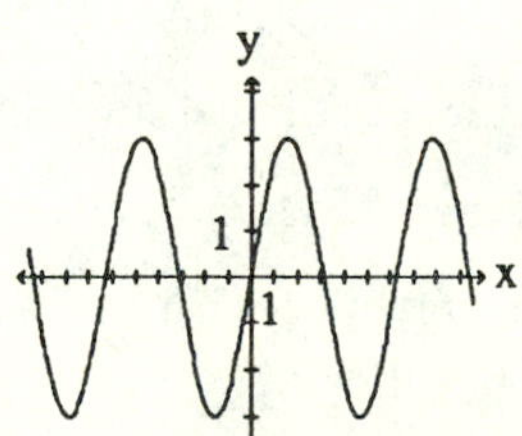

25.

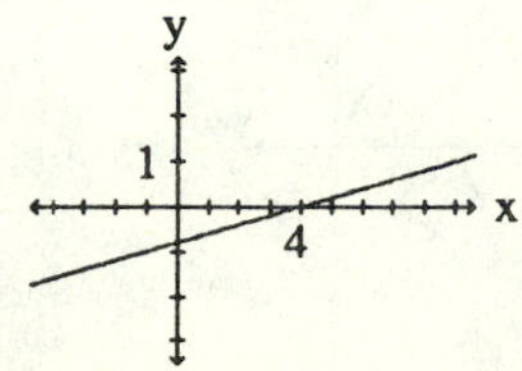

27.

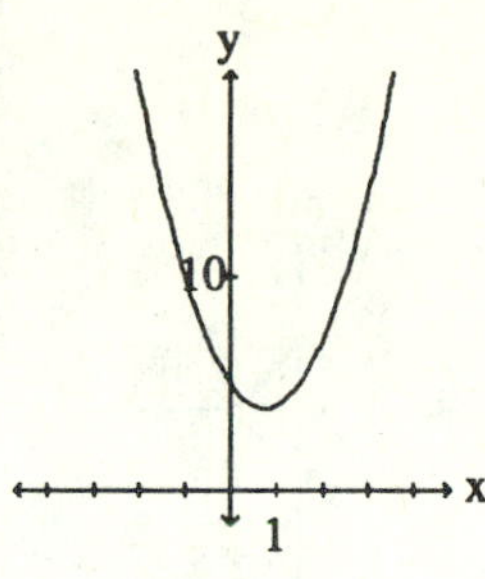

29.

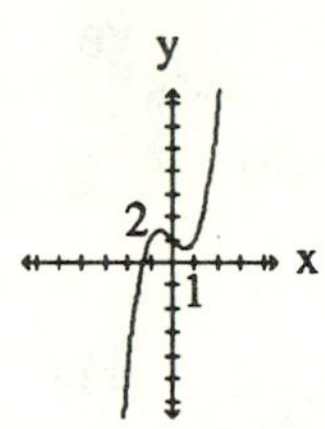

31. Dom: All real numbers.
Range: All real numbers.

33. Dom: All real numbers.
Range: All real numbers ≥ -27.

35. (a) $A(x) = 4x^2 + 21x$
(b)

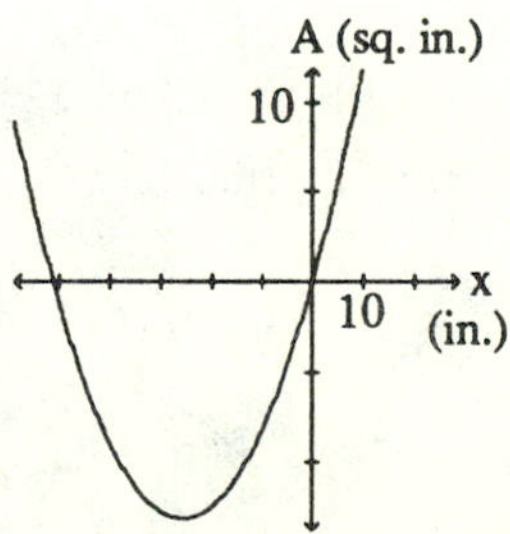

(c) Dom: All real numbers.
Range: All real numbers ≥ -27.56.

(d) All positive real numbers.

(e) $x = 0.84$ inches.

37.

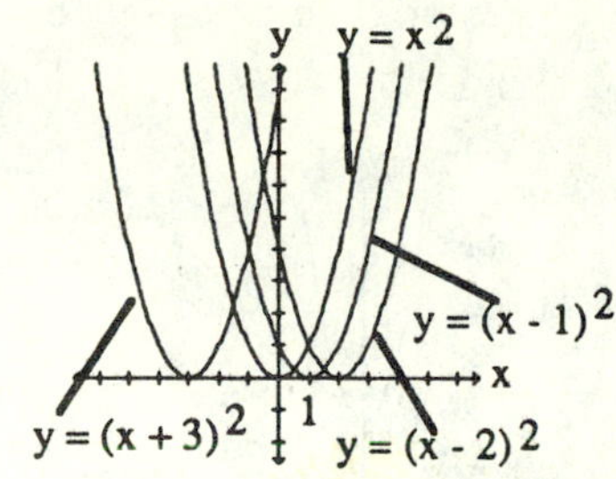

39. $y = x^2$
$y^2 = x$

41. If (a, b) and (a, c) are ordered pairs of $y = 3x - 1$, then $b = 3a - 1$ and $c = 3a - 1$. Thus, $b = c$ and $y = 3x - 1$ is a function.

Exercises 1.3

1. (a) $(2, -3)$
   (b) $(-2, 3)$
   (c) $(-2, -3)$

3. (a) (a, b)
   (b) $(-a, -b)$
   (c) $(-a, b)$

5. (a) none.
   (b) x-axis; (a, b) and $(a, -b)$ are points on the graph.
   (c) y-axis; (a, b) and $(-a, b)$ are points on the graph.

7. Dom: All real numbers.

Range: All real numbers greater than or equal to 6.

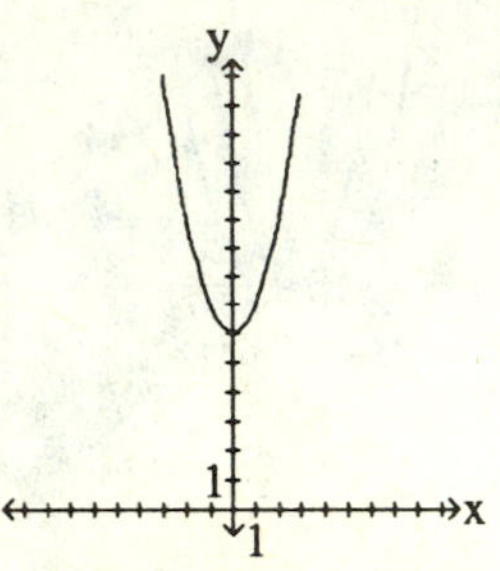

9. Dom: All real numbers.

   Range: All real numbers.

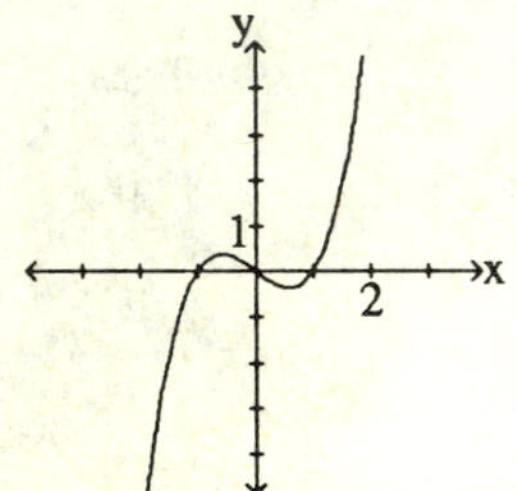

11. Dom: All real numbers.

    Range: All real numbers ≥ -2 and ≤ 2

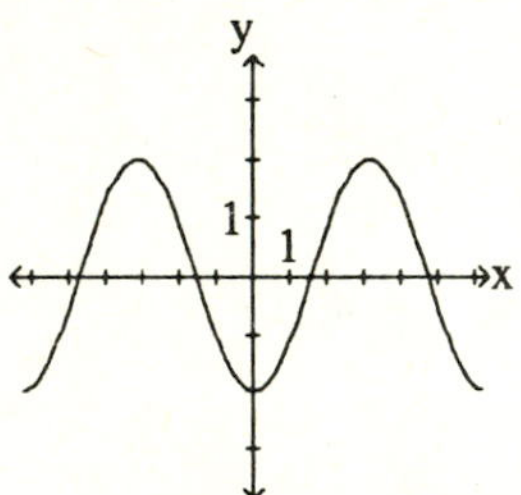

13. (a) Figure 1.4.21.

    (b) Cannot be symmetric with respect to the y-axis only.

    (c) Cannot be symmetric with respect to the origin only.

    (d)

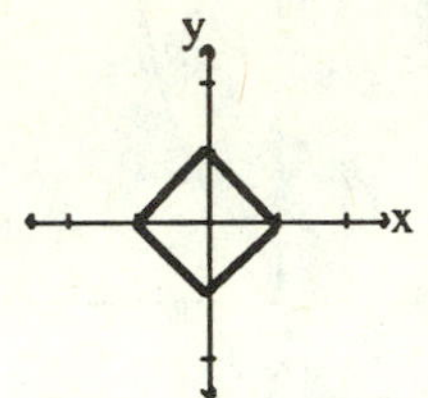

15. Symmetric with respect to x-axis

17. Symmetric with respect to y-axis

19. Not symmetric

21.

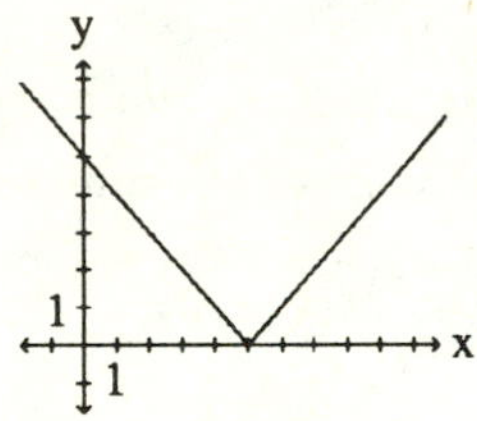

Dom: All real numbers.

Range: All nonnegative real numbers.

Line of Symmetry: $x = 5$.

y is a function of x.

23.

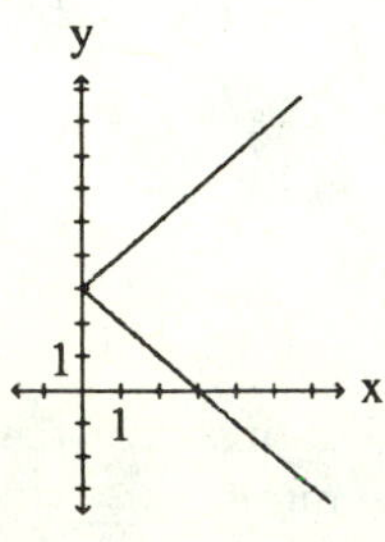

Dom: All nonnegative real numbers.

Range: All real numbers.

Line of Symmetry: $y = 3$.

25.

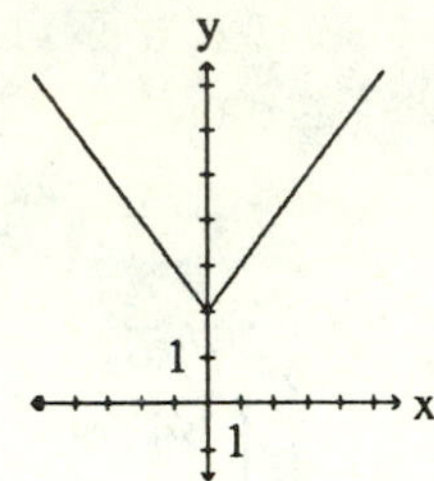

Dom: All real numbers.

Range: All real numbers ≥ 2.

Line of Symmetry: $x = 0$.

y is a function of x.

27.

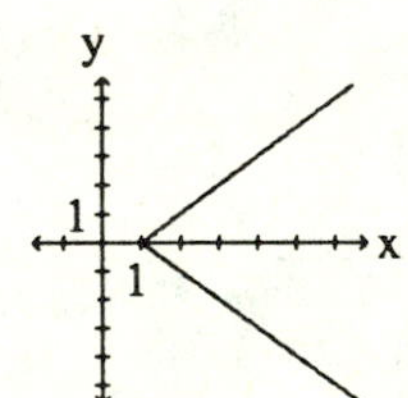

Dom: All real numbers ≥ 1.

Range: All real numbers.

Line of Symmetry: $y = 0$.

29. Symmetric with respect to origin

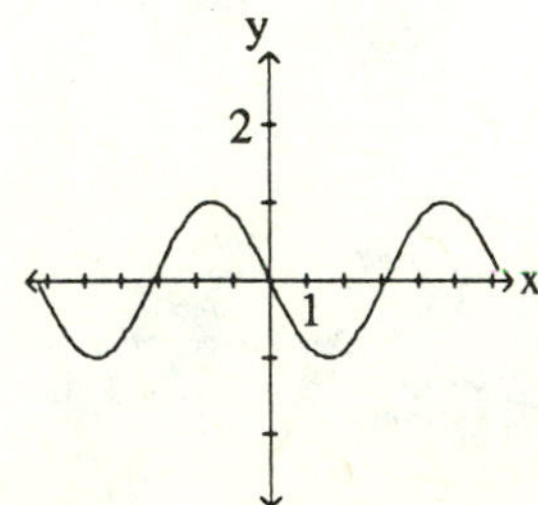

31. Graph $y = \sqrt{2x^2+1}$ and $y = -\sqrt{2x^2+1}$

33. (a) $A(x) = x(64 - x)$

(b) Domain: All real numbers.

Range: All real numbers less than or equal to 1024.

(c)

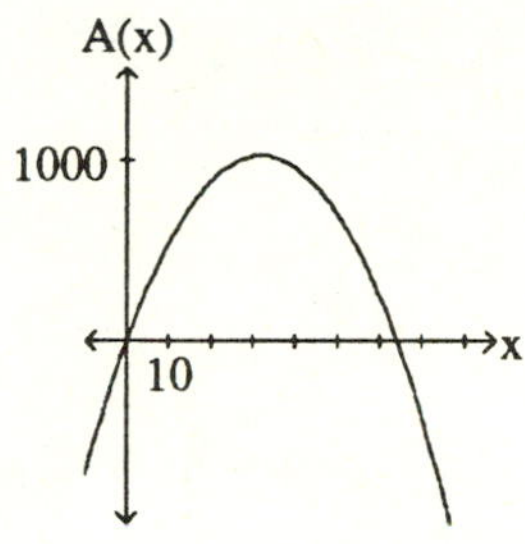

(d) Any real number between 0 and 64.

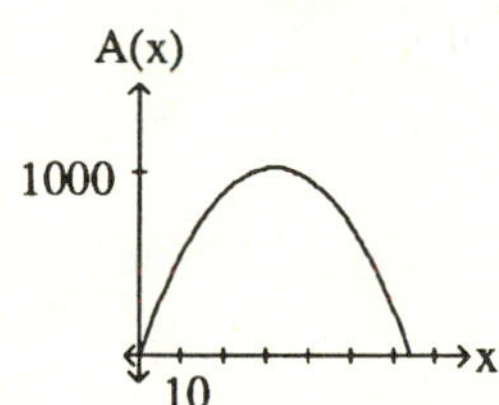

(e) 1024 square feet.

32 ft. by 32 ft.

35. (a) $P(x) = 5.25x - 83,000$

(b)

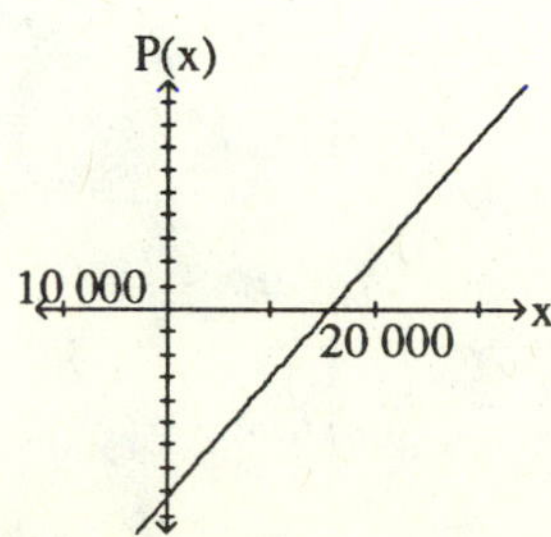

(c) $P = \$40,375$

37. Yes; the line of symmetry is $x = 2$.

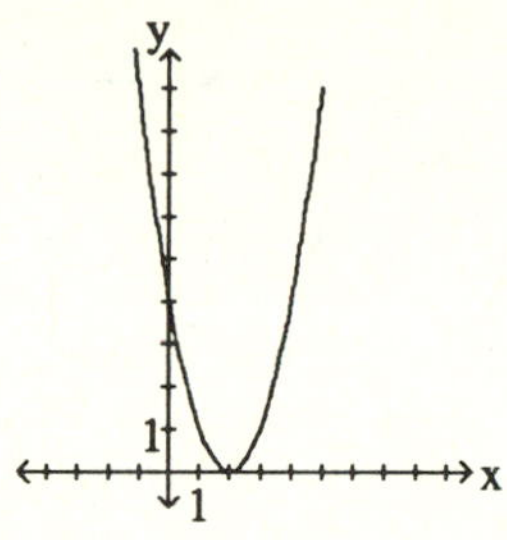

*39. Yes; the point is $(0, 2)$.

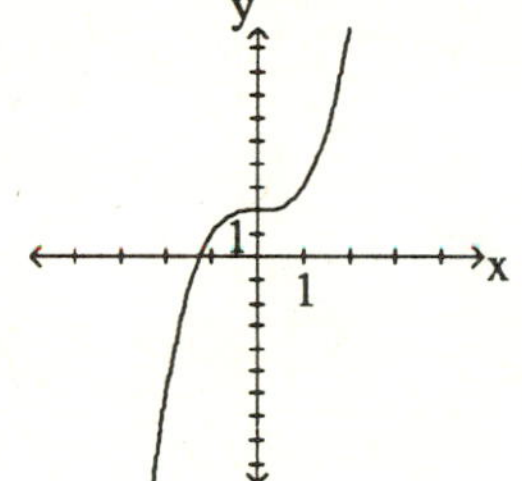

Exercises 1.4

1.

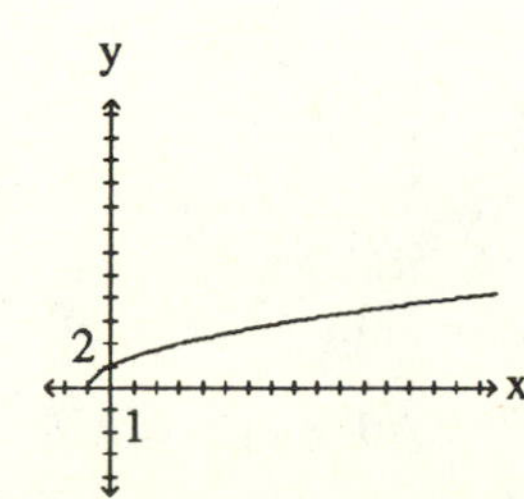

3.

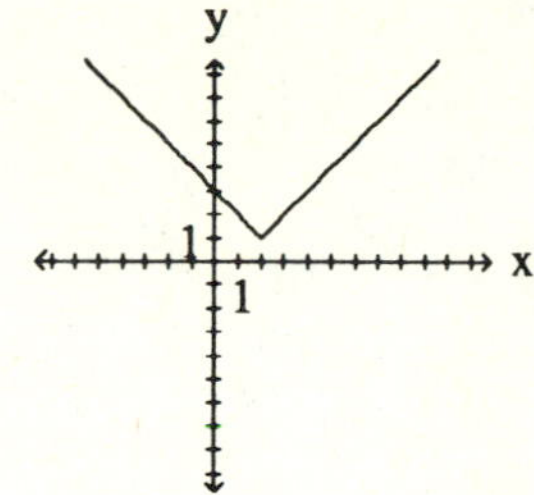

5.

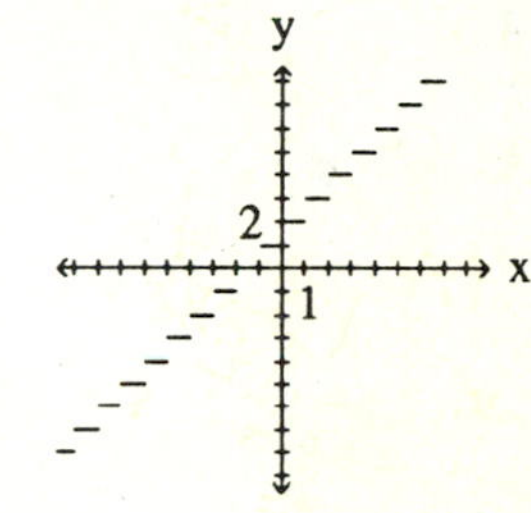

7.

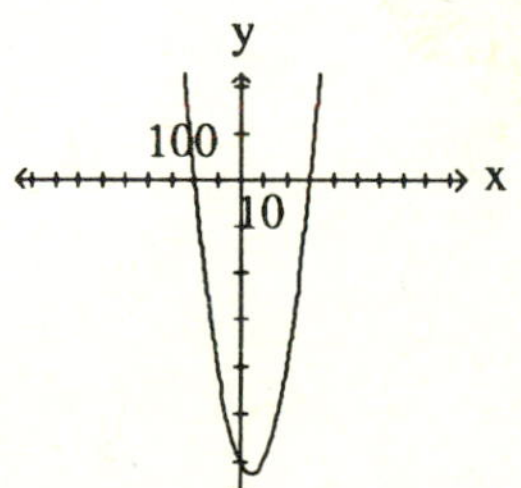

9.

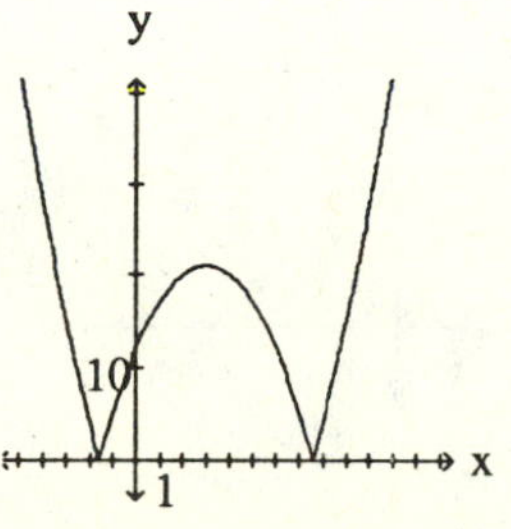

11.

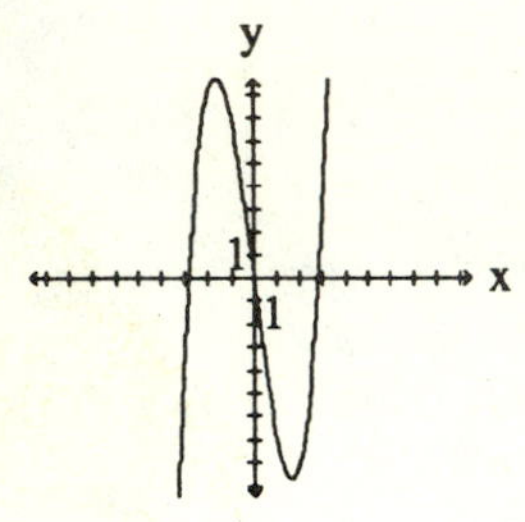

13. Dom: All real numbers ≥ -1.

Range: All nonnegative real numbers.

15. Dom: All real numbers.

17. Dom: All real numbers.

19. Dom: All real numbers ≥ 5.

21. The graphs are identical.

23. (e).

25. (c).

27. Dom: All real numbers.

Range: $[0, \infty)$

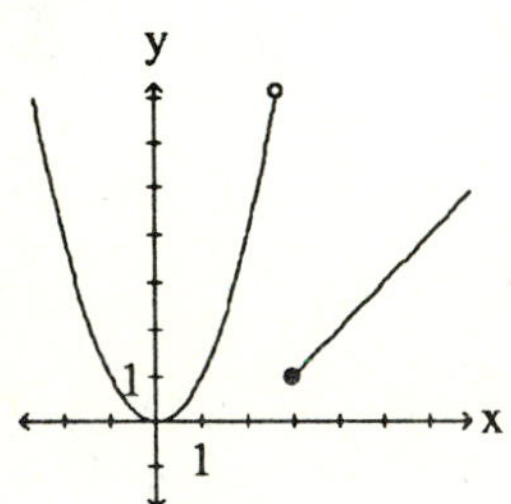

29. Dom: All real numbers.

Range: All real numbers ≤ -4 as well as 6

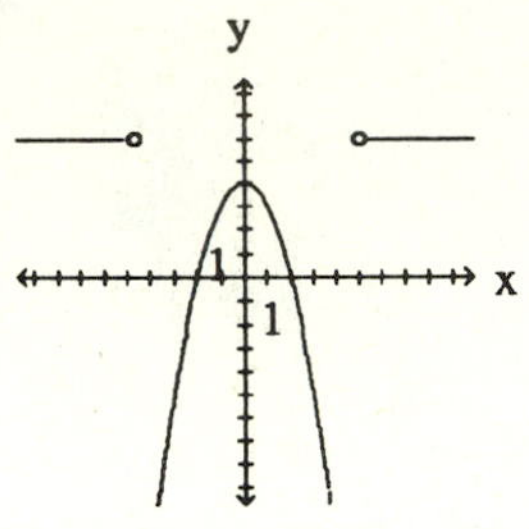

31. Dom: All real numbers

Range: All real numbers $\geq \frac{15}{4}$

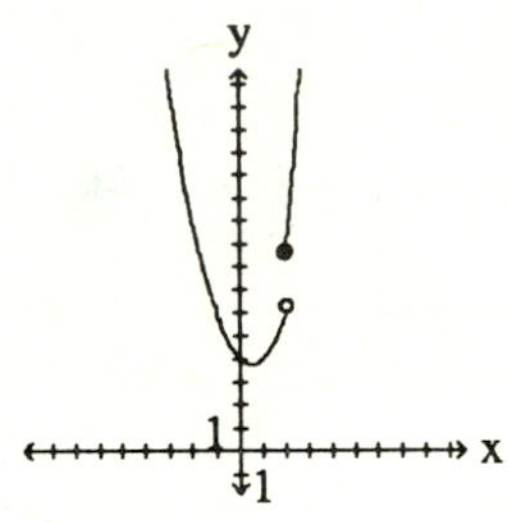

33. (a) $F = 3t + 5$

(b)

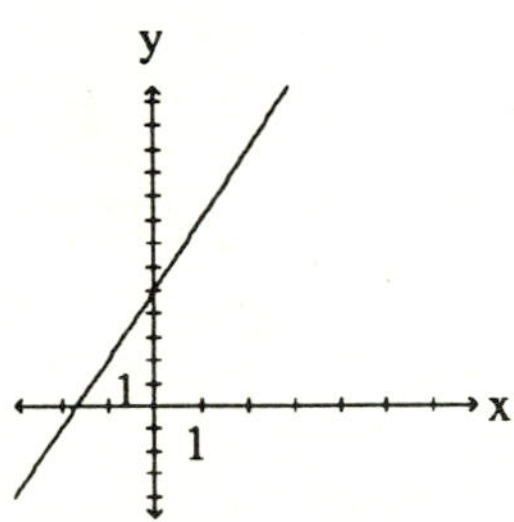

(c) Dom: All real numbers.

Range: All real numbers.

(d) All values of t greater than 0

represent the problem situation.

(e) $t = 3.2$ hours

35. (a) $P = 1.50x - [1.25x + 20,000] = .25x - 20,000$

(b)

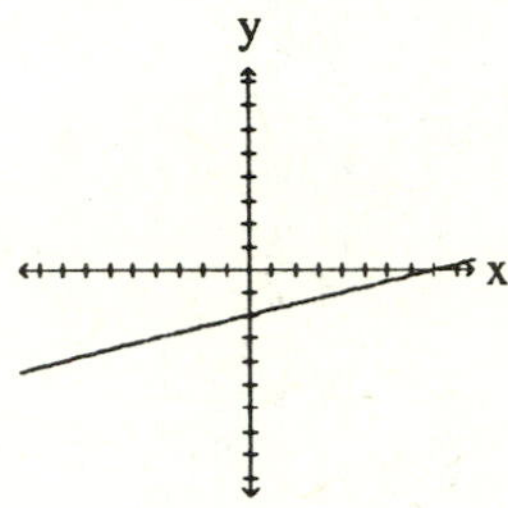

The scale is 10^4 units per tick mark.

(c) 80,000 lbs.

(d) 120,000 lbs.

(e) 80,120 lbs.

*37. (a) $C = .48 + .28(\text{ INT }(t))$ if t is *not* a whole number.

$C = .48 + .28(\text{ INT }(t-1))$ or simply $C = .48 + .28(t-1)$ if t is a whole number.

(b)

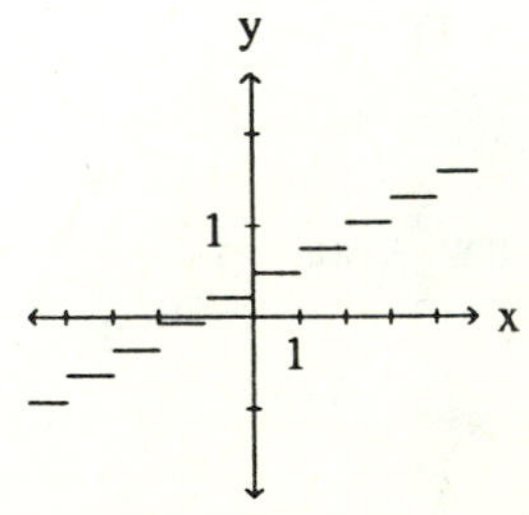

(c) Dom: All real numbers.

Range: $\{\ldots, -0.08, 0.20, 0.48, \ldots\}$

(d) All values of t greater than 0 represent the problem situation.

(e) Bill can talk to his girlfriend up to 6 minutes.

39. (a) $D = \sqrt{(3-x)^2 + (5-(1-x))^2}$

(b) Dom: All real numbers.

(c)

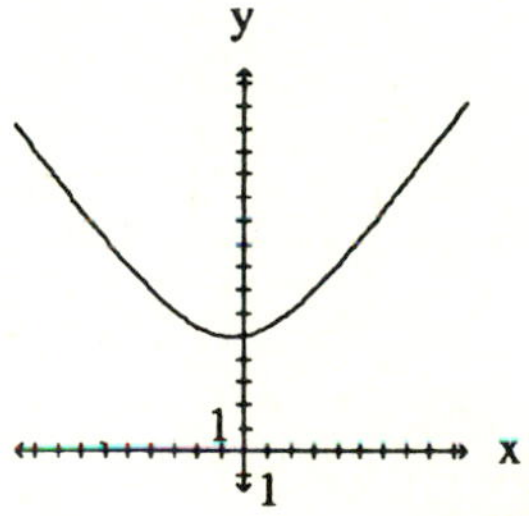

(d) $P: (6, -5)$ or $(-7, 8)$ (these are approximations).

41. a, c, f

43. a, d, e

45.

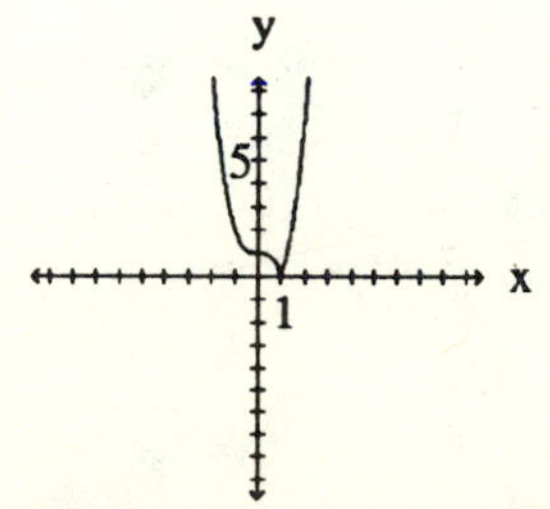

Dom: All real numbers

Range: All real numbers ≥ 0

47.

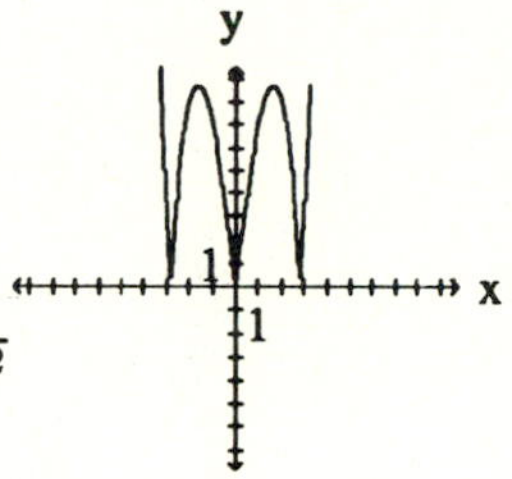

Dom: All real numbers

Range: All real numbers ≥ 0

Line of sym: $x = 0$

49.

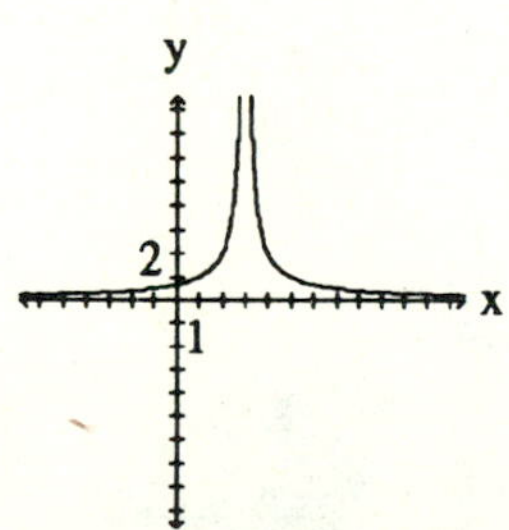

Dom: All real numbers except 3

Range: All real numbers > 0

Line of sym: $x = 3$

Exercises 1.5

1. (a)

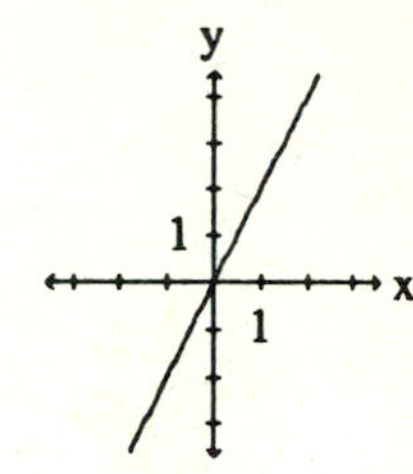

(b)

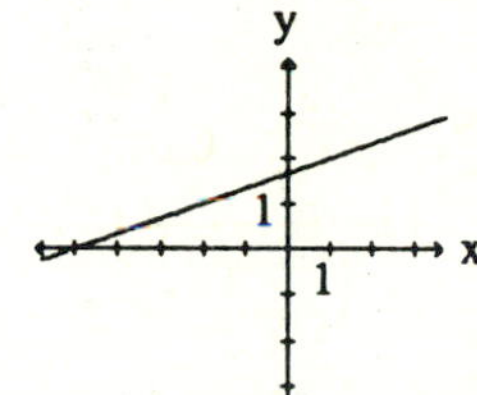

(c)

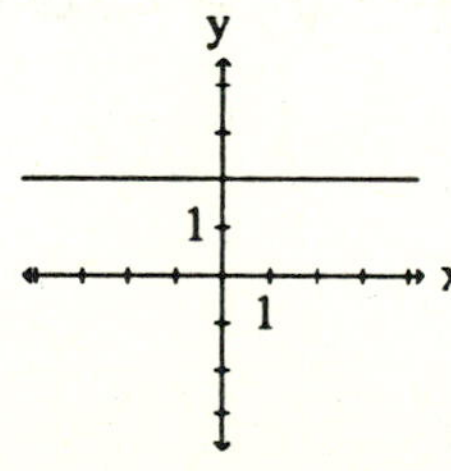

(d)

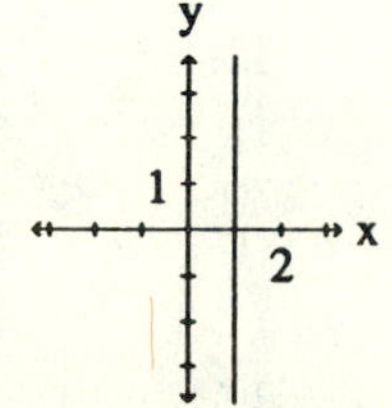

3.

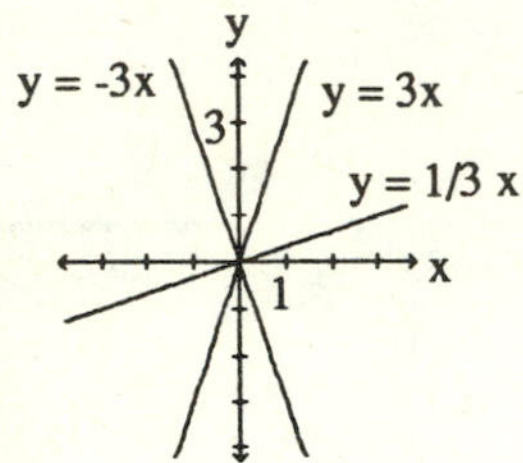

5.

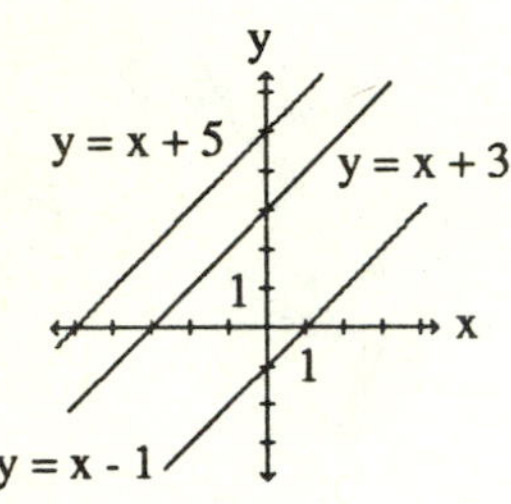

7.  $\frac{-8}{3}$

9.

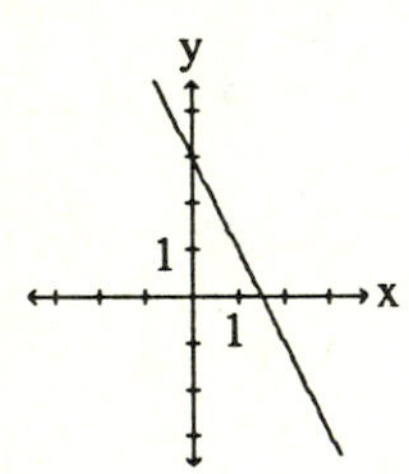

y intercept: 3

11.

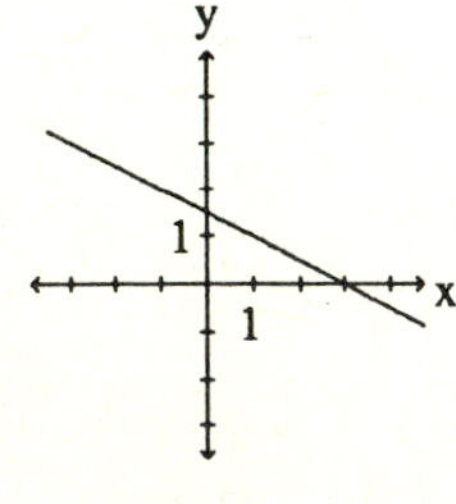

x intercept: 3

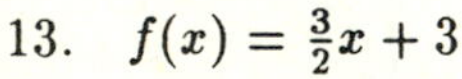

13. $f(x) = \frac{3}{2}x + 3$

15. $f(x) = 6.5x + 300$

17.

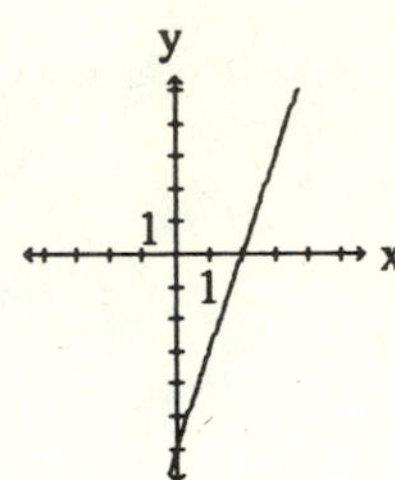

19.

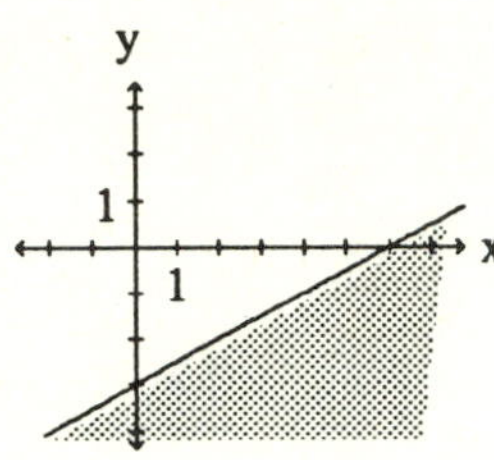
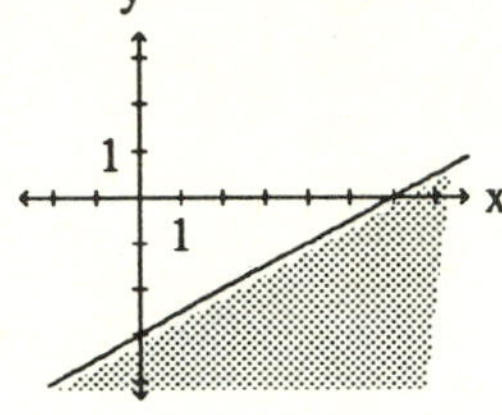

21.

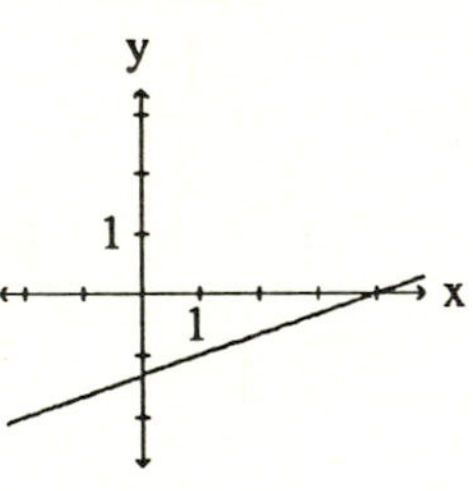

23.

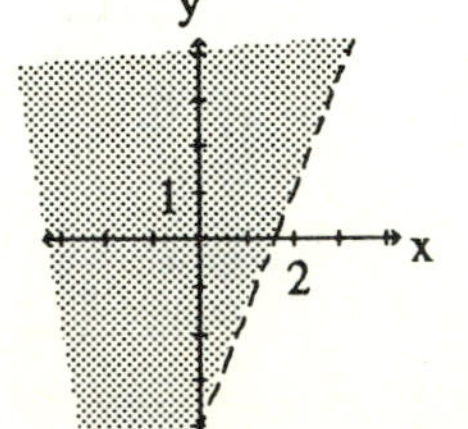

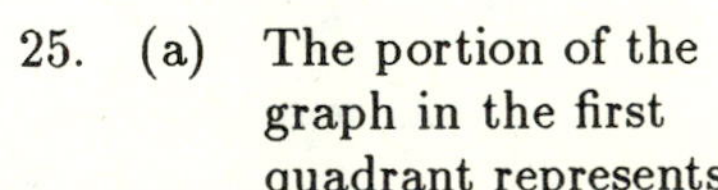

25. (a) The portion of the graph in the first quadrant represents the problem situation.

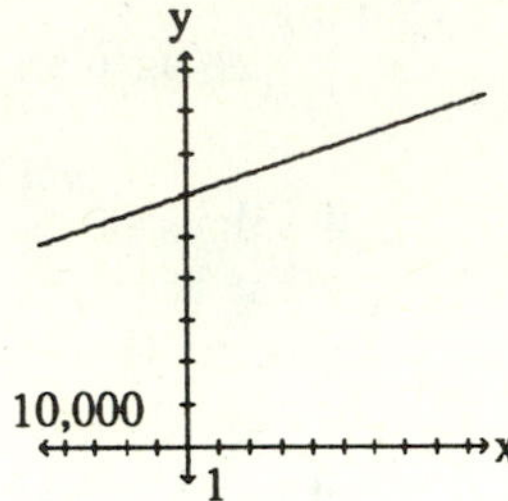

(b) $t = 4.5$ years

(c) $t \geq 8$ years

27. b represents the y-intercept and a represents the x-intercept.

29. (a) $V = 3187.5t + 42000$

(b) $t > 0$ for the problem situation.

V ($)

10,000

10 t (years)

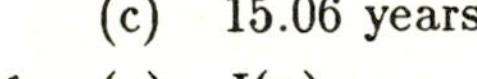

(c) 15.06 years

31. (a) $I(x) = .05(18,000-x)+.08x$

(b)

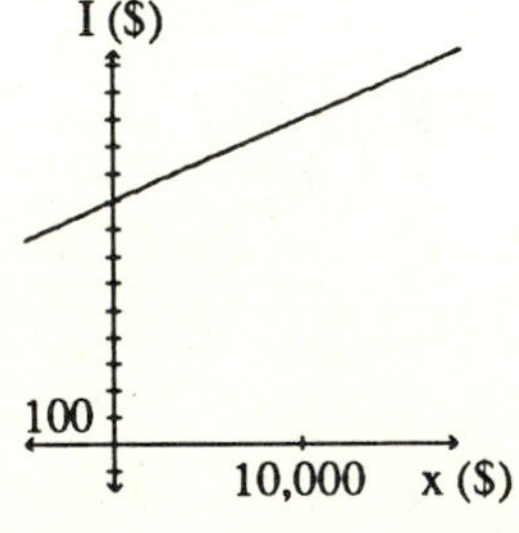

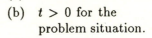

(c) The interest is a linear function of the amount invested at 8% with $I = 900 + .03x$

(d) $4000 invested at 8%

33.

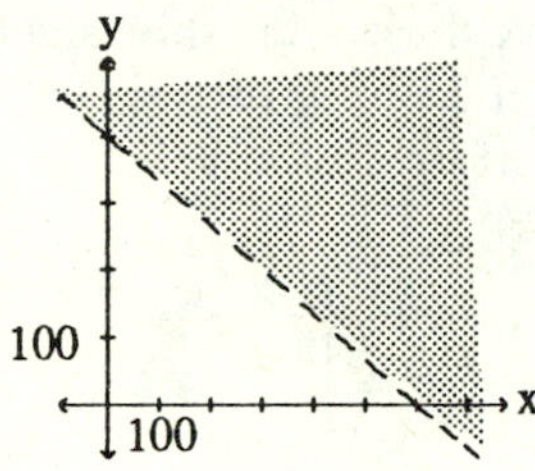

$2.50x + 3.75y > 1500$

35.

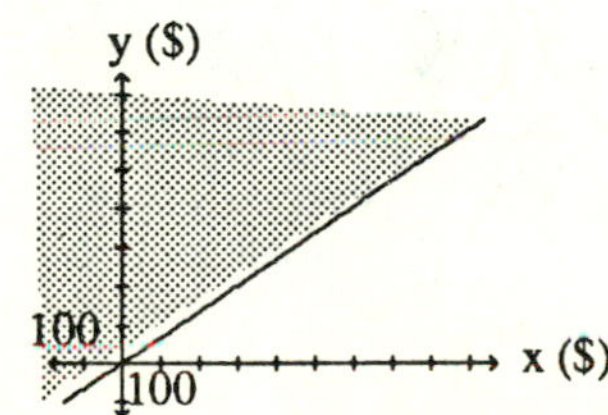

$0.105x + 0.167y \geq 0.13(x + y)$

37.

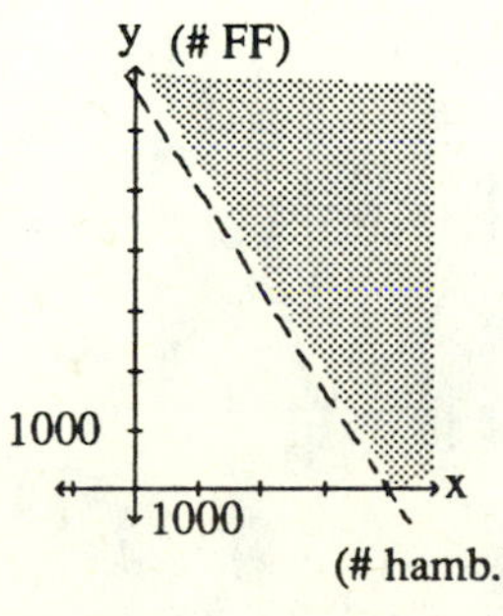

$0.25x + 0.15y - 600 > 400$

39. 1.37%: $I(t) = \frac{2500}{60000+(t-1)(2500)}$ or $I(t) = \frac{2500}{57500+2500t}$

41. No. It is unlikely that the value of the house would increase by a constant $2500 per year

43. Domain: $[0, \infty)$
Range: $\{0.72, 1.44, \ldots, 3.60\} \cup \{4.23, 4.86, \ldots, 9.90\} \cup \{9.90 + 0.51k | k \text{ is an integer} > 15\}$.

45. The model is called straight line depreciation because the graph of $B(t)$ is a straight line.

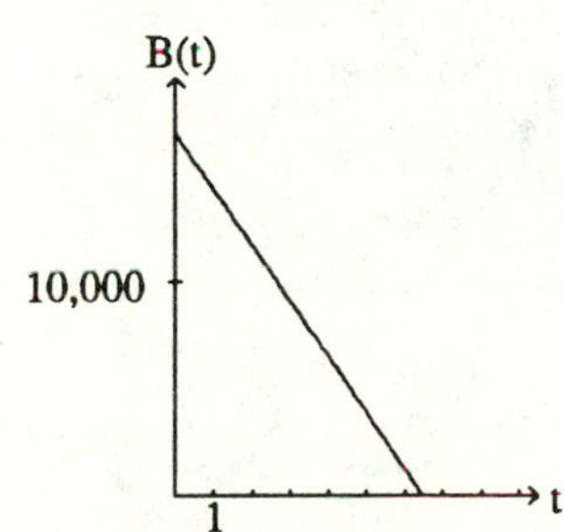

47.

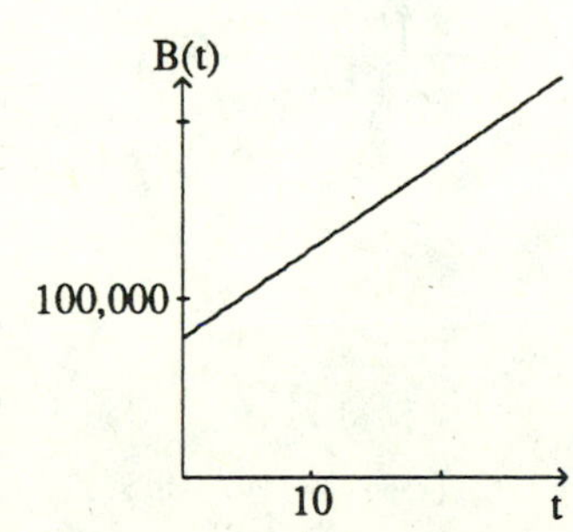

49.

$$B(t) = \begin{cases} 22600 + \frac{3800}{3}t & 0 < t \leq 3 \\ 10125 - 5425t & 3 < t \leq 7 \end{cases}$$

51. $B(5.5) = \$12{,}837.50$

Exercises 1.6

1.

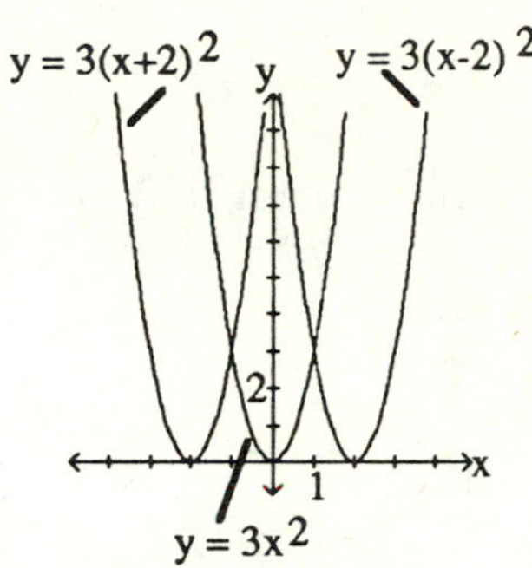

3.

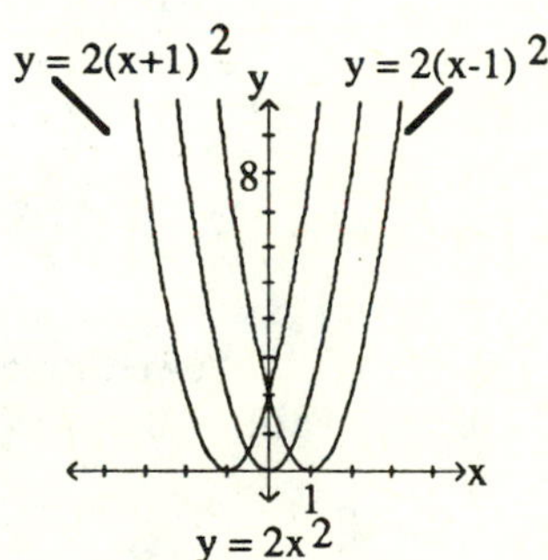

5.

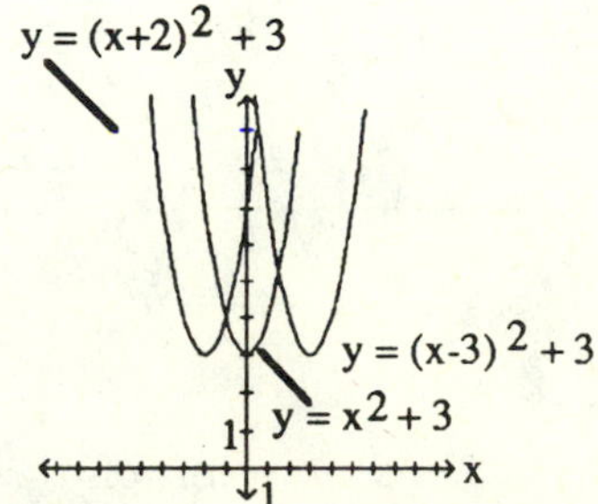

7. A vertical stretch of 4

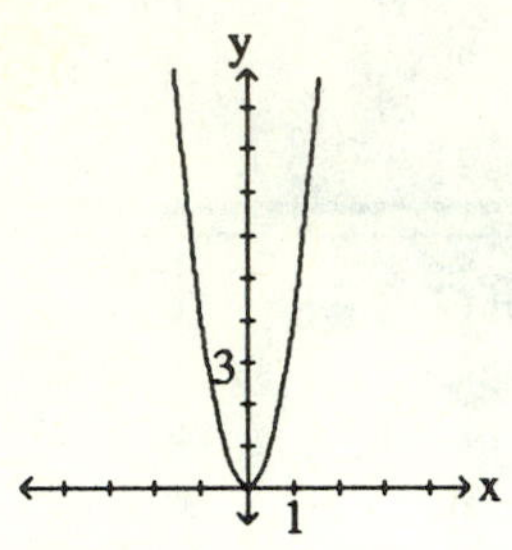

9. A horizontal shift of 5 units right

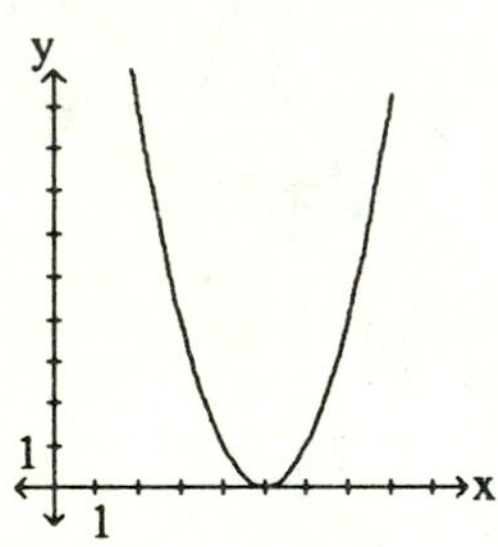

11. A vertical stretch by a factor of 2 followed by a vertical shift of 3 units down.

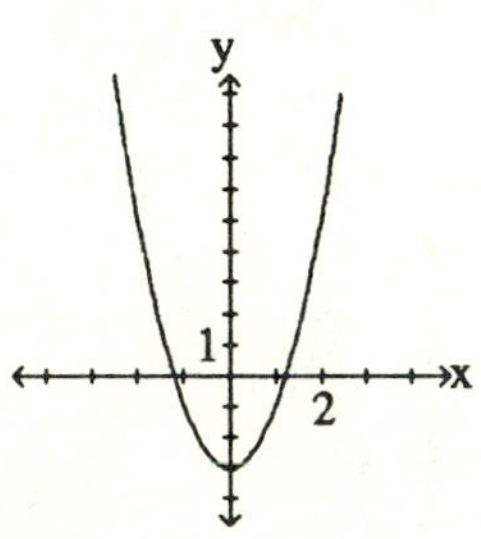

13. A horizontal shift of 4 units right followed by a vertical shift of 3 units up.

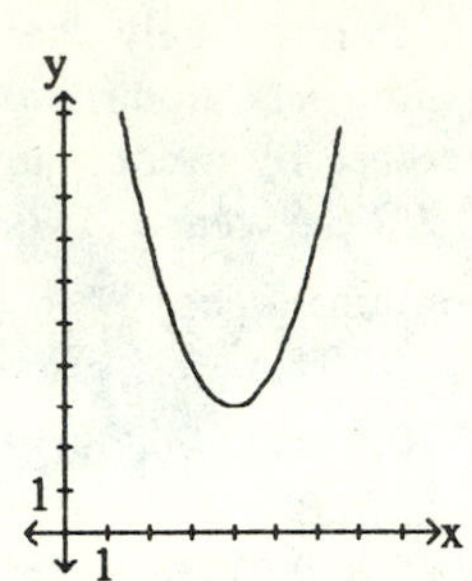

15. A horizontal shift right of 1 unit followed by a vertical stretch by a factor of 2, followed by a vertical shift of 3 units up.

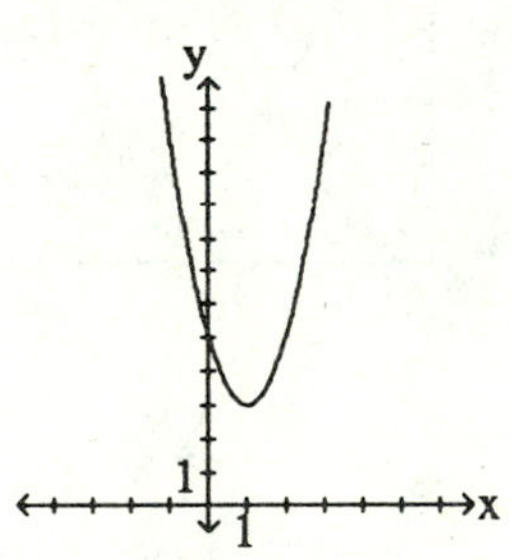

17.

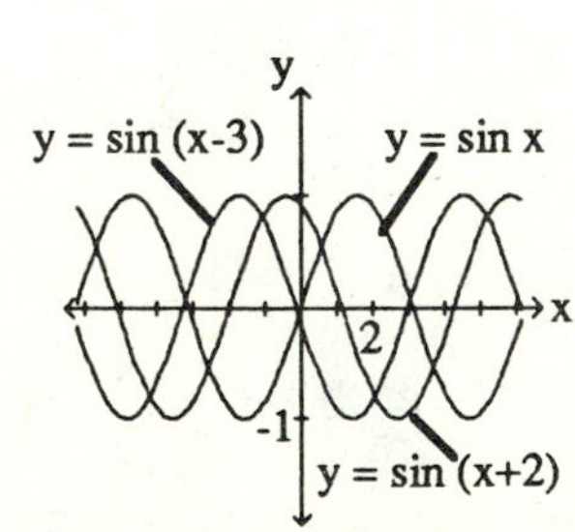

19.

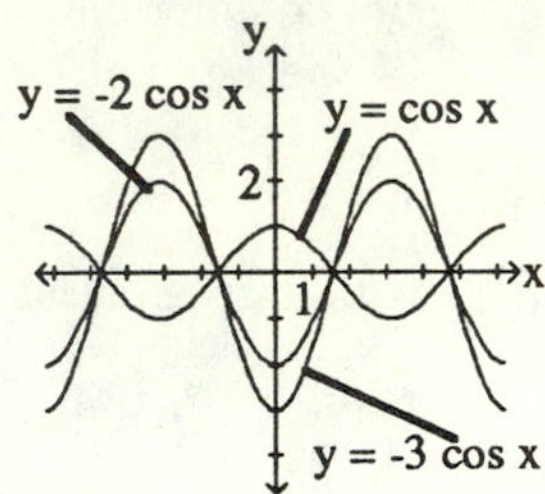

21. A horizontal shift right of 3 units, then a vertical stretch by a factor of 2.

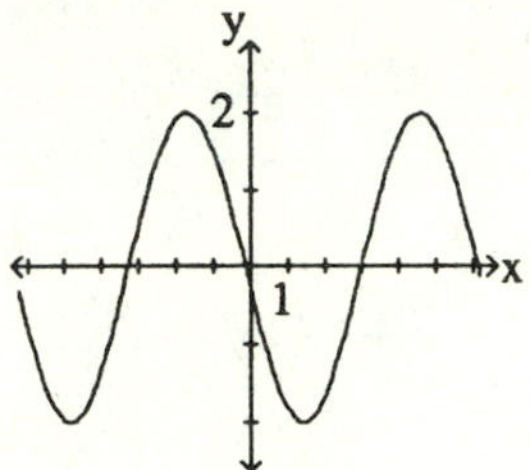

23. (a) $(3, 8)$
(b) $(3, 16)$
(c) $(3, 6)$
(d) $(3, -14)$

25. $b = 5$

27. (a) $y = 3(x - 4)^2$
(b) $y = 3(x - 4)^2$
(c) The graphs are the same.

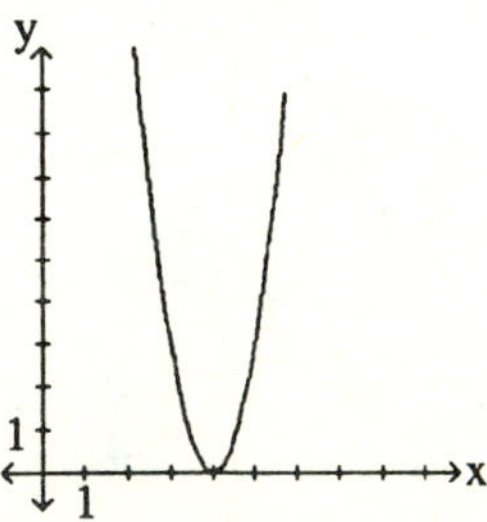

29. $y = 3(x + 2)^2 - 4$

31. $(-1, 10) \rightarrow (-1, 20) \rightarrow (2, 20)$

$(0,5) \to (0,10) \to (3,10)$
$(2,7) \to (2,14) \to (5,14)$

33.

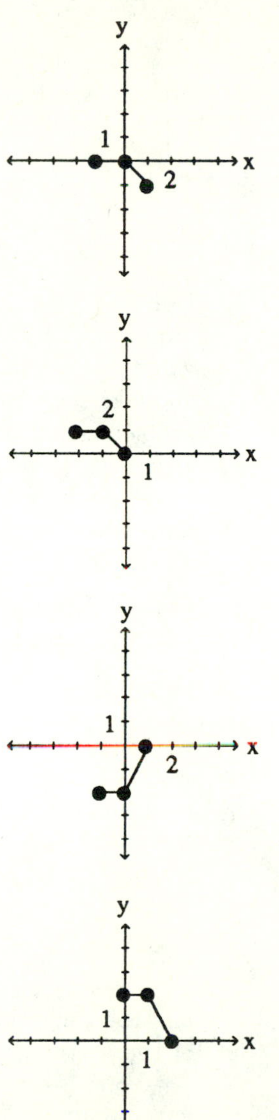

Exercises 1.7

1. A horizontal shift right 2 units followed by a vertical stretch of 3 (may be done in reverse order).

3. Vertical shift up 3 units.

5. Horizontal shift right 3 units followed by a vertical shift up 3 units (may be done in reverse order).

7. Horizontal shift of 8 units left followed by a reflection through the x-axis, followed by a vertical shift of 74 units (Steps 1 & 2 may be reversed but vertical shift must be last).

9.

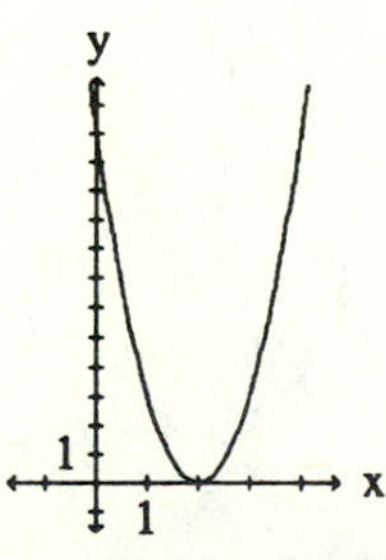

vertex: $(2,0)$

line of sym: $x = 2$

y-intercept: $(0,12)$

zeros: $x = 2$

11.

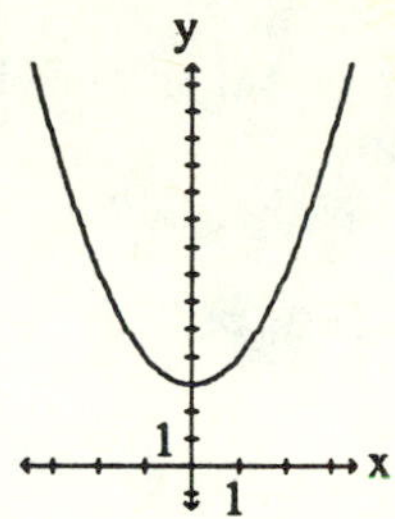

vertex: $(0,3)$
line of sym: $x = 0$
y-intercept: $(0,3)$
no real zeros

13.

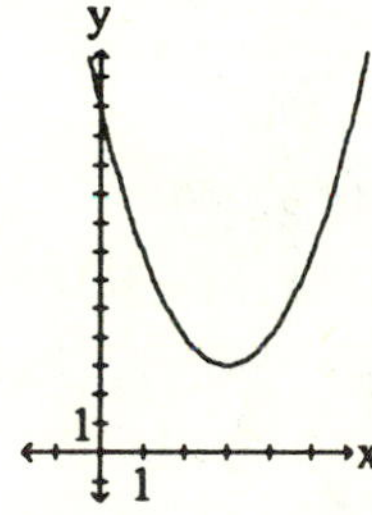

vertex: $(3,3)$
line of sym: $x = 3$
y-intercept: $(0,12)$
no real zeros

15.

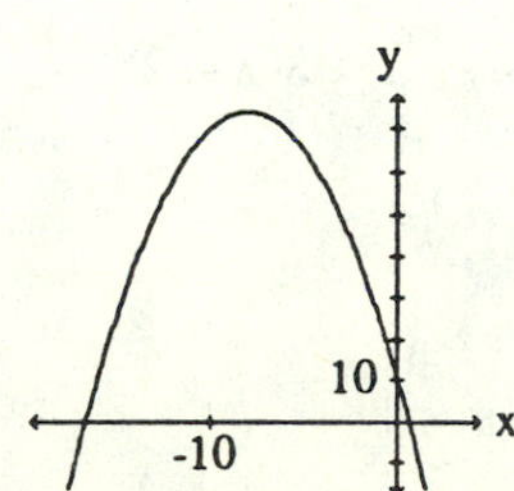

vertex: $(-8,74)$
line of sym: $x = -8$
y-intercept: $(0,10)$

zeros: $x = -8 \pm \sqrt{74}$

17. Horizontal shift right 3 units

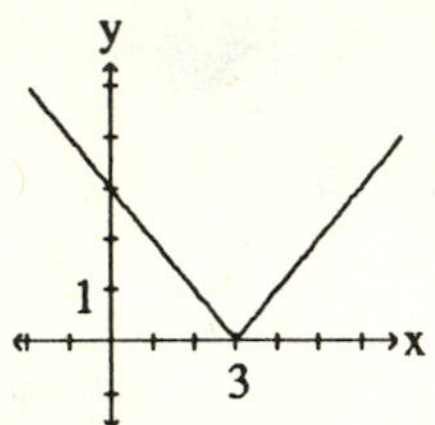

19. Horizontal shift right 4 units, followed by a vertical stretch by a factor of 3, then a vertical shift up 2 units.

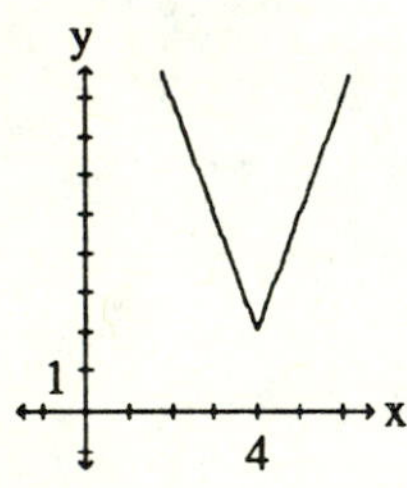

21. (a) $(3,-1)$, $(0,8)$, $(5,23)$

(b) $y = 3(x-2)^2 - 4$

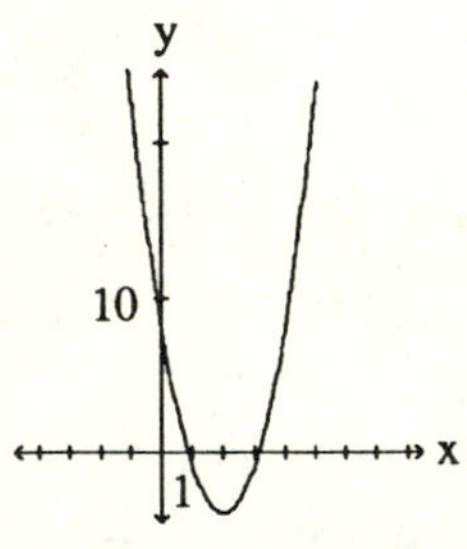

23. (a) $M : (3,0)$

Line of sym: $x = 3$

Vertex: $(3,-25)$

x-intercept: $(-2,0)$ and $(8,0)$

y-intercept: $(0,-16)$

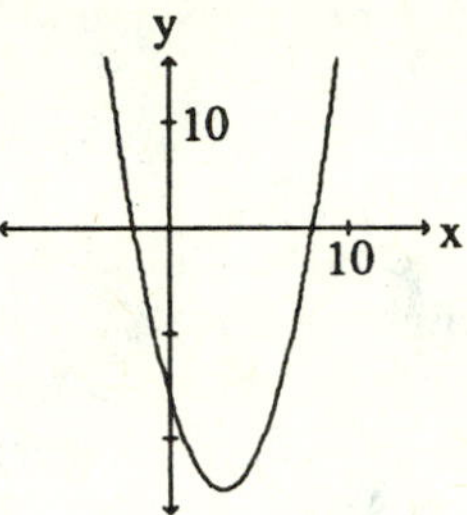

(b) $M : (-2.5,0)$

Line of sym: $x = -2.5$

Vertex: $(-2.5,-60.5)$

x-intercept: $(-8,0)$ and $(3,0)$

y-intercept: $(0,-48)$

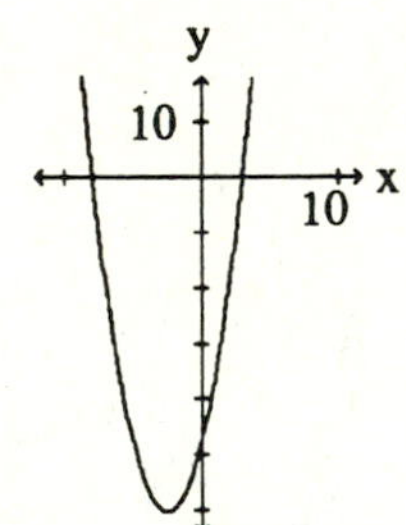

25. 11 inches ×7 inches

27. 8.5 feet × 10 feet

29. 11 feet & 9 feet

31. $H : 310$

$R : 24.45$

33. (a) $R = (300-5x)(800+20x) = 240{,}000 + 2000x - 100x^2$

(b)

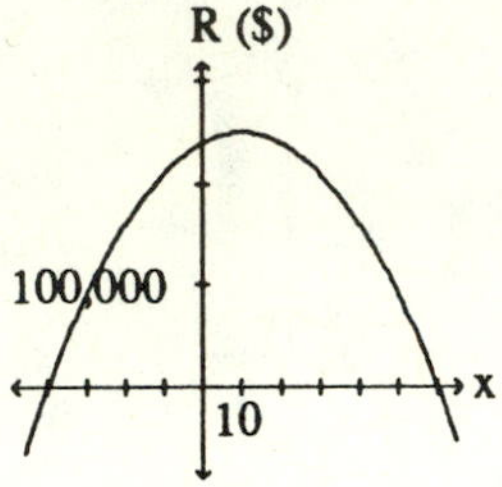

(c) Domain: All real numbers. All integer values of x from 0 to 40 make sense in the problem situation. (After 40 decreases in rent, all 1600 units are filled.)

(d) $250

35. $x = -0.5$

37. (a) $f(x) = (x - x_1)(x - x_2)$ since x_1, x_2 are real zeros.

Thus $f(x) = x^2 - (x_1+x_2)x + x_1x_2$ and the x-coordinate of the vertex is $-\frac{b}{2a} = \frac{x_1+x_2}{2}$ which is the midpoint of the 2 points x_1 and x_2.

(b) The line of symmetry is $x = -\frac{b}{2a} = \frac{x_1+x_2}{2}$.

39. (a)

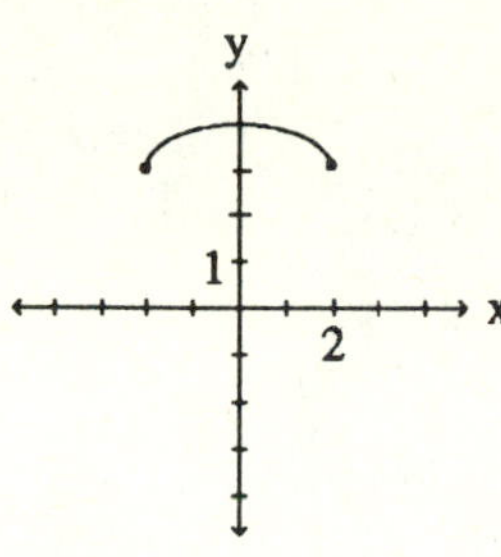

(b)

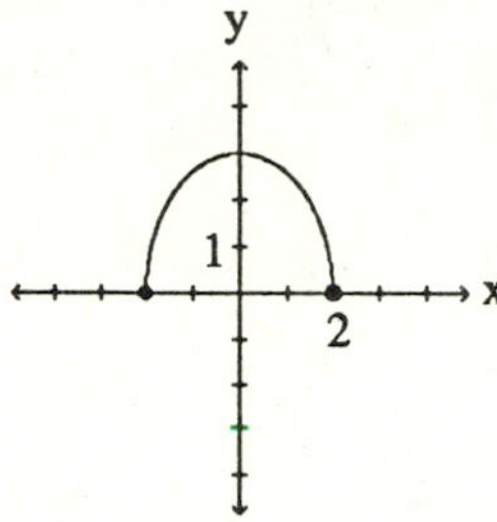

(c)

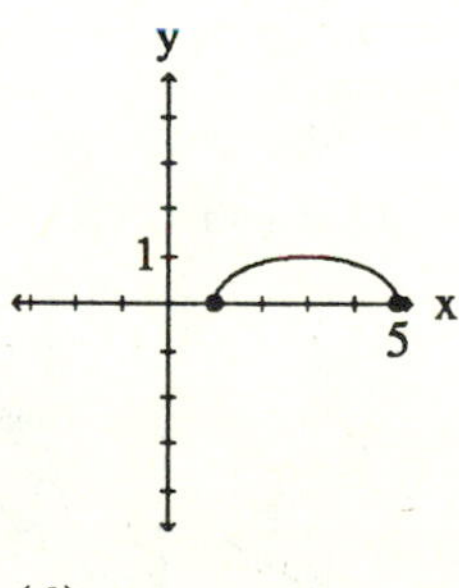

(d)

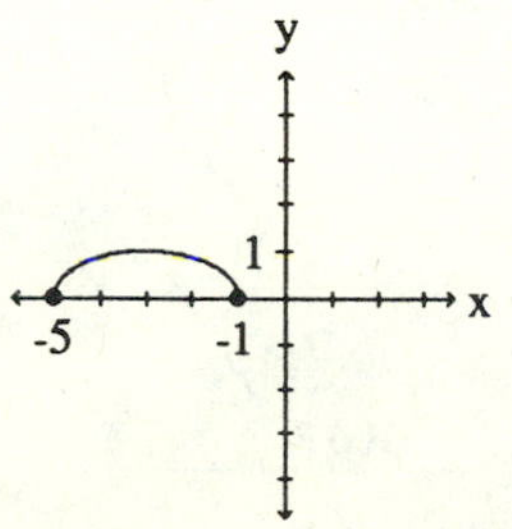

41. x-intercepts: $(h \pm \sqrt{\frac{-k}{a}}, 0)$, provided $\frac{-k}{a} > 0$

discriminant: $-4ak$

Exercises 1.8

1. $(f+g)(x) = x^2 + 2x - 1$
   Dom: $(-\infty, \infty)$
   $(f-g)(x) = 2x - 1 - x^2$
   Dom: $(-\infty, \infty)$
   $(fg)(x) = 2x^3 - x^2$
   Dom: $(-\infty, \infty)$
   $(\frac{f}{g})(x) = \frac{2x-1}{x^2}$
   Dom:
   $(-\infty, 0) \cup (0, \infty)$
3. $(f+g)(x) = x^2 + 2x$
   Dom: $(-\infty, \infty)$
   $(f-g)(x) = x^2 - 2x$
   Dom: $(-\infty, \infty)$
   $(fg)(x) = 2x^3$
   Dom: $(-\infty, \infty)$
   $(\frac{f}{g})(x) = \frac{x}{2}$
   Dom:
   $(-\infty, 0) \cup (0, \infty)$
5. $(f+g)(x) = \frac{5x+8}{3}$
   Dom: $(-\infty, \infty)$
   $(f-g)(x) = \frac{x+10}{3}$
   Dom: $(-\infty, \infty)$
   $(fg)(x) = \frac{(x+3)(2x-1)}{3}$
   Dom: $(-\infty, \infty)$
   $(\frac{f}{g})(x) = \frac{3x+9}{2x-1}$
   Dom:
   $(-\infty, \frac{1}{2}) \cup (\frac{1}{2}, \infty)$
7. $(f+g)(x)$: Range:
   $(-\infty, \infty)$
   $(f-g)(x)$: Range:
   $(-\infty, \infty)$
   $(fg)(x)$: Range:
   $(-2.04, \infty)$
   $(\frac{f}{g})(x)$: Range:
   $(-\infty, 1.5) \cup (1.5, \infty)$
9. The product of functions is a different operation than the composition of functions. Let $f(x) = x + 3$ and $g(x) = x - 1$. Then $fg(x) = (x+3)(x-1)$ and $f \circ g(x) = x + 2$.
11. Multiplication of polynomial functions is commutative, associative, has an identity $(f(x) = 1)$ and has an inverse $\left(\frac{1}{f(x)} \text{ where } f(x) \neq 0\right)$.
13.

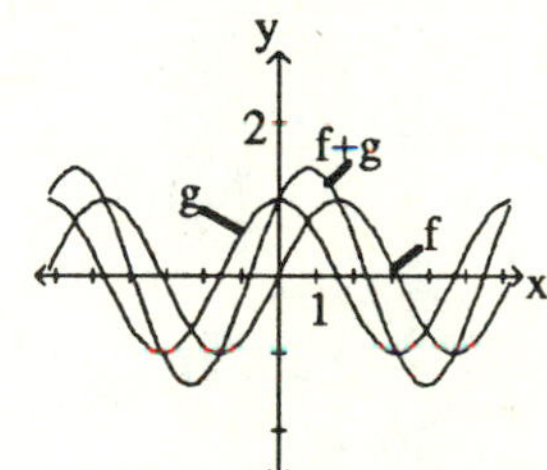

15.

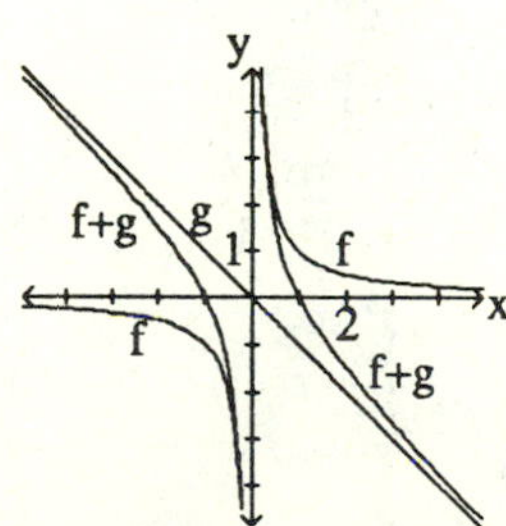

17. $f \circ g(3) = 5$
    $g \circ f(-2) = -6$
19. $f \circ g(3) = 8$
    $g \circ f(-2) = 3$
21. $f \circ g(x) = 3x - 1$

$g \circ f(x) = 3x + 1$

f: Dom: $(-\infty, \infty)$

Range: $(-\infty, \infty)$

g: Dom: $(-\infty, \infty)$

Range: $(-\infty, \infty)$

$f \circ g$: Dom: $(-\infty, \infty)$

Range: $(-\infty, \infty)$

$g \circ f$: Dom: $(-\infty, \infty)$

Range: $(-\infty, \infty)$

23. $f \circ g(x) = \frac{1}{(x-1)^2} - 1$

$g \circ f(x) = \frac{1}{x^2-2}$

f: Dom: $(-\infty, \infty)$

Range: $[-1, \infty)$

g: Dom: $(-\infty, 1) \cup (1, \infty)$

Range: $(-\infty, 0) \cup (0, \infty)$

$f \circ g$: Dom: $(-\infty, 1) \cup (1, \infty)$

Range: $(-1, \infty)$

$g \circ f$: Dom: $(-\infty, -\sqrt{2}) \cup (-\sqrt{2}, \sqrt{2}) \cup (\sqrt{2}, \infty)$

Range: $(-\infty, 0) \cup (0, \infty)$

25. $f \circ g(x) = \frac{1}{x(x+2)}$

$g \circ f(x) = \left(\frac{x}{x-1}\right)^2$

f: Dom: $(-\infty, 1) \cup (1, \infty)$

Range: $(-\infty, 0) \cup (0, \infty)$

g: Dom: $(-\infty, \infty)$

Range: $[0, \infty)$

$f \circ g$: Dom: $(-\infty, -2) \cup (-2, 0) \cup (0, \infty)$

Range: $(-\infty, -1] \cup (0, \infty)$

$g \circ f$: Dom: $(-\infty, 1) \cup (1, \infty)$

Range: $[0, \infty)$

27.

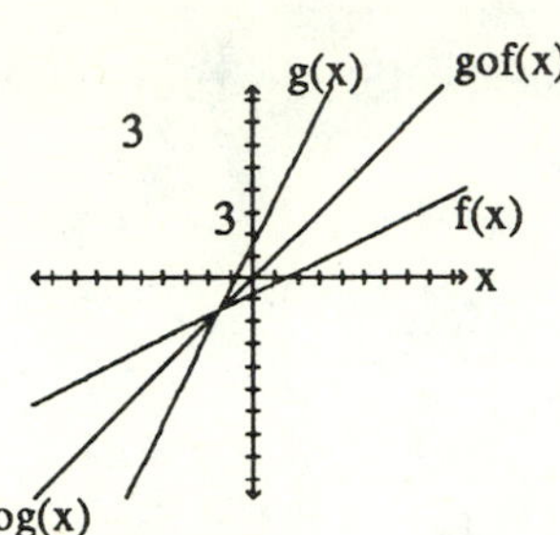

29. $g(x) = x^2$
$h(x) = x + 3$

31. $g(x) = x + 3$
$h(x) = x^2$

33. $g(x) = x - 2$
$h(x) = \sqrt[3]{x}$

35. Let $f_1(x) = \frac{1}{x}$, $f_2(x) = x^2$, $f_3(x) = x - 1$, and $f_4(x) = (x-1)^2$.

i) $f(x) = f_1 \circ f_2 \circ f_3(x)$

ii) $f(x) = f_1 \circ f_4(x)$

37. Let $f_1(x) = \frac{1}{x}$, $f_2(x) = x^2$, $f_3(x) = x + 1$, and $f_4(x) = x^2 + 2x + 1$.

i) $f(x) = f_1 \circ f_2 \circ f_3(x)$

ii) $f(x) = f_1 \circ f_4(x)$

39. $t = 3.63$ seconds.

41. (a) $V = \frac{4}{3}\pi t^3$

(b) $V = 36\pi t^3$

(c) $V = 3053.63$ cubic inches

(d) $t = 4.73$ sec.

43. (a) $S = f(r) = 4\pi r^2$; $r = g(t) = 1.6 - 0.0027t$; $S = f \circ g(t) = 4\pi(1.6 - 0.0027t)^2$

(b) $t = 592.59$ seconds or 9.88 minutes

45. $t = 1.62$ seconds

Review Exercises

1.

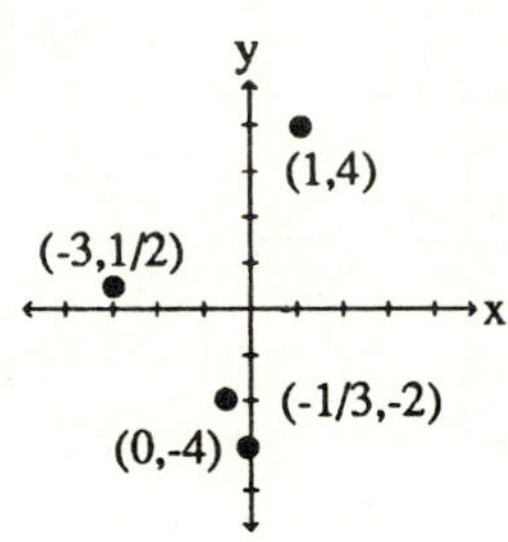

3. $a = 5$ or -5

5. Answers will vary

7.

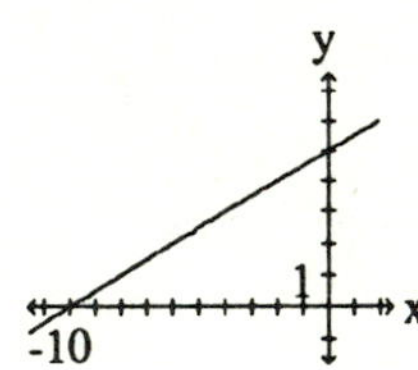

9.

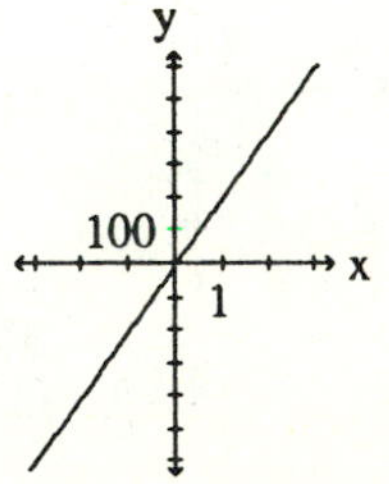

11.

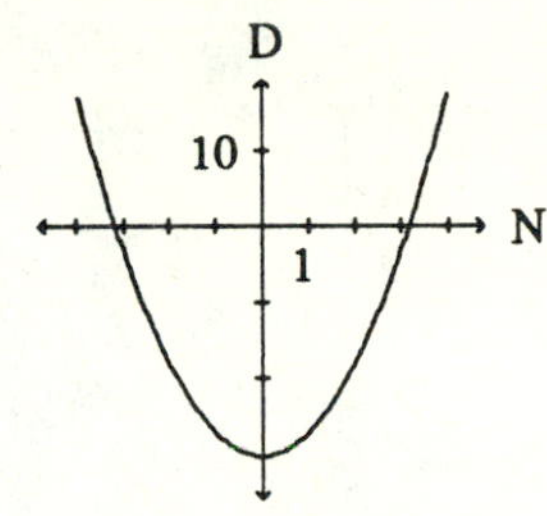

13. Dom: All real numbers.

Range: All real numbers.

15. (a) x-axis: if (a, b) is a point on the graph, so is $(a, -b)$.

(b) origin: if (a, b) is a point on the graph, so is $(-a, -b)$.

(c) y-axis: if (a, b) is a point on the graph, so is $(-a, b)$.

17. Dom: All real numbers ≥ -3.

Range: All real numbers.

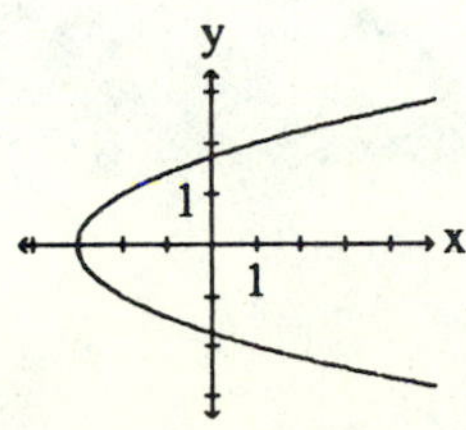

19. Symmetric with respect to the y-axis

21. Line of symmetry: $x = -5$.

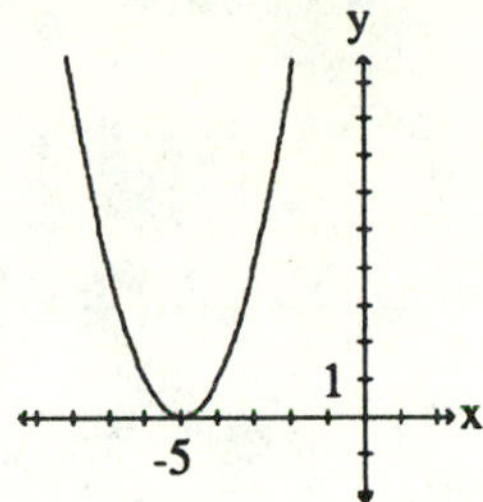

23. (a) $\sqrt{7} - 2.6$
(b) $\sqrt{7} - 2.6$
(c) $\pi - 3$
(d) $x - 5$

25. $\sqrt{13}$

27. (a) $\sqrt{4 + (y-3)^2}$
(b) $\sqrt{x^2 + 16}$

29. (a)

x	y
0	1
0, -2	1
-1	2
-2	1
-3	0

(b).

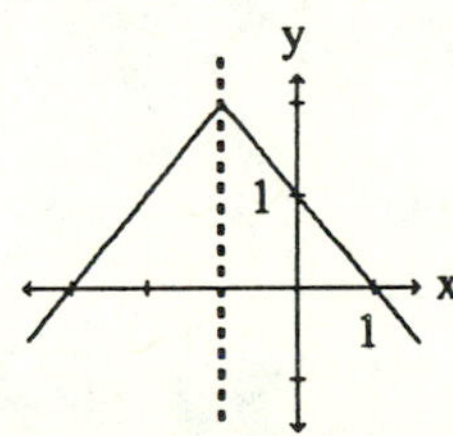

(c) Dom: All real numbers.

Range: All real numbers ≤ 2.

31. Dom: All real numbers.

Range: All real numbers ≤ 5.

Line of Symmetry: $x = 0$.

y defined as a function of x.

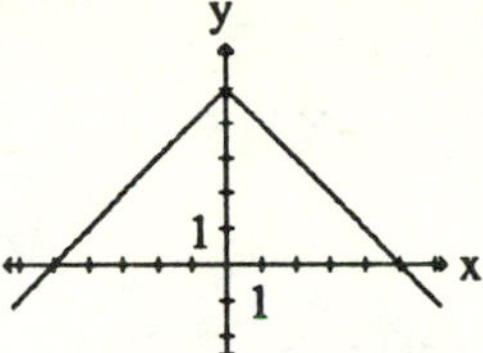

33. Dom: All real numbers.

Range: All real numbers≥ 0.

Line of Symmetry: $x = 0$.

y defined as a function of x.

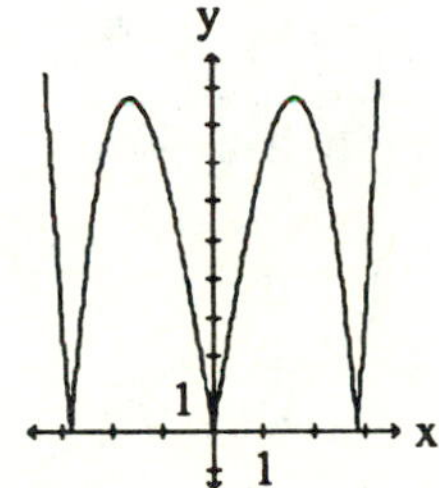

35. Dom: All real numbers $\neq 2$.

Range: All real numbers> 0.

Line of Symmetry: $x = 2$.

y defined as a function of x.

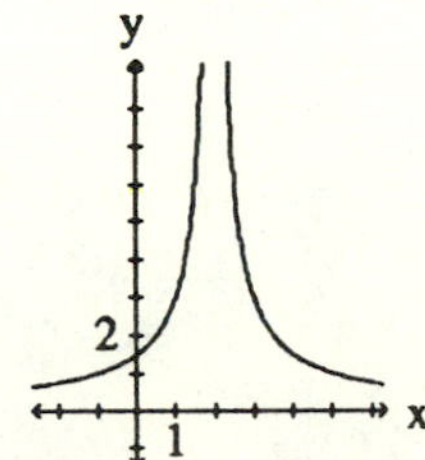

37. (a) (c)

39. No, it is not a function.

41. Yes, it is a function.

43. (a) $f(0) = 0$

$f(-1) = -2$

$f(2) = 4$

(b) $x = 0, 3$

(c) $x = 0.25, 2.75$

45. d

In problems 47–52 we know these are complete graphs because all critical portions are present and the end behavior is suggested.

47.

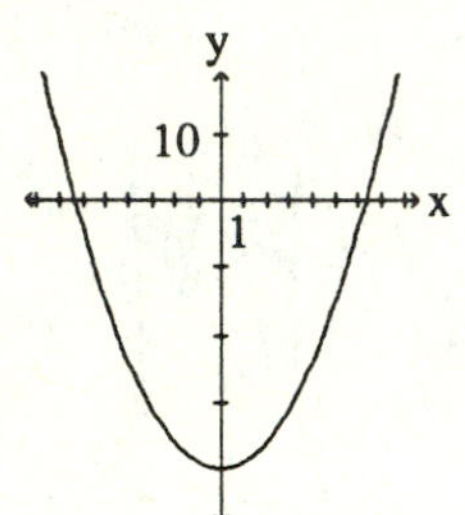

49.

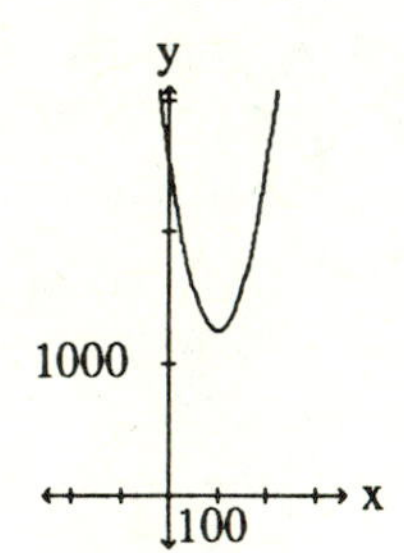

51.

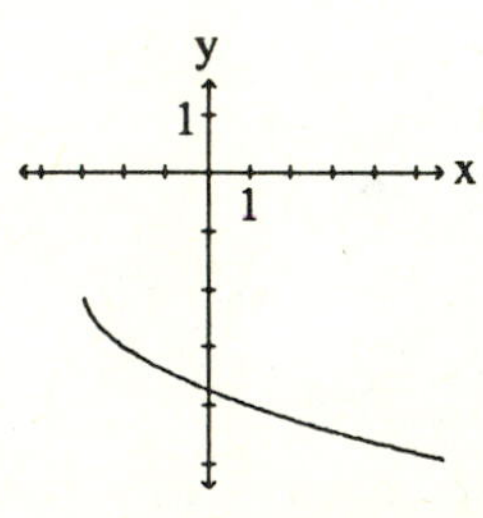

53. Dom: All real numbers.

Range: All real numbers ≥ -40.

55. Dom: All real numbers ≥ 3.

Range: All real numbers ≥ 0.

57. Dom: All real numbers.

Range: All real numbers.

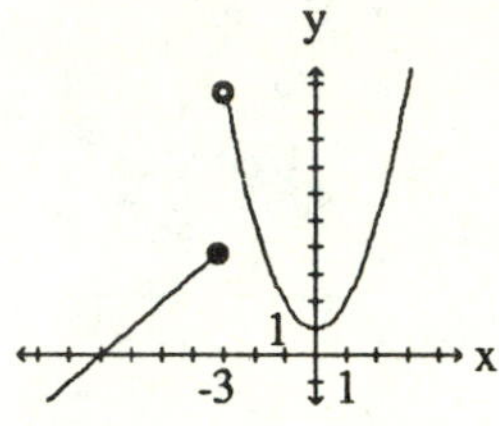

59.

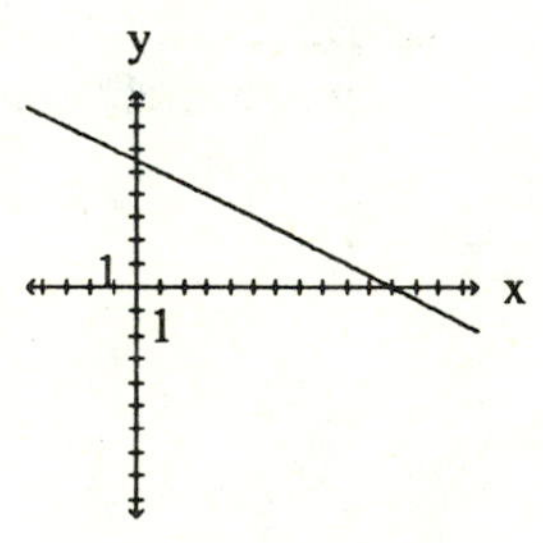

61.

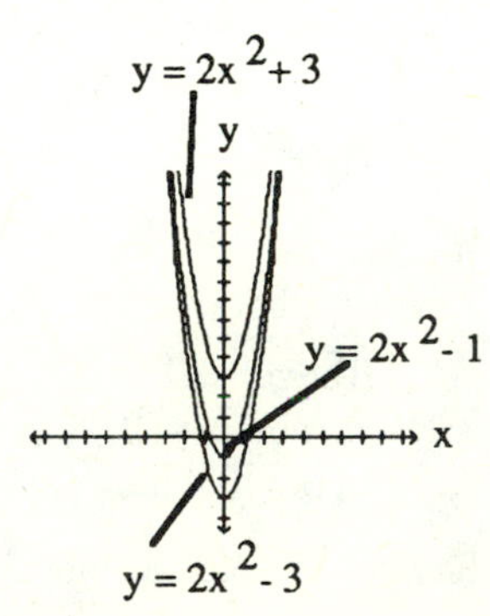

63. $(0, 4)$

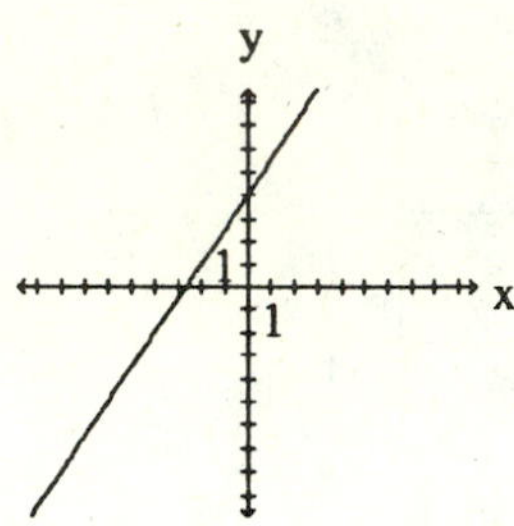

65. $f(x) = 5x + 250$

67.

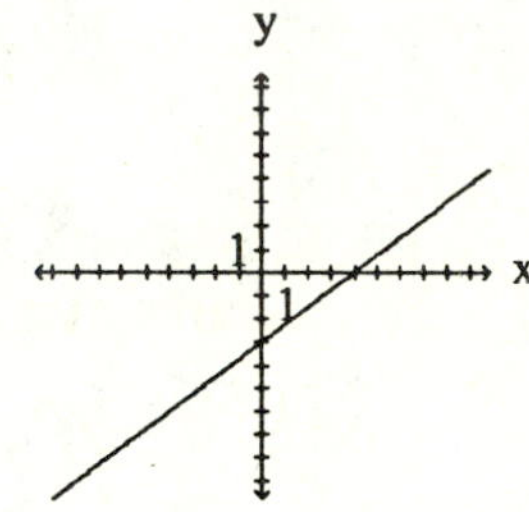

69.

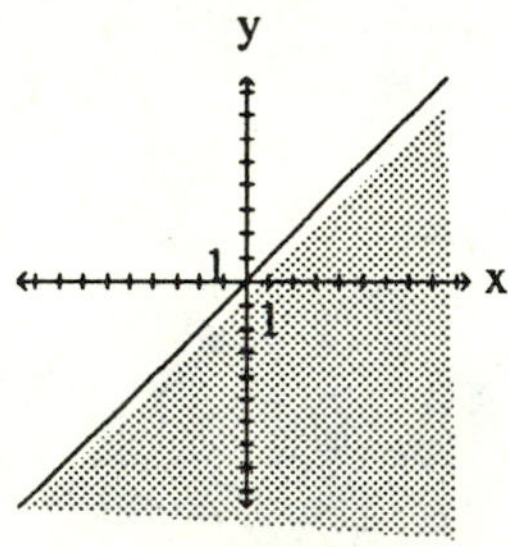

71. $y = \frac{3}{4}x + 1\frac{1}{4}$

73. (a) $x = 5$

(b) $y = 7$

75.

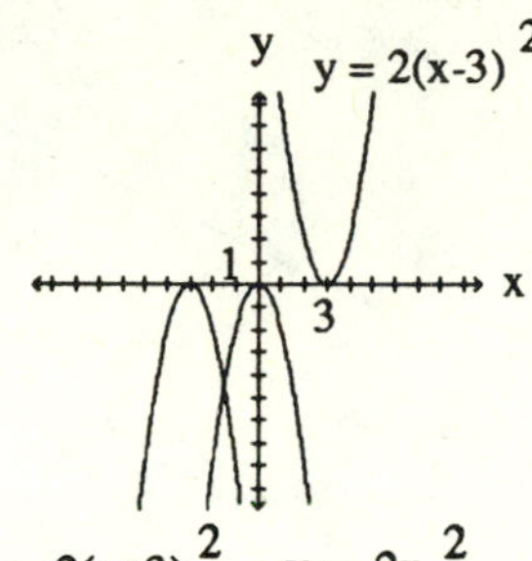

77. Horizontal shift right of 3 units

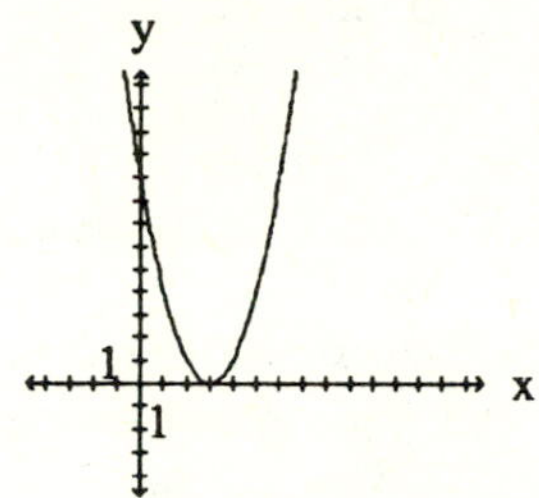

79. Horizontal shift left of 1 unit then vertical shift downward 4 units.

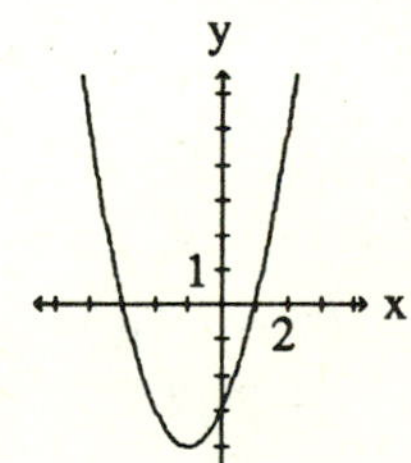

81.

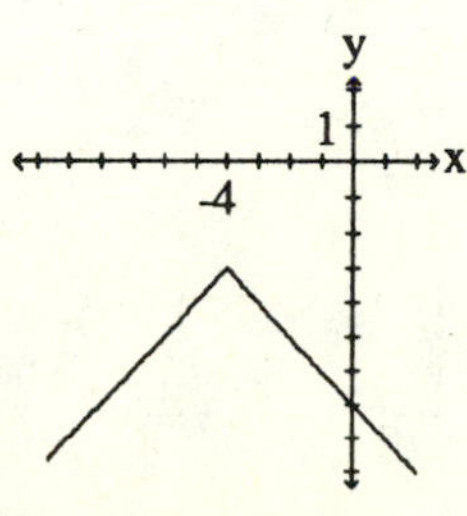

Horizontal shift left 4 units, then a reflection through the x-axis followed by a vertical shift down 3 units.

83. (a.) $b = 2$

(b.) $b = 0$

(c.) $b = -6$

(d.) $b = 2$

85. Vertical stretch by a factor of 2, then horizontal shift right 3 units, then vertical shift down 14 units. The vertical shift must come after the stretch.

87. vertex $(3, -14)$

line of symmetry $x = 3$

x-intercepts: $0.35, 5.65$

y-intercept 4

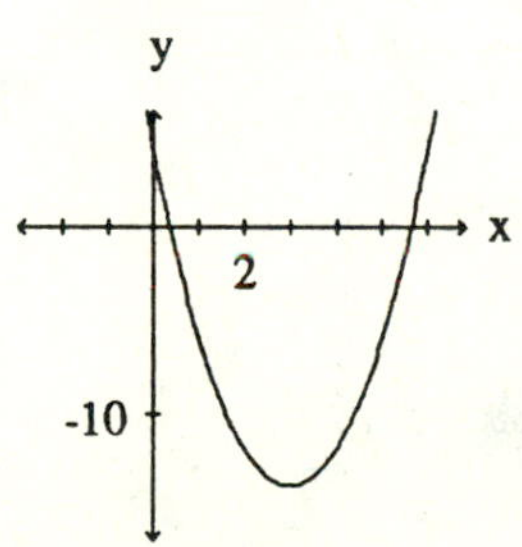

89. $x = 5.65$ and $x = 0.35$

91. No real solutions.

93. Shift horizontally left 2 units.

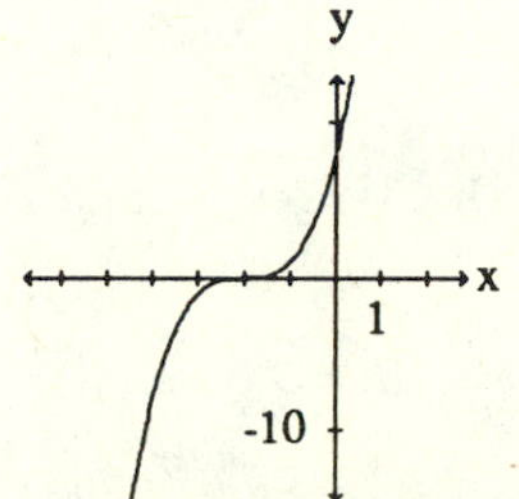

95. $f \circ g(-3) = -28$

$g \circ f(2) = 13$

97. $f \circ g(x) = \frac{x^2+6x+17}{4}$

$g \circ f(x) = \frac{x^2+5}{2}$

99. $f + g$: $\frac{x^3+3x^2-3}{x-1}$

Dom: $(-\infty, 1) \cup (1, \infty)$

$f - g$: $\frac{-x^3-3x^2+5}{x-1}$

Dom: $(-\infty, 1) \cup (1, \infty)$

fg: $\frac{(x+2)^2}{x-1}$

Dom: $(-\infty, 1) \cup (1, \infty)$

f/g: $\frac{1}{(x-1)(x+2)^2}$

Dom: $(-\infty, -2) \cup (-2, 1) \cup (1, \infty)$

101. Let $g(x) = (x-2)^2$ and $h(x) = \frac{3}{x}$

103.

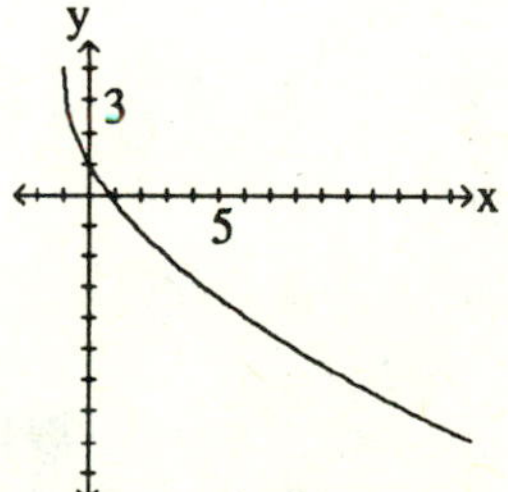

105. (a) $P = 0.50x - 18.25$

(b)

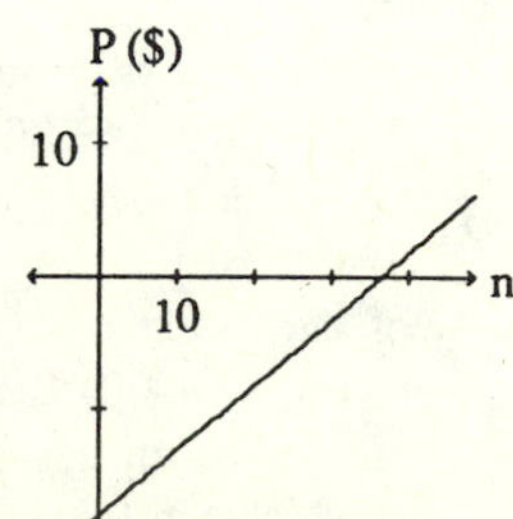

(c) Dom: All real numbers.

Range: All real numbers.

(d) The points (x, P), for nonnegative integers x, represent the problem situation.

(e) 37 tickets

107. (a) 10,000 square inches.

(b) Since $500 = 2l + 2w$, $250 = l + w$. If $l = x$ then $w = 250 - x$. Thus $A = lw = x(250-x)$.

(c)

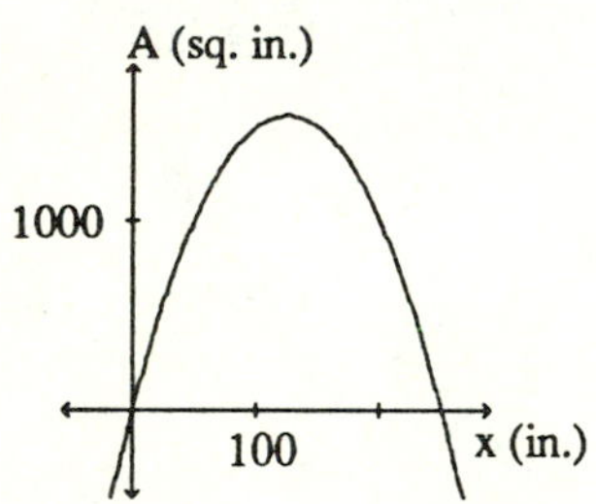

(d) If $l = 100$ then $w = 250 - 100 = 150$ and $A = 15000$. The coordinates of the point $(100, 15000)$ represent the length and area, respectively.

(e) If $l = -50$ then $w = 250 - (-50) = 300$ and $A = -15000$. The coordinates of the point $(-50, -15000)$ take on no meaning in this problem situation since length and area cannot take on negative values.

(f) All positive values of x less than 250 may be used in this problem situation, since length and width cannot have negative values.

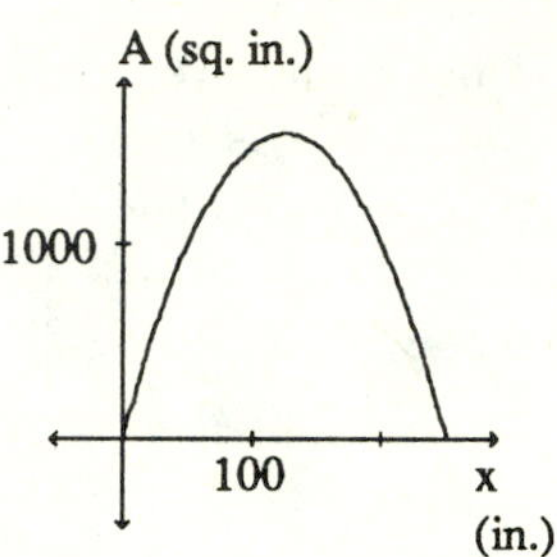

109. (a) $V = 40H$

(b)

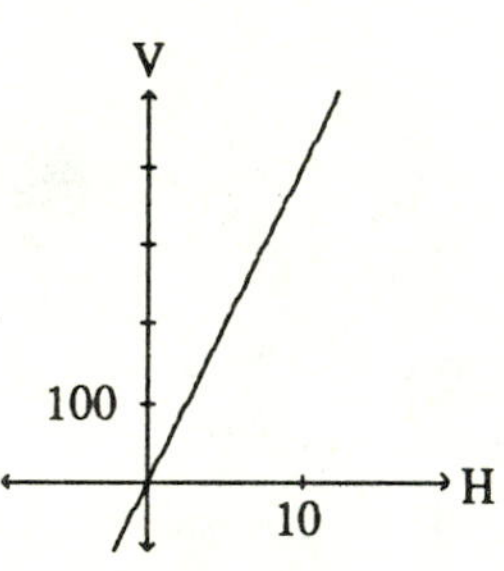

(c) Dom: All real numbers.

(d) Range: All real numbers.

(e) $H = 7.5$ units.

111. (a) $C = 85x + 75000$

(b)

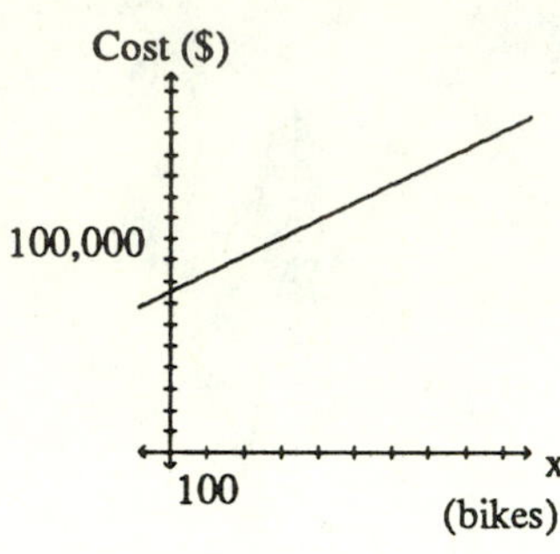

(c) Dom: All real numbers.

Range: All real numbers.

(d) The portion of the graph in the first quadrant represents possible total annual costs of producing bikes.

(e) 800 bikes.

(f) A represents the number of bikes made.

113. (a) $A = x\left(\frac{72-x}{2}\right)$

(b) Dom: All real numbers.

Range: All real numbers ≤ 648.

(c)

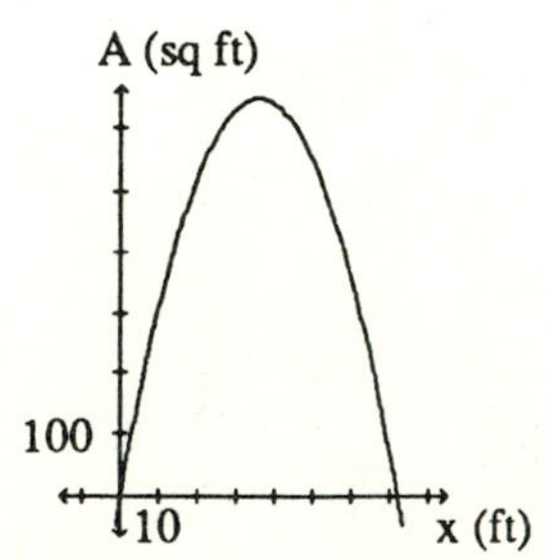

(d) All positive values of x less than 72 may be used in this problem situation.

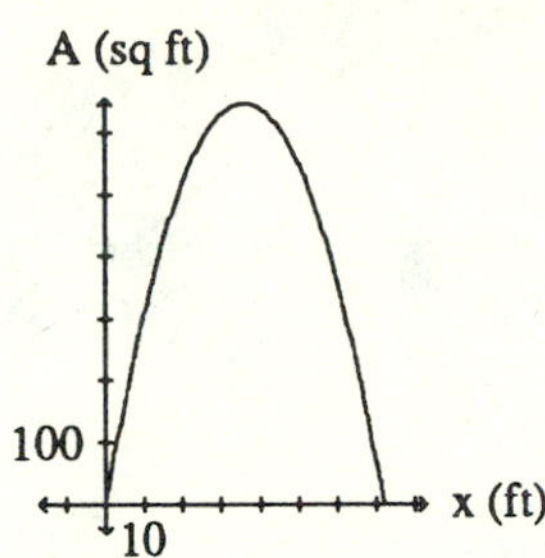

(e) $x = 65.94$ feet or $x = 6.06$ feet.

(f) Yes; 648 square feet.

115. (a) $D = 48t$

(b)

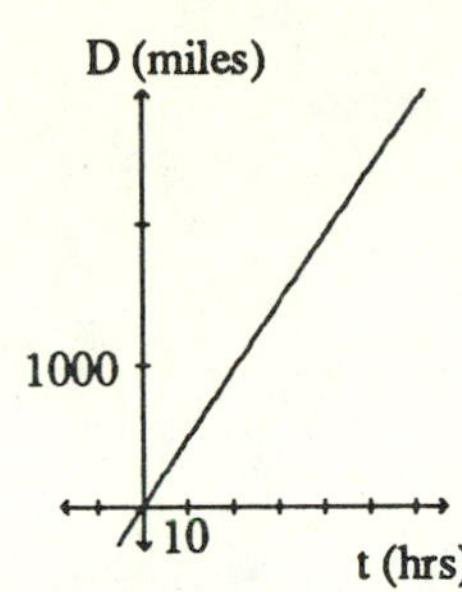

(c) Dom: All real numbers.
Range: All real numbers.

(d) All positive values of x may represent the problem situation.

(e) $t = 25$ hours.

117. (a) $s = -16t^2 + 250t + 30$

(b) The portion of the graph in the first quadrant represents the problem situation.

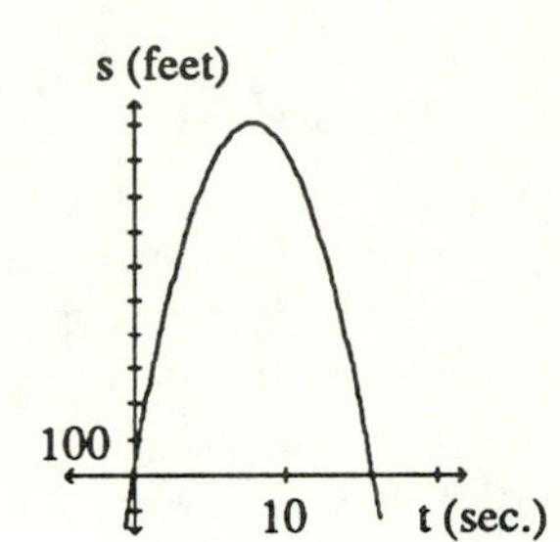

(c) $t = 2.47$ seconds, 13.15 seconds.

CHAPTER 2

Exercises 2.1

1. $x = -2, 5$

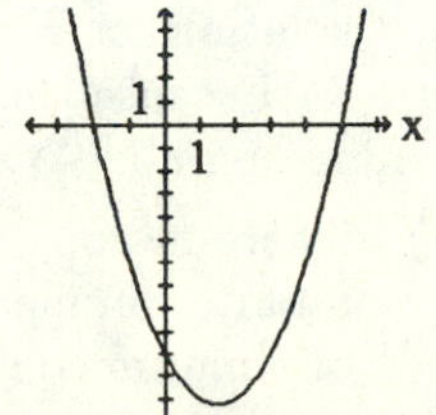

3. $x = -12, 4$

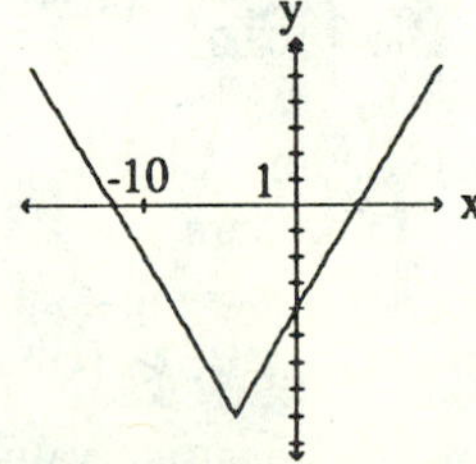

5. $x = \pm\sqrt{14}$

7. no real solutions

9. $x = -4, 8$

11. $x = 8, -3$

13. No; -1 and 2 are solutions. Another solution is $x = -3$. The x-intercepts of the graph are the solutions to the equation.

15. 3

17. $x = \frac{2}{3}$

19. Answers will vary. One possible solution is: $[-2, -1]$ by $[-1, 1]$, $[-1.9, -1.8]$ by $[-0.1, 0.1]$, $[-1.9, -1.89]$ by $[-0.01, 0.01]$, $[-1.894, -1.893]$ by $[-0.001, 0.001]$

21. 1.52 with error at most 0.06 for $f(x)$

23. 10.19 with error at most 0.09 for $f(x)$

25. $x = 0.61, 4.39$

27. $x = 5, 15, 40$

29. 0.70, 3.39

31. 1.86, 5, 8.14

33. Change $b_1x^3 + b_2x^2 + b_3x + b_4 = 0$, where $b_1 \neq 0$, to $x^3 + \frac{b_2}{b_1}x^2 + \frac{b_3}{b_1}x + \frac{b_4}{b_1} = 0$. Use the formula for the solution to $x^3 + a_1x^2 + a_2x + a_3 = 0$ substituting $\frac{b_2}{b_1}$ for a_1, $\frac{b_3}{b_1}$ for a_2, and $\frac{b_4}{b_1}$ for a_3.

35. (a) $y = \frac{216}{x}$

(b)

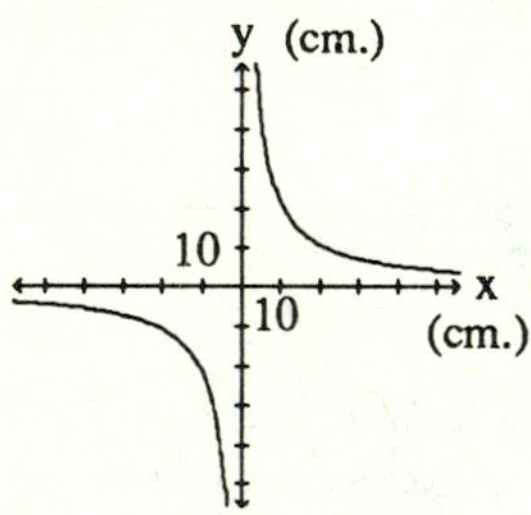

(c) The part of the graph in the first quadrant represents the problem situation.

(d) Width $= y = 9$ centimeters.

37. $C = \frac{5(F-32)}{9}$

39. (a) $n = \frac{1}{r}\left(\frac{S}{P} - 1\right)$

(b) 25 years

41. Side length = 11.98 units

43. (a)

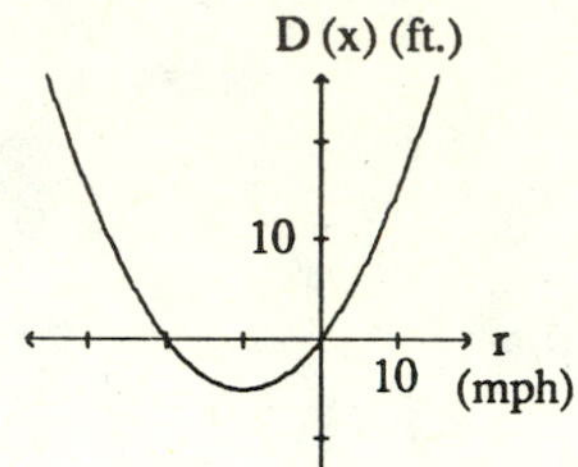

(b) Dom: All real numbers.

Range: All real numbers ≥ -4.96.

All positive values of r make sense in this problem, although, realistically, one is confined to a speed limit.

(c) 67.88 mph.

45. (a)

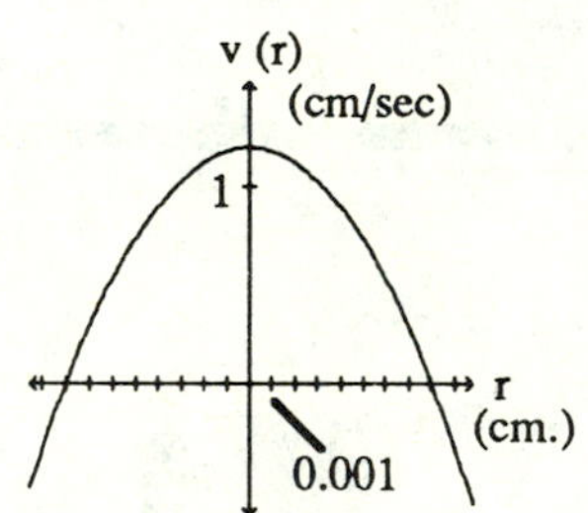

(b) Dom: All real numbers.

Range: All real numbers ≤ 1.19.

All positive values of $r < 0.008$ make sense in this problem situation.

(c) $r = 0.003$ centimeters.

47. $p = \$12.24$; $x = 553$ units (Integer answers given for problem situation.)

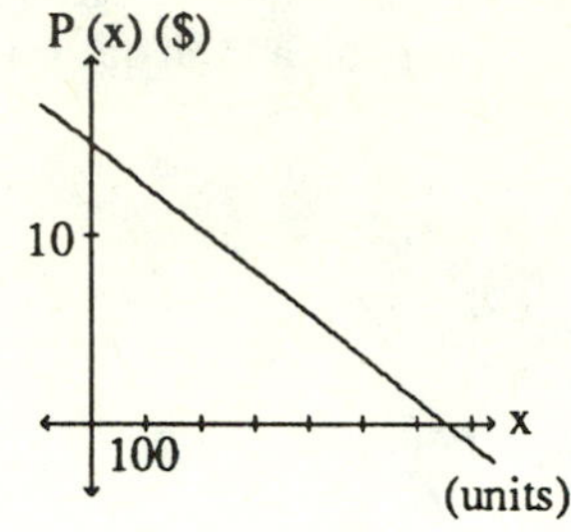

49.

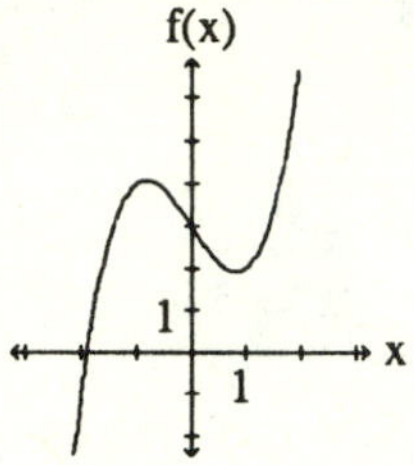

(a) 1

(b) 3

(c) $[-0.82, 0.82]$

51. $x = 3$

Exercises 2.2

1. $(-1, 7)$

3. $(8, -2)$

5. (a) There is not a simultaneous solution for E_1 & E_2 because the lines are parallel.

(b) There is an infinite number of simultaneous solutions because

E_1 & E_3 are the same line.

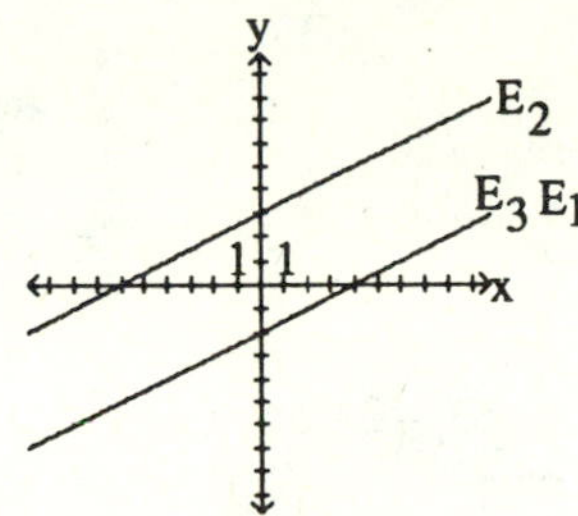

7. $(x, y) = (9, -5)$

9. $(x, y) =$ $(0,0); (\sqrt{2}, \sqrt{2}); (-\sqrt{2}, -\sqrt{2})$

11. $(-2.89, -2.36)$, $(2.49, -0.20)$

13. $(1.32, 2.27)$

15. $(-3.84, -4.76)$, $(0.62, 9.62)$, $(4.23, -7.86)$

17. $(-1.73, -0.99)$, $(1.06, 0.87)$

19. $(-1.32, -0.97)$, $(0, 0)$, $(1.32, 0.97)$

21. White sugar: 7.08 ounces

    Brown sugar: 15.92 ounces

23. Rowing speed: 3.56 mph

    Current speed: 1.06 mph

25. $a = \frac{1}{3}$, $b = \frac{-11}{3}$

27. 5%: $5494.50

    6%: $1098.90

    10%: $6593.40

29. 79

31. (a) $V(x) =$ $x(40 - 2x)(30 - 2x)$

    (b) $V(x) =$ $4x^3 - 140x^2 + 1200x$

    $V(x) = 1200$

    (c)

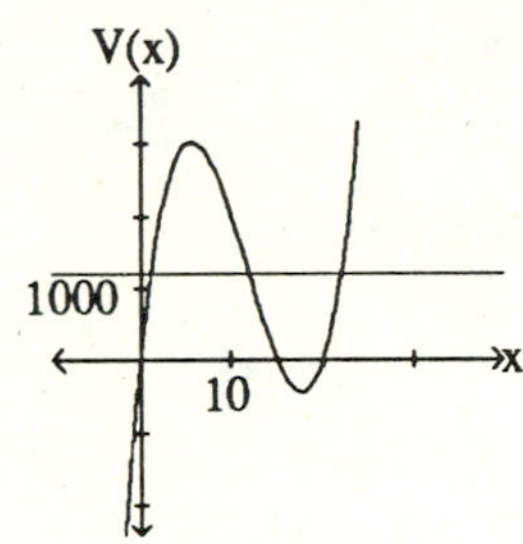

    (d) 3; the 2 solutions less than 20 are also solutions to the problem situation. The side lengths of the cut out squares can be no longer than 15 since one side of the box is 30 inches.

    (e) $x = 1.15$, maximum error in x is 0.0001.

    $x = 11.89$,maximum error in x is 0.0001.

33.

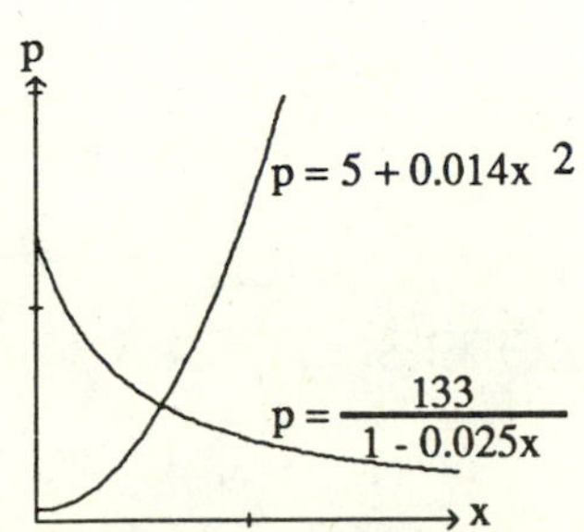

$53.74 ($x = 59$ units)

35. a

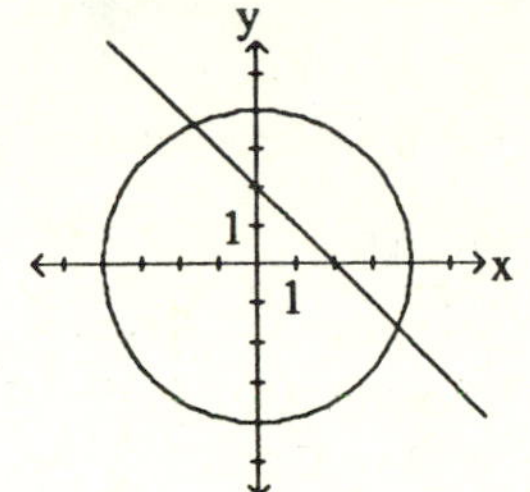

    (b) 2

    (c) $(-1.65, 3.65)$ & $(3.65, -1.65)$

    (d) see (c)

    (e)

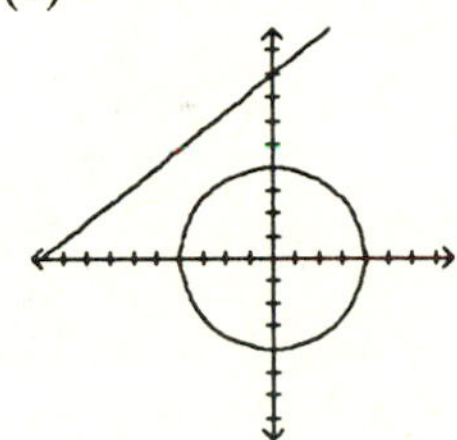

zero points of intersection

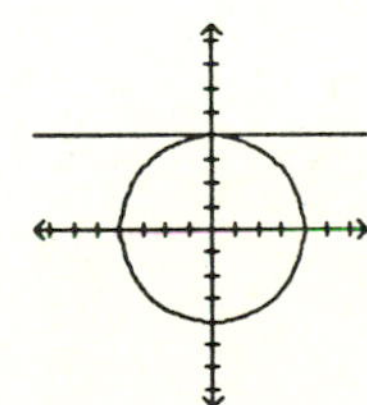

1 point of intersection

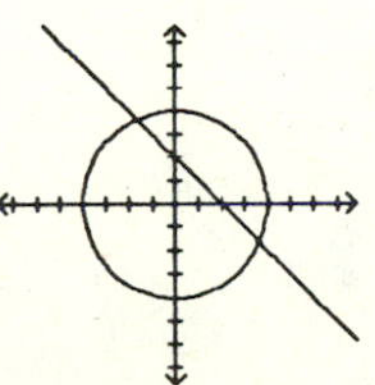

2 points of intersection

*37. $b = \pm\sqrt{32}$

Exercises 2.3

1. $(-1,1)$
3. $[-1,2)$
5. $(-\infty,4]$
7. $-11-(-7)$ must be negative for $-11<-7$. Since $-11-(-7)=-11+7=-4$,negative then $-11<-7$.
9. 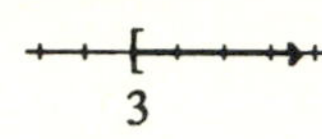
11. 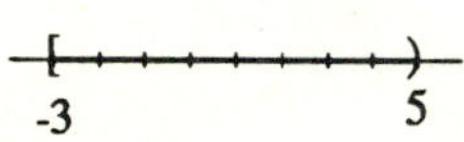
13. $(8,\infty)$;$x>8$; The set of all numbers greater than 8
15. $(-2,5)$
17. $(-2,5]$
19. (a) $f(-1)<2$
    (b) $g(2)<5$
21. (a) no
    (b) no
    (c) yes
23. $(-3,-1)$
25. $x<5$; $(-\infty,5)$

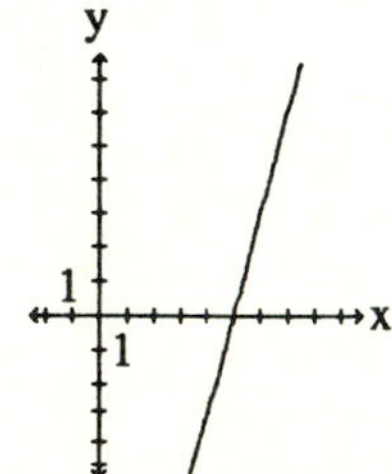

27. The graph of $y=3(x-1)+2-(5x+6)$ lies below the x-axis for values of $x\geq-3.5$ which is the solution set for Example 4.
29 Find where the graph of $y=3x-15$ lies below the x-axis.
31. $x\leq\frac{34}{7}$; $(-\infty,\frac{34}{7}]$
33. $x<\frac{33}{13}$; $(-\infty,\frac{33}{13})$
35. $-4\leq x<3$; $[-4,3)$
37. $-\frac{1}{2}\leq x\leq\frac{17}{2}$; $[-\frac{1}{2},\frac{17}{2}]$
39. $x>2$; $(2,\infty)$
41. $x>\frac{-33}{13}$;$(\frac{-33}{13},\infty)$
43. $x>\frac{-14}{3}$;$(\frac{-14}{3},\infty)$
45. She worked 4 hours or less.
47. $0<w<34$ inches

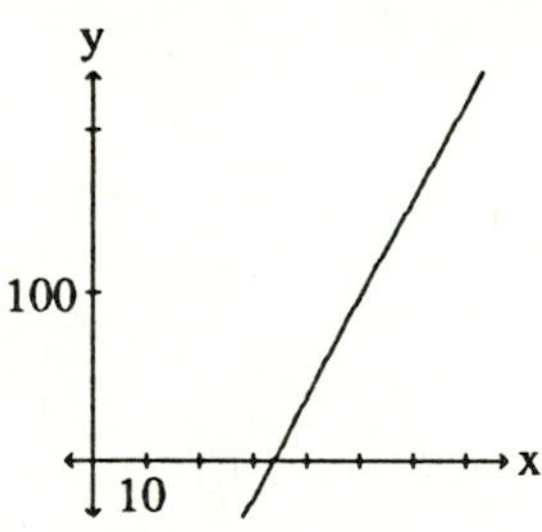

49. Counterexample:
    $-4<5$ and $-3<2$, but $12\not<10$. (Answers will vary)

*51. More than $100,000.

Exercises 2.4

1. $(-1,1)\cup(2,\infty)$
3. $[-1,0]\cup(1,\infty)$
5. 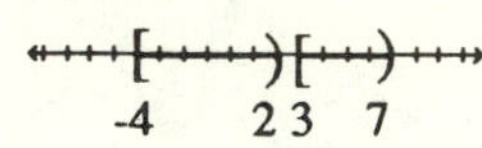
7. 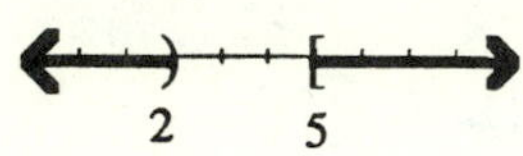
9. $(-\infty,-3)\cup(5,\infty)$; $x>5$ or $x<-3$; The set of all real numbers less than -3 or greater than 5.
11. $(-3,2)\cup(6,\infty)$; $-3<x<2$ or $x>6$; The set of all real numbers between -3 and 2 or greater than 6.
13. $(-1,7)$; $|x-3|<4$; The set of all real numbers less than 4 units from the point 3.
15. $(-\infty,1]\cup[6,\infty)$; $|x-3.5|\geq2.5$; The set of all real numbers at least 2.5 units from 3.5.
17. $(-\infty,-1)$, $(7,\infty)$

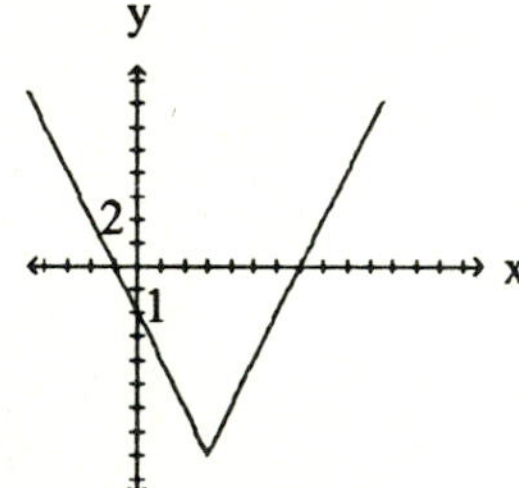

19. $(-\infty,\infty)$
21. $[-8,2]$
23. $(-\infty,-2]\cup[8,\infty)$
25. $(-6,14)$
27. $(-\infty,-\frac{4}{3})\cup(\frac{4}{3},\infty)$
29. $(-\infty,-2)\cup(-2,3)$

31. $|x| < 2$

33. $|x - 1| < 2$

35. (a) $f(2) < -2$
(b) yes
(c) $(0, 6)$

37. (a) $212 \leq \frac{9}{5}C + 32$
(b)

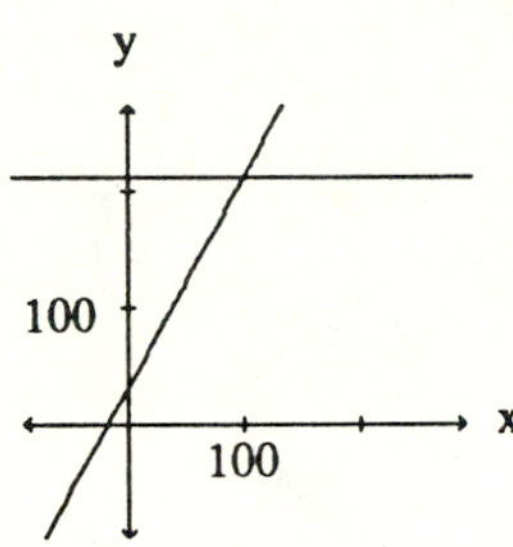

(c) $C \geq 100$

39. (a) $P = 0.42x - 20{,}000$
(b) $25{,}000 \leq 0.42x - 20{,}000$
(c)

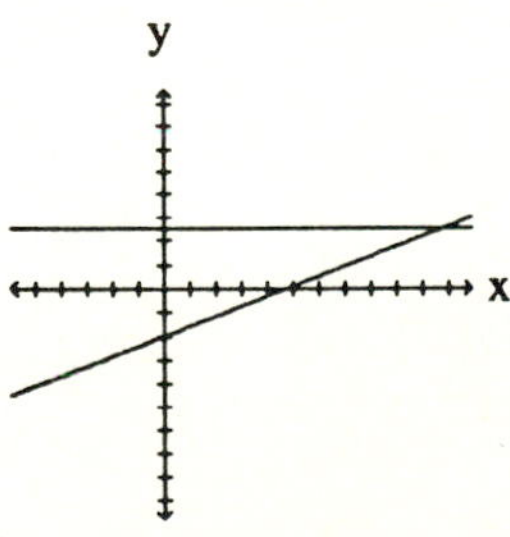

Scale: 10^4 per tick mark

(d) $x \geq 107{,}143$

41. $(4, \infty)$

43. $(-2.5, \infty)$

45. If $|x - 2| < d$ for $d < 0$, then $-d < x - 2 < d$ and $-d + 2 < x < d + 2$. So x is in the interval $(2 - d, 2 + d)$. Its length is $2 + d - (2 - d) = 2d$. The midpoint of the interval is 2 since $\frac{1}{2}$ the length of the interval is d and $(2 - d) + d = 2$.

*47. (a) $(0, 0.03)$
(b) $(0, 0.003)$
(c) $(0, 0.0003)$
(d) $(0, \frac{E+6}{3} - 2)$ or $(0, \frac{E}{3})$

Exercises 2.5

1. $(-2, 1)$

3. $(-\infty, 2] \cup [3, \infty)$

5. $(-\infty, 3] \cup [\frac{1}{2}, \infty)$

7. $(-1, 1)$

9. $(-\infty, -3] \cup (1, \infty)$

11. $(-\infty, -4) \cup (-1, 2)$

13. $(-\infty, 1] \cup [2, 2]$

15. $(3\frac{1}{3}, 4)$

17. $(-\infty, 0.70) \cup (4.30, \infty)$

19. $(-\infty, -1] \cup [0.67, 3]$

21. $[2.55, \infty)$

23. $[2.15, \infty)$

25. (a) $s = -16t^2 + 100t + 200$
(b)

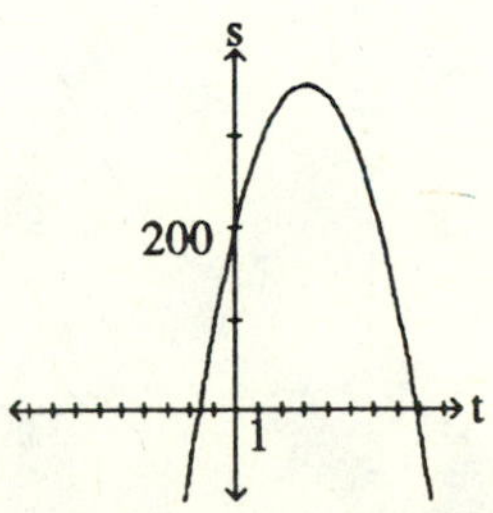

(c) Dom: $(-\infty, \infty)$
Range: $(-\infty, 356.25]$
(d) The portion of the graph in the first quadrant
(e) For $0 \leq t < 7.84$ seconds
(f) $t = 7.84$ seconds
(g) $s = 356.25$feet

27. i)

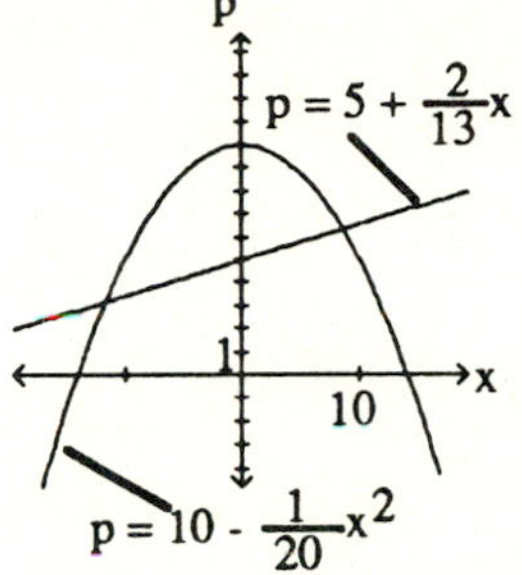

ii) The portion of the graphs in the first quadrant represent the problem situation.
iii) \$6.32
iv) 9

29. (a) $A(x) = 4x(x + 25)$
(b) $200 < 4x(x + 25) < 360$
(c) $(-28.19, -26.86) \cup (1.86, 3.19)$

Solution to problem situation:

$(1.86, 3.19)$

31. (a) $2000 > x(22 - 2x)(29 - 2x)$

(b)

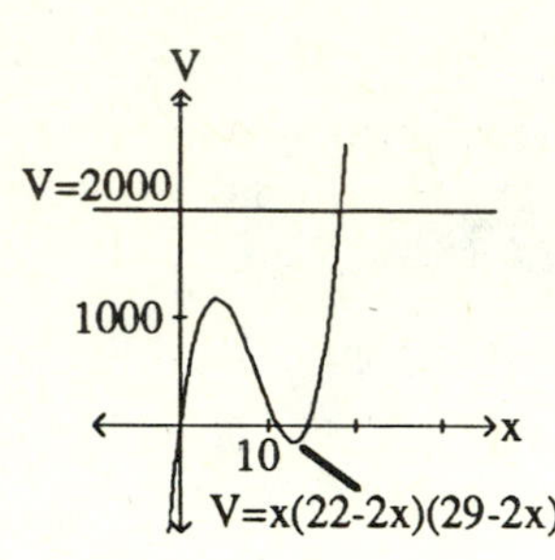

(c)

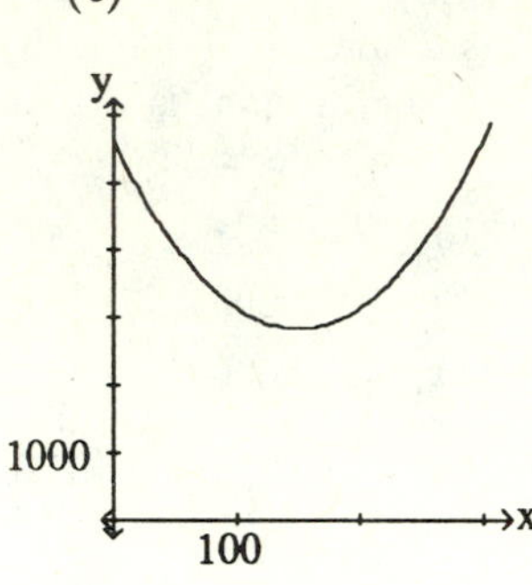

(d) $(52.53, 247.47)$

37. $[0.93, 3.64]$

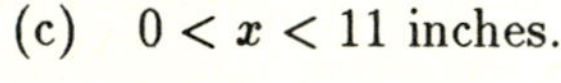

(c) $0 < x < 11$ inches.

*33. (a) Always true

(b) Counterexample:

$|-5+3| \neq |-5|+|3|$

(c) Always true

(d) Counterexample:

$|-5+3| \not\geq |-5|+|3|$

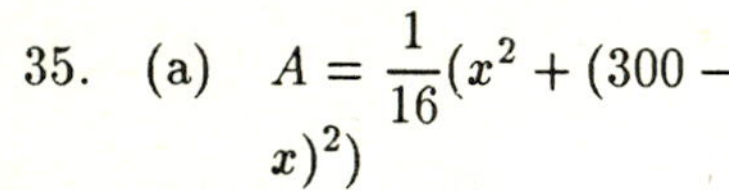

35. (a) $A = \frac{1}{16}(x^2 + (300 - x)^2)$

(b)

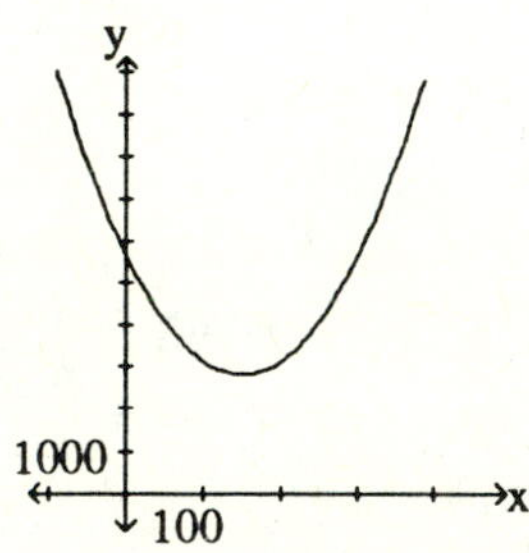

Exercises 2.6

1. $[-5, 5]$ by $[-2, 10]$

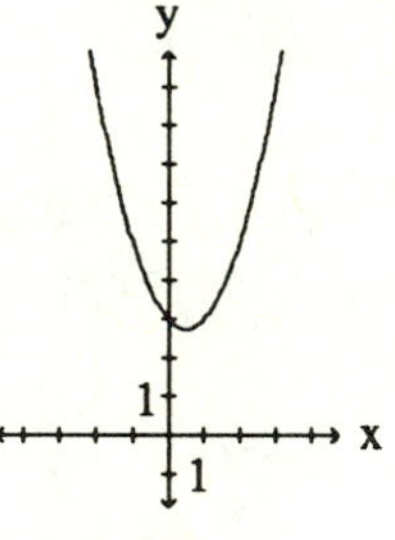

3. $[-100, 700]$ by $[-5000, 20000]$

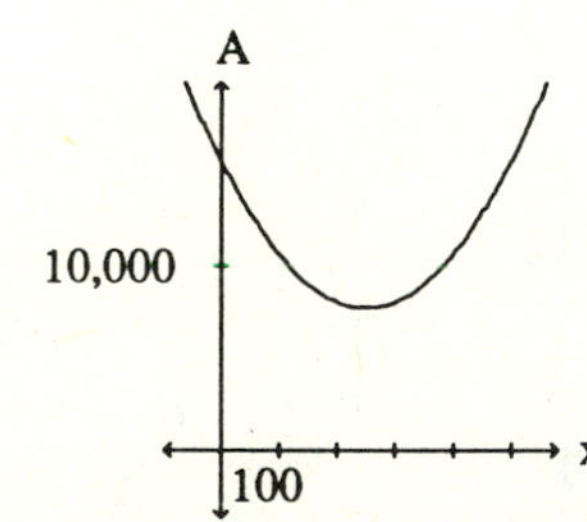

5. $[-5, 30]$ by $[-60000, 100000]$

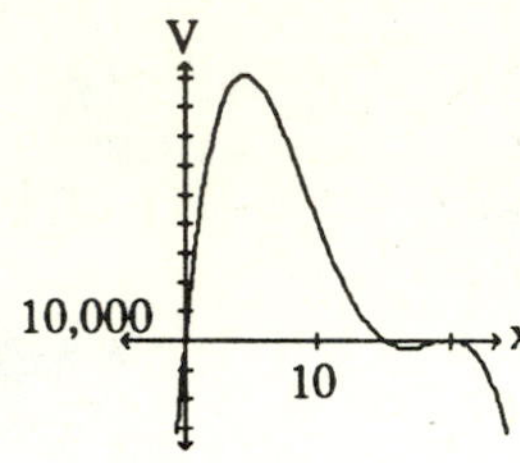

7. $[-1000, 3000]$ by $[-15000000, 2000000]$

9. local minimum at $(\frac{3}{14}, \frac{831}{28})$

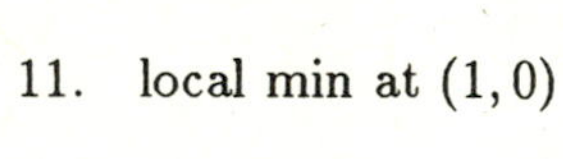

11. local min at $(1, 0)$

13.

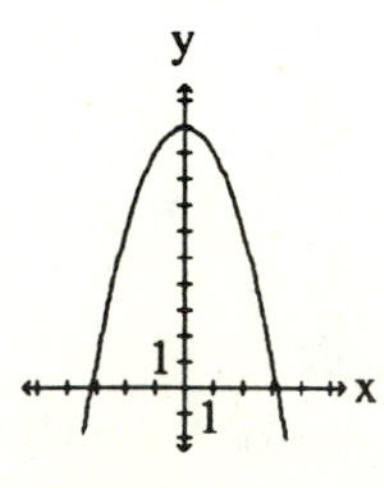

zeros: $x = \pm\sqrt{10}$

local maximum at $(0, 10)$

15.

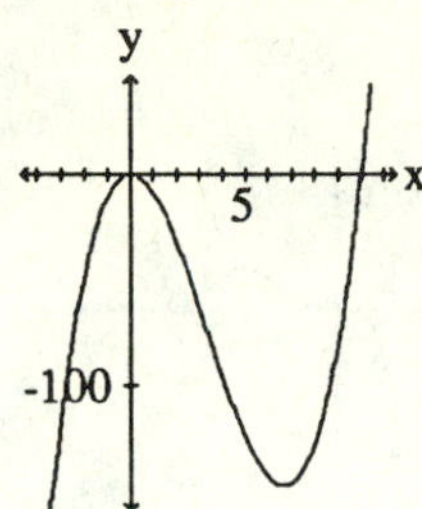

zeros: $x = 0, x = 10$

local max at $(0, 0)$

local min at $(6.67, -148.15)$

17.

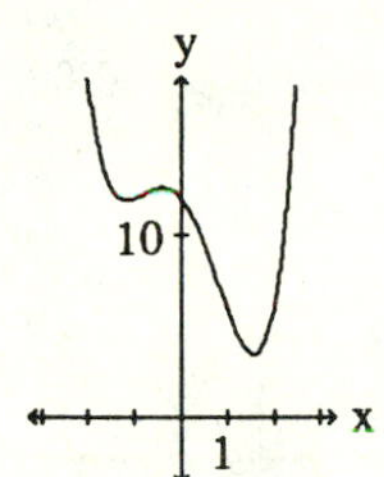

no real zeros

local max at $(-0.41, 12.59)$

local min at $(-1.16, 11.91)$ and $(1.57, 3.51)$

19.

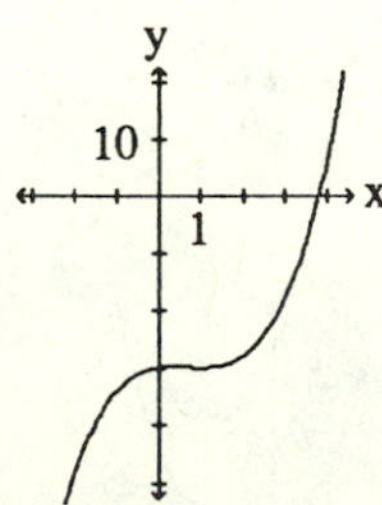

zeros: $x = 3.81$

Local maximum $(0.33, -29.85)$

Local minimum $(1, -30)$

21.

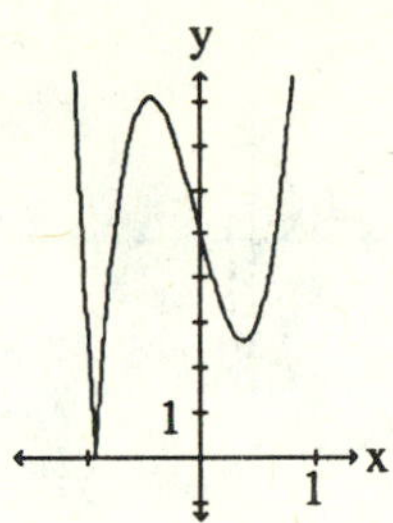

zeros: $x = -.93$

Local maximum $(-0.44, 8.08)$

Local minimum $(0.38, 2.59)$ and $(-0.93, 0)$

23. zeros: $x = 2.36$.

Local maximum $(0.79, 12.10)$.

Local minimum $(0.21, 11.90)$.

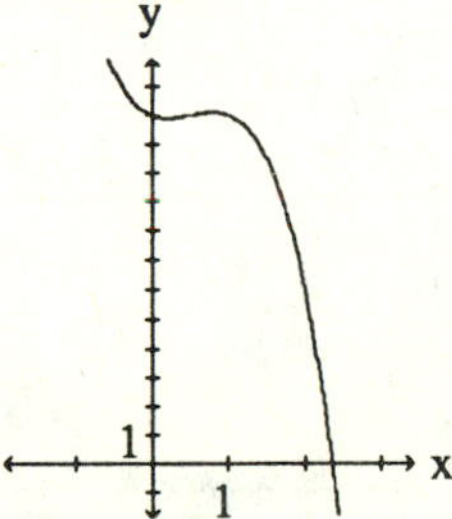

25. (a) $s(t) = -16t^2 + 40t + 300$

(b) $t = 5.76$ sec.

(c) maximum height of 325 ft. at 1.25 seconds.

27. 262.5 ft. by 525 sq. ft. with maximum area 137812.5 sq.ft.

29. (a) $R = x(2 + .002x - .0001x^2)$

(b)

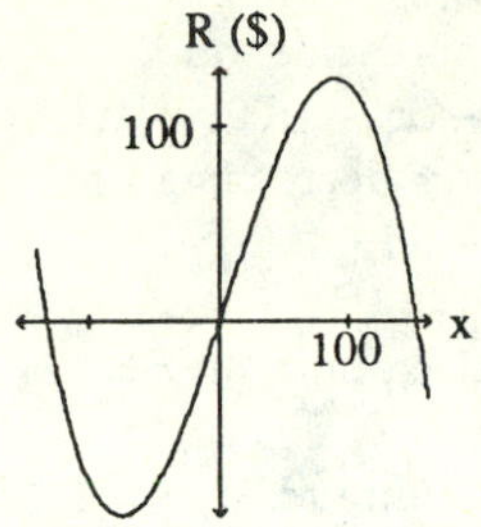

(c) See (b).

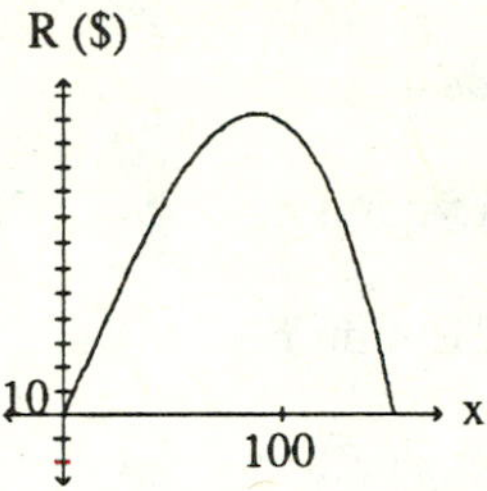

(d) $x = 89$, $R = \$123.35$

31. Length of sides are 10.93 in., 33.1 in., and 64.1 in. which gives maximum volume of 23231.91 cu. in.

Exercises 2.7

1. Vertical shift down 4 units.
3. Vertical stretch by a factor of 3, horizontal shift left 4 units then vertical shift down 5 units.
5. Reflect through the x-axis, horizontal shift right 2 units, then vertical shift down 2 units.
7. Increasing on $[1.5, \infty)$ and decreasing on $(-\infty, 1.5]$.

Minimum at $x = 1.5$

9. Increasing on $(-\infty, -1.79] \cup [1.12, \infty)$ and decreasing on $[-1.79, 1.12]$

Maximum at $x = -1.79$

Minimum at $x = 1.12$

11. Increasing on $(-\infty, 0.33] \cup [1, \infty)$ and decreasing on $[0.33, 1]$

Maximum at $x = 0.33$

Minimum at $x = 1$

13. g is not an increasing function nor a decreasing function on any interval.

15.

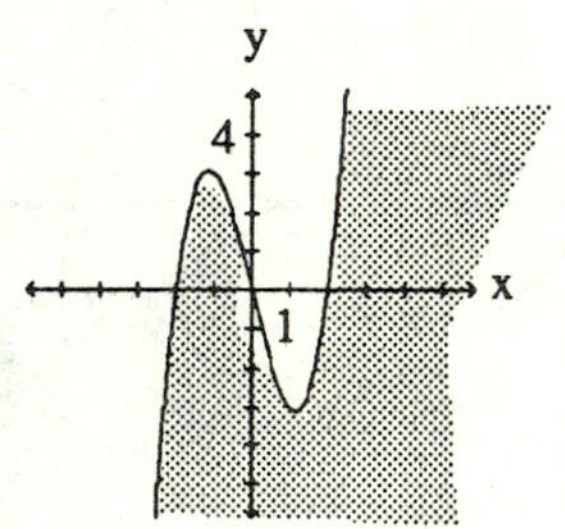

17.

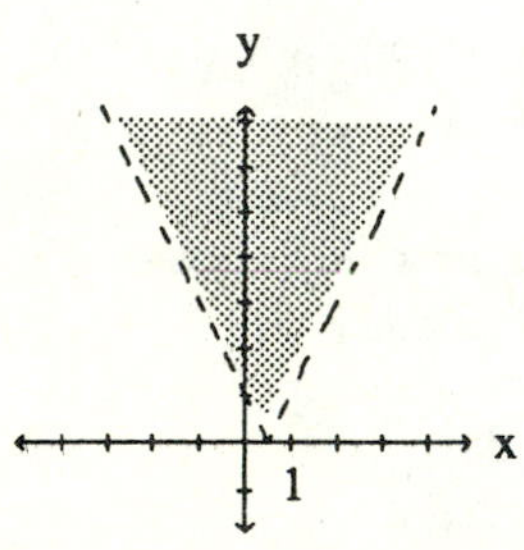

19. (a)

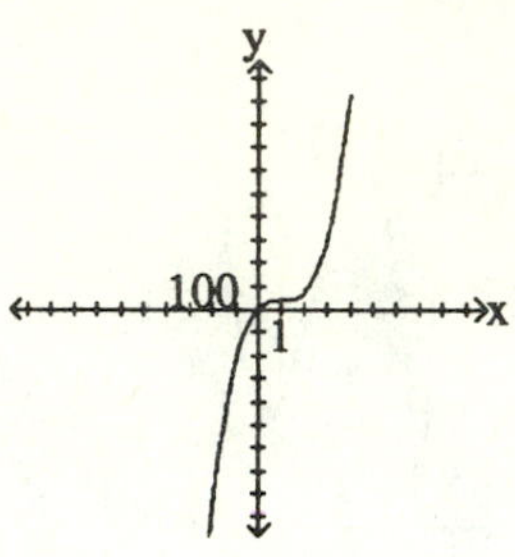

(b)

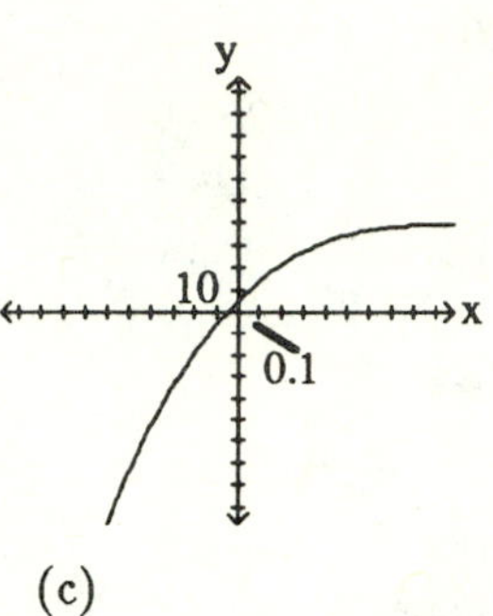

(c)

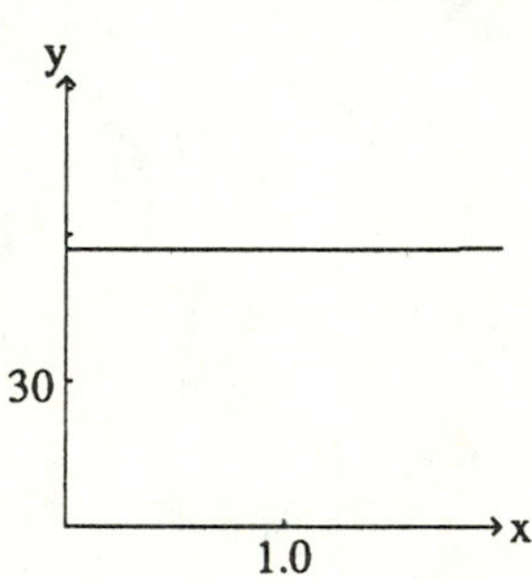

21. Expanded $a(x-b)^3 + c = ax^3 - 3abx^2 + 3ab^2x + (c - ab^3)$. If it is to equal $x^3 + x + 1$, then $-3ab = 0$ which implies that $a = 0$ or $b = 0$. If $a = 0$, then the x^3 term will also drop out. If $b = 0$, then the x term also drops out. Therefore, there can be no real numbers for a, b, and c.

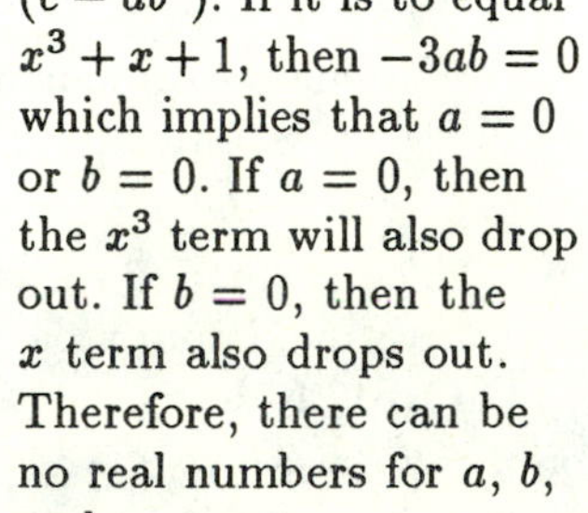

23.

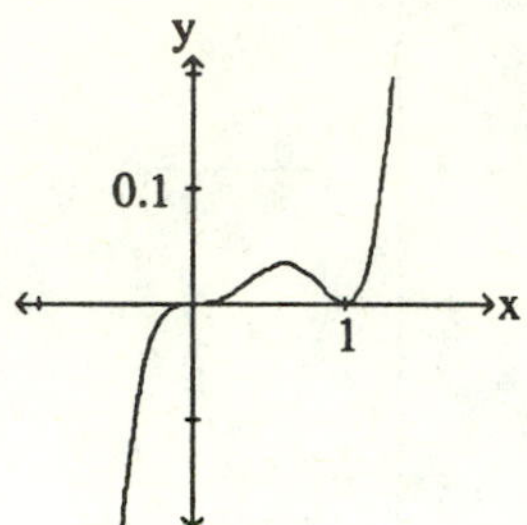

local maximum $(0.60, 0.03)$

local minimum $(1.00, 0)$

real zero: $x = 0,\ 1$

increasing on $(-\infty, 0.60]$, $[1.00, \infty)$

decreasing on $[0.60, 1.00]$

25.

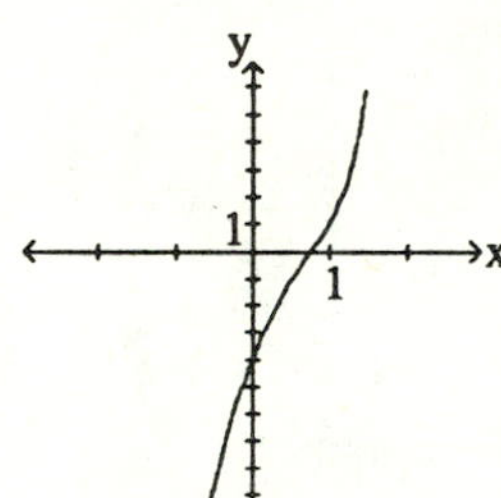

no local minimum or maximum

real zero: $x = 0.77$

increasing on $(-\infty, \infty)$

27. (a) $y = (1/4x)^2 + (75 - 1/4x)^2$

(b)

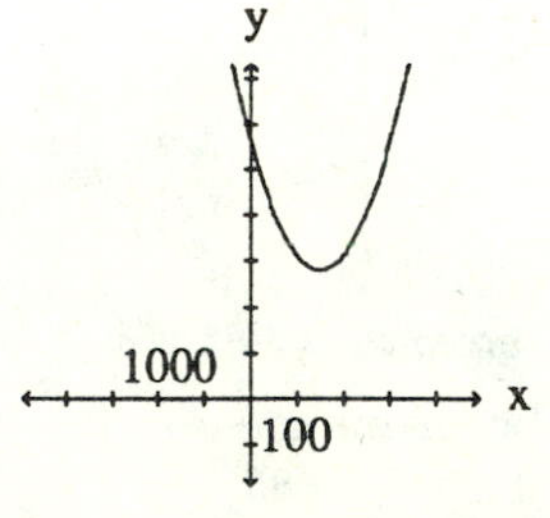

(c) The portion of the graph for which $0 < x < 300$ represents the problem situation.

(d) Increasing on $[150, \infty)$ and decreasing on $(-\infty, 150]$

29. (a) $V = x(50 - 2x)(85 - 2x)$

(b)

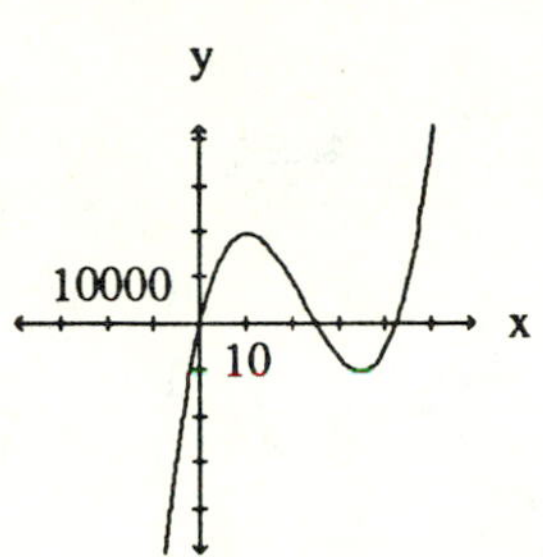

(c) The portion of the graph for which $0 < x < 25$ represents the problem situation.

(d) Increasing on $(-\infty, 10.17] \cup [34.83, \infty)$ and decreasing on $[10.17, 34.83]$

31. (a) $R = x(1.2x - 0.01x^2)$

(b)

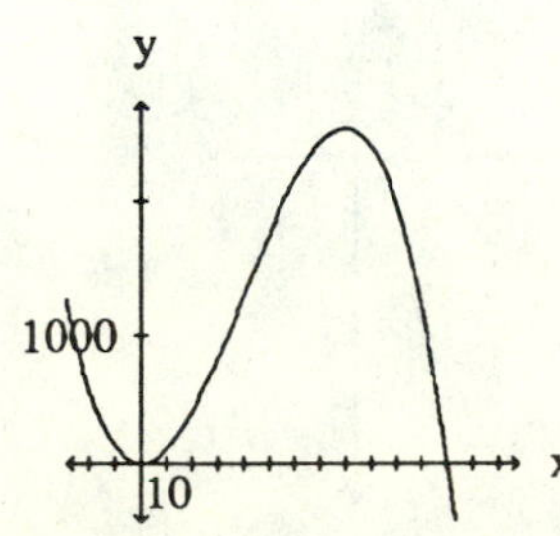

(c)

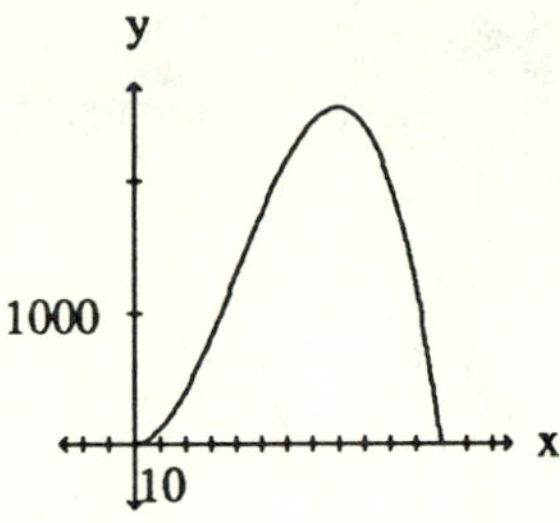

(d) Increasing on $[0, 80]$ and decreasing on $(-\infty, 0] \cup [80, \infty)$

Review Exercises

1. $x = \frac{27}{40}$
3. $x = 7, -3$
5. $x = \pm\sqrt{3}$
7. no real solutions
9. $x = -1.95$
11. No real solutions.
13. 1 solution
15. 1 solution
17. $(-0.67, 3.67)$ $(2.00, 9)$
19. $(-3.26, 5.36)$
21. $(-6.40, 49.68)$, $(0.40, 46.02)$, $(5.97, 47.61)$
23.

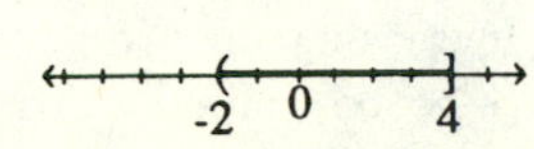

25. $-4 \leq x < 2$; The set of all real numbers less than 2 but greater than or equal to -4.
27. $(-4, 1) \cup (7, \infty)$; $x > 7$ or $-4 < x < 1$
29. (a) $f(-2) > g(-2)$
    (b) $f(2) < g(2)$
31. $(-\infty, -\frac{9}{2}]$
33. $[1, \frac{23}{4})$
35. No solution.
37. $(-\infty, -1] \cup [4, \infty)$
39. No real solutions.
41. $(-\infty, 2]$
43. $(-5, 2)$
45. $(-\infty, 0.23] \cup [1.43, \infty)$
47. $[-0.91, \infty)$
49. (a) $y = \frac{420}{x}$
    (b) 35 cm.
51. 8
53. (a) 16150 square inches.
    (b) If $x = l$ then $w = \frac{1}{2}P - x = \frac{1}{2} \cdot 530 - x = 265 - x$. The area of a rectangle is length times width thus $A = x(265 - x)$.
    (c)

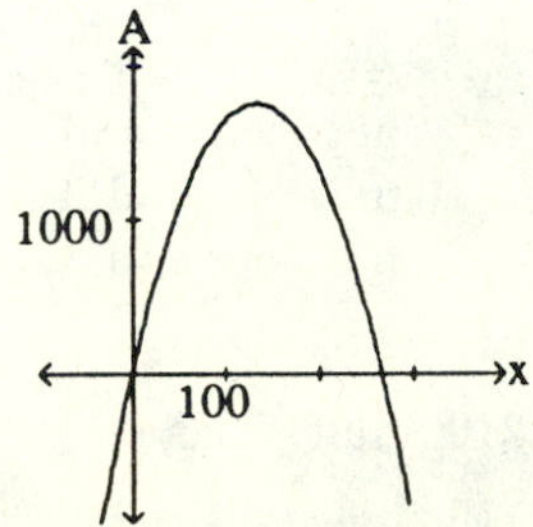

(d) If $x = 100$, then $A = 100(265 - 100) = 16500$. Thus $(100, 16500)$ is a point on the graph. This means if $l = 100$, then area is 16500.

(e) If $x = -100$ then $A = -100(265 + 100) = -36500$. Thus $(-100, -36500)$ is a point on the graph. These coordinates take on no meaning in this problem situation

55. (a) $A(x) = 320 - 72x + 4x^2$

(b)

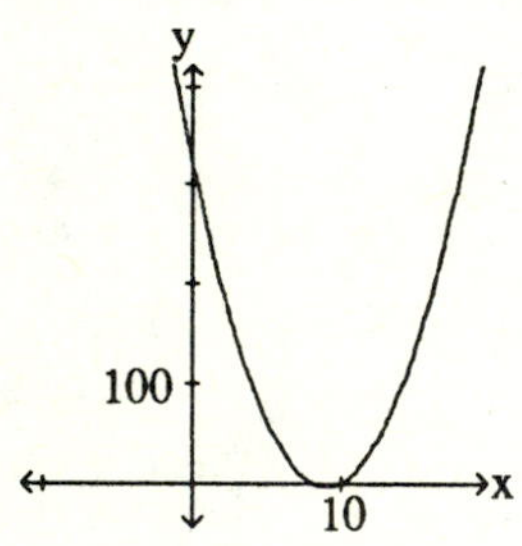

(c) Dom: All real numbers.

Range: All real numbers ≥ -4.

Any number for x between 0 and 8 may be used in this problem situation.

(d) 1.03 inches.

57. 210 adult tickets

115 children tickets

59. 6 pennies, 12 dimes, 5 nickels

61. $0 < w < 72.5$ inches

63. (a) $s = -16t^2 + 75t + 165$

(b)

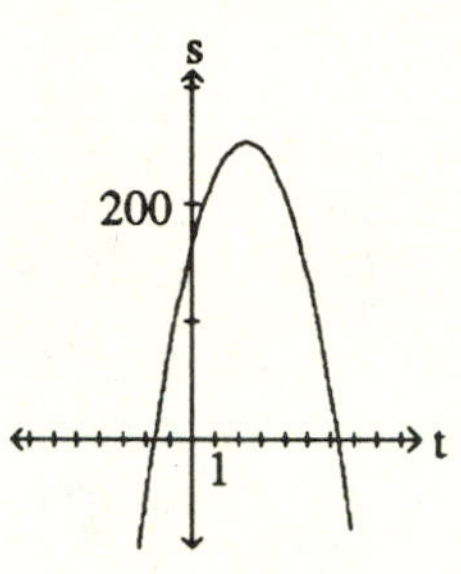

(c) Dom: All real numbers.

Range: All real numbers ≤ 252.89.

(d) The portion of the graph in the first quadrant.

(e) $0 \leq t < 6.32$

(f) $t = 6.32$ seconds

(g) $s = 252.89$feet

(h) $t = 2.34$ seconds

65. Increasing on $[3.5, \infty)$ and decreasing on $(-\infty, 3.5]$.

Real zeros: $x = 2, 5$

67. Increasing on $(-\infty, \infty)$. Real zeros: $x = 1.64$

69. Shift horizontally left 3 units then shift vertically up 2 units.

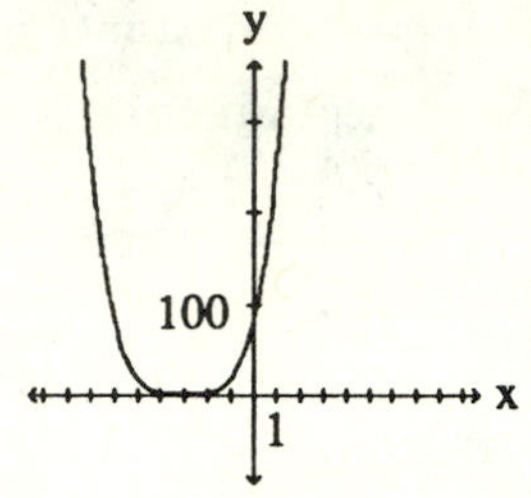

71.

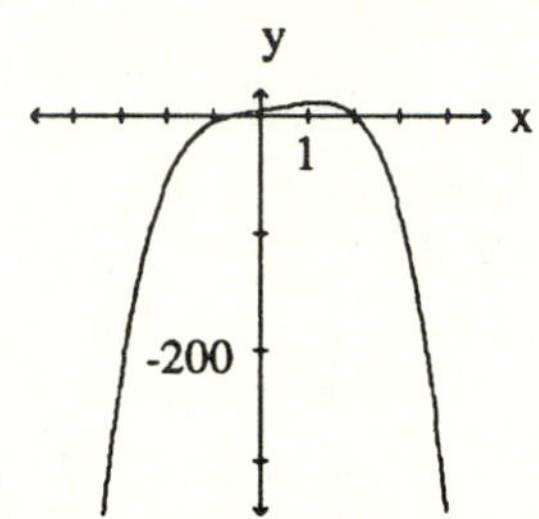

V.R. $[-5, 5]$ by $[-20, 20]$

Real zeros: $x = -0.67, 2.03$

73. Increasing on $(-\infty, 0.21] \cup [3.12, \infty)$, and decreasing on $[0.21, , 3.12]$

75. Local maximum at $(-1.33, -13.63)$

Local minimum at $(0, -16)$

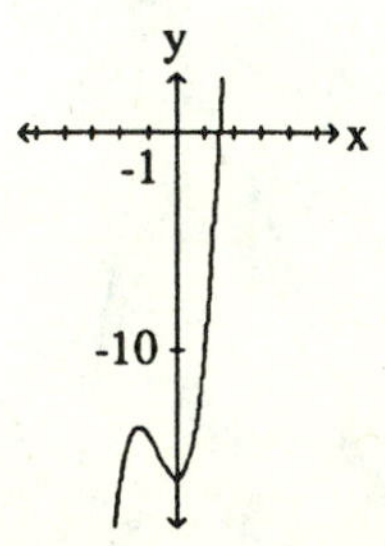

77.

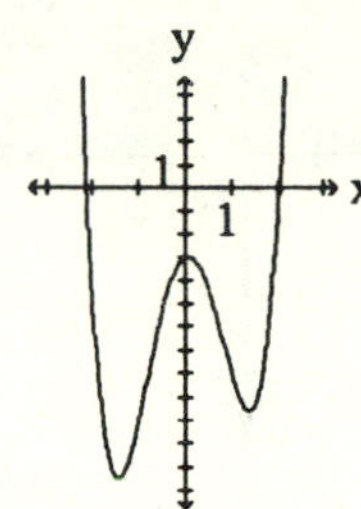

local maximum at $(0.06, -2.97)$

local minimum at $(-1.44, -12.43)$ and $(1.38, -9.60)$

79. $0.75x + 1.10y \leq 0.999(x + y)$

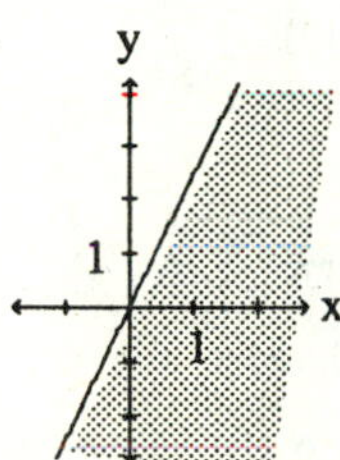

81. $W = 14$ft., $L = 16$ft.

83. (a) $s = -16t^2 + 70t + 200$

(b)

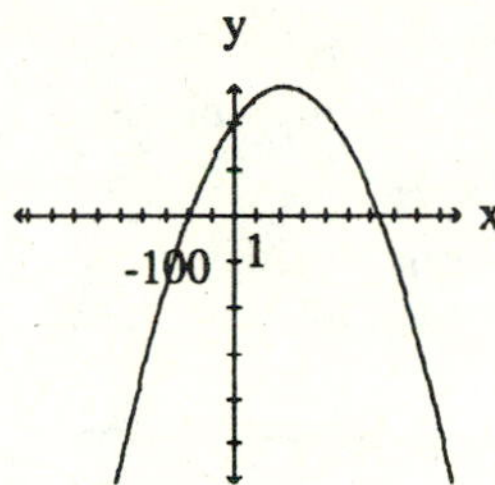

(c) The portion of the graph in the first quadrant represents the problem situation.

(d) $t = 0..39$ and $t = 3.98$

(e) $0.39 < t < 3.98$

85. (a) $R = x(1.5x - 0.03x^2)$

(b)

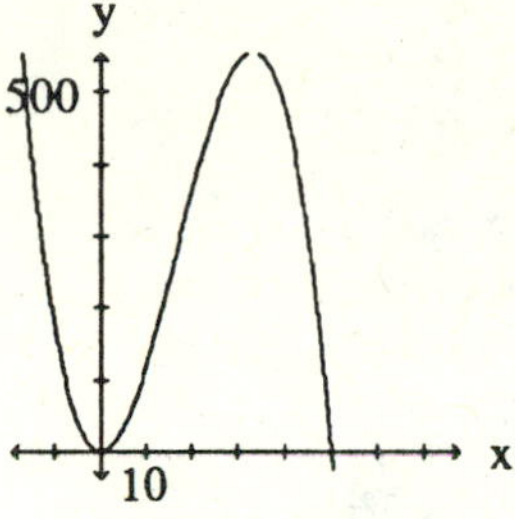

(c)

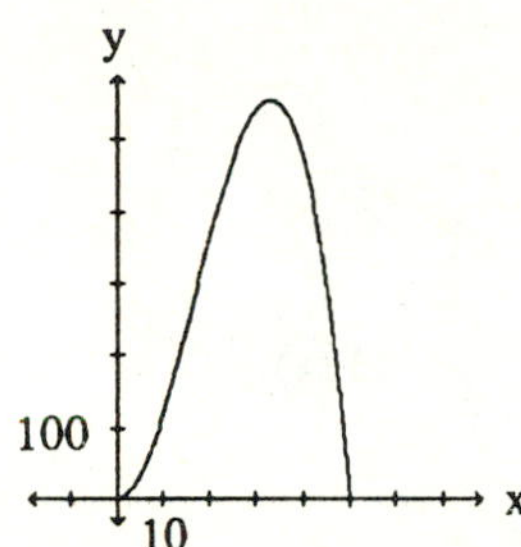

(d) $x = 33$ produces maximum daily revenue $R = \$555.56$ maximum daily revenue

• CHAPTER 3

Exercises 3.1

1. Dom: $(-\infty, \infty)$
   Range: $(-\infty, \infty)$
   no discontinuities
3. Dom: $(-\infty, \infty)$
   Range: $(2, \infty)$
   no discontinuities
5. Dom: $[-4, 2) \cup (2, \infty)$
   Range: $(-\infty, \infty)$
   discontinuous at $x = -4$
7. Dom: $(-\infty, 4) \cup (4, \infty)$
   Range: $[0, \infty)$
   discontinuous at $x = 2, x = 4$
9. Dom: $(-\infty, \infty)$
   Range: All integers.
   discontinuous at every integer value
11. Dom: $(-\infty, \infty)$
    Range: $(-\infty, \infty)$
    no discontinuities
13. Dom: $(-\infty, \infty)$
    Range: $(0, \infty)$
    discontinuous at $x = 0$
15. *(e)*
17. Answers will vary.
19. Answers will vary.
21. (a) $f(x) = -5x^4[1 - \frac{2}{5x} + \frac{1}{5x^3} - \frac{3}{5x^4}]$
    (b) Since each $r_i(x)$ approaches 0 as

$|x| \to \infty$, $f(x)$ approaches $-5x^4$.

23. $f(1) = -3$

    $f(2) = 1$

    Since f is continuous, it assumes every value between -3 and 1.

    $c = \frac{7}{4}$

25. $c = \frac{L-b}{a}$

27. $f(x) \to \infty$

29. $f(x) \to 0$

31. $f(x) \to -\infty$

33. (a) $[0, 676.52]$

    (b)

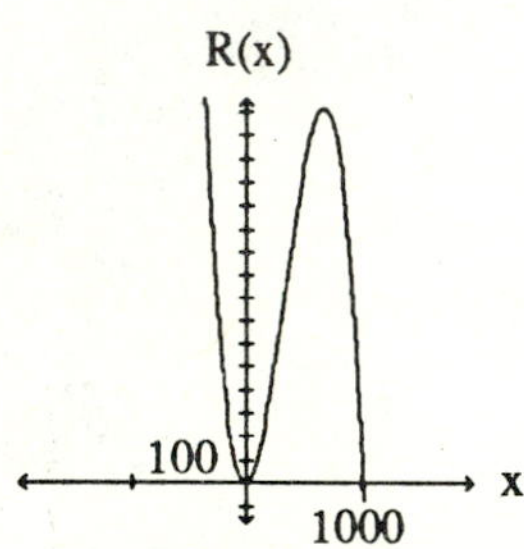

    (c) The portion of the graph in the first quadrant.

    (d) max revenue: $1615.01

    pounds of cookies: 676.52

*35. (a)

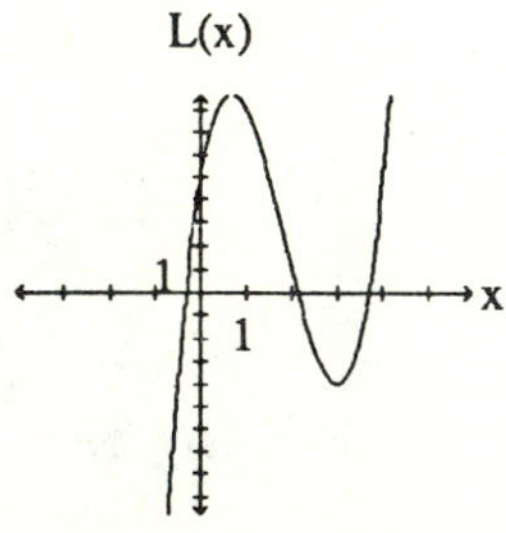

    (b) $L(-1) = -20$

    $L(0) = 5$

    $L(1) = 8$

    $L(2) = 1$

    (c) $L(1.5) = 5$

37. Triangles DEF and BDF are congruent by SSS.

    From Pythagorean Theorem, $y = \frac{9-x^2}{6}$.

    $A = \frac{1}{2}xy = \frac{1}{2}x\left(\frac{9-x^2}{6}\right) = \frac{9x-x^3}{12} = \frac{3}{4}x - \frac{1}{12}x^3$

Exercises 3.2

1. $q(x) = x - 1$

   $r(x) = 2$

3. $q(x) = 2x^2 - 5x + 3.5$

   $r(x) = -4.5$

5. 3

7. 5

9. yes

11. yes

13. $f(x) = (x-3)(x-2)$

    Real zeros: $x = 3, 2$

15. $f(x) = x(x+3)(x-3)$

    Real zeros: $0, 3, -3$

17. $T(x) = (2x^2 - 5)(x^2 + 3)$

    Real zeros: $x = \pm\sqrt{\frac{5}{2}}$

19. $f(x) = (5x - 17)(x + 3)$

    Real zeros: $x = -3, \frac{17}{5}$

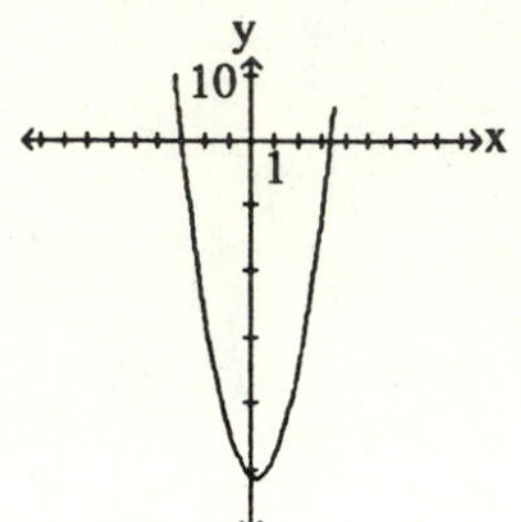

21. $f(x) = (x - 11)(x^2 + 1)$

    Real zero: $x = 11$

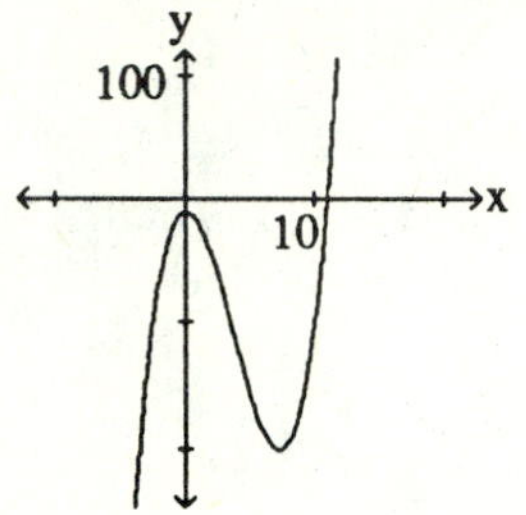

23. Real zeros: $-4.97, -0.10, 2.07$

25. Real zero: -0.002

27. 36 feet

29. (a) $t = 5.27$ seconds

    (b) $t = 1.09$ seconds, $s = 279.14$

31. $g(x) = a(x-2)^2$ where a is a nonzero real number.

33. $g(x) = a(x+1)(x^2+x+1)$ where a is a nonzero real number.

35. Equilibrium price: $62.53

    Production level: 27,960

37. $(x - 3)$

39. (a) Let x represent the base of the right triangle. Since $50 = \frac{1}{2}xh$ then $h = \frac{100}{x}$. From

the Pythagorean Theorem we know then $x^2 + (\frac{100}{x})^2 = (x+2)^2$. Simplifying this equation, we can find $10{,}000 - 4x^3 - 4x^2 = 0$.

(b) 13.25

Exercises 3.3

1. no real roots
3. $+\sqrt{3}, -\sqrt{3}$
   irrational
5. 0, integral; $\pm\sqrt{5}$, irrational
7. Synthetic division by $-\frac{1}{2}$ leaves a remainder of zero; $(2x-1)(x-3)$.
9.

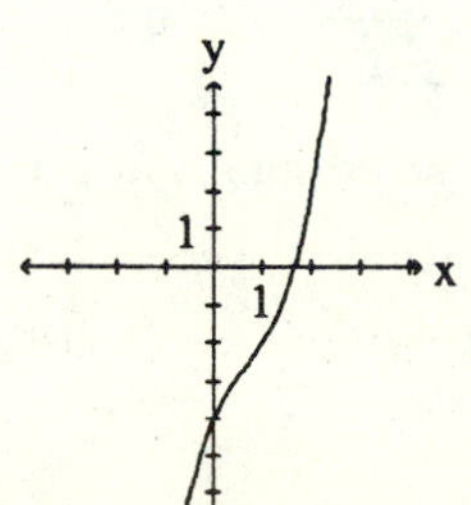

$[-5, 5]$ by $[-7, 5]$

Possible rational roots: ±1, ±2, ±4

rational zeros: none

Irrational zeros:1.65

11.

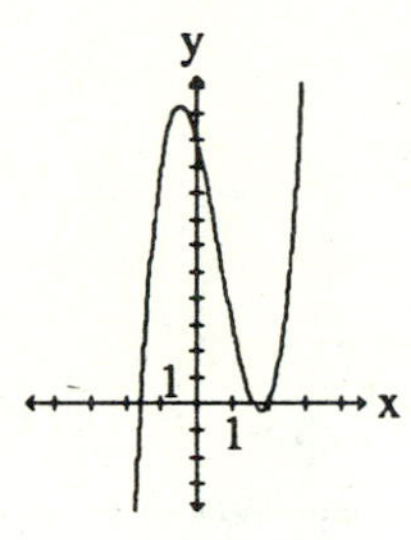

$[-5, 5]$ by $[-5, 12]$

Possible rational roots:$\pm\frac{1}{2}$, ±1, $\pm\frac{5}{2}$, ±2, ±5, ±10

rational zero: 2

Irrational zeros:-1.58, 1.58

13. quotient $x^2 - 3$; remainder 7
15. quotient $x^3 - 3x^2 + 4x - 5$; remainder 7
17. 1, rational

    $\frac{-5\pm\sqrt{21}}{2}$, irrational

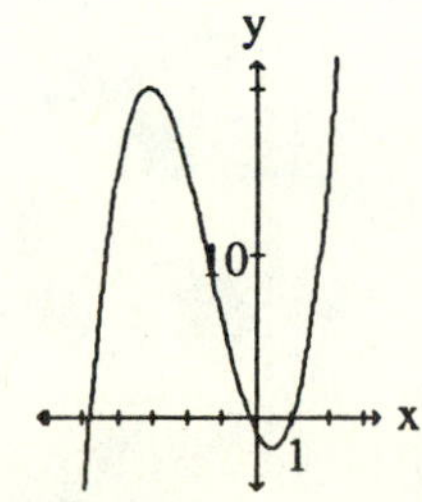

19. $\frac{7}{3}$, rational

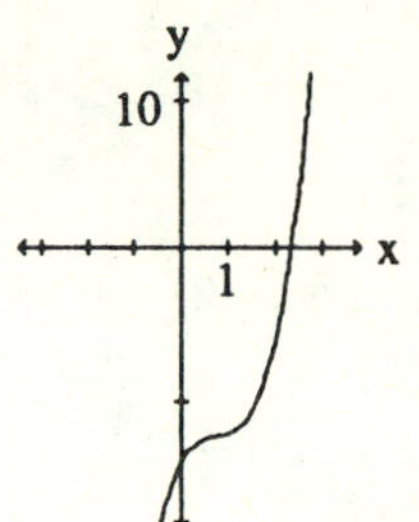

21. rational roots: $4, -\frac{1}{2}$
    no irrational roots
23. no rational roots
    irrational roots: -1.11, 0.86
25. $V(x) = 120x^2 - 2x^3$
27. 5.23ft $\times$5.23ft $\times$109.54ft
    59.58ft $\times$59.58ft $\times$0.84ft
29. $A(x) = 360x - 4x^2$
31. 5.23ft$\times$5.23ft$\times$109.54ft

Exercises 3.4

1. $8 - 3i$
3. $7 + 4i$
5. $1 + i$
7. $9 + 8i$
9. $2 + 3i$
11. 2
13. $\frac{2}{5} - \frac{1}{5}i$
15. $\frac{3}{5} + \frac{4}{5}i$
17. $\frac{1}{2} - \frac{7}{2}i$
19. i
21. 1
23. 2 non-real
25. 1 real zero,

2 non-real complex zeros

27. 2 real zeros,
2 non-real complex zeros

29. 1 is an integral zero, $-\frac{1}{2}+\frac{\sqrt{19}}{2}i$ and $-\frac{1}{2}-\frac{\sqrt{19}}{2}i$ are non-real complex zeros.

31. 1 and -1 are integral zeros, $-\frac{1}{2}+\frac{\sqrt{23}}{2}i$ and $-\frac{1}{2}-\frac{\sqrt{23}}{2}i$ are non-real complex zeros.

33. $x=1$ and $x=-1$

35. $a(x^2-4x+13)$, $a\neq 0$

37. $a(x^3-5x^2+8x-6)$, $a\neq 0$

39. (a) $a(x^3+6x^2+12x+8)$, $a\neq 0$
(b) no

41. $f(1+i)=3(1+i)^3-7(1+i)^2+8(1+i)-2=0$. By Theorem 3 we know $1-i$ is also a zero and $x=\frac{1}{3}$ is a zero.

43. $f(x)=(x-1)(x-i)(x+i)$

45. $f(x)=(x-1)(x+(\frac{1}{4}-\frac{\sqrt{31}}{4}i))(x+(\frac{1}{4}+\frac{\sqrt{31}}{4}i))$

47. $h=-2.98,\ 3.78,\ 14.21$

49. $h=-5.26$

51. $z+\bar{z}=a+bi+a-bi=2a$.

53. From Theorem 5 we know we can write a polynomial of degree n as a multiple of n linear factors. If the coefficients of these linear factors are not real, then using Theorem 4 we can write them as a quadratic.

55. $f(x)=(x-\sqrt{3})(x+\sqrt{3})(x^2-2x+2)$

57. $f(-i)=(-i)^3-i(-i)^2+2i(-i)+2=4+2i$. Theorem 3 states that the polynomial must have real coefficients.

Exercises 3.5

1.

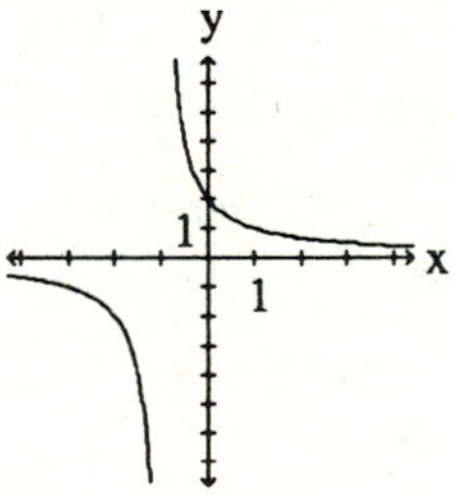

Vert. Asym.: $x=-1$

Horiz. Asym.: $y=0$

End Behavior Model: $y=0$

3.

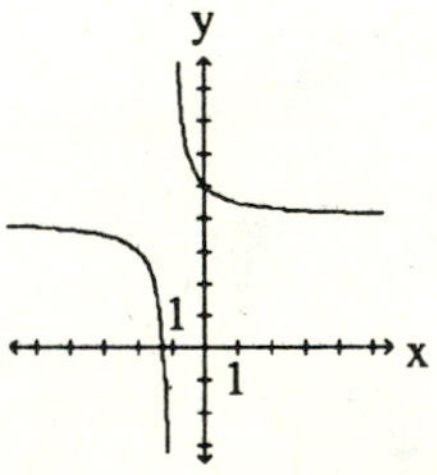

Vert. Asym.: $x=-1$

Horiz. Asym.: $y=4$

End Behavior Model: $y=4$

5.

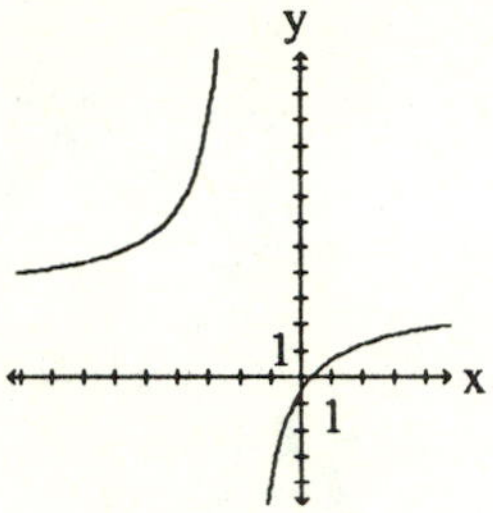

Vert. Asym.: $x=-2$

Horiz. Asym.: $y=3$

End Behavior Model: $y=3$

7.

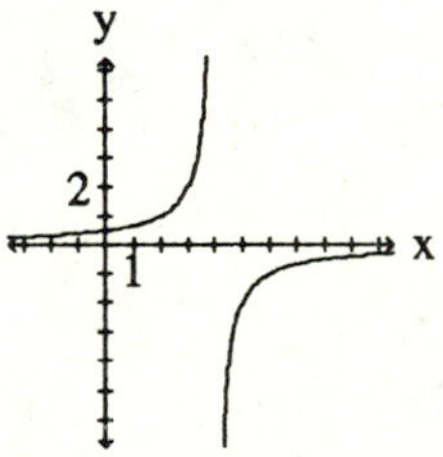

Vert. Asym.: $x=4$

Horiz. Asym.: $y=0$

End Behavior Model: $y=0$

9. $f(x)=\begin{cases}\frac{-x-1}{x+2} & x<0, x\neq -2\\ \frac{x-1}{x+2} & x\geq 0\end{cases}$
Discontinuous at $x=-2$

11. Domain: $(-\infty,2)\cup(2,\infty)$
Range: $(-\infty,0)\cup(0,\infty)$

13. $f(x)\rightarrow 0$

15. $g(x)\rightarrow 4$

17. Apply a horizontal shift 3 units right.

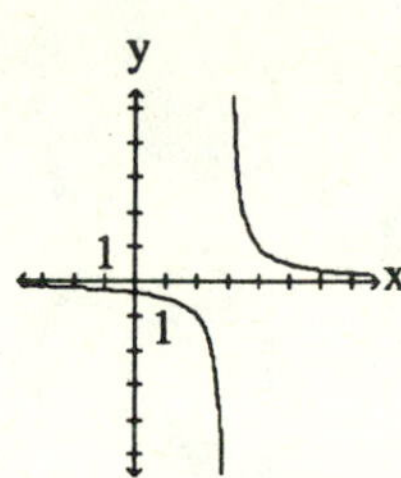

Vert. Asym.: $x = 3$

Horiz. Asym.: $y = 0$

19. Apply a horizontal shift left 2 units, followed by a vertical stretch of 2.

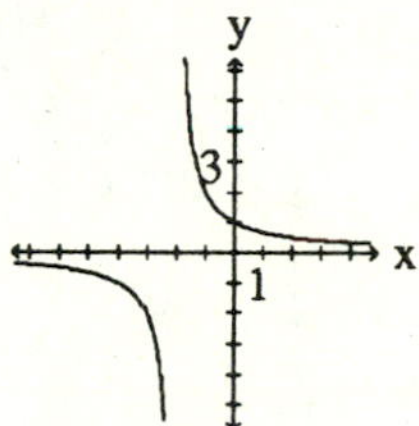

Vert. Asym.: $x = -2$

Horiz. Asym.: $y = 0$

21. $f = g \circ k$ where $k(x) = x - 3$

23. $f = h \circ g \circ k$ where $k(x) = x + 2$; $h(x) = 2x$

25. Dom: $(-\infty, 1) \cup (1, \infty)$

Range: $\frac{1}{2}$.

Vert. Asym.: none

Horiz. Asym.: $y = \frac{1}{2}$.

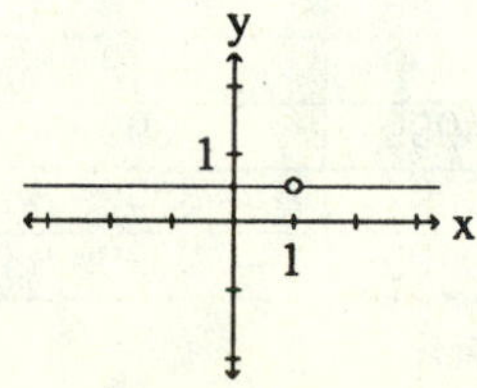

27.

x	f(x)
1	2.75
10	1.4
100	1.97
1,000	2.00
10,000	2.00
100,000	2.00
1,000,000	2.00

29. $f(x) \to 2$

31.

$f(x) \to \infty$

x	f(x)
4	5
4.9	32
4.99	302
4.999	3002
4.9999	30,002
4.99999	300,002
4.999999	3,000,002

33. (a) $\ell = \frac{300}{x}$

(b)

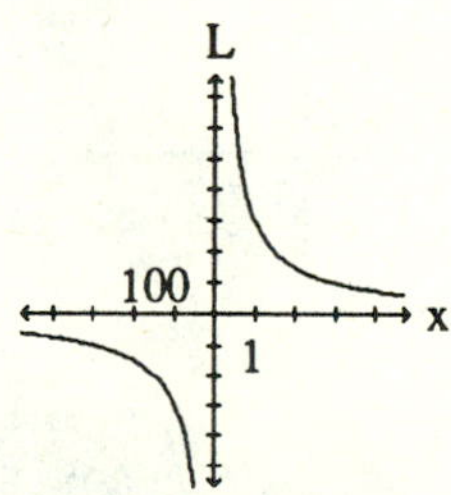

(c) $x = 0$; $y = 0$

(d) Domain: $(-\infty, 0) \cup (0, \infty)$

Range: $(-\infty, 0) \cup (0, \infty)$

(e) Positive values for x

(f) $x = 0.15$ units.

35. (a) $V = \frac{400}{p}$

$x = 0$; $y = 0$

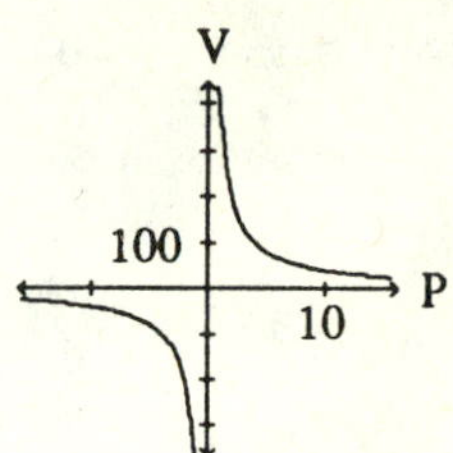

(b) $10 \leq p \leq 20$

37. 60 ounces of 60% solution and 40 ounces of 10% solution

39. Domain: all real numbers except $-\frac{d}{c}$

Range: all real numbers except $\frac{a}{c}$

Exercises 3.6

1. Domain: all real numbers except ± 1.

3. Domain: all real numbers except $2 \pm \sqrt{5}$.

5. Domain: all real numbers except $0, \pm 1$.

7.

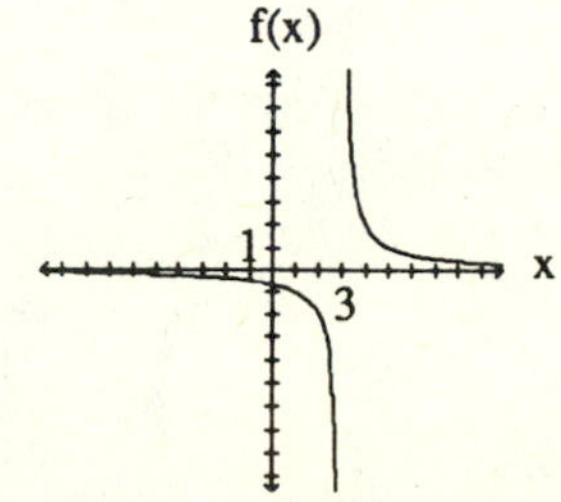

Vert asym: $x = 3$

Horiz asym: $y = 0$

End behavior model: $f(x) = \frac{1}{x}$

$f(x) \to 0$ as $|x| \to \infty$

9.

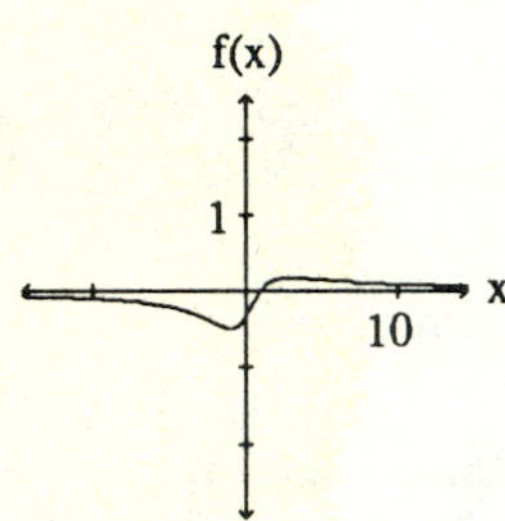

Horiz asym: $y = 0$

End behavior model: $f(x) = \frac{1}{x}$

$f(x) \to 0$ as $|x| \to \infty$

11.

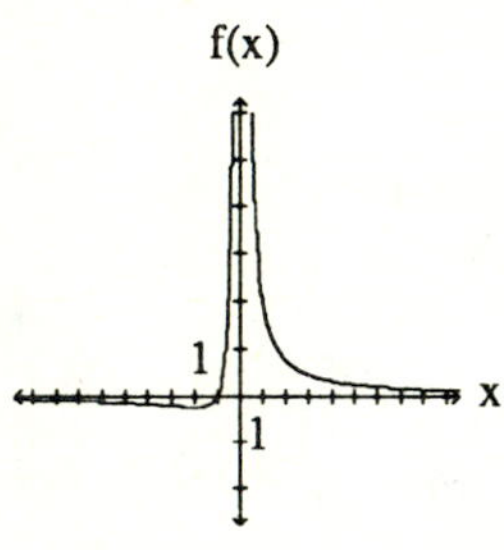

Dom: all real numbers except 0.

Range: all real numbers ≥ -0.25.

Vert asym: $x = 0$

Horiz asym: $y = 0$

$f(x) \to 0$ as $|x| \to \infty$

13.

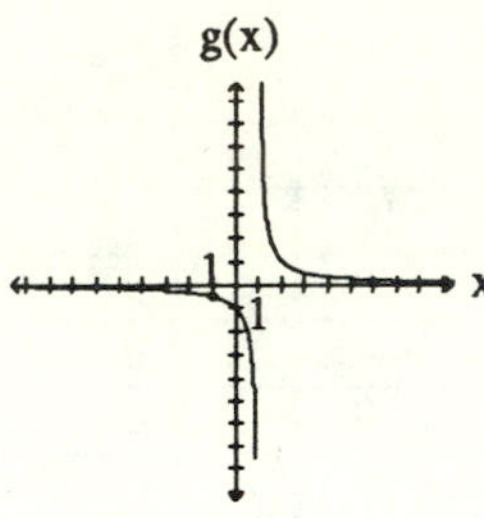

Dom: $(-\infty, -1) \cup (-1, 1) \cup (1, \infty)$.

Range: $(-\infty, -\frac{1}{2}) \cup (-\frac{1}{2}, 0) \cup (0, \infty)$

removable discontinuity at $x = -1$

Vert asym: $x = 1$

Horiz asym: $y = 0$

$f(x) \to 0$ as $|x| \to \infty$

15.

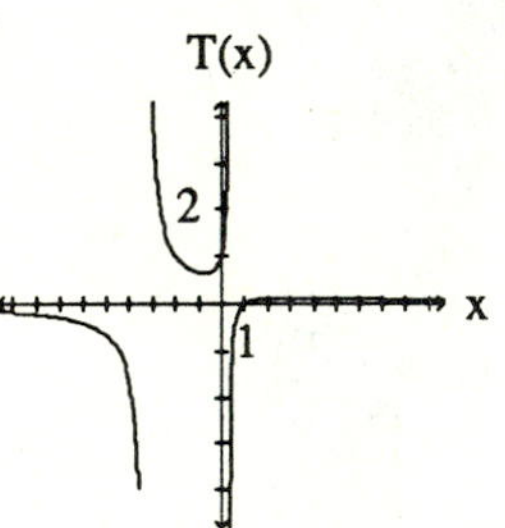

Dom: $(-\infty, -3.30) \cup (-3.30, 0.30) \cup (0.03, \infty)$

Range: $(-\infty, 0.12) \cup (0.65, \infty)$

Vert asym: $x = -3.30, 0.30$

Horiz asym: $y = 0$

$f(x) \to 0$ as $|x| \to \infty$

17. (a) 0

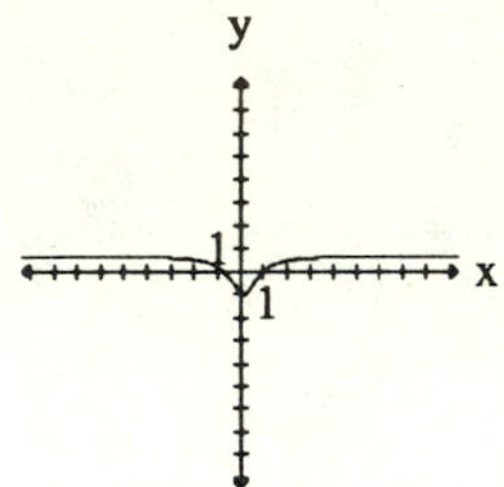

(b) yes; 1 asymptote appears.

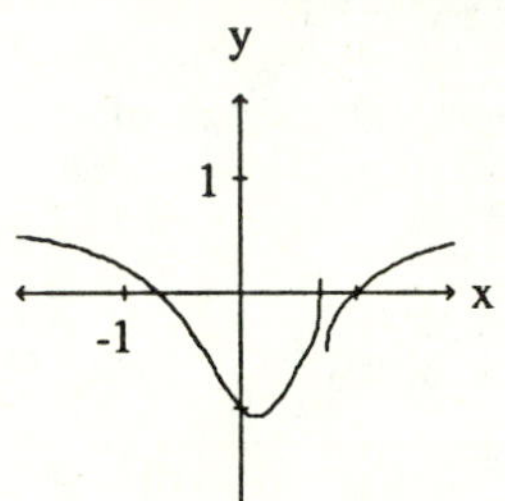

19. Increasing: $(-\infty, -0.62] \cup [0.69, \infty)$

Decreasing: $[-0.62, 0.69]$

21. Local minimum: $(-1.50, 3.12)$; $(5.83, 180.28)$

Local maximum: $(0.75, -0.92)$

23.

x	f(x)
3	undefined
2.9	−3.88
2.99	−39.88
2.999	−399.88
2.9999	−3999.88
2.99999	−39999.88
2.999999	−399999.88

25. $f(x) \to -\infty$

27. $L = \frac{860{,}672}{d^2}$; $L \approx 344.27$ decibels for $d = 50$ feet. As one gets very close to the band (d very small) the loudness increases.

29. (a) $C(x) = \frac{49.14+x}{x+78}$
(b) vert asym: $x = -78$
horiz asym: $y = 1$

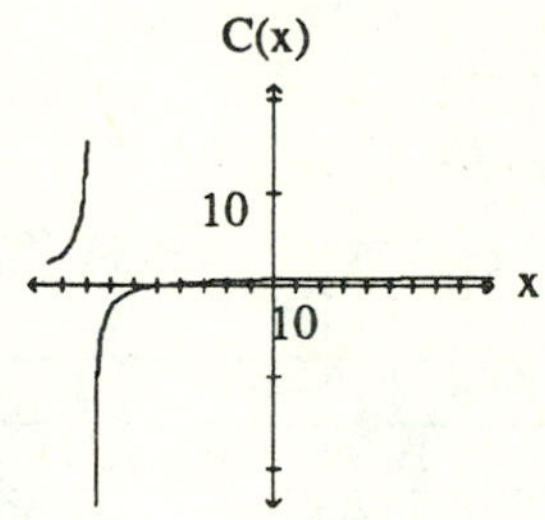

(c) All positive values of x.
(d) $x \geq 91.76$ ounces
(e) $x \geq 91.76$ ounces

31. (a) $C(x) = \frac{0.28x+26}{x+40}$
(b) vert asym: $x = -40$
horiz asym: $y = 0.28$

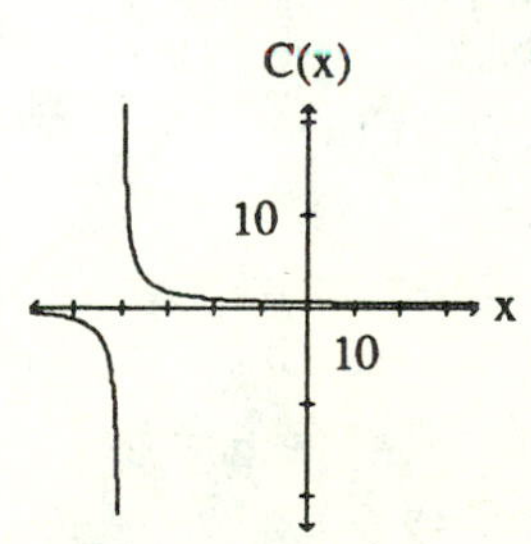

(c) All nonnegative values of x

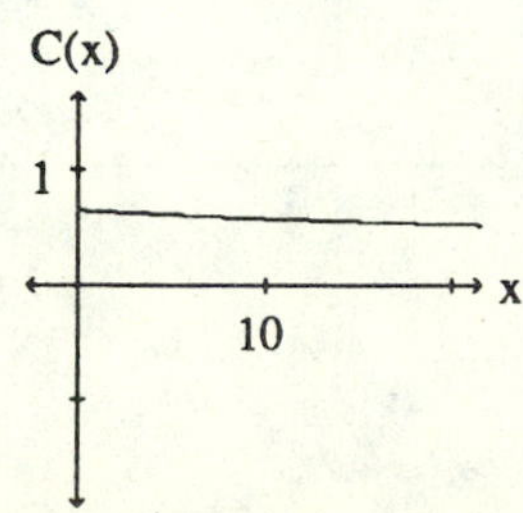

(d) 12.86 ounces

*33. If the power of $(x - a)$ in P is greater than or equal to the power of $(x - a)$ in h, then f has a removable discontinuity at $x = a$.

*35. A problem with a removable discontinuity at $x = a$ simplifies to to a problem which is continuous at $x = a$.

Exercises 3.7

1. $(-\infty, 0) \cup (0, \infty)$
3. $(-\infty, -1) \cup (-1, 1) \cup (1, \infty)$
5. $(-\infty, \infty)$
7. -3.5
9. $(-\infty, \frac{11}{4}) \cup (3, \infty)$
11.

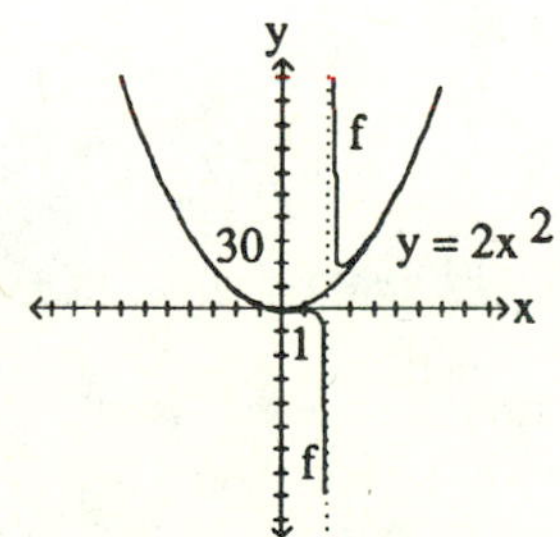

13. vert asym: $x = -2$
slant asym: $y = x - 4$

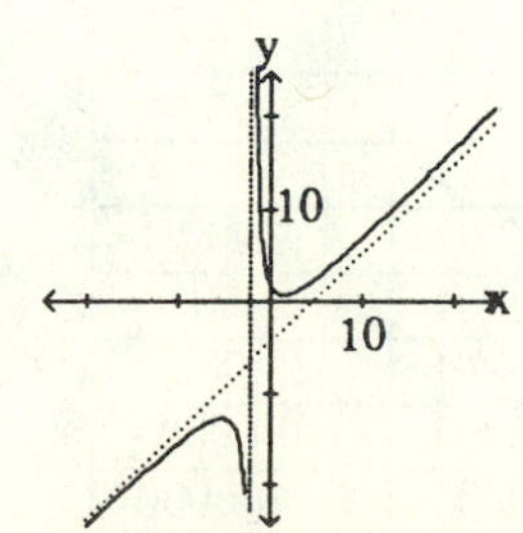

15. No asymptotes.

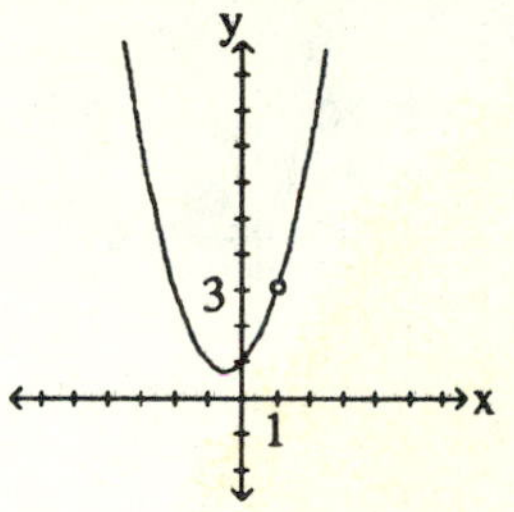

17. Vertical asymptote: $x = -3$

Slant asymptote: $y = x - 6$

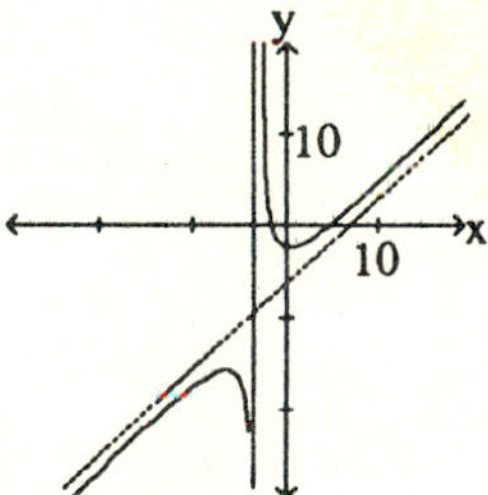

19. Vertical asymptote: $x = 2.5$

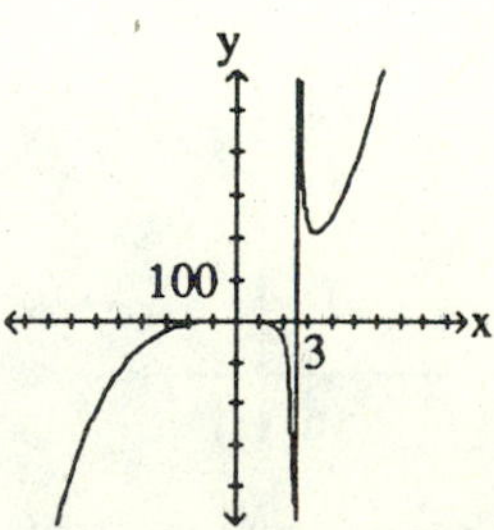

21. Vertical asymptote: $x = 3$

Horizontal asymptote:
$y = 2$

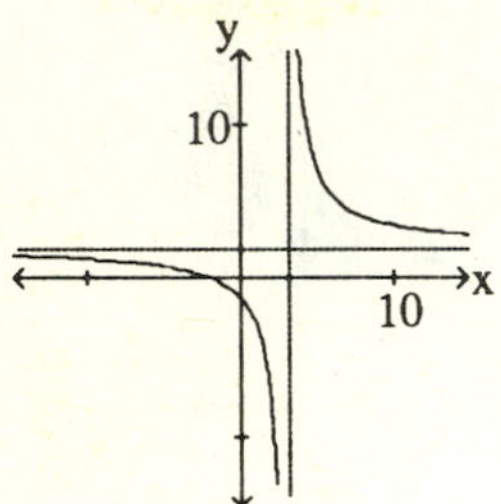

23. Vertical asymptote: $x = 2$

Domain: $(-\infty, 2) \cup (2, \infty)$

Range: $(-\infty, 0] \cup [4, \infty)$

Zero at $x = +1$

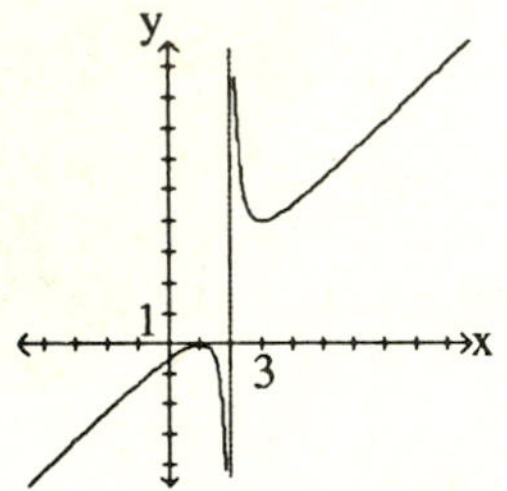

25. Vertical asymptote: $x = 2$

Domain: $(-\infty, 2) \cup (2, \infty)$

Range: (∞, ∞)

Zeros at $x = -1.62, 0.62, 1.00$

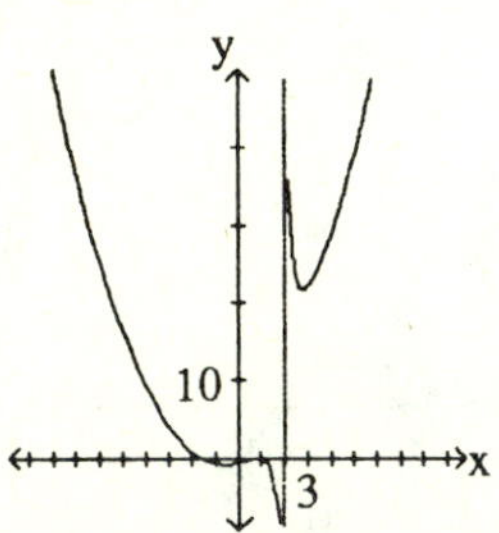

27. Vertical asymptotes: $x = 2$, $x = -2$

Domain: $(-\infty, -2) \cup (-2, 2) \cup (2, \infty)$

Range: $(-\infty, -0.33] \cup [11.78, \infty)$

No zeros

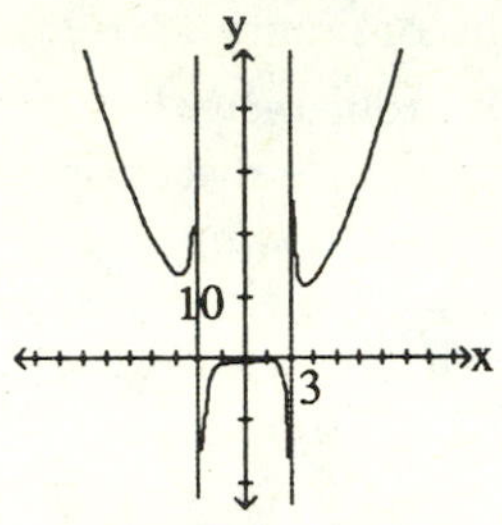

29. Decreasing $[2.94, \infty)$

Increasing $(-\infty, 2) \cup (2, 2.94]$

31. Local min: $(-2.62, -2.81)$

Local max: $(-5.40, -14.30)$

Increasing $(-\infty, -5.4) \cup (-2.62, \infty)$

Decreasing $(-5.4, -2.62)$

end behavior: $y = 2x - 3$

zero: $x = -3.46, -0.50$, and 3.46

33.

x	g(x)
3	3
2.1	30
2.01	300
2.001	3000
2.0001	30,000
2.00001	300,000
2.000001	3,000,000

x	f(x)
3	21
2.1	38.82
2.01	308.0802
2.001	3008.008002
2.0001	30008.0008
2.00001	300008.0001
2.000001	3000008.

35.

x	f(x)
1	−1
10	200.375
100	20,000.03061
1,000	2,000,000.003
10,000	200,000,000
100,000	2.0×10^{10}
1,000,000	2.0×10^{12}

37. $f(x) \to \infty$

39. $f(x) \to \infty$

41. If $(x - 2)$ is not a factor of $g(x)$.

43. (a) $P(x) = \frac{x+0.35(135)}{x+135}$

(b) $f(x) = \frac{x+0.35(135)}{x+135} - 0.63$

(c) 102.16 oz.

45. $f \circ f(x) = \frac{1}{\frac{1}{x}} = x$

$m = -1$

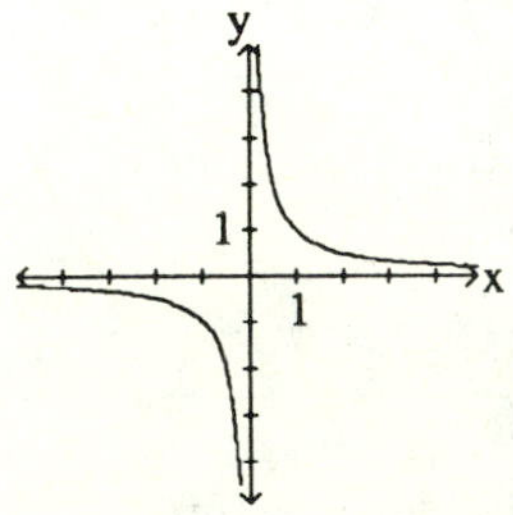

Exercises 3.8

1. (a)

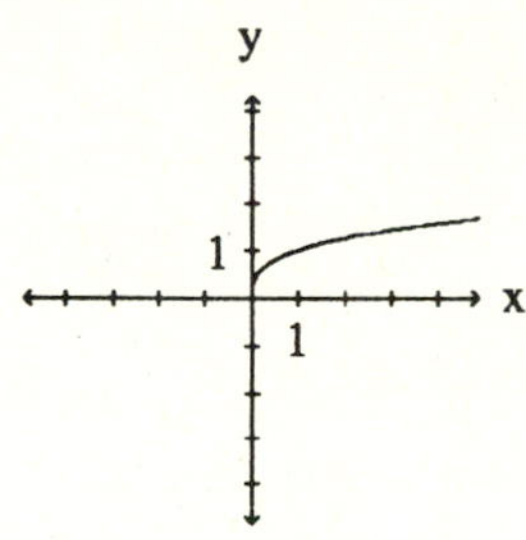

(b)

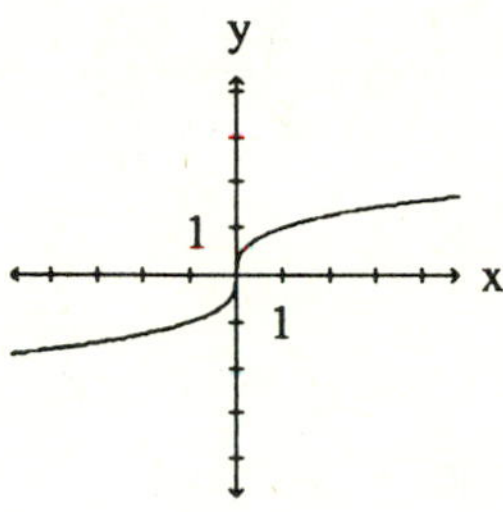

(c) b

(d) The function takes on values when x is less than0.

3. $(1, \infty)$; $(1, \infty)$
5. 2.5
7. no solution
9. 0.94 and -0.94
11. 5 and 10
13. 6
15. $(-\infty, -1.13)$
17. $(-1.30, 1.30)$
19. $(15.57, \infty)$
21.

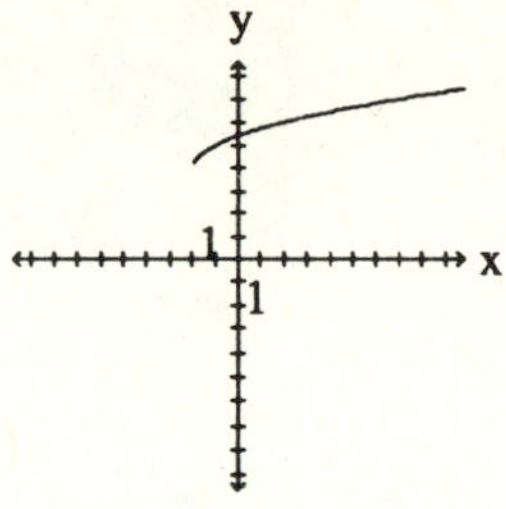

23.

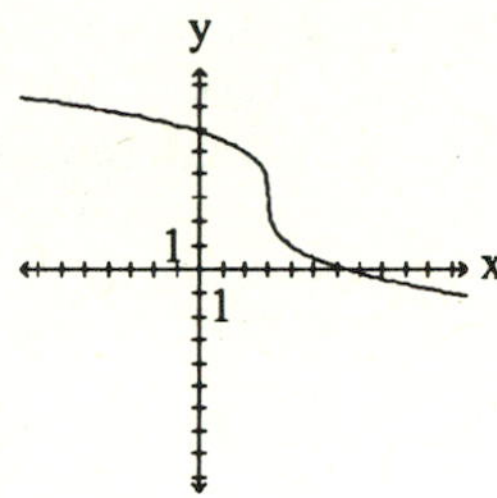

25.

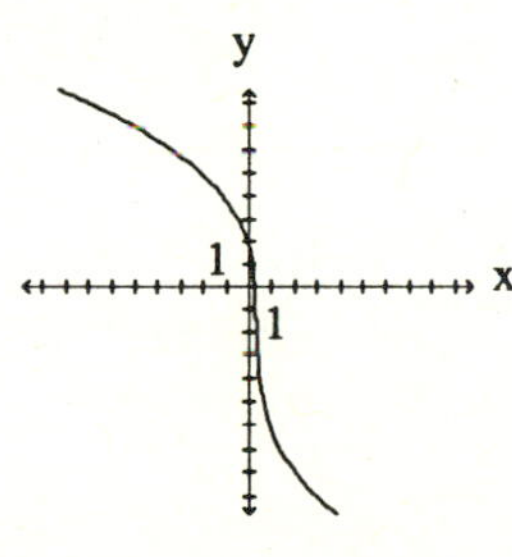

27.

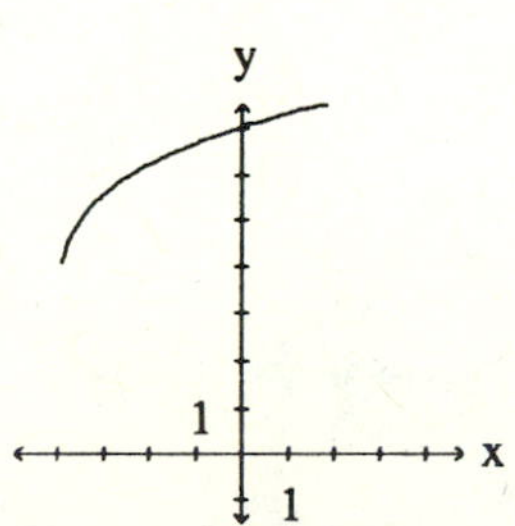

29. Apply a horizontal shift right of 3 units to the graph of $y = (x)^{1/3}$ to get $y = (x-3)^{1/3}$, then apply a vertical stretch of 2 followed by a reflection through the x-axis, then a vertical shift up of 3.
31.

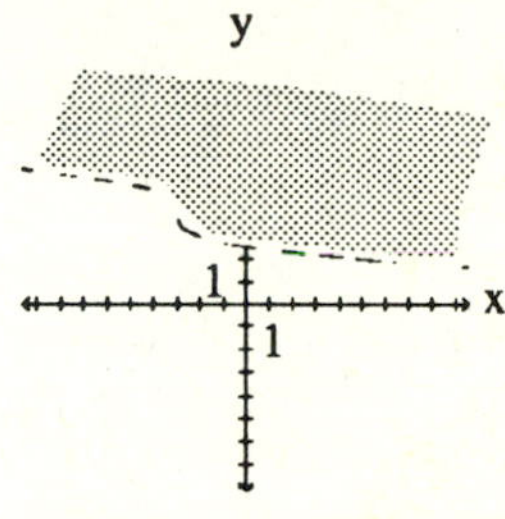

33. $x^3 - \sqrt{x+1} = 1$ can be written as $x^3 - 1 = \sqrt{x+1}$ or $f(x) = g(x)$. Hence the number of real solutions is the number of points of intersection of f and g. A quick sketch will reveal one solution.
35. End behavior model: $\sqrt{x^2}$ or $|x|$.
37. 0.54
39. 21.21
41. (a) $T(x) = \frac{\sqrt{x^2+400}}{30} + \frac{60-x}{50}$

    (b) $T(0) = 1.87$ hours. She rows to B. $T(60) = 2.11$ hours. She rows to C.

    (c)

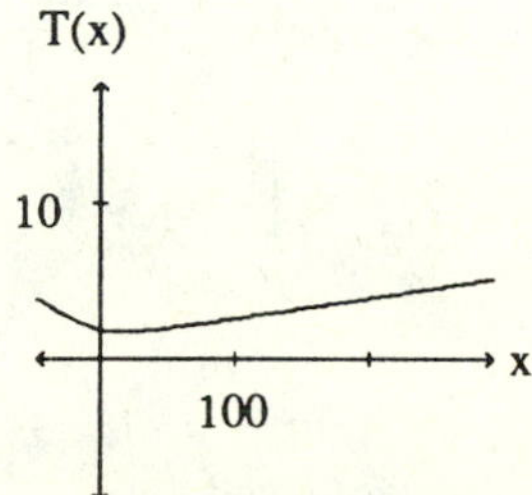

(d) $[0, 60]$

(e) $x = 14.98$ miles

(f) 1.73 hours

43. (a) Travel the road all the way.

(b) Yes, go to point B through the woods 2.77 miles from C then ride the bike to C.

45. Write as $y = -f(-(x-3))$

(a) Reflect through the y-axis.

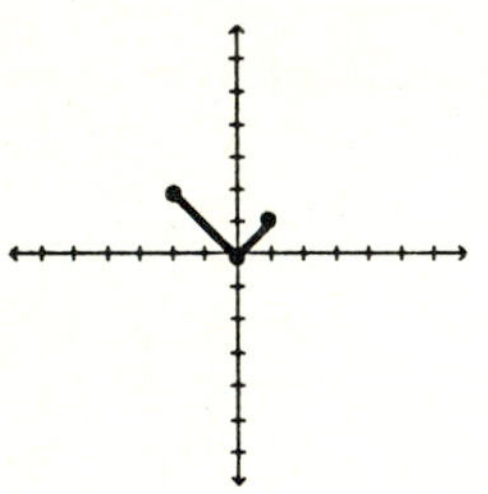

(b) Reflect through the x-axis.

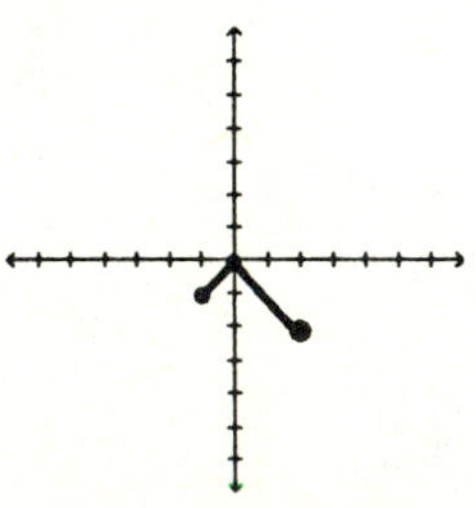

(c) Horizontal shift right 3 units, followed by a reflection through the x-axis.

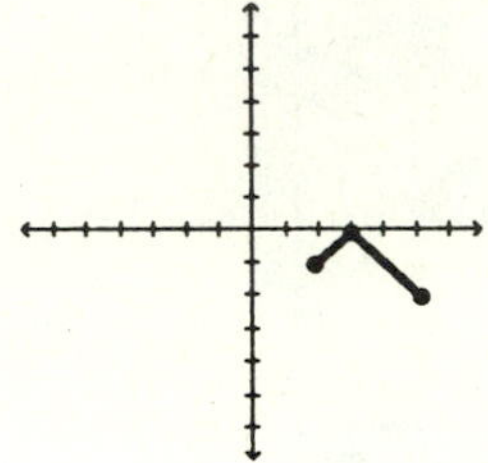

(d) Horizontal shift left 3 units, followed by a reflection through the y-axis, then a reflection through the x-axis.

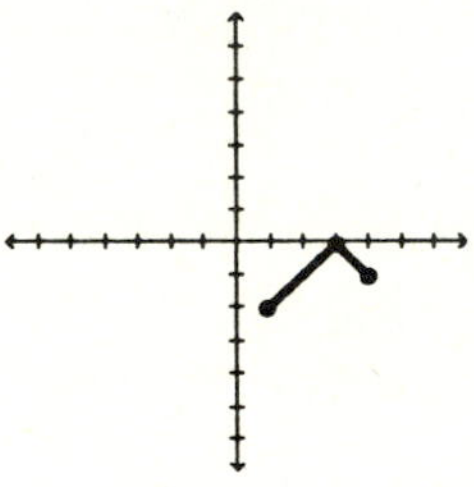

(e) Horizontal shift left of 1 unit, followed by a vertical stretch of 3, then a reflection through the x-axis, followed by a vertical shift up of 3.

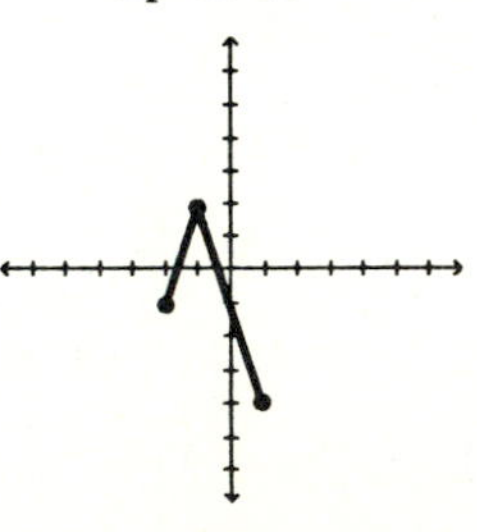

(f) Horizontal shift left of 1 unit, then reflect through the y-axis, followed by a vertical stretch of 2, then a vertical shift down of 1 unit.

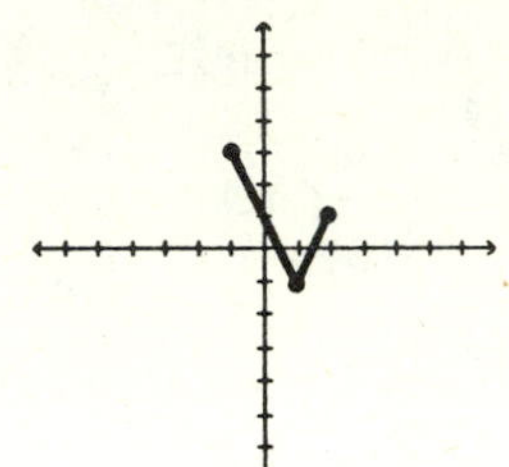

Review Exercises

1. Dom: $(-\infty, -\frac{1}{2}) \cup (-\frac{1}{2}, \infty)$
   Range: $(-\infty, \infty)$
   Discontinuous at x: $-\frac{1}{2}$
3. Dom: $(-\infty, \infty)$
   Range: $[-1, \infty)$
   Points of Discontinuity: none
5. Dom: $(-\infty, -5) \cup (-5, \infty)$
   Range: $(-\infty, 0) \cup (0, \infty)$
   Discontinuous at $x = -5$
7. V.R.: $[-12, 12]$ by $[-700, 700]$

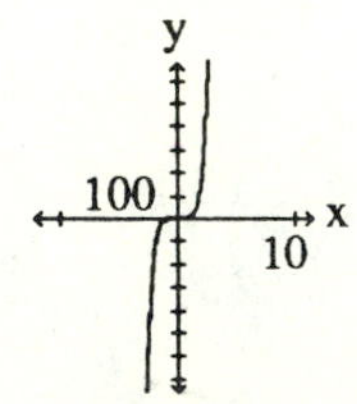

9. $f(x) \to \infty$

11. $f(x) \to 2$

13. $q(x) = 2x^2 - \frac{5x}{3} + \frac{37}{9}$
$r(x) = -\frac{65}{9}$

15. $r(x) = 27$

17. $(-2, -4)$

19. $x = \pm 2, \pm 3$

21. real zero: $x = 2$
complex zeros:
$x = -1 \pm \sqrt{2}i$

23. (a) 2
(b) 3
(c) $x = -1, -\frac{1}{40}, \frac{1}{20}$

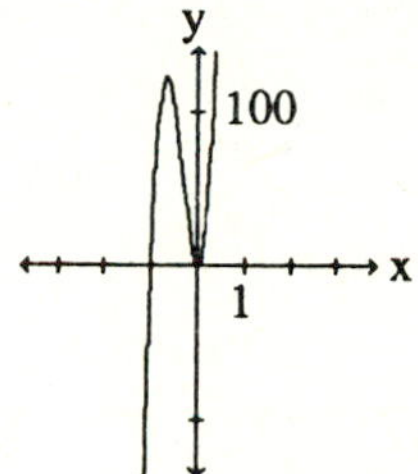

25. (a) $a = (x-2)(x+3)(x-c)$ for $a \neq 0$
(b) $a(x-2)^2(x+3)$ or $a(x-2)(x+3)^2$ for $a \neq 0$
(c) $a(x-2)(x^2-4x+5)$ for $a \neq 0$

27. $k = -2$

29. integral zeros: $x = \pm 3$
rational zero: $x = 1$
irrational zeros: $\frac{-3 \pm \sqrt{3}}{3}$

31. $\pm 1, \pm 2, \pm\frac{1}{3}, \pm\frac{2}{3}$

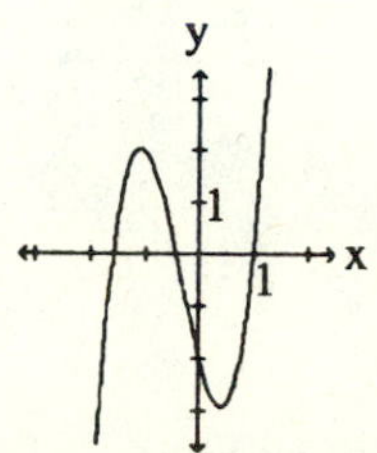

33. $x = -2$; rational
$x = \frac{-5 \pm \sqrt{41}}{2}$; irrational

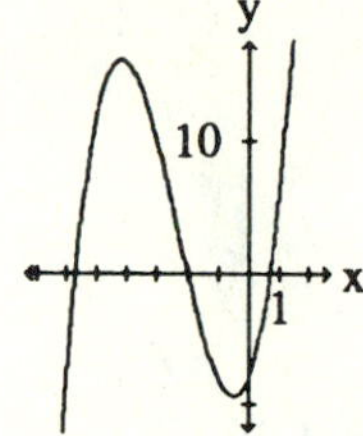

35. Vertical stretch by a factor of 2 to get $y = 2f(x)$; reflect through the y-axis to get $y = 2f(-x)$; Then, vertical shift up 1 unit to get $y = 1 + 2f(-x)$

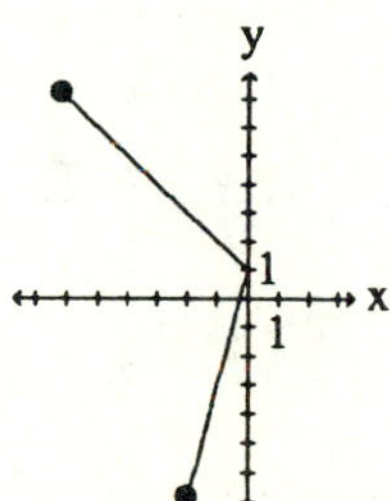

37. $3 - 7i$

39. $\frac{19}{41} - \frac{7}{41}i$

41. 1 real and 2 complex roots

43. integral roots: 4, −1
complex roots: $\pm\sqrt{2}i$

45. $f(x) = a(x^3 - 6x^2 + 37x - 58)$ where $a \neq 0$.

47. $f(2-3i) = (2-3i)^3 - 5(2-3i)^2 + 17(2-3i) - 13 = -46 - 9i - 5(-5-12i) + 17(2-3i) - 13 = 0$
$2 + 3i$ is another complex root
1 is an integral root

49. 4

51.

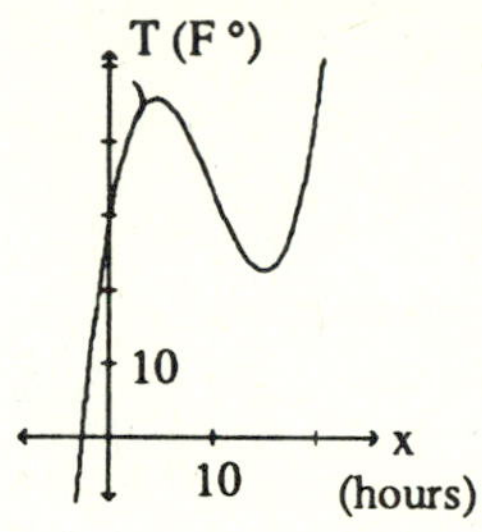

53. 45.80° at 10:26 a.m. (4.43 hours into the day).

55. (a) $(0, \infty)$
(b)

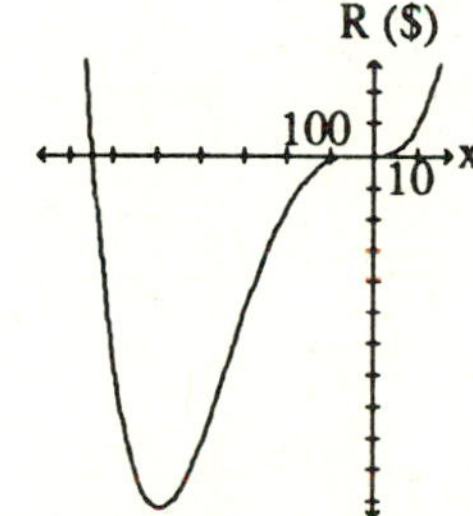

(c) $(0, \infty)$
(d) no maximum revenue

57.

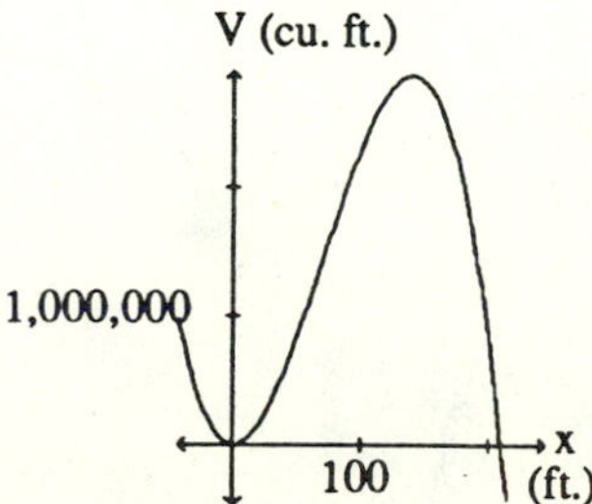

59. $x = 70$

61. 182 pheasants in 111.65 days

63. Answers will vary. One scenario could be that the population increases until there are too many pheasants for the food supply. They then begin to die off as the food disappears.

65.

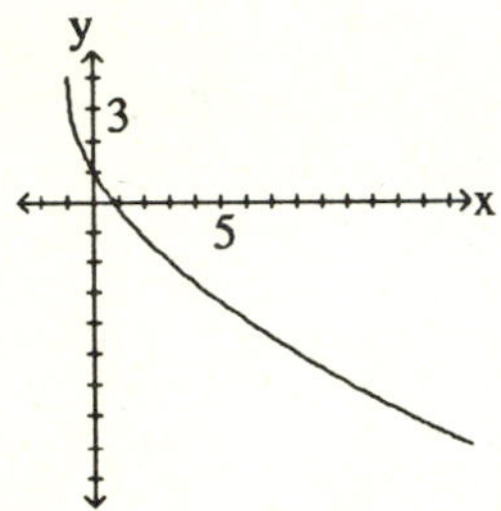

67. Vertical Asymptote: $x = -1$

Horizontal Asymptote: $y = 1$, $y = -1$

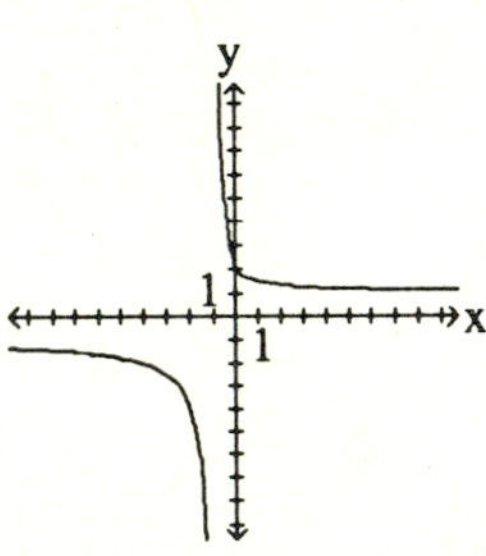

69. $f(x) \to 1$

71. (a) Let $g(x) = \frac{1}{x}$. Apply a horizontal shift right of 2 followed by a vertical stretch of 5.

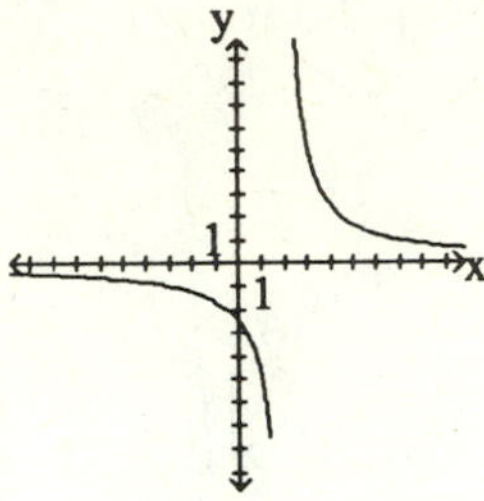

(b) Let $g(x) = \frac{1}{x}$. Apply a horizontal shift of 5 left, then a vertical stretch of 3, then a reflection through the x-axis, followed by a vertical shift up of 2.

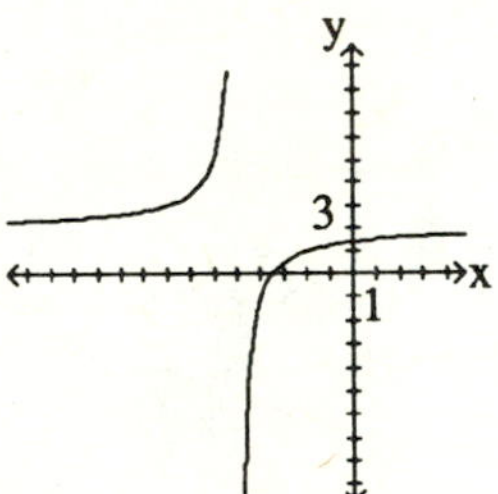

73. Domain: all real numbers except $\frac{3\pm\sqrt{5}}{2}$.

75. Vert. asym. at $x = -5$

Horiz. asym. at $y = 0$

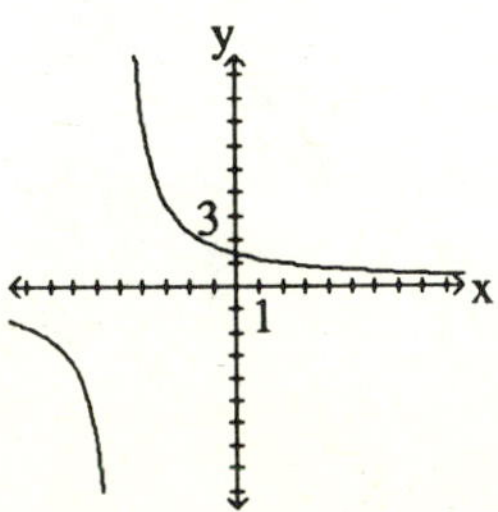

77. Domain: all real numbers except 5 and -4

Range: $(-\infty, 0) \cup (3, \infty)$

Zeros: $x = 0$

Vert. asym. at $x = 5, -4$

Horiz. asym. at $y = 3$

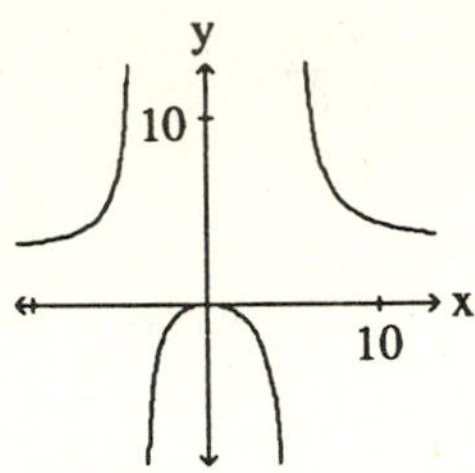

79. Dom: All real numbers except $-\sqrt[3]{2}$

Zeros: $x = -2.59, -0.16, 1.24$

Range: $(-\infty, \infty)$

Vert. asym. at $x = -\sqrt[3]{2}$

Horiz. asym. at $y = 2$

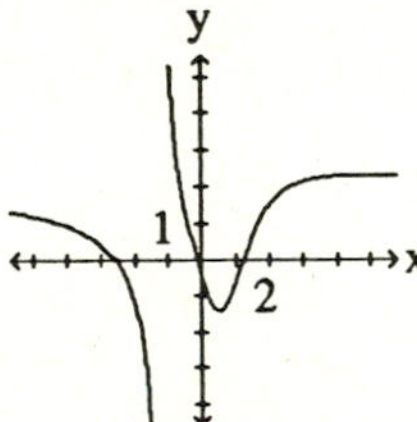

81. Increasing: $(-\infty, \infty)$

83. Minimum: $(-0.89, 3.47)$

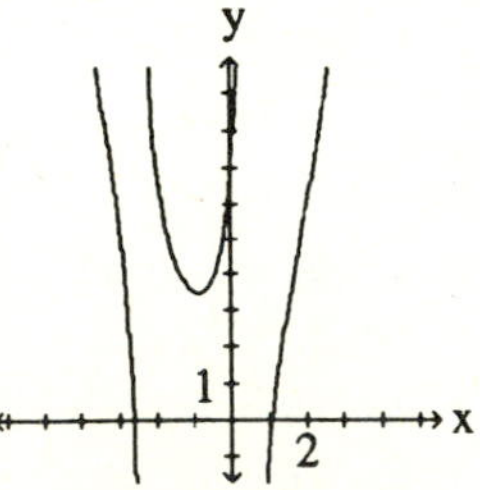

85. End behavior model: $y = x - 6$

Vert. asym. at $x = -2$

Zeros: none

minimum: $(3, 2)$

maximum: $(-7, -18)$

Increasing:
$(-\infty, -7] \cup [3, \infty)$

Decreasing:
$[-7, -2) \cup (-2, 3]$

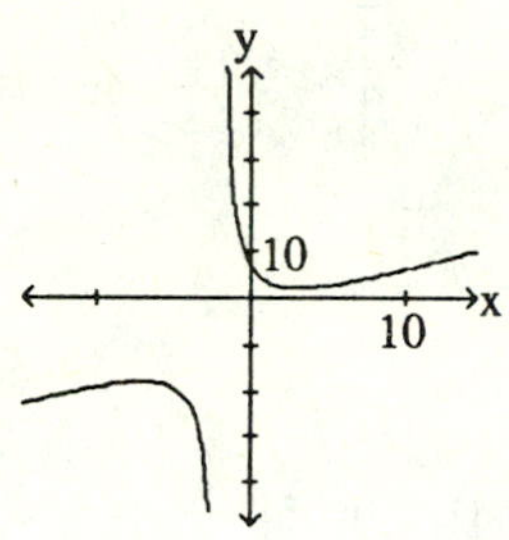

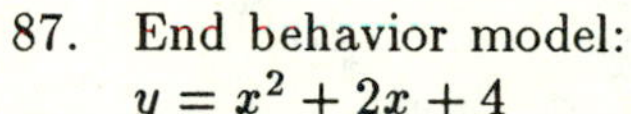
87. End behavior model:
$y = x^2 + 2x + 4$

No asymptotes but undefined at $x = 2$

89. End behavior model:
$x^3 - 3x^2 - 5x + 7$

Vert. asym. at $x = -1$

91. $f(x) \to \infty$ as $x \to -1^+$

93. 10

95. $(-5, 2)$

97. End behavior model:
$x^2 + 3x + 32$

y-intercept: 0.29

vert asymp $x = 7, -2$

zeros: $-0.93, 1.18$

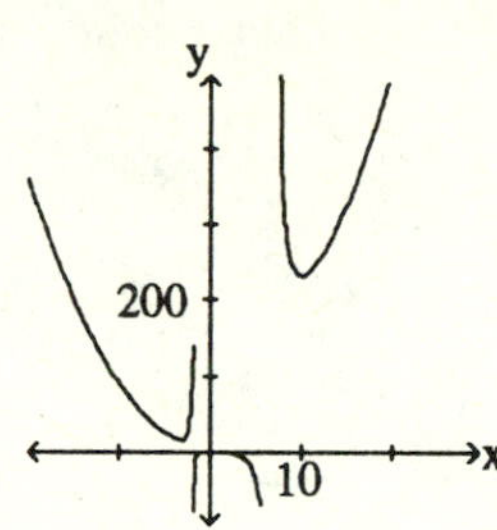

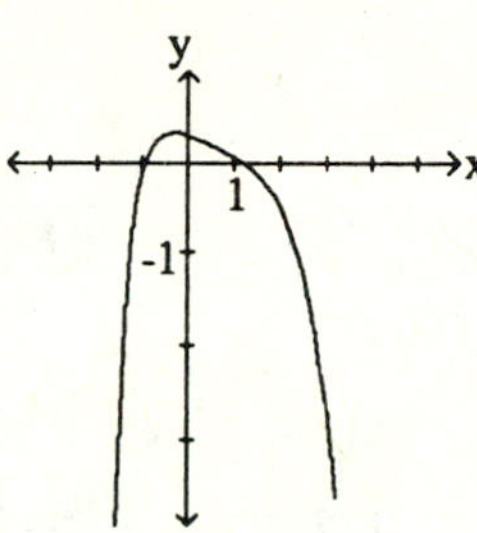

99. -4.54 and 1.54

101. -12.36 and 10.36

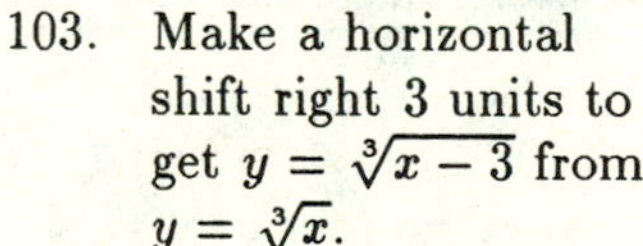
103. Make a horizontal shift right 3 units to get $y = \sqrt[3]{x-3}$ from $y = \sqrt[3]{x}$.

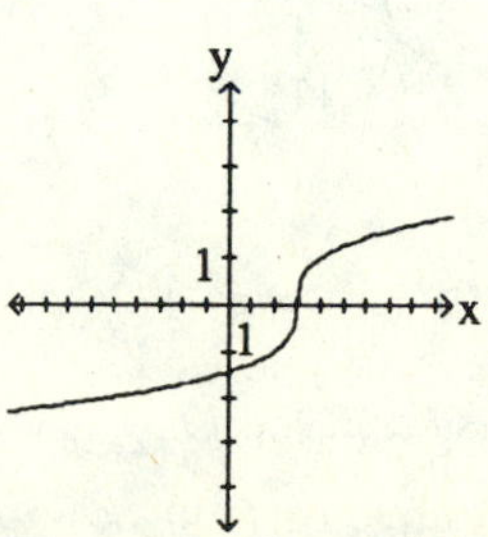

105.

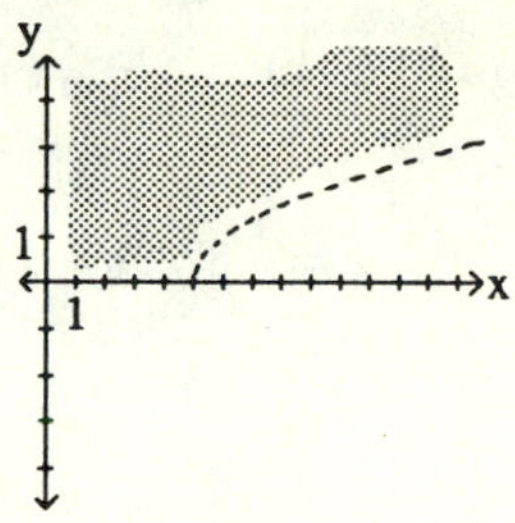

107. (a) $V(t) = 36\pi t^3$

(b) $V = 14137.17$ cubic inches

(c) $t = 5.10$ seconds

109. (a) $C(x) = \frac{75+x}{150+x}$

(b) $x = -150$, $y = 1$.

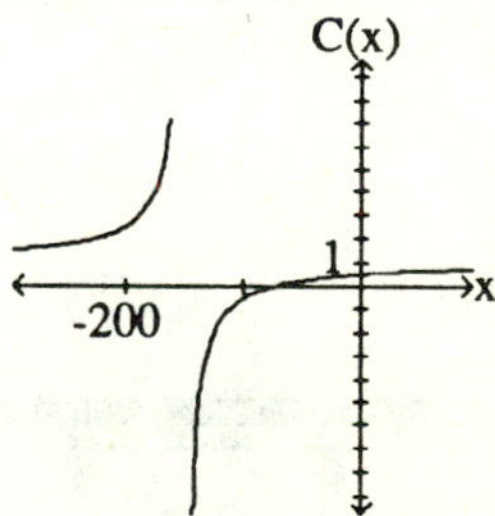

(c) All nonnegative values of x.

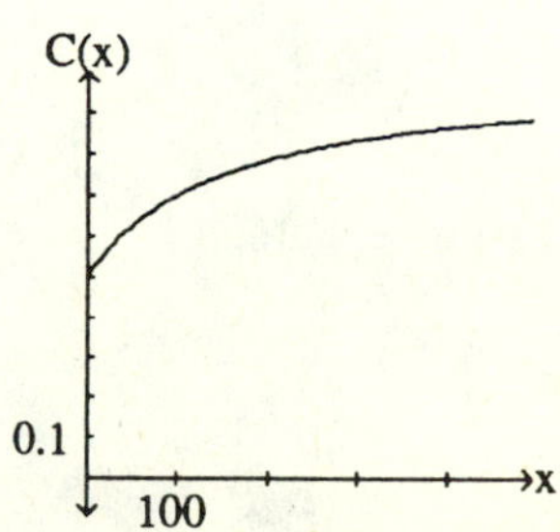

(d) $x = 190.91$ ounces

(e) $x = \frac{42}{0.22}$ ounces

111. (a) $C(x) = \frac{x+48.72}{x+87}$

(b) $x = -87$, $y = 1$.

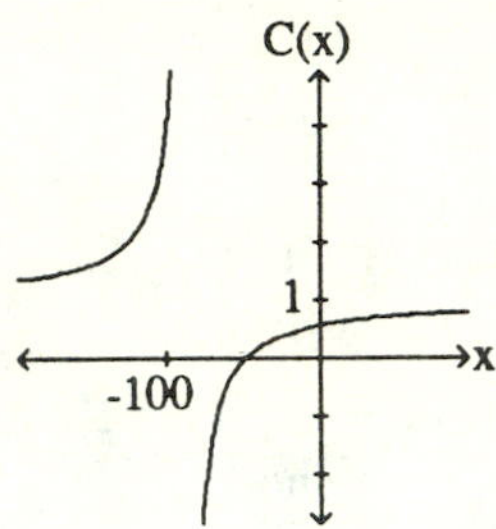

(c) All nonnegative values of x

(d) 87 ounces

(e) $x = \frac{19.14}{0.22}$ ounces

113. (a) $P(x) = \frac{2x^2+1000}{x}$

(b) vertical asymptote at $x = 0$ end behavior model is $y = 2x$

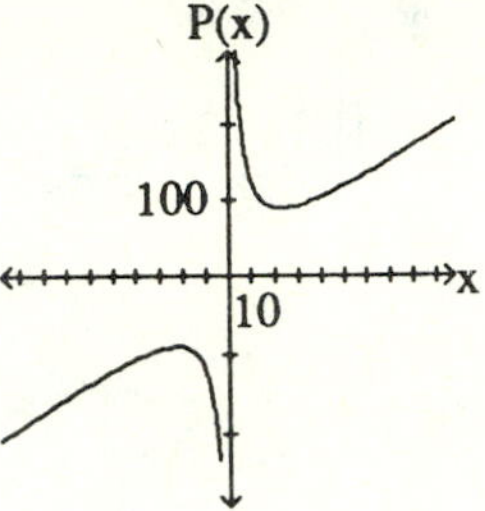

(c) $0 < x < 500$

(d) 22.36 ft by 22.36 ft

Perimeter = 89.44 ft

115. (a) $T(x) = \frac{\sqrt{x^2+15^2}}{25} + \frac{55-x}{40}$

(b) $T(0) = 1.975$ hours–row to B then drive to C

$T(55) = 2.28$ hours–row to C

(c)

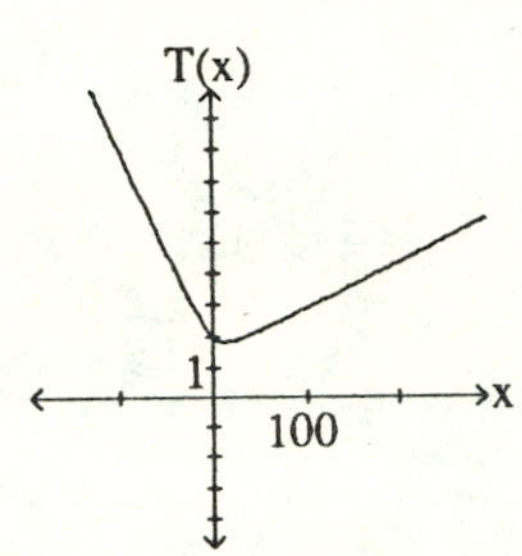

(d) $[0, 55]$

(e) To a point 12 miles from B.

(f) 1.84 hours

• CHAPTER 4

Exercises 4.1

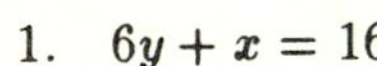

1. $6y + x = 16$

3. no

5. $(-4, -4)$

7.

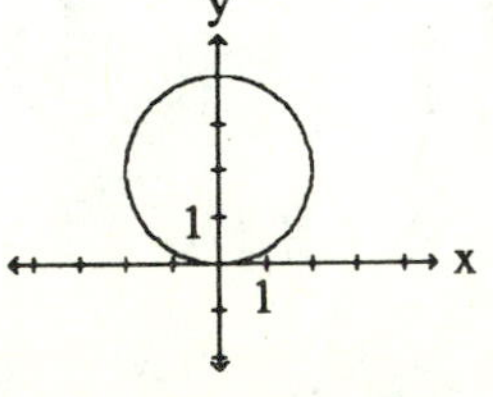

9. $f^{-1}(x) = x^2 + 2$ with $x \geq 0$, yes.
Domain of f: $[2, \infty)$
Range of f: $[0, \infty)$
Domain of f^{-1}: $(0, \infty)$
Range of f^{-1}: $[2, \infty)$

11.

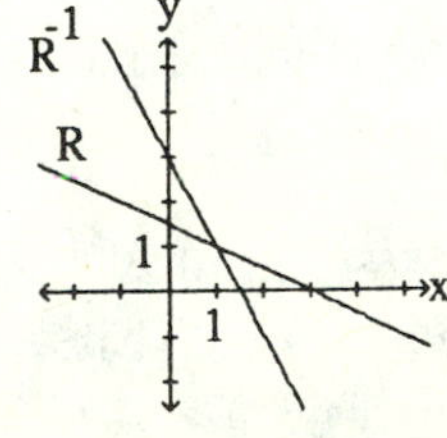

Inverse: $x = 3 - 2y$; function.

13.

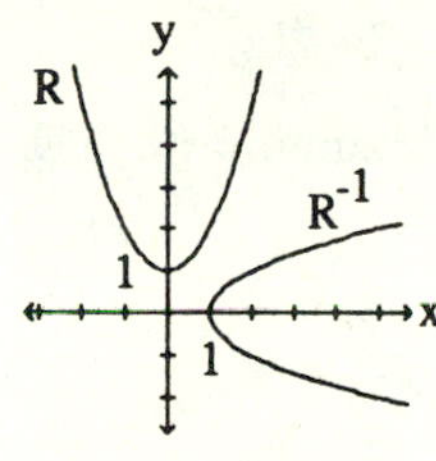

Inverse: $x = y^2 + 1$; not a function.

15.

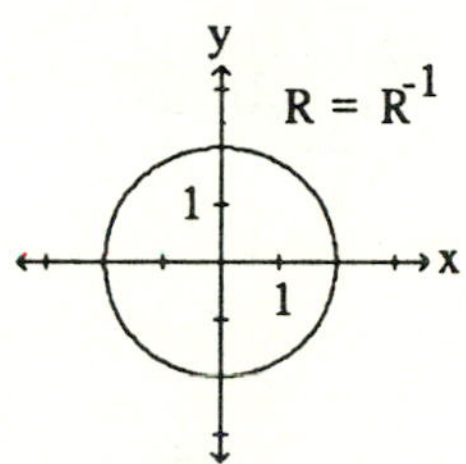

Inverse: $y^2 + x^2 = 4$; not a function.

17. f is one-to-one

$x = 3y - 6$; function;
$y = \frac{1}{3}x + 2$

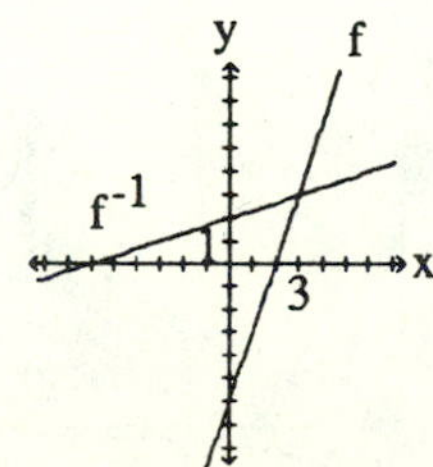

19. h is one-to-one

$x = \sqrt{y+2}$; function;
$y = x^2 - 2$ with $x \geq 0$

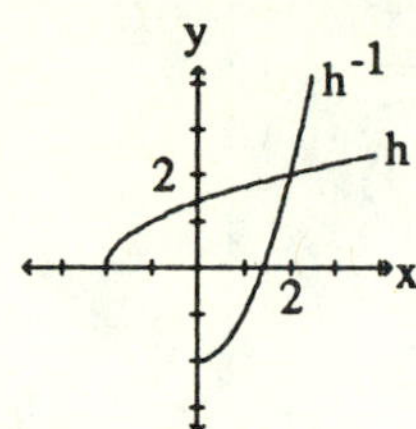

21. f is one-to-one
$x = \frac{y+3}{y-2}$; function;
$y = \frac{2x+3}{x-1}$

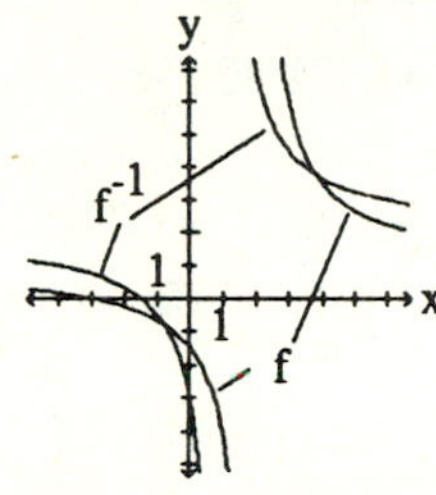

23. Domain of g: $(-\infty, \infty)$
Range of g: $[0, \infty)$
Domain of g^{-1}: $[0, \infty)$
Range of g^{-1}: $(-\infty, \infty)$

25. f is not one-to-one

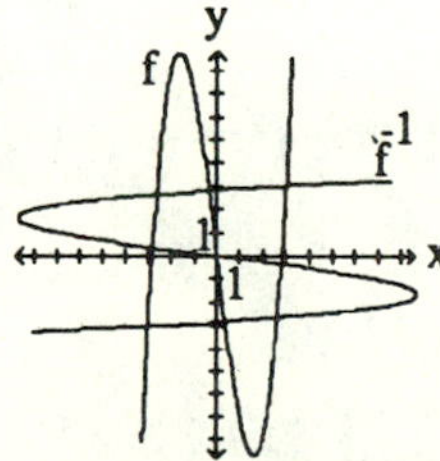

27. f is not one-to-one

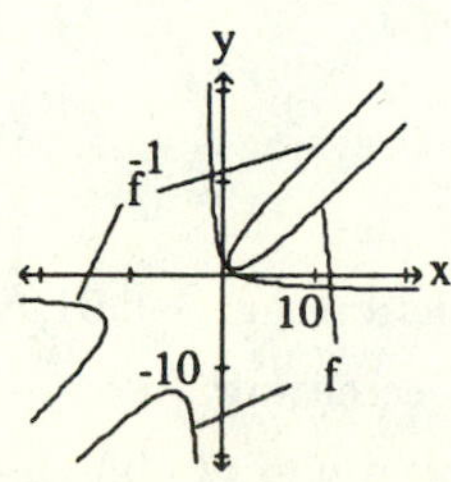

29. f is one-to-one

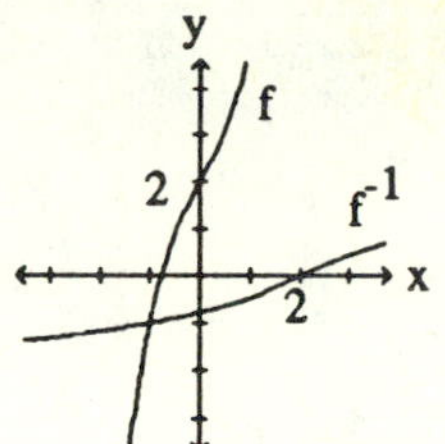

31. Domain of f: $(-\infty, \infty)$
Range of f: $(-\infty, \infty)$
Domain of f^{-1}: $(-\infty, \infty)$
Range of f^{-1}: $(-\infty, \infty)$

*33. Assume f is not one-to-one, then we can find two distinct ordered pairs, $(a, f(a))$ and $(b, f(b))$, where $f(a) = f(b)$ but $a \neq b$. But this is a contradiction to the statement "$f(x) = f(y)$ implies $x = y$ for all x and y in the domain of f." Hence, f is one-to-one.

35. $f(2\sqrt{2}) = 0$ and $f(-2\sqrt{2}) = 0$. Hence, f is not one-to-one, since we have found two distinct ordered pairs, $(2\sqrt{2}, 0)$ and $(-2\sqrt{2}, 0)$, where $f(2\sqrt{2}) = f(-2\sqrt{2})$, but $2\sqrt{2} \neq -2\sqrt{2}$.

37. $f(x) = x$; $f(x) = c - x$, c any constant; $f(x) = 1/x$

39. Domain $f : (-\infty, \infty)$
Range $f : (-\infty, 1) \cup [2, \infty)$
Domain f^{-1} : $(-\infty, 1) \cup [2, \infty)$
Domain $f^{-1} : (-\infty, \infty)$

$f^{-1}(x) =$
$\begin{cases} \frac{x}{x-1} & \text{for } x < 1 \\ \sqrt{x-1} & \text{for } x \geq 2 \end{cases}$

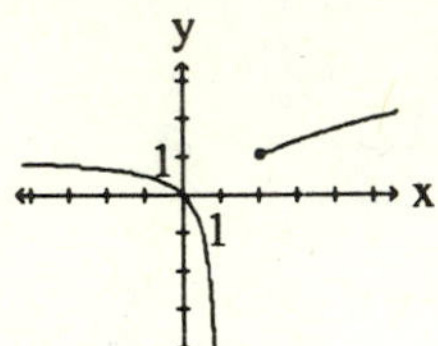

Exercises 4.2

1. The graphs are the same.

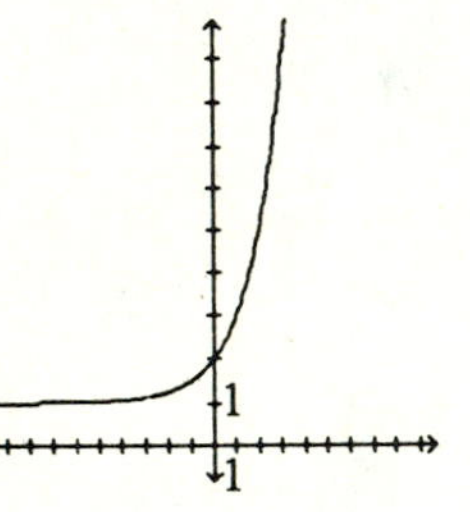

3.

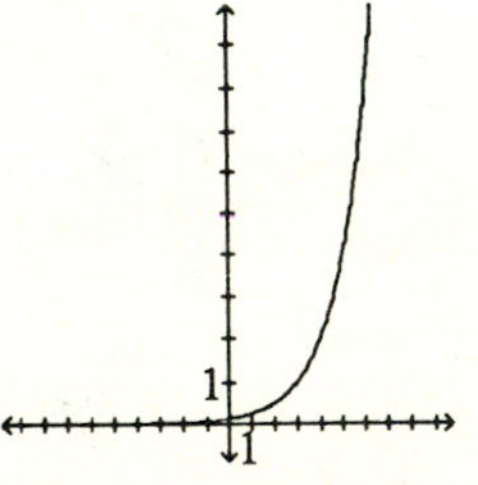

5.

7.

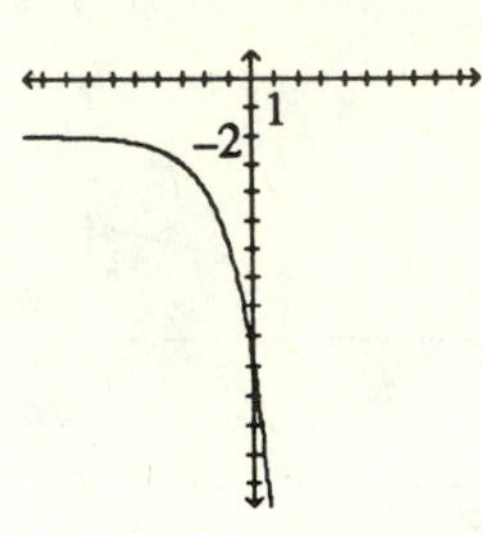

9. Domain: $(-\infty, \infty)$
 Range: $(0, \infty)$

11. $x = 4$

13. $x = -4$

15. $x = -79.80$

*17. This problem requires logarithms or a grapher to solve $e^{2x} = 3$ to get $x = 0.55$.

19. Domain: $(-\infty, \infty)$

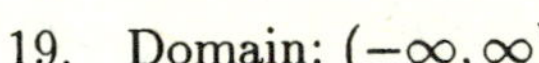

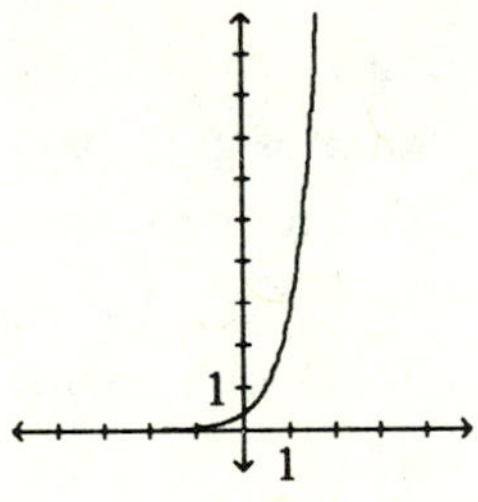

21. Domain: $(-\infty, \infty)$

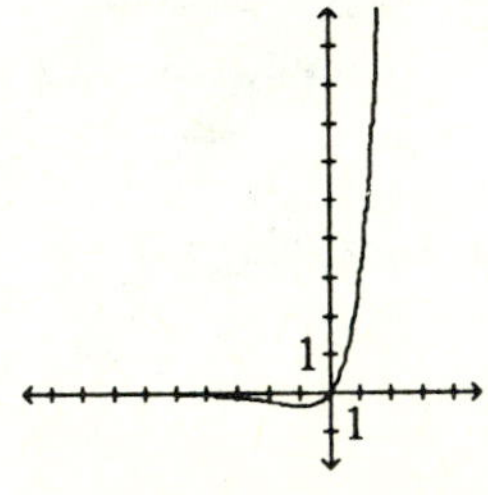

23. Increasing: $(-\infty, \infty)$
 Range: $(0, \infty)$

25. Increasing: $[-0.91, \infty)$
 Decreasing: $(-\infty, -0.91]$
 Minimum: $(-0.91, -0.33)$
 Range: $[-0.33, \infty)$

27. Increasing: $[-0.85, 0.85]$
 Decreasing: $(-\infty, -0.85] \cup [0.85, \infty)$
 Maximum: $(0.85, 0.52)$
 Minimum: $(-0.85, -0.52)$
 Range: $[-0.52, 0.52]$

29. From the horizontal line test, we see f is one-to-one.

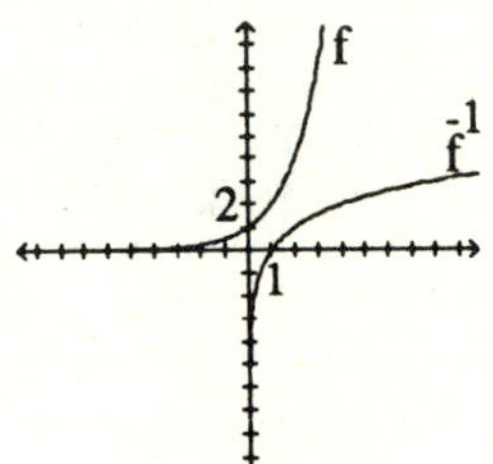

31.

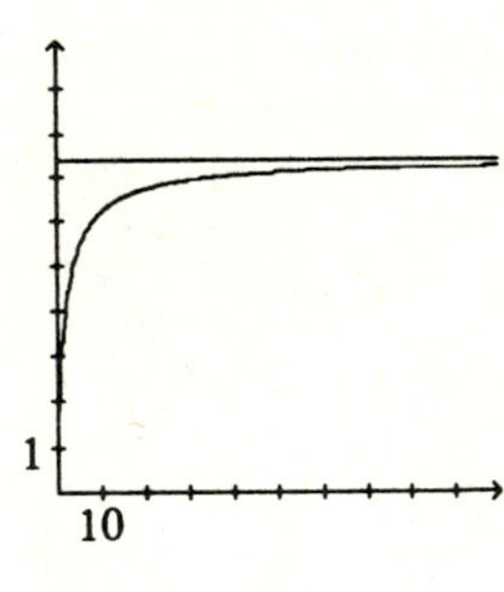

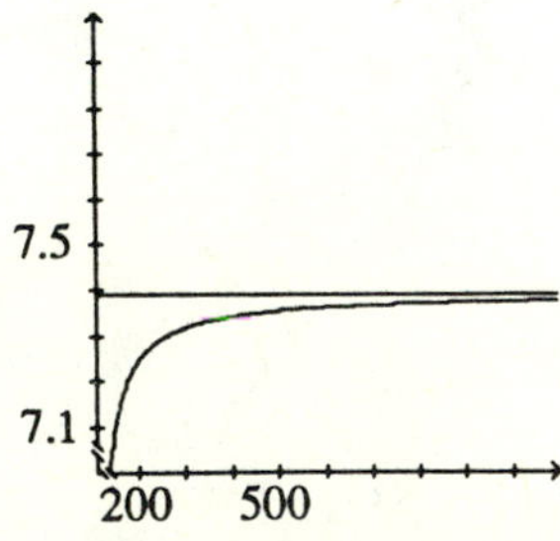

33.

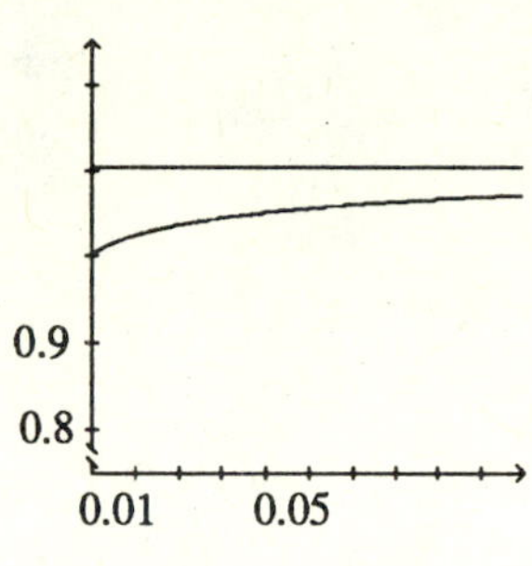

35.

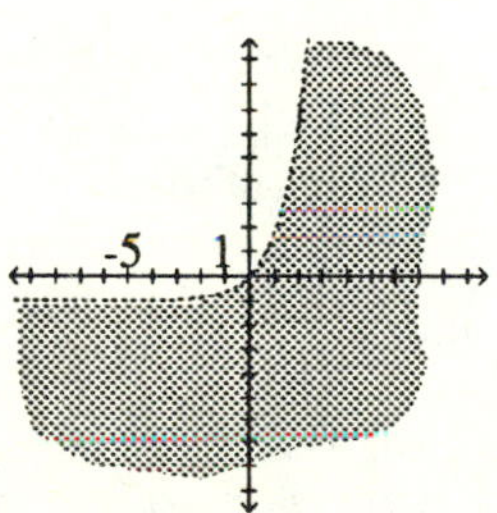

37. $x = -0.77, 2, 4$. For $x \in (-0.77, 2) \cup (4, \infty)$, $2^x > x^2$. For $x \in (-\infty, -0.77) \cup (2, 4)$, $x^2 > 2^x$.

39. (a) 10,000

(b)

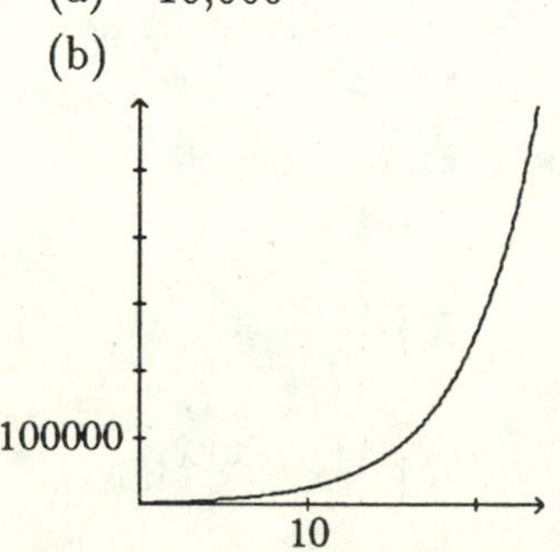

(c) 15.97 hours

41. (a) $P(t) = 475,000(1.0375)^t$

(b) The problem situation is represented by the portion of the graph in the first quadrant.

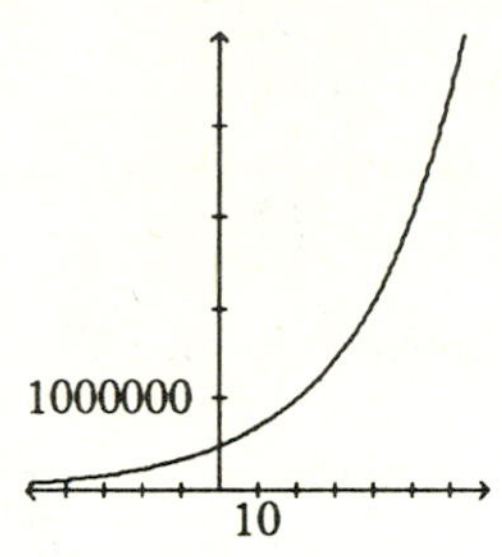

(c) after 20.22 years

43. $P(t) = 6250(1.0375)^t$

(b) The problem situation is represented by the portion of the graph in the first quadrant.

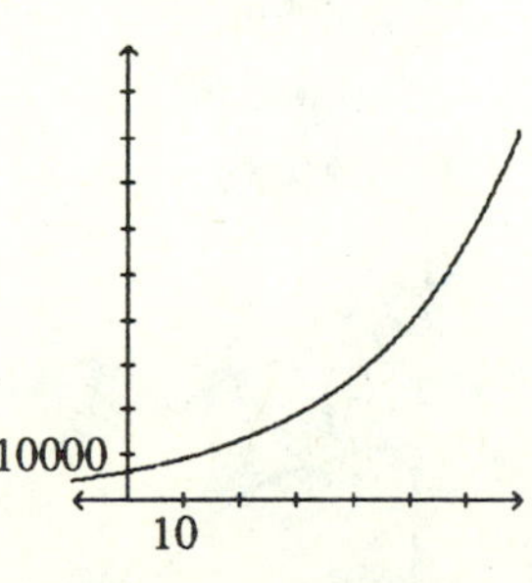

(c) in 1915–15,600 people

(d) in 1940–39,381 people

45. $A(t) = 3.5(0.5)^{\frac{t}{65}}$

(b)

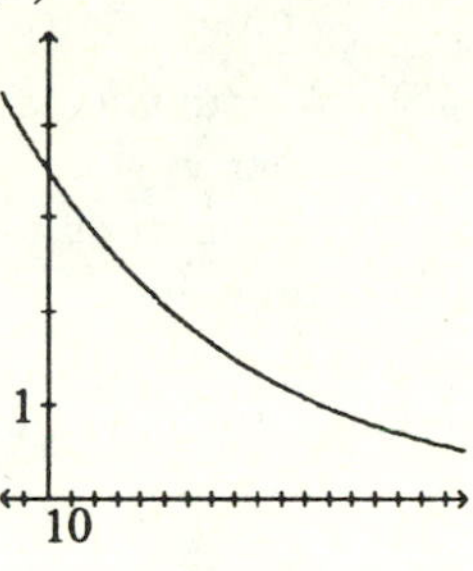

(c) $t = 117.49$

Exercises 4.3

1.

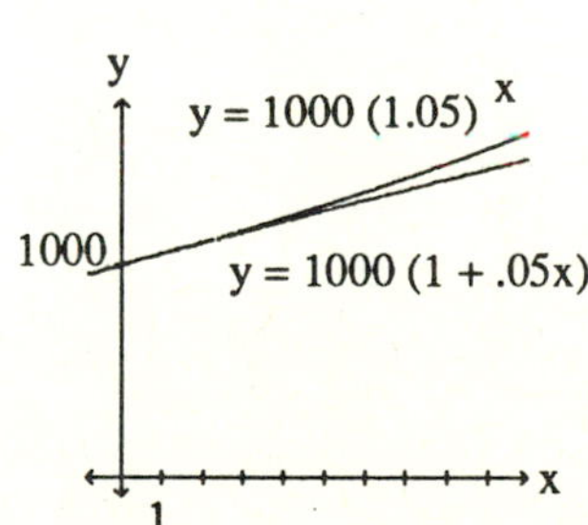

3.

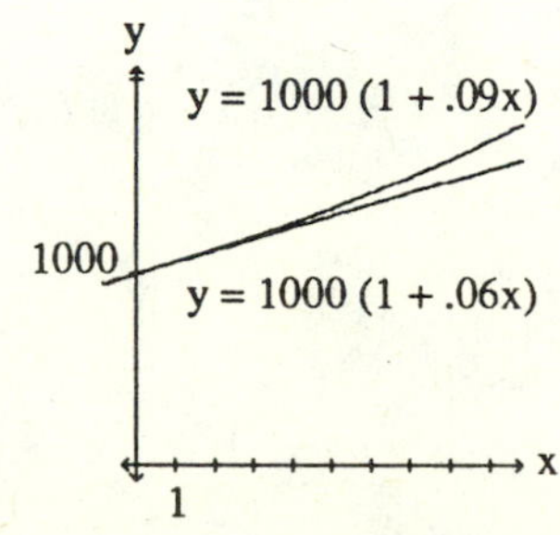

5. 1a: Amount in account after x years at 5% simple interest.
   1b: Amount in account after x years at 5% compounded annually.
   2a: Amount in account after x years at 8% simple interest.
   2b: Amount in account after x years at 8% compounded annually.
7. $r = 0.125$
9. $P = 1096.49$
11. $t = \frac{S-P}{rP}$
13. $r = 0.079$
15. $P = 729.73$
17. $P = \frac{S}{(1+r)^t}$
19. 11.55 years
21. 6.63 years
23. $r = 0.08$
25. 12.14 years
27. (a) \$669.11
    (b) \$673.43
    (c) \$674.43
    (d) \$674.91
    (e) \$674.93
29.

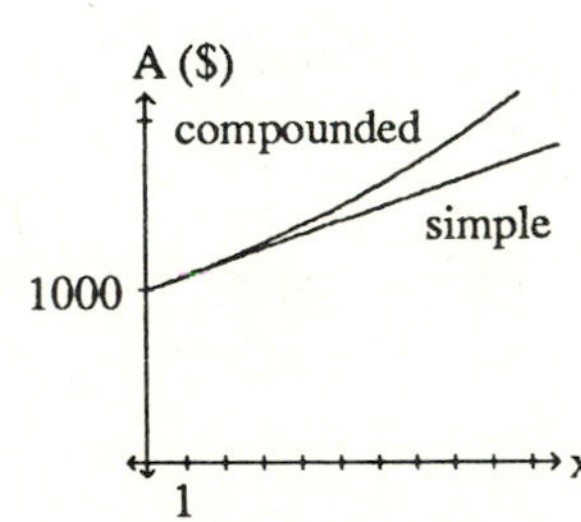

31. 6% compounded quarterly
33. 7.20% compounded daily
35.

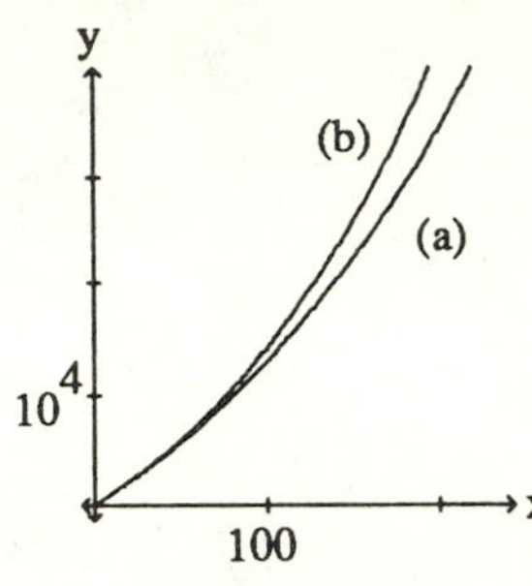

37. The values of y in 35(a) represent the future value of the annuity after x payments at 6% APR. The values of y in 35(b) represent the future value of the annuity after x payments at 8% APR.
39. \$187.71
41. \$36013.70
43. 14 years, 4 months
45. a) \$884.61
    b) \$905.77
    c) \$946.93
    d) \$1032.14
47. a)

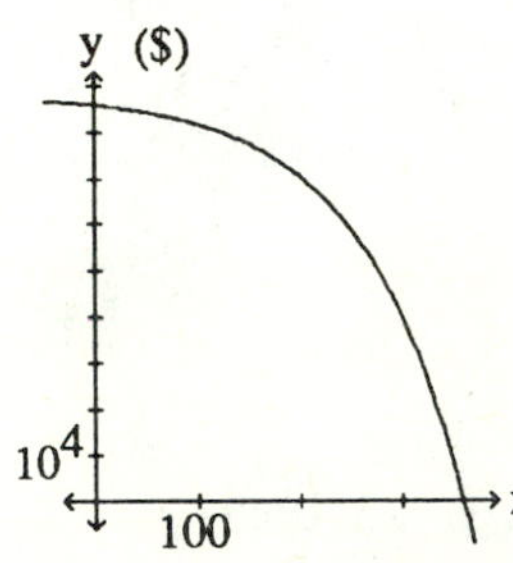

    b) The y value is the outstanding balance after n of the 360 payments have been made.
49. a) $R(x) = \frac{860}{1-(1.01)^{-x}}$
    b)

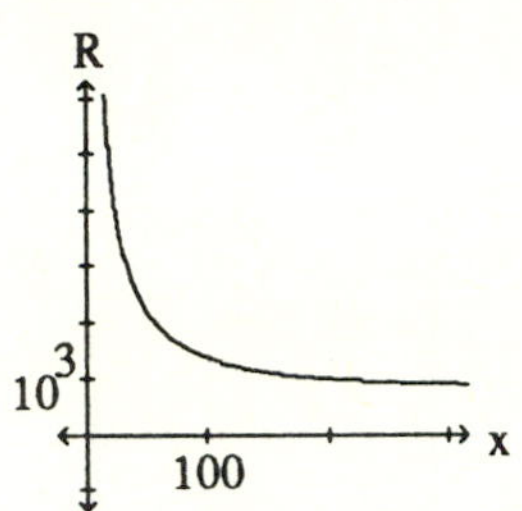

    c) R represents the monthly payment for a \$86,000 mortgage at 12% APR to be paid in x installments.
51. $i = 0.0125$; i is the monthly interest rate.

*53. Each $R(1+i)^k$ is the future value of one payment after k months. The number of months on deposit varies from $n-1$ for the first payment to 0 for the last payment. Summing the future values gives the desired result.

*55. From Exercise 52, we know $A(1+i)^n = S$. From Exercise 54 we know $S = R\left[\frac{(1+i)^n-1}{i}\right]$. Therefore, $A = S(1+i)^{-n} = R\left[\frac{(1+i)^n-1}{i}\right](1+i)^{-n} = R\left[\frac{1-(1+i)^{-n}}{i}\right]$

57.

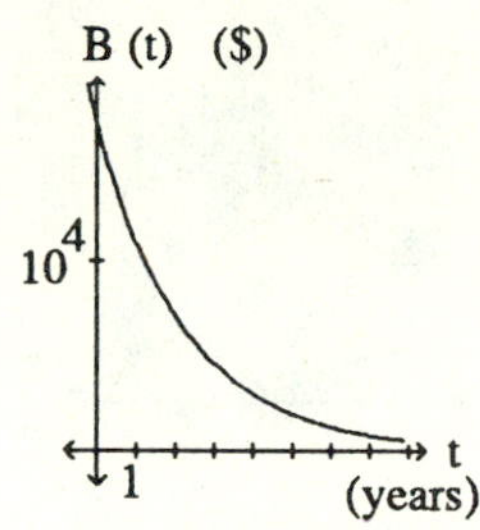

59. $B(4.25) = \$2599$

61. If $r < 0$, then $C(1-r)^n = C(1+|r|)^n$, where $|r| \geq 0$. But, this is equal to $C(1+r)^n = S$ when $r \geq 0$. $(1+r)^n$ will always increase as n gets larger (i.e. the length of the term increases), will always be positive, and will be greater than 1. So $S = C(1+r)^n > C$.

63. $S(7) = \$94{,}260.33$

65. $0.46

Exercises 4.4

1. 2
3. -4
5. No solution. Not defined $x < 0$.
7. 81
9. 5
11. 1.63
13. ± 3
15.

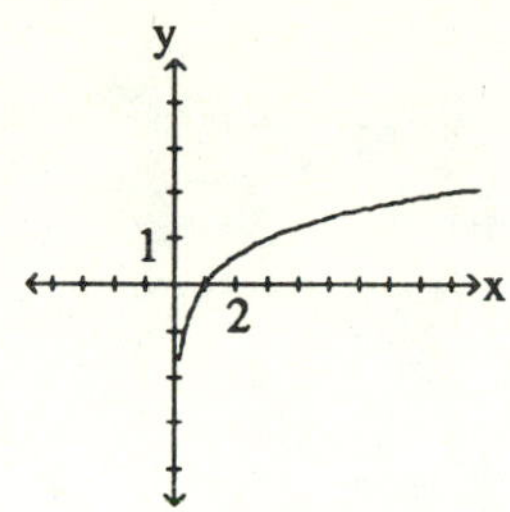

17.

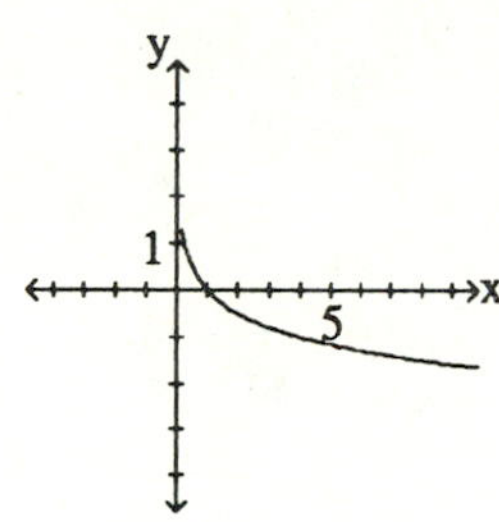

19. $\log_2(x+y) = 8$
21. $3^{-2} = \frac{x}{y}$
23. $\log 5000 + \log x + 360 \log(1+r)$
25. The domain of $f(x)$ is $(-\infty, 0) \cup (0, \infty)$, while the domain of $g(x)$ is $(0, \infty)$.
27. $f(x) \to 0$
29. $f(x) \to \infty$
31. $P(t) = 123{,}000(0.97625)^t$ in 37.45 years
33. $A(t) = 6.58(\frac{1}{2})^{t/14}$ when $t > 38.05$ days
35. Table 2 does not have a power rule since the points $(\ln x, \ln y)$ are not linear.

    Table 3 has a power rule since the points $(\ln x, \ln y)$ are linear.

    Table 4 does not have a power rule since the points $(\ln x, \ln y)$ are not linear.
37. Table 4 has an exponential algebraic representation since the points $(x, \ln y)$ are linear.
39. $t = 11.91$
41. $t = 15.52$
43. $\frac{\ln S - \ln p}{\ln(1+r)} = t$
45. 15.75 years
47. 40.75
49. $x = 0.01$
51.

$m = 1.51$

$b = -21.80$

	ln x	ln P
Earth	18.344	5.900
Mercury	17.399	4.477
Venus	18.022	5.416
Mars	18.769	6.532
Jupiter	19.996	8.373
Saturn	20.602	9.283

53. 43,504.80 days
55. $p = 49.90x^{.25}$

Exercises 4.5

1.

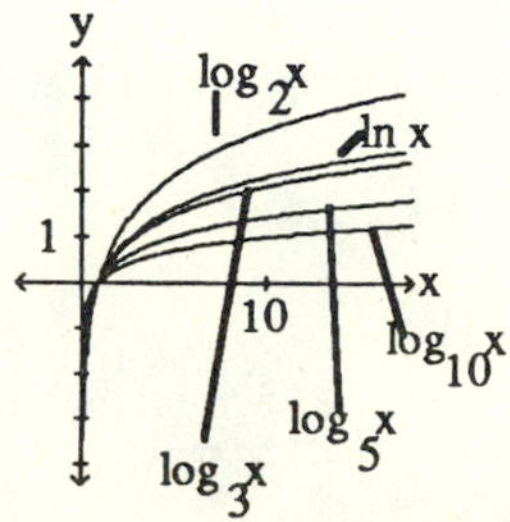

3. $(1, \infty)$

5. $(1, \infty)$

7.

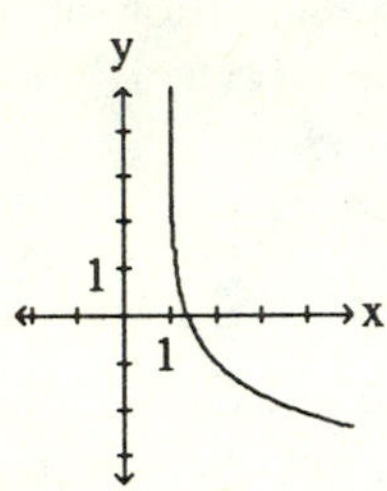

9.

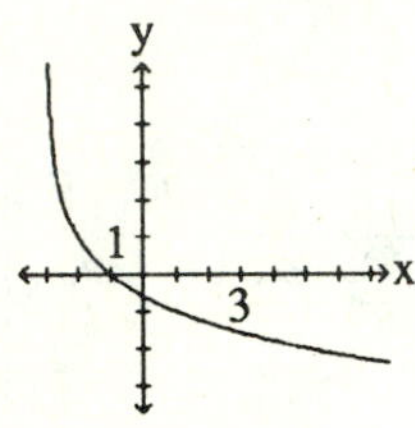

11. Domain: $(2, \infty)$

Range: $(-\infty, \infty)$

13. Domain: $(-\infty, -3) \cup (-3, \infty)$

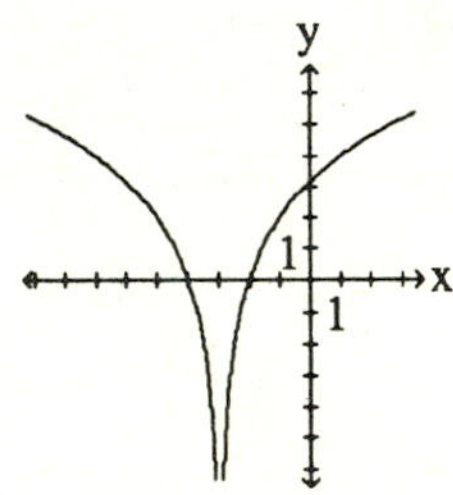

15. Domain: $(0, 1) \cup (1, \infty)$

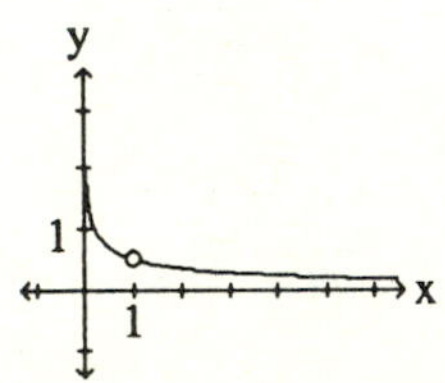

17. Domain: $(-\infty, \infty)$

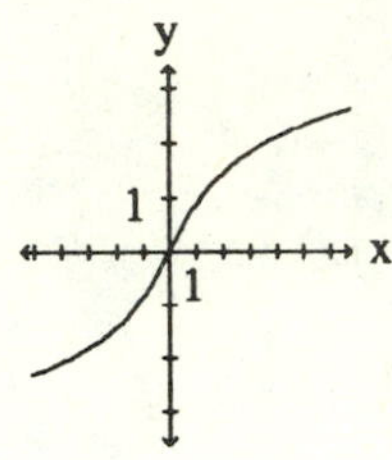

19. $f(x) \to \infty$ as $x \to \infty$

$f(x)$ not defined $x \in (-\infty, 0)$

21. $f(x) \to 0$ as $x \to \infty$

$f(x)$ not defined $x \in (-\infty, 0)$

23. local minimum: $(0.61, -0.18)$

decreasing on $(0, 0.61]$

increasing on $[0.61, \infty)$

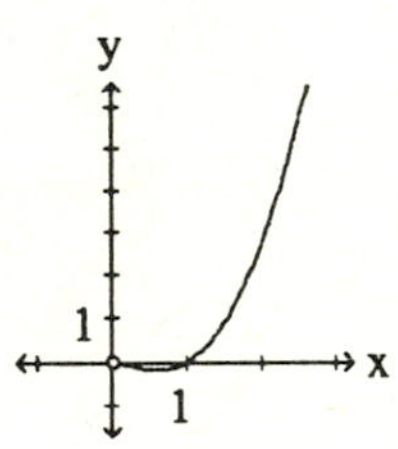

25. local minimum: $(2.72, 0.37)$

decreasing on $[2.72, \infty)$

increasing on $(0, 2.72]$

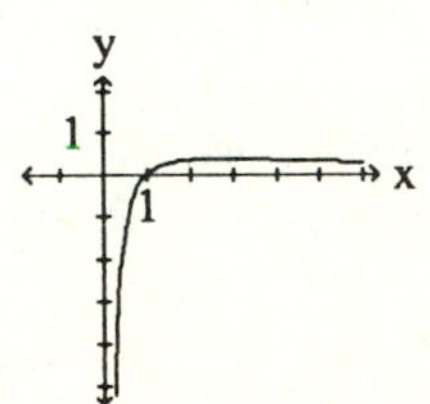

27.

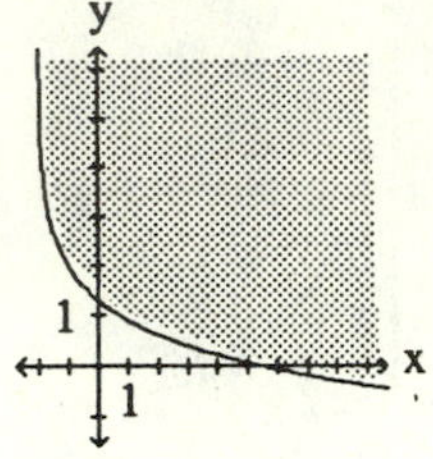

29. $(-2.02, 2.09)$ and $(2.23, 2.98)$

31. $f(x) > g(x)$ in $(-\infty, 2.48) \cup (3, \infty)$

$f(x) = g(x)$ at $(2.48, 15.22)$ and $(3, 27)$

$f(x) < g(x)$ in $(2.48, 3)$

33. $f(x) > g(x)$ in $(0, 10)$

$f(x) = g(x)$ at $(10, 10^{10})$ and $(0, 1)$

$f(x) < g(x)$ at $(10, \infty)$

$y = x^x$ is usually undefined for $x < 0$

35. $f(x) > g(x)$ in $(0, \infty)$

$g(x)$ is defined only in $(1, \infty)$

37. graph $y = \frac{\ln 4}{\ln x}$

39.

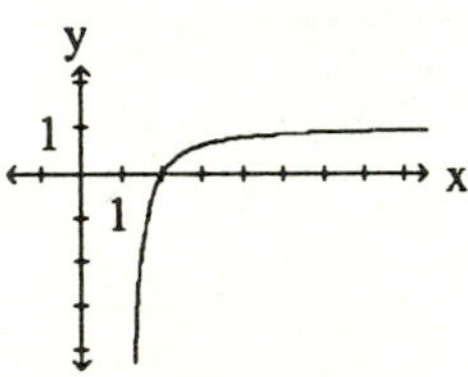

41. Domain: $(1, \infty)$

Range: $(-\infty, 1)$

43. Let a_1 be amplitude when 6 on Richter scale.

Let a_2 be amplitude when 4 on Richter scale.

Then $6 = \log(\frac{a_1}{T}) + B$ and $4 = \log(\frac{a_2}{T}) + B$ and

$$2 = \log(\frac{a_1}{T}) - \log(\frac{a_2}{T})$$

$$2 = \log(\frac{a_1}{T} \cdot \frac{T}{a_2})$$

$$2 = \log(\frac{a_1}{a_2})$$

$10^2 = \frac{a_1}{a_2}$ or $a_2 100 = a_1$

45. $I = 10^{(-0.00235x + \log 12)}$

$= \frac{12}{10^{(0.00235x)}}$

47. $I = 10.20$

49. $i(t) = \frac{10}{3}(1 - e^{(-\frac{3t}{0.03})})$

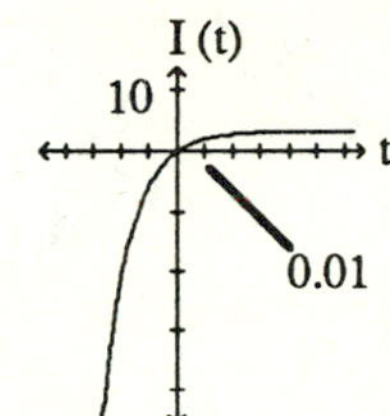

51. When $0 \le r \le 0.2$

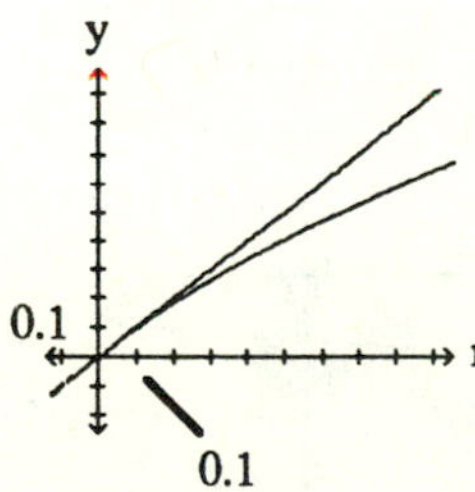

53.

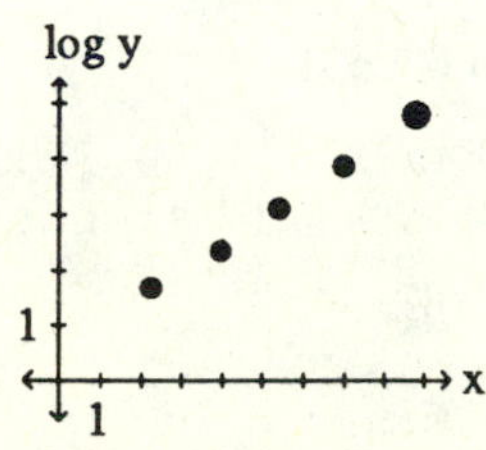

55. $f(x) = ab^x$

$\log(f(x)) = \log(ab^x)$

$\log(f(x)) = \log a + x \log b$

$\log(f(x)) = (\log b)x + \log a$

57. $\log(f(x)) = (\log b)x + \log a$

$\log f(x_1) = x_1 \log b + \log a$

$\log f(x_2) = x_2 \log b + \log a$

$\frac{\log f(x_2) - \log f(x_1)}{x_2 - x_1} = \log b$

59. (a)

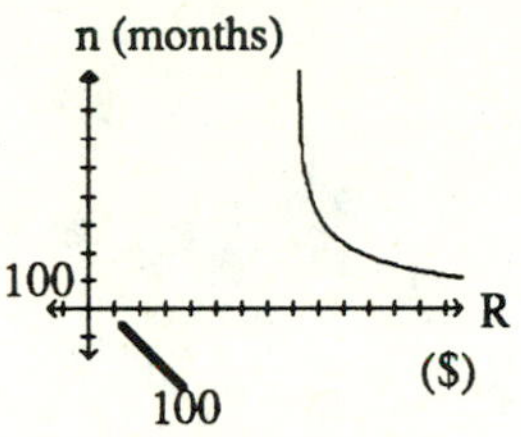

(b) \$923.17

(c) ~ 252 payments (21 years)

Exercises 4.6

1. 10,000
3. $5\frac{1}{4}$
5. 3
7. 10.04
9. 1
11. $y = 2 \log x$

Domain: $(0, \infty)$

Range: $(-\infty, \infty)$

$y = \log x^2$

Domain: $(-\infty, 0) \cup (0, \infty)$

Range: $(-\infty, \infty)$

13. $y = \log_5 \sqrt{x+6} - \log_5 x$

Domain: $(0, \infty)$

Range: $(-\infty, \infty)$

$y = \log_5 \frac{\sqrt{x+6}}{x}$

Domain: $(-\infty, 0) \cup (0, \infty)$

Range: $(-\infty, \infty)$

15. $x = \frac{\log 4.1}{\log 1.06}$

17. $x = \frac{\log\left(\frac{12+\sqrt{148}}{2}\right)}{\log 2}$

19. 24.22

21. 3.59

23. Domain: $(-\infty, \infty)$

Range: $[-0.37, \infty)$

Local minimum: $(-1.00, -0.37)$

Increasing in: $[-1, \infty)$

Decreasing in: $(-\infty, -1]$

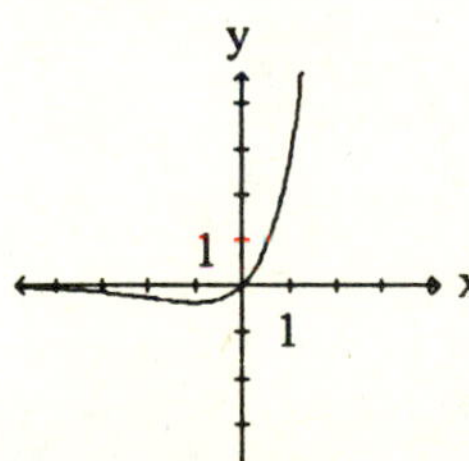

25. Domain: $(-\infty, \infty)$

Range: $(-\infty, \infty)$

Increasing in: $(-\infty, \infty)$

No local extrema

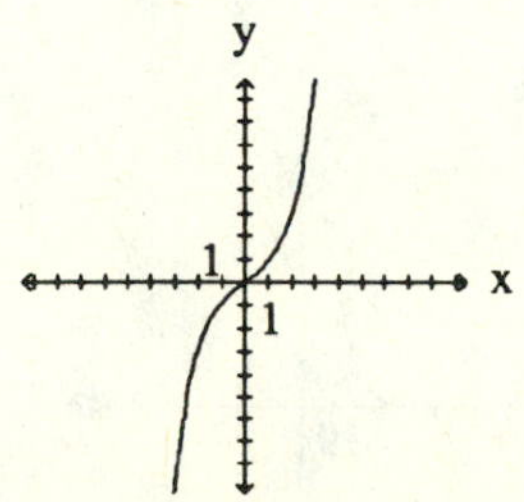

27. Domain: $(-\infty, -2) \cup (0, \infty)$
Range: $(-\infty, \infty)$
Increasing in: $(0, \infty)$
Decreasing in: $(-\infty, -2)$
No local extrema

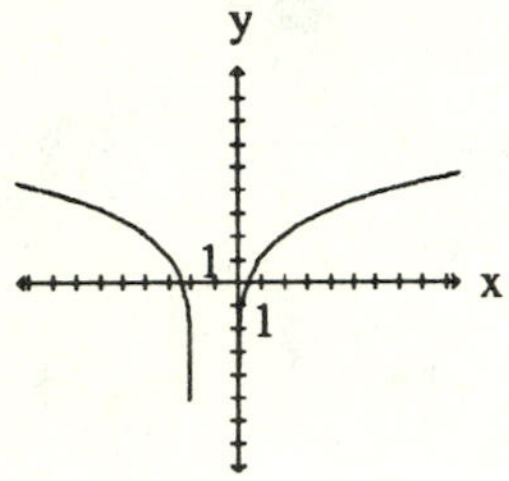

29. $(9, \infty)$
31. 1.31
33. $(0, 1.71)$
35. $1.08, 58.77$
37. $\frac{ln}{x} = \frac{1}{10}$ can be rewritten as $lnx = \frac{1}{10}xln2$ so $x = 2^{\frac{1}{10}x}$ or $x^{10} = 2^x$
39.

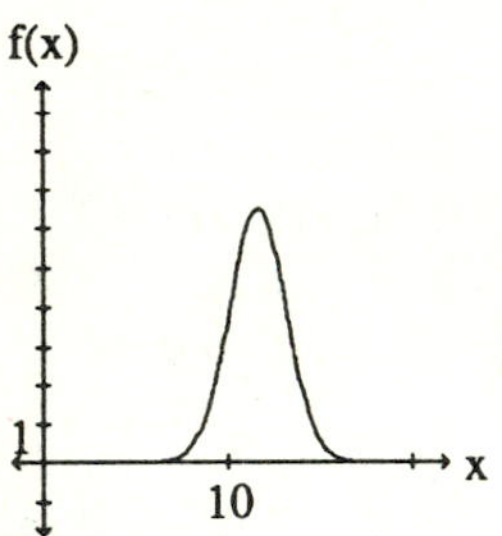

41.

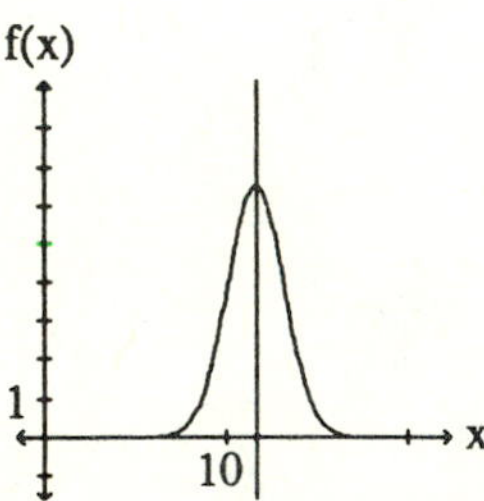

43. $f_2(x) = \frac{x^2}{2!} + x + 1 = \frac{x^2}{2!} + f_1(x)$
$f_3(x) = \frac{x^3}{3!} + \frac{x^2}{2!} + x + 1$
$= \frac{x^3}{3!} + f_2(x)$ ⋮
$f_n(x) = \frac{x^n}{n!} + f_{n-1}(x)$
45. $[-1.32, 1.17]$
47.

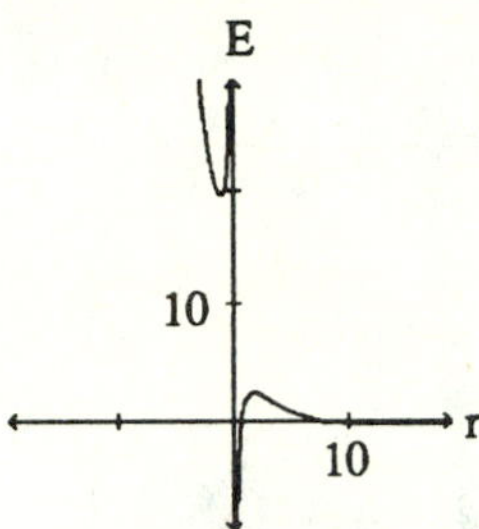

49. Relative maximum of 2.38 at $r = 1.73$
51. $B = \$6295.11$.
53. The straight line method allows for the greatest depreciation in the first two years.
55. ≈ 10.5 %

Review Exercises

1.

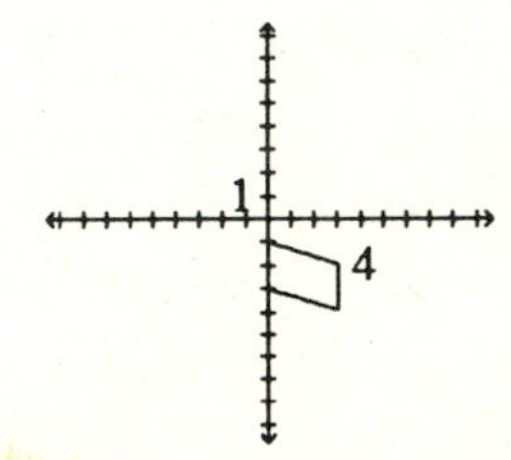

3. $x = y^2 - 2$; not a function

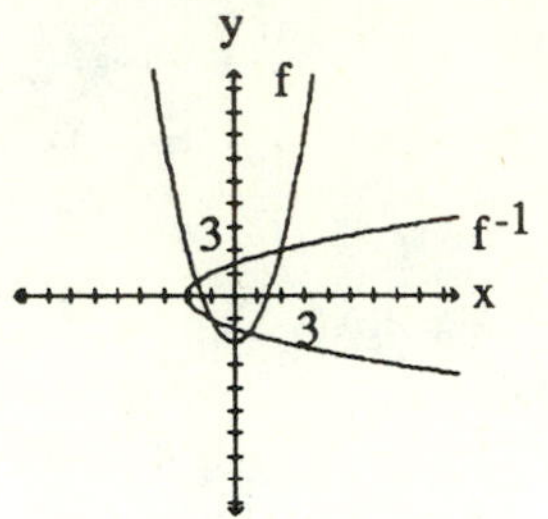

5. h is one-to-one
$y = \frac{x^2}{4} + 4$ with $x \geq 0$; function

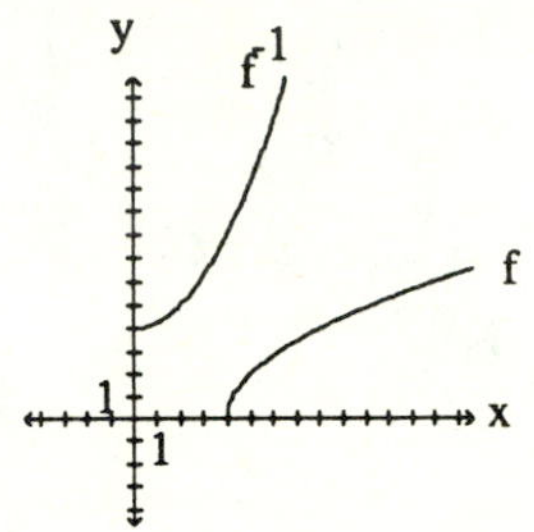

7. f is not one-to-one

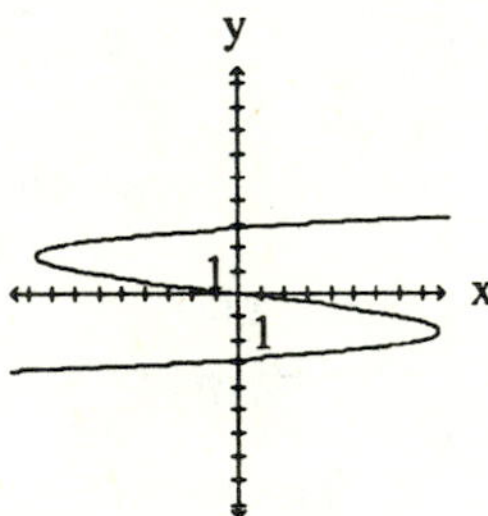

9. Dom f: $(-\infty, -2) \cup (-2, \infty)$
Range f: $(-\infty, -12.63) \cup (0.63, \infty)$
Dom f^{-1}: $(-\infty, -12.63) \cup (0.63, \infty)$
Range f^{-1}: $(-\infty, -2) \cup (-2, \infty)$
11. Domain: $(-\infty, \infty)$

Range: $(-\infty, 3)$

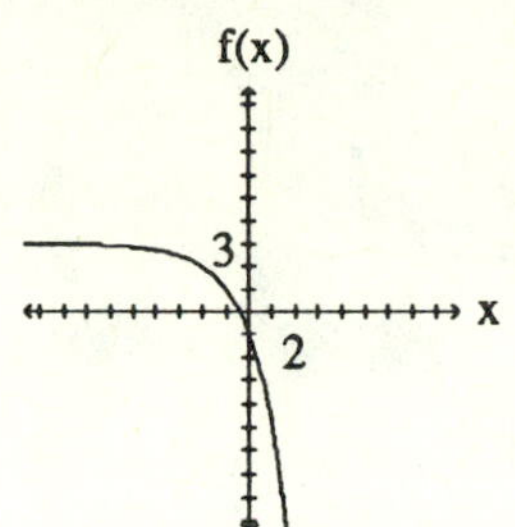

13.

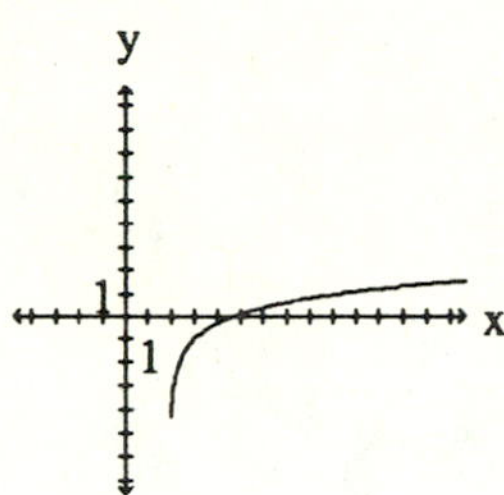

15.

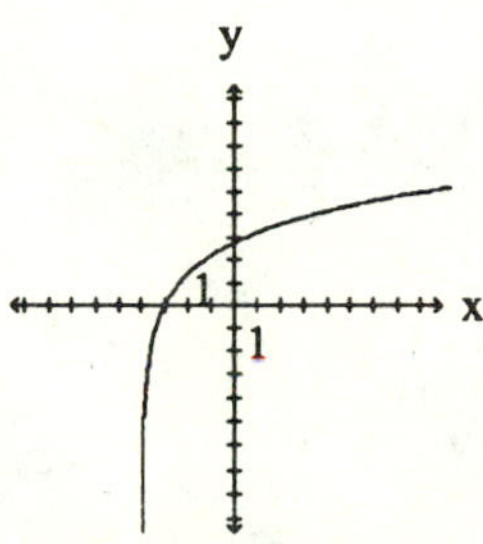

17. Domain: $(-\infty, \infty)$
Range: $(-\infty, 4.246)$

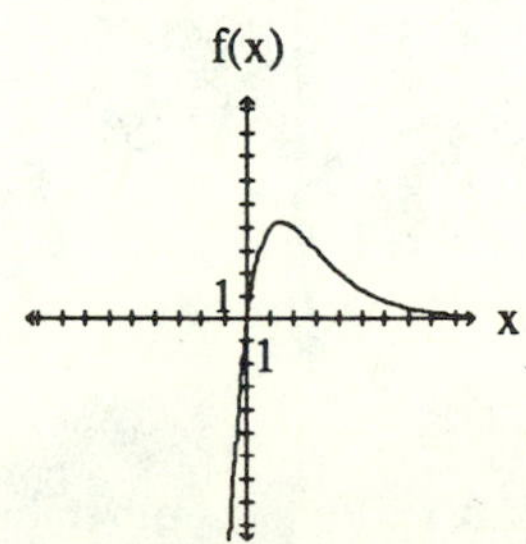

19. Increasing: $(-\infty, 1.44]$
Decreasing: $[1.44, \infty)$
Maximum: $(1.44, 4.25)$

21. no solution

23. 64

25. $\frac{\log z}{\log(23+x)}$

27. $g(x) \to \infty$ as $x \to 2^+$

29. $(0, 1)$

31. Domain: $(-\infty, 5) \cup (5, \infty)$

Range: $(-\infty, \infty)$

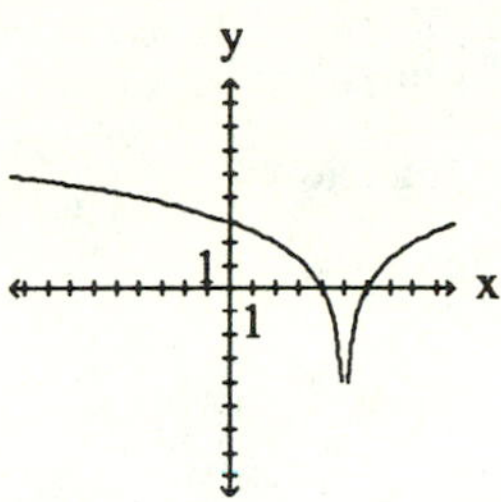

33. Decreasing: $(0, 0.72]$

Increasing: $[0.72, \infty)$

Local minimum: $(0.72, -0.12)$

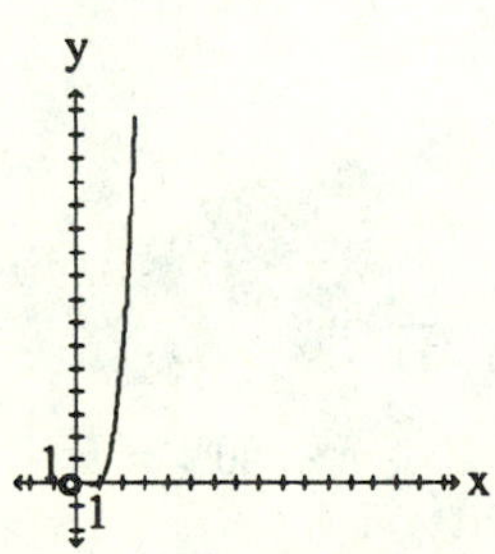

35. 5

37. $\frac{\log 0.90}{\log 1.5}$

39. Domain: $(-\infty, 0) \cup (3, \infty)$

Range: $(-\infty, 0) \cup (0, \infty)$

Decreasing in: $(-\infty, 0)$ and $(3, \infty)$

No local extrema

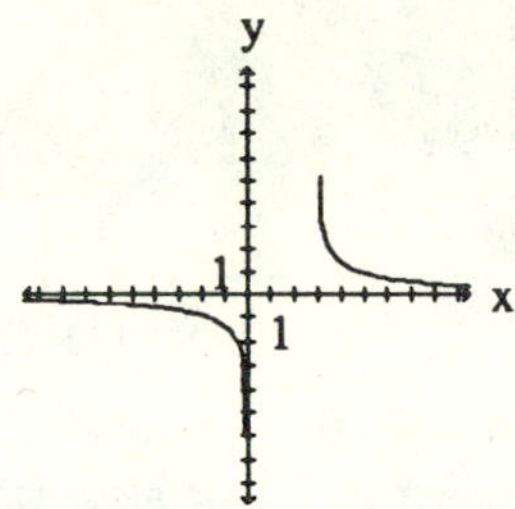

41. $(3.52, \infty)$

43. $(0.72, \infty)$

45. $(-\infty, 0.69]$

47. 6 years

49.

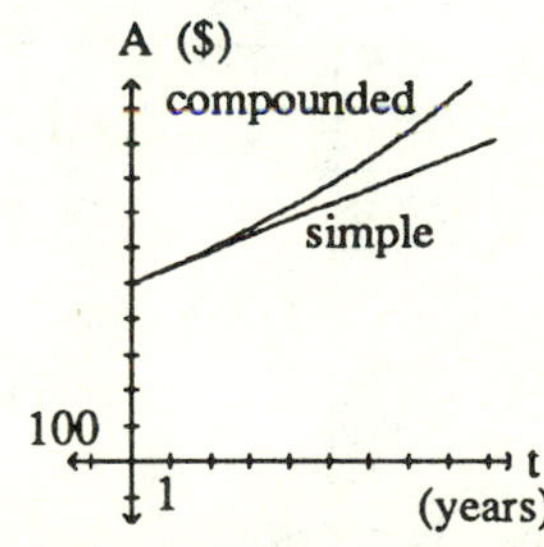

51. (a) $A(t) = 4.62\left(\frac{1}{2}\right)^{t/21}$

(b)

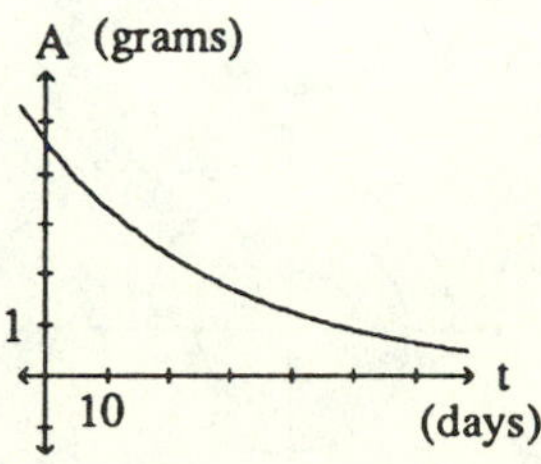

(c) after 46.37 days.

53. $a > 0$

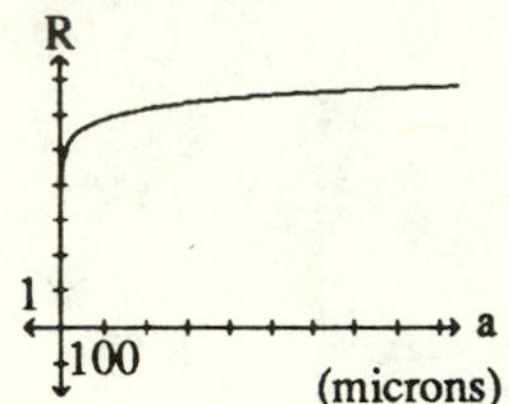

• CHAPTER 5

Exercises 5.1

1.

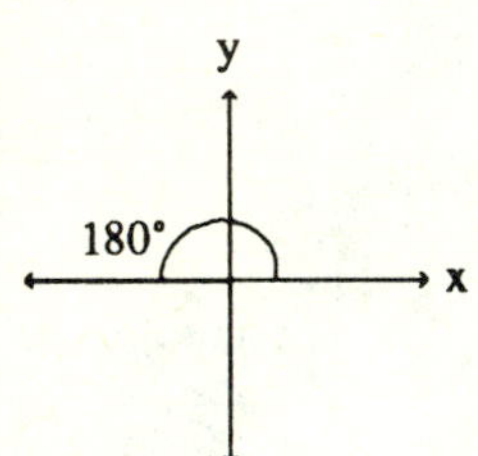

3.

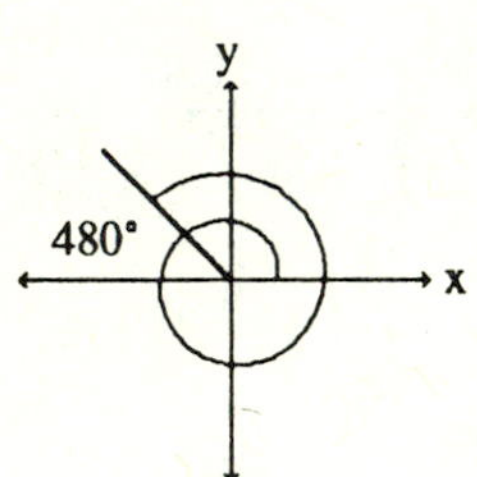

5.

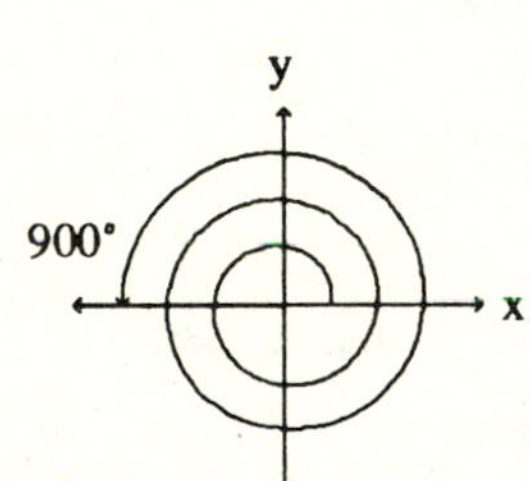

7. III Quadrant

9. II Quadrant

11. I Quadrant

13. 408°

15. 345°

17. 415°, 775°, −305°, −665° are possible answers.

19. 50°, 770°, −310°, −670° are possible answers.

21. 180°

23. 45°

25. 315°

27. $2\sqrt{2}$

29. $\sqrt{13}$

31. $2\sqrt{17}$

33. $37° + n \cdot 360°$, n any integer

35. $-72° + n \cdot 360°$, n any integer

37. $\frac{\pi}{3}$

39. 15.58

41. All four ratios are the same: $S = \frac{5}{18}\pi$

43. About 6107 feet

45.

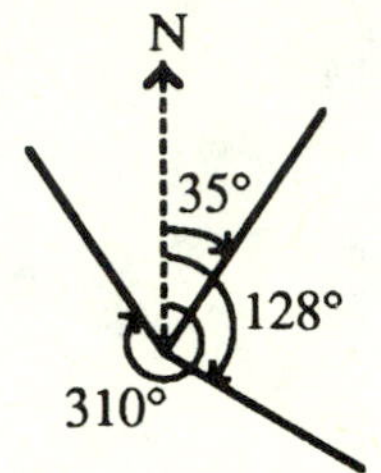

47.

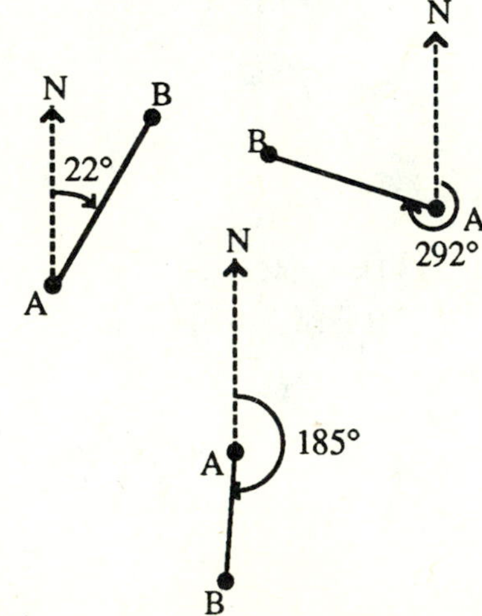

49. If a rational number has the form $\frac{a \cdot n}{b \cdot n}$ where a and b have no common factors, we divide out the common factor

n to get $\frac{a}{b}$ in lowest terms. If an angle has the form $\theta + n \cdot 360°$, with $0° \leq \theta \leq 360°$, we subtract $n \cdot 360°$ to get θ in "lowest terms."

Exercises 5.2

1. $\sin 23° = 0.39$
   $\cos 23° = 0.92$
   $\tan 23° = 0.42$
   $\csc 23° = 2.56$
   $\sec 23° = 1.09$
   $\cot 23° = 2.36$
3. $\sin 38° = 0.62$
   $\cos 38° = 0.79$
   $\tan 38° = 0.78$
   $\csc 38° = 1.62$
   $\sec 38° = 1.27$
   $\cot 38° = 1.28$
5. $\tan 30° = \frac{1}{\sqrt{3}}$
   $\cos 60° = \frac{1}{2}$
   $\sec 30° = \frac{2}{\sqrt{3}}$
   $\csc 60° = \frac{2}{\sqrt{3}}$
   $\cot 60° = \frac{1}{\sqrt{3}}$
7. For $\theta = 56.44269024$
   $\sin\theta = 0.8\overline{3}$
   $\cos\theta = 0.5528$
   $\tan\theta = 1.5076$
   $\csc\theta = 1.2$
   $\sec\theta = 1.809$
   $\cot\theta = 0.6633$
   These agree with the values in Example 1, to 4 decimal places.
9. $\sin\angle B = \frac{7}{\sqrt{113}}$
   $\cos\angle B = \frac{8}{\sqrt{113}}$
   $\tan\angle B = \frac{7}{8}$
   $\csc\angle B = \frac{\sqrt{113}}{7}$
   $\sec\angle B = \frac{\sqrt{113}}{8}$
   $\cot\angle B = \frac{8}{7}$
11. $\sin\angle B = \frac{2\sqrt{26}}{15}$
    $\cos\angle B = \frac{11}{15}$
    $\tan\angle B = \frac{2\sqrt{26}}{11}$
    $\csc\angle B = \frac{15}{2\sqrt{26}}$
    $\sec\angle B = \frac{15}{11}$
    $\cot\angle B = \frac{11}{2\sqrt{26}}$
13. $\sin\theta = \frac{1}{\sqrt{10}}$
    $\cos\theta = \frac{3}{\sqrt{10}}$
    $\tan\theta = \frac{1}{3}$
    $\csc\theta = \sqrt{10}$
    $\sec\theta = \frac{\sqrt{10}}{3}$
    $\cot\theta = 3$
15. 55.54°
17. 82.82°
19. 71.57°
21. 70.35°
23. $\frac{1}{\cot\theta} = \frac{1}{b/a} = \frac{a}{b} = \tan\theta$. Taking reciprocals, we get $\frac{1}{1/\cot\theta} = \frac{1}{\tan\theta}$ or $\cot\theta = \frac{1}{\tan\theta}$.
25. In a right triangle with a very small acute angle, the lengths a and b could differ by a great deal so that the ratio $\frac{a}{b} = \tan\theta$ is very small (for $a < b$) or very large (for $a > b$). For example, if $a = 1$ and $b = 1000$, $\tan\theta = \frac{1}{1000} = 0.0001$.
27. 74.78°
29. 28.07°
    Note: For $30 - 40$, $\angle C = 90°$
31. $\angle A = 36.87°$
    $\angle B = 53.13°$
    $c = 5$
33. $\angle B = 35°$
    $a = 22.25$
    $c = 27.16$
35. $\angle A = 23.46°$
    $\angle B = 66.54°$
    $b = 46.56$
37. $\angle A = 31°$
    $b = 8.32$
    $c = 9.71$
39. $\angle B = 79.8°$
    $a = 2.57$
    $b = 14.27$
41. $\sin^2\theta + \cos^2\theta = 1$ for each θ given.

Exercises 5.3

1. $D = 5$
3. $D = 9$
5. $D = 5\sqrt{2}$
7. 100 feet
9. 680.55 feet
11. 106.12 feet
13. 101.57 feet
15. 52.35 feet
17. 84.85 miles; bearing of 140°
19. 19°

21. 637.85 feet
23. 9.78 feet
25. 2931.09 feet
27. $\sin\theta = b$
$\cos\theta = a$
$\tan\theta = c$
$\csc\theta = f$
$\sec\theta = e$
$\cot\theta = d$
29. From the Pythagorean Theorem, we know the third side of the right triangle is $\sqrt{1-a^2}$. Therefore, $\sin\theta = \sqrt{1-a^2}$ and $\tan\theta = \frac{\sqrt{1-a^2}}{a}$.
31. $\sin\theta = \sqrt{1-a^2}$.

Exercises 5.4

1. $\sin(83°) = 0.99$
$\cos(83°) = 0.12$
$\tan(83°) = 8.14$
$\csc(83°) = 1.01$
$\sec(83°) = 8.21$
$\cot(83°) = 0.12$
3. $\sin(-400°) = -0.64$
$\cos(-400°) = 0.77$
$\tan(-400°) = -0.84$
$\csc(-400°) = -1.56$
$\sec(-400°) = 1.31$
$\cot(-400°) = -1.19$
5. (b) $P(0,1)$ $r = 1$
$\sin\theta = \frac{1}{1} = 1$
$\cos\theta = \frac{0}{1} = 0$
$\tan\theta$ (no value)
$\csc\theta = \frac{1}{1} = 1$
$\sec\theta$ (no value)
$\cot\theta = \frac{0}{1} = 0$
(c) $P(-1,0)$ $r = 1$
$\sin\theta = \frac{0}{1} = 0$
$\cos\theta = \frac{-1}{1} = -1$
$\tan\theta = \frac{0}{-1} = 0$
$\csc\theta$ (no value)
$\sec\theta = \frac{1}{-1} = -1$
$\cot\theta$ (no value)
7. In Quadrant I both x and y values of a point on the terminal side of an angle are positive. The radius is always positive. Thus, the ratios formed are positive for all six trig functions.
9. In Quadrant III both the x value and the y value are negative. Thus the only positive functions are those using both x and y (tan and cot). The rest are negative.
11. $\sin\theta = 0$
$\cos\theta = -1$
$\tan\theta = 0$
$\csc\theta$(no value)
$\sec\theta = -1$
$\cot\theta$ (no value)
$\sin\theta' = 0$
$\cos\theta' = 1$
$\tan\theta' = 0$
$\csc\theta'$ (no value)
$\sec\theta' = 1$
$\cot\theta'$ (no value)

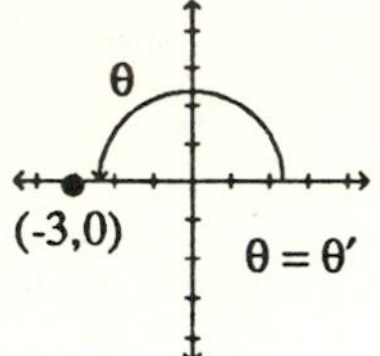

13. $\sin\theta = \frac{3}{5}$
$\cos\theta = \frac{4}{5}$
$\tan\theta = \frac{3}{4}$
$\csc\theta = \frac{5}{3}$
$\sec\theta = \frac{5}{4}$
$\cot\theta = \frac{4}{3}$
$\sin\theta' = \frac{3}{5}$
$\cos\theta' = \frac{4}{5}$
$\tan\theta' = \frac{3}{4}$
$\csc\theta' = \frac{5}{3}$
$\sec\theta' = \frac{5}{4}$
$\cot\theta' = \frac{4}{3}$

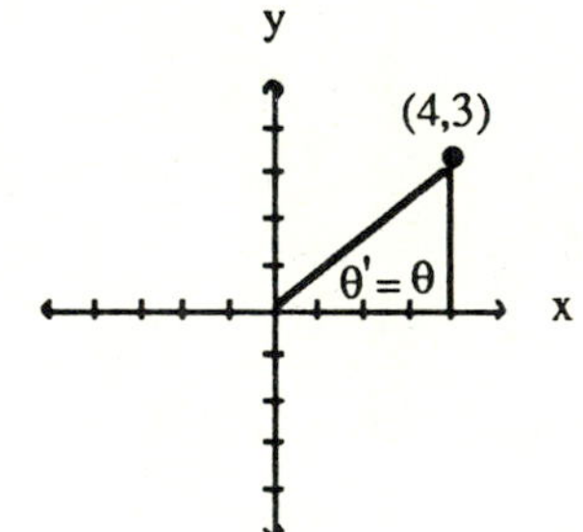

15. $\sin\theta = -\frac{3}{\sqrt{13}}$
$\cos\theta = -\frac{2}{\sqrt{13}}$
$\tan\theta = \frac{3}{2}$
$\csc\theta = -\frac{\sqrt{13}}{3}$
$\sec\theta = -\frac{\sqrt{13}}{2}$
$\cot\theta = \frac{2}{3}$
$\sin\theta' = \frac{3}{\sqrt{13}}$

$\cos\theta' = \frac{2}{\sqrt{13}}$

$\tan\theta' = \frac{3}{2}$

$\csc\theta' = \frac{\sqrt{13}}{3}$

$\sec\theta' = \frac{\sqrt{13}}{2}$

$\cot\theta' = \frac{2}{3}$

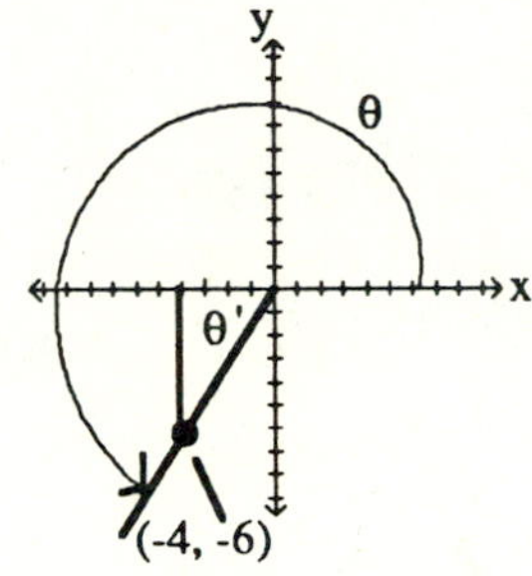

17. $\sin\theta = -\frac{1}{\sqrt{2}}$

$\cos\theta = \frac{1}{\sqrt{2}}$

$\tan\theta = -1$

$\csc\theta = -\sqrt{2}$

$\sec\theta = \sqrt{2}$

$\cot\theta = -1$

$\sin\theta' = \frac{1}{\sqrt{2}}$

$\cos\theta' = \frac{1}{\sqrt{2}}$

$\tan\theta' = 1$

$\csc\theta' = \sqrt{2}$

$\sec\theta' = \sqrt{2}$

$\cot\theta' = 1$

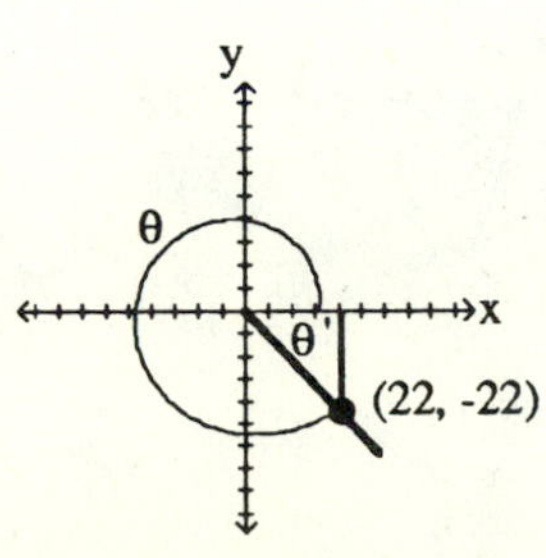

19.

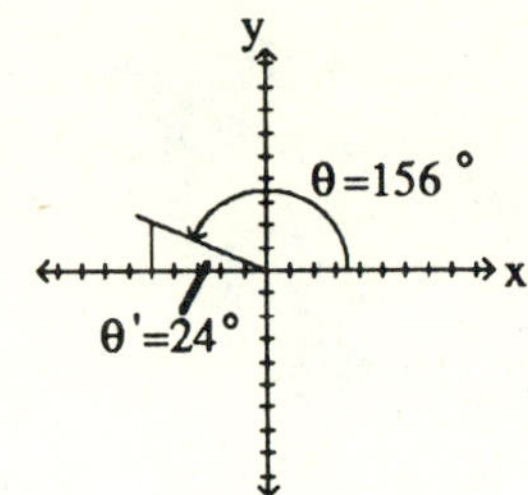

21.

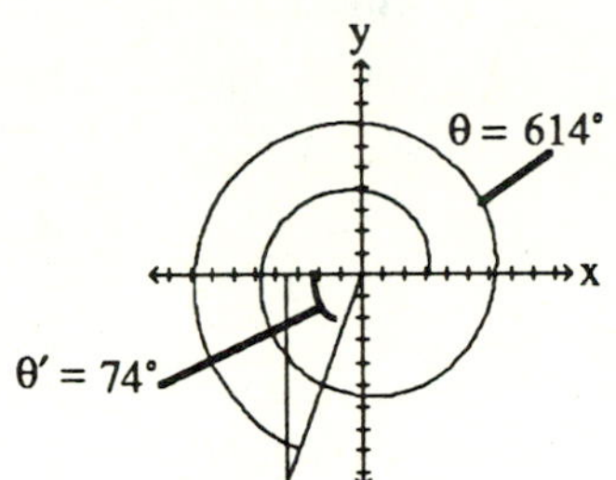

23. $\sin\theta = 0.41$

$\cos\theta = 0.91$

$\tan\theta = 0.45$

$\csc\theta = 2.46$

$\sec\theta = 1.09$

$\cot\theta = 2.25$

$\sin\theta' = 0.41$

$\cos\theta' = -0.91$

$\tan\theta' = -0.45$

$\csc\theta' = 2.46$

$\sec\theta' = -1.09$

$\cot\theta' = -2.25$

25. $\sin\theta = 0.96$

$\cos\theta = 0.28$

$\tan\theta = 3.49$

$\csc\theta = 1.04$

$\sec\theta = 3.63$

$\cot\theta = 2.90$

$\sin\theta' = -0.96$

$\cos\theta' = -0.28$

$\tan\theta' = 3.49$

$\csc\theta' = -1.04$

$\sec\theta' = -3.63$

$\cot\theta' = 2.90$

27. $\sin\theta = -\frac{1}{\sqrt{2}}$

$\cos\theta = -\frac{1}{\sqrt{2}}$

$\tan\theta = 1$

$\csc\theta = -\sqrt{2}$

$\sec\theta = -\sqrt{2}$

$\cot\theta = 1$

29. $\sin\theta = \frac{3}{5}$

$\cos\theta = \frac{4}{5}$

$\tan\theta = \frac{3}{4}$

$\csc\theta = \frac{5}{3}$

$\sec\theta = \frac{5}{4}$

$\cot\theta = \frac{4}{3}$

31. $\sin\theta = -\frac{2}{\sqrt{13}}$

$\cos\theta = -\frac{3}{\sqrt{13}}$

$\tan\theta = \frac{2}{3}$

$\csc\theta = -\frac{\sqrt{13}}{2}$

$\sec\theta = -\frac{\sqrt{13}}{3}$

$\cot\theta = \frac{3}{2}$

33. 74°

35. 23°

37. 34°

39. 63.43°

41. −71.57°

43. III quadrant

45. III quadrant

47. $\cot\theta\tan\theta = \left(\frac{1}{\tan\theta}\right)\tan\theta = 1$

49. $\sec^2\theta - \sin^2\theta = \frac{1}{\cos^2\theta} - (1-\cos^2\theta) = \frac{1-\cos^2\theta}{\cos^2\theta} + \cos^2\theta = \frac{\sin^2\theta}{\cos^2\theta} + \cos^2\theta = \tan^2\theta + \cos^2\theta$

51. Height of pole: 30.09 feet
Length of wire: 34.08 feet

53. 3290.53 feet

55. (a) $\sin 0° = 0$
$\cos 0° = 1$
$\tan 0° = 0$
$\cot 0° =$ no value
$\sec 0° = 1$
$\csc 0° =$ no value

(b) $\sin 30° = \frac{1}{2}$
$\cos 30° = \frac{\sqrt{3}}{2}$
$\tan 30° = \frac{1}{\sqrt{3}}$
$\cot 30° = \sqrt{3}$
$\sec 30° = \frac{2}{\sqrt{3}}$
$\csc 30° = 2$

(c) $\sin 45° = \frac{1}{\sqrt{2}}$
$\cos 45° = \frac{1}{\sqrt{2}}$
$\tan 45° = 1$
$\cot 45° = 1$
$\sec 45° = \sqrt{2}$
$\csc 45° = \sqrt{2}$

(d) $\sin 60° = \frac{\sqrt{3}}{2}$
$\cos 60° = \frac{1}{2}$
$\tan 60° = \sqrt{3}$
$\cot 60° = \frac{1}{\sqrt{3}}$
$\sec 60° = 2$
$\csc 60° = \frac{2}{\sqrt{3}}$

(e) $\sin 90° = 1$
$\cos 90° = 0$
$\tan 90° =$ no value
$\cot 90° = 0$
$\sec 90° =$ no value
$\csc 90° = 1$

(f) $\sin 180° = 0$
$\cos 180° - 1$
$\tan 180° = 0$
$\cot 180° =$ no value
$\sec 180° = -1$
$\csc 180° =$ no value

(g) $\sin 270° = -1$
$\cos 270° = 0$
$\tan 270° =$ no value
$\cot 270° = 0$
$\sec 270° =$ no value
$\csc 270° = -1$

(h) $\sin 360° = 0$
$\cos 360° = 1$
$\tan 360° = 0$
$\cot 360° =$ no value
$\sec 360° = 1$
$\csc 360° =$ no value

Exercises 5.5

1. $\frac{\pi}{180} = 0.02$
3. 0.52
5. −2.09
7. 57.30°
9. 135°
11. −131.78°
13. $\sin 200° = -0.34$
$\cos 200° = -0.94$
$\tan 200° = 0.36$
$\csc 200° = -2.92$
$\sec 200° = -1.06$
$\cot 200° = 2.75$
15. $\sin -2 = -0.91$
$\cos -2 = -0.42$
$\tan -2 = 2.19$
$\csc -2 = -1.10$
$\sec -2 = -2.40$
$\cot -2 = 0.46$
17. Domain: $(-\infty, \infty)$
Range: $[-1, 1]$

Period: $\frac{2\pi}{3}$

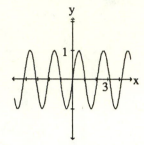

19. Domain: $(-\infty, \infty)$
Range: $[1, 3]$
Period: 2π

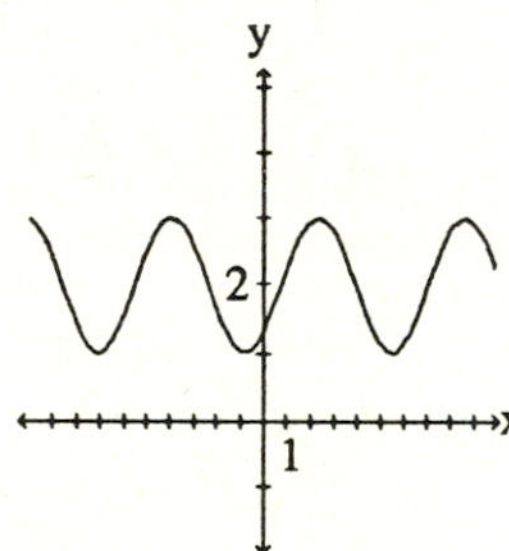

21. Domain: $(-\infty, \infty)$
Range: $[-4, 2]$
Period: 2π

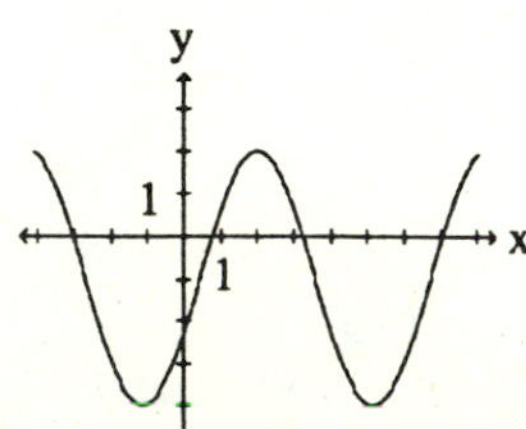

23. Domain: $(-\infty, \infty)$
Range: $[-4, 2]$

Period: 2π

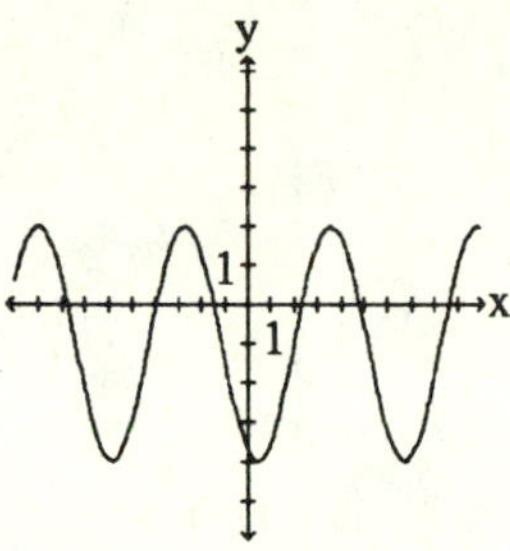

25. Horizontally shift right $180°$, then a vertical stretch of 2 followed by a reflection about the x-axis.

27. Let $-\theta$ be the angle determined by OP and the x-axis and θ be the angle in standard position. Then $\cos(-\theta) = \frac{a}{c}$ and $\cos\theta = \frac{a}{c}$ where $c = \sqrt{a^2+b^2}$. Since $\cos(-\theta) = \cos\theta$, cosine is an even function.

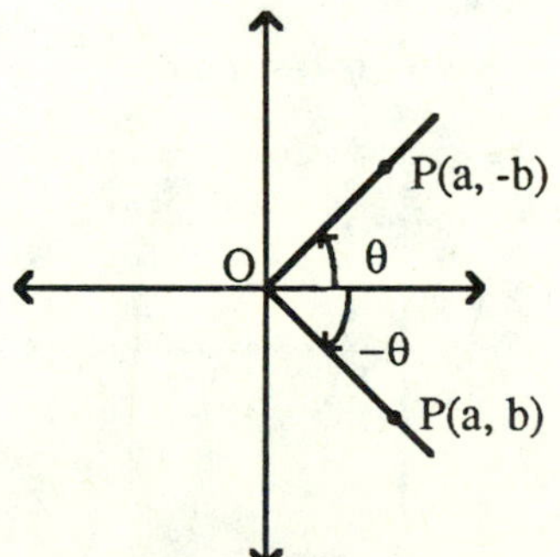

29. θ_1 measures 2 radians $\sin 2 = \sin(\pi - 2)$

    θ_2 measures 3 radians $\sin 3 = \sin(\pi - 3)$

    $\sin(\pi - 2) > \sin(\pi - 3)$ since an angle with measure $(\pi - 3)$ is smaller than angle with measure $(\pi - 2)$ and for angles less than $90°$ the sine values increase with increased angle measure. Therefore $\sin 2 > \sin 3$.

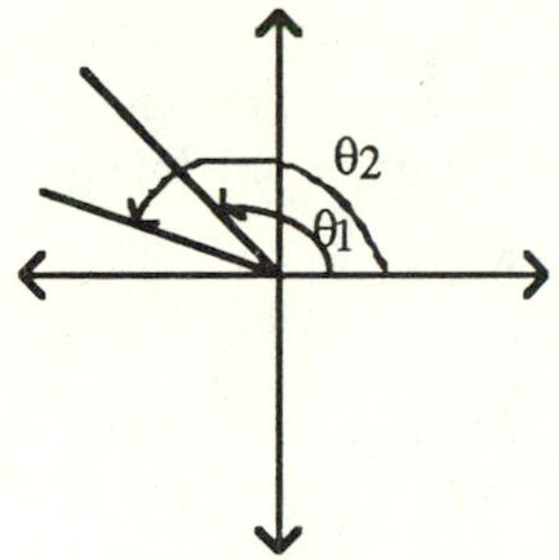

31. $\sin 10 = \sin(10 - 2\pi) = \sin(3.72)$. The angle with measure 3.72 lies in quadrant III, so the $\sin(3.72)$ would be less than zero.

33. $x = 0.52$, $x = 2.62$

35. $x = 0.93$, $x = 2.21$

37. $x = 0.93$, $x = 5.36$

39. $0 \le \theta \le 36.87°$ and $143.13° \le \theta \le 360°$

41. (a) True; the graphs of each side of the equation overlay each other.
    (b) True; the graphs of each side of the equation overlay each other.
    (c) False; the graphs of each side of the equation do **not** overlay each other.
    (d) True; the graphs of each side of the equation overlay each other.

43. 5.76 inches

45. 50.27 inches

47. 7539.82 radians per minute

49. 3769.91 radians per minute

51. Angular speed of wheel: 177,408 rads/hr.

    Angular speed of pedal sprocket: 55,440 rads/hr.

    Angular speed of wheel sprocket: 177,408 rads/hr.(same as wheel)

53. 18.71 inches

55.

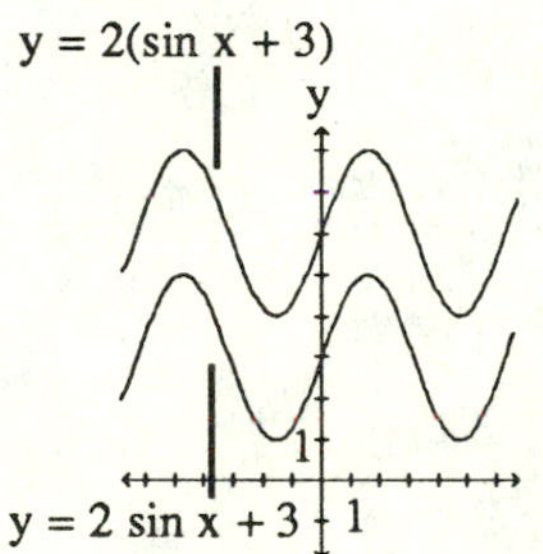

Exercises 5.6

1(a)

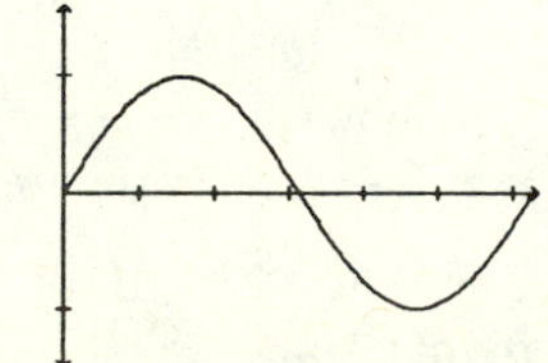

$[0, -2\pi]$ by $[-2, 2]$

(b)

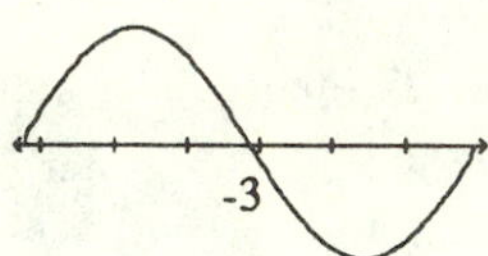

$[-2\pi, 0]$ by $[-2, 2]$

(c)

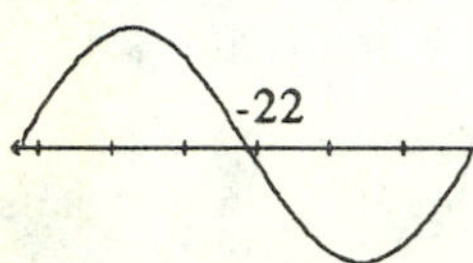

$[-8\pi, -6\pi]$ by $[-2, 2]$

(d)

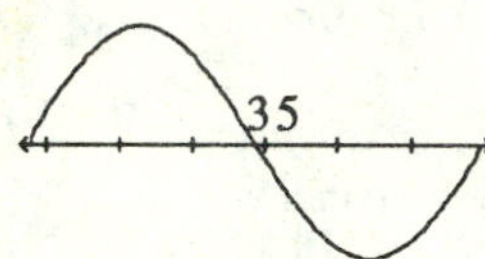

$[10\pi, 12\pi]$ by $[-2, 2]$

3. Domain: All reals except $x = \frac{\pi}{2} + k\pi$, k any integer

   Range: All reals

   Period: π

   Asymptotes: $x = \frac{\pi}{2} + k\pi$, k any integer

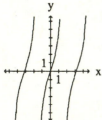

5. Domain: All reals except $x = \frac{\pi}{2} + k\pi$, k any integer

   Range: $(-\infty, -\frac{1}{2}] \cup [\frac{1}{2}, \infty)$

   Period: 2π

   Asymptotes: $x = \frac{\pi}{2} + k\pi$, k any integer

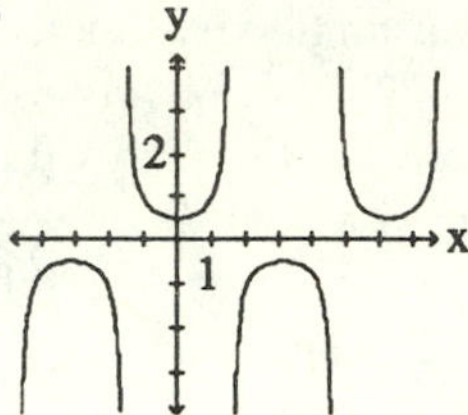

7. Domain: All reals except $x = k\pi$, k any integer

   Range: $(-\infty, -3] \cup [3, \infty)$

   Period: 2π

   Asymptotes: $x = k\pi$, k any integer

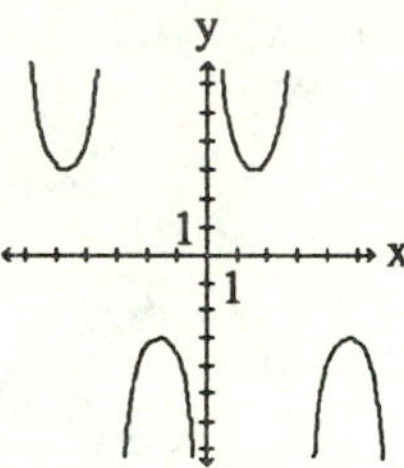

9. Domain: All reals

   Range: $[-3, 3]$

   Period: 2π

   Asymptotes: none

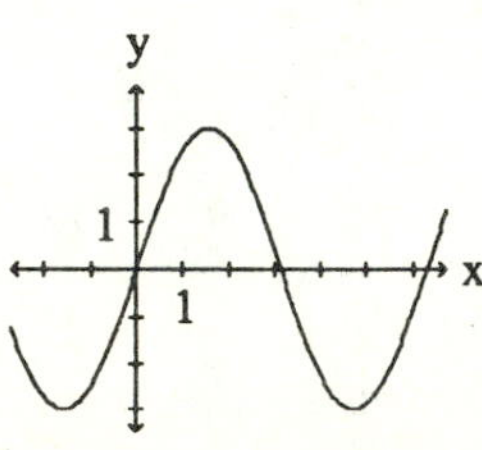

11. Domain: All reals except $x = \pi + 2k\pi$, k any integer

    Range: All reals

    Period: 2π

    Asymptotes: $x = \pi + 2k\pi$, k any integer

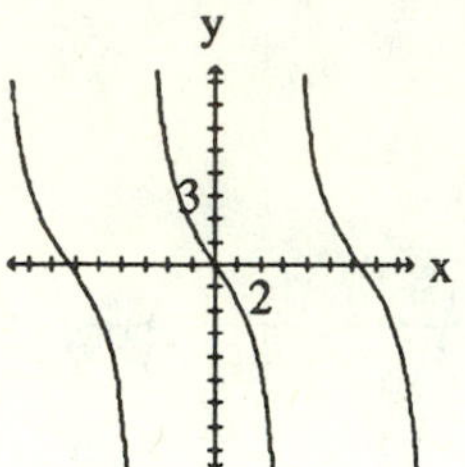

13. Domain: All reals except $x = k\pi, k$ any integer

    Range: $(-\infty, -2] \cup [2, \infty)$

    Period: 2π

    Asymptotes: $x = k\pi$, k any integer

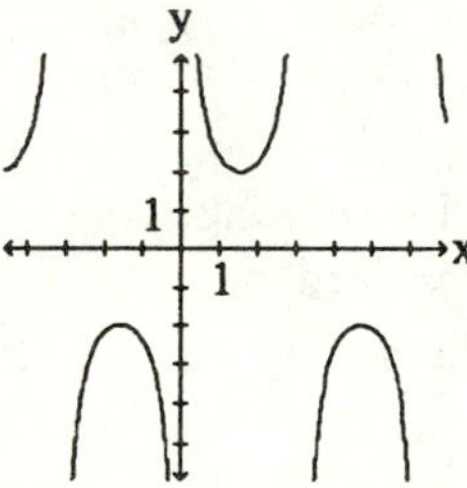

15. Domain: All reals

    Range: $[-1, 1]$

    Period: π

    Asymptotes: none

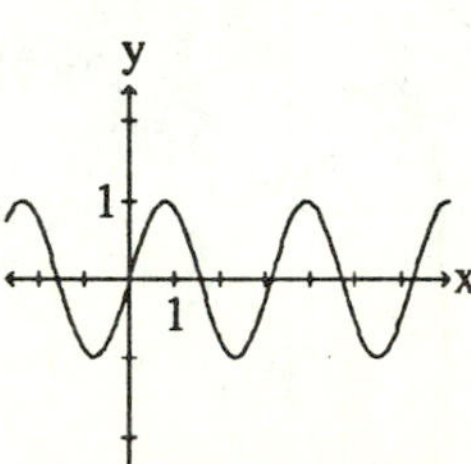

17. Domain: All reals except $x = \frac{\pi}{4} + k \cdot \frac{\pi}{2}$, k any integer

    Range: $(-\infty, -1] \cup [1, \infty)$

    Period: π

Asymptotes: $x = \frac{\pi}{4} + k \cdot \frac{\pi}{2}$, k any integer

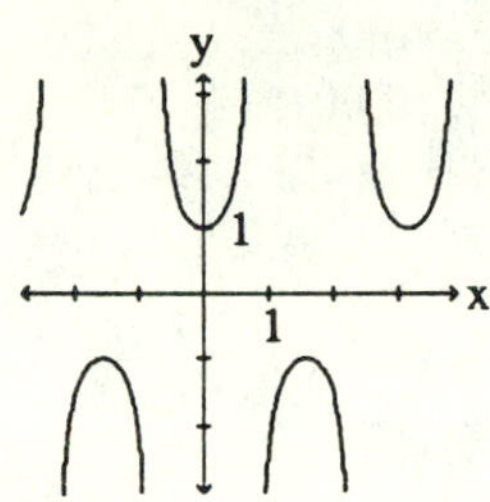

19. Domain: All reals

    Range: $[-1, 1]$

    Period: 4π

    Asymptotes: none

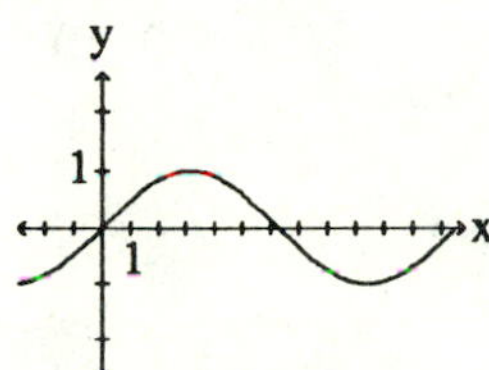

21. Domain: All reals

    Range: $[-2, 2]$

    Period: $\frac{2}{3}\pi$

    Asymptotes: none

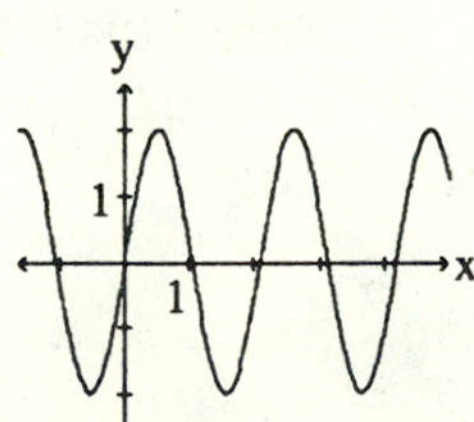

23. Domain: All reals

    Range: $[-3, 3]$

    Period: 4π

Asymptotes: none

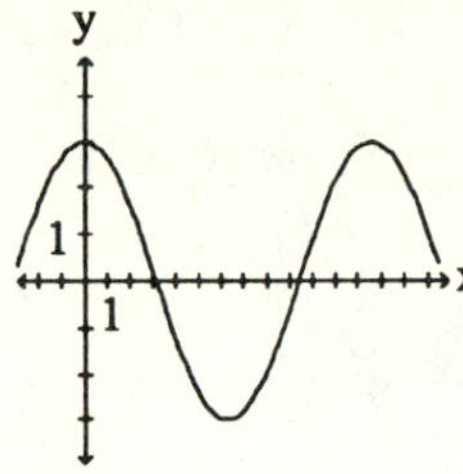

25. Domain: All reals

    Range: $[-2, 2]$

    Period: 2π

    Asymptotes: none

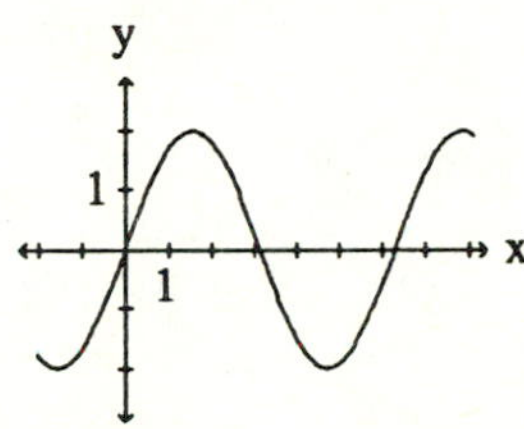

27. Domain: All reals except $x = \frac{1}{2} + k$, k any integer

    Range: All reals

    Period: 1

    Asymptotes: $x = \frac{1}{2} + k$, k any integer

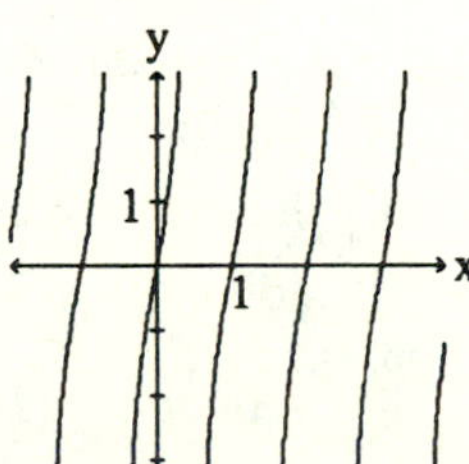

29. Domain: All reals except $x = \frac{\pi}{4} + k \cdot \frac{\pi}{2}$, k any integer

    Range: $(-\infty, -3] \cup [3, \infty)$

    Period: π

Asymptotes: $x = \frac{\pi}{4} + k \cdot \frac{\pi}{2}$, k any integer

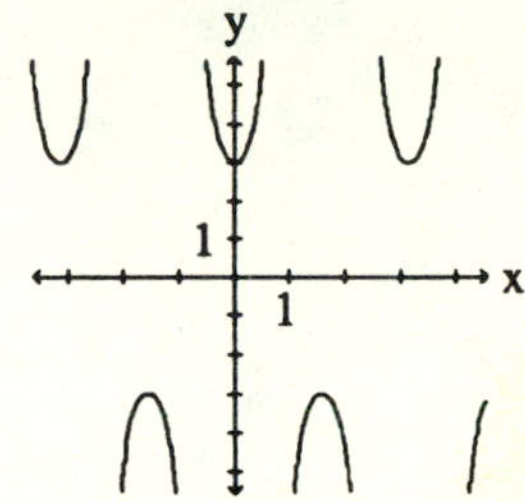

31. Horizontally shrink $y = \sin x$ by a factor of $\frac{1}{3}$ then apply a vertical stretch of 2.

33. Horizontally shrink $y = \cos x$ by a factor of $\frac{1}{\pi}$ then apply a vertical stretch of 3.

35. $x = 0.85, 2.29, -3.99, -5.43$

37. $x = 1.27, 4.41, -1.87, -5.01$

39. $(-5.82, -3.61) \cup (0.46, 2.68)$

41. $(-2\pi, -5.09) \cup (-\frac{3\pi}{2}, -1.95) \cup (-\frac{\pi}{2}, 1.19) \cup (\frac{\pi}{2}, 4.33) \cup (\frac{3\pi}{2}, 2\pi)$

43. $\cot(x + \pi) = \frac{1}{\tan(x+\pi)} = \frac{1}{\tan x} = \cot x$

45. $\csc(x + 2\pi) = \frac{1}{\sin(x+2\pi)} = \frac{1}{\sin x} = \csc x$

51. $x + 2k\pi$, k any integer, where $x = 0.25$ or 2.89

53. $x + 2k\pi$, k any integer, where $x = 1.14$ or 5.15

55. $x + 2k\pi$, k any integer, where $-3.29 < x < 0.15$

57. $x + \frac{2}{3}k\pi$, k any integer, where $x = 0.19$ or 0.85

59. $x + 4k\pi$, k any integer, where $0.51 \leq x \leq 5.78$ or $2\pi < x < 4\pi$

Exercises 5.7

1.

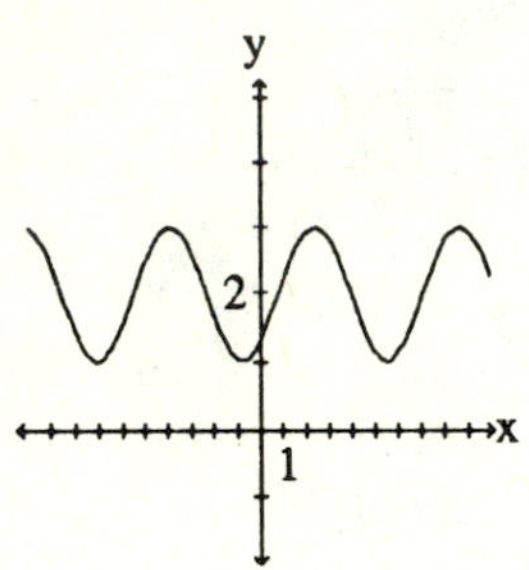

3.

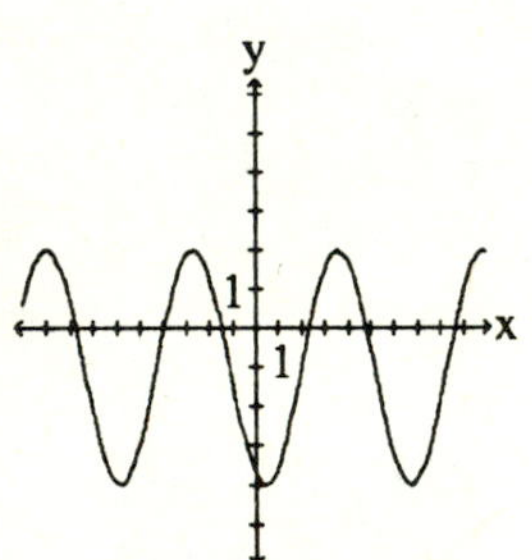

5.

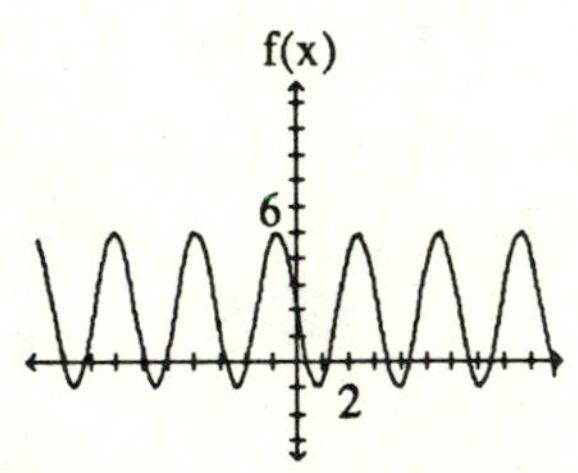

7.

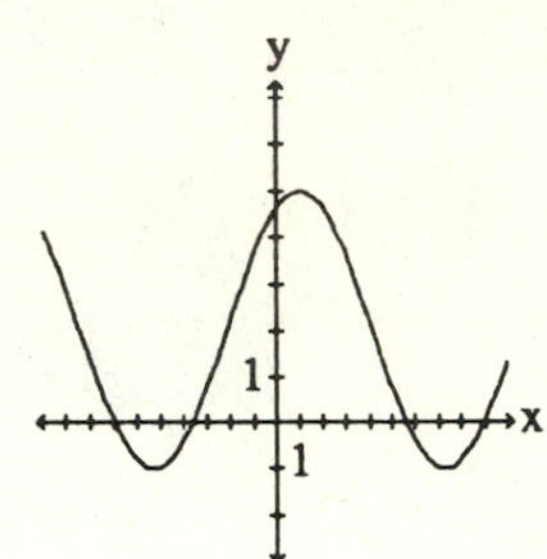

9.

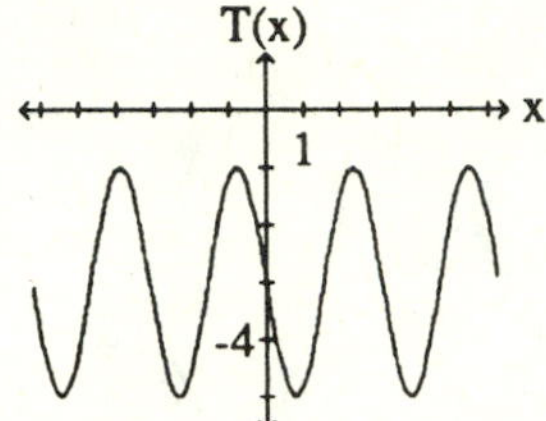

11. Dom: All real numbers.

    Range: $[-4, 2]$

    Period: 2π

    Asymptotes: none

13. Dom: All real numbers except $(2k+1)\pi$ where k is any integer.

    Range: $[-\infty, \infty]$

    Period: 2π

    Asymptotes: $(2k+1)\pi$ where k is any integer

15. Local maxima: $(0.79, 3.00)$, $(3.93, 3.00)$

    Local minima: $(2.36, -3.00)$, $(5.50, -3.00)$

    Increasing: $[0, 0.79] \cup [2.36, 3.93] \cup [5.50, 2\pi]$

    Decreasing: $[0.79, 2.36] \cup [3.93, 5.50]$

17. No extrema

    Increasing: $(0, \pi) \cup (\pi, 2\pi)$

    Asymptotes: $x = 0, \pi$, or 2π

19. Maxima at: $(\frac{\pi}{8}, 5.00)$, $(\frac{5\pi}{8}, 5.00)$, $(\frac{9\pi}{8}, 5.00)$, $(\frac{13\pi}{8}, 5.00)$

    Minima at: $(\frac{3\pi}{8}, -1.00)$, $(\frac{7\pi}{8}, -1.00)$, $(\frac{11\pi}{8}, -1.00)$, $(\frac{15\pi}{8}, -1.00)$

    Increasing in: $[0, \frac{\pi}{8}] \cup [\frac{3\pi}{8}, \frac{5\pi}{8}] \cup [\frac{7\pi}{8}, \frac{9\pi}{8}] \cup [\frac{11\pi}{8}, \frac{13\pi}{8}] \cup [\frac{15\pi}{8}, 2\pi]$

    Decreasing in: $[\frac{\pi}{8}, \frac{3\pi}{8}] \cup [\frac{5\pi}{8}, \frac{7\pi}{8}] \cup [\frac{9\pi}{8}, \frac{11\pi}{8}] \cup [\frac{13\pi}{8}, \frac{15\pi}{8}]$

21. $\frac{\pi}{18} + \frac{2k\pi}{3}$ with k any integer

    $\frac{5\pi}{18} + \frac{2k\pi}{3}$ with k any integer

23. $2.66 + 4k\pi$ with k any integer

    $-2.66 + 4k\pi$ with k any integer

25. $x = 0,\ 2.13$

27. In the interval $(0.69, 3.83)$

29. $y = 3\sin(2x - \pi)$

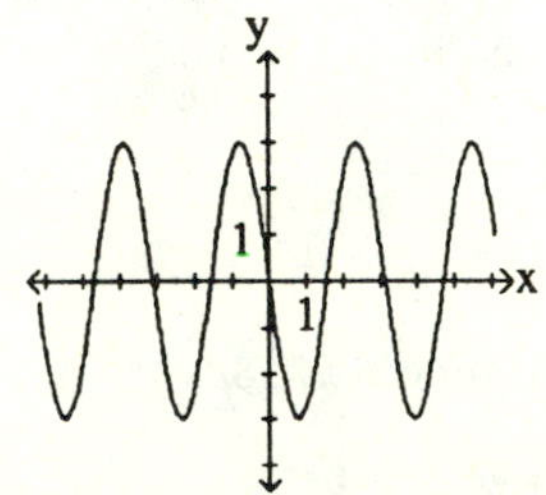

31. $y = 2\sin(\frac{1}{2}x + \frac{\pi}{8})$

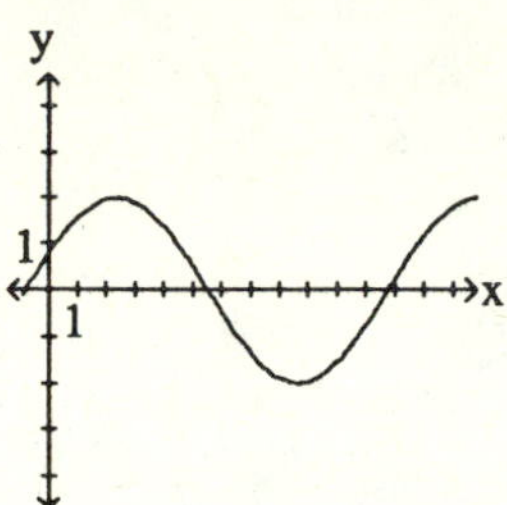

33. Vertical stretch by a factor of 4, a reflection through the x-axis, then horizontal shrink by a factor of 2,a horizontal shift right π units, then a vertical shift up 3 units.

35. A reflection through the x-axis, followed by a horizontal shift right π units, and a vertical shift up 2 units.

37. Period $= \frac{2\pi}{3}$

Frequency $= \frac{3}{2\pi}$

39. Period $= 4$

Frequency $= \frac{1}{4}$

41. Period $= 4$ seconds

$A = -15$

$B = \frac{\pi}{2}$

43. The part to the right of the y-axis

45. $\frac{1}{3}$ seconds

$A = -28$

$B = 6\pi$

47. for $x \geq 0$

49.

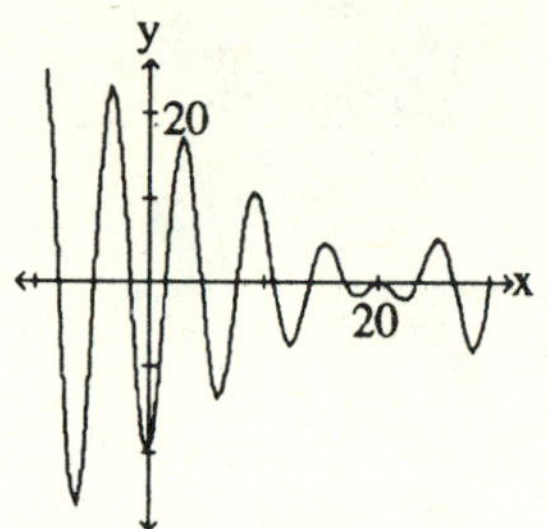

51. $d = 49.50$ inches

$\alpha = 1.43$ radians

$$\text{belt} = 2d + (2\pi - 2\alpha)22 + (2\pi - 2\alpha')15 = 264.36 \text{ inches}$$

(where α' is the angle betwen 50 and the dashed line)

53. 2.09 radians/second

55. 11.42 miles/hour

Review Exercises

1. $\sin 32° = 0.53$

$\cos 32° = 0.85$

$\tan 32° = 0.62$

$\csc 32° = 1.89$

$\sec 32° = 1.18$

$\cot 32° = 1.60$

3. $\sin 30° = \frac{1}{2}$

$\cos 270° = 0$

$\tan 135° = -1$

$\sin \frac{5\pi}{6} = \frac{1}{2}$

$\cos \frac{2\pi}{3} = -\frac{1}{2}$

$\cot 135° = -1$

$\csc \frac{\pi}{2} = 1$

$\sec 270° =$(no value)

$\cot \frac{\pi}{4} = 1$

$\csc 210° = -2$

$\sec 330° = \frac{2}{\sqrt{3}}$

$\cot \frac{5\pi}{4} = 1$

5. $\sin -\frac{\pi}{6} = -\frac{1}{2}$

$\cos -\frac{\pi}{6} = \frac{\sqrt{3}}{2}$

$\tan -\frac{\pi}{6} = -\frac{1}{\sqrt{3}}$

$\cot -\frac{\pi}{6} = -\sqrt{3}$

$\sec -\frac{\pi}{6} = \frac{2}{\sqrt{3}}$

$\csc -\frac{\pi}{6} = -2$

7. $\sin 60° = \frac{\sqrt{3}}{2}$

$\cos 60° = \frac{1}{2}$

$\tan 60° = \sqrt{3}$

$\cot 60° = \frac{1}{\sqrt{3}}$

$\sec 60° = 2$

$\csc 60° = \frac{2}{\sqrt{3}}$

9. $\sin 300° = -\frac{3}{\sqrt{2}}$

$\cos 300° = \frac{1}{2}$

$\tan 300° = -\sqrt{3}$

$\cot 300° = -\frac{1}{\sqrt{3}}$

$\sec 300° = 2$

$\csc 300° = -\frac{2}{\sqrt{3}}$

11. $\sin\theta = \frac{2\sqrt{6}}{7}$; $\csc\theta = \frac{7}{2\sqrt{6}}$

$\cos\theta = \frac{5}{7}$; $\sec\theta = \frac{7}{5}$

$\tan\theta = \frac{2\sqrt{6}}{5}$; $\cot\theta = \frac{5}{2\sqrt{6}}$

13. $\sin\theta = \frac{a}{c}$

$\cos\theta = \frac{b}{c}$

$\cot\theta = \frac{b}{a}$

$\frac{\cos\theta}{\sin\theta} = \frac{\frac{b}{c}}{\frac{a}{c}} = \frac{b}{a} = \cot\theta$

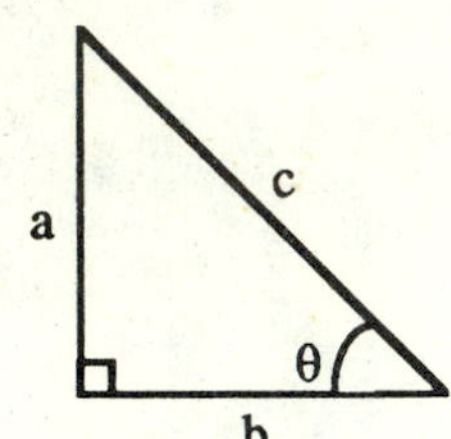

15. $\angle B = 55°$

$a = 8.60$

$b = 12.29$

17. 1.05 radians

19.

(a) 270°

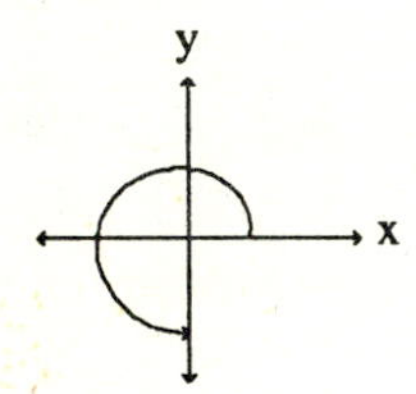

(b) 900°

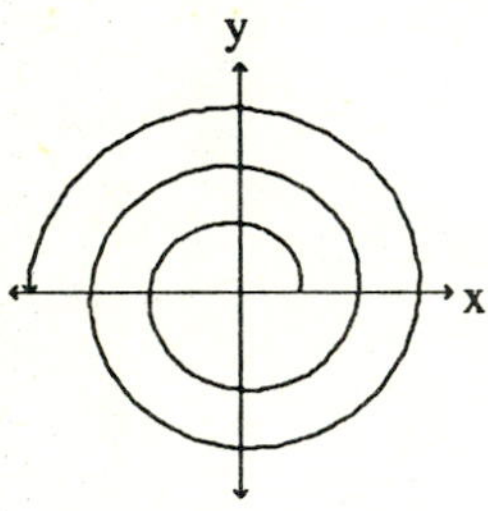

21. $\theta = 202.62°$ and $\theta = 22.62°$

23. $\sin\theta = \frac{4}{5}$

$\cos\theta = \frac{3}{5}$

$\tan\theta = \frac{4}{3}$

$\csc\theta = \frac{5}{4}$

$\sec\theta = \frac{5}{3}$

$\cot\theta = \frac{3}{4}$

25. $\frac{\sin^2\theta}{1-\cos\theta} = \frac{1-\cos^2\theta}{1-\cos\theta}$
$= \frac{(1+\cos\theta)(1-\cos\theta)}{1-\cos\theta}$
$= 1 + \cos\theta$

27. 15.95°

29. II

31.

(a) 394°

(b) $\frac{5\pi}{4}$

33. $x = \frac{\pi}{6} + 2k\pi$, k any integer; $x = \frac{11\pi}{6} + 2k\pi$

35.

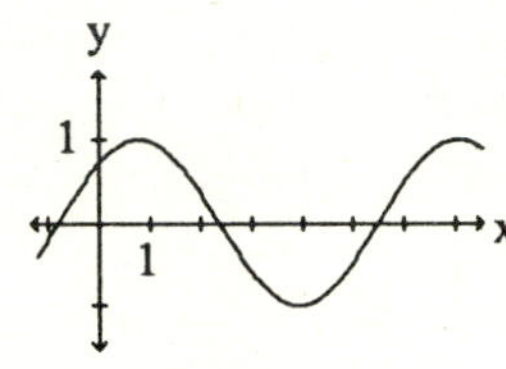

37.

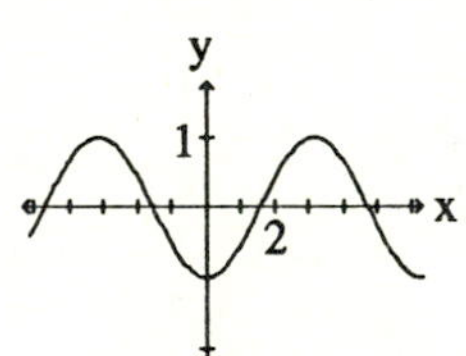

39. Domain: $(-\infty, \infty)$

Range: $[-3, 3]$

Period: π

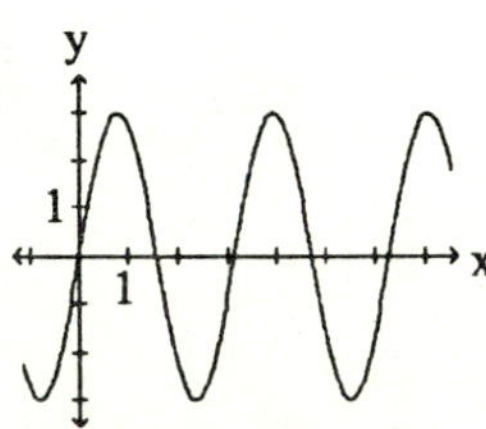

41. Domain: all real numbers different from $\frac{\pi}{6} + \frac{k\pi}{3}$, k any integer

Range: $(-\infty, -1] \cup [3, \infty)$

Period: $\frac{2\pi}{3}$

Asymptotes: $x = \frac{\pi}{6} + \frac{k\pi}{3}$, k any integer

43. Apply a horizontal stretch of 2 and a vertical shift down 5 units to the graph of $y = \tan x$.

45. $x = 0$

47. False; the graphs do not overlay each other.

49. II

51. 23.78 feet

53. 1.62 miles

55. 4669.58 feet

57. $\sin\theta' = 0.41$

$\cos\theta' = 0.91$

$\tan\theta' = 0.45$

$\csc\theta' = 2.46$

$\sec\theta' = 1.09$

$\cot\theta' = 2.25$

$\sin\theta = 0.41$

$\cos\theta = -0.91$

$\tan\theta = -0.45$

$\csc\theta = 2.46$

$\sec\theta = -1.09$

$\cot\theta = -2.25$

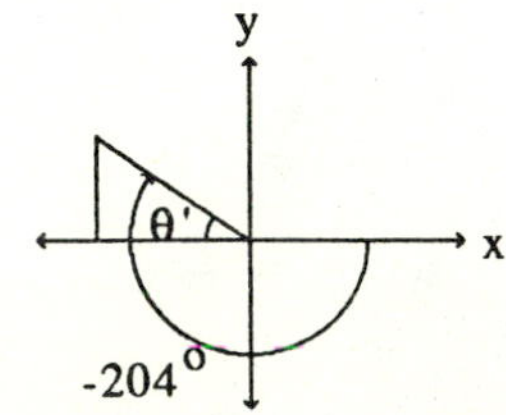

• CHAPTER 6

Exercises 6.1

1. $y = \sqrt{13}\sin(x + 0.59)$

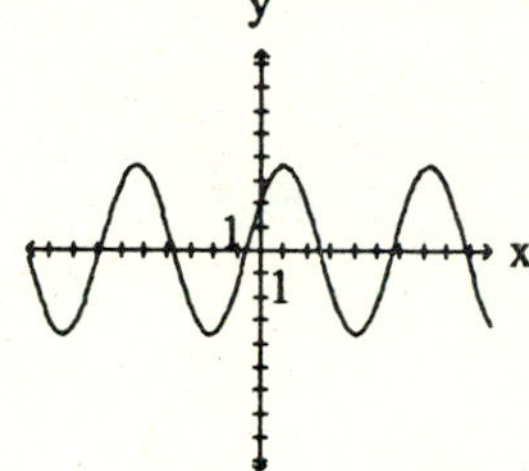

3. $y = \sqrt{2}\sin(x + \frac{\pi}{4})$

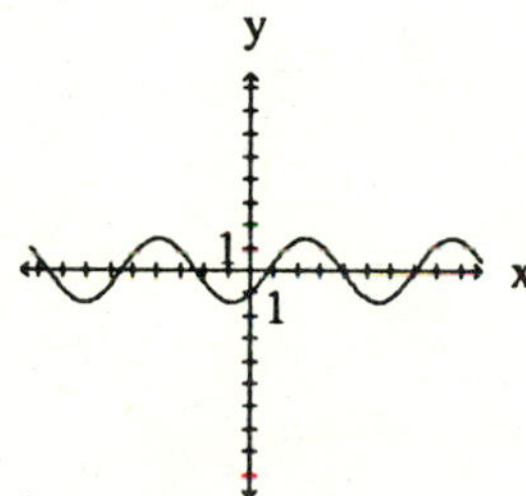

5. $A = 6.5$

7. Domain: $(-\infty, \infty)$
Range: $[-2, 1.12]$
Period: 2π
absolute maxima:
$(0.25, 1.12)$, $(2.89, 1.12)$
local minimum: $(1.57, 0)$
abs min: $(4.71, -2)$

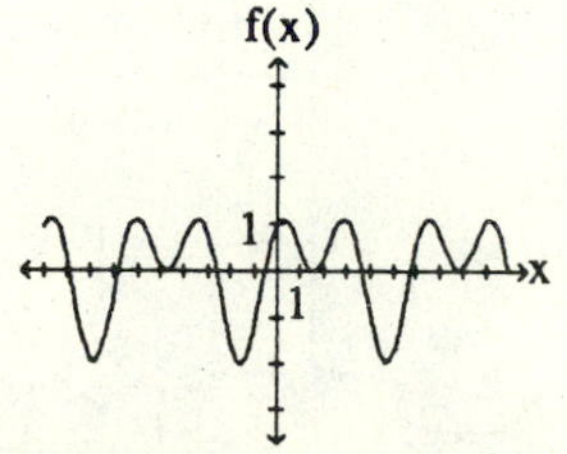

9. Domain: $(-\infty, \infty)$

Range: $[-1.88, 1.88]$

Period: 2π

local maxima:
$(2.56, 0.15)$, $(4.82, 1.06)$

absolute maximum:
$(0.47, 1.88)$

local minima:
$(1.68, -1.06)$,
$(5.70, -0.15)$

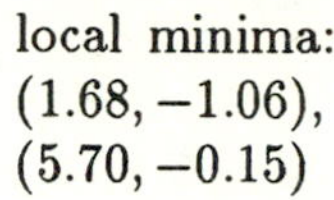

abs min: $(3.61, -1.88)$

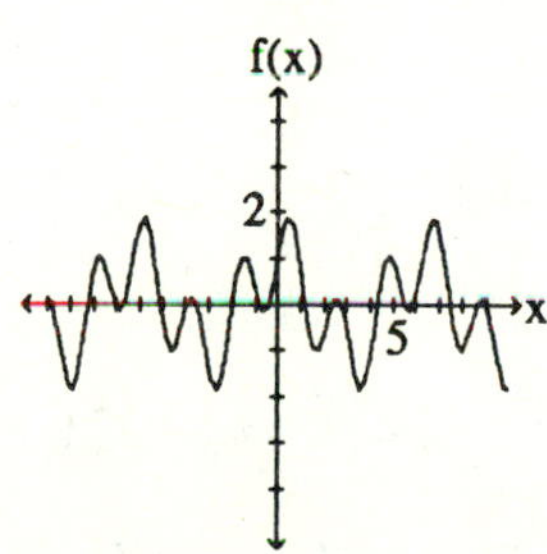

11. $y \to 0$ as $|x| \to \infty$

Undefined at $x = 0$

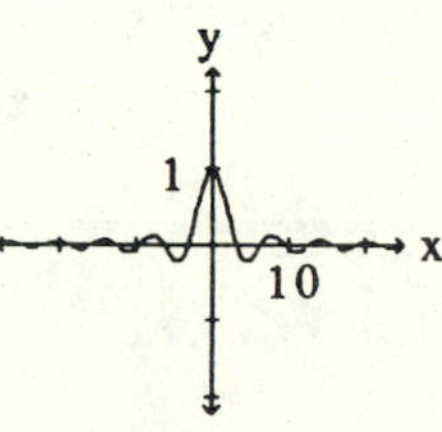

13. $y \to -1$ as $|x| \to \infty$

Undefined at $x = 0$

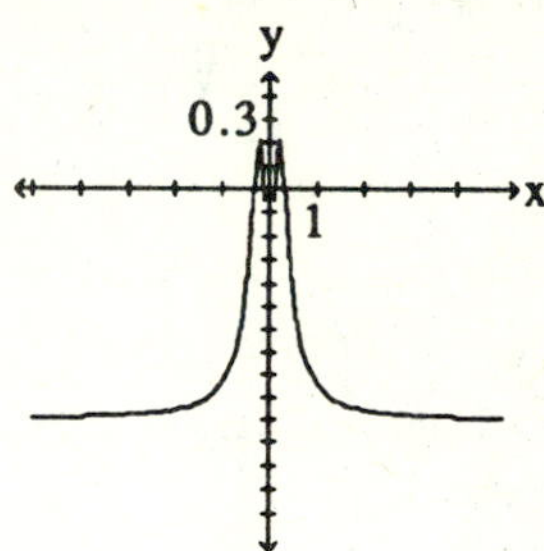

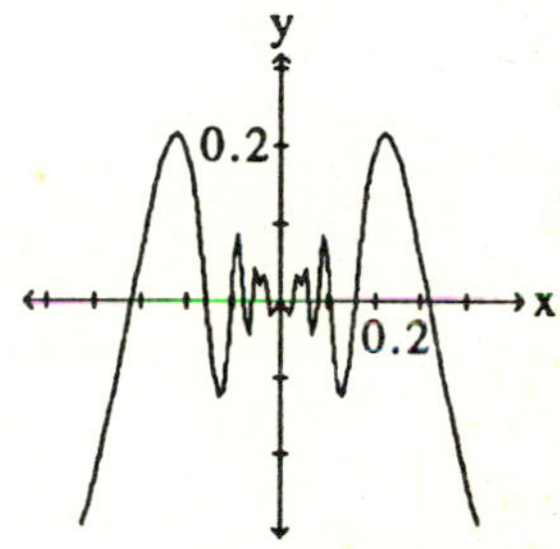

15. End behavior is periodic with period π

Vertical asymptote at $x = k\pi$, k any integer

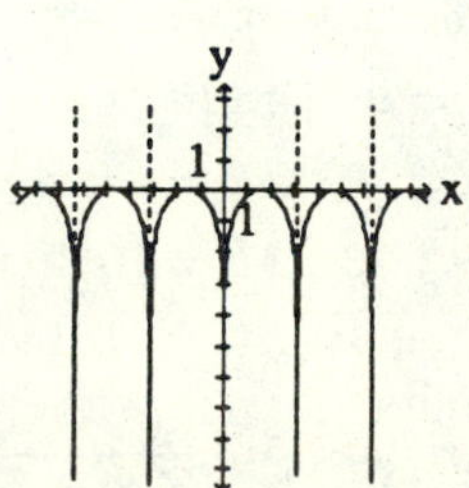

17. Symmetric with respect to the y-axis

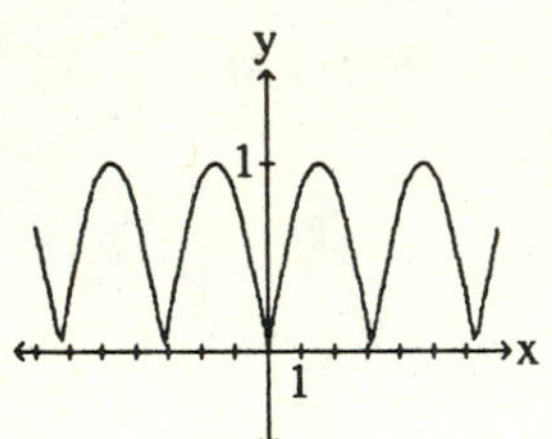

19. Symmetric with respect to the y-axis

Horizontal asymptote at $y = 1$

Undefined at $x = 0$

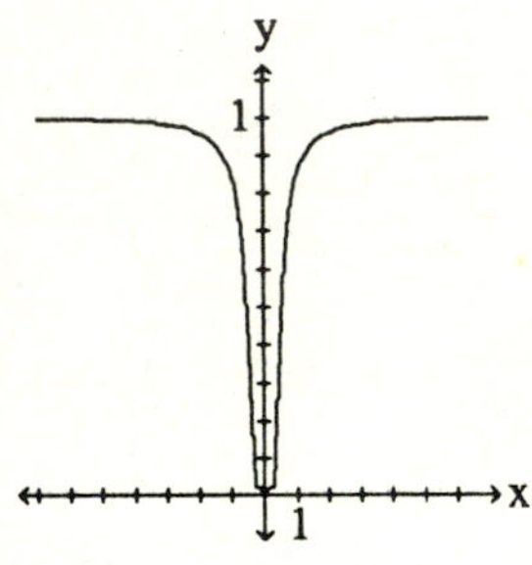

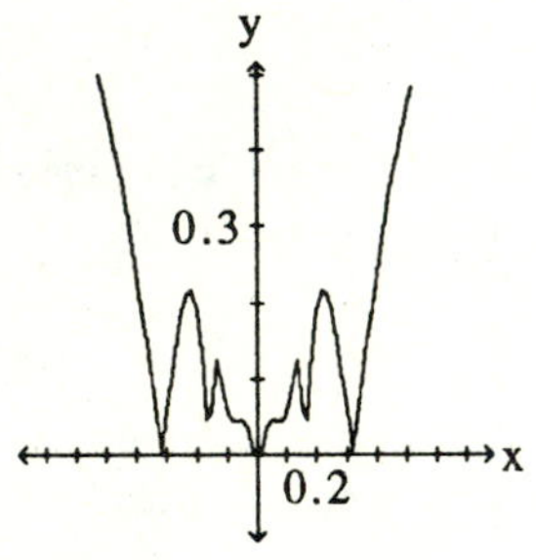

21. $V(\theta) = 16,000\pi \sin\theta \cos^2\theta$

23. $\theta = 0.62$ radians

$h = 40\sin\theta = 23.24$ units

$r = 20\cos\theta = 16.28$ units

$V = 19,347.18$ cubic units

25. $V(r) = \pi r^2\sqrt{400 - 4x^2}$

Max vol. 2418.40 cu. in.

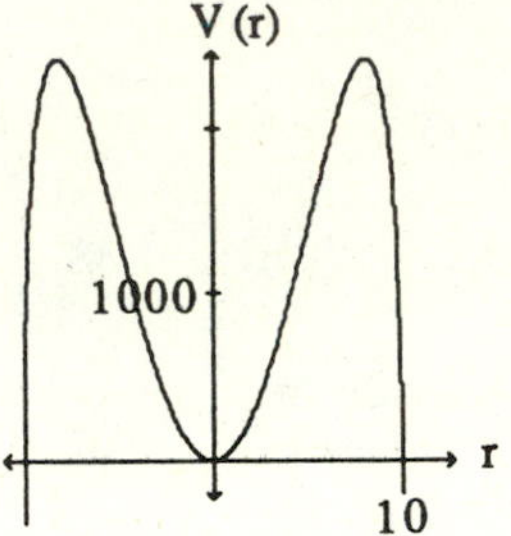

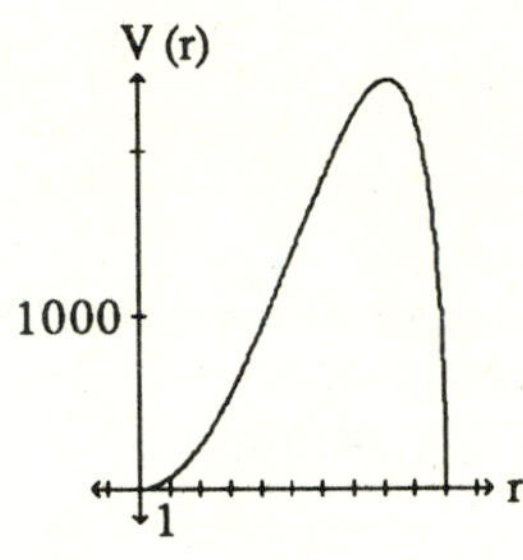

27.

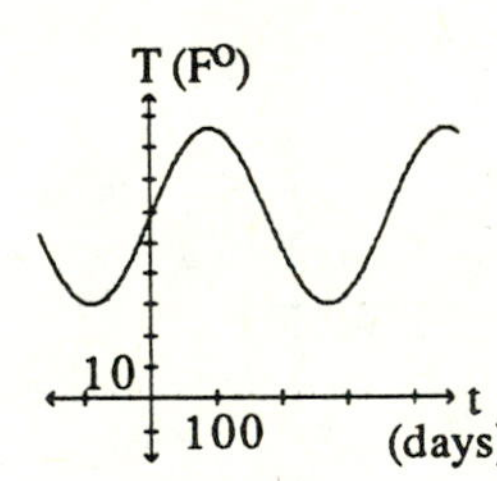

29. 90° on 86th day (June 26)

30° on 270th day (Dec. 26)

31. $f(x) \to 0.5$ as $x \to 0$

33. $f(x) \to 1$ as $x \to 0$

35. $\tan\theta = \frac{200t}{2055}$

Exercises 6.2

1. $\cos\theta$

3. $-\sin x$

5. $\tan x$

7. $\frac{\sin^2 x + \cos^2 x}{\cos x \cdot \sin x} = \frac{1}{\cos x \cdot \sin x} =$ $\sec x \cdot \csc x$

9. $\frac{\sin^2 x + \cos^2 x}{\cos x} = \frac{1}{\cos x} =$ $\sec x$

11. $\frac{1 + \cos^2 x - \sin^2 x}{2\sin x \cos x} =$ $\frac{2\cos^2 x}{2\sin x \cos x} = \frac{2\sin x \cos x}{2\sin^2 x} =$ $\frac{\sin 2x}{1 - \cos 2x}$

13. $\frac{\frac{\sin x + \cos x}{\cos x}}{\frac{\cos x + \sin x}{\sin x}} = \frac{\sin x}{\cos x} = \tan x$

15. true; graphs coincide

17. true; graphs coincide

19. $1 + \frac{\cos^2\theta}{\sin^2\theta} = \frac{\sin^2\theta + \cos^2\theta}{\sin^2\theta} =$ $\frac{1}{\sin^2\theta} = \csc^2\theta$

21. $\cos^2\theta + 1 =$ $\cos^2\theta + \sin^2\theta + \cos^2\theta =$ $2\cos^2\theta + \sin^2\theta$

23. $\frac{1+\sqrt{3}}{2\sqrt{2}}; \frac{1+\sqrt{3}}{2\sqrt{2}}$

25. $\cos x = \frac{\sqrt{5}}{3}; \csc x = \frac{3}{2}$

27. $\sec x = -2$

29. $\tan 2x = -\sqrt{3}$

31. Domain:x any real numbers different from $k \cdot \left(\sqrt{\frac{\pi}{2}}\right)$where k is any integer.

Range:$(-\infty, -1] \cup [1, \infty)$

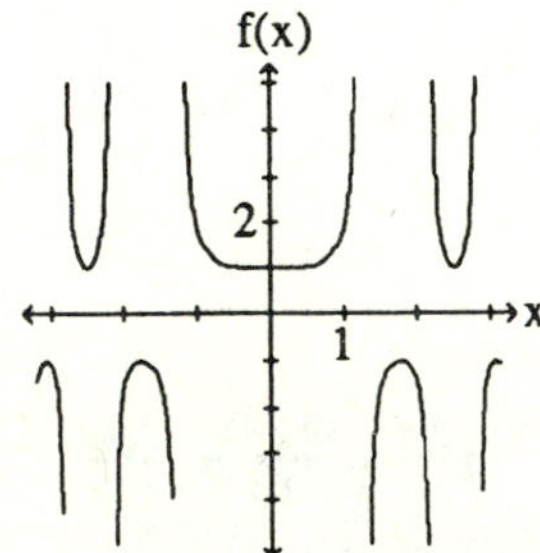

33. Domain: all real numbers

Range: $[0, 3]$

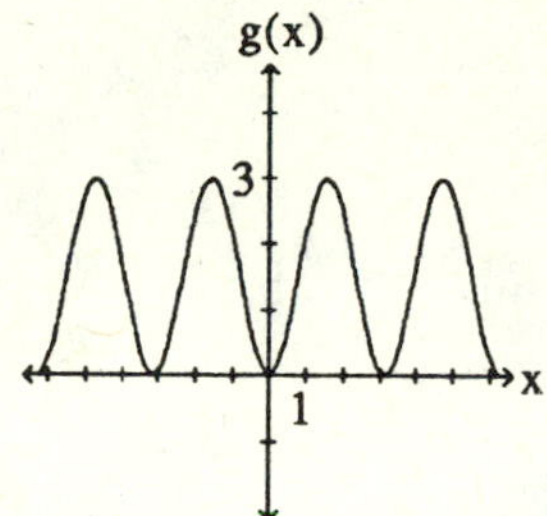

35. $\cos\theta(2\sin\theta + 1)$
37. $3\sin\theta\cos^2\theta + \cos^2\theta - \sin^3\theta - \sin^2\theta$
39. $\frac{\pi}{6}$ and $\frac{5\pi}{6}$
41. $\frac{\pi}{3}$ and $\frac{2\pi}{3}$
43. $\frac{\pi}{2}$ and $\frac{3\pi}{2}$
45. 0, π, and 2π
47. IV
49.

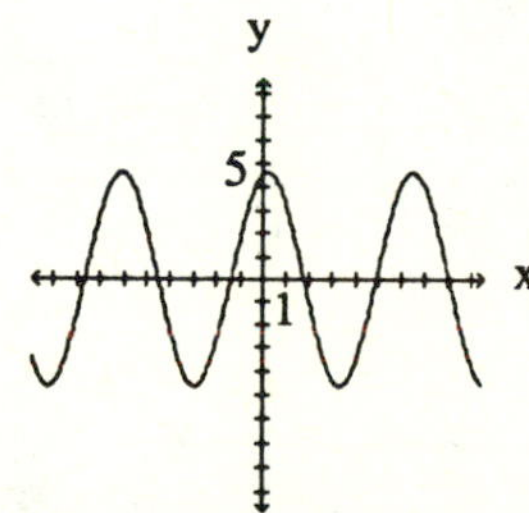

Period: 2π

Phase shift: 1.03

A=5.83

51. $A\sin(x + \alpha) = A(\sin x\cos\alpha + \cos x\sin\alpha) = A\sin x\cos\alpha + A\sin\alpha\cos x = (A\cos\alpha)\sin x + (A\sin\alpha)\cos x = a\sin x + b\cos x$

Exercises 6.3

1. 21.22°
3. 7.13°
5. 0.48
7. 1.17
9. $\frac{\pi}{2}$
11. $\frac{\pi}{4}$
13. $-\frac{\pi}{3}$
15. 0.36
17. 0.42
19. no value
21. no value
23. 0.74
25. 1.34
27. $\frac{\sqrt{3}}{2}$
29 $\frac{1}{2}$
31. 0.8
33. Domain: $[-1, 1]$
    Range: $[-0.57, 2.57]$

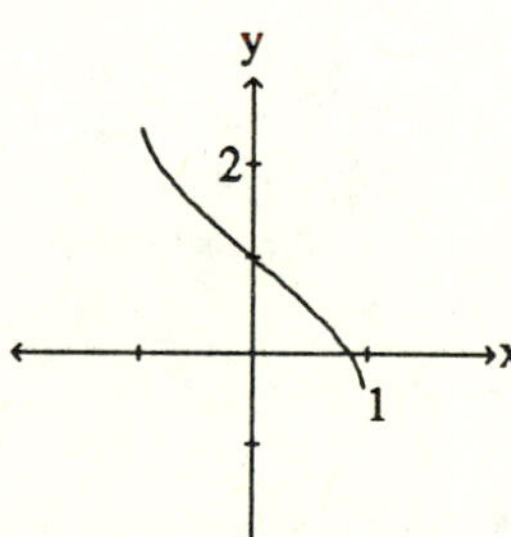

35. Domain: $(-\infty, \infty)$
    Range: $(-0.4, 0.4)$

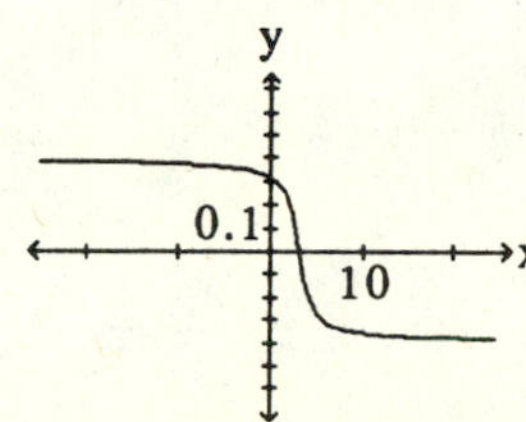

37. Domain: $[-1, 1]$

    Range: $[-1, 1]$

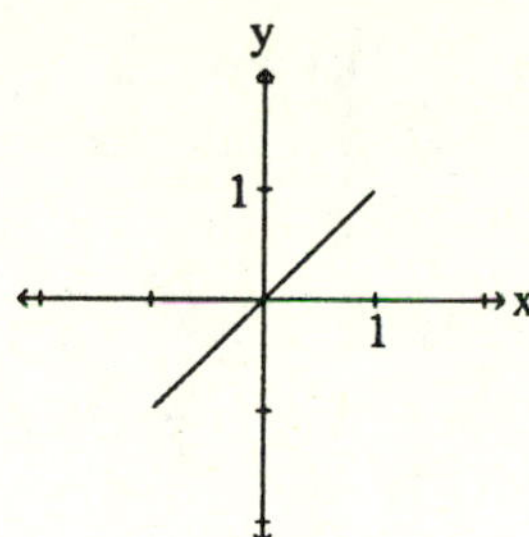

39. $[-1, 1]$
41. $\sin(.5)$
43. $\frac{x}{\sqrt{x^2+1}}$
45. $\frac{x}{\sqrt{1-x^2}}$
47. $1.16 + 2k\pi$ or $4.30 + 2k\pi$, where k is any integer.
49. $1.23 + 2k\pi$ or $5.05 + 2k\pi$, where k is any integer.
51. $f(x) \to 1$
53. $[-\pi, \pi]$
55. $\sin(-x) = -\sin x$ for $-\frac{\pi}{2} \le x \le \frac{\pi}{2}$, so for this interval $\sin^{-1}(-x) = -\sin^{-1}(x)$
57. Let $x = \arccos\theta$. Then $\cos x = \theta$, for $0° \le x \le 180°$. Since $\cos x = \sin(90° - x)$, $\theta = \sin(90° - x)$. So, $\arcsin\theta = 90° - x$. Since $x + (90° - x) = 90°$, $\arccos\theta + \arcsin\theta = 90°$.

59. $\theta = \tan^{-1}\left(\frac{50}{L}\right)$, $L \geq 0$

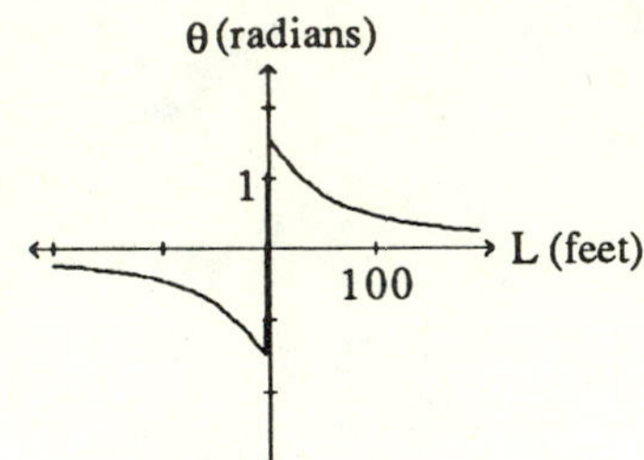

*61. $\sin\frac{\theta}{2} = \frac{\frac{c}{2}}{a} = \frac{c}{2a}$, then $c = 2a\sin\frac{\theta}{2}$.

$\cos\frac{\theta}{2} = \frac{h}{a}$ then $h = a\cos\frac{\theta}{2}$.

$A = \frac{1}{2}ch = \frac{1}{2}\left(2a\sin\frac{\theta}{2}\right)\left(a\cos\frac{\theta}{2}\right)$

$= a^2\sqrt{\frac{1-\cos\theta}{2}} \cdot \sqrt{\frac{1+\cos\theta}{2}}$

$A = \frac{1}{2}a^2\sin\theta$

63.

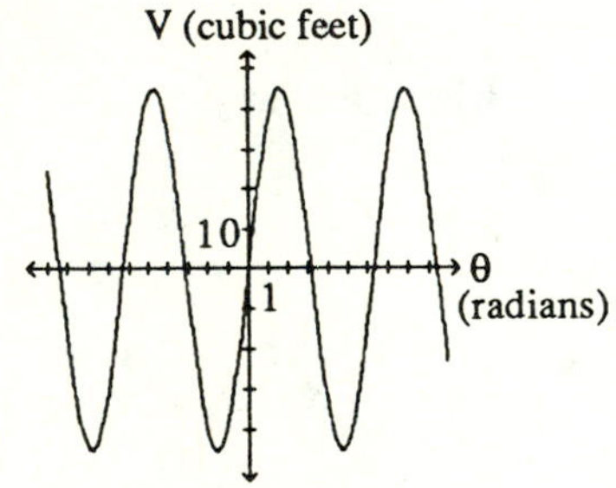

65. $V = 45\sin\theta$

67. $-1 \leq x \leq 1$

69. Compute $\tan^{-1}\frac{x}{\sqrt{1-x^2}}$.

71. $\sqrt{2}\cos\left(x - \frac{\pi}{4}\right)$

73. $5\cos(2x - 0.64)$

Exercises 6.4

1. $0.775 + 2k\pi$, $2.366 + 2k\pi$

3. No solution.

5. $\frac{\pi}{4}, \frac{5\pi}{4}$

7. $\frac{\pi}{6}, \frac{5\pi}{6}$

9. $\frac{1}{2}\arcsin(\frac{1}{3})$: 0.62, 2.53, 3.76, 5.66

11. $0, \pi, 2\pi$

13. $\frac{\pi}{2}, \frac{3\pi}{2}$

15. $k\pi$, where k is any integer

17. $\frac{\pi}{6} + 2k\pi$ and $\frac{5\pi}{6} + 2k\pi$, where k is any integer

19. $\frac{\pi}{6} + 2k\pi$ and $\frac{5\pi}{6} + 2k\pi$, where k is any integer.

21. $2k\pi \pm \frac{\pi}{3}$ and $(2k-1)\pi$, where k is any integer

23. $\frac{\pi}{2} + 2k\pi$, where k is any integer

25. $0 \leq x \leq \frac{\pi}{6}$, $\frac{5\pi}{6} < x \leq 2\pi$

27. $-\frac{3\pi}{8} + k\pi < x < \frac{\pi}{8} + k\pi$, where k is any integer.

29. -2.28, 0, 2.28

31. 0, 1.28

33. $1.57 + 2k\pi < x < 2.84 + 2k\pi$ and $4.71 + 2k\pi < x < 6.59 + 2k\pi$, where k is any integer.

35. $x = a + 2k\pi$ for $a = 1.18$, 2.36, 2.75, 4.32, 5.50, 5.89 where k is any integer

37. Local max: $(0, 5)$, $(12.56, 5)$

Local min: $(6.28, -1)$

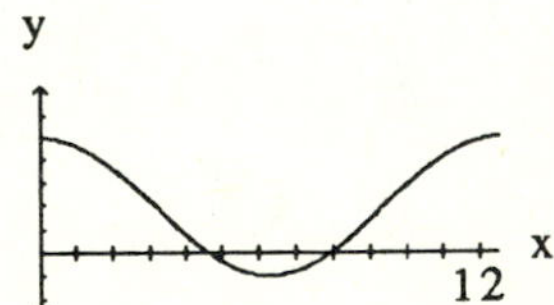

39. Local max: $(0.64, 1)$, $(2.64, 1)$

Local min: $(1.64, -3)$

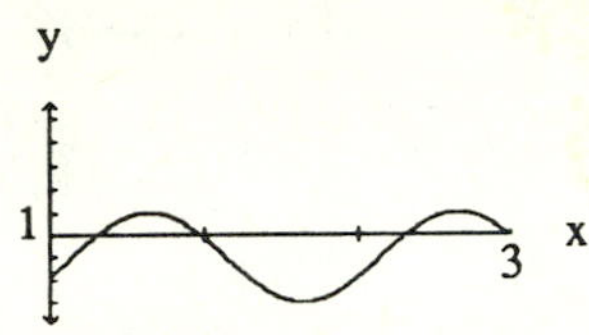

41. Local max: $(3.62, 1.91)$, $(26.97, 1.22)$, $(16.79, 0.22)$

Local min: $(10.72, -1.22)$, $(20.91, -0.22)$, $(34.08, -1.91)$

These extrema are in $[0, 12\pi]$ and will repeat every 12π which is the period of the function.

43. $A = 2$, $B = -3$, $C = -2$, and $D = \frac{3\pi}{2}$

45. $y = \frac{27}{2} + \frac{27}{2}\sin\left(\frac{2\pi}{40}\left(t - \frac{40}{4}\right)\right)$

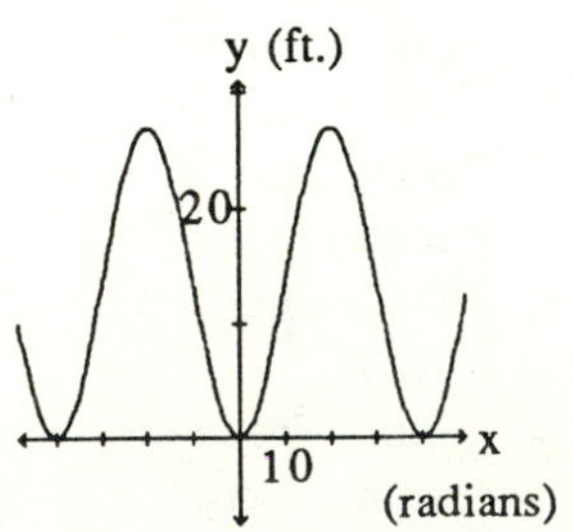

47. 0.15 radians ($\sim 9^\circ$)

49. 45,000.00 feet

51. a)

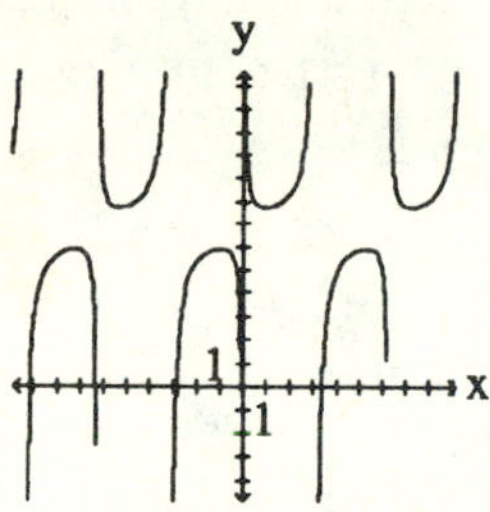

b) $0 < \theta < 180°$

c) $\theta = 54.72°$ or 0.96 radians

d) 7.72 sq. in.

53. $x = \frac{\pi}{2}$, $\sin^{-1}(\frac{-1+\sqrt{5}}{2})$, $\pi - \sin^{-1}(\frac{-1+\sqrt{5}}{2})$

55. $2k\pi < x < \frac{\pi}{2} + 2k\pi$ (otherwise at least one term is undefined.)

Exercises 6.5

1. Answers will vary, but once one side is chosen, the other 2 are determined. For example, $a = 11.2$cm then $b = 20.4$cm and $c = 20.2$cm.

3. $\gamma = 110°$, $a = 12.86$, $c = 18.79$

5. $\alpha = 90°$, $\gamma = 60°$, $c = 6\sqrt{3}$ or 10.39

7. $\gamma = 68°$, $a = 3.88$, $c = 6.61$

9. $\beta = 73.25°$, $\gamma = 56.75°$, $c = 4.37$

or

$\beta = 106.76°$, $\gamma = 23.24°$, $c = 2.06$

11. $\alpha = 55.17°$, $\gamma = 86.83°$, $c = 19.46$ or $\alpha = 124.83°$, $\gamma = 17.17°$, $c = 5.75$

13. no triangle possible.

15. $\alpha = 22.06°$, $\gamma = 5.94$, $c = 2.20$

17. (a) $6.69 < b < 10$

(b) $b = 6.69$ or $b \geq 10$

(c) $b < 6.69$

19. $a = 19.70$ miles

$b = 15.05$ miles

$h = 11.86$ miles

21. 0.72 miles

23. First, put in an altitude from B to AC to use in the solution of the triangle.

$c = 3.85$, $\beta = 128.89°$, $\alpha = 29.11°$

Exercises 6.6

1. no solution

3. $\alpha = 24.56°$, $\beta = 99.22°$, $\gamma = 56.22°$

5. $a = 9.83$, $\beta = 89.32°$, $\gamma = 35.68°$

7. no solution

9. no solution

11. (a) $6.78 < b < 8$

(b) $b = 6.78$ or $b \geq 8$

(c) $b < 6.78$

13. 17.55 square units

15. 110.34 square units

17. 5.56 square units

19. 841.22 feet

21. 42.50 feet

23. 41.56 feet

25. 893.64 feet, 1,123.56 feet

27. 61.73 feet

Review Exercises

1. $\gamma = 68°$, $b = 3.88$, $c = 6.61$

3. $\beta = 113.50°$, $b = 27.55$, $c = 18.16$

5. $\gamma = 72°$, $a = 2.94$, $b = 5.05$

7. $\alpha = 44.4°$, $\beta = 78.4°$, $\gamma = 57.2°$

9. i.) $5.63 < b < 12$

ii.) $b \geq 5.63$

iii.) $0 < b < 5.63$

11. 22.98 square units

13. $y = -5\sin(x - 0.64)$

15. $2\sin\theta\cos^3\theta + 2\sin^3\theta\cos\theta = 2\sin\theta(\cos\theta(1-\sin^2\theta)) + 2\sin\theta((1-\cos^2\theta)\cos\theta) = 2\sin\theta\cos\theta(1-\sin^2\theta+1-\cos^2\theta) = 2\sin\theta\cos\theta = \sin 2\theta$

17. Domain: $(-\infty, \infty)$

Range: $[-5.75, 7]$

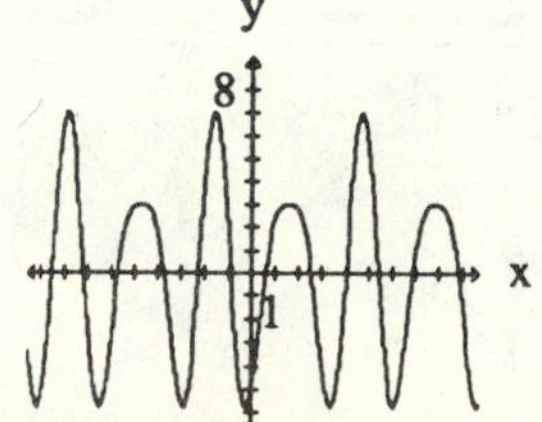

19. Domain: $(-\infty, \infty)$

Range: $(-\infty, \infty)$

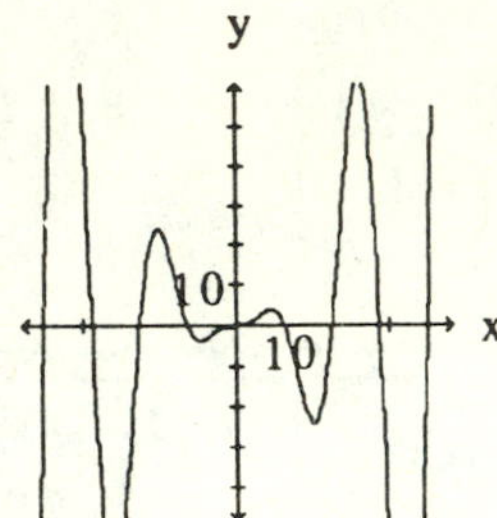

21. $y \to \infty$ as $|x| \to \infty$

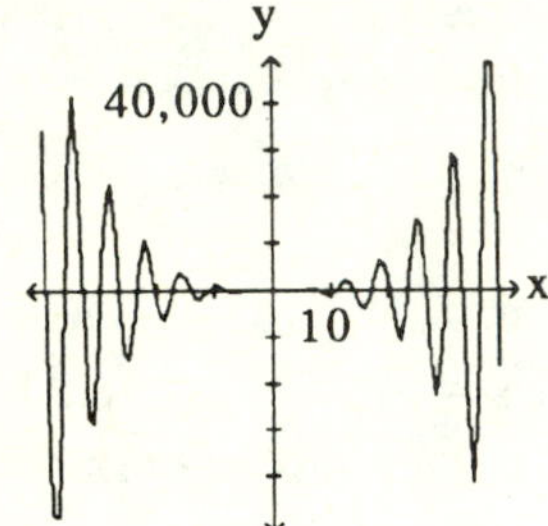

23. $|y| \to \infty$ as $|x| \to 0$

25. 0.37, 2.77

27. $\frac{\pi}{3}, \frac{2\pi}{3}$

29. $\frac{\tan\theta + \sin\theta}{2\tan\theta} = \frac{\tan\theta}{2\tan\theta} + \frac{\sin\theta}{2\frac{\sin\theta}{\cos\theta}} = \frac{1}{2} + \frac{\cos\theta}{2} = \frac{1+\cos\theta}{2} = \left(\sqrt{\frac{1+\cos\theta}{2}}\right)^2 = \cos^2\frac{\theta}{2}$

31. $\sin 3\theta = \sin(\theta + 2\theta) = \sin\theta\cos 2\theta + \cos\theta\sin 2\theta = \sin\theta(\cos^2\theta - \sin^2\theta) + (2\sin\theta\cos\theta)\cos\theta = \sin\theta\cos^2\theta - \sin^3\theta + 2\sin\theta\cos^2\theta = 3\sin\theta\cos^2\theta - \sin^3\theta$

33. Identity, the graphs coincide.

35. $\frac{\sqrt{2}+\sqrt{6}}{4}$

37. $\frac{2}{\sqrt{5}}$

39. $\sin 3\theta + \cos 3\theta = 3\sin\theta\cos^2\theta - \sin^3\theta + \cos^3\theta - 3\sin^2\theta\cos\theta$

41. $\frac{\pi}{4}, \frac{3\pi}{4}$

43. 0.50

45. $\frac{\pi}{4}$

47. 1.37

49. -1.43

51. Domain: $[-1, 1]$

Range: $[-1, \pi - 1]$

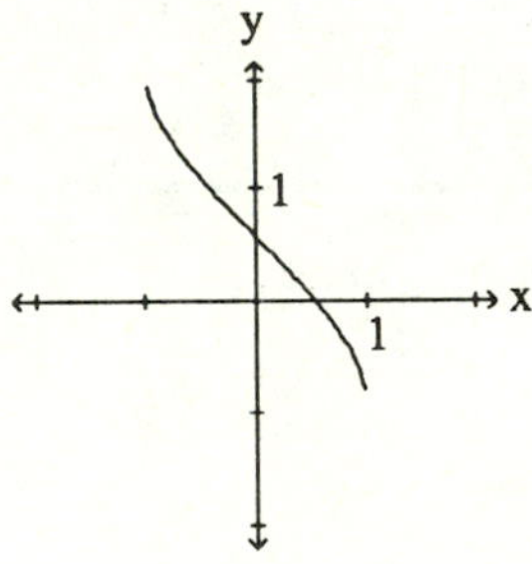

53. $\frac{\pi}{4}, \frac{3\pi}{4}$

55. $\frac{3\pi}{4} + 2k\pi$ and $\frac{5\pi}{4} + 2k\pi$, where k is any integer

57. 0.26, 1.31, 3.40, 4.45

59. $(2k-1)\cdot\frac{\pi}{4}$, where k is any integer

61. $2k\pi$ and $\frac{2k\pi}{3}$, where k is any integer

63. $0 \le x < \frac{\pi}{6}$, $\frac{5\pi}{6} < x < \frac{7\pi}{6}$, $\frac{11\pi}{6} < x \le 2\pi$

65. $\frac{4\sqrt{6}}{25}$

67. a.) 102.54 feet

b.) 96.35 feet

69. 849.77 feet

71. a.) $L = d(A,B) + d(B,C) = 3.5\sec\theta + 6\csc\theta$

b.)

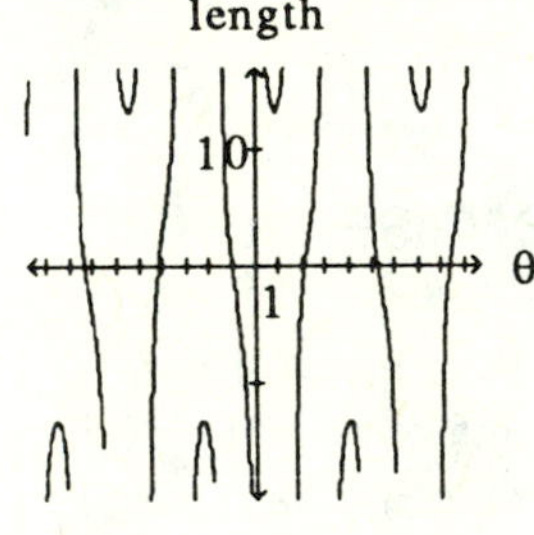

c.) $0 < \theta < \frac{\pi}{2}$

d.) $\theta = 0.87$ radians

• CHAPTER 7

Exercises 7.1

1. $2 + 7i$

3. $14 - 8i$

5. i

7.

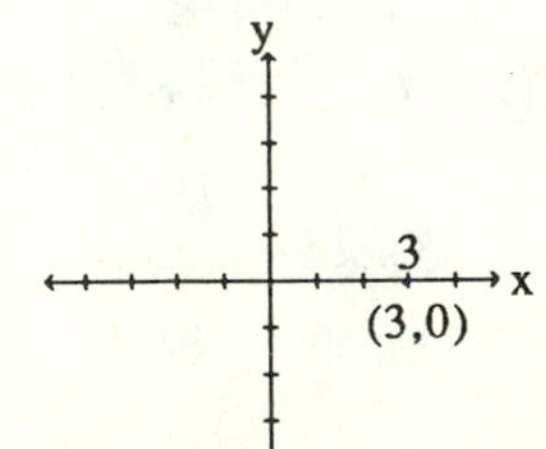

9.

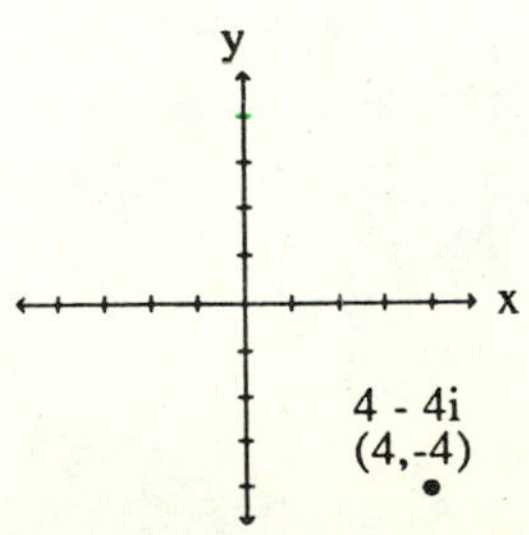

11.

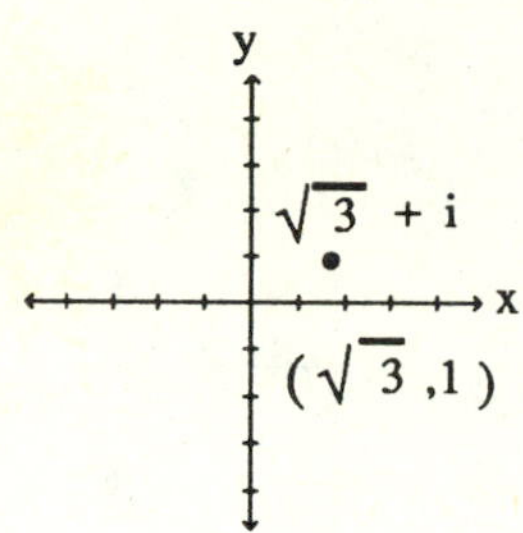

13. $a = 2$, $b = 2$, $|z_1| = 2\sqrt{2}$

15. $2\sqrt{2}(\cos\frac{\pi}{4} + i\sin\frac{\pi}{4})$

17. $1 + 7i$

19. $11 + 16i$

21. $5\sqrt{13}$

23. $5 + 7i$

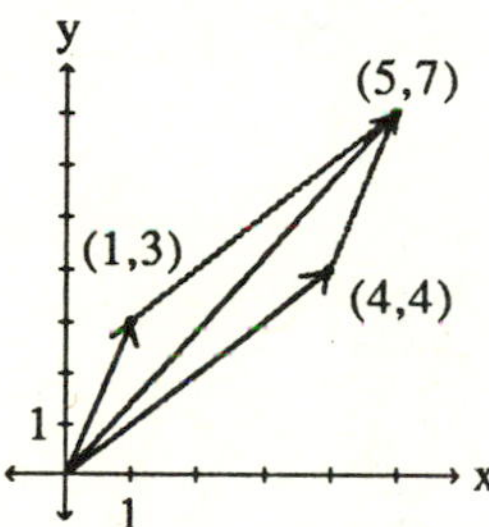

25. $1 + \sqrt{3}i$

27. $0.96 - 3.26i$

29. $\sqrt{13}(\cos(0.98) + i\sin(0.98))$ and $\sqrt{13}(\cos(0.98 + 2k\pi) + i\sin(0.98 + 2k\pi))$, where k is any integer.

31. $\sqrt{66}(\cos(4.89) + i\sin(4.89))$ and $\sqrt{66}(\cos(4.89 + 2k\pi) + i\sin(4.89 + 2k\pi))$, where k is any integer.

33. If $u = r\cos\theta = a + bi$, then $b = 0$ and $\bar{u} = a + bi = r\cos\theta - 0 \cdot i = r\cos\theta$. But $\cos\theta = \cos(-\theta)$. So $\bar{u} = r\cos\theta = r\cos(-\theta)$.

35. $uv = 0$ if and only if $|u \cdot v| = 0$. By Exercise 34, $|u \cdot v| = |u| \cdot |v|$. So $|u \cdot v| = |u| \cdot |v|$ if and only if $|u| = 0$ or $|v| = 0$; that is, if and only if $u = 0$ or $v = 0$.

37. $\frac{\sqrt{3}}{3} - \frac{1}{3}i$

39. $5\sqrt{13}\ cis\ 7.51 = 6 + 17i$

41. $2\,\text{cis}\,30° = 2(\frac{\sqrt{3}}{2} + \frac{1}{2}i) = \sqrt{3} + i$

$3\,\text{cis}\,60° = 3(\frac{1}{2} + \frac{\sqrt{3}}{2}i) = \frac{3}{2} + \frac{3\sqrt{3}}{2}i$

$\frac{\sqrt{3}+i}{\frac{3}{2}+\frac{3\sqrt{3}}{2}i}\frac{(\frac{3}{2}-\frac{3\sqrt{3}}{2}i)}{(\frac{3}{2}-\frac{3\sqrt{3}}{2}i)} == \frac{\sqrt{3}}{3} - \frac{i}{3}$

43. $6 + 17i$

Exercises 7.2

1. $5(\cos 5.36 + i\sin 5.36)$

3. $\sqrt{34}(\cos 0.54 + i\sin 0.54)$

5. $2(\cos\frac{3\pi}{2} + i\sin\frac{3\pi}{2})$

7. $6\,\text{cis}\,90° = 6i$

9. $(2\,\text{cis}\,30°)(3\,\text{cis}\,60°) = (\sqrt{3} + i)(\frac{3}{2} + \frac{3\sqrt{3}}{2}i) = 6i$

11. $\frac{-\sqrt{2}}{2} - \frac{\sqrt{2}}{2}i$

13. 64

15. Roots: $1, i, -1, -i$

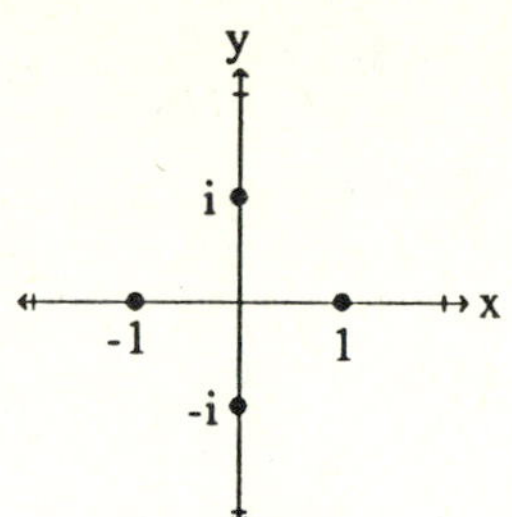

17. Roots: $\sqrt[8]{8}\,\text{cis}\,\frac{\pi}{16}$, $\sqrt[8]{8}\,\text{cis}\,\frac{9\pi}{16}$, $\sqrt[8]{8}\,\text{cis}\,\frac{17\pi}{16}$, $\sqrt[8]{8}\,\text{cis}\,\frac{25\pi}{16}$

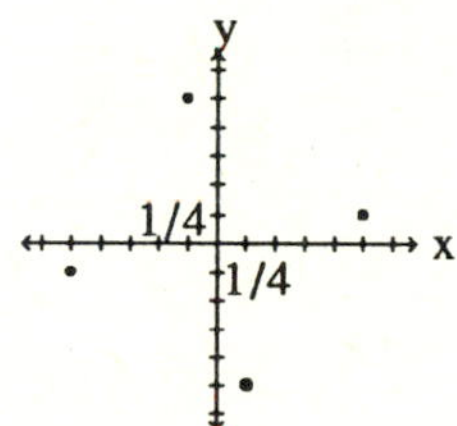

19. Roots: $\sqrt[12]{8}\,\text{cis}\,\frac{\pi}{8}$, $\sqrt[12]{8}\,\text{cis}\,\frac{11\pi}{24}$, $\sqrt[12]{8}\,\text{cis}\,\frac{9\pi}{8}$, $\sqrt[12]{8}\,\text{cis}\,\frac{19\pi}{24}$, $\sqrt[12]{8}\,\text{cis}\,\frac{35\pi}{24}$, $\sqrt[12]{8}\,\text{cis}\,\frac{43\pi}{24}$

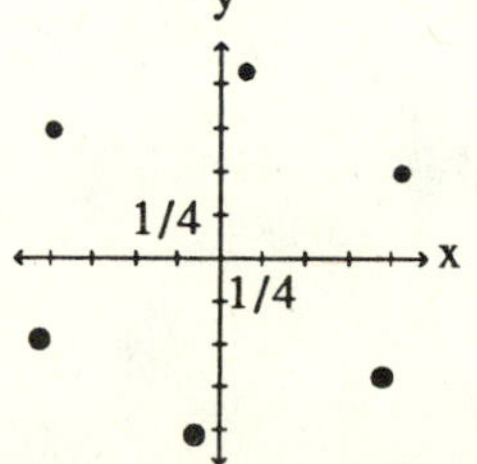

21. $-2,\ 1+\sqrt{3}i,\ 1-\sqrt{3}i$; $z=-8$

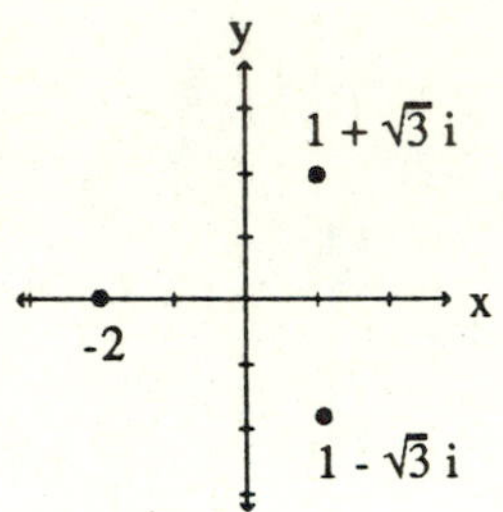

23. $\sqrt[8]{50}\operatorname{cis}\theta$ for $\theta = \frac{7\pi}{16}, \frac{15\pi}{16}, \frac{23\pi}{16}, \frac{31\pi}{16}$

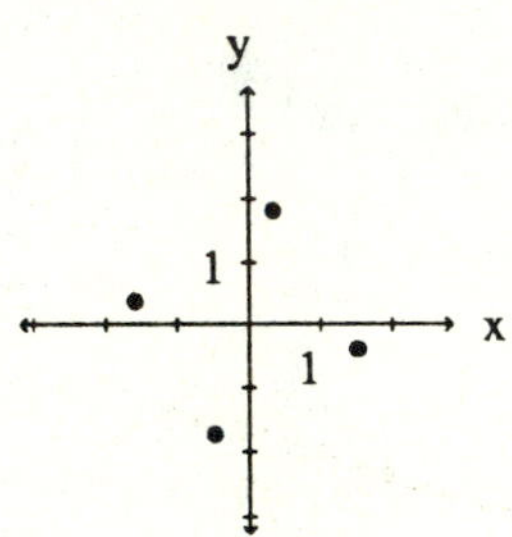

25. roots: $\sqrt{2}\operatorname{cis}\theta$ where $\theta = \frac{\pi}{12}, \frac{3\pi}{12}, \frac{5\pi}{12}, \frac{7\pi}{12}, \frac{9\pi}{12}, \frac{11\pi}{12}, \frac{13\pi}{12}, \frac{15\pi}{12}, \frac{17\pi}{12}, \frac{19\pi}{12}, \frac{21\pi}{12}, \frac{23\pi}{12}$

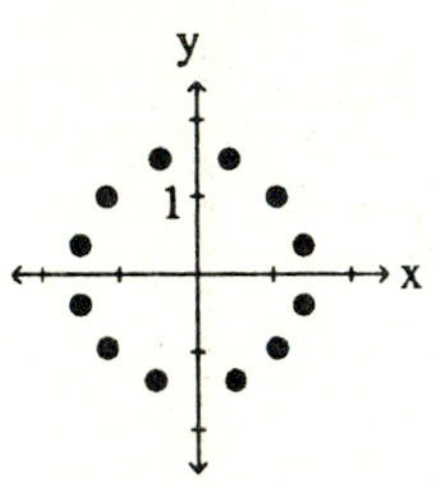

27. One possible pentagon has vertices $4\operatorname{cis}\frac{2\pi}{5}, 4\operatorname{cis}\frac{4\pi}{5}, 4\operatorname{cis}\frac{6\pi}{5}, 4\operatorname{cis}\frac{8\pi}{5}, 4\operatorname{cis}2\pi$

29. $(x+y)^6 = x^6 + 6x^5y + 15x^4y^2 + 20x^3y^3 + 15x^2y^4 + 6xy^5 + y^6$

31. $(x-y)^6 = x^6 - 6x^5y + 15x^4y^2 - 20x^3y^3 + 15x^2y^4 - 6xy^5 + y^6$

33. 56

35. The coefficient of $a^{n-k}b^k$ is $\frac{n(n-1)\cdots(n-k+1)}{1\cdot 2\cdots k} = \frac{n(n-1)\cdots(n-k+1)}{k!}\cdot \frac{(n-k)\cdots 2\cdot 1}{(n-k)\cdots 2\cdot 1} = \frac{n!}{k!(n-k)!}$

37. (a) $\binom{n}{k} = \frac{n!}{(n-k)!k!}$ and $\binom{n}{n-k} = \frac{n!}{(n-(n-k))!(n-k)!} = \frac{n!}{k!(n-k)!}$

(b) $\binom{n}{1} = \frac{n!}{(n-1)!1!} = \frac{n(n-1)\cdots 2\cdot 1}{(n-1)!1!} = \frac{n}{1!} = n$

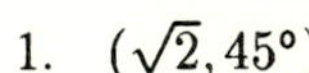

Exercises 7.3

1. $(\sqrt{2}, 45°)$

3. $(5, 126.87°)$

5. $(\sqrt{8}, \frac{\pi}{4} + 2k\pi)$, where k is any integer

7. $(0, 0)$

9. $(2.12, -2.12)$

11. 3.50

13.

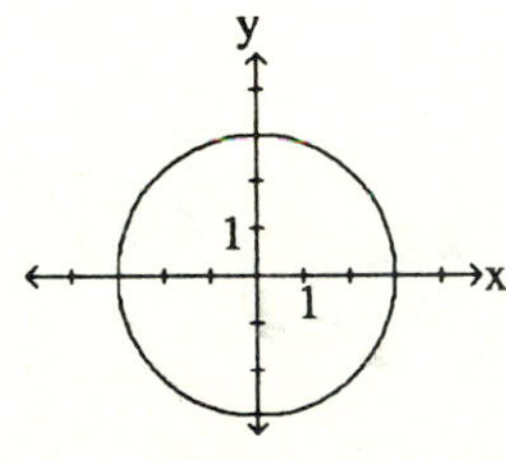

15. The graph is the y-axis.

17.

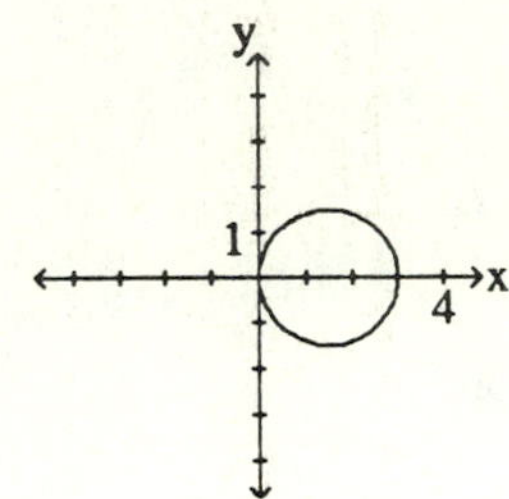

19.

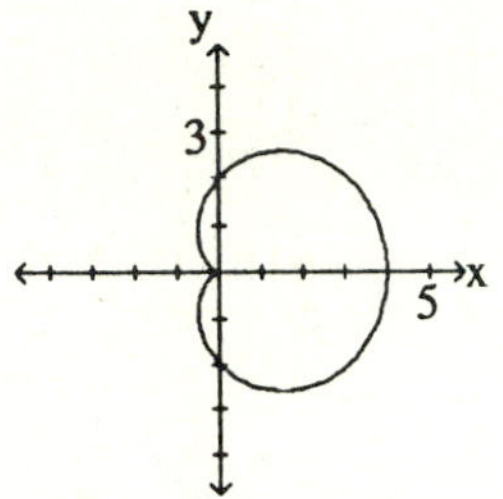

21.

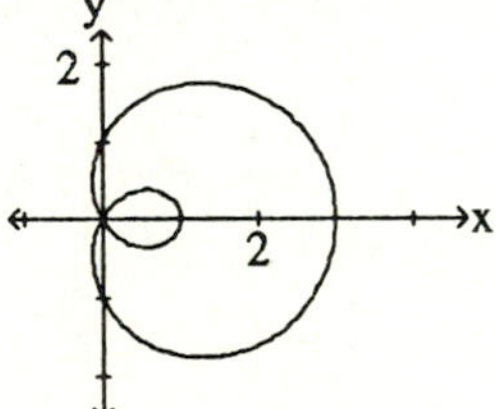

23. $0 \le \theta \le \frac{\pi}{2}$:

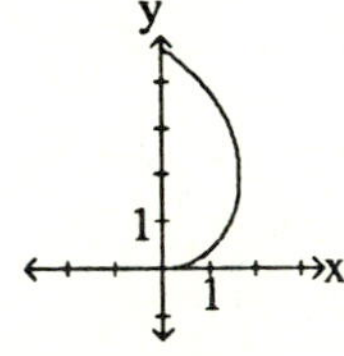

$0 \le \theta \le \pi$:

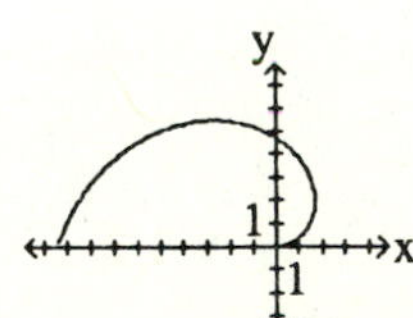

$0 \le \theta \le 2\pi$:

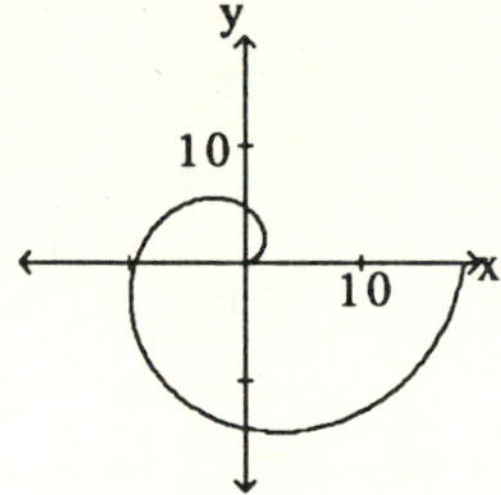

$-\pi \le \theta \le \pi$:

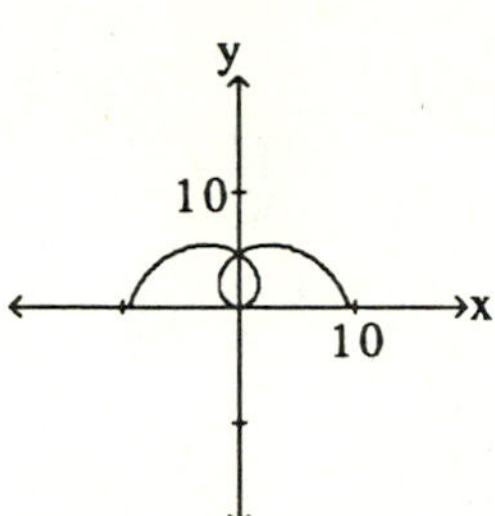

$0 \le \theta \le 4\pi$:

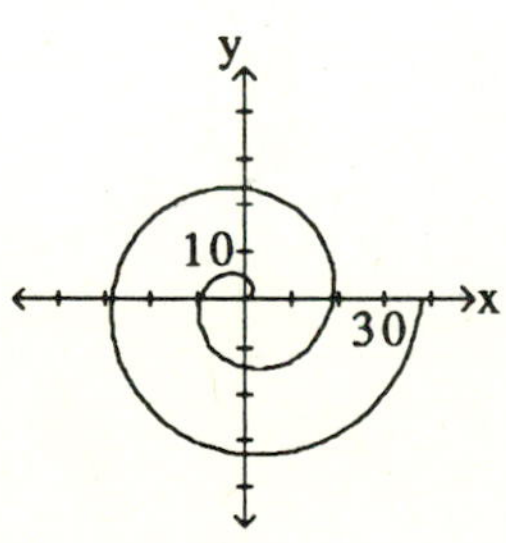

25. $0 \le \theta \le \frac{\pi}{2}$:

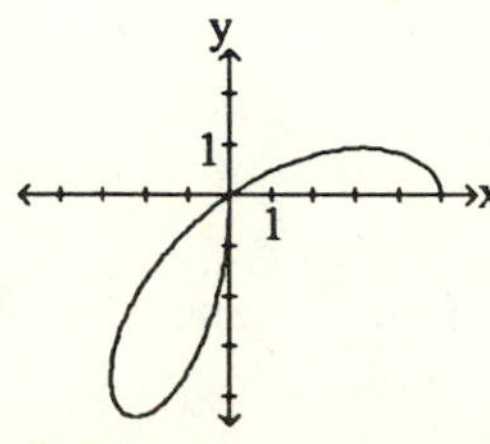

$0 \le \theta \le \pi$, $0 \le \theta \le 2\pi$, $-\pi \le \theta \le \pi$, $0 \le \theta \le 4\pi$:

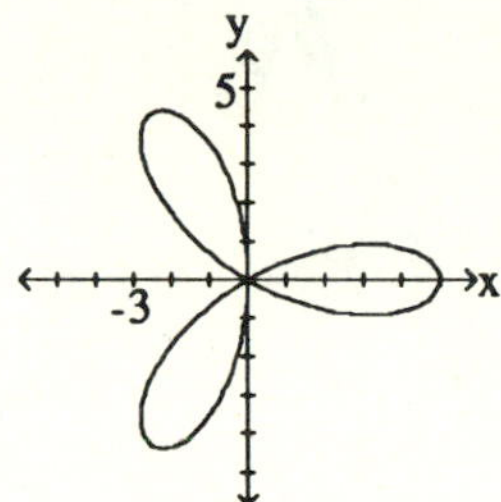

27. $-4\pi \le \theta \le 4\pi$:

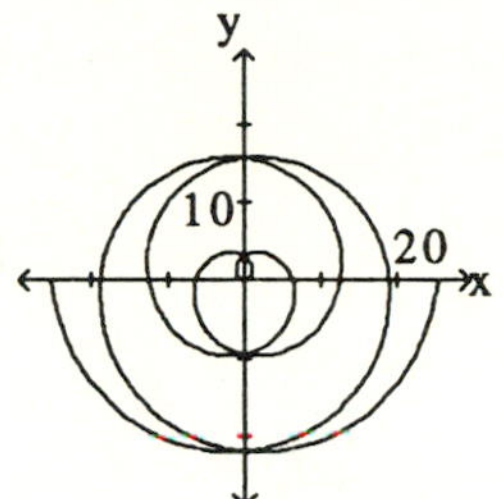

29. $0 \le \theta \le 2\pi$:

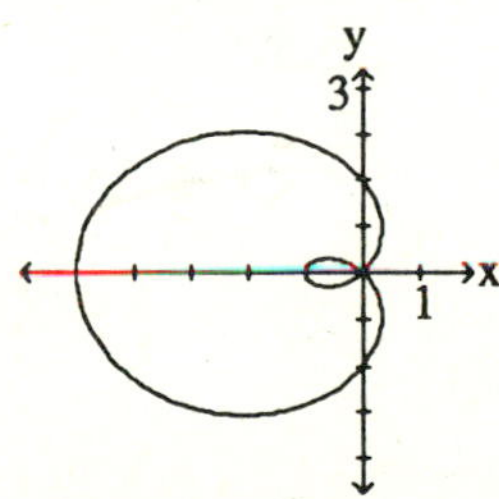

31. $0 \le \theta \le 2\pi$:

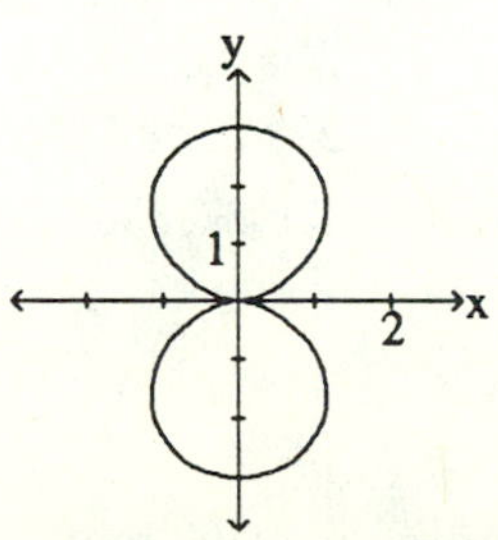

33. If n is odd, there are n petals. If n is even, there are $2n$ petals. $r = 5\sin\theta$ is a circle. $r = 5\sin 2\theta$ is a four-petal rose. $r = 5\sin 3\theta$ is a three-petal rose. $r = 5\sin 4\theta$ is a eight-petal rose. $r = 5\sin 5\theta$ is a five-petal rose. $r = 5\sin 6\theta$ is a twelve-petal rose.

35. The length of the petal is determined by a. All graphs are three-petal roses.

37. 5

39. 8

41.

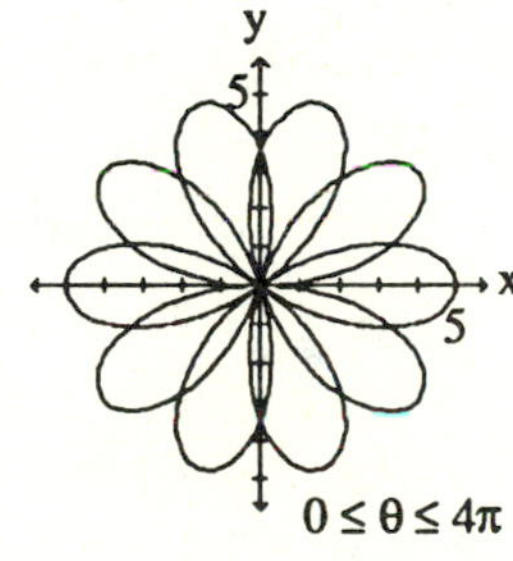

$0 \le \theta \le 4\pi$

43.

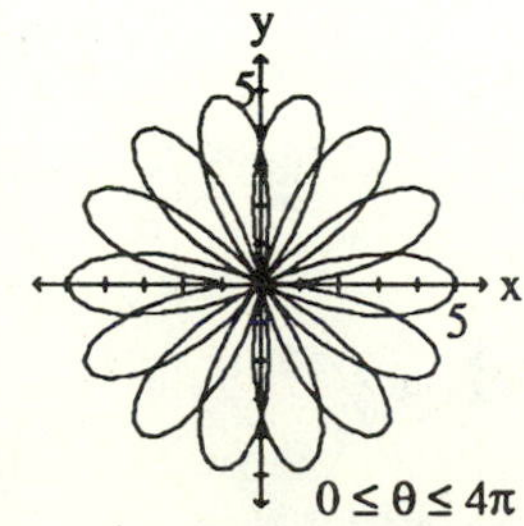

$0 \le \theta \le 4\pi$

45. $(\frac{a}{\sqrt{2}}, \theta)$ for $\theta = \frac{\pi}{4}, \frac{3\pi}{4}, \frac{5\pi}{4}, \frac{7\pi}{4}$

47.

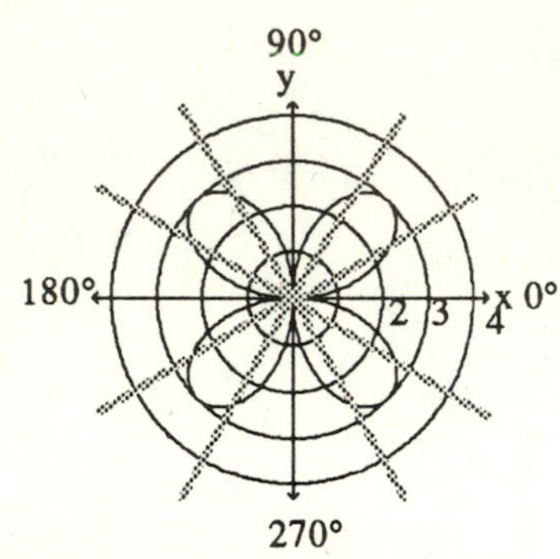

49. $(x, y) = (0, 0)$, $(2.5, 2.5)$, or $(-2.5, 2.5)$

51. $(x, y) = (0, 0)$, $(0.96, 0.16)$, $(0.96, -0.16)$, $(2.29, 1.43)$, $(2.29, -1.43)$, $(0.41, 1.47)$, $(0.41, -1.47)$, or $(-0.9, 0.73)$

Exercises 7.4

1. $(1, 7)$
3. $(-3, 8)$
5. $(4, -9)$
7. $\sqrt{50}$
9. $u + v = (4, 9)$

   $u - v = (-2, -3)$

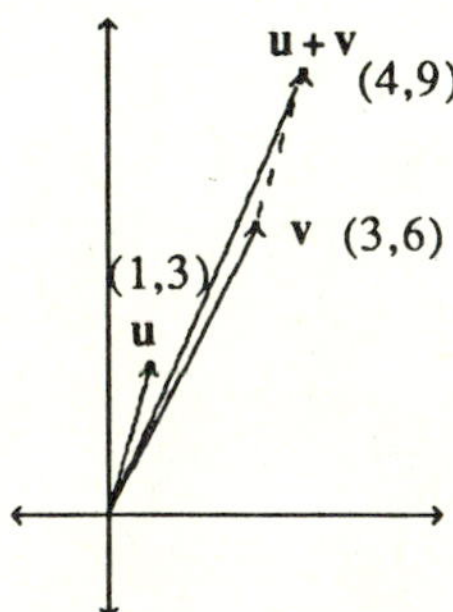

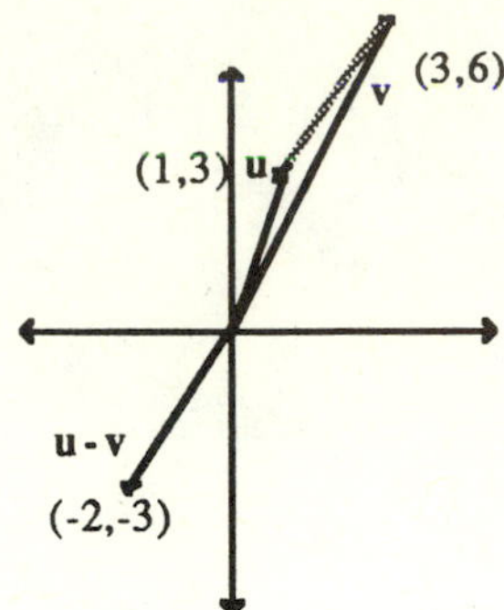

11. $(4, 2)$
13. $(12, 6)$
15. $(0, 15)$
17. $\sqrt{20}$
19. 0
21.

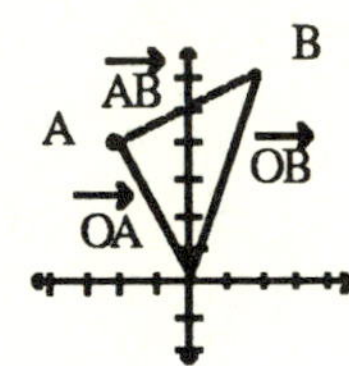

23. $3i - j$
25. $-i - 6j$
27. $u : 116.57°$; $v : 33.69°$; $\beta : 53.13°$
29. $u : 303.69°$; $v : 239.03°$; $\beta : 64.65°$
31. $(x, y) = (\frac{x+2y}{5})(1, 2) + (\frac{2x-y}{5})(2, -1)$
33. $v = (-223.99, 480.34)$
35. $v = (-136.87, 336.04)$;

    Speed = 362.84 mph;

    22.16° west of north
37. $v = (-17.95, -5.96)$;

    Speed = 18.92 mph;

    18.37° south of west
39. 758.91 mph

    in the direction

    1.05° east of north
41. $v = (0, -64)$; $B = (0, 156)$;

    $v = (0, -144)$; $B = (0, 76)$;

    $v = (0, -256)$; actually hits ground before $t = 4$.
43. $u + g = (36t \cos 70°, 36t \sin 70° - 16t^2)$

    $t = 1 : (12.31, 17.83)$

    $t = 2.5$: already on ground

    $t = 4$: already on ground
45. 17.88 feet high, 25.98 feet in horizontal direction (use exercise 43).

Exercises 7.5

1. Line segment from $(1, 0)$ to $(6, 5)$

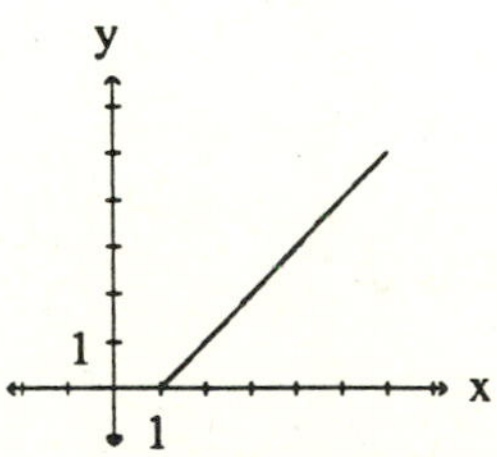

3. Line segment from $(-7, 8)$ to $(5, 4)$

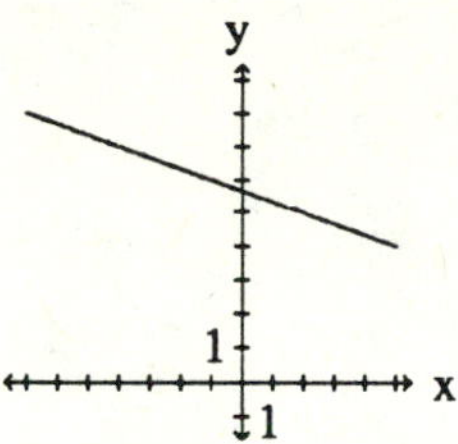

5. Part of one arm of a hyperbola with asymptote at $x = 0$

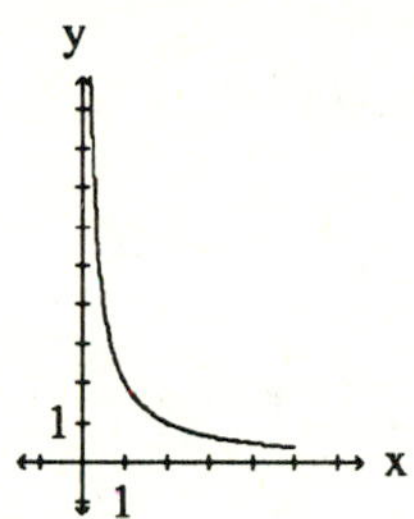

7. Circle with radius 3 centered at $(0, 0)$

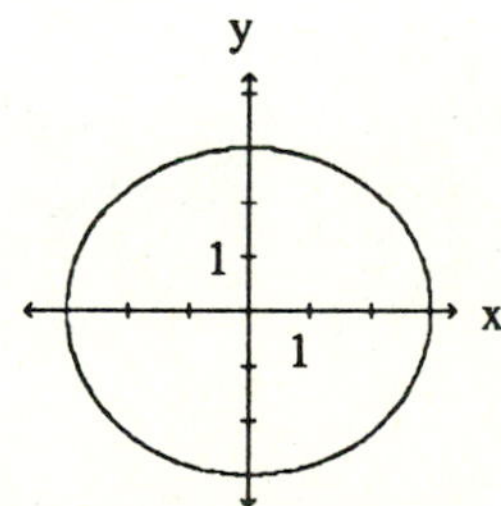

9. All except Exercises 7.

11. Domain: $[-2, 3]$;

    Range: $[-3, 2]$

13. Domain: $(0, 5]$;

    Range: $[\frac{2}{5}, \infty)$

15. $x(t) = 5\cos t$ and $y(t) = 5\sin t$ with t in $[0, 2\pi]$.

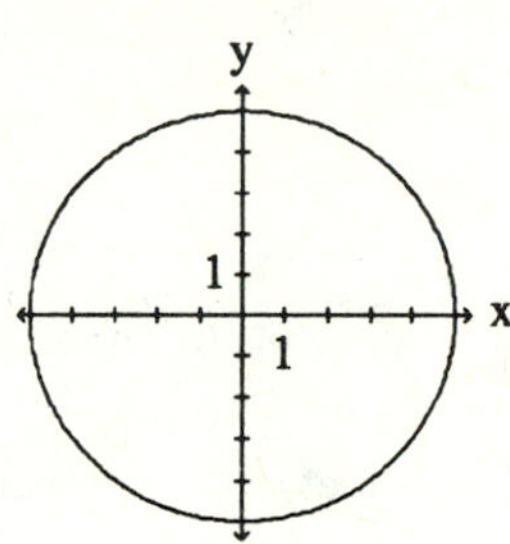

17. $x(t) = 6\cos t$ and $y = 10 + 6\sin t$ with t in $[0, 2\pi]$.

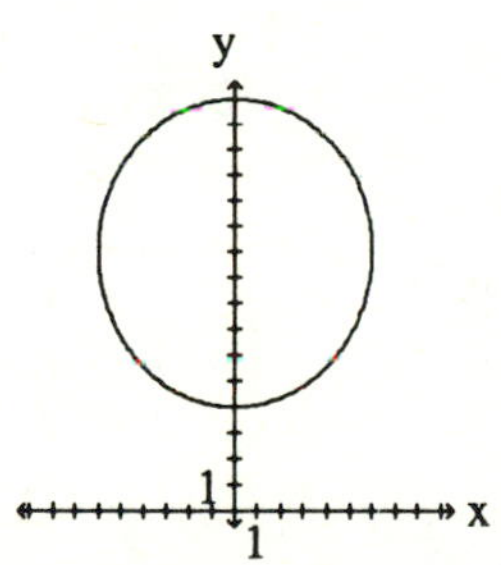

19.

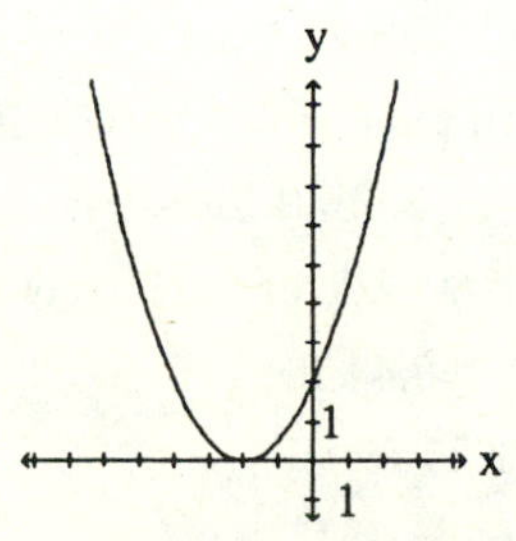

21.

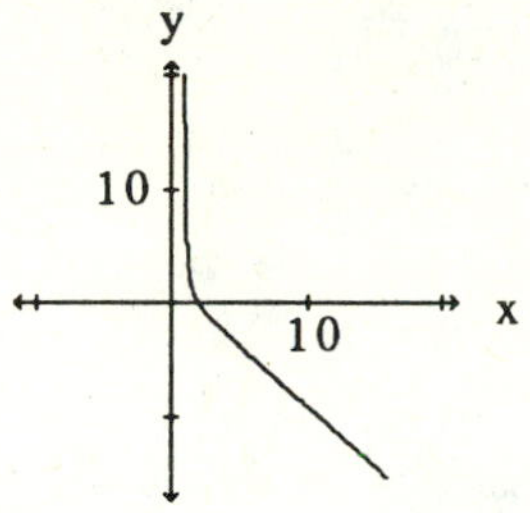

23.

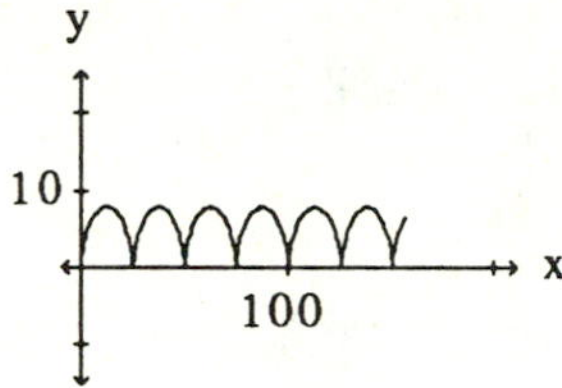

25.

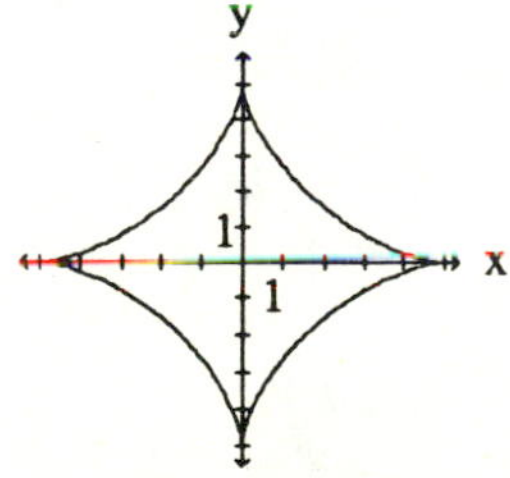

27. Domain: $[-5, 5]$

    Range: $[-5, 5]$

29. The function f is all ordered pairs $(x, f(x))$. The parametric curve defined by $x(t) = t$, $y(t) = f(t)$ is all ordered pairs $(x, y) = (t, f(t))$ for t in I. So the curve is the function f on I.

31. Maximum: $y(t) = 4$ when $t = 0$.

33. $x(t) = 5\sin t\cos t$ and $y(t) = 5\sin^2 t$ with t in $[-\pi, \pi]$

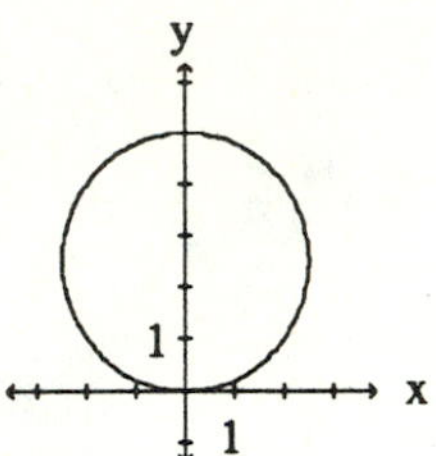

35. $x(t) = 5\sin 3t\cos t$ and $y(t) = 5\sin 3t\sin t$ with t in $[0, 2\pi]$

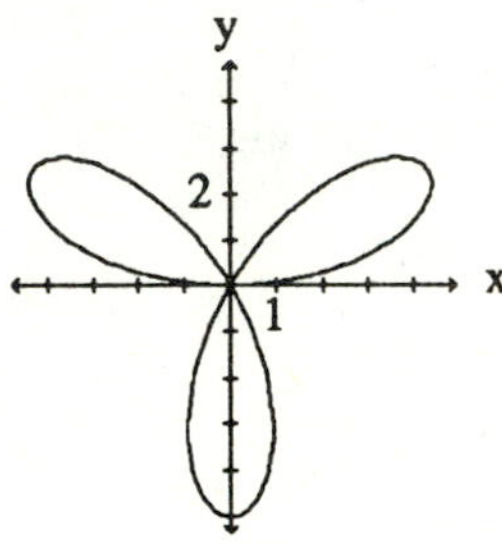

37. period $= 2a\pi$; max $y = 2a$

Exercises 7.6

1.

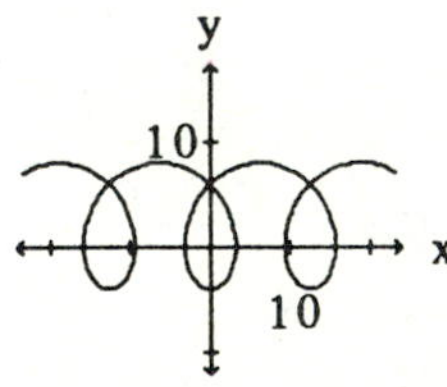

3.

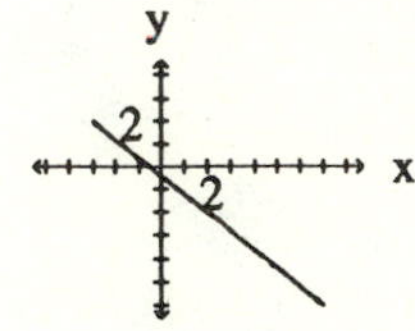

5. $x(t) = 58t\cos 41°$ and $y(t) = -16t^2 + 58t\sin 41°$
7. The portion in the first quadrant
9. 22.62 feet at 1.19 seconds
11. 9.52 seconds, 1306.09 feet
13. yes
15. Max. height: 88.74 feet,

    Distance: 248.55 feet
17.

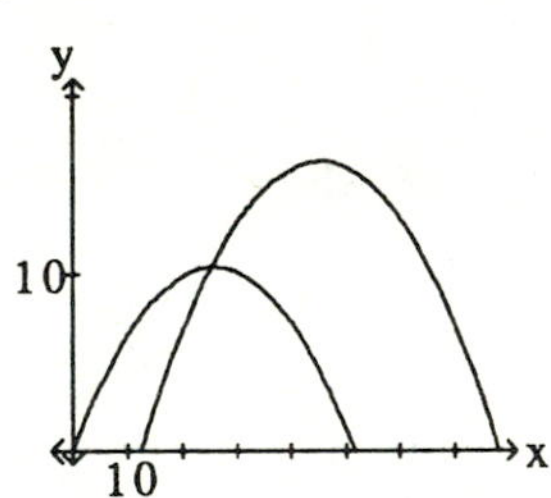

19. Chris: Max. height: 10.4 feet,

    Distance: 51.30 feet,

    Impact at 1.61 seconds

    Linda: Max. height: 15.27 feet,

    Distance: 63.12 feet,

    Impact at 1.95 seconds

    Chris' ball hits first.
21. 6.60 feet at 1.21 seconds
23. 2450.44 feet
25. 1.35 feet
27. about 51°
29. 209.34 feet, near the opponents' 15 yard line
31. 4.40 seconds
33. Home run:

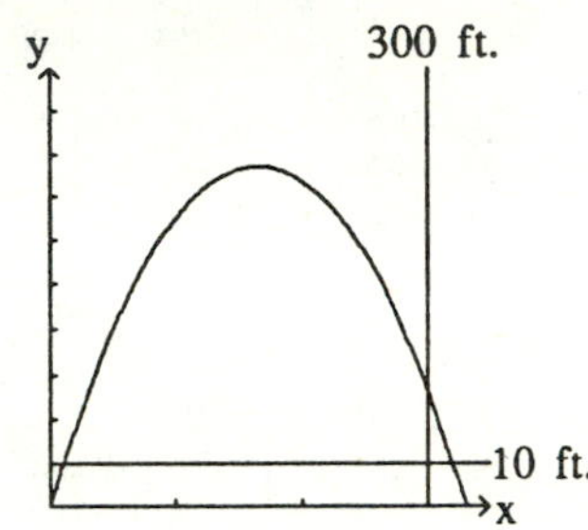

35. Home run:

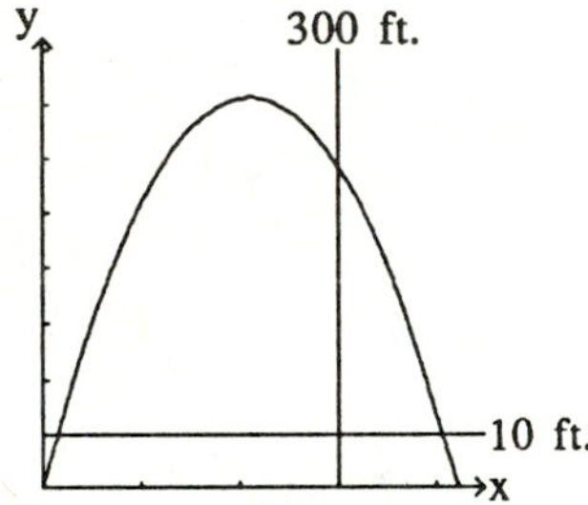

37. Not a home run:

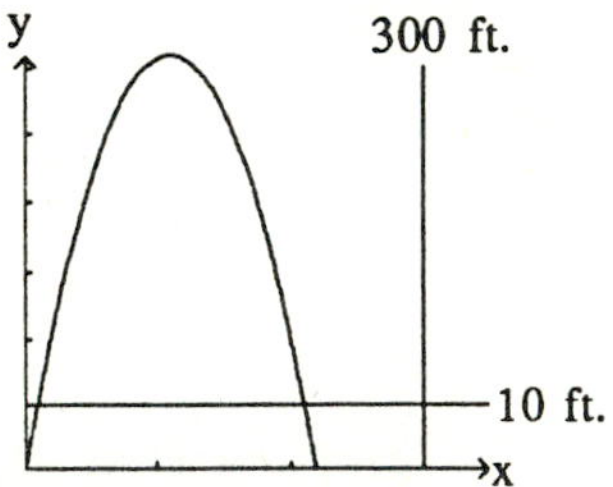

39. Point P is given by the coordinates $(a\theta - a\sin\theta, a - a\sin\theta)$. Thus, the parametric equations would be

$x(\theta) = a(\theta - \sin\theta)$ and
$y(\theta) = a(1 - \cos\theta)$

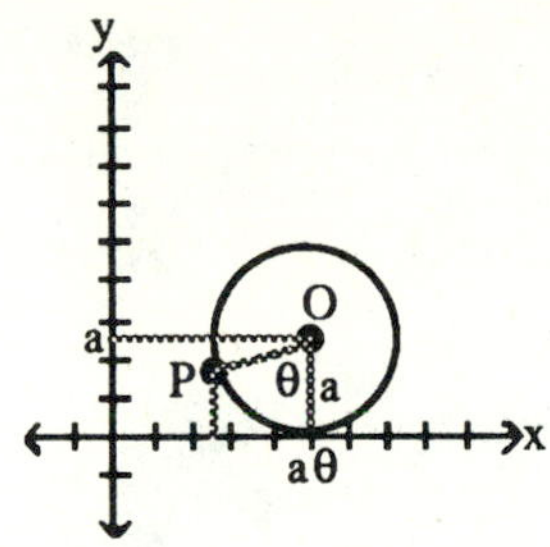

Review Exercises

1. $2 + 4i$
3. $-3 + i$
5.

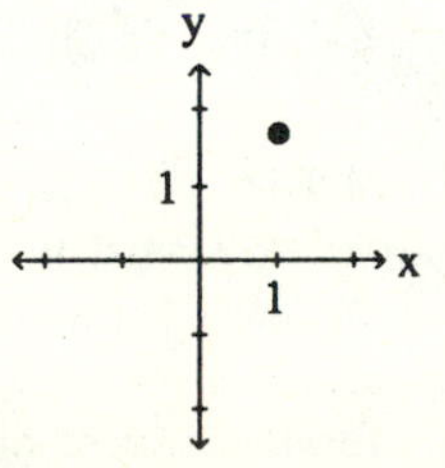

7. $5(\cos 5.35 + i\sin 5.35)$
9. $23 + 14i$
11. $3\sqrt{3} + 3i$
13. $3\sqrt{2}(\cos\frac{7\pi}{4} + i\sin\frac{7\pi}{4})$; $3\sqrt{2}(\cos\theta + i\sin\theta)$ for $\theta = \frac{7\pi}{4} + 2k\pi, k$ any integer
15. $\sqrt{34}(\cos 5.25 + i\sin 5.25)$; $\sqrt{34}(\cos\theta + i\sin\theta)$ for $\theta = 5.25 + 2k\pi, k$ any integer
17. $12\,\text{cis}\,90°$ or $12i$
19. $(3\,\text{cis}\,30°)(4\,\text{cis}\,60°) = 3(\frac{\sqrt{3}}{2} + \frac{1}{2}i)\cdot 4(\frac{1}{2} + \frac{\sqrt{3}}{2}i) = 3\sqrt{3} + 3i + 9i - 3\sqrt{3} = 12i$
21. $16\,\text{cis}\,14\pi = 16$
23. $\sqrt[8]{18}\,\text{cis}\,\theta$ for $\theta = \frac{\pi}{16}, \frac{9\pi}{16}, \frac{17\pi}{16}, \frac{25\pi}{16}$ 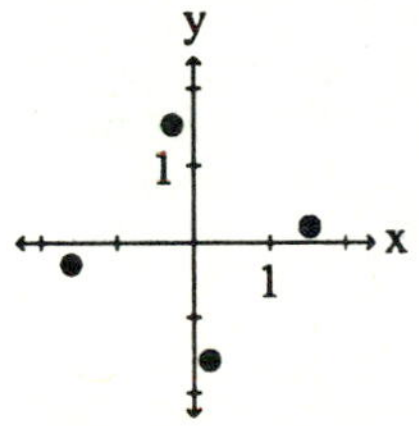
25. $z = -16 + 16i$; roots are $2\,\text{cis}\,\theta$ for $\theta = \frac{\pi}{4}, \frac{11\pi}{12}, \frac{19\pi}{12}$
27. $(\sqrt{41}, 2.25)$
29. $(-2\sqrt{2}, \frac{7\pi}{4} + 2k\pi)$ and $(2\sqrt{2}, \frac{3\pi}{4} + 2k\pi)$ where k is any integer
31. $(2\sqrt{2}, 2\sqrt{2})$
33. $d = 1$
35. 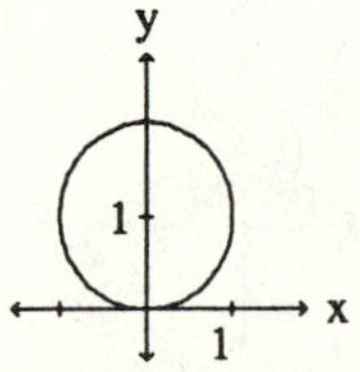
37.

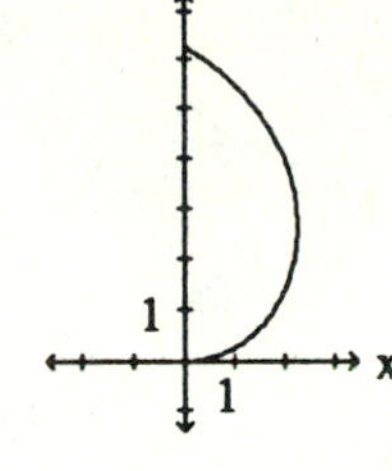

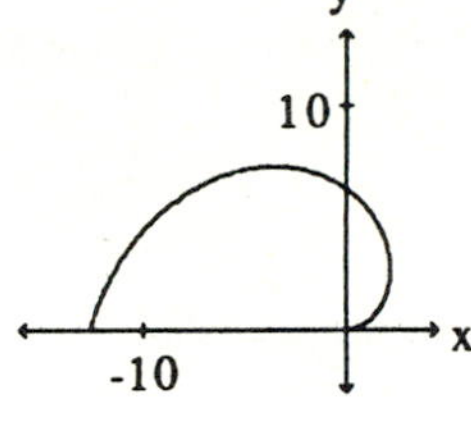

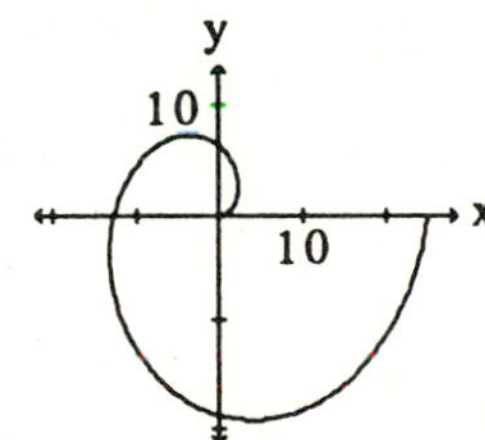

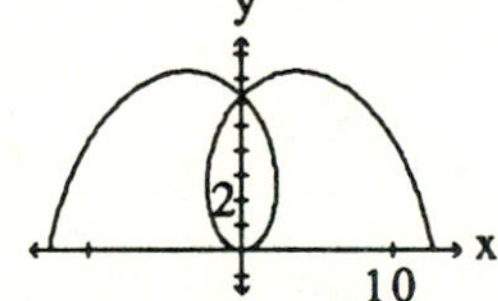

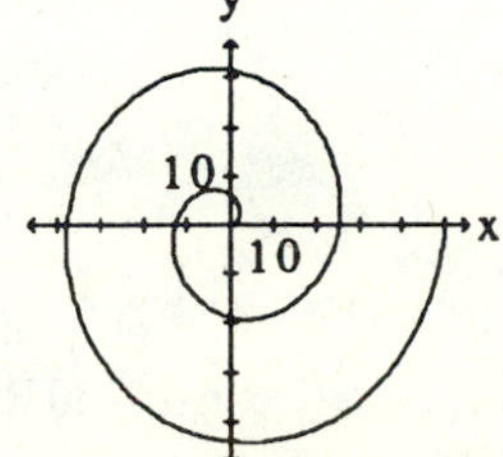

39. $0 \le \theta \le 2\pi$

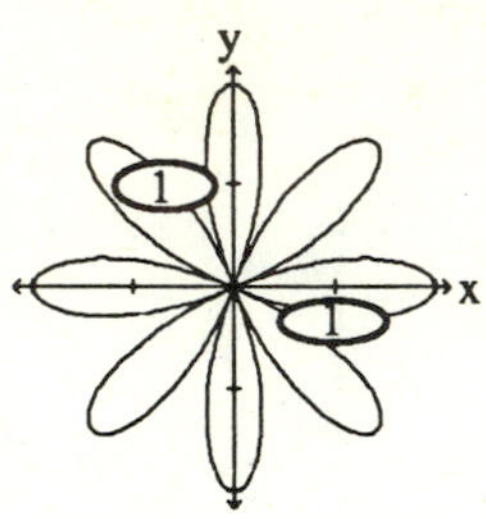

41. length $= 6$

43. $(x, y) = (-2, 0)$; $(x, y) = (0, -1)$

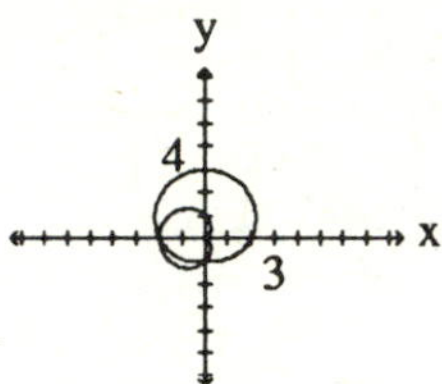

45. $(1, 7)$

47. $\boldsymbol{u} + \boldsymbol{v} = (1, 7)$

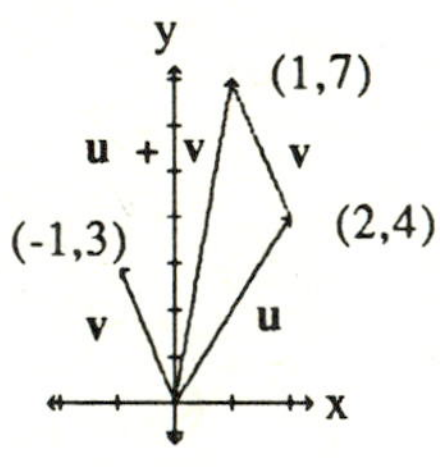

$\boldsymbol{u} - \boldsymbol{v} = (3, 1)$

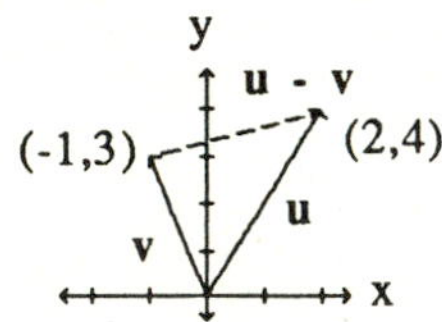

49. $(6, -5)$

51. $-2i + j$

53. u: 36.87°, v: 68.20°

angle between u and v is 31.33°

55. Let $(x, y) = ar + bs$ where $a = \frac{x+2y}{5}$ and $b = \frac{2x-y}{5}$. Then $\frac{x+2y}{5}(1, 2) + \frac{2x-y}{5}(2, -1) = (x, y)$

57. The curve is part of the hyperbola $y = \frac{6}{x}$.

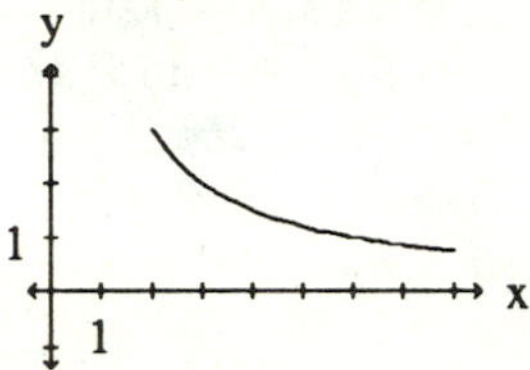

59. The curves in Exercises 56 and 57 are functions.

61. Domain: $[2, 8]$; Range: $[\frac{3}{4}, 3]$

63. The curves in Exercises 56 and 57 are one-to-one.

65.

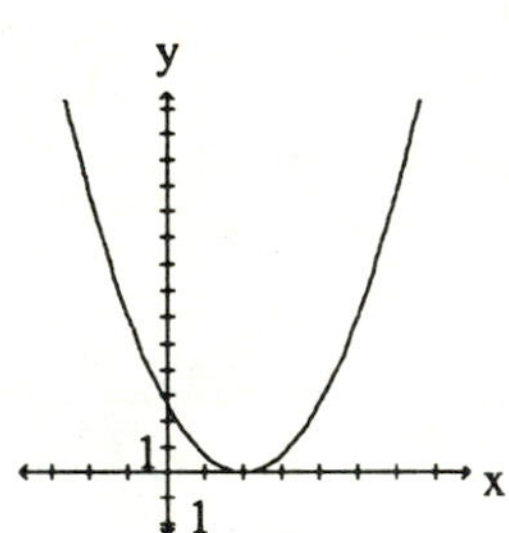

67.

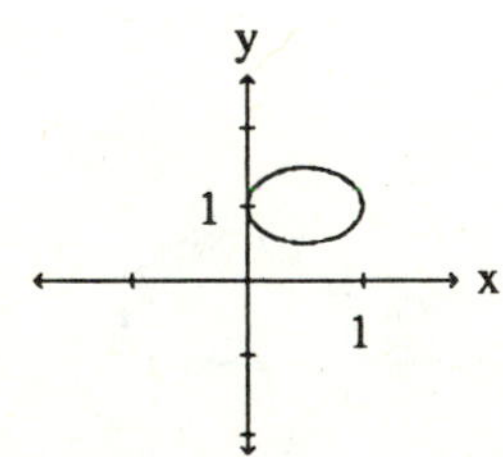

69.

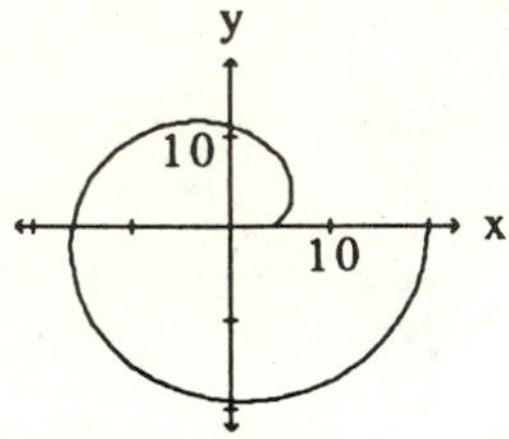

71. $x(t) = 2t \cos t$

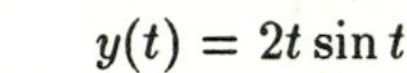
$y(t) = 2t \sin t$

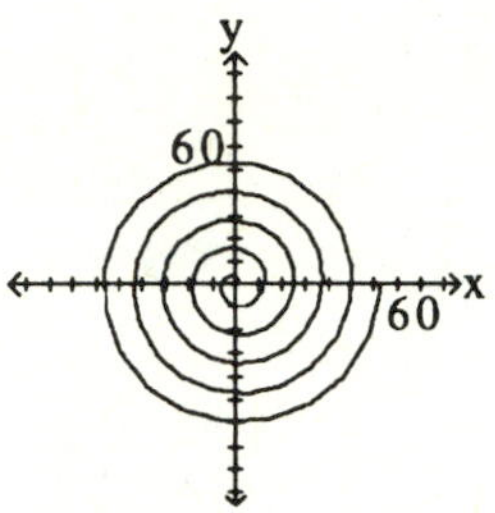

73. Velocity: $v = (90.145, -424.826)$

Speed: 434.28 mph

Direction: 11.98° east of south

75. $x(t) = (60 \cos 35°)t = 49.15t$

$y(t) = -16t^2 + (60 \sin 35°)t = -16t^2 + 34.41t$

77. The portion in the first quadrant.

79. Max. height = 18.5 feet when $t = 1.075$ seconds.

81. Max. height = 91.68 feet. The ball travels 307.73 feet.

83. 187.94 feet down field from the 20 yard line, around the opponents' 20 yard line. Note: this model ignores air

resistance, so the actual distance would be less.

85. 2.05 seconds.

• CHAPTER 8

Exercises 8.1

1. Std. form:
$(x-2)^2 = -8(y-1)$

Vertex: $(2,1)$

Line of Symmetry: $x = 2$

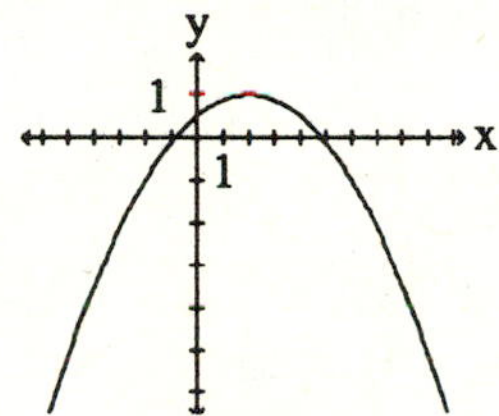

3. Std. form:
$(y+1)^2 = -2(x-\frac{5}{2})$

Vertex: $(\frac{5}{2},-1)$

Line of Symmetry:
$y = -1$

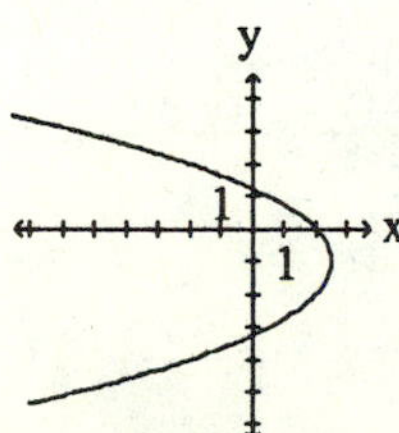

5. Std. form:
$(y-2)^2 = -8(x-3)$

Line of Symmetry: $y = 2$

Focus: $(1,2)$

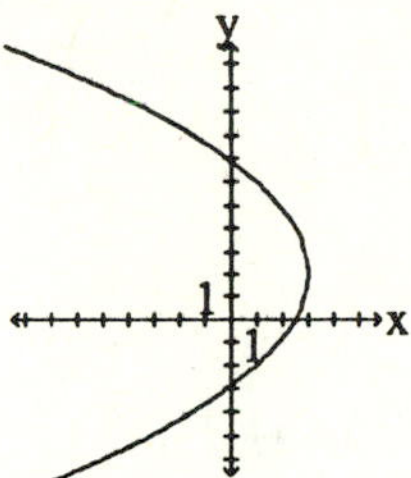

7. Std. form:
$(x-3)^2 = -12(y-2)$
Line of Symmetry: $x = 3$
Focus: $(3,-1)$

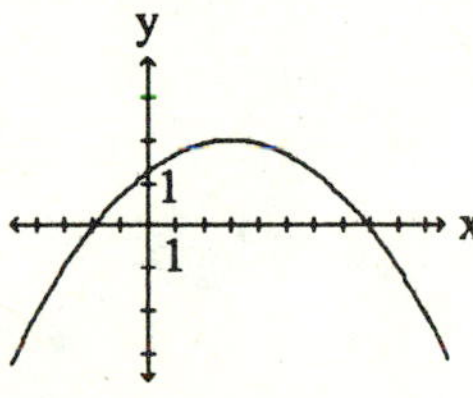

9. Std. form:
$(x+2)^2 = 24(y+6)$
Line of Symmetry:
$x = -2$
Directrix: $y = -12$

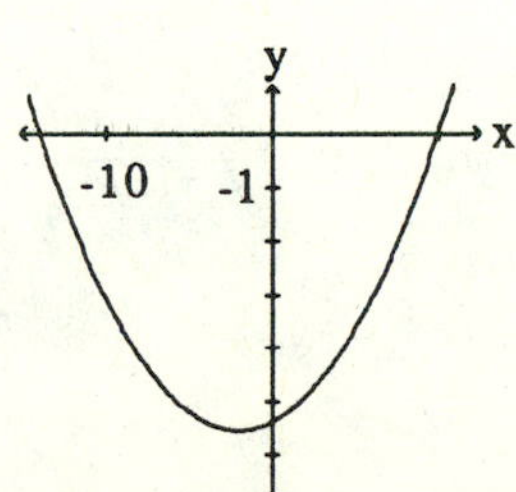

11. Std. form:
$(y-2)^2 = 16(x+1)$
Line of Symmetry: $y = 2$

Directrix: $x = -5$

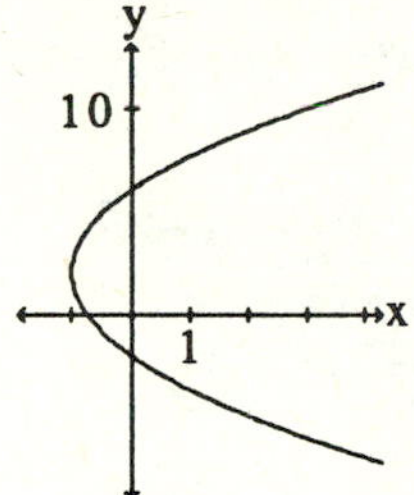

13. A horizontal stretch of factor 12, followed by a vertical shift down 1 unit, then a horizontal shift right 3 units.

15. A vertical shift down 1 unit, followed by a vertical stretch of factor 3, then a horizontal shift right 2 units.

17. Horizontal shrink by $\frac{1}{2}$ yields $y = (2x)^2 = 4x^2$. Vertical stretch by 4 yields $y = 4x^2$.

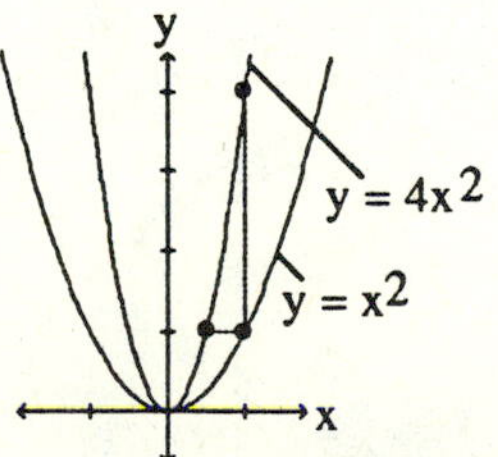

19. Put $y = Ax^2 + Bx + C$ in standard form by completing the squares to get $y = A(x^2 + \frac{B}{A}x + \frac{B^2}{A^2}) + C - \frac{B^2}{4A}$. Simplify to get

$\frac{1}{A}(y + \frac{B^2-4AC}{4A}) = (x + \frac{B}{2A})^2$. So $a = \frac{1}{4A}$, $h = \frac{-B}{2A}$, and $k = \frac{4AC-B^2+1}{4A}$. The focus is at $(h, k+a) = (\frac{-B}{2A}, \frac{4AC-B^2+1}{4A})$.

21. Std. form: $(x+1)^2 = y-2$

    Vertex: $(-1, 2)$

    Focus: $(-1, \frac{9}{4})$

    Directrix: $y = \frac{7}{4}$

    Line of Symmetry: $x = -1$

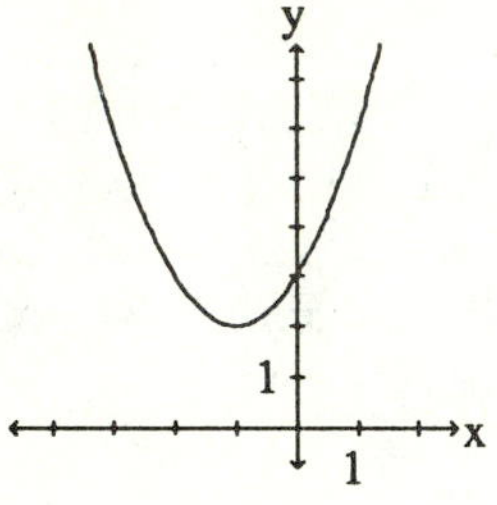

23. Std. form: $(y-1)^2 = -4(x - \frac{13}{4})$

    Vertex: $(\frac{13}{4}, 1)$

    Focus: $(\frac{9}{4}, 1)$

    Directrix: $x = \frac{17}{4}$

    Line of Symmetry: $y = 1$

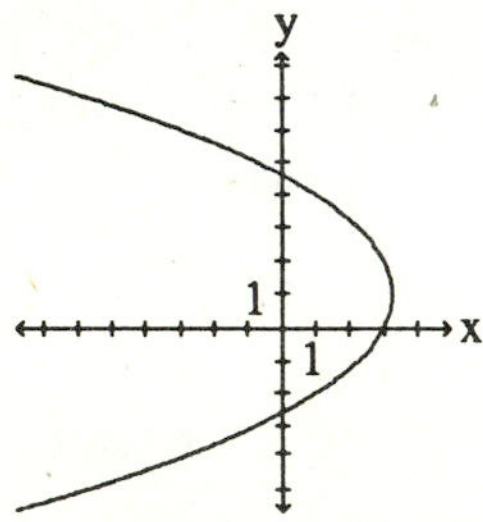

25. At the focus $(0, \frac{5}{2})$
27. At the focus $(3, 0)$
29. The equation is $x^2 = 40y$. Assume that the vertex is at $(0,0)$, the focus is $(0, 10)$, the directrix is $y = -10$ and the line of symmetry is $x = 0$.
31. No points of intersection.
33. $y = \frac{4}{3}x^2 + \frac{8}{3}$
35. $(-6, 4\sqrt{3})$ and $(-6, -4\sqrt{3})$
37.

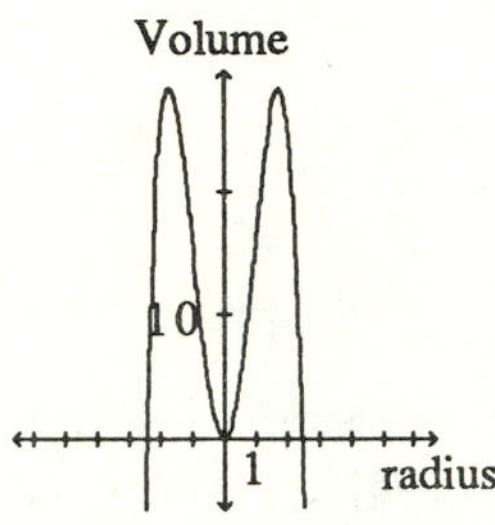

39. Radius 1.73, height 3.01, volume 28.27 cubic units.
41. a) $(\frac{3}{4}, \frac{9}{8})$

    b) $y = \frac{4}{3}x + \frac{1}{8}$

    c) 18.43°

    d) 18.43°

Exercises 8.2

1. Foci: $(\frac{-3\sqrt{7}}{2}, 0)$ and $(\frac{3\sqrt{7}}{2}, 0)$

   Std. form: $\frac{x^2}{36} + \frac{y^2}{9/4} = 1$

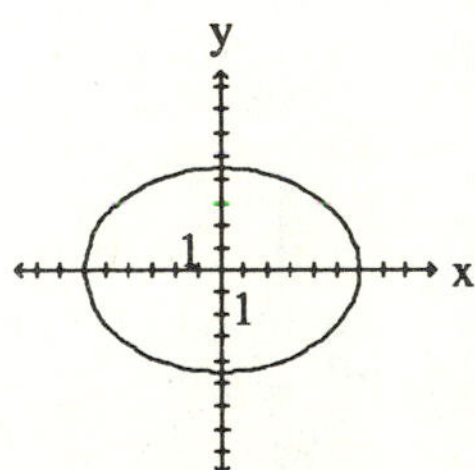

3. Endpts of major axis: $(-\sqrt{58}, 0)$ and $(\sqrt{58}, 0)$

   Endpts of minor axis $(0, -7)$ and $(0, 7)$

   Std. form: $\frac{x^2}{58} + \frac{y^2}{49} = 1$

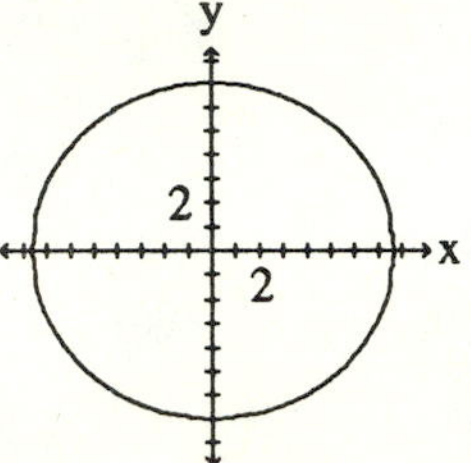

5. Endpts of major axis: $(-9/2, 0)$ and $(9/2, 0)$

   Endpoints of minor axis: $(0, -\frac{\sqrt{17}}{2})$ and $(0, \frac{\sqrt{17}}{2})$

   Std. form: $\frac{x^2}{81/4} + \frac{y^2}{17/4} = 1$

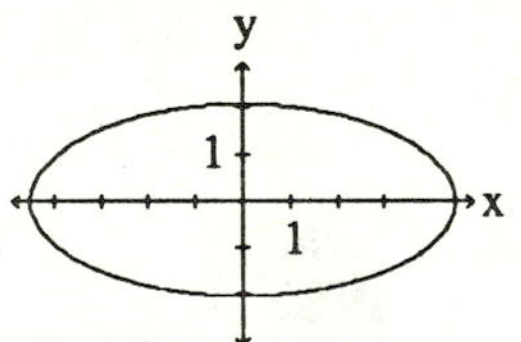

7. Endpts of major axis: $(1, 0)$ and $(1, 12)$

   Endpts of minor axis: $(1 - 3\sqrt{3}, 6)$ and $(1 + 3\sqrt{3}, 6)$

   Std. form: $\frac{(x-1)^2}{27} + \frac{(y-6)^2}{36} = 1$

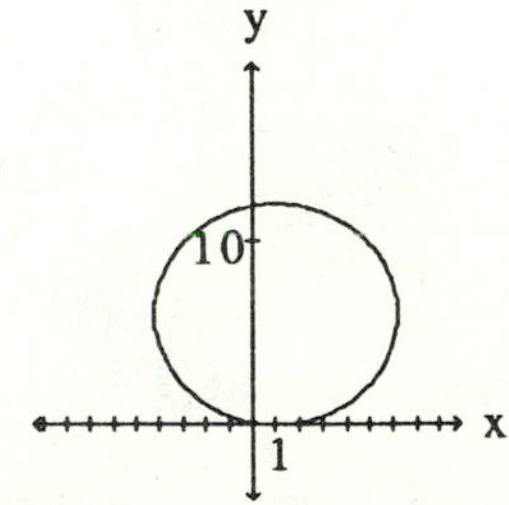

9. Endpts of major axis: $(1, -5)$ and $(1, -1)$

   Endpts of minor axis: $(1 - \sqrt{2}, -3)$ and $(1 + \sqrt{2}, -3)$

   Center: $(1, -3)$

   Foci: $(1, -3 - \sqrt{2})$ and $(1, -3 + \sqrt{2})$

   Lines of symmetry: $x = 1$ and $y = -3$

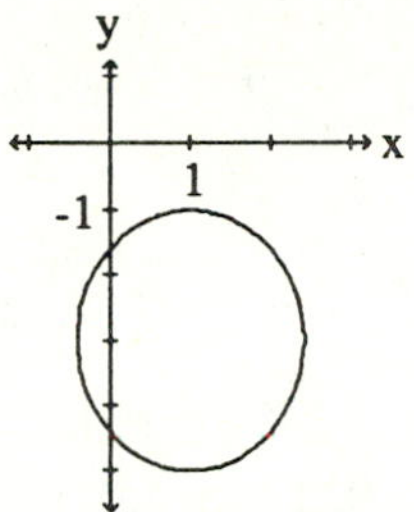

11. Endpts of major axis: $(-2 - \sqrt{5}, 1)$ and $(-2 + \sqrt{5}, 1)$

    Endpts of minor axis: $(-2, \frac{2-\sqrt{2}}{2})$ and $(-2, \frac{2+\sqrt{2}}{2})$

    Center: $(-2, 1)$

    Foci: $(\frac{-4-3\sqrt{2}}{2}, 1)$ and $(\frac{-4+3\sqrt{2}}{2}, 1)$

    Lines of symmetry: $x = -2$ and $y = 1$

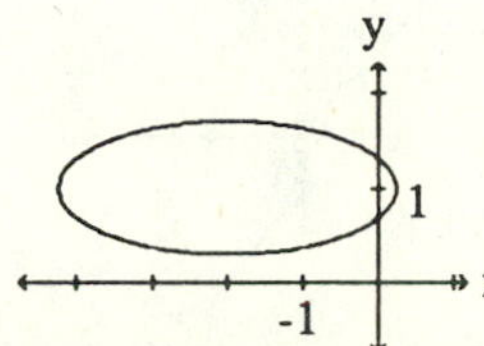

13.

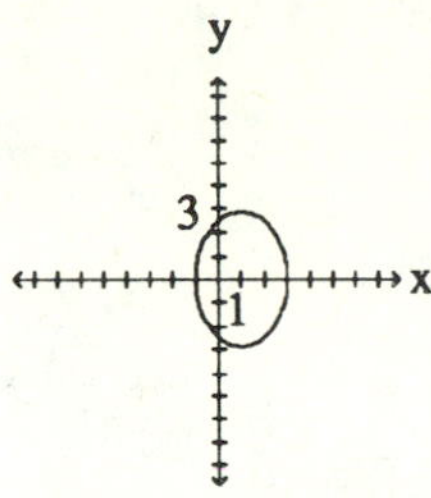

15. Std. form: $\frac{(x-1)^2}{4} + \frac{(y+1)^2}{9} = 1$

    Endpts of major axis: $(1, -4)$ and $(1, 2)$

    Endpts of minor axis: $(-1, -1)$ and $(3, -1)$

    Center: $(1, -1)$

    Foci: $(1, -1 - \sqrt{5})$ and $(1, -1 + \sqrt{5})$

    Lines of symmetry: $x = 1$ and $y = -1$

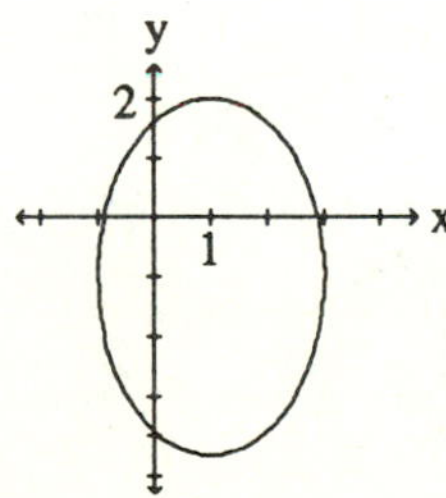

17. Std. form: $\frac{(x-4)^2}{1} + \frac{(y+8)^2}{4} = 1$

    Endpts of major axis: $(4, -10)$ and $(4, -6)$

    Endpts of minor axis: $(3, -8)$ and $(5, -8)$

    Center: $(4, -8)$

    Foci: $(4, -8 - \sqrt{3})$ and $(4, -8 + \sqrt{3})$

    Lines of symmetry: $x = 4$ and $y = -8$

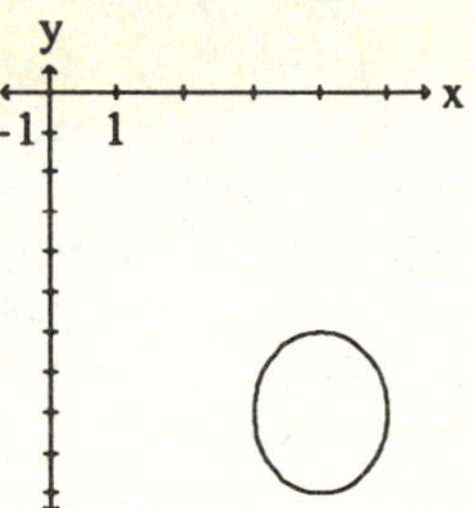

19. $5x^2 + y^2 + 5 = 0$ simplifies to $5x^2 + y^2 = -5$ which is impossible since $5x^2$ and y^2 are both positive. So this equation cannot be graphed. Also note that, if we convert to standard form, we get $\frac{x^2}{-1} + \frac{y^2}{-5} = 1$ but a^2 and b^2 must be positive so this is not an ellipse.

21. (a) The pool shark knows that any object passing through one focus of an ellipse will be reflected through the other focus (the pocket, in this case.)

    (b) Any ball passing through the other focus will bounce off the cushion into the pocket, if another ball is not in the way, no matter how it is hit. If we denote the center of the table $(0, 0)$, the equation for the elliptical cushion is $\frac{x^2}{9} + \frac{y^2}{4} = 1$, the pocket is at

$(-\sqrt{5}, 0)$, and the ball should pass through $(\sqrt{5}, 0)$.

23. If the center of the rotunda is $(0,0)$, the foci are at $(-20\sqrt{91}, 0)$ and $(20\sqrt{91}, 0)$. The senators stood along the major axis about 190.8 feet on either side of the center.

25. $e = 0.58$

27. $\frac{x^2}{16} + \frac{y^2}{15.84} = 1$

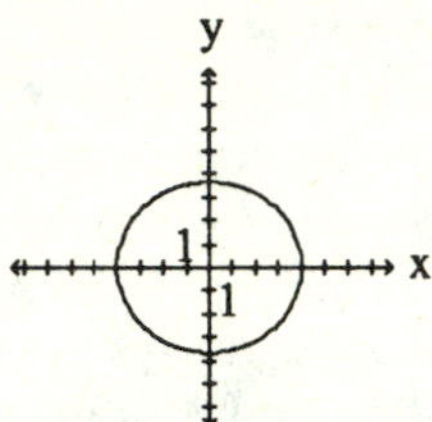

29. Since $a \geq b > 0$, we have $e = \frac{\sqrt{a^2-b^2}}{a} \geq 0$ because $a > 0$ and $\sqrt{a^2 - b^2} > 0$. Also $b = \sqrt{a^2(1-e^2)} > 0$ (from Exercise 10.2.26) and $a^2 > 0$, so $1 - e^2 > 0$. Thus, $e^2 < 1$ and $e < 1$. So we have $0 \leq e < 1$.

31. As e approaches 1, the distance from the focus to the center approaches the length of the major axis so the ellipse becomes more elongated.

33. $e = \frac{c}{a} = \frac{0}{a} = 0$ for any circle.

35. $\frac{x^2}{2.56\times10^{14}} + \frac{y^2}{2.15\times10^{14}} = 1$

37. Earth: $\frac{x^2}{8.649\times10^{15}} + \frac{y^2}{8.646\times10^{15}} = 1$

Mars: $\frac{x^2}{1.96\times10^{16}} + \frac{y^2}{1.94\times10^{16}} = 1$

Saturn: $\frac{x^2}{7.92\times10^{17}} + \frac{y^2}{5.94\times10^{17}} = 1$

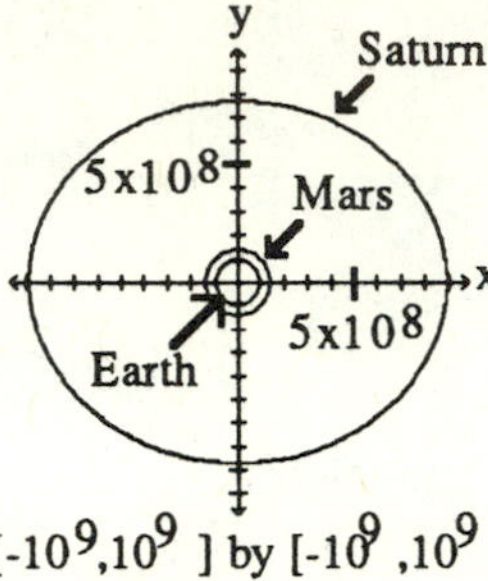

$[-10^9, 10^9]$ by $[-10^9, 10^9]$

Exercises 8.3

1. Std. form: $\frac{x^2}{4} - \frac{y^2}{5} = 1$

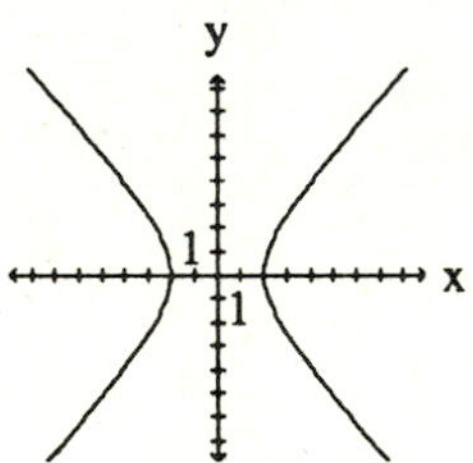

3. Std. form: $\frac{4x^2}{9} - \frac{4(y-2)^2}{27} = 1$

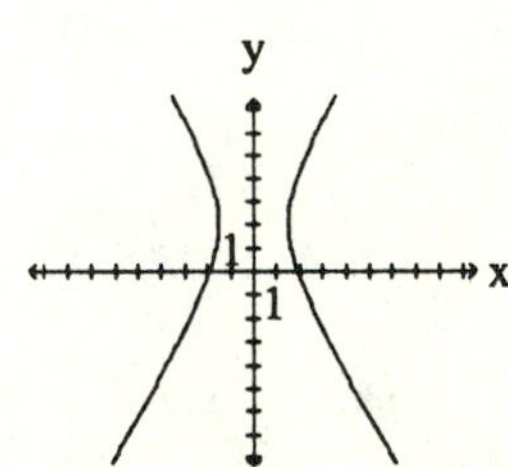

5. Center: $(0,3)$

Foci: $(3,3), (-3,3)$

Vertices: $(2,3), (-2,3)$

Lines of symmetry: $x = 0$, $y = 3$

Asymptotes: $y = \pm\frac{\sqrt{5}}{2}x + 3$

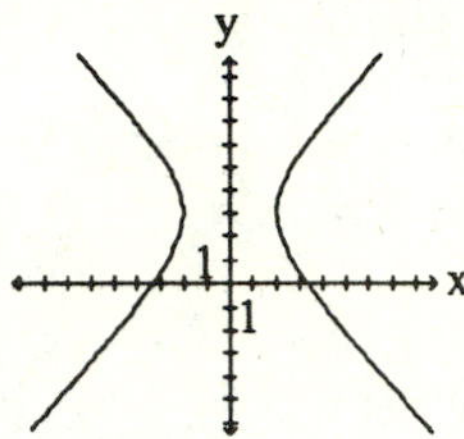

7. Center: $(3,1)$

Foci: $(3, 1+\sqrt{13}), (3, 1-\sqrt{13})$

Vertices: $(3,4), (3,-2)$

Lines of symmetry: $x = 3$, $y = 1$

Asymptotes: $y - 1 = \pm\frac{3}{2}(x-3)$

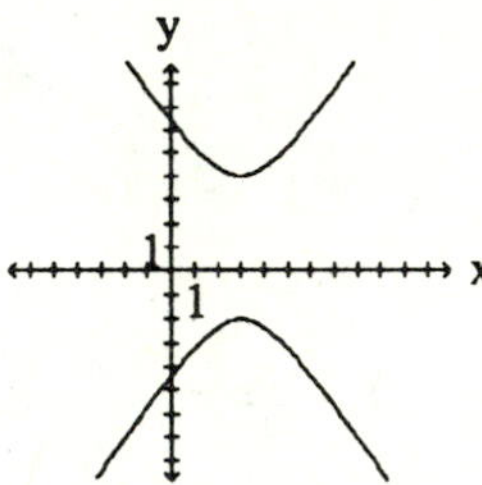

9.

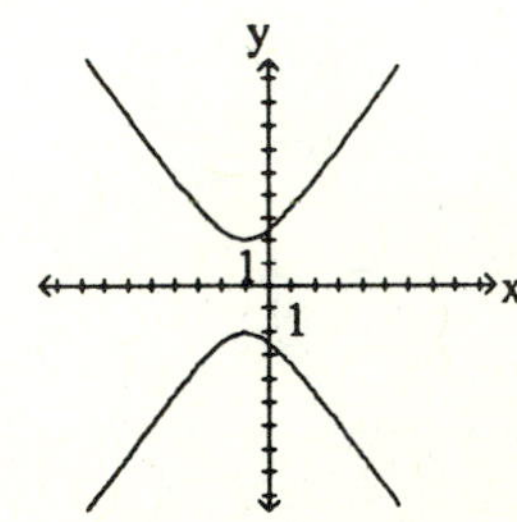

11. Hyperbola

Std. form: $\frac{(x-2)^2}{4} - \frac{(y-1)^2}{9} = 1$

Center: $(2, 1)$

Foci: $(2 + \sqrt{13}, 1), (2 - \sqrt{13}, 1)$

Vertices: $(0, 1), (4, 1)$

Lines of symmetry: $x = 2$, $y = 1$

Asymptotes: $y - 1 = \pm\frac{3}{2}(x - 2)$

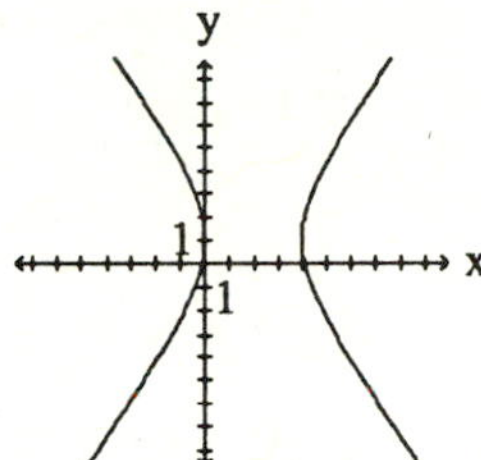

13. Hyperbola

Std. form: $\frac{(y-1)^2}{9} - \frac{(x+3)^2}{25} = 1$

Center: $(-3, 1)$

Foci: $(-3, 1 + \sqrt{34}), (-3, 1 - \sqrt{34})$

Vertices: $(-3, 4), (-3, -2)$

Lines of symmetry: $x = -3$, $y = 1$

Asymptotes: $y - 1 = \pm\frac{3}{5}(x + 3)$

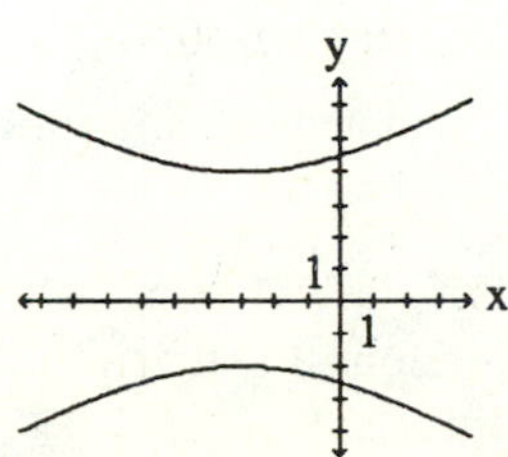

15. Hyperbola

Std. form: $\frac{(y-1)^2}{9} - \frac{(x+1)^2}{4} = 1$

Center: $(-1, 1)$

Foci: $(-1, 1 + \sqrt{13}), (-1, 1 - \sqrt{13})$

Vertices: $(-1, 4), (-1, -2)$

Lines of symmetry: $x = -1$, $y = 1$

Asymptotes: $y - 1 = \pm\frac{3}{2}(x + 1)$

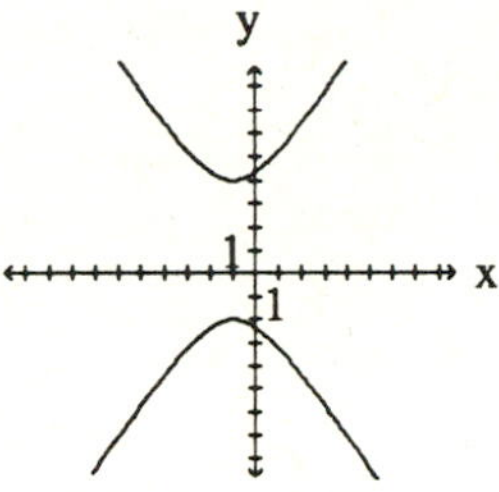

17. Parabola

Std. form: $(x - \frac{3}{2})^2 = -\frac{5}{2}(y - \frac{7}{2})$

Center: $(\frac{3}{2}, \frac{7}{2})$

Focus: $(\frac{3}{2}, \frac{23}{8})$

Line of symmetry: $x = \frac{3}{2}$

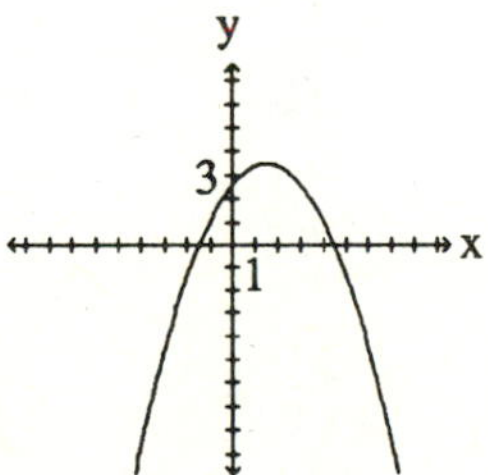

19. $e = \sqrt{5}$

21. Std. form: $\frac{x^2}{16} - \frac{y^2}{48} = 1$

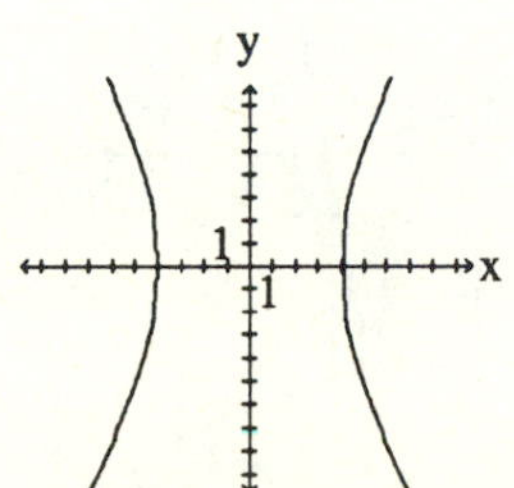

23. Std. form: $\frac{x^2}{16} - \frac{y^2}{159{,}984} = 1$

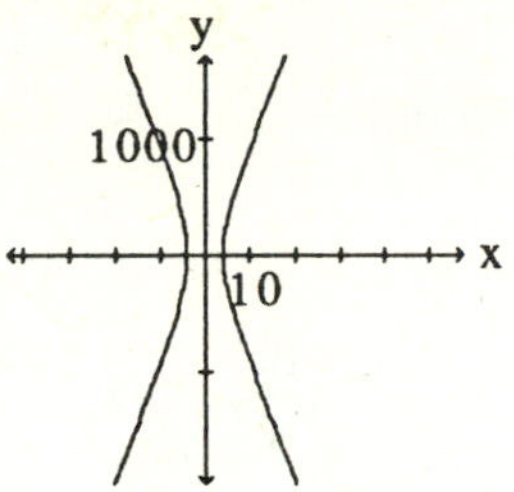

25. Eccentricity measures the ratio of the distance from the foci to the center versus the distance from the vertices to the center.

27. Std. form: $\frac{x^2}{8.649 \times 10^{15}} - \frac{y^2}{3.450951 \times 10^{18}} = 1$

29. Foci: $(6, 0)$, $(0, 0)$

31. Light reflects from the parabolic surface toward the focus, strikes the hyperbolic surface, and reflects back to the eye of the viewer (focus).

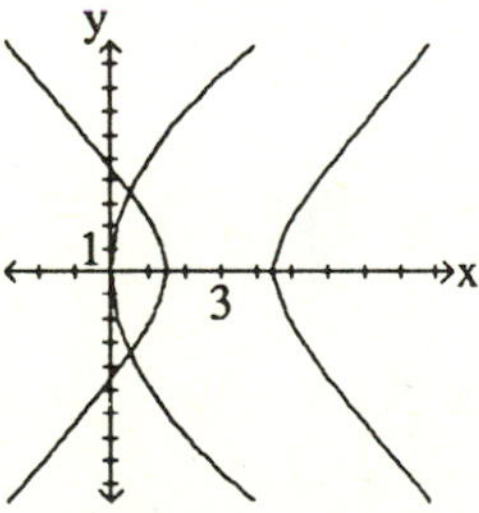

33. $\frac{(x-1.875)^2}{1.265625} - \frac{y^2}{2.25} = 1$

35. Std. form: $\frac{x^2}{2.5 \times 10^{13}} - \frac{y^2}{1.19 \times 10^{14}} = 1$

37. The graph would be one branch of a hyperbola.

Exercises 8.4

1. Ellipse

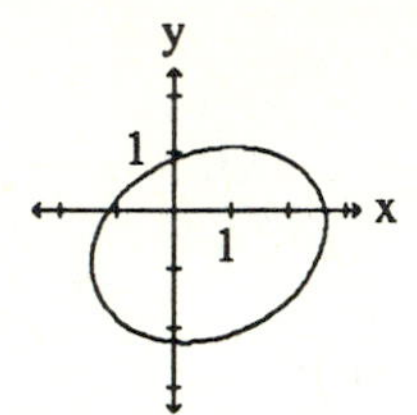

3. Ellipse

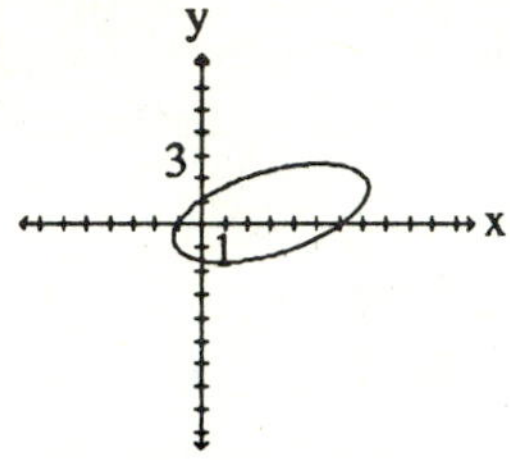

5. Degenerate conic; no graph

7. Hyperbola

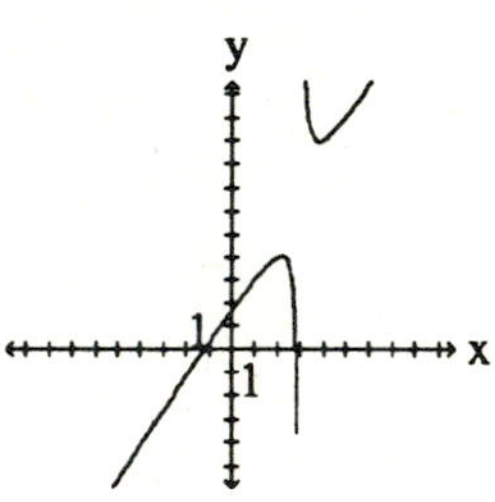

9. Hyperbola

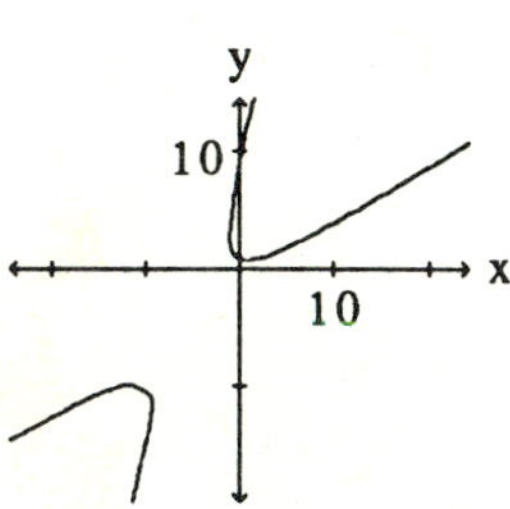

11. Major axis: $(1.87, -0.22)$ $(-3.30, -2.37)$

Minor axis: $(0.23, -2.98)$ $(-1.55, 0.82)$

Center: $(-0.59, -1.24)$

Line of Symmetry:
$y = 0.41x - 1$
$y = -2.41x - 2.5$

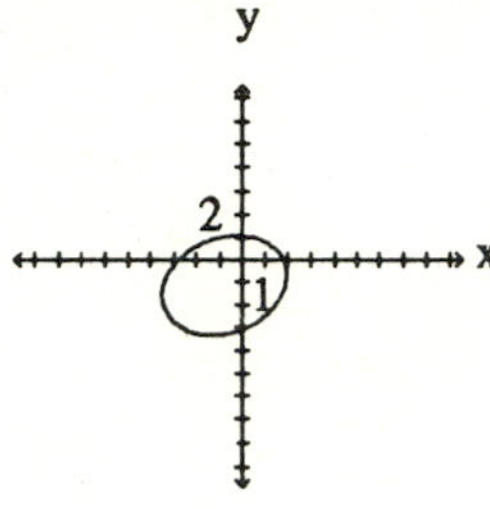

13. Center: $(0.83, 1.06)$

Asymptotes:
$y = 3.3x - 1.5$
$y = 0.58x + 0.6$

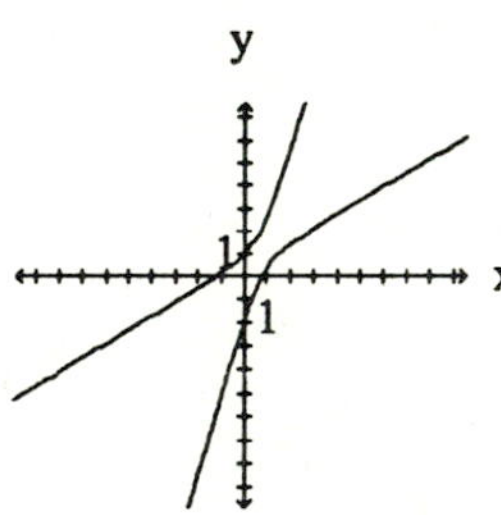

15. $y = 4.6x - 6$;

$y = -0.36x - 1.1$

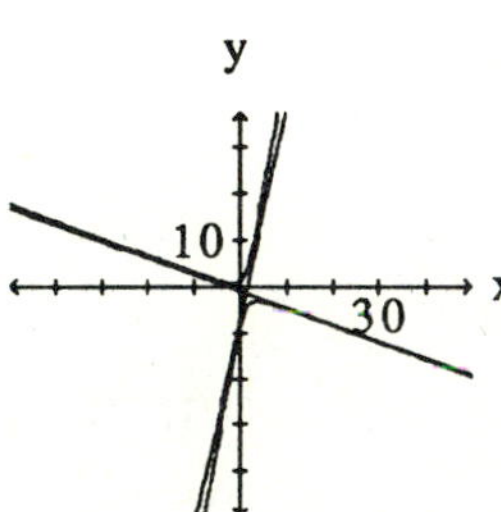

17. $y = 0.17x$;

$y = -6.8x - 5$

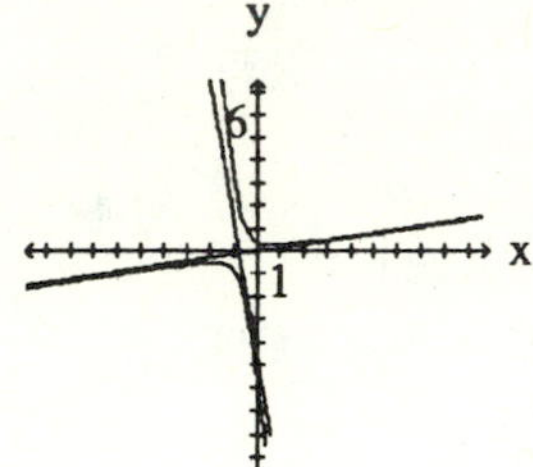

19.

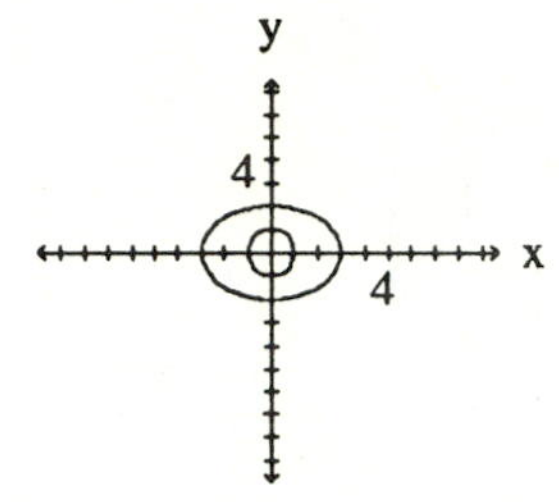

21. Vertical shift down 2 units, horizontal shift right 1 unit applied to $\frac{x^2}{9} + \frac{y^2}{4} = 1$

23. Rotation of 45° applied to $\frac{x^2}{4} - \frac{y^2}{4} = 1$

25. $20x^2 - 16xy + 32y^2 + 24x - 96y + 36 = 0$

27. $\frac{x^2}{8.978} + \frac{y^2}{3.9784} = 1$

29. $\theta = 45°$

31. Major axis: 8.78

Minor axis: 3.49

Center: $(2.92, 0.37)$

Transformations applied to

$\frac{x^2}{4.39^2} + \frac{y^2}{1.74^2} = 1$:

Horizontal shift right 2.92 units

Vertical shift up 0.37 units

Rotation of 16.85°

33. Final acidity percentage $= \frac{0.56(74)+x}{74+x} \times 100$

35. Domain: $(-\infty, -74) \cup (-74, \infty)$

    Range: $(-\infty, 100) \cup (100, \infty)$

37. Intersection of $y = \frac{0.56(74)+x}{74+x}(100)$ and $y = 8$

    Solution: $x = 96.37$ oz. acid

39. Hyperbola

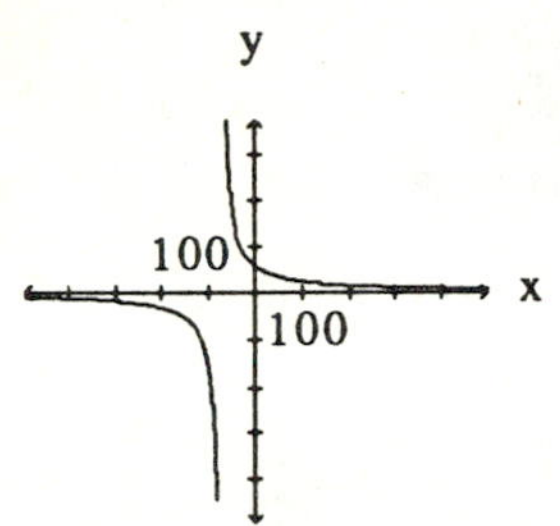

41. Positive values of x make sense for this problem

43. $B \neq 0$ and $A = C$; 45° rotation

Exercises 8.5

1. $(-1.48, 1.74)$ & $(2.92, -0.46)$

3. $(-3.10, 3.55)$ & $(2.10, 0.95)$

5. $(-2.20, 1.36)$ & $(2.20, 1.36)$ & $(-2.20, -1.36)$ & $(2.20, -1.36)$

7. $(-1.36, 0.70)$, $(1.36, 0.70)$, $(-1.03, -0.89)$, &$(1.03, -0.89)$

9. $(-2.32, 0.58)$ & $(-2.32, -0.58)$ & $(4.32, 1.91)$ & $(4.32, -1.91)$

11. $(0.21, 1.87)$ & $(1.87, 0.21)$ & $(-0.53, -2.75)$ & $(-2.75, 0.53)$

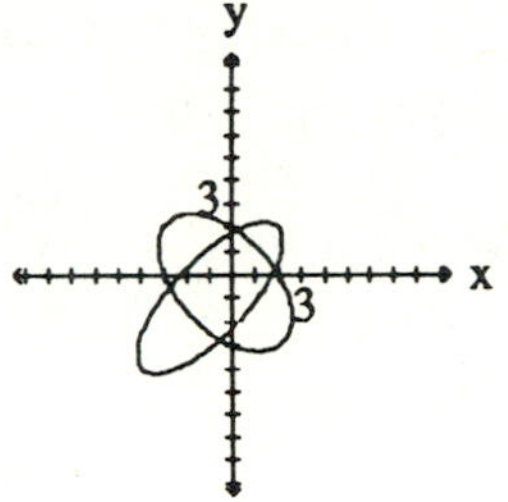

13. $(0.15, 1.84)$ & $(1.63, -0.24)$ & $(-2.08, 0.19)$ & $(-0.37, -2.61)$

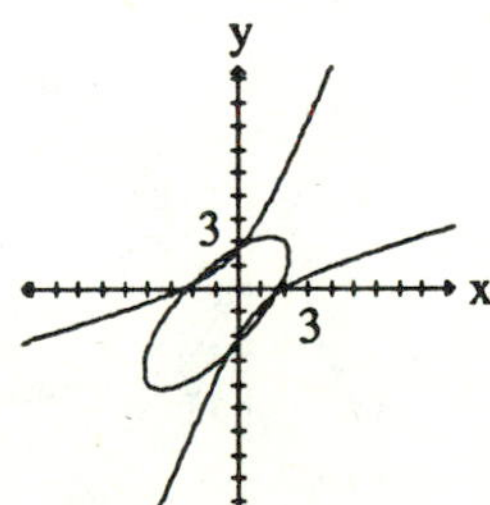

15.

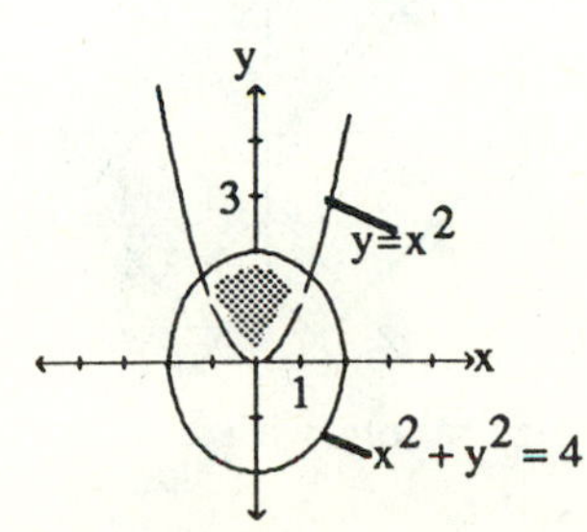

17.

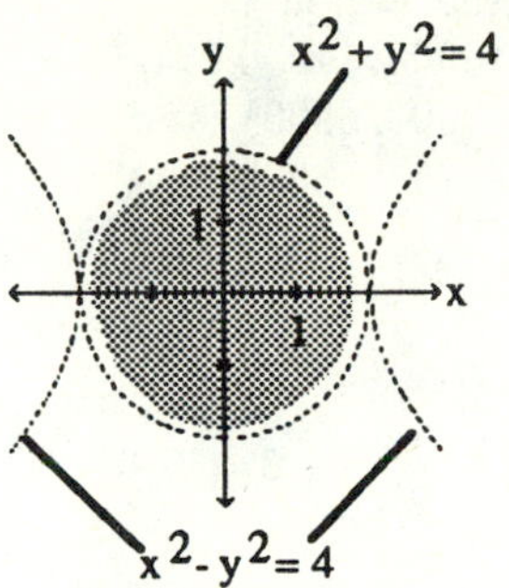

19. The shaded part of the graph is the area intersected by the right branch of the hyperbola and the ellipse.

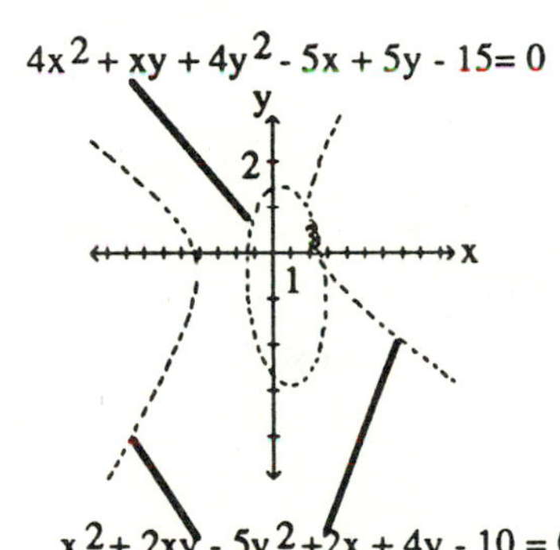

21. $f(x) = \frac{0.58(84)+x}{84+x}$

    $g(x) \geq 0.7$

    $h(x) \leq 0.8$

23. $x \geq 0$, $y \geq 0$

25. $\frac{(x-1000)^2}{346.5^2} - \frac{y^2}{879{,}937.75} = 1$

    $\frac{(y-3000)^2}{2233^2} - \frac{x^2}{4{,}013{,}711} = 1$

27. Closest solution is $(3953.20, 7939.73)$. Answers will vary widely due to roundoff error. The explosions are at about $(3950, 7950)$.

29. Vertices that define the region are $(3000, 14230.77)$, $(3000, 8600)$, and $(17640, 8600)$.

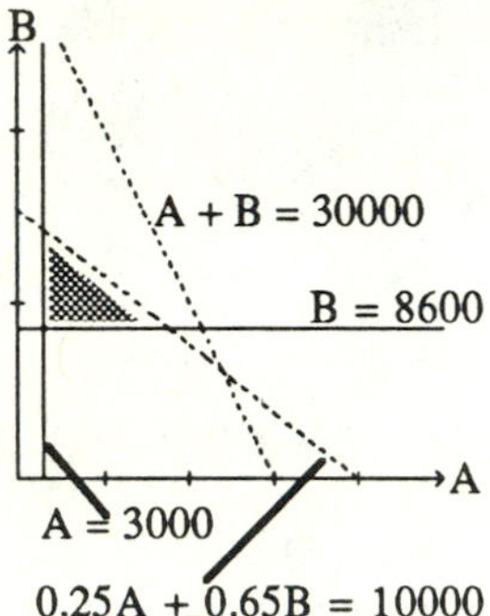

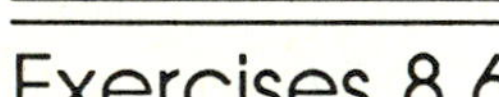

Exercises 8.6

1. $e = \frac{\sqrt{3}}{2}$

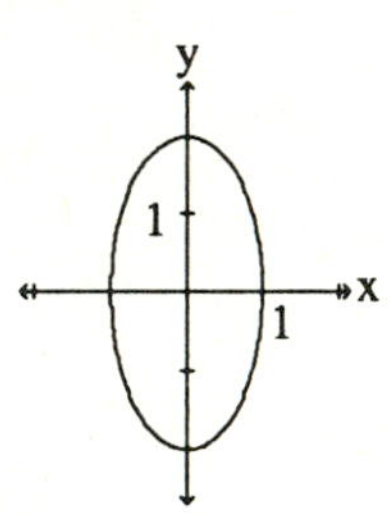

3. $e = \frac{5}{4}$

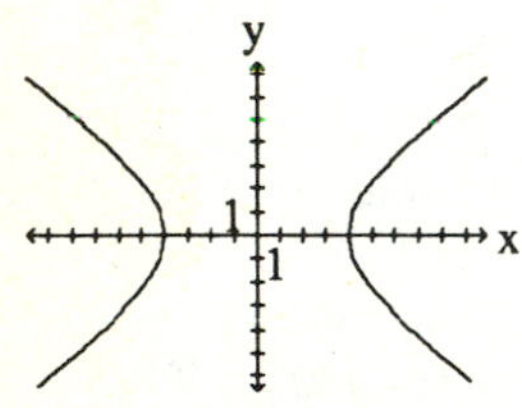

5. $e = \frac{\sqrt{15}}{3}$

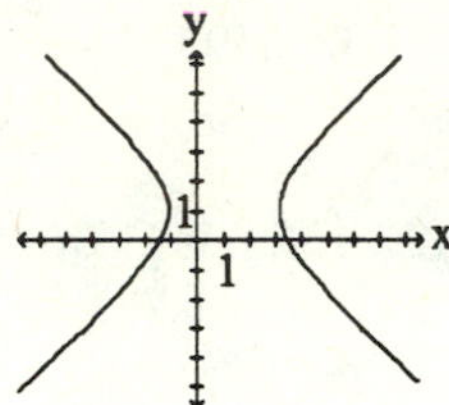

7. Center: $(1, 1)$
Transverse length: 4.5
Asym: $y = 1 \pm \frac{\sqrt{6}}{3}(x - 1)$

9. $e = 1.15$ (Use estimates of $a = 5.72$, $b = 3.24$)

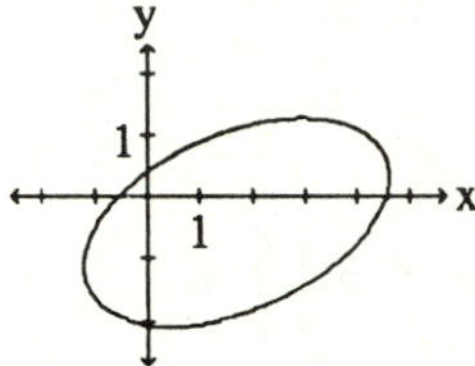

11.

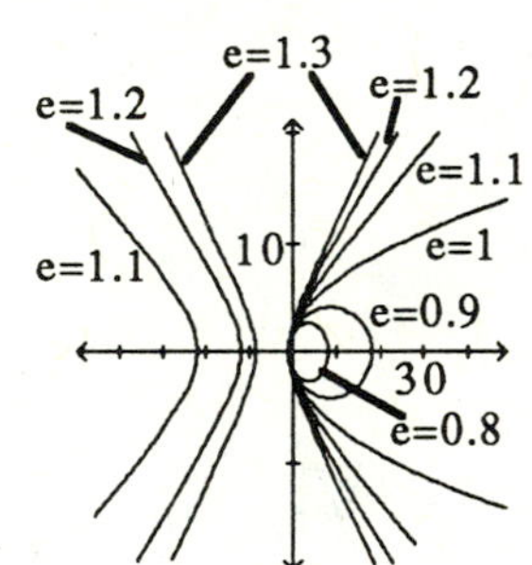

13.

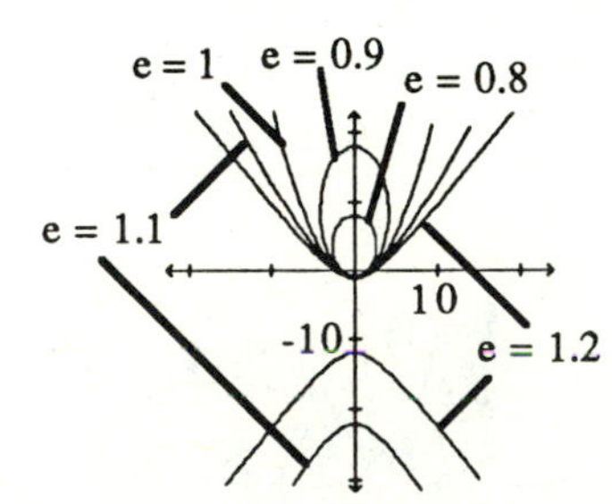

15. Parabola with $e = 1$, Directrix $x = -5$,

$0 \leq \theta \leq 2\pi$

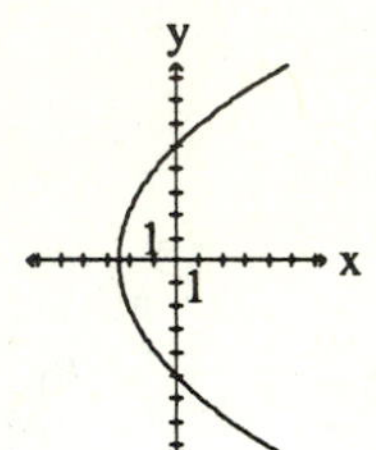

17. Ellipse with $e = \frac{1}{2}$,

Directrix $x = -\frac{3}{2}$,

$0 \leq \theta \leq 2\pi$

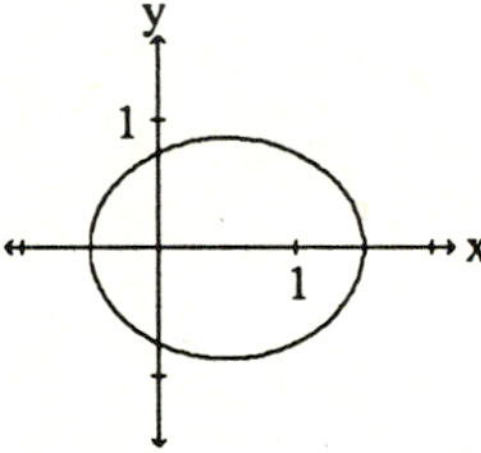

19. Ellipse with $e = \frac{1}{2}$,

Directrix $x = \frac{8}{3}$,

$0 \leq \theta \leq 2\pi$

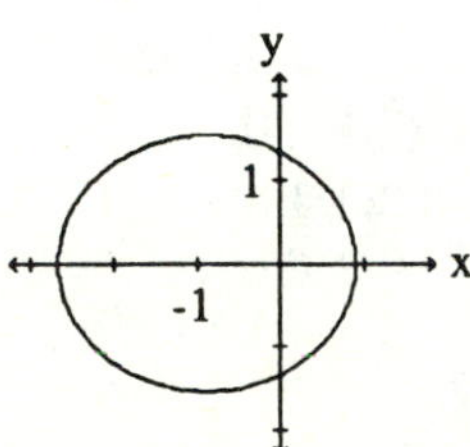

21. Ellipse with $e = \frac{3}{4}$

Std form: $\frac{(x+2.29)^2}{(6.09)^2} + \frac{y^2}{(5.64)^2} = 1$

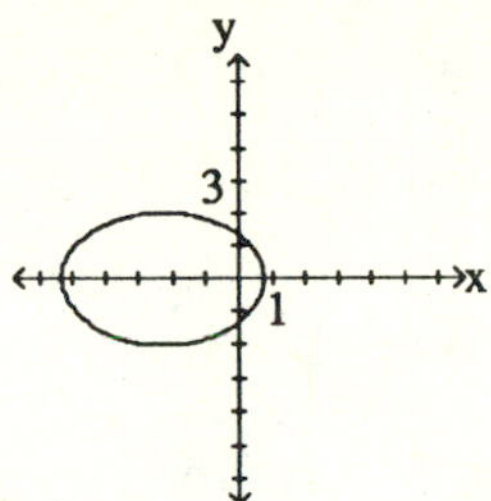

23. $e_1 = e_2 = \frac{1}{2}$. The graphs coincide.

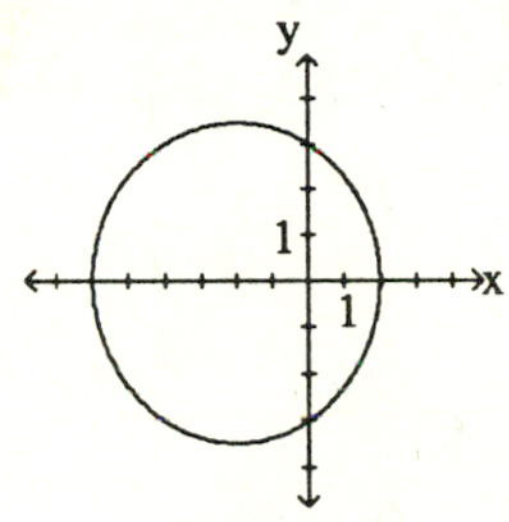

25. Yes, a hyperbola since $e > 1$

27. $e = \frac{1}{2}$. The equations in Exercises 27 and 29 yield the same graph.

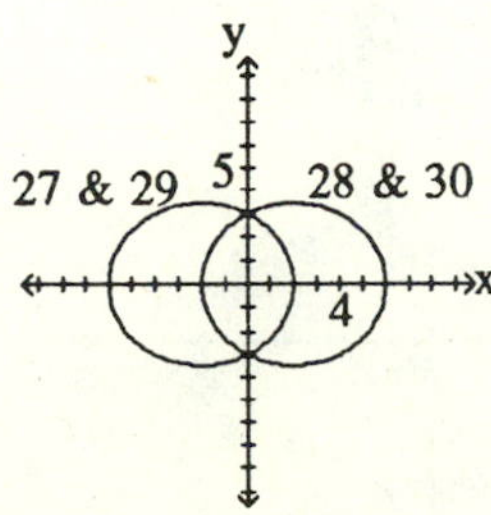

29. see 27.

31.

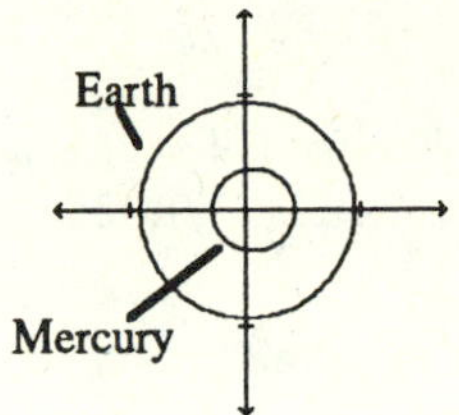

$[-2 \times 10^8, 2 \times 10^8]$ by $[-2 \times 10^8, 2 \times 10^8]$

33.

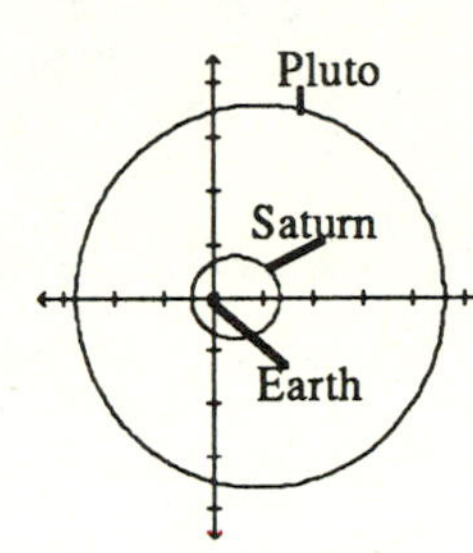

$[-4 \times 10^9, 6 \times 10^9]$ by $[-5 \times 10^9, 5 \times 10^9]$

35.

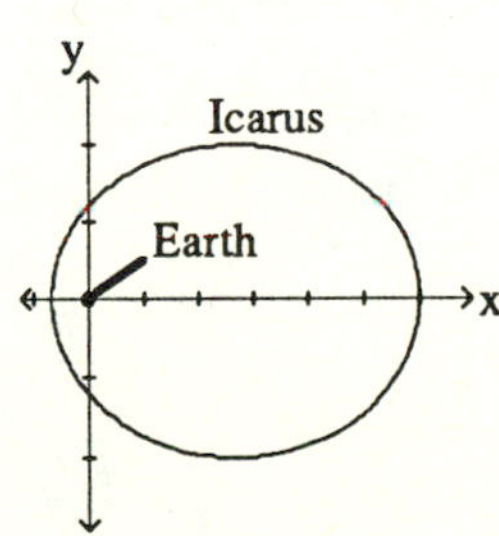

$[-1.25 \times 10^9, 7 \times 10^9]$ by $[-3 \times 10^9, 3 \times 10^9]$

37. They are similar ellipses.

39. $r = \frac{eh}{-1+e\cos\theta} = -\left(\frac{eh}{1-e\cos\theta}\right)$. The type of conic is determined by e; the sign of the directrix is reversed.

Review Exercises

Exercises

1. Std. form: $(x+2)^2 = 8(y+1)$
   Vertex: $(-2, -1)$
   Line of symmetry: $x = -2$

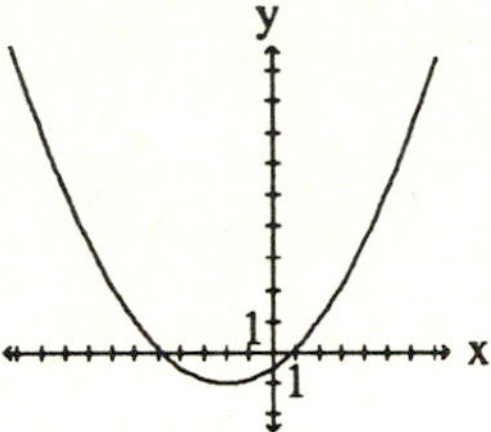

3. Std. form: $(y-4)^2 = -24(x-1)$
   Focus: $(-5, 4)$
   Line of symmetry: $y = 4$

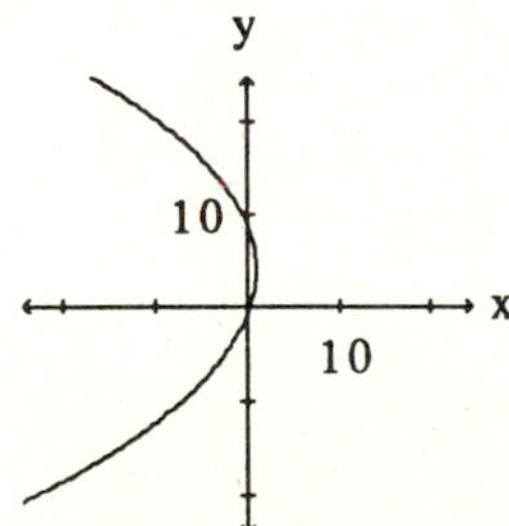

5. Horizontal shift right 3 units, followed by a vertical shrink by a factor of $\frac{1}{6}$, then a vertical shift up 1 unit.

7. Solve for y as a function of x: $y = 2 + \sqrt{x+2}$ or $y = 2 - \sqrt{x+2}$. Then graph each function on the same coordinate axis.

9. Std. form: $(y-1)^2 = 2(x - \frac{7}{6})$

Vertex: $(\frac{7}{6}, 1)$
Focus: $(\frac{5}{3}, 1)$
Directrix: $x = \frac{2}{3}$
Line of symmetry: $y = 1$

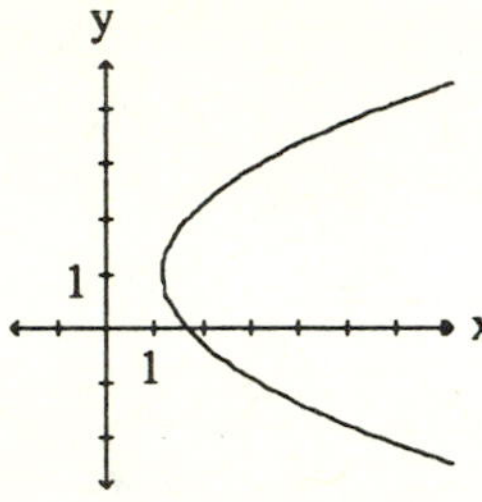

11. $y = \frac{4}{21}x^2 + \frac{68}{21}$

13. Std. form: $\frac{x^2}{64} + \frac{y^2}{39} = 1$
Endpts. of major axis: $(-8, 0)$, $(8, 0)$
Endpts. of minor axis: $(0, -\sqrt{39})$,$(0, \sqrt{39})$

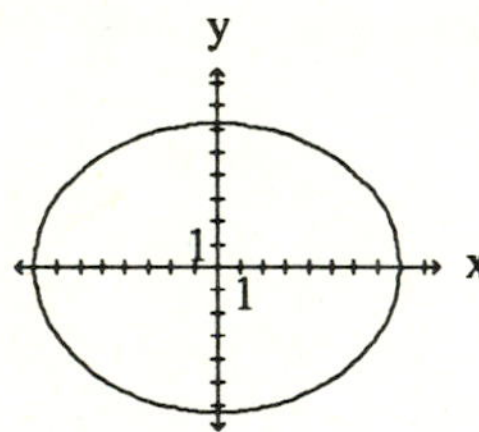

15. Std. form: $\frac{(x-2)^2}{27} + \frac{(y-6)^2}{36} = 1$
Endpts. of major axis: $(2, 0)$, $(2, 12)$
Endpts. of minor axis: $(2 - 3\sqrt{3}, 6)$,$(2 + 3\sqrt{3}, 6)$

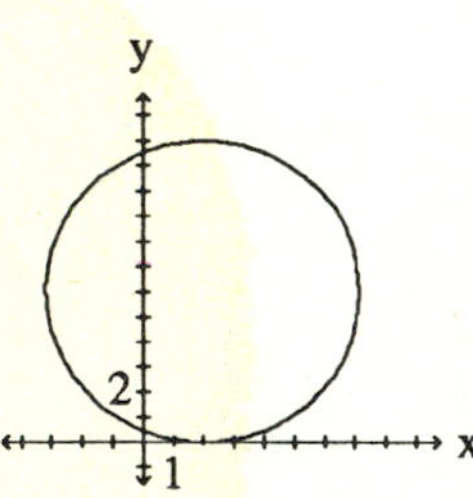

17. Endpts. of major axis: $(-3, -2)$, $(5, -2)$

Endpts. of minor axis: $(1, -2 - \sqrt{2})$, $(1, -2 + \sqrt{2})$

Center: $(1, -2)$

Foci: $(1 - \sqrt{14}, -2)$, $(1 + \sqrt{14}, -2)$

Lines of symmetry: $x = 1$, $y = -2$

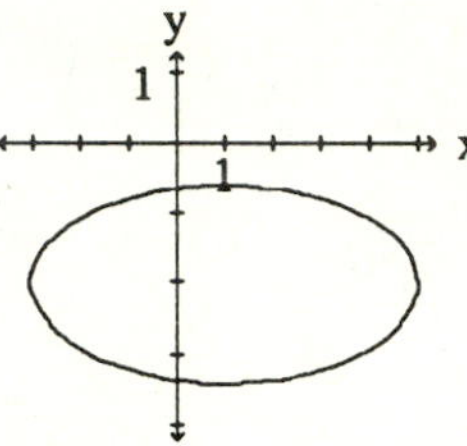

19. Std. form: $\frac{(x-3)^2}{5} + \frac{(y+1)^2}{4} = 1$

Endpts. of major axis: $(3 - \sqrt{5}, -1)$, $(3 + \sqrt{5}, -1)$

Endpts. of minor axis: $(3, -3)$, $(3, 1)$

Center: $(3, -1)$

Foci: $(2, -1)$, $(4, -1)$

Lines of symmetry: $x = 3$, $y = -1$

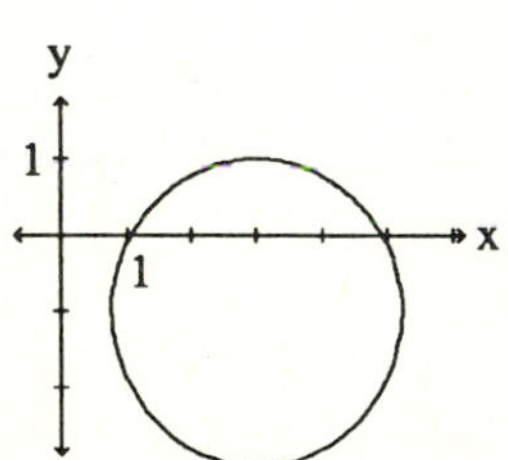

21. Std. form: $\frac{(y-4)^2}{25} - \frac{x^2}{11} = 1$

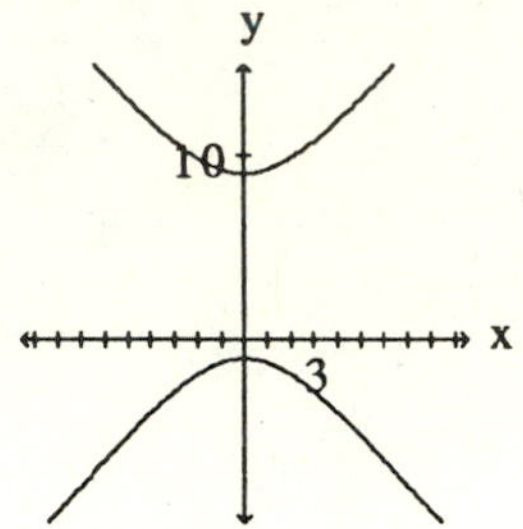

23. Center: $(-1, 3)$
Foci: $(-1, 3 - \sqrt{8})$, $(-1, 3 + \sqrt{8})$
Vertices: $(-1, 3 - \sqrt{5})$, $(-1, 3 + \sqrt{5})$
Lines of symmetry: $x = -1$, $y = 3$
Asymptotes: $y = 3 \pm \sqrt{\frac{5}{3}}(x + 1)$

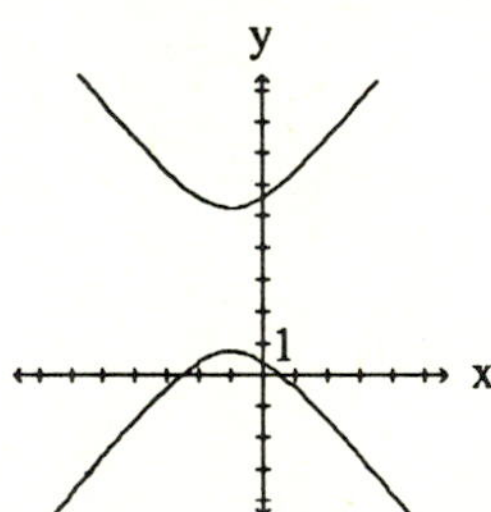

25. Parabola: $(y + 3)^2 = x$
Focus: $(\frac{1}{4}, -3)$
Vertex: $(0, -3)$
Line of symmetry: $y = -3$

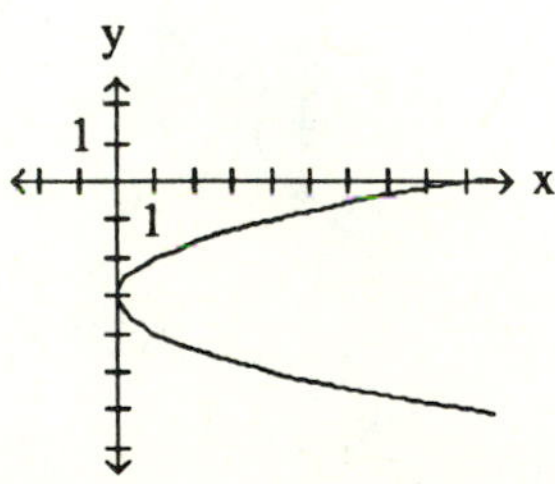

27. Hyperbola: $\frac{(y+3)^2}{\frac{1}{4}} - \frac{x^2}{\frac{1}{9}} = 1$
Center: $(0, -3)$
Foci: $(0, -3 - \frac{\sqrt{13}}{6})$, $(0, -3 + \frac{\sqrt{13}}{6})$
Vertices: $(0, -3.5)$, $(0, -2.5)$
Lines of symmetry: $x = 0$, $y = -3$
Asymptotes: $y = \pm\frac{3}{2}x + 3$

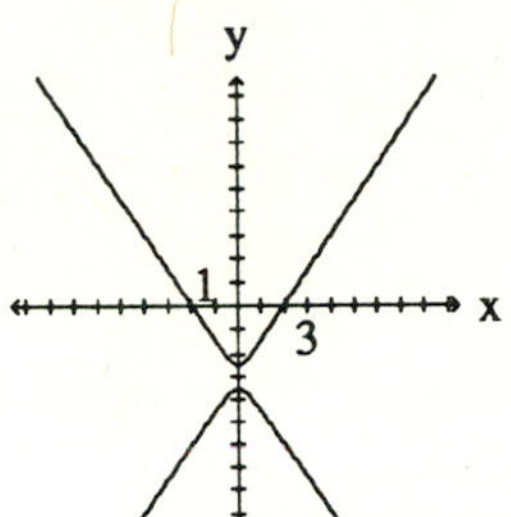

29. Hyperbola.

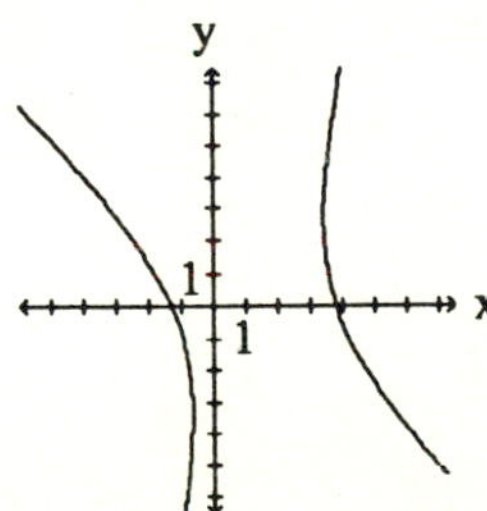

31. Ellipse.

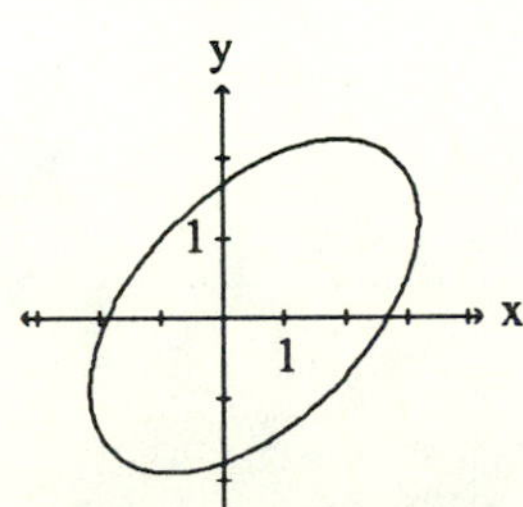

33. Asymptotes (est.): $y = 2x + 4$ and $y = x - 1$
Center (est.): $(-4.7, -5.6)$

35. $y = 3.3x - 3$ and $y = -0.44x - 1$

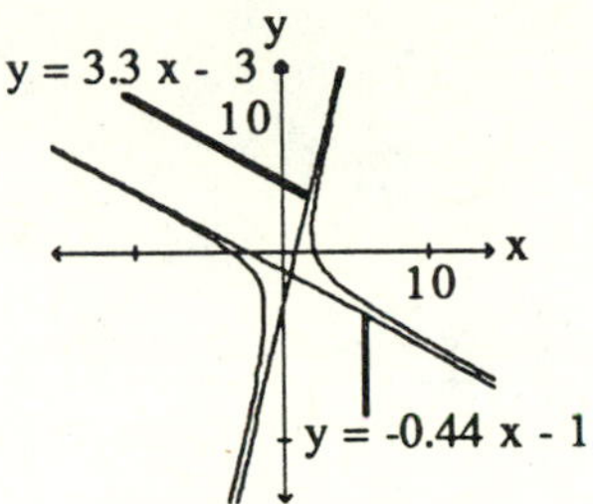

37. A horizontal shift left 2 units, a vertical shift up 1 unit, applied to $\frac{x^2}{9} + \frac{y^2}{4} = 1$

39. $(5.08, -0.54)$ and $(-23.26, 13.63)$

41. No solution.

43. 3 solutions.

45. $(-1.23, -1.34)$, $(2.48, 1.92)$

47.

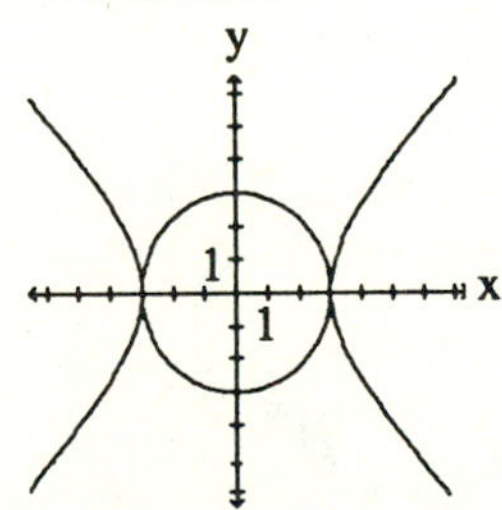

49. $e = 0.75$

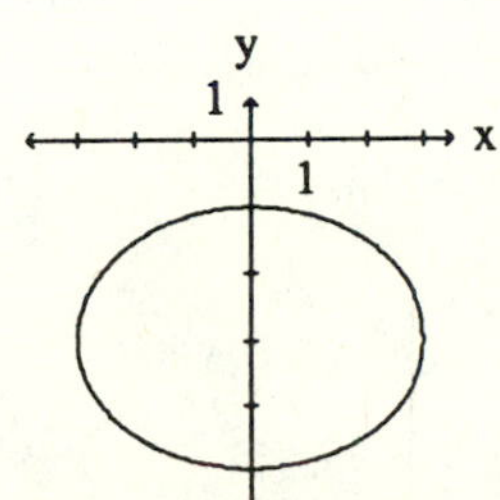

51.

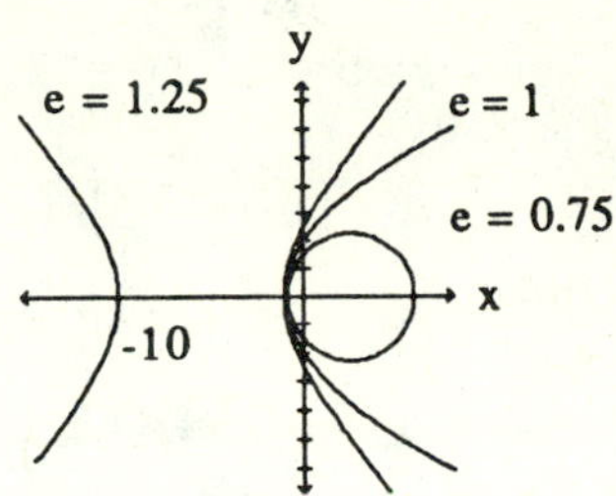

53. Hyperbola
Directrix: $x = -2$; $0 \leq \theta \leq 2\pi$

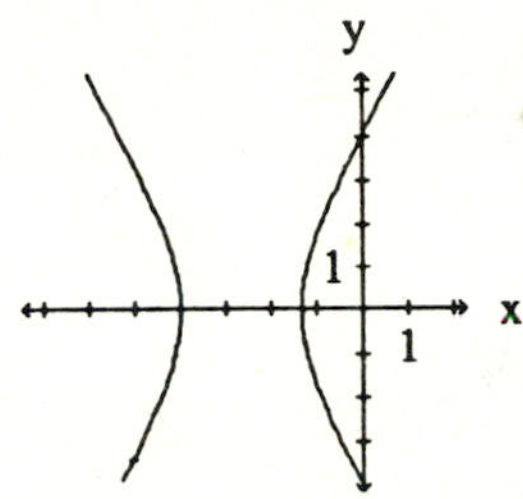

55. Directrix: $y = -2$; $0 \leq \theta \leq 2\pi$

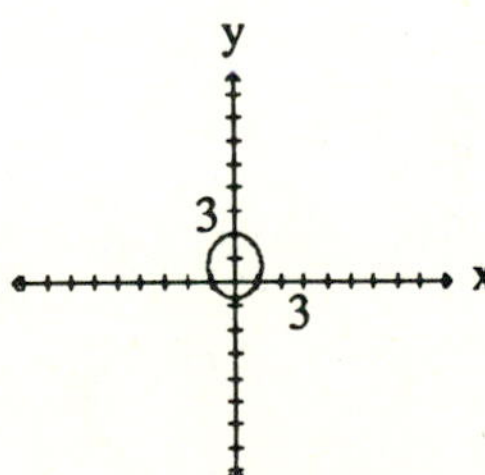

57. At the focus: $(\frac{5}{2}, 0)$

59. $\frac{x^2}{3.6\times10^{13}} - \frac{y^2}{1.08\times10^{14}} = 1$
(Assume the space station is at $(0, 0)$.)

61. $y = \frac{52+x}{80+x}(100)$

63. Domain: $(-\infty, -80) \cup (-80, \infty)$
Range: $(\infty, 100) \cup (100, \infty)$

65. The graphs of $y = \frac{52+x}{80+x}(100)$ and $y = 81$ intersect at 67.37.

67. $(-1635.17, 2721.88)$, $(-401.66, 1438.25)$, $(3022.11, 975.52)$, $(5645.91, 3705.94)$

Answers will vary due to round-off error

69. $A \geq 4000$
$B \geq 8000$
$A + B < 30,000$
$0.25A + 0.65B < 10,000$

71. 391.20 meters

72. $\frac{x^2}{52,441} + \frac{y^2}{38,260.95} = 1$

• CHAPTER 9

Exercises 9.1

1. $1, -1, 1, -1,\ a_{10} = -1$

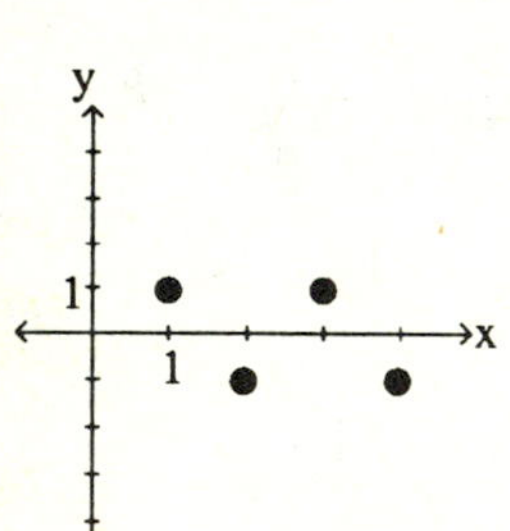

3. $2, \frac{3}{2}, \frac{4}{3}, \frac{5}{4},\ a_{10} = \frac{11}{10}$

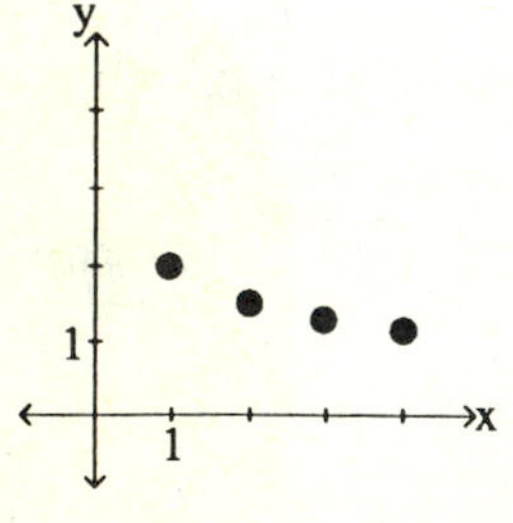

5. $0, \frac{7}{3}, \frac{13}{2}, \frac{63}{5},\ a_{10} = \frac{999}{11}$

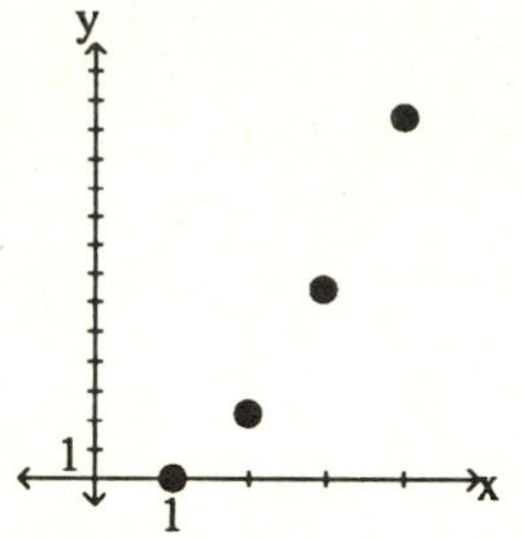

7. $3, 4, 5, 6,\ a_n = n + 2$

9. $2, 4, 8, 16,\ a_n = 2^n$

11. $d = 4,\ a_{10} = 42,\ a_n = 4n + 2$

13. $a_1 = -20,\ a_n = 4n - 24$

15. Geometric, $r = 2$, $a_n = 5(2)^{n-1}$

17. Arithmetic, $d = 7$, $a_n = 7n - 23$

19. Geometric, $r = -0.5$, $a_n = (-0.5)^{n-2}$

21. $S_n = 1000(.5)^n$, 500, 250, 125, 62.5, 31.25, 15.63, 7.81, 3.91, 1.95, 0.98, $r = 0.5$

23. No

25. About 28 months

33. 12, 15, $a_n = 3n - 6$, other a_n's are possible

35. 720, 5040, $a_n = n!$, other a_n's are possible

37. \$9216.56

39. $7\frac{1}{4}, 11\frac{1}{2}, 15\frac{3}{4}$

41. $5\frac{5}{6}, 8\frac{2}{3}, 11\frac{1}{2}, 14\frac{1}{3}, 17\frac{1}{6}$

43. 3, 5, 8, 13, 21

45. $a_n \to 31.42$ as $n \to \infty$ and $C = 31.42$

47. 1, 1, 2, 3, 5, 8, 13, 21, 34, the Fibonacci sequence

49. $K = 2$

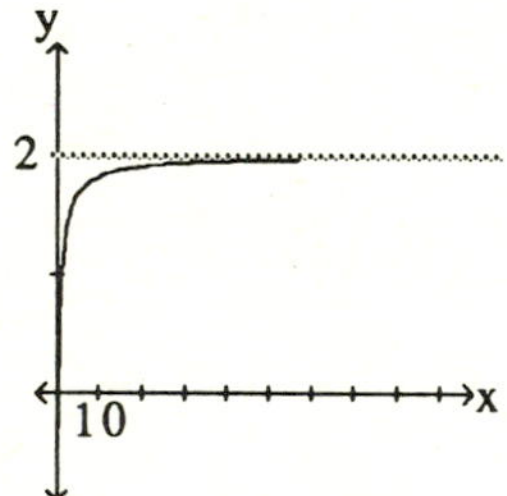

51. $K = 3$

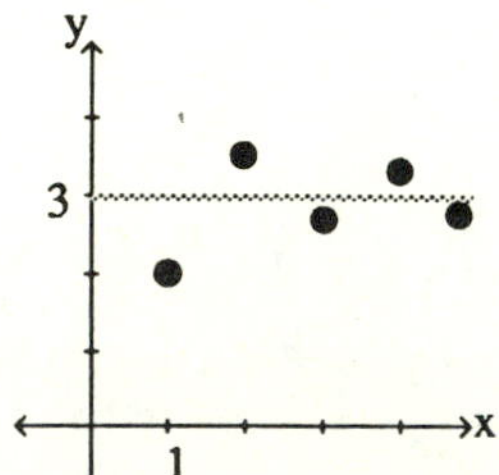

53. 1, 0.5, 0.71, 0.61, 0.65, 0.64, 0.64, 0.64

55. $x = 0.64$, same as iteration

57.

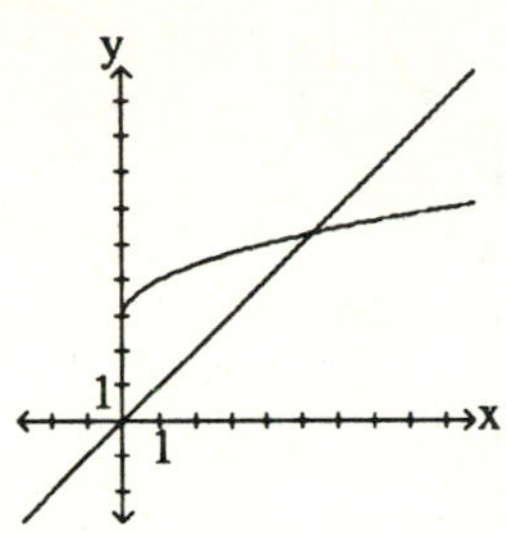

61. 10.57, 18.64, 32.84, 57.88

63. a)$1, 2, \frac{3}{2}, \frac{5}{3}, \frac{8}{5}, \frac{13}{8}, \frac{21}{13}, \frac{34}{21}, \frac{55}{34}, \frac{89}{55}$
b)$b_n \rightarrow \frac{1+\sqrt{5}}{2}$

Exercises 9.2

1. $\sum_{k=1}^{10}(3k-1)$

3. $\sum_{k=1}^{\infty}(4)^{-k}$

5. $S_n = \sum_{k=1}^{n}(2k-10)$

7. $S_n = \sum_{k=1}^{n} 3k(-1)^{k+1}$

9. Geometric,$1 + \frac{1}{2} + \frac{1}{4} + \dots + \left(\frac{1}{2}\right)^{n-1}$

11. Geometric,$-2 + 1 - \frac{1}{2} + \dots + (-2)^{2-n}$

13. 0.25

15. no sum

17. $\frac{1}{3}$

19. $0.9999... = 0.9 + 0.09 + 0.009 + ...$ By Theorem 3, as $n \rightarrow \infty$, $S_n \rightarrow \frac{0.9}{1-0.1} = 1$

21. As $n \rightarrow \infty$, $S_n \rightarrow 9$

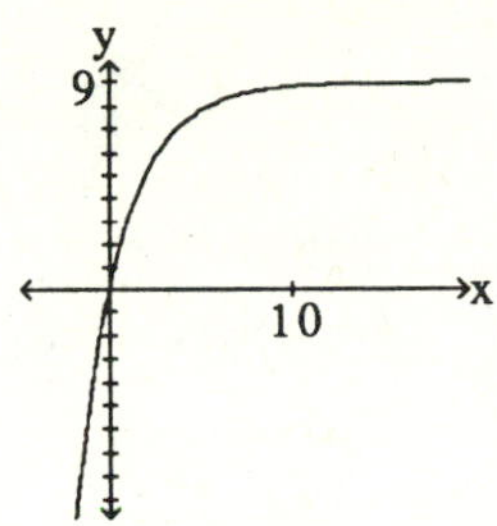

23. As $n \rightarrow \infty$, $S_n \rightarrow -3.75$

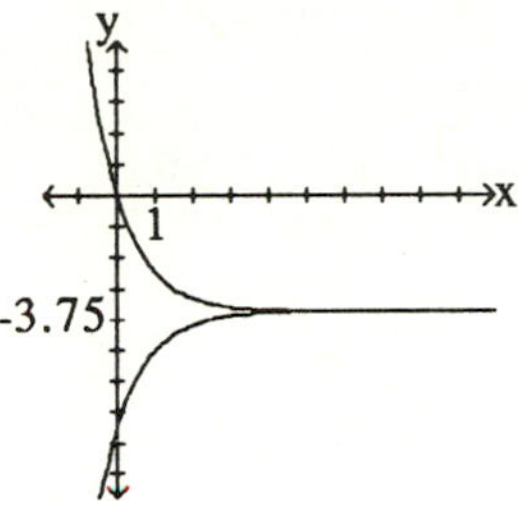

25. \$125, \$250.83, \$377.51, \$505.02, $S_n = 125\left(\frac{(1+\frac{.08}{12})^n - 1}{\frac{.08}{12}}\right)$, $r = 1 + \frac{.08}{12}$, $S_{180} = \$43,254.78$

27. \$222.77, \$443.34, \$661.72, \$877.94, $A_n = 225\left(\frac{1-(1+\frac{.12}{12})^{-n}}{\frac{.12}{12}}\right)$, $r = \left(1 + \frac{.12}{.12}\right)^{-1}$, $A_{48} = \$8544.14$

29. 1, 1.50, 1.83, 2.08, 2.28, 2.45, 2.59, 2.72, 2.83, 2.93, 3.02, 3.10, 3.18, 3.25, 3.32

31. $N = 4611$, $A = 9.01$ at 2 mins, $N = 23,050$, $A = 10.62$ at 10 minutes, $N = 69,269$, $A = 11.72$ at 30 minutes, $N = 138,620$, $A = 12.42$ at 1 hour, $N = 277,545$, $A = 13.11$ at 2 hours, $N = 833,692$, $A = 14.21$, overnight

33. Change the fourth line to: $(1 \div 2)x^y N + A \rightarrow A\Delta$

35. $\frac{5}{16}$

37. $\frac{63}{256}$

39. 0.5, 0.5^{50}

41. a)0.996, b)0.000007

44. It is an infinite series, but it has a finite sum.

Exercises 9.3

1. 1, 2, 2.5, 2.67, 2.7083, 2.7167, 2.7181, 2.71825, 2.718279, 2.718281527

3.

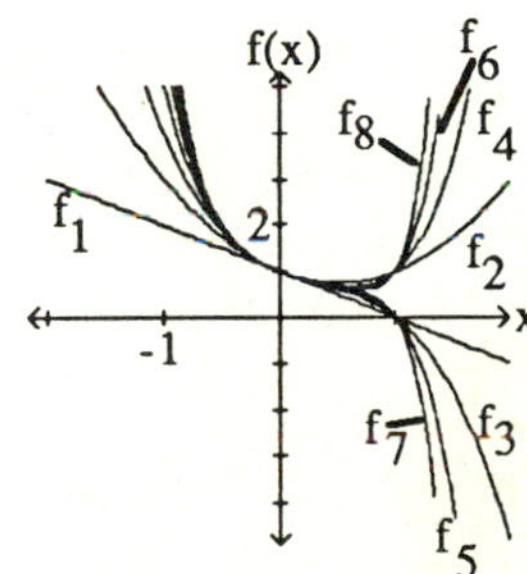

5. approx. (−.7, .7)

9. approx. 6.99

11. Underestimate

13. $|64 - 63| = 1$

15.

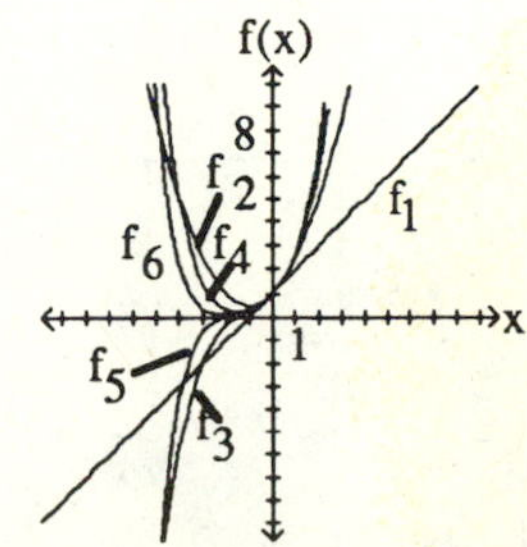

17. 7.09, 3.71, 1.69, 0.67, all underestimates

19.

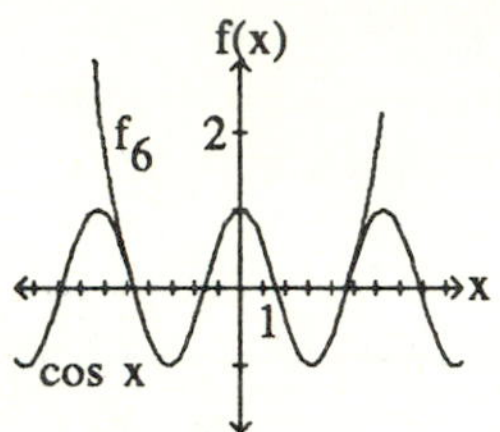

a) No. Error is 39.70.

b) Yes. Error is 5.29×10^{-5}.

21. $f_1 = 1 + \frac{1}{3}x$, $f_2 = 1 + \frac{1}{3}x - \frac{1}{9}x^2, \ldots,$ $f_6 = 1 + \frac{1}{3}x - \frac{1}{9}x^2 + \frac{5}{81}x^3 - \frac{10}{243}x^4 + \frac{22}{729}x^5 - \frac{154}{6561}x^6$

23. approx. $(-.7, 1)$

25. approx. 1.47×10^{-6}

27.

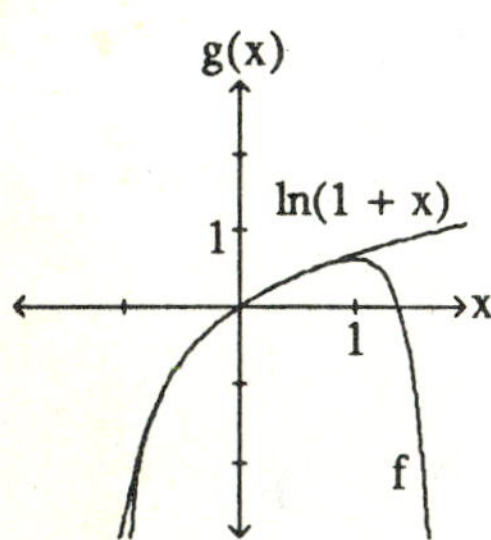

31. Three terms

33. Eight terms

Exercises 9.4

1. 3×4, 3 rows, 4 columns

3. a) $\begin{pmatrix} -1 & 2 & 3 & -1 \\ 3 & 2 & 4 & 3 \\ 2 & 6 & -2 & 3 \end{pmatrix}$

b) $\begin{pmatrix} -1 & 2 & 3 & -1 \\ 4 & 0 & 1 & 4 \\ 0 & 10 & 4 & 1 \end{pmatrix}$

c) $\begin{pmatrix} -1 & 2 & 3 & -1 \\ 0 & 8 & 13 & 0 \\ 2 & 6 & -2 & 3 \end{pmatrix}$

5. $\begin{pmatrix} 1 & 0 & 0 & \frac{50}{49} \\ 0 & 1 & 0 & \frac{13}{98} \\ 0 & 0 & 1 & -\frac{4}{49} \end{pmatrix}$

7. -9, 0, -1, 3. There is no a_{54} entry.

9. $\begin{pmatrix} 1 & 0 & -3 & \frac{2}{3} & \frac{1}{3} \\ 0 & 1 & 2 & -\frac{2}{3} & \frac{8}{3} \\ 0 & 0 & 0 & 0 & 0 \end{pmatrix}$

11. $\begin{pmatrix} 1 & 0 & 0 & -1 \\ 0 & 1 & 0 & 1 \\ 0 & 0 & 1 & 2 \end{pmatrix}$

13. $(9, 1)$

$\begin{pmatrix} 1 & -3 & 6 \\ 2 & 1 & 19 \end{pmatrix}$

15. no solution

$\begin{pmatrix} 1 & 1 & 3 \\ 1 & -1 & 5 \\ 2 & 1 & 4 \end{pmatrix}$

17. $(1, 2, 1)$

$\begin{pmatrix} 1 & -1 & 1 & 0 \\ 2 & 0 & -3 & -1 \\ -1 & -1 & 2 & -1 \end{pmatrix}$

19. no solution

$\begin{pmatrix} 1 & 1 & -2 & 2 \\ 3 & -1 & 1 & 4 \\ -2 & -2 & 4 & 6 \end{pmatrix}$

21. $(.75 + .25z, 1.25 + 1.75z, z)$

$\begin{pmatrix} 1 & 1 & -2 & 2 \\ 3 & -1 & 1 & 1 \\ -2 & -2 & 4 & -4 \end{pmatrix}$

23. $(2, 1, -1, 3)$

$\begin{pmatrix} 1 & 1 & -1 & 0 & 4 \\ 0 & 1 & 0 & 1 & 4 \\ 1 & -1 & 0 & 0 & 1 \\ 1 & 0 & 1 & 1 & 4 \end{pmatrix}$

25. $(-2 - w, .5 - z, z, w)$

$\begin{pmatrix} 2 & 1 & 1 & 2 & -3.5 \\ 1 & 1 & 1 & 1 & -1.5 \end{pmatrix}$

27. 724 children, 364 adults

29. \$7200 at 6%, \$3600 at 8%, \$9200 at 10%

31. $y = -x^2 + x + 10$

Exercises 9.5

1,2.

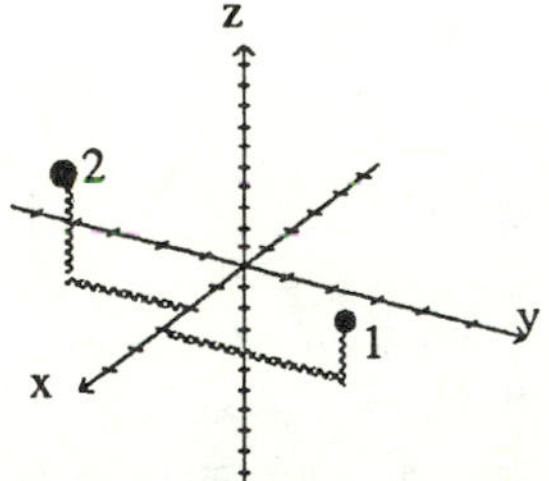

5. I

7. VIII

9.

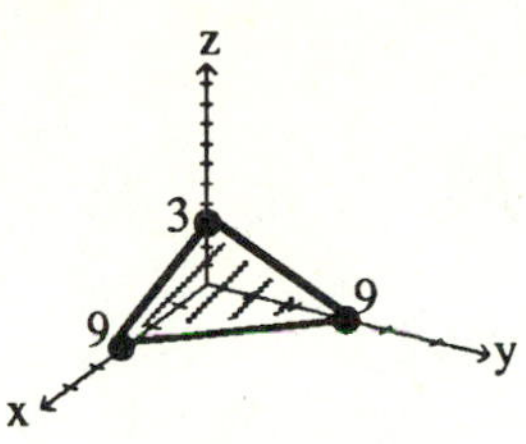

11.

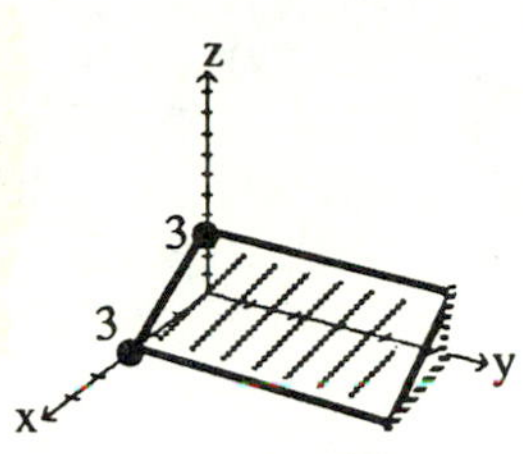

13.

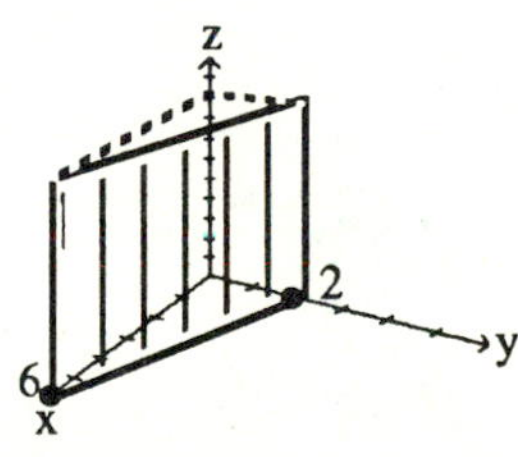

15.

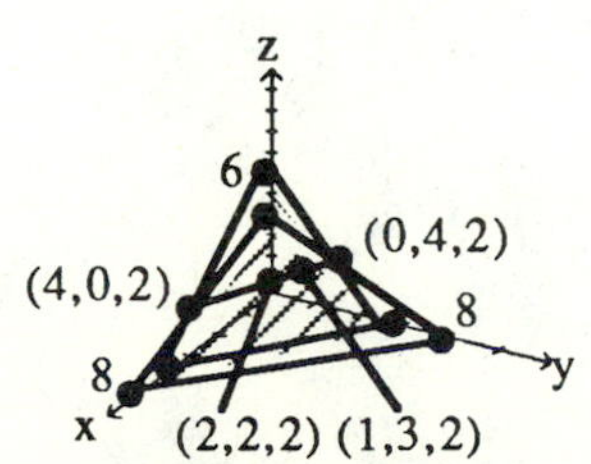

17. $\sqrt{53}$

19. $(1, -1, 5.5)$

21. $x = 4 + 2t$, $y = 6 + 4t$, $z = 1 + 3t$

23.

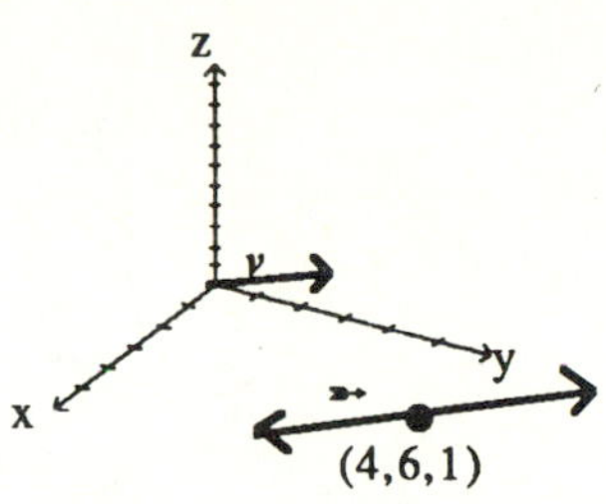

27. a) $\frac{x+1}{1} = \frac{y-2}{4} = \frac{z-4}{-7}$
b) $x = t - 1$, $y = 4t + 2$, $z = -7t + 4$

29. a) $\frac{x}{2} = \frac{y-6}{-10} = \frac{z+3}{4}$
b) $x = 2t$, $y = -10t + 6$, $z = 4t - 3$

31. Elliptic cylinder, center is z-axis

33. Ellipsoid

35. Elliptic paraboloid

37. Hyperbolic paraboloid

39. Cylinder, center is y-axis

41. $(4, 0°, 0)$

43. $(\sqrt{32}, 45°, 3)$

45. $(1, 90°, 0°)$

47. $(2, 45°, 45°)$

51. A plane $\perp$ to the xy-plane that cuts the plane $30°$ from the x-axis.

53. A sphere with radius 4, centered at $(0, 0, 0)$.

Exercises 9.6

1. Domain of f: R^2
Domain of g: R^2

3. Domain of z: $x^2 + y^2 \geq 9$, the circle $x^2 + y^2 = 9$ and its exterior

5.

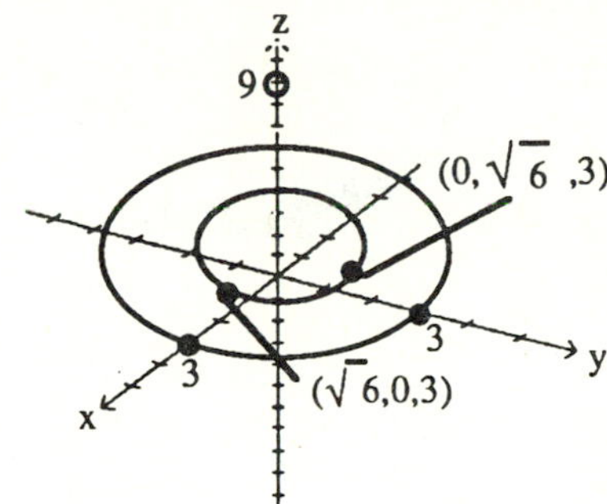

7.

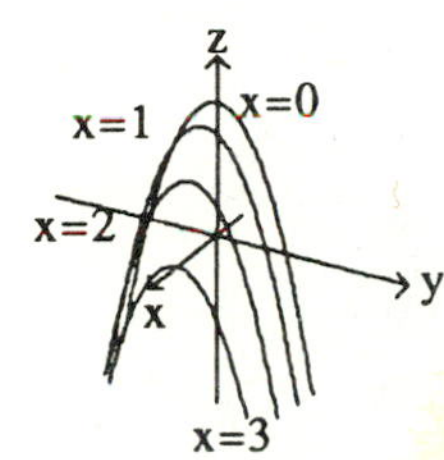

9.

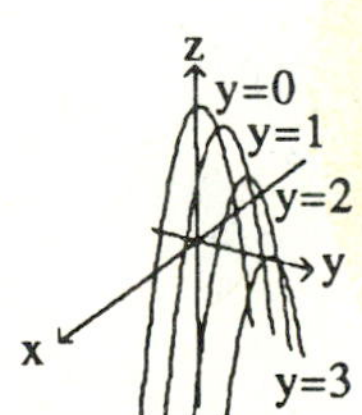

11. Max. at $(0, 0, 9)$

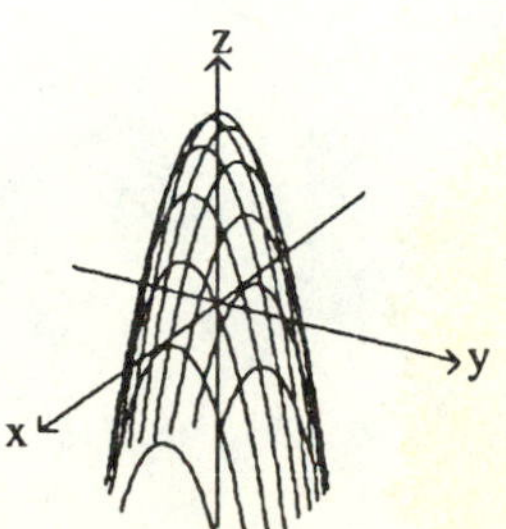

13. Domain of f: R^2

15. Max. at $(2, 3, 9)$

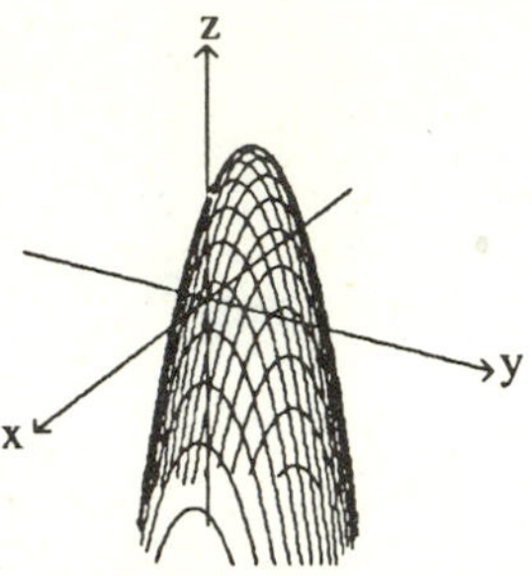

17. Domain: R^2, max. at $(0, 0, 25)$

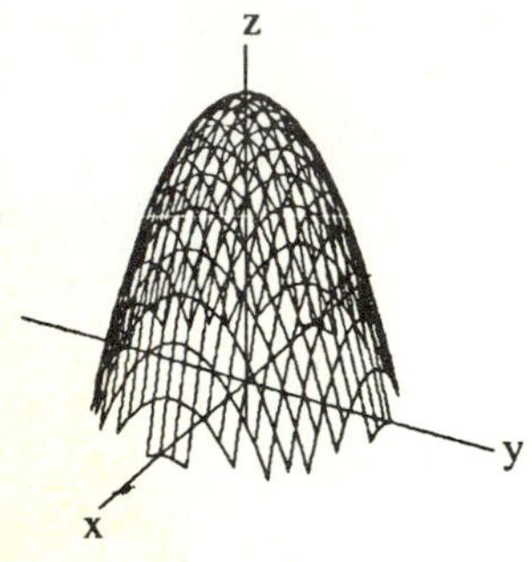

19. Domain: the ellipse $(x-2)^2+(y-4.5)^2 = 62.25$ and its interior, max. at $(2, 4.5, 7.9)$, min. at the ellipse

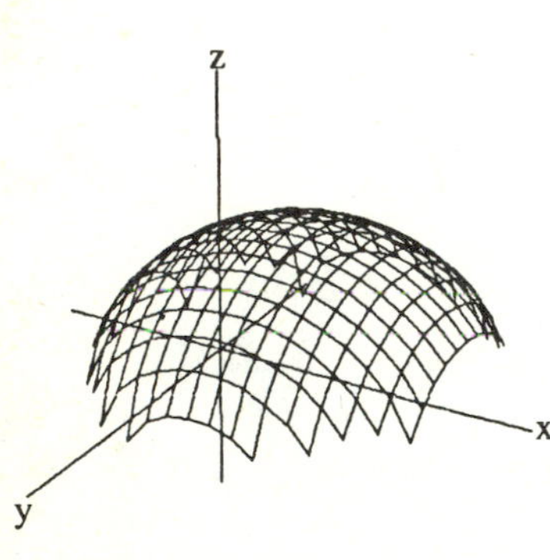

21. Domain: the interior of the circle $x^2 + y^2 = 9$, min. at $(0, 0, \frac{1}{3})$

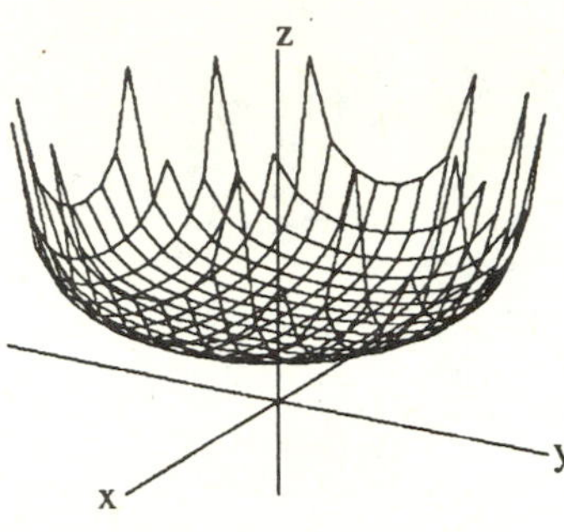

23. Domain: all R^2, saddle at $(0, 4, 0)$

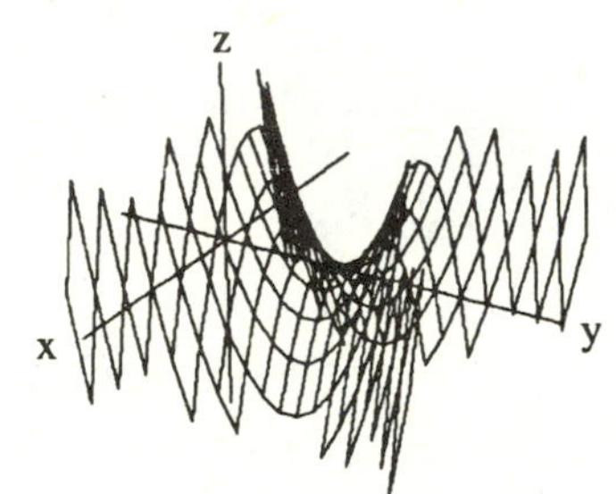

25. Domain: all R^2, max. at

$(0, 4.1, 231.1)$ and $(0, -2.9, 170.1)$, saddle at $(0, 0, 0)$

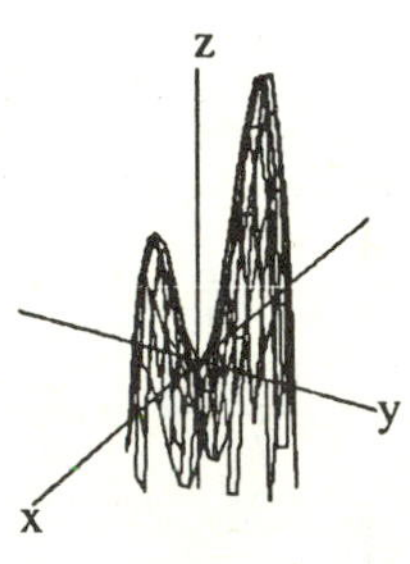

27. Domain: all R^2, max. at $(0, -2.8, 7.5)$ and $(0, 4.1, 11.7)$, saddle at $(0, 0, 5)$

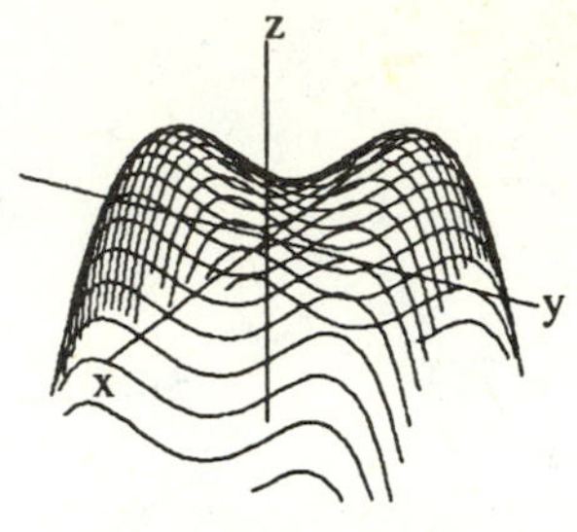

29. $(2, 1, 258)$, $(2, 3, 207.33)$, $(2, 7, 220.29)$, $(2, 10, 240)$, $(4, 1, 191)$, $(4, 3, 156.33)$, $(4, 7, 201.29)$, $(4, 10, 245)$, $(6, 1, 174)$, $(6, 3, 153.33)$, $(6, 7, 232.29)$, $(6, 10, 300)$, $(8, 1, 169.5)$, $(8, 3, 166.83)$, $(8, 7, 275.79)$, $(8, 10, 367.5)$

31.

33. $C = \frac{100}{y} + \frac{300}{x} + 4xy$

35. VC:[4,8] by [0,4] by [140,160]

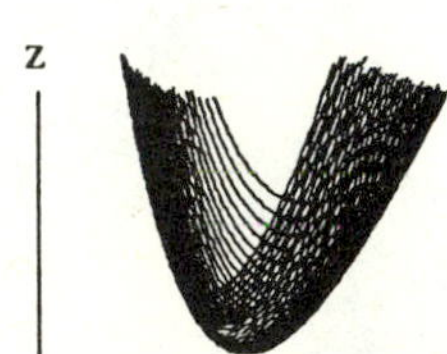

37. $S = \frac{1000}{y} + \frac{1000}{x} + 2xy$

39. VC:[4,12] by [4,12] by [360,385]

41. $S = \frac{1000}{y} + \frac{1000}{x} + xy$

43. VC:[7,13] by [7,13] by [297,303]

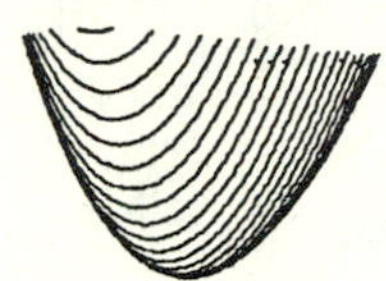

45. the plane $z = 0$

47. the plane $z = 2$

49.

r=2 at z=28
r=5 at z=7
r=3 at z=23
r=4 at z=16
r=$\sqrt{32}$ at z=0

Exercises 9.7

1. $-11 + 2i$
3. $6xy^2 - 2x^3 + (2y^3 - 6x^2y)i$
5. $\sqrt{(x^2 - y^2 - 1)^2 + (2xy)^2}$
7. $((x^2 - y^2 - 2x + 5)^2 + (2xy - 2y)^2)^{\frac{1}{2}}$
9. $((x^3 - 3xy^2 - 8)^2 + (3x^2y - y^3)^2)^{\frac{1}{2}}$
11. $1, -1$
13. $1 + 2i, 1 - 2i$
15. $2, -1 + \sqrt{3}i, -1 - \sqrt{3}i$
17. VC:[-2,2] by [-2,2] by [-4,4]

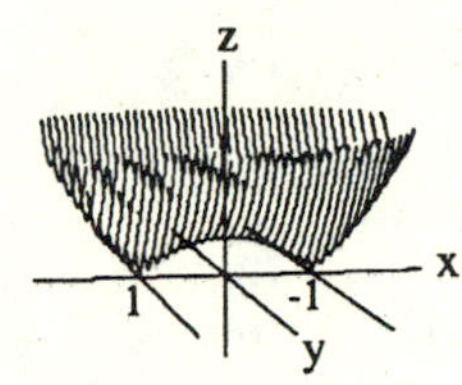

19. VC:[-2,2] by [-4,4] by [-6,6]

21. VC:[-4,4] by [-4,4] by [-10,10]

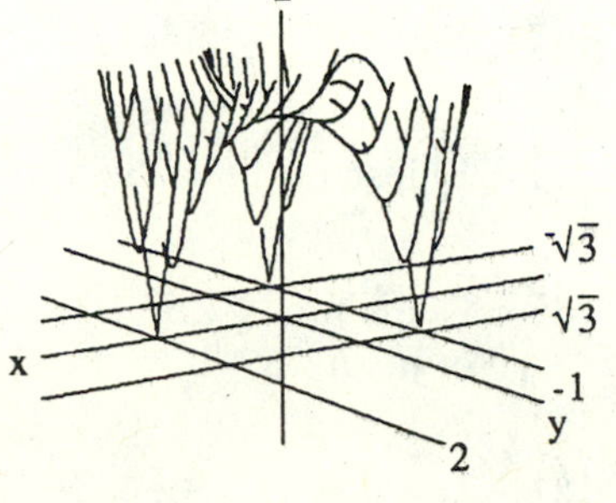

25. zeros: $i, -i, -1 + i, -1 - i$

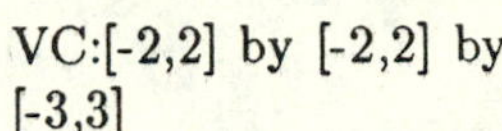
VC:[-2,2] by [-2,2] by [-3,3]

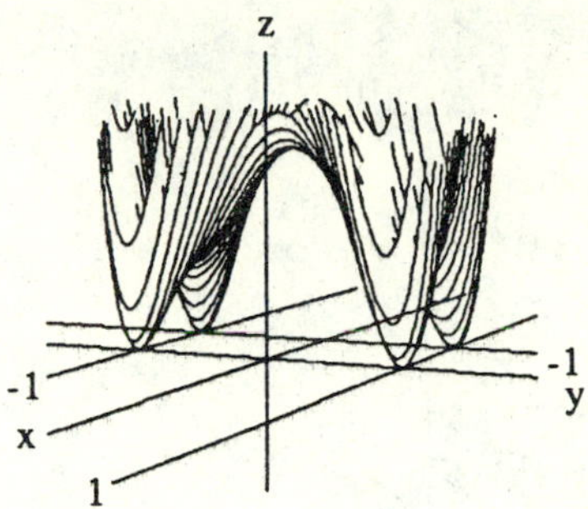

Review Exercises

1. $0, -1, 2, -3,\ a_{10} = -9$

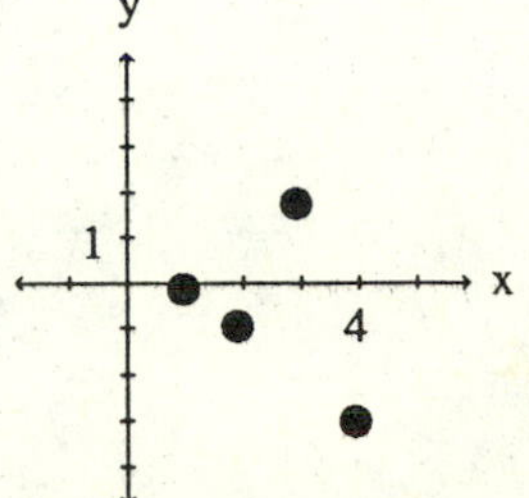

3. $a_1 = -32,\ a_n = -40 + 8n$

5. $\sum_{k=1}^{10} (1 - 3k)$

7. $\sum_{k=1}^{n} (10 - 2k)$

9. .125

11. As $n \to \infty,\ S_n \to 15$

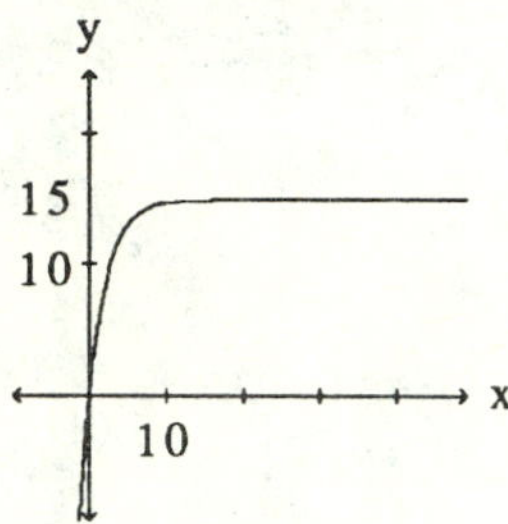

13. a)$|e^2 - f_6| = .0335$, yes

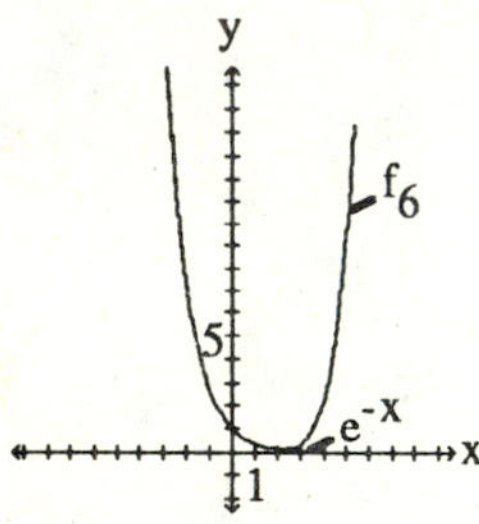

b)$|e^{-4} - f_6| = 1.03$, no

c)$(-\infty, 1.5)$

15. $(2,1)$

$$\begin{pmatrix} 1 & 1 & 3 \\ 1 & -1 & 1 \\ 2 & 1 & 5 \end{pmatrix}$$

17. $(2,2,0)$

$$\begin{pmatrix} 1 & -1 & 1 & 0 \\ 2 & 0 & -3 & 4 \\ -1 & -1 & 2 & -4 \end{pmatrix}$$

19. $(\frac{-11+z}{4}, \frac{7z-5}{4}, z)$

$$\begin{pmatrix} 1 & 1 & -2 & -4 \\ 3 & -1 & 1 & -7 \\ -2 & -2 & 4 & 8 \end{pmatrix}$$

21,22.

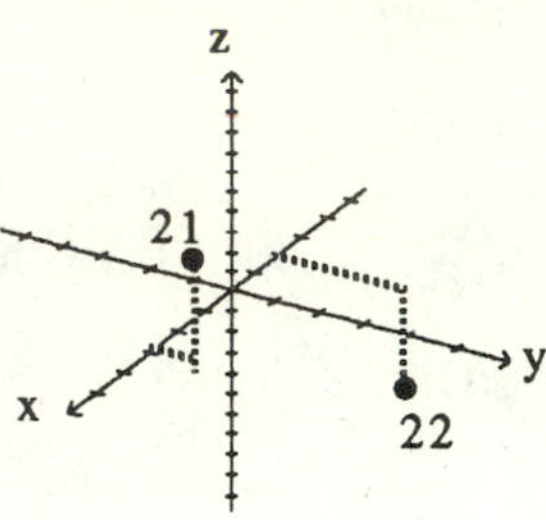

23. I

25.

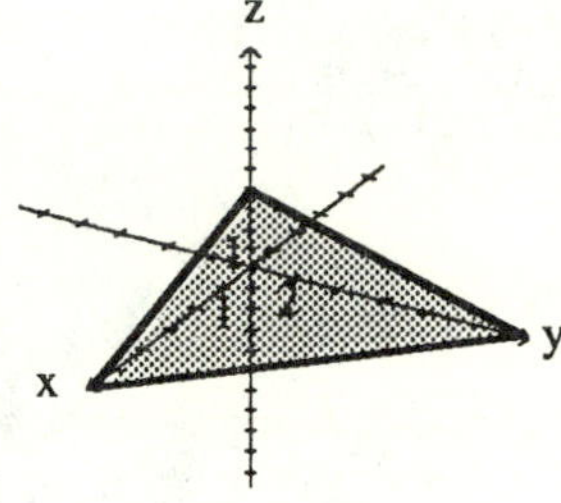

27. $\sqrt{141}$

29. a) $\frac{x-2}{4} = \frac{y-1}{3} = \frac{z-5}{3}$

b) $x = 4t + 2,$
$y = 3t + 1,\ z = 3t + 5$

31. Parabolic cylinder

33. Ellipsoid

35. Domain of $f : R^2$,Domain of $g : R^2$

37. Domain: the circle $x^2 + y^2 = 16$ and its exterior

39.

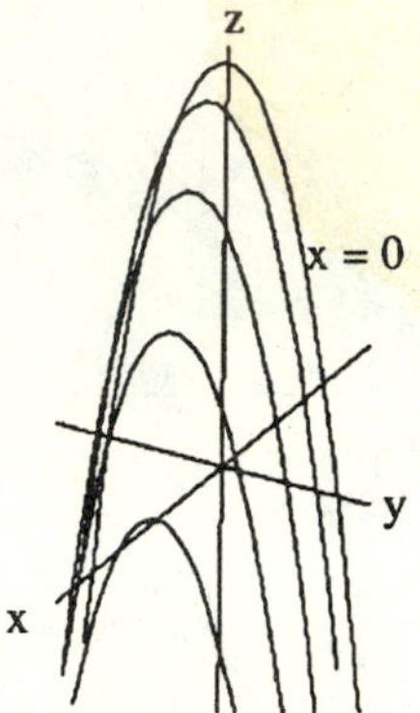

41. Domain: all R^2, max. at $(0,0,9)$

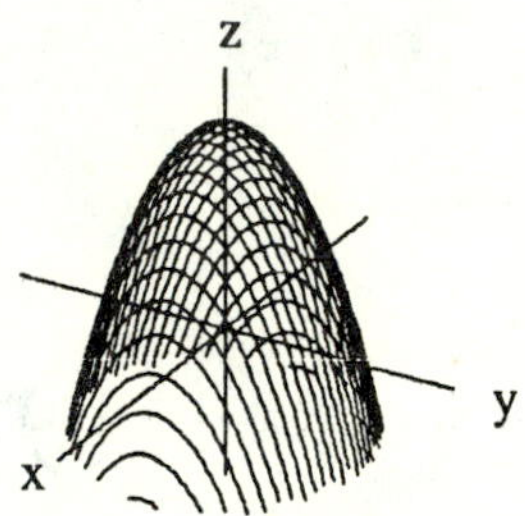

43. Domain: the hyperbola $(y-3)^2 + (x-1)^2 = 3$ and its interior, min. at the hyperbola

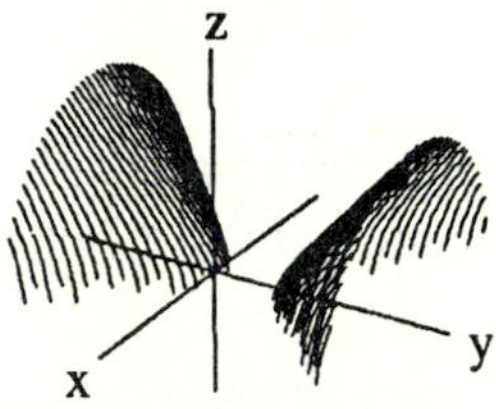

45. $\sqrt{(x^2 - y^2 - 4)^2 + (2xy)^2}$

47. $((x^2 - y^2 - 2x + 6)^2 + (2xy - 2y)^2)^{\frac{1}{2}}$

49. $-2, 2$

51. $1 + \sqrt{5}i, 1 - \sqrt{5}i$

53.

55.

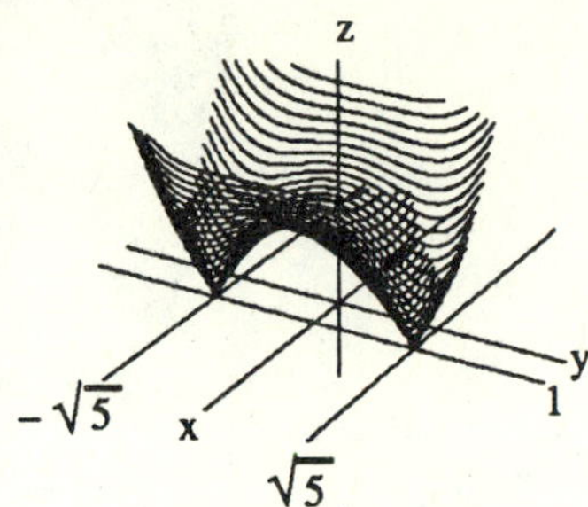

57. $a_n = 2000(\frac{1}{2})^n$,2000, 1000, 500, 250, 125, 62.5, 31.25, 15.625, 7.81, 3.91, 1.95, $r = \frac{1}{2}$

59. $2656.43

61. 67 children, 28 adults

63. $C = \frac{800}{y} + \frac{4000}{x} + 6xy$

65. VC:[10,20] by [0,12] by [799,812]

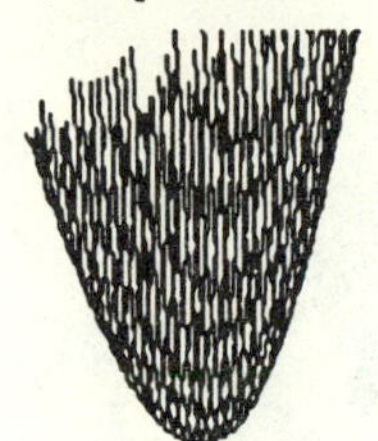

67. $S = \frac{1500}{y} + \frac{1500}{x} + 2xy$

69.

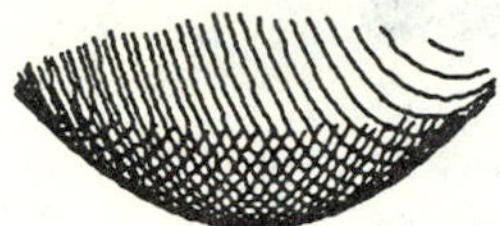

Index